T0180865

Lecture Notes in Computer Science 13699

More information about this series at https://link.springer.com/bookseries/558

Shai Avidan · Gabriel Brostow ·
Moustapha Cissé · Giovanni Maria Farinella ·
Tal Hassner (Eds.)

Computer Vision – ECCV 2022

17th European Conference
Tel Aviv, Israel, October 23–27, 2022
Proceedings, Part XXXIX

Editors
Shai Avidan
Tel Aviv University
Tel Aviv, Israel

Gabriel Brostow
University College London
London, UK

Moustapha Cissé
Google AI
Accra, Ghana

Giovanni Maria Farinella
University of Catania
Catania, Italy

Tal Hassner
Facebook (United States)
Menlo Park, CA, USA

ISSN 0302-9743 ISSN 1611-3349 (electronic)
Lecture Notes in Computer Science
ISBN 978-3-031-19841-0 ISBN 978-3-031-19842-7 (eBook)
https://doi.org/10.1007/978-3-031-19842-7

This Springer imprint is published by the registered company Springer Nature Switzerland AG
The registered company address is: Gewerbestrasse 11, 6330 Cham, Switzerland

Foreword

Organizing the European Conference on Computer Vision (ECCV 2022) in Tel-Aviv during a global pandemic was no easy feat. The uncertainty level was extremely high, and decisions had to be postponed to the last minute. Still, we managed to plan things just in time for ECCV 2022 to be held in person. Participation in physical events is crucial to stimulating collaborations and nurturing the culture of the Computer Vision community.

There were many people who worked hard to ensure attendees enjoyed the best science at the 16th edition of ECCV. We are grateful to the Program Chairs Gabriel Brostow and Tal Hassner, who went above and beyond to ensure the ECCV reviewing process ran smoothly. The scientific program includes dozens of workshops and tutorials in addition to the main conference and we would like to thank Leonid Karlinsky and Tomer Michaeli for their hard work. Finally, special thanks to the web chairs Lorenzo Baraldi and Kosta Derpanis, who put in extra hours to transfer information fast and efficiently to the ECCV community.

We would like to express gratitude to our generous sponsors and the Industry Chairs, Dimosthenis Karatzas and Chen Sagiv, who oversaw industry relations and proposed new ways for academia-industry collaboration and technology transfer. It's great to see so much industrial interest in what we're doing!

Authors' draft versions of the papers appeared online with open access on both the Computer Vision Foundation (CVF) and the European Computer Vision Association (ECVA) websites as with previous ECCVs. Springer, the publisher of the proceedings, has arranged for archival publication. The final version of the papers is hosted by SpringerLink, with active references and supplementary materials. It benefits all potential readers that we offer both a free and citeable version for all researchers, as well as an authoritative, citeable version for SpringerLink readers. Our thanks go to Ronan Nugent from Springer, who helped us negotiate this agreement. Last but not least, we wish to thank Eric Mortensen, our publication chair, whose expertise made the process smooth.

October 2022

Rita Cucchiara
Jiří Matas
Amnon Shashua
Lihi Zelnik-Manor

Preface

Welcome to the proceedings of the European Conference on Computer Vision (ECCV 2022). This was a hybrid edition of ECCV as we made our way out of the COVID-19 pandemic. The conference received 5804 valid paper submissions, compared to 5150 submissions to ECCV 2020 (a 12.7% increase) and 2439 in ECCV 2018. 1645 submissions were accepted for publication (28%) and, of those, 157 (2.7% overall) as orals.

846 of the submissions were desk-rejected for various reasons. Many of them because they revealed author identity, thus violating the double-blind policy. This violation came in many forms: some had author names with the title, others added acknowledgments to specific grants, yet others had links to their github account where their name was visible. Tampering with the LaTeX template was another reason for automatic desk rejection.

ECCV 2022 used the traditional CMT system to manage the entire double-blind reviewing process. Authors did not know the names of the reviewers and vice versa. Each paper received at least 3 reviews (except 6 papers that received only 2 reviews), totalling more than 15,000 reviews.

Handling the review process at this scale was a significant challenge. To ensure that each submission received as fair and high-quality reviews as possible, we recruited more than 4719 reviewers (in the end, 4719 reviewers did at least one review). Similarly we recruited more than 276 area chairs (eventually, only 276 area chairs handled a batch of papers). The area chairs were selected based on their technical expertise and reputation, largely among people who served as area chairs in previous top computer vision and machine learning conferences (ECCV, ICCV, CVPR, NeurIPS, etc.).

Reviewers were similarly invited from previous conferences, and also from the pool of authors. We also encouraged experienced area chairs to suggest additional chairs and reviewers in the initial phase of recruiting. The median reviewer load was five papers per reviewer, while the average load was about four papers, because of the emergency reviewers. The area chair load was 35 papers, on average.

Conflicts of interest between authors, area chairs, and reviewers were handled largely automatically by the CMT platform, with some manual help from the Program Chairs. Reviewers were allowed to describe themselves as senior reviewer (load of 8 papers to review) or junior reviewers (load of 4 papers). Papers were matched to area chairs based on a subject-area affinity score computed in CMT and an affinity score computed by the Toronto Paper Matching System (TPMS). TPMS is based on the paper's full text. An area chair handling each submission would bid for preferred expert reviewers, and we balanced load and prevented conflicts.

The assignment of submissions to area chairs was relatively smooth, as was the assignment of submissions to reviewers. A small percentage of reviewers were not happy with their assignments in terms of subjects and self-reported expertise. This is an area for improvement, although it's interesting that many of these cases were reviewers hand-picked by AC's. We made a later round of reviewer recruiting, targeted at the list of authors of papers submitted to the conference, and had an excellent response which helped provide enough emergency reviewers. In the end, all but six papers received at least 3 reviews.

The challenges of the reviewing process are in line with past experiences at ECCV 2020. As the community grows, and the number of submissions increases, it becomes ever more challenging to recruit enough reviewers and ensure a high enough quality of reviews. Enlisting authors by default as reviewers might be one step to address this challenge.

Authors were given a week to rebut the initial reviews, and address reviewers' concerns. Each rebuttal was limited to a single pdf page with a fixed template.

The Area Chairs then led discussions with the reviewers on the merits of each submission. The goal was to reach consensus, but, ultimately, it was up to the Area Chair to make a decision. The decision was then discussed with a buddy Area Chair to make sure decisions were fair and informative. The entire process was conducted virtually with no in-person meetings taking place.

The Program Chairs were informed in cases where the Area Chairs overturned a decisive consensus reached by the reviewers, and pushed for the meta-reviews to contain details that explained the reasoning for such decisions. Obviously these were the most contentious cases, where reviewer inexperience was the most common reported factor.

Once the list of accepted papers was finalized and released, we went through the laborious process of plagiarism (including self-plagiarism) detection. A total of 4 accepted papers were rejected because of that.

Finally, we would like to thank our Technical Program Chair, Pavel Lifshits, who did tremendous work behind the scenes, and we thank the tireless CMT team.

October 2022

Gabriel Brostow
Giovanni Maria Farinella
Moustapha Cissé
Shai Avidan
Tal Hassner

Organization

General Chairs

Rita Cucchiara	University of Modena and Reggio Emilia, Italy
Jiří Matas	Czech Technical University in Prague, Czech Republic
Amnon Shashua	Hebrew University of Jerusalem, Israel
Lihi Zelnik-Manor	Technion – Israel Institute of Technology, Israel

Program Chairs

Shai Avidan	Tel-Aviv University, Israel
Gabriel Brostow	University College London, UK
Moustapha Cissé	Google AI, Ghana
Giovanni Maria Farinella	University of Catania, Italy
Tal Hassner	Facebook AI, USA

Program Technical Chair

Pavel Lifshits	Technion – Israel Institute of Technology, Israel

Workshops Chairs

Leonid Karlinsky	IBM Research, Israel
Tomer Michaeli	Technion – Israel Institute of Technology, Israel
Ko Nishino	Kyoto University, Japan

Tutorial Chairs

Thomas Pock	Graz University of Technology, Austria
Natalia Neverova	Facebook AI Research, UK

Demo Chair

Bohyung Han	Seoul National University, Korea

Social and Student Activities Chairs

Tatiana Tommasi	Italian Institute of Technology, Italy
Sagie Benaim	University of Copenhagen, Denmark

Diversity and Inclusion Chairs

Xi Yin	Facebook AI Research, USA
Bryan Russell	Adobe, USA

Communications Chairs

Lorenzo Baraldi	University of Modena and Reggio Emilia, Italy
Kosta Derpanis	York University & Samsung AI Centre Toronto, Canada

Industrial Liaison Chairs

Dimosthenis Karatzas	Universitat Autònoma de Barcelona, Spain
Chen Sagiv	SagivTech, Israel

Finance Chair

Gerard Medioni	University of Southern California & Amazon, USA

Publication Chair

Eric Mortensen	MiCROTEC, USA

Area Chairs

Lourdes Agapito	University College London, UK
Zeynep Akata	University of Tübingen, Germany
Naveed Akhtar	University of Western Australia, Australia
Karteek Alahari	Inria Grenoble Rhône-Alpes, France
Alexandre Alahi	École polytechnique fédérale de Lausanne, Switzerland
Pablo Arbelaez	Universidad de Los Andes, Columbia
Antonis A. Argyros	University of Crete & Foundation for Research and Technology-Hellas, Crete
Yuki M. Asano	University of Amsterdam, The Netherlands
Kalle Åström	Lund University, Sweden
Hadar Averbuch-Elor	Cornell University, USA

Hossein Azizpour	KTH Royal Institute of Technology, Sweden
Vineeth N. Balasubramanian	Indian Institute of Technology, Hyderabad, India
Lamberto Ballan	University of Padova, Italy
Adrien Bartoli	Université Clermont Auvergne, France
Horst Bischof	Graz University of Technology, Austria
Matthew B. Blaschko	KU Leuven, Belgium
Federica Bogo	Meta Reality Labs Research, Switzerland
Katherine Bouman	California Institute of Technology, USA
Edmond Boyer	Inria Grenoble Rhône-Alpes, France
Michael S. Brown	York University, Canada
Vittorio Caggiano	Meta AI Research, USA
Neill Campbell	University of Bath, UK
Octavia Camps	Northeastern University, USA
Duygu Ceylan	Adobe Research, USA
Ayan Chakrabarti	Google Research, USA
Tat-Jen Cham	Nanyang Technological University, Singapore
Antoni Chan	City University of Hong Kong, Hong Kong, China
Manmohan Chandraker	NEC Labs America, USA
Xinlei Chen	Facebook AI Research, USA
Xilin Chen	Institute of Computing Technology, Chinese Academy of Sciences, China
Dongdong Chen	Microsoft Cloud AI, USA
Chen Chen	University of Central Florida, USA
Ondrej Chum	Vision Recognition Group, Czech Technical University in Prague, Czech Republic
John Collomosse	Adobe Research & University of Surrey, UK
Camille Couprie	Facebook, France
David Crandall	Indiana University, USA
Daniel Cremers	Technical University of Munich, Germany
Marco Cristani	University of Verona, Italy
Canton Cristian	Facebook AI Research, USA
Dengxin Dai	ETH Zurich, Switzerland
Dima Damen	University of Bristol, UK
Kostas Daniilidis	University of Pennsylvania, USA
Trevor Darrell	University of California, Berkeley, USA
Andrew Davison	Imperial College London, UK
Tali Dekel	Weizmann Institute of Science, Israel
Alessio Del Bue	Istituto Italiano di Tecnologia, Italy
Weihong Deng	Beijing University of Posts and Telecommunications, China
Konstantinos Derpanis	Ryerson University, Canada
Carl Doersch	DeepMind, UK

Matthijs Douze	Facebook AI Research, USA
Mohamed Elhoseiny	King Abdullah University of Science and Technology, Saudi Arabia
Sergio Escalera	University of Barcelona, Spain
Yi Fang	New York University, USA
Ryan Farrell	Brigham Young University, USA
Alireza Fathi	Google, USA
Christoph Feichtenhofer	Facebook AI Research, USA
Basura Fernando	Agency for Science, Technology and Research (A*STAR), Singapore
Vittorio Ferrari	Google Research, Switzerland
Andrew W. Fitzgibbon	Graphcore, UK
David J. Fleet	University of Toronto, Canada
David Forsyth	University of Illinois at Urbana-Champaign, USA
David Fouhey	University of Michigan, USA
Katerina Fragkiadaki	Carnegie Mellon University, USA
Friedrich Fraundorfer	Graz University of Technology, Austria
Oren Freifeld	Ben-Gurion University, Israel
Thomas Funkhouser	Google Research & Princeton University, USA
Yasutaka Furukawa	Simon Fraser University, Canada
Fabio Galasso	Sapienza University of Rome, Italy
Jürgen Gall	University of Bonn, Germany
Chuang Gan	Massachusetts Institute of Technology, USA
Zhe Gan	Microsoft, USA
Animesh Garg	University of Toronto, Vector Institute, Nvidia, Canada
Efstratios Gavves	University of Amsterdam, The Netherlands
Peter Gehler	Amazon, Germany
Theo Gevers	University of Amsterdam, The Netherlands
Bernard Ghanem	King Abdullah University of Science and Technology, Saudi Arabia
Ross B. Girshick	Facebook AI Research, USA
Georgia Gkioxari	Facebook AI Research, USA
Albert Gordo	Facebook, USA
Stephen Gould	Australian National University, Australia
Venu Madhav Govindu	Indian Institute of Science, India
Kristen Grauman	Facebook AI Research & UT Austin, USA
Abhinav Gupta	Carnegie Mellon University & Facebook AI Research, USA
Mohit Gupta	University of Wisconsin-Madison, USA
Hu Han	Institute of Computing Technology, Chinese Academy of Sciences, China

Vittorio Murino	Istituto Italiano di Tecnologia, Italy
P. J. Narayanan	International Institute of Information Technology, Hyderabad, India
Ram Nevatia	University of Southern California, USA
Natalia Neverova	Facebook AI Research, UK
Richard Newcombe	Facebook, USA
Cuong V. Nguyen	Florida International University, USA
Bingbing Ni	Shanghai Jiao Tong University, China
Juan Carlos Niebles	Salesforce & Stanford University, USA
Ko Nishino	Kyoto University, Japan
Jean-Marc Odobez	Idiap Research Institute, École polytechnique fédérale de Lausanne, Switzerland
Francesca Odone	University of Genova, Italy
Takayuki Okatani	Tohoku University & RIKEN Center for Advanced Intelligence Project, Japan
Manohar Paluri	Facebook, USA
Guan Pang	Facebook, USA
Maja Pantic	Imperial College London, UK
Sylvain Paris	Adobe Research, USA
Jaesik Park	Pohang University of Science and Technology, Korea
Hyun Soo Park	The University of Minnesota, USA
Omkar M. Parkhi	Facebook, USA
Deepak Pathak	Carnegie Mellon University, USA
Georgios Pavlakos	University of California, Berkeley, USA
Marcello Pelillo	University of Venice, Italy
Marc Pollefeys	ETH Zurich & Microsoft, Switzerland
Jean Ponce	Inria, France
Gerard Pons-Moll	University of Tübingen, Germany
Fatih Porikli	Qualcomm, USA
Victor Adrian Prisacariu	University of Oxford, UK
Petia Radeva	University of Barcelona, Spain
Ravi Ramamoorthi	University of California, San Diego, USA
Deva Ramanan	Carnegie Mellon University, USA
Vignesh Ramanathan	Facebook, USA
Nalini Ratha	State University of New York at Buffalo, USA
Tammy Riklin Raviv	Ben-Gurion University, Israel
Tobias Ritschel	University College London, UK
Emanuele Rodola	Sapienza University of Rome, Italy
Amit K. Roy-Chowdhury	University of California, Riverside, USA
Michael Rubinstein	Google, USA
Olga Russakovsky	Princeton University, USA

Mathieu Salzmann	École polytechnique fédérale de Lausanne, Switzerland
Dimitris Samaras	Stony Brook University, USA
Aswin Sankaranarayanan	Carnegie Mellon University, USA
Imari Sato	National Institute of Informatics, Japan
Yoichi Sato	University of Tokyo, Japan
Shin'ichi Satoh	National Institute of Informatics, Japan
Walter Scheirer	University of Notre Dame, USA
Bernt Schiele	Max Planck Institute for Informatics, Germany
Konrad Schindler	ETH Zurich, Switzerland
Cordelia Schmid	Inria & Google, France
Alexander Schwing	University of Illinois at Urbana-Champaign, USA
Nicu Sebe	University of Trento, Italy
Greg Shakhnarovich	Toyota Technological Institute at Chicago, USA
Eli Shechtman	Adobe Research, USA
Humphrey Shi	University of Oregon & University of Illinois at Urbana-Champaign & Picsart AI Research, USA
Jianbo Shi	University of Pennsylvania, USA
Roy Shilkrot	Massachusetts Institute of Technology, USA
Mike Zheng Shou	National University of Singapore, Singapore
Kaleem Siddiqi	McGill University, Canada
Richa Singh	Indian Institute of Technology Jodhpur, India
Greg Slabaugh	Queen Mary University of London, UK
Cees Snoek	University of Amsterdam, The Netherlands
Yale Song	Facebook AI Research, USA
Yi-Zhe Song	University of Surrey, UK
Bjorn Stenger	Rakuten Institute of Technology
Abby Stylianou	Saint Louis University, USA
Akihiro Sugimoto	National Institute of Informatics, Japan
Chen Sun	Brown University, USA
Deqing Sun	Google, USA
Kalyan Sunkavalli	Adobe Research, USA
Ying Tai	Tencent YouTu Lab, China
Ayellet Tal	Technion – Israel Institute of Technology, Israel
Ping Tan	Simon Fraser University, Canada
Siyu Tang	ETH Zurich, Switzerland
Chi-Keung Tang	Hong Kong University of Science and Technology, Hong Kong, China
Radu Timofte	University of Würzburg, Germany & ETH Zurich, Switzerland
Federico Tombari	Google, Switzerland & Technical University of Munich, Germany

James Tompkin	Brown University, USA
Lorenzo Torresani	Dartmouth College, USA
Alexander Toshev	Apple, USA
Du Tran	Facebook AI Research, USA
Anh T. Tran	VinAI, Vietnam
Zhuowen Tu	University of California, San Diego, USA
Georgios Tzimiropoulos	Queen Mary University of London, UK
Jasper Uijlings	Google Research, Switzerland
Jan C. van Gemert	Delft University of Technology, The Netherlands
Gul Varol	Ecole des Ponts ParisTech, France
Nuno Vasconcelos	University of California, San Diego, USA
Mayank Vatsa	Indian Institute of Technology Jodhpur, India
Ashok Veeraraghavan	Rice University, USA
Jakob Verbeek	Facebook AI Research, France
Carl Vondrick	Columbia University, USA
Ruiping Wang	Institute of Computing Technology, Chinese Academy of Sciences, China
Xinchao Wang	National University of Singapore, Singapore
Liwei Wang	The Chinese University of Hong Kong, Hong Kong, China
Chaohui Wang	Université Paris-Est, France
Xiaolong Wang	University of California, San Diego, USA
Christian Wolf	NAVER LABS Europe, France
Tao Xiang	University of Surrey, UK
Saining Xie	Facebook AI Research, USA
Cihang Xie	University of California, Santa Cruz, USA
Zeki Yalniz	Facebook, USA
Ming-Hsuan Yang	University of California, Merced, USA
Angela Yao	National University of Singapore, Singapore
Shaodi You	University of Amsterdam, The Netherlands
Stella X. Yu	University of California, Berkeley, USA
Junsong Yuan	State University of New York at Buffalo, USA
Stefanos Zafeiriou	Imperial College London, UK
Amir Zamir	École polytechnique fédérale de Lausanne, Switzerland
Lei Zhang	Alibaba & Hong Kong Polytechnic University, Hong Kong, China
Lei Zhang	International Digital Economy Academy (IDEA), China
Pengchuan Zhang	Meta AI, USA
Bolei Zhou	University of California, Los Angeles, USA
Yuke Zhu	University of Texas at Austin, USA

Todd Zickler | Harvard University, USA
Wangmeng Zuo | Harbin Institute of Technology, China

Technical Program Committee

Davide Abati
Soroush Abbasi Koohpayegani
Amos L. Abbott
Rameen Abdal
Rabab Abdelfattah
Sahar Abdelnabi
Hassan Abu Alhaija
Abulikemu Abuduweili
Ron Abutbul
Hanno Ackermann
Aikaterini Adam
Kamil Adamczewski
Ehsan Adeli
Vida Adeli
Donald Adjeroh
Arman Afrasiyabi
Akshay Agarwal
Sameer Agarwal
Abhinav Agarwalla
Vaibhav Aggarwal
Sara Aghajanzadeh
Susmit Agrawal
Antonio Agudo
Touqeer Ahmad
Sk Miraj Ahmed
Chaitanya Ahuja
Nilesh A. Ahuja
Abhishek Aich
Shubhra Aich
Noam Aigerman
Arash Akbarinia
Peri Akiva
Derya Akkaynak
Emre Aksan
Arjun R. Akula
Yuval Alaluf
Stephan Alaniz
Paul Albert
Cenek Albl
Filippo Aleotti
Konstantinos P. Alexandridis
Motasem Alfarra
Mohsen Ali
Thiemo Alldieck
Hadi Alzayer
Liang An
Shan An
Yi An
Zhulin An
Dongsheng An
Jie An
Xiang An
Saket Anand
Cosmin Ancuti
Juan Andrade-Cetto
Alexander Andreopoulos
Bjoern Andres
Jerone T. A. Andrews
Shivangi Aneja
Anelia Angelova
Dragomir Anguelov
Rushil Anirudh
Oron Anschel
Rao Muhammad Anwer
Djamila Aouada
Evlampios Apostolidis
Srikar Appalaraju
Nikita Araslanov
Andre Araujo
Eric Arazo
Dawit Mureja Argaw
Anurag Arnab
Aditya Arora
Chetan Arora
Sunpreet S. Arora
Alexey Artemov
Muhammad Asad
Kumar Ashutosh
Sinem Aslan
Vishal Asnani
Mahmoud Assran
Amir Atapour-Abarghouei
Nikos Athanasiou
Ali Athar
ShahRukh Athar
Sara Atito
Souhaib Attaiki
Matan Atzmon
Mathieu Aubry
Nicolas Audebert
Tristan T. Aumentado-Armstrong
Melinos Averkiou
Yannis Avrithis
Stephane Ayache
Mehmet Aygün
Seyed Mehdi Ayyoubzadeh
Hossein Azizpour
George Azzopardi
Mallikarjun B. R.
Yunhao Ba
Abhishek Badki
Seung-Hwan Bae
Seung-Hwan Baek
Seungryul Baek
Piyush Nitin Bagad
Shai Bagon
Gaetan Bahl
Shikhar Bahl
Sherwin Bahmani
Haoran Bai
Lei Bai
Jiawang Bai
Haoyue Bai
Jinbin Bai
Xiang Bai
Xuyang Bai

Yang Bai
Yuanchao Bai
Ziqian Bai
Sungyong Baik
Kevin Bailly
Max Bain
Federico Baldassarre
Wele Gedara Chaminda Bandara
Biplab Banerjee
Pratyay Banerjee
Sandipan Banerjee
Jihwan Bang
Antyanta Bangunharcana
Aayush Bansal
Ankan Bansal
Siddhant Bansal
Wentao Bao
Zhipeng Bao
Amir Bar
Manel Baradad Jurjo
Lorenzo Baraldi
Danny Barash
Daniel Barath
Connelly Barnes
Ioan Andrei Bârsan
Steven Basart
Dina Bashkirova
Chaim Baskin
Peyman Bateni
Anil Batra
Sebastiano Battiato
Ardhendu Behera
Harkirat Behl
Jens Behley
Vasileios Belagiannis
Boulbaba Ben Amor
Emanuel Ben Baruch
Abdessamad Ben Hamza
Gil Ben-Artzi
Assia Benbihi
Fabian Benitez-Quiroz
Guy Ben-Yosef
Philipp Benz
Alexander W. Bergman
Urs Bergmann
Jesus Bermudez-Cameo
Stefano Berretti
Gedas Bertasius
Zachary Bessinger
Petra Bevandić
Matthew Beveridge
Lucas Beyer
Yash Bhalgat
Suvaansh Bhambri
Samarth Bharadwaj
Gaurav Bharaj
Aparna Bharati
Bharat Lal Bhatnagar
Uttaran Bhattacharya
Apratim Bhattacharyya
Brojeshwar Bhowmick
Ankan Kumar Bhunia
Ayan Kumar Bhunia
Qi Bi
Sai Bi
Michael Bi Mi
Gui-Bin Bian
Jia-Wang Bian
Shaojun Bian
Pia Bideau
Mario Bijelic
Hakan Bilen
Guillaume-Alexandre Bilodeau
Alexander Binder
Tolga Birdal
Vighnesh N. Birodkar
Sandika Biswas
Andreas Blattmann
Janusz Bobulski
Giuseppe Boccignone
Vishnu Boddeti
Navaneeth Bodla
Moritz Böhle
Aleksei Bokhovkin
Sam Bond-Taylor
Vivek Boominathan
Shubhankar Borse
Mark Boss
Andrea Bottino
Adnane Boukhayma
Fadi Boutros
Nicolas C. Boutry
Richard S. Bowen
Ivaylo Boyadzhiev
Aidan Boyd
Yuri Boykov
Aljaz Bozic
Behzad Bozorgtabar
Eric Brachmann
Samarth Brahmbhatt
Gustav Bredell
Francois Bremond
Joel Brogan
Andrew Brown
Thomas Brox
Marcus A. Brubaker
Robert-Jan Bruintjes
Yuqi Bu
Anders G. Buch
Himanshu Buckchash
Mateusz Buda
Ignas Budvytis
José M. Buenaposada
Marcel C. Bühler
Tu Bui
Adrian Bulat
Hannah Bull
Evgeny Burnaev
Andrei Bursuc
Benjamin Busam
Sergey N. Buzykanov
Wonmin Byeon
Fabian Caba
Martin Cadik
Guanyu Cai
Minjie Cai
Qing Cai
Zhongang Cai
Qi Cai
Yancheng Cai
Shen Cai
Han Cai
Jiarui Cai

Bowen Cai
Mu Cai
Qin Cai
Ruojin Cai
Weidong Cai
Weiwei Cai
Yi Cai
Yujun Cai
Zhiping Cai
Akin Caliskan
Lilian Calvet
Baris Can Cam
Necati Cihan Camgoz
Tommaso Campari
Dylan Campbell
Ziang Cao
Ang Cao
Xu Cao
Zhiwen Cao
Shengcao Cao
Song Cao
Weipeng Cao
Xiangyong Cao
Xiaochun Cao
Yue Cao
Yunhao Cao
Zhangjie Cao
Jiale Cao
Yang Cao
Jiajiong Cao
Jie Cao
Jinkun Cao
Lele Cao
Yulong Cao
Zhiguo Cao
Chen Cao
Razvan Caramalau
Marlène Careil
Gustavo Carneiro
Joao Carreira
Dan Casas
Paola Cascante-Bonilla
Angela Castillo
Francisco M. Castro
Pedro Castro
Luca Cavalli
George J. Cazenavette
Oya Celiktutan
Hakan Cevikalp
Sri Harsha C. H.
Sungmin Cha
Geonho Cha
Menglei Chai
Lucy Chai
Yuning Chai
Zenghao Chai
Anirban Chakraborty
Deep Chakraborty
Rudrasis Chakraborty
Souradeep Chakraborty
Kelvin C. K. Chan
Chee Seng Chan
Paramanand Chandramouli
Arjun Chandrasekaran
Kenneth Chaney
Dongliang Chang
Huiwen Chang
Peng Chang
Xiaojun Chang
Jia-Ren Chang
Hyung Jin Chang
Hyun Sung Chang
Ju Yong Chang
Li-Jen Chang
Qi Chang
Wei-Yi Chang
Yi Chang
Nadine Chang
Hanqing Chao
Pradyumna Chari
Dibyadip Chatterjee
Chiranjoy Chattopadhyay
Siddhartha Chaudhuri
Zhengping Che
Gal Chechik
Lianggangxu Chen
Qi Alfred Chen
Brian Chen
Bor-Chun Chen
Bo-Hao Chen
Bohong Chen
Bin Chen
Ziliang Chen
Cheng Chen
Chen Chen
Chaofeng Chen
Xi Chen
Haoyu Chen
Xuanhong Chen
Wei Chen
Qiang Chen
Shi Chen
Xianyu Chen
Chang Chen
Changhuai Chen
Hao Chen
Jie Chen
Jianbo Chen
Jingjing Chen
Jun Chen
Kejiang Chen
Mingcai Chen
Nenglun Chen
Qifeng Chen
Ruoyu Chen
Shu-Yu Chen
Weidong Chen
Weijie Chen
Weikai Chen
Xiang Chen
Xiuyi Chen
Xingyu Chen
Yaofo Chen
Yueting Chen
Yu Chen
Yunjin Chen
Yuntao Chen
Yun Chen
Zhenfang Chen
Zhuangzhuang Chen
Chu-Song Chen
Xiangyu Chen
Zhuo Chen
Chaoqi Chen
Shizhe Chen

Xiaotong Chen
Xiaozhi Chen
Dian Chen
Defang Chen
Dingfan Chen
Ding-Jie Chen
Ee Heng Chen
Tao Chen
Yixin Chen
Wei-Ting Chen
Lin Chen
Guang Chen
Guangyi Chen
Guanying Chen
Guangyao Chen
Hwann-Tzong Chen
Junwen Chen
Jiacheng Chen
Jianxu Chen
Hui Chen
Kai Chen
Kan Chen
Kevin Chen
Kuan-Wen Chen
Weihua Chen
Zhang Chen
Liang-Chieh Chen
Lele Chen
Liang Chen
Fanglin Chen
Zehui Chen
Minghui Chen
Minghao Chen
Xiaokang Chen
Qian Chen
Jun-Cheng Chen
Qi Chen
Qingcai Chen
Richard J. Chen
Runnan Chen
Rui Chen
Shuo Chen
Sentao Chen
Shaoyu Chen
Shixing Chen
Shuai Chen
Shuya Chen
Sizhe Chen
Simin Chen
Shaoxiang Chen
Zitian Chen
Tianlong Chen
Tianshui Chen
Min-Hung Chen
Xiangning Chen
Xin Chen
Xinghao Chen
Xuejin Chen
Xu Chen
Xuxi Chen
Yunlu Chen
Yanbei Chen
Yuxiao Chen
Yun-Chun Chen
Yi-Ting Chen
Yi-Wen Chen
Yinbo Chen
Yiran Chen
Yuanhong Chen
Yubei Chen
Yuefeng Chen
Yuhua Chen
Yukang Chen
Zerui Chen
Zhaoyu Chen
Zhen Chen
Zhenyu Chen
Zhi Chen
Zhiwei Chen
Zhixiang Chen
Long Chen
Bowen Cheng
Jun Cheng
Yi Cheng
Jingchun Cheng
Lechao Cheng
Xi Cheng
Yuan Cheng
Ho Kei Cheng
Kevin Ho Man Cheng
Jiacheng Cheng
Kelvin B. Cheng
Li Cheng
Mengjun Cheng
Zhen Cheng
Qingrong Cheng
Tianheng Cheng
Harry Cheng
Yihua Cheng
Yu Cheng
Ziheng Cheng
Soon Yau Cheong
Anoop Cherian
Manuela Chessa
Zhixiang Chi
Naoki Chiba
Julian Chibane
Kashyap Chitta
Tai-Yin Chiu
Hsu-kuang Chiu
Wei-Chen Chiu
Sungmin Cho
Donghyeon Cho
Hyeon Cho
Yooshin Cho
Gyusang Cho
Jang Hyun Cho
Seungju Cho
Nam Ik Cho
Sunghyun Cho
Hanbyel Cho
Jaesung Choe
Jooyoung Choi
Chiho Choi
Changwoon Choi
Jongwon Choi
Myungsub Choi
Dooseop Choi
Jonghyun Choi
Jinwoo Choi
Jun Won Choi
Min-Kook Choi
Hongsuk Choi
Janghoon Choi
Yoon-Ho Choi

Yukyung Choi
Jaegul Choo
Ayush Chopra
Siddharth Choudhary
Subhabrata Choudhury
Vasileios Choutas
Ka-Ho Chow
Pinaki Nath Chowdhury
Sammy Christen
Anders Christensen
Grigorios Chrysos
Hang Chu
Wen-Hsuan Chu
Peng Chu
Qi Chu
Ruihang Chu
Wei-Ta Chu
Yung-Yu Chuang
Sanghyuk Chun
Se Young Chun
Antonio Cinà
Ramazan Gokberk Cinbis
Javier Civera
Albert Clapés
Ronald Clark
Brian S. Clipp
Felipe Codevilla
Daniel Coelho de Castro
Niv Cohen
Forrester Cole
Maxwell D. Collins
Robert T. Collins
Marc Comino Trinidad
Runmin Cong
Wenyan Cong
Maxime Cordy
Marcella Cornia
Enric Corona
Huseyin Coskun
Luca Cosmo
Dragos Costea
Davide Cozzolino
Arun C. S. Kumar
Aiyu Cui
Qiongjie Cui
Quan Cui
Shuhao Cui
Yiming Cui
Ying Cui
Zijun Cui
Jiali Cui
Jiequan Cui
Yawen Cui
Zhen Cui
Zhaopeng Cui
Jack Culpepper
Xiaodong Cun
Ross Cutler
Adam Czajka
Ali Dabouei
Konstantinos M. Dafnis
Manuel Dahnert
Tao Dai
Yuchao Dai
Bo Dai
Mengyu Dai
Hang Dai
Haixing Dai
Peng Dai
Pingyang Dai
Qi Dai
Qiyu Dai
Yutong Dai
Naser Damer
Zhiyuan Dang
Mohamed Daoudi
Ayan Das
Abir Das
Debasmit Das
Deepayan Das
Partha Das
Sagnik Das
Soumi Das
Srijan Das
Swagatam Das
Avijit Dasgupta
Jim Davis
Adrian K. Davison
Homa Davoudi
Laura Daza
Matthias De Lange
Shalini De Mello
Marco De Nadai
Christophe De Vleeschouwer
Alp Dener
Boyang Deng
Congyue Deng
Bailin Deng
Yong Deng
Ye Deng
Zhuo Deng
Zhijie Deng
Xiaoming Deng
Jiankang Deng
Jinhong Deng
Jingjing Deng
Liang-Jian Deng
Siqi Deng
Xiang Deng
Xueqing Deng
Zhongying Deng
Karan Desai
Jean-Emmanuel Deschaud
Aniket Anand Deshmukh
Neel Dey
Helisa Dhamo
Prithviraj Dhar
Amaya Dharmasiri
Yan Di
Xing Di
Ousmane A. Dia
Haiwen Diao
Xiaolei Diao
Gonçalo José Dias Pais
Abdallah Dib
Anastasios Dimou
Changxing Ding
Henghui Ding
Guodong Ding
Yaqing Ding
Shuangrui Ding
Yuhang Ding
Yikang Ding
Shouhong Ding

Haisong Ding
Hui Ding
Jiahao Ding
Jian Ding
Jian-Jiun Ding
Shuxiao Ding
Tianyu Ding
Wenhao Ding
Yuqi Ding
Yi Ding
Yuzhen Ding
Zhengming Ding
Tan Minh Dinh
Vu Dinh
Christos Diou
Mandar Dixit
Bao Gia Doan
Khoa D. Doan
Dzung Anh Doan
Debi Prosad Dogra
Nehal Doiphode
Chengdong Dong
Bowen Dong
Zhenxing Dong
Hang Dong
Xiaoyi Dong
Haoye Dong
Jiangxin Dong
Shichao Dong
Xuan Dong
Zhen Dong
Shuting Dong
Jing Dong
Li Dong
Ming Dong
Nanqing Dong
Qiulei Dong
Runpei Dong
Siyan Dong
Tian Dong
Wei Dong
Xiaomeng Dong
Xin Dong
Xingbo Dong
Yuan Dong
Samuel Dooley
Gianfranco Doretto
Michael Dorkenwald
Keval Doshi
Zhaopeng Dou
Xiaotian Dou
Hazel Doughty
Ahmad Droby
Iddo Drori
Jie Du
Yong Du
Dawei Du
Dong Du
Ruoyi Du
Yuntao Du
Xuefeng Du
Yilun Du
Yuming Du
Radhika Dua
Haodong Duan
Jiafei Duan
Kaiwen Duan
Peiqi Duan
Ye Duan
Haoran Duan
Jiali Duan
Amanda Duarte
Abhimanyu Dubey
Shiv Ram Dubey
Florian Dubost
Lukasz Dudziak
Shivam Duggal
Justin M. Dulay
Matteo Dunnhofer
Chi Nhan Duong
Thibaut Durand
Mihai Dusmanu
Ujjal Kr Dutta
Debidatta Dwibedi
Isht Dwivedi
Sai Kumar Dwivedi
Takeharu Eda
Mark Edmonds
Alexei A. Efros
Thibaud Ehret
Max Ehrlich
Mahsa Ehsanpour
Iván Eichhardt
Farshad Einabadi
Marvin Eisenberger
Hazim Kemal Ekenel
Mohamed El Banani
Ismail Elezi
Moshe Eliasof
Alaa El-Nouby
Ian Endres
Francis Engelmann
Deniz Engin
Chanho Eom
Dave Epstein
Maria C. Escobar
Victor A. Escorcia
Carlos Esteves
Sungmin Eum
Bernard J. E. Evans
Ivan Evtimov
Fevziye Irem Eyiokur Yaman
Matteo Fabbri
Sébastien Fabbro
Gabriele Facciolo
Masud Fahim
Bin Fan
Hehe Fan
Deng-Ping Fan
Aoxiang Fan
Chen-Chen Fan
Qi Fan
Zhaoxin Fan
Haoqi Fan
Heng Fan
Hongyi Fan
Linxi Fan
Baojie Fan
Jiayuan Fan
Lei Fan
Quanfu Fan
Yonghui Fan
Yingruo Fan
Zhiwen Fan

Zicong Fan
Sean Fanello
Jiansheng Fang
Chaowei Fang
Yuming Fang
Jianwu Fang
Jin Fang
Qi Fang
Shancheng Fang
Tian Fang
Xianyong Fang
Gongfan Fang
Zhen Fang
Hui Fang
Jiemin Fang
Le Fang
Pengfei Fang
Xiaolin Fang
Yuxin Fang
Zhaoyuan Fang
Ammarah Farooq
Azade Farshad
Zhengcong Fei
Michael Felsberg
Wei Feng
Chen Feng
Fan Feng
Andrew Feng
Xin Feng
Zheyun Feng
Ruicheng Feng
Mingtao Feng
Qianyu Feng
Shangbin Feng
Chun-Mei Feng
Zunlei Feng
Zhiyong Feng
Martin Fergie
Mustansar Fiaz
Marco Fiorucci
Michael Firman
Hamed Firooz
Volker Fischer
Corneliu O. Florea
Georgios Floros
Wolfgang Foerstner
Gianni Franchi
Jean-Sebastien Franco
Simone Frintrop
Anna Fruehstueck
Changhong Fu
Chaoyou Fu
Cheng-Yang Fu
Chi-Wing Fu
Deqing Fu
Huan Fu
Jun Fu
Kexue Fu
Ying Fu
Jianlong Fu
Jingjing Fu
Qichen Fu
Tsu-Jui Fu
Xueyang Fu
Yang Fu
Yanwei Fu
Yonggan Fu
Wolfgang Fuhl
Yasuhisa Fujii
Kent Fujiwara
Marco Fumero
Takuya Funatomi
Isabel Funke
Dario Fuoli
Antonino Furnari
Matheus A. Gadelha
Akshay Gadi Patil
Adrian Galdran
Guillermo Gallego
Silvano Galliani
Orazio Gallo
Leonardo Galteri
Matteo Gamba
Yiming Gan
Sujoy Ganguly
Harald Ganster
Boyan Gao
Changxin Gao
Daiheng Gao
Difei Gao
Chen Gao
Fei Gao
Lin Gao
Wei Gao
Yiming Gao
Junyu Gao
Guangyu Ryan Gao
Haichang Gao
Hongchang Gao
Jialin Gao
Jin Gao
Jun Gao
Katelyn Gao
Mingchen Gao
Mingfei Gao
Pan Gao
Shangqian Gao
Shanghua Gao
Xitong Gao
Yunhe Gao
Zhanning Gao
Elena Garces
Nuno Cruz Garcia
Noa Garcia
Guillermo Garcia-Hernando
Isha Garg
Rahul Garg
Sourav Garg
Quentin Garrido
Stefano Gasperini
Kent Gauen
Chandan Gautam
Shivam Gautam
Paul Gay
Chunjiang Ge
Shiming Ge
Wenhang Ge
Yanhao Ge
Zheng Ge
Songwei Ge
Weifeng Ge
Yixiao Ge
Yuying Ge
Shijie Geng

Zhengyang Geng
Kyle A. Genova
Georgios Georgakis
Markos Georgopoulos
Marcel Geppert
Shabnam Ghadar
Mina Ghadimi Atigh
Deepti Ghadiyaram
Maani Ghaffari Jadidi
Sedigh Ghamari
Zahra Gharaee
Michaël Gharbi
Golnaz Ghiasi
Reza Ghoddoosian
Soumya Suvra Ghosal
Adhiraj Ghosh
Arthita Ghosh
Pallabi Ghosh
Soumyadeep Ghosh
Andrew Gilbert
Igor Gilitschenski
Jhony H. Giraldo
Andreu Girbau Xalabarder
Rohit Girdhar
Sharath Girish
Xavier Giro-i-Nieto
Raja Giryes
Thomas Gittings
Nikolaos Gkanatsios
Ioannis Gkioulekas
Abhiram Gnanasambandam
Aurele T. Gnanha
Clement L. J. C. Godard
Arushi Goel
Vidit Goel
Shubham Goel
Zan Gojcic
Aaron K. Gokaslan
Tejas Gokhale
S. Alireza Golestaneh
Thiago L. Gomes
Nuno Goncalves
Boqing Gong
Chen Gong
Yuanhao Gong
Guoqiang Gong
Jingyu Gong
Rui Gong
Yu Gong
Mingming Gong
Neil Zhenqiang Gong
Xun Gong
Yunye Gong
Yihong Gong
Cristina I. González
Nithin Gopalakrishnan Nair
Gaurav Goswami
Jianping Gou
Shreyank N. Gowda
Ankit Goyal
Helmut Grabner
Patrick L. Grady
Ben Graham
Eric Granger
Douglas R. Gray
Matej Grcić
David Griffiths
Jinjin Gu
Yun Gu
Shuyang Gu
Jianyang Gu
Fuqiang Gu
Jiatao Gu
Jindong Gu
Jiaqi Gu
Jinwei Gu
Jiaxin Gu
Geonmo Gu
Xiao Gu
Xinqian Gu
Xiuye Gu
Yuming Gu
Zhangxuan Gu
Dayan Guan
Junfeng Guan
Qingji Guan
Tianrui Guan
Shanyan Guan
Denis A. Gudovskiy
Ricardo Guerrero
Pierre-Louis Guhur
Jie Gui
Liangyan Gui
Liangke Gui
Benoit Guillard
Erhan Gundogdu
Manuel Günther
Jingcai Guo
Yuanfang Guo
Junfeng Guo
Chenqi Guo
Dan Guo
Hongji Guo
Jia Guo
Jie Guo
Minghao Guo
Shi Guo
Yanhui Guo
Yangyang Guo
Yuan-Chen Guo
Yilu Guo
Yiluan Guo
Yong Guo
Guangyu Guo
Haiyun Guo
Jinyang Guo
Jianyuan Guo
Pengsheng Guo
Pengfei Guo
Shuxuan Guo
Song Guo
Tianyu Guo
Qing Guo
Qiushan Guo
Wen Guo
Xiefan Guo
Xiaohu Guo
Xiaoqing Guo
Yufei Guo
Yuhui Guo
Yuliang Guo
Yunhui Guo
Yanwen Guo

Akshita Gupta
Ankush Gupta
Kamal Gupta
Kartik Gupta
Ritwik Gupta
Rohit Gupta
Siddharth Gururani
Fredrik K. Gustafsson
Abner Guzman Rivera
Vladimir Guzov
Matthew A. Gwilliam
Jung-Woo Ha
Marc Habermann
Isma Hadji
Christian Haene
Martin Hahner
Levente Hajder
Alexandros Haliassos
Emanuela Haller
Bumsub Ham
Abdullah J. Hamdi
Shreyas Hampali
Dongyoon Han
Chunrui Han
Dong-Jun Han
Dong-Sig Han
Guangxing Han
Zhizhong Han
Ruize Han
Jiaming Han
Jin Han
Ligong Han
Xian-Hua Han
Xiaoguang Han
Yizeng Han
Zhi Han
Zhenjun Han
Zhongyi Han
Jungong Han
Junlin Han
Kai Han
Kun Han
Sungwon Han
Songfang Han
Wei Han
Xiao Han
Xintong Han
Xinzhe Han
Yahong Han
Yan Han
Zongbo Han
Nicolai Hani
Rana Hanocka
Niklas Hanselmann
Nicklas A. Hansen
Hong Hanyu
Fusheng Hao
Yanbin Hao
Shijie Hao
Udith Haputhanthri
Mehrtash Harandi
Josh Harguess
Adam Harley
David M. Hart
Atsushi Hashimoto
Ali Hassani
Mohammed Hassanin
Yana Hasson
Joakim Bruslund Haurum
Bo He
Kun He
Chen He
Xin He
Fazhi He
Gaoqi He
Hao He
Haoyu He
Jiangpeng He
Hongliang He
Qian He
Xiangteng He
Xuming He
Yannan He
Yuhang He
Yang He
Xiangyu He
Nanjun He
Pan He
Sen He
Shengfeng He
Songtao He
Tao He
Tong He
Wei He
Xuehai He
Xiaoxiao He
Ying He
Yisheng He
Ziwen He
Peter Hedman
Felix Heide
Yacov Hel-Or
Paul Henderson
Philipp Henzler
Byeongho Heo
Jae-Pil Heo
Miran Heo
Sachini A. Herath
Stephane Herbin
Pedro Hermosilla Casajus
Monica Hernandez
Charles Herrmann
Roei Herzig
Mauricio Hess-Flores
Carlos Hinojosa
Tobias Hinz
Tsubasa Hirakawa
Chih-Hui Ho
Lam Si Tung Ho
Jennifer Hobbs
Derek Hoiem
Yannick Hold-Geoffroy
Aleksander Holynski
Cheeun Hong
Fa-Ting Hong
Hanbin Hong
Guan Zhe Hong
Danfeng Hong
Lanqing Hong
Xiaopeng Hong
Xin Hong
Jie Hong
Seungbum Hong
Cheng-Yao Hong
Seunghoon Hong

Yi Hong
Yuan Hong
Yuchen Hong
Anthony Hoogs
Maxwell C. Horton
Kazuhiro Hotta
Qibin Hou
Tingbo Hou
Junhui Hou
Ji Hou
Qiqi Hou
Rui Hou
Ruibing Hou
Zhi Hou
Henry Howard-Jenkins
Lukas Hoyer
Wei-Lin Hsiao
Chiou-Ting Hsu
Anthony Hu
Brian Hu
Yusong Hu
Hexiang Hu
Haoji Hu
Di Hu
Hengtong Hu
Haigen Hu
Lianyu Hu
Hanzhe Hu
Jie Hu
Junlin Hu
Shizhe Hu
Jian Hu
Zhiming Hu
Juhua Hu
Peng Hu
Ping Hu
Ronghang Hu
MengShun Hu
Tao Hu
Vincent Tao Hu
Xiaoling Hu
Xinting Hu
Xiaolin Hu
Xuefeng Hu
Xiaowei Hu
Yang Hu
Yueyu Hu
Zeyu Hu
Zhongyun Hu
Binh-Son Hua
Guoliang Hua
Yi Hua
Linzhi Huang
Qiusheng Huang
Bo Huang
Chen Huang
Hsin-Ping Huang
Ye Huang
Shuangping Huang
Zeng Huang
Buzhen Huang
Cong Huang
Heng Huang
Hao Huang
Qidong Huang
Huaibo Huang
Chaoqin Huang
Feihu Huang
Jiahui Huang
Jingjia Huang
Kun Huang
Lei Huang
Sheng Huang
Shuaiyi Huang
Siyu Huang
Xiaoshui Huang
Xiaoyang Huang
Yan Huang
Yihao Huang
Ying Huang
Ziling Huang
Xiaoke Huang
Yifei Huang
Haiyang Huang
Zhewei Huang
Jin Huang
Haibin Huang
Jiaxing Huang
Junjie Huang
Keli Huang
Lang Huang
Lin Huang
Luojie Huang
Mingzhen Huang
Shijia Huang
Shengyu Huang
Siyuan Huang
He Huang
Xiuyu Huang
Lianghua Huang
Yue Huang
Yaping Huang
Yuge Huang
Zehao Huang
Zeyi Huang
Zhiqi Huang
Zhongzhan Huang
Zilong Huang
Ziyuan Huang
Tianrui Hui
Zhuo Hui
Le Hui
Jing Huo
Junhwa Hur
Shehzeen S. Hussain
Chuong Minh Huynh
Seunghyun Hwang
Jaehui Hwang
Jyh-Jing Hwang
Sukjun Hwang
Soonmin Hwang
Wonjun Hwang
Rakib Hyder
Sangeek Hyun
Sarah Ibrahimi
Tomoki Ichikawa
Yerlan Idelbayev
A. S. M. Iftekhar
Masaaki Iiyama
Satoshi Ikehata
Sunghoon Im
Atul N. Ingle
Eldar Insafutdinov
Yani A. Ioannou
Radu Tudor Ionescu

Umar Iqbal
Go Irie
Muhammad Zubair Irshad
Ahmet Iscen
Berivan Isik
Ashraful Islam
Md Amirul Islam
Syed Islam
Mariko Isogawa
Vamsi Krishna K. Ithapu
Boris Ivanovic
Darshan Iyer
Sarah Jabbour
Ayush Jain
Nishant Jain
Samyak Jain
Vidit Jain
Vineet Jain
Priyank Jaini
Tomas Jakab
Mohammad A. A. K. Jalwana
Muhammad Abdullah Jamal
Hadi Jamali-Rad
Stuart James
Varun Jampani
Young Kyun Jang
YeongJun Jang
Yunseok Jang
Ronnachai Jaroensri
Bhavan Jasani
Krishna Murthy Jatavallabhula
Mojan Javaheripi
Syed A. Javed
Guillaume Jeanneret
Pranav Jeevan
Herve Jegou
Rohit Jena
Tomas Jenicek
Porter Jenkins
Simon Jenni
Hae-Gon Jeon
Sangryul Jeon
Boseung Jeong
Yoonwoo Jeong
Seong-Gyun Jeong
Jisoo Jeong
Allan D. Jepson
Ankit Jha
Sumit K. Jha
I-Hong Jhuo
Ge-Peng Ji
Chaonan Ji
Deyi Ji
Jingwei Ji
Wei Ji
Zhong Ji
Jiayi Ji
Pengliang Ji
Hui Ji
Mingi Ji
Xiaopeng Ji
Yuzhu Ji
Baoxiong Jia
Songhao Jia
Dan Jia
Shan Jia
Xiaojun Jia
Xiuyi Jia
Xu Jia
Menglin Jia
Wenqi Jia
Boyuan Jiang
Wenhao Jiang
Huaizu Jiang
Hanwen Jiang
Haiyong Jiang
Hao Jiang
Huajie Jiang
Huiqin Jiang
Haojun Jiang
Haobo Jiang
Junjun Jiang
Xingyu Jiang
Yangbangyan Jiang
Yu Jiang
Jianmin Jiang
Jiaxi Jiang
Jing Jiang
Kui Jiang
Li Jiang
Liming Jiang
Chiyu Jiang
Meirui Jiang
Chen Jiang
Peng Jiang
Tai-Xiang Jiang
Wen Jiang
Xinyang Jiang
Yifan Jiang
Yuming Jiang
Yingying Jiang
Zeren Jiang
ZhengKai Jiang
Zhenyu Jiang
Shuming Jiao
Jianbo Jiao
Licheng Jiao
Dongkwon Jin
Yeying Jin
Cheng Jin
Linyi Jin
Qing Jin
Taisong Jin
Xiao Jin
Xin Jin
Sheng Jin
Kyong Hwan Jin
Ruibing Jin
SouYoung Jin
Yueming Jin
Chenchen Jing
Longlong Jing
Taotao Jing
Yongcheng Jing
Younghyun Jo
Joakim Johnander
Jeff Johnson
Michael J. Jones
R. Kenny Jones
Rico Jonschkowski
Ameya Joshi
Sunghun Joung

Felix Juefei-Xu
Claudio R. Jung
Steffen Jung
Hari Chandana K.
Rahul Vigneswaran K.
Prajwal K. R.
Abhishek Kadian
Jhony Kaesemodel Pontes
Kumara Kahatapitiya
Anmol Kalia
Sinan Kalkan
Tarun Kalluri
Jaewon Kam
Sandesh Kamath
Meina Kan
Menelaos Kanakis
Takuhiro Kaneko
Di Kang
Guoliang Kang
Hao Kang
Jaeyeon Kang
Kyoungkook Kang
Li-Wei Kang
MinGuk Kang
Suk-Ju Kang
Zhao Kang
Yash Mukund Kant
Yueying Kao
Aupendu Kar
Konstantinos Karantzalos
Sezer Karaoglu
Navid Kardan
Sanjay Kariyappa
Leonid Karlinsky
Animesh Karnewar
Shyamgopal Karthik
Hirak J. Kashyap
Marc A. Kastner
Hirokatsu Kataoka
Angelos Katharopoulos
Hiroharu Kato
Kai Katsumata
Manuel Kaufmann
Chaitanya Kaul
Prakhar Kaushik
Yuki Kawana
Lei Ke
Lipeng Ke
Tsung-Wei Ke
Wei Ke
Petr Kellnhofer
Aniruddha Kembhavi
John Kender
Corentin Kervadec
Leonid Keselman
Daniel Keysers
Nima Khademi Kalantari
Taras Khakhulin
Samir Khaki
Muhammad Haris Khan
Qadeer Khan
Salman Khan
Subash Khanal
Vaishnavi M. Khindkar
Rawal Khirodkar
Saeed Khorram
Pirazh Khorramshahi
Kourosh Khoshelham
Ansh Khurana
Benjamin Kiefer
Jae Myung Kim
Junho Kim
Boah Kim
Hyeonseong Kim
Dong-Jin Kim
Dongwan Kim
Donghyun Kim
Doyeon Kim
Yonghyun Kim
Hyung-Il Kim
Hyunwoo Kim
Hyeongwoo Kim
Hyo Jin Kim
Hyunwoo J. Kim
Taehoon Kim
Jaeha Kim
Jiwon Kim
Jung Uk Kim
Kangyeol Kim
Eunji Kim
Daeha Kim
Dongwon Kim
Kunhee Kim
Kyungmin Kim
Junsik Kim
Min H. Kim
Namil Kim
Kookhoi Kim
Sanghyun Kim
Seongyeop Kim
Seungryong Kim
Saehoon Kim
Euyoung Kim
Guisik Kim
Sungyeon Kim
Sunnie S. Y. Kim
Taehun Kim
Tae Oh Kim
Won Hwa Kim
Seungwook Kim
YoungBin Kim
Youngeun Kim
Akisato Kimura
Furkan Osman Kınlı
Zsolt Kira
Hedvig Kjellström
Florian Kleber
Jan P. Klopp
Florian Kluger
Laurent Kneip
Byungsoo Ko
Muhammed Kocabas
A. Sophia Koepke
Kevin Koeser
Nick Kolkin
Nikos Kolotouros
Wai-Kin Adams Kong
Deying Kong
Caihua Kong
Youyong Kong
Shuyu Kong
Shu Kong
Tao Kong
Yajing Kong
Yu Kong

Zishang Kong
Theodora Kontogianni
Anton S. Konushin
Julian F. P. Kooij
Bruno Korbar
Giorgos Kordopatis-Zilos
Jari Korhonen
Adam Kortylewski
Denis Korzhenkov
Divya Kothandaraman
Suraj Kothawade
Iuliia Kotseruba
Satwik Kottur
Shashank Kotyan
Alexandros Kouris
Petros Koutras
Anna Kreshuk
Ranjay Krishna
Dilip Krishnan
Andrey Kuehlkamp
Hilde Kuehne
Jason Kuen
David Kügler
Arjan Kuijper
Anna Kukleva
Sumith Kulal
Viveka Kulharia
Akshay R. Kulkarni
Nilesh Kulkarni
Dominik Kulon
Abhinav Kumar
Akash Kumar
Suryansh Kumar
B. V. K. Vijaya Kumar
Pulkit Kumar
Ratnesh Kumar
Sateesh Kumar
Satish Kumar
Vijay Kumar B. G.
Nupur Kumari
Sudhakar Kumawat
Jogendra Nath Kundu
Hsien-Kai Kuo
Meng-Yu Jennifer Kuo
Vinod Kumar Kurmi
Yusuke Kurose
Keerthy Kusumam
Alina Kuznetsova
Henry Kvinge
Ho Man Kwan
Hyeokjun Kweon
Heeseung Kwon
Gihyun Kwon
Myung-Joon Kwon
Taesung Kwon
YoungJoong Kwon
Christos Kyrkou
Jorma Laaksonen
Yann Labbe
Zorah Laehner
Florent Lafarge
Hamid Laga
Manuel Lagunas
Shenqi Lai
Jian-Huang Lai
Zihang Lai
Mohamed I. Lakhal
Mohit Lamba
Meng Lan
Loic Landrieu
Zhiqiang Lang
Natalie Lang
Dong Lao
Yizhen Lao
Yingjie Lao
Issam Hadj Laradji
Gustav Larsson
Viktor Larsson
Zakaria Laskar
Stéphane Lathuilière
Chun Pong Lau
Rynson W. H. Lau
Hei Law
Justin Lazarow
Verica Lazova
Eric-Tuan Le
Hieu Le
Trung-Nghia Le
Mathias Lechner
Byeong-Uk Lee
Chen-Yu Lee
Che-Rung Lee
Chul Lee
Hong Joo Lee
Dongsoo Lee
Jiyoung Lee
Eugene Eu Tzuan Lee
Daeun Lee
Saehyung Lee
Jewook Lee
Hyungtae Lee
Hyunmin Lee
Jungbeom Lee
Joon-Young Lee
Jong-Seok Lee
Joonseok Lee
Junha Lee
Kibok Lee
Byung-Kwan Lee
Jangwon Lee
Jinho Lee
Jongmin Lee
Seunghyun Lee
Sohyun Lee
Minsik Lee
Dogyoon Lee
Seungmin Lee
Min Jun Lee
Sangho Lee
Sangmin Lee
Seungeun Lee
Seon-Ho Lee
Sungmin Lee
Sungho Lee
Sangyoun Lee
Vincent C. S. S. Lee
Jaeseong Lee
Yong Jae Lee
Chenyang Lei
Chenyi Lei
Jiahui Lei
Xinyu Lei
Yinjie Lei
Jiaxu Leng
Luziwei Leng

Jan E. Lenssen
Vincent Lepetit
Thomas Leung
María Leyva-Vallina
Xin Li
Yikang Li
Baoxin Li
Bin Li
Bing Li
Bowen Li
Changlin Li
Chao Li
Chongyi Li
Guanyue Li
Shuai Li
Jin Li
Dingquan Li
Dongxu Li
Yiting Li
Gang Li
Dian Li
Guohao Li
Haoang Li
Haoliang Li
Haoran Li
Hengduo Li
Huafeng Li
Xiaoming Li
Hanao Li
Hongwei Li
Ziqiang Li
Jisheng Li
Jiacheng Li
Jia Li
Jiachen Li
Jiahao Li
Jianwei Li
Jiazhi Li
Jie Li
Jing Li
Jingjing Li
Jingtao Li
Jun Li
Junxuan Li
Kai Li
Kailin Li
Kenneth Li
Kun Li
Kunpeng Li
Aoxue Li
Chenglong Li
Chenglin Li
Changsheng Li
Zhichao Li
Qiang Li
Yanyu Li
Zuoyue Li
Xiang Li
Xuelong Li
Fangda Li
Ailin Li
Liang Li
Chun-Guang Li
Daiqing Li
Dong Li
Guanbin Li
Guorong Li
Haifeng Li
Jianan Li
Jianing Li
Jiaxin Li
Ke Li
Lei Li
Lincheng Li
Liulei Li
Lujun Li
Linjie Li
Lin Li
Pengyu Li
Ping Li
Qiufu Li
Qingyong Li
Rui Li
Siyuan Li
Wei Li
Wenbin Li
Xiangyang Li
Xinyu Li
Xiujun Li
Xiu Li
Xu Li
Ya-Li Li
Yao Li
Yongjie Li
Yijun Li
Yiming Li
Yuezun Li
Yu Li
Yunheng Li
Yuqi Li
Zhe Li
Zeming Li
Zhen Li
Zhengqin Li
Zhimin Li
Jiefeng Li
Jinpeng Li
Chengze Li
Jianwu Li
Lerenhan Li
Shan Li
Suichan Li
Xiangtai Li
Yanjie Li
Yandong Li
Zhuoling Li
Zhenqiang Li
Manyi Li
Maosen Li
Ji Li
Minjun Li
Mingrui Li
Mengtian Li
Junyi Li
Nianyi Li
Bo Li
Xiao Li
Peihua Li
Peike Li
Peizhao Li
Peiliang Li
Qi Li
Ren Li
Runze Li
Shile Li

Sheng Li
Shigang Li
Shiyu Li
Shuang Li
Shasha Li
Shichao Li
Tianye Li
Yuexiang Li
Wei-Hong Li
Wanhua Li
Weihao Li
Weiming Li
Weixin Li
Wenbo Li
Wenshuo Li
Weijian Li
Yunan Li
Xirong Li
Xianhang Li
Xiaoyu Li
Xueqian Li
Xuanlin Li
Xianzhi Li
Yunqiang Li
Yanjing Li
Yansheng Li
Yawei Li
Yi Li
Yong Li
Yong-Lu Li
Yuhang Li
Yu-Jhe Li
Yuxi Li
Yunsheng Li
Yanwei Li
Zechao Li
Zejian Li
Zeju Li
Zekun Li
Zhaowen Li
Zheng Li
Zhenyu Li
Zhiheng Li
Zhi Li
Zhong Li
Zhuowei Li
Zhuowan Li
Zhuohang Li
Zizhang Li
Chen Li
Yuan-Fang Li
Dongze Lian
Xiaochen Lian
Zhouhui Lian
Long Lian
Qing Lian
Jin Lianbao
Jinxiu S. Liang
Dingkang Liang
Jiahao Liang
Jianming Liang
Jingyun Liang
Kevin J. Liang
Kaizhao Liang
Chen Liang
Jie Liang
Senwei Liang
Ding Liang
Jiajun Liang
Jian Liang
Kongming Liang
Siyuan Liang
Yuanzhi Liang
Zhengfa Liang
Mingfu Liang
Xiaodan Liang
Xuefeng Liang
Yuxuan Liang
Kang Liao
Liang Liao
Hong-Yuan Mark Liao
Wentong Liao
Haofu Liao
Yue Liao
Minghui Liao
Shengcai Liao
Ting-Hsuan Liao
Xin Liao
Yinghong Liao
Teck Yian Lim
Che-Tsung Lin
Chung-Ching Lin
Chen-Hsuan Lin
Cheng Lin
Chuming Lin
Chunyu Lin
Dahua Lin
Wei Lin
Zheng Lin
Huaijia Lin
Jason Lin
Jierui Lin
Jiaying Lin
Jie Lin
Kai-En Lin
Kevin Lin
Guangfeng Lin
Jiehong Lin
Feng Lin
Hang Lin
Kwan-Yee Lin
Ke Lin
Luojun Lin
Qinghong Lin
Xiangbo Lin
Yi Lin
Zudi Lin
Shijie Lin
Yiqun Lin
Tzu-Heng Lin
Ming Lin
Shaohui Lin
SongNan Lin
Ji Lin
Tsung-Yu Lin
Xudong Lin
Yancong Lin
Yen-Chen Lin
Yiming Lin
Yuewei Lin
Zhiqiu Lin
Zinan Lin
Zhe Lin
David B. Lindell
Zhixin Ling

Zhan Ling
Alexander Liniger
Venice Erin B. Liong
Joey Litalien
Or Litany
Roee Litman
Ron Litman
Jim Little
Dor Litvak
Shaoteng Liu
Shuaicheng Liu
Andrew Liu
Xian Liu
Shaohui Liu
Bei Liu
Bo Liu
Yong Liu
Ming Liu
Yanbin Liu
Chenxi Liu
Daqi Liu
Di Liu
Difan Liu
Dong Liu
Dongfang Liu
Daizong Liu
Xiao Liu
Fangyi Liu
Fengbei Liu
Fenglin Liu
Bin Liu
Yuang Liu
Ao Liu
Hong Liu
Hongfu Liu
Huidong Liu
Ziyi Liu
Feng Liu
Hao Liu
Jie Liu
Jialun Liu
Jiang Liu
Jing Liu
Jingya Liu
Jiaming Liu
Jun Liu
Juncheng Liu
Jiawei Liu
Hongyu Liu
Chuanbin Liu
Haotian Liu
Lingqiao Liu
Chang Liu
Han Liu
Liu Liu
Min Liu
Yingqi Liu
Aishan Liu
Bingyu Liu
Benlin Liu
Boxiao Liu
Chenchen Liu
Chuanjian Liu
Daqing Liu
Huan Liu
Haozhe Liu
Jiaheng Liu
Wei Liu
Jingzhou Liu
Jiyuan Liu
Lingbo Liu
Nian Liu
Peiye Liu
Qiankun Liu
Shenglan Liu
Shilong Liu
Wen Liu
Wenyu Liu
Weifeng Liu
Wu Liu
Xiaolong Liu
Yang Liu
Yanwei Liu
Yingcheng Liu
Yongfei Liu
Yihao Liu
Yu Liu
Yunze Liu
Ze Liu
Zhenhua Liu
Zhenguang Liu
Lin Liu
Lihao Liu
Pengju Liu
Xinhai Liu
Yunfei Liu
Meng Liu
Minghua Liu
Mingyuan Liu
Miao Liu
Peirong Liu
Ping Liu
Qingjie Liu
Ruoshi Liu
Risheng Liu
Songtao Liu
Xing Liu
Shikun Liu
Shuming Liu
Sheng Liu
Songhua Liu
Tongliang Liu
Weibo Liu
Weide Liu
Weizhe Liu
Wenxi Liu
Weiyang Liu
Xin Liu
Xiaobin Liu
Xudong Liu
Xiaoyi Liu
Xihui Liu
Xinchen Liu
Xingtong Liu
Xinpeng Liu
Xinyu Liu
Xianpeng Liu
Xu Liu
Xingyu Liu
Yongtuo Liu
Yahui Liu
Yangxin Liu
Yaoyao Liu
Yaojie Liu
Yuliang Liu

Yongcheng Liu
Yuan Liu
Yufan Liu
Yu-Lun Liu
Yun Liu
Yunfan Liu
Yuanzhong Liu
Zhuoran Liu
Zhen Liu
Zheng Liu
Zhijian Liu
Zhisong Liu
Ziquan Liu
Ziyu Liu
Zhihua Liu
Zechun Liu
Zhaoyang Liu
Zhengzhe Liu
Stephan Liwicki
Shao-Yuan Lo
Sylvain Lobry
Suhas Lohit
Vishnu Suresh Lokhande
Vincenzo Lomonaco
Chengjiang Long
Guodong Long
Fuchen Long
Shangbang Long
Yang Long
Zijun Long
Vasco Lopes
Antonio M. Lopez
Roberto Javier Lopez-Sastre
Tobias Lorenz
Javier Lorenzo-Navarro
Yujing Lou
Qian Lou
Xiankai Lu
Changsheng Lu
Huimin Lu
Yongxi Lu
Hao Lu
Hong Lu
Jiasen Lu
Juwei Lu
Fan Lu
Guangming Lu
Jiwen Lu
Shun Lu
Tao Lu
Xiaonan Lu
Yang Lu
Yao Lu
Yongchun Lu
Zhiwu Lu
Cheng Lu
Liying Lu
Guo Lu
Xuequan Lu
Yanye Lu
Yantao Lu
Yuhang Lu
Fujun Luan
Jonathon Luiten
Jovita Lukasik
Alan Lukezic
Jonathan Samuel Lumentut
Mayank Lunayach
Ao Luo
Canjie Luo
Chong Luo
Xu Luo
Grace Luo
Jun Luo
Katie Z. Luo
Tao Luo
Cheng Luo
Fangzhou Luo
Gen Luo
Lei Luo
Sihui Luo
Weixin Luo
Yan Luo
Xiaoyan Luo
Yong Luo
Yadan Luo
Hao Luo
Ruotian Luo
Mi Luo
Tiange Luo
Wenjie Luo
Wenhan Luo
Xiao Luo
Zhiming Luo
Zhipeng Luo
Zhengyi Luo
Diogo C. Luvizon
Zhaoyang Lv
Gengyu Lyu
Lingjuan Lyu
Jun Lyu
Yuanyuan Lyu
Youwei Lyu
Yueming Lyu
Bingpeng Ma
Chao Ma
Chongyang Ma
Congbo Ma
Chih-Yao Ma
Fan Ma
Lin Ma
Haoyu Ma
Hengbo Ma
Jianqi Ma
Jiawei Ma
Jiayi Ma
Kede Ma
Kai Ma
Lingni Ma
Lei Ma
Xu Ma
Ning Ma
Benteng Ma
Cheng Ma
Andy J. Ma
Long Ma
Zhanyu Ma
Zhiheng Ma
Qianli Ma
Shiqiang Ma
Sizhuo Ma
Shiqing Ma
Xiaolong Ma
Xinzhu Ma

Gautam B. Machiraju
Spandan Madan
Mathew Magimai-Doss
Luca Magri
Behrooz Mahasseni
Upal Mahbub
Siddharth Mahendran
Paridhi Maheshwari
Rishabh Maheshwary
Mohammed Mahmoud
Shishira R. R. Maiya
Sylwia Majchrowska
Arjun Majumdar
Puspita Majumdar
Orchid Majumder
Sagnik Majumder
Ilya Makarov
Farkhod F. Makhmudkhujaev
Yasushi Makihara
Ankur Mali
Mateusz Malinowski
Utkarsh Mall
Srikanth Malla
Clement Mallet
Dimitrios Mallis
Yunze Man
Dipu Manandhar
Massimiliano Mancini
Murari Mandal
Raunak Manekar
Karttikeya Mangalam
Puneet Mangla
Fabian Manhardt
Sivabalan Manivasagam
Fahim Mannan
Chengzhi Mao
Hanzi Mao
Jiayuan Mao
Junhua Mao
Zhiyuan Mao
Jiageng Mao
Yunyao Mao
Zhendong Mao
Alberto Marchisio
Diego Marcos
Riccardo Marin
Aram Markosyan
Renaud Marlet
Ricardo Marques
Miquel Martí i Rabadán
Diego Martin Arroyo
Niki Martinel
Brais Martinez
Julieta Martinez
Marc Masana
Tomohiro Mashita
Timothée Masquelier
Minesh Mathew
Tetsu Matsukawa
Marwan Mattar
Bruce A. Maxwell
Christoph Mayer
Mantas Mazeika
Pratik Mazumder
Scott McCloskey
Steven McDonagh
Ishit Mehta
Jie Mei
Kangfu Mei
Jieru Mei
Xiaoguang Mei
Givi Meishvili
Luke Melas-Kyriazi
Iaroslav Melekhov
Andres Mendez-Vazquez
Heydi Mendez-Vazquez
Matias Mendieta
Ricardo A. Mendoza-León
Chenlin Meng
Depu Meng
Rang Meng
Zibo Meng
Qingjie Meng
Qier Meng
Yanda Meng
Zihang Meng
Thomas Mensink
Fabian Mentzer
Christopher Metzler
Gregory P. Meyer
Vasileios Mezaris
Liang Mi
Lu Mi
Bo Miao
Changtao Miao
Zichen Miao
Qiguang Miao
Xin Miao
Zhongqi Miao
Frank Michel
Simone Milani
Ben Mildenhall
Roy V. Miles
Juhong Min
Kyle Min
Hyun-Seok Min
Weiqing Min
Yuecong Min
Zhixiang Min
Qi Ming
David Minnen
Aymen Mir
Deepak Mishra
Anand Mishra
Shlok K. Mishra
Niluthpol Mithun
Gaurav Mittal
Trisha Mittal
Daisuke Miyazaki
Kaichun Mo
Hong Mo
Zhipeng Mo
Davide Modolo
Abduallah A. Mohamed
Mohamed Afham Mohamed Aflal
Ron Mokady
Pavlo Molchanov
Davide Moltisanti
Liliane Momeni
Gianluca Monaci
Pascal Monasse
Ajoy Mondal
Tom Monnier

Aron Monszpart
Gyeongsik Moon
Suhong Moon
Taesup Moon
Sean Moran
Daniel Moreira
Pietro Morerio
Alexandre Morgand
Lia Morra
Ali Mosleh
Inbar Mosseri
Sayed Mohammad Mostafavi Isfahani
Saman Motamed
Ramy A. Mounir
Fangzhou Mu
Jiteng Mu
Norman Mu
Yasuhiro Mukaigawa
Ryan Mukherjee
Tanmoy Mukherjee
Yusuke Mukuta
Ravi Teja Mullapudi
Lea Müller
Matthias Müller
Martin Mundt
Nils Murrugarra-Llerena
Damien Muselet
Armin Mustafa
Muhammad Ferjad Naeem
Sauradip Nag
Hajime Nagahara
Pravin Nagar
Rajendra Nagar
Naveen Shankar Nagaraja
Varun Nagaraja
Tushar Nagarajan
Seungjun Nah
Gaku Nakano
Yuta Nakashima
Giljoo Nam
Seonghyeon Nam
Liangliang Nan
Yuesong Nan
Yeshwanth Napolean
Dinesh Reddy Narapureddy
Medhini Narasimhan
Supreeth Narasimhaswamy
Sriram Narayanan
Erickson R. Nascimento
Varun Nasery
K. L. Navaneet
Pablo Navarrete Michelini
Shant Navasardyan
Shah Nawaz
Nihal Nayak
Farhood Negin
Lukáš Neumann
Alejandro Newell
Evonne Ng
Kam Woh Ng
Tony Ng
Anh Nguyen
Tuan Anh Nguyen
Cuong Cao Nguyen
Ngoc Cuong Nguyen
Thanh Nguyen
Khoi Nguyen
Phi Le Nguyen
Phong Ha Nguyen
Tam Nguyen
Truong Nguyen
Anh Tuan Nguyen
Rang Nguyen
Thao Thi Phuong Nguyen
Van Nguyen Nguyen
Zhen-Liang Ni
Yao Ni
Shijie Nie
Xuecheng Nie
Yongwei Nie
Weizhi Nie
Ying Nie
Yinyu Nie
Kshitij N. Nikhal
Simon Niklaus
Xuefei Ning
Jifeng Ning
Yotam Nitzan
Di Niu
Shuaicheng Niu
Li Niu
Wei Niu
Yulei Niu
Zhenxing Niu
Albert No
Shohei Nobuhara
Nicoletta Noceti
Junhyug Noh
Sotiris Nousias
Slawomir Nowaczyk
Ewa M. Nowara
Valsamis Ntouskos
Gilberto Ochoa-Ruiz
Ferda Ofli
Jihyong Oh
Sangyun Oh
Youngtaek Oh
Hiroki Ohashi
Takahiro Okabe
Kemal Oksuz
Fumio Okura
Daniel Olmeda Reino
Matthew Olson
Carl Olsson
Roy Or-El
Alessandro Ortis
Guillermo Ortiz-Jimenez
Magnus Oskarsson
Ahmed A. A. Osman
Martin R. Oswald
Mayu Otani
Naima Otberdout
Cheng Ouyang
Jiahong Ouyang
Wanli Ouyang
Andrew Owens
Poojan B. Oza
Mete Ozay
A. Cengiz Oztireli
Gautam Pai
Tomas Pajdla
Umapada Pal

Simone Palazzo
Luca Palmieri
Bowen Pan
Hao Pan
Lili Pan
Tai-Yu Pan
Liang Pan
Chengwei Pan
Yingwei Pan
Xuran Pan
Jinshan Pan
Xinyu Pan
Liyuan Pan
Xingang Pan
Xingjia Pan
Zhihong Pan
Zizheng Pan
Priyadarshini Panda
Rameswar Panda
Rohit Pandey
Kaiyue Pang
Bo Pang
Guansong Pang
Jiangmiao Pang
Meng Pang
Tianyu Pang
Ziqi Pang
Omiros Pantazis
Andreas Panteli
Maja Pantic
Marina Paolanti
Joao P. Papa
Samuele Papa
Mike Papadakis
Dim P. Papadopoulos
George Papandreou
Constantin Pape
Toufiq Parag
Chethan Parameshwara
Shaifali Parashar
Alejandro Pardo
Rishubh Parihar
Sarah Parisot
JaeYoo Park
Gyeong-Moon Park
Hyojin Park
Hyoungseob Park
Jongchan Park
Jae Sung Park
Kiru Park
Chunghyun Park
Kwanyong Park
Sunghyun Park
Sungrae Park
Seongsik Park
Sanghyun Park
Sungjune Park
Taesung Park
Gaurav Parmar
Paritosh Parmar
Alvaro Parra
Despoina Paschalidou
Or Patashnik
Shivansh Patel
Pushpak Pati
Prashant W. Patil
Vaishakh Patil
Suvam Patra
Jay Patravali
Badri Narayana Patro
Angshuman Paul
Sudipta Paul
Rémi Pautrat
Nick E. Pears
Adithya Pediredla
Wenjie Pei
Shmuel Peleg
Latha Pemula
Bo Peng
Houwen Peng
Yue Peng
Liangzu Peng
Baoyun Peng
Jun Peng
Pai Peng
Sida Peng
Xi Peng
Yuxin Peng
Songyou Peng
Wei Peng
Weiqi Peng
Wen-Hsiao Peng
Pramuditha Perera
Juan C. Perez
Eduardo Pérez Pellitero
Juan-Manuel Perez-Rua
Federico Pernici
Marco Pesavento
Stavros Petridis
Ilya A. Petrov
Vladan Petrovic
Mathis Petrovich
Suzanne Petryk
Hieu Pham
Quang Pham
Khoi Pham
Tung Pham
Huy Phan
Stephen Phillips
Cheng Perng Phoo
David Picard
Marco Piccirilli
Georg Pichler
A. J. Piergiovanni
Vipin Pillai
Silvia L. Pintea
Giovanni Pintore
Robinson Piramuthu
Fiora Pirri
Theodoros Pissas
Fabio Pizzati
Benjamin Planche
Bryan Plummer
Matteo Poggi
Ashwini Pokle
Georgy E. Ponimatkin
Adrian Popescu
Stefan Popov
Nikola Popović
Ronald Poppe
Angelo Porrello
Michael Potter
Charalambos Poullis
Hadi Pouransari
Omid Poursaeed

Shraman Pramanick
Mantini Pranav
Dilip K. Prasad
Meghshyam Prasad
B. H. Pawan Prasad
Shitala Prasad
Prateek Prasanna
Ekta Prashnani
Derek S. Prijatelj
Luke Y. Prince
Véronique Prinet
Victor Adrian Prisacariu
James Pritts
Thomas Probst
Sergey Prokudin
Rita Pucci
Chi-Man Pun
Matthew Purri
Haozhi Qi
Lu Qi
Lei Qi
Xianbiao Qi
Yonggang Qi
Yuankai Qi
Siyuan Qi
Guocheng Qian
Hangwei Qian
Qi Qian
Deheng Qian
Shengsheng Qian
Wen Qian
Rui Qian
Yiming Qian
Shengju Qian
Shengyi Qian
Xuelin Qian
Zhenxing Qian
Nan Qiao
Xiaotian Qiao
Jing Qin
Can Qin
Siyang Qin
Hongwei Qin
Jie Qin
Minghai Qin
Yipeng Qin
Yongqiang Qin
Wenda Qin
Xuebin Qin
Yuzhe Qin
Yao Qin
Zhenyue Qin
Zhiwu Qing
Heqian Qiu
Jiayan Qiu
Jielin Qiu
Yue Qiu
Jiaxiong Qiu
Zhongxi Qiu
Shi Qiu
Zhaofan Qiu
Zhongnan Qu
Yanyun Qu
Kha Gia Quach
Yuhui Quan
Ruijie Quan
Mike Rabbat
Rahul Shekhar Rade
Filip Radenovic
Gorjan Radevski
Bogdan Raducanu
Francesco Ragusa
Shafin Rahman
Md Mahfuzur Rahman Siddiquee
Hossein Rahmani
Kiran Raja
Sivaramakrishnan Rajaraman
Jathushan Rajasegaran
Adnan Siraj Rakin
Michaël Ramamonjisoa
Chirag A. Raman
Shanmuganathan Raman
Vignesh Ramanathan
Vasili Ramanishka
Vikram V. Ramaswamy
Merey Ramazanova
Jason Rambach
Sai Saketh Rambhatla
Clément Rambour
Ashwin Ramesh Babu
Adín Ramírez Rivera
Arianna Rampini
Haoxi Ran
Aakanksha Rana
Aayush Jung Bahadur Rana
Kanchana N. Ranasinghe
Aneesh Rangnekar
Samrudhdhi B. Rangrej
Harsh Rangwani
Viresh Ranjan
Anyi Rao
Yongming Rao
Carolina Raposo
Michalis Raptis
Amir Rasouli
Vivek Rathod
Adepu Ravi Sankar
Avinash Ravichandran
Bharadwaj Ravichandran
Dripta S. Raychaudhuri
Adria Recasens
Simon Reiß
Davis Rempe
Daxuan Ren
Jiawei Ren
Jimmy Ren
Sucheng Ren
Dayong Ren
Zhile Ren
Dongwei Ren
Qibing Ren
Pengfei Ren
Zhenwen Ren
Xuqian Ren
Yixuan Ren
Zhongzheng Ren
Ambareesh Revanur
Hamed Rezazadegan Tavakoli
Rafael S. Rezende
Wonjong Rhee
Alexander Richard

Christian Richardt
Stephan R. Richter
Benjamin Riggan
Dominik Rivoir
Mamshad Nayeem Rizve
Joshua D. Robinson
Joseph Robinson
Chris Rockwell
Ranga Rodrigo
Andres C. Rodriguez
Carlos Rodriguez-Pardo
Marcus Rohrbach
Gemma Roig
Yu Rong
David A. Ross
Mohammad Rostami
Edward Rosten
Karsten Roth
Anirban Roy
Debaditya Roy
Shuvendu Roy
Ahana Roy Choudhury
Aruni Roy Chowdhury
Denys Rozumnyi
Shulan Ruan
Wenjie Ruan
Patrick Ruhkamp
Danila Rukhovich
Anian Ruoss
Chris Russell
Dan Ruta
Dawid Damian Rymarczyk
DongHun Ryu
Hyeonggon Ryu
Kwonyoung Ryu
Balasubramanian S.
Alexandre Sablayrolles
Mohammad Sabokrou
Arka Sadhu
Aniruddha Saha
Oindrila Saha
Pritish Sahu
Aneeshan Sain
Nirat Saini
Saurabh Saini
Takeshi Saitoh
Christos Sakaridis
Fumihiko Sakaue
Dimitrios Sakkos
Ken Sakurada
Parikshit V. Sakurikar
Rohit Saluja
Nermin Samet
Leo Sampaio Ferraz Ribeiro
Jorge Sanchez
Enrique Sanchez
Shengtian Sang
Anush Sankaran
Soubhik Sanyal
Nikolaos Sarafianos
Vishwanath Saragadam
István Sárándi
Saquib Sarfraz
Mert Bulent Sariyildiz
Anindya Sarkar
Pritam Sarkar
Paul-Edouard Sarlin
Hiroshi Sasaki
Takami Sato
Torsten Sattler
Ravi Kumar Satzoda
Axel Sauer
Stefano Savian
Artem Savkin
Manolis Savva
Gerald Schaefer
Simone Schaub-Meyer
Yoni Schirris
Samuel Schulter
Katja Schwarz
Jesse Scott
Sinisa Segvic
Constantin Marc Seibold
Lorenzo Seidenari
Matan Sela
Fadime Sener
Paul Hongsuck Seo
Kwanggyoon Seo
Hongje Seong
Dario Serez
Francesco Setti
Bryan Seybold
Mohamad Shahbazi
Shima Shahfar
Xinxin Shan
Caifeng Shan
Dandan Shan
Shawn Shan
Wei Shang
Jinghuan Shang
Jiaxiang Shang
Lei Shang
Sukrit Shankar
Ken Shao
Rui Shao
Jie Shao
Mingwen Shao
Aashish Sharma
Gaurav Sharma
Vivek Sharma
Abhishek Sharma
Yoli Shavit
Shashank Shekhar
Sumit Shekhar
Zhijie Shen
Fengyi Shen
Furao Shen
Jialie Shen
Jingjing Shen
Ziyi Shen
Linlin Shen
Guangyu Shen
Biluo Shen
Falong Shen
Jiajun Shen
Qiu Shen
Qiuhong Shen
Shuai Shen
Wang Shen
Yiqing Shen
Yunhang Shen
Siqi Shen
Bin Shen
Tianwei Shen

Xi Shen
Yilin Shen
Yuming Shen
Yucong Shen
Zhiqiang Shen
Lu Sheng
Yichen Sheng
Shivanand Venkanna Sheshappanavar
Shelly Sheynin
Baifeng Shi
Ruoxi Shi
Botian Shi
Hailin Shi
Jia Shi
Jing Shi
Shaoshuai Shi
Baoguang Shi
Boxin Shi
Hengcan Shi
Tianyang Shi
Xiaodan Shi
Yongjie Shi
Zhensheng Shi
Yinghuan Shi
Weiqi Shi
Wu Shi
Xuepeng Shi
Xiaoshuang Shi
Yujiao Shi
Zenglin Shi
Zhenmei Shi
Takashi Shibata
Meng-Li Shih
Yichang Shih
Hyunjung Shim
Dongseok Shim
Soshi Shimada
Inkyu Shin
Jinwoo Shin
Seungjoo Shin
Seungjae Shin
Koichi Shinoda
Suprosanna Shit
Palaiahnakote Shivakumara
Eli Shlizerman
Gaurav Shrivastava
Xiao Shu
Xiangbo Shu
Xiujun Shu
Yang Shu
Tianmin Shu
Jun Shu
Zhixin Shu
Bing Shuai
Maria Shugrina
Ivan Shugurov
Satya Narayan Shukla
Pranjay Shyam
Jianlou Si
Yawar Siddiqui
Alberto Signoroni
Pedro Silva
Jae-Young Sim
Oriane Siméoni
Martin Simon
Andrea Simonelli
Abhishek Singh
Ashish Singh
Dinesh Singh
Gurkirt Singh
Krishna Kumar Singh
Mannat Singh
Pravendra Singh
Rajat Vikram Singh
Utkarsh Singhal
Dipika Singhania
Vasu Singla
Harsh Sinha
Sudipta Sinha
Josef Sivic
Elena Sizikova
Geri Skenderi
Ivan Skorokhodov
Dmitriy Smirnov
Cameron Y. Smith
James S. Smith
Patrick Snape
Mattia Soldan
Hyeongseok Son
Sanghyun Son
Chuanbiao Song
Chen Song
Chunfeng Song
Dan Song
Dongjin Song
Hwanjun Song
Guoxian Song
Jiaming Song
Jie Song
Liangchen Song
Ran Song
Luchuan Song
Xibin Song
Li Song
Fenglong Song
Guoli Song
Guanglu Song
Zhenbo Song
Lin Song
Xinhang Song
Yang Song
Yibing Song
Rajiv Soundararajan
Hossein Souri
Cristovao Sousa
Riccardo Spezialetti
Leonidas Spinoulas
Michael W. Spratling
Deepak Sridhar
Srinath Sridhar
Gaurang Sriramanan
Vinkle Kumar Srivastav
Themos Stafylakis
Serban Stan
Anastasis Stathopoulos
Markus Steinberger
Jan Steinbrener
Sinisa Stekovic
Alexandros Stergiou
Gleb Sterkin
Rainer Stiefelhagen
Pierre Stock

Ombretta Strafforello
Julian Straub
Yannick Strümpler
Joerg Stueckler
Hang Su
Weijie Su
Jong-Chyi Su
Bing Su
Haisheng Su
Jinming Su
Yiyang Su
Yukun Su
Yuxin Su
Zhuo Su
Zhaoqi Su
Xiu Su
Yu-Chuan Su
Zhixun Su
Arulkumar Subramaniam
Akshayvarun Subramanya
A. Subramanyam
Swathikiran Sudhakaran
Yusuke Sugano
Masanori Suganuma
Yumin Suh
Yang Sui
Baochen Sun
Cheng Sun
Long Sun
Guolei Sun
Haoliang Sun
Haomiao Sun
He Sun
Hanqing Sun
Hao Sun
Lichao Sun
Jiachen Sun
Jiaming Sun
Jian Sun
Jin Sun
Jennifer J. Sun
Tiancheng Sun
Libo Sun
Peize Sun
Qianru Sun
Shanlin Sun
Yu Sun
Zhun Sun
Che Sun
Lin Sun
Tao Sun
Yiyou Sun
Chunyi Sun
Chong Sun
Weiwei Sun
Weixuan Sun
Xiuyu Sun
Yanan Sun
Zeren Sun
Zhaodong Sun
Zhiqing Sun
Minhyuk Sung
Jinli Suo
Simon Suo
Abhijit Suprem
Anshuman Suri
Saksham Suri
Joshua M. Susskind
Roman Suvorov
Gurumurthy Swaminathan
Robin Swanson
Paul Swoboda
Tabish A. Syed
Richard Szeliski
Fariborz Taherkhani
Yu-Wing Tai
Keita Takahashi
Walter Talbott
Gary Tam
Masato Tamura
Feitong Tan
Fuwen Tan
Shuhan Tan
Andong Tan
Bin Tan
Cheng Tan
Jianchao Tan
Lei Tan
Mingxing Tan
Xin Tan
Zichang Tan
Zhentao Tan
Kenichiro Tanaka
Masayuki Tanaka
Yushun Tang
Hao Tang
Jingqun Tang
Jinhui Tang
Kaihua Tang
Luming Tang
Lv Tang
Sheyang Tang
Shitao Tang
Siliang Tang
Shixiang Tang
Yansong Tang
Keke Tang
Chang Tang
Chenwei Tang
Jie Tang
Junshu Tang
Ming Tang
Peng Tang
Xu Tang
Yao Tang
Chen Tang
Fan Tang
Haoran Tang
Shengeng Tang
Yehui Tang
Zhipeng Tang
Ugo Tanielian
Chaofan Tao
Jiale Tao
Junli Tao
Renshuai Tao
An Tao
Guanhong Tao
Zhiqiang Tao
Makarand Tapaswi
Jean-Philippe G. Tarel
Juan J. Tarrio
Enzo Tartaglione
Keisuke Tateno
Zachary Teed

Ajinkya B. Tejankar
Bugra Tekin
Purva Tendulkar
Damien Teney
Minggui Teng
Chris Tensmeyer
Andrew Beng Jin Teoh
Philipp Terhörst
Kartik Thakral
Nupur Thakur
Kevin Thandiackal
Spyridon Thermos
Diego Thomas
William Thong
Yuesong Tian
Guanzhong Tian
Lin Tian
Shiqi Tian
Kai Tian
Meng Tian
Tai-Peng Tian
Zhuotao Tian
Shangxuan Tian
Tian Tian
Yapeng Tian
Yu Tian
Yuxin Tian
Leslie Ching Ow Tiong
Praveen Tirupattur
Garvita Tiwari
George Toderici
Antoine Toisoul
Aysim Toker
Tatiana Tommasi
Zhan Tong
Alessio Tonioni
Alessandro Torcinovich
Fabio Tosi
Matteo Toso
Hugo Touvron
Quan Hung Tran
Son Tran
Hung Tran
Ngoc-Trung Tran
Vinh Tran
Phong Tran
Giovanni Trappolini
Edith Tretschk
Subarna Tripathi
Shubhendu Trivedi
Eduard Trulls
Prune Truong
Thanh-Dat Truong
Tomasz Trzcinski
Sam Tsai
Yi-Hsuan Tsai
Ethan Tseng
Yu-Chee Tseng
Shahar Tsiper
Stavros Tsogkas
Shikui Tu
Zhigang Tu
Zhengzhong Tu
Richard Tucker
Sergey Tulyakov
Cigdem Turan
Daniyar Turmukhambetov
Victor G. Turrisi da Costa
Bartlomiej Twardowski
Christopher D. Twigg
Radim Tylecek
Mostofa Rafid Uddin
Md. Zasim Uddin
Kohei Uehara
Nicolas Ugrinovic
Youngjung Uh
Norimichi Ukita
Anwaar Ulhaq
Devesh Upadhyay
Paul Upchurch
Yoshitaka Ushiku
Yuzuko Utsumi
Mikaela Angelina Uy
Mohit Vaishnav
Pratik Vaishnavi
Jeya Maria Jose Valanarasu
Matias A. Valdenegro Toro
Diego Valsesia
Wouter Van Gansbeke
Nanne van Noord
Simon Vandenhende
Farshid Varno
Cristina Vasconcelos
Francisco Vasconcelos
Alex Vasilescu
Subeesh Vasu
Arun Balajee Vasudevan
Kanav Vats
Vaibhav S. Vavilala
Sagar Vaze
Javier Vazquez-Corral
Andrea Vedaldi
Olga Veksler
Andreas Velten
Sai H. Vemprala
Raviteja Vemulapalli
Shashanka Venkataramanan
Dor Verbin
Luisa Verdoliva
Manisha Verma
Yashaswi Verma
Constantin Vertan
Eli Verwimp
Deepak Vijaykeerthy
Pablo Villanueva
Ruben Villegas
Markus Vincze
Vibhav Vineet
Minh P. Vo
Huy V. Vo
Duc Minh Vo
Tomas Vojir
Igor Vozniak
Nicholas Vretos
Vibashan VS
Tuan-Anh Vu
Thang Vu
Mårten Wadenbäck
Neal Wadhwa
Aaron T. Walsman
Steven Walton
Jin Wan
Alvin Wan
Jia Wan

Jun Wan
Xiaoyue Wan
Fang Wan
Guowei Wan
Renjie Wan
Zhiqiang Wan
Ziyu Wan
Bastian Wandt
Dongdong Wang
Limin Wang
Haiyang Wang
Xiaobing Wang
Angtian Wang
Angelina Wang
Bing Wang
Bo Wang
Boyu Wang
Binghui Wang
Chen Wang
Chien-Yi Wang
Congli Wang
Qi Wang
Chengrui Wang
Rui Wang
Yiqun Wang
Cong Wang
Wenjing Wang
Dongkai Wang
Di Wang
Xiaogang Wang
Kai Wang
Zhizhong Wang
Fangjinhua Wang
Feng Wang
Hang Wang
Gaoang Wang
Guoqing Wang
Guangcong Wang
Guangzhi Wang
Hanqing Wang
Hao Wang
Haohan Wang
Haoran Wang
Hong Wang
Haotao Wang
Hu Wang
Huan Wang
Hua Wang
Hui-Po Wang
Hengli Wang
Hanyu Wang
Hongxing Wang
Jingwen Wang
Jialiang Wang
Jian Wang
Jianyi Wang
Jiashun Wang
Jiahao Wang
Tsun-Hsuan Wang
Xiaoqian Wang
Jinqiao Wang
Jun Wang
Jianzong Wang
Kaihong Wang
Ke Wang
Lei Wang
Lingjing Wang
Linnan Wang
Lin Wang
Liansheng Wang
Mengjiao Wang
Manning Wang
Nannan Wang
Peihao Wang
Jiayun Wang
Pu Wang
Qiang Wang
Qiufeng Wang
Qilong Wang
Qiangchang Wang
Qin Wang
Qing Wang
Ruocheng Wang
Ruibin Wang
Ruisheng Wang
Ruizhe Wang
Runqi Wang
Runzhong Wang
Wenxuan Wang
Sen Wang
Shangfei Wang
Shaofei Wang
Shijie Wang
Shiqi Wang
Zhibo Wang
Song Wang
Xinjiang Wang
Tai Wang
Tao Wang
Teng Wang
Xiang Wang
Tianren Wang
Tiantian Wang
Tianyi Wang
Fengjiao Wang
Wei Wang
Miaohui Wang
Suchen Wang
Siyue Wang
Yaoming Wang
Xiao Wang
Ze Wang
Biao Wang
Chaofei Wang
Dong Wang
Gu Wang
Guangrun Wang
Guangming Wang
Guo-Hua Wang
Haoqing Wang
Hesheng Wang
Huafeng Wang
Jinghua Wang
Jingdong Wang
Jingjing Wang
Jingya Wang
Jingkang Wang
Jiakai Wang
Junke Wang
Kuo Wang
Lichen Wang
Lizhi Wang
Longguang Wang
Mang Wang
Mei Wang

Min Wang
Peng-Shuai Wang
Run Wang
Shaoru Wang
Shuhui Wang
Tan Wang
Tiancai Wang
Tianqi Wang
Wenhai Wang
Wenzhe Wang
Xiaobo Wang
Xiudong Wang
Xu Wang
Yajie Wang
Yan Wang
Yuan-Gen Wang
Yingqian Wang
Yizhi Wang
Yulin Wang
Yu Wang
Yujie Wang
Yunhe Wang
Yuxi Wang
Yaowei Wang
Yiwei Wang
Zezheng Wang
Hongzhi Wang
Zhiqiang Wang
Ziteng Wang
Ziwei Wang
Zheng Wang
Zhenyu Wang
Binglu Wang
Zhongdao Wang
Ce Wang
Weining Wang
Weiyao Wang
Wenbin Wang
Wenguan Wang
Guangting Wang
Haolin Wang
Haiyan Wang
Huiyu Wang
Naiyan Wang
Jingbo Wang
Jinpeng Wang
Jiaqi Wang
Liyuan Wang
Lizhen Wang
Ning Wang
Wenqian Wang
Sheng-Yu Wang
Weimin Wang
Xiaohan Wang
Yifan Wang
Yi Wang
Yongtao Wang
Yizhou Wang
Zhuo Wang
Zhe Wang
Xudong Wang
Xiaofang Wang
Xinggang Wang
Xiaosen Wang
Xiaosong Wang
Xiaoyang Wang
Lijun Wang
Xinlong Wang
Xuan Wang
Xue Wang
Yangang Wang
Yaohui Wang
Yu-Chiang Frank Wang
Yida Wang
Yilin Wang
Yi Ru Wang
Yali Wang
Yinglong Wang
Yufu Wang
Yujiang Wang
Yuwang Wang
Yuting Wang
Yang Wang
Yu-Xiong Wang
Yixu Wang
Ziqi Wang
Zhicheng Wang
Zeyu Wang
Zhaowen Wang
Zhenyi Wang
Zhenzhi Wang
Zhijie Wang
Zhiyong Wang
Zhongling Wang
Zhuowei Wang
Zian Wang
Zifu Wang
Zihao Wang
Zirui Wang
Ziyan Wang
Wenxiao Wang
Zhen Wang
Zhepeng Wang
Zi Wang
Zihao W. Wang
Steven L. Waslander
Olivia Watkins
Daniel Watson
Silvan Weder
Dongyoon Wee
Dongming Wei
Tianyi Wei
Jia Wei
Dong Wei
Fangyun Wei
Longhui Wei
Mingqiang Wei
Xinyue Wei
Chen Wei
Donglai Wei
Pengxu Wei
Xing Wei
Xiu-Shen Wei
Wenqi Wei
Guoqiang Wei
Wei Wei
XingKui Wei
Xian Wei
Xingxing Wei
Yake Wei
Yuxiang Wei
Yi Wei
Luca Weihs
Michael Weinmann
Martin Weinmann

Congcong Wen
Chuan Wen
Jie Wen
Sijia Wen
Song Wen
Chao Wen
Xiang Wen
Zeyi Wen
Xin Wen
Yilin Wen
Yijia Weng
Shuchen Weng
Junwu Weng
Wenming Weng
Renliang Weng
Zhenyu Weng
Xinshuo Weng
Nicholas J. Westlake
Gordon Wetzstein
Lena M. Widin Klasén
Rick Wildes
Bryan M. Williams
Williem Williem
Ole Winther
Scott Wisdom
Alex Wong
Chau-Wai Wong
Kwan-Yee K. Wong
Yongkang Wong
Scott Workman
Marcel Worring
Michael Wray
Safwan Wshah
Xiang Wu
Aming Wu
Chongruo Wu
Cho-Ying Wu
Chunpeng Wu
Chenyan Wu
Ziyi Wu
Fuxiang Wu
Gang Wu
Haiping Wu
Huisi Wu
Jane Wu
Jialian Wu
Jing Wu
Jinjian Wu
Jianlong Wu
Xian Wu
Lifang Wu
Lifan Wu
Minye Wu
Qianyi Wu
Rongliang Wu
Rui Wu
Shiqian Wu
Shuzhe Wu
Shangzhe Wu
Tsung-Han Wu
Tz-Ying Wu
Ting-Wei Wu
Jiannan Wu
Zhiliang Wu
Yu Wu
Chenyun Wu
Dayan Wu
Dongxian Wu
Fei Wu
Hefeng Wu
Jianxin Wu
Weibin Wu
Wenxuan Wu
Wenhao Wu
Xiao Wu
Yicheng Wu
Yuanwei Wu
Yu-Huan Wu
Zhenxin Wu
Zhenyu Wu
Wei Wu
Peng Wu
Xiaohe Wu
Xindi Wu
Xinxing Wu
Xinyi Wu
Xingjiao Wu
Xiongwei Wu
Yangzheng Wu
Yanzhao Wu
Yawen Wu
Yong Wu
Yi Wu
Ying Nian Wu
Zhenyao Wu
Zhonghua Wu
Zongze Wu
Zuxuan Wu
Stefanie Wuhrer
Teng Xi
Jianing Xi
Fei Xia
Haifeng Xia
Menghan Xia
Yuanqing Xia
Zhihua Xia
Xiaobo Xia
Weihao Xia
Shihong Xia
Yan Xia
Yong Xia
Zhaoyang Xia
Zhihao Xia
Chuhua Xian
Yongqin Xian
Wangmeng Xiang
Fanbo Xiang
Tiange Xiang
Tao Xiang
Liuyu Xiang
Xiaoyu Xiang
Zhiyu Xiang
Aoran Xiao
Chunxia Xiao
Fanyi Xiao
Jimin Xiao
Jun Xiao
Taihong Xiao
Anqi Xiao
Junfei Xiao
Jing Xiao
Liang Xiao
Yang Xiao
Yuting Xiao
Yijun Xiao

Yao Xiao
Zeyu Xiao
Zhisheng Xiao
Zihao Xiao
Binhui Xie
Christopher Xie
Haozhe Xie
Jin Xie
Guo-Sen Xie
Hongtao Xie
Ming-Kun Xie
Tingting Xie
Chaohao Xie
Weicheng Xie
Xudong Xie
Jiyang Xie
Xiaohua Xie
Yuan Xie
Zhenyu Xie
Ning Xie
Xianghui Xie
Xiufeng Xie
You Xie
Yutong Xie
Fuyong Xing
Yifan Xing
Zhen Xing
Yuanjun Xiong
Jinhui Xiong
Weihua Xiong
Hongkai Xiong
Zhitong Xiong
Yuanhao Xiong
Yunyang Xiong
Yuwen Xiong
Zhiwei Xiong
Yuliang Xiu
An Xu
Chang Xu
Chenliang Xu
Chengming Xu
Chenshu Xu
Xiang Xu
Huijuan Xu
Zhe Xu
Jie Xu
Jingyi Xu
Jiarui Xu
Yinghao Xu
Kele Xu
Ke Xu
Li Xu
Linchuan Xu
Linning Xu
Mengde Xu
Mengmeng Frost Xu
Min Xu
Mingye Xu
Jun Xu
Ning Xu
Peng Xu
Runsheng Xu
Sheng Xu
Wenqiang Xu
Xiaogang Xu
Renzhe Xu
Kaidi Xu
Yi Xu
Chi Xu
Qiuling Xu
Baobei Xu
Feng Xu
Haohang Xu
Haofei Xu
Lan Xu
Mingze Xu
Songcen Xu
Weipeng Xu
Wenjia Xu
Wenju Xu
Xiangyu Xu
Xin Xu
Yinshuang Xu
Yixing Xu
Yuting Xu
Yanyu Xu
Zhenbo Xu
Zhiliang Xu
Zhiyuan Xu
Xiaohao Xu
Yanwu Xu
Yan Xu
Yiran Xu
Yifan Xu
Yufei Xu
Yong Xu
Zichuan Xu
Zenglin Xu
Zexiang Xu
Zhan Xu
Zheng Xu
Zhiwei Xu
Ziyue Xu
Shiyu Xuan
Hanyu Xuan
Fei Xue
Jianru Xue
Mingfu Xue
Qinghan Xue
Tianfan Xue
Chao Xue
Chuhui Xue
Nan Xue
Zhou Xue
Xiangyang Xue
Yuan Xue
Abhay Yadav
Ravindra Yadav
Kota Yamaguchi
Toshihiko Yamasaki
Kohei Yamashita
Chaochao Yan
Feng Yan
Kun Yan
Qingsen Yan
Qixin Yan
Rui Yan
Siming Yan
Xinchen Yan
Yaping Yan
Bin Yan
Qingan Yan
Shen Yan
Shipeng Yan
Xu Yan

Yan Yan
Yichao Yan
Zhaoyi Yan
Zike Yan
Zhiqiang Yan
Hongliang Yan
Zizheng Yan
Jiewen Yang
Anqi Joyce Yang
Shan Yang
Anqi Yang
Antoine Yang
Bo Yang
Baoyao Yang
Chenhongyi Yang
Dingkang Yang
De-Nian Yang
Dong Yang
David Yang
Fan Yang
Fengyu Yang
Fengting Yang
Fei Yang
Gengshan Yang
Heng Yang
Han Yang
Huan Yang
Yibo Yang
Jiancheng Yang
Jihan Yang
Jiawei Yang
Jiayu Yang
Jie Yang
Jinfa Yang
Jingkang Yang
Jinyu Yang
Cheng-Fu Yang
Ji Yang
Jianyu Yang
Kailun Yang
Tian Yang
Luyu Yang
Liang Yang
Li Yang
Michael Ying Yang
Yang Yang
Muli Yang
Le Yang
Qiushi Yang
Ren Yang
Ruihan Yang
Shuang Yang
Siyuan Yang
Su Yang
Shiqi Yang
Taojiannan Yang
Tianyu Yang
Lei Yang
Wanzhao Yang
Shuai Yang
William Yang
Wei Yang
Xiaofeng Yang
Xiaoshan Yang
Xin Yang
Xuan Yang
Xu Yang
Xingyi Yang
Xitong Yang
Jing Yang
Yanchao Yang
Wenming Yang
Yujiu Yang
Herb Yang
Jianfei Yang
Jinhui Yang
Chuanguang Yang
Guanglei Yang
Haitao Yang
Kewei Yang
Linlin Yang
Lijin Yang
Longrong Yang
Meng Yang
MingKun Yang
Sibei Yang
Shicai Yang
Tong Yang
Wen Yang
Xi Yang
Xiaolong Yang
Xue Yang
Yubin Yang
Ze Yang
Ziyi Yang
Yi Yang
Linjie Yang
Yuzhe Yang
Yiding Yang
Zhenpei Yang
Zhaohui Yang
Zhengyuan Yang
Zhibo Yang
Zongxin Yang
Hantao Yao
Mingde Yao
Rui Yao
Taiping Yao
Ting Yao
Cong Yao
Qingsong Yao
Quanming Yao
Xu Yao
Yuan Yao
Yao Yao
Yazhou Yao
Jiawen Yao
Shunyu Yao
Pew-Thian Yap
Sudhir Yarram
Rajeev Yasarla
Peng Ye
Botao Ye
Mao Ye
Fei Ye
Hanrong Ye
Jingwen Ye
Jinwei Ye
Jiarong Ye
Mang Ye
Meng Ye
Qi Ye
Qian Ye
Qixiang Ye
Junjie Ye

Sheng Ye
Nanyang Ye
Yufei Ye
Xiaoqing Ye
Ruolin Ye
Yousef Yeganeh
Chun-Hsiao Yeh
Raymond A. Yeh
Yu-Ying Yeh
Kai Yi
Chang Yi
Renjiao Yi
Xinping Yi
Peng Yi
Alper Yilmaz
Junho Yim
Hui Yin
Bangjie Yin
Jia-Li Yin
Miao Yin
Wenzhe Yin
Xuwang Yin
Ming Yin
Yu Yin
Aoxiong Yin
Kangxue Yin
Tianwei Yin
Wei Yin
Xianghua Ying
Rio Yokota
Tatsuya Yokota
Naoto Yokoya
Ryo Yonetani
Ki Yoon Yoo
Jinsu Yoo
Sunjae Yoon
Jae Shin Yoon
Jihun Yoon
Sung-Hoon Yoon
Ryota Yoshihashi
Yusuke Yoshiyasu
Chenyu You
Haoran You
Haoxuan You
Yang You
Quanzeng You
Tackgeun You
Kaichao You
Shan You
Xinge You
Yurong You
Baosheng Yu
Bei Yu
Haichao Yu
Hao Yu
Chaohui Yu
Fisher Yu
Jin-Gang Yu
Jiyang Yu
Jason J. Yu
Jiashuo Yu
Hong-Xing Yu
Lei Yu
Mulin Yu
Ning Yu
Peilin Yu
Qi Yu
Qian Yu
Rui Yu
Shuzhi Yu
Gang Yu
Tan Yu
Weijiang Yu
Xin Yu
Bingyao Yu
Ye Yu
Hanchao Yu
Yingchen Yu
Tao Yu
Xiaotian Yu
Qing Yu
Houjian Yu
Changqian Yu
Jing Yu
Jun Yu
Shujian Yu
Xiang Yu
Zhaofei Yu
Zhenbo Yu
Yinfeng Yu
Zhuoran Yu
Zitong Yu
Bo Yuan
Jiangbo Yuan
Liangzhe Yuan
Weihao Yuan
Jianbo Yuan
Xiaoyun Yuan
Ye Yuan
Li Yuan
Geng Yuan
Jialin Yuan
Maoxun Yuan
Peng Yuan
Xin Yuan
Yuan Yuan
Yuhui Yuan
Yixuan Yuan
Zheng Yuan
Mehmet Kerim Yücel
Kaiyu Yue
Haixiao Yue
Heeseung Yun
Sangdoo Yun
Tian Yun
Mahmut Yurt
Ekim Yurtsever
Ahmet Yüzügüler
Edouard Yvinec
Eloi Zablocki
Christopher Zach
Muhammad Zaigham Zaheer
Pierluigi Zama Ramirez
Yuhang Zang
Pietro Zanuttigh
Alexey Zaytsev
Bernhard Zeisl
Haitian Zeng
Pengpeng Zeng
Jiabei Zeng
Runhao Zeng
Wei Zeng
Yawen Zeng
Yi Zeng

Yiming Zeng
Tieyong Zeng
Huanqiang Zeng
Dan Zeng
Yu Zeng
Wei Zhai
Yuanhao Zhai
Fangneng Zhan
Kun Zhan
Xiong Zhang
Jingdong Zhang
Jiangning Zhang
Zhilu Zhang
Gengwei Zhang
Dongsu Zhang
Hui Zhang
Binjie Zhang
Bo Zhang
Tianhao Zhang
Cecilia Zhang
Jing Zhang
Chaoning Zhang
Chenxu Zhang
Chi Zhang
Chris Zhang
Yabin Zhang
Zhao Zhang
Rufeng Zhang
Chaoyi Zhang
Zheng Zhang
Da Zhang
Yi Zhang
Edward Zhang
Xin Zhang
Feifei Zhang
Feilong Zhang
Yuqi Zhang
GuiXuan Zhang
Hanlin Zhang
Hanwang Zhang
Hanzhen Zhang
Haotian Zhang
He Zhang
Haokui Zhang
Hongyuan Zhang
Hengrui Zhang
Hongming Zhang
Mingfang Zhang
Jianpeng Zhang
Jiaming Zhang
Jichao Zhang
Jie Zhang
Jingfeng Zhang
Jingyi Zhang
Jinnian Zhang
David Junhao Zhang
Junjie Zhang
Junzhe Zhang
Jiawan Zhang
Jingyang Zhang
Kai Zhang
Lei Zhang
Lihua Zhang
Lu Zhang
Miao Zhang
Minjia Zhang
Mingjin Zhang
Qi Zhang
Qian Zhang
Qilong Zhang
Qiming Zhang
Qiang Zhang
Richard Zhang
Ruimao Zhang
Ruisi Zhang
Ruixin Zhang
Runze Zhang
Qilin Zhang
Shan Zhang
Shanshan Zhang
Xi Sheryl Zhang
Song-Hai Zhang
Chongyang Zhang
Kaihao Zhang
Songyang Zhang
Shu Zhang
Siwei Zhang
Shujian Zhang
Tianyun Zhang
Tong Zhang
Tao Zhang
Wenwei Zhang
Wenqiang Zhang
Wen Zhang
Xiaolin Zhang
Xingchen Zhang
Xingxuan Zhang
Xiuming Zhang
Xiaoshuai Zhang
Xuanmeng Zhang
Xuanyang Zhang
Xucong Zhang
Xingxing Zhang
Xikun Zhang
Xiaohan Zhang
Yahui Zhang
Yunhua Zhang
Yan Zhang
Yanghao Zhang
Yifei Zhang
Yifan Zhang
Yi-Fan Zhang
Yihao Zhang
Yingliang Zhang
Youshan Zhang
Yulun Zhang
Yushu Zhang
Yixiao Zhang
Yide Zhang
Zhongwen Zhang
Bowen Zhang
Chen-Lin Zhang
Zehua Zhang
Zekun Zhang
Zeyu Zhang
Xiaowei Zhang
Yifeng Zhang
Cheng Zhang
Hongguang Zhang
Yuexi Zhang
Fa Zhang
Guofeng Zhang
Hao Zhang
Haofeng Zhang
Hongwen Zhang

Hua Zhang
Jiaxin Zhang
Zhenyu Zhang
Jian Zhang
Jianfeng Zhang
Jiao Zhang
Jiakai Zhang
Lefei Zhang
Le Zhang
Mi Zhang
Min Zhang
Ning Zhang
Pan Zhang
Pu Zhang
Qing Zhang
Renrui Zhang
Shifeng Zhang
Shuo Zhang
Shaoxiong Zhang
Weizhong Zhang
Xi Zhang
Xiaomei Zhang
Xinyu Zhang
Yin Zhang
Zicheng Zhang
Zihao Zhang
Ziqi Zhang
Zhaoxiang Zhang
Zhen Zhang
Zhipeng Zhang
Zhixing Zhang
Zhizheng Zhang
Jiawei Zhang
Zhong Zhang
Pingping Zhang
Yixin Zhang
Kui Zhang
Lingzhi Zhang
Huaiwen Zhang
Quanshi Zhang
Zhoutong Zhang
Yuhang Zhang
Yuting Zhang
Zhang Zhang
Ziming Zhang
Zhizhong Zhang
Qilong Zhangli
Bingyin Zhao
Bin Zhao
Chenglong Zhao
Lei Zhao
Feng Zhao
Gangming Zhao
Haiyan Zhao
Hao Zhao
Handong Zhao
Hengshuang Zhao
Yinan Zhao
Jiaojiao Zhao
Jiaqi Zhao
Jing Zhao
Kaili Zhao
Haojie Zhao
Yucheng Zhao
Longjiao Zhao
Long Zhao
Qingsong Zhao
Qingyu Zhao
Rui Zhao
Rui-Wei Zhao
Sicheng Zhao
Shuang Zhao
Siyan Zhao
Zelin Zhao
Shiyu Zhao
Wang Zhao
Tiesong Zhao
Qian Zhao
Wangbo Zhao
Xi-Le Zhao
Xu Zhao
Yajie Zhao
Yang Zhao
Ying Zhao
Yin Zhao
Yizhou Zhao
Yunhan Zhao
Yuyang Zhao
Yue Zhao
Yuzhi Zhao
Bowen Zhao
Pu Zhao
Bingchen Zhao
Borui Zhao
Fuqiang Zhao
Hanbin Zhao
Jian Zhao
Mingyang Zhao
Na Zhao
Rongchang Zhao
Ruiqi Zhao
Shuai Zhao
Wenda Zhao
Wenliang Zhao
Xiangyun Zhao
Yifan Zhao
Yaping Zhao
Zhou Zhao
He Zhao
Jie Zhao
Xibin Zhao
Xiaoqi Zhao
Zhengyu Zhao
Jin Zhe
Chuanxia Zheng
Huan Zheng
Hao Zheng
Jia Zheng
Jian-Qing Zheng
Shuai Zheng
Meng Zheng
Mingkai Zheng
Qian Zheng
Qi Zheng
Wu Zheng
Yinqiang Zheng
Yufeng Zheng
Yutong Zheng
Yalin Zheng
Yu Zheng
Feng Zheng
Zhaoheng Zheng
Haitian Zheng
Kang Zheng
Bolun Zheng

Haiyong Zheng
Mingwu Zheng
Sipeng Zheng
Tu Zheng
Wenzhao Zheng
Xiawu Zheng
Yinglin Zheng
Zhuo Zheng
Zilong Zheng
Kecheng Zheng
Zerong Zheng
Shuaifeng Zhi
Tiancheng Zhi
Jia-Xing Zhong
Yiwu Zhong
Fangwei Zhong
Zhihang Zhong
Yaoyao Zhong
Yiran Zhong
Zhun Zhong
Zichun Zhong
Bo Zhou
Boyao Zhou
Brady Zhou
Mo Zhou
Chunluan Zhou
Dingfu Zhou
Fan Zhou
Jingkai Zhou
Honglu Zhou
Jiaming Zhou
Jiahuan Zhou
Jun Zhou
Kaiyang Zhou
Keyang Zhou
Kuangqi Zhou
Lei Zhou
Lihua Zhou
Man Zhou
Mingyi Zhou
Mingyuan Zhou
Ning Zhou
Peng Zhou
Penghao Zhou
Qianyi Zhou
Shuigeng Zhou
Shangchen Zhou
Huayi Zhou
Zhize Zhou
Sanping Zhou
Qin Zhou
Tao Zhou
Wenbo Zhou
Xiangdong Zhou
Xiao-Yun Zhou
Xiao Zhou
Yang Zhou
Yipin Zhou
Zhenyu Zhou
Hao Zhou
Chu Zhou
Daquan Zhou
Da-Wei Zhou
Hang Zhou
Kang Zhou
Qianyu Zhou
Sheng Zhou
Wenhui Zhou
Xingyi Zhou
Yan-Jie Zhou
Yiyi Zhou
Yu Zhou
Yuan Zhou
Yuqian Zhou
Yuxuan Zhou
Zixiang Zhou
Wengang Zhou
Shuchang Zhou
Tianfei Zhou
Yichao Zhou
Alex Zhu
Chenchen Zhu
Deyao Zhu
Xiatian Zhu
Guibo Zhu
Haidong Zhu
Hao Zhu
Hongzi Zhu
Rui Zhu
Jing Zhu
Jianke Zhu
Junchen Zhu
Lei Zhu
Lingyu Zhu
Luyang Zhu
Menglong Zhu
Peihao Zhu
Hui Zhu
Xiaofeng Zhu
Tyler (Lixuan) Zhu
Wentao Zhu
Xiangyu Zhu
Xinqi Zhu
Xinxin Zhu
Xinliang Zhu
Yangguang Zhu
Yichen Zhu
Yixin Zhu
Yanjun Zhu
Yousong Zhu
Yuhao Zhu
Ye Zhu
Feng Zhu
Zhen Zhu
Fangrui Zhu
Jinjing Zhu
Linchao Zhu
Pengfei Zhu
Sijie Zhu
Xiaobin Zhu
Xiaoguang Zhu
Zezhou Zhu
Zhenyao Zhu
Kai Zhu
Pengkai Zhu
Bingbing Zhuang
Chengyuan Zhuang
Liansheng Zhuang
Peiye Zhuang
Yixin Zhuang
Yihong Zhuang
Junbao Zhuo
Andrea Ziani
Bartosz Zieliński
Primo Zingaretti

Nikolaos Zioulis
Andrew Zisserman
Yael Ziv
Liu Ziyin
Xingxing Zou
Danping Zou
Qi Zou
Shihao Zou
Xueyan Zou
Yang Zou
Yuliang Zou
Zihang Zou
Chuhang Zou
Dongqing Zou
Xu Zou
Zhiming Zou
Maria A. Zuluaga
Xinxin Zuo
Zhiwen Zuo
Reyer Zwiggelaar

Contents – Part XXXIX

Lane Detection Transformer Based on Multi-frame Horizontal and Vertical Attention and Visual Transformer Module

Han Zhang, Yunchao Gu(✉), Xinliang Wang, Junjun Pan, and Minghui Wang

Beihang University, XueYuan Road No. 37, Haidian District, Beijing, China
{allenzhang,guyunchao,wangxinliang,pan_junjun,minghuiw}@buaa.edu.cn

Abstract. Lane detection requires adequate global information due to the simplicity of lane line features and changeable road scenes. In this paper, we propose a novel lane detection Transformer based on multi-frame input to regress the parameters of lanes under a lane shape modeling. We design a Multi-frame Horizontal and Vertical Attention (MHVA) module to obtain more global features and use Visual Transformer (VT) module to get "lane tokens" with interaction information of lane instances. Extensive experiments on two public datasets show that our model can achieve state-of-art results on VIL-100 dataset and comparable performance on TuSimple dataset. In addition, our model runs at 46 fps on multi-frame data while using few parameters, indicating the feasibility and practicability in real-time self-driving applications of our proposed method.

Keywords: Autonomous driving · Lane detection · Transformer

1 Introduction

Autonomous driving has been developing rapidly in recent years and has received full attention from academia and industry. In perception task—the "eyes" of autonomous driving, lane detection plays a significant role in understanding road environment. It relates to the self-positioning, observance of traffic rules, and subsequent decisions on control of autonomous vehicles.

Lane detection is challenging mainly due to two reasons. Firstly, the lanes[1] are slender in shape, single in structure, and rare in appearance clues. In addition, lanes may disappear due to various reasons such as wear and tear, shadow occlusion, road congestion, terrible weather conditions, or dazzling light. However, humans can easily determine the location of lane lines based on vehicle alignment, road shape, visible local lane lines information, etc. Inspired by the human visual system, people found that the global information containing additional visual clues is preferable to detecting the lane lines.

[1] For academic consistency, we use "lane" to denote lane line in this paper.

S. Avidan et al. (Eds.): ECCV 2022, LNCS 13699, pp. 1–16, 2022.
https://doi.org/10.1007/978-3-031-19842-7_1

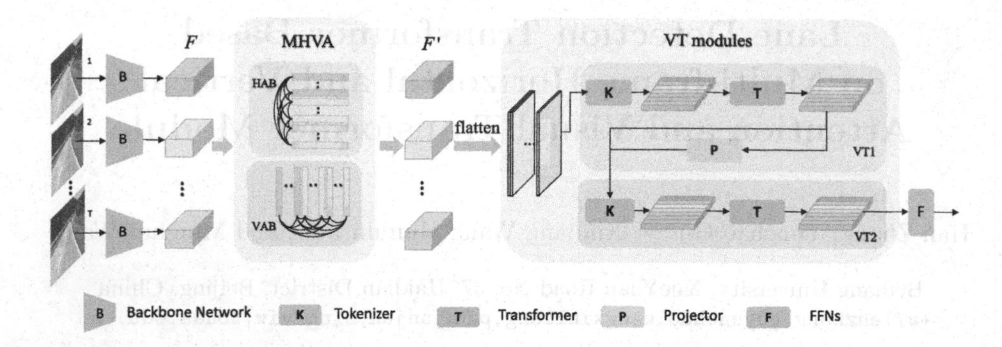

Fig. 1. Overall Framework. Continuous frames are compressed to 1/8 of origin through the backbone network. Then the compressed features are fed into the MHVA to pass messages between rows or columns. The VT modules are used to generate "lane tokens" and enable interaction between lane instances. Finally, the parameters are obtained through FFNs with the tokens as input

Many methods [2,7,12,16,24] focus on obtaining sufficient global features from current frame. Among them, LSTR [7] proposes to use the most conventional Transformer to capture global information, but it does not take full advantage of the structural features of lane lines and the pixel-wise attention mechanism is computationally intensive. SCNN [12], on the other hand, designs a special CNN to capture spatial relationships of pixels across rows and columns. However, using the convolution operation which is good at processing local features, it needs a large number of local information transfers(e.g. slice-by-slice transfer, multiple iterations of convolution modules) to achieve global information interaction. That is the reason why it is computationally intensive and inefficient in terms of information interaction. The RESA [24] on top of it does not solve the problem fundamentally.

The importance of global features to lane detection can not be ignored, but it is obviously unrealistic and ill-considered to expect deep learning models which only use individual frames as input and completely ignore more visual information of previous frames in dynamic driving can infer complete and accurate detection results. Although there are some lane methods [22,23,25] use multi-frame input data, they are all segmentation-based. These methods need to classify each pixel, which are indirect and lead to higher computational cost. Besides, they also need extra post-processing steps to extract lane information.

To solve the above problems, we propose a novel end-to-end model using multiple frames as input, having the customized structure suitable for lane detection based on Transformer. To extract more global information from frames, we design a Transformer-Based structure called Multi-frame Horizontal and Vertical Attention (MHVA) module. It can solve the dilemma of SCNN, due to its advantage of establishing efficient and direct information interaction via attention module across multiple frames. Here are the two main differences from previous work. First, MHVA builds the interaction among row and column features separately via direct attention, which adopts the divide-and-conquer idea to reduce the computation and fully makes use of horizontal and vertical features. Second, MHVA can be easily extended to detection based on multiple frames.

Besides, we use the VT module in our method. We notice a certain continuity between lanes, such as they are parallel and always intersect at vanishing points. Inspired by [19], we introduce the concept of "lane token" into lane detection to get lane instance features. By utilizing the VT module, we can assign the features obtained from MHVA to specific lanes, and establish relationship between lanes to realize information interaction. Finally, the regression task is built on tokens to get the parameters of each lane.

The main contributions of our method can be summarized as follows:

- We propose a novel curve-fitting lane detection method using multi-frame information and achieve instance-level lane detection.
- We design a lane detection Transformer based on MHVA and VT modules capturing more global information for parametric regression.
- The experiments on two public datasets verify the effectiveness of our method. In addition, we achieve the state-of-the-art results on the VIL-100 dataset and competitive performance on the TuSimple dataset, both with real-time speed.

2 Related Work

Early lane detection approaches are based on traditional computer vision methods. However, traditional methods have poor robustness in complex scenarios due to their dependence on highly-specialized and hand-crafted features. Recently, with the spring breeze of deep learning, the lane detection approaches based on it have sprung up. They become the mainstream solutions relying on powerful learning ability. We taxonomize these methods into four categories.

Segmentation-Based Method. This kind of method regards lane detection as a segmentation problem. Some semantic segmentation methods [1,10,25] only distinguish the pixels of lanes from background, while others [2,8,9,12,13,24] regard each lane as an individual classification category by detecting a fixed number of lanes. Methods [3,11,21] based on instance segmentation take each lane as an instance of a lane category.

Based on multi-frame input, [22,25] only segment lanes from background and both need post-processing to distinguish lanes. Different from them, MMANet [23] is an instance lane detection method using multi-frame input. But it is inefficient and requires a redundant extraction step. Instead of using segmentation-based paradigm, we choose to regress the parameters of curves.

Row-Wise-Classification-Based Method. This kind of method is based on the prior domain knowledge that the number of intersection points between a horizontal line and each other lanes is most one. Thus, it formulates lane detection as a row-wise classification problem by selecting the cell position for intersection points. Finally, a complete lane can be composed of these intersection points. This divide-and-conquer idea reduces the calculation cost, but they also need a post-processing step to extract lanes. [20] and [14] are representative models of this method.

Anchor-Based Method. Inspired by Faster-RCNN [15], Line-CNN [6] proposes the first anchor-based lane detection method. Its core design LPU (line proposal unit) predicts lines from the original straight proposal line. However, due to the lack of ability to capture global information, this method is inefficient. Based on [6], LaneATT [16] proposes an anchor-based attention mechanism to capture global information.

Curve-Fitting-Based Method. Lane detection is regarded as a parametric regression problem which use polynomial curves to express lanes in curve-fitting-based method. PolyLaneNet [17] connects a regression part after backbone network. Although it is efficient, the lack of global information leads to a gap with other methods in terms of performance. On this basis, R. Liu *et al.* [7] change the lane modeling method and use Transformer to learn more global features. They are both based on single-image input and discard the visual information in previous frames. Using continuous multi-frame input, we propose a more powerful lane detection Transformer to capture more global information.

3 Approach

We propose a novel lane detection method based on curve fitting, whose framework is shown in Fig. 1. It receives several time-ordered RGB images taken from a camera mounted in the vehicle as input and outputs the parameters of the predicted lanes. It consists of a backbone network, a MHVA module, several VT modules, and feed-forward networks (FFNs) for parametric regression. Given several continuous images in order, the backbone network first extracts a high-level feature map F. Then, the feature map F and positional embedding E are fed into the MHVA module to get the enhanced feature map F'. After received "lane tokens" through VT modules, FFNs will regress the parameters on them. Hungarian fitting loss are used to train our network.

3.1 Lane Shape Model

Following the lane shape model in LSTR [7], we regress the cubic polynomial parameters related to the internal and external parameters of the camera and the angle between the camera with the ground plane. In the image coordinate system, the modeling formula is as follows:

$$x = \frac{k}{(y-f)^2} + \frac{m}{y-f} + n + b \times y - b', \tag{1}$$

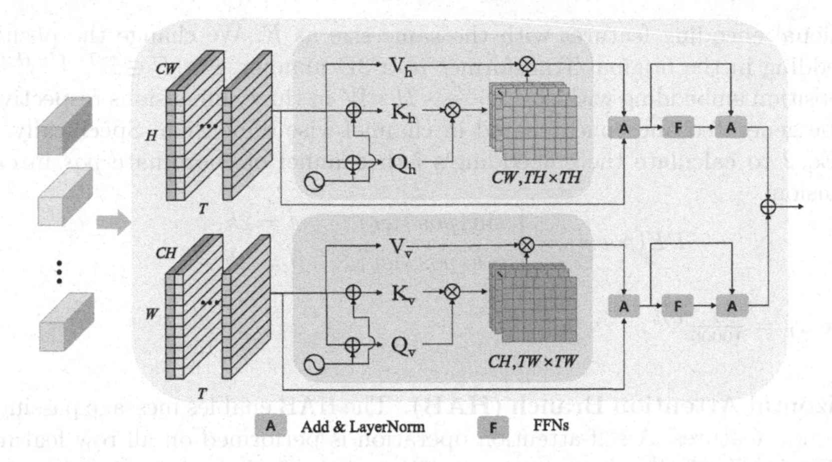

Fig. 2. MHVA. It has a HAB branch for interaction among rows and a VHB for interaction among columns. The final output is the sum of the results of the two branches

where (x, y) is position coordinate, n and b' are constants terms that cannot be integrated, and k, f, m, b are variable parameters.

For one lane l, it relates to eight parameters $(k, f, m, n, b_l, b_l^{'}, \alpha_l, \beta_l)$. The first four parameters are shared parameters among all lanes, and b_l and $b_l^{'}$ are particular parameters of lane l. α_l and β_l are vertical starting offset and ending offset respectively. Kindly refer to [7] for more details.

3.2 The MHVA Module

Transformer is popular in computer vision due to its effectiveness in capturing long distance dependence. Therefore, to capture global content information from a multi-frame feature map, the intuitive idea is to build attention connection at pixel level by Transformer just as in [18]. However, when only focusing on lanes, we construct a more efficient module called Multi-frame Horizontal and Vertical Attention (MHVA) based on the prior knowledge of lane structure and frame continuity.

When we input T frames, the feature map $F \in \mathbb{R}^{T \times C \times H \times W}$ is obtained from the backbone network. Instead of building the pixel level connections, MHVA models the relationships among all row features and all column features respectively. It decomposes the $THW \times THW$ connection into $TH \times TH$ and $TW \times TW$ connections. The two branches of MHVA are Horizontal Attention Branch (HAB) and Vertical Attention Branch (VAB). Figure 2 shows the architecture of MHVA.

Position Embedding. To supplement precise position information in three dimensions (temporal, horizontal, and vertical dimensions), we generate fixed positional encoding features with the same size as F. We change the position embedding in the original Transformer to a 3D manner. For $F \in \mathbb{R}^{T \times C \times H \times W}$, the position embedding with size $T \times \frac{C}{3} \times H \times W$ in three dimensions respectively will be generated and concatenated in channel-wise dimension. Specifically, we use Eq. 2 to calculate the embedding's i-th channel of coordinate pos in each dimension:

$$PE(pos, i) = \begin{cases} \sin(pos \cdot \omega_k), & i = 2k, \\ \cos(pos \cdot \omega_k), & i = 2k+1, \end{cases} \tag{2}$$

where $\omega_k = \frac{1}{10000}^{\frac{2k}{C/3}}$.

Horizontal Attention Branch (HAB). The HAB enables message passing of horizontal features. A self-attention operation is performed on all row features. For $F \in \mathbb{R}^{T \times C \times H \times W}$, the row num is TH and the feature channel size is CW. Afterward, the feature map F and positional embedding E will be integrated to $TH \times CW$ and fed into attention module, with query, key, and value tensors (Q_h, K_h, V_h) of the same dimension $TH \times d\_model$ as output. Then, a $TH \times TH$ attention matrix A_h is obtained as Eq. 3:

$$A_h = softmax(\frac{Q_h K_h^T}{\sqrt{d\_model}}). \tag{3}$$

where A_{ij} represents the similarity between features of frame $[i/h]$, row $[i \bmod h]$ and frame $[j/h]$, row $[j \bmod h]$. Intuitively, the value of adjacent rows or corresponding rows in adjacent frames will be higher. Finally, we obtain a refined feature map using the attention matrix A_h and V_h tensor as follows:

$$F_h = A_h V_h. \tag{4}$$

Vertical Attention Branch (VAB). The VAB models the relationships between column features similar to that of HAB's. Differently, the feature map F and positional embedding E are resized to $TW \times CH$, and the attention matrix A_v is $TW \times TW$. Finally, F_v is given by the multiplication of A_v and V_v.

F_h and F_v will be resized to $T \times C \times H \times W$ and added together to get the final feature map F'.

3.3 The VT Module

Inspired by [19], we define the "lane tokens" to represent the lane instance, and use the Visual Transformer (VT) module to learn the compact lane instance features from the high-level features. A VT module involves three steps. The first step is to group pixels into lane instances to generate a serial of corresponding lane tokens. Then, a transformer module is performed to the lane tokens to model the relationships between them. Finally, we reproject the lane tokens to pixel level to obtain an augmented feature map. If multiple VT modules are used continuously, the module VT_i will build on the reprojected feature map of the previous VT module VT_{i-1}. In last VT module, the third step is omitted directly to regress parameters on lane tokens.

Tokenizer. The function of tokenizer is to convert feature maps into compact sets of lane tokens. Formally, we resize feature map F' to $THW \times C$ and denote it as input by $X \in \mathbb{R}^{N \times C} (N = THW)$. The lane tokens are denoted as $T \in \mathbb{R}^{L \times C}$, where L is the fixed maximum number of lanes. The core mechanism for calculating tokens is attention. Firstly, an attention map $A \in \mathbb{R}^{N \times L}$ is generated where A_{ij} represents the contribution of one pixel $p_i \in \mathbb{R}^C$ to the lane instance j. Next, we compute the weighted averages of pixels in X to form lane tokens by multiplying A and X. Formally,

$$T = A^T X. \tag{5}$$

There are two ways to generate the attention map A for modules in different locations as shown in Fig. 3. For the first VT module, we generate it as follows:

$$A = \mathit{softmax}_N (X W_A), \tag{6}$$

the $W_A \in \mathbb{R}^{C \times L}$ in Eq. 6 forms lane instance groups from X, and the *softmax* operation converts the values into normalized attention values.

While for the subsequent VT modules, we use the output lane tokens T_{last} of its previous module to guide the extraction of current new lane tokens. Formally,

$$\begin{aligned} W_R &= T_{last} W_{T \to R}, \\ A &= \mathit{softmax}_N (X W_R), \end{aligned} \tag{7}$$

in which $W_{T \to R} \in \mathbb{R}^{C \times C}$, $W_R \in \mathbb{R}^{C \times L}$.

Transformer. After Tokenizer, we need to establish the relationship between lanes to realize information interaction. To dynamically model interactions, a standard transformer with minor changes is used, in which self-attention generates the input-dependent weights.

$$\begin{aligned} T'_{out} &= T_{in} + \mathit{softmax}_L (T_{in} W_K (T_{in} W_Q)^T) T_{in}, \\ T_{out} &= T'_{out} + \delta(T'_{out} F_1) F_2, \end{aligned} \tag{8}$$

where T_{in}, T'_{out}, $T_{out} \in \mathbb{R}^{L \times C}$ are lane tokens, W_K and W_Q are key weight and query weight with size $C \times C$, and F_1, $F_2 \in \mathbb{R}^{C \times C}$ are linear weights. $\delta(\cdot)$ is the *ReLu* function.

Projector. When we overlay multiple VT modules, the role of the projector in the current VT is to generate the input feature map for the next VT module.

When using only one VT module, there is no need to generate feature maps for the subsequent modules, parametric regression will be operated at lane tokens.

$$X_{out} = X_{in} + softmax_{_L}((X_{in}W_Q)(TW_K)^T)T, \tag{9}$$

where X_{in}, $X_{out} \in \mathbb{R}^{N\times C}$. W_Q, $W_K \in \mathbb{R}^{C\times C}$ are the weights for query and key tensors. The product of key and query determines how to project information encoded in lane tokens to the pixel level feature map.

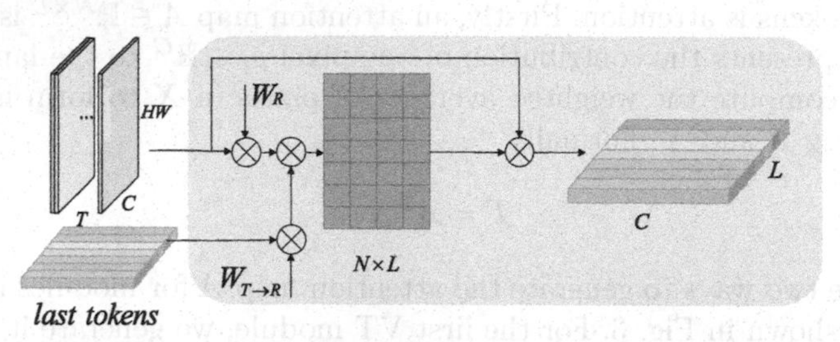

(a) Tokenizer without previous tokens

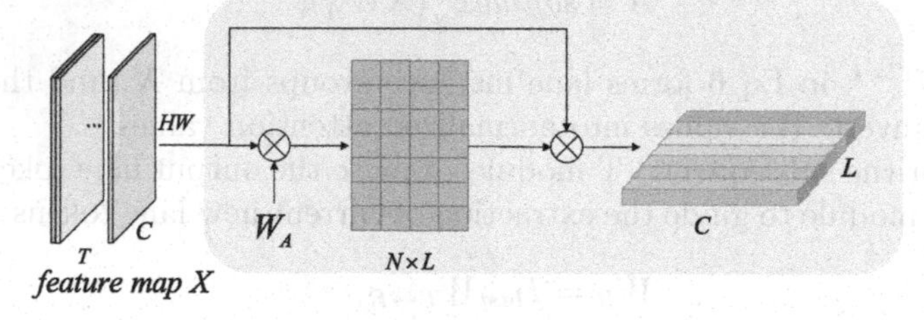

(b) Tokenizer with previous tokens

Fig. 3. The two kinds of Tokenizer. The N is equal to THW

3.4 FFNs

Given the lane tokens $T \in \mathbb{R}^{L\times C}$, we use three branches to predict the parameters. In the classification branch, we use a linear operation to generate $C \in \mathbb{R}^{L\times 2}$, which indicates the probability of lane instances represented by each lane token. On another two branches, we use three-layer perceptrons to predict four shared parameters and four lane-specific parameters respectively. Those shared parameters will be averaged between lanes.

3.5 Loss Function

Our proposed method infers the predicted parameters of lane instances, which are stochastic in order. Thus we match them with the ground truth lanes first by utilizing the Hungarian algorithm. The regression loss is optimized based on the above matching result.

The L predicted curves are denoted as $\mathcal{H} = \{h_i | h_i = (c_i, m_i)\}_{i=1}^{L}$, where m is the set of eight parameters and $c \in \{0, 1\}$ represents the possibility of existence of lanes. The ground truth labels of lanes are represented by a set of key points $s = (x, y)_{r=1}^{R}$, where $y_{i+1} > y_i$, and the ground truth curves are denoted as $\mathcal{L} = \{l_i | l_i = (c_i', s_i')\}_{i=1}^{L}$. Since L is larger than the number of ground truth lanes, $\mathcal{L}$ will be padded with non-lane values.

Bipartite Matching. In order to find a bipartite matching between the predicted parameters and the ground truth lanes, a permutation of L elements δ is searched with the lowest cost:

$$\hat{\delta} = \arg\min_{\delta} \sum_{i}^{L} \mathcal{D}(l_i, h_{\delta(i)}), \tag{10}$$

where $\mathcal{D}$ is a pair-wise matching cost between the *i-th* ground truth lane and a predicted curve with index $\delta(i)$.

For the prediction curve with index δ_i, the probability of class c_i' is denoted as $p_{\delta(i)}(c_i')$, and the prediction key points $s_{\delta(i)}$ can be calculated by Eq. 1 based on parameters $m_{\delta(i)}$. The matching cost $\mathcal{D}$ is defined as:

$$\begin{aligned} \mathcal{D} = &- \omega_1 p_{\delta(i)}(c_i') + \mathbb{1}(c_i' = 1)\omega_2 L_1(s_i', s_{\delta(i)}) \\ &+ \mathbb{1}(c_i' = 1)\omega_3 L_1(\alpha_i', \alpha_{\delta(i)}, \beta_i', \beta_{\delta(i)}), \end{aligned} \tag{11}$$

where $\mathbb{1}$ is the indicate function, ω_1, ω_2 and ω_3 are the effect parameters, and L_1 is the mean absolute error.

Regression Loss. After getting the matching results, we calculate the regression loss as follows:

$$\begin{aligned} L = \sum_{i=1}^{L} &-\omega_1 p_{\delta(i)}(c_i') + \mathbb{1}(c_i' = 1)\omega_2 L_1(s_i', s_{\delta(i)}) \\ &+ \mathbb{1}(c_i' = 1)\omega_3 L_1(\alpha_i', \alpha_{\delta(i)}, \beta_i', \beta_{\delta(i)}). \end{aligned} \tag{12}$$

4 Experiments

In this section, we evaluate the performance of our proposed method with the widely-used metrics on two public datasets TuSimple[2] and VIL-100 [23]. Afterward, we provide a detailed ablation study to prove the rationality of our design and the selected hyperparameters.

[2] https://github.com/TuSimple/TuSimple-benchmark.

4.1 Dataset

VIL-100. VIL-100 is the first video instance-level lane detection dataset with all frames annotated. It contains 100 videos, 100 frames per video within 10 s, in total 10000 labeled frames. Among 100 videos, 97 are collected by the monocular forward-facing camera, and the remaining three videos are from the Internet. Besides, it contains one normal scenario and nine challenging scenarios. The training set has 80 videos and the rest belongs to the testing set.

TuSimple. The TuSimple dataset is a widely-used dataset collected on America's highway with constant illumination and weather in the daytime. It consists of 6408 sequences, each of which contains 20 continuous frames collected in one second. We take the annotated 3626 clips for training and adopt the remaining 2782 clips for testing.

4.2 Evaluation Metrics

On the TuSimple dataset, we use its three official metrics *Accuracy*, *FP* and *FN*. *Accuracy* refers to the average number of correct points per image. The standard for "correct points" is that when the vertical coordinate is the same, the horizontal distance between the predicted point and the ground truth point is smaller than 20 pixels. *Accuracy* can be defined in the following way:

$$Accuracy = \sum_{i=1}^{n} \frac{C_i}{S_i}, \tag{13}$$

where C_i is the number of correct points and S_i is the total number of ground-truth points in frame i, n is the number of total frames.

The false positive (*FP*) and false negative (*FN*) are computed as:

$$FP = \frac{F_{pred}}{N_{pred}}, \tag{14}$$

$$FN = \frac{M_{pred}}{N_{gt}}, \tag{15}$$

where F_{pred} is the wrongly predicted lanes, M_{pred} represents the number of missed lanes, N_{pred} is the total number of predicted lanes, N_{gt} is the total number of ground-truth lanes.

On the VIL-100 dataset, in addition to the above three metrics, we also use the *F1* metric. *F1* is a region-based metric based on the intersection-over-union (IoU). For one lane instance, we can get the predicted key points using the predicted parameters. A binary mask can be obtained by connecting these points with a line of 30 pixels in width. Then, if the IoU between the predicted mask and the ground truth label is larger than 0.5, the predicted lane can be considered as true positive (tp), otherwise, it is a false positive (fp). The missed lanes are denoted as fn. The metric $F1$ is formulated as:

$$F1 = \frac{2 \times precision \times recall}{precision + recall}, \tag{16}$$

$$precision = \frac{tp}{tp + fp}, \tag{17}$$

$$recall = \frac{tp}{tp + fn}. \tag{18}$$

4.3 Implementation Details

The input data for our model are 5 continuous frames with size 360 × 640. Same data augmentation methods are used for these continuous frames, such as shadow, flipping horizontally, rotating, and color jittering.

The backbone network we use is a reduced ResNet18 which is the same as [7]. The output channels of four blocks are changed from "64, 128, 256, 512" to "16, 32, 64, 128" and the downsampling factor is set to 8. We use the multi-head self-attention in all attention modules and the head number is set to 2. The number of VT modules is 2. The fixed number of predicted curves L is set to 7 and the loss coefficients $\omega_1, \omega_2, \omega_3$ are set to 3, 5 and 2 respectively.

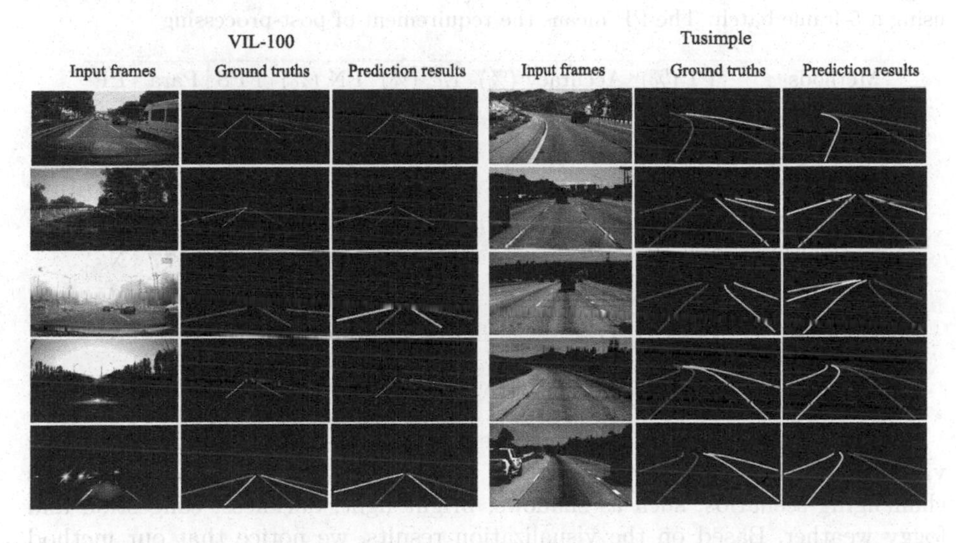

Fig. 4. Visualization results for VIL-100 and TuSimple dataset. The colors are just to distinguish different lane instances

We train our model on one TitanX GPU with batch size 8 in PyTorch. For both datasets, we use Adam optimizer with the base learning rate of 0.0001. The learning rate policy is StepLR with stepsize 300. The numbers of total training epochs are 1000 for TuSimple and 1200 for VIL-100 dataset separately.

4.4 Results

In this section, we treat LSTR as the baseline model since it is also a curve-fitting method based on Transformer. To show our performance, we compare our method with other state-of-the-art methods. Apart from the video instance lane detection method MMANet [23], the input of other methods are almost one image. However, because MMANet has not been trained on TuSimple, we only compare with it on VIL-100 dataset. Besides the above-mentioned four metrics, the FPS and the total number of parameters are also compared.

VIL-100. Table 1 shows the performance on VIL-100 dataset. Our proposed method outperforms the mentioned state-of-the-art lane detectors on four above-mentioned metrics. Compared with baseline method LSTR, ours shows significant improvement on four metrics, especially the *FP* is reduced by half. Compared with MMANet, the performance of our model raises 0.7% on *F1*, 0.5% on *Accuracy*, 2.9% on *FP*, and 1.0% on *FN* with 15 × fewer parameters and 5 × faster speed. The large improvement on the challenging dataset fully demonstrates the effectiveness of our proposed method.

Table 1. Quantitative comparisons on VIL-100 dataset. The FPS is evaluated using a single batch of inputs on the TitanX GPU, which means that our method is performed using a 5-frame batch. The PP means the requirement of post-processing

Methods	F1 (%)	Accuracy (%)	FP (%)	FN (%)	FPS	Para	PP
LaneNet [11]	72.1	85.8	12.2	20.7	36	1.48	Y
SCNN [12]	49.1	90.7	12.8	11.0	16	20.72	Y
ENet-SAD [2]	75.5	88.6	17.0	15.2	14	0.98	Y
UFSA [14]	31.0	85.2	11.5	21.5	–	–	Y
LSTR [7]	70.3	88.4	16.3	14.8	**48**	**0.77**	N
MMA [23]	83.9	91.0	11.1	10.5	9	57.91	Y
Ours	**84.6**	**91.5**	**8.2**	**9.5**	46	3.87	N

Besides quantitative comparisons, we show our results qualitatively through visualization in Fig. 4. We visualize the performance of our method in various challenging scenarios, such as shadows, bright light, darkness, congestion and foggy weather. Based on the visualization results, we notice that our method can adequately capture global features and detect lanes well even when the visual information of lanes is very scarce.

TuSimple. The results on TuSimple are shown as Table 2. It can be noted that our method achieves comparable results compared with other methods. The visualization results are shown in the Fig. 4. We can notice that our method performs well, even can detect lanes not labeled in the ground truth (shown in row 3).

Compared to LSTR, our method does not improve significantly. We guess that the root cause may lie in the limitations of the dataset itself. Firstly, as mentioned in [5,6,16], it is a relatively simple dataset because the highway scenario is easier than street scenes and not congested. Thus, it's hard to expose the model's advantages in realistic complex scenes and has more saturated results. Besides, the road scene with no congestion and stable illumination results in the high similarity of multi-frame input data, which makes the use of multi-frame input meaningless. Combined with the results of the VIL-100 dataset, the results on the TuSimple dataset confirm the effectiveness of multi-frame input for global content information in the complex real driving environment.

Table 2. Quantitative comparisons on TuSimple dataset. *means our re-implemented results on TuSimple

Method	Accuracy (%)	FP(%)	FN (%)
SCNN [12]	96.53	6.17	1.80
Enet-SAD [2]	96.64	4.67	5.18
UFSA [14]	95.82	19.05	3.92
Line-CNN [6]	**96.87**	4.42	**1.97**
PINet [4]	96.75	**3.10**	2.50
LaneATT [16]	95.57	3.56	3.10
PolyLaneNet [17]	93.36	9.42	9.33
LSTR* [7]	96.03	3.26	3.44
Ours	96.17	3.50	3.38

4.5 Ablation Study

Because the TuSimple dataset cannot clearly reflect the difference among methods, we conduct ablation experiments on the VIL-100 dataset.

Effectiveness of the Multi-frame Mechanism. To verify the effectiveness of multiple frames, we improve the Transformer in LSTR to adapt to multi-frame feature maps. Specifically, at the core self-attention part, the attention between pixels in a single frame is replaced by the relationship modeling among all pixels in multiple frames, which is similar to the video instance segmentation method [18]. In addition to this, single-frame input experiments are also conducted on our method.

As shown in Table 3, compared with original LSTR (method A), method B with multi-frame input outperforms it by 0.43% on *F1*, 0.89% on *Accuracy*,

3.49% on *FP*. On our model, the multi-frame continuous input brings an improvement of 5.49% on *F1*, 2.33% improvement on *Accuracy*, a reduction of 3.43% on *FP*, and a reduction of 4% on *FN*.

To exclude the influence of complexity, we experiment on our method with 5 same frames as input and the results are shown as method D. We found that compared to single frame input, all metrics are worse. This result excludes the influence of complexity and fully demonstrates the effectiveness of multi-frame input.

Effectiveness of Our Model. Based on method B, we generate the method C, replacing the Transformer encoder part with our MHVA module. As shown in Table 3, we find that method C outperforms B by 9.81% on *F1*, 1.1% on *Accuracy*, 3.53% on *FP*, and 3.97% on *FN*. It shows that the strategic information passing is more effective than blind full-pixel information interaction, which indicates the rationality and effectiveness of our MHVA design.

The difference between method C and our method is the decode part. We replace the Transformer decoder with VT modules. Compared with method C, we can see that the addition of the VT module in method F has largely improved the values of metrics. The value of *F1* increases 4.1%, and *Accuracy* increases 1.1%, while *FN* decreases 2.07%, and *FP* decreases 1.9%.

Besides, based on single frame input, the comparison of method D and method A can prove the superiority of our model. We can notice that our model can bring an improvement of 4.05% on *F1*, 0.81% improvement on *Accuracy*, a reduction of 4.5% on *FP*, and a reduction of 0.76% on *FN*.

Table 3. Quantitative results of different models with two kinds of input

Methods	Model	Input	F1 (%)	Accuracy (%)	FP (%)	FN (%)
A	LSTR	Single	70.30	88.40	16.30	14.30
B		5 frames	70.73	89.31	13.93	14.37
C	+MHVA	5 frames	80.54	90.43	10.44	11.44
D	our	Single	79.51	89.21	11.80	13.54
E		5 same	76.85	89.05	14.17	14.34
F		5 frames	84.59	91.54	8.37	9.54

Table 4. Quantitative results of our method with different sampling numbers

Frames	F1 (%)	Accuracy (%)	FP (%)	FN (%)
3 (frames)	83.68	90.68	9.47	10.82
5 (frames)	**84.59**	91.54	**8.37**	9.54
7 (frames)	84.34	**91.80**	8.41	**9.34**

Selection of Sampling Number. To test the influence of the choice of different sampling numbers, we experiment with different frame sampling numbers 3, 5 and 7. The results are shown in Table 4. It can be found that the result of 5 frames outperforms that of 3 frames significantly, and is comparable with the result of 7 frames under smaller GPU memory and less inference time. Therefore, we choose 5 frames as the final sampling number of input data.

5 Conclusions

In conclusion, we propose a novel lane detection Transformer using multiple frames as input. Based on curve fitting, it can detect lanes directly and efficiently. Besides, the customized MHVA can capture more global information in two directions, and the VT modules are very effectual in improving detection result. Our method can achieve real-time results despite the use of multi-frame information, which enables the deployment in practical applications.

References

1. Ghafoorian, M., Nugteren, C., Baka, N., Booij, O., Hofmann, M.: EL-GAN: embedding loss driven generative adversarial networks for lane detection. In: Leal-Taixé, L., Roth, S. (eds.) ECCV 2018. LNCS, vol. 11129, pp. 256–272. Springer, Cham (2019). https://doi.org/10.1007/978-3-030-11009-3_15
2. Hou, Y., Ma, Z., Liu, C., Loy, C.C.: Learning lightweight lane detection CNNs by self attention distillation. In: Proceedings of the IEEE/CVF International Conference on Computer Vision, pp. 1013–1021 (2019)
3. Jung, S., Choi, S., Khan, M.A., Choo, J.: Towards lightweight lane detection by optimizing spatial embedding. arXiv preprint arXiv:2008.08311 (2020)
4. Ko, Y., Lee, Y., Azam, S., Munir, F., Jeon, M., Pedrycz, W.: Key points estimation and point instance segmentation approach for lane detection. IEEE Trans. Intell. Transp. Syst. **23**, 8949–8958 (2021)
5. Loo, M., Lee, J., Lee, D., Kim, W., Hwang, S., Lee, S.: Robust lane detection via expanded self attention. In: Proceedings of the IEEE/CVF Winter Conference on Applications of Computer Vision, pp. 533–542 (2022)
6. Li, X., Li, J., Hu, X., Yang, J.: Line-CNN: end-to-end traffic line detection with line proposal unit. IEEE Trans. Intell. Transp. Syst. **21**(1), 248–258 (2019)
7. Liu, R., Yuan, Z., Liu, T., Xiong, Z.: End-to-end lane shape prediction with transformers. In: Proceedings of the IEEE/CVF Winter Conference on Applications of Computer Vision, pp. 3694–3702 (2021)
8. Liu, Y.B., Zeng, M., Meng, Q.H.: Heatmap-based vanishing point boosts lane detection (2020)
9. Lo, S.Y., Hang, H.M., Chan, S.W., Lin, J.J.: Multi-class lane semantic segmentation using efficient convolutional networks. In: 2019 IEEE 21st International Workshop on Multimedia Signal Processing (MMSP), pp. 1–6. IEEE (2019)
10. Mamidala, R.S., Uthkota, U., Shankar, M.B., Antony, A.J., Narasimhadhan, A.: Dynamic approach for lane detection using google street view and CNN. In: TENCON 2019–2019 IEEE Region 10 Conference (TENCON), pp. 2454–2459. IEEE (2019)

11. Neven, D., De Brabandere, B., Georgoulis, S., Proesmans, M., Van Gool, L.: Towards end-to-end lane detection: an instance segmentation approach. In: 2018 IEEE Intelligent Vehicles Symposium (IV), pp. 286–291. IEEE (2018)
12. Pan, X., Shi, J., Luo, P., Wang, X., Tang, X.: Spatial as deep: spatial CNN for traffic scene understanding. In: Proceedings of the AAAI Conference on Artificial Intelligence, vol. 32 (2018)
13. Pizzati, F., Allodi, M., Barrera, A., García, F.: Lane detection and classification using cascaded CNNs. In: Moreno-Díaz, R., Pichler, F., Quesada-Arencibia, A. (eds.) EUROCAST 2019. LNCS, vol. 12014, pp. 95–103. Springer, Cham (2020). https://doi.org/10.1007/978-3-030-45096-0_12
14. Qin, Z., Wang, H., Li, X.: Ultra fast structure-aware deep lane detection. In: Vedaldi, A., Bischof, H., Brox, T., Frahm, J.-M. (eds.) ECCV 2020. LNCS, vol. 12369, pp. 276–291. Springer, Cham (2020). https://doi.org/10.1007/978-3-030-58586-0_17
15. Ren, S., He, K., Girshick, R., Sun, J.: Faster R-CNN: towards real-time object detection with region proposal networks. In: Advances in Neural Information Processing Systems, vol. 28 (2015)
16. Tabelini, L., Berriel, R., Paixao, T.M., Badue, C., De Souza, A.F., Oliveira-Santos, T.: Keep your eyes on the lane: real-time attention-guided lane detection. In: Proceedings of the IEEE/CVF Conference on Computer Vision and Pattern Recognition, pp. 294–302 (2021)
17. Tabelini, L., Berriel, R., Paixao, T.M., Badue, C., De Souza, A.F., Oliveira-Santos, T.: PolylaneNet: lane estimation via deep polynomial regression. In: 2020 25th International Conference on Pattern Recognition (ICPR), pp. 6150–6156. IEEE (2021)
18. Wang, Y., et al.: End-to-end video instance segmentation with transformers. In: Proceedings of the IEEE/CVF Conference on Computer Vision and Pattern Recognition, pp. 8741–8750 (2021)
19. Wu, B., et al.: Visual transformers: token-based image representation and processing for computer vision. arXiv preprint arXiv:2006.03677 (2020)
20. Yoo, S, et al.: End-to-end lane marker detection via row-wise classification. In: Proceedings of the IEEE/CVF Conference on Computer Vision and Pattern Recognition Workshops, pp. 1006–1007 (2020)
21. Zhang, H., Gu, Y., Wang, X., Wang, M., Pan, J.: SololaneNet: instance segmentation-based lane detection method using locations. In: 2021 IEEE International Intelligent Transportation Systems Conference (ITSC), pp. 2725–2731. IEEE (2021)
22. Zhang, J., Deng, T., Yan, F., Liu, W.: Lane detection model based on spatio-temporal network with double convolutional gated recurrent units. IEEE Trans. Intell. Transp. Syst., 1–13 (2021). https://doi.org/10.1109/tits.2021.3060258
23. Zhang, Y., et al.: VIL-100: a new dataset and a baseline model for video instance lane detection. In: Proceedings of the IEEE/CVF International Conference on Computer Vision, pp. 15681–15690 (2021)
24. Zheng, T., et al.: RESA: recurrent feature-shift aggregator for lane detection. In: Proceedings of the AAAI Conference on Artificial Intelligence, vol. 35, pp. 3547–3554 (2021)
25. Zou, Q., Jiang, H., Dai, Q., Yue, Y., Chen, L., Wang, Q.: Robust lane detection from continuous driving scenes using deep neural networks. IEEE Trans. Veh. Technol. **69**(1), 41–54 (2019)

ProposalContrast: Unsupervised Pre-training for LiDAR-Based 3D Object Detection

Junbo Yin[1], Dingfu Zhou[2,3], Liangjun Zhang[2,3], Jin Fang[2,3,4], Cheng-Zhong Xu[4], Jianbing Shen[4(✉)], and Wenguan Wang[5(✉)]

[1] School of Computer Science, Beijing Institute of Technology, Beijing, China
[2] Baidu Research, Beijing, China
[3] National Engineering Laboratory of Deep Learning Technology and Application, Beijing, China
[4] SKL-IOTSC, CIS, University of Macau, Zhuhai, China
shenjianbingcg@gmail.com
[5] ReLER, AAII, University of Technology Sydney, Ultimo, Australia
wenguanwang.ai@gmail.com
https://github.com/yinjunbo/ProposalContrast

Abstract. Existing approaches for unsupervised point cloud pre-training are constrained to either scene-level or point/voxel-level instance discrimination. Scene-level methods tend to lose local details that are crucial for recognizing the road objects, while point/voxel-level methods inherently suffer from limited receptive field that is incapable of perceiving large objects or context environments. Considering region-level representations are more suitable for 3D object detection, we devise a new unsupervised point cloud pre-training framework, called *ProposalContrast*, that learns robust 3D representations by contrasting region proposals. Specifically, with an exhaustive set of region proposals sampled from each point cloud, geometric point relations within each proposal are modeled for creating expressive proposal representations. To better accommodate 3D detection properties, *ProposalContrast* optimizes with both inter-cluster and inter-proposal separation, *i.e.*, sharpening the discriminativeness of proposal representations across semantic classes and object instances. The generalizability and transferability of *ProposalContrast* are verified on various 3D detectors (*i.e.*, PV-RCNN, CenterPoint, PointPillars and PointRCNN) and datasets (*i.e.*, KITTI, Waymo and ONCE).

Keywords: 3D object detection · Unsupervised point cloud pre-training

1 Introduction

3D object detection from LiDAR point clouds has received great interest in recent years due to its significance to self-driving vehicles. Most existing 3D

J. Yin—Work done during an internship at Baidu Research.

S. Avidan et al. (Eds.): ECCV 2022, LNCS 13699, pp. 17–33, 2022.
https://doi.org/10.1007/978-3-031-19842-7_2

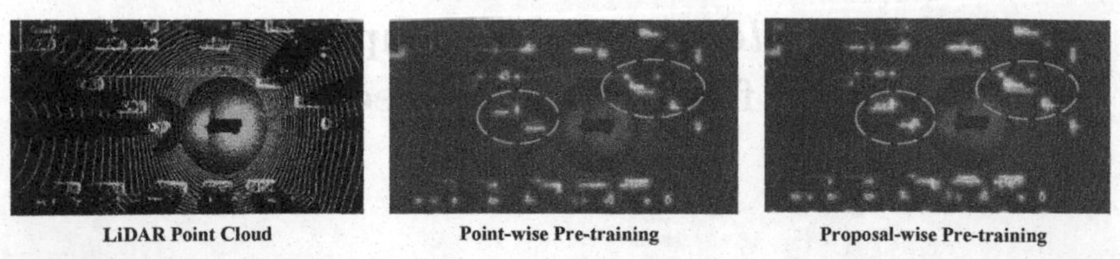

(a) Compared with our proposal-level pre-training, point-level pre-training tends to produce incomplete object representations (as indicated by the dotted circle).

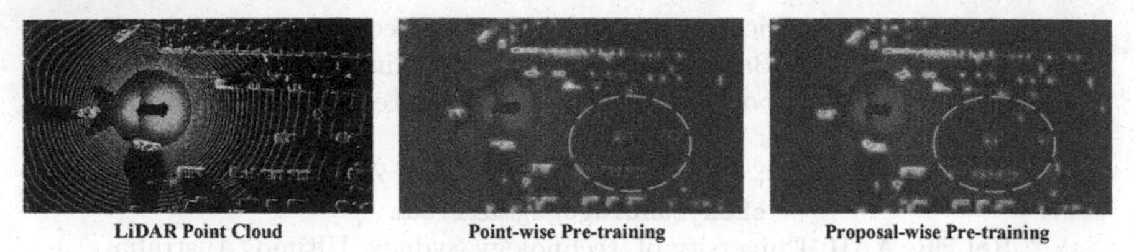

(b) By exploring larger context, our proposal-level pre-training successfully perceives objects that are missed by the point-wise method (as indicated by the dotted circle).

Fig. 1. Comparison of VoxelNet representations learned from PointContrast [38] and our *ProposalContrast.*

detectors are trained on massive labeled data. However, annotating point clouds is expensive and time-consuming. On the other hand, unlabeled point cloud data can be easily generated by self-driving vehicles: it is estimated that a self-driving car could collect 200k point cloud frames within only 8 working hours.

Self-supervised learning (SSL) [5,12,13,20,46] provides a feasible way to make use of unlabeled data. SSL methods typically define a pretext task, where free supervision signals can be derived from the data itself for representation learning. With a few labeled data of downstream tasks, the learnt representations can be fine-tuned and show excellent performance. Recent advances in SSL can be largely ascribed to contrastive learning [5,6,13,14]. Contrastive learning explores the pretext task of instance discrimination, *i.e.*, maximizing the agreement of the feature embeddings between two differently augmented views of the same data instance, and minimizing the agreement between different instances. The instances are often defined as images in the 2D domain. Intuitively, the instance discrimination pretext relies on an *object-centric* assumption [4,48], where the object of interest should lie in the center of an image such that different augmentations can be applied to achieve cross-view consistency. Although this is hold for some image datasets like ImageNet [9], it is suboptimal for LiDAR point cloud datasets; taking Waymo [33] as an example, objects are of smaller sizes (*e.g.*, $4 \times 2\,\mathrm{m}^2$ for car), and are unevenly distributed across a considerably wide range (*e.g.*, $150 \times 150\,\mathrm{m}^2$ for a point cloud scene).

Therefore, how to define the instances is crucial for the adoption of contrastive SSL techniques in point clouds. However, previous approaches for point cloud SSL either directly contrast different views of a whole point cloud scene [15,47], or merely focus on point-/voxel-level instance discrimination [21,38].

The scene-level methods struggle to describe the locality of the road objects, while the point-/voxel-level methods overemphasize fine-grained details, lacking object-level characteristics. Each point cloud consists of several objects and the objects such as vehicles typically contain numerous points and span several voxels. Thus previous approaches take little consideration of the properties of point data, hurting the utility of the learned representations in downstream tasks, such as 3D object detection. In light of the analysis above, we believe learning point representations on the proposal-level is more desired (see Fig. 1 for more evidence).

In this work, we propose a proposal-level point cloud SSL framework, named *ProposalContrast*, that conducts proposal-wise contrastive pre-training for 3D detection-aligned representation learning. In particular, for each point cloud region proposal, its representation is designed to explicitly encode the geometrical relations of the points inside the proposal. This is achieved by a cross-attention based encoding module, which attends each point to its neighbors [22]. To better align proposal-level pre-training with the nature of 3D detection, we propose to jointly optimize two pretext tasks, *i.e.*, inter-proposal discrimination (IPD) and inter-cluster separation (ICS). IPD is for instance-discriminative representation learning. Through minimizing a proposal-wise contrastive loss, proposal representations are encouraged to gather instance-specific characteristics, eventually benefiting the localization of objects. Differently, ICS is for class-discriminative representation learning. Since class label is not available during pre-training, ICS conducts cluster-based contrastive learning. It groups proposals into clusters and enforces consistency between cluster predictions (*i.e.*, pseudo and soft class labels) of different views of each proposal. In this way, ICS encourages the proposal representations to abstract instance-invariant, common patterns, hence facilitating the semantic recognition of objects.

We comprehensively demonstrate the generalizability of our *ProposalContrast* on prevalent 3D detection network architectures, *i.e.*, PV-RCNN [30], CenterPoint [45], PointPillars [19], and PointRCNN [31], as well as empirically validate the transferability on three popular self-driving point cloud datasets, *i.e.*, Waymo [33], KITTI [11], and ONCE [24]. Moreover, *ProposalContrast* can significantly save the annotation cost, *e.g.*, with only half annotations, our pre-trained PV-RCNN outperforms the scratch model trained with full annotations.

2 Related Work

3D Object Detection. Existing solutions for 3D object detection can be broadly grouped into two classes in terms of point representations, *i.e.*, grid-based and point-based. Specifically, grid-based methods [19,39,43,44,49] typically first discretize the raw point clouds into regular girds (*e.g.*, voxels [39] or pillars [19]), which can then be processed by 3D or 2D convolutional networks. Point-based methods [25,26,28,31,40] directly extract features and predict 3D objects from raw point clouds. In general, grid-based approaches are efficient but suffer from information loss in the quantification process. Point-based methods yield impressive results, but the computational cost is high for large-scale point clouds. Hence some recent detectors [30], built upon the hybrid of voxel- and point-based architectures, are developed to enjoy the best of both worlds. In this work, we show that a wide range of modern 3D detectors can benefit from our self-supervised pre-training algorithm, which can learn meaningful and transferable representations from large-scale unlabeled LiDAR point clouds.

Self-supervised Learning (SSL) in Point Cloud. SSL [14,16,18,27,41,46] is to learn expressive feature representation without manual annotations. Recently, contrastive learning based SSL algorithms [5,13,14,36,42] proved impressive results on various downstream tasks, even surpassing the supervised alternatives. In this article, we follow this paradigm with a proposal-level pretraining method specifically designed for the task of point cloud object detection. Concurrent to our study, PointContrast [38], DepthContrast [47], GCC-3D [21], and STRL [15] also exploit the potential of contrastive SSL in point cloud pretraining. Though impressive, these methods have a few limitations. First, [15,47] takes the whole point cloud scene as the instance, neglecting the underlying *object-centric* assumption [4,48], since self-driving point cloud scenes typically comprise of multiple object instances. Second, other methods like [21,38] only take into account instance discrimination at the point-/voxel-level. Thus they are hard to acquire object-level representations that are compatible with 3D object detection. Third, [15,38,47] overlook the semantic relations among instances, focusing on modeling low-level characteristics instead of high-level informative patterns. In [21], although an extra self-clustering strategy is introduced for capturing semantic properties, it only provides supervisory signals on moving voxels which are too sparse to cover potential object candidates, and leads to a complicated two-stage pipeline that trains the 3D encoder and 2D encoder separately. Differently, our point cloud pretraining method encourages discrimination between proposal instances and clusters, hence comprehensively capturing the properties of point cloud data and well aligning with 3D detection. Although RandomRooms [29] also considers region-level representations, it refers to synthetic CAD objects in indoor environments. In contrast, we automatically mine potential object instances from raw point data in self-driving scenes.

3 Approach

In Sect. 3.1, we first briefly describe our proposal-level pre-training method specifically designed for 3D object detection. Then, we detail the crucial components, *i.e.*, region proposal encoding module (in Sect. 3.2) and joint optimization of inter-proposal discrimination and inter-cluster separation (in Sect. 3.3).

3.1 Overview of *ProposalContrast*

Due to the specific characteristics of 3D LiDAR point cloud data, such as the irregular and sparse structures across large perception ranges, simply applying 2D pre-training techniques to point clouds cannot get satisfactory results. This calls for better adapting existing pre-training techniques to the inherent

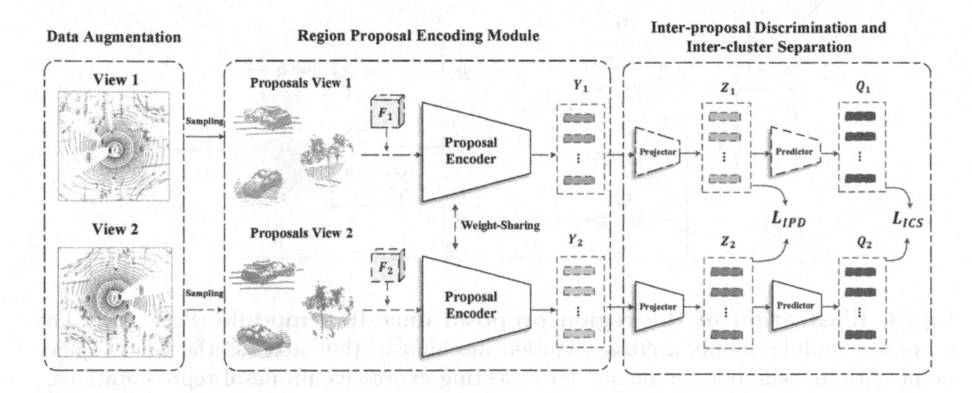

Fig. 2. Illustration of our *ProposalContrast* framework. Given augmented point cloud with different views, we first sample paired region proposals and then extract the features with a region proposal encoding module. After that, inter-proposal discrimination and inter-cluster separation are enforced to optimize the whole network.

structures of point cloud data. Rather than current point cloud pretraining methods investigating the unsupervised representation learning on the scene-/point-/voxel-level [15,38,38,47], our *ProposalContrast* learns representations by contrasting directly on region proposals.

As shown in Fig. 2, to achieve proposal-level pre-training, *ProposalContrast* has five core components: data augmentation, correspondence mapping, region proposal generation, region proposal encoding, as well as a joint optimization of inter-proposal discrimination (IPD) and inter-cluster separation (ICS).

Data Augmentation. Let $X_0 \in \mathbb{R}^{L_0 \times 3}$ denote an input point cloud with L_0 points (here we describe point cloud with 3D coordinates for simplicity). We apply two different data augmentation operators $\mathcal{T}_1, \mathcal{T}_2$ on X_0 to produce two augmented views X_1, X_2:

$$X_1 = \mathcal{T}_1(X_0) \in \mathbb{R}^{L_1 \times 3}, \quad X_2 = \mathcal{T}_2(X_0) \in \mathbb{R}^{L_2 \times 3}, \tag{1}$$

where L_1 and L_2 are number of points of X_1 and X_2. The data augmentation strategies include random rotation, flip, scaling and random point drop out (see Sect. 4.1 for detailed definition). Note that random rotation is only applied on the upright axis since we are aware of self-driving scenarios.

Correspondence Mapping. Before generating proposals, we first get the correspondence mapping M between point sets in X_1 and X_2. This can be easily achieved by recording the index of each point in the original view X_0. M is later used for sampling and grouping region proposals.

Region Proposal Generation. Some SSL methods for 2D representation learning make use of image proposals in the form of 2D bounding boxes [4]. For point clouds, straightforwardly representing proposals as 3D bounding boxes, however, is not a feasible choice, due to the significantly enlarged space of object candidates in 3D scenarios. In addition, generating a dense set of 3D bounding boxes will lead to unacceptable high computational cost. These considerations motivate

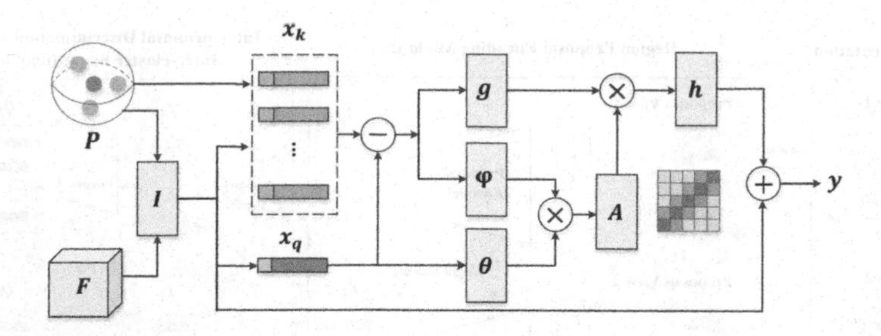

Fig. 3. Illustration of the region proposal encoding module (Sect. 3.2). The encoding module adopts a cross-attention mechanism that attends the center query point with its neighbor key points, for collecting expressive proposal representations.

us to adopt spherical proposals, instead of 3D bounding box proposals. Specifically, given the input point cloud X_0, we first abandon the road plane points so as to sample less from the background [1]. Then we perform farthest point sampling (FPS) [28] on X_0 to sample a total of N points as the centers of N spherical proposals (the corresponding samples in X_1 and X_2 can also be identified according to the correspondence M). FPS encourages the sampled points to be away from each other and thus guarantees the diversity of the sampled proposals. Next, spherical proposals are generated by searching K points around each sampled center point within a pre-defined radius r. Finally, we get two sets of spherical proposals $P_1, P_2 \in \mathbb{R}^{N\times K\times 3}$ from the two views, *i.e.*, $P_1 \in X_1, P_2 \in X_2$.

Region Proposal Encoding. An encoding module is further adopted to extract expressive proposal representations, by considering the geometric relations of the points inside each proposal. Given the scene-level representations, *i.e.*, $\boldsymbol{F}_1$ and $\boldsymbol{F}_2$, of X_1 and X_2, extracted by the backbone network, the proposal encoding module outputs geometry-aware proposal representations $\boldsymbol{Y}_1, \boldsymbol{Y}_2 \in \mathbb{R}^{N\times C}$ for P_1 and P_2:

$$\boldsymbol{Y}_1 = f_{\text{En}}(P_1, \boldsymbol{F}_1), \quad \boldsymbol{Y}_2 = f_{\text{En}}(P_2, \boldsymbol{F}_2), \tag{2}$$

where the encoding module f_{En} is achieved by a neural attention mechanism, which will be detailed in Sect. 3.2.

Joint Optimization. The proposal features $\boldsymbol{Y}_1$, $\boldsymbol{Y}_2$ from different views are learned by enforcing both cluster-based class consistency and instance-wise discrimination. This endows $\boldsymbol{Y}_1$, $\boldsymbol{Y}_2$ with desired properties for 3D object detection. The detailed design is presented in Sect. 3.3.

3.2 Region Proposal Encoding Module

The relations among points in a proposal provide crucial geometry information for describing the proposal. To extract better proposal representations, we leverage the cross-attention [10] to capture the geometric relations between points.

For a point cloud scene X and corresponding proposals P, we extract its global scene-wise representation through a backbone network, *i.e.*, $\boldsymbol{F} =$

$f_{\text{Bbone}}(X)$. Popular 3D backbones like VoxelNet [39] and PointNet++ [28] can be used to instantiate f_{Bbone}.

Then we obtain initial representation $\boldsymbol{P} \in \mathbb{R}^{N\times K\times C}$ for all the proposals P by applying bilinear interpolation function $I(\cdot)$ over $\boldsymbol{F}$, *i.e.*, $\boldsymbol{P} = I(P, \boldsymbol{F})$, where N is the proposal number of one view, K is the point number inside a proposal and C is the backbone channel number.

After that, we capture the geometrical relation between points via cross attention, *i.e.*, attending a query point with other key points. Let $\boldsymbol{p} \in \boldsymbol{P}$ with $K \times C$ size denote the initial representation of a proposal $p \in P$. As illustrated in Fig. 3, we set the center point feature $\boldsymbol{x}_q \in \mathbb{R}^{1\times C}$ of proposal p, *i.e.*, $\boldsymbol{x}_q \in \boldsymbol{p}$ as the query, since the center point is more informative [45]. Next, we get the neighbor features $\boldsymbol{x}_k \in \mathbb{R}^{K\times C}$, where $\boldsymbol{x}_k \in \boldsymbol{p}$, and use the difference between $\boldsymbol{x}_k$ and $\boldsymbol{x}_q$ as keys to encode the asymmetric geometry relation as recommended by [37]. The point coordinate of $\boldsymbol{x}_q$ and $\boldsymbol{x}_k$ are also integrated to provide explicit position information.

Formally, the $\boldsymbol{x}_q$ and $\boldsymbol{x}_k$ are first projected to query, key, and value vectors through:

$$\boldsymbol{w}_q = \theta(\boldsymbol{x}_q), \quad \boldsymbol{w}_k = \phi(\boldsymbol{x}_k - \boldsymbol{x}_q), \quad \boldsymbol{w}_v = g(\boldsymbol{x}_k - \boldsymbol{x}_q), \tag{3}$$

where $\theta(\cdot)$, $\phi(\cdot)$, and $g(\cdot)$ are linear layers. Afterwards, we compute the attention weight $A \in [0, 1]^K$ based on normalized similarities between the query $\boldsymbol{w}_q$ and each key $\boldsymbol{w}_k$. The attention weight is then applied on the value vector $\boldsymbol{w}_v$ to aggregate information from all the keys:

$$\boldsymbol{w}_o = \sum\nolimits_{\boldsymbol{w}_k} A(\boldsymbol{w}_q, \boldsymbol{w}_k) \cdot \boldsymbol{w}_v, \quad A(\boldsymbol{w}_q, \boldsymbol{w}_k) = \frac{\boldsymbol{w}_q^\top \boldsymbol{w}_k}{\sum_{\boldsymbol{w}_k} \boldsymbol{w}_q^\top \boldsymbol{w}_k}. \tag{4}$$

As seen, the attention weight encodes the geometry relation of each point pair under a metric space, which makes the center point aware of the informative neighborhoods and thus encourages effective information exchange between points inside a proposal. Finally, the representation of proposal p is given as:

$$\boldsymbol{y} = \boldsymbol{x}_q + h(\boldsymbol{w}_o), \tag{5}$$

where $h(\cdot)$ is a linear layer. In this way, by applying Eqs. (3–5) on each the region proposal in X_1 and X_2, we got all the proposal representations, *i.e.*, $\boldsymbol{Y}_1, \boldsymbol{Y}_2 \in \mathbb{R}^{N\times C}$, which will be trained by joint optimization of inter-proposal discrimination and inter-cluster separation.

3.3 Joint Optimization of Inter-proposal Discrimination and Inter-cluster Separation

A good point cloud representation for 3D detection is desired to be discriminative across both instances and classes; object instance-sensitive representation benefits the localization of 3D objects, while class-discriminative representation is crucial for the recognition of object categories. We therefore propose to optimize with both inter-proposal discrimination (IPD) and inter-cluster separation (ICS) simultaneously. Because annotations for object bounding box and semantics labels are not given during pretraining, IPD is designed to contrast the representations on the proposal-level, while ICS discovers stable and informative 3D patterns through grouping the proposals into clusters.

Inter-proposal Discrimination (IPD). For object instance-sensitive representation learning, we conduct proposal-level contrastive learning [5]. Specifically, given the proposal representations, *i.e.*, $\boldsymbol{Y}_1, \boldsymbol{Y}_2 \in \mathbb{R}^{N\times C}$, from the two different views X_1, X_2. We aim to pull positive pairs (*e.g.*, the same proposal with different views) close, as well as push negative pairs (*e.g.*, the different proposals) apart.

Following the common practice in contrastive learning, we first adopt a projection layer $f_{\text{Proj}}(\cdot)$ to project each proposal representation $\boldsymbol{y} \in \{\boldsymbol{Y}_1, \boldsymbol{Y}_2\}$ to a ℓ_2-normalized embedding space:

$$\boldsymbol{z} = \frac{f_{\text{Proj}}(\boldsymbol{y})}{||f_{\text{Proj}}(\boldsymbol{y})||_2} \in [0,1]^C. \tag{6}$$

Given the sets of normalized proposal embeddings $\boldsymbol{Z}_1, \boldsymbol{Z}_2 \in [0,1]^{N\times C}$, the IPD loss is designed in a form of the InfoNCE loss [27]:

$$\mathcal{L}_{\text{IPD}} = \frac{1}{N}\sum_{\boldsymbol{z}_1^n \in \boldsymbol{Z}_1} -\log \frac{\exp(\boldsymbol{z}_1^{n\top} \cdot \boldsymbol{z}_2^{n'}/\tau)}{\sum_{\boldsymbol{z}_2^m \in \boldsymbol{Z}_2} \exp(\boldsymbol{z}_1^{n\top} \cdot \boldsymbol{z}_2^{m}/\tau)} + \frac{1}{N}\sum_{\boldsymbol{z}_2^m \in \boldsymbol{Z}_2} -\log \frac{\exp(\boldsymbol{z}_2^{m\top} \cdot \boldsymbol{z}_1^{m'}/\tau)}{\sum_{\boldsymbol{z}_1^n \in \boldsymbol{Z}_1} \exp(\boldsymbol{z}_2^{m\top} \cdot \boldsymbol{z}_1^{n}/\tau)}, \tag{7}$$

where τ is a temperature hyper-parameter. For $\boldsymbol{z}_1^n$, $\boldsymbol{z}_2^{n'}$ refers to a positive sample; $\boldsymbol{z}_1^n$ and $\boldsymbol{z}_2^{n'}$ correspond to a same proposal in X_0. Similarly, $\boldsymbol{z}_2^m$ is a negative sample for $\boldsymbol{z}_1^n$, *i.e.*, $\boldsymbol{z}_2^m \neq \boldsymbol{z}_2^{n'}$.

Inter-cluster Separation (ICS). For class-discriminative representation learning, we group the spherical proposals into different clusters, which can be viewed as pseudo class labels. Inspired by recent clustering-based contrastive SSL [2], ICS is designed to maximize the agreement between cross-view cluster assignments.

For each normalized proposal feature $\boldsymbol{z}_1^n \in \boldsymbol{Z}_1$ from view X_1, we first apply a predictor $f_{\text{Pred}}(\cdot)$ to map it to vector $\boldsymbol{q}_1^n \in \mathbb{R}^O$ that represents its pseudo-class embedding, such that $\boldsymbol{q}_1^n = f_{\text{Pred}}(\boldsymbol{z}_1^n)$, where the output channel dimension O of $\boldsymbol{q}_1^n$ refers to the number of pseudo classes. To get cluster assignment, *i.e.*, $\hat{\boldsymbol{q}}_1^n \in [0,1]^O$, we stop gradient on $\boldsymbol{q}_1^n$ and adopt Sinkhorn-Knopp algorithm [7] to group all the proposals in each training batch into O clusters. Then the training target of ICS is defined as:

$$\mathcal{L}_{\text{ICS}} = \frac{1}{N}\sum\nolimits_{\boldsymbol{q}_1^n} -\hat{\boldsymbol{q}}_1^n \log \sigma(\boldsymbol{q}_2^{n'}) + \frac{1}{N}\sum\nolimits_{\boldsymbol{q}_2^m} -\hat{\boldsymbol{q}}_2^m \log \sigma(\boldsymbol{q}_1^{m'}), \tag{8}$$

where $\sigma(\cdot)$ refers to the softmax function that maps the pseudo-class embedding into class probability distribution. $\mathcal{L}_{\text{ICS}}$ encourages the cross-view clustering consistency of each proposal, *i.e.*, use the cluster assignment $\hat{\boldsymbol{q}}_1^n$ (resp. $\hat{\boldsymbol{q}}_2^m$) of view X_1 (resp. X_2) as pseudo grouptruth to supervise the class probability distribution $\sigma(\boldsymbol{q}_2^{n'})$ (resp. $\sigma(\boldsymbol{q}_1^{m'})$). Note that $\hat{\boldsymbol{q}}_1^n$ and $\boldsymbol{q}_2^{n'}$ correspond to a same proposal in X_0.

Finally, the overall self-supervised learning loss is defined as:

$$\mathcal{L} = \alpha \mathcal{L}_{\text{IPD}} + \beta \mathcal{L}_{\text{ICS}}, \tag{9}$$

where α and β are the balancing coefficients, respectively. In Sect. 4.3, we provide analysis on the effectiveness of the two optimization targets.

4 Experimental Results

4.1 Pre-training Settings

Datasets. We adopt the common experimental protocol of SSL, *i.e.*, first pre-training a backbone network with large-scale unlabeled data and then fine-tuning it on downstream tasks with much fewer labeled data. Some previous 3D SSL methods make use of ShapeNet [3] and ScanNet [8] datasets to pre-train the 3D backbones, thus they only focus on the indoor setting and suffer from large domain gap when transferring to the self-driving setting. In our experiments, we adopt Waymo Open Dataset [33] for the self-supervised pre-training. The Waymo dataset contains 798 scenes (158,361 frames) for training and 202 scenes (40,077 frames) for validation; it is 20× larger than KITTI [11]. We adopt the whole training set to pre-train various 3D backbones without using the labels.

Network Architectures. For thorough examination of the efficacy and versatility of our approach, we investigate the performance of *ProposalContrast* on diverse 3D backbone architectures, including grid-based, *i.e.*, VoxelNet [39] and PointPillars [19], as well as point-based, *i.e.*, PointNet++ [28]. The projection layer, $f_{\text{Proj}}(\cdot)$, in Sect. 3.3 is implemented as two linear layers, with the first layer followed by a batch normalization (BN) layer and a ReLU. The channel dimension of the output is set as $C = 128$. The predictor $f_{\text{Pred}}(\cdot)$, implemented as a linear layer, outputs a 128-d vector as the pseudo-class embedding, *i.e.*, $O = 128$. All the functions in the proposal encoding module are instantiated by linear layers with 128 channels. The attention linear head $h(\cdot)$ transforms the attended features to the original backbone channels.

Implementation Details. We empirically consider four types of data augmentations to generate different views, including random rotation ($[-180°, 180°]$), random scaling ([0.8, 1.2]), random flipping along X-axis or Y-axis, and random point drop out. For random point drop out, we sample 100k points from the original point cloud for each of the two augmented views. 20k points are chosen from the same indexes to ensure a 20% overlap for the two augmented views, while the other 80k points are randomly sampled from the remained point clouds. We sample $N = 2048$ spherical proposals for every point cloud frame; each proposal contains $K = 16$ points within $r = 1.0$ m radius. The parameters for the VoxelNet [39] backbone are the same as the corresponding 3D object detectors. The temperature parameter τ in the IPD loss $\mathcal{L}_{\text{IPD}}$ (Eq. 7) is set to 0.1. The coefficients α and γ in Eq. 9 are both set to 1 empirically. We pre-train the models for 36 epochs, and use Adam optimizer [17] to optimize the network. Cosine learning rate schedule [23] is adopted with warmup strategy in the first 5 epochs. The maximum learning rate is set to 0.003.

Table 1. Data-efficient 3D Object Detection on KITTI. We pre-train the backbones of PointRCNN [31] and PV-RCNN [30] on Waymo and transfer to KITTI 3D object detection with different label configurations. Consistent improvements are obtained under each setting. Our approach outperforms all the concurrent self-supervised learning methods, *i.e.*, DepthContrast [47], PointContrast [38], GCC-3D [21], and STRL [15].

Fine-tuning with various label ratios	Detector	Pre-train. Schedule	mAP (Mod.)	Car Easy	Car Mod.	Car Hard	Pedestrian Easy	Pedestrian Mod.	Pedestrian Hard	Cyclist Easy	Cyclist Mod.	Cyclist Hard
20% (∼ 0.7k frames)	PointRCNN	Scratch	63.51	88.64	75.23	72.47	55.49	48.90	42.23	85.41	66.39	61.74
		Ours	**66.20**$_{+2.69}$	88.52	**77.02**$_{+1.79}$	72.56	58.66	**51.90**$_{+3.00}$	44.98	90.27	**69.67**$_{+3.28}$	65.05
	PV-RCNN	Scratch	66.71	91.81	82.52	80.11	58.78	53.33	47.61	86.74	64.28	59.53
		Ours	**68.13**$_{+1.42}$	91.96	**82.65**$_{+0.13}$	80.15	62.58	**55.05**$_{+1.72}$	50.06	88.58	**66.68**$_{+2.40}$	62.32
50% (∼ 1.8k frames)	PointRCNN	Scratch	66.73	89.12	77.85	75.36	61.82	**54.58**	47.90	86.30	67.76	63.26
		Ours	**69.23**$_{+2.50}$	89.32	**79.97**$_{+2.12}$	77.39	62.19	54.47$_{-0.11}$	46.49	92.26	**73.25**$_{+5.69}$	68.51
	PV-RCNN	Scratch	69.63	91.77	82.68	81.9	63.70	57.10	52.77	89.77	69.12	64.61
		Ours	**71.76**$_{+2.13}$	92.29	**82.92**$_{+0.24}$	82.09	65.82	**59.92**$_{+2.82}$	55.06	91.87	**72.45**$_{+3.33}$	67.53
100% (∼ 3.7k frames)	PointRCNN	Scratch	69.45	90.02	**80.56**	78.02	62.59	55.66	48.69	89.87	72.12	67.52
		DepthCon. [47]	70.26$_{+0.81}$	89.38	80.32$_{-0.24}$	77.92	65.55	57.62$_{+1.96}$	50.98	90.52	72.84$_{+0.72}$	68.22
		Ours	**70.71**$_{+1.26}$	89.51	80.23$_{-0.33}$	77.96	66.15	**58.82**$_{+3.16}$	52.00	91.28	**73.08**$_{+0.96}$	68.45
	PV-RCNN	Scratch	70.57	–	84.50	–	–	57.06	–	–	70.14	–
		GCC-3D [21]	71.26$_{+0.69}$	–	–	–	–	–	–	–	–	–
		STRL [15]	71.46$_{+0.89}$	–	84.70$_{+0.20}$	–	–	57.80$_{+0.74}$	–	–	71.88$_{+1.74}$	–
		PointCon. [38]	71.55$_{+0.98}$	91.40	84.18$_{-0.32}$	82.25	65.73	57.74$_{+0.68}$	52.46	91.47	72.72$_{+2.58}$	67.95
		Ours	**72.92**$_{+2.35}$	92.45	**84.72**$_{+0.22}$	82.47	68.43	**60.36**$_{+3.30}$	55.01	92.77	**73.69**$_{+3.55}$	69.51

4.2 Transfer Learning Settings and Results

In this paper, we investigate self-supervised pre-training in an autonomous driving setting. We evaluate our approach on several widely used LiDAR point cloud datasets, *i.e.*, KITTI [11], Waymo [33] and ONCE [24]. Specifically, we compare our *ProposalContrast* with other pre-training methods by fine-tuning the detection models. Different amounts of labeled data are used for fine-tuning to show the data-efficient ability. Besides, various modern 3D object detectors are involved to demonstrate the generalizability of our pre-trained models.

KITTI Dataset. KITTI 3D object benchmark [11] has been widely used for 3D object detection from LiDAR point cloud. It contains 7,481 labeled samples, which are divided into two groups, *i.e.*, a training set (3,712 samples) and a validation set (3,769 samples). Mean Average Precision (mAP) with 40 recall positions are usually adopted to evaluate the detection performance, with a 3D IoU thresholds of 0.7 for cars and 0.5 for pedestrians and cyclists.

We assess the transferability of our pre-trained model by pre-training on Waymo then fine-tuning on KITTI. Two typical 3D object detectors, *i.e.*, PointRCNN [31] and PV-RCNN [30], are used as the baselines. The two detectors are based on different 3D backbones (*i.e.*, point-wise or voxel-wise networks), covering most cases of mainstream 3D detectors. A crucial advantage of self-supervised pre-training is to improve data efficiency for the downstream task with limited annotated data. To this end, we evaluate the data-efficient 3D object detection. In particular, the training samples are split into three groups, with each containing 20% (0.7k), 50% (1.8k) and 100% (3.7k) labeled samples, respectively. The experimental results are shown in Table 1. On both 3D detectors, our self-supervised pre-trained model effectively improves the performance

Table 2. Comparisons between our model and other self-supervised learning methods on Waymo. All the detectors are trained by 20% training samples following the OpenPCDet [34] configuration and evaluated on the validation set. Both PV-RCNN [30] and CenterPoint [45] are used as beseline detectors.

3D Object Detector	Transfer Paradigm	Overall AP/APH	Vehicle AP/APH	Pedestrian AP/APH	Cyclist AP/APH
SECOND [39]	Scratch	55.08/51.32	59.57/59.04	53.00/43.56	52.67/51.37
Part-A^2-Anchor [32]	Scratch	60.39/57.43	64.33/63.82	54.24/47.11	62.61/61.35
PV-RCNN [30]	Scratch	59.84/56.23	64.99/64.38	53.80/45.14	60.72/59.18
GCC-3D (PV-RCNN) [21]	Fine-tuning	61.30/58.18$_{(+1.46/+1.95)}$	65.65/65.10	55.54/48.02	62.72/61.43
Ours (PV-RCNN)	Fine-tuning	**62.62/59.28**$_{(+2.78/+3.05)}$	66.04/65.47	57.58/49.51	64.23/62.86
CenterPoint [45]	Scratch	63.46/60.95	61.81/61.30	63.62/57.79	64.96/63.77
GCC-3D (CenterPoint) [21]	Fine-tuning	65.29/62.79$_{(+1.83/+1.84)}$	63.97/63.47	64.23/58.47	67.68/66.44
Ours (CenterPoint)	Fine-tuning	**66.42/63.85**$_{(+2.96/+2.90)}$	64.94/64.42	66.13/60.11	68.19 67.01
CenterPoint-Stage2 [45]	Scratch	65.29/62.47	64.70/64.11	63.26/58.46	65.93/64.85
GCC-3D (CenterPoint-Stage2) [21]	Fine-tuning	67.29/64.95$_{(+2.00/+2.48)}$	66.45/65.93	66.82/61.47	68.61/67.46
Ours (CenterPoint-Stage2)	Fine-tuning	**68.06/65.69**$_{(+2.77/+3.22)}$	66.98/66.48	68.15/62.61	69.04/67.97

in comparison to the model trained from scratch. Our model also outperforms several concurrent works. For example, based on PV-RCNN detector, our model exceeds STRL [15] and GCC-3D [21] by 1.46% and 1.66%, respectively. Our model also surpasses DepthContrast [47] and PointContrast [38], thanks to the proposal-level representation. Besides, the classes with fewer labeled instances (*e.g.*, pedestrian and cyclist) are improved a lot, showcasing the ability to address imbalanced class distribution. More importantly, with our pre-trained model, PointRCNN and PV-RCNN using half annotation achieve comparative performance compared with the counterparts with full annotations. This also suggests the potential of our approach in reducing the heavy annotation burden.

Waymo Open Dataset. Waymo dataset [33] contains three classes: vehicles, pedestrians, and cyclists. 3D Average Precision (AP) and Average Precision with Heading (APH) are defined as the evaluation metrics for all classes. The AP and APH are based on IoU thresholds of 0.7 for vehicles and 0.5 for pedestrians and cyclists. Two difficulty levels, *i.e.*, LEVEL_1 and LEVEL_2 are defined according to the points number in the bounding boxes, where we mainly consider the LEVEL_2 metric.

We follow the schedule of OpenPCDet [34] to fine-tune the detectors on 20% training samples for 30 epochs. In particular, we first report the training-from-scratch results of SECOND [39], Part-A^2-Anchor [32], CenterPoint [45] (VoxelNet version) and PV-RCNN [30], in terms of GCC3D [21]. After that, we apply our *ProposalContrast* on two strong baselines, CenterPoint [45] and PV-RCNN [30], to verify our model. According to Table 2, with our self-supervised pre-training, the performances of popular 3D detectors are substantially improved. For PV-RCNN [30], we improve the model training from scratch by 3.05% APH, as well as outperform GCC-3D [21] by 1.10% APH on average. We further evaluate our pre-trained model on CenterPoint with VoxelNet backbone. The experimental results show that our approach improves 2.9% and 1.06% APH, compared with training from scratch and GCC-3D [21], respectively. Furthermore, based on the two-stage CenterPoint, our model reaches 65.69% APH, improving the model trained from scratch by 3.22%.

Table 3. Data-efficient 3D object detection on Waymo dataset. Our *ProposalContrast* consistently improves the performance of modern 3D object detectors, especially when only limited labeled data are available. In each row, we present the results trained from scratch on the top and show the fine-tuning results at the bottom.

Fine-tuning with various label ratios	Detector	Relative Gain	3D AP/APH (LEVEL 2)			
			Overall	Vehicle	Pedestrian	Cyclist
1% (~ 0.8k frames)	PointPillars [19]	**+8.60/+8.26**	23.05/18.08	27.15/26.17	30.31/18.79	11.68/9.28
			31.65/26.34	35.88/35.08	37.61/25.22	21.47/18.73
	VoxelNet [39]	**+17.48/+16.95**	20.88/17.83	21.95/21.45	27.98/20.52	12.70/11.53
			38.36/34.78	37.60/36.91	39.74/31.70	37.74/35.73
10% (~ 8k frames)	PointPillars [19]	**+2.33/+2.85**	51.75/46.58	54.94/54.32	54.01/41.53	46.31/43.88
			54.08/49.43	57.54/56.93	56.97/45.25	47.74/46.1
	VoxelNet [39]	**+4.96/+5.06**	54.04/51.24	54.37/53.74	51.45/45.05	56.30/54.93
			59.00/56.30	58.83/58.23	57.75/51.75	60.42/58.91
50% (~ 40k frames)	PointPillars [19]	**+1.06/+1.08**	59.77/55.58	61.89/61.32	61.89/51.26	55.54/54.16
			60.83/56.66	63.01/62.44	62.57/52.02	56.91/55.53
	VoxelNet [39]	**+1.14/+1.05**	63.51/61.05	63.18/62.66	63.35/57.67	63.99/62.82
			64.65/62.10	64.07/63.54	64.64/58.71	65.24/64.04
100% (~ 80k frames)	PointPillars [19]	**+0.54/+0.40**	61.68/57.92	63.95/63.39	62.91/53.38	58.17/57.01
			62.22/58.32	64.05/63.85	63.51/53.69	58.66/57.41
	VoxelNet [39]	**+0.55/+0.61**	64.84/62.29	64.38/63.86	66.05/60.06	64.09/62.95
			65.39/62.90	64.67/64.16	66.52/60.65	64.97/63.88

Table 4. 3D Object Detection Performance on ONCE validation set. The improved CenterPoint achieves the best performance among these SOTA detectors.

Methods	mAP	Orientation-aware AP		
		Vehicle	Pedestrian	Cyclist
PointPillars [19]	44.34	68.57	17.63	46.81
SECOND [39]	51.89	71.19	26.44	58.04
PV-RCNN [30]	53.55	77.77	23.50	59.37
CenterPoint [30]	60.05	66.79	49.90	63.45
PointPainting [35]	57.78	66.17	44.84	62.34
CenterPoint* [45]	64.24	75.26	51.65	65.79
Ours (CenterPoint*)	$\mathbf{66.24}_{+2.00}$	$\mathbf{78.00}_{+2.74}$	$\mathbf{52.56}_{+0.91}$	$\mathbf{68.17}_{+2.38}$

We also evaluate our model under a data-efficient 3D object detection setting. To be specific, we split the Waymo training set into two groups, with each group containing 399 scenes (~80k frames). We first conduct pre-training on one group without using the labels, and then fine-tune the pre-trained model with labels on another group. During fine-tuning, various fractions of training data are uniformly sampled: 1% (0.8k frames), 10% (8k frames), 50% (40k frames) and 100% (80k frames). Two different backbones, *i.e.*, PointPillars [19] and VoxelNet [39] based on CenterPoint, are involved to measure the performance of our pre-trained model. The detection model trained from random initialization is viewed as the baseline. The advantage of our pre-trained model over the baseline is presented in Table 3. In essence, our pre-trained model can consistently

Table 5. Ablation studies on different granularities of instance features.

Methods	Initialization	Overall (mAP/mAPH)	Vehicle	Pedestrian	Cyclist
Baseline	Random initialization	59.63/56.96	58.96/58.40	56.72/50.77	63.20/61.70
PointContrast [38]	Point-level pre-train	60.32/57.75$_{(+0.69/+0.79)}$	60.47/59.85	56.67/50.78	63.82/62.63
DepthContrast [47]	Scene-level pre-train	60.71/57.96$_{(+1.08/+1.00)}$	60.11/59.62	58.42/52.27	63.59/62.00
ProposalContrast	Proposal-level pre-train	**62.75/60.13**$_{(+3.12/+3.17)}$	61.20/60.64	60.75/54.83	66.29/64.93

Table 6. Ablation studies on each module of *ProposalContrast*.

Modules	Aspect	Param.	mAP/mAPH
Baseline	Random Init.	-	59.63/56.96
MaxPooling	N=2048, r=1.0	-	62.09/59.32
Attentive Proposal Encoder	.5Proposal .5Number (N)	1024	62.10/59.47
		2048	**62.75/60.13**
		4096	62.54/59.91
	Spherical Radius (r)	0.5	62.43/59.81
		1.0	**62.75/60.13**
		2.0	62.36/59.55

Modules	Aspect	Param.	mAP/mAPH
Baseline	Random Init.	-	59.63/56.96
IPD task	*w/o* ICS	-	62.17/59.56
.5ICS task	#Cluster (*w/o* IPD)	64	61.47/58.87
		128	61.77/59.16
		256	61.36/58.91
	w/o SKC	128	57.29/54.52
IPD + ICS	#Cluster	128	**62.75/60.13**

promote detection performance on both backbones, especially when the labled data is scarce, *i.e.*, improving 8.26% and 16.95% APH with 1% labeled data. Our model also outperforms the baselines under all label settings.

ONCE Dataset. ONCE [24] is a newly released dataset for 3D object detection in autonomous driving. It involves 5k frames for training or fine-tuning and 3k frames for validation. An orientation-aware AP is introduced to account for objects with opposite orientations. ONCE evaluates 3 categories for 3D object detection: vehicle, pedestrian and cyclist. The official ONCE benchmark provides some results from popular 3D detectors on the validation set. Since our implemented CenterPoint* achieves much better results than the official version, we use it as the baseline. We pre-train the VoxelNet backbone of CenterPoint* on Waymo with our *ProposalContrast* and fine-tune it with the ONCE training set. Then we give a performance comparison with several 3D detectors on the validation set. As shown in Table 4, our model improves the baseline by 2.00% mAP, achieving better performance on the validation set. This gives another evidence of the transferability of our pre-trained model.

4.3 Ablation Study

In this section, we examine our *ProposalContrast* model in depth. We conduct each group experiment by pre-training the VoxelNet backbone on the full Waymo training set in an unsupervised manner, and evaluate the performance by fine-tuning the detector on Waymo 20% training data. CenterPoint trained from random initialization is viewed as the baseline. 1× schedule (12 epochs) in [45] is used to save computation on the large-scale Waymo.

Comparison to Point-/Scene-level Pre-training. The main contribution of this work is to propose a proposal-wise pre-training paradigm. To show its merits over previous contrastive learning methods in point cloud, we re-implement the pioneer works like PointContrast [38] and DepthContrast [47] based on the VoxelNet backbone of CenterPoint. As seen in Table 5, *ProposalContrast* achieves much better results when transferring to 3D object detection in large-scale LiDAR point clouds, thanks to the more suitable representation.

Effectiveness of the Proposal Encoding Module. Next, we ablate the design choices in the proposal encoding module. We first apply a heuristic method (MaxPool) that directly pools the point features inside a proposal and uses liner layers for embedding. As shown in Table 6, this has obtained improvement over the baseline due to operating on more informative candidate proposals. After being equipped with the attentive proposal encoder, better results are achieved (+0.81 mAPH). This shows the importance of modeling the geometry structures of proposals. After that, we also search the proposal number N and proposal radius r. It turns out that $N = 2048$ and $r = 1.0$ m give better results. We infer that more proposals or a larger proposal radius will cause overlaps between neighbor proposals, which may confuse the instance discrimination process.

Effectiveness of the Joint Optimization Module. We further investigate the effect of the two learning targets in Table 6. We first consider only the inter-proposal discrimination (IPD) task for self-supervised learning, which improves the baseline by 2.60% mAPH. Then, we evaluate the pre-training model with only the inter-cluster separation (ICS) task and check the effect of class (cluster) number, *i.e.*, the output dimensions of the predictor. After that, we examine the importance of Sinkhorn-Knopp clustering (SKC). As seen, the performance drops a lot (−2.44 mAPH) without SKC. Finally, the joint learning of the two tasks further improves the performance.

5 Conclusion

This paper presented *ProposalContrast*, a proposal-wise pre-training framework for LiDAR-based 3D object detection. Despite the previous works for scene-level or point/voxel-level instance discrimination, we argue that the proposal-level representation is more suitable for 3D object detection in the large-scale LiDAR point cloud, which is not well addressed by previous works. To achieve proposal-wise contrastive learning, we carefully designed a proposal generation module, a region proposal encoding module and a joint optimization module. In particular, the proposal generation module samples dense and diverse spherical proposals with different augmented views. The proposal encoding module abstracts proposal features to model the intrinsic geometry structure of proposals by considering relationships of points inside proposals. To further build a comprehensive representation, we proposed to jointly optimize an inter-proposal discrimination task and an inter-cluster separation task. Extensive experiments on diverse prevalent 3D object detectors and datasets show the effectiveness of our model. We expect this work will encourage the community to explore the unsupervised pre-training paradigm in driving scenarios.

Acknowledgements. This work was partially supported by Zhejiang Lab's International Talent Fund for Young Professionals (ZJ2020GZ023), ARC DECRA DE220101390, FDCT under grant 0015/2019/AKP, and the Start-up Research Grant (SRG) of University of Macau.

References

1. Bogoslavskyi, I., Stachniss, C.: Efficient online segmentation for sparse 3D laser scans. J. Photogrammetry Remote Sens. Geoinf. Sci. **85**(1), 41–52 (2017)
2. Caron, M., Misra, I., Mairal, J., Goyal, P., Bojanowski, P., Joulin, A.: Unsupervised learning of visual features by contrasting cluster assignments. In: NeurIPS (2020)
3. Chang, A.X., et al.: ShapeNet: an information-rich 3d model repository. arXiv preprint (2015)
4. Chen, K., Hong, L., Xu, H., Li, Z., Yeung, D.Y.: MultiSiam: self-supervised multi-instance Siamese representation learning for autonomous driving. In: ICCV (2021)
5. Chen, T., Kornblith, S., Norouzi, M., Hinton, G.: A simple framework for contrastive learning of visual representations. In: ICML (2020)
6. Chen, X., He, K.: Exploring simple Siamese representation learning. In: CVPR (2021)
7. Cuturi, M.: Sinkhorn distances: lightspeed computation of optimal transport. In: NeurIPS (2013)
8. Dai, A., Chang, A.X., Savva, M., Halber, M., Funkhouser, T., Nießner, M.: ScanNet: richly-annotated 3d reconstructions of indoor scenes. In: CVPR (2017)
9. Deng, J., Dong, W., Socher, R., Li, L.J., Li, K., Fei-Fei, L.: ImageNet: a large-scale hierarchical image database. In: CVPR (2009)
10. Dosovitskiy, A., et al.: An image is worth 16×16 words: transformers for image recognition at scale. In: ICLR (2021)
11. Geiger, A., Lenz, P., Urtasun, R.: Are we ready for autonomous driving? The KITTI vision benchmark suite. In: CVPR (2012)
12. Gidaris, S., Bursuc, A., Komodakis, N., Pérez, P., Cord, M.: Learning representations by predicting bags of visual words. In: CVPR (2020)
13. He, K., Fan, H., Wu, Y., Xie, S., Girshick, R.: Momentum contrast for unsupervised visual representation learning. In: CVPR (2020)
14. Hjelm, R.D., et al.: Learning deep representations by mutual information estimation and maximization. In: ICLR (2019)
15. Huang, S., Xie, Y., Zhu, S.C., Zhu, Y.: Spatio-temporal self-supervised representation learning for 3D point clouds. In: ICCV (2021)
16. Jing, L., Tian, Y.: Self-supervised visual feature learning with deep neural networks: a survey. TPAMI **43**(11), 4037–4058 (2020)
17. Kingma, D.P., Ba, J.: Adam: a method for stochastic optimization. In: ICLR (2015)
18. Kingma, D.P., Mohamed, S., Jimenez Rezende, D., Welling, M.: Semi-supervised learning with deep generative models. In: NeurIPS (2014)
19. Lang, A.H., Vora, S., Caesar, H., Zhou, L., Yang, J., Beijbom, O.: PointPillars: fast encoders for object detection from point clouds. In: CVPR (2019)
20. Li, X., Liu, S., De Mello, S., Wang, X., Kautz, J., Yang, M.H.: Joint-task self-supervised learning for temporal correspondence. In: NeurIPS (2019)
21. Liang, H., et al.: Exploring geometry-aware contrast and clustering harmonization for self-supervised 3D object detection. In: ICCV (2021)

22. Liu, Z., et al.: Swin transformer: hierarchical vision transformer using shifted windows. In: ICCV (2021)
23. Loshchilov, I., Hutter, F.: SGDR: stochastic gradient descent with warm restarts. In: ICLR (2017)
24. Mao, J., et al.: One million scenes for autonomous driving: once dataset. In: NeurIPS Datasets and Benchmarks (2021)
25. Meng, Q., Wang, W., Zhou, T., Shen, J., Jia, Y., Van Gool, L.: Towards a weakly supervised framework for 3D point cloud object detection and annotation. TPAMI **44**, 4454–4468 (2021)
26. Meng, Q., Wang, W., Zhou, T., Shen, J., Van Gool, L., Dai, D.: Weakly supervised 3D object detection from lidar point cloud. In: Vedaldi, A., Bischof, H., Brox, T., Frahm, J.-M. (eds.) ECCV 2020. LNCS, vol. 12358, pp. 515–531. Springer, Cham (2020). https://doi.org/10.1007/978-3-030-58601-0_31
27. Oord, A.V.D., Li, Y., Vinyals, O.: Representation learning with contrastive predictive coding. arXiv preprint (2018)
28. Qi, C.R., Yi, L., Su, H., Guibas, L.J.: PointNet++: deep hierarchical feature learning on point sets in a metric space. In: NeurIPS (2017)
29. Rao, Y., Liu, B., Wei, Y., Lu, J., Hsieh, C.J., Zhou, J.: RandomRooms: unsupervised pre-training from synthetic shapes and randomized layouts for 3D object detection. In: ICCV (2021)
30. Shi, S., Guo, C., Jiang, L., Wang, Z., Shi, J., Wang, X., Li, H.: PV-RCNN: point-voxel feature set abstraction for 3D object detection. In: CVPR (2020)
31. Shi, S., Wang, X., Li, H.: PoinTRCNN: 3D object proposal generation and detection from point cloud. In: CVPR (2019)
32. Shi, S., Wang, Z., Shi, J., Wang, X., Li, H.: From points to parts: 3D object detection from point cloud with part-aware and part-aggregation network. TPAMI **43**(8), 2647–2664 (2020)
33. Sun, P., et al.: Scalability in perception for autonomous driving: waymo open dataset. In: CVPR (2020)
34. Team, O.D.: OpenPCDet: an open-source toolbox for 3D object detection from point clouds (2020). https://github.com/open-mmlab/OpenPCDet
35. Vora, S., Lang, A.H., Helou, B., Beijbom, O.: PointPainting: sequential fusion for 3D object detection. In: CVPR (2020)
36. Wang, W., Zhou, T., Yu, F., Dai, J., Konukoglu, E., Van Gool, L.: Exploring cross-image pixel contrast for semantic segmentation. In: ICCV (2021)
37. Wang, Y., Sun, Y., Liu, Z., Sarma, S.E., Bronstein, M.M., Solomon, J.M.: Dynamic graph CNN for learning on point clouds. ACM Trans. Graph. **38**(5), 1–12 (2019)
38. Xie, S., Gu, J., Guo, D., Qi, C.R., Guibas, L., Litany, O.: PointContrast: unsupervised pre-training for 3D point cloud understanding. In: Vedaldi, A., Bischof, H., Brox, T., Frahm, J.-M. (eds.) ECCV 2020. LNCS, vol. 12348, pp. 574–591. Springer, Cham (2020). https://doi.org/10.1007/978-3-030-58580-8_34
39. Yan, Y., Mao, Y., Li, B.: Second: sparsely embedded convolutional detection. Sensors **18**(10), 3337 (2018)
40. Yang, Z., Sun, Y., Liu, S., Jia, J.: 3DSSD: point-based 3D single stage object detector. In: CVPR (2020)
41. Ye, M., Shen, J.: Probabilistic structural latent representation for unsupervised embedding. In: CVPR (2020)
42. Yin, J., et al.: Semi-supervised 3D object detection with proficient teachers. In: ECCV (2022)

43. Yin, J., Shen, J., Gao, X., Crandall, D., Yang, R.: Graph neural network and spatiotemporal transformer attention for 3d video object detection from point clouds. TPAMI (2021)
44. Yin, J., Shen, J., Guan, C., Zhou, D., Yang, R.: Lidar-based online 3D video object detection with graph-based message passing and spatiotemporal transformer attention. In: CVPR (2020)
45. Yin, T., Zhou, X., Krahenbuhl, P.: Center-based 3D object detection and tracking. In: CVPR (2021)
46. Zhai, X., Oliver, A., Kolesnikov, A., Beyer, L.: S4L: self-supervised semi-supervised learning. In: CVPR (2019)
47. Zhang, Z., Girdhar, R., Joulin, A., Misra, I.: Self-supervised pretraining of 3D features on any point-cloud. In: ICCV (2021)
48. Zhao, Y., Wang, G., Luo, C., Zeng, W., Zha, Z.J.: Self-supervised visual representations learning by contrastive mask prediction. In: ICCV (2021)
49. Zhou, D., et al.: Joint 3D instance segmentation and object detection for autonomous driving. In: CVPR (2020)

PreTraM: Self-supervised Pre-training via Connecting Trajectory and Map

Chenfeng Xu[1], Tian Li[2], Chen Tang[1(✉)], Lingfeng Sun[1], Kurt Keutzer[1], Masayoshi Tomizuka[1], Alireza Fathi[3], and Wei Zhan[1]

[1] University of California, Berkeley, USA
chen_tang@berkeley.edu
[2] University of California, San Diego, USA
[3] Google Research, Mountain View, CA, USA

Abstract. Deep learning has recently achieved significant progress in trajectory forecasting. However, the scarcity of trajectory data inhibits the data-hungry deep-learning models from learning good representations. While pre-training methods for representation learning exist in computer vision and natural language processing, they still require large-scale data. It is hard to replicate their success in trajectory forecasting due to the inadequate trajectory data (e.g., 34K samples in the nuScenes dataset). To work around the scarcity of trajectory data, we resort to another data modality closely related to trajectories—HD-maps, which is abundantly provided in existing datasets. In this paper, we propose *PreTraM*, a self-supervised **Pre**-training scheme via connecting **Tra**jectories and **M**aps for trajectory forecasting. PreTraM consists of two parts: 1) Trajectory-Map Contrastive Learning, where we project trajectories and maps to a shared embedding space with cross-modal contrastive learning, 2) Map Contrastive Learning, where we enhance map representation with contrastive learning on large quantities of HD-maps. On top of popular baselines such as AgentFormer and Trajectron++, PreTraM reduces their errors by 5.5% and 6.9% relatively on the nuScenes dataset. We show that PreTraM improves data efficiency and scales well with model size. Our code and pre-trained models will be released at https://github.com/chenfengxu714/PreTraM.

Keywords: Trajectory forecasting · Self-supervised learning · Pre-training · Contrastive learning · Multi-modality

1 Introduction

Trajectory forecasting is a challenging task in autonomous driving, which aims at predicting the future trajectory conditioned on past trajectories and surrounding

C. Xu and T. Li—Equal contribution.

Supplementary Information The online version contains supplementary material available at https://doi.org/10.1007/978-3-031-19842-7_3.

S. Avidan et al. (Eds.): ECCV 2022, LNCS 13699, pp. 34–50, 2022.
https://doi.org/10.1007/978-3-031-19842-7_3

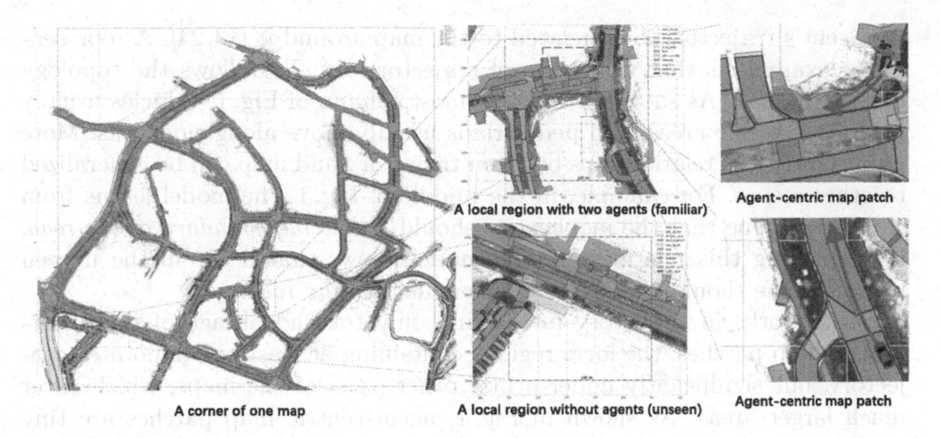

Fig. 1. We have two key observation about maps and trajectories: 1) As shown in the rightmost column, vehicles usually move in drivable areas and pedestrians usually move along sidewalks. And the relationship learnt from familiar scenes can generalize to unseen scenes. (Please zoom in for better view.) 2) Agent-centric map patches are taken from a local region of the map, which is just a tiny part of the whole map.

scenes. Current deep learning models have dominated trajectory forecasting by data-driven supervised learning. However, both the collection and the annotation of trajectory data are extremely difficult and costly. Trajectory data is collected by vehicles with sophisticated sensor systems. Then annotators need to label the objects, associate their positions, generate and smoothen trajectories. This complex procedure limits the scale of the data. The popular open-sourced trajectory forecasting dataset nuScenes [3] has only 34K samples, much fewer than that of the elementary small-scale image dataset MNIST (60K samples) [10]. The scarcity of trajectory data prohibits the models from learning good trajectory representation, and thus restrains their performance.

In the Natural Language Processing (NLP) and computer vision (CV) communities, it was found effective to use self-supervised pre-training on vast unlabeled datasets to learn language/visual representations. The classic methods, such as autoregressive language modeling [2], masked autoencoding [12], and contrastive learning [6,18], are conceptually simple, but require billions of training data. Although recent results from CLIP [27] show that cross-modal contrastive learning requires much fewer pre-training data (4x fewer), the amount of data used is still far more than available trajectory data. Unlike NLP and CV, where large-scale unlabeled datasets exist, the bottleneck for scaling trajectory datasets lies in data collection and annotation. It poses the critical challenge for trajectory forecasting to benefit from existing pre-training schemes. And to the best of our knowledge, few efforts in trajectory forecasting have explored pre-training.

To work around the scarcity of trajectories, we resort to another modality of data that is closely related to trajectories—HD-maps. In fact, we observe two important facts about maps:

- An agent's trajectory is correlated to the map around it [13,24]. A representative example is that the shape of trajectory usually follows the topology of the HD-map. As shown in the rightmost column of Fig. 1, vehicles usually move in drivable areas, and pedestrians usually move along sidewalks. More importantly, the relationships between trajectory and map can be generalized to other scenes. For example, in the middle of Fig. 1, the model learns from the upper scene that the moving car should *follow the boundary of the road.* By capturing this relationship, the model knows that a car in the unseen bottom scene should also *follow the boundary of the road.*
- Existing works in trajectory forecasting only take advantage of the agent-centric map patches, the local regions containing at least one annotated trajectory, but significantly under-utilize other parts of the maps, which cover much larger areas. As shown in Fig. 1, agent-centric map patches are tiny compared with the leftmost global map.

Based on the above observations, we propose PreTraM, a self-supervised **pre**-training scheme via connecting **tra**jectories and **m**aps for trajectory forecasting. Specifically, we jointly pre-train the trajectory encoder and the map encoder of a model in two ways: 1) *Trajectory-Map Contrastive Learning (TMCL)*: Inspired by CLIP [27], we contrast trajectories with corresponding map patches to enforce the model to capture their relationship. 2) *Map Contrastive Learning (MCL)*: We train a stronger map encoder with contrastive learning on large quantities of trajectory-decoupled map patches, which outnumber the agent-centric ones by 782x. In short, PreTraM is a synergy of TMCL and MCL: a better trajectory representation is learned via bridging the map and trajectory representations with TMCL, so that the trajectory encoder benefits from the map representation enhanced by MCL.

Our method reduces the prediction error of a variety of popular prediction models including AgentFormer [34] and Trajectron++ [28] by 5.5% and 6.9% relatively on the nuScenes dataset [3]. More importantly, we find that PreTraM is able to achieve larger performance gain when less data is available. Impressively, using only 70 % of the trajectory data, PreTraM on top of AgentFormer show superior performance than AgentFormer trained on 100 % trajectory data. This demonstrates the proposed pre-training scheme brings strong data efficiency. Furthermore, we apply PreTraM to larger versions of AgentFormer and observe it consistently improves prediction accuracy when the model scales up. We also conduct sufficient ablation studies and shed light on how PreTraM works.

In summary, our key contributions are as follows:

- We propose PreTraM, a novel self-supervised pre-training scheme for trajectory forecasting by connecting trajectories and maps, which consists of trajectory-map contrastive learning and map contrastive learning.
- We show with experiments that PreTraM achieves up to 6.9 % relative improvement in FDE-10 upon popular baselines.
- PreTraM enhances the data efficiency of prediction models, using 70% training data but beating the baseline with 100% training data, and generalizes to models of larger scales.

- Through ablation studies and analysis, we demonstrate the efficacy of TMCL and MCL respectively, and shed light on how PreTraM works.

2 Background

2.1 Problem Formulation of Trajectory Forecasting

In trajectory forecasting, we aim to predict the future trajectories of multiple target agents in a scene. Typically, a set of history states x for all agents and the surrounding HD-map patches M are input to the model f_ω and the model predicts the future trajectories of each agent $y = f_\omega(x, M)$.

The HD-map contains rich semantic information (e.g., drivable area, stop line, and traffic light) [3]. In this work, we employ rasterized top-down semantic images around each of the agents as the input HD-map patches M, i.e., $M = \{m_i\}_{i \in \{1,\ldots,A\}}, m_i \in \mathbb{R}^{C \times C \times 3}$, where C is the context size and 3 denotes the RGB channels. Note that each color has its specific semantic meaning in HD-maps.

As for the history states, denoting the number of agents in the scene as A, and the history time span as T, then $x = s_{1,\ldots,A}^{(-T:0)} \in \mathbb{R}^{T \times A \times D}$, where s_i is the history states of agent i, and 0 denotes the current timestamp. D is the dimension of features that generally contain the agent's 2D or 3D coordinates, as well as other information such as its heading and its speed.

2.2 Contrastive Learning

Contrastive learning is a powerful method for self-supervised representation learning that was made popular by [6–8,18]. Using instance discrimination as the pretext task, they pull the semantically-close neighbors together and push away non-neighbors [14]. For example, in SimCLR [6], given a mini-batch of inputs, each input x_i is transformed into a positive sample x_i^+. Let h_i, h_i^+ denote the hidden representation of x_i, x_i^+. Then on a mini-batch of N pairs of (x_i, x_i^+), it adopts the InfoNCE loss [25] as its training objective.

In particular, we are interested in one specific work that explored contrastive learning in NLP: SimCSE [14]. Instead of using word replacement or deletion as augmentation, it uses different dropout masks in the model as the minimal augmentation for the positive samples. This simple approach turns out to be very effective in that it fully preserves the semantic of the text, compared with other augmentation operations. To preserve the semantic of HD-map, we also adopt dropout for the positive samples in map contrastive learning.

More recently, CLIP [27] demonstrated the power of cross-modal contrastive learning conditioned on huge amounts of data. It collects paired images and captions from the Internet and asks the model to pair an image with the corresponding text, using large batches. For a mini-batch of N pairs of images I_i and texts T_i, denoting their hidden representations as (h_i^I, h_i^T), it applies cross-entropy loss on the $N \times N$ similarity matrix over all pairs of images and texts, stated as follows:

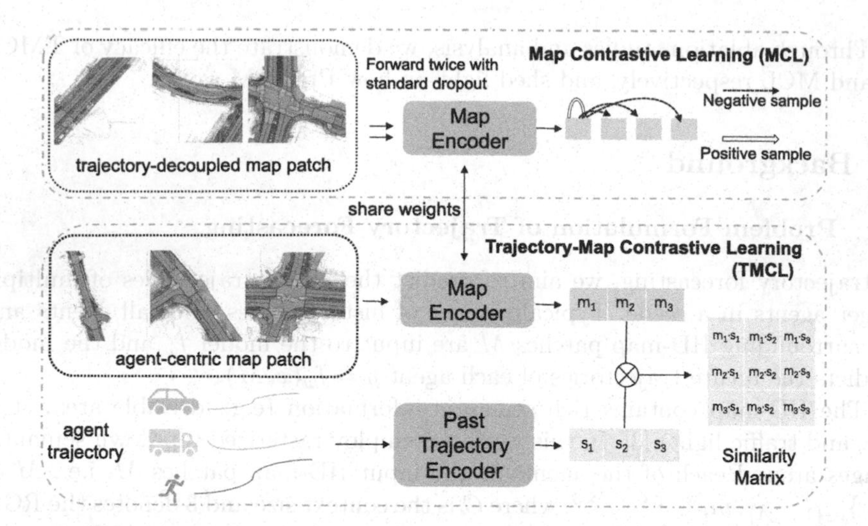

Fig. 2. Top: Map Contrastive Learning (MCL). On the contrary to agent-centric map patches, the trajectory-decoupled ones do not necessarily contain agent trajectories. During training, we randomly crop those patches from the whole map around positions on the road. Bottom: Trajectory-Map Contrastive Learning (TMCL).

$$l_i = -\log \frac{e^{\text{sim}(h_i^I, h_i^T)/\tau}}{\sum_{j=0}^{N} e^{\text{sim}(h_i^I, h_j^T)/\tau}} \tag{1}$$

where $\text{sim}(\cdot, \cdot)$ is a measurement of similarity, typically the cosine similarity, and τ is the temperature parameter. Note that it can be seen as the InfoNCE loss using the corresponding text as the positive sample of an image.

Intuitively, using natural language as supervision of images, CLIP puts image and text in a shared embedding space. Besides, Eq. (refeq1) enforces similarity between the correct pair of images and text, and thus learns the pattern of image-text relationship. Following this intuition, we design a trajectory-map contrastive learning objective to capture the relationship between them.

3 Method

We propose a novel self-supervised **pre**-training scheme by connecting **tra**jectory and **m**ap (PreTraM) to enhance the trajectory and map representations when there are small-scale trajectory data, but large-scale map data. We jointly pre-train a trajectory encoder and a map encoder to obtain good trajectory representation by encoding the trajectory-map relationship into the representation.

As illustrated in Fig. 2, the proposed PreTraM is composed of two parts: 1) A simple trajectory-map contrastive learning (TMCL) that is conducted between map encoder and trajectory encoder, using limited number of trajectories and the paired map patches. 2) A simple map contrastive learning (MCL) that is conducted on map encoder using large batch size on trajectory-decoupled map patches, where there are not necessarily agent trajectories. After pre-training, we load the pre-trained weights and finetune under the prediction objective with the same training schedules as the original models.

3.1 Trajectory-Map Contrastive Learning (TMCL)

We propose to use a cross-modal contrastive learning method that facilitates both trajectory encoder and map encoder. Specifically, given a mini-batch of scenes, for all the input history states x, we split them into single agent trajectories and treat them *independently*, i.e., $S = \{s_i | s_i \in x, \forall x \in B\}$, where B denotes the mini-batch. For each agent, we crop an agent-centric map patch around its current position from the HD-map, and then we have $N_{\text{traj}} = |S|$ pairs of correlated trajectories and HD-map patches (s_i, m_i). We also rotate the map with respect to the orientation of the agent following the common practice [28,34]. The model is required to match each trajectory s_i with the paired map patch m_i among all the map patches in the mini-batch and vice versa. As shown in bottom of Fig. 2, we input the trajectories and maps into the corresponding encoders to obtain the features $\{h_i^{\text{traj}}\}, \{h_i^{\text{map}}\}$. Then we compute a similarity matrix across all pairs of trajectories and maps in the mini-batch. Note that we apply a linear projection layer [27] after the map encoder and the trajectory encoder for the hidden representations h but omit it above for the sake of simplicity. It is the same case in Sect. 3.2 for MCL. Finally, we optimize a symmetric cross-entropy loss over these similarity scores as follows [27]:

$$l_i^{\text{TMCL}} = -\log \frac{e^{\text{sim}(h_i^{\text{traj}}, h_i^{\text{map}})/\tau_{\text{traj}}}}{\sum_{j=1}^{N_{\text{traj}}} e^{\text{sim}(h_i^{\text{traj}}, h_j^{\text{map}})/\tau_{\text{traj}}}} \tag{2}$$

Through this objective, the similarities of the correct pairs of the trajectories and maps are maximized and those of the other pairs are minimized. It results in a shared embedding space of trajectories and maps. We find that a prediction model that fuses map and trajectories to make prediction benefits from such a shared embedding space. It agrees with the finding in [20] that models for vision-language tasks benefit from an aligned embedding space for the visual and language inputs. The TMCL objective teaches the model to encode the relationship between maps and trajectories into the representation. By capturing the relationship, the trajectory embedding contains the information of the underlying map conditioned on the input trajectory, which implies the geometric and routing information of the future trajectories for the predictor.

3.2 Map Contrastive Learning (MCL)

To further facilitate learning the trajectory-map relationship, we learn a general map representation by map contrastive learning. At each training iteration, we randomly crop N_{map} map patches from a random subset of HD-Maps in the dataset. Note that N_{map} is much greater than the agent number A.

In addition to using a large number of map patches, the key ingredient to get MCL to work effectively is using the exact identical instance as its positive sample, i.e., $m_i^+ = m_i$, and apply dropout in the map encoder [14]. Denote the hidden representation $h_i^z = g_\theta(m_i, z)$ where g_θ is the map encoder and z is a random mask for dropout. As shown in the top part of Fig. 2, we feed the same map patch to the encoder in two independent forward passes with different dropout masks z, z', which gives two representation $h_i^z, h_i^{z'}$ for each m_i. Thanks to dropout, h_i^z and $h_i^{z'}$ are different, but still encode the same topology and semantic. In contrast, regular augmentation operations in CV such as random rotation, flip, gaussian noise or color jitter do not work here. Gaussian noise and color jitter transform the semantics of HD-maps, while flip and rotation change their topologies. Instead, dropout serves as a minimal augmentation for the positive sample and turns out to be effective through experiments. We provide comparison experiments on dropout against other augmentations in Sect. 2 of the supplementary material. Formally, the training objective of MCL is:

$$l_i^{\mathrm{MCL}} = -\log \frac{e^{\mathrm{sim}(h_i^{z_i}, h_i^{z_i'})/\tau_{\mathrm{map}}}}{\sum_{j=1}^{N_{\mathrm{map}}} e^{\mathrm{sim}(h_i^{z_i}, h_j^{z_j'})/\tau_{\mathrm{map}}}} \tag{3}$$

It is worth noting that MCL is a novel use of HD-maps not only because we make use of every piece of the HD-map, but also because we design a customized training objective to make better use of HD-maps.

3.3 Training Objective

The PreTraM scheme is complete with the joint of TMCL and MCL. The overall objective function combines their objectives, given by:

$$\mathcal{L} = \sum_{i=1}^{N_{\mathrm{traj}}} l_i^{\mathrm{TMCL}} + \lambda \sum_{i=1}^{N_{\mathrm{map}}} l_i^{\mathrm{MCL}} \tag{4}$$

4 Experiments

4.1 Dataset and Implementation Details

Dataset. nuScenes is a recent large-scale autonomous driving dataset collected from Boston and Singapore. It consists of 1000 driving scenes with each scene annotated 2 Hz, and the driving routes are carefully chosen to capture challenging scenarios. The nuScenes dataset provides HD semantic maps from Boston Seaport together with Singapore's One North, Queenstown and Holland Village districts, with 11 semantic classes. It is split into 700 scenes for training, 150 scenes for validation, and 150 scenes for testing.

Our main experiments follow the split used in AgentFormer [34], in which the original training set is split into two parts: 500 scenes for training, and 200 scenes for validation. The original validation set is used for testing our model.

Baseline. We performed experiments with PreTraM on two models, AgentFormer [34] and Trajectron++ [28]. Both of them are CVAE models including a past trajectory encoder, a map encoder, a future trajectory encoder, and a future trajectory decoder. We reproduced AgentFormer to support parallel training. Compared with the original code, our reproduced code trains 17.1x faster than the official code (4.5 h vs. 77 h on one V100 GPU), and its performance is competitive—0.029 better than the official implementation on ADE-5. Note that AgentFormer separately trains DLow [33] for better sampling. We did not reproduce this part since we focus on representation learning and want a precise quantitative evaluation on the benefit of PreTraM to the model itself. Plus, Trajectron++ does not use DLow while applicable. We want to keep the setting consistent, so that it is meaningful to compare the performance gains between different models. As for Trajectron++ [28], we use their official implementation but re-train it using the data split in AgentFormer to ensure fair comparison. In the following sections, we denote AgentFormer/Trajectron++ pre-trained with PreTraM as PreTraM-AgentFormer/PreTraM-Trajectron++, or PreTraM when the model is clear in the context.

Pre-training and Finetuning. Our pre-training is applied to the *past* trajectory encoder and map encoder. To train TMCL, we pair the historical trajectories of last 2 s and map patches of context size 100×100. We randomly rotate the trajectories and maps simultaneously for data augmentation. For MCL, we collect the trajectory-decoupled map patches dynamically at training. For each instance in the mini-batch, we crop 120 map patches centered at random positions along the road in the HD-map. We pre-train the encoders with the PreTraM objective function for 20 epochs using batch size 32 (which means 3440 map patches for MCL in one iteration). Throughout the pre-training phase, we use 28.8M map patches to train our map encoder, which is 782x more than agent-centric map patches. The pre-training phase is fast—only *30* min on one V100 GPU for AgentFormer. More details are in Sect. 3 of the Supplementary material.

Recall that we use dropout for positive samples in MCL. In shallow map encoders, such as Map-CNN used in AgentFormer and Trajectron++, we place the dropout at post-activation of each convolution. For relatively deeper map encoder such as ResNet family, we place two dropout masks on each residual block. The mask ratio of dropout is default as $p = 0.1$.

At finetuning phase, we use the same training recipes as AgentFormer and Trajectron++. The prediction horizon is 6 s and we use the ground-truth future trajectories to supervise the training.

Metric. The main metrics are Average Displacement Error (ADE) and Final Displacement Error (FDE). We follow previous works [28,34] to sample k trajectories during inference and pick the minimum of the error, denoting as ADE-k and FDE-k. Apart from the sampling based metrics above, we also use a deterministic metric meanFDE, which is the FDE of the trajectory that the model deems as the most likely.

Table 1. Comparison experiments based on AgentFormer [34] and Trajectron++ [28]. Note that the reported AgentFormer is removed of DLow. The AgentFormer* denotes our reproduced implementation. *Lower* number is better.

Method	ADE-5	FDE-5	ADE-10	FDE-10
MTP [9]	2.93	–	–	–
AgentFormer [34]	2.517	5.459	1.852	3.869
MultiPath [5]	2.32	–	1.96	–
DLow-AF [33]	2.11	4.70	1.78	3.58
DSF-AF [23]	2.06	4.67	1.66	3.71
CoverNet [26]	1.96	–	1.48	–
AgentFormer*	2.488	5.420	1.893	3.902
PreTraM-AgentFormer*	**2.391**(−0.097)	**5.177**(−0.243)	**1.796**(−0.097)	**3.687**(−0.215)
Trajectron++ [28]	1.772	4.150	1.405	3.221
PreTraM-Trajectron++	**1.698**(−0.074)	**3.963**(−0.197)	**1.348**(−0.057)	**3.040**(−0.181)

Table 2. Experimental evaluation on meanFDE, KDE NLL, and Boundary violation rate (B. Viol.) provided by Trajectron++ [28]. *Lower* number is better.

Method	meanFDE	KDE NLL	B. Viol. (%)
Trajectron++	8.242	2.487	23.7
PreTraM-Trajectron++	**8.212**(−0.030)	**2.380**(−0.107)	**21.9**(−1.8)

We also leverage the metrics including Kernel Density Estimate-based Negative Log Likelihood (KDE NLL) [28] and boundary violation rate. The former measures the NLL of the ground truth trajectory under a distribution created by fitting a kernel density estimate on trajectory samples, which shows the likelihood of the ground truth trajectory given the sampled trajectory predictions. The latter is the ratio of the predicted trajectories that hit road boundaries.

4.2 Comparison Experiments

The results compared with the baselines and the other prior-arts are shown in Table 1. Observe that using PreTraM improves the performance by 0.097 (*resp. 0.074*) ADE-5, 0.243 (*resp. 0.197*) FDE-5, 0.097 (*resp. 0.057*) ADE-10, 0.215 (*resp. 0.181*) FDE-10, on top of AgentFormer (*resp.* Trajectron++). This is up to *4.1%* relative improvement on ADE-5 and *6.9%* on FDE-10. Remarkably, we achieve it with a simple pre-training scheme. PreTraM does not rely on long pre-training epochs or huge quantities of external data as said in Sect. 4.1. The HD-map we use during pre-training is inherently provided by the dataset. Besides, PreTraM is plug-and-play, and can be easily applied to most prediction model that fuses HD-map and trajectory. In conclusion, these results demonstrate that PreTraM indeed facilitates the models in representation learning. Note that our reproduced AgentFormer does not include DLow as stated in Sect. 4.1.

We also evaluate the results on the metrics provided by Trajectron++ to show the advantage of PreTraM. PreTraM-Trajectron++ improves baseline by 0.107 KDE NLL and 1.8% boundary violation rate (Table 2). The improvements on these two metrics show that our pre-training scheme not only improves prediction accuracy, but also improves stability and safety.

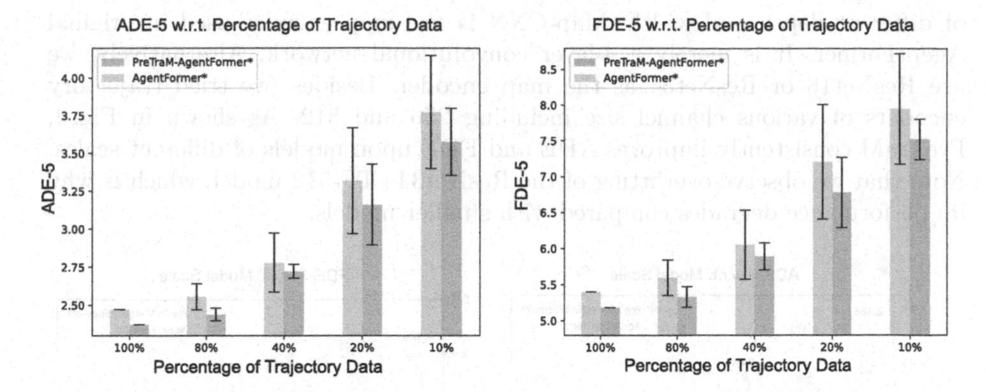

Fig. 3. Experiments with part of the trajectory data. Left: ADE-5 results. Right: FDE-5 results. We repeat the experiments with 3 different random seeds and report the mean performance. The error bars are 3 times the standard deviation. As the percentage of trajectory data becomes lower, the improvements of PreTraM are larger. Moreover, the std of PreTraM is much smaller than the baseline over all the settings.

4.3 Data Efficiency

In this section, we explore whether the learned representations of trajectory and map can improve data efficiency. To investigate this we evaluate PreTraM-AgentFormer on a fraction of the dataset, comparing its result with baseline AgentFormer. In this set of experiments to best demonstrate our strength, we use ResNet18 as a substitute for the 4-layer map encoder, Map-CNN, in the original AgentFormer. This is due to the intuition that larger models are better at representation learning [6,18,27]. We randomly sample 80%, 40%, 20% and 10% trajectories from the dataset, but keep all the HD-maps available. For each setting, we repeat the experiments with 3 different random seeds and report the mean and the standard deviation in Fig. 3. We observe that PreTraM-AgentFormer outperforms the baseline in all settings. More importantly, the performance gain of PreTraM gets larger as the percentage of data goes smaller. With *10%* of data, i.e., around 1200 samples, PreTraM surpasses the baseline by *0.32* on ADE-5 and *0.77* on FDE-5. Moreover, the std of PreTraM is much smaller than the baseline. The std of ADE-5 of the baseline are 0.035 (80%), 0.079 (40%), 0.143 (20%), and 0.154 (10%) respectively, while those of PreTraM are 0.017, 0.018, 0.108, 0.091. It is the same case in terms of FDE-5.

In addition, we observe that training on 70% of data, PreTraM-AgentFormer still outperforms the baseline with 100 % of data (2.470 ADE-5 vs. 2.472 ADE-5). More results are shown in Sect. 1 of the supplementary material.

4.4 Scalability Analysis

A good representation learning method is able to scale with the model size [6, 18]. Therefore, we evaluate PreTraM on map encoders and trajectory encoders of different depth and width. Map-CNN is the map encoder used in original AgentFormer. It is merely a 4-layer convolutional network. Alternatively, we use ResNet18 or ResNet34 as the map encoder. Besides, we tried trajectory encoders of various channel size including 256 and 512. As shown in Fig. 4, PreTraM consistently improves ADE and FDE upon models of different scales. Note that we observe overfitting of the ResNet34+TF-512 model, which is why its performance degrades compared with smaller models.

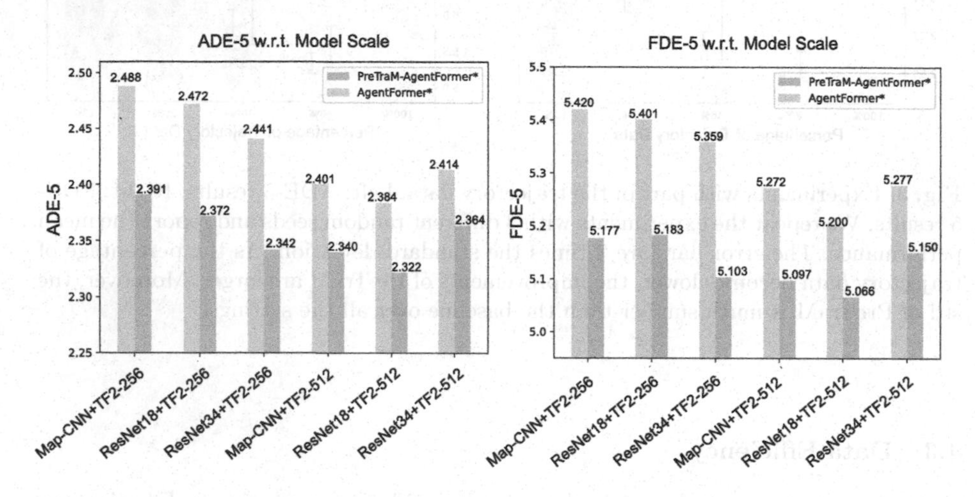

Fig. 4. Experiment with models at different scales. As the model gets deeper and wider, PreTraM consistently improves AgentFormer by a large margin.

4.5 Analysis

It is natural that PreTraM enhances the map representation since we utilize 28.8M map samples for pre-training the map encoder, but as we proposed in previous sections, another important goal is to further enhance the trajectory representation. Therefore, we conduct experiments and delve deep into the function of PreTraM to discuss how it enhances the trajectory representation. All the experiments are completed upon AgentFormer with Map-CNN.

Does PreTraM Indeed Improve Trajectory Representation? In fact, we can quantitatively demonstrate this by loading one of the pre-trained encoders, trajectory encoder (TE) and map encoder (ME), at finetuning phase. As shown in Table 3, we can first observe that loading only the pre-trained map encoder improves prediction performance. More importantly, we observe that just loading TE pre-trained weights is able to give almost the same result as loading both of ME and TE. This means the learnt trajectory representation is strong, and that the major benefit of PreTraM owes to the trajectory representation.

Table 3. Comparison with loading one of the pretrained models when finetuning. ME: map encoder, TE: past trajectory encoder.

Method	ADE-5	FDE-5
Baseline AgentFormer	2.488	5.420
Finetune with both pretrained weights (PreTraM)	2.391(−0.097)	5.177(−0.243)
Finetune with only TE pretrained weights	2.399(−0.089)	5.277(−0.143)
Finetune with only ME pretrained weights	2.454(−0.034)	5.372(−0.048)

Table 4. Comparison with different pre-training strategies. MTM means masked trajectory modeling, recovering the masked trajectories during pre-training, which is a mimic of masked language modeling in NLP [12].

Method	ADE-5	FDE-5
Baseline AgentFormer	2.488	5.420
Pre-training with both TMCL and MCL (PreTraM)	2.391(−0.097)	5.177(−0.243)
Pre-training with MTM and MCL	2.431(−0.057)	5.322(−0.098)
Pre-training with only MCL	2.442(−0.046)	5.373(−0.057)
Pre-training with only TMCL	2.451(−0.037)	5.369(−0.051)

So our answer is "Yes, PreTraM indeed improves trajectory representation."

Is TMCL crucial for improving trajectory representation? To examine the contribution of TMCL, we experiment with an alternative to TMCL as the objective function for pre-training trajectory representation. Inspired by Masked Language Modeling (MLM) [12] for sequence modeling in NLP, we randomly mask out part of the input history states and ask the trajectory encoder to recover the masked part. Denoting this task as Masked Trajectory Modeling (MTM), we jointly pre-train the model on the objective of MTM and MCL. For variable controlling, we also pre-train the model solely on MCL. As shown in Table 4 we find that MTM plus MCL does improve from the baseline but is almost comparable to pre-training with only MCL. It shows the important role of TMCL in trajectory representation learning as it learns trajectory-map relationship and bridges the trajectory and map embedding space.

So our answer is "Yes, TMCL is crucial to improve trajectory representation."

Is MCL Crucial for Improving Trajectory Representation? Indeed, as shown in Table 4, when pre-training only with MCL, the improvement is 0.046. This makes sense in that HD-map is an important prior to prediction and thus better map representation in general can improve prediction. But is MCL also helpful to trajectory representation? To examine this, we only pre-train with TMCL. We find that without MCL, TMCL brings limited improvements compared with PreTraM, *e.g.,* 0.037 ADE-5 vs. 0.097 ADE-5 with PreTraM (Table 4). This demonstrates that although map and trajectory are totally different modalities, PreTraM makes use of much more maps to enhance trajectory representation under the situation that the trajectory data is limited.

So our answer is "Yes, MCL is crucial to improve trajectory representation."

5 Related Works

Given a trajectory forecasting model, the applicable pre-training schemes largely depend on the adopted scene representation. In this section, we first give a concise summary of the literature from the perspective of scene representation. Then, we review several works related to self-supervised learning for trajectory forecasting.

5.1 Scene Representation in Trajectory Forecasting

In complex urban traffic scenarios, it is crucial to utilize the semantic information of the scene to make accurate predictions. A widely-adopted approach is to employ rasterized top-down semantic images around the target agents as input and use CNNs to encode the context [5,9,26,28,34]. The past trajectories of the predicted agents are encoded separately and then aggregated with the context embedding. Our proposed PreTraM can be directly applied to pretrain models with this scene representation. The image-based representation has constant input size regardless of the complexity of the scene, which makes encoding simple and unified. However, some argued recently that rich semantic and structured information (e.g., relations between road segments) of the maps is lost through rasterization [13,21]. To this end, they proposed to represent scenes as graphs that naturally inherit the structured information. Graph neural networks (GNN) [1,21] and Transformers [13,32] were then adopted to encode the context information from the scene graphs. Many graph-based models have then achieved state-of-the-art performance on multiple benchmarks [11,15–17,31,35].

5.2 Self-supervised Learning in Trajectory Forecasting

Pre-training and, in a broader sense, self-supervised learning are under-explored for trajectory forecasting. There are only a few recent works investigating their applications in trajectory forecasting. Inspired by similar methods in NLP, an auxiliary graph completion task was proposed in [13] to enhance the node representation, including both road elements and agents. However, the graph completion objective was jointly optimized with the prediction task. Moreover, the auxiliary task was applied to their Transformer-based encoder for the scene graphs, which limits the amount of data for self-supervised training to the size of prediction datasets. In contrast, our PreTraM framework lets trajectory encoder benefit from the large number of map patches that are not associated with agents. In [22], SimCLR was adopted to pre-train the representation of rasterized maps and agent relations. They deliberately introduced assumptions on semantic invariant operations based on domain knowledge. In our MCL, we follow [14] to avoid any assumptions on semantic invariant operations. Moreover, [22] focuses on contrastive learning within the same modality of data, and in the broader community of autonomous driving, [4] adopts single-modal contrastive learning on maps to improve sample efficiency, whereas our PreTraM framework leverages single-modal *and* cross-modal contrastive learning to jointly train trajectory and map representations. As shown in Sect. 4.5, PreTraM has clear performance advantage over single-modal contrastive learning within each data modality.

6 Discussion and Limitations

Our experiments demonstrated that PreTraM is effective for prediction models based on rasterized map representation. In principle, PreTraM is *not* limited to image-based map encoders. We can also apply PreTraM to those popular graph-based methods reviewed in Sect. 5.1, as long as we can obtain separate map and trajectory embeddings from the pipeline. For instance, some works adopted a two-stage graph encoding scheme, where the map graph was encoded before being fused with trajectory embeddings [15,16,21]. We want to point out that GNNs may behave differently from CNNs during pre-training, and we are interested in extending PreTraM to graph-based methods as future study. Meanwhile, other works integrate the road elements and agents into a single graph before aggregation [11,13,17,31,35]. PreTraM cannot be applied to these models, as there are no matching pairs of map and trajectory embeddings in their pipelines. We are interested in exploring alternative pre-training methods for these models. Besides, PreTraM may also benefit end-to-end methods that predict trajectories directly from raw sensor inputs [19,29]. Cross-modal contrastive learning can still be applied to enhance trajectory representation with sensor inputs such as 3D point-clouds or images. Another interesting extension is to contrast multi-agent trajectory embeddings with maps to enhance interaction modeling [30].

7 Conclusion

In this paper, we propose PreTraM, a novel self-supervised pre-training scheme for trajectory forecasting. We design Trajectory-Map Contrastive Learning (TMCL) to help models capture the relationship between agents and the surrounding HD-map, and Map Contrastive Learning (MCL) to enhance map representation via a large number of augmented map patches that are not associated with the agents. With PreTraM, we reduce the error of Trajectron++ and AgentFormer by 5.5% and 6.9% relatively. Furthermore, PreTraM promotes data efficiency of the models. We also demonstrate that our method can consistently improve performance when the model size scales up. Through ablation studies and analysis, we show PreTraM indeed enhances map and trajectory representations. In particular, a better trajectory representation is learned via bridging the map and trajectory representations with TMCL, so that the trajectory encoder can benefit from the map representation enhanced by MCL. Therefore the performance improvement is attributed to the coherent integration of MCL and TMCL in our framework.

Acknowledgements. We sincerely appreciate Boris Ivanovic and Rowan McAllister for providing help on the experiments related to Trajectron++. This work was sponsored by Google-BAIR Commons program. Google also provided a generous donation of cloud compute credits through the Google-BAIR Commons program.

References

1. Battaglia, P.W., et al.: Relational inductive biases, deep learning, and graph networks. arXiv preprint arXiv:1806.01261 (2018)
2. Brown, T., et al.: Language models are few-shot learners. In: Advances in Neural Information Processing Systems, vol. 33, pp. 1877–1901 (2020)
3. Caesar, H., et al.: nuScenes: a multimodal dataset for autonomous driving. In: Proceedings of the IEEE/CVF Conference on Computer Vision and Pattern Recognition, pp. 11621–11631 (2020)
4. Cai, P., Wang, S., Wang, H., Liu, M.: Carl-lead: lidar-based end-to-end autonomous driving with contrastive deep reinforcement learning. arXiv preprint arXiv:2109.08473 (2021)
5. Chai, Y., Sapp, B., Bansal, M., Anguelov, D.: MultiPath: multiple probabilistic anchor trajectory hypotheses for behavior prediction. In: CoRL (2019)
6. Chen, T., Kornblith, S., Norouzi, M., Hinton, G.: A simple framework for contrastive learning of visual representations. In: International Conference on Machine Learning, pp. 1597–1607. PMLR (2020)
7. Chen, T., Kornblith, S., Swersky, K., Norouzi, M., Hinton, G.E.: Big self-supervised models are strong semi-supervised learners. In: Advances in Neural Information Processing Systems, vol. 33, pp. 22243–22255 (2020)
8. Chen, X., Fan, H., Girshick, R., He, K.: Improved baselines with momentum contrastive learning. arXiv preprint arXiv:2003.04297 (2020)
9. Cui, H., et al.: Multimodal trajectory predictions for autonomous driving using deep convolutional networks. In: 2019 International Conference on Robotics and Automation (ICRA), pp. 2090–2096. IEEE (2019)
10. Deng, L.: The MNIST database of handwritten digit images for machine learning research. IEEE Signal Process. Mag. **29**(6), 141–142 (2012)
11. Deo, N., Wolff, E., Beijbom, O.: Multimodal trajectory prediction conditioned on lane-graph traversals. In: Conference on Robot Learning, pp. 203–212. PMLR (2022)
12. Devlin, J., Chang, M.W., Lee, K., Toutanova, K.: BERT: pre-training of deep bidirectional transformers for language understanding. In: Proceedings of the 2019 Conference of the North American Chapter of the Association for Computational Linguistics: Human Language Technologies, Volume 1 (Long and Short Papers), pp. 4171–4186. Association for Computational Linguistics, Minneapolis, June 2019. https://doi.org/10.18653/v1/N19-1423
13. Gao, J., et al.: VectorNet: encoding HD maps and agent dynamics from vectorized representation. In: 2020 IEEE/CVF Conference on Computer Vision and Pattern Recognition (CVPR) (2020)
14. Gao, T., Yao, X., Chen, D.: SimCSE: simple contrastive learning of sentence embeddings. In: Proceedings of the 2021 Conference on Empirical Methods in Natural Language Processing, pp. 6894–6910. Association for Computational Linguistics, Online and Punta Cana, Dominican Republic, November 2021. https://doi.org/10.18653/v1/2021.emnlp-main.552
15. Gilles, T., Sabatini, S., Tsishkou, D., Stanciulescu, B., Moutarde, F.: Home: heatmap output for future motion estimation. In: 2021 IEEE International Intelligent Transportation Systems Conference (ITSC), pp. 500–507 (2021). https://doi.org/10.1109/ITSC48978.2021.9564944

16. Gilles, T., Sabatini, S., Tsishkou, D., Stanciulescu, B., Moutarde, F.: THOMAS: trajectory heatmap output with learned multi-agent sampling. In: International Conference on Learning Representations (2022). https://openreview.net/forum?id=QDdJhACYrlX
17. Gu, J., Sun, C., Zhao, H.: DenseTNT: end-to-end trajectory prediction from dense goal sets. In: Proceedings of the IEEE/CVF International Conference on Computer Vision, pp. 15303–15312 (2021)
18. He, K., Fan, H., Wu, Y., Xie, S., Girshick, R.: Momentum contrast for unsupervised visual representation learning. In: Proceedings of the IEEE/CVF Conference on Computer Vision and Pattern Recognition, pp. 9729–9738 (2020)
19. Laddha, A.G., Gautam, S., Palombo, S., Pandey, S., Vallespi-Gonzalez, C.: MVFuseNet: improving end-to-end object detection and motion forecasting through multi-view fusion of lidar data. In: 2021 IEEE/CVF Conference on Computer Vision and Pattern Recognition Workshops (CVPRW), pp. 2859–2868 (2021)
20. Li, J., Selvaraju, R.R., Gotmare, A.D., Joty, S., Xiong, C., Hoi, S.: Align before fuse: vision and language representation learning with momentum distillation. In: NeurIPS (2021)
21. Liang, M., et al.: Learning lane graph representations for motion forecasting. In: Vedaldi, A., Bischof, H., Brox, T., Frahm, J.-M. (eds.) ECCV 2020. LNCS, vol. 12347, pp. 541–556. Springer, Cham (2020). https://doi.org/10.1007/978-3-030-58536-5_32
22. Ma, H., Sun, Y., Li, J., Tomizuka, M.: Multi-agent driving behavior prediction across different scenarios with self-supervised domain knowledge. In: 2021 IEEE International Intelligent Transportation Systems Conference (ITSC) (2021)
23. Ma, Y.J., Inala, J.P., Jayaraman, D., Bastani, O.: Likelihood-based diverse sampling for trajectory forecasting. In: Proceedings of the IEEE/CVF International Conference on Computer Vision, pp. 13279–13288 (2021)
24. Ngiam, J., et al.: Scene transformer: a unified architecture for predicting future trajectories of multiple agents. In: International Conference on Learning Representations (2022). https://openreview.net/forum?id=Wm3EA5OlHsG
25. Van den Oord, A., Li, Y., Vinyals, O.: Representation learning with contrastive predictive coding. arXiv e-prints. arXiv-1807 (2018)
26. Phan-Minh, T., Grigore, E.C., Boulton, F.A., Beijbom, O., Wolff, E.M.: CoverNet: multimodal behavior prediction using trajectory sets. In: Proceedings of the IEEE/CVF Conference on Computer Vision and Pattern Recognition, pp. 14074–14083 (2020)
27. Radford, A., et al.: Learning transferable visual models from natural language supervision. In: International Conference on Machine Learning, pp. 8748–8763. PMLR (2021)
28. Salzmann, T., Ivanovic, B., Chakravarty, P., Pavone, M.: Trajectron++: dynamically-feasible trajectory forecasting with heterogeneous data. In: Vedaldi, A., Bischof, H., Brox, T., Frahm, J.-M. (eds.) ECCV 2020. LNCS, vol. 12363, pp. 683–700. Springer, Cham (2020). https://doi.org/10.1007/978-3-030-58523-5_40
29. Shah, M., et al.: LiRaNet: end-to-end trajectory prediction using spatio-temporal radar fusion. In: CoRL (2020)
30. Tang, C., Zhan, W., Tomizuka, M.: Exploring social posterior collapse in variational autoencoder for interaction modeling. Adv. Neural. Inf. Process. Syst. **34**, 8481–8494 (2021)

31. Varadarajan, B., et al.: Multipath++: efficient information fusion and trajectory aggregation for behavior prediction. In: 2022 International Conference on Robotics and Automation (ICRA), pp. 7814–7821 (2022). https://doi.org/10.1109/ICRA46639.2022.9812107
32. Vaswani, A., et al.: Attention is all you need. In: Advances in Neural Information Processing Systems, vol. 30 (2017)
33. Yuan, Y., Kitani, K.: DLow: diversifying latent flows for diverse human motion prediction. In: Vedaldi, A., Bischof, H., Brox, T., Frahm, J.-M. (eds.) ECCV 2020. LNCS, vol. 12354, pp. 346–364. Springer, Cham (2020). https://doi.org/10.1007/978-3-030-58545-7_20
34. Yuan, Y., Weng, X., Ou, Y., Kitani, K.: AgentFormer: agent-aware transformers for socio-temporal multi-agent forecasting. In: Proceedings of the IEEE/CVF International Conference on Computer Vision (ICCV) (2021)
35. Zhao, H., et al.: TNT: target-driven trajectory prediction. In: Kober, J., Ramos, F., Tomlin, C. (eds.) Proceedings of the 2020 Conference on Robot Learning. Proceedings of Machine Learning Research, vol. 155, pp. 895–904. PMLR, 16–18 November 2021. https://proceedings.mlr.press/v155/zhao21b.html

Master of All: Simultaneous Generalization of Urban-Scene Segmentation to All Adverse Weather Conditions

Nikhil Reddy[1](✉), Abhinav Singhal[2], Abhishek Kumar[2], Mahsa Baktashmotlagh[3], and Chetan Arora[2]

[1] University of Queensland – IIT Delhi Academy of Research (UQIDAR), Delhi, India
nikhil.jangamreddy@uqidar.iitd.ac.in
[2] Indian Institute of Technology Delhi, Delhi, India
[3] The University of Queensland, Brisbane, Australia
https://mall-iitd.github.io/

Abstract. Computer vision systems for autonomous navigation must generalize well in adverse weather and illumination conditions expected in the real world. However, semantic segmentation of images captured in such conditions remains a challenging task for current state-of-the-art (SOTA) methods trained on broad daylight images, due to the associated distribution shift. On the other hand, domain adaptation techniques developed for the purpose rely on the availability of the source data, (un)labeled target data and/or its auxiliary information (e.g., GPS). Even then, they typically adapt to a single(specific) target domain(s). To remedy this, we propose a novel, fully test time, adaptation technique, named *Master of ALL* (MALL), for simultaneous generalization to multiple target domains. MALL learns to generalize on unseen adverse weather images from multiple target domains directly at the inference time. More specifically, given a pre-trained model and its parameters, MALL enforces edge consistency prior at the inference stage and updates the model based on (a) a single test sample at a time (MALL-sample), or (b) continuously for the whole test domain (MALL-domain). Not only the target data, MALL also does not need access to the source data and thus, can be used with any pre-trained model. Using a simple model pre-trained on daylight images, MALL outperforms specially designed adverse weather semantic segmentation methods, both in domain generalization and test-time adaptation settings. Our experiments on foggy, snow, night, cloudy, overcast, and rainy conditions demonstrate the target domain-agnostic effectiveness of our approach. We further show that MALL can improve the performance of a model on an adverse weather condition, even when the model is already pre-trained for the specific condition.

Supplementary Information The online version contains supplementary material available at https://doi.org/10.1007/978-3-031-19842-7_4.

S. Avidan et al. (Eds.): ECCV 2022, LNCS 13699, pp. 51–69, 2022.
https://doi.org/10.1007/978-3-031-19842-7_4

Keywords: Multi-target domain generalization · All weather urban-scene segmentation · Test-time adaptation · Normalized cuts

1 Introduction

Out-of-domain generalization plays a pivotal role in the success of deep neural networks (DNNs) for safety-critical applications such as autonomous navigation. Although DNN based architectures have achieved tremendous success in many computer vision tasks [12,13,23], they don't perform well when the test data comes from a domain with different sample distribution, referred to as the domain shift [28]. Domain shift can be caused by different factors such as input corruption, adverse weather (rain, snow etc.), illumination changes (e.g. night time), adversarial attacks, sensor malfunction etc. Ensuring robust performance on unseen target domains is critical for the real-world applicability of DNNs (Table 1).

Table 1. Comparing different problem settings, the proposed technique fits in as a *fully test-time adaptation* paradigm, which aims to adapt a pre-trained model to an unseen target domain.

Problem Setting	Source Data	Target Label (y^t)	Train Loss	Test Loss	Open Targets
Fine-tuning	–	✓	$\mathcal{L}(x^t, y^t)$	✗	✗
Unsupervised domain adaptation[51]	x^s, y^s	✗	$\mathcal{L}(x^s, y^s) + \mathcal{L}(x^t, x^s)$	✗	✗
Source-free domain adaptation[18]	✗	✗	$\mathcal{L}(x^t)$	✗	✗
Domain Generalization[6]	x^s, y^s	✗	$\mathcal{L}(x^s, y^s)$	✗	✓
Test-time training[41]	x^s, y^s	✗	$\mathcal{L}(x^s, y^s) + \mathcal{L}(x^s)$	$\mathcal{L}(x^t)$	✓
Fully test-time adaptation[46]	–	✗	✗	$\mathcal{L}(x^t)$	✓

Despite the success of DNNs for the task of semantic segmentation in sunny, daytime conditions [3,4], these models suffer from severe performance degradation when applied on night, or adverse weather images due to low illumination, shadows, motion blur, glare/or and overexposure. There is scarcity of labeled ground truth data in night-time and adverse weather conditions for training a DNN model. Further, semantic segmentation models are required to exhibit robust performance and reliable operation in a wide range of changing environments, and expecting availability of the data corresponding to all such conditions, and their combinations (e.g. rainy night-time) is unrealistic, and impractical.

To mitigate the aforementioned issues, Domain Generalization (DG) approaches, such as RobustNet [6], aim to improve robustness on the unseen target domains. However the performance of DG methods is still quite low on adverse weather conditions (see Table 4). On the other hand, unsupervised domain adaptation (UDA) techniques for semantic segmentation [5,14,17,20,24,37,43–45,52,53,56,56,57] typically focus on synthetic-to-real domain adaptation, and are not relevant to our focus. UDA techniques specially designed for adapting from normal to adverse weather, require access to target dataset, and are limited to single target domain only, such as nighttime [31,34,47] or fog [8,32]. This limits their real life applicability which requires simultaneous adaptation to all weather conditions.

While effective, it might not be practical in a real-world scenario to collect (even unlabeled) target data corresponding to all adverse visual conditions or their exponential number of combinations. Moreover, there might be a shift in the target data itself due to the dynamic nature of real-world processes. This necessitates designing a model which can adapt on-the-fly when the data is being received during inference. In the light of the above discussion, in this work, we propose MALL for semantic segmentation of adverse weather images. MALL adapts a pre-trained model to unseen weather deteriorated images by enforcing an edge consistency prior. The prior is enforced by aligning the edges in the input RGB image with the edges in the output predicted label image using a loss inspired from the normalized cuts [39]. Moreover, motivated by the recent fully-test time adaptation paradigm [46], we enforce the prior only during the inference (and not at the train time). This has a significant advantage in terms of removing our reliance on the availability of unlabeled target data during training, and capability to use any pre-trained model for adaptation during inference. The key contributions of our work include:

- We propose to enforce edge consistency prior during inference for unsupervised test time adaptation of semantic segmentation techniques for adverse weather images. Taking inspiration from [39], we propose a *Weighted Log Multi-class Normalized Cut* loss for the unsupervised test time training.
- We propose a new framework named MALL in two different settings: MALL-sample adapts a pre-trained model to a single image at a time, whereas MALL-domain adapts a model to the target domain at the test time.
- We perform rigorous experiments to demonstrate that MALL outperforms all SOTA techniques in multiple settings: domain generalization, and unsupervised domain adaptation.
- In a one of its kind experiment, we demonstrate the simultaneous, and domain-agnostic effectiveness of the MALL on six different adverse weather conditions: night, overcast, cloudy, snow, fog, and rain (Fig. 1).

2 Related Work

Test-Time Adaptation: In a test-time training framework [41], in the first step, one trains a DNN with the main task like classification and an auxiliary task like rotation prediction of an image. During inference, the auxiliary task is used to adapt network parameters to handle distribution shifts. TENT [46] has proposed a fully test-time adaptation framework assuming no access to source data and without altering training. It adapts the affine parameters of a batch normalization layer to minimize entropy-based loss for a particular test sample. We also propose a fully test-time adaptation framework specifically for unseen

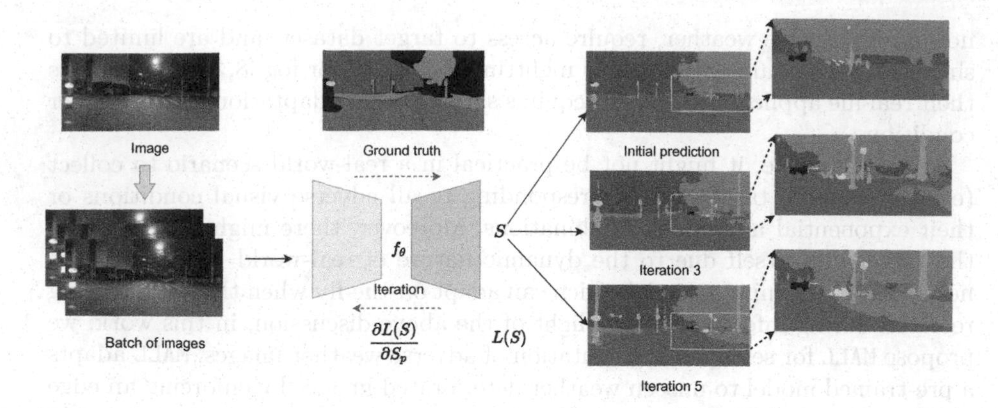

Fig. 1. Our technique improves the prediction of a pre-trained segmentation model by penalizing the loss if the edges in the predicted label image, S, do not align with the visual edges in the input image.

adverse weather conditions such as nighttime, rain, etc. However, unlike TENT, which uses entropy-based loss, we enforce edge prior to achieve our objective.

Domain Generalization (DG): These approaches aim to improve generalization on the unseen target domains. While popular for image classification [29,54,55], very few DG approaches have been proposed for semantic segmentation. IBNNet [25] combines the advantage of instance normalization (to prevent over-fitting on the training data) and batch normalization (learning discriminative intermediate feature representations) to improve generalization in semantic segmentation. Switchable whitening [26] de-correlates intermediate feature maps to remove domain-specific style information. RobustNet [6] extends whitening approach to selectively remove higher values in channel-wise covariance matrix. MALL improves generalization and complement above DG methods by directly adapting to unseen adverse weather images during inference.

UDA Methods for Night-Time Image Segmentation: GCMA [31] proposes the use of intermediate twilight domain to adapt to nighttime images gradually from day to twilight, and twilight to the night-time domain. MGCDA [34] extends it using self-training with curriculum learning to adapt gradually to night images. Similarly [9,30,40], rely on image transfer models like cycleGAN as a pre-processing step. However, their performance depend significantly on the quality of images generated in the pre-processing step. To mitigate this, DANNet [47] proposes a one-stage adaptation network without relying on an intermediate domain or image transfer model. DANNet uses reweighing strategy to handle misalignment between day-night image pairs, and adversarial training accompanied by pseudo-label supervision to enhance segmentation predictions. DANNet also uses additional supervision from GPS to enhance semantic label predictions. DANNet generates noisy predictions for moving objects like cars, and small objects like traffic lights. Our proposed method, MALL, does not need any extra information from the target domain dataset.

Adaptation Without Target Data: In a real-world scenario, it is often impractical to collect large labeled or unlabeled target data covering all possible illumination scenarios. Recent work Zeroshot-DayNight [19] assumes no access to unlabeled target night images and uses the color invariance idea [11], to propose a new layer in a DNN, referred to as color invariant convolution (CIConv). The method claims to reduce the day-night distribution shift for intermediate feature maps, but requires retraining the entire network from scratch using CIConv, thus restricting its applicability. On the other hand, the proposed MALL technique does not do any retraining of a model, but adapts the model on-the-fly to unseen weather images during inference. MALL improves current SOTA methods in both the scenarios, when having access to unlabeled night images during training, as well as without access to them.

Normalized Cuts: Normalized cuts [39] is a popular graph partitioning algorithm proposed for binary image segmentation. Tang *et al.* [42] has proposed to use the normalized cut loss as a regularizer to improve performance in weakly-supervised segmentation tasks. It uses cross-entropy loss for labeled pixels and continuous normalized cut-based loss for weakly-labeled (scribbles) pixels to improve segmentation predictions. We extend the normalized cut loss as edge consistency prior for unsupervised domain generalization setting to penalize the segmentation predictions if the label edges do not align with the visual edges in the input.

3 Methodology

In this section, we go through the formulation of the proposed MALL technique. Specifically, we propose to optimize the model parameters directly during inference using *Weighted Log Multi-class Normalized Cut* (WL-MNC) loss. The loss is designed to penalize incorrect semantic label predictions when the consistency of visual edges present at the boundaries of classes in an image is not preserved. Note that, our assumption in the proposed MALL framework is that we do not need access to the source data. The MALL technique can work in two different settings, namely, MALL-sample and MALL-domain. The former updates the pre-trained model based on a single nighttime image at a time, whereas the later adapts the model continuously to all samples of the target domain as they are presented in a streaming mode.

3.1 Softmax Multi-class Normalized Cut Loss

In this section, we extend normalized cut [39] as a loss function for an unsupervised, multi-class setting that can handle class imbalance in the target dataset. Normalized Cut [39] is a popular graph partitioning algorithm for binary image segmentation task. For the binary image segmentation task, we define a graph node corresponding to every pixel in the image. Let $\mathbf{I}$ be an image, and $\mathbf{A}$ be the affinity matrix, with $\mathbf{A} = [\mathbf{A}_{ij}]$ being the similarity between pixel i and pixel j in the image. Let $\mathbf{d}$ be the degree vector defined as $\mathbf{d} = \mathbf{A}\mathbf{1}$, where $\mathbf{1}$ denotes a vector of all ones. For a binary image segmentation task, where pixels are labeled either as a foreground or background pixel, Normalized Cut loss for partitioning pixels of an image into foreground ($\mathcal{F}$) and background pixels ($\mathcal{B}$) is defined as:

$$\mathrm{cut}(\mathcal{F},\mathcal{B}) = \sum_{x\in\mathcal{F},y\in\mathcal{B}} \mathbf{A}(x,y) \tag{1}$$

$$\mathrm{Ncut}(\mathcal{F},\mathcal{B}) = \frac{\mathrm{cut}(\mathcal{F},\mathcal{B})}{\mathrm{assoc}(\mathcal{F},\mathcal{V})} + \frac{\mathrm{cut}(\mathcal{F},\mathcal{B})}{\mathrm{assoc}(\mathcal{B},\mathcal{V})} \tag{2}$$

where $\mathcal{V} = \mathcal{F} \cup \mathcal{B}$, $\mathrm{assoc}(\mathcal{F},\mathcal{V}) = \sum_{p\in\mathcal{F},q\in\mathcal{V}} \mathbf{A}_{pq}$. If we denote binary image segmentation prediction by a matrix $\mathbf{S}$, then Eq. (2) can be rewritten as:

$$\mathrm{Ncut} = \frac{\mathbf{S}^\top \mathbf{A}(\mathbf{1}-\mathbf{S})}{\mathbf{d}^\top \mathbf{S}} \tag{3}$$

Further, Normalized Cut loss can be extended to a multi-class setting as follows. Let c be the total number of classes, and p denote a particular class. Let $\mathbf{S}_p$ be the segmentation label prediction matrix corresponding to class p, such that $\mathbf{S}_p[i][j] = 1$, if the predicted label of pixel (i,j) is p, and $\mathbf{S}_p[i][j] = 0$, otherwise. Then, Multi-class Normalized Cut loss is a non-differentiable loss as shown below:

$$\mathrm{MultiClassNormalizedCut} = \sum_{p=0}^{c-1} \frac{(\mathbf{S}_p)^\top \mathbf{A}(\mathbf{1}-\mathbf{S}_p)}{\mathbf{d}^\top \mathbf{S}_p} \tag{4}$$

To make a network end-to-end trainable, one can relax Multi-class Normalized Cut loss to Softmax Multi-class Normalized Cut loss [42], by relaxing the value of $S_p[i][j]$ to predicted probability of pixel (i,j) taking label p.

3.2 Class Imbalance Re-weighting

We observe that most benchmark datasets (e.g., Cityscapes [7]) have highly skewed ratio corresponding to the pixels of minority to majority classes. This leads to reduced accuracy for the pixels belonging to the minority class. To handle the class imbalance, we introduce weighted log class probability weighting into the Multi-class Normalized Cut loss function. Let r_p denote the normalized frequency of class p in the source domain, i.e., number of pixels of class p divided by total number of pixels[1]. We define the weight, w_p, as:

$$w_p = \frac{\log(r_p)}{\sum_{i=0}^{c-1}\log(r_i)}. \tag{5}$$

We use w_p to define a new loss function, Weighted Log Multi-class Normalized Cut (`WL-MNC`) loss, to handle the class imbalance as follows:

$$\mathcal{L}_{\texttt{WL-MNC}} = \sum_{p=0}^{c-1} w_p \frac{(\mathbf{S}_p)^\top \mathbf{A}(\mathbf{1}-\mathbf{S}_p)}{\mathbf{d}^\top \mathbf{S}_p}. \tag{6}$$

[1] Note that we do not assume the availability of source data as our method does not need any training for the chosen backbone. However, we do assume the availability of the statistics for the source dataset on which a given model is pre-trained on.

3.3 Sample Importance Weighting

Given an image at the test time, we first create a mini-batch by multiple distortions (called samples) of the input image. Then, we assign an importance weight to a sample k in the mini-batch based on the entropy, e^k, of the sample. We compute the entropy of a sample as:

$$e^k = -\frac{1}{w \cdot h} \sum_{i=0}^{h-1} \sum_{j=0}^{w-1} \sum_{p=0}^{c-1} \mathbf{S}^k[i][j][p] \log\left(\mathbf{S}^k[i][j][p]\right), \tag{7}$$

where $\mathbf{S}^k$ denotes the segmentation output for sample k, and w, and h denotes width and height of the image respectively. If the entropy of a sample is high, we give more weight to the `WL-MNC` loss of the corresponding sample thus increasing the importance of edge consistency for the sample. Importance weighting is based on the softmax normalization of unsupervised Shannon entropy [38] loss. Importance weight w_{imp}^k of kth sample in the mini-batch of size b is defined as:

$$w_{\text{imp}}^k = \frac{\exp(e^k)}{\sum_{i=0}^{b-1} \exp(e^i)}. \tag{8}$$

Overall loss for a mini-batch is computed as:

$$\mathcal{L}_{\text{total}} = \sum_{i=0}^{b-1} w_{\text{imp}}^k \mathcal{L}_{\texttt{WL-MNC}}(\mathbf{S}^k) \tag{9}$$

For a pixel (i, j) in an image, we define affinity matrix $\mathbf{A}$ using a Gaussian kernel in 5D (`RGBXY`) space. Where `RGB` corresponds to R, G, and B color values of the pixel, and `XY` corresponds to the pixel location (i, j) in the image. For two pixels with similar color and location, similarity or affinity is high. In order to facilitate efficient calculation of loss and its gradient, we use a fast bilateral filtering based technique [1].

3.4 MALL-Sample

In the `MALL-sample` method, a pre-trained `DNN` model f_θ, with parameters, θ is adapted to a single image. Given an image x_t, we create a mini-batch of size b, and batch size b, we generate a batch of images by applying $(b-1)$ data augmentation transformations to x_t. During forward pass, the generated batch of images is passed into the network, and loss is back-propagated. We conduct multiple forward and back-propagation iterations over the same batch (and update the weights) until a pre-defined termination criterion (explained below) is met. Once termination criteria are met, we save segmentation label predictions of the image, and weights of the network revert back to the initial model parameters θ. The advantage of reverting back weights of the network for each image is that we do not suffer performance degradation on the source domain (e.g. daylight images). $\mathbf{S}$ denotes segmentation predictions from the network.

$$\min_\theta \mathcal{L}(x_t, \theta, \mathbf{S}). \tag{10}$$

For the `MALL-sample` setting, as a termination criterion, we consider the number of iterations per image as 5, with an early stopping criteria when the difference of loss between two consecutive iterations is less than a particular threshold.

3.5 MALL-Domain

In `MALL-domain` method, a pre-trained model f_θ is adapted to a different target domain directly during inference. Similar to `MALL-sample`, for an unseen target domain image x_t, we pass a batch of images to the network in the forward pass, and `WL-MNC` loss is similarly back-propagated. We also conduct multiple iterations in the same fashion. However, in this case, we do not revert back the weights of network parameters at the end of iterations. Further, no augmentations of the input image are performed to generate a mini-batch in the `MALL-domain` method. The compilation of a batch of images X_t in `MALL-domain` is based on maintaining a buffer of size b; once the buffer is full, we forward pass the batch of images into the network and empty the buffer. We use X_t to denote the batch of images at time t, The benefit of the `MALL-domain` method over unsupervised domain adaptation methods is the ability to adapt on-the-fly to unseen adverse weather images (e.g. rainy night images) directly during inference.

$$\min_{\theta} \mathcal{L}\left(X_t, \theta, \mathbf{S}\right). \tag{11}$$

For the `MALL-domain` setting, we consider the early stopping criteria such that the difference of loss of the network between two consecutive iterations is less than the threshold. The threshold is set to 10^{-10} based on empirical experiments.

4 Experiments

4.1 Datasets and Evaluation Criteria

Nighttime Driving (ND) [9]: The Nighttime Driving-test dataset consists of 50 nighttime images and their corresponding pixel-level coarse ground truth annotations. Each image is of resolution 1920 × 1080 labeled with 19 classes.

Dark Zurich (DZ) [31]: It is a collection of 8779 images with a resolution of 1920 × 1080 captured during the daytime, twilight, and nighttime. For each image, corresponding `GPS` coordinates of the camera are also provided. Dark Zurich-val corresponds to 50 nighttime images used for validation. Dark Zurich-test corresponds to 151 nighttime images used for testing. Dark Zurich-test does not provide ground truth pixel label information to users, and an online evaluation server has been provided to evaluate the performance.

ACDC [35]: The dataset is a collection of 4006 adverse visual condition images with a resolution of 1920×1080 pixels. It contains images from adverse visual conditions of fog, night, snow, and rain. Each visual domain (fog, night, snow, and rain) contains 400 train images along with ground truth semantic labels, 100 validation images (106 validation images for night domain) along with ground truth semantic labels, and 500 unlabeled test images. ACDC Night-val corresponds to results reported on ACDC night validation images.

Foggy Driving (FDD, FD) [33]: It is a collection of 101 real-world fog images with a resolution of 960 $\times$1280 pixels. It contains images captured with fog ranging from moderate to dense fog. Foggy Driving dense is a subset of Foggy Driving dataset consisting of 21 images with dense fog.

Foggy Zurich (FZ) [32]: It contains 3808 real-world fog images collected in Zurich city with a resolution of 1920 $\times$ 1080 pixels. Foggy Zurich-test consists of 40 real fog images with corresponding ground truth semantic labels used for evaluation.

C-Driving [22]: The dataset is collection of four adverse weather conditions compiled from BDD100K dataset [50] consisting of four weather conditions such as cloudy, rain, snow and overcast conditions.

Cityscapes (Day) [7]: It is a collection of 5000 broad daylight images of resolution 2048 $\times$ 1024 pixels. For evaluation,

Evaluation Criteria: Mean Intersection over Union (`mIoU`) is considered as an evaluation criterion. Higher `mIoU` indicates better segmentation label predictions.

Implementation Details: We implement the proposed `MALL` technique using Pytorch [27], and train it on an NVIDIA 32GB V100 GPU. During training, the network parameters are updated using Stochastic Gradient Descent(SGD)

Table 2. Results of `MALL` on pre-trained daytime models. Dataset descriptions are provided in Subsect. 4.1.

Model	ND	DZ	FDD	FD	FZ	ACDC				C-Driving				Day
Daytime (Cityscapes) →						Fog	Rain	Snow	Night	Cloud	Rain	Snow	Overcast	
DeepLabv3+ mobilenet [4]	28.5	11.9	25.2	36.1	26.4	47.4	37.8	30.3	15.8	34.1	27.2	27.0	37.1	61.6
with TENT [46]	33.3	17.4	23.7	39.3	31.4	52.3	42.0	41.3	21.9	36.6	29.5	30.2	38.6	61.6
with `MALL-sample`	36.8	19.8	26.6	40.5	32.6	54.9	43.4	42.7	23.9	37.9	30.8	31.4	39.7	61.6
with `MALL-domain`	38.4	21.7	27.6	40.3	32.3	55.3	42.8	41.9	24.4	37.7	30.4	30.9	39.8	59.6
DeepLabv3+ Resnet101 [4]	38.2	20.2	37.9	44.4	31.2	64.1	48.3	44.0	23.5	42.1	34.6	35.7	44.9	78.5
with TENT [46]	39.3	21.6	37.4	45.8	33.6	63.5	49.4	47.6	25.2	42.6	35.9	36.9	45.9	78.5
with `MALL-sample`	42.1	22.3	38.8	47.2	34.3	64.8	49.9	48.8	26.8	43.8	36.6	38.4	47.4	78.5
with `MALL-domain`	42.7	22.8	38.7	46.7	34.1	65.4	49.7	49.5	26.9	43.7	37.4	38.9	47.7	75.7
RefineNet [21]	33.5	17.1	25.1	35.6	24.9	55.9	42.6	44.2	21.5	41.1	34.6	35.9	44.7	71.4
with TENT [46]	34.0	18.3	26.8	37.3	29.0	57.5	43.1	45.8	22.8	41.0	34.2	37.3	44.8	71.4
with `MALL-sample`	35.5	19.7	30.5	39.0	30.8	59.3	44.5	47.3	23.9	42.5	35.7	38.3	46.2	71.4
with `MALL-domain`	36.3	21.9	31.8	39.9	30.4	59.6	44.2	47.4	23.6	42.8	36.4	38.9	47.3	69.2

optimizer with a learning rate of 1×10^{-3}, momentum of 0.9, and weight decay of 5×10^{-4}. σ_{rgb} and σ_{xy} is set to 15, 80 respectively when reporting our results. Augmentation used for `MALL` is horizontal flip, gaussian blur and color jitter. For `MALL-sample` and `MALL-domain` methods, the mini-batch size is set to 12, images are resized to the resolution of 1024×512 pixels during test-time adaptation.

Comparison with TENT [46]: `TENT` is designed for single image adaptation during inference. We perform semantic segmentation task on the adverse weather datasets discussed in Sect. 4.1. For a fair comparison, we report the results of `MALL-sample` using three pre-trained models namely, DeepLabv3+ mobilenet [4], DeepLabv3+ resnet101 [4] and RefineNet [21]. The mIOU of pre-trained models on the cityscapes-val dataset is 61.6%, 78.5% and 71.4% respectively. Results are presented in Table 2. The `MALL` framework significantly improves the pre-trained DeepLabv3+ mobilenet, DeepLabv3+ resnet101 and RefineNet models to adapt to unseen adverse weather images directly during inference. We have observed that `MALL-sample` and `MALL-domain` methods achieve 29%, 35%, 66% and 82% better mIOU over pretrained DeepLabv3+ mobilenet model on the Night-time Driving test and Dark Zurich validation dataset respectively. Similarly `MALL-sample` and `MALL-domain` methods achieve 10.2%, 11.8%, 10.4% and 12.8% better mIOU over pretrained DeepLabv3+ resnet101 model on the Night-time Driving test and Dark Zurich validation dataset respectively. Results signify that `MALL-sample` outperforms `TENT`. `MALL` is able to adapt to unseen nighttime images using 50 images in night test datasets without any additional target domain data like `GPS` information.

Impact on Daytime Performance: For `MALL-sample` there is no drop in daytime performance as we revert to initial weights after adapting the model to a single image. For `MALL-domain` on nighttime driving dataset, we observed that a decrease in daytime mIOU performance for DeepLabv3+ mobilenet, DeepLabv3+ resnet101 and RefineNet by 3.2%, 3.6% and 3.1% respectively. Adapting to a single image using `MALL` directly during inference can enhance generalization capability without any impact on daytime performance.

Inference Time: For a batch size of 12, with each image resolution of 1024×512, `MALL-sample` takes 1312 ms, 1564 ms and 1578 ms per iteration on DeepLabv3+ mobilenet, DeepLabv3+ resnet101 and RefineNet respectively. `MALL-domain` takes 1180 ms, 1219 ms and 1282 ms per iteration on DeepLabv3+ mobilenet, DeepLabv3+ resnet101 and RefineNet respectively.

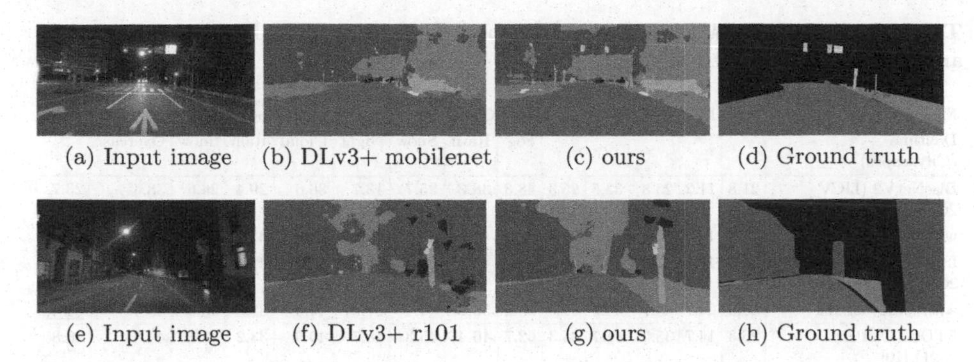

(a) Input image (b) DLv3+ mobilenet (c) ours (d) Ground truth

(e) Input image (f) DLv3+ r101 (g) ours (h) Ground truth

Fig. 2. Qualitative visual comparison of our proposed MALL framework on pre-trained daylight models: Deeplabv3+ mobilenet, Deeplabv3+ resnet101

4.2 Improvement on SOTA Daytime Models

We apply the proposed MALL framework on the SOTA daytime pre-trained models. We consider pre-trained models namely, BiseNetV2 [49], STDC [10], ISANet [16], LRASPP [15], SegFormer-B0 [48], GCNet [2] and Mobilenet V2 [36]. Results are presented in Table 3. Average mIoU is reported to demonstrate overall improvement in segmentation performance. We observed significant improvements with MALL across multiple weather conditions. We report that MALL-domain improves STDC [10] performance on Nighttime driving by 66% in terms of mIoU. MALL also performs well with transformer-based models such as Segformer [48]. MALL-domain improves Mobilenet V2 [36] performance on Dark Zurich by 419%. In terms of average mIoU we report an improvement of 60% for Mobilenet V2. Performance gains by the MALL technique on various baselines validates the generalization ability of proposed approach for adverse weather and visual conditions.

4.3 Improvement on Domain Generalization Models

In Table 4, we demonstrate the effectiveness of the MALL framework over the state-of-the-art domain generalization methods for semantic segmentation IBN-Net [25], Switchable whitening (SW) [26], and RobustNet [6]. We consider pre-trained models of the above-mentioned methods and apply our MALL-domain methods. We report that MALL-domain on IBNNet, RobustNet-Resnet101 shows an increase of 6.1%, 5.2% mIoU on the Nighttime Driving dataset. This shows that MALL can complement domain generalization approaches to adapt to unseen adverse weather images without altering training or requiring unlabeled target data during training. Visual results are shown in Fig. 3.

Table 3. Results of `MALL` on `SOTA` pre-trained daytime models. Dataset descriptions are provided in Subsect. 4.1.

Model Daytime (Cityscapes) →	ND	DZ	FDD	FD	FZ	ACDC				C-Driving				Avg
						Fog	Rain	Snow	Night	Cloud	Rain	Snow	Overcast	
BiseNetV2 (IJCV 2021) [49]	21.8	11.2	23.8	33.5	25.3	48.3	38.3	35.7	13.7	36.6	29.4	26.6	38.6	23.7
with `MALL-domain`	27.7	16.8	33.6	35.6	26.4	52.3	40.2	36.1	19.8	38.8	31.6	31.0	41.2	**30.4**
ISANet (IJCV 2021) [16]	28.1	15.8	37.3	38.6	39.7	62.2	46.8	44.9	18.8	42.2	32.6	35.0	43.9	37.3
with `MALL-domain`	38.3	24.7	38.7	44.9	41.5	63.2	52.8	50.4	27.6	46.5	38.6	39.5	48.3	**42.7**
STDC (CVPR 2021) [10]	26.3	14.7	35.5	41.7	35.3	62.7	46.4	45.3	18.7	40.9	33.2	33.1	44.6	36.8
with `MALL-domain`	43.6	24.1	42.4	44.7	42.8	67.0	48.9	50.4	23.5	44.4	36.4	37.5	47.3	**42.5**
SegFormer (NeurIPS 2021) [48]	32.3	18.9	35.1	40.8	31.4	63.4	48.2	46.4	21.5	39.7	32.7	32.6	42.6	37.3
with `MALL-domain`	34.8	20.2	36.0	41.5	33.8	65.3	49.2	48.6	23.2	40.6	33.0	33.2	43.4	**38.7**
GCNet (TPAMI 2020) [2]	26.5	16.8	40.7	45.8	35.3	62.7	47.9	48.1	19.4	42.6	33.3	35.0	46.0	38.5
with `MALL-domain`	40.2	25.3	42.1	46.8	40.0	64.8	52.0	52.0	25.8	46.4	39.3	39.1	49.7	**43.4**
LRASPP (ICCV 2019) [15]	24.1	12.7	28.6	30.5	21.4	51.3	37.4	38.2	15.4	33.3	26.4	24.9	35.4	29.2
with `MALL-domain`	33.4	16.9	31.3	36.0	28.5	54.1	41.6	39.7	18.9	36.5	29.5	29.3	39.6	**33.5**
Mobilenet V2 (CVPR 2019) [36]	9.7	3.9	23.4	30.1	20.4	41.2	38.3	26.1	3.7	34.6	25.3	22.0	36.6	24.2
with `MALL-domain`	33.8	20.3	30.0	43.9	32.8	61.2	49.0	43.6	26.5	42.9	35.9	36.2	45.1	**38.6**

Table 4. Results of `MALL` framework on `SOTA` Domain Generalization methods. Dataset descriptions are provided in Subsect. 4.1.

Model Daytime (Cityscapes) →	ND	DZ	FDD	FD	FZ	ACDC				C-Driving				Avg
						Fog	Rain	Snow	Night	Cloud	Rain	Snow	Overcast	
IBNNet [25]	33.4	21.7	31.7	41.9	34.5	63.6	50.4	50.2	25.9	43.8	35.8	35.0	48.0	39.6
with `MALL-domain`	39.5	24.8	32.5	42.7	35.2	64.7	51.5	51.2	28.9	44.9	36.4	36.2	48.9	**41.3**
SW [26]	34.0	15.6	22.2	38.2	28.6	54.8	40.2	45.6	20.7	39.2	30.9	31.6	41.7	34.1
with `MALL-domain`	37.2	21.1	30.6	39.8	30.5	59.8	44.5	49.1	26.2	40.5	33.6	34.8	44.4	**37.9**
RobustNet-Resnet50 [6]	35.8	20.9	37.0	43.0	34.5	60.6	46.9	50.1	26.1	42.7	36.5	34.7	45.7	39.5
with `MALL-domain`	41.0	22.5	39.3	43.8	35.7	61.9	47.7	51.2	28.7	43.5	37.6	36.1	45.9	**41.2**
RobustNet-Resnet101 [6]	37.7	21.8	38.8	43.5	33.5	59.3	46.5	45.2	26.8	41.6	34.4	33.8	42.7	38.8
with `MALL-domain`	42.9	24.8	39.7	44.7	37.4	64.2	49.9	51.1	30.7	42.0	35.8	36.4	47.3	**42.0**

4.4 Improvement on Unsupervised Domain Adaptation Models

In this section, we consider SOTA UDA methods for night image segmentation. We evaluate the performance in two scenarios: (1) where the pre-trained model has not seen night images during training, and (2) where the model is trained on nighttime images. For the first category of models trained on source data (daytime) only, we experiment with Zeroshot-DayNight [19]. For the second category, when the models are trained on source data (daytime) and target data (nighttime), we consider DANNet (PSPNet) [47]. Table 5, Table 6 shows the results. We observe that MALL-domain improves Zeroshot-DayNight [19] by 2.6% and 1.7% in terms of mIoU on Nighttime Driving-test and Dark Zurich-test respectively. MALL-domain method on DANNet [47] and MGCDA [34] increases mIOU by 1.1% and 0.3%, 0.5% and 0.7% on the Nighttime Driving test, Dark Zurich test datasets, respectively. Furthermore, MALL can significantly improve DANNet performance in other weather conditions, as demonstrated in Table 7. For instance MALL-domain on DANNet improves mIoU on Foggy Zurich dataset by 3.6%. Significant improvement in the performance using MALL-domain demonstrates the ability of our framework to enhance the generalization ability of SOTA nighttime image segmentation methods in both the scenarios: with and without access to unlabeled target data during training. Figure 2 shows qualitative results for improvement over models trained on daytime images. Whereas in Fig. 4 we show qualitative results for models which are trained on both daytime and nighttime images.

Table 5. Results of MALL framework on state-of-the-art methods in night image segmentation, zeroshot-DN: Zeroshot DayNight [19].

Method	ND-test	DZ-test
Trained on source data only		
Zeroshot-DN [19]	41.2	34.5
with MALL-domain	**43.8**	**36.2**
Trained on source and target data		
ADVENT [45]	34.7	29.7
BDL [20]	34.7	30.8
AdaptSegNet [43]	34.5	30.4
DMAda [9]	41.6	32.1
Day2Night [40]	45.1	–
GCMA [31]	45.6	42.0
MGCDA [34]	49.4	42.5
with MALL-domain	**49.9**	43.2
DANNet [47]	47.7	45.2
with MALL-sample	48.3	45.3
with MALL-domain	48.8	**45.5**

MALL qualitative visual results for the night image datasets for the DANNet pre-trained model with MALL-domain are reported in Fig. 4. In Fig. 4 we see that DANNet fails to label moving object (car), or detect traffic light, whereas by using MALL during inference it can detect traffic light, and car correctly. Results on Dark Zurich-val, MGCDA segmentation outputs, class wise mIoU values on Dark Zurich-test dataset are reported in the supplementary material.

Table 6. Results of MALL framework on state-of-the-art methods in night image segmentation

Method	DZ-val
GCMA [31]	26.6
MGCDA [34]	26.1
with MALL-domain (ours)	**26.8**
DANNet [47]	36.7
with MALL-sample (ours)	**37.1**
with MALL-domain (ours)	**37.3**

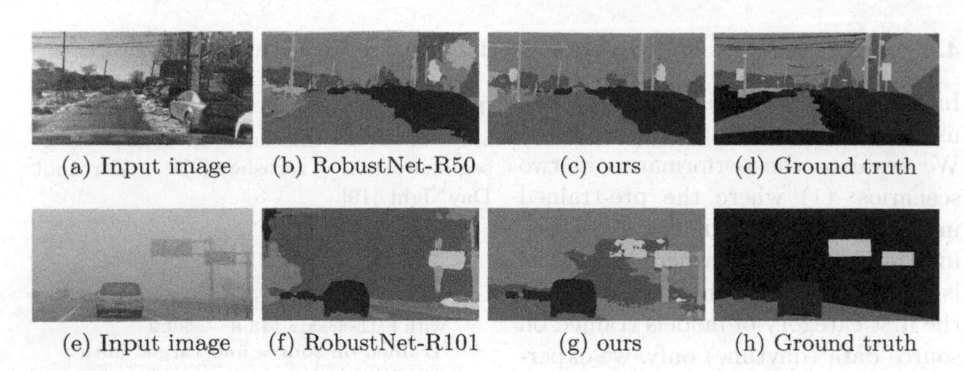

(a) Input image (b) RobustNet-R50 (c) ours (d) Ground truth

(e) Input image (f) RobustNet-R101 (g) ours (h) Ground truth

Fig. 3. Qualitative visual comparison of our proposed `MALL` framework on pre-trained Domain generalization models: Robustnet-Resnet50, Robustnet-Resnet101 ours: `MALL-domain` method.

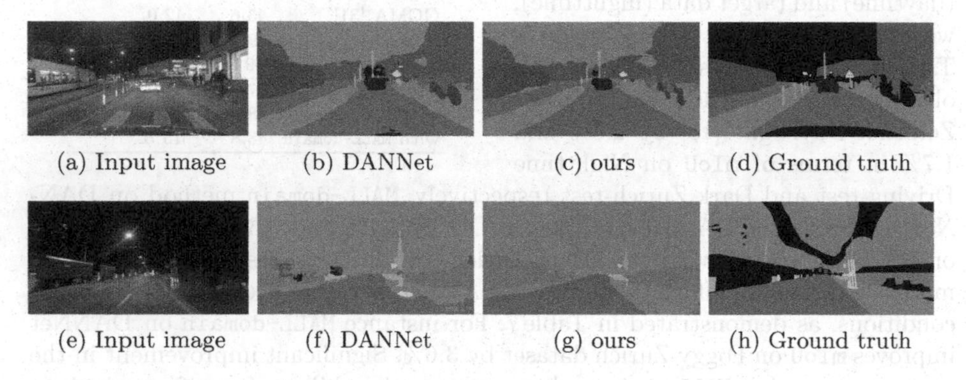

(a) Input image (b) DANNet (c) ours (d) Ground truth

(e) Input image (f) DANNet (g) ours (h) Ground truth

Fig. 4. Qualitative visual comparison of our proposed `MALL` framework on DANNet pre-trained model on two images from night image datasets. DANNet with `MALL-domain` further improve the results.

Table 7. Results of `MALL` framework on DANNet [47]. Dataset descriptions are provided in Subsect. 4.1.

Model Nighttime (DZ) →	ND	DZ	FDD	FD	FZ	ACDC				C-Driving				Avg
						Fog	Rain	Snow	Night	Cloud	Rain	Snow	Overcast	
DANNet [47]	47.7	36.7	33.9	35.4	31.6	52.5	44.3	47.0	39.8	42.0	34.1	35.5	42.7	40.2
with `MALL-domain`	48.8	37.3	36.4	37.8	35.2	53.8	45.8	48.9	41.5	43.5	36.7	37.8	43.9	**42.1**

Ablation Study: In order to demonstrate the effectiveness of class imbalance re-weighting and sample importance weighting, we perform an ablation study on loss formulation considering a pre-trained model, DeepLabv3+ mobilenet [4]. Results are shown in Table 8.

Table 8. Ablation study on several loss variants using DeepLabv3+ mobilenet [4]

Method	ND-test	DZ-val
DeepLabv3+ mobilenet [4]	28.5	11.9
with MALL-domain (ours)	38.4	21.7
w/o class imbalance re-weighting	37.4	20.6
w/o sample importance weighting	36.1	19.2

Conclusion: This paper introduces a novel weighted log multi-class normalized cut loss to enforce edge consistency prior for improving semantic segmentation predictions in adverse weather and visual conditions. The proposed framework improves segmentation performance of SOTA models trained on daytime images as well as the ones which have seen adverse weather images during training. We show that our framework can be used in conjunction with SOTA domain generalization approaches to further improve their performance for adverse weather images. Our experiments indicate that edge consistency prior could also be effective in multiple adverse weather conditions, such as rain, snow, night, cloudy, overcast and fog.

Acknowledgment. We thank UQ-IIT Delhi Research Academy (UQIDAR) for providing contingency grant. This work has also been partly supported by the funding received from DST through the IMPRINT program (IMP/2019/000250). We acknowledge National Super-computing Mission (NSM) for providing computing resources of 'PARAM Siddhi-AI', under National PARAM Super-computing Facility (NPSF), C-DAC, Pune, and supported by the Ministry of Electronics and Information Technology (MeitY) and Department of Science and Technology (DST), Government of India.

References

1. Adams, A., Baek, J., Davis, M.A.: Fast high-dimensional filtering using the permutohedral lattice. Comput. Graph. Forum **29**, 753–762 (2010)
2. Cao, Y., Xu, J., Lin, S., Wei, F., Hu, H.: Global context networks. IEEE Trans. Pattern Anal. Mach. Intell., 1 (2020). https://doi.org/10.1109/TPAMI.2020.3047209
3. Chen, L.C., Papandreou, G., Kokkinos, I., Murphy, K., Yuille, A.L.: DeepLab: semantic image segmentation with deep convolutional nets, Atrous convolution, and fully connected CRFs. IEEE Trans. Pattern Anal. Mach. Intell. **40**(4), 834–848 (2017)
4. Chen, L.-C., Zhu, Y., Papandreou, G., Schroff, F., Adam, H.: Encoder-Decoder with atrous separable convolution for semantic image segmentation. In: Ferrari, V., Hebert, M., Sminchisescu, C., Weiss, Y. (eds.) ECCV 2018. LNCS, vol. 11211, pp. 833–851. Springer, Cham (2018). https://doi.org/10.1007/978-3-030-01234-2_49
5. Chen, Y., Li, W., Van Gool, L.: Road: reality oriented adaptation for semantic segmentation of urban scenes. In: Proceedings of the IEEE Conference on Computer Vision and Pattern Recognition, pp. 7892–7901 (2018)

6. Choi, S., Jung, S., Yun, H., Kim, J.T., Kim, S., Choo, J.: RobustNet: improving domain generalization in urban-scene segmentation via instance selective whitening. In: Proceedings of the IEEE/CVF Conference on Computer Vision and Pattern Recognition, pp. 11580–11590 (2021)
7. Cordts, M., et al.: The cityscapes dataset for semantic urban scene understanding. In: Proceedings of the IEEE Conference on Computer Vision and Pattern Recognition, pp. 3213–3223 (2016)
8. Dai, D., Sakaridis, C., Hecker, S., Van Gool, L.: Curriculum model adaptation with synthetic and real data for semantic foggy scene understanding. Int. J. Comput. Vis. **128**(5), 1182–1204 (2020). https://doi.org/10.1007/s11263-019-01182-4
9. Dai, D., Van Gool, L.: Dark model adaptation: semantic image segmentation from daytime to nighttime. In: 2018 21st International Conference on Intelligent Transportation Systems (ITSC), pp. 3819–3824. IEEE (2018)
10. Fan, M., et al.: Rethinking BiSeNet for real-time semantic segmentation. In: Proceedings of the IEEE/CVF Conference on Computer Vision and Pattern Recognition, pp. 9716–9725 (2021)
11. Geusebroek, J.M., Van den Boomgaard, R., Smeulders, A.W.M., Geerts, H.: Color invariance. IEEE Trans. Pattern Anal. Mach. Intell. **23**(12), 1338–1350 (2001)
12. Girshick, R.: Fast R-CNN. In: Proceedings of the IEEE International Conference on Computer Vision, pp. 1440–1448. IEEE, Piscataway (2015)
13. He, K., Zhang, X., Ren, S., Sun, J.: Deep residual learning for image recognition. In: Proceedings of the IEEE Conference on Computer Vision and Pattern Recognition, pp. 770–778 (2016)
14. Hoffman, J., et al.: CyCADA: cycle-consistent adversarial domain adaptation. In: International Conference on Machine Learning, pp. 1989–1998. PMLR (2018)
15. Howard, A., et al.: Searching for MobileNetV3. In: The IEEE International Conference on Computer Vision (ICCV), pp. 1314–1324, October 2019. https://doi.org/10.1109/ICCV.2019.00140
16. Huang, L., Yuan, Y., Guo, J., Zhang, C., Chen, X., Wang, J.: Interlaced sparse self-attention for semantic segmentation. arXiv preprint arXiv:1907.12273 (2019)
17. Kim, M., Byun, H.: Learning texture invariant representation for domain adaptation of semantic segmentation. In: Proceedings of the IEEE/CVF Conference on Computer Vision and Pattern Recognition, pp. 12975–12984 (2020)
18. Kundu, J.N., Kulkarni, A., Singh, A., Jampani, V., Babu, R.V.: Generalize then adapt: source-free domain adaptive semantic segmentation. In: Proceedings of the IEEE/CVF International Conference on Computer Vision (ICCV), pp. 7046–7056, October 2021
19. Lengyel, A., Garg, S., Milford, M., van Gemert, J.C.: Zero-shot day-night domain adaptation with a physics prior. In: Proceedings of the IEEE/CVF International Conference on Computer Vision, pp. 4399–4409 (2021)
20. Li, Y., Yuan, L., Vasconcelos, N.: Bidirectional learning for domain adaptation of semantic segmentation. In: Proceedings of the IEEE/CVF Conference on Computer Vision and Pattern Recognition, pp. 6936–6945 (2019)
21. Lin, G., Milan, A., Shen, C., Reid, I.: RefineNet: multi-path refinement networks for high-resolution semantic segmentation. In: Proceedings of the IEEE Conference on Computer Vision and Pattern Recognition, pp. 1925–1934 (2017)
22. Liu, Z., et al.: Open compound domain adaptation. In: IEEE Conference on Computer Vision and Pattern Recognition (CVPR) (2020)
23. Long, J., Shelhamer, E., Darrell, T.: Fully convolutional networks for semantic segmentation. In: Proceedings of the IEEE Conference on Computer Vision and Pattern Recognition, pp. 3431–3440 (2015)

24. Luo, Y., Zheng, L., Guan, T., Yu, J., Yang, Y.: Taking a closer look at domain shift: category-level adversaries for semantics consistent domain adaptation. In: Proceedings of the IEEE/CVF Conference on Computer Vision and Pattern Recognition, pp. 2507–2516 (2019)
25. Pan, X., Luo, P., Shi, J., Tang, X.: Two at once: enhancing learning and generalization capacities via IBN-Net. In: Ferrari, V., Hebert, M., Sminchisescu, C., Weiss, Y. (eds.) ECCV 2018. LNCS, vol. 11208, pp. 484–500. Springer, Cham (2018). https://doi.org/10.1007/978-3-030-01225-0_29
26. Pan, X., Zhan, X., Shi, J., Tang, X., Luo, P.: Switchable whitening for deep representation learning. In: Proceedings of the IEEE/CVF International Conference on Computer Vision, pp. 1863–1871 (2019)
27. Paszke, A., et al.: PyTorch: an imperative style, high-performance deep learning library. Adv. Neural. Inf. Process. Syst. **32**, 8026–8037 (2019)
28. Quiñonero-Candela, J., Sugiyama, M., Lawrence, N.D., Schwaighofer, A.: Dataset Shift in Machine Learning. MIT Press (2009)
29. Robey, A., Pappas, G.J., Hassani, H.: Model-based domain generalization. arXiv preprint arXiv:2102.11436 (2021)
30. Romera, E., Bergasa, L.M., Yang, K., Alvarez, J.M., Barea, R.: Bridging the day and night domain gap for semantic segmentation. In: 2019 IEEE Intelligent Vehicles Symposium (IV), pp. 1312–1318. IEEE (2019)
31. Sakaridis, C., Dai, D., Gool, L.V.: Guided curriculum model adaptation and uncertainty-aware evaluation for semantic nighttime image segmentation. In: Proceedings of the IEEE/CVF International Conference on Computer Vision, pp. 7374–7383 (2019)
32. Sakaridis, C., Dai, D., Hecker, S., Van Gool, L.: Model adaptation with synthetic and real data for semantic dense foggy scene understanding. In: Ferrari, V., Hebert, M., Sminchisescu, C., Weiss, Y. (eds.) ECCV 2018. LNCS, vol. 11217, pp. 707–724. Springer, Cham (2018). https://doi.org/10.1007/978-3-030-01261-8_42
33. Sakaridis, C., Dai, D., Van Gool, L.: Semantic foggy scene understanding with synthetic data. Int. J. Comput. Vision **126**(9), 973–992 (2018)
34. Sakaridis, C., Dai, D., Van Gool, L.: Map-guided curriculum domain adaptation and uncertainty-aware evaluation for semantic nighttime image segmentation. IEEE Trans. Pattern Anal. Mach. Intell. (2020)
35. Sakaridis, C., Dai, D., Van Gool, L.: ACDC: the adverse conditions dataset with correspondences for semantic driving scene understanding. arXiv preprint arXiv:2104.13395 (2021)
36. Sandler, M., Howard, A., Zhu, M., Zhmoginov, A., Chen, L.C.: MobileNetV 2: inverted residuals and linear bottlenecks. In: Proceedings of the IEEE Conference on Computer Vision and Pattern Recognition, pp. 4510–4520 (2018)
37. Sankaranarayanan, S., Balaji, Y., Jain, A., Lim, S.N., Chellappa, R.: Learning from synthetic data: addressing domain shift for semantic segmentation. In: Proceedings of the IEEE Conference on Computer Vision and Pattern Recognition, pp. 3752–3761 (2018)
38. Shannon, C.E.: A mathematical theory of communication. Bell Syst. Tech. J. **27**(3), 379–423 (1948)
39. Shi, J., Malik, J.: Normalized cuts and image segmentation. IEEE Trans. Pattern Anal. Mach. Intell. **22**(8), 888–905 (2000)
40. Sun, L., Wang, K., Yang, K., Xiang, K.: See clearer at night: towards robust nighttime semantic segmentation through day-night image conversion. In: Artificial Intelligence and Machine Learning in Defense Applications, vol. 11169, p. 111690A. International Society for Optics and Photonics (2019)

41. Sun, Y., Wang, X., Zhuang, L., Miller, J., Hardt, M., Efros, A.A.: Test-time training with self-supervision for generalization under distribution shifts. In: ICML (2020)
42. Tang, M., Djelouah, A., Perazzi, F., Boykov, Y., Schroers, C.: Normalized cut loss for weakly-supervised CNN segmentation. In: Proceedings of the IEEE Conference on Computer Vision and Pattern Recognition, pp. 1818–1827 (2018)
43. Tsai, Y.H., Hung, W.C., Schulter, S., Sohn, K., Yang, M.H., Chandraker, M.: Learning to adapt structured output space for semantic segmentation. In: Proceedings of the IEEE Conference on Computer Vision and Pattern Recognition, pp. 7472–7481 (2018)
44. Tsai, Y.H., Sohn, K., Schulter, S., Chandraker, M.: Domain adaptation for structured output via discriminative patch representations. In: Proceedings of the IEEE/CVF International Conference on Computer Vision, pp. 1456–1465 (2019)
45. Vu, T.H., Jain, H., Bucher, M., Cord, M., Pérez, P.: Advent: adversarial entropy minimization for domain adaptation in semantic segmentation. In: Proceedings of the IEEE/CVF Conference on Computer Vision and Pattern Recognition, pp. 2517–2526 (2019)
46. Wang, D., Shelhamer, E., Liu, S., Olshausen, B., Darrell, T.: Tent: fully test-time adaptation by entropy minimization. arXiv preprint arXiv:2006.10726 (2020)
47. Wu, X., Wu, Z., Guo, H., Ju, L., Wang, S.: DanNet: a one-stage domain adaptation network for unsupervised nighttime semantic segmentation. In: Proceedings of the IEEE/CVF Conference on Computer Vision and Pattern Recognition, pp. 15769–15778 (2021)
48. Xie, E., Wang, W., Yu, Z., Anandkumar, A., Alvarez, J.M., Luo, P.: SegFormer: simple and efficient design for semantic segmentation with transformers. arXiv preprint arXiv:2105.15203 (2021)
49. Yu, C., Gao, C., Wang, J., Yu, G., Shen, C., Sang, N.: BiSeNet V2: bilateral network with guided aggregation for real-time semantic segmentation. Int. J. Comput. Vision **129**(11), 3051–3068 (2021)
50. Yu, F., et al.: BDD100K: a diverse driving dataset for heterogeneous multitask learning. In: IEEE/CVF Conference on Computer Vision and Pattern Recognition (CVPR), June 2020
51. Zhang, P., Zhang, B., Zhang, T., Chen, D., Wang, Y., Wen, F.: Prototypical pseudo label denoising and target structure learning for domain adaptive semantic segmentation. arXiv preprint arXiv:2101.10979 (2021)
52. Zhang, Y., Qiu, Z., Yao, T., Liu, D., Mei, T.: Fully convolutional adaptation networks for semantic segmentation. In: Proceedings of the IEEE Conference on Computer Vision and Pattern Recognition, pp. 6810–6818 (2018)
53. Zheng, Z., Yang, Y.: Rectifying pseudo label learning via uncertainty estimation for domain adaptive semantic segmentation. Int. J. Comput. Vision **129**(4), 1106–1120 (2021)
54. Zhou, K., Yang, Y., Hospedales, T., Xiang, T.: Deep domain-adversarial image generation for domain generalisation. In: Proceedings of the AAAI Conference on Artificial Intelligence, vol. 34, pp. 13025–13032 (2020)
55. Zhou, K., Yang, Y., Hospedales, T., Xiang, T.: Learning to generate novel domains for domain generalization. In: Vedaldi, A., Bischof, H., Brox, T., Frahm, J.-M. (eds.) ECCV 2020. LNCS, vol. 12361, pp. 561–578. Springer, Cham (2020). https://doi.org/10.1007/978-3-030-58517-4_33

56. Zou, Y., Yu, Z., Vijaya Kumar, B.V.K., Wang, J.: Unsupervised domain adaptation for semantic segmentation via class-balanced self-training. In: Ferrari, V., Hebert, M., Sminchisescu, C., Weiss, Y. (eds.) ECCV 2018. LNCS, vol. 11207, pp. 297–313. Springer, Cham (2018). https://doi.org/10.1007/978-3-030-01219-9_18
57. Zou, Y., Yu, Z., Liu, X., Kumar, B., Wang, J.: Confidence regularized self-training. In: Proceedings of the IEEE/CVF International Conference on Computer Vision, pp. 5982–5991 (2019)

LESS: Label-Efficient Semantic Segmentation for LiDAR Point Clouds

Minghua Liu[1], Yin Zhou[2(✉)], Charles R. Qi[2], Boqing Gong[3], Hao Su[1], and Dragomir Anguelov[2]

[1] UC San Diego, San Diego, USA
[2] Waymo, Mountain View, USA
yinzhou@waymo.com
[3] Google, Washington, D.C., USA

Abstract. Semantic segmentation of LiDAR point clouds is an important task in autonomous driving. However, training deep models via conventional supervised methods requires large datasets which are costly to label. It is critical to have label-efficient segmentation approaches to scale up the model to new operational domains or to improve performance on rare cases. While most prior works focus on indoor scenes, we are one of the first to propose a label-efficient semantic segmentation pipeline for outdoor scenes with LiDAR point clouds. Our method co-designs an efficient labeling process with semi/weakly supervised learning and is applicable to nearly any 3D semantic segmentation backbones. Specifically, we leverage geometry patterns in outdoor scenes to have a heuristic pre-segmentation to reduce the manual labeling and jointly design the learning targets with the labeling process. In the learning step, we leverage prototype learning to get more descriptive point embeddings and use multi-scan distillation to exploit richer semantics from temporally aggregated point clouds to boost the performance of single-scan models. Evaluated on the SemanticKITTI and the nuScenes datasets, we show that our proposed method outperforms existing label-efficient methods. With extremely limited human annotations (*e.g.*, 0.1% point labels), our proposed method is even highly competitive compared to the fully supervised counterpart with 100% labels.

1 Introduction

Light detection and ranging (LiDAR) sensors have become a necessity for most autonomous vehicles. They capture more precise depth measurements and are more robust against various lighting conditions compared to visual cameras. Semantic segmentation for LiDAR point clouds is an indispensable technology as

M. Liu—Work done during internship at Waymo LLC.

Supplementary Information The online version contains supplementary material available at https://doi.org/10.1007/978-3-031-19842-7_5.

S. Avidan et al. (Eds.): ECCV 2022, LNCS 13699, pp. 70–89, 2022.
https://doi.org/10.1007/978-3-031-19842-7_5

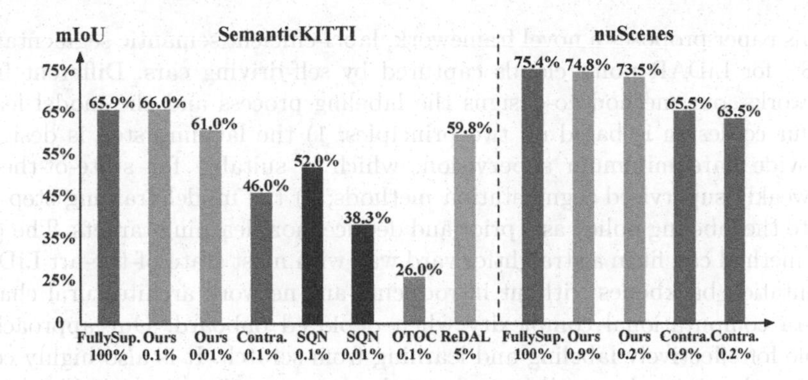

Fig. 1. We compare LESS with Cylinder3D [68] (our fully-supervised counterpart), ContrastiveSceneContext [23], SQN [24], OneThingOneClick [33], and ReDAL [55] on the SemanticKITTI [4] and nuScenes [5] validation sets. The ratio between labels used and all points is listed below each bar. Please note that all competing label-efficient methods mainly focus on indoor settings and are not specially designed for outdoor LiDAR segmentation.

it provides fine-grained scene understanding, complementary to object detection. For example, semantic segmentation help self-driving cars distinguish drivable and non-drivable road surfaces and reason about their functionalities, like parking areas and sidewalks, which is beyond the scope of modern object detectors.

Based on large-scale public driving-scene datasets [4,5], several LiDAR semantic segmentation approaches have recently been developed [9,50,59,62,68]. Typically, these methods require fully labeled point clouds during training. Since a LiDAR sensor may perceive millions of points per second, exhaustively labeling all points is extremely laborious and time-consuming. Moreover, it may fail to scale when we extend the operational domain (*e.g.*, various cities and weather conditions) and seek to cover more rare cases. Therefore, to scale up the system, it is critical to have label-efficient approaches for LiDAR semantic segmentation, whose goal is to minimize the quantity of human annotations while still achieving high performance.

While there are some prior works studying label-efficient semantic segmentation, they mostly focus on indoor scenes [3,11] or 3D object parts [6], which are quite different in point cloud appearance and object type distribution, compared to the outdoor driving scenes (*e.g.*, significant variances in point density, extremely unbalanced point counts between common types, like ground and vehicles, and less common ones, such as cyclists and pedestrians). Besides, most prior explorations tend to address the problem from two independent perspectives, which may be less effective in our outdoor setting. Specifically, one perspective is improving labeling efficiency, where the methods resort to active learning [34,47,55], weak labels [44,54], and 2D supervision [53] to reduce labeling efforts. The other perspective focuses on training, where the efforts assume the partial labels are given and design semi/weakly supervised learning algorithms to exploit the limited labels and strive for better performance [20,33,34,44,60,61,66].

This paper proposes a novel framework, label-efficient semantic segmentation (LESS), for LiDAR point clouds captured by self-driving cars. Different from prior works, our method co-designs the labeling process and the model learning. Our co-design is based on two principles: 1) the labeling step is designed to provide bare minimum supervision, which is suitable for state-of-the-art semi/weakly supervised segmentation methods; 2) the model training step can tap into the labeling policy as a prior and deduce more learning targets. The proposed method can fit in a straightforward way with most state-of-the-art LiDAR segmentation backbones without introducing any network architectural change or extra computational complexity when deployed onboard. Our approach is suitable for effectively labeling and learning from scratch. It is also highly compatible with mining long-tail instances, where, in practice, we mainly want to identify and annotate rare cases based on trained models.

Specifically, we leverage a philosophy that outdoor-scene objects are often well-separated when isolating ground points and design a heuristic approach to pre-segment an outdoor scene into a set of connected components. The component proposals are of high purity (*i.e.*, only contain one or a few classes) and cover most of the points. Then, instead of meticulously labeling all points, the annotators are only required to label one point per class for each component. In the model learning process, we train the backbone segmentation network with the sparse labels directly annotated by humans as well as the derived labels based on component proposals. To encourage a more descriptive embedding space, we employ contrastive prototype learning [18,29,33,48,63], which increases intra-class similarity and inter-class separation. We also leverage a multi-scan teacher model to exploit richer semantics within the temporally fused point clouds and distill the knowledge to boost the performance of the single-scan model.

We evaluate the proposed method on two large-scale autonomous driving datasets, SemanticKITTI [4] and nuScenes [5]. We show that our method significantly outperforms existing label-efficient methods (see Fig. 1). With extremely limited human annotations, such as 0.1% labeled points, the approach achieves highly competitive performance compared to the fully supervised counterpart, demonstrating the potential of practical deployment.

In summary, our contribution mainly includes:

- Analyze how label-efficient segmentation of outdoor LiDAR point clouds differs from the indoor settings, and show that the unbalanced category distribution is one of the main challenges.
- Leverage the unique geometric structure of LiDAR point clouds and design a heuristic algorithm to pre-segment input points into high-purity connected components. A customized labeling policy is then proposed to exploit the components with tailored labels and losses.
- Adapt beneficial components into label-efficient LiDAR segmentation and carefully design a network-agnostic pipeline that achieves on-par performance with the fully supervised counterpart.
- Evaluate the proposed pipeline on two large-scale autonomous driving datasets and extensively ablate each module.

2 Related Work

2.1 Segmentation Networks for LiDAR Point Clouds

In contrast to indoor-scene point clouds, outdoor LiDAR point clouds' large scale, varying density, and sparsity require the segmentation networks to be more efficient. Many works project the 3D point clouds from spherical view [2,10,12,27,30,39,43,58] (*i.e.*, range images) or bird's-eye-view [45,65] onto 2D images, or try to fuse different views [1,19,31]. There are also some works directly consuming point clouds [8,14,25,52]. They aim to structure the irregular data more efficiently. Zhu *et al.* [68] employ the voxel-based representation and alleviate the high computational burden by leveraging cylindrical partition and sparse asymmetrical convolution. Recent works also try to fuse the point and voxel representations [9,50,62,64], and even with range images [59]. All of these works can serve as the backbone network in our label-efficient framework.

Table 1. Point distribution across the most common and rarest categories of SemanticKITTI [4] **and nuScenes** [5]. Numbers are normalized by the sample quantity of bicycles.

Dataset	Vegetation	Road	Building	Car	Motorcycle	Person	Bicycle
SemanticKITTI	1606	1197	799	257	2	2	1
nuScenes	867	2242	1261	270	3	16	1

2.2 Label-Efficient 3D Semantic Segmentation

Label-efficient 3D semantic segmentation has recently received lots of attention [17]. Previous explorations are mainly two-fold: labeling and training.

As for labeling, several approaches seek active learning [34,47,55], which iteratively selects and requests points to be labeled during the network training. Hou *et al.* [23] utilize features from unsupervised pre-training to choose points for labeling. Wang *et al.* [53] project the point clouds to 2D and leverage 2D supervision signals. Some works utilize scene-level or sub-cloud-level weak labels [44,54]. There are also several approaches using rule-based heuristics or handcrafted features to help annotation [20,37,51].

As for training, Xie *et al.* [23,57] utilize contrastive learning for unsupervised pre-training. Some approaches employ self-training to generate pseudo-labels [33, 44,60]. Lots of works use Conditional Random Fields (CRFs) [20,33,34,61] or random walk [66] to propagate labels. Moreover, there are also works that utilize prototype learning [33,66], siamese learning [44,61], temporal constraints [36], smoothness constraints [44,47], attention [54,66], cross task consistency [44], and synthetic data [56] to help training.

However, most recent works mainly focus on indoor scenes [3,11] or 3D object parts [6], while outdoor scenarios are largely under-explored.

3 Method

In this section, we present our LESS framework. Since existing label-efficient segmentation works typically address domains other than autonomous driving, we first conduct a pilot study to understand the challenges in this novel setting and introduce motivations behind LESS (Sect. 3.1). After briefly going over our LESS framework (Sect. 3.2), we dive into the details of each part (Sects. 3.3 to 3.6).

3.1 Pilot Study: What Should We Pay Attention To?

Previous works [23,33,44,47,53,54,61,66] on label-efficient 3D semantic segmentation mainly focused on indoor datasets, such as ScanNet-v2 [11] and S3DIS [3]. In these datasets, input points are sampled from high-quality reconstructed meshes and are thus densely and uniformly distributed. Also, objects in indoor scenarios typically share similar sizes and have a relatively balanced class distribution. However, in outdoor settings, input point clouds demonstrate substantially higher complexity due to the varying point density and the ubiquitous occlusions throughout the scene. Moreover, in outdoor driving scenes, the sample distribution across different categories is highly unbalanced due to factors including occurring frequency and object size. Table 1 shows the point distribution over two autonomous driving datasets, where the numbers of road points are 1,197 and 2,242 times larger than that of bicycle points, respectively. The extremely unbalanced distribution adds extra difficulty for label-efficient segmentation, whose goal is to only label a tiny portion of points (Fig. 2).

We conduct a pilot study to further examine this challenge. Specifically, we train a state-of-the-art semantic segmentation network, Cylinder3D [68], on the SemanticKITTI dataset with three intuitive setups: (a) 100% labels, (b) randomly annotating 0.1% points per scan, and (c) randomly selecting 0.1% scans and annotating all points for the selected scans. The results are shown in Sect. 4.2. Without any special efforts, "0.1% random points" can already achieve a mean IoU of 48.0%, compared to 65.9% by the fully supervised version. On common categories, such as car, road, building, and vegetation, the performances of the "0.1% label" models are close to the fully supervised model. However, on the underrepresented categories, such as bicycle, person, and motorcycle, we observe substantial performance gaps compared to the fully supervised

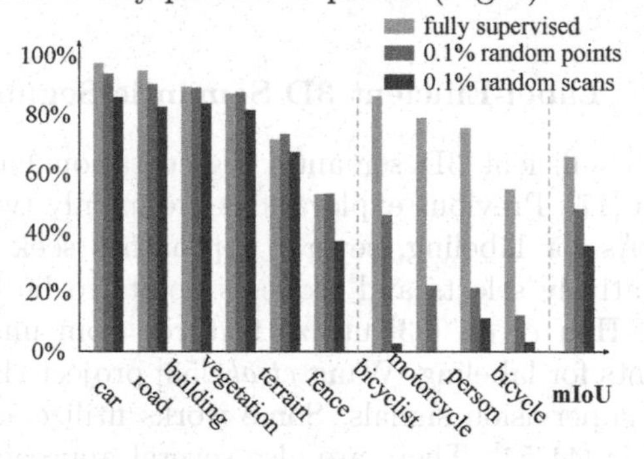

Fig. 2. Pilot study: performances (IoU) of the most common and rarest categories. Models are trained with 100% of labels and 0.1% of labels (in terms of points or scans) on SemanticKITTI [4].

model. These categories tend to have small sizes, appear less frequently, and are thus more vulnerable when reducing the annotation budget. However, they are still critical for many applications such as autonomous driving. Moreover, we find that "0.1% random points" outperforms "0.1% random scans" by a large margin, mainly due to its label diversity.

These observations inspire us to rethink the existing paradigm of label-efficient segmentation. While prior works typically focus on either efficient labeling or improving training approaches, we argue that it can be more effective to address the problem by co-designing both. By integrating the two parts, we may cover more underrepresented instances with a limited labeling budget, and exploit the labeling efforts more effectively during network training.

3.2 Overview

Our LESS framework integrates pre-segmentation, labeling, and network training. It can work with most existing LiDAR segmentation backbones without changing their network architectures or inference latency. As shown in Fig. 3, our pipeline takes raw LiDAR sequences as input. It first employs a heuristic method to partition the point clouds into a set of high-purity components (Sect. 3.3). Instead of exhaustively labeling all points, annotators only need to quickly label a few points for each component proposal (*e.g.*, one point label for each class that appears). Besides the human-annotated sparse labels, we derive other types of labels so as to train the network with more context information (Sect. 3.4). During the network training, we employ contrastive prototype learning to realize a more descriptive embedding space (Sect. 3.5). We also boost the single-scan model by distilling the knowledge from a multi-scan teacher, which exploits richer semantics within the temporally fused point clouds (Sect. 3.6).

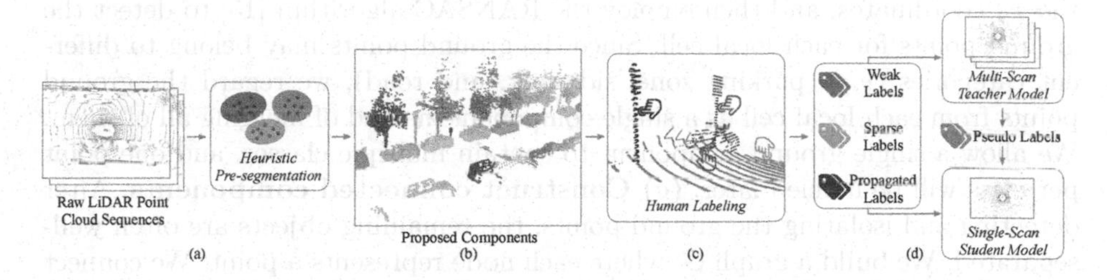

Fig. 3. Overview of our LESS pipeline. (a) We first utilize a heuristic algorithm to pre-segment each LiDAR sequence into a set of connected components. (b) Examples of the proposed components. Different colors indicate different components. For clear visualization, components of ground points are not shown. (c) Human annotators only need to coarsely label each component. Each color denotes a proposed component, and each click icon indicates a labeled point. Only sparse labels are directly annotated by humans. (d) We then train the network to digest various labels and utilize multi-scan distillation to exploit richer semantics in the temporally fused point clouds. (Color figure online)

3.3 Pre-segmentation

We design a heuristic pre-segmentation to subdivide the point cloud into a collection of components. Each resulting component proposal is of high purity, containing only one or a few categories, which facilitates annotators to coarsely label all the proposals, *i.e.*, one point label per class (Sect. 3.4). In this way, we can derive dense supervision by disseminating the sparse point-wise annotations to the whole components. Since modern networks can learn the semantics of homogeneous neighborhoods from sparse annotations, spending lots of annotation budgets on large objects may be futile. Our component-wise coarse annotation is agnostic to the object size, which benefits underrepresented small objects.

For indoor scenarios, many prior arts [20,33,47] leverage the surface normal and color information to generate super voxels and assume that the points within each super voxel share the same category. These approaches, however, might not generalize to outdoor LiDAR point clouds, where the surface can be noisy and color information is not available. Since the homogeneity assumption is hard to hold, we instead propose to lift this constraint and allow each component to contain more than one category.

Unlike indoor scenarios, objects in outdoor scans are often well-separated after detecting and isolating the ground points. Inspired by this philosophy, we design an intuitive approach to pre-segment each LiDAR sequence, which includes four steps: **(a) Fuse overlapping scans.** We first split a LiDAR sequence into sub-sequences, each containing t consecutive scans. We then fuse the scans of each sub-sequence based on the provided ego-poses. In this way, we can label the same instance across overlapping scans at one click. **(b) Detect ground points.** While the ground surface may not be flat at the full-scene scale, we assume for each local region (*e.g.*, $5\,\text{m} \times 5\,\text{m}$), the ground points can be fitted by a plane. We thus partition the whole scene into a uniform grid according to the xy coordinates, and then employ the RANSAC algorithm [15] to detect the ground points for each local cell. Since the ground points may belong to different categories (*e.g.*, parking zone, sidewalk, and road), we regard the ground points from each local cell as a single component instead of merging all of them. We allow a single ground component to contain multiple classes, and one point per class will be labeled later. **(c) Construct connected components.** After detecting and isolating the ground points, the remaining objects are often well-separated. We build a graph G, where each node represents a point. We connect every pair of points (u, v) in the graph, whose Euclidean distance is smaller than a threshold τ. We then divide the points into groups by calculating the connected components for the graph G. Due to the non-uniform point density distribution of the LiDAR point clouds, it is hard to use a fixed threshold across different ranges. We thus propose an adaptive threshold $\tau(u, v) = \max(r_u, r_v) \times d$ to compensate for the varying density, where r_u and r_v are the distances between the points and the sensor centers, and d is a pre-defined hyper-parameter. **(d) Subdivide large components.** After step (c), there usually exist some connected components covering an enormous area (*e.g.*, buildings and vegetation), which are prone to include some small objects. To keep each component of high purity and facilitate network training, we subdivide oversized components to ensure each component is bounded within a fixed size. Also, we ignore small components with only a few points, which tend to be noisy and can lead to excessive component proposals.

In practice, we find our pre-segmentation generates a small number of components for each sequence. The component proposals cover most of the points, and each component tends to have high purity. These open up the possibility of quickly bootstrapping the labeling from scratch. Moreover, unlike other methods [20,33,47] relying on various handcrafted features, our method only utilizes the simple geometrical connectivity, allowing it to generalize to various scenarios without tuning lots of hyper-parameters. Please refer to Sect. 4.4 for statistics of the pre-segmentation results and the supplementary material for more details.

3.4 Annotation Policy and Training Labels

Instead of meticulously labeling every point, we propose to coarsely annotate the component proposals. Specifically, for each component proposal, an annotator needs to first skim through the component and then label only one point for each identified category. Figure 3 (c) illustrates an example where the pre-segmentation yields three components colored in red, blue, and green, respectively. Because the blue component only has traffic-sign points, the annotator only needs to randomly select one point to label. The green component is similar, as it only contains road points. In the red component, there is a bicycle lying against a traffic sign, and the annotator needs to select one point for each class to label. By coarsely labeling all components, we are unlikely to miss any underrepresented instances, as the proposed components cover the majority of points. Moreover, since the number of components is orders of magnitude smaller than that of points and our coarse annotation policy frees annotators from carefully labeling instance boundaries (required in the labeling process to build SemanticKITTI [4] dataset), we are thus able to reduce manual labeling costs.

Based on the component proposals, we can obtain three types of labels. **Sparse labels:** points directly labeled by annotators. Although only a tiny subset of points are labeled, sparse labels provide the most accurate and diverse supervision. **Weak labels:** classes that appear in each component. Weak labels are derived based on human-annotated sparse labels within each component. In the example of Fig. 3 (c), all red points can only be either bicycles or traffic signs. We disseminate weak labels from each component to the points therein. The multi-category weak labels provide weak but dense supervision and cover most points. **Propagated labels:** for the pure components (*i.e.*, only one category appears), we can propagate the label to the entire component. Given the effectiveness of our pre-segmentation approach, the propagated labels also cover a wide range of points. However, since some categories may be easier to be separated and prone to form pure components, the distribution of the propagated labels may be biased and less diverse than the sparse labels.

We formulate a joint loss function by exploiting the three types of labels: $\mathcal{L} = \mathcal{L}_{\text{sparse}} + \mathcal{L}_{\text{propagated}} + \mathcal{L}_{\text{weak}}$, where $\mathcal{L}_{\text{sparse}}$ and $\mathcal{L}_{\text{propagated}}$ are weighted cross-entropy loss with respect to the sparse labels and propagated labels, respectively. We utilize inverse square root of label frequency [35,38,69] as category weights to emphasize underrepresented categories. Here, we calculate a cross-entropy loss for each label type separately, because propagated labels significantly outnumber sparse labels while sparse labels provide more diverse supervision.

Denote the weak labels as binary masks l_{ij} for point i and category j. $l_{ij} = 1$ when point i belongs to a component that contains category j. We exploit the multi-category weak labels by penalizing the impossible predictions:

$$\mathcal{L}_{\text{weak}} = -\frac{1}{n}\sum_{i=1}^{n} \log(1 - \sum_{l_{ij}=0} p_{ij}) \tag{1}$$

where p_{ij} is the predicted probability of point i, and n is the number of points. Prior approaches [44,54] aggregate per-point predictions into component-level predictions and then utilize the multiple-instance learning loss (MIL) [41,42] to supervise the learning. Here, we only penalize the negative predictions without encouraging the positive ones. This is because our network takes a single-scan point cloud as input, but the labels are collected and derived over the temporally fused point clouds. Hence, a positive instance may not always appear in each individual scan, due to occlusions or limited sensor coverage.

3.5 Contrastive Prototype Learning

Besides the great success in self-supervised representation learning [7,21,40], contrastive learning has also shown effectiveness in supervised learning and few-shot learning [18,26,46,49]. It can overcome shortcomings of the cross-entropy loss, such as poor margins [13,26,32,63], and construct a more descriptive embedding space. Following [18,29,33,48,63], we exploit the limited annotations by learning distinctive class prototypes (*i.e.*, class centroids in the feature space). Without pre-training, a contrastive prototype loss $\mathcal{L}_{\text{proto}}$ is added to Sect. 3.4 as an auxiliary loss. Due to the limited annotations and unbalanced label distribution, only using samples within each batch to determine class prototypes may lead to unstable results. Inspired by the idea of momentum contrast [21], we instead learn the class prototypes $\mathbf{P}_c$ by using a moving average over iterations:

$$\mathbf{P}_c \leftarrow m\mathbf{P}_c + (1-m)\frac{1}{n_c}\sum_{y_i=c} \text{stopgrad}(h(f(x_i))) \tag{2}$$

where $f(x_i)$ is the embedding of point x_i, h is a linear projection head with vector normalization, stopgrad denotes the stop gradient operation, y_i is the label of x_i, n_c is the number of points with label c in a batch, and m is a momentum coefficient. In the beginning, $\mathbf{P}_c$ are initialized randomly.

The prototype loss $\mathcal{L}_{\text{proto}}$ is calculated for the points with sparse labels and propagated labels within each batch:

$$\mathcal{L}_{\text{proto}} = \frac{1}{n} \sum_{i}^{n} -w_{y_i} \log \frac{\exp(h(f(x_i)) \cdot \mathbf{P}_{y_i}/\tau)}{\sum_c \exp(h(f(x_i)) \cdot \mathbf{P}_c/\tau)} \tag{3}$$

where $h(f(x_i)) \cdot \mathbf{P}_{y_i}$ indicates the cosine similarity between the projected embedding and the prototype, τ is a temperature hyper-parameter, n is the number of points, and w_{y_i} is the inverse square root weight of category y_i. $\mathcal{L}_{\text{proto}}$ aims to learn a better embedding space by increasing intra-class compactness and inter-class separability.

3.6 Multi-scan Distillation

We aim to learn a segmentation network that takes a single LiDAR scan as input and can be deployed in real-time onboard applications. During our label-efficient training, we can train a multi-scan network as a teacher model. It applies temporal fusion of multiple scans and takes the densified point cloud as input, compensating for the sparsity and incompleteness within a single scan. The teacher model is thus expected to exploit the richer semantics and perform better than a single-scan model. Especially, it may improve the performance for those underrepresented categories, which tend to be small and sparse. After that, we distill the knowledge from the multi-scan teacher model to boost the performance of the single-scan student model.

Specifically, for a scan at time t, we fuse the point clouds of neighboring scans at time $\{t + i\Delta; i \in [-2, 2]\}$ (Δ is a time interval) using the ego-poses of the LiDAR sensor. To enable a large batch size, we use voxel subsampling [67] to normalize the fused point cloud to a fixed size. Labels are then fused accordingly. Besides the spatial coordinates, we also concatenate an additional channel indicating the time index i of each point. The teacher model is trained using the loss functions introduced in Sects. 3.4 and 3.5.

The student model shares the same backbone network and is first trained from scratch in the same way as the teacher model except for the single-scan input. We then fine-tune it by incorporating an additional distillation loss $\mathcal{L}_{\text{dis}}$. Specifically, following [22], we match student predictions with the soft pseudo-labels generated by the teacher model via a cross-entropy loss:

$$\mathcal{L}_{\text{dis}} = -\frac{T^2}{n} \sum_{i}^{n} \sum_{c} \frac{\exp(u_{ic}/T)}{\sum_{c'} \exp(u_{ic'}/T)} \log \left(\frac{\exp(v_{ic}/T)}{\sum_{c'} \exp(v_{ic'}/T)} \right) \tag{4}$$

where u_{ic} and v_{ic} are the predicted logits for point i and category c by the teacher and student models respectively, and T is a temperature hyper-parameter. A higher temperature is typically used so that the probability distribution across classes is smoother, and the distillation is thus encouraged to match the negative logits, which also contain rich information. The cross-entropy is multiplied by T^2 to align the magnitudes of the gradients with existing other losses [22].

Please note that the idea of multi-scan distillation may only be beneficial for our label-efficient LiDAR segmentation setting. For the fully supervised setting, all labels are already available and accurate, and there is no need to leverage the pseudo labels. For the indoor setting, all points are sampled from high-quality reconstructed meshes, and there is no need for a multi-scan teacher model.

4 Experiments

We employ Cylinder3D [68], a recent state-of-the-art method for LiDAR semantic segmentation, as our backbone network. We utilize ground truth labels to mimic the obtained human annotations, and no extra noise is added. Please refer to the supplementary material for more implementation and training details.

We evaluate the proposed method on two large-scale autonomous driving datasets, SemanticKITTI [4] and nuScenes [5]. **SemanticKITTI** [4] is collected in Germany with 64-beam LiDAR sensors. The (sensor) capture and annotation frequency 10 Hz. It contains 10 training sequences (19k scans), 1 validation sequence (4k scans), and 11 testing sequences (20k scans). 19 classes are used for segmentation. **nuScenes** [5] is collected in Boston and Singapore with 32-beam LiDAR sensors. Although the (sensor) capture frequency 20 Hz, the annotation frequency is 2 Hz. It contains 700 training sequences (28k scans), 150 validation sequences (6k scans), and 150 testing sequences (6k scans). 16 classes are used for segmentation. For both datasets, we follow the official guidance [4,5] to use mean intersection-over-union (mIoU) as the evaluation metric.

4.1 Comparison on SemanticKITTI

We compare the proposed method with both label-efficient [23,24,33,55] and fully supervised [1,12,16,27,31,50,58,65,68] methods. Please note that all competing label-efficient methods mainly focus on indoor settings and are not specially designed for outdoor LiDAR segmentation. Among them, ContrastiveSC [23] employs contrastive learning as unsupervised pre-training and uses the learned features for active labeling, ReDAL [55] also employs active labeling, OneThingOneClick [33] proposes a self-training approach and iteratively propagate the labels, and SQN [24] presents a network by leveraging the similarity between neighboring points. We report the results on the validation set. Since ContrastiveSC [23] and OneThingOneClick [33] are only tested on indoor datasets in the original paper, we adapt the source code published by the authors and train their models on SemanticKITTI [4]. For other methods, the results are either obtained from the literature or correspondences with the authors.

Table 2 lists the results, where our method outperforms existing label-efficient methods by a large margin. With only 0.1% sparse labels (as defined in Sect. 3.4), it even completely match the performance of the fully supervised baseline Cylinder3D [68], which demonstrates the potential of deployment into real applications. By checking the breakdown results, we find that the differences between methods mainly come from the underrepresented categories, such as bicycle,

Table 2. Comparison on the SemanticKITTI validation set. Cylinder3D [68] is our fully supervised counterpart. Cylinder3D* is our re-trained version with our proposed prototype learning and multi-scan distillation.

Method	Annot.	mIoU	Car	Bicycle	Motorcycle	Truck	Other-vehicle	Person	Bicyclist	Motorcyclist	Road	Parking	Sidewalk	Other-ground	Building	Fence	Vegetation	Trunk	Terrain	Pole	Traffic-sign
SqueezeSegV3 [58]	100%	52.7	86	31	48	51	42	52	52	0	95	47	82	0	80	47	83	53	72	42	38
PolarNet [65]		53.6	92	31	39	46	24	54	62	0	92	47	78	2	89	46	85	60	72	58	42
MPF [1]		57.0	94	28	55	62	36	57	74	0	95	47	81	1	88	53	86	54	73	57	42
S-BKI [16]		57.4	94	34	57	45	27	53	72	0	94	50	84	0	89	60	87	63	75	64	45
TemporalLidarSeg [12]		61.3	92	43	54	84	61	64	68	0	95	44	83	1	89	60	85	64	71	59	47
KPRNet [27]		63.1	95	43	60	76	51	75	81	0	96	51	84	0	90	60	88	66	76	63	43
SPVNAS [50]		64.7	97	35	72	81	66	71	86	0	94	48	81	0	92	67	88	65	74	64	49
AMVNet [31]		65.2	96	49	65	89	55	71	86	0	96	54	83	0	91	62	88	67	74	65	49
Cylinder3D [68]		65.9	97	55	79	80	67	75	86	1	95	46	82	1	89	53	87	71	71	66	53
Cylinder3D*		66.2	97	48	72	94	67	74	91	0	93	44	79	3	91	60	88	70	72	63	53
ReDAL [55]	5%	59.8	95	30	59	63	50	63	84	1	92	39	78	1	89	54	87	62	74	64	50
OneThingOneClick [33]	0.1%	26.0	77	0	0	2	1	0	2	0	63	0	38	0	73	44	78	39	53	25	0
ContrastiveSC [23]	0.1%	46.0	93	0	0	62	45	28	0	0	90	39	71	6	90	42	89	57	75	54	34
SQN [24]	0.1%	52.0	93	8	35	59	46	41	59	0	91	37	76	1	89	51	85	61	73	53	35
LESS (Ours)	0.1%	**66.0**	97	50	73	94	67	76	92	0	93	40	79	3	91	60	87	68	71	62	51
SQN [24]	0.01%	38.3	83	0	22	12	17	15	47	0	85	21	65	0	79	37	77	46	67	44	12
LESS (Ours)	0.01%	**61.0**	96	33	61	73	59	68	87	0	92	38	76	5	89	52	87	67	71	59	46

motorcycle, person, and bicyclist. Existing label-efficient methods, which are mainly designed for indoor settings, suffer a lot from the highly unbalanced sample distribution, while our method is remarkably competitive in those underrepresented classes. See Fig. 4 for further demonstration. OneThingOneClick [33] fails to produce decent results, which is partially due to its pure super-voxel assumption that does not always hold in outdoor scenes. As for the 0.01% annotations setting, the performance of SQN [24] drops drastically to 38.3%, whereas our proposed method can still achieve a high mIoU of 61.0%. For completeness, we also re-train Cylinder3D [68] with our proposed prototype learning and multi-scan distillation. We find that the two strategies provide marginal gain in the fully-supervised setting, where all labels are available and accurate.

4.2 Comparison on nuScenes

We also compare the proposed method with existing approaches on the nuScenes [5] dataset and report the results on the validation set. Since the author-released model of Cylinder3D [68] utilizes SemanticKITTI for pre-training, here, we report its result based on training the model from scratch for a fair comparison.

Fig. 4. Qualitative examples on the SemanticKITTI [4] **(first row) and nuScenes** [5] **(second row) validation sets.** Please zoom in for the details. Red rectangles highlight the wrong predictions. Our results are similar to the fully supervised counterpart, while ContrastiveSceneContext [23] produces worse results on underrepresented categories (see persons and bicycles). Please note that, points in two datasets (with different density) are visualized in different point size for better visualization. (Color figure online)

Table 3. Comparison on nuScenes validation set. Cylinder3D [68] is our fully supervised counterpart.

Method	Anno.	mIOU(%)
(AF)2-S3Net [9]	100%	62.2
SPVNAS [50]		74.8
Cylinder3D [68]		75.4
AMVNet [31]		77.2
RPVNet [59]		77.6
ContrastiveSC [23]	0.2%	63.5
LESS (Ours)	0.2%	**73.5**
ContrastiveSC [23]	0.9%	65.5
LESS (Ours)	0.9%	**74.8**

Table 4. Ablation study on the SemanticKITTI validation set. All variants use 0.1% sparse labels.

Pre-seg.	Weak labels	Propa. labels	Proto. learning	Multi-scan distillation	mIoU (%)
✗	✗	✗	✗	✗	48.1
✓	✗	✗	✗	✗	59.3
✓	✓	✗	✗	✗	61.6
✓	✗	✓	✗	✗	62.2
✓	✓	✓	✗	✗	63.5
✓	✓	✓	✓	✗	64.9
✓	✓	✓	✓	✓	66.0

For other fully-supervised methods [9,31,50,59], the results are either obtained from the literature or correspondences with the authors. Since no prior label-efficient work is tested on the nuScenes [5] dataset, we adapt the source code published by the authors to train ContrastiveSceneContext [23] from scratch.

We want to point out that points in the nuScenes dataset are much sparser than those in SemanticKITTI. In nuScenes, only 2 scans per second are labeled, while in SemanticKITTI, 10 scans per second are labeled. Due to the difference of sensors (32-beam vs. 64-beam), the number of points per scan in nuScenes is also much smaller (26k vs. 120k). See the right inset for the comparison of two datasets (fused points for 0.5 s). Considering the sparsity of the original ground truth labels, here we report the 0.2% and 0.9% annotation settings.

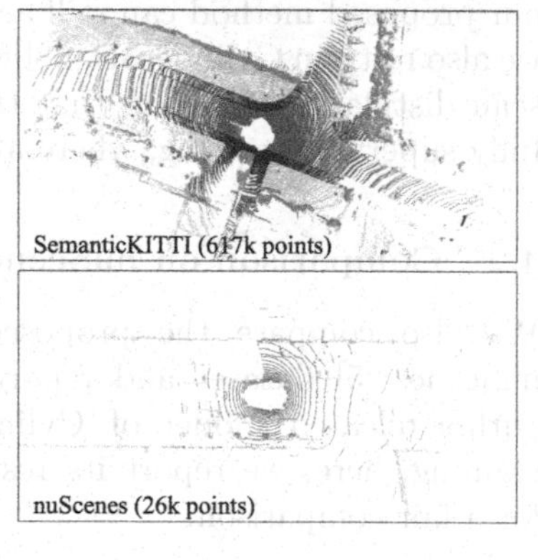

Table 3 shows the results, where our proposed method outperforms ContrastiveSceneContext [23] by a large margin. With only 0.2% sparse labels, our result is also highly competitive with the fully-supervised counterpart [68].

Table 5. Statistics of the pre-segmentation and labeling. Only sparse labels are directly annotated by humans.

Statistics	SemanticKITTI	nuScenes
One-category components	68.6%	80.6%
Two-category components	23.8%	14.9%
Components with more than two categories	7.6%	4.5%
Average number of categories per component	1.40	1.25
Coverage of sparse labels	0.1%	0.2%
Coverage of propagated labels	42.0%	53.6%
Coverage of weak labels	95.5%	99.0%

Table 6. Comparison of various annotation policies on SemanticKITTI. All methods utilize 0.1% annotations and the same backbone network [68]. The fourth column indicates the number of sparse labels (and propagated labels) for an underrepresented category (*i.e.*, motorcycle). Multi-scan distillation is not utilized here. The IoU results are calculated on the validation set.

Annotation policy	Pre-segmentation	mIoU (%)	#labels for motorcycle	IoU (%) of motorcycle
Randomly sample points	✗	48.1	943	22.2
Randomly sample scans	✗	35.6	548	0.0
Active labeling [23]	✗	54.2	456	36.7
Uniform grid partition	✓	61.4	1024 (76k)	61.3
Geometric partition [28,33]	✓	61.9	1190 (294k)	64.0
LESS (Ours)	✓	64.9	1146 (933k)	72.3

4.3 Ablation Study

Table 4 shows the ablation study of each component. The first row is the result of training with 0.1% random point labels. By incorporating the pre-segmentation, we spend the limited annotation budget on more underrepresented instances, thereby significantly increasing mIoU from 48.1% to 59.3%. Derived from the component proposals, weak labels and propagated labels complement the human-annotated sparse labels and provide dense supervision. Compared to multi-category weak labels, propagated labels provide more accurate supervision and thus lead to a slightly higher gain. Both contrastive prototype learning and multi-scan distillation further boost the performance and finally close the gap between LESS and the fully-supervised counterpart in terms of mIoU.

4.4 Analysis of Pre-segmentation and Labeling

By leveraging the unique geometric structure and a careful design, our pre-segmentation works well for outdoor LiDAR point clouds. Table 5 summarizes some statistics of the pre-segmentation and labeling results. For both datasets,

Fig. 5. Improving segmentation with multi-scan distillation. The multi-scan teacher leverages the richer semantics via temporal fusion to accurately segment the bicycle and ground, which provides high-quality supervision to enhance the single-scan model.

Table 7. Results of the multi-scan distillation on the SemanticKITTI validation set. 0.1% annotations are used.

Single-scan (before)		Multi-scan teacher		Single-scan (after)	
mIoU	Bicycle	mIoU	Bicycle	mIoU	Bicycle
64.9%	45.6%	66.8%	51.5%	66.0%	49.9%

only less than 10% of the components contain more than two categories, which validates that our pre-segmentation generates high-purity components. The high "coverage of propagated labels" indicates that we thus deduce a good amount of "free" supervision from the pure components. The low "coverage of sparse labels" shows that annotators indeed only need to label a tiny portion of points, thus reducing human effort. The "coverage of weak labels" confirms that the proposed components can faithfully cover most points. Furthermore, the consistent results across two distinct datasets verify that our method generalizes well in practice.

Table 6 shows the comparison of different annotation policies (*i.e.*, how to use the labeling budget). The first two baselines are introduced in Sect. 3.1, "active labeling" utilizes the features from contrastive pre-training to actively select points [23], "uniform grid partition" uniformly divides the fused point clouds into a grid according to the xy coordinates and treats each cell as a component, "geometric partition" extracts handcrafted geometric features and solves a minimal partition problem [28,68]. All of them are trained with the same backbone Cylinder3D [68]. The first three methods employ no pre-segmentation and are trained with $\mathcal{L}_{\text{sparse}}$ only. The other approaches utilize our labeling policy (*i.e.*, one label per class for each component) and are trained with additional $\mathcal{L}_{\text{propagated}}$, $\mathcal{L}_{\text{weak}}$, and $\mathcal{L}_{\text{proto}}$. As a result, their performances are much higher than the first three methods. We also report the number of labels and the IoU for an underrepresented category. We see that our policy leads to more useful supervisions and higher IoUs for underrepresented categories.

4.5 Analysis of Multi-scan Distillation

Table 7 and Fig. 5 show the results of multi-scan distillation. The teacher model exploits the densified point clouds via temporal fusion and thus performs better than the single-scan model (even compared to the fully supervised single-scan model). Through knowledge distillation from the teacher model, the student model improves a lot in the underrepresented classes and completely matches the fully supervised model in mIoU.

5 Conclusion and Future Work

We study label-efficient LiDAR point cloud semantic segmentation and propose a pipeline that co-designs the labeling and the model learning and can work with most 3D segmentation backbones. We show that our method can utilize bare minimum human annotations to achieve highly competitive performance.

We have shown LESS is an effective approach for bootstrapping labeling and learning from scratch. In addition, LESS is also highly compatible for efficiently improving a performant model. With the predictions of an existing model, the proposed pipeline can be used for annotators to pick and label component proposals of high-values, such as underrepresented classes, long-tail instances, classes with most failures, *etc.* We leave this for future exploration.

References

1. Alnaggar, Y.A., Afifi, M., Amer, K., ElHelw, M.: Multi projection fusion for real-time semantic segmentation of 3D lidar point clouds. In: Proceedings of the IEEE/CVF Winter Conference on Applications of Computer Vision, pp. 1800–1809 (2021)
2. Alonso, I., Riazuelo, L., Montesano, L., Murillo, A.C.: 3D-MiniNet: learning a 2D representation from point clouds for fast and efficient 3D LIDAR semantic segmentation. IEEE Rob. Autom. Lett. **5**(4), 5432–5439 (2020)
3. Armeni, I., Sax, S., Zamir, A.R., Savarese, S.: Joint 2D–3D-semantic data for indoor scene understanding. arXiv preprint arXiv:1702.01105 (2017)
4. Behley, J., et al.: SemanticKITTI: a dataset for semantic scene understanding of lidar sequences. In: Proceedings of the IEEE International Conference on Computer Vision (ICCV), pp. 9297–9307 (2019)
5. Caesar, H., et al.: nuScenes: a multimodal dataset for autonomous driving. In: Proceedings of the IEEE/CVF Conference on Computer Vision and Pattern Recognition (CVPR), pp. 11621–11631 (2020)
6. Chang, A.X., et al.: ShapeNet: an information-rich 3D model repository. arXiv preprint arXiv:1512.03012 (2015)
7. Chen, T., Kornblith, S., Norouzi, M., Hinton, G.: A simple framework for contrastive learning of visual representations. In: Proceedings of the International Conference on Machine Learning (ICML), pp. 1597–1607. PMLR (2020)
8. Cheng, M., Hui, L., Xie, J., Yang, J., Kong, H.: Cascaded non-local neural network for point cloud semantic segmentation. In: 2020 IEEE/RSJ International Conference on Intelligent Robots and Systems (IROS), pp. 8447–8452. IEEE (2020)
9. Cheng, R., Razani, R., Taghavi, E., Li, E., Liu, B.: AF2-S3Net: attentive feature fusion with adaptive feature selection for sparse semantic segmentation network. In: Proceedings of the IEEE/CVF Conference on Computer Vision and Pattern Recognition (CVPR), pp. 12547–12556 (2021)

10. Cortinhal, T., Tzelepis, G., Erdal Aksoy, E.: SalsaNext: fast, uncertainty-aware semantic segmentation of LiDAR point clouds. In: Bebis, G., et al. (eds.) ISVC 2020. LNCS, vol. 12510, pp. 207–222. Springer, Cham (2020). https://doi.org/10.1007/978-3-030-64559-5_16
11. Dai, A., Chang, A.X., Savva, M., Halber, M., Funkhouser, T., Nießner, M.: ScanNet: richly-annotated 3D reconstructions of indoor scenes. In: Proceedings of the IEEE/CVF Conference on Computer Vision and Pattern Recognition (CVPR), pp. 5828–5839 (2017)
12. Duerr, F., Pfaller, M., Weigel, H., Beyerer, J.: LiDAR-based recurrent 3D semantic segmentation with temporal memory alignment. In: Proceedings of the International Conference on 3D Vision (3DV), pp. 781–790. IEEE (2020)
13. Elsayed, G.F., Krishnan, D., Mobahi, H., Regan, K., Bengio, S.: Large margin deep networks for classification. In: Advances in Neural Information Processing Systems (NeurIPS) (2018)
14. Fang, Y., Xu, C., Cui, Z., Zong, Y., Yang, J.: Spatial transformer point convolution. arXiv preprint arXiv:2009.01427 (2020)
15. Fischler, M.A., Bolles, R.C.: Random sample consensus: a paradigm for model fitting with applications to image analysis and automated cartography. Commun. ACM **24**(6), 381–395 (1981)
16. Gan, L., Zhang, R., Grizzle, J.W., Eustice, R.M., Ghaffari, M.: Bayesian spatial kernel smoothing for scalable dense semantic mapping. IEEE Rob. Autom. Lett. **5**(2), 790–797 (2020)
17. Gao, B., Pan, Y., Li, C., Geng, S., Zhao, H.: Are we hungry for 3d lidar data for semantic segmentation? ArXiv abs/2006.04307 3, 20 (2020)
18. Gao, Y., Fei, N., Liu, G., Lu, Z., Xiang, T., Huang, S.: Contrastive prototype learning with augmented embeddings for few-shot learning. arXiv preprint arXiv:2101.09499 (2021)
19. Gerdzhev, M., Razani, R., Taghavi, E., Bingbing, L.: TORNADO-Net: mulTiview tOtal vaRiatioN semantic segmentAtion with diamond inceptiOn module. In: Proceedings of the IEEE International Conference on Robotics and Automation (ICRA), pp. 9543–9549. IEEE (2021)
20. Guinard, S., Landrieu, L.: Weakly supervised segmentation-aided classification of urban scenes from 3D LiDAR point clouds. In: ISPRS Workshop 2017 (2017)
21. He, K., Fan, H., Wu, Y., Xie, S., Girshick, R.: Momentum contrast for unsupervised visual representation learning. In: Proceedings of the IEEE/CVF Conference on Computer Vision and Pattern Recognition (CVPR), pp. 9729–9738 (2020)
22. Hinton, G., Vinyals, O., Dean, J.: Distilling the knowledge in a neural network. In: Advances in Neural Information Processing Systems (NeurIPS) (2015)
23. Hou, J., Graham, B., Nießner, M., Xie, S.: Exploring data-efficient 3D scene understanding with contrastive scene contexts. In: Proceedings of the IEEE/CVF Conference on Computer Vision and Pattern Recognition (CVPR), pp. 15587–15597 (2021)
24. Hu, Q., et al.: SQN: weakly-supervised semantic segmentation of large-scale 3D point clouds with 1000x fewer labels. arXiv preprint arXiv:2104.04891 (2021)
25. Hu, Q., et al.: RandLA-Net: efficient semantic segmentation of large-scale point clouds. In: Proceedings of the IEEE/CVF Conference on Computer Vision and Pattern Recognition (CVPR), pp. 11108–11117 (2020)
26. Khosla, P., et al.: Supervised contrastive learning. In: Advances in Neural Information Processing Systems (NeurIPS) (2020)
27. Kochanov, D., Nejadasl, F.K., Booij, O.: KPRNet: improving projection-based lidar semantic segmentation. arXiv preprint arXiv:2007.12668 (2020)

28. Landrieu, L., Simonovsky, M.: Large-scale point cloud semantic segmentation with superpoint graphs. In: Proceedings of the IEEE/CVF Conference on Computer Vision and Pattern Recognition (CVPR), pp. 4558–4567 (2018)
29. Li, J., Zhou, P., Xiong, C., Hoi, S.C.: Prototypical contrastive learning of unsupervised representations. In: Proceedings of the International Conference on Learning Representations (ICLR) (2020)
30. Li, S., Chen, X., Liu, Y., Dai, D., Stachniss, C., Gall, J.: Multi-scale interaction for real-time lidar data segmentation on an embedded platform. arXiv preprint arXiv:2008.09162 (2020)
31. Liong, V.E., Nguyen, T.N.T., Widjaja, S., Sharma, D., Chong, Z.J.: AMVNet: assertion-based multi-view fusion network for lidar semantic segmentation. arXiv preprint arXiv:2012.04934 (2020)
32. Liu, W., Wen, Y., Yu, Z., Yang, M.: Large-margin softmax loss for convolutional neural networks. In: Proceedings of the International Conference on Machine Learning (ICML), vol. 2, p. 7 (2016)
33. Liu, Z., Qi, X., Fu, C.W.: One thing one click: a self-training approach for weakly supervised 3D semantic segmentation. In: Proceedings of the IEEE/CVF Conference on Computer Vision and Pattern Recognition (CVPR), pp. 1726–1736 (2021)
34. Luo, H., et al.: Semantic labeling of mobile lidar point clouds via active learning and higher order MRF. IEEE Trans. Geosci. Remote Sens. **56**(7), 3631–3644 (2018)
35. Mahajan, D., et al.: Exploring the limits of weakly supervised pretraining. In: Ferrari, V., Hebert, M., Sminchisescu, C., Weiss, Y. (eds.) ECCV 2018. LNCS, vol. 11206, pp. 185–201. Springer, Cham (2018). https://doi.org/10.1007/978-3-030-01216-8_12
36. Mei, J., Gao, B., Xu, D., Yao, W., Zhao, X., Zhao, H.: Semantic segmentation of 3D lidar data in dynamic scene using semi-supervised learning. IEEE Trans. Intell. Transp. Syst. **21**(6), 2496–2509 (2019)
37. Mei, J., Zhao, H.: Incorporating human domain knowledge in 3-D LiDAR-based semantic segmentation. IEEE Transa. Intell. Veh. **5**(2), 178–187 (2019)
38. Mikolov, T., Sutskever, I., Chen, K., Corrado, G.S., Dean, J.: Distributed representations of words and phrases and their compositionality. In: Advances in Neural Information Processing Systems (NeurIPS), pp. 3111–3119 (2013)
39. Milioto, A., Vizzo, I., Behley, J., Stachniss, C.: RangeNet++: fast and accurate lidar semantic segmentation. In: 2019 IEEE/RSJ International Conference on Intelligent Robots and Systems (IROS), pp. 4213–4220. IEEE (2019)
40. Oord, A.V.D., Li, Y., Vinyals, O.: Representation learning with contrastive predictive coding. arXiv preprint arXiv:1807.03748 (2018)
41. Pathak, D., Shelhamer, E., Long, J., Darrell, T.: Fully convolutional multi-class multiple instance learning. arXiv preprint arXiv:1412.7144 (2014)
42. Pinheiro, P.O., Collobert, R.: From image-level to pixel-level labeling with convolutional networks. In: Proceedings of the IEEE/CVF Conference on Computer Vision and Pattern Recognition (CVPR), pp. 1713–1721 (2015)
43. Razani, R., Cheng, R., Taghavi, E., Bingbing, L.: Lite-HDSeg: LiDAR semantic segmentation using lite harmonic dense convolutions. arXiv preprint arXiv:2103.08852 (2021)
44. Ren, Z., Misra, I., Schwing, A.G., Girdhar, R.: 3D spatial recognition without spatially labeled 3D. In: Proceedings of the IEEE/CVF Conference on Computer Vision and Pattern Recognition (CVPR), pp. 13204–13213 (2021)
45. Rist, C.B., Schmidt, D., Enzweiler, M., Gavrila, D.M.: SCSSNet: learning spatially-conditioned scene segmentation on LiDAR point clouds. In: 2020 IEEE Intelligent Vehicles Symposium (IV), pp. 1086–1093. IEEE (2020)

46. Schroff, F., Kalenichenko, D., Philbin, J.: FaceNet: a unified embedding for face recognition and clustering. In: Proceedings of the IEEE/CVF Conference on Computer Vision and Pattern Recognition (CVPR), pp. 815–823 (2015)
47. Shi, X., Xu, X., Chen, K., Cai, L., Foo, C.S., Jia, K.: Label-efficient point cloud semantic segmentation: an active learning approach. arXiv preprint arXiv:2101.06931 (2021)
48. Snell, J., Swersky, K., Zemel, R.S.: Prototypical networks for few-shot learning. In: Advances in Neural Information Processing Systems (NeurIPS) (2017)
49. Sohn, K.: Improved deep metric learning with multi-class N-pair loss objective. In: Advances in Neural Information Processing Systems (NeurIPS), pp. 1857–1865 (2016)
50. Tang, H., et al.: Searching efficient 3D architectures with sparse point-voxel convolution. In: Vedaldi, A., Bischof, H., Brox, T., Frahm, J.-M. (eds.) ECCV 2020. LNCS, vol. 12373, pp. 685–702. Springer, Cham (2020). https://doi.org/10.1007/978-3-030-58604-1_41
51. Thomas, H., Agro, B., Gridseth, M., Zhang, J., Barfoot, T.D.: Self-supervised learning of lidar segmentation for autonomous indoor navigation. In: Proceedings of the IEEE International Conference on Robotics and Automation (ICRA), pp. 14047–14053. IEEE (2021)
52. Thomas, H., Qi, C.R., Deschaud, J.E., Marcotegui, B., Goulette, F., Guibas, L.J.: KPConv: flexible and deformable convolution for point clouds. In: Proceedings of the IEEE International Conference on Computer Vision (ICCV), pp. 6411–6420 (2019)
53. Wang, H., Rong, X., Yang, L., Feng, J., Xiao, J., Tian, Y.: Weakly supervised semantic segmentation in 3D graph-structured point clouds of wild scenes. arXiv preprint arXiv:2004.12498 (2020)
54. Wei, J., Lin, G., Yap, K.H., Hung, T.Y., Xie, L.: Multi-path region mining for weakly supervised 3D semantic segmentation on point clouds. In: Proceedings of the IEEE/CVF Conference on Computer Vision and Pattern Recognition (CVPR), pp. 4384–4393 (2020)
55. Wu, T.H., et al.: ReDAL: region-based and diversity-aware active learning for point cloud semantic segmentation. In: Proceedings of the IEEE International Conference on Computer Vision (ICCV), pp. 15510–15519 (2021)
56. Xiao, A., Huang, J., Guan, D., Zhan, F., Lu, S.: SynLiDAR: learning from synthetic LiDAR sequential point cloud for semantic segmentation. arXiv preprint arXiv:2107.05399 (2021)
57. Xie, S., Gu, J., Guo, D., Qi, C.R., Guibas, L., Litany, O.: PointContrast: unsupervised pre-training for 3D point cloud understanding. In: Vedaldi, A., Bischof, H., Brox, T., Frahm, J.-M. (eds.) ECCV 2020. LNCS, vol. 12348, pp. 574–591. Springer, Cham (2020). https://doi.org/10.1007/978-3-030-58580-8_34
58. Xu, C., et al.: SqueezeSegV3: spatially-adaptive convolution for efficient point-cloud segmentation. In: Vedaldi, A., Bischof, H., Brox, T., Frahm, J.-M. (eds.) ECCV 2020. LNCS, vol. 12373, pp. 1–19. Springer, Cham (2020). https://doi.org/10.1007/978-3-030-58604-1_1
59. Xu, J., Zhang, R., Dou, J., Zhu, Y., Sun, J., Pu, S.: RpvNet: a deep and efficient range-point-voxel fusion network for LiDAR point cloud segmentation. arXiv preprint arXiv:2103.12978 (2021)
60. Xu, K., Yao, Y., Murasaki, K., Ando, S., Sagata, A.: Semantic segmentation of sparsely annotated 3D point clouds by pseudo-labelling. In: Proceedings of the International Conference on 3D Vision (3DV), pp. 463–471. IEEE (2019)

61. Xu, X., Lee, G.H.: Weakly supervised semantic point cloud segmentation: towards 10x fewer labels. In: Proceedings of the IEEE/CVF Conference on Computer Vision and Pattern Recognition (CVPR), pp. 13706–13715 (2020)
62. Yan, X., et al.: Sparse single sweep LiDAR point cloud segmentation via learning contextual shape priors from scene completion. In: Proceedings of the AAAI Conference on Artificial Intelligence (AAAI) (2020)
63. Yang, H.M., Zhang, X.Y., Yin, F., Liu, C.L.: Robust classification with convolutional prototype learning. In: Proceedings of the IEEE/CVF Conference on Computer Vision and Pattern Recognition (CVPR), pp. 3474–3482 (2018)
64. Zhang, F., Fang, J., Wah, B., Torr, P.: Deep FusionNet for point cloud semantic segmentation. In: Vedaldi, A., Bischof, H., Brox, T., Frahm, J.-M. (eds.) ECCV 2020. LNCS, vol. 12369, pp. 644–663. Springer, Cham (2020). https://doi.org/10.1007/978-3-030-58586-0_38
65. Zhang, Y., et al.: PolarNet: an improved grid representation for online lidar point clouds semantic segmentation. In: Proceedings of the IEEE/CVF Conference on Computer Vision and Pattern Recognition (CVPR), pp. 9601–9610 (2020)
66. Zhao, N., Chua, T.S., Lee, G.H.: Few-shot 3D point cloud semantic segmentation. In: Proceedings of the IEEE/CVF Conference on Computer Vision and Pattern Recognition (CVPR), pp. 8873–8882 (2021)
67. Zhou, Y., Tuzel, O.: VoxelNet: end-to-end learning for point cloud based 3D object detection. In: Proceedings of the IEEE/CVF Conference on Computer Vision and Pattern Recognition (CVPR), June 2018
68. Zhu, X., et al.: Cylindrical and asymmetrical 3D convolution networks for LiDAR segmentation. In: Proceedings of the IEEE/CVF Conference on Computer Vision and Pattern Recognition (CVPR), pp. 9939–9948 (2021)
69. Zou, Y., Weinacker, H., Koch, B.: Towards urban scene semantic segmentation with deep learning from LiDAR point clouds: a case study in Baden-Württemberg, Germany. Remote Sens. **13**(16), 3220 (2021)

Visual Cross-View Metric Localization with Dense Uncertainty Estimates

Zimin Xia[1(✉)], Olaf Booij[2], Marco Manfredi[2], and Julian F. P. Kooij[1]

[1] Intelligent Vehicles Group, Technical University Delft, Delft, The Netherlands
{z.xia,j.f.p.kooij}@tudelft.nl
[2] TomTom, Amsterdam, The Netherlands
{olaf.booij,marco.manfredi}@tomtom.com

Abstract. This work addresses visual cross-view metric localization for outdoor robotics. Given a ground-level color image and a satellite patch that contains the local surroundings, the task is to identify the location of the ground camera within the satellite patch. Related work addressed this task for range-sensors (LiDAR, Radar), but for vision, only as a secondary regression step after an initial cross-view image retrieval step. Since the local satellite patch could also be retrieved through any rough localization prior (e.g. from GPS/GNSS, temporal filtering), we drop the image retrieval objective and focus on the metric localization only. We devise a novel network architecture with denser satellite descriptors, similarity matching at the bottleneck (rather than at the output as in image retrieval), and a dense spatial distribution as output to capture multi-modal localization ambiguities. We compare against a state-of-the-art regression baseline that uses global image descriptors. Quantitative and qualitative experimental results on the recently proposed VIGOR and the Oxford RobotCar datasets validate our design. The produced probabilities are correlated with localization accuracy, and can even be used to roughly estimate the ground camera's heading when its orientation is unknown. Overall, our method reduces the median metric localization error by 51%, 37%, and 28% compared to the state-of-the-art when generalizing respectively in the same area, across areas, and across time.

1 Introduction

Ground-to-aerial/satellite image matching, also known as cross-view image matching, has shown notable performance in large-scale geolocalization [8,15,16,25,34,37,40,48]. Usually, this global localization task is formulated as image retrieval. For each ground-level query image the system retrieves the most similar geo-tagged aerial/satellite patch in the database and uses the location of the center pixel in that patch as the location of the query. In practice, global localization can also be obtained by other means in outdoor robotics, such as temporal filtering or coarse GPS/GNSS [31,41,42], but can still have errors of

Supplementary Information The online version contains supplementary material available at https://doi.org/10.1007/978-3-031-19842-7_6.

S. Avidan et al. (Eds.): ECCV 2022, LNCS 13699, pp. 90–106, 2022.
https://doi.org/10.1007/978-3-031-19842-7_6

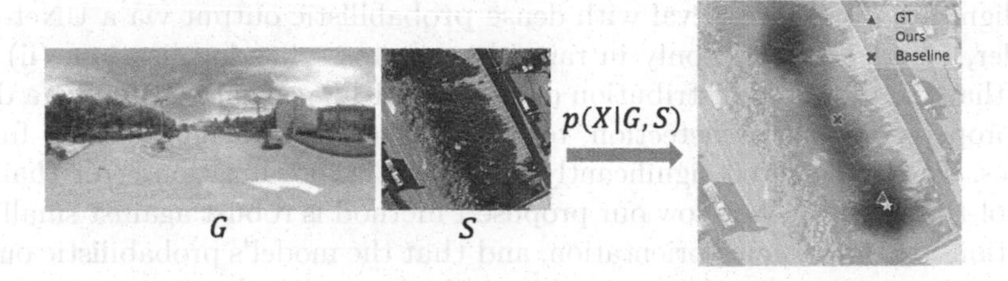

Fig. 1. Example of visual cross-view metric localization. Given a ground-level image G (left), and a satellite patch S (middle) with its local area, we aim to identify the location X within S where G was taken. Our method estimates a dense probability distribution over the satellite image. The resulting (log) probability heat map is overlayed in red on top of the satellite patch (right). Compared to the regression-based baseline that tends to roughly regress to the midpoint among multiple modes, our method captures the underlying multi-modal distribution. Our final predicted location, $\text{argmax}(p(X|G, S))$, is closer to the ground truth. (Color figure online)

tens of meters [4,41,42]. In this work, we therefore follow [31,41,42] by exploiting a coarse location estimate, and zoom into fine-grained *metric localization* within a known satellite image, i.e. to identify which image coordinates in the satellite patch correspond to the location of ground measurement. We adopt the common assumption [16,25,27,34,48] of known orientation, e.g. the center of a ground panorama points north, though we will seek to loosen this restriction in our experiments and roughly estimate the camera's heading too.

In vision, even though ground-to-ground metric localization is a well-studied task [1,6,13], so far in the cross-view setting, the only end-to-end approach that considers metric localization is the regression-based approach proposed in [48], which we will refer here to as *Cross-View Regression* (CVR) for simplicity. CVR tries to solve both the global coarse localization and local metric localization. As a result, its metric localization regressor is built on top of global image descriptors and might miss fine-grained scene information from the satellite image.

Rather than formulating visual cross-view metric localization as a regression task, we propose to produce a dense multi-modal distribution to capture localization ambiguities, and avoid regressing to the midpoint between multiple visually similar places, see Fig. 1. To capture more spatial information, we compute multiple local satellite image descriptors rather than a single global one, and train these in a locally discriminative manner. We note that dense uncertainty output for localization was shown to be successful with range-sensing modalities, like LiDAR and Radar, for localization within top-down maps [3,38,44]. However, these methods are not directly applicable to monocular vision, as they rely on highly accurate depth information which images lack.

Unlike existing literature [8,16,25,27,34,37,40,48], we address local metric localization as a standalone task in visual cross-view matching, and make the following contributions: (i) We propose to predict a dense multi-modal distribution for localization, which can represent localization ambiguity. For this, we propose a new Siamese-like network that exploits multiple local satellite descriptors and uses similarity matching in the fusion bottleneck. It combines the metric learning paradigm from image retrieval with dense probabilistic output via a UNet-style decoder, found previously only in range-based cross-view localization. (ii) We show that the produced distribution correlates with localization quality, a desirable property for outlier detection, temporal filtering, and multi-sensor fusion. Besides, we also achieved significantly lower median localization error than the state-of-the-art. (iii) We show our proposed method is robust against small perturbations on the assumed orientation, and that the model's probabilistic output can even be used to classify a ground image orientation when it is unknown.

Our experiments use the recent large-scale VIGOR dataset for standalone cross-view metric localization to test generalization to new locations in both known and unknown areas. We also collect and stitch additional satellite data for data augmentation and metric localization on the Oxford RobotCar dataset, testing generalization to new measurements along the same route across time.[1]

2 Related Work

We here review the works most related to visual cross-view metric localization.

Cross-view Image Retrieval is a special case of image retrieval. For place recognition [17], a majority of works [5,35,36] construct a reference database using ground-level images, but it is infeasible to guarantee the coverage of the images everywhere. Alternatively, satellite images provide continuous coverage over the world and are publicly available. Given this advantage, a series of approaches [15,37,40] have been proposed to solve large-scale geolocalization using ground-to-satellite cross-view image retrieval. CVM-Net [8] adopts the powerful image descriptor NetVLAD [2] to summarize the view-point invariant information for the cross-view image retrieval. In [16], the authors encode the azimuth and altitude of the pixels in the ground-level query to guide the ground-to-satellite matching. To explicitly minimize the visual difference between satellite and ground domains, various improvements have been proposed. SAFA [25] proposes to use a polar transformation to warp the satellite patch towards the ground-level panorama and uses attention modules to extract the specific features that are visible from both views. In [21,34], a conditional GAN [10] is used to generate synthetic satellite images from the ground-level panorama or to synthesize the panoramic street view from the satellite image to direct the cross-view matching. Instead of constructing a visually similar input, CVFT [27] tries to transport the features from the ground domain towards the satellite domain inside an end-to-end network. Some works [26,37,47] jointly estimate the orientation of the ground query during retrieval without any metric localization. Recently, transformers [43,46] are also used in cross-view image retrieval.

Limitations in Cross-view Image Retrieval are also evident despite its increasing popularity for geolocalization. Recently, [48] points out that cross-view image retrieval methods assume that query ground images correspond to the center of satellite patches in the database, and this assumption is not valid

[1] Models and code, plus extended data are available at https://github.com/tudelft-iv/CrossViewMetricLocalization.

during test time. To break this assumption, [48] introduces a new cross-view matching benchmark VIGOR in which the ground images are not aligned with the center of satellite patches. Another limitation of retrieval is the trade-off between localization accuracy and computation or dataset density. To acquire meter-level localization accuracy, reference satellite patches often have a large overlap with each other, such as sampling the patch every 5m as done in [9,41].

Range Sensing Sensors-to-satellite Metric Localization received more attention than its visual counterpart. RSL-Net [31] localizes Radar scans on a known satellite image. This task is formulated as generating a top-down Radar scan conditioned on the satellite image using [10], and then comparing the online scan to synthetic scan for pose estimation. Later, this idea is extended to self-supervised learning [30]. In [29], the top-down representation of a LiDAR scan is compared to UNet [22] encoded satellite features for metric localization. The range information is crucial in representing the measurement in a top-down view.

LiDAR-to-BEV Map Metric Localization is another frontier that benefited from the range sensing. Dense pixel-to-pixel matchable LiDAR and bird's eye view (BEV) map embeddings can be learned by a deep network [3]. Localization becomes finding the position that has the maximum cross-correlation between two embeddings. Later work [38] shows that it is possible to localize the online LiDAR sweep on HD maps in a similar manner. Those works deliver a dense probabilistic output by formulating the localization task as a classification problem. This property is ideal in probabilistic robot localization [32], as it enables multi-sensor fusion and temporal filtering.

Visual Ground-to-satellite Metric Localization cannot directly reuse the same architecture used to localize LiDAR scans in a BEV map, since an RGB ground image does not provide reliable depth information. Hence pixel-level dense comparison, such as cross-correlation, cannot be leveraged. [45] predicts ground-view semantics from aerial imagery for orientation estimation, and shows only qualitatively that metric localization is possible by comparing the predicted semantics across viewpoints. To the best of our knowledge, CVR [48] is the only end-to-end approach in the vision domain that attempts metric localization on a satellite patch. Given a ground-level query, it first retrieves the matched satellite patch and then regresses the offset between the ground image and satellite patch center. However, its offset regression is based on global feature descriptors, which might cause the regression head to miss detailed scene layout information, and it limits the output to uni-modal estimates. Plus, CVR lacks dense uncertainty estimation to identify ambiguous locations, or a way to filter out unreliable results. A concurrent work [24] perform unimodal localization and orientation estimation by warping features across views and solving an iterative optimization.

3 Methodology

In our work, we assume that a rough prior localization estimate is available, e.g. through GPS/GNSS, odometry, or some other robot-localization techniques [31,41,42]. Given a ground-level image G and a top-down $L \times L$ satellite

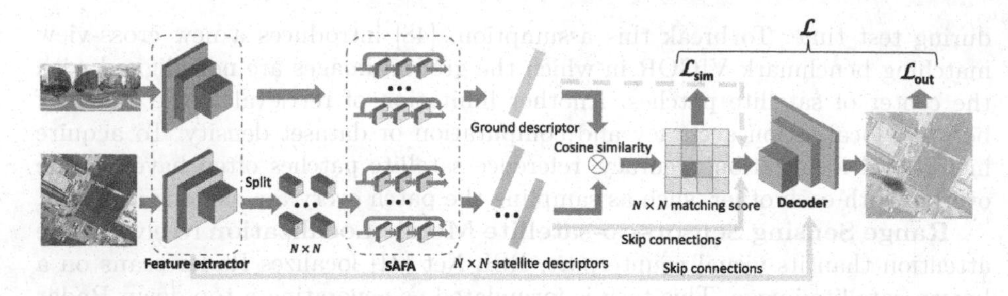

Fig. 2. An overview of the proposed cross-view metric localization architecture (trainable parts in bold). Dashed skip connection is optional, see ablation study. We overlay an exemplar output heat map on top of the input satellite image for intuition.

image S that represents the local area where G was taken, our metric localization objective is to estimate the 2D image coordinates $X \in [0,1]^2$ within S that correspond to the ground location of the camera of G. Moreover, we aim for a dense probabilistic output to benefit a downstream sensor fusion task, similar to [3]. Note that in practice, G and S are often provided with their heading prealigned [16,48], such that the center vertical line of G points in the up direction of S.

Both the baseline CVR [48] and our proposed method adapt a common cross-view image retrieval architecture [25]. This basic backbone is a Siamese-like architecture without weight-sharing. Both the ground and satellite input branches consist of a VGG [28] feature extractor. E.g. for the satellite branch, these features form a $L' \times L' \times 512$ volume. On the feature volume 8 Spatial-Aware Feature Aggregation (SAFA) modules [25] are applied, each generating a 512-dimensional vector, which is all concatenated. Each branch thus yields a single global $1 \times 1 \times 4096$-dimensional descriptor. In an image retrieval task, this network would be trained through metric learning such that descriptors of matching (S, G) pairs are close together in the 4906-dimensional space.

Importantly, our proposed architecture and CVR make distinct choices on (1) the used descriptor representation for S, (2) how the descriptors are fused, (3) how the output head represents the localization result, and (4) consequently, the losses. We explain these choices for both methods in turn.

3.1 Baseline Cross-View Regression

The CVR method in [48] uses a single architecture for a two-step approach. First global localization is done through image retrieval by comparing descriptor G to descriptors of all known satellite patches. After retrieving satellite patch S, metric localization is performed using the already computed descriptors of both G and S. We employ CVR here for the metric localization task only, and therefore keep its proposed architecture, but will not train it for image retrieval. Focusing on metric localization only, our CVR baseline makes the following design choices:

Feature Descriptors: CVR follows the image-retrieval concept of encoding the satellite and ground image each into a single image-global 4096-dimensional descriptor. Both descriptors are fed as-is to the fusion step.

Fusion: CVR simply concatenates the two feature descriptors into a single 8192-dimensional vector.

Output Head: A multi-layer perceptron is used on the fused descriptors which outputs the relative 2D offset ΔX between G's true location within S and the center $X_S = (0.5, 0.5)$ of the satellite patch, s.t. $X = X_S + \Delta X$.

Loss: The standard L2 regression loss is used on the predicted offset and true offset.

We note that most of these choices follow from the need to use a single global descriptor for a whole satellite patch, as such descriptors are necessary for image retrieval. Our argument is however that if a localization prior is already available and global image retrieval is not necessary, this state-of-the-art architecture is sub-optimal for metric localization only compared to our proposed approach.

3.2 Proposed Method

Our proposed architecture starts with a mostly similar Siamese-like backbone. The method overview is shown in Fig. 2. It differs from CVR as follows:

Feature Descriptors: Instead of building one image-global descriptor to represent S, we increase the top-down spatial resolution by splitting the satellite $L' \times L' \times 512$ feature volume along spatial directions into $N \times N$ sub-volumes, where N is a hyper-parameter. Now the 8 SAFA [25] modules are applied to each $L'/N \times L'/N \times 512$ sub-volume in parallel, resulting in an $N \times N \times 4096$ descriptor $g(S)$ for the satellite branch, shown as the green vectors in Fig. 2. Let $g(S)^{ij}$ denote the i-th row j-th column of the satellite descriptor, $1 \leq i, j < N$. The ground image is still encoded as a single global 4096-dimensional descriptor $f(G)$, shown as the blue vector in Fig. 2.

Fusion: To help distinguish different satellite image sub-regions, we compute the cosine similarity between $f(G)$ and each $g(S)^{ij}$, and use this similarity as a feature itself at this fusion bottleneck. This similarity computation results in a $N \times N \times 1$ matching score map M, thus $M^{ij} = \text{sim}(f(G), g(S)^{ij})$. To complete our fusion step, the M is concatenated to the satellite descriptors $g(S)$ through a skip connection, shown as the upper yellow solid arrow in Fig. 2. Optionally, one could also concatenate $f(G)$ again into the fused descriptor (yellow dashed arrow), similar to CVR; we explore this in our experiments.

Output Head: Rather than treating metric localization as a regression task, we seek to generate a dense distribution over the image coordinates. Such output enables us to represent localization ambiguities and estimate the (un)certainty of our prediction. Towards this, we feed the fusion volume to a decoder which can progressively up-sample the $N \times N$ matching map to higher resolutions. Akin to the UNet architecture [22], skip connections between satellite encoder and decoder are used to pass the fine-grained scene layout information to guide the decoding. Finally, a softmax activation function is applied on the last layer, and outputs a $L \times L \times 1$ heat map H, where each pixel $H^{u,v} = p(X \in c(u, v)|G, S)$ represents the probability of G being located within pixel area $c(u, v)$. This heat map is useful by itself, e.g. in a sensor fusion framework. For a single frame estimate, we simply output the center image coordinates $\overline{c}[\cdot]$ of the most probable pixel, i.e. $X = \overline{c}[\text{argmax}_{(u,v)} H^{u,v}]$.

Losses: A benefit of our framework is that we can add losses on both the final output and the fusion bottleneck. The full loss $\mathcal{L} = \mathcal{L}_{\text{out}} + \beta \times \mathcal{L}_{\text{sim}}$ is thus a weighted sum of the output loss, $\mathcal{L}_{\text{out}}$, and the bottleneck loss, $\mathcal{L}_{\text{sim}}$, where β is a hyper-parameter. We discuss each term next.

Since the output H is a discrete probability distribution that sums to one, we treat our task as a multi-class classification problem. $\mathcal{L}_{\text{out}}$ is simply a cross-entropy loss over the $L \times L$ output cells. The ground truth is one-hot encoded as a heat map with the same $L \times L$ resolution and label 1 at the true location and 0 elsewhere, In practice, we will apply Gaussian label smoothing to the one-hot encoding of the output head, and tune the smoothing σ as part of the hyperparameter optimization.

To guide the model to already learn locally discriminative satellite descriptors at the fusion bottleneck, we apply the infoNCE loss [20] from contrastive representation learning [11], which can be seen as a generalized version of triplet loss [23] used in image retrieval in the case of multiple negative samples are presented at the same time,

$$\mathcal{L}'(ij^+) = -\log \frac{\exp(\text{sim}(f(G), g(S)^{ij^+})/\tau)}{\sum_{i,j} \exp(\text{sim}(f(G), g(S)^{ij})/\tau)}. \tag{1}$$

Here τ is a hyper-parameter introduced by [20], and its role is similar to the margin between positive and negative samples in triplet loss, and (ij^+) is the cell index of the positive satellite descriptor w.r.t. the ground descriptor.

We reuse the smoothed one-hot encoding from the output loss to allow multiple soft positives if the true location is near a cell border. We max-pool the $L \times L$ target map to the $N \times N$ resolution and renormalize it to generate 'positiveness' weights w_{ij}^+ for each cell $1 \leq i, j \leq N$. Our bottleneck loss is simply a weighted version of Eq. (1), $\mathcal{L}_{\text{sim}} = \sum_{i,j} w_{ij}^+ \mathcal{L}'(ij)$.

4 Experiments

In this section, we first introduce the two datasets and evaluation metrics for our experiments. Then we motivate each of our design choices and provide a detailed ablation study. Finally, our model is compared to the baseline CVR approach [48] to show our advantage in metric localization in generalizing to new measurements in the same area, across areas, and across time.

4.1 Datasets

The first used dataset, **VIGOR** [48], contains geo-tagged ground-level panoramic images and satellite images collected in four cities in the US. Unlike previous cross-view image retrieval datasets [14,16,33,45], the satellite patches in VIGOR seamlessly cover the target area. Importantly, the ground-level panoramas are not located at the center of satellite patches. Each satellite patch corresponds to 72.96×72.96 m ground area with a ground resolution of 0.114 m. The orientation of the satellite patch and ground panorama are aligned in a way that the vertical line at the center of the panorama corresponds to the north direction in the satellite patch. Typically, each patch has ~50% overlap with its neighboring patch in the North, South, East, and West direction. This means every ground image is covered by 4 satellite patches. If the ground image is at the center 1/4 area of a satellite patch, the patch is denoted as "positive", otherwise "semi-positive". In practice, "positive" samples simulate the case that the global localization prior is more accurate, e.g. error $< \sqrt{2} \times 18.24$ m in the case of VIGOR. Similarly, "positive + semi-positive" samples would be a result of a coarser localization prior, e.g. error $< \sqrt{2} \times 36.48$ m. During training, we include both "positives" and "semi-positives" samples. Our main evaluation will be based on positive samples, since it is representative for most real-world situations, e.g. localization prior from GNSS positioning in an open area or temporal filtering. For completeness, we also evaluate on "positive + semi-positive", to showcase how the methods behave with a less certain localization prior, e.g. GNSS positioning in an urban canyon. We adopt the "same-area" and "cross-area" splits from [48] to test the model's generalization in the same cities and across different cities. To find one set of hyper-parameters for both "same-area" and "cross-area", we create a subset of the shared training data from New York as a smaller "tuning" split with 11108/2777 training/validation samples.

The second dataset, **Oxford RobotCar** [18,19] contains multi-sensor measurements from multiple traversals over a consistent route through Oxford collected over a year. The original dataset does not contain satellite images. To enable cross-view metric localization, we stitch the satellite patches provided by [41,42] and our additionally collected ones to create a continuous satellite map that covers the target area. We follow the same data split as in [41] to test how our method generalizes to new ground images collected at different time. In total, there are 17067, 1698, and 5089 ground-level front-viewing images in the training, validation, and test set respectively. The test set contains 3 traversals collected at later times of day than the training recordings. Benefiting from a full continuous satellite map, we randomly and uniformly sample satellite patches around the ground image locations during training for data augmentation. Each patch is rotationally aligned with the view direction of the ground image and has a resolution of $800 pixels \times 800 pixels$, which corresponds to 73.92×73.92 m on the ground. A fixed set of satellite patches is used for validation and testing. Each patch has a 50% area overlap with the closest neighboring patches. We pair each ground image with the patch at the closest center location to allow for mutual information between the satellite and ground front-facing views.

Similarly, during training, we also control the sampled locations to make sure the ground image locates inside the central area of the sampled satellite patch.

4.2 Evaluation Metrics

To measure the localization error, we report the mean and median distance between the predicted location and ground truth location in meters over all samples. Note that the mean error can be biased by a few samples with large error, and including the median error provides a measurement more robust w.r.t. outliers. In practice, a localization method that operates on a single image frame can be extended to process a sequence of data using a Bayesian filter [9,41]. In such a setting, the estimated probability at the ground truth location plays an important role in accurately localizing over the whole path. Motivated by this, we include the probability at the ground truth pixel area as an additional metric. The baseline CVR method does not have any probability estimation on its output. Hence, we post-process the baseline output by assuming the regressed location is the mean of an isotropic Gaussian distribution, and we estimate the standard deviation of this Gauss on the validation set.

4.3 Hyper-parameters and Ablation Study

We first discuss our hyper-parameter choices, then investigate the main components in our proposed architecture. The weight β in our loss function is set to 10^4, and the τ in infoNCE loss is set to 0.1, as done in [20]. The loss is optimized by Adam optimizer [12] with a learning rate of 1×10^{-5}, and the VGG feature extractors are pre-trained on ImageNet [7]. Our main model variations and hyper-parameters are now compared on the VIGOR "tuning" split.

We initially set $N = 8$ for the matching at the bottleneck. The infoNCE loss at the bottleneck is key to improve the final model output. The model trained with it achieves a much better mean error, 14.30 m, than the model trained without it, 19.25 m. Label smoothing with $\sigma = 4$ pixels further reduces the mean error to 13.39 m (we tested $\sigma = 1, 2, 4, 8$). We use both in all future experiments.

Next, we study the influence of different resolutions $N \times N$ at the model bottleneck. When $N = 1, 2, 4, \mathbf{8}, 16$, the mean error is $19.62, 15.98, 15.23, \mathbf{13.39}, 15.04$ meters respectively (best in bold). With $N = 1$ no infoNCE loss is applied, and the decoder receives a single matching score concatenated with features from the satellite branch. Increasing N improves the spatial resolution at the model bottleneck. However, with larger N the decoder also operates on larger inputs but with fewer upsampling layers. We observe a balance at $N = 8$.

To further explore the role of metric learning at the bottleneck and the feature concatenation, we create four extra model variations in Table 1. Directly concatenating the ground with a single global satellite descriptor (see "1,S+G") is akin to CVR's fusion with a decoder head instead of regression, but this change alone does not perform well. The model does not work when the decoder

Table 1. Fusion bottleneck, error on tuning split: "S" stands for satellite descriptors $g(S)$, "G" for ground descriptors $f(G)$, "M" for cosine similarity feature. Best in bold

N, descriptors	1,S+G	8,M	8,S+G	8,S+M+G	8,S+M
Mean error (m)	18.62	24.37	18.35	13.83	**13.39**

Table 2. Localization error on VIGOR. Best in bold. "Center-only" denotes using satellite patch center as the prediction. The term "Positives" stands for evaluation on positive satellite patches. "Pos.+semi-pos." takes the mean over the results from the positive satellite patches and all semi-positive satellite patches

	Same-area				*Cross-area*			
	Positives		Pos.+semi-pos.		Positives		Pos.+semi-pos.	
Error(m)	Mean	Median	Mean	Median	Mean	Median	Mean	Median
Center-only	14.15	14.82	27.78	28.85	14.07	14.07	27.80	28.89
CVR [48]	10.55	9.31	16.64	13.82	**11.26**	10.02	18.66	16.73
Ours	**9.86**	**4.58**	**13.45**	**5.39**	13.06	**6.31**	**17.13**	**7.78**

operates on only a single channel map ("8,M") without any context from the satellite patch. Increasing the satellite resolution is also still insufficient with only ground descriptors ("8,S+G"), the descriptors must also be trained to be locally discriminative. Interestingly, we do not observe any benefit from also concatenating the ground descriptor ("8,S+M+G") to our default of satellite descriptors with matching scores ("8,S+M"). In all next experiments, we fuse only the satellite descriptors and the matching score.

We note that to forward-pass an input pair from VIGOR on a Tesla V100 GPU, CVR uses 0.020s, and our best-performing model 0.034s (i.e. $\sim$ 30 FPS).

4.4 Generalization in the Same Area/Across Areas

We now compare our method to CVR on the VIGOR splits for generalizing to unseen ground images inside the same area, and across areas. When tested on the "same-area" correctly retrieved samples, CVR trained for only regression has a better mean (-1.5 m) and median (-1.1 m) localization error than CVR trained for both retrieval and regression, as expected. From now on, we will always train CVR model for regression only as the baseline.

Metric Error: The quantitative comparison against CVR on both VIGOR splits is summarized in Table 2. To highlight the value of conducting cross-view metric localization, we also include a "center-only" prediction which always outputs $X = (0.5, 0.5)$. Note that retrieval-only methods typically assume that the center of a satellite patch is representative of the true location.

When the ground image is compared to the positive-only satellite patches, our model reduces the median error by 51% over CVR when generalizing within the

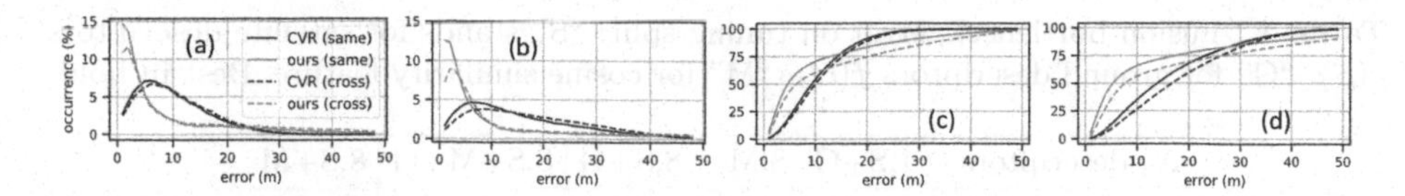

Fig. 3. Error distributions (plots a,b: regular, c,d: cumulative, a,c: positives, b,d: positives and semi-positives) on VIGOR, for same-area and cross-area experiments.

same area (4.58 m vs 9.31m), and by 37% when generalizing across areas (6.31 m vs 10.02 m). Generally, our model improves over the baselines, but across areas, our mean error is higher than that of CVR. The error (cumulative) distribution in Fig. 3 confirms that this is due to a few large-error outliers in our prediction. These outliers are a result of selecting a wrong mode, or of large uncertainty in our multi-modal output, whereas regression might pick an averaged location in the middle resulting in neither small nor very large errors. We will show below that our location's probability can be used to detect such potential large error cases. This would aid an external sensor fusion module, which can also directly integrate the distribution to reduce the uncertainty through other measurements.

When ground images could be located further from the center, as in the "positive+semi-positive" test cases, there is less matchable visual information between the two views. In this case, the performance of both CVR and our model somewhat degenerates, though our model suffers less than CVR. Our mean and median errors are lower than CVR's, both within the same area and across areas. Moreover, our method's advantage in the median error further increases. In the Sup. Mat. we furthermore investigate the effect of using a CVR-like regression layer and loss on top of our dense output but find it hurts median performance.

Qualitative Results: To intuitively understand where our advantage comes from, we provide qualitative examples of success and failure cases in Fig. 4. In the context of image-based cross-view metric localization, there can exist multiple visually similar locations on the satellite image given a ground image. In such cases, it is important for the model to have the capability to express the underlying localization uncertainty. In our model, the uncertainty is already present at the model bottleneck, see Fig. 4 top row. This distribution is upsampled by the decoder and aligned with the observed environmental features, such as roads and crossings, resulting in the dense multi-modal uncertainty map. We emphasize that no explicit semantic map information, e.g. on road layout, was used during training. Since the regression-based baseline method forces the output to be a single location, it risks 'averaging' multiple similar locations and provides a wrong final estimate without any uncertainty information.

We argue that in practice our outliers are still more acceptable than CVR's errors. When our model is uncertain about the exact location, our output heat map can be rather homogeneous. As shown in Fig. 4 example 3, given a ground image taken on the road, our model assigns high probability to roads in the center and on the left. In this situation, the distance between our predicted location and the ground truth can be large. Instead, CVR tends to output the

Fig. 4. Top: input ground images and matching score maps at the model bottleneck, bottom: input satellite image overlayed with outputs from CVR and our method. From left to right: 1: VIGOR, same-area, 2,3: VIGOR: cross-area, 4: Oxford RobotCar.

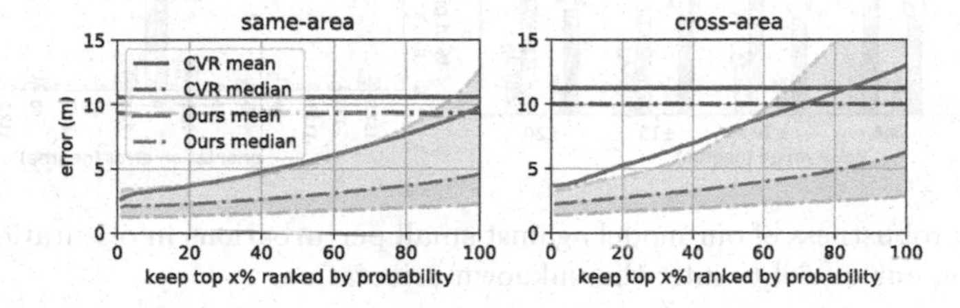

Fig. 5. Ranking the predictions using their probabilities on VIGOR "positives". Red lines show our error statistics over the top $x\%$. Cyan/orange: error between median and 25%/75% quantile line. As x decreases, only the more probable predictions are kept. Blue lines: CVR cannot rank predictions this way. (Color figure online)

average between the visually similar areas, which can result in a location near the center but that is intuitively unreasonable, e.g. within some vegetation, even though it may have a smaller distance to the ground truth location.

Probability Evaluation: Apart from the metric localization error, we will compare how well each model can predict the probability at the ground truth location. For CVR we estimate this assuming a fixed Gaussian error distribution, see Sect. 4.2. Table 3 reports both mean and median probabilities at the ground truth pixel. Our multi-modal approach outperforms the fixed error distribution for CVR. Importantly, the probability at our predicted location (the maximum in H) is correlated to its localization error. If we apply a rejection threshold to only keep the top $x\%$, predictions, we can reduce the expected error. See Fig. 5 with the statistics over the top-ranked estimates. These properties are beneficial when the single frame localization results are temporally filtered or fused with other sensors.

Orientation: Till now, we have relied on a known orientation during test time, e.g. estimated in the preceding retrieval step [26,37,47] or by the sensor stack [39]. We here study our model's robustness against orientation perturbations and ability to infer orientation when it is unknown *without retraining.* To

Table 3. Probabilities at the ground truth pixel on VIGOR. Best in bold. The magnitude of the probabilities is low due to the normalization over the 512 × 512 grid. "Uniform" shows for reference the prob. at GT for a homogeneous map, 1/(512 × 512)

Prob. at GT,	Same-area		Cross-area	
Positives	Mean	Median	Mean	Median
Uniform	3.81×10^{-6}	3.81×10^{-6}	3.81×10^{-6}	3.81×10^{-6}
CVR [48]	1.55×10^{-5}	1.70×10^{-5}	1.57×10^{-5}	1.72×10^{-5}
Ours	$\mathbf{2.93 \times 10^{-4}}$	$\mathbf{1.17 \times 10^{-4}}$	$\mathbf{1.54 \times 10^{-4}}$	$\mathbf{7.06 \times 10^{-5}}$

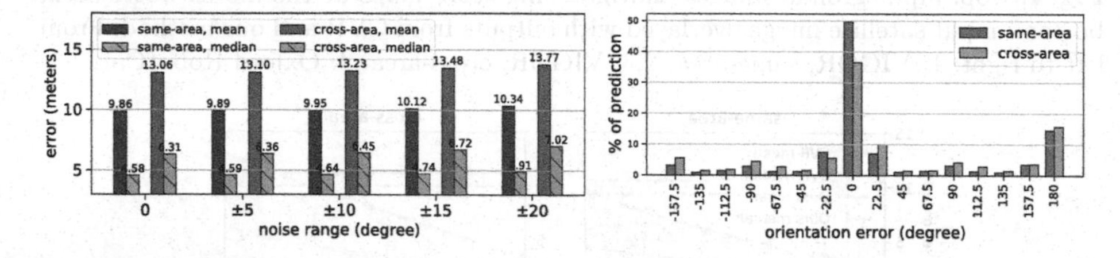

Fig. 6. Left: robustness of our model against small perturbations in orientation. Right: directly using our model to infer the unknown orientation.

test robustness, we uniformly sample angular noise from a range up to ±20° [27] to horizontally shift the ground panoramas (i.e. "rotate" the heading) at test time. As shown in Fig. 6 left, the predicted location of our model remains stable under such noise.

Still, the model's confidence is not invariant to orientation shift, as our prediction confidence can help classify a ground panorama's unknown orientation. We rotate the ground panorama by multiples of 22.5° up to 360°, apply our model to each rotated panorama with the satellite patch, and collect all 16 activation maps before the final softmax operation. The classification output is the orientation of the map with the highest activation. As shown in Fig. 6 right, for same and across areas, our model correctly classifies 50% and 37% samples into the true orientation class out of 16 classes. Most erroneous predictions have an error of 180°, corresponding to the opposite driving direction.

4.5 Generalization Across Time

Finally, we test how our method generalizes to new measurements collected at different times and days on the Oxford RobotCar dataset. For a comparison to cross-view image retrieval, we include "GeolocalRetrieval", which was previously also proposed on the Oxford RobotCar dataset [41]. While regular image retrieval is trained to be globally discriminative, this method learns descriptors that are only discriminative for nearby satellite patches within a 50 m radius, and thus assumes a localization prior during both training and testing, similar to our task. To increase its localization accuracy, we feed it a larger idealized dataset

Table 4. Localization error on Oxford RobotCar. Shown are the average ± standard deviation of 'mean' and 'median' errors over 3 test traversals. Best results in bold. ⋆: uses the same training and test ground images, but more overlapping satellite patches to obtain finer localization through image retrieval only

Error (meters)	Mean	Median
Center-only	12.09 ± 0.02	12.65 ± 0.01
GeolocalRetrieval⋆ [41]	6.01 ± 0.68	4.62 ± 0.49
CVR [48]	2.29 ± 0.31	1.72 ± 0.21
Ours	**1.77** ± 0.25	**1.24** ± 0.10

of satellite patches at more densely sampled locations (200+ patches in a 50 m radius) including even patches centered on the actual test locations. Therefore GeolocalRetrieval could obtain zero meter error if it correctly retrieves the exact satellite patch at each test image location.

Table 4 shows the localization error among all included methods. As expected, with the idealized satellite patches, GeolocalRetrieval delivers lower error than "center-only". However, metric localization methods (CVR and ours) show a clear advantage over GeolocalRetrieval. This highlights the benefit of conducting metric localization over simply densifying the dataset for retrieval. Moreover, using our model for metric localization reduces the mean error by 23% and the median error by 28% compared to using CVR. Qualitatively, we again observe the benefit of expressing multi-modal distribution over the CVR's regression even without the use of panoramic ground images, see Fig. 4 example 4. The probability evaluation also aligns with the findings on VIGOR. Our probability at the ground truth pixel are consistently higher than that under CVR with its estimated error distribution over three test traversals. Averaged over three test traversals, the mean/median probability at the ground truth pixel for CVR are $1.67 \times 10^{-4}/1.89 \times 10^{-4}$, and for ours are $1.54 \times 10^{-3}/1.38 \times 10^{-3}$.

We also test classification of the orientation on this non-panoramic dataset. Instead of shifting the ground image, we now rotate the satellite patch 16 times with 22.5°, starting at 0° where north points in the vertical up direction. The orientation of a ground image is inferred by selecting the peak probability as we did for VIGOR. On three test traversals, 72.3%, 70.7%, and 70.6% of the test samples are predicted with the correct orientation out of the 16 possible directions. More details on orientation classification and the localization results with unknown orientation can be found in the Supplementary Material.

To summarize, also for test images at new days our method shows all-round superiority, similar to generalization within the same and across areas.

5 Conclusion

In this work, we focused on visual cross-view metric localization on a known satellite image, a relatively unexplored task. In contrast to the state-of-the-art regression-based baseline, our method provides a dense multi-modal spatial distribution. We studied the architectural design differences, and showed generalization to new measurements in the same area, across areas, and generalizing across time on two state-of-the-art datasets. Our method surpasses the state-of-the-art by 51%, 37%, and 28% respectively in the median localization error. In a few cases the multi-modal output yields higher distance errors, e.g. when an incorrect mode is deemed more probable. Still, our probabilities can be used to filter such large errors and have less risk of excluding the true location. We show that our method is robust against small orientation noise, and is capable to roughly classify the orientation from its prediction confidence. Future work will address temporal filtering and fine-grained orientation estimation.

Acknowledgements. This work is part of the research programme Efficient Deep Learning (EDL) with project number P16-25, which is (partly) financed by the Dutch Research Council (NWO).

References

1. Agarwal, P., Burgard, W., Spinello, L.: Metric localization using google street view. In: IEEE/RSJ IROS, pp. 3111–3118 (2015)
2. Arandjelovic, R., Gronat, P., Torii, A., Pajdla, T., Sivic, J.: NetVLAD: CNN architecture for weakly supervised place recognition. In: Proceedings of IEEE/CVF CVPR, pp. 5297–5307 (2016)
3. Barsan, I.A., Wang, S., Pokrovsky, A., Urtasun, R.: Learning to localize using a lidar intensity map. In: CoRL (10 2018)
4. Ben-Moshe, B., Elkin, E., et al.: Improving accuracy of gnss devices in urban canyons. In: CCCG, pp. 511–515 (2011)
5. Chen, D.M., et al.: City-scale landmark identification on mobile devices. In: Proceedings of IEEE/CVF CVPR, pp. 737–744 (2011)
6. Clement, L., Gridseth, M., Tomasi, J., Kelly, J.: Learning matchable image transformations for long-term metric visual localization. IEEE Robot. Autom. Lett. **5**(2), 1492–1499 (2020). https://doi.org/10.1109/LRA.2020.2967659
7. Deng, J., Dong, W., Socher, R., et al.: Imagenet: A large-scale hierarchical image database. In: Proceedings of IEEE/CVF CVPR, pp. 248–255 (2009)
8. Hu, S., Feng, M., Nguyen, R.M., Hee Lee, G.: CVM-net: Cross-view matching network for image-based ground-to-aerial geo-localization. In: Proceedings of IEEE/CVF CVPR, pp. 7258–7267 (2018)
9. Hu, S., Lee, G.H.: Image-based geo-localization using satellite imagery. IJCV, pp. 1–15 (2019)
10. Isola, P., Zhu, J.Y., Zhou, T., Efros, A.A.: Image-to-image translation with conditional adversarial networks. In: Proceedings of IEEE/CVF CVPR, pp. 1125–1134 (2017)
11. Khosla, P., et al.: Supervised contrastive learning. arXiv preprint arXiv:2004.11362 (2020)
12. Kingma, D.P., Ba, J.: Adam: A method for stochastic optimization. ICLR (2014)
13. Lategahn, H., Stiller, C.: Vision-only localization. IEEE Trans. Intell. Transport. Syst. **15**(3), 1246–1257 (2014). https://doi.org/10.1109/TITS.2014.2298492

14. Lin, T.Y., Belongie, S., Hays, J.: Cross-view image geolocalization. In: Proceedings of IEEE/CVF CVPR, pp. 891–898 (2013)
15. Lin, T.Y., Cui, Y., Belongie, S., Hays, J.: Learning deep representations for ground-to-aerial geolocalization. In: Proceedings of IEEE/CVF CVPR, pp. 5007–5015 (2015)
16. Liu, L., Li, H.: Lending orientation to neural networks for cross-view geo-localization. In: Proceedings of IEEE/CVF CVPR, pp. 5624–5633 (2019)
17. Lowry, S., et al.: Visual place recognition: a survey. IEEE Trans. Robot. **32**(1), 1–19 (2015)
18. Maddern, W., Pascoe, G., Linegar, C., Newman, P.: 1 year, 1000 km: The oxford robotcar dataset. IJRR **36**(1), 3–15 (2017)
19. Maddern, W., Pascoe, G., et al.: Real-time kinematic ground truth for the oxford robotcar dataset. arXiv preprint: 2002.10152 (2020)
20. Oord, A.v.d., Li, Y., Vinyals, O.: Representation learning with contrastive predictive coding. arXiv preprint arXiv:1807.03748 (2018)
21. Regmi, K., Shah, M.: Bridging the domain gap for ground-to-aerial image matching. In: Proc. of IEEE/CVF ICCV, pp. 470–479 (2019)
22. Ronneberger, O., Fischer, P., Brox, T.: U-Net: convolutional networks for biomedical image segmentation. In: Navab, N., Hornegger, J., Wells, W.M., Frangi, A.F. (eds.) MICCAI 2015. LNCS, vol. 9351, pp. 234–241. Springer, Cham (2015). https://doi.org/10.1007/978-3-319-24574-4_28
23. Schroff, F., Kalenichenko, D., Philbin, J.: Facenet: A unified embedding for face recognition and clustering. In: Proceedings of IEEE/CVF CVPR, pp. 815–823 (2015)
24. Shi, Y., Li, H.: Beyond cross-view image retrieval: Highly accurate vehicle localization using satellite image. In: Proceedings of the IEEE/CVF CVPR (2022)
25. Shi, Y., Liu, L., Yu, X., Li, H.: Spatial-aware feature aggregation for image based cross-view geo-localization. In: NeurIPS, pp. 10090–10100 (2019)
26. Shi, Y., Yu, X., Campbell, D., Li, H.: Where am i looking at? joint location and orientation estimation by cross-view matching. In: Proceedings of IEEE/CVF CVPR, pp. 4064–4072 (2020)
27. Shi, Y., Yu, X., Liu, L., et al.: Optimal feature transport for cross-view image geo-localization. In: Proceedings of AAAI, pp. 11990–11997 (2020)
28. Simonyan, K., Zisserman, A.: Very deep convolutional networks for large-scale image recognition. In: ICLR (2015)
29. Tang, T.Y., De Martini, D., Newman, P.: Get to the point: Learning lidar place recognition and metric localisation using overhead imagery. Robotics: Science and Systems (2021)
30. Tang, T.Y., De Martini, D., Wu, S., Newman, P.: Self-supervised learning for using overhead imagery as maps in outdoor range sensor localization. IJRR **40**(12–14), 1488–1509 (2021)
31. Tang, T.Y., De Martini, D., Barnes, D., Newman, P.: Rsl-net: Localising in satellite images from a radar on the ground. IEEE Robot. Autom. Lett. **5**(2), 1087–1094 (2020)
32. Thrun, S., Burgard, W., Fox, D.: Probabilistic robotics. MIT press (2005)
33. Tian, Y., Chen, C., Shah, M.: Cross-view image matching for geo-localization in urban environments. In: Proceeidngs of IEEE/CVF CVPR, pp. 3608–3616 (2017)
34. Toker, A., Zhou, Q., Maximov, M., Leal-Taixe, L.: Coming down to earth: Satellite-to-street view synthesis for geo-localization. In: Proc. of IEEE/CVF CVPR. pp. 6488–6497 (June 2021)

35. Torii, A., Arandjelovic, R., Sivic, J., Okutomi, M., Pajdla, T.: 24/7 place recognition by view synthesis. In: Proceedings of IEEE/CVF CVPR, pp. 1808–1817 (2015)
36. Torii, A., Sivic, J., Okutomi, M., Pajdla, T.: Visual place recognition with repetitive structures. IEEE Trans. Pattern Anal. Mach. Intell. **37**(11), 2346–2359 (2015). https://doi.org/10.1109/TPAMI.2015.2409868
37. Vo, Nam N.., Hays, James: Localizing and orienting street views using overhead imagery. In: Leibe, Bastian, Matas, Jiri, Sebe, Nicu, Welling, Max (eds.) ECCV 2016. LNCS, vol. 9905, pp. 494–509. Springer, Cham (2016). https://doi.org/10.1007/978-3-319-46448-0_30
38. Wei, X., Bârsan, I.A., Wang, S., Martinez, J., Urtasun, R.: Learning to localize through compressed binary maps. In: Proceedings of IEEE/CVF CVPR, pp. 10316–10324 (2019)
39. Won, D., et al.: Performance improvement of inertial navigation system by using magnetometer with vehicle dynamic constraints. J. Sensors, 1–11 (2015)
40. Workman, S., Souvenir, R., Jacobs, N.: Wide-area image geolocalization with aerial reference imagery. In: Proceedings of IEEE/CVF ICCV, pp. 3961–3969 (2015)
41. Xia, Z., Booij, O., Manfredi, M., Kooij, J.F.P.: Cross-view matching for vehicle localization by learning geographically local representations. IEEE Robot. Autom. Lett. **6**(3), 5921–5928 (2021). https://doi.org/10.1109/LRA.2021.3088076
42. Xia, Z., Booij, O., Manfredi, M., Kooij, J.F.P.: Geographically local representation learning with a spatial prior for visual localization. In: Bartoli, A., Fusiello, A. (eds.) ECCV 2020. LNCS, vol. 12536, pp. 557–573. Springer, Cham (2020). https://doi.org/10.1007/978-3-030-66096-3_38
43. Yang, H., Lu, X., Zhu, Y.: Cross-view geo-localization with layer-to-layer transformer. In: NeurIPS. pp. 29009–29020 (2021)
44. Yin, H., Chen, R., Wang, Y., Xiong, R.: Rall: end-to-end radar localization on lidar map using differentiable measurement model. IEEE Transactions on Intelligent Transportation Systems (2021)
45. Zhai, M., Bessinger, Z., Workman, S., Jacobs, N.: Predicting ground-level scene layout from aerial imagery. In: Proceedings of IEEE/CVF CVPR, pp. 867–875 (2017)
46. Zhu, S., Shah, M., Chen, C.: Transgeo: Transformer is all you need for cross-view image geo-localization. In: Proceedings of the IEEE/CVF CVPR, pp. 1162–1171 (2022)
47. Zhu, S., Yang, T., Chen, C.: Revisiting street-to-aerial view image geo-localization and orientation estimation. In: Proceedings of IEEE/CVF WACV, pp. 756–765 (2021)
48. Zhu, S., Yang, T., Chen, C.: Vigor: Cross-view image geo-localization beyond one-to-one retrieval. In: Proceedings of IEEE/CVF CVPR, pp. 3640–3649 (2021)

V2X-ViT: Vehicle-to-Everything Cooperative Perception with Vision Transformer

Runsheng Xu[1], Hao Xiang[1], Zhengzhong Tu[2], Xin Xia[1], Ming-Hsuan Yang[3,4], and Jiaqi Ma[1](✉)

[1] University of California, Los Angeles, Los Angeles, USA
jiaqima@ucla.edu
[2] University of Texas at Austin, Austin, USA
[3] Google Research, New York, USA
[4] University of California, Merced, Merced, USA

Abstract. In this paper, we investigate the application of Vehicle-to-Everything (V2X) communication to improve the perception performance of autonomous vehicles. We present a robust cooperative perception framework with V2X communication using a novel vision Transformer. Specifically, we build a holistic attention model, namely V2X-ViT, to effectively fuse information across on-road agents (i.e., vehicles and infrastructure). V2X-ViT consists of alternating layers of heterogeneous multi-agent self-attention and multi-scale window self-attention, which captures inter-agent interaction and per-agent spatial relationships. These key modules are designed in a unified Transformer architecture to handle common V2X challenges, including asynchronous information sharing, pose errors, and heterogeneity of V2X components. To validate our approach, we create a large-scale V2X perception dataset using CARLA and OpenCDA. Extensive experimental results demonstrate that V2X-ViT sets new state-of-the-art performance for 3D object detection and achieves robust performance even under harsh, noisy environments. The code is available at https://github.com/DerrickXuNu/v2x-vit.

Keywords: V2X · Vehicle-to-Everything · Cooperative perception · Autonomous driving · Transformer

1 Introduction

Perceiving the complex driving environment precisely is crucial to the safety of autonomous vehicles (AVs). With recent advancements of deep learning, the robustness of single-vehicle perception systems has demonstrated significant

R. Xu, H. Xiang and Z. Tu—Equal contribution.

Supplementary Information The online version contains supplementary material available at https://doi.org/10.1007/978-3-031-19842-7_7.

S. Avidan et al. (Eds.): ECCV 2022, LNCS 13699, pp. 107–124, 2022.
https://doi.org/10.1007/978-3-031-19842-7_7

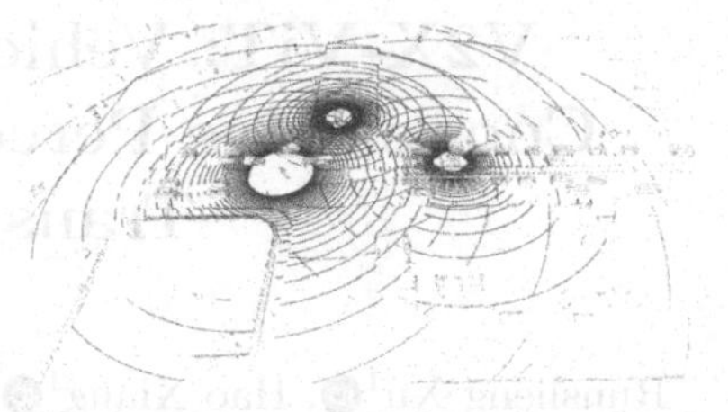

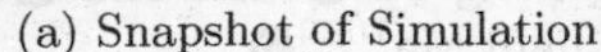

(a) Snapshot of Simulation (b) Aggregated LiDAR point cloud

Fig. 1. A data sample from the proposed V2XSet. (a) A simulated scenario in CARLA where two AVs and infrastructure are located at different sides of a busy intersection. (b) The aggregated LiDAR point clouds of these three agents.

improvement in several tasks such as semantic segmentation [10,32], depth estimation [11,54], and object detection and tracking [13,20,23,48]. Despite recent advancements, challenges remain. Single-agent perception system tends to suffer from occlusion and sparse sensor observation at a far distance, which can potentially cause catastrophic consequences [50]. The cause of such a problem is that an individual vehicle can only perceive the environment from a single perspective with limited sight-of-view. To address these issues, recent studies [4,5,39,44] leverage the advantages of multiple viewpoints of the same scene by investigating Vehicle-to-Vehicle (V2V) collaboration, where visual information (*e.g.*, detection outputs, raw sensory information, intermediate deep learning features, details see Sect. 2) from multiple nearby AVs are shared for a complete and accurate understanding of the environment.

Although V2V technologies have the prospect to revolutionize the mobility industry, it ignores a critical collaborator – roadside infrastructure. The presence of AVs is usually unpredictable, whereas the infrastructure can always provide supports once installed in key scenes such as intersections and crosswalks. Moreover, infrastructure equipped with sensors on an elevated position has a broader sight-of-view and potentially less occlusion. Despite these advantages, including infrastructure to deploy a robust V2X perception system is non-trivial. Unlike V2V collaboration, where all agents are homogeneous, V2X systems often involve a heterogeneous graph formed by infrastructure and AVs. The configuration discrepancies between infrastructure and vehicle sensors, such as types, noise levels, installation height, and even sensor attributes and modality, make the design of a V2X perception system challenging. Moreover, the GPS localization noises and the asynchronous sensor measurements of AVs and infrastructure can introduce inaccurate coordinate transformation and lagged sensing information. Failing to properly handle these challenges will make the system vulnerable.

In this paper, we introduce a unified fusion framework, namely V2X Vision Transformer or **V2X-ViT**, for V2X perception, that can jointly handle these challenges. Figure 2 illustrates the entire system. AVs and infrastructure capture, encode, compress, and send intermediate visual features with each other, while the ego vehicle (*i.e.*, receiver) employs V2X-Transformer to perform information fusion for object detection. We propose two novel attention modules to

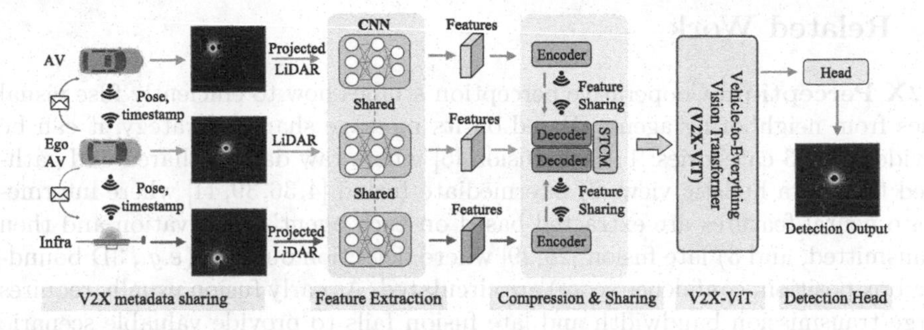

Fig. 2. Overview of our proposed V2X perception system. It consists of five sequential steps: V2X metadata sharing, feature extraction, compression & sharing, V2X-ViT, and the detection head. The details of each individual component are illustrated in Sect. 3.1.

accommodate V2X challenges: 1) a customized heterogeneous multi-agent self-attention module that explicitly considers agent types (vehicles and infrastructure) and their connections when performing attentive fusion; 2) a multi-scale window attention module that can handle localization errors by using multi-resolution windows in parallel. These two modules will adaptively fuse visual features in an iterative fashion to capture inter-agent interaction and per-agent spatial relationship, correcting the feature misalignment caused by localization error and time delay. Moreover, we also integrate a delay-aware positional encoding to further handle the time delay uncertainty. Notably, all these modules are incorporated in a single transformer that learns to address these challenges end-to-end.

To evaluate our approach, we collect a new large-scale open dataset, namely V2XSet, that explicitly considers real-world noises during V2X communication using the high-fidelity simulator CARLA [9], and a cooperative driving automation simulation tool OpenCDA. Figure 1 shows a data sample in the collected dataset. Experiments show that our proposed V2X-ViT significantly advances the performance on V2X LiDAR-based 3D object detection, achieving a 21.2% gain of AP compared to single-agent baseline and performing favorably against leading intermediate fusion methods by at least 7.3%. Our contributions are:

- We present the first unified transformer architecture (V2X-ViT) for V2X perception, which can capture the heterogeneity nature of V2X systems with strong robustness against various noises. Moreover, the proposed model achieves state-of-the-art performance on the challenging cooperative detection task.
- We propose a novel heterogeneous multi-agent attention module (HMSA) tailored for adaptive information fusion between heterogeneous agents.
- We present a new multi-scale window attention module (MSWin) that simultaneously captures local and global spatial feature interactions in parallel.
- We construct V2XSet, a new large-scale open simulation dataset for V2X perception, which explicitly accounts for imperfect real-world conditions.

2 Related Work

V2X Perception. Cooperative perception studies how to efficiently fuse visual cues from neighboring agents. Based on its message sharing strategy, it can be divided into 3 categories: 1) early fusion [5] where raw data is shared and gathered to form a holistic view, 2) intermediate fusion [4,36,39,44] where intermediate neural features are extracted based on each agent's observation and then transmitted, and 3) late fusion [28,29] where detection outputs (*e.g.*, 3D bounding box position, confidence score) are circulated. As early fusion usually requires large transmission bandwidth and late fusion fails to provide valuable scenario context [39], intermediate fusion has attracted increasing attention because of its good balance between accuracy and transmission bandwidth. Several intermediate fusion methods have been proposed for V2V perception recently. OPV2V [44] implements a single-head self-attention module to fuse features, while F-Cooper employs *maxout* [14] fusion operation. V2VNet [39] proposes a spatial-aware message passing mechanism to jointly reason detection and prediction. To attenuate outlier messages, [36] regresses vehicles' localization errors with consistent pose constraints. DiscoNet [21] leverages knowledge distillation to enhance training by constraining the corresponding features to the ones from the network for early fusion. However, intermediate fusion for V2X is still in its infancy. Most V2X methods explored late fusion strategies to aggregate information from infrastructure and vehicles. For example, a late fusion two-level Kalman filter is proposed by [26] for roadside infrastructure failure conditions. Xiangmo *et al.* [51] propose fusing the lane mark detection from infrastructure and vehicle sensors, leveraging Dempster-Shafer theory to model the uncertainty.

LiDAR-Based 3D Object Detection. Numerous methods have been explored to extract features from raw points, voxels, bird-eye-view (BEV) images, and their mixtures. PointRCNN [31] proposes a two-stage strategy based on raw point clouds, which learns rough estimation in the first stage and then refines it with semantic attributes. The authors of [45,53] propose to split the space into voxels and produce features per voxel. Despite having high accuracy, their inference speed and memory consumption are difficult to optimize due to reliance on 3D convolutions. To avoid computationally expensive 3D convolutions, [20,46] propose an efficient BEV representation. To satisfy both computational and flexible receptive field requirements, [30,47,52] combine voxel-based and point-based approaches to detect 3D objects.

Transformers in Vision. The Transformer [38] is first proposed for machine translation [38], where multi-head self-attention and feed-forward layers are stacked to capture long-range interactions between words. Dosovitskiy *et al.* [8] present a Vision Transformer (ViT) for image recognition by regarding image patches as visual words and directly applying self-attention. The full self-attention in ViT [8,12,38], despite having global interaction, suffers from heavy computational complexity and does not scale to long-range sequences or high-resolution images. To ameliorate this issue, numerous methods have introduced locality into self-attention, such as Swin [25], CSwin [7], Twins [6], window [34,40], and sparse attention [35,37,43]. A hierarchical architecture is usually adopted to progressively increase the receptive fields for capturing longer dependencies.

While these vision transformers have proven efficient in modeling homogeneous structured data, their efficacy to represent heterogeneous graphs has been less studied. One of the developments related to our work is the heterogeneous graph transformer (HGT) [16]. HGT was originally designed for web-scale Open Academic Graph where the nodes are text and attributes. Inspired by HGT, we build a customized heterogeneous multi-head self-attention module tailored for graph attribute-aware multi-agent 3D visual feature fusion, which is able to capture the heterogeneity of V2X systems.

3 Methodology

In this paper, we consider V2X perception as a heterogeneous multi-agent perception system, where different types of agents (*i.e.*, smart infrastructure and AVs) perceive the surrounding environment and communicate with each other. To simulate real-world scenarios, we assume that all the agents have imperfect localization and time delay exists during feature transmission. Given this, our goal is to develop a robust fusion system to enhance the vehicle's perception capability and handle these aforementioned challenges in a unified end-to-end fashion. The overall architecture of our framework is illustrated in Fig. 2, which includes five major components: 1) metadata sharing, 2) feature extraction, 3) compression and sharing, 4) V2X vision Transformer, and 5) a detection head.

3.1 Main Architecture Design

V2X Metadata Sharing. During the early stage of collaboration, every agent $i \in \{1 \dots N\}$ within the communication networks shares metadata such as poses, extrinsics, and agent type $c_i \in \{I, V\}$ (meaning infrastructure or vehicle) with each other. We select one of the connected AVs as the ego vehicle (e) to construct a V2X graph around it where the nodes are either AVs or infrastructure and the edges represent directional V2X communication channels. To be more specific, we assume the transmission of metadata is well-synchronized, which means each agent i can receive ego pose $x_e^{t_i}$ at the time t_i. Upon receiving the pose of the ego vehicle, all the other connected agents nearby will project their own LiDAR point clouds to the ego-vehicle's coordinate frame before feature extraction.

Feature Extraction. We leverage the anchor-based PointPillar method [20] to extract visual features from point clouds because of its low inference latency and optimized memory usage [44]. The raw point clouds will be converted to a stacked pillar tensor, then scattered to a 2D pseudo-image and fed to the PointPillar backbone. The backbone extracts informative feature maps $\mathbf{F}_i^{t_i} \in \mathbb{R}^{H \times W \times C}$, denoting agent i's feature at time t_i with height H, width W, and channels C.

Compression and Sharing. To reduce the required transmission bandwidth, we utilize a series of 1×1 convolutions to progressively compress the feature maps

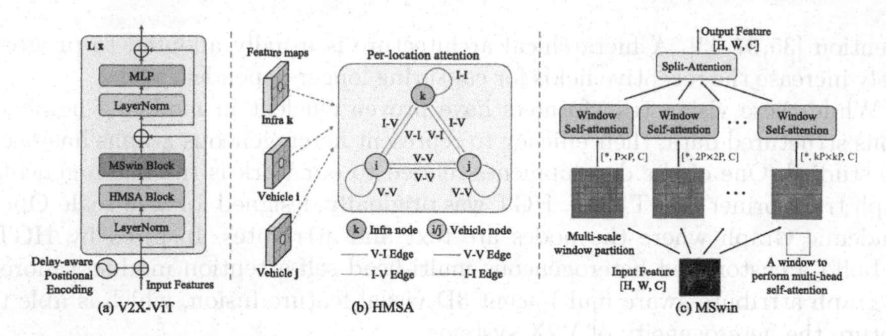

Fig. 3. V2X-ViT architecture. (a) The architecture of our proposed V2X-ViT model. (b) Heterogeneous multi-agent self-attention (HMSA) presented in Sect. 3.2. (c) Multi-scale window attention module (MSwin) illustrated in Sect. 3.2.

along the channel dimension. The compressed features with the size (H, W, C') (where $C' \ll C$) are then transmitted to the ego vehicle (e), on which the features are projected back to (H, W, C) using 1×1 convolutions.

There exists an inevitable time gap between the time when the LiDAR data is captured by connected agents and when the extracted features are received by the ego vehicle (details see appendix). Thus, features collected from surrounding agents are often temporally misaligned with the features captured on the ego vehicle. To correct this delay-induced global spatial misalignment, we need to transform (*i.e.*, rotate and translate) the received features to the current ego-vehicle's pose. Thus, we leverage a spatial-temporal correction module (STCM), which employs a differential transformation and sampling operator $\mathbf{\Gamma}_\xi$ to spatially warp the feature maps [17]. An ROI mask is also calculated to prevent the network from paying attention to the padded zeros caused by the spatial warp.

V2X-ViT. The intermediate features $\mathbf{H}_i = \mathbf{\Gamma}_\xi\left(\mathbf{F}_i^{t_i}\right) \in \mathbb{R}^{H \times W \times C}$ aggregated from connected agents are fed into the major component of our framework *i.e.*, V2X-ViT to conduct an iterative inter-agent and intra-agent feature fusion using self-attention mechanisms. We maintain the feature maps in the same level of high resolution throughout the entire Transformer as we have observed that the absence of high-definition features greatly harms the objection detection performance. The details of our proposed V2X-ViT will be unfolded in Sect. 3.2.

Detection Head. After receiving the final fused feature maps, we apply two 1×1 convolution layers for box regression and classification. The regression output is $(x, y, z, w, l, h, \theta)$, denoting the position, size, and yaw angle of the predefined anchor boxes, respectively. The classification output is the confidence score of being an object or background for each anchor box. We use the smooth ℓ_1 loss for regression and a focal loss [24] for classification.

3.2 V2X-Vision Transformer

Our goal is to design a customized vision Transformer that can jointly handle the common V2X challenges. Firstly, to effectively capture the heterogeneous graph representation between infrastructure and AVs, we build a heterogeneous multi-agent self-attention module that learns different relationships based on node and edge types. Moreover, we propose a novel spatial attention module, namely multi-scale window attention (MSwin), that captures long-range interactions at various scales. MSwin uses multiple window sizes to aggregate spatial information, which greatly improves the detection robustness against localization errors. Lastly, these two attention modules are integrated into a single V2X-ViT block in a factorized manner (illustrated in Fig. 3a), enabling us to maintain high-resolution features throughout the entire process. We stack a series of V2X-ViT blocks to iteratively learn inter-agent interaction and per-agent spatial attention, leading to a robust aggregated feature representation for detection.

Heterogeneous Multi-agent Self-attention. The sensor measurements captured by infrastructure and AVs possibly have distinct characteristics. The infrastructure's LiDAR is often installed at a higher position with less occlusion and different view angles. In addition, the sensors may have different levels of sensor noise due to maintenance frequency, hardware quality *etc.* To encode this heterogeneity, we build a novel heterogeneous multi-agent self-attention (HMSA) where we attach types to both nodes and edges in the directed graph. To simplify the graph structure, we assume the sensor setups among the same category of agents are identical. As shown in Fig. 3b, we have two types of nodes and four types of edges, *i.e.*, node type $c_i \in \{I, V\}$ and edge type $\phi(e_{ij}) \in \{V\!-\!V, V\!-\!I, I\!-\!V, I\!-\!I\}$. Note that unlike traditional attention where the node features are treated as a vector, we only reason the interaction of features *in the same spatial position* from different agents to preserve spatial cues. Formally, HSMA is expressed as:

$$\mathbf{H}_i = \underset{\forall j \in N(i)}{\mathsf{Dense}_{c_i}} \left(\mathbf{ATT}(i,j) \cdot \mathbf{MSG}(i,j)\right) \tag{1}$$

which contains 3 operators: a linear aggregator Dense_{c_i}, attention weights estimator **ATT**, and message aggregator **MSG**. The Dense is a set of linear projectors indexed by the node type c_i, aggregating multi-head information. **ATT** calculates the importance weights between pairs of nodes conditioned on the associated node and edge types:

$$\mathbf{ATT}(i,j) = \underset{\forall j \in N(i)}{\mathsf{softmax}} \left(\underset{m \in [1,h]}{\Big\Vert} \mathrm{head}^m_{\mathrm{ATT}}(i,j) \right) \tag{2}$$

$$\mathrm{head}^m_{\mathrm{ATT}}(i,j) = \left(\mathbf{K}^m(j)\, \mathbf{W}^{m,\mathrm{ATT}}_{\phi(e_{ij})} \mathbf{Q}^m(i)^T \right) \frac{1}{\sqrt{C}} \tag{3}$$

$$\mathbf{K}^m(j) = \mathsf{Dense}^m_{c_j}(\mathbf{H}_j) \tag{4}$$

$$\mathbf{Q}^m(i) = \mathsf{Dense}^m_{c_i}(\mathbf{H}_i) \tag{5}$$

where $\|$ denotes concatenation, m is the current head number and h is the total number of heads. Notice that Dense here is indexed by both node type $c_{i/j}$, and head number m. The linear layers in $\mathbf{K}$ and $\mathbf{Q}$ have distinct parameters. To incorporate the semantic meaning of edges, we calculate the dot product between Query and Key vectors weighted by a matrix $\mathbf{W}^{m,\text{ATT}}_{\phi(e_{ij})} \in \mathbb{R}^{C\times C}$. Similarly, when parsing messages from the neighboring agent, we embed infrastructure and vehicle's features separately via $\text{Dense}^m_{c_j}$. A matrix $\mathbf{W}^{m,\text{MSG}}_{\phi(e_{ij})}$ is used to project the features based on the edge type between source node and target node:

$$\mathbf{MSG}(i,j) = \underset{m\in[1,h]}{\|} \text{head}^m_{\text{MSG}}(i,j) \tag{6}$$

$$\text{head}^m_{\text{MSG}}(i,j) = \text{Dense}^m_{c_j}(\mathbf{H}_j)\,\mathbf{W}^{m,\text{MSG}}_{\phi(e_{ij})}. \tag{7}$$

Multi-scale Window Attention. We present a new type of attention mechanism tailored for efficient long-range spatial interaction on high-resolution detection, called multi-scale window attention (MSwin). It uses a pyramid of windows, each of which caps a different attention range, as illustrated in Fig. 3c. The usage of variable window sizes can greatly improve the detection robustness of V2X-ViT against localization errors (see ablation study in Fig. 5b). Attention performed within larger windows can capture long-range visual cues to compensate for large localization errors, whereas smaller window branches perform attention at finer scales to preserve local context. Afterward, the split-attention module [49] is used to adaptively fuse information coming from multiple branches, empowering MSwin to handle a range of pose errors. Note that MSwin is applied on each agent independently without considering any inter-agent fusion; therefore we omit the agent subscript in this subsection for simplicity.

Formally, let $\mathbf{H} \in \mathbb{R}^{H\times W\times C}$ be an input feature map of a single agent. In branch j out of k parallel branches, $\mathbf{H}$ is partitioned using window size $P_j \times P_j$, into a tensor of shape $(\frac{H}{P_j} \times \frac{W}{P_j}, P_j \times P_j, C)$, which represents a $\frac{H}{P_j} \times \frac{W}{P_j}$ grid of non-overlapping patches each with size $P_j \times P_j$. We use h_j number of heads to improve the attention power at j-th branch. More detailed formulation can be found in Appendix. Following [15,25], we also consider an additional relative positional encoding $\mathbf{B}$ that acts as a bias term added to the attention map. As the relative position along each axis lies in the range $[-P_j+1, P_j-1]$, we take $\mathbf{B}$ from a parameterized matrix $\hat{\mathbf{B}} \in \mathbb{R}^{(2P_j-1)\times(2P_j-1)}$.

To attain per-agent multi-range spatial relationship, each branch partitions input tensor $\mathbf{H}$ with different window sizes *i.e.* $\{P_j\}_{j=1}^k = \{P, 2P, ..., kP\}$. We progressively decrease the number of heads when using a larger window size to save memory usage. Finally, we fuse the features from all the branches by a Split-Attention module [49], yielding the output feature $\mathbf{Y}$. The complexity of the proposed MSwin is *linear* to image size HW, while enjoying long-range multi-scale receptive fields and adaptively fuses both local and (sub)-global visual hints in parallel. Notably, unlike Swin Transformer [25], our multi-scale window approach requires no masking, padding, or cyclic-shifting, making it more efficient in implementations while having larger-scale spatial interactions.

Delay-Aware Positional Encoding. Although the global misalignment is captured by the spatial warping matrix Γ_ξ, another type of local misalignment, arising from object motions during the delay-induced time lag, also needs to be considered. To encode this temporal information, we leverage an adaptive delay-aware positional encoding (DPE), composed of a linear projection and a learnable embedding. We initialize it with sinusoid functions conditioned on time delay Δt_i and channel $c \in [1, C]$:

$$\mathbf{p}_c(\Delta t_i) = \begin{cases} \sin\left(\Delta t_i / 10000^{\frac{2c}{C}}\right), & c = 2k \\ \cos\left(\Delta t_i / 10000^{\frac{2c}{C}}\right), & c = 2k+1 \end{cases} \tag{8}$$

A linear projection $f : \mathbb{R}^C \rightarrow \mathbb{R}^C$ will further warp the learnable embedding so it can generalize better for unseen time delay [16]. We add this projected embedding to each agents' feature $\mathbf{H}_i$ before feeding into the Transformer so that the features are temporally aligned beforehand.

$$\mathrm{DPE}(\Delta t_i) = f(\mathbf{p}(\Delta t_i)) \tag{9}$$

$$\mathbf{H}_i = \mathbf{H}_i + \mathrm{DPE}(\Delta t_i) \tag{10}$$

4 Experiments

4.1 V2XSet: An Open Dataset for V2X Cooperative Perception

To the best of our knowledge, there exists no fully public V2X perception dataset suitable for investigating common V2X challenges such as localization error and transmission time delay. DAIR-V2X [2] is a large-scale real-world V2I dataset without V2V cooperation. V2X-Sim [21] is an open V2X simulated dataset but does not simulate noisy settings and only contains a single road type. OPV2V [44] contains more road types but are restricted to V2V cooperation. To this end, we collect a new large-scale dataset for V2X perception that explicitly considers these real-world noises during V2X collaboration using CARLA [9] and OpenCDA [42] together. In total, there are 11,447 frames in our dataset (33,081 samples if we count frames per agent in the same scene), and the train/validation/test splits are 6,694/1,920/2,833, respectively. Compared with existing datasets, V2XSet incorporates both V2X cooperation and realistic noise simulation. Please refer to the supplementary material for more details.

4.2 Experimental Setup

Th evaluation range in x and y direction are $[-140, 140]$ m and $[-40, 40]$ m respectively. We assess models under two settings: 1) *Perfect Setting*, where the pose is accurate, and everything is synchronized across agents; 2) *Noisy Setting*, where pose error and time delay are both considered. In the *Noisy Setting*, the positional and heading noises of the transmitter are drawn from a Gaussian distribution with a default standard deviation of 0.2 m and 0.2° respectively,

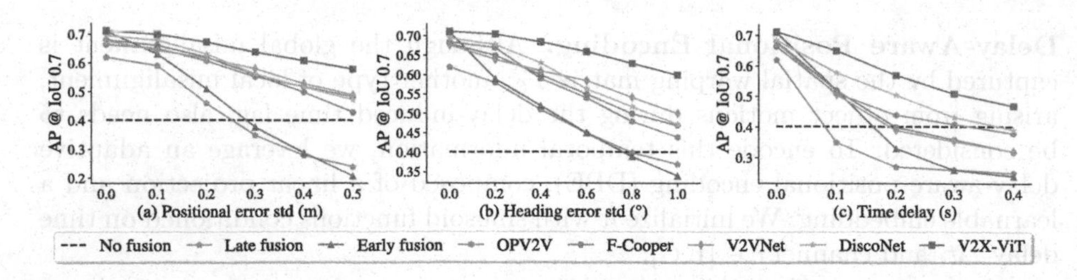

Fig. 4. Robustness assessment on positional and heading errors.

Table 1. 3D detection performance comparison on V2XSet. We show Average Precision (AP) at IoU=0.5, 0.7 on *Perfect* and *Noisy* settings, respectively.

	Perfect		Noisy	
Models	AP0.5	AP0.7	AP0.5	AP0.7
No Fusion	0.606	0.402	0.606	0.402
Late Fusion	0.727	0.620	0.549	0.307
Early Fusion	0.819	0.710	0.720	0.384
F-Cooper [4]	0.840	0.680	0.715	0.469
OPV2V [44]	0.807	0.664	0.709	0.487
V2VNet [39]	0.845	0.677	0.791	0.493
DiscoNet [21]	0.844	0.695	0.798	0.541
V2X-ViT (Ours)	**0.882**	**0.712**	**0.836**	**0.614**

following the real-world noise levels [1,22,41]. The time delay is set to 100 ms for all the evaluated models to have a fair comparison of their robustness against asynchronous message propagation.

Evaluation Metrics. The detection performance is measured with Average Precisions (AP) at Intersection-over-Union (IoU) thresholds of 0.5 and 0.7. In this work, we focus on LiDAR-based vehicle detection. Vehicles hit by at least one LiDAR point from any connected agent will be included as evaluation targets.

Implementation Details. During training, a random AV is selected as the ego vehicle, while during testing, we evaluate on a fixed ego vehicle for all the compared models. The communication range of each agent is set as 70 m based on [18], whereas all the agents out of this broadcasting radius of ego vehicle is ignored. For the PointPillar backbone, we set the voxel resolution to 0.4 m for both height and width. The default compression rate is 32 for all intermediate fusion methods. Our V2X-ViT has 3 encoder layers with 3 window sizes in MSwin: 4, 8, and 16. We first train the model under the *Perfect Setting*, then fine-tune it while fixing the backbone for *Noisy Setting*. We adopt Adam optimizer [19] with an initial learning rate of 10^{-3} and steadily decay it every 10 epochs using a factor of 0.1. All models are trained on Tesla V100 with 10^5 iterations.

Compared Methods. We consider *No Fusion* as our baseline, which only uses ego-vehicle's LiDAR point clouds. We also compare with *Late Fusion*, which gathers all detected outputs from agents and applies Non-maximum suppression to produce the final results, and *Early Fusion*, which directly aggregates raw LiDAR point clouds from nearby agents. For *intermediate fusion* strategy, we evaluate four state-of-the-art approaches: OPV2V [44], F-Cooper [4] V2VNet [39], and DiscoNet [21]. For a fair comparison, all the models use PointPillar as the backbone, and every compared V2V methods also receive infrastructure data, but they do not distinguish between infrastructure and vehicles.

4.3 Quantitative Evaluation

Main Performance Comparison. Table 1 shows the performance comparisons on both *Perfect* and *Noisy Setting.* Under the *Perfect Setting*, all the cooperative methods significantly outperform *No Fusion* baseline. Our proposed V2X-ViT outperforms SOTA intermediate fusion methods by 3.8%/1.7% for AP@0.5/0.7. It is even higher than the ideal *Early fusion* by 0.2% AP@0.7, which receives complete raw information. Under noisy setting, when localization error and time delay are considered, the performance of *Early Fusion* and *Late Fusion* drastically drop to 38.4% and 30.7% in AP@0.7, even worse than single-agent baseline *No Fusion.* Although OPV2V [44], F-Cooper [4] V2VNet [39], and DiscoNet [21] are still higher than *No fusion*, their performance decrease by 17.7%, 21.1%, 18.4% and 15.4% in AP@0.7, respectively. In contrast, V2X-ViT performs favorably against the *No fusion* method by a large margin, *i.e.* 23% and 21.2% higher in AP@0.5 and AP@0.7. Moreover, when compared to the *Perfect Setting*, V2X-ViT only drops by less than 5% and 10% in AP@0.5 and AP@0.7 under *Noisy Setting*, demonstrating its robustness against normal V2X noises. The real-time performance of V2X-ViT is also shown in Table 4. The inference time of V2X-ViT is 57 ms, and by using only 1 encoder layer, V2X-ViT$_S$ can still beat DiscoNet while reaching only 28 ms inference time, which achieves real-time performance.

Sensitivity to Localization Error. To assess the models' sensitivity to pose error, we sample noises from Gaussian distribution with standard deviation $\sigma_{xyz} \in [0, 0.5]$ m, $\sigma_{\text{heading}} \in [0°, 1.0°]$. As Fig. 4 depicts, when the positional and heading errors stay within a normal range (*i.e.*, $\sigma_{xyz} \leq 0.2m, \sigma_{heading} \leq 0.4°$ [1,22,41]), the performance of V2X-ViT only drops by less than 3%, whereas other intermediate fusion methods decrease at least 6%. Moreover, the accuracy of *Early Fusion* and *Late Fusion* degrade by nearly 20% in AP@0.7. When the noise is massive (*e.g.*, 0.5 m and 1° std), V2X-ViT can still stay around 60% detection accuracy while the performance of other methods significantly degrades, showing the robustness of V2X-ViT against pose errors.

Time Delay Analysis. We further investigate the impact of time delay with range [0, 400] ms. As Fig. 4c shows, the AP of *Late Fusion* drops dramatically below *No Fusion* with only 100 ms delay. *Early Fusion* and other intermediate fusion methods are relatively less sensitive, but they still drop rapidly when delay keeps increasing and are all below the baseline after 400 ms. Our V2X-ViT, in

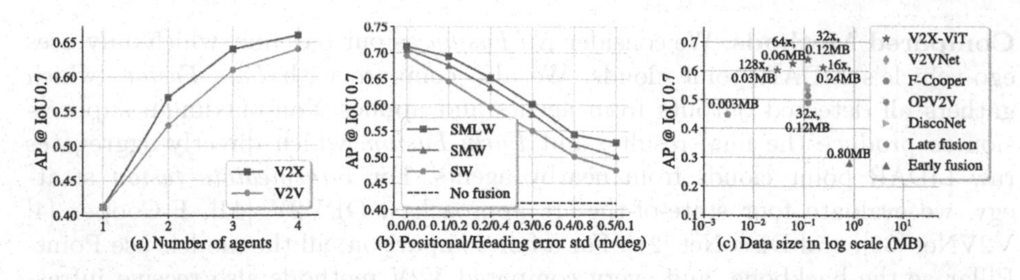

Fig. 5. Ablation studies. (a) AP *vs.* number of agents. (b) MSwin for localization error with window sizes: 4^2 (S), 8^2 (M), 16^2 (L). (c) AP *vs.* data size.

Table 2. Component ablation study. MSwin, SpAttn, HMSA, DPE represent adding i) multi-scale window attention, ii) split attention, iii) heterogeneous multi-agent self-attention, and iv) delay-aware positional encoding, respectively.

MSwin	SpAttn	HMSA	DPE	AP0.5/AP0.7
				0.719/0.478
✓				0.748/0.519
✓	✓			0.786/0.548
✓	✓	✓		0.823/0.601
✓	✓	✓	✓	**0.836/0.614**

Table 3. Effect of DPE w.r.t. time delay on AP@0.7.

Delay/Model	w/o DPE	w/DPE
100 ms	0.639	0.650
200 ms	0.558	0.572
300 ms	0.496	0.514
400 ms	0.458	0.478

Table 4. Inference time measured on GPU Tesla V100.

Model	Time	AP0.7(prf/nsy)
V2X-ViT$_S$	28 ms	0.696/0.591
V2X-ViT	57 ms	0.712/0.614

contrast, exceeds *No Fusion* by 6.8% in AP@0.7 even under 400 ms delay, which is much larger than usual transmission delay in real-world system [33]. This clearly demonstrates its great robustness against time delay.

Infrastructure *vs.* Vehicles. To analyze the effect of infrastructure in the V2X system, we evaluate the performance between V2V, where only vehicles can share information, and V2X, where infrastructure can also transmit messages. We denote the number of agents as the total number of infrastructure and vehicles that can share information. As shown in Fig. 5a, both V2V and V2X have better performance when the number of agents increases. The V2X system has better APs compared with V2V in our collected scenes. We argue this is due to the better sight-of-view and less occlusion of infrastructure sensors, leading to more informative features for reasoning the environmental context.

Effects of Transmission Size. The size of the transmitted message can significantly affect the transmission delay, thereby affecting the detection performance. Here we study the model's detection performance with respect to transmitted data size. The data transmission time is calculated by $t_c = f_s/v$, where f_s denotes the feature size and transmission rate v is set to 27 Mbps [3]. Following [27], we also include another system-wise asynchronous delay that follows a uniform distribution between 0 and 200 ms. See supplementary materials for more details. From Fig. 5c, we can observe: 1) Large bandwidth requirement can eliminate the advantages of cooperative perception quickly, *e.g.*, *Early Fusion* drops to 28%, indicating the necessity of compression; 2) With the default compression rate (32x), our V2X-ViT outperforms other intermediate fusion methods substantially; 3) V2X-ViT is insensitive to large compression rate. Even under a 128x compression rate, our model can still maintain high performance.

4.4 Qualitative Evaluation

Detection Visualization. Figure 6 shows the detection visualization of OPV2V, V2VNet, DiscoNet, and V2X-ViT in two challenging scenarios under *Noisy setting*. Our model predicts highly accurate bounding boxes which are well-aligned with ground truths, while other approaches exhibit larger displacements. More importantly, V2X-ViT can identify more dynamic objects (more ground-truth bounding boxes have matches), which proves its capability of effectively fusing all sensing information from nearby agents. Please see Appendix for more results.

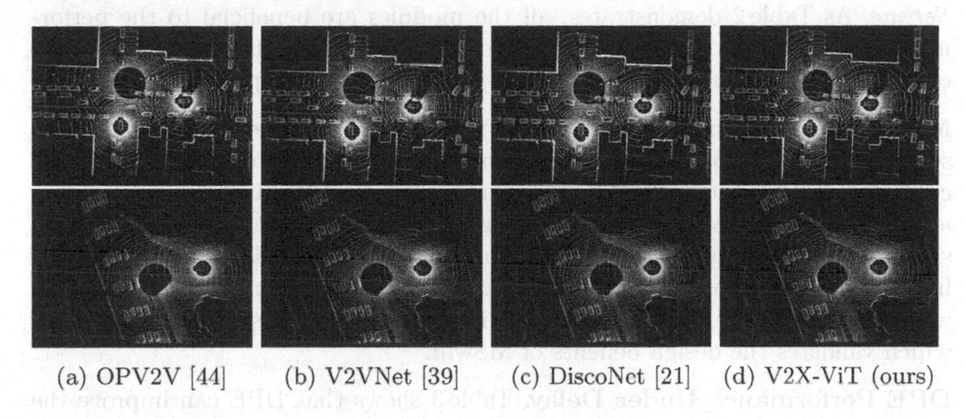

(a) OPV2V [44] (b) V2VNet [39] (c) DiscoNet [21] (d) V2X-ViT (ours)

Fig. 6. Qualitative comparison in a congested intersection and a highway entrance ramp. Green and red 3D bounding boxes represent the ground truth and prediction respectively. Our method yields more accurate detection results. More visual examples are provided in the supplementary materials. (Color figure online)

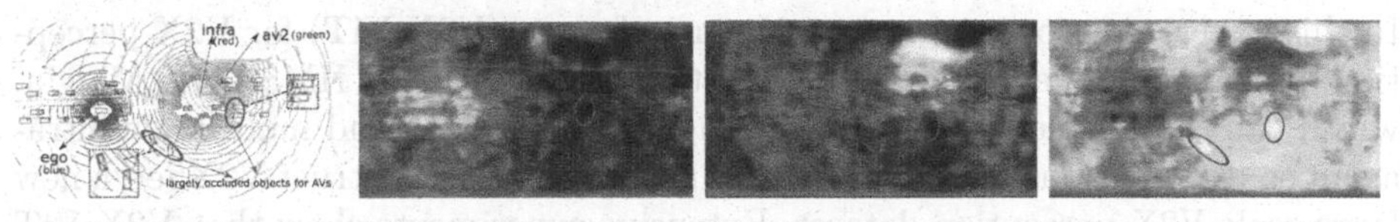

(a) LiDAR points (better zoom-in) (b) attention weights ego paid to ego (c) attention weights ego paid to av2 (d) attention weights ego paid to infra

Fig. 7. Aggregated LiDAR points and attention maps for ego. Several objects are occluded (blue circle) from both AV's perspectives, whereas infra can still capture rich point clouds. V2X-ViT learned to pay more attention to infra on occluded areas, shown in (d). We provide more visualizations in Appendix. (Color figure online)

Attention Map Visualization. To understand the importance of infra, we also visualize the learned attention maps in Fig. 7, where brighter color means more attention ego pays. As shown in Fig. 7a, several objects are largely occluded (circled in blue) from both ego and AV2's perspectives, whereas infrastructure can still capture rich point clouds. Therefore, V2X-ViT pays much more attention to infra on occluded areas (Fig. 7d) than other agents (Figs. 7b and 7c), demonstrating the critical role of infra on occlusions. Moreover, the attention map for infra is generally brighter than the vehicles, indicating more importance on infra seen by the trained V2X-ViT model.

4.5 Ablation Studies

Contribution of Major Components in V2X-ViT. Now we investigate the effectiveness of individual components in V2X-ViT. Our base model is PointPillars with naive multi-agent self-attention fusion, which treats vehicles and infrastructure equally. We evaluate the impact of each component by progressively adding i) MSwin, ii) split attention, iii) HMSA, and iv) DPE on the *Noisy Setting*. As Table 2 demonstrates, all the modules are beneficial to the performance gains, while our proposed MSwin and HMSA have the most significant contributions by increasing the AP@0.7 4.1% and 6.6%, respectively.

MSwin for Localization Error. To validate the effectiveness of the multi-scale design in MSwin on localization error, we compare three different window configurations: i) using a single small window branch (SW), ii) using a small and a middle window (SMW), and iii) using all three window branches (SMLW). We simulate the localization error by combining different levels of positional and heading noises. From Fig. 5b, we can clearly observe that using a large and small window in parallel remarkably increased its robustness against localization error, which validates the design benefits of MSwin.

DPE Performance Under Delay. Table 3 shows that DPE can improve the performance under various time delays. The AP gain increases as delay increases.

5 Conclusion

In this paper, we propose a new vision transformer (V2X-ViT) for V2X perception. Its key components are two novel attention modules *i.e.* HMSA and MSwin, which can capture heterogeneous inter-agent interactions and multi-scale intra-agent spatial relationship. To evaluate our approach, we construct V2XSet, a new large-scale V2X perception dataset. Extensive experiments show that V2X-ViT can significantly boost cooperative 3D object detection under both perfect and noisy settings. This work focuses on LiDAR-based cooperative 3D vehicle detection, limited to single sensor modality and vehicle detection task. Our future work involves multi-sensor fusion for joint V2X perception and prediction.

Broader Impacts and Limitations. The proposed model can be deployed to improve the performance and robustness of autonomous driving systems by incorporating V2X communication using a novel vision Transformer. However, for models trained on simulated datasets, there are known issues on data bias and generalization ability to real-world scenarios. Furthermore, although the design choice of our communication approach (i.e., project LiDAR to others at the beginning) has an advantage of accuracy (see supplementary for details), its scalability is limited. In addition, new concerns around privacy and adversarial robustness may arise during data capturing and sharing, which has not received much attention. This work facilitates future research on fairness, privacy, and robustness in visual learning systems for autonomous vehicles.

References

1. Rt3000. https://www.oxts.com/products/rt3000-v3. Accessed 11 Nov 2021
2. Institue for AI Industry Research (AIR), T.U.: Vehicle-infrastructure cooperative autonomous driving: DAIR-V2X dataset (2021)
3. Arena, F., Pau, G.: An overview of vehicular communications. Future Internet **11**(2), 27 (2019)
4. Chen, Q., Ma, X., Tang, S., Guo, J., Yang, Q., Fu, S.: F-Cooper: feature based cooperative perception for autonomous vehicle edge computing system using 3D point clouds. In: Proceedings of the 4th ACM/IEEE Symposium on Edge Computing, pp. 88–100 (2019)
5. Chen, Q., Tang, S., Yang, Q., Fu, S.: Cooper: cooperative perception for connected autonomous vehicles based on 3D point clouds. In: 2019 IEEE 39th International Conference on Distributed Computing Systems (ICDCS), pp. 514–524. OPTorganization (2019)
6. Chu, X., et al.: Twins: revisiting the design of spatial attention in vision transformers. arXiv preprint arXiv:2104.13840 **1**(2), 3 (2021)
7. Dong, X., et al.: CSWin transformer: a general vision transformer backbone with cross-shaped windows. arXiv preprint arXiv:2107.00652 (2021)
8. Dosovitskiy, A., et al.: An image is worth 16×16 words: transformers for image recognition at scale. arXiv preprint arXiv:2010.11929 (2020)
9. Dosovitskiy, A., Ros, G., Codevilla, F., Lopez, A., Koltun, V.: CARLA: an open urban driving simulator. In: Proceedings of the 1st Annual Conference on Robot Learning, pp. 1–16 (2017)
10. El Madawi, K., Rashed, H., El Sallab, A., Nasr, O., Kamel, H., Yogamani, S.: RGB and LiDAR fusion based 3D semantic segmentation for autonomous driving. In: 2019 IEEE Intelligent Transportation Systems Conference (ITSC), pp. 7–12. OPTorganization (2019)
11. Fan, X., Zhou, Z., Shi, P., Xin, Y., Zhou, X.: RAFM: recurrent atrous feature modulation for accurate monocular depth estimating. IEEE Signal Process. Lett., 1–5 (2022). https://doi.org/10.1109/LSP.2022.3189597
12. Fan, Z., Song, Z., Liu, H., Lu, Z., He, J., Du, X.: SVT-Net: super light-weight sparse voxel transformer for large scale place recognition. In: AAAI (2022)
13. Fan, Z., Zhu, Y., He, Y., Sun, Q., Liu, H., He, J.: Deep learning on monocular object pose detection and tracking: a comprehensive overview. ACM Comput. Surv. (CSUR) (2021)

14. Goodfellow, I., Warde-Farley, D., Mirza, M., Courville, A., Bengio, Y.: Maxout networks. In: International Conference on Machine Learning, pp. 1319–1327. PMLR (2013)
15. Hu, H., Zhang, Z., Xie, Z., Lin, S.: Local relation networks for image recognition. In: ICCV, pp. 3464–3473 (2019)
16. Hu, Z., Dong, Y., Wang, K., Sun, Y.: Heterogeneous graph transformer. In: Proceedings of The Web Conference 2020, pp. 2704–2710 (2020)
17. Jaderberg, M., Simonyan, K., Zisserman, A., et al.: Spatial transformer networks. In: NeurIPS (2015)
18. Kenney, J.B.: Dedicated short-range communications (DSRC) standards in the united states. Proc. IEEE **99**(7), 1162–1182 (2011)
19. Kingma, D.P., Ba, J.: Adam: a method for stochastic optimization. arXiv preprint arXiv:1412.6980 (2014)
20. Lang, A.H., et al: Fast encoders for object detection from point clouds. In: CVPR, pp. 12697–12705 (2019)
21. Li, Y., Ren, S., Wu, P., Chen, S., Feng, C., Zhang, W.: Learning distilled collaboration graph for multi-agent perception. In: NeurIPS 34 (2021)
22. Li, Y., et al.: Toward location-enabled IoT (LE-IoT): IoT positioning techniques, error sources, and error mitigation. IEEE Internet Things J. **8**(6), 4035–4062 (2020)
23. Liang, M., Yang, B., Wang, S., Urtasun, R.: Deep continuous fusion for multi-sensor 3D object detection. In: Ferrari, V., Hebert, M., Sminchisescu, C., Weiss, Y. (eds.) ECCV 2018. LNCS, vol. 11220, pp. 663–678. Springer, Cham (2018). https://doi.org/10.1007/978-3-030-01270-0_39
24. Lin, T.Y., Goyal, P., Girshick, R., He, K., Dollár, P.: Focal loss for dense object detection. In: ICCV, pp. 2980–2988 (2017)
25. Liu, Z., et al.: Swin transformer: Hierarchical vision transformer using shifted windows. arXiv preprint arXiv:2103.14030 (2021)
26. Mo, Y., Zhang, P., Chen, Z., Ran, B.: A method of vehicle-infrastructure cooperative perception based vehicle state information fusion using improved Kalman filter. Multimedia Tools Appl., 1–18 (2021). https://doi.org/10.1007/s11042-020-10488-2
27. Rauch, A., Klanner, F., Dietmayer, K.: Analysis of V2X communication parameters for the development of a fusion architecture for cooperative perception systems. In: 2011 IEEE Intelligent Vehicles Symposium (IV), pp. 685–690. OPTorganization (2011)
28. Rauch, A., Klanner, F., Rasshofer, R., Dietmayer, K.: Car2X-based perception in a high-level fusion architecture for cooperative perception systems. In: 2012 IEEE Intelligent Vehicles Symposium, pp. 270–275. OPTorganization (2012)
29. Rawashdeh, Z.Y., Wang, Z.: Collaborative automated driving: a machine learning-based method to enhance the accuracy of shared information. In: 2018 21st International Conference on Intelligent Transportation Systems (ITSC), pp. 3961–3966. OPTorganization (2018)
30. Shi, S., et al.: PV-RCNN: point-voxel feature set abstraction for 3D object detection. In: CVPR, pp. 10529–10538 (2020)
31. Shi, S., Wang, X., Li, H.: PointRCNN: 3D object proposal generation and detection from point cloud. In: CVPR, pp. 770–779 (2019)
32. Treml, M., et al.: Speeding up semantic segmentation for autonomous driving. In: NeurIPS Workshop MLITS (2016)

33. Tsukada, M., Oi, T., Ito, A., Hirata, M., Esaki, H.: AutoC2X: open-source software to realize V2X cooperative perception among autonomous vehicles. In: 2020 IEEE 92nd Vehicular Technology Conference (VTC2020-Fall), pp. 1–6. OPTorganization (2020)
34. Tu, Z., Talebi, H., Zhang, H., Yang, F., Milanfar, P., Bovik, A., Li, Y.: MAXIM: multi-axis MLP for image processing. In: Proceedings of the IEEE/CVF Conference on Computer Vision and Pattern Recognition, pp. 5769–5780 (2022)
35. Tu, Z., et al.: MaxViT: multi-axis vision transformer. arXiv preprint arXiv:2204.01697 (2022)
36. Vadivelu, N., Ren, M., Tu, J., Wang, J., Urtasun, R.: Learning to communicate and correct pose errors. arXiv preprint arXiv:2011.05289 (2020)
37. Vaswani, A., Ramachandran, P., Srinivas, A., Parmar, N., Hechtman, B., Shlens, J.: Scaling local self-attention for parameter efficient visual backbones. In: CVPR, pp. 12894–12904 (2021)
38. Vaswani, A., et al.: Attention is all you need. In: NeurIPS, pp. 5998–6008 (2017)
39. Wang, T.-H., Manivasagam, S., Liang, M., Yang, B., Zeng, W., Urtasun, R.: V2VNet: vehicle-to-vehicle communication for joint perception and prediction. In: Vedaldi, A., Bischof, H., Brox, T., Frahm, J.-M. (eds.) ECCV 2020. LNCS, vol. 12347, pp. 605–621. Springer, Cham (2020). https://doi.org/10.1007/978-3-030-58536-5_36
40. Wang, Z., Cun, X., Bao, J., Liu, J.: Uformer: a general U-shaped transformer for image restoration. arXiv preprint arXiv:2106.03106 (2021)
41. Xia, X., Hang, P., Xu, N., Huang, Y., Xiong, L., Yu, Z.: Advancing estimation accuracy of sideslip angle by fusing vehicle kinematics and dynamics information with fuzzy logic. IEEE Trans. Veh. Technol. **70**, 6577–6590 (2021)
42. Xu, R., Guo, Y., Han, X., Xia, X., Xiang, H., Ma, J.: OpenCDA: an open cooperative driving automation framework integrated with co-simulation. In: 2021 IEEE International Intelligent Transportation Systems Conference (ITSC), pp. 1155–1162. OPTorganization (2021)
43. Xu, R., Tu, Z., Xiang, H., Shao, W., Zhou, B., Ma, J.: CoBEVT: cooperative bird's eye view semantic segmentation with sparse transformers. arXiv preprint arXiv:2207.02202 (2022)
44. Xu, R., Xiang, H., Xia, X., Han, X., Liu, J., Ma, J.: OPV2V: an open benchmark dataset and fusion pipeline for perception with vehicle-to-vehicle communication. arXiv preprint arXiv:2109.07644 (2021)
45. Yan, Y., Mao, Y., Li, B.: SECOND: sparsely embedded convolutional detection. Sensors **18**(10), 3337 (2018)
46. Yang, B., Luo, W., Urtasun, R.: PIXOR: real-time 3D object detection from point clouds. In: CVPR, pp. 7652–7660 (2018)
47. Yang, Z., Sun, Y., Liu, S., Shen, X., Jia, J.: STD: sparse-to-dense 3D object detector for point cloud. In: CVPR, pp. 1951–1960 (2019)
48. Zelin, Z., Ze, W., Yueqing, Z., Boxun, L., Jiaya, J.: Tracking objects as pixel-wise distributions. arXiv preprint arXiv:2207.05518 (2022)
49. Zhang, H., et al.: ResNeSt: split-attention networks. arXiv preprint arXiv:2004.08955 (2020)
50. Zhang, Z., Fisac, J.F.: Safe occlusion-aware autonomous driving via game-theoretic active perception. arXiv preprint arXiv:2105.08169 (2021)
51. Zhao, X., Mu, K., Hui, F., Prehofer, C.: A cooperative vehicle-infrastructure based urban driving environment perception method using a DS theory-based credibility map. Optik **138**, 407–415 (2017)

52. Zhong, Y., Zhu, M., Peng, H.: VIN: voxel-based implicit network for joint 3D object detection and segmentation for lidars. arXiv preprint arXiv:2107.02980 (2021)
53. Zhou, Y., Tuzel, O.: VoxelNet: end-to-end learning for point cloud based 3D object detection. In: CVPR, pp. 4490–4499 (2018)
54. Zhou, Z., Fan, X., Shi, P., Xin, Y.: R-MSFM: recurrent multi-scale feature modulation for monocular depth estimating. In: Proceedings of the IEEE/CVF International Conference on Computer Vision, pp. 12777–12786 (2021)

DevNet: Self-supervised Monocular Depth Learning via Density Volume Construction

Kaichen Zhou[1], Lanqing Hong[2(✉)], Changhao Chen[1], Hang Xu[2], Chaoqiang Ye[2], Qingyong Hu[1], and Zhenguo Li[2]

[1] University of Oxford, Oxford, UK
[2] Huawei Noah's Ark Lab, Shatin, Hong Kong
{honglanqing,xu.hang,yechaoqiang,li.zhenguo}@huawei.com

Abstract. Self-supervised depth learning from monocular images normally relies on the 2D pixel-wise photometric relation between temporally adjacent image frames. However, they neither fully exploit the 3D point-wise geometric correspondences, nor effectively tackle the ambiguities in the photometric warping caused by occlusions or illumination inconsistency. To address these problems, this work proposes Density Volume Construction Network (DevNet), a novel self-supervised monocular depth learning framework, that can consider 3D spatial information, and exploit stronger geometric constraints among adjacent camera frustums. Instead of directly regressing the pixel value from a single image, our DevNet divides the camera frustum into multiple parallel planes and predicts the pointwise occlusion probability density on each plane. The final depth map is generated by integrating the density along corresponding rays. During the training process, novel regularization strategies and loss functions are introduced to mitigate photometric ambiguities and overfitting. Without obviously enlarging model parameters size or running time, DevNet outperforms several representative baselines on both the KITTI-2015 outdoor dataset and NYU-V2 indoor dataset. In particular, the root mean-square-deviation is reduced by around 4% with DevNet on both KITTI-2015 and NYU-V2 in the task of depth estimation.

Keywords: Depth estimation · Monocular camera · Self-supervised learning · Occlusion probability density · Volume rendering

1 Introduction

Vision-based depth estimation (VDE) attracts attentions due to its significance in understanding the geometry of a 3D scene. It is the basis of higher-level 3D tasks, e.g., scene reconstruction [25] and object detection [38], and supports a number of cutting-edge applications, from autonomous driving [16,47] to augmented reality [21].

S. Avidan et al. (Eds.): ECCV 2022, LNCS 13699, pp. 125–142, 2022.
https://doi.org/10.1007/978-3-031-19842-7_8

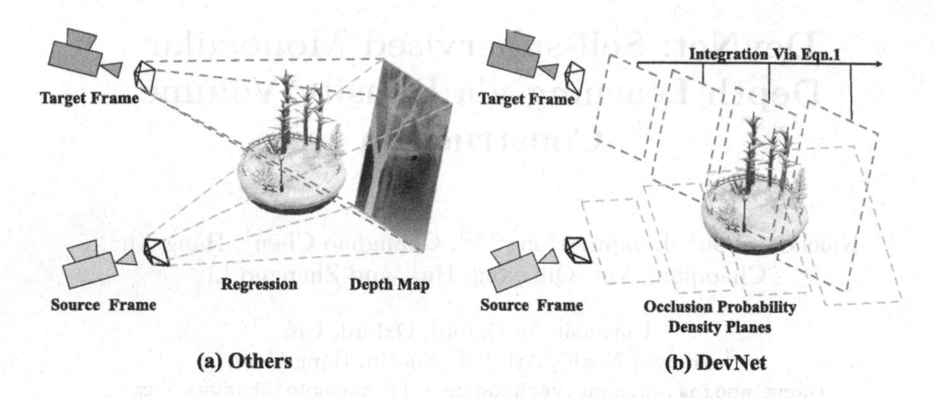

Fig. 1. A comparison between our proposed DevNet and other depth learning approaches [13,14,32,36,51]. **(a)** Most existing depth learning methods predict depth maps directly from input images [10,11], and the correspondences among adjacent frames are built only with 2D pixel-wise depth maps. **(b)** Our DevNet renders depth maps by predicting the density information of parallel planes, and the correspondences among adjacent frames are built with 3D frustums. The pyramid denotes the frustum.

Recently learning-based VDE becomes a focus, that can be generally divided into two categories – supervised [1,4] and self-supervised approaches [46,50,52]. Supervised VDE usually requires high-precision ground-truth depth as training labels, whose process would be costly and time-consuming. Self-supervised methods learn depth maps from monocular images [50] or stereo image pairs [52]. Self-supervised monocular VDE [11,36,51], exploits the photometric loss based on the warping relation between temporally adjacent image frames [50]. Along with per-pixel depth predictions, it also produces ego-motion estimation. Though self-supervised monocular VDE has seen great progresses, there still exists a large performance gap between self-supervised and supervised methods [31]. This gap comes from the fact that it is unreliable to optimize the framework only based on the photometric loss without understanding the geometry of the whole scene. The assumption that lower photometric loss denotes higher depth accuracy does not hold in many cases, e.g., the occlusions, illumination inconsistency, non-Lambertian surfaces, and textureless regions.

To tackle these problems, we propose DevNet, a novel density volume construction based monocular depth learning framework. *By exploiting volume rendering* [20], *3D spatial information along the ray can be integrated into pixel-wise depth estimation, which contributes to building stronger correspondences among adjacent camera frustums.* Instead of directly producing a depth map from a single RGB image, DevNet discretizes the scene into multiple parallel planes, as shown in Fig. 1(b), and predicts the occlusion probability density information of each point on planes. Then, the pixelwise depth value is computed by integrating the probability density values of intersections between the ray and planes. During the training phase, our DevNet mitigates the limitations of photometric loss by adopting an occlusion regularization and a brightness regularization. Finally, a depth consistency loss based on occupancy probability density volume is introduced to maintain the geometric consistency between temporally adjacent frames. Extensive experiments conducted on both the KITTI-2015 [9] and NYU-V2 [37] datasets demonstrate the effectiveness of our proposed DevNet.

Our main contributions are summarized as below:

- We introduce DevNet, a novel self-supervised depth learning framework that renders depth maps via predicting the occlusion probability density of sampled points in the scene.
- We present a novel depth consistency loss using the predicted density volume, to ensure the geometric consistency between the depth maps of adjacent frames, which can further contribute to performance improvement.
- An occlusion regularization is introduced to detect and reduce the photometric ambiguity caused by camera motion and dynamic objects.

2 Related Works

2.1 Supervised Monocular Depth Learning

In monocular depth learning, supervised approaches [4,22,31,49] usually take a single image as input and use depth data measured with range sensors (e.g. RGB-D camera or LIDAR) as ground truth labels during training. [35] first proposed a learning-based method to learn depth from single monocular images in an end-to-end manner and the final algorithm could recover even accurate prediction for unstructured scenes. With the development of deep learning, [4] constructs a monocular depth estimator based on the convolutional architecture to infer the corresponding depth with a global estimation layer, followed by a local refinement layer. [34] further improved neural network-based depth estimation by using a predicted categorical depth distribution for each pixel. Although the depth estimation network trained with ground-truth labels can achieve high accuracy in this task, obtaining high-resolution ground-truth labels from different domain/scenes is costly, time-consuming and not always applicable, which limits the potentials of these supervised methods.

2.2 Self-supervised Depth Learning

Self-supervised methods realize depth learning based on a series of unlabeled images. [8] firstly made use of a pair of images to estimate corresponding by minimizing a photometric loss between the left input image and the warped right image as the target of optimization. [50] stated that only based on image reconstruction loss would lead to a poor result. They further improved the performance of depth estimation by proposing a new network architecture and a novel left-right consistency loss. However, due to the convenience of self-supervised monocular depth learning, it attracts more attention. [50] designed an end-to-end self-supervised monocular depth estimation network for depth estimation and motion estimation of monocular unlabeled video. By using the estimated pose, the loss is constructed based on warping nearby views to the target view. [44] further improved the accuracy of the self-supervised depth estimation task by jointly optimizing depth estimation, optic flow, and camera ego-motion. Predictions from three modules are used to construct the target view and form the loss function for both static background and dynamic components. [11] proposes lots of useful strategies such as a robust reprojection loss, multi-scale sampling, and an automasking loss to further improve the accuracy of monocular depth estimation. To improve the performance at non-contiguous regions and motion boundaries. [2] found that image reconstruction loss ignores the influence of dynamic components and proposes to exploit the geometric consistency among different frames for realizing scale-consistency prediction results. [19] further introduces the self-attention mechanism and takes the discrete disparity prediction into the framework for monocular depth estimation, which could learn more general contextual information and generate a more robust prediction.

2.3 Neural Rendering

Rendering represents the process of generating 2D images from a 2D/3D model. Classic rendering contains two common approaches: Rasterization and Raytracing. Recent neural rendering generally denotes the combination of the recent proposed generative model with traditional computer graphics rendering techniques, which generates images of 3D scenarios from different camera views based on learned implicit representation. Based on this conception, NeRF [30] is proposed to encode both the appearance and the geometry information of a 3D scenario with a Multilayer Perceptron (MLP), which realizes the view synthesis of real-life scenes. After that, numerous studies further adapt NERF to various scenarios, including images in wild [27], dynamic scenes with monocular or stereo videos [3,24,33], post estimation [43], and depth estimation [41]. Despite diverse applications of neural rendering, we are the first to integrate this technique in the monocular depth estimation process. Considering the characteristics of this task and neural rendering, we further propose a novel occlusion regularization and depth consistency loss function to improve its performance in depth learning.

3 Density Volume Rendering Based Depth Learning

Most previous self-supervised monocular VDEs [13,14,32,36,51] adopt the encoder-decoder structure to predict depth map directly from a single RGB image. However, it is not always reliable to optimize these models only by minimizing the photometric loss constructed by warping adjacent frames as in Fig. 1(a) without considering the 3D spatial constraints. DevNet aims to build stronger geometric constraints among adjacent frames by learning the density volume of corresponding camera frustums. The framework of DevNet consists of a rendering-based depth module and a pose module, as introduced in Sect. 3.2. To reduce the influences of occlusions and illumination inconsistency, an occlusion regularization and a brightness regularization are introduced in Sect. 3.3.

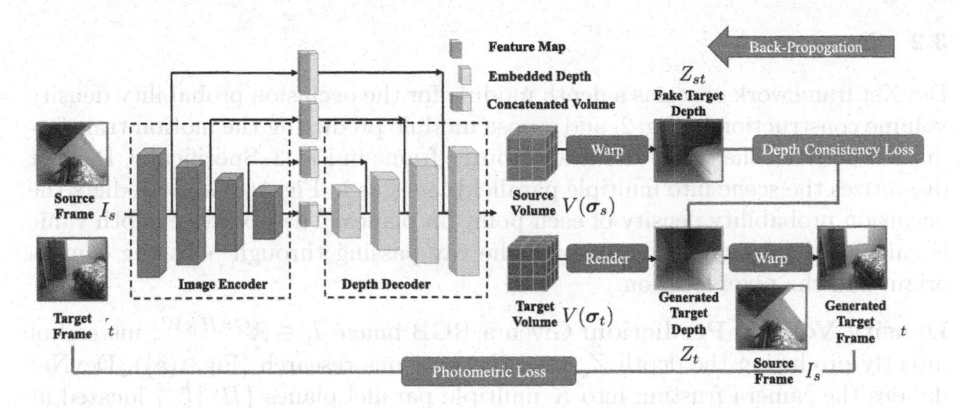

Fig. 2. The illustration for the Rendering-Based Depth Module. This module is designed for estimating a density volume from an RGB frame and the density volume would be used to render the depth map. During the training phase, the photometric loss and depth consistency loss are used to realize self-supervised learning.

We achieve the self-supervised depth estimation via using multiple loss functions in Sect. 3.4. Finally, Sect. 3.5 discusses the differences between proposed DevNet and several similar works.

3.1 Preliminary

Volume rendering has been widely used in computing 2D RGB projections based on discrete 3D samples. Based on [28], to compute the color information along the ray $\boldsymbol{r} = \boldsymbol{o} + \boldsymbol{d} * s$ within the range $[s_n, s_f]$, we divide it into K intervals with the length δ_k for each interval. The color information can be approximated as

$$C = \sum_{k=1}^{K} T_k(1 - \exp(-\sigma_k \delta_k))c_k, \quad \text{where} \quad T_k = \exp(-\sum_{k'=1}^{k-1} \sigma_{k'} \delta_{k'}), \tag{1}$$

where σ_k denotes occlusion probability density at interval δ_k; the occlusion probability within δ_k is written as $\sigma_k \delta_k = 1 - \exp(-\sigma_k \delta_k)$, when δ_k is small enough. Equation (1) can be regarded expectation of color information $c(s)$ in the occlusion distribution σ_k. Similar to color information, the depth is computed as

$$Z = \sum_{k=1}^{K} T_k(1 - \exp(-\sigma_k \delta_k))z_k, \tag{2}$$

where z_k is the depth value for point s_k with respect to image plane. With Eq. (2), the depth computation can exploit the spatial correspondence among different frames. This advantage has already been proven in [30,41].

3.2 Framework

DevNet framework contains a depth module for the occlusion probability density volume construction in Fig. 2, and a pose module predicting the motion transformation between the target frame and source frame in Fig. 3. Specifically, DevNet discretizes the scene into multiple parallel planes as in Fig. 1(b) and predicts the occlusion probability density of each point on planes. The pixelwise depth value is calculated through rendering over the ray passing through both the camera origin and the pixel location.

Density Volume Prediction: Given a RGB image $I_i \in \mathbb{R}^{C\times H\times W}$, instead of directly producing the depth Z_i as in the previous research (Fig. 1(a)), DevNet divides the camera frustum into K multiple parallel planes $\{\Pi_i^k\}_{k=1}^K$ located at the depth $\{z_i^k\}_{k=1}^K$ (Fig. 1(b)). Our depth module produces the occlusion probability density of planes $V(\boldsymbol{\sigma}_i) = \{\boldsymbol{\sigma}_i^k\}_{k=1}^K, \boldsymbol{\sigma}_i^k \in \mathbb{R}^{1\times H\times W}$ based on I_i. Finally, the Z_i map is predicted through volume rendering Eq. (2).

Rendering-Based Depth Estimation Module: The framework of depth module is shown in Fig. 2, including an image encoder and a depth decoder. The image encoder is based on the ResNet structure [15] which takes I_i as input and generates hierarchical feature maps $\{\boldsymbol{f}_i^j\}_{j=1}^N, \boldsymbol{f}_i^j \in \mathbb{R}^{C^j\times\frac{H}{2^j}\times\frac{W}{2^j}}$, where we adopt $N = 4$ in our experiment. The embedding strategy [30] is introduced to process depth information of each plane, written as

$$e(z_i^k) = (\sin(2^0\pi z_i^k), \cos(2^0\pi z_i^k), ..., \sin(2^{(\frac{E}{2}-1)}\pi z_i^k), \cos(2^{(\frac{E}{2}-1)}\pi z_i^k)), \quad (3)$$

where E is dimension of embedded depth. After that, feature maps are concatenated with the corresponding embedded depth to obtain multi-scale feature maps $\{\text{cat}(e(z_i)^j, \boldsymbol{f}_i^j)\}_{j=1}^N$. These feature maps are skip-connected with the decoder to generate the probability density volume $V(\boldsymbol{\sigma}_i)$. During the training phase, DevNet outputs multi-scale density volumes $V(\boldsymbol{\sigma}_i) = \{V^j(\boldsymbol{\sigma}_i)\}_{j=1}^N$ all of which are used in constructing the training loss. During the inference stage, DevNet only uses the biggest layer density volume.

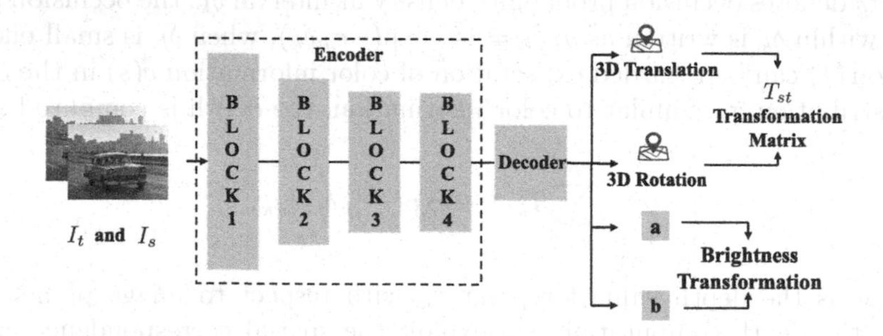

Fig. 3. The architecture of the Pose Module. This module would take concatenated two frames as input and output the pose transformation and the brightness transformation parameters.

Pose Estimation Module: This module also adopts encoder-decoder structure as in Fig. 3. The pose encoder adopts the ResNet structure and takes concatenated target frame I_t and source frame I_s as input. The pose decoder takes the feature map from the last layer of the encoder as input. It outputs 3-dimensional translation and 3-dimensional rotation vectors, which are used to construct the transformation matrix T_{ts}, and the 2-dimensional brightness transformation parameters presented in Sect. 3.3.

3.3 Regularizations

The self-supervised depth learning algorithms [10,11,36,42] generally rely on the warping relation between the pixel $P_s^m(x_s^m, y_s^m, 1)$ in the source frame and its counterpart $P_{st}^m(x_t^m, y_t^m, 1)$ in the target frame, which can be written as

$$P_{st}^m = Z_t(P_{st}^m)^{-1} K_t T_{st} Z_s(P_s^m) K_s^{-1} P_s^m, \tag{4}$$

where K_t, K_s are the intrinsic matrices for the target frame and the source frame; T_{st} is the transformation matrix from the source frame to the target frame; $Z_s(P_s^m)$ is the depth of pixel P_s^m in source camera frustum and $Z_t(P_{ts}^m)$ is the depth of pixel P_{ts}^m with the respect to the target camera frustum. However, this relation is destroyed in the following cases, such as moving objects, the camera motion, the illumination inconsistency, non-Lambertian surfaces, etc. To mitigate these influences of aforementioned occasions, we propose the following regularization strategies.

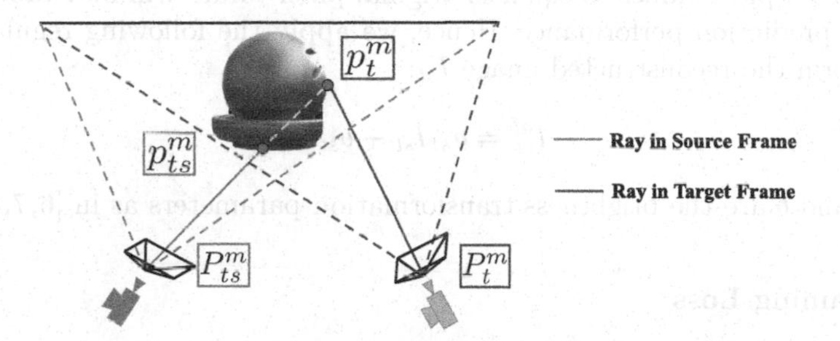

Fig. 4. The illustration of the occlusion regularization. This figure shows the intuition behind proposed occlusion regularization where camera is moving from left to right.

Regularization for Occlusion: Due to dynamic objects and camera motion, occlusions are inevitable in application scenarios. These occlusions would cause ambiguities in the photometric warping between source and target frames. To minimize its influence, we design a novel occlusion regularization strategy to mitigate this problem as illustrated in Fig. 4. During the training phase, after predicting the depth map Z_t of the target frame, the intersection $p_t^m(x, y, z)$

between the rays of pixel $P_t^m(X,Y,1)$ and the surfaces in the target camera frustum can be formulated as

$$p_t^m = Z_t(P_t^m)K_t^{-1}P_t^m, \tag{5}$$

whose depth with respect to the source frame is denoted as $Z_s(p_t^m)$. Then the corresponding pixel P_{ts}^m in the source pixel frame is:

$$P_{ts}^m = \frac{1}{Z_s(p_t^m)}K_t T_{ts} p_t^m, \tag{6}$$

whose depth is denoted as $Z_s(P_{ts}^m)$ and corresponding points p_{ts}^m in the source frame could also be found. Due to the movement of object or the motion of camera, point p_{ts}^m would be different from point p_t^m. This difference could be used to compute the occlusion mask, which could be written as:

$$\mathcal{M}_o = (|Z_s(p_t^m) - Z_s(P_{ts}^m)| < \gamma), \tag{7}$$

where γ is the threshold in terms of the depth difference. Details could be found in our provided code.

Regularization for Brightness: With the help of the Eq. (4), we project the source frame I_s to the target frame and get the reconstructed target image I_{st}, which are used in photometric loss function construction introduced in Sect. 3.4. However, despite the proposed occlusion regularization, the adjacent frames of outdoor scenarios often suffer from illumination inconsistency. Directly forcing reconstructed pixel values to equal to original pixel values will inevitably deteriorate the prediction performance. Hence, we apply the following regularization to transform the reconstructed image I_{st}:

$$I_{st}^{ab} = a_{st}I_{st} + b_{st}, \tag{8}$$

where a and b are the brightness transformation parameters as in [6,7,18,42].

3.4 Training Loss

Our training loss function consists of four parts, including a photometric loss $\mathcal{L}_p$, a smoothness loss $\mathcal{L}_s$, a depth synthesis loss $\mathcal{L}_d$ and a regularization loss $\mathcal{L}_r$:

$$\mathcal{L} = \mathcal{L}_p + \alpha\mathcal{L}_s + \beta\mathcal{L}_d + \eta\mathcal{L}_r, \tag{9}$$

where α, β and η are corresponding weights to balance these losses.

The photometric loss function consists of L1 and SSIM [10,11]:

$$\begin{aligned}\mathcal{L}_p = 0.15 * \sum_p \mathcal{M}_o \odot \mathcal{M}_i \odot |I_t - I_{st}^{ab}| \\ + 0.85 * \frac{1 - \mathrm{SSIM}(\mathcal{M}_o \odot \mathcal{M}_i \odot I_t, \mathcal{M}_o \odot \mathcal{M}_i \odot I_{st}^{ab})}{2},\end{aligned} \tag{10}$$

Fig. 5. The qualitative results of depth estimation on the KITTI-2015. The results generated by DevNet are shown in the last row, which are sharper than results estimated by listed methods. This comparison is consistent with quantitative results in Table 1.

where SSIM is used to quantify the quality degradation in the image processing [40] and $\mathcal{M}_i$ is the identity mask provided in MonoDepth2 [11].
The smoothness loss function is applied to the target depth map Z_t as:

$$\mathcal{L}_s = |\partial_x Z_t^*| e^{-|\partial_x I_t|} + |\partial_y Z_t^*| e^{-|\partial_y I_t|}, \tag{11}$$

where $Z_t^* = Z_t / \overline{Z_t}$ and $\overline{Z_t}$ is the average inverse depth of target frame [39]. This loss function is used to perverse the edge in the depth reconstruction.

The Depth Consistency Loss. Based on this Eq. (4), our DevNet reconstructs the depth map of target frame Z_{st} based on the density volume of source frame $V^j(\boldsymbol{\sigma}_s)$, as the density of each point $\boldsymbol{\sigma}_t(p_t)$ in target frame can be found by querying the density information of its counterpart in source frame $\boldsymbol{\sigma}_s(T_{st}p_t)$. The depth consistency loss is formulated as:

$$\mathcal{L}_d = \frac{1}{M'} \sum_m \left(\frac{|\mathcal{M}_o \odot Z_{st}^m - \mathcal{M}_o \odot Z_t^m|}{\mathcal{M}_o \odot Z_{st}^m + \mathcal{M}_o \odot Z_t^m} \right), \tag{12}$$

where M' is the number of estimated valid depth values.
The regularization loss $\mathcal{L}_r$ regularizes brightness transformation parameters:

$$\mathcal{L}_r = (a-1)^2 + b^2, \tag{13}$$

which is proposed under the assumption that there is no significant illumination inconsistency between adjacent frames.

Table 1. The quantitative results of depth estimation on the KITTI-2015. Comparison between DevNet and other depth estimation methods on KITTI-2015 with the Eigen split. Best results are in bold. M: trained with monocular sequence. MS: trained with both monocular sequence and stereo images. MSup: supervised trained with monocular sequence.

Methods	Type	Lower is better				Larger is better		
		Abs Rel	Sq Rel	RMSE	RMSE log	$\delta < 1.25$	$\delta < 1.25^2$	$\delta < 1.25^3$
SIGNet [29]	M	0.133	0.905	5.181	0.208	0.825	0.947	0.981
LearnK [13]	M	0.128	0.959	5.230	0.212	0.845	0.947	0.976
DualNet [49]	M	0.121	0.837	4.945	0.197	0.853	0.955	0.982
MonoDepth2(R50) [11]	M	0.110	0.831	4.642	0.187	0.883	0.962	0.982
Johnston [19]	M	0.106	0.861	4.699	0.185	0.889	0.962	0.982
FeatDepth(R50) [36]	M	0.104	0.729	4.481	0.179	**0.893**	0.965	0.984
PackNet-SfM [14]	M	0.107	0.802	4.538	0.186	0.889	0.962	0.981
R-MSFM6 [51]	M	0.108	0.748	4.470	0.185	0.889	0.963	0.982
w/o Pretrain(R50)(P24)	M	0.114	0.836	4.888	0.191	0.877	0.958	0.981
DevNet(R50)(P16)	M	0.103	0.713	4.459	0.177	0.890	0.965	0.982
DevNet(R50)(P24)	M	**0.100**	**0.699**	**4.412**	**0.174**	**0.893**	**0.966**	**0.985**
DFR [46]	MS	0.135	1.132	5.585	0.229	0.820	0.933	0.971
EPC++ [26]	MS	0.128	0.935	5.011	0.209	0.831	0.945	0.979
MonoDepth2(R50) [11]	MS	0.106	0.818	4.75	0.196	0.874	0.957	0.979
FeatDepth(R50) [36]	MS	0.099	0.697	4.427	0.184	0.889	0.963	0.982
R-MSFM6 [51]	MS	0.108	0.753	4.469	0.185	0.888	0.963	0.982
w/o Pretrain(R50)(P24)	MS	0.107	0.751	4.461	0.189	0.883	0.963	0.981
DevNet(R50)(P24)	MS	**0.095**	**0.671**	**4.365**	**0.174**	**0.895**	**0.970**	**0.988**
NeWCRFs [45]	MSup	0.052	0.155	2.129	0.079	0.974	0.997	0.999

3.5 Discussions

The Relation with NeRF [30] **Similarity:** Both of us reconstruct the color image or the depth map by discretizing the scenario into thousands of points. The color or depth value of each pixel is calculated by integrating along with its rays. **Difference:** (1) Instead of running the neural network *thousands of times* to obtain point-wise information for the whole scene, DevNet only passes through the network *once* to acquire the density information of the whole scene. (2) DevNet concentrates on self-supervised depth estimation and this network can be used in multiple scenarios, while NeRF focuses on learning an implicit representation in a certain scenario.

The Relation with FLA [12] **Similarity:** Both of us adapt the auto-encoder structure and output multi-channel geometry information used for depth map construction. **Difference:** (1) FLA outputs multi-channel disparity logit volume which constructs depth through *SoftMax* calculation. DevNet outputs multi-channel density volume which generates depth through *integration*. (2) FLA pre-defines the position of each channel and does not consider the position information. At each time, DevNet randomly takes samples from a uniform distribution, and the position is used to construct embedded depth volume V_i^j. (3) Unlike FLA, DevNet does not need the two-stage training strategy and realizes end-to-end training. (4) DevNet focuses on self-supervised *monocular* depth learning, while FLA focuses on self-supervised *stereo* depth learning.

4 Experiments

Our experiments are mainly conducted on the KITTI-2015 [9] and NYU-V2 [37]. For KITTI-2015, the Eigen split [4] is used to train and test our model. For a fair comparison, following the same data selection strategy in previous research [50,51], we remove all static images from the training set, after which 39,810 training images and 4,424 test images are generated. We adapt the camera matrix setting in [11], which uses the same intrinsic matrix for all images. The length focal is set as the average value. For NYU-V2, we adapt the official train and test split used in [48], where the train set contains 20000 images and the test set consists of 654 labeled images. In this section, the performance of DevNet is firstly tested in terms of depth prediction. Depth prediction results are compared with following widely used evaluation metrics [5,11]:

- Abs Rel $= \frac{1}{|M|}\sum_{m=1}^{M}\frac{|\hat{Z}^m - Z^m|}{\hat{Z}^m}$;
- Sq Rel $= \frac{1}{|M|}\sum_{m=1}^{M}\frac{||\hat{Z}^m - Z^m||_2}{\hat{Z}^m}$;
- RMSE $= \sqrt{\frac{1}{|M|}\sum_{m=1}^{M}||\hat{Z}^m - Z^m||_2}$;
- RMSE log $= \sqrt{\frac{1}{|M|}\sum_{m=1}^{M}||log(\hat{Z}^m) - log(Z^m)||_2}$;
- $\frac{1}{M}|\{\delta = \max(\frac{\hat{Z}^m}{Z^m}, \frac{Z^m}{\hat{Z}^m},) < 1.25^t, m = 1, ..., M\}|$;

where $\hat{Z}^m$ is the ground truth; Z^m is predicted depth value and $t = 1, 2, 3$. Then, to evaluate each component in DevNet, ablation studies are also conducted by varying the sample number, removing mask and loss functions, and changing the backbone. Besides the depth estimation, the DevNet is capable of achieving visual odometry estimation, which is compared to baselines. Finally, the generalization of DevNet trained on KITTI-2015 is directly tested on the NYU-V2 [37].

Table 2. The quantitative results of depth estimation on the NYU-V2. Comparison between DevNet and other depth estimation methods on NYU-V2 with the official split. Best results are in bold.

Model	Type	Abs Rel	Sq Rel	RMSE	RMSE log	$\delta < 1.25$	$\delta < 1.25^2$	$\delta < 1.25^3$
MonoDepth2(R18) [11]	M	0.160	0.152	0.601	0.186	0.767	0.949	0.988
Monoindoor(R18) [17]	M	0.134	–	0.526	–	0.823	0.958	0.989
DevNet(R18)(P24)	M	**0.129**	**0.098**	**0.492**	**0.153**	**0.843**	**0.965**	**0.992**

4.1 Depth Estimation

KITTI-2015: In this experiment, the depth encoder adapts ResNet50(R50), while the pose encoder adapts ResNet18(R18) as base model. During the training phase, our DevNet takes the image with the resolution of 320×1024 as input. When testing our DevNet, we set $80m$ as the max depth value and $0.1m$ as the minimum depth value [11,18]. Same as in [50], the median value of the ground-truth image is used to calculate the scale factor. The comparison between the performance of DevNet and that of state-of-art algorithms is demonstrated in Table 1. It is clear to see that DevNet generally performs better no matter using monocular images, or a combination of monocular and stereo images. Compared to [19] that also learns a disparity volume used to generate depth map through SoftMax, DevNet outperforms them largely by 5.7% in terms of Abs Rel. FeatDepth [36] proposes to pre-train an feature encoder for low-texture region learning. DevNet outperforms them by 3.8% in terms of Abs Rel without pretraining the feature auto-encoder. A qualitative comparison of depth predictions is presented in Fig. 5, showing that DevNet detects the details of moving objects and thin objects, highlighted by red circles. One thing worth mentioning, even though DevNet could achieve good performance with unsupervised setting, supervised algorithms, e.g., NeWCRFs [45], still largely outperform unsupervised algorithms. *Size&Speed* : One thing worth mentioning is that DevNet slightly increases the parameter size and the running time only, as DevNet only modifies skip-connects and the last layer channel number of MonoDepth2 (R50) backbone. While DevNet significantly improves the performance.

NYU-V2: For this indoor dataset, during the training phase, DevNet is fed with a resolution of 288×384. We set $10.0m$ as the max depth value and $0.1m$ as the minimum depth value. Considering the lower resolution of NYU-V2, we adapt the ResNet18 backbone for both depth and pose modules. The result in Table 2 shows that DevNet also outperforms its counterparts in the indoor

Table 3. The quantitative results of odometry estimation on the KITTI odometry dataset. Comparison between DevNet and others in terms of Average translational root mean square error drift and average rotational root mean square error drift.

Algorithm	Seq9		Seq10	
	Translation	Rotation	Translation	Rotation
DFR [46]	11.93	3.91	12.45	3.46
MonoDepth2 [11]	10.85	2.86	11.60	5.72
NeuralBundler [23]	8.10	2.81	12.90	**3.71**
SC-SfMlearner [50]	8.24	2.19	10.70	4.58
FeatDepth [36]	8.75	2.11	10.67	4.91
DevNet(R50)(P24)	**8.09**	**2.05**	**10.05**	4.32

dataset. Specifically, DevNet outperforms Monoindoor by around 4% in terms of Abs Rel.

4.2 Odometry Estimation

To test the performance of odometry estimation, we adapt the split strategy used in [46,50]. This strategy uses the Sequences 00–08 of the KITTI odometry dataset as the training set and the Sequences 09–10 as the test set, as shown in Table 3. The averaged translational and rotational root mean square error (RMSE) are reported. Our DevNet outperforms other representative learning-based methods, e.g. MonoDepth2 and FeatDepth. This is because our DevNet focuses on learning the geometric correspondences between adjacent frames and its accurate depth estimation benefits odometry estimation when jointly optimizing both the odometry module and depth estimation module.

4.3 Ablation Study

In this section, we analyze the performance of DevNet by studying the influence of different components and training strategies on the depth prediction result.

Training Strategies: To evaluate the contribution of each training strategy towards our DevNet framework, we conducted ablation studies under several settings, including the DevNet without embedding strategy applied on position information, without color regularization, without occlusion mask, and without depth loss. The baseline algorithm denotes the DevNet with only photometric loss, smooth loss, and general min-reprojection mask. Based on the analysis

Table 4. The ablation study on depth estimation above the KITTI-2015. The ablation studies concern the influences of different components, training strategies, and different sample number of planes.

Regularization	Type	Abs Rel	Sq Rel	RMSE	$\delta < 1.25$	$\delta < 1.25^2$	$\delta < 1.25^3$
Baseline(R50)(P24)	M	0.118	0.881	4.897	0.872	0.950	0.971
w/o Embedding(R50)(P24)	M	0.102	0.711	4.439	0.889	0.963	0.983
w/o Color Reg(R50)(P24)	M	0.103	0.719	4.462	0.889	0.962	0.979
w/o Occlusion Mask(R50)(P24)	M	0.106	0.791	4.547	0.882	0.959	0.978
w/o Depth Loss(R50)(P24)	M	0.107	0.784	4.487	0.880	0.959	0.967
DevNet(R18)(P24)	M	0.108	0.801	4.712	0.874	0.951	0.973
DevNet(R50)(P08)	M	0.108	0.746	4.471	0.884	0.949	0.969
DevNet(R50)(P16)	M	0.103	0.713	4.459	0.890	0.965	0.982
DevNet(R50)(P24)	M	0.100	0.699	4.412	**0.893**	0.966	0.985
DevNet(R50)(P32)	M	**0.100**	**0.697**	**4.408**	0.891	**0.968**	**0.988**

in Sect. 3, adding the proposed components in the training process is useful to improving estimation accuracy, which is also verified in Table 4. Hence, the main concern here is which component contributes largest towards the model. By comparing the results of DevNet without certain components, it indicates that the position embedding strategy and the color regularization are less influential compared with the occlusion mask and depth loss. The performances of DevNet w/o occlusion mask and DevNet w/o depth mask are similar, as both of two terms focus on the spatial correspondence. However, the highest performance achieved by DevNet shows that the occlusion mask and the depth loss are beneficial towards each other. **Small Backbone:** Resnet50 is used as our backbone. To study the influence of backbone with different size, we also test the performance DevNet with Resnet18 in Table 4. We could notice that replacing the backbone R50 with the R18 could decline DevNet's performance, while DevNet still outperforms the presented methods on KITTI-2015. **Sampled Planes Number:** DevNet splits the camera frustum into multiple parallel planes, therefore we are also interested in the influence of the number of plane samples towards the performance of DevNet. As demonstrated in Table 4, increasing the number of samples improves the estimation accuracy, which is noticed by comparing the performance of DevNet(P08), DevNet(P16), DevNet(P24) and DevNet(P32). However, it can be observed that when the number of samples is already large enough, increasing the number of samples will not result in a high margin in terms of prediction accuracy.

4.4 Generalization

To test the generalization ability of our DevNet, we adapt the DevNet model trained on the KITTI-2015, to directly evaluate it above the NYU-V2 [37] without fine-tuning shown in Table 5 and Fig. 6. Despite huge differences, indoor and outdoor scenes still share some features, e.g., color inconsistency probably indicates depth difference, which makes this generalization possible. The network is fed with images with a resolution of 288×384. The comparison among FeatDepth, R-MSFM3, and R-MSFM6 indicates that better performance on the KITTI-2015

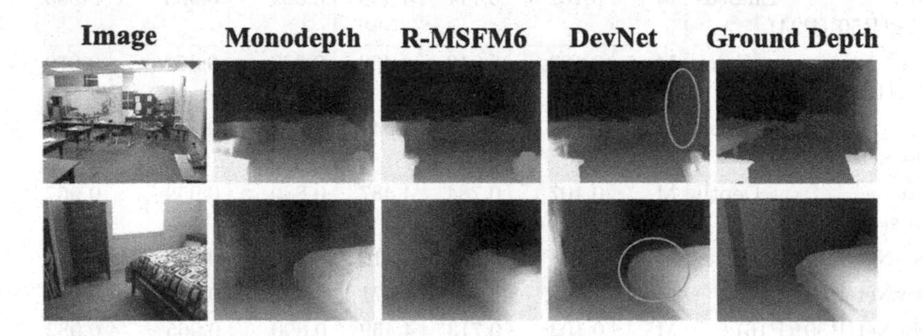

Fig. 6. The qualitative results of depth estimation on the NYU-V2. All methods are trained on KITTI-2015 and tested on NYU-V2.

Table 5. The quantitative results of depth estimation on NYU-V2. Algorithms are trained on monocular video of KITTI-2015 and tested on NYU-V2.

Algorithm	Type	Abs Rel	Sq Rel	RMSE	$\delta < 1.25$	$\delta < 1.25^2$	$\delta < 1.25^3$
MonoDepth2 [11]	M	0.391	0.898	1.458	0.415	0.724	0.874
FeatDepth [36]	M	0.353	0.701	1.231	0.516	0.776	0.904
R-MSFM3 [51]	M	0.355	0.672	1.276	0.476	0.755	0.898
R-MSFM6 [51]	M	0.372	0.803	1.344	0.494	0.746	0.884
DevNet(R50)(P24)	M	**0.333**	**0.605**	**1.142**	**0.541**	**0.790**	**0.911**

doesn't necessarily mean better performance on the NYU-V2. It also could be noticed that our model outperforms other monocular self-supervised methods in all evaluation metrics, which is a shred of strong evidence to support that our model has better generalization capability and could effectively learn the geometric information based on appearance information.

5 Conclusion

In this work, we propose DevNet, an effective self-supervised monocular depth learning framework, that can accurately predict depth maps by constructing the density volume of the scene. To mitigate the influences of occlusions and illumination inconsistency, a novel occlusion mask and a brightness regularization are designed to benefit the training process of this framework. Moreover, the depth consistency loss function is designed to maintain the consistency of density volume for successive frames in the temporal domain. We show that using the training strategy consisting of them provides a simple and efficient model to produce monocular depth and odometry estimation accurately for both indoor and outdoor scenarios, which also possesses high generalization ability.

Acknowledgements. This work was partially supported by the Huawei UK AI Fellowship. We gratefully acknowledge the support of MindSpore, CANN(Compute Architecture for Neural Networks), and Ascend AI Processor used for this research.

References

1. Bhat, S.F., Alhashim, I., Wonka, P.: AdaBins: depth estimation using adaptive bins. In: CVPR (2021)
2. Bian, J., et al.: Unsupervised scale-consistent depth and ego-motion learning from monocular video. In: NeurIPS (2019)
3. Du, Y., Zhang, Y., Yu, H.X., Tenenbaum, J.B., Wu, J.: Neural radiance flow for 4D view synthesis and video processing. In: ICCV (2021)
4. Eigen, D., Fergus, R.: Predicting depth, surface normals and semantic labels with a common multi-scale convolutional architecture. In: ICCV (2015)
5. Eigen, D., Puhrsch, C., Fergus, R.: Depth map prediction from a single image using a multi-scale deep network. arXiv preprint arXiv:1406.2283 (2014)

6. Engel, J., Koltun, V., Cremers, D.: Direct sparse odometry. TPAMI **40**(3), 611–625 (2017)
7. Engel, J., Stückler, J., Cremers, D.: Large-scale direct slam with stereo cameras. In: IROS (2015)
8. Garg, R., B.G., V.K., Carneiro, G., Reid, I.: Unsupervised CNN for single view depth estimation: geometry to the rescue. In: Leibe, B., Matas, J., Sebe, N., Welling, M. (eds.) ECCV 2016. LNCS, vol. 9912, pp. 740–756. Springer, Cham (2016). https://doi.org/10.1007/978-3-319-46484-8_45
9. Geiger, A., Lenz, P., Urtasun, R.: Are we ready for autonomous driving? The KITTI vision benchmark suite. In: CVPR (2012)
10. Godard, C., Mac Aodha, O., Brostow, G.J.: Unsupervised monocular depth estimation with left-right consistency. In: CVPR (2017)
11. Godard, C., Mac Aodha, O., Firman, M., Brostow, G.J.: Digging into self-supervised monocular depth estimation. In: ICCV (2019)
12. GonzalezBello, J.L., Kim, M.: Forget about the LiDAR: self-supervised depth estimators with med probability volumes. In: NeurIPS (2020)
13. Gordon, A., Li, H., Jonschkowski, R., Angelova, A.: Depth from videos in the wild: Unsupervised monocular depth learning from unknown cameras. In: ICCV (2019)
14. Guizilini, V., Ambrus, R., Pillai, S., Raventos, A., Gaidon, A.: 3D packing for self-supervised monocular depth estimation. In: CVPR (2020)
15. He, K., Zhang, X., Ren, S., Sun, J.: Deep residual learning for image recognition. In: CVPR (2016)
16. Huo, Y., et al.: Learning depth-guided convolutions for monocular 3D object detection. In: CVPR Workshop (2020)
17. Ji, P., Li, R., Bhanu, B., Xu, Y.: MonoIndoor: towards good practice of self-supervised monocular depth estimation for indoor environments. In: ICCV (2021)
18. Jin, H., Favaro, P., Soatto, S.: Real-time feature tracking and outlier rejection with changes in illumination. In: ICCV (2001)
19. Johnston, A., Carneiro, G.: Self-supervised monocular trained depth estimation using self-attention and discrete disparity volume. In: CVPR (2020)
20. Kajiya, J.T.: The rendering equation. In: Annual Conference on Computer Graphics and Interactive Techniques (1986)
21. Kalia, M., Navab, N., Salcudean, T.: A real-time interactive augmented reality depth estimation technique for surgical robotics. In: ICRA (2019)
22. Lee, J.H., Han, M.K., Ko, D.W., Suh, I.H.: From big to small: multi-scale local planar guidance for monocular depth estimation. arXiv preprint arXiv:1907.10326 (2019)
23. Li, Y., Ushiku, Y., Harada, T.: Pose graph optimization for unsupervised monocular visual odometry. In: ICRA (2019)
24. Li, Z., Niklaus, S., Snavely, N., Wang, O.: Neural scene flow fields for space-time view synthesis of dynamic scenes. In: CVPR (2021)
25. Liu, H., Tang, X., Shen, S.: Depth-map completion for large indoor scene reconstruction. Pattern Recogn. **99**, 107112 (2020)
26. Luo, C., et al.: Every pixel counts++: joint learning of geometry and motion with 3D holistic understanding. TPAMI **42**(10), 2624–2641 (2019)
27. Martin-Brualla, R., Radwan, N., Sajjadi, M., Barron, J.T., Dosovitskiy, A., Duckworth, D.: NeRF in the wild: neural radiance fields for unconstrained photo collections. arXiv preprint arXiv:2008.02268 (2020)
28. Max, N.: Optical models for direct volume rendering. IEEE Trans. Visual. Comput. Graph. **1**(2), 99–108 (1995)

29. Meng, Y., et al.: SigNet: semantic instance aided unsupervised 3D geometry perception. In: CVPR (2019)
30. Mildenhall, B., Srinivasan, P.P., Tancik, M., Barron, J.T., Ramamoorthi, R., Ng, R.: NeRF: representing scenes as neural radiance fields for view synthesis. In: Vedaldi, A., Bischof, H., Brox, T., Frahm, J.-M. (eds.) ECCV 2020. LNCS, vol. 12346, pp. 405–421. Springer, Cham (2020). https://doi.org/10.1007/978-3-030-58452-8_24
31. Peng, R., Wang, R., Lai, Y., Tang, L., Cai, Y.: Excavating the potential capacity of self-supervised monocular depth estimation. In: ICCV (2021)
32. Pillai, S., Ambruş, R., Gaidon, A.: SuperDepth: self-supervised, super-resolved monocular depth estimation. In: ICRA (2019)
33. Pumarola, A., Corona, E., Pons-Moll, G., Moreno-Noguer, F.: D-NeRF: neural radiance fields for dynamic scenes. In: CVPR (2021)
34. Reading, C., Harakeh, A., Chae, J., Waslander, S.L.: Categorical depth distribution network for monocular 3d object detection. In: CVPR (2021)
35. Saxena, A., Chung, S.H., Ng, A.Y., et al.: Learning depth from single monocular images. In: NeurIPS (2005)
36. Shu, C., Yu, K., Duan, Z., Yang, K.: Feature-metric loss for self-supervised learning of depth and egomotion. In: Vedaldi, A., Bischof, H., Brox, T., Frahm, J.-M. (eds.) ECCV 2020. LNCS, vol. 12364, pp. 572–588. Springer, Cham (2020). https://doi.org/10.1007/978-3-030-58529-7_34
37. Silberman, N., Hoiem, D., Kohli, P., Fergus, R.: Indoor segmentation and support inference from RGBD images. In: Fitzgibbon, A., Lazebnik, S., Perona, P., Sato, Y., Schmid, C. (eds.) ECCV 2012. LNCS, vol. 7576, pp. 746–760. Springer, Heidelberg (2012). https://doi.org/10.1007/978-3-642-33715-4_54
38. Sun, J., et al.: Disp R-CNN: Stereo 3D object detection via shape prior guided instance disparity estimation. In: CVPR (2020)
39. Wang, C., Buenaposada, J.M., Zhu, R., Lucey, S.: Learning depth from monocular videos using direct methods. In: CVPR (2018)
40. Wang, Z., Bovik, A.C., Sheikh, H.R., Simoncelli, E.P.: Image quality assessment: from error visibility to structural similarity. IEEE Trans. Image Process. **13**(4), 600–612 (2004)
41. Wei, Y., Liu, S., Rao, Y., Zhao, W., Lu, J., Zhou, J.: NerfingMVS: guided optimization of neural radiance fields for indoor multi-view stereo. In: ICCV (2021)
42. Yang, N., Stumberg, L.v., Wang, R., Cremers, D.: D3VO: deep depth, deep pose and deep uncertainty for monocular visual odometry. In: CVPR (2020)
43. Yen-Chen, L., Florence, P., Barron, J.T., Rodriguez, A., Isola, P., Lin, T.Y.: iNeRF: inverting neural radiance fields for pose estimation. arXiv preprint arXiv:2012.05877 (2020)
44. Yin, Z., Shi, J.: GeoNet: unsupervised learning of dense depth, optical flow and camera pose. In: CVPR, pp. 1983–1992 (2018)
45. Yuan, W., Gu, X., Dai, Z., Zhu, S., Tan, P.: New CRFs: neural window fully-connected CRFs for monocular depth estimation. arXiv preprint arXiv:2203.01502 (2022)
46. Zhan, H., Garg, R., Weerasekera, C.S., Li, K., Agarwal, H., Reid, I.: Unsupervised learning of monocular depth estimation and visual odometry with deep feature reconstruction. In: CVPR (2018)
47. Zhang, D., Lo, F.P.W., Zheng, J.Q., Bai, W., Yang, G.Z., Lo, B.: Data-driven microscopic pose and depth estimation for optical microrobot manipulation. ACS Photonics **7**(11), 3003–3014 (2020)

48. Zhao, W., Liu, S., Shu, Y., Liu, Y.J.: Towards better generalization: joint depth-pose learning without PoseNet. In: CVPR (2020)
49. Zhou, J., Wang, Y., Qin, K., Zeng, W.: Unsupervised high-resolution depth learning from videos with dual networks. In: ICCV (2019)
50. Zhou, T., Brown, M., Snavely, N., Lowe, D.G.: Unsupervised learning of depth and ego-motion from video. In: CVPR (2017)
51. Zhou, Z., Fan, X., Shi, P., Xin, Y.: R-MSFM: recurrent multi-scale feature modulation for monocular depth estimating. In: ICCV (2021)
52. Zhu, S., Brazil, G., Liu, X.: The edge of depth: explicit constraints between segmentation and depth. In: CVPR (2020)

Action-Based Contrastive Learning for Trajectory Prediction

Marah Halawa(✉), Olaf Hellwich, and Pia Bideau

Excellence Cluster Science of Intelligence, Technische Universitat Berlin, Berlin, Germany
{marah.halawa,olaf.hellwich,p.bideau}@tu-berlin.de

Abstract. Trajectory prediction is an essential task for successful human-robot interaction, such as in autonomous driving. In this work, we address the problem of predicting future pedestrian trajectories in a first-person view setting with a moving camera. To that end, we propose a novel action-based contrastive learning loss, that utilizes pedestrian action information to improve the learned trajectory embeddings. The fundamental idea behind this new loss is that trajectories of pedestrians performing the same action should be closer to each other in the feature space than the trajectories of pedestrians with significantly different actions. In other words, we argue that behavioral information about pedestrian action influences their future trajectory. Furthermore, we introduce a novel sampling strategy for trajectories that is able to effectively increase negative and positive contrastive samples. Additional synthetic trajectory samples are generated using a trained Conditional Variational Autoencoder (CVAE), which is at the core of several models developed for trajectory prediction. Results show that our proposed contrastive framework employs contextual information about pedestrian behavior, *i.e.* action, effectively, and it learns a better trajectory representation. Thus, integrating the proposed contrastive framework within a trajectory prediction model improves its results and outperforms state-of-the-art methods on three trajectory prediction benchmarks.

1 Introduction

Predicting the future trajectories of pedestrians is an important task in many applications, such as in social robot interaction and autonomous driving. Typically, the future trajectory of an agent/pedestrian is predicted based on its own past movement history [33]. Nonetheless, integrating additional information is possible, such as the trajectories of surrounding agents [1,9], or visual scene data [34]. When the surrounding agents in the scene are cars or robots, modeling the motion information based on past trajectories only is a reasonable way to solve the task. However, in this work, we argue that when other agents in the scene are pedestrians, then limiting the information used for prediction to past trajectories is not sufficient. In those cases additional information about pedestrian behavior (*e.g.* action) plays an important role for predicting their future trajectory. For

S. Avidan et al. (Eds.): ECCV 2022, LNCS 13699, pp. 143–159, 2022.
https://doi.org/10.1007/978-3-031-19842-7_9

example, the future trajectories of a pedestrian who is walking while texting on a phone could be different from a pedestrian carrying an object or pushing a baby stroller even if they have the same previous observed trajectories, and the same end goal.

In this work, we study the influence of observed pedestrians' actions on their predicted trajectories. We propose a novel contrastive learning loss called *Action-based Contrastive Loss*. This novel loss is employed as a regularizer to the main trajectory prediction loss. The action-based contrastive loss encourages the trajectory embeddings of agents performing the same action (called positive samples) to come closer to each other in the feature space, and the embeddings of trajectories observed while performing different actions (called negative samples) far away from each other. For instance, the representations of trajectories of walking pedestrians are encouraged to become closer in the feature space, but farther from the representations of trajectories of pedestrians riding bikes or standing, as illustrated in Fig. 1.

Contrastive learning losses, including ours (action-based contrastive loss), utilize a mechanism called negative sampling/mining, which aims to choose the samples that are deemed different and therefore their corresponding features are driven farther in the embedding space. In our case, the negatives are trajectories of pedestrians that have different actions. Commonly used negative sampling techniques include choosing all other samples from the same mini-batch [5] or from a fixed-size memory bank [14]. Nevertheless, while these mechanisms prove effective on natural imaging datasets, we find they do not provide similarly high gains on trajectory datasets. We conjecture that this is due to the higher variation in visual data compared to trajectory data, and most importantly, to the larger sizes of imaging datasets, *e.g.* Imagenet [8] contains 1.6M images compared to PIE [31] that contains 738,970 trajectory samples. This results in limited numbers of negative samples, an issue that becomes more evident when conditioning samples by class information, *e.g.* action or behavior. Few works attempt to address this issue via designing special heuristics for negative mining [13,23,38]. Alternatively, in this work, we propose to utilize the data distribution learned by a Conditional Variational Auto-Encoder (CVAE) [35]. This avoids designing special heuristics for negative mining. While this form of sampling may be utilized to create negative samples only, we employ it to create both positive and negative samples. This is possible due to the different definition of our contrastive loss compared to the traditional Noise Contrastive Estimation loss (NCE loss); the notion of positive/negative in our case is tied to the different classes of action in the data. As explained above, the samples that belong to the same action class are positives from the point of view of this class, and other samples are negative.

Contributions. Our main contributions in this paper are as follows:

- A novel contrastive loss, called action-based contrastive loss, which provides the model with additional information about the action of an agent by guiding the development of the embedding space for trajectories during learning.

- A novel sampling/mining technique that utilizes the latent trajectory distributions learned by CVAEs, circumventing the need to design special mechanisms based on heuristics.

Our proposed contrastive learning framework improves the performance results on three first-person view trajectory prediction benchmarks. It also provides evidence that utilizing agent behavior information, in the form of action type in this case, is beneficial for trajectory prediction, aligning with [26]. However, our proposed learning framework requires action information only during training.

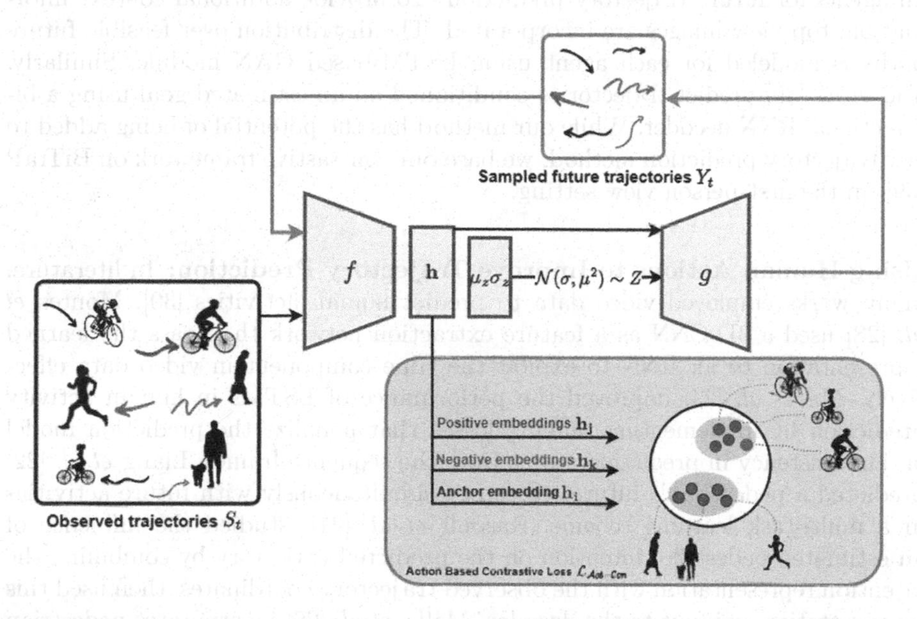

Fig. 1. Overview of our action-based contrastive learning framework during training phase. The contrastive loss $\mathcal{L}_{\text{Act-Con}}$ gets as input both positive (green) and negative (red) embeddings h for an anchor (blue). The positive and negative samples are the samples other than the anchor in the batch, as well as the synthetic samples from the CVAE. The parts shown in yellow refer to our novel action-based contrastive learning framework. It is worth mentioning that the action-based contrastive loss illustrated in this figure updates only the weights of encoder f, and it is jointly optimized with $\mathcal{L}_{traj}$ that updates both encoder f and decoder g. $\mathcal{L}_{traj}$ is not shown in the figure. (Color figure online)

2 Related Work

Multi-modal Trajectory Prediction: A human can reach a desired location following many possible trajectories. Therefore, multiple works utilize multi-modal trajectory models, instead of predicting a single-path solution. Lee *et al.* [21] proposed multi-modal trajectory model by incorporating samples from the Gaussian distribution of a trained conditional variational autoencoder (CVAE) into a long short-term memory encoder-decoder (LSTMED) model. Mangalam *et al.* [27] predict the multi-modal trajectory of an agent by modeling three factors: the desired endpoint goal, the social interaction with other agents in the scene, and the planned trajectories with respect to the environmental constraints in the scene. Similarly, their model is based on CVAE, which takes as input both the encodings of the past trajectory and of the endpoint goal. Sadeghian *et al.* [34] additionally include the past/observed trajectories of all agents for future trajectory prediction. To provide additional context information, top view images are incorporated. The distribution over feasible future paths is modeled for each agent using LSTM-based GAN module. Similarly, Yao *et al.* [39] predict trajectories conditioned on an estimated goal using a bi-directional RNN decoder. While our method has the potential of being added to any trajectory prediction method, we base our contrastive framework on BiTraP [39], in the first-person view setting.

Using Human Actions to Improve Trajectory Prediction: In literature, many works employed video data to predict human activities [30]. Montes *et al.* [28] used a 3D-CNN as a feature extraction network then pass the learned representation to an RNN to exploit the time component in video data effectively. Ma *et al.* [24] improved the performance of LSTMs in human activity prediction by implementing ranking losses that penalize the prediction model on inconsistency in prediction scores from the sequence frames. Liang *et al.* [22] predicted a pedestrian's future trajectory simultaneously with future activities in a multi-task learning scheme. Rasouli *et al.* [31], studied the influence of an estimated pedestrian intention on the predicted trajectory by combining the intention representation with the observed trajectory coordinates, then used this representation as input to the decoder. Malla *et al.* [26] incorporates pedestrian action information with a trajectory prediction model. They require this information as prior information and learn a joint representation for both observed trajectory and pedestrian action. In this work, we also highlight the importance of analyzing the pedestrian's behavior and action in the prediction of their future trajectory. However, we propose to incorporate action information only during training using a novel action-based contrastive loss.

Contrastive Trajectory Prediction: Contrastive learning is a representation learning approach, first proposed by [29]. This approach encourages similar high-dimensional input vectors to be mapped closely to each other in a lower-dimensional embedding manifold, and the dissimilar ones are mapped far away from each other. Contrastive learning has been applied in several unsupervised [4–6,10,11,14,15,17,29,41] and supervised [18] representation learning methods. Recently, only few works applied contrastive learning to trajectory prediction in a multi-agent setting. The flexibility of defining a contrastive loss by using positive and negative samples addresses the shortage problem in critical and challenging scenarios in training datasets. Such rare scenarios are necessary for the model, as the agent could face these in the real-world. Makansi *et al.* [25] utilize this idea by separating the hard and critical samples in the feature space that do not satisfy some certain favorable criterion far away from the positive easy samples. Liu *et al.* [23] proposed a social sampling strategy that relies on augmenting negative samples with prior knowledge about undesired scenarios in the multi-agent setting. Both methods use the contrastive loss as a weighted combination to the future trajectory forecasting loss, which may be the mean squared error (MSE) or negative log-likelihood (NLL). Our method follows this family of algorithms, and uses a novel action-based contrastive loss to add context information about pedestrian actions to the trajectory prediction model.

Supervised Contrastive Loss: Khosla *et al.* [18] proposed a supervised contrastive loss that is a generalization of the Triplet loss [16]. In this supervised contrastive loss for each anchor there are more than one positive sample, in addition to many negative samples. There are two major differences between our proposed action-based contrastive loss and the supervised contrastive loss used in [18]. First, they employ the supervised contrastive loss to *replace* the cross-entropy loss for training the image classifier using image labels. However, we utilize the contrastive loss to *regularize* the trajectory prediction loss, which may be MSE or NLL. Second, due to the differences between the nature of datasets we use in this paper and the image data used in [18], it is simpler to extract many positive and negative samples from a large dataset, such as ImageNet [8]. However, in first-person view trajectory prediction datasets, the number of pedestrians with same actions is limited, therefore we address this with a novel sampling process from a CVAE trained to predict trajectories based on observing a short past trajectory. This CVAE predictive model ensures consistency between observed and predicted trajectories. Thus, it allows sampling additional positive and negative samples that belong to specific actions. Using this novel sampling technique avoids designing hard negative mining techniques, which use heuristics, as in [19,36] for domain adaptation.

3 Methodology

In this section, we present our method for the task of pedestrian trajectory prediction, that focuses on integrating contextual information such as actions for more reliable future predictions. We address this by employing an action-based contrastive loss that enhances the trajectory prediction model with action information.

3.1 Problem Formulation

For each pedestrian we have an observed past trajectory $S_t = [s_1, ..., s_{t-1}, s_t]$ at time t, and we predict a future trajectory $Y_t = [y_{t+1}, y_{t+2}, ..., y_T]$, where s and y are bounding box coordinates for the observed and predicted trajectories, respectively. T is the maximum predicted trajectory time length in the future. In addition, we also have for each trajectory the action class information a, where the set of available actions $a \in \{a_1, a_2, ..\}$ may vary across different datasets. Then in the training data, we assume there are N different training samples, where for each sample $i \in [1, .., N]$, we know S^i, Y^i, and a^i. Finally, we process the dataset samples in mini-batches, where each batch contains B samples.

3.2 Multi-modal Trajectory Prediction

We follow the commonly used approach of an encoder-decoder prediction model, where an encoder f learns the representation h given an observed trajectory S_t as an input, then a decoder g uses the representation h together with a sampled latent variable z to predict the future trajectory Y_t. We employ a standard long-short term encoder-decoder model (LSTMED) [21]. In fact, we extend on the bi-directional version of LSTMED, proposed in Yao *et al.*'s BiTraP [39]. The possibility to draw multiple future trajectories for each observed trajectory is achieved with a CVAE, which is a non-parametric model, that learns the distribution of target trajectory through a stochastic latent variable. The distribution learned by the CVAE is essential for our proposed contrastive framework, which we explain below. As a trajectory prediction loss function $\mathcal{L}_{traj}$, the Best-of-Many (BoM) L2-loss [3] between predicted and target trajectory is used. It is noteworthy that we do not restrict our proposed framework, explained below, to these choices of model architectures or loss functions; we adopt standard and effective techniques to study its influence on predicted trajectories. The essential factor for our learning framework is that the predicting future trajectory model is based on CVAE, similar to trajectory prediction models in [27,39].

3.3 Action-Based Contrastive Learning Framework

In order to enhance the model with contextual information about the pedestrian actions, we propose a novel loss that is called action-based contrastive loss, which acts as a regularizer for the trajectory prediction loss, and they jointly train the trajectory prediction model. The proposed action-based contrastive loss is based on a novel action-based sampling strategy shown in Fig. 1. We first describe the proposed contrastive loss in the simple case, without including additional samples from the CVAE distribution, and we generalize it later.

Action-Based Contrastive Loss Let B be the number of samples within a batch. For each observed past trajectory S^i where $i \in \{1, .., B\}$, called the anchor, there exists multiple positive and negative samples. The positive samples S^{i+} are the trajectories that have the same action class as the anchor, which are denoted by $S^{i''}$. Moreover, we also add an augmented version $S^{i'}$ of the anchor trajectory as a positive sample, following [23], which is created by adding small white noise ϵ to the bounding box coordinates of the anchor trajectory.

Formally:

$$S^{i'} = \{S^i + \epsilon\}$$
$$S^{i''} = \{S^j\}\,; \text{where } 0 < j < B, a^j = a^i, i \neq j$$
$$S^{i+} = S^{i'} \cup S^{i''}$$

Negative samples S^{i-} are trajectories belonging to a different action class than the anchor.

$$S^{i-} = \{S^k\}\,; \text{where } 0 < k < B, a^k \neq a^i, i \neq k$$

Afterwards, all batch samples $\{S^i\}_{i=1}^{B}$ are processed by the model encoder f to produce their hidden representations $\{\mathbf{h}^i\}_{i=1}^{B}$. Assuming M positive samples and K negative samples in the batch, with $B = M + K$. The proposed loss is calculated as follows:

$$\begin{aligned}\ell_{\text{Act-Con}} &= -\frac{1}{B}\sum_{i=1}^{B} \log \frac{\sum_{j=1, j\neq i, a^j=a^i}^{M} \exp(sim(\mathbf{h}^i, \mathbf{h}^j)/\tau)}{\sum_{k=1, k\neq i}^{K} \exp(sim(\mathbf{h}^i, \mathbf{h}^k)/\tau)} \\ \mathcal{L}_{\text{Act-Con}} &= \frac{1}{N/B}\sum^{N/B} \ell_{\text{Act-Con}}\end{aligned} \tag{1}$$

where sim is the similarity between the vector representations of the samples, for which we use the dot-product. τ is the temperature hyperparameter. The above loss function encourages the embeddings $\mathbf{h}^i$ of positive sample trajectories to be closer to each other in the embedding space, and far away from the embeddings of negative samples. The complete loss function sums both the trajectory prediction loss $\mathcal{L}_{traj}$ and the action-based contrastive loss $\mathcal{L}_{\text{Act-Con}}$:

$$\mathcal{L}_{final} = \mathcal{L}_{traj} + \beta \mathcal{L}_{\text{Act-Con}} \tag{2}$$

where β is a hyper-parameter that controls the contribution of action-based contrastive loss. It is worth mentioning that additional behavioral information such as pedestrian's action class is only needed during training. However, during inference, the model only takes the observed trajectory as input to predict the future trajectory.

Action-Based Synthetic Trajectory Sampling The above loss formulation assumes no additional synthetic samples, *i.e.* it considers observed trajectories in the batch only. However, due to the relatively limited sizes of trajectory datasets, and the shortage of diversity in action classes in captured scenes, commonly used negative sampling techniques may not be sufficient. Those include sampling from the same mini-batch [5] or from a fixed-size memory bank [14]. More comprehensive negative and positive samples, from various behavior scenarios are rather needed. Therefore, we extend training samples by drawing trajectories from the distribution learned by the generative Conditional Variational Autoencoder (CVAE) model. CVAE is a generative model that introduces a stochastic latent variable Z in order to learn the distribution of target future trajectory $P\left(Y^i|S^i, Z\right)$. This distribution is conditioned on the input observed trajectories S^i, and the stochastic latent variable Z. Thus, the model is able to predict *multiple* feasible trajectories Y^i given the input S^i. We assume the latent variable following a Gaussian distribution $Z \sim N\left(\mu_Z, \sigma_Z^2\right)$, and we train the CVAE to capture this distribution. Afterwards, the training dataset is extended by sampling from the Gaussian latent space multiple times, and passing samples through the decoder g to effectively predict different feasible future trajectories conditioned on an observed trajectory. The conditioning on the observed trajectory ensures a consistent behavior in the predicted future trajectory. This behavior is captured in both the continuity of the trajectory as well as the identical action class in both observed and future trajectories. We employ the same encoder-decoder trajectory prediction model explained above in Sect. 3.2, which is a CVAE that predicts multiple feasible future trajectories, as the example in Fig. 2 shows. This sampling strategy is illustrated in Fig. 1, and it has the advantage that it avoids designing heuristics for negative sample mining techniques, as mentioned before. The intuition behind this sampling strategy is that the encoder-decoder CVAE model is capable of generating future trajectories with the same behavior of the observed trajectory. Since it is trained to predict the future trajectory of an observed trajectory, then it captures the characteristics of the observed trajectory.

Let $\left\{Y^{i,l}\right\}_{l=1}^{L}$ be the multiple predicted trajectories for an observed trajectory S^i. Here, $Y^{i,l}$ is sampled from $P\left(Y^{i,l}|S^i, Z\right)$, and L is the number of times we sample a different Z from the normal distribution $N\left(\mu_Z, \sigma_Z^2\right)$. Given these synthetic trajectory samples, the set of positive samples for trajectory S^i are then reformulated as follows:

$$S_{1:t}^{i+} = \left\{S^{i'}_{1:t}\right\} \cup \left\{S^{i''}_{1:t}\right\} \cup \left\{Y_{t+1:T}^{j,l}\right\}_{l=0}^{L}$$

where $i, j \in 1, .., B$ and $a^l = a^i$ and $a^j = a^i$. And the negative samples are reformulated as follows:

$$S_{1:t}^{i-} = \left\{S_{1:t}^{k}\right\} \cup \left\{Y_{t+1:T}^{k,l}\right\}_{l=0}^{L}$$

where $i, k \in 1, .., B$ and $i \neq k$ and $a^i \neq a^k$ and $a^i \neq a^l$ and $a^k = a^l$. In words, the synthetic samples created for sample S^k, which we denote $Y^{k,l}$, are considered negative from the point of view of sample S^i. These synthetic samples have the same action class of sample S^k, hence denoted $a^k = a^l$.

The described action-based synthetic trajectory sampling strategy changes the sets of positive and negative samples used in creating training batches. However, the proposed contrastive loss equation Eq. 1 remains the same, only M and K are affected.

4 Experiments

In this section, we present the evaluation results of our method on three first-person-view trajectory prediction datasets [26,31,32]. First, we describe the used

Fig. 2. Examples from TITAN dataset showing the multi-modality in the trajectory prediction space. CVAE is able to predict multiple feasible future trajectories (Green bounding boxes), conditioned on previously observed trajectories (Blue bounding boxes). The red bounding boxes refer to the ground truth future trajectories. (Color figure online)

datasets. Then, we provide an overview of the experimental setup and used evaluation metrics. Finally, we discuss our results and findings.

4.1 Datasets

We evaluate our method on first-person view datasets. In this domain, the Pedestrian Intention Estimation (PIE) [31] and the Joint Attention for Autonomous Driving (JAAD) [32] datasets are the most commonly used benchmarks in literature. The PIE dataset provides 293,437 annotated frames, containing 1,842 pedestrians with behavior annotations such as walking, standing, crossing, looking, etc. Since a pedestrian could be "walking" and "looking" at the same time, for example, then a pedestrian could have multiple behavior labels in a single frame. Therefore, we only use two classes "walking" and "standing", which are exclusive. We use the same train and test splits in [31]. On the other hand, the JAAD dataset provides 82,032 annotated frames, containing 2,786 pedestrians, 686 of them have behavior annotations. Similar to the PIE dataset, we only use for JAAD dataset two classes "walking" and "standing", which are exclusive. We use the same train and test splits in [32].

We also use a third dataset named TITAN [26], which contains more action classes compared to PIE and JAAD. TITAN provides 75,262 frames with 395,770 pedestrians with multiple action labels organized in five hierarchical contextual activities, such as individual atomic actions, simple scene contextual actions, complex contextual actions, transportive actions, and communicative actions. For the same reason of not having multiple labels for each pedestrian, we use individual atomic actions labels for the TITAN dataset. The atomic action labels describe the primitive action, and are categorized into 9 labels (sitting, standing, walking, running, bending, kneeling, squatting, jumping, laying down).

Table 1. The quantitative results on **PIE** and **JAAD** datasets. The evaluation metrics are reported for different prediction lengths in *squared pixels*. ABC+ is our proposed action-based contrastive framework with sampling from a learned CVAE. BiTraP is a baseline trajectory prediction model without adding any contrastive loss. The other baseline results are obtained from [39]. Lower is better.

Method	**PIE**					**JAAD**				
	ADE			C-ADE	C-FDE	ADE			C-ADE	C-FDE
	0.5	1.0	1.5	1.5		0.5	1.0	1.5	1.5	
Linear [31]	123	477	1365	950	3983	233	857	2303	1565	6111
LSTM [31]	172	330	911	837	3352	289	569	1558	1473	5766
B-LSTM [2]	101	296	855	811	3259	159	539	1535	1447	5615
FOL-X [40]	47	183	584	546	2303	147	484	1374	1290	4924
PIE_{traj} [31]	58	200	636	596	2477	110	399	1280	1183	4780
BiTraP [39]	23	48	102	81	261	**38**	94	222	177	565
ABC+	**16**	**38**	**87**	**65**	**191**	40	**89**	**189**	**145**	**409**

4.2 Experimental Setup

We use the same setup for all datasets, where we observe 0.5 s and predict 0.5, 1.0, and 1.5 s, following [31,39]. The predicted trajectories have two forms: bounding boxes coordinates and centers, that are evaluated separately. PIE and JAAD datasets are both annotated 30 Hz frequency, therefore we observe 15 frames and predict 45 frames. However, TITAN dataset is annotated at 10 HZ sampling frequency. Thus, we observe 5 frames and predict 15 frames.

Implementation Details. We use 256 as the size for all hidden layers in the encoder-decoder model that is detailed in Sect. 3.2. It is noteworthy that we implement the loss in Eq. 1 using the efficient matrix-form (especially on GPU machines), instead of performing expensive pairwise computations. We train the model on all datasets with Adam optimizer [20] using a batch size of 128 and a learning rate of 0.001. On training datasets, we perform hyper-parameter tuning for β (Eq. 2). We achieve our best results using $\beta = 0.75$ for all datasets.

Evaluation Metrics. Following the commonly used evaluation protocols in literature [31,37,39], we use the following evaluation metrics: i) Bounding box Average Displacement Error (ADE), ii) Bounding box Center ADE (C-ADE), iii) Bounding box Final Displacement Error (FDE), and iv) Bounding box Center FDE (C-FDE). All are computed in squared pixels. The bounding box ADE is the mean square error (MSE) for all predicted trajectories and ground-truth future trajectories. This error is calculated using the bounding box upper-left and lower-right coordinates. However, in C-ADE, otherwise called C-MSE, the error is calculated using the centers of the bounding boxes. Bounding box FDE, otherwise called FMSE, is the distance between the destination point of the predicted trajectory and of the ground truth at the last time step. FDE is also calculated using the bounding boxes coordinates. Finally, C-FDE or C-FMSE is the mean squared error between the centers of final destination bounding boxes.

Table 2. The quantitative results on **TITAN** dataset. The evaluation metrics are reported for observing 15 time steps and predicting 45 time steps of trajectories in *squared pixels*. ABC+ is our proposed action-based contrastive framework with sampling from a learned CVAE. BiTraP is a baseline trajectory prediction model without adding any contrastive loss. Lower is better.

Method	ADE			C-ADE	C-FDE
	0.5	1.0	1.5	1.5	
BiTraP [39]	194	352	658	498	989
ABC+	**165**	**302**	**575**	**434**	**843**

Table 3. The quantitative results on **TITAN** dataset. The evaluation metrics are reported for observing 10 time steps and predicting 20 time steps of trajectories in *pixels*. ABC+ is our proposed action-based contrastive framework with sampling from a learned CVAE. The other baseline results are obtained from [26]. Lower is better.

Method	ADE	FDE
Social-LSTM [1]	37.01	66.78
Social-GAN [12]	35.41	69.41
Titan-vanilla [26]	38.56	72.42
Titan-AP [26]	33.54	55.80
ABC+	**30.52**	**46.84**

Baselines. The trajectory prediction model trained with our proposed Action-Based Contrastive framework (loss and sampling strategy) is indicated by (ABC+). First, we evaluate the performance of our action-based contrastive framework by comparing its results to the original BiTraP trajectory prediction model [39] on all datasets, *i.e.* without adding our contrastive loss. This baseline aims to highlight the gains obtained by our proposed contrastive framework. BiTraP had previously achieved state-of-the-art on PIE and JAAD datasets. Additionally, for PIE and JAAD datasets, we compare our results with PIEtraj [31], FOL-X [40], B-LSTM [2], LSTM [31], and Linear [31] trajectory prediction models. On the TITAN dataset, we first report the evaluation results compared to BiTraP using observed and predicted lengths equal to PIE and JAAD. However, to fairly compare our results on the TITAN dataset with prior work of Malla *et al.* [26], Social-LSTM [1], and Social-GAN [12], we follow the same experimental setup used in [26]. To that end, we retrain both the BiTraP baseline model, and our proposed model (ABC+) to predict 20 frames, after observing 10 frames, and we report our results using ADE and FDE in pixels, not in squared pixels.

4.3 Trajectory Prediction Results

The evaluation results are shown in Table 1 for PIE and JAAD. For TITAN, in Table 2 we show the evaluation results when observing 10 frames and predicting 20 frames, similar to [26], and in Table 3 we compare to BiTraP when observing 5 frames and predicting 10 frames. As the tables show, our method (ABC+) achieves superior performance compared to the baseline BiTraP, which does not use our action-based contrastive loss. This result highlights the effectiveness of adding the proposed contrastive objective and sampling strategy. Our proposed method also outperforms other baseline methods, with significant margins.

These evaluation results confirm the gains obtained by using our proposed action-based contrastive loss and sampling strategy. Utilizing action information with our contrastive approach exhibits improved performance across all evaluated benchmarks. In the TITAN dataset, particularly, the performance benefits appear larger. We believe this is due to TITAN's more comprehensive action class structure, compared to PIE or JAAD. In other words, a more diverse set of pedestrian action classes improves the learned embedding space by our action-based contrastive loss. Nevertheless, our method improves the obtained results even with a simpler binary action class structure in PIE and JAAD.

Another significant result is that our method (ABC+) outperforms the baseline Titan-AP [26] on TITAN, which incorporates the same action class informa-

Table 4. Ablation results on **PIE, JAAD**, and **TITAN** datasets. ABC+ is our proposed action-based contrastive framework with sampling from a learned CVAE. ABC uses our action-based contrastive loss but without sampling from a learned CVAE. SimCLR uses a normal batch contrastive loss instead of our proposed action-based contrastive loss. Lower is better.

	Method	ADE			C-ADE	C-FDE
		0.5	1.0	1.5	1.5	
PIE	SimCLR	26	67	163	125	399
	ABC	16	40	93	69	213
	ABC+	**16**	**38**	**87**	**65**	**191**
JAAD	SimCLR	50	124	273	211	608
	ABC	41	93	201	150	425
	ABC+	**40**	**89**	**189**	**145**	**409**
TITAN	SimCLR	255	506	999	773	1805
	ABC	188	345	634	488	951
	ABC+	**165**	**302**	**575**	**434**	**843**

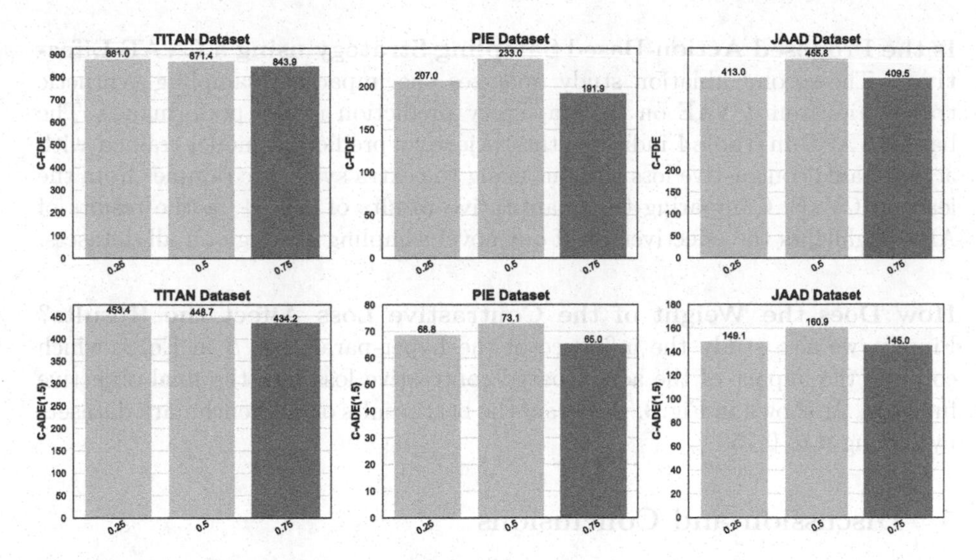

Fig. 3. -FDE results (top) and C-ADE(1.5) (bottom) of trajectory prediction model by applying different β values $0.25, 0.5, 0.75$ in Eq. 2. The results are reported for TITAN, PIE, and JAAD datasets. Lower values are better.

tion with observed trajectory information, and produces a combined embedding to predict the future trajectory. This indicates that our approach of supporting the trajectory prediction model with behavioral context information by using action-based contrastive loss is more effective than encoding the action classes in the embedding space representation.

4.4 Ablation Study

In this section, we present the ablation studies to provide further insights into our proposed action-based contrastive loss. Similar to the evaluation results shown above, we refer to our proposed action-based contrastive framework by ABC+, where we use the action-based contrastive loss as a regularizer to the trajectory prediction loss, and we also increase negative and positive samples during training by sampling synthetic trajectories from CVAE.

Does Action Information Improve the Contrastive Loss? The first ablation study examines how the action-based contrastive loss Eq. 1 compares to the batch contrastive loss, namely SimCLR [5], shown in Table 4. For this baseline, we replace the action-based contrastive loss with SimCLR contrastive loss, and we measure the trajectory prediction performance. The results demonstrate the impact of utilizing contextual information in form of action on the future trajectory prediction model.

Is the Proposed Action-Based Sampling Strategy using a CVAE Effective? The second ablation study analyzes the impact of sampling synthetic trajectories from CVAE on the trajectory prediction model performance. The baseline ABC in Table 4 indicates the trajectory prediction model trained with action-based contrastive loss without using the extra synthetic samples from the learned CVAE. Comparing the quantitative results of ABC+ to the results of ABC highlights the effectiveness of our novel sampling strategy on all datasets.

How Does the Weight of the Contrastive Loss Affect the Results? Finally, we also study the influence of the hyper-parameter β in Eq. 2, which controls the impact of the action-based contrastive loss into the final objective function. As shown in Fig. 3, we obtain the best results on all benchmark datasets by setting β to 0.75.

5 Discussion and Conclusions

We presented a contrastive framework for learning behavior-aware pedestrian trajectory representations. Our proposed framework consists of an action-based contrastive loss, and a novel trajectory sampling technique from a learned distribution of a C-VAE model. The proposed framework significantly improves the performance of trajectory prediction models on three different first-person view benchmarks. Our evaluation results provide evidence that including pedestrian behavior information, in the form of action or activity class in this case, is beneficial for trajectory prediction. Moreover, our results also confirm that our action-based contrastive loss, in conjunction with our sampling strategy, is superior to alternative approaches that also utilize action class information.

This work comes with a number of strengths. First, we ensure our proposed contrastive loss can be easily integrated with commonly used trajectory prediction models. Second, our proposed sampling strategy utilizes readily learned distributions by generative models, such as CVAEs, and it avoids designing data-specific heuristics. This allows for wider range of applications, such as on animal trajectory data. Finally, contrastive learning in general, and our proposed action-based framework, in particular, allow for enhancing the quantities of underrepresented action classes in the data. This line of work may help address the shortage of necessary edge-cases in training datasets, which may be encountered in real-world scenarios.

This work also comes with a limitation. While effective, our proposed action-based contrastive framework requires pedestrian action labels during the training phase only. However, this requirement is mitigated by our new trajectory sampling technique from CVAE, which does not require action labels for generated samples. Making the training scheme semi-supervised in our model. In addition, many modern trajectory datasets are increasingly providing action information. Action prediction tasks from video data achieve high performances [7], and hence can be performed efficiently and reliably as a pre-processing step for our trajectory prediction framework. We deem evaluating such idea as future work.

Acknowledgment. This work was funded by the Deutsche Forschungsgemeinschaft (DFG, German Research Foundation) under Germany's Excellence Strategy - EXC 2002/1 "Science of Intelligence" - project number 390523135.

References

1. Alahi, A., Goel, K., Ramanathan, V., Robicquet, A., Fei-Fei, L., Savarese, S.: Social LSTM: human trajectory prediction in crowded spaces. In: 2016 IEEE Conference on Computer Vision and Pattern Recognition (CVPR), pp. 961–971 (2016). https://doi.org/10.1109/CVPR.2016.110
2. Bhattacharyya, A., Fritz, M., Schiele, B.: Long-term on-board prediction of people in traffic scenes under uncertainty. In: 2018 IEEE/CVF Conference on Computer Vision and Pattern Recognition, pp. 4194–4202 (2018)
3. Bhattacharyya, A., Schiele, B., Fritz, M.: Accurate and diverse sampling of sequences based on a "best of many" sample objective. In: Proceedings of the IEEE Conference on Computer Vision and Pattern Recognition, pp. 8485–8493 (2018)
4. Caron, M., Misra, I., Mairal, J., Goyal, P., Bojanowski, P., Joulin, A.: Unsupervised learning of visual features by contrasting cluster assignments. Adv. Neural. Inf. Process. Syst. **33**, 9912–9924 (2020)
5. Chen, T., Kornblith, S., Norouzi, M., Hinton, G.: A simple framework for contrastive learning of visual representations. In: International Conference on Machine Learning, pp. 1597–1607. PMLR (2020)
6. Chen, X., He, K.: Exploring simple Siamese representation learning. In: IEEE Conference on Computer Vision and Pattern Recognition (CVPR), pp. 15750–15758 (2021)
7. Degardin, B., Proença, H.: Human behavior analysis: a survey on action recognition. Appl. Sci. **11**(18), 8324 (2021). https://doi.org/10.3390/app11188324. www.mdpi.com/2076-3417/11/18/8324
8. Deng, J., Dong, W., Socher, R., Li, L.J., Li, K., Fei-Fei, L.: ImageNet: a large-scale hierarchical image database. In: 2009 IEEE Conference on Computer Vision and Pattern Recognition, pp. 248–255. IEEE (2009)
9. Deo, N., Trivedi, M.M.: Convolutional social pooling for vehicle trajectory prediction. In: 2018 IEEE/CVF Conference on Computer Vision and Pattern Recognition Workshops (CVPRW), pp. 1549–15498 (2018). https://doi.org/10.1109/CVPRW.2018.00196
10. Dwibedi, D., Aytar, Y., Tompson, J., Sermanet, P., Zisserman, A.: With a little help from my friends: nearest-neighbor contrastive learning of visual representations. In: International Conference on Computer Vision (ICCV), pp. 9588–9597 (2021)
11. Grill, J.B., et al.: Bootstrap your own latent: a new approach to self-supervised learning (2020)
12. Gupta, A., Johnson, J., Fei-Fei, L., Savarese, S., Alahi, A.: Social GAN: socially acceptable trajectories with generative adversarial networks. In: 2018 IEEE/CVF Conference on Computer Vision and Pattern Recognition, pp. 2255–2264 (2018)
13. Harwood, B., Kumar, B.G.V., Carneiro, G., Reid, I., Drummond, T.: Smart mining for deep metric learning. In: Proceedings of the IEEE International Conference on Computer Vision (ICCV) (2017)

14. He, K., Fan, H., Wu, Y., Xie, S., Girshick, R.: Momentum contrast for unsupervised visual representation learning. In: Proceedings of the IEEE/CVF Conference on Computer Vision and Pattern Recognition, pp. 9729–9738 (2020)
15. Henaff, O.: Data-efficient image recognition with contrastive predictive coding. In: International Conference on Machine Learning, pp. 4182–4192. PMLR (2020)
16. Hoffer, E., Ailon, N.: Deep metric learning using triplet network. In: Feragen, A., Pelillo, M., Loog, M. (eds.) SIMBAD 2015. LNCS, vol. 9370, pp. 84–92. Springer, Cham (2015). https://doi.org/10.1007/978-3-319-24261-3_7
17. Jahanian, A., Puig, X., Tian, Y., Isola, P.: Generative models as a data source for multiview representation learning. In: International Conference on Learning Representations (2022). https://openreview.net/forum?id=qhAeZjs7dCL
18. Khosla, P., et al.: Supervised contrastive learning. In: Larochelle, H., Ranzato, M., Hadsell, R., Balcan, M.F., Lin, H. (eds.) Advances in Neural Information Processing Systems., vol. 33, pp. 18661–18673. Curran Associates, Inc. (2020)
19. Kim, D., Yoo, Y., Park, S., Kim, J., Lee, J.: Selfreg: self-supervised contrastive regularization for domain generalization. In: Proceedings of the IEEE/CVF International Conference on Computer Vision, pp. 9619–9628 (2021)
20. Kingma, D.P., Ba, J.: Adam: a method for stochastic optimization. arXiv preprint arXiv:1412.6980 (2014)
21. Lee, N., Choi, W., Vernaza, P., Choy, C.B., Torr, P.H., Chandraker, M.: Desire: distant future prediction in dynamic scenes with interacting agents. In: Proceedings of the IEEE Conference on Computer Vision and Pattern Recognition, pp. 336–345 (2017)
22. Liang, J., Jiang, L., Niebles, J.C., Hauptmann, A.G., Fei-Fei, L.: Peeking into the future: predicting future person activities and locations in videos. In: Proceedings of the IEEE Conference on Computer Vision and Pattern Recognition, pp. 5725–5734 (2019)
23. Liu, Y., Yan, Q., Alahi, A.: Social NCE: contrastive learning of socially-aware motion representations. In: Proceedings of the IEEE/CVF International Conference on Computer Vision (ICCV), pp. 15118–15129 (2021)
24. Ma, S., Sigal, L., Sclaroff, S.: Learning activity progression in LSTMs for activity detection and early detection. In: 2016 IEEE Conference on Computer Vision and Pattern Recognition (CVPR), pp. 1942–1950 (2016)
25. Makansi, O., Çiçek, O., Marrakchi, Y., Brox, T.: On exposing the challenging long tail in future prediction of traffic actors. In: Proceedings of the IEEE/CVF International Conference on Computer Vision (ICCV), pp. 13147–13157 (2021)
26. Malla, S., Dariush, B., Choi, C.: Titan: future forecast using action priors. In: Proceedings of the IEEE/CVF Conference on Computer Vision and Pattern Recognition, pp. 11186–11196 (2020)
27. Mangalam, K., et al.: It is not the journey but the destination: endpoint conditioned trajectory prediction. In: Vedaldi, A., Bischof, H., Brox, T., Frahm, J.-M. (eds.) ECCV 2020. LNCS, vol. 12347, pp. 759–776. Springer, Cham (2020). https://doi.org/10.1007/978-3-030-58536-5_45
28. Montes, A., Salvador, A., Pascual, S., Giro-i Nieto, X.: Temporal activity detection in untrimmed videos with recurrent neural networks. In: 1st NIPS Workshop on Large Scale Computer Vision Systems (2016)
29. Van den Oord, A., Li, Y., Vinyals, O.: Representation learning with contrastive predictive coding. arXiv:1807.03748 (2018)
30. Pareek, P., Thakkar, A.: A survey on video-based human action recognition: recent updates, datasets, challenges, and applications. Artif. Intell. Rev. **54**(3), 2259–2322 (2021)

31. Rasouli, A., Kotseruba, I., Kunic, T., Tsotsos, J.K.: Pie: a large-scale dataset and models for pedestrian intention estimation and trajectory prediction. In: International Conference on Computer Vision (ICCV) (2019)
32. Rasouli, A., Kotseruba, I., Tsotsos, J.K.: Are they going to cross? a benchmark dataset and baseline for pedestrian crosswalk behavior. In: Proceedings of the IEEE International Conference on Computer Vision Workshops, pp. 206–213 (2017)
33. Rudenko, A., Palmieri, L., Herman, M., Kitani, K.M., Gavrila, D.M., Arras, K.O.: Human motion trajectory prediction: a survey. Int. J. Robot. Res. **39**(8), 895–935 (2020). https://doi.org/10.1177/0278364920917446
34. Sadeghian, A., Kosaraju, V., Sadeghian, A., Hirose, N., Rezatofighi, H., Savarese, S.: Sophie: An attentive GAN for predicting paths compliant to social and physical constraints. In: Proceedings of the IEEE/CVF Conference on Computer Vision and Pattern Recognition (CVPR) (2019)
35. Sohn, K., Lee, H., Yan, X.: Learning structured output representation using deep conditional generative models. In: Cortes, C., Lawrence, N., Lee, D., Sugiyama, M., Garnett, R. (eds.) Advances in Neural Information Processing Systems, vol. 28. Curran Associates, Inc. (2015)
36. Sun, Y., Tzeng, E., Darrell, T., Efros, A.A.: Unsupervised domain adaptation through self-supervision. CoRR abs/1909.11825 (2019). http://arxiv.org/abs/1909.11825
37. Wang, C., Wang, Y., Xu, M., Crandall, D.J.: Stepwise goal-driven networks for trajectory prediction. IEEE Robot. Autom. Lett. **7**(2), 2716–2723 (2022). https://doi.org/10.1109/LRA.2022.3145090
38. Wu, C.Y., Manmatha, R., Smola, A.J., Krahenbuhl, P.: Sampling matters in deep embedding learning. In: Proceedings of the IEEE International Conference on Computer Vision (ICCV) (2017)
39. Yao, Y., Atkins, E., Johnson-Roberson, M., Vasudevan, R., Du, X.: Bitrap: bi-directional pedestrian trajectory prediction with multi-modal goal estimation. IEEE Robot. Autom. Lett. (RA-L) **6**(2), 1463–1470 (2021)
40. Yao, Y., Xu, M., Choi, C., Crandall, D.J., Atkins, E.M., Dariush, B.: Egocentric vision-based future vehicle localization for intelligent driving assistance systems. In: 2019 International Conference on Robotics and Automation (ICRA), pp. 9711–9717 (2019)
41. Zbontar, J., Jing, L., Misra, I., LeCun, Y., Deny, S.: Barlow twins: self-supervised learning via redundancy reduction. In: International Conference on Machine Learning, pp. 12310–12320. PMLR (2021)

Radatron: Accurate Detection Using Multi-resolution Cascaded MIMO Radar

Sohrab Madani[1(✉)], Jayden Guan[1], Waleed Ahmed[1], Saurabh Gupta[1], and Haitham Hassanieh[2]

[1] University of Illinois Urbana-Champaign, Urbana, USA
smadani2@illinois.edu

[2] EPFL, Ecublens, Switzerland

Abstract. Millimeter wave (mmWave) radars are becoming a more popular sensing modality in self-driving cars due to their favorable characteristics in adverse weather. Yet, they currently lack sufficient spatial resolution for semantic scene understanding. In this paper, we present Radatron, a system capable of accurate object detection using mmWave radar as a stand-alone sensor. To enable Radatron, we introduce a first-of-its-kind, high-resolution automotive radar dataset collected with a cascaded MIMO (Multiple Input Multiple Output) radar. Our radar achieves 5 cm range resolution and 1.2° angular resolution, 10× finer than other publicly available datasets. We also develop a novel hybrid radar processing and deep learning approach to achieve high vehicle detection accuracy. We train and extensively evaluate Radatron to show it achieves 92.6% AP_{50} and 56.3% AP_{75} accuracy in 2D bounding box detection, an 8% and 15.9% improvement over prior art respectively. Code and dataset is available on https://jguan.page/Radatron/.

1 Introduction

Recently, there has been a significant amount of work, from both academia [2, 14, 43, 49] and industry [3, 26, 30, 35], on leveraging millimeter wave (mmWave) radars for imaging and object detection in autonomous vehicles. Millimeter wave radars are relatively cheap and can operate in adverse weather conditions such as fog, smog, snowstorms, and sandstorms where today's sensory modalities like cameras and LiDAR fail [38, 46]. Despite that, today's commercial use of mmWave automotive radars remains limited to unidirectional ranging in tasks like adaptive cruise control and parking assistance. This is mainly due to the fact that radar's angular resolution is extremely low, 100× lower than LiDAR as shown in Fig. 1(b, c), making it difficult to use radar for object detection. As

S. Madani, J. Guan and W. Ahmed—Indicates equal contribution.

Supplementary Information The online version contains supplementary material available at https://doi.org/10.1007/978-3-031-19842-7_10.

S. Avidan et al. (Eds.): ECCV 2022, LNCS 13699, pp. 160–178, 2022.
https://doi.org/10.1007/978-3-031-19842-7_10

a result, prior work aiming to gain semantic understanding directly from low resolution radar heatmaps is only able to coarsely localize objects [8,11,48] or must fuse radar with LiDAR or cameras to enable object detection [3,37]. In this paper, we focus on exploring how well radar performs in object detection tasks and devise techniques to improve its performance.

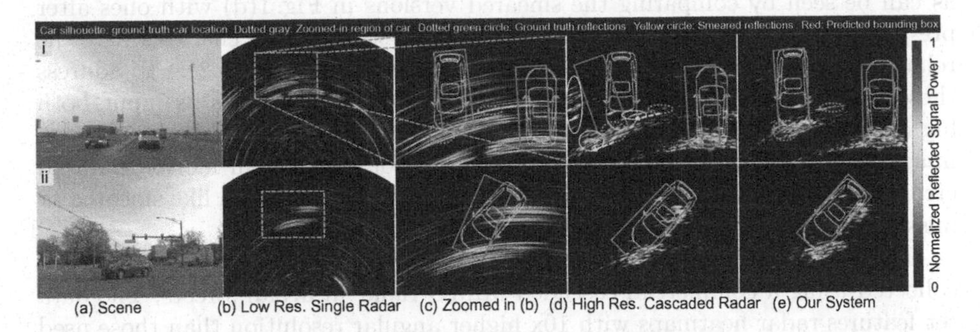

Fig. 1. The low resolution of millimeter wave radar makes it difficult to perform accurate bounding box detection in (c). High resolution cascaded MIMO radars can improve the resolution but suffer from motion smearing in (d). Radatron delivers accurate detection in (e) by combining motion compensation with a two stream deep learning architecture that takes low and high resolution radar images as input.

Improving the angular resolution of conventional radar sensors is challenging. This is because in principle, radar's angular resolution is inversely proportional to the size of the radar antenna aperture [11]. For example, in order to achieve $0.1°$ angular resolution similar to LiDAR [12], we require a 10 m-long aperture consisting of an array of 3000 antennas. The cost, power, and large form factor make such a design prohibitively expensive. An alternative cheaper solution is to use a cascaded MIMO (Multiple Input Multiple Output) radar in which multiple radars are combined to emulate a much larger radar aperture [45,47]. The radars take turns transmitting to avoid interference between the transmitters. Signals from multiple transmitters and receivers are then combined coherently to generate a high resolution image as shown in Fig. 1(d) (for primer on radar, see Sect. 3). This design, however, cannot work well for dynamic scenes like self-driving cars where the different radar transmitters capture snapshots of the scene at slight timing offsets. In vision, such a problem leads to motion blur which can be addressed using a higher frame rate or deblurring techniques [6,42]. Radar, on the other hand, uses mmWave RF signals that travel as sine/cosine waves with millimeter scale wavelength. As a result, even a slight motion of few millimeters can completely change the sign of signal across transmitters which can destructively combine to smear, defocus and even eliminate the object especially as the number of radar transmitters increases. Fig. 1(d.i) shows this effect: reflections in the moving scene get smeared and appear in different locations than where they really are, which leads to inaccurate bounding boxes prediction.

In this paper, we present Radatron, a mmWave radar-based object detection system that can detect precise bounding boxes of vehicles using a cascaded MIMO radar. Radatron overcomes the above challenge by combining a novel radar data pre-processing method with a new deep learning framework. First, we show how to compensate for motion induced errors in pre-processing the raw radar data from a large cascaded MIMO radar. This alleviates most errors, as can be seen by comparing the smeared versions in Fig. 1(d) with ones after pre-processing in Fig. 1(e). The remaining errors stem from scenarios where the relative speed of the cars is high (e.g. incoming cars, see Sect. 4.2). To address these cases, we design a two stream neural network that takes as input both high and low resolution versions of the radar image. Since the low resolution image uses a single radar transmitter, it does not suffer from motion induced errors which allows the network to correct for faulty information like smeared or missed cars that might be mistaken as noise and artifacts.

The paper also introduces a first-of-its-kind high resolution radar data set collected using a commercial cascaded MIMO radar in urban streets. The data set features radar heatmaps with 10x higher angular resolution than those used in prior work [1,11,48], resulting in rich geometric information of objects in the scene, i.e. boundaries and sizes. The data set also includes stereo-camera images which are used for extracting the ground truth and annotating the data. The data set includes 152 k frames representing 4.2 h of driving over 12 days. We also leverage data augmentation to generate significantly more data especially for less common cases (e.g. oriented cars).

We train and extensively evaluate Radatron using our self-collected dataset. Our results show that Radatron improves overall detection accuracy by 8% for AP_{50} and 15.9% for AP_{75} compared to low resolution radars used in prior work [1,11,48]. For hard cases like oriented and incoming cars, Radatron improves overall detection accuracy by upto 14.8% for AP_{50} and 33.1% for AP_{75} compared to low resolution radars, and by upto 13.8% for AP_{50} and 25.2% for AP_{75} compared to a cascaded MIMO Radar without Radatron's pre-processing and two stream network. Besides, we also conducted controlled experiments to qualitatively evaluate Radatron's performance in fog.

Finally, this paper makes the following contributions. First, we demonstrate the ability of achieving accurate vehicle detection using radar by leveraging the high resolution heatmaps captured by cascaded MIMO radars. Second, we propose a network architecture leveraging multi-resolution radar data along with a motion compensation pre-processing algorithm. Third, we collect a high resolution automotive radar dataset with real-world driving scenarios on urban streets using cascaded MIMO radar, which we plan to release once the paper is accepted.

2 Related Work

A. Radar-Based Datasets . Several radar datasets have recently been introduced using single TI chips [10,33,35,49,54], the Navtech CTS350-X radar device [2,8,43], or other low resolution and 1D radar device [9,30]. Unlike these datasets, Radatron uses the cascaded MIMO TI radar which provides an angular resolution of 1.18° in azimuth, 18° in elevation and a range resolution of 5 cm enabling accurate object detection. Additional details of our dataset can be found in Sect. 5. We summarize and compare our data set to other publicly available datasets in Table 1. [2,43] are the closest in terms of resolution but use a mechanically rotating horn antenna which results in a low frame rate 4 Hz, motion smearing that cannot be corrected in pre-processing, and inability to compute velocity from Doppler information in the radar signals.

Low-cost radar has been used with deep learning in applications such as hand-gesture recognition [55], imaging and tracking of the human body [20,56–58], as well as indoor mapping [25]. We focus on using radar for autonomous driving where prior work comprises two groups:

Table 1. Publicly available radar datasets. We only include publicly available data sets with more than 500 frames that provide 2D and 3D radar heatmaps. Hence, data sets like [3,8,14,26,28] are not included. N/A: Not Applicable. N/R: Not Reported.

Dataset	Dim.	Resolution			#Total	#Labeled	Frame	Size	Ground	Radar
		Azi.	Ele	Range	Frames	Frames	Rate		Truth	
Nuscenes [4]	1D/2D	N/R	N/A	N/R	1.3 M	40 K	13 fps	5.5 hrs	LiDAR	N/R
CARRADA [35]	2D[1]	15°	N/A	20 cm	12.7 K	7.2 K	10 fps	21 mins	Camera	AWR1642
CRUW [49]	2D[1]	15°	N/A	23 cm	400 K	N/R[4]	30 fps	3.5 hrs	Camera	AWR1843
OXFORD [2]	2D	1.8°	N/A	17 cm	240 K	0	4 fps	280 km[3]	N/A	CTS350-X
RADIATE [43]	2D	1.8°	N/A	17 cm	200 K	44 K	4 fps	3 hrs	Camera	CTS350-X
Zendar [30]	2D	30°	N/A	18 cm	400 K	11 K	10 fps	11 hrs	LiDAR	N/R
SCORP [33]	3D	15°	30°	12 cm	4 K	4 K	10 fps	6.6 mins	Camera	AWR1843
RADDet [54]	3D	15°	30°	20 cm	10 K	10 K	N/R	Static[2]	Camera	AWR1843
Radatron	3D	1.2°	18°	5 cm	152 K	16 K	10 fps	4.22 hrs	Camera	MMWCAS

[1] The radar in [35,49] can provide 3D data with 30° resolution in elevation. However, the data sets provided are 2D.
[2] The radar is mounted on the side of the road rather than on a moving car.
[3] Driving for 280 km which can correspond to 3 to 10 hrs.
[4] Report 260 K objects but only the center is annotated, not the bounding box.

1. Radar Point Clouds: Learning radar data in the format of point clouds is widely studied [1,7,39,40]. [39,40] demonstrated a semantic segmentation network on radar point clouds while [7] adjusts PointNets [36] for radar data to perform 2D object detection. Pointillism [1] performs 3D bounding box by combining point clouds from multiple spatially separated radars. However, to get point clouds, filtering and thresholding are performed to remove sensor leakage, background clutter, and noise. These hard-coded filtering algorithms lead to the loss of useful information and result in point clouds that are 10 to 100 times sparser than LiDARs [29].

2. Radar Heatmaps: To avoid loss of information, radar data can be processed as heatmaps with range-angle-Doppler tensors [11,26,29,34,54]. In order to learn the 3D radar tensors, past methods collapse the 3D radar tensor onto each dimension separately to extract features, and then concatenate the resulting multi-view feature maps for semantic segmentation [34], object classification and center point detection [11], as well as 2D bounding box detection [26]. Other work feeds the 2D BEV range-angle heatmap into the network as an image [8]. Note that while [8,26] achieved relatively accurate 2D bounding box detection results, their datasets were collected on highways and are not publicly available. Compared to highway driving scenarios, where cars are all moving in the same direction and with similar speeds, our dataset is on urban and suburban streets with more complicated traffic intersections, parked cars on the curbside, and various clutters. In [54], dataset is available but places the radar on the side of the street for traffic monitoring which leads to a poor accuracy of 51.6% AP_{50}. In addition to CNN-based networks, [29] uses graph neural network to achieve a 69% AP_{50} but their data and code are not available. Complementary features of multi-sensor data along with the added redundancy has encouraged previous work to combine different sensors. In particular, Radar and LiDAR fusion has been studied in [37,41,53] while radar and monocular camera fusion has also been studied in [5,18,19,21,24,31]. In this work, we focus on radar as a standalone sensor and aim to show the capabilities of high resolution radar in detecting objects with high accuracy, even in urban and dynamic scenarios.

3 Background on mmWave MIMO Radar

Millimeter wave radars transmit FMCW (Frequency Modulated Continuous Wave) chirps to sense the environment. The chirps emitted from transmitter antenna (TX) reflect off objects in the scene which are then captured by the receiver antenna (RX). By comparing the transmitted and received chirp, we can estimate the round-trip Time-of-Flight (ToF) τ, and hence the ranges of the reflectors $\rho = \tau c/2$ (c denotes the speed of light) in the scene. This is the technique used in today's commercial vehicles that perform radar ranging. Ranging alone, however, is not sufficient to localize objects. One step further is to use a radar with multiple RX antennas that all receive the reflected chirp. The minute ToF differences $\Delta\tau_{ij} = \tau_i - \tau_j$ between these received versions can be exploited to estimate the angle from which the reflections arrive (denoted by ϕ) [17]. The pair (ρ, ϕ) creates a radar heatmap that localizes objects in the 2D polar coordinate.

For this technique to be viable for applications such as semantic scene understanding and object detection, we need to consider the resolution of the radar, which is closely tied to hardware configuration: the range resolution is proportional to the bandwidth of the FMCW chirp, while the angular resolution is proportional to the number of RX antennas. Thanks to the high bandwidth in the mmWave band, mmWave radars achieve cm-level ranging resolution, which is sufficient for most applications. However, reaching an acceptable angular resolution is much more difficult. For instance, to achieve the same angular resolution as a commercial LiDAR, we would need to build a radar with hundreds of RX antennas, which is simply impractical due to the hardware complexity, cost, and power consumption. A much more scalable solution is to use multiple TX as well as multiple RX antennas, a technique referred to as MIMO radar. In MIMO, each of the N TX antennas take turns to transmit one FMCW chirp, which is then received by all M RX antennas, thereby emulating N×M total *virtual* antennas, while using only N+M *physical* antennas [45]. The received chirps from all N·M virtual antennas are then combined to create the (ρ, ϕ) heatmap of the scene.

While MIMO enables higher angular resolution, it comes at the cost of unique challenges. To understand these challenges, we reiterate that in MIMO, TX antennas each transmit one chirp, and all these chirps jointly contribute to the radar heatmap. As TX antennas need to take turns transmitting, there will be a slight time offset δt_{ij} between when the ith and j^{th} chirp are transmitted. For stationary scenes, such time offsets are harmless since they will not affect the ToF difference $\Delta\tau_{ij}$ between different virtual antennas. However, if the scene moves even by as much as 1 mm ($\sim \frac{\lambda}{4}$ at 77 GHz) during the transmitting interval δt_{ij}, the angle estimation and overall radar heatmap can be significantly distorted. This is because the movement of reflections within δt_{ij} contaminates the ToF differences $\Delta\tau_{ij}$ between different virtual antennas as follows:

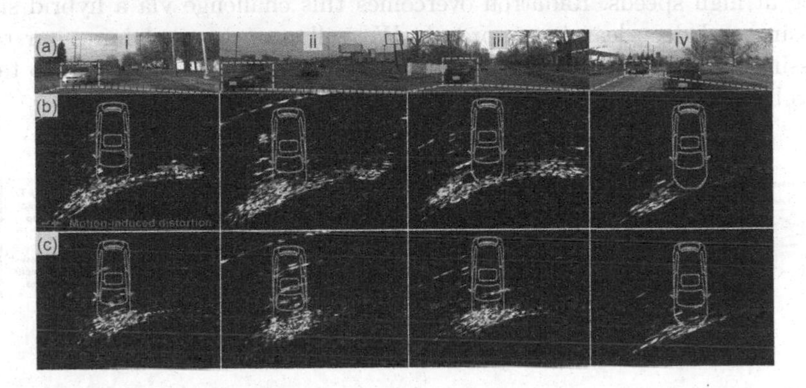

Fig. 2. Motion induced distortion and Radatron's compensation algorithm. (a) Original scene. (b) Bird's-eye view radar heatmap under motion-induced distortion. (c) Processed heatmap after applying Radatron's motion compensation algorithm.

$$\Delta\tau'_{ij} = \tau_i - \tau_j + \delta t_{ij}\frac{2v}{c} = \Delta\tau_{ij} + \delta t_{ij}\frac{2v}{c}, \tag{1}$$

where v is the relative speed of the object in the scene, and c is the speed of light. Note that the motion induced ToF change $\delta t_{ij}\frac{2v}{c}$ cannot be isolated from the angle of arrival dependent ToF difference $\Delta\tau_{ij}$. Consequently, object reflections can get smeared in the radar heatmap, moved into another location, or split into multiple less prominent reflections at different angles. We note that the effect of the error term increases with the speed of the object v, making the problem even more severe for high speed objects. We call this effect the *motion-induced distortion* of the MIMO radar. Fig. 2(b) shows the impact of *motion-induced distortion* in selected range-azimuth radar heatmaps where there is a car moving towards the radar, and we zoom into the region of the incoming car. As one can see, reflections of the car got smeared along ϕ axis, and even split into multiple less prominent reflections appearing at wrong locations away from the car.

4 Method

Our goal is to design a system that can leverage the high resolution cascaded radar as a stand-alone sensor and perform accurate object detection. While the radar heatmaps created using cascaded radar benefit from high angular and range resolution, they come with a set of unique challenges as laid out in Sects. 1 and 3. On the one hand, if we cascade multiple TX antennas to emulate a virtual array with more antenna elements, we can maximize the angular resolution and minimize leakages due to sparsity in the antenna array. However, the transmit time offsets between different TX antennas can cause *motion-induced distortion* (Sect. 3), and the resulting radar heatmap will be smeared. This issue is particularly severe for automotive radars since both the radar and the scene are moving at high speeds. Radatron overcomes this challenge via a hybrid signal processing and deep learning approach. We will start by explaining our radar processing solution, and then proceed to describe our network design to tackle this problem.

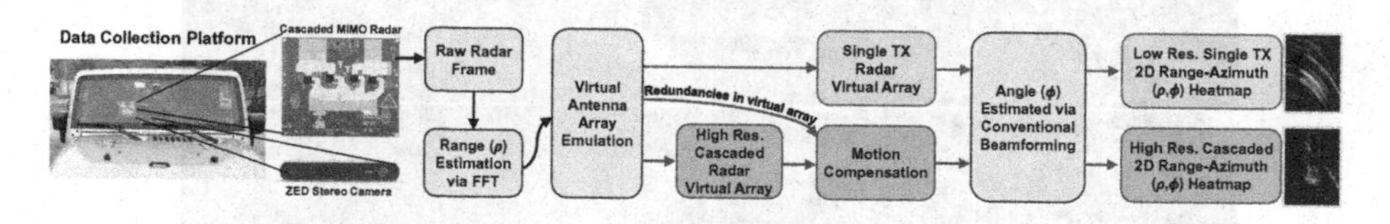

Fig. 3. Radatron's data pre-processing pipeline.

4.1 Radar Signal Processing

On the signal processing end, we design a *motion compensation* algorithm and integrate it into our radar processing pipeline as shown in Fig. 3. It takes the raw radar signal samples as input, and first applies a standard fast Fourier transform to the time-domain signal, which estimates the reflected power from different ranges. Then, before estimating the angles of reflections to localize the objects, we first compensate for the motion-induced distortion. To do so, we leverage the fact that the emulated virtual antenna array has some *redundancies*; that is, there are some co-located virtual antennas pairs. For the co-located virtual antennas i and i', the estimated ToF difference becomes $\tau_i - \tau_i' + \delta t_{ii'} \frac{2v}{c}$, where $\delta t_{ii'}$ represents the TX interval between co-located virtual antenna pairs. Note that $\tau_i = \tau_i'$ for co-located antennas and they cancel out. Therefore, the measured ToF difference between antenna i and i' is the motion-induced ToF variance: $\delta t_{ii'} \frac{2v}{c}$. As the only unknown in this equation is the speed of the object v, we can estimate v, and therefore the motion-induced variance. We then compensate for the estimated motion-induced variance by adding opposite values to all TX antennas. We explain our algorithm in more detail in the supplementary material. Fig. 2(c) shows the intermediate motion compensation results, where the smearing artifacts are mostly corrected, and the reflections overlap well with the ground truth location of the car. After compensating for the motion-induced variance, we then utilize the corrected $\delta\tau$ among non-overlapping virtual antennas to extract the angular information of the reflections. We use the Conventional Beamforming algorithm [27] that outputs a 2D range-azimuth (RA) radar heatmap of the scene in the polar coordinates, where the pixel values represent the reflected signal power. We use this radar signal processing pipeline to create two types of inputs for the network:

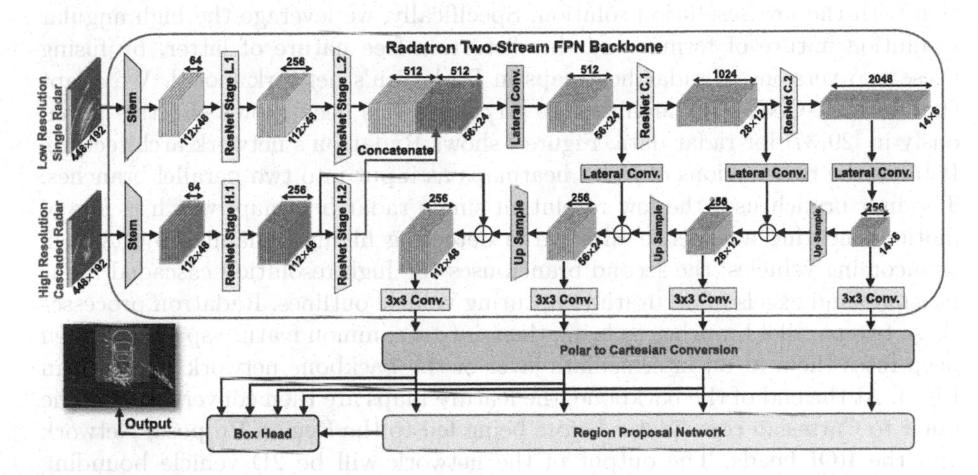

Fig. 4. Radatron's network architecture. We combine two branches of high resolution and low resolution radar data in an intermediate layer. For each feature map the number of channels and dimensions is indicated above and below it respectively.

High Resolution Cascaded Radar: The radar heatmap is created using a uniform 86×1 virtual antenna array, emulated with multiple TX antennas.[1] It features the high azimuth resolution achieved using our cascaded MIMO radar.

Low Resolution Single Radar: Instead of using multiple TX antennas, here we only use one TX antenna with all the RX antennas to emulate a non-uniform 16×1 virtual antenna array, so motion compensation is not needed and hence skipped. This processing pipeline approximately reduces the angle resolution by half and introduces leakage artifacts.

[1] We describe the virtual antenna array emulation in the supp. material.

4.2 Radatron's Network Design

Although our motion compensation algorithm can alleviate the motion-induced distortions to some extent, it is not perfect. Specifically, the algorithm fails in cases of high speed incoming cars, and there will be residual distortions even after applying the motion compensation algorithm. For example, in Fig. 2(c.iv), although after compensation the reflection is centered at the location of the car, it's still smeared across a wider range of angles. To deal with these residual distortions, one potential solution would be to cascade M RX antennas with a single TX antenna. As we use only one TX here, the radar heatmap does not suffer from any motion-induced distortion. However, the virtual antennas in the low resolution version are a sparse subset of the complete N·M virtual array. This results in a heatmap with lower resolution and more leakages, as shown in Fig. 3. Using this heatmap alone as a solution is therefore not sufficient.

In order to get the best of both worlds, Radatron combines the high resolution with the low resolution solution. Specifically, we leverage the high angular resolution nature of former and the distortion-free nature of latter, by fusing these two versions of radar heatmaps in Radatron's network model. We adapt the Faster R-CNN FPN architecture [51] which has been shown effective previously in [29,37] for radar data. Figure 4 shows Radatron's network architecture. It takes the two versions of radar heatmaps as input into two parallel branches: The first branch uses the low resolution single radar heatmap, which is free of motion-smearing and hence effective in detecting highly dynamic objects such as incoming vehicles; the second branch uses the high resolution cascaded radar heatmap and excels in accurately capturing vehicle outlines. Radatron processes these two parallel branches to bring them into a common feature space and then deep-fuses them at an intermediate layer of the backbone network as shown in Fig. 4. At the end of the backbone, the feature maps are then converted from the polar to Cartesian coordinates before being fed to the Region Proposal Network and the ROI heads. The output of the network will be 2D vehicle bounding boxes. We will now explain each part of Radatron's network in more detail.

Radatron's Backbone: For the backbone, we adapt an FPN-based architecture. We process the two input heatmaps to have the same dimension, and feed them into two identical branches. Each of the two branches first goes through a stem layer which consists of a 7×7 Conv. layer, ReLU non-linearity [32] and BatchNorm [16]. Each branch then goes through two ResNet stages, which are the same ones used as the building blocks of ResNet50 [15]. We then combine the two branches by concatenating their feature maps of the same dimension across channels, and fuse them by applying a 3×3 Conv. layer. We further encode the feature maps by passing them through ResNet stages, and combine them to create the feature maps similar to [51].

Coordinate Conversion: Compared to the Cartesian coordinate, the polar coordinate is more natural to radar data as radar has uniform resolution across range and angle. It is also easier for a convolutional network to learn radar artifacts like side lobe leakages in the polar coordinates as they appear parallel to the range and angle coordinates, but extend in a circular fashion in the Cartesian coordinates. On the other hand, bounding boxes work naturally with Cartesian coordinates. We therefore feed in the radar data in the polar coordinates to Radatron's backbone network, and at the end of the backbone explicitly map the features from polar to Cartesian coordinates using bilinear interpolation and before feeding it to the RPN and ROI heads.

RPN and ROI Head: As described earlier, the output feature maps of the backbone are converted from polar to Cartesian coordinates before being fed into the network. We adopt the RPN and ROI architecture in [51] and add oriented boxes. Implementation details can be found in Sect. 6.

Data Augmentation: We applied two forms of data augmentations in training:

A. Flipping in Angle. The input heatmap is flipped along the angle axis. In normal driving scenarios, most incoming cars appear on only one side of the ego vehicle, and flipping azimuth angles eliminates such inherent bias in the dataset.

B. Translation in Angle. We translate the input heatmap along the angle axis. This transformation is similar to one in [11], with the difference that we perform circular shift in angle; i.e., the angles outside the field of view wrap around and fill in the resulting blank space after translation. As most other vehicles appear straight with respect to the ego vehicle, this helps create more oriented cars.

5 Radatron Dataset

Data Collection Platform: Our data collection platform consists of a TI-MMWCAS cascaded MIMO radar [45] and a ZED stereo camera [44] as shown in Fig. 3. Our radar data features high resolution in both range and angle. Our hardware cascades four TI radar chips, with 3 TX and 4 RX antennas each similar to the ones used in prior work [1,11,48], into a 12 TX and 16 RX MIMO radar system. This cascaded MIMO radar can emulate a large virtual antenna array with up to 192 antenna elements, which provides us with 1.2° azimuth resolution and and 18° elevation resolution. We transmit FMCW radar signals at 77 GHz with 3 GHz bandwidth, yielding a range resolution of 5 cm. We show more details on our radar hardware in the supplementary material.

We drove with our data collection platform in diverse scenarios including campus road, our local urban streets, and downtown area of a nearby major city over 12 days. Each day, we conducted four 20 m data collection sessions, during which we streamed data with a frame rate of 10 FPS. Then we further refined the data and filtered out empty frames with no objects. Our final dataset consists of 152 K frames translating into a duration of 4.2 h. Note that although

Radatron's network only takes 2D range-azimuth heatmap as the input, the raw radar data in our dataset also contains elevation and Doppler information. For operator safety and numerical evaluation need, our dataset was collected in clear weather, but we expect the results to hold in tough weather, as vast prior work have shown that radar works well in fog, rain, and snow [2,50,52]. As a initial verification, we conducted controlled fog experiments to qualitatively evaluate Radatron's performance in fog.

6 Evaluation and Experiments

Evaluation Metrics . We use Average Precision (AP) as our main metric to evaluate Radatron's detection performance, following recent work [29,37] in radar object detection, using Intersection over Union (IoU) thresholds values of 0.5, and 0.75. We also use the mean AP (mAP) of IOU values from 0.5 to 0.95 with 0.05 steps. We follow the COCO framework [23] to evaluate Radatron.

Baselines . We compare with the following baselines:

A. Radar used in Prior Work: We implement a virtual array equivalent to the radar used in recent radar datasets [10,11,22,33,35,43,49].

B. Stand-alone Single Radar TX: We trim Radatron's network to parse one TX antenna only, which is equivalent to having stand-alone top stream in Fig. 4.

C. Stand-alone Cascaded Radar: We process the Cascaded radar data with high resolution but bypass our motion compensation algorithm, and feed it into stand-alone bottom stream in Fig. 4.

Table 2. Performance against baselines. Best performing model is boldfaced. Str. stands for straight. Ori. stands for oriented. Inc. stands for incoming.

Eval Metric	AP 50 (%)				AP 75 (%)				mAP (%)			
Model / Split	str.	ori.	inc.	overall	str.	ori.	inc.	overall	str.	ori.	inc.	Overall
Radar in Prior work	88.6	73.9	69.4	84.6	45.0	24.0	24.6	40.4	47.3	34.4	31.2	44.2
Stand-alone single-TX	92.4	77.6	74.3	88.9	50.2	31.6	33.6	46.4	51.4	36.6	37.6	48.4
Stand-alone cascaded	87.7	80.9	65.9	84.6	42.9	31.9	26.2	39.8	45.5	38.1	30.9	43.2
Radatron (multi-res)	**95.6**	**88.7**	**79.7**	**92.6**	**56.3**	**57.1**	**38.2**	**56.3**	**53.8**	**53.1**	**41.4**	**53.8**

D. MVDNet: We train the radar-only version of MVDNet [37] using our dataset. We implement this baseline once with and once again without our compensation algorithm. As MVDNet only accepts one input, we compare it with the "high-res only" version of Radatron.

Radatron Variants. We implement three different variants of Radatron:

A. Radatron (No Compensation): We remove the motion compensation algorithm (4.1) from the signal processing pipeline.

B. Radatron (High-res Only): We remove the *top branch* from Fig. 4 and only feed in the high-resolution processed radar data through the bottom branch.

C. Radatron(Multi-res): We perform the motion compensation algorithm and use both branches with high- and low-resolution processed radar data in Fig. 4.

Dataset Split. Out of 152K overall frames, we manually annotate 16K frames following Sect. 5. We split the dataset into train and test sets by a 3 to 1 ratio. The set of days from which train and test frames were chosen were disjoint.

Test Set Split . To show Radatron's performance under different difficulty scenarios following Sects. 4.1 and 4.2, we split vehicles of the test set into 3 categories:

1. *straight*: Any vehicle on the same lane with an orientation within $\pm 5^{\circ}$.
2. *oriented*: Any vehicle whose orientation is out of the $\pm 5^{\circ}$ range.
3. *incoming*: Any vehicle on the opposite lane, moving towards the ego vehicle.

The *straight* vehicles are relatively easy to detect even using low resolution radars. However, for *oriented* vehicles, high resolution radar is required to accurately detect their angle with respect to the ego vehicle. Finally, *incoming* vehicles tend to get missed by the high resolution heatmap due to the motion induced distortions, as explained in Sect. 4. Instead, our partial cascade radar will pick up the incoming cars when the high resolution heatmap fails. Our test set includes 2854 straight, 327 oriented, and 512 incoming cars.

6.1 Performance Against Baselines

We first compare Radatron with the prior work radar baseline which uses radar heatmaps used by previous art. As seen in Table 2 , Radatron outperforms the prior work radar baseline consistently across all evaluation metrics. This proves empirically that the higher angular resolution of our radar data indeed improves the vehicle detection task. We highlight that while their difference in the overall AP_{50} is around 8%, for the harder cases of oriented cars, Radatron outperforms the baseline by as much as 14.8% in the AP_{50} metric. The gap in performance becomes even more prominent for AP_{75}, where Radatron outperforms the prior work radar baseline by as much as 15.9% overall and 33.1% for oriented cars. The same trend is also seen using the mAP metric. We attribute this performance gap to our motion compensation algorithm, multi-resolution network, and high angular resolution of our dataset. For example, as shown in Fig. 1, one can visually make out the outline of a vehicle by only looking at the radar heatmaps of Radatron, while the prior work radar baseline only roughly localizes the car.

This also explains increased performance gap for the harder cases of oriented cars, and for the higher IoU thresholds.

We next compare Radatron with the other two baselines to show the impact of the our compensation algorithm (Sect. 4.1) as well as our fusion network (Sect. 4.2) on Radatron's performance. We state few points. First, in AP_{50}, Radatron outperforms the single-TX and cascaded baseline baselines by 3.7% and 8% respectively. For AP_{75}, the margin jumps to 9.9% and 16.5% respectively. This indicates that Radatron is better able to capture the harder cases compared to the two baselines. Second, Radatron outperforms the single-TX baseline in the oriented cars significantly, by 11.1% and 25.5% in AP_{50} and AP_{75} respectively. This is in line with our expectation from Sect. 4.2, as the low-resolution and high leakage of single-TX makes it difficult to find the vehicle orientation. Finally, for the incoming cars, Radatron outperforms the cascaded baseline by large margins of 13.8% and 12% for AP_{50} and AP_{75} respectively. This confirms our hypothesis in Sect. 4.2 and 4.1, as the lack of motion compensation algorithm severely distorts the cascaded baseline, as shown in Fig. 2(b).

Finally, we compare Radatron with a radar-only version of MVDNet in Table 4. As seen, Radatron outperforms MVDNet by large margins. We note, however, that our motion compensation algorithm helps improve MVDNet's performance, especially for incoming cars by 38.2%. However, Radatron still outperforms MVDNet by 4.7%, 9.5%, and 31.2% for the three categories respectively in AP50. We attribute this improvement to the use of the ResNet FPN backbone and the polar to cartesian conversion in Radatron.

Table 3. Performance of Radatron's variants. Best performing model is boldfaced. Str. stands for straight. Ori. stands for oriented. Inc. stands for incoming.

Eval metric		AP 50 (%)				AP 75 (%)				mAP (%)			
Model	Split	str.	ori.	inc.	overall	str.	ori.	inc.	overall	str.	ori.	inc.	Overall
Radatron (no comp.)		93.3	84.6	78.9	91.1	49.9	40.4	37.3	46.9	51.3	43.9	40.6	49.1
Radatron (high-res only)		94.7	**90.7**	73.1	92.4	**61.4**	56.3	34.6	**57.1**	**56.6**	52.3	37.6	**53.9**
Radatron (multi-res)		**95.6**	88.7	**79.7**	**92.6**	56.3	**57.1**	**38.2**	56.3	53.8	**53.1**	**41.4**	53.8

Table 4. Comparison with prior work with best results boldfaced.

Eval metric		AP 50 (%)			AP 75 (%)		
Ablation	Split	str	ori	inc	str	ori	inc.
MVDNet (w/o our comp.)		62.5	62.9	3.7	18.1	22.0	0.6
MVDNet (w/ our comp.)		90.5	81.2	41.9	43.8	39.1	8.0
Radatron (high-res only)		**94.7**	**90.7**	**73.1**	**61.4**	**56.3**	**34.6**

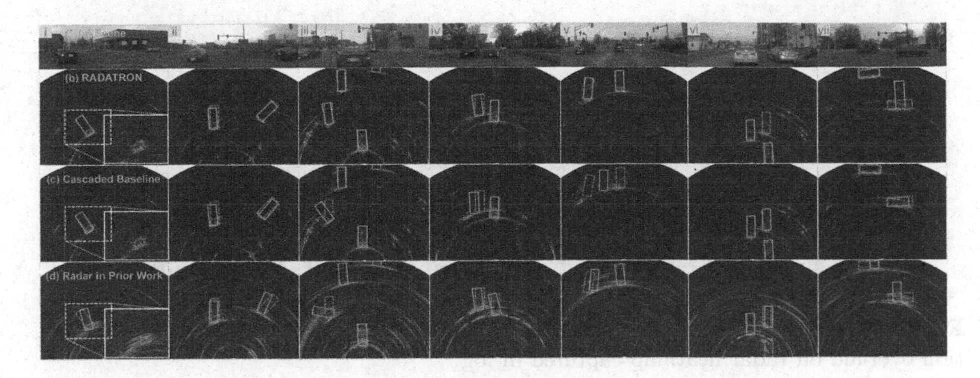

Fig. 5. Examples from our test set. Ground truth marked in green and predictions in red. (a) Original scene. Row (b) shows Radatron's performance overlaid on distortion compensated radar heatmaps. Row (c) and (d) show the performances of stand-alone cascaded and radar in prior work baselines along with their input heatmaps respectively.(Color figure online)

6.2 Radatron's Performance

We now analyze the performance of three different variants of Radatron defined earlier in this section. The results are shown in Table 3. The *multi-resolution* model outperforms the *no compensation* model by 1.5% and 9.4% in AP_{50} and AP_{75} respectively, which means that the multi-res architecture alone without the motion compensation algorithm will not perform well enough, especially for the harder cases, like high-speed incoming cars. On the other hand, the *multi-resolution* model also outperforms *high-resolution only* for incoming cars by 6.6% and 3.6% respectively, which further shows that the motion compensation algorithm alone is not sufficient and can be improved upon using the multi-res network. We note, however, that *multi-resolution*'s performance improvement for the high speed incoming vehicles comes with a slight decrease in performance for oriented cars compared to the *high-resolution only* network. We envision that one could come up with smart combination of high-res and multi-res variants of Radatron to improve the results on all metrics.

Ablation Studies: We also perform extensive ablation studies to better study each component of Radatron. Specifically, we perform ablation studies on the impact of data augmentation, using different coordinate systems, fusing the high- and low-res versions at different stages, and using the Doppler information. The detailed results are presentend in the supp. material.

6.3 Qualitative Results

We show example qualitative results from our test set in Fig. 5, by overlaying the predictions (in solid red line) and ground truth bounding boxes (dotted green

Fig. 6. Controlled fog experiment. (a) Original scene. (b) Scene in fog. (c) Prediction overlaid on radar heatmap captured in fog.

line) on top of Radatron's high-resolution input radar heatmaps in row (b). We also compare Radatron's performance against other baselines, and summarize our observations as follows. As the resolution of the radar heatmap improves, the predictions also become more accurate especially for oriented cars. However, even with the same resolution as Radatron's heatmap, the cascaded baseline suffers when the targets are moving with a high relative speed to the radar, e.g. the incoming cars in Fig. 5(c.iii-vi), due to motion-induced distortion as we described in Sect. 3. Through distortion compensation and fusion network, Radatron is able to overcome this challenge and accurate predict incoming cars. We also noticed some typical failure cases for Radatron, which we show in Fig. 5(b.vi-vii). These cases are likely caused by the fusion network falsely trusting the low-resolution branch and trying to resolve non-existing motion distortion. We show more results and failure mode analysis in supplementary material.

Controlled Fog Experiment . Figure 6 shows Radatron's performance in realistic fog emulated using a fog machine with high-density water-based fog fluid, following past work [13,52]. As depicted in the figure, while the cars are not be visible in the RGB image, Radatron can accurately detect cars in the scene.

7 Limitations and Discussion

First, the maximum range of Radatron's radar was configured to 25m to match that of our stereo camera [44]. Hence, our dataset does not include cars beyond 25m. Second, Radatron does not leverage the 3D nature of its high resolution datasets, which could potentially be used to detect 3D bounding boxes. Third, Radatron was trained and tested using data collected in the same country and may not work as well in other locations. Finally, Radatron currently only detecs vehicles but could be expanded to more objects like pedestrians and bikes by annotating these classes. Addressing these limitations is left for future work. Besides, our cascaded radar also provides Doppler information. We have conducted some initial experiments on leveraging this Doppler information. The implementation and results are presented in the supplementary material.

References

1. Bansal, K., Rungta, K., Zhu, S., Bharadia, D.: Pointillism: accurate 3D bounding box estimation with multi-radars. In: Proceedings of the 18th Conference on Embedded Networked Sensor Systems SenSys 2020, pp. 340–353 (2020)
2. Barnes, D., Gadd, M., Murcutt, P., Newman, P., Posner, I.: The oxford radar robotcar dataset: a radar extension to the oxford robotcar dataset. In: 2020 IEEE International Conference on Robotics and Automation (ICRA), pp. 6433–6438. IEEE (2020)
3. Bijelic, M., et al.: Seeing through fog without seeing fog: deep multimodal sensor fusion in unseen adverse weather. In: Proceedings of the IEEE/CVF Conference on Computer Vision and Pattern Recognition, pp. 11682–11692 (2020)
4. Caesar, H., et al.: Nuscenes: a multimodal dataset for autonomous driving. In: Proceedings of the IEEE/CVF Conference on Computer Vision and Pattern Recognition, pp. 11621–11631 (2020)
5. Chadwick, S., Maddern, W., Newman, P.: Distant vehicle detection using radar and vision (2019)
6. Cho, S., Lee, S.: Fast motion deblurring. In: ACM SIGGRAPH Asia 2009 Papers, pp. 1–8 (2009)
7. Danzer, A., Griebel, T., Bach, M., Dietmayer, K.: 2D car detection in radar data with pointNets. In: 2019 IEEE Intelligent Transportation Systems Conference (ITSC), pp. 61–66 (2019)
8. Dong, X., Wang, P., Zhang, P., Liu, L.: Probabilistic oriented object detection in automotive radar. In: Proceedings of the IEEE/CVF Conference on Computer Vision and Pattern Recognition Workshops, pp. 102–103 (2020)
9. Feng, D., et al.: Deep multi-modal object detection and semantic segmentation for autonomous driving: datasets, methods, and challenges. IEEE Trans. Intell. Transp. Syst. **22**(3), 1341–1360 (2020)
10. Gao, X., Xing, G., Roy, S., Liu, H.: Experiments with mmWave automotive radar test-bed. In: 2019 53rd Asilomar Conference on Signals, Systems, and Computers, pp. 1–6. IEEE (2019)
11. Gao, X., Xing, G., Roy, S., Liu, H.: Ramp-CNN: a novel neural network for enhanced automotive radar object recognition. IEEE Sensors J. **21**(4), 5119–5132 (2021)
12. Geiger, A., Lenz, P., Urtasun, R.: Are we ready for autonomous driving? the kitti vision benchmark suite. In: 2012 IEEE Conference on Computer Vision and Pattern Recognition, pp. 3354–3361. IEEE (2012)
13. Golovachev, Y., Etinger, A., Pinhasi, G., Pinhasi, Y.: Propagation properties of sub-millimeter waves in foggy conditions. J. Appl. Phys. **125**(15), 151612 (2019)
14. Guan, J., Madani, S., Jog, S., Gupta, S., Hassanieh, H.: Through fog high-resolution imaging using millimeter wave radar. In: Proceedings of the IEEE/CVF Conference on Computer Vision and Pattern Recognition (CVPR) (2020)
15. He, K., Zhang, X., Ren, S., Sun, J.: Deep residual learning for image recognition. In: Proceedings of the IEEE Conference on Computer Vision and Pattern Recognition, pp. 770–778 (2016)
16. Ioffe, S., Szegedy, C.: Batch normalization: accelerating deep network training by reducing internal covariate shift. In: International Conference on Machine Learning, pp. 448–456. PMLR (2015)
17. Iovescu, C., Rao, S.: The fundamentals of millimeter wave sensors. In: Texas Instruments, pp. 1–8 (2017)

18. Kim, J., Kim, Y., Kum, D.: Low-level sensor fusion network for 3d vehicle detection using radar range-azimuth heatmap and monocular image. In: Proceedings of the Asian Conference on Computer Vision (ACCV) (2020)
19. Kim, Y., Choi, J.W., Kum, D.: Grif Net: gated region of interest fusion network for robust 3D object detection from radar point cloud and monocular image. In: 2020 IEEE/RSJ International Conference on Intelligent Robots and Systems (IROS), pp. 10857–10864. IEEE (2020)
20. Li, T., Fan, L., Zhao, M., Liu, Y., Katabi, D.: Making the invisible visible: action recognition through walls and occlusions. In: Proceedings of the IEEE/CVF International Conference on Computer Vision, pp. 872–881 (2019)
21. Lim, T.Y., et al.: Radar and camera early fusion for vehicle detection in advanced driver assistance systems. In: NeurIPS Machine Learning for Autonomous Driving Workshop (2019)
22. Lim, T.Y., Markowitz, S.A., Do, M.N.: Radical: a synchronized FMCW radar, depth, IMU and RGB camera data dataset with low-level FMCW radar signals. IEEE J. Select. Topics Signal Process. **15**(4), 941–953 (2021)
23. Lin, T.-Y., Maire, M., Belongie, S., Hays, J., Perona, P., Ramanan, D., Dollár, P., Zitnick, C.L.: Microsoft COCO: common objects in context. In: Fleet, D., Pajdla, T., Schiele, B., Tuytelaars, T. (eds.) ECCV 2014. LNCS, vol. 8693, pp. 740–755. Springer, Cham (2014). https://doi.org/10.1007/978-3-319-10602-1_48
24. Long, Y., Morris, D., Liu, X., Castro, M., Chakravarty, P., Narayanan, P.: Radar-camera pixel depth association for depth completion (2021)
25. Lu, C.X., et al.: See through smoke: robust indoor mapping with low-cost mmWave radar. In: Proceedings of the 18th International Conference on Mobile Systems, Applications, and Services, MobiSys 2020, pp. 14–27. Association for Computing Machinery, New York, USA (2020)
26. Major, B., et al.: Vehicle detection with automotive radar using deep learning on range-azimuth-doppler tensors. In: 2019 IEEE/CVF International Conference on Computer Vision Workshop (ICCVW), pp. 924–932 (2019)
27. Manikas, A.: Beamforming: Sensor Signal Processing for Defence Applications, vol. 5. World Scientific (2015)
28. Meyer, M., Kuschk, G.: Automotive radar dataset for deep learning based 3D object detection. In: 2019 16th European Radar Conference (EuRAD), pp. 129–132. IEEE (2019)
29. Meyer, M., Kuschk, G., Tomforde, S.: Graph convolutional networks for 3D object detection on radar data. In: Proceedings of the IEEE/CVF International Conference on Computer Vision, pp. 3060–3069 (2021)
30. Mostajabi, M., Wang, C.M., Ranjan, D., Hsyu, G.: High-resolution radar dataset for semi-supervised learning of dynamic objects. In: Proceedings of the IEEE/CVF Conference on Computer Vision and Pattern Recognition Workshops, pp. 100–101 (2020)
31. Nabati, R., Qi, H.: Centerfusion: center-based radar and camera fusion for 3D object detection. In: Proceedings of the IEEE/CVF Winter Conference on Applications of Computer Vision (WACV), pp. 1527–1536 (2021)
32. Nair, V., Hinton, G.E.: Rectified linear units improve restricted boltzmann machines. In: ICML (2010)
33. Nowruzi, F.E., et al.: Deep open space segmentation using automotive radar. In: 2020 IEEE MTT-S International Conference on Microwaves for Intelligent Mobility (ICMIM), pp. 1–4. IEEE (2020)

34. Ouaknine, A., Newson, A., Perez, P., Tupin, F., Rebut, J.: Multi-view radar semantic segmentation. In: Proceedings of the IEEE/CVF International Conference on Computer Vision (ICCV), pp. 15671–15680 (2021)
35. Ouaknine, A., Newson, A., Rebut, J., Tupin, F., Pérez, P.: Carrada dataset: camera and automotive radar with range-angle-doppler annotations. In: 2020 25th International Conference on Pattern Recognition (ICPR), pp. 5068–5075. IEEE (2021)
36. Qi, C.R., Su, H., Mo, K., Guibas, L.J.: PointNet: deep learning on point sets for 3D classification and segmentation. In: Proceedings of the IEEE Conference on Computer Vision and Pattern Recognition, pp. 652–660 (2017)
37. Qian, K., Zhu, S., Zhang, X., Li, L.E.: Robust multimodal vehicle detection in foggy weather using complementary lidar and radar signals. In: Proceedings of the IEEE/CVF Conference on Computer Vision and Pattern Recognition (CVPR), pp. 444–453 (2021)
38. Satat, G., Tancik, M., Raskar, R.: Towards photography through realistic fog. In: 2018 IEEE International Conference on Computational Photography (ICCP), pp. 1–10. IEEE (2018)
39. Schumann, O., Hahn, M., Dickmann, J., Wöhler, C.: Semantic segmentation on radar point clouds. In: 2018 21st International Conference on Information Fusion (FUSION), pp. 2179–2186 (2018)
40. Schumann, O., Wöhler, C., Hahn, M., Dickmann, J.: Comparison of random forest and long short-term memory network performances in classification tasks using radar. In: 2017 Sensor Data Fusion: Trends, Solutions, Applications (SDF), pp. 1–6 (2017). https://doi.org/10.1109/SDF.2017.8126350
41. Shah, M., et al.: LiRaNet: end-to-end trajectory prediction using Spatio-temporal radar fusion (2020)
42. Shan, Q., Jia, J., Agarwala, A.: High-quality motion deblurring from a single image. ACM Trans. Graph. (TOG) **27**(3), 1–10 (2008)
43. Sheeny, M., De Pellegrin, E., Mukherjee, S., Ahrabian, A., Wang, S., Wallace, A.: Radiate: a radar dataset for automotive perception. arXiv preprint arXiv:2010.09076 (2020)
44. Stereolabs Inc.: Zed stereo camera (2022). https://www.stereolabs.com/zed/ [Online; Accessed 7 Mar 2022]
45. Texas Instruments Inc.: mmWave cascade imaging radar RF evaluation module (2022). https://www.ti.com/tool/MMWCAS-RF-EVM [Online; Accessed 7 Mar 2022]
46. Times, N.Y.: 5 things that give self-driving cars headaches (2016). https://www.nytimes.com/interactive/2016/06/06/automobiles/autonomous-cars-problems.html
47. Uhnder Inc.: Uhnder - digital automotive radar (2022). https://www.uhnder.com/ [Online; Accessed 7 Mar 2022]
48. Wang, Y., Jiang, Z., Gao, X., Hwang, J.N., Xing, G., Liu, H.: RodNet: radar object detection using cross-modal supervision. In: Proceedings of the IEEE/CVF Winter Conference on Applications of Computer Vision (WACV), pp. 504–513 (2021)
49. Wang, Y., Wang, G., Hsu, H.M., Liu, H., Hwang, J.N.: Rethinking of radar's role: a camera-radar dataset and systematic annotator via coordinate alignment. In: Proceedings of the IEEE/CVF Conference on Computer Vision and Pattern Recognition, pp. 2815–2824 (2021)
50. Waymo: A fog blog (2021). https://blog.waymo.com/2021/11/a-fog-blog.html
51. Wu, Y., Kirillov, A., Massa, F., Lo, W.Y., Girshick, R.: Detectron2 (2019). https://github.com/facebookresearch/detectron2

52. Golovachev, Y., et al.: Millimeter wave high resolution radar accuracy in fog conditions-theory and experimental verification. Sensors **18**(7), 2148 (2018)
53. Yang, B., Guo, R., Liang, M., Casas, S., Urtasun, R.: RadarNet: exploiting radar for robust perception of dynamic objects. In: Vedaldi, A., Bischof, H., Brox, T., Frahm, J.-M. (eds.) ECCV 2020. LNCS, vol. 12363, pp. 496–512. Springer, Cham (2020). https://doi.org/10.1007/978-3-030-58523-5_29
54. Zhang, A., Nowruzi, F.E., Laganiere, R.: Raddet: range-azimuth-doppler based radar object detection for dynamic road users. arXiv preprint arXiv:2105.00363 (2021)
55. Zhang, Z., Tian, Z., Zhou, M.: Latern: dynamic continuous hand gesture recognition using FMCW radar sensor. IEEE Sens. J. **18**(8), 3278–3289 (2018). https://doi.org/10.1109/JSEN.2018.2808688
56. Zhao, M., et al.: Through-wall human pose estimation using radio signals. In: 2018 IEEE/CVF Conference on Computer Vision and Pattern Recognition, pp. 7356–7365 (2018)
57. Zhao, M., et al.: Through-wall human mesh recovery using radio signals. In: 2019 IEEE/CVF International Conference on Computer Vision (ICCV), pp. 10112–10121 (2019)
58. Zhao, M., et al.: Rf-based 3D skeletons. In: Proceedings of the 2018 Conference of the ACM Special Interest Group on Data Communication, SIGCOMM 2018, pp. 267–281. Association for Computing Machinery, New York, USA (2018)

LiDAR Distillation: Bridging the Beam-Induced Domain Gap for 3D Object Detection

Yi Wei[1,2], Zibu Wei[1], Yongming Rao[1,2], Jiaxin Li[3], Jie Zhou[1,2], and Jiwen Lu[1,2](✉)

[1] Department of Automation, Tsinghua University, Beijing, China
{y-wei19,weizb18}@mails.tsinghua.edu.cn, {jzhou,lujiwen}@tsinghua.edu.cn
[2] Beijing National Research Center for Information Science and Technology, Beijing, China
[3] Gaussian Robotics, Shanghai, China

Abstract. In this paper, we propose the LiDAR Distillation to bridge the domain gap induced by different LiDAR beams for 3D object detection. In many real-world applications, the LiDAR points used by mass-produced robots and vehicles usually have fewer beams than that in large-scale public datasets. Moreover, as the LiDARs are upgraded to other product models with different beam amount, it becomes challenging to utilize the labeled data captured by previous versions' high-resolution sensors. Despite the recent progress on domain adaptive 3D detection, most methods struggle to eliminate the beam-induced domain gap. We find that it is essential to align the point cloud density of the source domain with that of the target domain during the training process. Inspired by this discovery, we propose a progressive framework to mitigate the beam-induced domain shift. In each iteration, we first generate low-beam pseudo LiDAR by downsampling the high-beam point clouds. Then the teacher-student framework is employed to distill rich information from the data with more beams. Extensive experiments on Waymo, nuScenes and KITTI datasets with three different LiDAR-based detectors demonstrate the effectiveness of our LiDAR Distillation. Notably, our approach does not increase any additional computation cost for inference. Code is available at https://github.com/weiyithu/LiDAR-Distillation.

Keywords: Unsupervised domain adaptation · 3D object detection · Knowledge distillation

1 Introduction

Recently, LiDAR-based 3D object detection [20,33–35,47,58] has attracted more and more attention due to its crucial role in autonomous driving. While the great

Supplementary Information The online version contains supplementary material available at https://doi.org/10.1007/978-3-031-19842-7_11.

S. Avidan et al. (Eds.): ECCV 2022, LNCS 13699, pp. 179–195, 2022.
https://doi.org/10.1007/978-3-031-19842-7_11

improvement has been made, most of the existing works focus on the single domain, where the training and testing data are captured with the same LiDAR sensors. In many real-world scenarios, we can only get low-beam (*e.g.* 16 beams) point clouds since high-resolution LiDAR (*e.g.* 64 beams) is prohibitively expensive. For example, the price of a 64-beam Velodyne LiDAR is even higher than a cleaning robot or an unmanned delivery vehicle. However, since 3D detectors [20,33,47] cannot simply transfer the knowledge from the high-beam data to the low-beam one, it is challenging to make use of large-scale datasets [9,18,36] collected by high-resolution sensors. Moreover, with the update of product models, the robots or vehicles will be equipped with the LiDAR sensors with different beams. It is unrealistic to collect and annotate massive data for each kind of product. One solution is to use the highest-beam LiDAR to collect informative training data and adapt them to the low-beam LiDARs. Thus, it is an important direction to bridge the domain gap induced by different LiDAR beams.

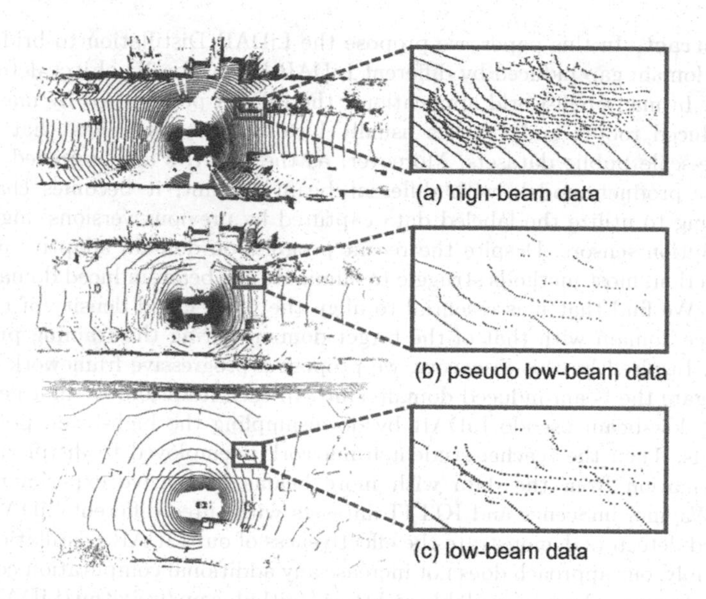

Fig. 1. The beam-induced domain gap. The high-beam data (source domain) and the low-beam data (target domain) are from Waymo [36] and nuScenes [2] datasets respectively. We generate pseudo low-beam data by downsampling high-beam data to align the point cloud density of the source domain to the target domain. The models are trained on this synthetic data with a teacher-student framework.

Since beam-induced domain gap directly comes from the sensors, it is specific and important among many domain-variant factors. Different with research-guided datasets, the training data in industry tends to have similar environments with that during inference, where the environmental domain gaps such as object sizes, weather conditions are marginal. The main gap comes from the beam difference of LiDARs used during training and inference. Moreover, although LiDAR Distillation is designed for beam-induced domain gap, we can easily

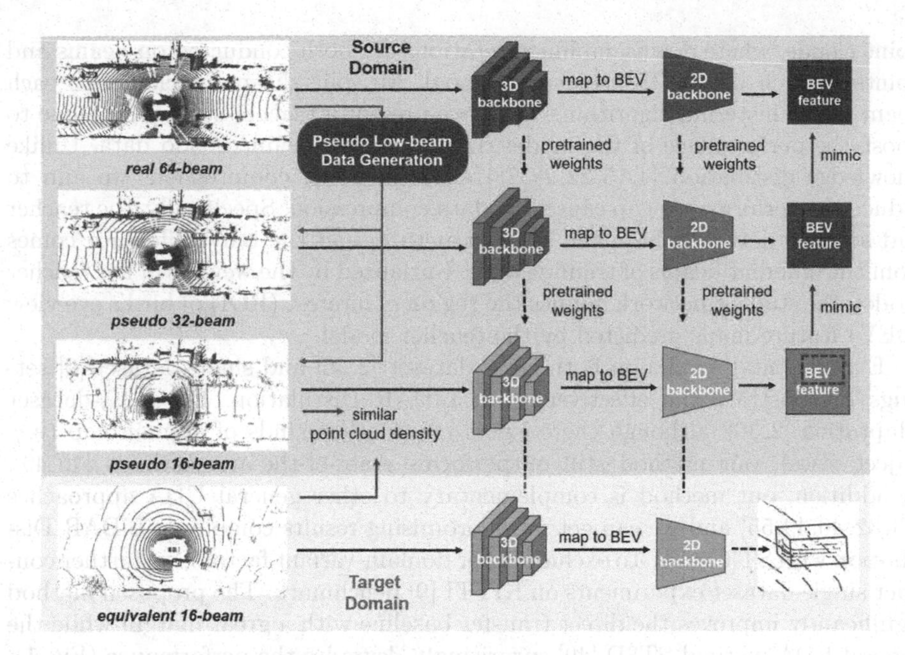

Fig. 2. An overview of LiDAR Distillation. Our method aligns the point cloud density of the source domain with that of the target domain in an progressive way. In each iteration, we first generate pseudo low-beam data by downsampling high-beam point clouds. Then we employ a teacher-student framework to improve the model trained on synthetic low-beam data. Specifically, teacher and student networks are trained on high-beam and low-beam data respectively, which have the same architecture. The student network is initialized with the pretrained weights of the teacher model. As summarized in [37], many 3D detectors employ 3D backbones and 2D backbones, and predict dense BEV features. Based on this fact, we conduct the mimicking operation on the ROI regions of BEV feature maps.

combine our method with other 3D unsupervised domain adaptation methods to solve general domain gaps.

While there are several works [23,32,40,45,49,55] that concentrate on the unsupervised domain adaptation (UDA) task in 3D object detection, most of them aim to address general UDA issues. However, due to the fact that most 3D detectors are sensitive to point cloud densities, the beam-induced domain gap cannot be handled well with these methods. As mentioned in ST3D [49], it is difficult for their UDA method to adapt detectors from the data with more beams to the data with fewer beams. Moreover, since the structure of point clouds is totally different from images, the UDA methods [3,6,21,29–31,44,56,59,60] designed for image tasks are not suitable for 3D object detection.

To address the issue, we propose the LiDAR Distillation to bridge the beam-induced domain gap in 3D object detection task. The key insight is to align the point cloud density and distill the rich knowledge from the high-beam data in a progressive way. First, we downsample high-beam data to pseudo low-beam point clouds, where downsampling operations are both conducted on beams and points in each beam. To achieve this goal, we split the point clouds to each beam via a clustering algorithm. Then, we present a teacher-student pipeline to boost the performance of the model trained on low-beam pseudo data. Unlike knowledge distillation [4,15,22,28,28] used in model compression, we aim to reduce the performance gap caused by data compression. Specifically, the teacher and student networks have the same structure, and the only difference comes from the different beams of training data. Initialized by the weights of the teacher model, the student network mimics the region of interest (ROI) of bird's eye view (BEV) feature maps predicted by the teacher model.

Experimental results on both cross-dataset [2,36] and single-dataset [9] settings demonstrate the effectiveness of LiDAR Distillation. For cross-dataset adaptation [2,36], although there exists many other kinds of domain gap (*e.g.* object sizes), our method still outperforms state-of-the-art methods [40,49]. In addition, our method is complementary to other general UDA approaches [23,32,40,49,55] and we can get more promising results combining LiDAR Distillation with ST3D [49]. To exclude other domain-variant factors, we further conduct single-dataset experiments on KITTI [9] benchmark. The proposed method significantly improves the direct transfer baseline with a great margin while the general UDA method ST3D [49] surprisingly degrades the performance (Fig. 1).

2 Related Work

LiDAR-based 3D Object Detection: LiDAR-based 3D object detection [20,25,33–35,47,50,58] focuses on the problem of localizing and classifying objects in 3D space. It has attracted increasing attention in recent works due to its eager demand in computer vision and robotics. As the pioneer works, [5,19,48] directly project point clouds to 2D BEV maps and adopt 2D detection methods to extract features and localize objects. Beyond these works, as a commonly used 3D detector in the industry due to its good trade-off of efficiency and accuracy, PointPillars [20] leverages an encoder to learn the representation of point clouds organized in pillars. Recently, voxel-based methods [8,33,47,58] achieve remarkable performances thanks to the development of 3D sparse convolution [7,10]. As a classical backbone, SECOND [47] investigates an improved sparse convolution method embedded with voxel feature encoding layers. Combining voxel-based networks with point-based set abstraction via a two-step strategy, PV-RCNN [33] becomes one of the state-of-the-art methods. In our work, we adopt PointPillars [20], SECOND [47] and PV-RCNN [33] to demonstrate the effectiveness of our LiDAR Distillation.

Unsupervised Domain Adaptation on Point Clouds: The goal of unsupervised domain adaptation [3,6,21,29–31,44,52,56,59,60] is to transfer the model trained on source domain to unlabeled target domains. Recently, some works begin to concentrate on the UDA problem in 3D tasks. [26,57] explore object-level feature alignment between global features and local features. Among the

UDA methods for 3D semantic segmentation [17,43,51], [51] tries to solve the domain shift induced by different LiDAR sensors. However, their method adopts 3D convolutions for scene completion, which will bring huge additional computation cost. Moreover, the complete scenes can be easily used in 3D segmentation but they are not suitable for 3D object detection. Compared with great development in 3D object detection, only a few methods [23,32,40,45,49,55] have been proposed for the UDA of 3D detection. Wang *et al.* [40] conclude that the key factor of the domain gap is the difference in car size and they propose SN to normalize the object sizes of the source and target domains. Yang *et al.* [49] propose ST3D which redesigns the self-training pipeline in 2D UDA methods for 3D object detection task. While most of these works target at general UDA methods, we find that they struggle to bridge the beam-induced domain gap.

Knowledge Distillation: Knowledge distillation (KD) [4,15,22,28] has become one of the most effective techniques for model compression. KD leverages the teacher-student framework and distills the knowledge from teacher models to improve the performance of student models. As one of the pioneer works, Hinton *et al.* [15] transfer the knowledge through soft targets. Compared with the KD in classification task [1,14,38,53,54], compressing object detectors [4,11,22,39,41] is more challenging. Chen *et al.* [4] improve the students by mimicking all components, including intermediate features and soft outputs. Beyond this work, Li *et al.* [22] only mimic the areas sampled from region proposals. Recently, Guo *et al.* [11] point out that features from background regions are also important. Different from above methods, we utilize knowledge distillation for data compression (high-beam data to low-beam data) instead of model compression. In addition, few works study the problem of knowledge distillation in 3D object detection. Due to the differences in backbones, it is not trivial to directly apply 2D methods to 3D tasks. In our method, we adopt feature imitation on bird's eye view (BEV) feature maps.

3 Approach

3.1 Problem Statement

The goal of unsupervised domain adaptation is to mitigate the domain shift between the source domain and the target domain and maximize the performance on the target domain. Specifically, in this work, we aim to transfer a 3D object detector trained on high-beam source labeled data $\{(X_i^s, Y_i^s)\}_{i=1}^{N_s}$ to low-beam unlabeled target domain $\{X_i^t\}_{i=1}^{N_t}$, where s and t represent source and target domains respectively. The N_s is the samples number of the source domain while X_i^s and Y_i^s mean the ith source point cloud and its corresponding label. The 3D bounding box label Y_i^s is parameterized by its center location of (c_x, c_y, c_z), the size (d_x, d_y, d_z), and the orientation θ.

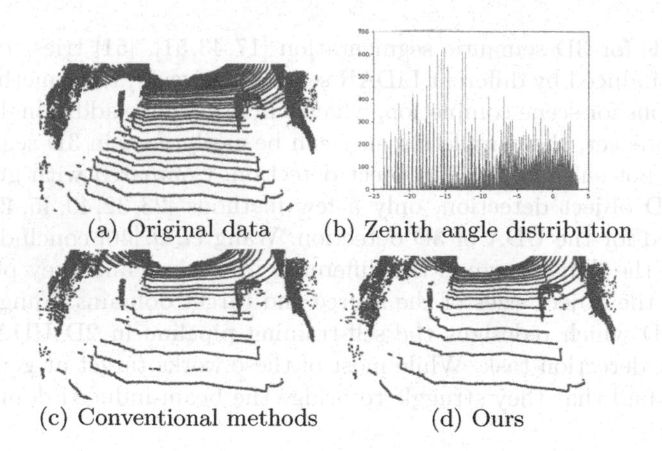

Fig. 3. Pseudo low-beam data generation. **Top row:** (a) original 64-beam LiDAR points. (b) the distribution of zenith angles θ. **Bottom row:** (c) pseudo 32-beam data generated by conventional methods [24,42]. (d) pseudo 32-beam data generated by our clustering method. Although we can see the distinct peaks in the distribution of θ, there also exists some noise. Conventional methods will produce short broken lines with the wrong beam labels while our method can generate more realistic data.

3.2 Overview

We propose the LiDAR Distillation to bridge the domain gap induced by LiDAR beam difference between source and target domains. Figure 2 shows the pipeline of our method. A key factor of our approach is to progressively generate low-beam pseudo LiDAR $\{X_i^m\}_{i=1}^{N_s}$ in the source domain, which has the similar density with the point clouds of the target domain. The density indicates the density of beams and point number per beam. Section 3.3 introduces this part in details. Then we leverage the teacher-student framework to improve the performance of the model trained on pseudo low-beam data, which is discussed in Sect. 3.4. Finally, Sect. 3.5 shows how we progressively distill knowledge from the informative high-beam real point clouds.

We find that the general UDA method [49] will surprisingly degrade the performance when dealing with the domain shift only caused by LiDAR beams. The observation indicates that it is hard to directly transfer the knowledge from the high-beam source domain to the low-beam target domain. Although subsampling high-beam source domain data to low-beam ones will lose rich data knowledge, it guarantees the point cloud density of the source domain is similar to that of the target domain, which is more important compared with data information gain. In this work, we focus on the margin between real high-beam data and generated low-beam data without leveraging the training set in the target domain. Therefore, our approach can be easily integrated with other UDA methods by substituting the source domain with the generated middle domain.

3.3 Pseudo Low-beam Data Generation

The objective of this stage is to close the gap of input point cloud densities between two domains. An alternative solution is to increase the beams of the point clouds in the target domain. This can be easily implemented by upsampling the range images converted from point clouds. However, our experiments show that the upsampled point clouds have many noisy points that heavily destroy objects' features. In this work, we propose to downsample the real high-beam data in the source domain to realize point cloud density's alignment. To this end, we first need to split point clouds into different beams. Although some datasets have beam annotations, many LiDAR points do not have these labels. Here we present a clustering algorithm to get the beam label of each point. We assume the beam number and mean point number in each beam of source and target domains are B_s, P_s and B_t, P_t respectively.

We first transfer cartesian coordinates (x, y, z) to spherical coordinates:

$$\begin{aligned}\theta &= \arctan \frac{z}{\sqrt{x^2+y^2}} \\ \phi &= \arcsin \frac{y}{\sqrt{x^2+y^2}}\end{aligned} \tag{1}$$

where ϕ and θ are azimuth and zenith angles. Figure 3 shows one case in KITTI dataset [9] and the subfigure (b) is the distribution of zenith angles θ. Conventional methods [24,42] assume that beam angles are uniformly distributed in a range (*e.g.* $[-23.6°, 3.2°]$ in KITTI dataset) and assign a beam label to each point according to the distance between its zenith angle and beam angles. However, this does not always hold true due to the noise in raw data. To tackle the problem, we apply K-Means algorithm on zenith angles to find B_s cluster centers as beam angles. As shown in Fig. 3 (c)(d), compared to the conventional methods, our method is more robust to the noise and can generate more realistic pseudo data. Moreover, from Table 2, we can see that the vertical field of view (the range of beam angles) of datasets are different. Intuitively aligning B_s beams to B_t beams is not correct, which will cause different beam densities. We define the equivalent beams B_t' to the source domain as follows:

$$B_t' = [\frac{\alpha_s - \beta_s}{\alpha_t - \beta_t} B_t] \tag{2}$$

where $[\alpha_s, \beta_s]$ and $[\alpha_t, \beta_t]$ represent the vertical field of view in source and target domains. With the beam label of each point, we can easily downsample B_s beams to B_t' beams. Moreover, we should also align P_s to P_t by sampling points in each beam. Instead of random sampling, we sort points according to their azimuth angles ϕ and subsample them orderly.

3.4 Distillation from High-beam Data

Knowledge distillation usually employs a teacher-student framework, where the teacher network is a large model while the student network is a small model. The goal is to improve the performance of the student network by learning the prediction of teacher network. There are two kinds of mimic targets: classification logits and intermediate feature maps. In object detection task, since feature maps are crucial for both object classification and localization, they are the commonly used mimic targets. Formally, regression loss is applied between the feature maps predicted by student and teacher networks:

$$L_m = \frac{1}{N} \sum_i ||f_t^i - r(f_s^i)||_2 \tag{3}$$

where f_t^i and f_s^i are the teacher and student networks' feature maps of the ith sample. r is a transformation layer to align the dimensions of feature maps. During the mimic training, the weights of teacher are fixed and we only train the student network.

Different from knowledge distillation methods aiming at model compression, our goal is to distill the knowledge from the high-beam data $\{X_i^s\}_{i=1}^{N_s}$ to the pseudo low-beam data $\{X_i^m\}_{i=1}^{N_s}$, which can be regarded as a data compression problem. Unlike model compression, our student and teacher models employ the same network architecture. The only difference between them is that the input to the student and teacher models are X^m and X^s respectively. To get higher accuracy, we use the pretrained teacher model's parameters as the initial weights of the student network.

However, compared with 2D detectors, the architectures of 3D detectors are more diverse. To the best of our knowledge, few works study the mimicking methods for 3D object detection. Moreover, it is not trivial to mimic sparse 3D features since the occupied voxels of student and teacher networks are different. We observe that many 3D object detection methods [5,8,19,20,33,47,48] project 3D features to bird's eye view (BEV). Based on this fact, we conduct mimicking operation on dense BEV features.

As mentioned in [22,41], the dimension of feature maps can be million magnitude and it is difficult to perform regression of two high-dimension features. This phenomenon also exists in BEV maps. In addition, many regions on feature maps have weak responses and not all elements are crucial for final prediction. The student network should pay more attention to the features in important regions. Here, we propose to mimic the BEV features sampled from the region of interest (ROI). The ROIs contain both positive and negative samples with different ratios and sizes, which are informative and have significant guidance effects. These ROIs generated by the teacher model will force the student network to learn more representative features. For the 3D detectors which do not use ROIs (*e.g.* PointPillars [20] and SECOND [47]), we can add a light-weight ROI head assigning negative and positive targets during training and remove it during inference.

The loss function that the student network intends to minimize is defined as follows:

$$\begin{aligned} L &= L_{gt} + \lambda L_m^{3D} \\ L_m^{3D} &= \frac{1}{N} \sum_i \frac{1}{M_i} \sum_j ||b_t^{ij} - b_s^{ij}||_2 \end{aligned} \tag{4}$$

where b^{ij} indicates the jth ROI of the ith sample's BEV feature maps. N and M_i represent sample number and ROI number in the ith sample. L_{gt} is the original objective function for 3D object detection and λ is the mimic loss weight. Note that there is no need for transformation layer r since student and teacher networks have the same architecture.

3.5 Progressive Knowledge Transfer

If the beam difference between real high-beam data and pseudo low-beam data is too large (*e.g.* 128 vs 16), the student network cannot learn from the teacher model effectively. To solve the issue, we further present a progressive pipeline to distill the knowledge step by step, which is described in Algorithm 1. For example, if the source domain is 64-beam and the target domain is 16-beam, we first need to generate pseudo 32-beam data and train a student model on it. This student model will be used as the teacher network in the next iteration. Then we downsample both the beams and points per beam of 32-beam data to 16-beam data. We utilize the student model trained on the generated 16-beam data as the final model for the target domain. We downsample beams for 2 times in each iteration since lots of LiDAR data contains 2^k beams.

4 Experiments

4.1 Experimental Setup

Datasets: We conduct experiments on three popular autonomous driving datasets: KITTI [9], Waymo [36] and nuScenes [2]. Specifically, cross-dataset experiments are conducted from Waymo to nuScenes while single-dataset adaptation uses KITTI benchmark. Table 2 shows an overview of three datasets. Following [40,49], we adopt the KITTI evaluation metric for all datasets on the commonly used car category (the vehicle category in Waymo). We report the average precision (AP) over 40 recall positions. The IoU thresholds are 0.7 for both the BEV IoUs and 3D IoUs evaluation. The models are evaluated on the validation set of each dataset.

Implementation Details: We evaluate our LiDAR Distillation on three commonly used 3D detectors: PointPillars [20], SECOND [47] and PV-RCNN [33]. Following [49], we also add an extra IoU head to SECOND, which is named as SECOND-IoU. To guarantee the same input format in each dataset, we only use (x, y, z) as the raw points' features. To fairly compare with ST3D [49], we adopt same data augmentation without using GT sampling. The data augmentation is performed simultaneously in a low-beam and high-beam point clouds pair. We utilize 128 ROIs to sample the features on the BEV feature maps and the ratio of positive and negative samples of ROIs was 1:1. The mimic loss weight is set to 1 to balance multiple objective functions. We run full training epochs for each iteration in the progressive knowledge transfer pipeline, *i.e.* , 80 epochs for KITTI and 30 epochs for Waymo. Our work is built upon the 3D object detection codebase OpenPCDet [37].

Algorithm 1. Progressive Knowledge Transfer

Require: Source domain labeled data $\{(X_i^s, Y_i^s)\}_{i=1}^{N_s}$. The (equivalent) beam number and mean point number in each beam of source and target domains are B_s, P_s and B_t', P_t respectively.

Ensure: The 3D object detection model for target domain.

1: Calculate the iteration n for progressive knowledge transfer: $n = \lfloor \log_2(B_s/B_t') \rfloor$

2: Pretrain 3D detector D_0 on $\{(X_i^s, Y_i^s)\}_{i=1}^{N_s}$.

3: **for** $j = 1$ to n: **do**

4: Use the method introduced in Sect. 3.3 to downsample $\{X_i^s\}_{i=1}^{N_s}$ to $\{X_i^{m_j}\}_{i=1}^{N_s}$ which have $\lceil B_s/2^j \rceil$ beams. If $j == n$, also downsample the mean point number per beam P_s to P_t.

5: Adopt D_{j-1} as the teacher model to train the student model D_j on pseudo low-beam data $\{(X_i^{m_j}, Y_i^s)\}_{i=1}^{N_s}$ with the distillation method described in Sect. 3.4.

6: **end for**

7: D_n is the final model for the target domain.

Table 1. Results of cross-dataset adaptation experiments. **Direct Transfer** indicates that the model trained on the Waymo dataset is directly tested on nuScenes. We report AP_{BEV} and AP_{3D} over 40 recall positions of the car category at IoU = 0.7.

Task	Method	SECOND-IoU	PV-RCNN	PointPillars
		AP_{BEV} / AP_{3D}	AP_{BEV} / AP_{3D}	AP_{BEV} / AP_{3D}
Waymo → nuScenes	Direct Transfer	32.91 / 17.24	34.50 / 21.47	27.8 / 12.1
	SN [40]	33.23 / 18.57	34.22 / 22.29	28.31 / 12.98
	ST3D	35.92 / 20.19	36.42 / 22.99	30.6 / 15.6
	Hegde *et al.*[12]	- / 20.47	- / -	- / -
	UMT [13]	35.10 / 21.05	- / -	- / -
	3D-CoCo [52]	- / -	- / -	33.1 / 20.7
	Ours	40.66 / 22.86	43.31 / 25.63	40.23 / 19.12
	Ours (w/ ST3D)	**42.04 / 24.50**	**44.08 / 26.37**	**40.83/ 20.97**

4.2 Cross-Dataset Adaptation

Table 1 shows the results of cross-dataset experiments. We compare our method with two state-of-the-art general UDA methods SN [40] and ST3D [49]. The reported numbers of these two methods are from [49]. Although our LiDAR Distillation only aims to bridge the domain gap caused by LiDAR beams and there exist many other domain-variant factors (*e.g.* car sizes, geography and weather) in two datasets, our method still outperforms other general UDA methods for a large margin. The improvements over Direct Transfer baseline are also significant. Moreover, our LiDAR Distillation does not use the training set of the target domain and is complementary to other general UDA methods. Take the ST3D as an example: we can simply replace the model pretrained on the source domain with the model pretrained on pseudo low-beam data. From Table 1, we can see that our method greatly boosts the performance of ST3D (AP_{BEV} : 35.92 → 42.04), which demonstrates that point cloud density is the key factor among many domain-variant factors.

Table 2. Datasets overview. We use **version 1.0** of Waymo Open Dataset. **VFOV** means vertical field of view (the range of beam angles). Statistical information is computed from whole dataset.

Dataset	LiDAR Type	VFOV	Points Per Beam
Waymo [36]	64-beam	[-17.6°, 2.4°]	2258
KITTI [9]	64-beam	[-23.6°, 3.2°]	1863
nuScenes [2]	32-beam	[-30.0°, 10.0°]	1084

4.3 Single-Dataset Adaptation

To decouple the influence of other factors, we further conduct single-dataset adaptation experiments, where the domain gap is only induced by LiDAR beams. Different with research-guided datasets, the training data in industry tends to have similar environments with that during inference, where the environmental domain gaps such as object sizes are marginal. The main gap is the beam difference of LiDARs used during training and inference. Thus, it is valuable to investigate the performance drop only caused by this certain domain gap. Unfortunately, the popular autonomous driving datasets [2,9,36] are captured by only one type of LiDAR sensor. To solve the problem, we build the synthetic low-beam target domain by uniformly downsampling the validation set of KITTI. We adopt widely used 3D detector PointPillars [20] in this experiment.

Table 3 shows the results of single-dataset adaptation on KITTI dataset. 32* and 16* mean that we not only reduce LiDAR beams but also subsample 1/2 points in each beam. We do not compare with SN since car sizes are the same in training and validation sets. We can see that our LiDAR Distillation significantly improves Direct Transfer baseline. As the difference between point cloud densities increases, the performance gap becomes larger. Surprisingly, we find that the general UDA method ST3D [49] will degrade the performance. This is mainly due to different point cloud densities in source and target domains, which are not friendly to the self-training framework.

4.4 Ablation Studies

In this section, we conduct sufficient experiments to verify the effectiveness of each proposed component. To exclude the effects of other domain variables, all experiments are conducted under the single-dataset setting with PointPillars [20] backbones.

Component Analysis of Teacher-student Framework: The experiment uses 32-beam data as the target domain. As shown in Table 4, all components contribute to the final performance. While high-beam data is informative and we can get high-performance model, it is hard to directly transfer the knowledge to the low-beam domain. With the alignment of the point cloud density, 3D detectors can learn more robust domain-invariant features in the source domain training set. The distillation from real high-beam data will introduce rich knowledge,

Table 3. Results of single-dataset adaptation on KITTI dataset [9]. **Direct Transfer** indicates that the model trained on the original KITTI training set is directly evaluated on the low-beam validation set. For **32*** and **16***, we not only reduce LiDAR beams but also subsample 1/2 points in each beam. We report AP_{BEV} and AP_{3D} over 40 recall positions of the car category at IoU = 0.7. The improvement is calculated according to Direct Transfer baseline. We mark the best result of each target domain in **bold**.

Target Domain Beams	Method	PointPillars					
		AP_{BEV}			AP_{3D}		
		Easy	Moderate	Hard	Easy	Moderate	Hard
32	Direct Transfer	91.19	79.81	76.53	80.79	65.91	61.09
	ST3D [49]	86.65	71.29	66.77	75.36	57.57	52.88
	Ours	**91.59**	**82.22**	**79.57**	**86.00**	**70.15**	**66.86**
	Improvement	+0.40	+2.41	+3.04	+5.21	+4.24	+5.77
32*	Direct Transfer	88.39	73.56	68.22	74.59	57.77	51.45
	ST3D [49]	83.81	67.08	62.57	71.09	53.30	48.34
	Ours	**90.60**	**79.47**	**76.58**	**82.83**	**66.96**	**62.51**
	Improvement	+2.21	+5.91	+8.36	+8.24	+9.19	+11.06
16	Direct Transfer	83.11	64.91	58.32	67.64	47.48	41.41
	ST3D [49]	75.49	57.58	52.33	60.36	42.40	37.93
	Ours	**89.98**	**74.32**	**71.54**	**80.21**	**59.87**	**55.32**
	Improvement	+6.87	+9.41	+13.22	+12.57	+12.39	+13.91
16*	Direct Transfer	77.22	56.32	49.51	57.36	38.75	32.88
	ST3D [49]	76.04	55.63	49.18	55.17	37.02	31.84
	Ours	**87.44**	**70.43**	**66.35**	**75.35**	**55.24**	**50.96**
	Improvement	+10.22	+14.11	+16.84	+17.99	+16.49	+18.08

which can be maximumly preserved in teacher pretrained weights. To better understand the effects of point cloud density alignment and knowledge distillation, we further conduct experiments of all synthetic target domains in supplementary material, where we find that knowledge distillation plays an important role when the domain gap is small while density alignment is more crucial when the domain gap is large. The experiments also show that our method significantly outperforms point density alignment baseline, i.e. taking the low-beam point clouds as training data, which verify the effectiveness of the proposed teacher-student pipeline.

Effectiveness of Progressive Knowledge Transfer: We further investigate the effectiveness of progressive knowledge transfer. From Table 5, we can see that if the teacher model is trained on original 64-beam data, it cannot provide effective guidance for the student model since there exist large gaps between 64-beam and 16-beam data. Moreover, compared with the teacher model directly trained on 32-beam data, the teacher network obtained from the proposed progressive pipeline gets better results.

Analysis for Mimicking Regions: The experimental result is shown in Table 6. We find that it is not effective to mimic all elements of BEV feature maps. We also try mimicking the regions inside groundtruth bounding boxes.

Table 4. The ablation study of the teacher-student framework. **Density alignment** means that we use generated low-beam data to train the network. **Pretrained** indicates that we employ pretrained teacher weights as the initial weights for the student network. 32-beam data is set as the target domain.

density alignment	knowledge distillation	pretrained	AP_{3D}		
			Easy	Moderate	Hard
			80.79	65.91	61.09
✓			82.80	67.01	63.82
✓		✓	83.05	68.66	64.79
✓	✓		83.71	69.01	64.97
✓	✓	✓	**86.00**	**70.15**	**66.86**

Table 5. The ablation study of progressive knowledge transfer. **64** and **32** represent that the teacher model is trained without progressive distillation. The teacher model obtained from our method is named as **32 progressive**. The result is evaluated on 16-beam data.

teacher model	AP_{3D}		
	Easy	Moderate	Hard
64	78.75	58.80	54.47
32	78.31	58.74	**55.62**
32 progressive	**80.21**	**59.87**	55.32

Table 6. The ablation study of mimicking regions. **all** means that we perform mimicking operation on the whole BEV feature maps while **groundtruth** indicates that we only mimic the regions inside groundtruth bounding boxes.

mimicking regions	AP_{3D}		
	Easy	Moderate	Hard
all	83.83	69.36	64.98
groundtruth	84.46	69.99	65.41
ROI	**86.00**	**70.15**	**66.86**

However, the results are not promising since some background features are also useful. With a good balance of foreground and background information, mimicking ROI regions performs best.

4.5 Further Analysis

The Necessity of Point Cloud Density Alignment: In this part, we study whether the point cloud density is a key factor to bridge the domain gap among many domain-variant factors. Although the data in nuScenes [2] is 32-beam, according to the VFOV in Table 2 and Eq. 2, its equivalent beam in the source domain is 16. In addition, the points per beam of Waymo data is about twice as much as that of nuScenes data. Therefore, 16* beam is the most similar point cloud density compared with target domain data. Experimental results are shown in Table 7, where evaluation results on target and source domain are illustrated in the first and second columns respectively. We observe that we can get better results when point cloud densities of source and target domain become closer. Indeed, both the beams of Waymo and nuScenes obtain non-uniform distribution and our method does not require them to strictly align. While the alignment of the point cloud density will sacrifice the performance on the source domain due to the information loss, it will bring huge improvements on the target domain.

Table 7. The Necessity of Point Cloud Density Alignment. We downsample Waymo [36] data to different beams and evaluate models on nuScenes [2] (target domain) and Waymo [36] (source domain). Notice that the equivalent beam of nuScenes is 16*. The experiment is conducted with SECOND-IoU backbone.

Source Domain	AP_{BEV}/AP_{3D}	
Beams	nuScenes	Waymo
64	32.91/17.24	**67.72/54.01**
32	37.35/20.66	64.67/51.18
16	40.08/21.51	57.04/43.41
16*	**40.66/22.86**	53.23/39.97

Table 8. Comparison with LiDAR upsampling methods. We first convert point clouds of the target domain to range images. Then we upsample range images with bilinear interpolation and a light-weight super-resolution network [16], and convert them back to point clouds. 32-beam KITTI data is used as the target domain.

Method	AP_{3D}		
	Easy	Moderate	Hard
interpolation	74.43	54.56	49.72
super resolution	79.78	62.58	58.04
Ours	**86.00**	**70.15**	**66.86**

Comparison with LiDAR Upsampling Methods: An alternative solution for point cloud density alignment is to upsample the low-beam data in the target domain. Different from scene completion methods [27,46] which utilize 3D convolutions, we do not need to complete the whole scenes but need to process point clouds efficiently. To solve the problem, we first convert point clouds to range images. Then we upsample range images by naive non-parameter bilinear interpolation and a super-resolution network [16]. As shown in Table 8, these two upsampling methods fail to boost the accuracy, which demonstrates that it is not trivial to accurately upsample LiDAR points with high efficiency. This is mainly because there exist some noisy points, which will lead to incorrect object shapes.

5 Conclusion

In this work, we propose the LiDAR Distillation to bridge the domain gap caused by different LiDAR beams in 3D object detection task. Inspired by the discovery that point cloud density is an important factor for this problem, we first generate pseudo low-beam data by downsampling real high-beam LiDAR points. Then knowledge distillation technique is used to progressively boost the performance of the model trained on synthetic low-beam point clouds. Experimental results on three popular autonomous driving datasets demonstrate the effectiveness of our method.

Acknowledgments. This work was supported in part by the National Key Research and Development Program of China under Grant 2017YFA0700802, in part by the National Natural Science Foundation of China under Grant 62125603 and Grant U1813218, in part by a grant from the Beijing Academy of Artificial Intelligence (BAAI).

References

1. Anil, R., Pereyra, G., Passos, A., Ormandi, R., Dahl, G.E., Hinton, G.E.: Large scale distributed neural network training through online distillation. arXiv preprint arXiv:1804.03235 (2018)
2. Caesar, H., et al.: nuscenes: a multimodal dataset for autonomous driving. In: CVPR, pp. 11621–11631 (2020)
3. Chen, C., Zheng, Z., Ding, X., Huang, Y., Dou, Q.: Harmonizing transferability and discriminability for adapting object detectors. In: CVPR, pp. 8869–8878 (2020)
4. Chen, G., Choi, W., Yu, X., Han, T., Chandraker, M.: Learning efficient object detection models with knowledge distillation. In: NeurIPS (2017)
5. Chen, X., Ma, H., Wan, J., Li, B., Xia, T.: Multi-view 3D object detection network for autonomous driving. In: CVPR, pp. 1907–1915 (2017)
6. Chen, Y., Li, W., Sakaridis, C., Dai, D., Van Gool, L.: Domain adaptive faster R-CNN for object detection in the wild. In: CVPR, pp. 3339–3348 (2018)
7. Choy, C., Gwak, J., Savarese, S.: 4D spatio-temporal convnets: Minkowski convolutional neural networks. In: CVPR, pp. 3075–3084 (2019)
8. Deng, J., Shi, S., Li, P., Zhou, W., Zhang, Y., Li, H.: Voxel R-CNN: towards high performance voxel-based 3D object detection. In: AAAI, pp. 1201–1209 (2021)
9. Geiger, A., Lenz, P., Urtasun, R.: Are we ready for autonomous driving? the kitti vision benchmark suite. In: CVPR (2012)
10. Graham, B., Engelcke, M., Van Der Maaten, L.: 3D semantic segmentation with submanifold sparse convolutional networks. In: CVPR, pp. 9224–9232 (2018)
11. Guo, J., et al.: Distilling object detectors via decoupled features. In: CVPR, pp. 2154–2164 (2021)
12. Hegde, D., Patel, V.: Attentive prototypes for source-free unsupervised domain adaptive 3D object detection. arXiv preprint arXiv:2111.15656 (2021)
13. Hegde, D., Sindagi, V., Kilic, V., Cooper, A.B., Foster, M., Patel, V.: Uncertainty-aware mean teacher for source-free unsupervised domain adaptive 3D object detection. arXiv preprint arXiv:2109.14651 (2021)
14. Heo, B., Kim, J., Yun, S., Park, H., Kwak, N., Choi, J.Y.: A comprehensive overhaul of feature distillation. In: ICCV, pp. 1921–1930 (2019)
15. Hinton, G., Vinyals, O., Dean, J.: Distilling the knowledge in a neural network. arXiv preprint arXiv:1503.02531 (2015)
16. Hui, Z., Gao, X., Yang, Y., Wang, X.: Lightweight image super-resolution with information multi-distillation network. In: ACMMM, pp. 2024–2032 (2019)
17. Jaritz, M., Vu, T.H., Charette, R.D., Wirbel, E., Pérez, P.: xMUDA: cross-modal unsupervised domain adaptation for 3D semantic segmentation. In: CVPR, pp. 12605–12614 (2020)
18. Kesten, R., et al.: Lyft level 5 perception dataset 2020 (2019). https://level5.lyft.com/dataset/
19. Ku, J., Mozifian, M., Lee, J., Harakeh, A., Waslander, S.L.: Joint 3D proposal generation and object detection from view aggregation. In: IROS, pp. 1–8 (2018)
20. Lang, A.H., Vora, S., Caesar, H., Zhou, L., Yang, J., Beijbom, O.: Pointpillars: fast encoders for object detection from point clouds. In: CVPR, pp. 12697–12705 (2019)
21. Li, C., et al.: Spatial attention pyramid network for unsupervised domain adaptation. In: ECCV, pp. 481–497 (2020)
22. Li, Q., Jin, S., Yan, J.: Mimicking very efficient network for object detection. In: CVPR, pp. 6356–6364 (2017)

23. Luo, Z., et al.: Unsupervised domain adaptive 3D detection with multi-level consistency. In: ICCV, pp. 8866–8875 (2021)
24. Milioto, A., Vizzo, I., Behley, J., Stachniss, C.: Rangenet++: fast and accurate lidar semantic segmentation. In: 2019 IEEE/RSJ International Conference on Intelligent Robots and Systems (IROS), pp. 4213–4220. IEEE (2019)
25. Qi, C.R., Liu, W., Wu, C., Su, H., Guibas, L.J.: Frustum pointnets for 3D object detection from RGB-D data. In: CVPR, pp. 918–927 (2018)
26. Qin, C., You, H., Wang, L., Kuo, C.C.J., Fu, Y.: Pointdan: a multi-scale 3D domain adaption network for point cloud representation. In: NeurIPS, pp. 7192–7203 (2019)
27. Roldão, L., de Charette, R., Verroust-Blondet, A.: LMSCNet: lightweight multiscale 3D semantic completion. In: 3DV (2020)
28. Romero, A., Ballas, N., Kahou, S.E., Chassang, A., Gatta, C., Bengio, Y.: Fitnets: hints for thin deep nets. arXiv preprint arXiv:1412.6550 (2014)
29. Saito, K., Ushiku, Y., Harada, T.: Asymmetric tri-training for unsupervised domain adaptation. In: ICML, pp. 2988–2997 (2017)
30. Saito, K., Ushiku, Y., Harada, T., Saenko, K.: Strong-weak distribution alignment for adaptive object detection. In: CVPR, pp. 6956–6965 (2019)
31. Saito, K., Watanabe, K., Ushiku, Y., Harada, T.: Maximum classifier discrepancy for unsupervised domain adaptation. In: CVPR, pp. 3723–3732 (2018)
32. Saltori, C., Lathuilière, S., Sebe, N., Ricci, E., Galasso, F.: SF-UDA 3D: source-free unsupervised domain adaptation for LiDAR-based 3D object detection. In: 3DV, pp. 771–780 (2020)
33. Shi, S., et al.: PV-RCNN: point-voxel feature set abstraction for 3D object detection. In: CVPR, pp. 10529–10538 (2020)
34. Shi, S., Wang, X., Li, H.: PointRCNN: 3D object proposal generation and detection from point cloud. In: CVPR, pp. 770–779 (2019)
35. Shi, S., Wang, Z., Shi, J., Wang, X., Li, H.: From points to parts: 3D object detection from point cloud with part-aware and part-aggregation network. arXiv preprint arXiv:1907.03670 (2019)
36. Sun, P., et al.: Scalability in perception for autonomous driving: waymo open dataset. In: CVPR, pp. 2446–2454 (2020)
37. Team, O.D.: OpenPCDet: an open-source toolbox for 3D object detection from point clouds (2020). https://github.com/open-mmlab/OpenPCDet
38. Tung, F., Mori, G.: Similarity-preserving knowledge distillation. In: ICCV, pp. 1365–1374 (2019)
39. Wang, T., Yuan, L., Zhang, X., Feng, J.: Distilling object detectors with fine-grained feature imitation. In: CVPR, pp. 4933–4942 (2019)
40. Wang, Y., et al.: Train in Germany, test in the USA: making 3D object detectors generalize. In: CVPR, pp. 11713–11723 (2020)
41. Wei, Y., Pan, X., Qin, H., Ouyang, W., Yan, J.: Quantization mimic: towards very tiny CNN for object detection. In: Ferrari, V., Hebert, M., Sminchisescu, C., Weiss, Y. (eds.) ECCV 2018. LNCS, vol. 11212, pp. 274–290. Springer, Cham (2018). https://doi.org/10.1007/978-3-030-01237-3_17
42. Wu, B., Wan, A., Yue, X., Keutzer, K.: Squeezeseg: convolutional neural nets with recurrent CRF for real-time road-object segmentation from 3D lidar point cloud. In: 2018 IEEE International Conference on Robotics and Automation (ICRA), pp. 1887–1893. IEEE (2018)
43. Wu, B., Zhou, X., Zhao, S., Yue, X., Keutzer, K.: Squeezesegv 2: improved model structure and unsupervised domain adaptation for road-object segmentation from a lidar point cloud. In: ICRA, pp. 4376–4382 (2019)

44. Xu, C.D., Zhao, X.R., Jin, X., Wei, X.S.: Exploring categorical regularization for domain adaptive object detection. In: CVPR, pp. 11724–11733 (2020)
45. Xu, Q., Zhou, Y., Wang, W., Qi, C.R., Anguelov, D.: SPG: unsupervised domain adaptation for 3D object detection via semantic point generation. In: ICCV, pp. 15446–15456 (2021)
46. Yan, X., Gao, J., Li, J., Zhang, R., Li, Z., Huang, R., Cui, S.: Sparse single sweep lidar point cloud segmentation via learning contextual shape priors from scene completion. In: AAAI, pp. 3101–3109 (2021)
47. Yan, Y., Mao, Y., Li, B.: Second: Sparsely embedded convolutional detection. Sensors **18**(10), 3337 (2018)
48. Yang, B., Luo, W., Urtasun, R.: Pixor: real-time 3D object detection from point clouds. In: CVPR, pp. 7652–7660 (2018)
49. Yang, J., Shi, S., Wang, Z., Li, H., Qi, X.: ST3D: self-training for unsupervised domain adaptation on 3D object detection. In: CVPR, pp. 10368–10378 (2021)
50. Yang, Z., Sun, Y., Liu, S., Shen, X., Jia, J.: STD: sparse-to-dense 3D object detector for point cloud. In: ICCV, pp. 1951–1960 (2019)
51. Yi, L., Gong, B., Funkhouser, T.: Complete and label: a domain adaptation approach to semantic segmentation of lidar point clouds. In: CVPR, pp. 15363–15373 (2021)
52. Yihan, Z., et al.: Learning transferable features for point cloud detection via 3D contrastive co-training. In: NeurIPS, vol. 34 (2021)
53. Yim, J., Joo, D., Bae, J., Kim, J.: A gift from knowledge distillation: fast optimization, network minimization and transfer learning. In: CVPR, pp. 4133–4141 (2017)
54. You, S., Xu, C., Xu, C., Tao, D.: Learning from multiple teacher networks. In: SIGKDD, pp. 1285–1294 (2017)
55. Zhang, W., Li, W., Xu, D.: SRDAN: scale-aware and range-aware domain adaptation network for cross-dataset 3D object detection. In: CVPR, pp. 6769–6779 (2021)
56. Zheng, Y., Huang, D., Liu, S., Wang, Y.: Cross-domain object detection through coarse-to-fine feature adaptation. In: CVPR, pp. 13766–13775 (2020)
57. Zhou, X., Karpur, A., Gan, C., Luo, L., Huang, Q.: Unsupervised domain adaptation for 3D keypoint estimation via view consistency. In: ECCV, pp. 137–153 (2018)
58. Zhou, Y., Tuzel, O.: VoxelNet: end-to-end learning for point cloud based 3D object detection. In: Proceedings of the IEEE Conference on Computer Vision and Pattern Recognition, pp. 4490–4499 (2018)
59. Zhu, X., Pang, J., Yang, C., Shi, J., Lin, D.: Adapting object detectors via selective cross-domain alignment. In: CVPR, pp. 687–696 (2019)
60. Zou, Y., Yu, Z., Vijaya Kumar, B.V.K., Wang, J.: Unsupervised domain adaptation for semantic segmentation via class-balanced self-training. In: Ferrari, V., Hebert, M., Sminchisescu, C., Weiss, Y. (eds.) ECCV 2018. LNCS, vol. 11207, pp. 297–313. Springer, Cham (2018). https://doi.org/10.1007/978-3-030-01219-9_18

Efficient Point Cloud Segmentation with Geometry-Aware Sparse Networks

Maosheng Ye[1], Rui Wan[2], Shuangjie Xu[1], Tongyi Cao[2], and Qifeng Chen[1(✉)]

[1] HKUST, Hong Kong, China
{myeag,sxubj}@connect.ust.hk, cqf@ust.hk
[2] DeepRoute.Ai, Shenzhen, China
{ruiwan,tongyicao}@deeproute.ai

Abstract. In point cloud learning, sparsity and geometry are two core properties. Recently, many approaches have been proposed through single or multiple representations to improve the performance of point cloud semantic segmentation. However, these works fail to maintain the balance among performance, efficiency, and memory consumption, showing incapability to integrate sparsity and geometry appropriately. To address these issues, we propose the Geometry-aware Sparse Networks (GASN) by utilizing the sparsity and geometry of a point cloud in a single voxel representation. GASN mainly consists of two modules, namely Sparse Feature Encoder and Sparse Geometry Feature Enhancement. The Sparse Feature Encoder extracts the local context information, and the Sparse Geometry Feature Enhancement enhances the geometric properties of a sparse point cloud to improve both efficiency and performance. In addition, we propose deep sparse supervision in the training phase to help convergence and alleviate the memory consumption problem. Our GASN achieves state-of-the-art performance on both SemanticKITTI and Nuscenes datasets while running significantly faster and consuming less memory.

1 Introduction

Large-scale outdoor point cloud segmentation has been a crucial task for autonomous driving systems and has demanding requirements for efficiency, performance, and memory consumption. PointNet [29] and PointNet++ [30] are the pioneering works that directly operate on the raw point cloud to maintain and utilize the pointwise geometry (accurate measurement information), which is one of the core properties of a point cloud. However, it is hard to apply these approaches in outdoor scenarios due to memory consumption and runtime efficiency. RandLA [14] applies a random sampling strategy to reduce the number of points to improve efficiency, which leads to some information loss. With the popularity of sparse convolutions [10,50], there has been some progress in utilizing the sparse voxel-based representation (e.g., AF2S3Net [6] and Cylinder3D [59]), which is a kind of representation that preserves the metric space. Compared with point representation, the merits of the sparse voxel-based representation

S. Avidan et al. (Eds.): ECCV 2022, LNCS 13699, pp. 196–212, 2022.
https://doi.org/10.1007/978-3-031-19842-7_12

lie in the effectiveness and efficiency of quickly expanding the receptive fields based on sparsity, another important property of a point cloud. Furthermore, sparse voxel-based representation, which aggregates point features within the local neighborhood, can significantly reduce memory usage. Moreover, traditional or current popular convolutional neural networks (CNNs) can be directly applied to extract better context information.

Recently, several works recognize the limitation of a single representation and explore richer information by combining multiple representations. PVCNN [21] fuses point-based and voxel-based representations with MLP layers and dense 3D convolution layers but does not take point cloud sparsity into consideration. SPVCNN [35] and DRINet [52] design the sparse convolution layers and pointwise operational layers to fuse features considering sparsity and geometry. Furthermore, RPVNet [47] combines the range-view, point, and voxel representations for point cloud segmentation. The general framework of current multi-representation learning is to utilize sparse convolutions for locality and sparsity and pointwise operations for geometry learning, aiming to combine sparsity and geometry for better performance and efficiency. While these approaches lead to some performance improvements, they are not efficient enough to meet the need of a real-time system due to the extra computation cost brought by the additional views or representations. Meanwhile, according to the results of the experiments in these approaches, the voxel-based representation is still the dominant one, with which these methods have already achieved decent performance. Inspired by these observations, we propose the Geometry-aware Sparse Networks to explore extra geometric properties based on a single sparse voxel-based representation. Our Geometry-aware Sparse Networks incorporate sparsity and geometry in a single representation without introducing extra computation costs from multi-representation fusion. Our Geometry-aware Sparse Networks mainly

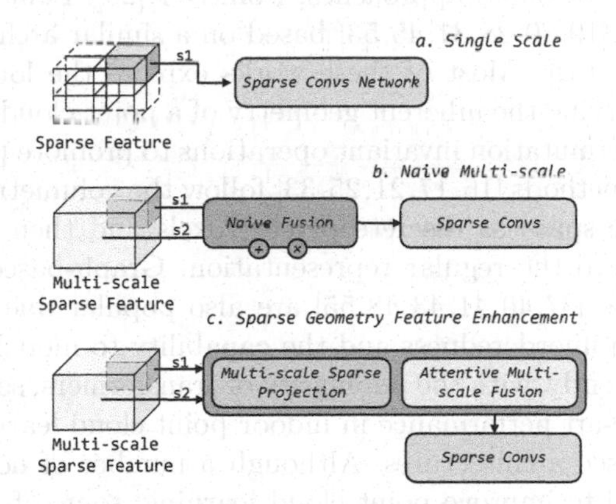

Fig. 1. Comparison between two common ways to deal with sparse features and our proposed method.

have two modules: Sparse Feature Encoder (SFE) and Sparse Geometric Feature Enhancement (SGFE). Each module takes the output of the other module as input to fully explore the sparsity and geometry of a point cloud at a low computation cost and memory usage. In SGFE (shown in Fig. 1), we propose a novel multi-scale sparse projection layer to explore more geometry and Attentive Scale Selection for multi-scale feature selection. Apart from that, we apply deep sparse supervision compared with the most common dense manner to alleviate the pressure of memory consumption.

Our contributions are summarized as follows:

- We propose a novel network architecture to fully exploit the sparsity and geometry properties. Multi-scale sparse projection layer and Attentive Scale Selection in the Sparse Geometric Feature Enhancement are proposed to enhance the geometry for feature learning.
- Deep Sparse Supervision is proposed to design the supervision in a sparse manner to reduce the memory cost.
- We evaluate our proposed approach on large-scale outdoor scenario datasets including SemanticKITTI [1] and nuScenes-lidarseg [3] to demonstrate the effectiveness of our method. We achieve state-of-the-art performance on both datasets with a runtime speed of 59 ms on average on an Nvidia RTX 2080 Ti GPU.

2 Related Work

Indoor Point Cloud Segmentation. The point cloud from an indoor scene often has closely positioned points with a small range. The existing indoor point cloud segmentation approaches can be classified according to their model representations. For point-based approaches, PointNet [29], PointNet++ [29], and their related works [19,20,28,31,45,53] based on a similar architecture are popular models in this task. Most of these works explore the local neighborhood context while preserving the inherent geometry of a point cloud. They use different grouping and permutation invariant operations to promote performance. The other mainstream methods [16,17,21,25,33] follow the volumetric representation by partitioning the space as discrete pixels/voxels and then applying 2D/3D CNN architectures to the regular representation. Graph-based approaches for point cloud learning [37,40,41,43,48,55] are also popular due to the nature of graphs to deal with unorderedness and the capability to model the relationship among points. Currently, with the popularity of transformers, some works [11,56] achieve state-of-the-art performance in indoor point cloud learning by introducing transformer-based architectures. Although a number of novel architectures have been proposed to improve point cloud learning, some of them fail to generalize to outdoor scenarios with thousands of hundreds of points.

Outdoor Point Cloud Segmentation. Compared with indoor point cloud segmentation, the sparsity and larger number of points pose great challenges for existing approaches. Point-based methods such as KPConv [38] and RandLA [14] extend the architecture of PointNet [29] or PointNet++ [30] and adopt sampling strategies to alleviate these problems but lead to extra information loss. KPConv [38] introduces the kernel point selection process to generate high-quality sampling points. Range-view-based approaches [8,44,46] project the point cloud into range views or spherical representations and apply efficient CNN architectures. However, the range view cannot maintain the metric space and introduces distortions, which potentially leads to performance degradation. Some other approaches [6,7,49,54,59] quantize a point cloud into some predefined space or representations (e.g., polar grids, 2D grids, and sparse 3D grids) and then apply regular convolution neural networks or sparse convolutions [10,42,50] to achieve the balance between efficiency and performance. A line of works integrates the multiple representations, including range views, voxel representations, and point representation, to exploit the potential of different representations deeply [21,35,47,52]. These works utilize different architectures for different representations and propose various fusion strategies and show strong performance gain compared to single-representation-based methods, at the cost of extra running time.

Image Segmentation to Point Cloud Segmentation. The fully convolutional network (FCN) [22] is one of the pioneering works for image segmentation with deep learning. Based on FCN and existing prevalent CNN architecture, DeepLab [4], PSPNet [57] and their following works [5,51] are proposed with multi-scale or multiple dilation rate strategies to explore more hierarchical local context information. Furtherly, HRNet [34] fuses different resolution heatmaps in a single framework and keeps the high resolution to improve the performance. Considering the great process achieved in image segmentation, lots of works [26,39,52,54,59] have applied these tricks, including hierarchy learning, attention mechanism, or backbones into point cloud segmentation. Some works [7,59] are built on U-net [32] with sparse convolution acceleration.

3 Approach

The overall network architecture of our approach consists of two modules: 1) Sparse Feature Encoder and 2) Sparse Geometry Feature Enhancement. Sparse Feature Encoder serves as a basic block for fast local context aggregation. While Sparse Geometry Feature Enhancement, which takes the output of Sparse Feature Encoder as input, enhances the geometric information by multi-scale sparse projection and attentive scale selection layer. Both modules interact in sparse space, saving the computation cost and increasing the runtime efficiency. Moreover, we apply Deep Sparse Supervision at the voxel level to alleviate the memory issues resulting from dense supervision. Figure 2 demonstrates the overall framework of our proposed approach.

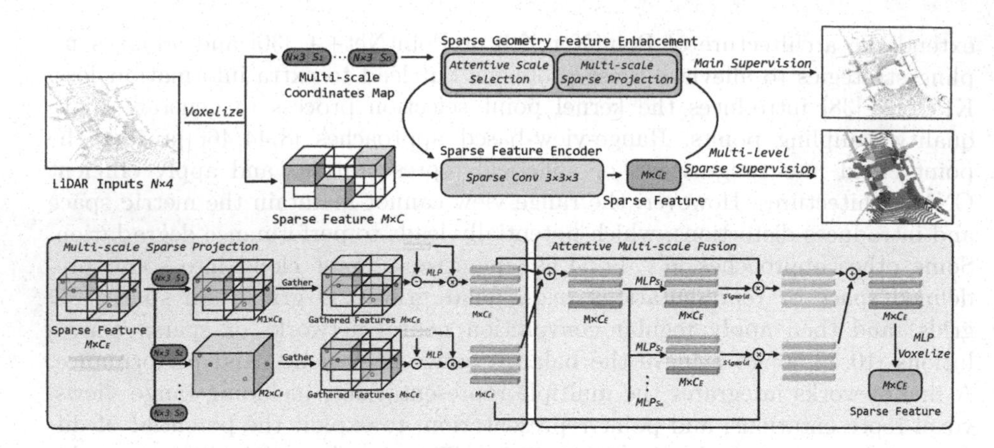

Fig. 2. The overall structure of our GASN. In the top half of the figure, LiDAR input is firstly voxelized as sparse features. Then the sparse feature encoder utilizes sparse convolutions to process the sparse features. Furthermore, sparse geometry feature enhancement will enhance the features by multi-scale sparse projection and attentive scale selection layer to generate the input of sparse feature encoder at the next stage. Sparse supervision will be attached to the output of the sparse feature encoder as an auxiliary loss. The bottom line describes the details of multi-scale sparse projection and attentive scale selection. N is the number of points, M_i is the number of voxels, C_E is the channel dimension.

3.1 Prerequisite

Voxelization. Given a grid size s, voxelization is the process that discretizes a point $p_i = (x_i, y_i, z_i)$ to its voxel index V_i by the following equation:

$$V_i = (\lfloor x_i/s \rfloor, \lfloor y_i/s \rfloor, \lfloor z_i/s \rfloor), \tag{1}$$

where $\lfloor \cdot \rfloor$ is a floor function. V_i is represented as a scalar of the voxel index for the i-th point p_i, and. Since each voxel could contains multiples points or multiple fine-grained sub-voxels, we define a scatter function Ψ and gather function Φ which is widely used in [52,58], where the former performs the clustering process to cluster point features or sub-voxel features $\mathcal{S} \in R^{N_\mathcal{S} \times C}$ to voxel features $\mathcal{V} \in R^{N_\mathcal{V} \times C}$ under larger voxel scale while the latter reverses from voxel features $\mathcal{V}$ to point features or sub-voxel features $\mathcal{S}$:

$$\mathcal{V} = \Psi(\mathcal{S}, V_\mathcal{S}) = \left\{ \sum_j \mathcal{S}_j \left| V_{\mathcal{S}_j} \in V_{\mathcal{V}_i}, i = 1, \ldots, N_\mathcal{V} \right. \right\}, \tag{2}$$

$$\mathcal{S} = \Phi(\mathcal{V}, V_\mathcal{V}) = \left\{ \mathcal{V}_j | V_{\mathcal{S}_i} \in V_{\mathcal{V}_j}, i = 1, \ldots, N_\mathcal{S} \right\}, \tag{3}$$

where C is the channel number, $V_\mathcal{V}$ and $V_\mathcal{S}$ is the voxel index for point clouds, $V_{\mathcal{V}_j}$ means the voxel index for the voxel feature $\mathcal{V}_j$, which is the same as $V_{\mathcal{S}_i}$. The Eq. 2 and Eq. 3 mainly reveal the mutual transformation between point

Algorithm 1. Multi-scale Sparse Projection

Input: Input sparse voxel features F_v with corresponding voxel index V and pre-defined scale set S
Output: Multi-scale features O

1: $L = []$
2: **for** each $s \in S$ **do**
3: $\mathcal{V}^s = (\lfloor \mathcal{V}_x/s \rfloor, \lfloor \mathcal{V}_y/s \rfloor, \lfloor \mathcal{V}_z/s \rfloor)$
4: $F_v^s = \Psi(F_v, \mathcal{V}^s)$
5: $G^s = \text{MLP}(\Phi(F_v^s, \mathcal{V}))$
6: $O^s = G^s * F_v$
7: $L.\text{append}(O^s)$
8: **end for**
9: $O \leftarrow \text{Stack}(L)$
10: return O

features and voxel features: Scatter and Gather. To convert a raw point cloud into voxelwise features, we use the similar approach used in DRINet [52] and Cylinder3D [59] namely **GAFE**.

3.2 Sparse Feature Encoder

The sparse voxel representation makes it easy to apply standard convolution operations to extract local context information. Thus, in order to keep high runtime efficiency and explore more locality, we utilize sparse convolution layers [10,42,50] instead of dense convolution [21]. One of the good merits of sparse convolution lies in the sparsity, with which the convolution operation only considers the non-empty voxels. Based on this observation, we build our sparse feature encoder (SFE) with sparse convolution to quickly expand the receptive field with less computation cost. We adopt the ResNet BottleNeck [12] while replacing the ReLU activation with Leaky ReLU activation [24]. In order to keep a high efficiency, all the channel number for sparse convolution is set to 64.

3.3 Sparse Geometry Feature Enhancement

After obtaining the sparse voxelwise features from the sparse feature encoder, we aim to enhance the voxelwise features with more geometric guidance by our Sparse Geometry Feature Enhancement (SGFE).

Multi-scale Sparse Projection Layer. Inspired by previous works [30,52,57] which focus on multi-scale features aggregation, we notice that hierarchical context information helps enhance the capability of feature extraction, especially for point cloud, which has inherent scale invariance and geometry. Multi-scale features bring point cloud learning more geometric enhancement since each voxel scale reflects one specific physical dimension property. However, it is not applicable to directly apply Pyramid Pooling Module from PSPNet [57] due to the sparsity of point clouds. Also, DRINet [52] proposes points pooling at point level by scattering operation while introducing extra huge memory cost when considering large-scale scenarios. As a result, we propose **Multi-scale Sparse Projection Layer** to exploit the multi-scale features at a sparse voxel level with a much lower memory cost.

Given the input voxelwise features $F_v \in R^{N_{\mathcal{V}} \times C}$ with corresponding voxel index V and pre-defined pooling scale set S, where $N_{\mathcal{V}}$ is the voxel number and C is the channel number. Multi-scale Sparse Projection is described in Algorithm 1. Different scales contain different geometry prior since the scale semantics in the point cloud are proportional to the real dimension, reflecting the physical metric space. For each pooling scale s, we first re-calculate the voxel index under the scale s to ensure the features within the same pooling window will have the same voxel index. Then, we apply *Scatter* operation Ψ to perform the sparse pooling in order to obtain feature mean with local geometric priors $F_v^s \in R^{N_{\mathcal{V}_s} \times C}$ within each pooled region, where $N_{\mathcal{V}_s}$ is the voxel number under scale s. The embedding features $G^s \in R^{N_{\mathcal{V}} \times C}$ are based on the normalized features with learnable MLP layers. We then use tensor elementwise multiplication to obtain projection features O^s at scale s. Finally, features that reveal the geometry at different scales are stacked together.

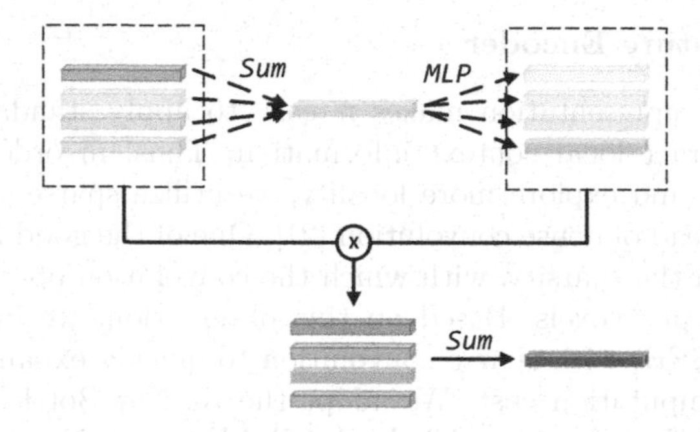

Fig. 3. An example that demonstrates the attentive scale selection. ⊗ indicates tensor element-wise multiplication.

Attentive Scale Selection. After obtaining the multi-scale features from the multi-scale sparse projection layer, the naive way of multi-scale features fusion is to apply tensor concatenation or tensor summation. As such, all the features from different scales share the same weights and are treated equally. A common consensus in the point cloud is that features from different scales have different geometry prior and focus on scene understanding. From this point of view and motivated by the SENet [13] and SKNet [18], a better approach to fusing multi-scale features is to apply a re-weighting strategy along scale dimension for each feature channel to re-distribute the importance of each scale.

We first sum over all the input tensors $\{O^s|s \in S\}$ for the first stage fusion to collect all the information from different scales as follows:

$$\mathcal{O} = \sum_s O^s, \tag{4}$$

where $\mathcal{O} \in R^{N_\mathcal{V} \times C}$ in the summed features. Inspired by recent popular attention works [18], we apply a scale-wise MLP layer with sigmoid activation for the results to get the attentive embedding for each scale. Finally, tensor multiplication between attention weight tensor and multi-scale features is applied, followed by the scale dimension's tensor sum:

$$\mathcal{A}\_\mathcal{O} = \sum_s \text{Sigmoid}(\text{MLP}^s(\mathcal{O})) * O^s, \tag{5}$$

where $\mathcal{A}\_\mathcal{O} \in R^{N_\mathcal{V} \times C}$ is the output attentive features for the SGFE module. Fig. 3 demonstrates the overall process.

3.4 Deep Sparse Supervision

Dense supervision on each pixel/voxel is a popular way in both 2D and 3D semantic segmentation tasks. Previous methods [54,59] generate dense feature maps with dense supervision. Although these works have considered the sparsity with their network architectures, they ignore this property when designing the loss, one of the key differences between 2D and 3D data. In fact, dense supervision with fine-grained feature maps brings a significant overload on memory usage. For example, when using a grid size of 0.2 m, the memory consumption (about 500 Mb) for a single dense feature map with 20 classes could be problematic. Based on these observations and inspired by [23], we propose a novel **Deep Sparse Supervision (DSS)** to deal with supervision in a deep sparse style.

Specifically, we generate voxel semantic labels at different scales in the training stage, and the supervision only acts on valid voxels rather than whole dense feature maps. Each voxel label is assigned with the major vote point labels within the voxel. Since we stack multiple blocks of SFE which generate sparse voxelwise features at different scales, we apply sparse supervision to the output voxelwise features stage by stage as auxiliary loss, one of the useful optimization techniques. We also apply sparse supervision for the main final prediction branch. The auxiliary loss helps optimize the training, while the main branch loss accounts for the most gradients.

In the testing stages, all the auxiliary branches are disabled to keep the runtime efficiency. This kind of training strategy has proven its effectiveness in image-based segmentation [22]. We consider the sparsity of point clouds and apply it in a sparse manner to save memory consumption.

3.5 Final Prediction

For the final semantic prediction, we fuse the multi-stage features from the output of each Sparse Geometry Feature Enhancement Layer by gathering operations to the most fine-grained scale of the voxel. To obtain pointwise results,

Algorithm 2. Geometry-aware Sparse Networks

Input: Input sparse voxel features F, the number of blocks B, and voxel index V^s at the target scale s

Output: Semantic prediction P

1: $L = []$
2: $Loss = 0$
3: **for** $i = 1$ to B **do**
4: $\mathcal{V} \leftarrow$ SFE(F)
5: $F \leftarrow$ SGFE($\mathcal{V}$)
6: $O^s \leftarrow \Phi(F, \mathcal{V}^s)$
7: $Loss$ += LossFunc($\mathcal{V}$)
8: L.append(O^s)
9: **end for**
10: $L \leftarrow$ Stack(L)
11: P = Softmax(MLP(L))
12: return P

we also apply the gathering strategy, in which each point is attached with the semantic features from the voxel that it lies in. The whole algorithm for our framework is illustrated in Algorithm 2.

4 Experiments

We conduct experiments on large-scale outdoor scenarios dataset SemanticKITTI [1] and Nuscenes [3] to show the effectiveness of our proposed approaches. Besides that, we also conduct ablation studies to validate proposed components.

4.1 Datasets

SemanticKITTI. SemanticKITTI [1] generated from KITTI odometry dataset [9] contains 22 sequences which involve the most common scenes for autonomous driving. Each scan in the dataset has more than 100K points on average, with pointwise annotation labels for 20 classes. According to the official settings, sequences from 00 to 10 except 08 are the training split, sequence 08 is the validation split, and the rest sequences from 11 to 21 are treated as test split.

Nuscenes. Nuscenes [3] has a total of 40,000 scans collected by 32 beams Lidar sensor. Compared with SemanticKITTI [1], it contains fewer points and annotations classes(16 classes).

For both datasets, they use **mIoU** as one evaluation metric, which is one of the most popular criteria in point cloud semantic segmentation. Apart from that, SemanticKITTI provides Acc, the average accuracy metric, and Nuscenes uses **fwIoU** as additional criteria, which is the weighted sum of IoU for each class based on the point-level frequency.

Table 1. The per-class mIoU results on the SemanticKITTI test set.

Methods	Road	Sidewalk	Parking	Other ground	Building	Car	Truck	Bicycle	Motorcycle	Other vehicle	Vegetation	Trunk	Terrain	Person	Bicyclist	Motorcyclist	Fence	Pole	Traffic sign	mIoU	Speed (ms)
PointNet [29]	61.6	35.7	15.8	1.4	41.4	46.3	0.1	1.3	0.3	0.8	31.0	4.6	17.6	0.2	0.2	0.0	12.9	2.4	3.7	14.6	500
PointNet++ [30]	72.0	41.8	18.7	5.6	62.3	53.7	0.9	1.9	0.2	0.2	46.5	13.8	30.0	0.9	1.0	0.0	16.9	6.0	8.9	20.1	5900
KPConv [38]	88.8	72.7	61.3	31.6	90.5	96.0	33.4	30.2	42.5	44.3	84.8	69.2	69.1	61.5	61.6	11.8	64.2	56.4	47.4	58.8	–
RandLA [14]	90.7	73.7	60.2	20.4	86.9	94.2	40.1	26.0	25.8	38.9	81.4	66.8	49.2	49.2	48.2	7.2	56.3	47.7	38.1	53.9	880
SqueezeSegV3 [46]	91.7	74.8	63.4	26.4	89.0	92.5	29.6	38.7	36.5	33.0	82.0	58.7	65.4	45.6	46.2	20.1	59.4	49.6	58.9	55.9	238
RangeNet++ [27]	91.8	75.2	65.0	27.8	87.4	91.4	25.7	25.7	34.4	23.0	80.5	55.1	64.6	38.3	38.8	4.8	58.6	47.9	55.9	52.2	83.3
TangentConv [36]	83.9	63.9	33.4	15.4	83.4	90.8	15.2	2.7	16.5	12.1	79.5	49.3	58.1	23.0	28.4	8.1	49.0	35.8	28.5	35.9	3000
SPVCNN [35]	90.2	75.4	67.6	21.8	91.6	97.2	56.6	50.6	50.4	58.0	86.1	73.4	71.0	67.4	67.1	50.3	66.9	64.3	67.3	67.0	259
PolarNet [54]	90.8	74.4	61.7	21.7	90.0	93.8	22.9	40.3	30.1	28.5	84.0	65.5	67.8	43.2	40.2	5.6	61.3	51.8	57.5	54.3	62
DASS [39]	92.8	71.0	31.7	0.0	82.1	91.4	**66.7**	25.8	31.0	43.8	83.5	56.6	69.6	47.7	70.8	0.0	39.1	45.5	35.1	51.8	90
DRINet [52]	90.7	75.2	65.0	26.2	91.5	96.9	43.3	57.0	56.0	54.5	85.2	72.6	68.8	69.4	75.1	58.9	67.3	63.5	66.0	67.5	62
Cylinder3D [59]	92.0	70.0	65.0	32.3	90.7	97.1	50.8	67.6	63.8	58.5	85.6	72.5	69.8	73.7	69.2	48.0	66.5	62.4	66.2	68.9	131
AF2S3Net [6]	92.0	76.2	66.8	**45.8**	92.5	94.3	40.2	63.0	**81.4**	40.0	78.6	68.0	63.1	76.4	81.7	**77.7**	69.6	64.0	**73.3**	**70.8**	-
RPVNet [47]	**93.4**	**80.7**	**70.3**	33.3	**93.5**	**97.6**	44.2	**68.4**	68.7	**61.1**	86.5	**75.1**	71.7	75.9	74.4	43.4	**72.1**	64.8	61.4	70.3	168
Ours	89.8	74.6	66.2	30.1	92.3	96.9	59.3	65.8	58.0	61.0	**87.3**	73.0	**72.5**	**80.4**	**82.7**	46.3	69.6	**66.1**	71.6	70.7	**59**

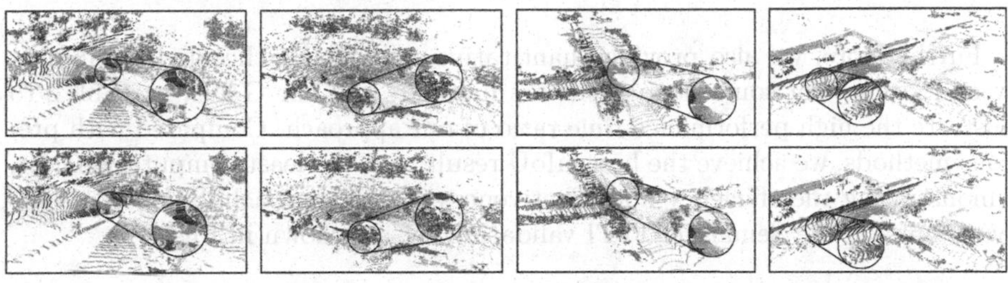

Fig. 4. The results on the SemanticKITTI validation set. The top row is the ground truth, and the bottom row is the predictions by our Geometry-aware Sparse Networks.

Network Details. We use the same settings for both datasets. We quantize the point cloud with a voxel scale 0.2 m along xyz dimensions to generate the initial sparse voxel features. We design our Geometry-aware Sparse Networks with four blocks of Sparse Feature Encoder and Sparse Geometry Feature Enhancement. In the ablation study, we also evaluate the effect of block number choice between performance and efficiency. As for the multi-scale sparse projection layer, we adopt kernel sizes and strides with [2, 4, 6, 8], which could cover the coarse and the fine pooling regions. Similar to previous works, we apply random flipping, random points dropout, random scale, and global rotation in the training stage. For the loss design, we combine the Lovasz loss [2] and cross-entropy loss as supervision. Adam optimizer [15] is employed with an initial learning rate of 0.0002 at batch size 4 for 50 epochs. The learning rate decays in a ratio of 0.1 for every 15 epochs.

4.2 Results On SemanticKITTI

SemanticKITTI Single Scan: We provide the detailed per-class quantitative results of our Geometry-aware Sparse Networks as well as the other state-of-the-art methods in Table 1. Compared with previous methods, our GASN achieves state-of-the-art performance while maintaining a real-time inference efficiency. Although our Geometry-aware Sparse Networks fails to achieve the best result for every class, it achieves balanced results among all the classes. Even compared to multiple representation fusion approaches [35,47,52], our method still surpasses by a considerable margin.

Table 2. Quantitative results of model complexity with performance. The performance is obtained from the leaderboard of SemanticKITTI. Statistics of the number of parameters and Macs are from the corresponding papers. The running times are all evaluated on a single Nvidia RTX 2080Ti GPU. ∗ donates the statistics from our reproduction.

Method	Param (M)	Macs (G)	Speed	Memory	mIoU
SPVCNN	12.5	73.8	120 ms	2.4 Gb	67.0
DRINet	3.5	14.58	62* ms	2.1* Gb	67.5
Cylind3D	53.3	64.3	131 ms	3.0 Gb	68.9
RPVNet	24.8	119.5	168* ms	2.7* Gb	70.3
Ours	**2.2**	**12.1**	**59 ms**	**1.6 Gb**	**70.7**

Furthermore, we also provide quantitative analysis for the model complexity and latency for some state-of-the-art methods and our GASN in Table 2 to illustrate the high performance-time ratio of our approach. Compared with previous methods, we achieve the best mIoU result with the least computation cost, demonstrating the efficiency and effectiveness of our approach. Some quantitative results on the SemanticKITTI validation set are shown in Fig. 4.

Table 3. Comparison to the state-of-the-art methods on the test set of SemanticKITTI multiple scans challenge. ∗ donates the moving category.

Approach	Road	Sidewalk	Parking	Other ground	Building	Car	Car*	Truck	Truck*	Bicycle	Motorcycle	Other vehicle	Other vehicle *	Vegetation	Trunk	Terrain	Person	Person*	Bicyclist	Bicyclist*	Motorcyclist	Motorcyclist*	Fence	Pole	Traffic sign	mIoU
Cylinder3D	90.7	74.5	65.0	32.3	92.6	94.6	74.9	41.3	0.0	**67.6**	63.8	38.8	0.1	85.8	72.0	68.9	12.5	65.7	1.7	68.3	0.2	11.9	66.0	63.1	61.4	52.5
TemporalLidarSeg	91.8	75.8	59.6	23.2	89.8	92.1	68.2	39.2	2.1	47.7	40.9	35.0	12.4	82.3	62.5	64.7	14.4	40.4	0.0	42.8	0.0	12.9	63.8	52.6	60.4	47.0
KPConv	86.5	70.5	58.4	26.7	90.8	93.7	69.4	42.5	5.8	44.9	47.2	38.6	4.7	84.6	70.3	66.0	21.6	67.5	0.0	67.4	0.0	47.2	64.5	57.0	53.9	51.2
AF2S3Net	91.3	72.5	68.8	**53.5**	87.9	91.8	65.3	15.7	5.6	65.4	**86.8**	27.5	3.9	75.1	64.6	57.4	16.4	67.6	**15.1**	66.4	**67.1**	59.6	63.2	62.6	71.0	56.9
Ours	**92.3**	**79.1**	**69.6**	30.9	**93.7**	**97.2**	**85.4**	**46.8**	**15.9**	63.6	53.3	**64.6**	**26.3**	**86.8**	**75.8**	**71.2**	**30.5**	**84.8**	0.0	**73.1**	0.0	**76.9**	**72.8**	**68.0**	**73.5**	**61.3**

SemanticKITTI Multiple Scans: Meanwhile, we also conduct experiments on SemanticKITTI multiple scans challenge to verify the effectiveness of our GASN. In this task, we directly stack multiple aligned scans as input without any temporal fusion algorithm, then generate the pointwise output prediction according to Algorithm 2. We do not apply any post-processing for refinement. As shown in Table 3, our Geometry-aware Sparse Network shows a significant improvement compared with previous methods in terms of both metrics. Compared with AF2S3Net [6], which is also a voxel-based approach, the proposed method brings a very competitive gain (about 4.4%). Our method surpasses the existing approaches in nearly all categories, demonstrating the effectiveness and generalization capability of our method. To this end, GASN can serve as an efficient and strong backbone for the large-scale point cloud semantic segmentation task.

4.3 Results On Nuscenes

In order to verify the generalization ability of our approach, we also report the results on Nuscenes-lidarSeg [3] task on the test set. As shown in Table 4, GASN achieves highly strong results, with 10 out of 16 categories surpassing

Table 4. The per-class mIoU results on the Nuscenes test set.

Method	Barrier	Bicycle	Bus	Car	Construction	Motorcycle	Pedestrian	Traffic cone	Trailer	Truck	Driveable	Other_flat	Sidewalk	Terrain	Manmade	Vegetation	FW mIoU	mIoU
AF2S3Net [6]	78.9	**52.2**	89.9	84.2	**77.4**	74.3	77.3	72.0	83.9	73.8	97.1	66.5	77.5	74.0	87.7	86.8	88.5	78.3
Cylinder3D [59]	82.8	29.8	84.3	89.4	63.0	79.3	77.2	**73.4**	**84.6**	69.1	**97.7**	70.2	80.3	75.5	90.4	87.6	89.9	77.2
SPVCNN [35]	80.0	30.0	**91.9**	90.8	64.7	79.0	75.6	70.9	81.0	74.6	97.4	69.2	80.0	76.1	89.3	87.1	89.7	77.4
PolarNet [54]	72.2	16.8	77.0	86.5	51.1	69.7	64.8	54.1	69.7	63.5	96.6	67.1	77.7	72.1	87.1	84.5	87.4	69.4
Ours	**85.5**	43.2	90.5	**92.1**	64.7	**86.0**	**83.0**	73.3	83.9	**75.8**	97.0	**71.0**	**81.0**	**77.7**	**91.6**	**90.2**	**91.0**	**80.4**

the other approaches. There is a noticeable improvement for classes with small sizes, such as pedestrians and motorcycles, demonstrating the effectiveness of our sparse geometry feature enhancement, which is designed to capture and integrate multi-scale context information through hierarchical feature learning and attentive scale selection.

4.4 Ablation Study

Components Study. Our baseline model, shown in Table 5, which only uses a sparse feature encoder, has achieved decent performance. It reveals that the voxel representation with sparse convolution has a strong capability for feature extraction and context information learning in the outdoor point cloud semantic segmentation task. The multi-scale sparse projection layer brings 1.8% improvement with its ability to capture hierarchical geometry information. Furtherly, we apply the attentive scale selection to re-distribute the importance of each scale for each channel. This strategy enables the network the focus on a more significant scale and has a performance gain of about 1.9%. Next, adding the deep sparse supervision in the training stage improves the mIoU to 69.4%, an increase of 1.7%. More importantly, deep sparse supervision will be disabled when inference, bringing no extra computation cost for deployment.

Table 5. Ablation study on the SemanticKITTI validation set. MSP refers to Multi-scale Sparse Projection. ASS refers to Attentive Scale Selection. DSS refers to Deep Sparse Supervision

SFE	SGFE		DSS	mIoU (%)
	MSP	ASS		
✓				64.0
✓	✓			65.8
✓	✓	✓		67.7
✓	✓	✓	✓	69.4

Number of Blocks. The block number choice plays a crucial role in our network. Fewer blocks may lead to underfitting problems, while more blocks mean more computation cost and memory consumption, indicating inefficiency. In order to find a better block number, we conduct the experiment varying this parameter. With block number ranging from 1 to 5, the performance is 50.8%,

61.3%, 67.4%, 69.4% and 69.9%, while the speed will be 26 ms, 33 ms, 41 ms, 59 ms and 70 ms respectively. The performance increases from 50.8% to 70.2% when the block number rises to 5, while the runtime speed nearly doubles. Furthermore, adding one more block when the block number is 4 introduces extra 11 ms latency while only improving the mIoU by about 0.5%. Therefore, we choose block number as 4 in our experiment.

Table 6. Ablation study for fusion strategy. TC refers to concatenation and TS refers to Tensor sum.

Method	mIoU (%)	Memory	Speed
ASS	**69.4**	1.6 Gb	59 ms
TS	67.7	**1.45 Gb**	**57 ms**
TC	68.7	1.48 Gb	**57 ms**

Scale Features Fusion Strategy. We analyze the effectiveness of the proposed ASS by comparing it with different fusion strategies. We use tensor concatenation and summation, which are commonly used in feature fusion. For this work, we enable all other proposed modules for a fair comparison. As shown in Table 6, the ASS layer serves as a better approach for fusing the multi-scale features, which leads to a 2.0% increase compared with the tensor summation while only bringing 2 ms cost. Table 7 also shows that other models that utilize multiple representations could benefit from the ASS strategy. Compared with DRINet [52], the original SPVCNN [35] does not have hierarchical learning in the pointwise branches, and then its performance will improve when ASS is enabled.

Table 7. Ablation study for DSS and ASS on the SemanticKITTI [1] validation set with different models. The statistics are from our reproduction.

Method	Original (%)	DSS (%)	ASS (%)
SPVCNN [35]	64.7	66.1 (**+1.4**)	66.7 (**+2.0**)
DRINet [52]	67.3	68.1 (**+0.8**)	67.9 (**+0.6**)
Cylinder3D [59]	66.5	67.8 (**+1.3**)	–

Sparse or Dense Supervision. Deep Sparse supervision is another characteristic of our GASN. In this experiment, we compare the sparse and dense supervision in terms of memory cost and performance. We remove the deep auxiliary loss branch for simplicity, with only the main supervision left. The memory consumption for both sparse and dense supervision only includes prediction tensor and label tensor without the consumption used by gradient tensors. The memory footprint in sparse supervision (6 Mb) only accounts for about one percent of that in dense supervision (552 Mb), and the results are close between the methods, which are 69.4% and 69.5% respectively. By incorporating sparse supervision, we can use a larger batch size for training, leading to efficient training.

Deep Sparse Supervision. Our Deep Sparse Supervision can be a general component in the point cloud semantic segmentation task, and we incorporate this strategy with other popular models to verify its effectiveness. As shown in Table 7, Deep Sparse Supervision could help the performance of popular models without any extra computation cost for inference.

Table 8. Ablation study on SemanticKITTI validation set for voxel size choice.

Voxel size (m)	mIoU (%)	Memory	Speed (ms)
0.2	**69.4**	1.6 Gb	59
0.3	68.3	1.42 Gb	47
0.4	67.1	1.31 Gb	42
0.5	64.0	**1.25 Gb**	**39**

Voxel Size. To choose the best voxel size for the experiments, we also conduct experiments to verify the effect. As shown in Table 8, the runtime speed, memory consumption, and performance drop significantly with a larger voxel size. As a result, we choose voxel size as 0.2 m in our experiments.

5 Conclusion

In this paper, we propose our Geometry-aware Sparse Networks that serve as an efficient network architecture for point cloud segmentation. Our GASN deals with point cloud segmentation by fully utilizing the sparsity and geometry in a single sparse voxel representation to maintain performance and efficiency. GASN consists of Sparse Feature Encoder (SFE) and Sparse Geometry Feature Enhancement (SGFE). SFE helps extract local context information, while SGFE enhances the geometry with multi-scale sparse projection and attentive scale selection. Moreover, we apply deep sparse supervision to accelerate convergence with a lower memory cost. The experiments on large-scale outdoor scenarios datasets demonstrate that our approach achieves state-of-the-art performance with impressive runtime efficiency.

References

1. Behley, J., et al.: Semantickitti: a dataset for semantic scene understanding of lidar sequences. In: ICCV, pp. 9297–9307 (2019)
2. Berman, M., Rannen Triki, A., Blaschko, M.B.: The lovász-softmax loss: a tractable surrogate for the optimization of the intersection-over-union measure in neural networks. In: CVPR, pp. 4413–4421 (2018)

3. Caesar, H., et al.: nuscenes: a multimodal dataset for autonomous driving. In: CVPR, pp. 11621–11631 (2020)
4. Chen, L.C., Papandreou, G., Kokkinos, I., Murphy, K., Yuille, A.L.: Deeplab: semantic image segmentation with deep convolutional nets, atrous convolution, and fully connected crfs. IEEE Trans. Patt. Anal. Mach. Intell. **40**(4), 834–848 (2017)
5. Chen, L.C., Papandreou, G., Schroff, F., Adam, H.: Rethinking atrous convolution for semantic image segmentation. arXiv preprint arXiv:1706.05587 (2017)
6. Cheng, R., Razani, R., Taghavi, E., Li, E., Liu, B.: 2–s3net: attentive feature fusion with adaptive feature selection for sparse semantic segmentation network. In: CVPR, pp. 12547–12556 (2021)
7. Choy, C., Gwak, J., Savarese, S.: 4D spatio-temporal convnets: minkowski convolutional neural networks. In: CVPR, pp. 3075–3084 (2019)
8. Cortinhal, T., Tzelepis, G., Erdal Aksoy, E.: SalsaNext: fast, uncertainty-aware semantic segmentation of LiDAR point clouds. In: Bebis, G., et al. (eds.) ISVC 2020. LNCS, vol. 12510, pp. 207–222. Springer, Cham (2020). https://doi.org/10.1007/978-3-030-64559-5_16
9. Geiger, A., Lenz, P., Stiller, C., Urtasun, R.: Vision meets robotics: the kitti dataset. Int. J. Robot. Res. **32**(11), 1231–1237 (2013)
10. Graham, B., Engelcke, M., Maaten, L.V.D.: 3D semantic segmentation with submanifold sparse convolutional networks. In: CVPR (2018)
11. Guo, M.H., Cai, J.X., Liu, Z.N., Mu, T.J., Martin, R.R., Hu, S.M.: PCT: point cloud transformer. arXiv preprint arXiv:2012.09688 (2020)
12. He, K., Zhang, X., Ren, S., Sun, J.: Deep residual learning for image recognition. In: CVPR, pp. 770–778 (2016)
13. Hu, J., Shen, L., Sun, G.: Squeeze-and-excitation networks. In: CVPR, pp. 7132–7141 (2018)
14. Hu, Q., et al.: Randla-net: efficient semantic segmentation of large-scale point clouds. In: CVPR, pp. 11108–11117 (2020)
15. Kingma, D.P., Ba, J.: Adam: a method for stochastic optimization. arXiv preprint arXiv:1412.6980 (2014)
16. Kumawat, S., Raman, S.: LP-3DCNN: unveiling local phase in 3D convolutional neural networks. In: CVPR, pp. 4903–4912 (2019)
17. Le, T., Duan, Y.: Pointgrid: a deep network for 3D shape understanding. In: CVPR, pp. 9204–9214 (2018)
18. Li, X., Wang, W., Hu, X., Yang, J.: Selective kernel networks. In: CVPR, pp. 510–519 (2019)
19. Li, Y., Bu, R., Sun, M., Wu, W., Di, X., Chen, B.: Pointcnn: convolution on x-transformed points. Adv. Neural. Inf. Process. Syst. **31**, 820–830 (2018)
20. Liu, Z., Hu, H., Cao, Y., Zhang, Z., Tong, X.: A closer look at local aggregation operators in point cloud analysis. In: Vedaldi, A., Bischof, H., Brox, T., Frahm, J.-M. (eds.) ECCV 2020. LNCS, vol. 12368, pp. 326–342. Springer, Cham (2020). https://doi.org/10.1007/978-3-030-58592-1_20
21. Liu, Z., Tang, H., Lin, Y., Han, S.: Point-voxel CNN for efficient 3D deep learning. In: Advances in Neural Information Processing Systems, pp. 965–975 (2019)
22. Long, J., Shelhamer, E., Darrell, T.: Fully convolutional networks for semantic segmentation. In: CVPR, pp. 3431–3440 (2015)
23. Loquercio, A., Dosovitskiy, A., Scaramuzza, D.: Learning depth with very sparse supervision. IEEE Robot. Autom. Lett. **5**(4), 5542–5549 (2020)
24. Maas, A.L., Hannun, A.Y., Ng, A.Y., et al.: Rectifier nonlinearities improve neural network acoustic models. In: Proc. icml, vol. 30, p. 3. Citeseer (2013)

25. Maturana, D., Scherer, S.: Voxnet: a 3D convolutional neural network for real-time object recognition. In: IROS, pp. 922–928. IEEE (2015)
26. Milioto, A., Vizzo, I., Behley, J., Stachniss, C.: Rangenet++: fast and accurate lidar semantic segmentation. In: IROS, pp. 4213–4220. IEEE (2019)
27. Milioto, A., Vizzo, I., Behley, J., Stachniss, C.: Rangenet++: fast and accurate lidar semantic segmentation. In: IROS, pp. 4213–4220. IEEE (2019)
28. Pham, Q.H., Nguyen, T., Hua, B.S., Roig, G., Yeung, S.K.: Jsis3d: joint semantic-instance segmentation of 3D point clouds with multi-task pointwise networks and multi-value conditional random fields. In: CVPR, pp. 8827–8836 (2019)
29. Qi, C.R., Su, H., Mo, K., Guibas, L.J.: Pointnet: deep learning on point sets for 3D classification and segmentation. In: CVPR, pp. 652–660 (2017)
30. Qi, C.R., Yi, L., Su, H., Guibas, L.J.: PointNet++: deep hierarchical feature learning on point sets in a metric space. In: Advances in Neural Information Processing Systems 30, vol. 30, pp. 5099–5108. Curran Associates, Inc. (2017)
31. Qi, X., Liao, R., Jia, J., Fidler, S., Urtasun, R.: 3D graph neural networks for RGBD semantic segmentation. In: ICCV, pp. 5199–5208 (2017)
32. Ronneberger, O., Fischer, P., Brox, T.: U-Net: convolutional networks for biomedical image segmentation. In: Navab, N., Hornegger, J., Wells, W.M., Frangi, A.F. (eds.) MICCAI 2015. LNCS, vol. 9351, pp. 234–241. Springer, Cham (2015). https://doi.org/10.1007/978-3-319-24574-4_28
33. Su, H., Maji, S., Kalogerakis, E., Learned-Miller, E.: Multi-view convolutional neural networks for 3D shape recognition. In: ICCV, pp. 945–953 (2015)
34. Sun, K., Xiao, B., Liu, D., Wang, J.: Deep high-resolution representation learning for human pose estimation. In: CVPR, pp. 5693–5703 (2019)
35. Tang, H., et al.: Searching efficient 3D architectures with sparse point-voxel convolution. In: Vedaldi, A., Bischof, H., Brox, T., Frahm, J.-M. (eds.) ECCV 2020. LNCS, vol. 12373, pp. 685–702. Springer, Cham (2020). https://doi.org/10.1007/978-3-030-58604-1_41
36. Tatarchenko, M., Park, J., Koltun, V., Zhou, Q.Y.: Tangent convolutions for dense prediction in 3D. In: CVPR, pp. 3887–3896 (2018)
37. Te, G., Hu, W., Zheng, A., Guo, Z.: Rgcnn: regularized graph CNN for point cloud segmentation. In: Proceedings of the 26th ACM International Conference on Multimedia, pp. 746–754 (2018)
38. Thomas, H., Qi, C.R., Deschaud, J.E., Marcotegui, B., Goulette, F., Guibas, L.J.: Kpconv: flexible and deformable convolution for point clouds. In: ICCV, pp. 6411–6420 (2019)
39. Unal, O., Van Gool, L., Dai, D.: Improving point cloud semantic segmentation by learning 3D object detection. In: WACV, pp. 2950–2959 (2021)
40. Wang, C., Samari, B., Siddiqi, K.: Local spectral graph convolution for point set feature learning. In: ECCV, pp. 52–66 (2018)
41. Wang, L., Huang, Y., Hou, Y., Zhang, S., Shan, J.: Graph attention convolution for point cloud semantic segmentation. In: CVPR, pp. 10296–10305 (2019)
42. Wang, P.S., Liu, Y., Guo, Y.X., Sun, C.Y., Tong, X.: O-cnn: octree-based convolutional neural networks for 3D shape analysis. ACM Trans. Graph. (TOG) **36**(4), 1–11 (2017)
43. Wang, Y., Sun, Y., Liu, Z., Sarma, S.E., Bronstein, M.M., Solomon, J.M.: Dynamic graph CNN for learning on point clouds. ACM Trans. Graph. (ToG) **38**(5), 1–12 (2019)
44. Wu, B., Wan, A., Yue, X., Keutzer, K.: Squeezeseg: convolutional neural nets with recurrent CRF for real-time road-object segmentation from 3D lidar point cloud. In: ICRA, pp. 1887–1893. IEEE (2018)

45. Wu, W., Qi, Z., Fuxin, L.: Pointconv: deep convolutional networks on 3D point clouds. In: CVPR, pp. 9621–9630 (2019)
46. Xu, C., et al.: SqueezeSegV3: spatially-adaptive convolution for efficient point-cloud segmentation. In: Vedaldi, A., Bischof, H., Brox, T., Frahm, J.-M. (eds.) ECCV 2020. LNCS, vol. 12373, pp. 1–19. Springer, Cham (2020). https://doi.org/10.1007/978-3-030-58604-1_1
47. Xu, J., Zhang, R., Dou, J., Zhu, Y., Sun, J., Pu, S.: Rpvnet: a deep and efficient range-point-voxel fusion network for lidar point cloud segmentation. arXiv preprint arXiv:2103.12978 (2021)
48. Xu, Q., Sun, X., Wu, C.Y., Wang, P., Neumann, U.: Grid-GCN for fast and scalable point cloud learning. In: CVPR, pp. 5661–5670 (2020)
49. Yan, X., et al.: Sparse single sweep lidar point cloud segmentation via learning contextual shape priors from scene completion. arXiv preprint arXiv:2012.03762 (2020)
50. Yan, Y., Mao, Y., Li, B.: Second: sparsely embedded convolutional detection. Sensors **18**(10), 3337 (2018)
51. Yang, M., Yu, K., Zhang, C., Li, Z., Yang, K.: Denseaspp for semantic segmentation in street scenes. In: CVPR, pp. 3684–3692 (2018)
52. Ye, M., Xu, S., Cao, T., Chen, Q.: Drinet: a dual-representation iterative learning network for point cloud segmentation. In: ICCV (2021)
53. Zhang, K., Hao, M., Wang, J., de Silva, C.W., Fu, C.: Linked dynamic graph CNN: learning on point cloud via linking hierarchical features. arXiv preprint arXiv:1904.10014 (2019)
54. Zhang, Y., et al.: Polarnet: an improved grid representation for online lidar point clouds semantic segmentation. In: CVPR, pp. 9601–9610 (2020)
55. Zhang, Y., Rabbat, M.: A graph-cnn for 3D point cloud classification. In: 2018 IEEE International Conference on Acoustics, Speech and Signal Processing (ICASSP), pp. 6279–6283. IEEE (2018)
56. Zhao, H., Jiang, L., Jia, J., Torr, P., Koltun, V.: Point transformer. arXiv preprint arXiv:2012.09164 (2020)
57. Zhao, H., Shi, J., Qi, X., Wang, X., Jia, J.: Pyramid scene parsing network. In: CVPR, pp. 2881–2890 (2017)
58. Zhou, Y., et al.: End-to-end multi-view fusion for 3D object detection in lidar point clouds. In: Conference on Robot Learning, pp. 923–932. PMLR (2020)
59. Zhu, X., et al.: Cylindrical and asymmetrical 3D convolution networks for lidar segmentation. In: CVPR, pp. 9939–9948 (2021)

FH-Net: A Fast Hierarchical Network for Scene Flow Estimation on Real-World Point Clouds

Lihe Ding, Shaocong Dong, Tingfa Xu(✉), Xinli Xu, Jie Wang, and Jianan Li(✉)

Beijing Institute of Technology, Beijing, China
dean.dinglihe@gmail.com, dscdyc101029579@gmail.com,
xxlbigbrother@gmail.com, {ciom_xtf1,lijianan}@bit.edu.cn,
jwang123bit@gmail.com

Abstract. Estimating scene flow from real-world point clouds is a fundamental task for practical 3D vision. Previous methods often rely on deep models to first extract expensive per-point features at full resolution, and then get the flow either from complex matching mechanism or feature decoding, suffering high computational cost and latency. In this work, we propose a fast hierarchical network, FH-Net, which directly gets the key points flow through a lightweight Trans-flow layer utilizing the reliable local geometry prior, and optionally back-propagates the computed sparse flows through an inverse Trans-up layer to obtain hierarchical flows at different resolutions. To focus more on challenging dynamic objects, we also provide a new copy-and-paste data augmentation technique based on dynamic object pairs generation. Moreover, to alleviate the chronic shortage of real-world training data, we establish two new large-scale datasets to this field by collecting lidar-scanned point clouds from public autonomous driving datasets and annotating the collected data through novel pseudo-labeling. Extensive experiments on both public and proposed datasets show that our method outperforms prior state-of-the-arts while running at least **7×** faster at **113** FPS. Code and data are released at https://github.com/pigtigger/FH-Net.

Keywords: Scene flow · Real-world point cloud · Transformer · Copy-and-paste

1 Introduction

Scene flow estimation from point clouds, which accurately measures point movement between consecutive frames, serves as an fundamental step for downstream tasks like 3D motion segmentation [32] and tracking [45], and thus plays an increasingly important role in robotics [3] and autonomous driving [46]. Previous works [10,18] for scene flow estimation, however, mainly focus on synthetic datasets [20] with CAD models and pseudo lidar points [21] obtained from disparity images where point clouds are dense and under direct correspondence, totally different from the lidar-scanned and leading to catastrophic performance when move to real world.

Conventional scene flow estimation paradigm Fig. 1a [10,18] uses a U-Net [31] style deep model to obtain per-point fused feature from two consecutive frames at full

L. Ding and S. Dong—Equal contribution.

S. Avidan et al. (Eds.): ECCV 2022, LNCS 13699, pp. 213–229, 2022.
https://doi.org/10.1007/978-3-031-19842-7_13

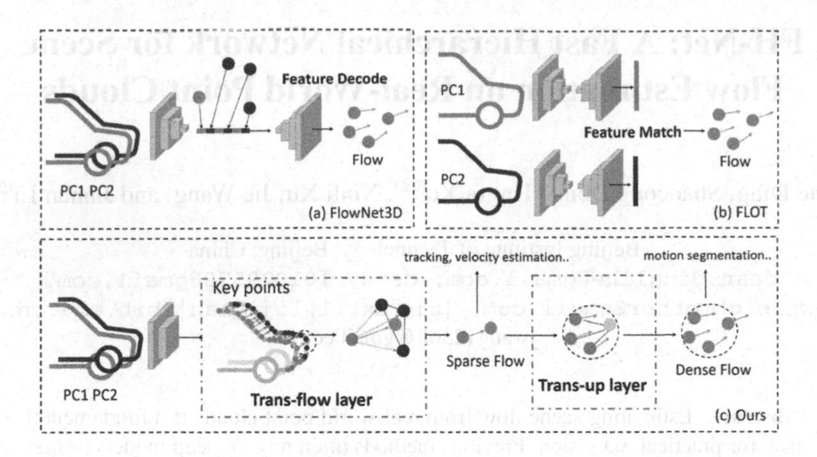

Fig. 1. Most previous works (a) fuse two consecutive point clouds in deep feature space from which to decode the flow, or (b) match the features extracted from two point cloud separately by a deep network to predict the flow. Our approach (c) directly computes key points flow (green) by aggregating the spatial offsets from the source point (red) to its neighboring target points (blue) via the well-designed local relationship modeling then use trans-up layer to get the hierarchical flow. (Color figure online)

resolution, and then decode flow from the fused feature, one other line of works Fig. 1b [27] try to extract point feature without temporal fusion and apply complex feature matching algorithm to compute the flow. However, both of these methods need to get the full resolution deep point feature at first then predict the flow, which not only can not well meet the various demand of downstream tasks (i.e., velocity estimation needs real-time instance-level flow while motion segmentation needs higher resolution flow) but also brings high computational cost and latency thus hard to be deployed on real-world autonomous vehicle and robots. To deal with these problems, we propose a fast hierarchical network Fig. 1c, FH-Net, which can output different resolution flows from different layers while running much faster than real-time.

Specifically, we first extract the key points' geometric features for each frame through shallow hierarchical set abstraction layers [28], then utilize a novel transformer-based Trans-flow layer to directly compute key points' flow by applying point-wise attention in the local region between two consecutive frames. We take the offset from source point to every target point as the so-called value while get query and key from the feature of source and target point respectively. Followed by a well-designed subtraction relation mechanism, we can obtain the source key point flow directly by leveraging the strong guide from local relative position instead of decoding from deep feature or applying expensive feature matching mechanism after fully up-sampling like previous methods as shown in Fig. 1.

Through Trans-flow layer, we can get the sparse keypoint flows which well meet the need of object level flow in scenarios like 3D tracking. However, dense prediction tasks like segmentation, requires more fine-grained flows at higher resolution. So optionally, we need to up-sample the predicted flow from Trans-flow layer. A trial solution is interpolation up-sampling, which however only weights flow by distance and suffers large propagation error when close points have different flow. To address this problem, we propose Trans-up layer sharing the same mechanism as Trans-flow layer to up sample the flow. For arbitrary high resolution point, by attending each of its neighbor points in low resolution point set both in spatial and feature space, we can dynamically weight the sparse flows and aggregate them to get the dense flow. Benefit from this design, we can get up-sampled flow at each resolution through hierarchical Trans-up layers and fed it into downstream tasks with higher flexibility and efficiency. Benefit from the hierarchical flow outputs, we further present a Hierarchical loss which calculates the EPE (every point error) of intermediate flows in each up-sampling layer to better constrain the flow accuracy and speed up the convergence.

Based on the proposed fundamental Trans-flow and Trans-up layers, we build our FH-Net with series version, FH-L, FH-R and FH-S with flexible combinations of the above modules to trade off between complexity and efficiency, which can adapt to various tasks. Additionally, dynamic objects is commonly of small number in a single frame, which poses challenges to learn the prediction of more important foreground flow. To tackle this problem, inspired by [8,44], we propose a new data augmentation strategy which during training randomly select objects from pre-built dynamic objects database and augment them with different strategy, i.e., dropout, down-sampling, to generate object pairs then paste them in two frames respectively.

Furthermore, to mitigate the problem of lacking real-world training data for scene flow estimation, we build two new point cloud scene flow datasets named SF-KITTI and SF-Waymo by annotating general autonomous driving dataset i.e. KITTI [7], Waymo [34] with pseudo sceneflow labels. Specifically, we compute every foreground point's label from its gt box's pose transform between two frames and annotate background point via ego-motion obtained from IMU. We note that the recent released dataset [15] shares similar ideas. The key difference are it only focuses on foreground point and different processing strategies for ground point.

Extensive experiments show our FH-Net achieves state-of-the-art lidar-scanned point cloud scene flow estimation results both on public Lidar-KITTI [7,9] and our more challenging large scale dataset SF-Waymo with faster inference speed and significant reduced parameters compared with previous methods [9,18,43]. Furthermore, we also achieve competitive results on no-lidar-scanned dataset FlyingThings3D [20] and Stereo-KITTI [21], which shows the great generality of our method.

The key contributions of this work are as follows:

- We propose a fast hierarchical architecture, which outputs accurate hierarchical scene flow on real-world point clouds with high efficiency.
- We introduce a new data augmentation strategy which improves predicted flow's accuracy especially on more challenging dynamic objects.
- We construct two lidar-scanned point cloud scene flow datasets to the community, which can facilitate the research on real-world scene flow estimation.
- We establish new state-of-the-art results on Lidar-KITTI and SF-Waymo.

2 Related Works

Scene Flow on Point Clouds. While there is extensive literature on traditional 3D scene flow [13,14,25,36–38,42], we pay more attention to recent learning-based methods [6,20,24,26,30,40] which have shown advantages in many aspects. Pioneer work

FlowNet3D [18] directly consumed point cloud and estimated scene flow through an encoder-decoder architecture. FlowNet++ [41] further used geometric constraints to get more accurate flow. HPLFlowNet [10] projected point clouds into permutohedral lattices then got flow from Bilateral Convolutional Layers. Inspired by PWC-Net [33], PointPWCNet [43] estimated scene flow in a coarse-to-fine fashion. FLOT [27] estimated scene flow through optimal transport and Meteor-Net [19] improved the accuracy of the inferred flow utilizing multiple temporal information. Recently, Gojcic et al. [9] estimated foreground flow and background flow respectively with weakly supervision. Other works [16,23] also explored the unsupervised/self-supervised sceneflow estimation because of the absence of annotated real-world scene flow dataset.

Transformer and Attention. Transformer architectures with attention mechanisms have recently revolutionized computer vision research. For 2D image, Dosovitskiy et al. [5] treat images as sequences of patches. Hu et al. [12] and Ramachandran et al. [29] applied scalar dot-product self-attention within local image patches and Zhao et al. [47] developed a family of vector self-attention operators. Recently, Zhao et al. [48] and Guo et al. [11] introduced vector and dot-product transformer into 3D point cloud to better extract geometric features in a single frame. While in this work we use attention mechanism to better model the local relationship across consecutive frames.

3 Methodology

Given two consecutive point clouds, $\boldsymbol{P}_0$ at frame t and $\boldsymbol{Q}_0$ at frame $t+1$, the task of scene flow estimation aims to predict the translation of every point in $\boldsymbol{P}_0$ from frame t to frame $t+1$.

3.1 Network Architectures

Figure 2 presents the overall workflow of our FH-Net, which comprises four key steps: *i) Feature embedding:* get key-points with shallow features for each frame; *ii) Trans-flow layer:* compute every key point's flow by directly aggregating the spatial vectors between source point and target point set; *iii) Trans-up layer:* upsample previous flow based on the similarity between sparse and dense points both in spatial and feature space; *iv) Flow propagation:* propagate the flow to raw resolution.

Feature Embedding. This step aims to sample a subset of keypoints from the input point cloud, and enrich each of the keypoints with local geometric features. We apply two sequential set abstractions [28] to $\boldsymbol{P}_0$ and $\boldsymbol{Q}_0$, outputting two keypoint sets, $\boldsymbol{P}_2 = \{(\boldsymbol{x}_i, \boldsymbol{u}_i)\}_{i=1}^{N_2}$ and $\boldsymbol{Q}_2 = \{(\boldsymbol{y}_i, \boldsymbol{v}_i)\}_{i=1}^{N_2}$, respectively, where $\boldsymbol{x}_i, \boldsymbol{y}_i \in \mathbb{N}^3$ denote *xyz* coordinates and $\boldsymbol{u}_i, \boldsymbol{v}_i \in \mathbb{R}^C$ denote embedded local features.

Trans-flow Layer. The trans-flow layer outputs the scene flow $\boldsymbol{F}_2 = \left\{\boldsymbol{f}_i^2 \in \mathbb{R}^3\right\}_{i=1}^{N_2}$ for $\boldsymbol{P}_2$. It computes each flow element as a weighted sum of the spatial offsets from a source keypoint in $\boldsymbol{P}_2$ to its neighboring target keypoints in $\boldsymbol{Q}_2$ using transformer attention mechanism, through the following two steps:

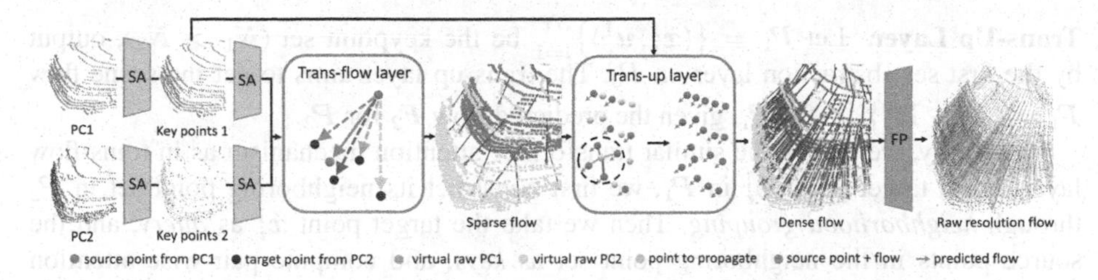

Fig. 2. Overall architecture. FH-Net first obtains keypoints from set abstraction layers (SA), then uses Trans-flow layer to compute each key point's flow by directly aggregating the spatial vectors between source point and target neighbor point set through local region relationship modeling. The following trans-up layer is adopted to dynamically upsample the key points' flow which are then propagated to the raw resolution by Flow propagation module (FP).

Neighborhood grouping: For source keypoint $\boldsymbol{p}_i = (\boldsymbol{x}_i, \boldsymbol{u}_i)$ in $\boldsymbol{P}_2$, we first use ball query to group its spatially neighboring points in $\boldsymbol{Q}_2$, forming a target candidate set:

$$\boldsymbol{Q}_2^i = \{\boldsymbol{q}_j = (\boldsymbol{y}_j, \boldsymbol{v}_j) \mid \| \boldsymbol{y}_j - \boldsymbol{x}_i \|_2 < R\}_{j=1}^{M}, \tag{1}$$

where R is a preset radius and M is the number of neighboring points.

Cross-attention: We take $\boldsymbol{p}_i$ as *query* point and $\{\boldsymbol{q}_j\}_{j=1}^{M}$ as *key* points, and compute pair-wise attention weight as:

$$\alpha_{i,j} = \gamma(\varphi(\boldsymbol{u}_i) - \psi(\boldsymbol{v}_j) + \sigma(\boldsymbol{y}_j - \boldsymbol{x}_i)), \tag{2}$$

where φ and ψ are learnable point-wise linear transformations, σ is a relative position encoding. γ is a mapping function, learned by a MLP, that produces an attention weight by considering both feature and spatial relations (Fig. 3).

Attentional Aggregation: Given the attention weight, we directly take the positional offset between $\boldsymbol{p}_i$ and $\boldsymbol{q}_j$ as *value*, and compute the translation $\boldsymbol{f}_i^2$ for $\boldsymbol{p}_i$ as a weighted sum of the positional offsets with all the *key* points:

$$\boldsymbol{f}_i^2 = \sum_{j=1}^{M} \rho(\alpha_{i,j})(\boldsymbol{y}_j - \boldsymbol{x}_i), \tag{3}$$

where ρ is a normalization function, *softmax* hereby.

Cross-frame Feature Enhancement: After obtaining the low-resolution flow, it will inevitably introduce noise if we upsample the flow rely on structure feature from single frame only. Therefore, in addition to the aggregation of the flow, the Trans-flow layer also fuse the cross-frame features for further usage. Specifically, for source keypoint $\boldsymbol{p}_i$, we integrate its corresponding local context from $\boldsymbol{Q}_2$:

$$\boldsymbol{u}_i' = \underset{j=1,\ldots,M}{\text{MAX}} \{h\left(\mathcal{C}\left(\boldsymbol{u}_i, \boldsymbol{v}_j, \sigma\left(\boldsymbol{y}_j - \boldsymbol{x}_i\right)\right)\right)\}, \tag{4}$$

where $\boldsymbol{u}_i'$ denotes enhanced features, $\mathcal{C}$ denotes concatenation in channel dimension, h denotes learnable linear transformation, MAX is element-wise max pooling.

Trans-Up Layer. Let $P_1 = \{(x_i^1, u_i^1)\}_{i=1}^{N_1}$ be the keypoint set ($N_1 > N_2$), output by the first set abstraction layer on P_0. The trans-up layer aims to get the scene flow $F_1 = \{f_i^1 \in \mathbb{R}^3\}_{i=1}^{N_1}$ for P_1, given the predicted flow F_2 for P_2.

Generally, we follow the similar transformer attention mechanism as in trans-flow layer. For a target point x_i^1 in P_1, we first construct its neighboring point set in P_2 through *neighborhood grouping*. Then we take the target point x_i^1 as *query*, and the source points in the neighboring point set as *keys*, and compute pair-wise attention weights through *cross attention*. Finally, we take the given flows of these source points as *values*, and compute the flow f_i^1 for x_i^1 through *attentional aggregation*. Optionally, we update the feature u_i^1 by applying Eq. (4) to P_1 and P_2 for further flow refinement.

Figure 4 shows that benefited from Trans-up layer, the intermediate flow not only becomes denser but also gets refined by considering the integrity and correlation of local regions which can conquer the influence of outliers.

Flow Propagation Layer. We simply use 3D interpolations to upsample scene flow F_1 to the final output F_0 instead of Trans-up layer since there are no extracted features corresponded to the raw resolution points. Optionally, to mitigate the propagation error, a flow refinement residual learned from the last layers' feature can be added to the output flow. Specifically, we upsample the last layers' feature to the raw resolution by 3D interpolations and feed it into a MLP layer then added to the interpolation result:

$$f_j^0 = \sum_{x_i^1 \in \chi} \frac{1}{\| x_i^1 - x_j^0 \|} f_i^1 + MLP\Big(\sum_{x_i^1 \in \chi} \frac{1}{\| x_i^1 - x_j^0 \|} u_i^1\Big), \tag{5}$$

where χ is the local region satisfies $\| x_i^1 - x_j^0 \|_2 < R$.

3.2 Hierarchical Loss

Our FH-Net can predict hierarchical scene flows $\{\{f_i^k \mid_{i=1,\dots,N_k}\}\}_{k=0}^{2}$, which enables us to add supervisions on each flow hierarchy during optimization, to improve predictive accuracy and speed up convergence:

$$\mathcal{L}_h = \sum_{k=0}^{2} \lambda_k \frac{1}{N_k} \sum_{i=1}^{N_k} \beta_i \left\| f_i^k - \hat{f}_i^k \right\|_2^2, \tag{6}$$

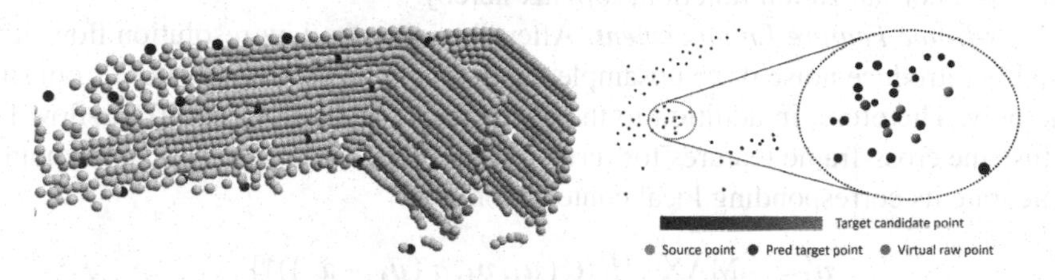

Fig. 3. Attention visualization of Trans-flow layer. Brighter green points have higher attention weights and prediction is obtained by aggregating the offset from source point to these four activated point (we draw raw points in gray here for auxiliary understanding). (Color figure online)

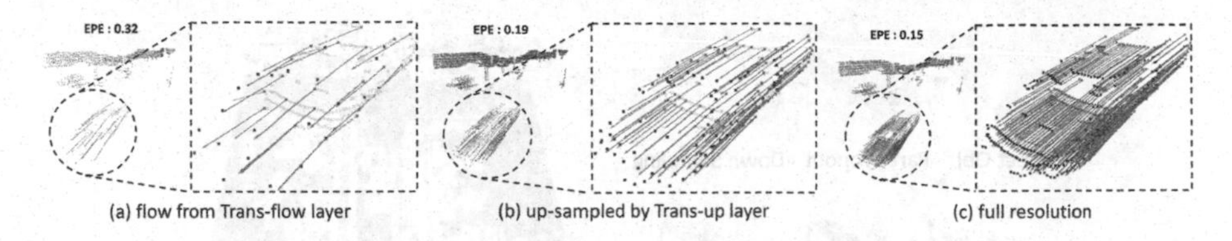

Fig. 4. Hirachical flow visualization in Lidar-KITTI. red denotes source key point in a certain resolution, blue denotes the ground truth target points and green arrows denotes the flow. **gray** points are raw $\boldsymbol{P}_0$ points for reader to better understanding local structure. (Color figure online)

where $\hat{\boldsymbol{f}}_i^k$ denotes the ground-truth flow corresponding to $\boldsymbol{f}_i^k$, λ_k is the balance weight for hierarchy k, and β_i is set to 5 and 0.5 for foreground and background points, respectively.

3.3 FH-Net Families

Based on the above fundamental layers, we provide three versions of FH-Net: **FH-R** is a standard version whose architecture specifications are listed in Table 1 **FH-L** is a lite version where residual refinement process is excluded during flow propagation, hence with less parameters and higher speed; **FH-S** is an advanced venison equipped with an extra segmentation module [4,35] to divide the point cloud into foreground and background points [1,9], whose flows are predicted by our model and by using the ego-motion from IMU sensors, respectively.

Table 1. Architecture specs of FH-R. SA: set abstraction.

Layer type	R (m)	Sampling rate	Group num	MLP width
SA layer	0.5	0.5×	16	[32, 32, 64]
SA layer	1	0.25×	16	[64, 64, 128]
Trans-flow	10	1×	64	[128, 256], [3, 64, 128], [128, 512, 1]
Trans-up	0.6	2×	64	[64, 64], [256, 128, 64], [64, 256, 1], [3, 64, 64]
Flow propagation	0.5	4×	3	[256, 256, 128, 3]

3.4 Data Augmentation

Our proposed data augmentation method first generates dynamic object pairs from a established object database then pastes them to the well designed safe area.

Dynamic Object Pairs Generation. We first establish a dynamic object database by cropping a large number of independent dynamic object point clouds from all training data. During training, we take out the point cloud of one or more dynamic objects, then for each object we apply random downsampling and part dropout strategy to generate a pair of objects which is closer to the lidar scanned pattern as shown in Fig. 5 (a). Then we paste one of the object pairs to the current frame with random pose $\boldsymbol{T}_1$, and paste the

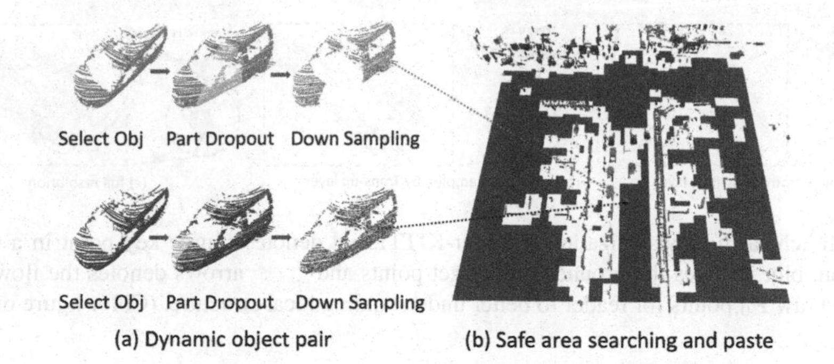

Fig. 5. Data augmentation pipeline. green denotes safe area, red and blue denote points from two frames respectively. (Color figure online)

other one with random pose T_2 in the next frame. At the same time, we must ensure that the transformation from T_1 to T_2 conforms to the motion law of real object. In this way, a dynamic object pair is established, and the scene flow can be calculated as follows:

$$F_{1\to2} = T_2 P_1 - T_1 P_1, \tag{7}$$

where $P_1 \in \mathbb{R}^4$ denotes the homogeneous points of first object in box coordinate and $F_{1\to2} \in \mathbb{R}^4$ is the homogeneous flow.

Safe Area Searching for Object Paste. Importantly, we need to prevent the pasted objects from overlapping with existing scenes in the raw frame, which does not happen in the real world. Inspired by [17], we first pillarize the points of current frame, and divide them into $h * w$ pillars in BEV (bird eye view). The pillars with points less than threshold is noted as *P-empty*, the safe area where object can be pasted. Both box_1 and box_2 should be located in the safe area. We further carry out morphological corrosion on the 2D *P-empty* image from BEV perspective to get stricter safe area of current frame as shown in Fig. 5 (b). As we first paste box_1 then paste box_2 according to the motion law, which is more convenient to implement, it can be ensured to great extent that box_2 is also located in safe area as long as the constraints of motion law are met.

4 Real-World Dataset Creation

We build a large-scale 360° point cloud scene flow dataset named SF-Waymo as well as a smaller one SF-KITTI, which are more in line with the real scene based on the general automatic driving dataset, Waymo and KITTI.

Data Collection. We collect real-world point cloud pairs from Waymo and KITTI as well as 3D gt boxes and ego-motion annotations, then process them by removing the ground points with preset height threshold as ground points are useless for understanding scene motion. For KITTI, we also crop the point cloud within the view of front camera where annotations like 3D boxes are available. In general, our SF-Waymo contains 100 scenes, total 20k pairs and SF-KITTI has 20 scenes, total 7k pairs.

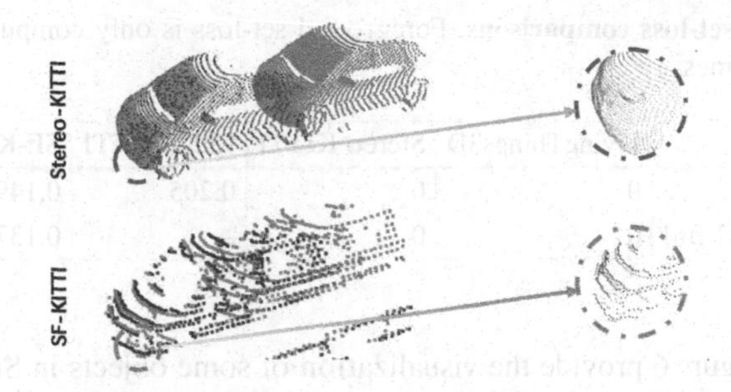

Fig. 6. Comparisons of point density and correspondence between our SF-KITTI and previous Stereo-KITTI dataset. Red denotes source point, green denotes gt flow plus source point and blue denotes target point. (Color figure online)

Table 2 provides more details and specific comparisons between our new proposed datasets with existing ones. Our datasets are closer to real scene with obvious advantages in scale, temporal frames, providing bidirectional flow and points semantic label.

Pseudo Label. Observing that the scene flow of background is mainly generated by ego motion, while foreground flow is the combination of ego motion and dynamic object's movement, we divide points into background and foreground using gt boxes and compute label respectively. In addition, we also compute bidirectional flow between two frames. For background points P_1^b at frame t, the forward flow from time t to time $t+1$ is

$$F_{for}^b = T_{21}P_1^b - P_1^b, \tag{8}$$

where T_{21} denotes the pose transform from PC_1 to PC_2. For foreground dynamic object points P_1^f cropped by gt box, the forward flow can be written as:

$$F_{for}^f = T_{box,2}T_{box,1}^{-1}P_1^f - P_1^f, \tag{9}$$

where $T_{box,2}$ and $T_{box,1}$ respectively denote the bbox's posture in PC_1 and PC_2 with vehicle coordinate, box_1 and box_2 can be matched by object ID. The backword flow can be calculated symmetrically:

$$F_{back}^b = T_{12}P_2^b - P_2^b, \tag{10}$$

$$F_{back}^f = T_{box,1}T_{box,2}^{-1}P_2^f - P_2^f. \tag{11}$$

Table 2. Statistics of Different Datasets. DC: Direct point-to-point correspondence.

Dataset	Total	Train	Test	Pattern	Temporal	DC	FOV
FlyingThings3D	23k	19k	3824	Synthetic	✗	✓	360°
Stereo KITTI	150	✗	150	Disparity projection	✗	✓	90°
Lidar-KITTI	142	✗	142	Lidar-scanned	✗	✗	90°
SF-KITTI	7k	6400	600	Lidar-scanned	✓	✗	90°
SF-Waymo	20k	17k	3k	Lidar-scanned	✓	✗	360°

Table 3. Datasets set-loss comparisons. Foreground set-loss is only computed by foreground points from two frames.

Dataset	FlyingThings3D	Stereo KITTI	Lidar-KITTI	SF-KITTI	SF-Waymo
Set-loss↓ [m]	0	0	0.205	0.149	**0.117**
Foreground set-loss↓ [m]	0	0	–	0.137	**0.077**

Visualization. Figure 6 provide the visualization of some objects in Stereo-KITTI and our SF-KITTI. It can be seen that the points in Stereo-KITTI are in direct correspondence: for every source point in the first frame, we can find its matching point in the second frame, thus the green points and blue points almost coincide with each other in the first row of the figure. In addition, Stereo-KITTI is obtained from disparity images, where the points are dense and regular, it also has a large gap compared with the real-world. While in our SF-KITTI, since points are scanned by Lidar, they are sparse and not in direct correspondence, which is more inline with the real-world.

Label Accuracy. We further verify that our pseudo labels are also accurate compared with Lidar-KITTI (labels obtained from carefully annotated KITTI-Sceneflow dataset). As GT flow is unavailable in real-world, we introduce **Set-loss** as a metric to evaluate the accuracy of flow annotations. Set-loss measures the degree of overlap between two point clouds pc_1, pc_2 (with n_1 and n_2 points, respectively) by averaging the point-to-set distance for every point in pc_1 and pc_2:

$$\frac{1}{n_1+n_2}\Big(\sum_{\boldsymbol{x}_i \in pc_1} \min_{\boldsymbol{x}_j \in pc_2} (\| \boldsymbol{x}_i - \boldsymbol{x}_j \|_2) + \sum_{\boldsymbol{x}_j \in pc_2} \min_{\boldsymbol{x}_i \in pc_1} (\| \boldsymbol{x}_j - \boldsymbol{x}_i \|_2)\Big). \tag{12}$$

Given two consecutive point cloud frames $\boldsymbol{P}$ and $\boldsymbol{Q}$ as well as the flow labels $\boldsymbol{f}$, we then compute the set loss between $\boldsymbol{P}+\boldsymbol{f}$ and $\boldsymbol{Q}$ to evaluate the accuracy of flow annotations. Ideally, for synthetic datasets FlyingThings3D and Stereo-KITTI (KITTI-Sceneflow), where points are in one to one correspondence, the set loss is **zero**. For real-world datasets like Lidar-KITTI, SF-KITTI and SF-Waymo, more accurate flow labels should also have smaller set loss. Significantly, the labels of Lidar-KITTI are computed by projecting points to image plane then assigning them the carefully annotated labels from KITTI-Sceneflow, which have been used widely. As shown in Table 3, the set-loss of SF-Waymo and SF-KITTI is lower than Lidar-KITTI and becomes smaller on foreground objects, which proves that the annotations of SF-Waymo and SF-KITTI, though generated from bounding boxes, are satisfactorily accurate and even better than those of existing real-world dataset.

5 Experiments

To evaluate the behavior of our FH-Net, we conduct comprehensive experiments on Lidar-KITTI [7,9], Stereo-KITTI and FlyingThings3D and the proposed SF-Waymo datasets. First, we describe settings of experiments, and then extend into results of our method. Further, we conduct extensive ablation studies to verify the effectiveness of FH-Net in detail.

Table 4. Evaluation results on Lidar-KITTI.

Method	Training set	EPE3D↓ [m]	Acc3DS↑	Acc3DR↑	Outliers↓	Speed (Hz)	Parameters
FlowNet3D [18]	FT3D	0.722	0.03	0.122	0.965	15.6	4,800k
PointPWC-Net [43]	FT3D	**0.39**	**0.387**	**0.55**	**0.653**	13.2	30,000k
FLOT [27]	FT3D	0.653	0.155	0.313	0.837	0.9	**440k**
Rigid3DSceneFlow [9]	FT3D	0.535	0.262	0.437	0.742	7.2	7,700k
FH-L	FT3D	0.531	0.20	0.397	0.825	**112.6**	680k
FlowNet3D [18]	SF-KITTI	0.289	0.107	0.334	0.749	15.6	4,800k
PointPWC-Net [43]	SF-KITTI	0.275	0.151	0.405	0.737	13.2	30,000k
FLOT [27]	SF-KITTI	0.271	0.133	0.424	0.725	0.9	**440k**
FH-L	SF-KITTI	0.255	0.241	0.538	0.683	**112.6**	680k
FH-R	SF-KITTI	**0.156**	**0.341**	**0.636**	**0.612**	57.3	1,500k
Rigid3DSceneFlow [9]	FT3D + SemanticKITTI	0.15	0.521	0.744	0.45	7.2	7,700k
FH-S	SF-KITTI + SemantcKITTI	**0.096**	**0.698**	**0.812**	**0.367**	**33.8**	**5,300k**

Table 5. Evaluation results on SF-Waymo.

Method	Training set	EPE3D↓ [m]	Acc3DS↑	Acc3DR↑	Outliers↓	Speed (Hz)	Parameters
FlowNet3D [18]	SF-Waymo	0.225	0.230	0.486	0.779	9.3	4,800k
PointPWC-Net [43]	SF-Waymo	0.307	0.103	0.231	0.786	1.0	30,000k
FESTA [39]	SF-Waymo	0.223	0.245	0.272	0.765	6.2	5,400k
FH-L	SF-Waymo	0.243	0.212	0.508	0.652	**66.7**	**680k**
FH-R	SF-Waymo	0.211	0.209	0.522	0.621	33.6	1,500k
FH-S	SF-Waymo	**0.175**	**0.358**	**0.674**	**0.603**	24.4	5,300k

5.1 Experimental Settings

Implementation Details. In our experiments, we set epochs to 150, batch size to 8, 4 for experiments on SF-KITTI and SF-Waymo, respectively. We train FH-Net by Adam optimizer with learning rate of 0.001. For hierarchical loss in Eq. (6), we set $\alpha_0 = 0.2$, $\alpha_1 = 0.2$ and $\alpha_2 = 0.6$, especially. In terms of data augmentation strategy, we paste 3 cars, 1–5 pedestrian and 1–4 cyclists with 0.4 probability of random down-sampling, as well as dropout. Notably, all models are trained end-to-end from scratch.

Evaluation Metrics. We take 3D end-point-error (EPE3D) as the main evaluation metric, defined as the mean L_2 distance between the ground truth and predicted scene flow. In addition, strict accuracy (Acc3DS), relaxed accuracy (Acc3DR) and Outliers are used to describe the performance of models.

5.2 Results on Lidar-KITTI

Setup. Lidar-KITTI [7,9] is a lidar-scanned point cloud scene flow dataset with ground truth from back projecting 3D points to Stereo-KITTI [21,22]. We train FH-Net on SF-KITTI with data-augmentation strategy referred in Sect. 3.4 and compare it with current SoTA methods [9,18,27,43], on Lidar-KITTI dataset. The results of previous methods trained on FlyingThings3D are from [9] and we also re-train them on SF-KITTI with the same settings as FH-Net for fair comparision. We train the segmentation branch of

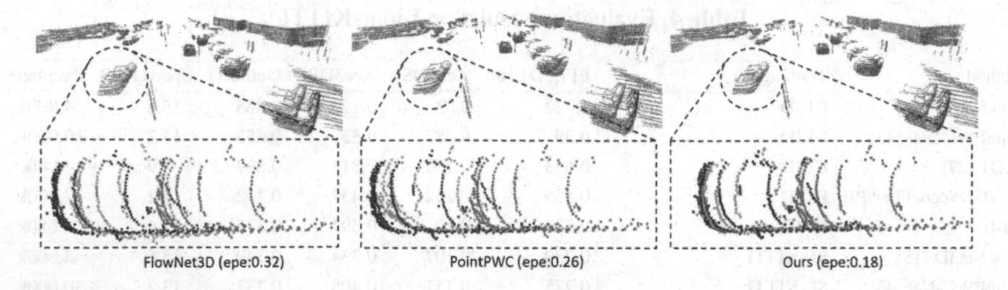

Fig. 7. Qualitative results on Lidar-KITTI. red denotes PC1 points, green denotes predicted points and blue points are the ground truth. (Color figure online)

FH-S with semantic-KITTI [2] dataset. In addition, following [9,10,18], we remove the ground points by a preset height threshold, cause they are meaningless to scene motion understanding. All methods are measured on a single RTX 3090 GPU.

Results. Table 4 shows our FH-Net achieves an EPE3D accuracy of 0.096, superior to SoTA method Rigid3DSceneFlow [9], as well as obvious advantages in terms of speed (about 5×). In addition, the elite version FH-L achieves competitive performance with fewer parameters (0.68M) and higher inference speed (112.6 fps). Compared with [9], the *params* of FH-L are reduced by 91.2% and the *speedup* is measured 15×.

Speed. Compared with current methods, FH-Net shows tremendous advantages in terms of speed. All three versions of FH-Net (FH-L, FH-R, FH-S) outperform all other methods in inference speed, achieving an extremely outstanding performance. With elite version (FH-L), we further speed up the inference to an amazing level of 112.6 fps at a sacrifice of little accuracy drop, while maintaining high-performance 0.255 EPE3D.

Visualization. We select some representative scenarios to visualize our performance and give some analysis. As shown in Fig. 7, previous works fail to fit the wide range object movement for lacking local position information, while ours predicts behave well with the help of strong local relative position.

5.3 Results on SF-Waymo

Setup. We train all methods on SF-waymo training set and evaluate them on SF-Waymo validation set. Other settings are the same as Sect. 5.1.

Table 6. Evaluation results on no-lidar-scanned dataset

Method	Training set	EPE3D
FlowNet3D [18]	FT3D	0.177
HPLFlowNet [10]	FT3D	0.117
Rigid3DSceneflow [9]	FT3D	**0.042**
FH-R	FT3D	0.098
FH-S	SF-KITTI + FT3D	0.072

(a) **Results on Stereo-KITTI**

Method	EPE3D
FlowNet3D [18]	0.114
PointPWC-Net [43]	0.059
FLOT [27]	**0.052**
Rigid3DSceneflow [9]	**0.052**
FH-R	0.054

(b) **Results on FlyingThings3D**

Results. Empirically, our method surpasses all other methods and achieves the state-of-the-art EPE3D accuracy of **0.175** with higher inference speed and fewer parameters. For example, we outperform FESTA [39] by 0.048 EPE3ED and 4× speedup in inference efficiency. Notably, the lite version FH-L achieves amazing speed (**66.7 Hz**) with slight drop in accuracy, Table 5 illustrates all results on SF-Waymo.

Speed. Due to the lightweight design of architecture, our method outperforms all other methods in inference speed. Compared with FESTA ($6.2fps$), our FH-Net still performs much better ($24.4fps, 33.6fps$ and $66.7fps$). Moreover, we notice even in the large-scale SF-Waymo dataset, our FH-Net still exceeds the requirement of real-time scene flow prediction, showing great possibilities for autonomous driving.

5.4 Results on No-lidar-scanned Dataset

We also report the results on no-lidar scanned datasets such as Stereo-KITTI and FlyingThings3D. Results in Table 6 indicate that our well-designed architecture for real-world scene flow estimation can well generalize to human-made data. Notably, on such datasets our method does not perform as well as on real-world datasets like Lidar-KITTI and SF-Waymo. Cause points in Stereo-KITTI and FlyingThings3D datasets always have direct correspondence, our method that estimates target points has poorer performance than Rigid3DSceneflow [9] that adopts matching strategy.

5.5 Ablation Studies

Table 7 shows the ablation results based on FH-L. Models are trained on SF-KITTI and evaluated on Lidar-KITTI, unless otherwise specified, analyzed as follows:

Effect of Trans-Flow Layer. We replace Trans-flow layer with other operators such as taking the average of candidate flows or directly using the nearest point's offset as flow. Table 7d shows that Trans-flow layer has obvious superiority.

Effect of Trans-Up Layer. Table 7b demonstrates that upsampling the sparse flow through Trans-up layer can better modeling the local region's relationship and get more accurate flow than simply linear interpolation. We also observe that the interacted features is beneficial in the Trans-up layer.

Effect of Position Encoding. Table 7h shows that considering the semantic feature similarity as well as the spatial similarity by adding positional encoding in trans-flow layer leads to apparent improvement, which indicates that the local geometry distribution plays an important role when predict the flow.

Refinement in Flow Propagation. Table 7f illustrates that adding residual refinement in flow propagation module can help mitigate some propagation errors as we expected. We think that the refinement module plays an important role in case of occlusion.

Effect of Hierarchical Loss. Table 7c shows that intermediate flow supervision are beneficial, which is consistent with our assumptions that the more accurate intermediate flows also lead to better performance of the final prediction. We also find that expanding foreground weight by five times significantly boost the performance.

Table 7. Ablations studies

Car	Ped	Cyc	EPE3D*	EPE3D
✗	✗	✗	0.278	0.308
0	1-5	1-4	0.276	0.301
3	1-5	1-4	0.127	**0.255**
6	1-5	1-4	**0.106**	0.259

(a) **Data augmentation:** * denotes the error of foreground.

Linear	Trans	F.I	EPE3D
✓	✗	✗	0.459
✗	✓	✗	0.314
✗	✓	✓	**0.255**

(b) **Trans-up layer:** F.I denotes feature interaction.

$[\lambda_0, \lambda_1, \lambda_2]$	$[\beta_0, \beta_1]$	EPE3D
✗	✗	0.306
0.5, 0.5, 0	✗	0.298
0.6, 0.2, 0.2	1, 1	0.265
0.6, 0.2, 0.2	5, 0.5	**0.255**

(c) **Hierarchical loss:** λ and β are hyperparameters of loss.

TF	Avg	N	EPE3D
✗	✓	✗	1.384
✗	✗	✓	1.025
✓	✗	✗	**0.255**

(d) **Trans-flow layer:** TF,Avg,N denote trans-flow, average, nearest.

Method	Train set	Validation set	EPE3D
FlowNet3D	FT3D	SF-Waymo/eval	0.834
FlowNet3D	SF-Waymo/train	SF-Waymo/eval	**0.225**
FH-L	FT3D	SF-Waymo/eval	0.673
FH-L	SF-Waymo/train	SF-Waymo/eval	**0.243**

(e) **Training data**: We train baseline model and FH-L on commonly used training set FT3D and SF-Waymo.

FR	EPE3D
✗	0.255
✓	**0.223**

(f) **FR**: res- idual flow refinement.

Method	Pre-train set	Fine-tune set	Validation set	EPE3D
FlowNet3D	FT3D	Stereo-KITTI	Stereo-KITTI	0.144
FH-L	FT3D	Stereo-KITTI	Stereo-KITTI	0.110
FlowNet3D	SF-KITTI	Stereo-KITTI	Stereo-KITTI	**0.114**
FH-L	SF-KITTI	Stereo-KITTI	Stereo-KITTI	**0.079**

(e) **Pre-train and fine-tune:** We pre-train the baseline and our FH-L on our dataset then fine-tune on others.

Pos.enc	EPE3D
✗	0.278
✓	**0.255**

(h) **Position encoding**

Effect of Data Augmentation. Table 7a shows the effectiveness of our new proposed data augmentation method with different number of pasted dynamic objects. Furthermore, we also evaluate the foreground EPE3D to investigate the performance gain from foreground dynamic object pair paste strategy. The result on Table 7a is consistent with our assumptions that increasing the number and diversity of dynamic objects can significantly boost the performance especially on more challenging foreground objects.

Effect of Training Data. In Table 7e and Table 4, we analyze the effect of our new proposed SF-KITTI and SF-Waymo by re-training models on them and compare the results with those trained on FT3D. It can be seen that the EPE3D is obviously reduced both on Lidar-KITTI and SF-Waymo, which proves that our new dataset is more friendly to the real-world sceneflow esitamtion. Furthermore, we evaluate the effectiveness of our datasets by using it to pretrain the model then fine-tune on other public datasets, Table 7g shows that the network can achieve better performance by learning more challenging real-world scenes in our new proposed dataset.

6 Conclusion

In this paper, we present a fast hierarchical framework named FH-Net which can output hierarchical flows and maintains high inference speed. We also propose a new copy-and-paste data augmentation method to focus more on challenging dynamic objects and establish two real-world sceneflow dataset for real-world scene flow estimation.

Acknowledgements. This work was financially supported by the National Natural Science Foundation of China (No. 62101032), the Postdoctoral Science Foundation of China (No. 2021M690015), and Beijing Institute of Technology Research Fund Program for Young Scholars (No. 3040011182111).

References

1. Behl, A., Paschalidou, D., Donné, S., Geiger, A.: Pointflownet: learning representations for rigid motion estimation from point clouds. In: CVPR (2019)
2. Behley, J., Garbade, M., Milioto, A., Quenzel, J., Behnke, S., Stachniss, C., Gall, J.: Semantickitti: a dataset for semantic scene understanding of lidar sequences. In: ICCV (2019)
3. Chen, S., Li, Y., Kwok, N.M.: Active vision in robotic systems: a survey of recent developments. In: IJRR (2011)
4. Choy, C., Gwak, J., Savarese, S.: 4D spatio-temporal convnets: minkowski convolutional neural networks. In: CVPR (2019)
5. Dosovitskiy, A., et al.: An image is worth 16 × 16 words: transformers for image recognition at scale. arXiv preprint arXiv:2010.11929 (2020)
6. Fan, H., Yang, Y.: PointRNN: point recurrent neural network for moving point cloud processing. arXiv preprint arXiv:1910.08287 (2019)
7. Geiger, A., Lenz, P., Urtasun, R.: Are we ready for autonomous driving? the kitti vision benchmark suite. In: CVPR (2012)
8. Ghiasi, G., et al.: Simple copy-paste is a strong data augmentation method for instance segmentation. In: CVPR (2021)
9. Gojcic, Z., Litany, O., Wieser, A., Guibas, L.J., Birdal, T.: Weakly supervised learning of rigid 3D scene flow. In: Proceedings of the IEEE/CVF Conference on Computer Vision and Pattern Recognition, pp. 5692–5703 (2021)
10. Gu, X., Wang, Y., Wu, C., Lee, Y.J., Wang, P.: Hplflownet: hierarchical permutohedral lattice flownet for scene flow estimation on large-scale point clouds. In: CVPR (2019)
11. Guo, M.H., Cai, J.X., Liu, Z.N., Mu, T.J., Martin, R.R., Hu, S.M.: PCT: point cloud transformer. Comput. Vis. Media 7(2), 187–199 (2021)
12. Hu, H., Zhang, Z., Xie, Z., Lin, S.: Local relation networks for image recognition. In: ICCV (2019)
13. Huguet, F., Devernay, F.: A variational method for scene flow estimation from stereo sequences. In: ICCV (2007)
14. Jaimez, M., Souiai, M., Gonzalez-Jimenez, J., Cremers, D.: A primal-dual framework for real-time dense RGB-D scene flow. In: ICRA (2015)
15. Jund, P., Sweeney, C., Abdo, N., Chen, Z., Shlens, J.: Scalable scene flow from point clouds in the real world. IEEE Robot. Autom. Lett. (2021)
16. Kittenplon, Y., Eldar, Y.C., Raviv, D.: Flowstep3D: model unrolling for self-supervised scene flow estimation. In: CVPR (2021)
17. Lang, A.H., Vora, S., Caesar, H., Zhou, L., Yang, J., Beijbom, O.: Pointpillars: fast encoders for object detection from point clouds. In: CVPR (2019)
18. Liu, X., Qi, C.R., Guibas, L.J.: Flownet3D: learning scene flow in 3D point clouds. In: CVPR (2019)
19. Liu, X., Yan, M., Bohg, J.: Meteornet: deep learning on dynamic 3D point cloud sequences. In: ICCV (2019)
20. Mayer, N., Ilg, E., Hausser, P., Fischer, P., Cremers, D., Dosovitskiy, A., Brox, T.: A large dataset to train convolutional networks for disparity, optical flow, and scene flow estimation. In: CVPR (2016)

21. Menze, M., Heipke, C., Geiger, A.: Joint 3D estimation of vehicles and scene flow. ISPRS (2015)
22. Menze, M., Heipke, C., Geiger, A.: Object scene flow. ISPRS (2018)
23. Mittal, H., Okorn, B., Held, D.: Just go with the flow: self-supervised scene flow estimation. In: CVPR (2020)
24. Mustafa, A., Hilton, A.: Semantically coherent 4D scene flow of dynamic scenes. IJCV (2020)
25. Newcombe, R.A., Fox, D., Seitz, S.M.: Dynamicfusion: reconstruction and tracking of non-rigid scenes in real-time. In: CVPR (2015)
26. Niemeyer, M., Mescheder, L., Oechsle, M., Geiger, A.: Occupancy flow: 4D reconstruction by learning particle dynamics. In: ICCV (2019)
27. Puy, G., Boulch, A., Marlet, R.: Flot: scene flow on point clouds guided by optimal transport. In: ECCV (2020)
28. Qi, C.R., Yi, L., Su, H., Guibas, L.J.: Pointnet++: deep hierarchical feature learning on point sets in a metric space. arXiv preprint arXiv:1706.02413 (2017)
29. Ramachandran, P., Parmar, N., Vaswani, A., Bello, I., Levskaya, A., Shlens, J.: Stand-alone self-attention in vision models. arXiv preprint arXiv:1906.05909 (2019)
30. Rempe, D., Birdal, T., Zhao, Y., Gojcic, Z., Sridhar, S., Guibas, L.J.: Caspr: learning canonical spatiotemporal point cloud representations. arXiv preprint arXiv:2008.02792 (2020)
31. Ronneberger, O., Fischer, P., Brox, T.: U-Net: convolutional networks for biomedical image segmentation. In: Navab, N., Hornegger, J., Wells, W.M., Frangi, A.F. (eds.) MICCAI 2015. LNCS, vol. 9351, pp. 234–241. Springer, Cham (2015). https://doi.org/10.1007/978-3-319-24574-4_28
32. Shao, L., Shah, P., Dwaracherla, V., Bohg, J.: Motion-based object segmentation based on dense RGB-D scene flow. IEEE Robot. Autom. Lett. **3**(4), 3797–3804 (2018)
33. Sun, D., Yang, X., Liu, M.Y., Kautz, J.: Pwc-net: CNNs for optical flow using pyramid, warping, and cost volume. In: CVPR (2018)
34. Sun, P., et al.: Scalability in perception for autonomous driving: waymo open dataset. In: CVPR (2020)
35. Sun, P., et al.: RSN: range sparse net for efficient, accurate lidar 3D object detection. In: CVPR (2021)
36. Tanzmeister, G., Thomas, J., Wollherr, D., Buss, M.: Grid-based mapping and tracking in dynamic environments using a uniform evidential environment representation. In: ICRA (2014)
37. Ushani, A.K., Wolcott, R.W., Walls, J.M., Eustice, R.M.: A learning approach for real-time temporal scene flow estimation from lidar data. In: ICRA (2017)
38. Vedula, S., Baker, S., Rander, P., Collins, R., Kanade, T.: Three-dimensional scene flow. In: ICCV (1999)
39. Wang, H., Pang, J., Lodhi, M.A., Tian, Y., Tian, D.: Festa: flow estimation via spatial-temporal attention for scene point clouds. In: CVPR (2021)
40. Wang, S., Suo, S., Ma, W.C., Pokrovsky, A., Urtasun, R.: Deep parametric continuous convolutional neural networks. In: CVPR (2018)
41. Wang, Z., Li, S., Howard-Jenkins, H., Prisacariu, V., Chen, M.: Flownet3D++: geometric losses for deep scene flow estimation. In: WACV (2020)
42. Wedel, A., Rabe, C., Vaudrey, T., Brox, T., Franke, U., Cremers, D.: Efficient dense scene flow from sparse or dense stereo data. In: ECCV (2008)
43. Wu, W., Wang, Z., Li, Z., Liu, W., Fuxin, L.: Pointpwc-net: a coarse-to-fine network for supervised and self-supervised scene flow estimation on 3D point clouds. arXiv preprint arXiv:1911.12408 (2019)
44. Yan, Y., Mao, Y., Li, B.: Second: sparsely embedded convolutional detection. Sensors **18**(10), 3337 (2018)

45. Yin, T., Zhou, X., Krahenbuhl, P.: Center-based 3D object detection and tracking. In: CVPR (2021)
46. Yurtsever, E., Lambert, J., Carballo, A., Takeda, K.: A survey of autonomous driving: common practices and emerging technologies. IEEE Access **8**, 58443–58469 (2020)
47. Zhao, H., Jia, J., Koltun, V.: Exploring self-attention for image recognition. In: CVPR (2020)
48. Zhao, H., Jiang, L., Jia, J., Torr, P.H., Koltun, V.: Point transformer. In: ICCV (2021)

SpatialDETR: Robust Scalable Transformer-Based 3D Object Detection From Multi-view Camera Images With Global Cross-Sensor Attention

Simon Doll[1,3](✉), Richard Schulz[1], Lukas Schneider[1], Viviane Benzin[1], Markus Enzweiler[2], and Hendrik P. A. Lensch[3]

[1] Mercedes -Benz, Stuttgart, Germany
simon.doll@mercedes-benz.com
[2] Esslingen University of Applied Sciences, Stuttgart, Germany
[3] University of Tübingen, Tübingen, Germany
simon.doll@student.uni-tuebingen.de

Abstract. Based on the key idea of DETR this paper introduces an object-centric 3D object detection framework that operates on a limited number of 3D object queries instead of dense bounding box proposals followed by non-maximum suppression. After image feature extraction a decoder-only transformer architecture is trained on a set-based loss. SpatialDETR infers the classification and bounding box estimates based on attention both spatially within each image and across the different views. To fuse the multi-view information in the attention block we introduce a novel geometric positional encoding that incorporates the view ray geometry to explicitly consider the extrinsic and intrinsic camera setup. This way, the spatially-aware cross-view attention exploits arbitrary receptive fields to integrate cross-sensor data and therefore global context. Extensive experiments on the nuScenes benchmark demonstrate the potential of global attention and result in state-of-the-art performance. Code available at https://github.com/cgtuebingen/SpatialDETR.

Keywords: 3D object detection · Cross-sensor attention · Autonomous driving

1 Introduction

To achieve the goal of autonomous driving it is necessary to perform 3D scene understanding and to detect objects such as other cars or pedestrians. The required 3D object detection task is typically either performed using a LiDAR sensor [10,30], multi-view cameras [27,29] or by using multiple sensor modalities [25,31]. While LiDAR-based methods have the advantage of precise information about the 3D structure of objects, camera-only approaches come with higher sensor-refresh rates and scale also to low-cost autonomous vehicles.

The problem of multi-view 3D object detection solely from camera images is challenging due to the fact that the images only contain a 2D projection of the

S. Avidan et al. (Eds.): ECCV 2022, LNCS 13699, pp. 230–245, 2022.
https://doi.org/10.1007/978-3-031-19842-7_14

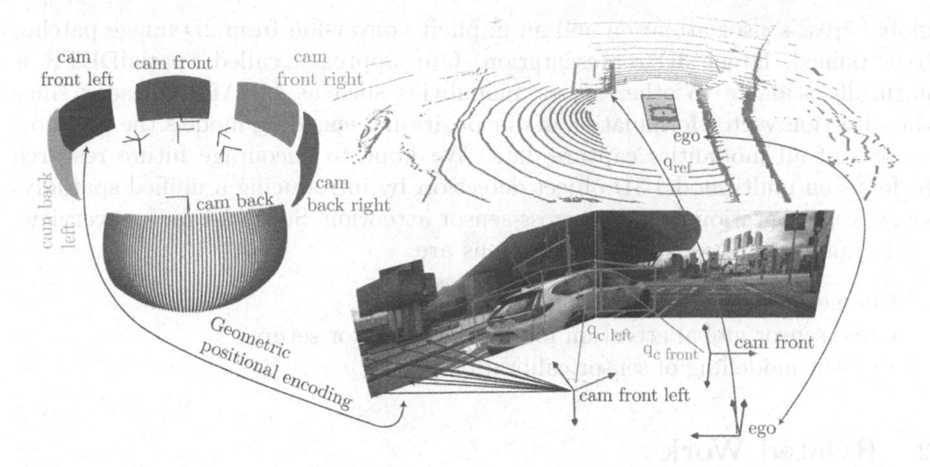

Fig. 1. Visualization of our spatially-aware global cross-sensor attention. Fusing the 3D information from the individual images is supported by a positional encoding (left) that incorporates the global 3D view direction of each pixel

3D scene. Furthermore, the position of the camera sensors needs to be taken into account, since some cameras might have overlapping field of views. FCOS3D [27] tackles this issue by extending concepts from the 2D object detection to the 3D domain. This typically results in the need for post-processing steps to filter out duplicate detections in overlapping regions since the 3D structure and the positioning of the sensors is not taken into account. Pseudo-LiDAR-based methods [28] solve this by converting the images to a 3D point cloud [16] and achieve state-of-the-art performance on the nuScenes benchmark [1]. Using multi-head dot-product attention as introduced in [24], DETR [2] proposed a new object-centric approach to 2D object detection. Recently, DETR3D [29] pioneered in extending this concept to the multi-view 3D object detection domain. But in contrast to DETR where global attention between a query and each feature patch is used to select pixels corresponding to an object, DETR3D only uses pixels that correspond to the query center and therefore supposably misses feature patches corresponding to a large object in the scene.

To allow for global, cross-sensor attention and to therefore enable an object query to select all feature patches that correspond to an object in the scene the DETR3D formulation is not sufficient. Global, cross-sensor attention requires a way to compare an object query with keys that correspond to feature patches in the different camera images.

Motivated by the findings in [16] that the success of pseudo-LiDAR methods results from the explicit modelling of the sensor positions, we propose a new geometric positional encoding to exploit the spatial and geometric structure of object detection in 3D. We furthermore introduce a spatially-aware cross-attention block to account for different coordinate spaces and to model query-key similarity in a sensor-specific fashion. The resulting combination enables global cross-sensor attention and an implicit conversion from 2D image patches to a unified, latent 3D representation. Our approach called SpatialDETR is naturally scalable to other sensor modalities such as LiDAR or Radar since the direction vector formulation of our positional encoding models the way how sensors of all modalities capture data. We hope to encourage future research to focus on multi-modal 3D object detection by introducing a unified spatially-aware representation for global cross-sensor attention. See Fig. 1 for an overview.

Summarizing, our main contributions are:

- Geometric positional encoding
- Cross-sensor global attention for arbitrary sensor setups
- Explicit modelling of sensor calibrations

2 Related Work

2.1 Transformer-Based Object Detection

Modern 2D object detectors often predict a high amount of *oriented bounding boxes* (OBB)s [12,20,23] by using anchor-boxes or a dense prediction scheme. As a result, such methods rely on post-processing steps such as non-maximum suppression to filter out highly overlapping OBB proposals. DETR [2] mitigates this issue by viewing object detection as a set-prediction problem. Using a transformer encoder-decoder architecture and learnable object queries as input to the decoder results in a small number of object proposals which allow for a set-based loss via bipartite matching with the ground truth. To improve the slow convergence of DETR various methods have been proposed such as introducing sparsity to keys [33] or queries [21] or by using an encoder-only version of DETR [22].

To perform object detection in 3D from monocular camera images, two-stage methods convert the images to a pseudo-LiDAR point cloud [8,28] using dense depth-estimation. Single stage methods either leverage an object detection network that was pretrained on a large-scale monocular depth estimation task [18] or multi-task learning [27].

DETR3D [29] proposes a DETR-based decoder-only approach that is capable of 3D object detection via 3D to 2D queries without dense depth prediction. For each object query the object center is extracted via a *feed-forward network* (FFN) and projected to the different image planes to extract multi-view features from the backbone. Similar to DeformableDETR, the attention weights are not image-feature dependent via query-key similarity as in the original attention formulation [24] but are instead directly inferred from the object queries with a FFN. As a result, the attended feature patches correspond to the object centers only, which requires deformable convolutions in the backbone to allow for larger receptive fields since no global attention is performed. In contrast to this, we propose a geometric positional encoding that explicitly models the sensor calibration and allows to incorporate global context across multiple sensors via global cross-attention.

2.2 Prior-Guided Attention

As stated in [33] attention-based methods like DETR require long training schedules since they do not introduce inductive biases like convolutional architectures and learn a fully flexible receptive field over a sequence of inputs [24]. Deformable DETR and DETR3D mitigate this issue by attending to a small number of patches only or by even using only the object center. Spatial-Gaussian priors are introduced in [6] to increase the attention weights of feature patches close to the objects and to therefore account for the local nature of objects. In TransFusion [15] the epipolar-line and epipolar field are used in a dual camera setup as geometric loss function to guide the attention weights during training. In our work we avoid hard constraints, guide attention through explicit modeling of the scene geometry and model geometric dependencies between an object query and a feature patch directly in the cross-attention blocks using sensor-relative query projections. Due to the geometric interpretation of our encoding it is still possible to introduce task or sensor-set specific biases, e.g. by only using queries that lie in front of the image plane to speed up the convergence.

2.3 Positional Encodings

Since transformers are invariant to input permutations, positional encodings as introduced in [24] and [5] are used to incorporate positional information to the feature patches. Typically, either learned or fixed encodings are used, whilst the latter lead to a slightly higher performance in DETR [2]. Fixed encodings consisting of alternating sine and cosine functions [2] might have the benefit of better extrapolation behavior. Although those encodings are well-suited for grid-based data like images the encoding of multi-view data from cameras mounted at arbitrary positions results in a complex graph of pixel directions. Additionally, small calibration changes, e.g. due to different vehicles in a fleet of cars might result in changed pixel directions. Our proposed encoding models pixel directions as a function of the sensor extrinsics and is scalable to varying sensor-modalities and arbitrary sensor calibrations.

3 Method

3.1 Overall Architecture

An overview of the proposed architecture is presented in Fig. 2. The input consists of multi-view, RGB camera images which are utilized to predict a set of 3D bounding boxes in the scene.

A shared backbone for all cameras is applied to compress the images to a latent representation. Following the design of DETR and DETR3D a transformer decoder is used to generate OBB proposals. Our proposed geometric positional encoding is used to incorporate the extrinsics and intrinsics of the camera setup and scene geometry into the sequence of input features supporting global attention across all cameras. The so-called *object queries* are learnable vectors that are refined iteratively using a stack of decoder layers. Each

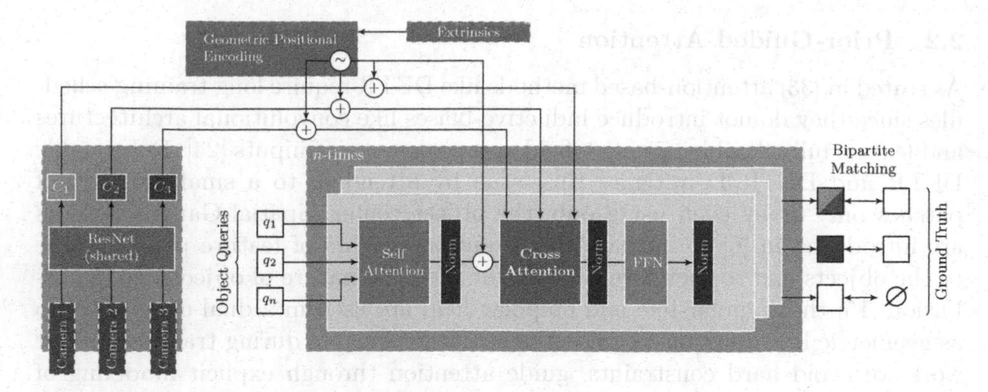

Fig. 2. SpatialDETR architecture. The camera images are encoded using a shared backbone. Afterwards object proposals are iteratively refined with self- and cross-attention in n decoder layers. The newly proposed geometric positional encoding is used in each spatially-aware cross-attention block for query-key similarity

decoder layer consists of a self-attention block between the object queries and a cross-attention block between object queries and image feature maps. In contrast to DETR our decoder-only approach scales linearly with the image resolution since no self-attention between feature patches is performed as motivated in [9]. As in DETR3D [29], the final bounding box predictions are generated by a FFN converting each latent object query to an OBB described as $(x, y, z, w, h, l, \theta_{cos}, \theta_{sin}, v_x, v_y)$ where (x, y, z) correspond to the object center, (w, h, l) to the extend (width, length and height) of the bounding box, v_x and v_y to the object velocity and $\theta = \arctan\left(\frac{\theta_{sin}}{\theta_{cos}}\right)$ to the object heading. Following [29], the bounding box center coordinates are predicted in normalized coordinates using a sigmoid function and are afterwards scaled to the desired object detection range to increase the numeric stability. During training each decoder layer computes bounding box predictions to improve the gradient flow, during inference only the output of the last decoder layer is used.

After each decoder layer the output object centers $\mathbf{o}_c^{n+1}$ are interpreted as relative offsets to the object centers of the last decoder layer before the sigmoid which results in the following update of a query center $\mathbf{q}_c^{n+1}$ in decoder layer $n+1$ as introduced in [29].

$$\mathbf{q}_c^{n+1} = \text{sigmoid}\left(\text{sigmoid}^{-1}(\mathbf{q}_c^n) + \mathbf{o}_c^{n+1}\right) \tag{1}$$

3.2 Revisiting Attention In DETR

Vaswani et al. introduced the concept of dot-product attention in [24] as

$$Attn(\mathbf{Q}, \mathbf{K}, \mathbf{V}) = \text{softmax}(\frac{\mathbf{Q} \cdot \mathbf{K}^T}{\sqrt{d_k}}) \cdot \mathbf{V} \tag{2}$$

For each value $\mathbf{v} \in \mathbf{V}$, an attention weight is computed as the scaled dot product between a query $\mathbf{q} \in \mathbf{Q}$ and the corresponding key vector $\mathbf{k}$ (see [24]). This query-key similarity is responsible for selecting the relevant values in the input. Afterwards, the resulting weighted sum of values is the output of an attention head. In the case of cross-attention for object detection the values and keys correspond to linearly projected feature patches while the queries correspond to objects in the scene. To adapt this concept to multi-sensor cross-attention, we need to solve two main challenges in our framework:

- **Position information:** The spatial location of each feature patch from all cameras needs to be incorporated, e.g. by using a positional encoding.
- **Query-key similarity:** The query-key similarity via the dot-product in Eq. (2) needs to take the different coordinate spaces into account, since object queries are defined with respect to a reference system while feature patches encode camera-relative information.

As a solution we propose a novel geometric positional encoding in Sect. 3.3 to model spatial dependencies between feature pixels. Furthermore, we introduce a spatially-aware cross-attention block that resolves the issues related with different coordinate spaces in Sect. 3.4.

3.3 Geometric Positional Encoding

As shown in DETR [2] global spatial attention between a query that represents an object and feature patches from the images is a simple yet effective way to allow for flexible receptive fields and therefore global context. The selected image patches are used to update the query to allow for a precise localization and classification of the object. Furthermore, object detection based on attention removes the need for post-processing steps like non-maximum suppression or prior knowledge such as anchor boxes while still achieving state of the art performance since the learnable object queries specialize to learned priors during training [2].

To utilize dot-product attention as defined in Eq. (2) the keys need a positional encoding to incorporate position information of an image patch, since the general attention mechanism is permutation invariant with respect to the input sequence [2,24]. Modelling this with a 1D or 2D fixed *sine-cosine* encoding [5] or a learnable positional encoding [2] loses the advantage of explicitly modelling the sensor extrinsics as discussed beforehand. Additionally, in contrast to encodings for a single image in which the pixels follow a grid-like structure, the introduction of multi-view images results in a complex graph of pixel directions (see Fig. 1). A simple 1D or 2D enumeration of pixel positions as done in [5] or [2] is therefore not applicable. Taking the positioning of the different sensors with respect to a reference frame into account is required to preserve robustness and scalability to a fleet of cars with varying extrinsic calibrations or even different sensor sets. Our approach is in contrast to DETR3D where the 3D query centers are projected into the individual images via the extrinsic and intrinsic calibration, followed by a simple lookup of the image feature at that position without any spatial attention. The subsequent cross-attention across the different views does therefore not need image patch information. As a result, a query in DETR3D only aggregates features corresponding to object centers across multiple-feature levels and different views without a flexible receptive field and global context or key to query similarity.

Having the possibility to use global attention is required to improve the detection of large objects such as trucks or busses and is the key step towards a unified representation for multi-task learning since tasks like behavior prediction or planning rely on global context as shown at the TESLA-AI-Day[1] and in [19].

To allow for robust global attention we propose a sensor-relative, spatially-aware encoding. From the camera intrinsic parameters for a camera c which are given from the dataset we can directly compute normalized direction vectors $\mathbf{d}_c^p$ per pixel p. This view-ray vectors are then element-wise added to the latent key vectors via a fully learnable network *dir2latent* that maps the three dimensional inputs to the latent feature dimension which is shared for all key and query vectors (see Fig. 3).

This results in the following update for the latent key vectors:

$$\hat{\mathbf{k}}_c^p = \mathbf{k}_c^p + \mathit{dir2latent}(\mathbf{d}_c^p) \tag{3}$$

Therefore, no hyper parameter for scaling the encoding, to control the amount of shift that is applied to the values, is required and the model learns how to merge semantic and geometric information during training.

It is noteworthy that this formulation results in sensor-relative encodings, since the direction vectors are defined with respect to the corresponding camera frames. A transformation of the key encodings to reference coordinates is not applicable without an explicit depth estimate per feature pixel due to the translation component. Similar to fixed sine-cosine encodings our geometric positional encoding can model relative offsets with a linear operator [24] by using a three-dimensional rotation matrix as operator.

3.4 Spatially-Aware Attention

To use the sensor-relative encodings and to perform similarity matching within the same coordinate space while allowing to explicitly model the sensor extrinsics, the queries in a cross-attention block need to be camera specific as they are used in the dot-product with sensor specific keys (see Eq. (2)).

To tackle this each object query $\mathbf{q}$ is projected to all camera frames using

$$\mathbf{q}_{\text{ref}}^{3d} = \text{center}(\mathit{query2box}(\mathbf{q})) \tag{4}$$

$$\mathbf{q}_c^{3d} = {}^{c}\mathbf{T}_{\text{ref}} \cdot \mathbf{q}_{\text{ref}}^{3d} \tag{5}$$

$$\mathbf{q}_c = \mathbf{q} + \mathit{dir2latent}\left(\mathbf{q}_c^{3d} \,/\, ||\mathbf{q}_c^{3d}||_2\right) \tag{6}$$

[1] TESLA-AI-Day: https://www.youtube.com/watch?v=j0z4FweCy4M.

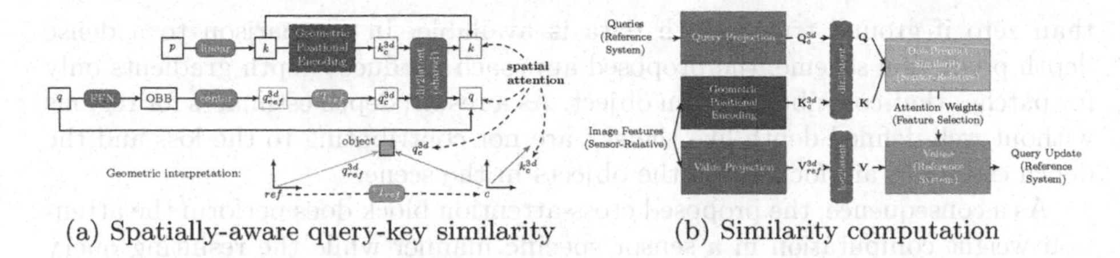

(a) Spatially-aware query-key similarity (b) Similarity computation

Fig. 3. Visualization of the proposed spatial attention module. Each query $\mathbf{q}$ gets a sensor-relative update and each key $\mathbf{k}$ a geometric update. The resulting spatially-aware latent vectors can then be compared using a dot-product to perform spatial attention

The FFN *query2box* projects the latent query vector to an OBB as defined in Sect. 3.1. The query center is then projected to all camera frames. Transformations ${}^{b}\mathbf{T}_{a}$ from one frame a to another frame b are modelled via homogenous matrices

$$ {}^{b}\mathbf{T}_{a} = \begin{bmatrix} \mathbf{R} & \mathbf{t} \\ 0 & 1 \end{bmatrix}. \tag{7} $$

A visualization of the camera-relative queries and their geometric similarity with keys from a camera image is given in Fig. 3.

As a result, the queries get a camera-relative geometric update that enforces a latent encoding which is similar to the geometric-aware keys by sharing the network for direction to latent projection. The resulting queries $\mathbf{q}_c$ and keys $\hat{\mathbf{k}}_c^p$ are then used for the attention weight computation.

In contrast to this sensor-relative space, the resulting sum of weighted values is used to perform a query update in reference coordinates. This is due to the fact, that patches from different cameras may contribute to the same object query. This results in the need for values that are encoded with respect to the reference frame to decouple sensor extrinsics from the feature encodings. Analogously to the query projection this can be done by computing a latent update to transform a value to the reference system. This is achieved by computing an explicit depth estimate v_{depth} for each latent value vector $\mathbf{v}_c^p$ utilizing a FFN:

$$ v_{\text{depth}} = \mathit{FFN}(\mathbf{v}_c^p) \tag{8} $$

$$ \mathbf{v}_c^{3d} = v_{depth} \cdot \mathbf{d}_c^p \tag{9} $$

$$ \mathbf{v}_{\text{ref}}^{3d} = {}^{\text{ref}}\mathbf{T}_{\text{c}} \cdot \mathbf{v}_{\text{c}}^{3d} \tag{10} $$

$$ \mathbf{v}_{\text{ref}}^{p} = \mathbf{v}_c^p + \mathit{loc2latent}(\mathbf{v}_{\text{ref}}^{3d}). \tag{11} $$

The direction vector $\mathbf{d}_c^p$ is scaled according to the computed factor which results in a sensor-relative depth estimate. Finally, the latent value update is applied using the FFN *loc2latent* to convert the 3D global position $\mathbf{v}_{\text{ref}}^{3d}$ to a latent vector, resulting in the spatially-aware latent value vector $\mathbf{v}_{\text{ref}}^{p}$.

This formulation also directly integrates multi-task learning into the framework by introducing a depth loss for all values with an attention weight greater than zero if ground truth depth data is available. In comparison to a dense depth prediction scheme, the proposed approach produces depth gradients only for patches that contributed to an object. As a result, depth estimates for regions without well-defined depth like the sky are not contributing to the loss and the depth estimates are focused on the objects in the scene.

As a consequence, the proposed cross-attention block does perform the attention weight computation in a sensor specific manner while the resulting query updates from weighted values are performed in reference coordinates. We expect this formulation to help the model to decouple the sensor positioning from the feature encoding and to scale to small calibration changes or online calibration.

4 Experiments

We evaluate the performance of SpatialDETR on the well-established nuScenes benchmark in the setting of camera-only single-shot 3D object detection. In addition, we provide qualitative results and several ablation studies to evaluate the effects of the proposed encoding, the spatial attention block and decoder layer configuration.

4.1 Experimental Setup

Dataset. All experiments are performed on the nuScenes dataset [1] for autonomous driving. Data was collected using cars that feature a setup of six cameras mounted on the roof of the car to capture images in different directions as well as other sensors. No radar or LiDAR data is used during training or evaluation. The dataset consists of 1000 scenes with a length of 20 s each, containing annotations of 23 classes from which 10 are used to compute benchmark metrics [17]. Training is performed utilizing the officially defined `train` split. Since ground-truth annotations are only provided for the training and validation set, all ablation experiments are performed on the validation set.

Metrics. For evaluation we report metrics according to the official nuScenes standard [1] which include the mean Average Precision (mAP), true-positive metrics such as Average Translation Error (ATE), Average Scale Error (ASE) and Average Orientation Error (AOE), Average Velocity Error (AVE), Average Attribute Error (AAE) and the summarizing nuScenes Detection Score (NDS). For a detailed metric definition, the mAP computation and class-specific detection ranges we refer the reader to [1].

Model Configuration. All model configurations use a shared ResNet-101 [7] backbone which is initialized from a FCOS3D [27] checkpoint as in [4]. The attached SpatialDETR head uses the output of the last backbone stage as input. As done in DETR3D [29] a stack of $d = 6$ decoder layers is then applied to produce the object bounding box proposals for $q = 900$ object queries. We closely follow the decoder configuration proposed in DETR3D [4] to increase the comparability. Each decoder layer uses a latent dimension $d_e = 256$ for the attention computation which is then split to $h = 8$ heads of parallel multi-head-attention modules with a latent dimension $d_h = \frac{d_e}{h} = 32$ in sequential query-to-query self-attention and query-to-patch cross-attention blocks. All FFNs used in the spatial attention blocks utilize two hidden layers consisting of a linear projection of dimensionality d_e, Layer-norm and a ReLU activation function. The two sub-networks that predict the OBB and a corresponding class label are chosen as proposed in [29]. The benchmark configurations are trained with shared decoder weights and attend only to camera images in which the object center is visible, see Sect. 4.3 for details.

Training Parameters. As proposed in [4] we train all models for 24 epochs, a total batch size of 8 on 8 NVIDIA A100 GPUs with 40GB RAM and a backbone in which the first block is frozen. The ablation models are trained with a batch size of 4 on 4 NVIDIA V100 GPUs with 16GB RAM and a fully frozen backbone to reduce the memory footprint. All models are trained using the settings proposed in DETR3D [4], consisting of an AdamW [14] optimizer and a cosine-annealing lr-schedule with an initial learning rate of $2e^{-4}$. We did not perform hyperparameter tuning for the used learning rate schedule, optimizer, batch size or used backbone and use no test-time augmentation or stochastic weight averaging. Our implementation is based on DETR3D [4] and MMDetection3D [3].

4.2 Comparison To Existing Works

We compare SpatialDETR to previous state-of-the-art methods for multi-view 3D object-detection. While FCOS3D [27], PGD [26] and BEVDet [8] follow a pseudo-LiDAR approach, DETR3D [29] and SpatialDETR use a transformer-based object-centric paradigm. As shown in Table 1 our method achieves the highest performance in terms of mean Average Precision and performs equally with DETR3D in the nuScenes detection score using the same training configuration as DETR3D.

It is noteworthy that the transformer-based methods gain a large performance increase when using a feature extractor that was initialized from a pseudo-LiDAR method. We suspect this to result from the dense-depth estimation and the depth-based loss in pseudo-LiDAR methods which might result in a faster convergence and spatially-aware backbone features while transformer-based methods, have a slower convergence behavior due to the uniformly distributed attention in early training stages [33]. Our model does not use multi-scale feature maps with a FPN [11] via multi-scale attention as proposed in [6] which might be the reason for the slightly less precise bounding boxes as compared to DETR3D.

Without any bells and whistles like test-time augmentation, model ensembling, finetuning or complex data augmentation as in BEVDet [8], SpatialDETR ranks within the Top-3 on the official nuScenes benchmark [17] and clearly outperforms DETR3D (see Table 2). As shown in Table 3 our global attention approach increases the performance by up to 3% on large objects such as trucks

Table 1. Comparison of state-of-the-art methods with different configurations on the nuScenes `val` set. PT flags models that use a pretrained backbone initialized from a FCOS3D checkpoint, MS indicates methods that utilize multi-scale features. ‡ marks methods using class-balanced grouping and sampling (*CBGS*) [32], + stands for methods using a Swin-Transformer-Tiny backbone [13], ¶ for results reported in [8] and † for results reported in [29]

Name	PT	MS	mAP↑	ATE↓	ASE↓	AOE↓	AVE↓	AAE↓	NDS↑
FCOS3D [27]†	✗	✓	0.299	0.785	0.268	0.557	1.396	**0.154**	0.373
FCOS3D [27]†	✓	✓	0.321	0.746	0.265	0.503	1.351	0.160	0.393
PGD [26]¶	✗	✓	0.335	0.732	**0.263**	0.423	1.285	0.172	0.409
BEVDET [8]‡	✗	✓	0.317	0.704	0.273	0.531	0.940	0.250	0.389
BEVDET [8]+‡	✗	✓	0.349	**0.637**	0.269	0.490	0.914	0.268	0.417
DETR3D [29]	✗	✓	0.303	0.860	0.278	0.437	0.967	0.235	0.374
DETR3D [29]	✓	✓	0.346	0.773	0.268	**0.383**	**0.842**	0.216	**0.425**
SpatialDETR	✗	✗	0.303	0.849	0.282	0.522	0.941	0.229	0.369
SpatialDETR	✓	✗	**0.351**	0.772	0.274	0.395	0.847	0.217	**0.425**

and busses in comparison to DETR3D. This supports the assumption that our attention allows for global context and arbitrary receptive fields that can capture large objects across the different cameras.

Table 2. Comparison of state-of-the-art methods with different configurations on the nuScenes `test` set. Our model was trained with CBGS [32] on the training and validation set as proposed by DETR3D [4]

Name	mAP↑	ATE↓	ASE↓	AOE↓	AVE↓	AAE↓	NDS↑
BEVDet [8]	**0.424**	**0.524**	**0.242**	**0.373**	0.950	0.148	**0.488**
DETR3D [29]	0.412	0.641	0.255	0.394	**0.845**	0.133	0.479
SpatialDETR	**0.424**	0.613	0.253	0.402	0.857	**0.131**	0.486

Since we do not use multi-scale feature maps, the performance on small objects such as pedestrians is slightly reduced due to the low-resolution backbone feature map which is in accordance to the findings in DETR [2]. As for the 2D case this can be tackled by using multi-scale attention as proposed in [33] or [6]. Additionally, our global attention formulation does not significantly increase the runtime since the entire global attention computation and feature projections can be computed in parallel.

4.3 Ablations And Analysis

Qualitative results for the performance of SpatialDETR on the nuScenes `val` set are shown in Fig. 4. The visualized attention maps in Fig. 5 show how the model

Table 3. Comparison of transformer-based methods in terms of average precision for different classes on the `val` set and runtime on a single NVIDIA V100 GPU. Both methods use a ResNet101 initialized from a FCOS3D checkpoint. Note how our spatially-aware attention improves the localization error particularly for extended objects

Method	Truck↑	Bus↑	Trailer↑	Pedestrian↑	Traffic cone↑	FPS↑
DETR3D	0.286	0.347	0.167	**0.424**	**0.529**	**2.5**
SpatialDETR	**0.302**	**0.378**	**0.175**	0.418	0.514	2.4

uses the global attention mechanism to focus on feature patches that belong to an object even across image borders as shown for query q^{93}.

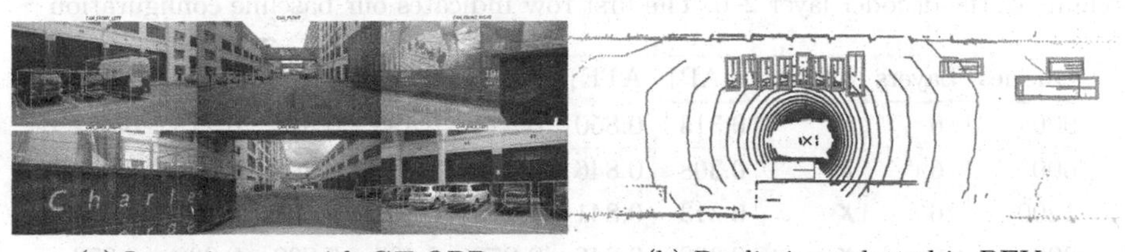

(a) Input images with GT-OBBs (b) Predictions plotted in BEV

Fig. 4. Qualitative detection results. a) input images and ground truth OBBs, b) detection results in BEV (blue) and ground truth (orange) (Color figure online)

The effect of different configurations of the proposed positional encoding and the proposed spatial-attention is analyzed in Table 4. Here, a configuration without sensor-specific queries reduces the baseline performance where the queries use a sensor specific, geometric update $P(Q)$ as defined in Eq. (6). This is likely due to the fact that the translation of the cameras with respect to the reference system is not accounted for in the global formulation. Introducing the geometric constraint that the predicted object center of an object query has to be in front of the image plane increases the performance which proves the effectiveness of our proposed geometric encoding since it allows to model various task-specific geometric constraints. A more complex computation that does not only take the object center but instead the full shape estimate of the bounding box into account might improve the performance even further.

Interpreting values directly as global features without the proposed explicit value projection update, increases the performance compared to the baseline but might result in a coupling between backbone features and sensor calibration. Since we directly predict a depth value per feature patch with a FFN this might cause numeric instabilities and could therefore require a customized scaling dependent on the desired object detection range. Furthermore, depth-supervision either by ground-truth depth maps or LiDAR hits might be needed to increase the performance. Table 5 shows the effect of different decoder settings. Changing

Table 4. Effect of the different components of SpatialDETR. $P(Q)$ indicates the projection of queries in sensor-relative coordinates, $P(V)$ the projection of values to global coordinates. Models with $C(Q)$ only attend to feature patches of an image if the query center is visible within that image

$P(Q)$	$P(V)$	$C(Q)$	MAP↑	ATE↓	ASE↓	AOE↓	AVE↓	AAE↓	NDS↑
✓	✓	✗	0.313	0.850	0.274	0.494	0.814	0.213	0.392
✗	✗	✗	0.248	0.993	0.284	0.506	**0.771**	0.202	0.348
✓	✗	✗	0.317	**0.824**	0.276	0.470	0.797	0.200	**0.402**
✓	✗	✓	0.318	0.835	0.277	**0.460**	0.837	0.210	0.397
✓	✓	✓	**0.319**	**0.824**	**0.272**	0.502	0.813	**0.192**	0.399

Table 5. Analysis of different decoder configurations. Models flagged as shared, share weights of the decoder layer 2–6. The first row indicates our baseline configuration

Queries	Layers	Shared	mAP↑	ATE↓	ASE↓	AOE↓	AVE↓	AAE↓	NDS↑
900	6	✗	0.313	0.850	0.274	0.494	0.814	0.213	0.392
600	6	✗	0.308	0.846	0.279	0.521	0.809	0.209	0.388
1200	6	✗	0.313	0.841	0.274	0.501	**0.798**	**0.207**	0.394
900	2	✗	0.308	0.846	0.278	0.493	0.899	0.230	0.379
900	10	✗	0.311	0.838	0.277	0.483	0.810	0.223	0.392
900	6	✓	**0.314**	**0.823**	**0.273**	**0.472**	0.822	0.224	**0.395**

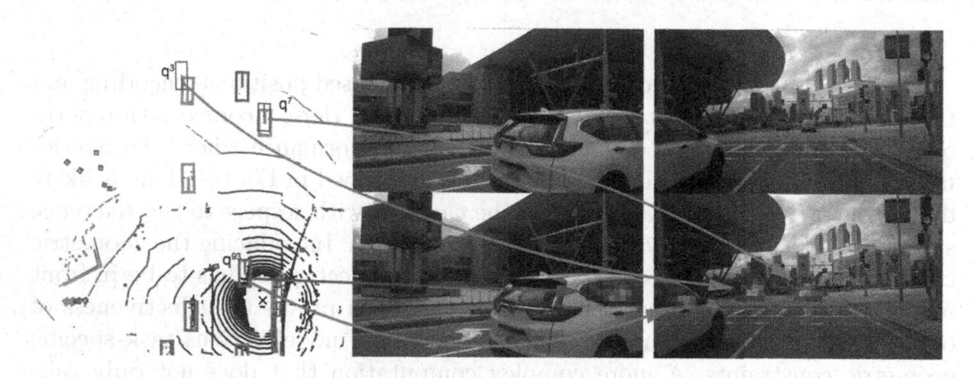

Fig. 5. Visualization of the spatially-aware global cross-sensor attention. Left: predictions (blue) and ground truth (orange). For three queries ($\mathbf{q}^3, \mathbf{q}^7, \mathbf{q}^{93}$) we jointly visualize the query-key attention maps for the sensor-relative queries (bottom images). Green arrows indicate queries relative to the front camera, red arrows queries relative to the front-left camera (Color figure online)

the number of decoder layers as well as the number of used object queries has a similar effect to DETR [2] and DETR3D [4]. Decreasing the amount of object queries or decoder layers (see Table 5) decreases the performance, the performance saturates close to the default values of six decoder layers and 900 object queries as proposed in [29]. Sharing the weights between the decoder layers and therefore treating the full decoder like a RNN with the same input for each layer and a cross-attention block as proposed in [9] results in a performance gain, reduction of trainable parameters and more precise bounding box estimates.

5 Conclusion

This paper presents SpatialDETR, a novel spatially and geometrically motivated transformer-based 3D object detection framework for multi-view data. It explicitly models the sensor calibration and features linear complexity with respect to the input images size due to a decoder-only attention architecture. The introduced geometric positional encoding and spatially-aware attention block enable us to use the global, cross-sensor context for arbitrary sensor positions and achieve state-of-the-art performance while significantly improving the performance for large objects without task-specific, hand-crafted constraints or post-processing steps. Similar to DETR, the detection of small objects is challenging and might be tackled in future work as done in the 2D case already. To prove the effectiveness of spatially-aware features that are decoupled from the sensor calibration, new data sets that model a fleet of cars, contain varying sensor calibrations and ground truth depth maps might be required. We are looking forward to apply the presented approach to multi-modal sensor data and to investigate its unified latent cross-sensor representation with multi-task learning.

Acknowledgement. The research leading to these results is funded by the German Federal Ministry for Economic Affairs and Climate Action within the project "KI Delta Learning"(Förderkennzeichen 19A19013A). The authors would like to thank the consortium for the successful cooperation."

References

1. Caesar, H., et al.: nuScenes: a multimodal dataset for autonomous driving. In: IEEE/CVF Conference on Computer Vision and Pattern Recognition (CVPR), pp. 11621–11631 (2020)
2. Carion, N., Massa, F., Synnaeve, G., Usunier, N., Kirillov, A., Zagoruyko, S.: End-to-end object detection with transformers. In: European Conference on Computer Vision (ECCV), pp. 213–229 (2020)
3. Contributors, M.: MMDetection3D: openMMLab next-generation platform for general 3D object detection (2020). https://github.com/open-mmlab/mmdetection3d. Accessed 07 Mar 2022
4. DETR3D Github-Repository. https://github.com/WangYueFt/detr3d. Accessed 07 Mar 2022
5. Dosovitskiy, A., et al.: An image is worth 16×16 words: transformers for image recognition at scale. arXiv preprint arXiv:2010.11929 (2020)
6. Gao, P., Zheng, M., Wang, X., Dai, J., Li, H.: Fast convergence of DETR with spatially modulated co-attention. In: IEEE/CVF International Conference on Computer Vision (ICCV), pp. 3621–3630 (2021)

7. He, K., Zhang, X., Ren, S., Sun, J.: Deep residual learning for image recognition. In: IEEE/CVF Conference on Computer Vision and Pattern Recognition (CVPR), pp. 770–778 (2016)
8. Huang, J., Huang, G., Zhu, Z., Du, D.: Bevdet: high-performance multi-camera 3D object detection in bird-eye-view. arXiv preprint arXiv:2112.11790 (2021)
9. Jaegle, A., Gimeno, F., Brock, A., Vinyals, O., Zisserman, A., Carreira, J.: Perceiver: general perception with iterative attention. In: International Conference on Machine Learning (ICML), pp. 4651–4664 (2021)
10. Lang, A.H., Vora, S., Caesar, H., Zhou, L., Yang, J., Beijbom, O.: Pointpillars: fast encoders for object detection from point clouds. In: IEEE/CVF Conference on Computer Vision and Pattern Recognition (CVPR), pp. 12697–12705 (2019)
11. Lin, T.Y., Dollár, P., Girshick, R., He, K., Hariharan, B., Belongie, S.: Feature pyramid networks for object detection. In: IEEE/CVF Conference on Computer Vision and Pattern Recognition (CVPR), pp. 2117–2125 (2017)
12. Lin, T.Y., Goyal, P., Girshick, R., He, K., Dollár, P.: Focal loss for dense object detection. In: IEEE/CVF International Conference on Computer Vision (ICCV), pp. 2980–2988 (2017)
13. Liu, Z., Lin, Y., Cao, Y., Hu, H., Wei, Y., Zhang, Z., Lin, S., Guo, B.: Swin transformer: hierarchical vision transformer using shifted windows. In: IEEE/CVF International Conference on Computer Vision (ICCV), pp. 10012–10022 (2021)
14. Loshchilov, I., Hutter, F.: Decoupled weight decay regularization. arXiv preprint arXiv:1711.05101 (2017)
15. Ma, H., et al.: Transfusion: cross-view fusion with transformer for 3D human pose estimation. arXiv preprint arXiv:2110.09554 (2021)
16. Ma, X., Liu, S., Xia, Z., Zhang, H., Zeng, X., Ouyang, W.: Rethinking pseudo-lidar representation. In: European Conference on Computer Vision (ECCV), pp. 311–327 (2020)
17. nuScenes Detection Task. https://www.nuscenes.org/object-detection/. Accessed 07 Mar 2022
18. Park, D., Ambrus, R., Guizilini, V., Li, J., Gaidon, A.: Is pseudo-lidar needed for monocular 3D object detection? In: IEEE/CVF International Conference on Computer Vision (ICCV), pp. 3142–3152 (2021)
19. Prakash, A., Chitta, K., Geiger, A.: Multi-modal fusion transformer for end-to-end autonomous driving. In: IEEE/CVF Conference on Computer Vision and Pattern Recognition (CVPR), pp. 7077–7087 (2021)
20. Ren, S., He, K., Girshick, R., Sun, J.: Faster r-cnn: towards real-time object detection with region proposal networks. Advances in neural information processing systems 28 (2015)
21. Roh, B., Shin, J., Shin, W., Kim, S.: Sparse DETR: efficient end-to-end object detection with learnable sparsity. arXiv preprint arXiv:2111.14330 (2021)
22. Sun, Z., Cao, S., Yang, Y., Kitani, K.M.: Rethinking transformer-based set prediction for object detection. In: IEEE/CVF International Conference on Computer Vision (ICCV), pp. 3611–3620 (2021)
23. Tian, Z., Shen, C., Chen, H., He, T.: Fcos: fully convolutional one-stage object detection. In: IEEE/CVF International Conference on Computer Vision (ICCV), pp. 9627–9636 (2019)
24. Vaswani, A., Shazeer, N., Parmar, N., Uszkoreit, J., Jones, L., Gomez, A.N., Kaiser, Ł., Polosukhin, I.: Attention is all you need. Advances in neural information processing systems 30 (2017)

25. Wang, C., Ma, C., Zhu, M., Yang, X.: Pointaugmenting: cross-modal augmentation for 3D object detection. In: IEEE/CVF Conference on Computer Vision and Pattern Recognition (CVPR), pp. 11794–11803 (2021)
26. Wang, T., Xinge, Z., Pang, J., Lin, D.: Probabilistic and geometric depth: detecting objects in perspective. In: Conference on Robot Learning (CORL), pp. 1475–1485 (2022)
27. Wang, T., Zhu, X., Pang, J., Lin, D.: Fcos3d: fully convolutional one-stage monocular 3d object detection. In: IEEE/CVF International Conference on Computer Vision (ICCV), pp. 913–922 (2021)
28. Wang, Y., Chao, W.L., Garg, D., Hariharan, B., Campbell, M., Weinberger, K.Q.: Pseudo-lidar from visual depth estimation: bridging the gap in 3D object detection for autonomous driving. In: IEEE/CVF Conference on Computer Vision and Pattern Recognition (CVPR), pp. 8445–8453 (2019)
29. Wang, Y., Guizilini, V.C., Zhang, T., Wang, Y., Zhao, H., Solomon, J.: Detr3D: 3D object detection from multi-view images via 3D-to-2D queries. In: Conference on Robot Learning (CORL), pp. 180–191 (2022)
30. Yin, T., Zhou, X., Krahenbuhl, P.: Center-based 3D object detection and tracking. In: IEEE/CVF Conference on Computer Vision and Pattern Recognition (CVPR), pp. 11784–11793 (2021)
31. Zhang, W., Wang, Z., Change Loy, C.: Multi-modality cut and paste for 3D object detection. arXiv e-prints .arXiv-2012 (2020)
32. Zhu, B., Jiang, Z., Zhou, X., Li, Z., Yu, G.: Class-balanced grouping and sampling for point cloud 3D object detection. arXiv preprint arXiv:1908.09492 (2019)
33. Zhu, X., Su, W., Lu, L., Li, B., Wang, X., Dai, J.: Deformable DETR: deformable transformers for end-to-end object detection. arXiv preprint arXiv:2010.04159 (2020)

Pixel-Wise Energy-Biased Abstention Learning for Anomaly Segmentation on Complex Urban Driving Scenes

Yu Tian[1], Yuyuan Liu[1], Guansong Pang[2](✉), Fengbei Liu[1], Yuanhong Chen[1], and Gustavo Carneiro[1]

[1] Australian Institute for Machine Learning, University of Adelaide, Adelaide, Australia

[2] School of Computing and Information Systems, Singapore Management University, Singapore, Singapore

gspang@smu.edu.sg

Abstract. State-of-the-art (SOTA) anomaly segmentation approaches on complex urban driving scenes explore pixel-wise classification uncertainty learned from outlier exposure, or external reconstruction models. However, previous uncertainty approaches that directly associate high uncertainty to anomaly may sometimes lead to incorrect anomaly predictions, and external reconstruction models tend to be too inefficient for real-time self-driving embedded systems. In this paper, we propose a new anomaly segmentation method, named pixel-wise energy-biased abstention learning (PEBAL), that explores pixel-wise abstention learning (AL) with a model that learns an adaptive pixel-level anomaly class, and an energy-based model (EBM) that learns inlier pixel distribution. More specifically, PEBAL is based on a non-trivial joint training of EBM and AL, where EBM is trained to output high-energy for anomaly pixels (from outlier exposure) and AL is trained such that these high-energy pixels receive adaptive low penalty for being included to the anomaly class. We extensively evaluate PEBAL against the SOTA and show that it achieves the best performance across four benchmarks. Code is available at https://github.com/tianyu0207/PEBAL.

1 Introduction

Recent advances in semantic segmentation have shown tremendous improvements on complex urban driving scenes [23]. Despite the accurate predictions on the inlier classes, the model fails to properly recognise anomalous objects that deviate from the training inlier distribution (col. 2 of Fig. 1). Addressing

Y. Tian and Y. Liu—Contributed equally to this work.

Supplementary Information The online version contains supplementary material available at https://doi.org/10.1007/978-3-031-19842-7_15.

S. Avidan et al. (Eds.): ECCV 2022, LNCS 13699, pp. 246–263, 2022.
https://doi.org/10.1007/978-3-031-19842-7_15

such failure cases is crucial to road safety for autonomous driving vehicles. For example, anomalies can be represented by unexpected objects in the middle of the road, such as a large rock or an unexpected animal that can be incorrectly predicted as a part of the road class, leading to potentially fatal traffic collisions.

Current methods [2,4,6,11,20,30,34,40] to detect and segment anomalous objects in complex urban driving scenes tend to depend on classification uncertainty or image reconstruction. The association of high classification uncertainty with anomaly is intuitive, but it has a few caveats. For instance, classification uncertainty happens when samples are close to classification decision boundaries, but there is no guarantee that all anomalies will be close to classification boundaries. Furthermore, samples close to classification boundaries may not be anomalies at all, but just hard inlier samples. Hence, these uncertainty based methods may detect a large number of false positive and false negative anomalies. For example, Fig. 1 shows that the previous SOTA Meta-OoD [6] misses important anomalous pixels (all rows), while misclassifying anomalies (e.g., vegetation in rows 1, 2, 3), even with the use of the outlier exposure (OE) strategy [19]. In fact, the OE strategy maximises the uncertainty for proxy anomalies, which can cause the model to be more uncertain for all inlier classes and detect false positive anomalies (e.g., Meta-OoD mis-classifies trees or bush with high anomaly scores – Fig. 1 col 4). Reconstruction methods [11,40] add an extra network to reconstruct the input images from the estimated segmentation, where differences are assumed to be anomalous. Not only does this approach depend on accurate segmentation results for precise reconstruction, but they also require an extra reconstruction network that is hard to train and inefficient to run in real-time self-driving embedded systems. Moreover, reconstruction methods that rely on a discrepancy module require re-training whenever the inlier segmentation model changes due to input distribution shift [11], limiting their applicability in real-world systems. Furthermore, previous approaches [2,6,11,15,20,30] ignore a couple of important constraints for anomaly segmentation, namely smoothness (e.g., Meta-OoD fails to classify neighbouring anomaly pixels in Fig. 1, rows 1, 4) and sparsity (e.g., Meta-OoD incorrectly detects a large number of anomalous pixels–see yellow and red regions in Fig. 1, rows 1, 2, 3). Another common issue shared by previous methods [2,6,30] is that they usually rely on the re-training of the entire network for OE, which is inefficient and can also bias the classification towards outliers.

In this paper, we propose a new anomaly segmentation method, the pixel-wise energy-biased abstention learning (PEBAL), that directly learns a pixel-level anomaly class, in addition to the pre-defined inlier classes, to reject/abstain anomalous pixels that are dissimilar to any of the inlier classes. It is achieved by a joint optimisation of a novel pixel-wise anomaly abstention learning (PAL) and an energy based model (EBM) [14,24,31]. Particularly, abstention learning (AL) [32] was originally developed to learn an image-level anomaly class, which is significantly challenged by the pixel-wise anomaly segmentation task that requires pixel-level anomaly class learning. This is because the original AL model treats all pixel inputs equally with a single pre-defined fixed penalty factor to regularise the classification of anomalous pixels, while adaptive penal-

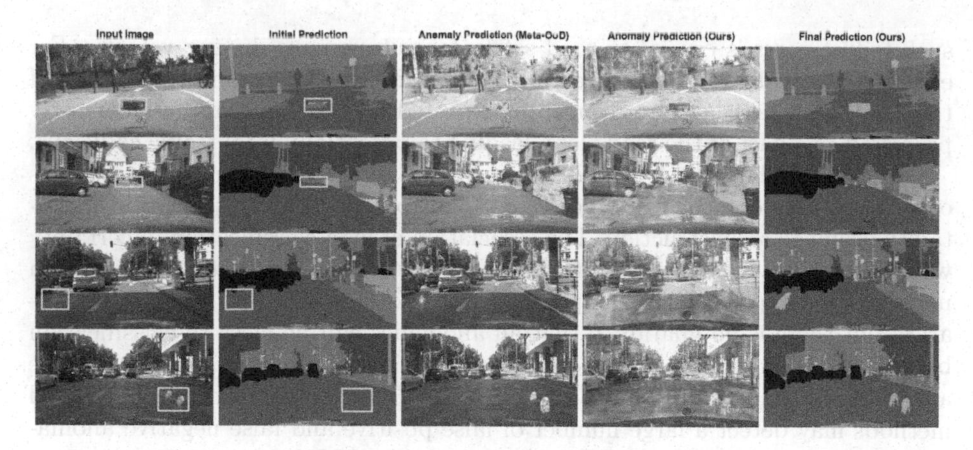

Fig. 1. Anomaly segmentation overview. From the **input image** (anomaly highlighted with a yellow box), the **initial prediction** shows the original segmentation results with anomalies classified as a one of the pre-defined inlier classes. **Anomaly predictions** by the previous SOTA **Meta-OoD** [6] and **our** method show an anomaly map with high scores (in yellow and red) for anomalous pixels, where our approach shows less false positive and false negative detections. Consequently, our method can detect small and distant anomalies (row 2) and blurry/unclear anomalies (rows 1, 3, 4) more accurately than Meta-OoD [6]. In our **final prediction**, anomalous pixels are coloured as cyan. **Some anomalies are small and blurred (e.g., row 2), so please zoom in the PDF for better visualisation.** (Color figure online)

ties are typically required for different pixels in a complex driving scene, e.g., pixels in small (distant) objects vs. large (near) objects, or centred pixels vs fringe pixels of objects. PEBAL is designed to address this issue by learning adaptive pixel-wise energy-based penalties, which automatically decreases the penalty for pixels that are likely to be anomalies. Hence, our model does not explore previously proposed uncertainty measures (e.g., entropy or softmax criteria) or image reconstruction, and instead, for the first time, explicitly learns a new pixel-wise anomaly class. The learned penalty factors are jointly optimised with EBM, resulting in a mutually beneficial optimisation of anomaly and inlier segmentation. Additionally, we impose smoothness and sparsity constraints to the learning of the anomaly segmentation by PEBAL, incorporating local and global dependencies into the pixel-wise penalty estimation and anomaly score learning. Finally, the training of PEBAL is efficient given that we only need to fine-tune the last block of the segmentation model to achieve accurate inference. To summarise, our contributions are the following:

- We propose the pixel-wise energy-biased abstention learning (PEBAL) that jointly optimises a novel pixel-wise anomaly abstention learning (PAL) and energy-based models (EBM) to learn adaptive pixel-level anomalies. PEBAL mutually reinforces PAL and EBM in detecting anomalies, enabling accurate segmentation of anomalous pixels without compromising the segmentation of inlier pixels (cols. 4,5 of Fig. 1).

- We introduce a new pixel-wise energy-biased penalty estimation, which can learn adaptive energy-based penalties to highly varying pixels in a complex driving scene, allowing a robust detection of small/distant and blurry anomalous objects (Fig. 1 row 2).
- We further refine our PEBAL training, using a novel smoothness and sparsity regularisation on anomaly scores to consider the local and global dependencies of the pixels, enabling the reduction of false positive/negative anomaly predictions.

We validate our approach on Fishyscapes leaderboard [4], and achieve SOTA classification accuracy on all relevant benchmarks. We also achieve the best classification results on LostandFound [36] and Road Anomaly [30] test sets, significantly surpassing other competing methods.

2 Related Work

Uncertainty-Based Anomaly Segmentation. Early uncertainty-based methods [18,25,28] focused on the estimation of image-level anomalies, but they tended to misclassify object boundaries as anomalies [20]. Jung et al. [20] mitigate this issue by iteratively replacing false anomalous boundary pixels with neighbouring non-boundary pixels that have low anomaly score. In [21,22,34], the boundary issue was tackled with a pixel-wise uncertainty estimated with MC dropout, but they showed a low pixel-wise anomaly detection accuracy [30]. Without fine-tuning using a proxy outlier dataset, uncertainty estimation may not be accurate enough to detect anomalies and can predict high uncertainty for challenging inliers or low uncertainty for outliers due to overconfident misclassification.

Reconstruction-Based Anomaly Segmentation. Anomalies can also be segmented from the errors between the input image and its reconstruction obtained from its predicted segmentation map [1,8,10,11,16,30,39,40]. Those approaches are challenged by the dependence on an accurate segmentation prediction, by the complexity of reconstruction models that usually require long training and inference processes, and also by the low quality of the reconstructed images.

Anomaly Segmentation via Outlier Exposure. Hendrycks et al. [19] propose the outlier exposure (OE) strategy that uses an auxiliary dataset of outliers that do not overlap with the real outliers/anomalies to improve the anomaly detection performance. This OE strategy uses outliers from ImageNet [2,3,38], void class of Cityscape [11] or COCO [6], where the expectation is that the model can generalise to unseen outliers. Maximising uncertainty for outliers using the OE strategy can lead to a deterioration of the segmentation of inliers [3,38]. Another major drawback of OE methods is that they are trained using outlier images or objects without considering the fact that outliers are rare events that appear around inliers. Hence, the training contains a disproportionately high amount of outliers [6] that can bias the segmentation toward the anomaly class. We address this issue by respecting the anomaly detection assumption, where anomalous objects are rare, contribute to a small proportion of the training set, and appear around inliers.

Abstention Learning. The abstention learning mechanism [12] adds a "reserve" (i.e., anomaly) class that is predicted when the classification predictions for all inlier classes are not high enough. This method shows good performance in learning holistic image-level anomaly class with a single predefined penalty factor for the whole training set, but it fails to learn fine-grained pixel-level anomaly class as an adaptive pixel-wise penalty is required for highly varying pixel-level anomalies (see Table 5). We address this issue by learning a novel pixel-wise energy-biased penalty estimator that is jointly trained with fine-grained abstention learning. It is worth noting that differently from uncertainty-based methods [4,6,17,20] that assume anomaly even when the model is uncertain but confident, abstention learning requires all classes to have low confidence to predict the anomaly class.

Energy-Based Models. EBM is trained such that inlier training samples have low energy, whereas non-training outlier samples (i.e., anomalies) are expected to have high energy [24]. This energy value can then be used to compute the probability of a sample to belong to the inlier distribution. Recently, EBMs are being implemented with deep learning models [14,31,35], and to learn them, it is necessary to compute the partition function, which is generally estimated with Markov Chain Monte Carlo (MCMC) [14], but this estimation cannot generate accurate high-resolution images. Hence, we follow the simpler idea of estimating the energy score with the *logsumexp* operator [14,31], where we minimise the energy of inliers and use an OE strategy [19] to maximise the energy of outliers. Hence, we do not need to compute the partition function.

3 Method

We present our PEBAL in this section (see Fig. 2), where we first describe the dataset, then introduce abstention learning and EBM. Next, we present the loss function to train the model, followed by the training and inference procedures.

3.1 Training Set

We assume to have a set of inlier training images and annotations $\mathcal{D}^{in} = \{(\mathbf{x}_i, \mathbf{y}_i^{in})\}_{i=1}^{|\mathcal{D}^{in}|}$, where $\mathbf{x} \in \mathcal{X} \subset \mathbb{R}^{H \times W \times C}$ denotes an image with C colour channels, and $\mathbf{y}^{in} \in \mathcal{Y}^{in} \subset \{0,1\}^{H \times W \times Y}$ denotes the inlier pixel level labels that can belong to Y classes. We also have a set of outlier images and annotations $\mathcal{D}^{out} = \{(\mathbf{x}_i, \mathbf{y}_i^{out})\}_{i=1}^{|\mathcal{D}^{out}|}$, where $\mathbf{y}^{out} \in \mathcal{Y}^{out} \subset \{0,1\}^{H \times W \times (Y+1)}$ denotes the outlier pixel-level labels, with the class $Y+1$ reserved for pixels belonging to the anomaly class. Note that similarly to previous papers [6], the types of anomalies in training set $\mathcal{D}^{out}$ do not overlap with the anomalies to be found in the testing set.

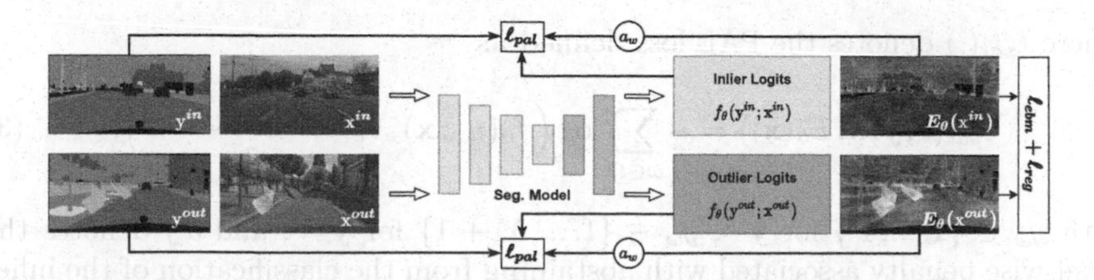

Fig. 2. PEBAL. The pixel-wise anomaly abstention (PAL) loss ℓ_{pal} learns to abstain the prediction of outlier pixels from $\mathbf{x}^{out}$ containing OE objects (i.e., cyan coloured masks) and calibrate the logit of inlier classes (i.e., reduction of the inlier logits) from both inlier image $\mathbf{x}^{in}$ and outlier image $\mathbf{x}^{out}$. The EBM loss ℓ_{ebm} pushes the free energy E_θ to low values for inlier pixels and pulls that to high values for outlier pixels, where a regularisation loss ℓ_{reg} enforces the smoothness and sparsity constraints on the energy maps. Such EBM learning reduces the logit of inlier classes to share similar values at the same time, facilitating the ℓ_{pal} learning. Then, the pixel-wise penalty a_ω associated with the abstention class at position ω is estimated to bias the penalty to be low for outlier pixels and high for inlier pixels, which in turn encourages high free energy for anomalies and enforces ℓ_{pal} to abstain the anomalous pixels.

3.2 Pixel-Wise Energy-Biased Abstention Learning (PEBAL)

The PEBAL model is denoted by

$$p_\theta(y|\mathbf{x})_\omega = \frac{\exp(f_\theta(y;\mathbf{x})_\omega)}{\sum_{y' \in \{1,...,Y+1\}} \exp(f_\theta(y';\mathbf{x})_\omega)}, \tag{1}$$

where θ is the model parameter, ω indexes a pixel in the image lattice Ω, $p_\theta(y|\mathbf{x})_\omega$ represents the probability of labelling pixel ω with $y \in \{1, ..., Y+1\}$, and $f_\theta(y;\mathbf{x})_\omega$ is the logit for class y at pixel ω.

To train the model in (1), we formulate a cost function that jointly trains PAL and EBM to classify anomalous pixels. An important training hyper-parameter for PAL is the penalty to abstain from the classification into one of the inlier classes in $\{1, ..., Y\}$–this penalty is generally tuned to a single value for all training samples through model selection (e.g., cross validation) [32]. Instead of treating this as a tunable hyper-parameter, we propose the use of EBM (defined below in (4)) to automatically estimate this penalty during the training process for each pixel within each training image. More specifically, the cost function to train the PEBAL model in (1) is:

$$\begin{aligned} \ell(\mathcal{D}^{in}, \mathcal{D}^{out}, \theta) = & \\ & \sum_{(\mathbf{x},\mathbf{y}^{in})\in\mathcal{D}^{in}} \left(\ell_{pal}(\theta, \mathbf{y}^{in}, \mathbf{x}, E_\theta(\mathbf{x})) + \lambda \ell_{ebm}^{in}(E_\theta(\mathbf{x})) + \ell_{reg}(E_\theta(\mathbf{x}))\right) + \\ & \sum_{(\mathbf{x},\mathbf{y}^{out})\in\mathcal{D}^{out}} \left(\ell_{pal}(\theta, \mathbf{y}^{out}, \mathbf{x}, E_\theta(\mathbf{x})) + \lambda \ell_{ebm}^{out}(E_\theta(\mathbf{x})) + \ell_{reg}(E_\theta(\mathbf{x}))\right). \end{aligned} \tag{2}$$

where $\ell_{pal}(.)$ denotes the PAL loss defined as

$$\ell_{pal}(\theta, \mathbf{y}, \mathbf{x}, E_\theta(\mathbf{x})) = -\sum_{\omega \in \Omega} \log \Big(f_\theta(y_\omega; \mathbf{x})_\omega + \frac{f_\theta(Y+1; \mathbf{x})_\omega}{a_\omega} \Big), \tag{3}$$

with $y_\omega \in \{1, ..., Y\}$ for $\mathbf{y}^{in}$, $y_\omega \in \{1, ..., Y+1\}$ for $\mathbf{y}^{out}$, and a_ω denotes the pixel-wise penalty associated with abstaining from the classification of the inlier classes. The minimisation of the loss in (3) will abstain from classifying outlier pixels into one of the inlier classes, where a pixel is estimated to be an outlier with a_ω. Before formulating a_ω, let us define the inlier free energy at pixel ω, which is denoted by $E_\theta(\mathbf{x})_\omega$ and computed with the *logsumexp* operator as follows [14,24,31]:

$$E_\theta(\mathbf{x})_\omega = -\log \sum_{y \in \{1, ..., Y\}} \exp(f_\theta(y; \mathbf{x})_\omega). \tag{4}$$

The pixel-wise penalty associated with abstaining from the classification of the inlier classes is defined by

$$a_\omega = (-E_\theta(\mathbf{x})_\omega)^2, \tag{5}$$

which means that the larger the a_ω (i.e., low inlier free energy, so the sample is an inlier), the higher the loss to abstain from classifying into one of the Y classes, and low value of a_ω (i.e., high free inlier energy, which means an outlier sample) implies a lower loss to abstain from classifying one of the Y classes. Also in (2), $\ell^{in}_{ebm}(.)$ (weighted by hyper-parameter λ) represents the EBM loss that pushes the inlier free energy in (4) for samples in $\mathcal{D}^{in}$ to low values, with

$$\ell^{in}_{ebm}(E_\theta(\mathbf{x})) = \sum_{\omega \in \Omega} \big(\max(0, E_\theta(\mathbf{x})_\omega - m_{in}) \big)^2, \tag{6}$$

representing the loss of having inlier samples with free energy larger than threshold m_{in}, and

$$\ell^{out}_{ebm}(E_\theta(\mathbf{x})) = \sum_{\omega \in \Omega} \big(\max(0, m_{out} - E_\theta(\mathbf{x})_\omega) \big)^2, \tag{7}$$

denoting the loss of having outlier samples with inlier free energy smaller than threshold m_{out}, where the margin losses in (6) and (7) effectively create an energy gap between normal and abnormal pixels. The last term to define in (2) is the inlier free energy regularisation loss to enforce that anomalous pixels are sparse and pixel anomaly classification is smooth (i.e., anomalous pixels tend to have anomalous neighbouring pixels), which is defined as

$$\ell_{reg}(E_\theta(\mathbf{x})) = \sum_{\omega \in \Omega} \beta_1 |E_\theta(\mathbf{x})_\omega - E_\theta(\mathbf{x})_{\mathcal{N}(\omega)}| + \beta_2 |E_\theta(\mathbf{x})_\omega|, \tag{8}$$

where β_1 and β_2 are hyper-parameters that weight the contributions of the smoothness and sparsity and sparsity regularisations, and $\mathcal{N}(\omega)$ denotes neighbouring pixels in horizontal and vertical directions.

3.3 Training and Inference

Training. An important point of the training process is how to setup the inlier and outlier datasets $\mathcal{D}^{in}$ and $\mathcal{D}^{out}$. A recently published paper [6] carefully selects images to be included in $\mathcal{D}^{out}$ by making sure that the segmentation labels presented in those images do not overlaps with the inlier labels. In particular for [6], $\mathcal{D}^{in}$ has images and annotations from Cityscape and $\mathcal{D}^{out}$ has images and annotations from COCO [29]. We argue that there are two issues with this strategy to form $\mathcal{D}^{out}$, which are: 1) the selected COCO images generally only contain anomalous pixel labels, leading to unstable training of the outlier losses (i.e., second summation in (2)) given the exclusive presence of the anomaly class (in effect, this becomes a one-class segmentation problem); 2) re-training the model with images containing only anomalous pixels removes the semantic context of inlier pixels when training for the outlier losses, which can deteriorate the segmentation accuracy of the inlier labels.

To mitigate these issues, we form $\mathcal{D}^{out}$ using a novel extension based on CutMix and CutPaste [27,41], which we refer to as AnomalyMix. AnomalyMix cuts the anomalous objects from an outlier dataset (e.g., COCO) using its labelled masks and paste them into the images of the inlier dataset (e.g., CitySpace), where we label the pixels of the anomalous object with the class $Y+1$ – these images are then inserted into $\mathcal{D}^{out}$. AnomalyMix addresses the two issues above because the outlier images now contain a combination of inlier and outlier pixels, allowing a balanced learning and keeping the visual context of inlier labels when training for the outlier losses. Furthermore, AnomalyMix can form a potentially infinite number of training images for $\mathcal{D}^{out}$ given the range of transformations to be applied to the cut objects and the locations of the inlier images that the objects can be pasted. Previous papers [11,20] argue that re-training the whole segmentation model can jeopardise the segmentation accuracy for the inlier classes. Furthermore, such re-training requires a long training time, leading to inefficient optimisation. In this work, we propose to **fine-tune only the final classification block** using the loss in (2), instead of re-training the whole segmentation model. Besides being efficient, this fast fine-tuning keeps the segmentation accuracy of the model in the original dataset used for pre-training the model. Furthermore, an interesting side-effect of our training is that the cost function in (2) will calibrate the segmentation prediction for the inlier classes. This happens because the terms $\ell_{pal}(.)$, $\ell_{ebm}^{in}(.)$ and $\ell_{ebm}^{out}(.)$ jointly constrain the maximisation of logits and naturally calibrate classification confidence (See supplementary material).

Inference. During inference, pixel-wise anomaly detection is performed by computing the inlier free energy score $\mathbb{E}_\theta(\mathbf{x})_\omega$ from (4) for each pixel position ω given a test image $\mathbf{x}$ and inlier segmentation is obtained from the inlier classes from the PEBAL model in (1). Following [20], we also apply a Gaussian smoothing kernel to produce the final energy map.

4 Experiment

4.1 Datasets

LostAndFound [36] is one of the first publicly available urban driving scene anomaly detection datasets containing real-world anomalous objects. The dataset has an official testing set containing 1,203 images with small obstacles in front of the cars, collecting from 13 different street scenes, featuring 37 different types of anomalous objects with various sizes and material.
Fishyscapes [4] is a high-resolution dataset for anomaly estimation in semantic segmentation for urban driving scenes. The benchmark has an online testing set that is entirely unknown to the methods. The dataset is composed by two data sources: Fishyscapes LostAndFound that contains a set of real road anomalous objects [36] and a blending-based Fishyscapes Static dataset. The Fishyscapes LostAndFound validation set consists of 100 images from the aforementioned LostAndFound dataset with refined labels and the Fishyscapes Static validation set contains 30 images with the blended anomalous objects from Pascal VOC [13]. For all datasets, we select the checkpoints based on the results on the public validation sets, but submitted our code and checkpoints to the benchmark website to be evaluated on their hidden test sets.
Road Anomaly [30] contains real-world road anomalies in front of the vehicles. The dataset has 60 images from the Internet, containing unexpected animals rocks, cones and obstacles.Unlike the LostAndFound and Fishyscapes, this dataset contains abnormal objects with various scales and sizes, making it even more challenging.

4.2 Implementation Details

Following [5,6], we use DeepLabv3+ [7] with WideResnet38 trained by Nvidia [42] and ResNet101 from [20] as the backbone of our segmentation models. The training details of those models can be found in their original papers or our supplementary material. The models are trained on Cityscapes [9] training set. For our PEBAL fine-tuning, we empirically set the m_{in} and m_{out} in Eq. 6 and Eq. 7 as -12 and -6, respectively. The weights β_1 and β_2 in Eq. 8 are set to $5e-4$ and $3e-6$ [37], λ in Eq. 2 to 0.1, and the weight of ℓ_{ebm} to 0.1, respectively. Note that those hyper-parameters are selected at the first training epoch to normalise loss values to a similar scale. We also show our model can obtain consistently SOTA results regardless of the selection of hyper-parameters in the supplementary material. Our training consists of fine-tuning the final classification block of the model for 20 epochs. We use the same resolution of random crop as in [42], and use Adam with a learning rate of $1e^{-5}$. The batch size is set to 16. Following [6], for our AnomalyMix augmentation, we randomly sample 297 images as training data from the remaining COCO images that do not contain objects in Cityscapes or our anomaly validation/testing sets and randomly apply AnomalyMix to mix them into the Cityscape training images, following Chan et al. [6].

4.3 Evaluation Measures

Following [4,6,11,20], we compute the the area under receiver operating characteristics (AUROC), average precision (AP), and the false positive rate at a true positive rate of 95% (FPR95) to validate our approach. For Fishyscapes public leaderboard, we use AP and FPR95 to compare with other methods, same as their website.

4.4 Comparison on Anomaly Segmentation Benchmarks

Table 1. Anomaly segmentation results on **LostAndFound** testing set, with **WideResnet38** backbone. All methods use the same segmentation models. * indicate that the model requires additional learnable parameters. † indicates that the results are obtained from the official code with our WideResnet38 backbone.

Methods	AUC ↑	AP ↑	FPR_{95} ↓
MSP [17]	85.49	38.20	18.56
Mahalanobis [26]	79.53	42.56	24.51
Max Logit [18]	94.52	65.45	15.56
Entropy [18]	86.52	50.66	16.95
Energy [31]	94.45	66.37	15.69
Meta-OoD [6]	97.95	71.23	5.95
†SML [20]	88.05	25.89	44.48
†SynBoost* [11]	98.38	70.43	4.89
Deep Gambler [32]	98.67	72.73	3.81
Ours	**99.76**	**78.29**	**0.81**

Comparison on LostAndFound. Table 1 shows the result on the testing set of LostAndFound. Notably, our approach surpasses the previous baseline approaches (i.e., MSP [17], Mahalanobis [26], Max Logit [18] and Entropy [18]) by 10% to 40% AP, and 13% to 22% FPR95, respectively. When compared with previous SOTA approaches such as SynBoost [11], SML [20] and Meta-OoD [6], we improve the AP performance by a large margin (15% to 40%), and decrease the FPR95 by about 5% to 70%. This illustrates the robustness and effectiveness on detecting small and distant anomalous objects given that the dataset contains mostly real-world small objects. Our PEBAL also improves the EBM baseline [31] and the AL baseline based on Deep Gambler [32]. This demonstrates that a simple adaptation of AL and EBM is not enough to enable accurate pixel-wise anomaly detection. Previous SOTA SML [20] aims to balance the inlier class-wise discrepancy on prediction scores, which is disadvantageous for measuring performance on LostAndFound test set since there may be no classes in the evaluation other than the road class (i.e., most of the inlier classes within LF test set is road class), thus leading to significant performance variations between LostAndFound and Fishyscapes. It is worth noting that our approach achieves 1.03% FPR95, significantly reducing the false positive pixels, improving the chances of applying it to real-world applications.

Comparison on Fishyscapes Leaderboard. Table 2 shows the leaderboard results on the test set of Fishyscapes LostAndFound and Fishyscapes Static. Following [20], we compared the methods based on whether they require re-training of the entire segmentation network, adding the extra network, or utilising the OoD data. We achieve the SOTA performance by a large margin on Fishyscapes leaderboard when compared with the previous methods except [2] (Static) that rely on an inefficient re-training segmentation model, extra learnable parameters, and extra OoD training data. Without re-training the entire network or adding extra learnable parameters, our approach can work efficiently to surpass previous SOTA competing approaches that fall into the same category by about 13% to 42% on LostAndFound and 40% to 50% AP on Static. Such significant improvements indicate the generalisation ability of our proposed PEPAL on detecting a wide variety of unseen abnormalities (i.e., of different size, type, scene, and distance) substantially reducing false negative and positive pixels. Moreover, it is worth noting that PEBAL reduces the amount of false positive pixels to 7.58 and 1.73 FPR on the two datasets. This result is publicly available on the Fishyscapes website.

Table 2. Comparison with previous approaches on **Fishyscapes Leaderboard**. We achieve a new state-of-the-art performance among the approaches that require extra OoD data, and without re-training the segmentation networks and extra networks on Fishyscapes Leaderboard.

Models	re-training	Extra Network	OoD Data	FS LostAndFound		FS Static	
				AP ↑	FPR95 ↓	AP ↑	FPR95 ↓
Discriminative Outlier Detection Head [2]	✓	✓	✓	31.31	19.02	96.76	0.29
MSP [17]	✖	✖	✖	1.77	44.85	12.88	39.83
Entropy [18]	✖	✖	✖	2.93	44.83	15.41	39.75
SML [20]	✖	✖	✖	31.05	21.52	53.11	19.64
kNN Embedding - density [4]	✖	✖	✖	3.55	30.02	44.03	20.25
Bayesian Deeplab [34]	✓	✖	✖	9.81	38.46	48.70	15.05
Density - Single-layer NLL [4]	✖	✓	✖	3.01	32.9	40.86	21.29
Density - Minimum NLL [4]	✖	✓	✖	4.25	47.15	62.14	17.43
Image Resynthesis [30]	✖	✓	✖	5.70	48.05	29.6	27.13
OoD Training - Void Class	✓	✖	✓	10.29	22.11	45.00	19.40
Dirichlet Deeplab [33]	✓	✖	✓	34.28	47.43	31.30	84.60
Density - Logistic Regression [4]	✖	✓	✓	4.65	24.36	57.16	13.39
SynBoost [11]	✖	✓	✓	43.22	15.79	72.59	18.75
Ours	✖	✖	✓	**44.17**	**7.58**	**92.38**	**1.73**

Table 3. Anomaly segmentation results on **Fishyscapes validation sets** (LostAndFound and Static), and the **Road Anomaly testing set**, with **WideResnet38** backbone. * indicate that the model requires additional learnable parameters. † indicates that the results are obtained from the official code with our WideResnet38 backbone. Best and second best results in bold.

Methods	FS LostAndFound			FS Static			Road Anomaly		
	AUC ↑	AP ↑	FPR_{95} ↓	AUC ↑	AP ↑	FPR_{95} ↓	AUC ↑	AP ↑	FPR_{95} ↓
MSP [17]	89.29	4.59	40.59	92.36	19.09	23.99	67.53	15.72	71.38
Max Logit [17]	93.41	14.59	42.21	95.66	38.64	18.26	72.78	18.98	70.48
Entropy [18]	90.82	10.36	40.34	93.14	26.77	23.31	68.80	16.97	71.10
Energy [31]	93.72	16.05	41.78	95.90	41.68	17.78	73.35	19.54	70.17
Mahalanobis [26]	96.75	56.57	11.24	96.76	27.37	11.7	62.85	14.37	81.09
Meta-OoD [11]	93.06	41.31	37.69	97.56	72.91	13.57	–	–	–
†Synboost* [11]	96.21	**60.58**	31.02	95.87	66.44	25.59	**81.91**	**38.21**	**64.75**
†SML [20]	94.97	22.74	33.49	97.25	66.72	12.14	75.16	17.52	70.70
Deep Gambler [32]	**97.82**	31.34	**10.16**	**98.88**	**84.57**	**3.39**	78.29	23.26	65.12
Ours	**98.96**	**58.81**	**4.76**	**99.61**	**92.08**	**1.52**	**87.63**	**45.10**	**44.58**

Table 4. Anomaly segmentation results on **Fishyscapes validation sets** (LostAndFound and Static), and the **Road Anomaly testing set**, with **Resnet101** backbone. * indicate that the model requires additional learnable parameters. † indicates that the results are obtained from the official code with our Resnet101 backbone. Best and second best results in bold.

Methods	FS LostAndFound			FS Static			Road Anomaly		
	AUC ↑	AP ↑	FPR_{95} ↓	AUC ↑	AP ↑	FPR_{95} ↓	AUC ↑	AP ↑	FPR_{95} ↓
MSP [17]	86.99	6.02	45.63	88.94	14.24	34.10	73.76	20.59	68.44
Max Logit [17]	92.00	18.77	38.13	92.80	27.99	28.50	77.97	24.44	64.85
Entropy [18]	88.32	13.91	44.85	89.99	21.78	33.74	75.12	22.38	68.15
Energy [31]	93.50	25.79	32.26	91.28	31.66	37.32	78.13	24.44	63.36
†SynthCP* [40]	88.34	6.54	45.95	89.9	23.22	34.02	76.08	24.86	64.69
†Synboost* [11]	94.89	**40.99**	34.47	92.03	48.44	47.71	85.23	**41.83**	59.72
SML [20]	96.88	36.55	14.53	96.69	48.67	16.75	81.96	25.82	49.74
Deep Gambler [32]	**97.19**	39.77	**12.41**	**97.51**	**67.69**	**15.39**	**85.45**	31.45	**48.79**
Ours	**99.09**	**59.83**	**6.49**	**99.23**	**82.73**	**6.81**	**92.51**	**62.37**	**28.29**

Comparison on Fishyscapes Validation Sets and Road Anomaly. In Tables 3 and 4, we compare our approach on the Fishyscapes validation sets and Road Anomaly using two different backbones. Our model outperforms the previous methods by a large margin on all three benchmarks, regardless of the backbones and their segmentation accuracy. To verify the applicability of our method, except for the modern WideResnet38 backbone, we use a ResNet101 DeepLabv3+ to investigate the performance in terms of the size of the architecture and its inlier segmentation accuracy. The results demonstrate that our approach is applicable to a wide-range of segmentation models, indicating the effectiveness of PEBAL to adapt to real-world systems.

Moreover, our fine-tuning sacrifices only marginally the inlier segmentation accuracy (i.e., 0.2%–0.7% mIoU on Cityscapes) for both backbones, achieving good performance on both inlier and anomaly segmentation. We present details of all inlier segmentation models (i.e., Cityscapes training setup and mIoU), and include more experimental results of other DeepLabv3+ checkpoints in supplementary material.

Remarks – Superior Performance on Challenging Benchmarks. Each dataset has different challenges. For example, the LostAndFound testing set considers only drivable areas with homogeneous normal scenes (i.e., road) and limited categories of abnormalities (i.e., road obstacles), leading to a relatively less challenging benchmark on which most methods can obtain good AUC performance, as shown in Tables 1, 3 and 4. On the contrary, Fishyscapes and RoadAnomaly contain large number of heterogeneous inlier and outlier pixels from diverse classes, leading to significantly more difficult testbeds than the LostAndFound testing set. Furthermore, Fishyscapes and RoadAnomaly contain domain shift compared with Cityscapes (e.g., both datasets contain different scenes than Cityscapes) and have different types/sizes of OoD objects. Most existing SOTA methods work ineffectively on these two datasets due to those challenges, while our adaptive pixel-level anomaly class learning helps our model effectively detect these challenging inlier and outlier pixels in the aforementioned heterogeneous and domain-shifted scenes, yielding substantial improvements (i.e., 20% to 50%) to previous approaches, as shown in Tables 2, 3 and 4.

4.5 Ablation Study

Table 5 shows the contribution of each component of our PEBAL on the LostAndFound testing set. All modules are trained with COCO OE images using AnomalyMix. Adding an extra OoD class to learn the OE training samples with entropy maximisation (**EM**) is our baseline (first row). To justify the effectiveness of our proposed joint training, we show the results using energy-based models (ℓ_{ebm} without ℓ_{pal}) and pixel-wise abstention (ℓ_{pal} with pre-defined fixed penalty). Both outperform the baselines (AP=70.2, FPR=8.9 and AP=72.7, FPR=3.8 vs. AP=69, FPR=8.03), while our proposed joint training (ℓ_{ebm} + ℓ_{pal}) obtains 77.19% of AP and 1.19% of FPR, improving over each module by 4% to 7%. This indicates the effectiveness of our joint training and the significance of our proposed PAL with learnable adaptive energy-based penalties a_{ω}. Finally, the smoothness and sparsity regularisation losses stabilise the training and further improve the performance.

Table 5. Ablation studies for anomaly segmentation on **LostAndFound**, with **WideResnet38** backbone, where all proposed modules are trained with COCO OE images with AnomalyMix. EM denotes the baseline method that adds an extra OoD class to learn the OE training samples with entropy maximisation (first row).

EM	ℓ_{ebm}	ℓ_{pal}	ℓ_{reg}	AUC ↑	AP ↑	FPR_{95} ↓
✓				96.88	69.02	8.03
	✓			97.88	70.24	8.92
		✓		98.67	72.73	3.81
	✓	✓		99.63	77.19	1.19
	✓	✓	✓	**99.76**	**78.29**	**0.81**

Table 6. The performance comparison of our approach on Fishyscapes benchmark w.r.t different **diversity of OE classes** (mean results over six random seeds), in terms of AP and FPR95.

Class Per.	FS LostAndFound		FS Static	
	AP ↑	FPR_{95} ↓	AP ↑	FPR_{95} ↓
1%	53.57 ± 3.74	6.97 ± 1.98	85.84 ± 1.01	3.05 ± 0.97
5%	52.16 ± 3.88	6.58 ± 1.95	90.57 ± 1.75	1.93 ± 0.52
10%	55.14 ± 3.02	5.78 ± 1.59	91.37 ± 1.28	1.64 ± 0.58
25%	55.48 ± 3.32	5.98 ± 1.27	91.28 ± 1.94	1.77 ± 0.18
50%	56.69 ± 2.57	5.32 ± 1.16	91.88 ± 0.71	1.62 ± 0.05
75%	57.86 ± 2.83	5.11 ± 1.69	91.85 ± 0.56	1.63 ± 0.09

4.6 Outlier Samples and Computational Efficiency

Outlier Diversity and Efficiency. In Table 6, we randomly select 1%, 5%, 10%, 25% 50%, and 75% of COCO classes as the OE data during training and compute the mean results over six different random seeds. We achieve consistent AP and FPR performance regardless of the number of COCO classes used during the training on Fishyscapes. It is also worth noting that our approach can effectively learn the PEBAL model using **only one class** (1% in Table 6) of outlier data, which selects some of the irrelevant classes of COCO objects that are not possible to be found on road in real life (e.g., dining table, laptop, and clock). The results indicate that our model can consistently achieve SOTA performance on Fishyscapes without a careful selection of OE classes, demonstrating the robustness of our approach under diverse outlier classes. We also investigate the outlier sample efficiency of our model w.r.t smaller OE training sets with a fixed 100% COCO classes (80 classes) on Fishyscapes in Table 7, and we achieve consistently good performance regardless the number of outlier training samples. All those experiments show the applicability of our PEBAL to real-world autonomous driving systems.

Computational Efficiency. We compare the computational efficiency of our PEBAL with previous SOTA Meta-OoD [6] and Synboost [11] in terms of the trainable parameters, training time and mean inference time per image, on an NVIDIA3090. As PEBAL requires the fine-tuning of the final classification block, it has only 1.3M parameters and each training epoch takes about 12 min, which is significantly less than the re-training approach Meta-OoD that has 137.1M parameters and each training epoch takes about 26 min, and the reconstruction based approach Synboost that takes about 33 min to train a epoch of its re-synthesis and dissimilarity networks with 157.3M parameters. Moreover, our method also has a much faster mean inference time of 0.55 s compared to 0.85 s of Meta-OoD and 1.95 s of Synboost. Those results suggest the practicability of our model in real-world self-driving systems.

Table 7. The performance comparison of our approach on Fishyscapes benchmark w.r.t different **amount of OE training samples** (mean results over six random seeds), in terms of AP and FPR95.

Train Size	FS LostAndFound		FS Static	
	AP ↑	FPR_{95} ↓	AP ↑	FPR_{95} ↓
5%	54.32 ± 1.89	5.77 ± 2.38	89.11 ± 1.52	2.23 ± 0.65
10%	56.28 ± 1.05	4.66 ± 1.36	90.02 ± 0.57	1.67 ± 0.28
25%	56.18 ± 1.69	4.81 ± 1.44	91.23 ± 0.95	1.63 ± 0.22
50%	57.34 ± 1.19	4.75 ± 1.32	91.29 ± 0.92	1.67 ± 0.17

5 Conclusions and Discussions

We proposed a simple yet effective approach, named Pixel-wise Energy-biased Abstention Learning (PEBAL), to fine-tune the last block of a segmentation model to detect unexpected road anomalies. The approach introduces a non-trivial training that jointly optimises a novel pixel-wise abstention learning and an energy-based model to learn an adaptive pixel-wise anomaly class, in which a new pixel-wise energy-biased penalty estimation method is proposed to improve the precision and robustness to detect small and distant anomalous objects. The resulting model significantly reduces the false positive and false negative detected anomalies, compared with previous SOTA methods. The results on four benchmarks demonstrate the accuracy and robustness of our approach to detect anomalous objects regardless of the amount or diversity of exposed training outliers. Despite the remarkable performance on most datasets, PEBAL is not as effective on the most challenging dataset, Road Anomaly, that contains significantly more diverse and realistic anomalous objects. We plan to further enhance the generalisation of our model to accurately detect more unknown, diverse anomalies.[1]

[1] Supported by Australian Research Council through grants DP180103232 and FT190100525.

References

1. Baur, C., Wiestler, B., Albarqouni, S., Navab, N.: Deep autoencoding models for unsupervised anomaly segmentation in brain MR images. In: Crimi, A., Bakas, S., Kuijf, H., Keyvan, F., Reyes, M., van Walsum, T. (eds.) BrainLes 2018. LNCS, vol. 11383, pp. 161–169. Springer, Cham (2019). https://doi.org/10.1007/978-3-030-11723-8_16
2. Bevandić, P., Krešo, I., Oršić, M., Šegvić, S.: Simultaneous semantic segmentation and outlier detection in presence of domain shift. In: Fink, G.A., Frintrop, S., Jiang, X. (eds.) DAGM GCPR 2019. LNCS, vol. 11824, pp. 33–47. Springer, Cham (2019). https://doi.org/10.1007/978-3-030-33676-9_3
3. Bevandić, P., Krešo, I., Oršić, M., Šegvić, S.: Discriminative out-of-distribution detection for semantic segmentation. arXiv preprint arXiv:1808.07703 (2018)
4. Blum, H., Sarlin, P.E., Nieto, J., Siegwart, R., Cadena, C.: The fishyscapes benchmark: measuring blind spots in semantic segmentation. arXiv preprint arXiv:1904.03215 (2019)
5. Chan, R., et al.: Segmentmeifyoucan: a benchmark for anomaly segmentation. NeurIPS (2021)
6. Chan, R., Rottmann, M., Gottschalk, H.: Entropy maximization and meta classification for out-of-distribution detection in semantic segmentation. In: Proceedings of the IEEE/CVF International Conference on Computer Vision, pp. 5128–5137 (2021)
7. Chen, L.C., Zhu, Y., Papandreou, G., Schroff, F., Adam, H.: Encoder-decoder with atrous separable convolution for semantic image segmentation. In: Proceedings of the European conference on computer vision (ECCV), pp. 801–818 (2018)
8. Chen, Y., Tian, Y., Pang, G., Carneiro, G.: Deep one-class classification via interpolated gaussian descriptor (2021)
9. Cordts, M., et al.: The cityscapes dataset for semantic urban scene understanding. In: Proceedings of the IEEE Conference on Computer Vision and Pattern Recognition, pp. 3213–3223 (2016)
10. Creusot, C., Munawar, A.: Real-time small obstacle detection on highways using compressive RBM road reconstruction. In: 2015 IEEE Intelligent Vehicles Symposium (IV), pp. 162–167. IEEE (2015)
11. Di Biase, G., Blum, H., Siegwart, R., Cadena, C.: Pixel-wise anomaly detection in complex driving scenes. In: Proceedings of the IEEE/CVF Conference on Computer Vision and Pattern Recognition, pp. 16918–16927 (2021)
12. El-Yaniv, R., et al.: On the foundations of noise-free selective classification. J. Mach. Learn. Res. **11**(5) (2010)
13. Everingham, M., Van Gool, L., Williams, C.K., Winn, J., Zisserman, A.: The pascal visual object classes (VOC) challenge. Int. J. Comput. Vis. **88**(2), 303–338 (2010)
14. Grathwohl, W., Wang, K.C., Jacobsen, J.H., Duvenaud, D., Norouzi, M., Swersky, K.: Your classifier is secretly an energy based model and you should treat it like one. arXiv preprint arXiv:1912.03263 (2019)
15. Grcić, M., Bevandić, P., Šegvić, S.: Dense anomaly detection by robust learning on synthetic negative data. arXiv preprint arXiv:2112.12833 (2021)
16. Haldimann, D., Blum, H., Siegwart, R., Cadena, C.: This is not what i imagined: error detection for semantic segmentation through visual dissimilarity. arXiv preprint arXiv:1909.00676 (2019)
17. Hendrycks, D., Basart, S., Mazeika, M., Mostajabi, M., Steinhardt, J., Song, D.: Scaling out-of-distribution detection for real-world settings. arXiv preprint arXiv:1911.11132 (2019)

18. Hendrycks, D., Gimpel, K.: A baseline for detecting misclassified and out-of-distribution examples in neural networks. arXiv preprint arXiv:1610.02136 (2016)
19. Hendrycks, D., Mazeika, M., Dietterich, T.: Deep anomaly detection with outlier exposure. arXiv preprint arXiv:1812.04606 (2018)
20. Jung, S., Lee, J., Gwak, D., Choi, S., Choo, J.: Standardized max logits: a simple yet effective approach for identifying unexpected road obstacles in urban-scene segmentation. In: Proceedings of the IEEE/CVF International Conference on Computer Vision, pp. 15425–15434 (2021)
21. Kendall, A., Gal, Y.: What uncertainties do we need in bayesian deep learning for computer vision? arXiv preprint arXiv:1703.04977 (2017)
22. Lakshminarayanan, B., Pritzel, A., Blundell, C.: Simple and scalable predictive uncertainty estimation using deep ensembles. arXiv preprint arXiv:1612.01474 (2016)
23. Lateef, F., Ruichek, Y.: Survey on semantic segmentation using deep learning techniques. Neurocomputing **338**, 321–348 (2019)
24. LeCun, Y., Chopra, S., Hadsell, R., Ranzato, M., Huang, F.: A tutorial on energy-based learning. Predicting Structured Data **1**(0) (2006)
25. Lee, K., Lee, H., Lee, K., Shin, J.: Training confidence-calibrated classifiers for detecting out-of-distribution samples. arXiv preprint arXiv:1711.09325 (2017)
26. Lee, K., Lee, K., Lee, H., Shin, J.: A simple unified framework for detecting out-of-distribution samples and adversarial attacks. Advances in Neural Information Processing Systems 31 (2018)
27. Li, C.L., Sohn, K., Yoon, J., Pfister, T.: Cutpaste: self-supervised learning for anomaly detection and localization. In: Proceedings of the IEEE/CVF Conference on Computer Vision and Pattern Recognition, pp. 9664–9674 (2021)
28. Liang, S., Li, Y., Srikant, R.: Enhancing the reliability of out-of-distribution image detection in neural networks. arXiv preprint arXiv:1706.02690 (2017)
29. Lin, T.-Y., Maire, M., Belongie, S., Hays, J., Perona, P., Ramanan, D., Dollár, P., Zitnick, C.L.: Microsoft COCO: common objects in context. In: Fleet, D., Pajdla, T., Schiele, B., Tuytelaars, T. (eds.) ECCV 2014. LNCS, vol. 8693, pp. 740–755. Springer, Cham (2014). https://doi.org/10.1007/978-3-319-10602-1_48
30. Lis, K., Nakka, K., Fua, P., Salzmann, M.: Detecting the unexpected via image resynthesis. In: Proceedings of the IEEE/CVF International Conference on Computer Vision, pp. 2152–2161 (2019)
31. Liu, W., Wang, X., Owens, J.D., Li, Y.: Energy-based out-of-distribution detection. arXiv preprint arXiv:2010.03759 (2020)
32. Liu, Z., Wang, Z., Liang, P.P., Salakhutdinov, R.R., Morency, L.P., Ueda, M.: Deep gamblers: learning to abstain with portfolio theory. Adv. Neural. Inf. Process. Syst. **32**, 10623–10633 (2019)
33. Malinin, A., Gales, M.: Predictive uncertainty estimation via prior networks. arXiv preprint arXiv:1802.10501 (2018)
34. Mukhoti, J., Gal, Y.: Evaluating bayesian deep learning methods for semantic segmentation. arXiv preprint arXiv:1811.12709 (2018)
35. Nijkamp, E., Hill, M., Zhu, S.C., Wu, Y.N.: Learning non-convergent non-persistent short-run MCMC toward energy-based model. arXiv preprint arXiv:1904.09770 (2019)
36. Pinggera, P., Ramos, S., Gehrig, S., Franke, U., Rother, C., Mester, R.: Lost and found: detecting small road hazards for self-driving vehicles. In: 2016 IEEE/RSJ International Conference on Intelligent Robots and Systems (IROS), pp. 1099–1106. IEEE (2016)

37. Sultani, W., Chen, C., Shah, M.: Real-world anomaly detection in surveillance videos. In: Proceedings of the IEEE Conference on Computer Vision and Pattern Recognition, pp. 6479–6488 (2018)
38. Vandenhende, S., Georgoulis, S., Proesmans, M., Dai, D., Van Gool, L.: Revisiting multi-task learning in the deep learning era. arXiv preprint arXiv:2004.13379 2 (2020)
39. Vojir, T., Šipka, T., Aljundi, R., Chumerin, N., Reino, D.O., Matas, J.: Road anomaly detection by partial image reconstruction with segmentation coupling. In: Proceedings of the IEEE/CVF International Conference on Computer Vision, pp. 15651–15660 (2021)
40. Xia, Y., Zhang, Y., Liu, F., Shen, W., Yuille, A.L.: Synthesize then compare: detecting failures and anomalies for semantic segmentation. In: Vedaldi, A., Bischof, H., Brox, T., Frahm, J.-M. (eds.) ECCV 2020. LNCS, vol. 12346, pp. 145–161. Springer, Cham (2020). https://doi.org/10.1007/978-3-030-58452-8_9
41. Yun, S., Han, D., Oh, S.J., Chun, S., Choe, J., Yoo, Y.: Cutmix: regularization strategy to train strong classifiers with localizable features. In: Proceedings of the IEEE/CVF International Conference on Computer Vision, pp. 6023–6032 (2019)
42. Zhu, Y., Sapra, K., Reda, F.A., Shih, K.J., Newsam, S., Tao, A., Catanzaro, B.: Improving semantic segmentation via video propagation and label relaxation. In: Proceedings of the IEEE/CVF Conference on Computer Vision and Pattern Recognition, pp. 8856–8865 (2019)

Rethinking Closed-Loop Training for Autonomous Driving

Chris Zhang[1,2(✉)], Runsheng Guo[3], Wenyuan Zeng[1,2], Yuwen Xiong[1,2], Binbin Dai[1], Rui Hu[1], Mengye Ren[4], and Raquel Urtasun[1,2]

[1] Waabi, Toronto, Canada
{czhang,wzeng,yxiong,bdai,rhu,urtasun}@waabi.ai
[2] University of Toronto, Toronto, Canada
[3] University of Waterloo, Waterloo, Canada
r9guo@uwaterloo.ca
[4] New York University, New York, USA
mengye@nyu.edu

Abstract. Recent advances in high-fidelity simulators [22,44,82] have enabled closed-loop training of autonomous driving agents, potentially solving the distribution shift in training v.s. deployment and allowing training to be scaled both safely and cheaply. However, there is a lack of understanding of how to build effective training benchmarks for closed-loop training. In this work, we present the first empirical study which analyzes the effects of different training benchmark designs on the success of learning agents, such as how to design traffic scenarios and scale training environments. Furthermore, we show that many popular RL algorithms cannot achieve satisfactory performance in the context of autonomous driving, as they lack long-term planning and take an extremely long time to train. To address these issues, we propose trajectory value learning (TRAVL), an RL-based driving agent that performs planning with multistep look-ahead and exploits cheaply generated imagined data for efficient learning. Our experiments show that TRAVL can learn much faster and produce safer maneuvers compared to all the baselines.

Keywords: Closed-loop learning · Autonomous driving · RL

1 Introduction

Self-driving vehicles require complex decision-making processes that guarantee safety while maximizing comfort and progress towards the destination. Most

C. Zhang, R. Guo and W. Zeng—Equal contribution.
R. Guo and M. Ren—Work done during affiliation with Waabi.

Supplementary Information The online version contains supplementary material available at https://doi.org/10.1007/978-3-031-19842-7_16.

S. Avidan et al. (Eds.): ECCV 2022, LNCS 13699, pp. 264–282, 2022.
https://doi.org/10.1007/978-3-031-19842-7_16

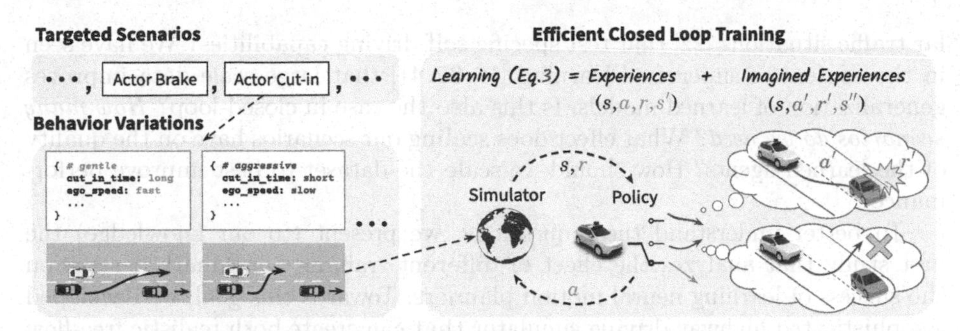

Fig. 1. We use behavioral variations on top of scenarios designed to target specific traffic interactions as the basis for learning. TRAVL efficiently learns to plan trajectories from both real and imagined experience.

approaches have relied on hand-engineered planners that are built on top of perception and motion forecasting modules. However, a robust decision process has proven elusive, failing to handle the complexity of the real world.

In recent years, several approaches have been proposed, aiming at exploiting machine learning to learn to drive. Supervised learning approaches such as behavior cloning [3,16,17,47,56,62] that learn from human demonstrations are amongst the most popular, as large amounts of driving data can be readily obtained by instrumenting a vehicle to collect data while a human is driving. However, learning in this open-loop manner leads to distribution shift between training and deployment [17,18,59], as the model does not understand the closed-loop effects of its actions when passively learning from the expert data distribution.

Closed-loop training is one principled way to tackle this, by enabling the agent to continuously interact with the environment and thus learn to recover from its own mistakes. However, closing the loop while driving in the real world comes with many safety concerns as it is dangerous to update the software on the fly without proper safety verification. Furthermore, it is unethical to expose the self-driving vehicle (SDV) to safety-critical situations, which is necessary for learning to handle them. Finally, rare scenarios can take extremely long to capture in the wild and is impractical to scale. An appealing alternative is to learn to drive in a virtual environment by exploiting simulation systems [22,82]. While encouraging results have been demonstrated [39,62,73], there is still a lack of understanding of how to build training benchmarks for effective closed-loop training.

In this paper we aim to shed some light on this by studying the following questions: *What type of scenarios do we need to learn to drive safely?* Simple scenarios used in previous works such as car racing [38,53,78] and empty-lanes [33] are insufficient in capturing the full complexity of driving. Should we instead simulate complex free-flow traffic[1] [27,29] or design scenarios targeting particu-

[1] This is similar to how we encounter random events when collecting data.

lar traffic situations [22] that test specific self-driving capabilities? We have seen in the context of supervised learning [19,32,41] that large scale data improves generalization of learned models. Is this also the case in closed-loop? *How many scenarios do we need?* What effect does scaling our scenarios have on the quality of the learned agents? How should we scale the dataset to best improve performance?

To better understand these questions, we present (to our knowledge) the first study that analyzes the effect of different training benchmark designs on the success of learning neural motion planners. Towards this goal, we developed a sophisticated highway driving simulator that can create both realistic free-flow traffic as well as targeted scenarios capable of testing specific self-driving capabilities. In particular, the latter is achieved by exploiting procedural modeling which composes variations of unique traffic patterns (*e.g.*, lead actor breaking, merging in from an on ramp). Since each of these patterns is parameterized, we can sample diverse variations and generate a large set of scenarios automatically. Under this benchmark, we show:

1. Scenarios designed for specific traffic interactions provide a richer learning signal than generic free-flow traffic simulations. This is likely because the former ensures the presence of interesting and safety-critical interactions.
2. Training on smaller scenario variations leads to more unsafe driving behaviors. This suggests that crafting more variations of traffic situations is key when building training benchmarks.
3. Existing RL-based approaches have difficulty learning the intricacies of many scenarios. This is likely because they typically learn a direct mapping from observations to control signals (*e.g.*, throttle, steering). Thus, they regrettably lack multi-step lookahead reasoning into the future, which is necessary to handle complex scenarios such as merging into crowded lanes. Furthermore, learning these agents in a model-free manner can be extremely slow, as the agents have to learn with trial and error through costly simulations.

To address the struggles of current RL approaches, we propose trajectory value learning (TRAVL), a method which learns *long-term reasoning efficiently in closed-loop* (Fig. 1). Instead of myopically outputting control commands independently at each timestep, TRAVL can perform decision-making with explicit multistep look-ahead by *planning* in trajectory space. Our model learns a deep feature map representation of the state which can be fused with trajectory features to directly predict the Q-value of following that trajectory. Inference amounts to selecting the maximum value trajectory plan from a sampled trajectory set. Unlike conventional model-based planning, this bypasses the need to explicitly model and predict all state variables and transitions, as not all of them are equally important (*e.g.*, a far away vehicle is of less interest in driving). Furthermore, our trajectory-based formulation allows us to cheaply produce additional *imagined* (*i.e.*, counterfactual) experiences, resulting in significantly better learning efficiency compared to model-free methods which need to rely solely on interacting in the environment.

Summary of Contributions: In this paper, we present an in-depth empirical study on how various design choices of training data generation can affect the success of learning driving policies in closed-loop. This allows us to identify a number of guidelines for building effective closed-loop training benchmarks. We further propose a new algorithm for efficient and effective learning of long horizon driving policies, that better handle complex scenarios which mimic the complexity of the real-world. We believe our work can serve as a starting point to rethink how we shall conduct closed-loop training for autonomous driving.

2 Related Work

Open-Loop Training: In open-loop training, the agent does not take any actions and instead learns passively by observing expert states and actions. ALVINN [55] first explored behavior cloning as an open-loop training method for end-to-end visuomotor self-driving. Since then, several advances in data augmentation [3,16], neural network architecture [16,47,56] and auxiliary task design [17,62] have been made in order to handle more complex environments. Additionally, margin-based learning approaches [60,61,79,80] incorporate structured output spaces, while offline RL [37,68] exploits reward functions. The primary challenge in open-loop training is the distribution shift encountered when the predicted actions are rolled out in closed-loop. While techniques such as cleverly augmenting the training data [3] partially alleviate this issue, challenges remain.

Closed-Loop Training: The most popular paradigm for closed-loop training is reinforcement learning (RL). In contrast to open-loop learning, online RL approaches [40,65,66] do not require pre-collected expert data, but instead learn through interacting with the environment. However, such methods have prohibitively low sample efficiency [13] and can take several weeks to train a single model [73]. To address this issue, auxiliary tasks have been used as additional sources of supervision, such as predicting affordances [12,62,73], scene reconstruction [33] or imitation-based pre-training [39]. Note that our learning approach is orthogonal to these tasks and thus can be easily combined. Model-based RL approaches are more sample efficient. They assume access to a world model, which provides a cheaper way to generate training data [8,23,72] in addition to the simulator. It can also be used during inference for planning with multi-step lookaheads [15,25,26,48,51,64,69,70]. Yet, not many works have been explored in the context of self-driving. [13] uses an on-rails world model to generate data for training and empirically shows better efficiency. [52] uses a learned semantic predictor to perform multistep look-ahead during inference. Our work enjoys both benefits with a unified trajectory-based formulation.

When an expert can be queried online, other closed-loop training techniques such as DAgger style approaches [14,53,59] can be applied. When the expert is only available during offline data collection, one can apply closed-loop imitation learning methods [31,36]. However, building an expert can be expensive and difficult, limiting the applications of these methods.

Closed-Loop Benchmarking: Directly closing the loop in the real world [33] provides the best realism but is unsafe. Thus, leveraging simulation has been a dominant approach for autonomous driving. Environments focused on simple car racing [7,78] are useful in prototyping quickly but can be over-simplified for real-world driving applications. More complex traffic simulators [2,4,11,43] typically use heuristic-based actors to simulate general traffic flow. Learning-based traffic models have also been explored recently [5,71]. However, we show that while useful for evaluating the SDV in nominal conditions, general traffic flow provides limited learning signal as interesting interactions are not guaranteed to happen frequently. As the majority of traffic accidents can be categorized into a few number of situations [49], recent works focus on crafting traffic scenarios specifically targeting these situations for more efficient coverage [1,21,22,58,67,81]. However, visual diversity is often more stressed than behavioral diversity. For example, the CARLA benchmark [22] has 10 different scenario types and uses geolocation for variation, resulting in visually diverse backgrounds but fixed actor policies[2]. Other approaches include mining real data [9], or using adversarial approaches [20,24,35,75]. Multi-agent approaches that control different actors with different policies [6,82] have also been proposed. However, these works have not studied the effects on learning in detail.

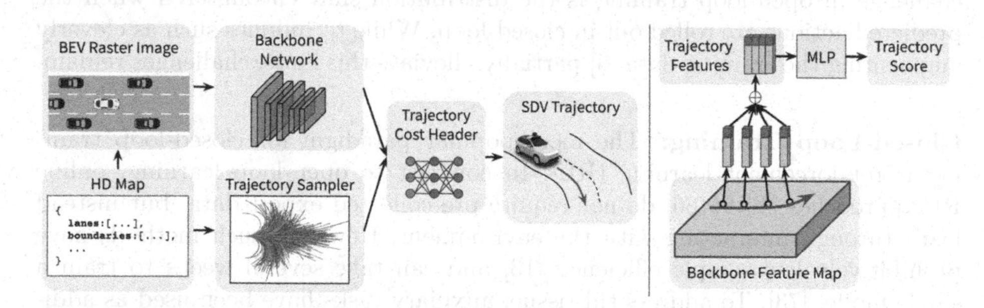

Fig. 2. TRAVL leverages rich backbone features to predict the cost of following a trajectory. The lowest costed trajectory is selected as the SDV's plan

3 Learning Neural Planners in Closed-Loop

Most model-free RL-based self-driving approaches parametrize the action space as instant control signals (e.g., throttle, steering), which are directly predicted from state observations. While simple, this parameterization hampers the ability to perform long-term reasoning into the future, which is necessary in order to handle complex driving situations. Furthermore, this approach can lead to inefficient learning as it relies solely on experiences collected from the environment, which may contain only sparse supervisory signals. Model-based approaches address these issues by explicitly a predictive model of the world. However,

[2] https://github.com/carla-simulator/scenario_runner.

performing explicit model rollouts online during inference can be prohibitively expensive especially if the number of potential future trajectories considered is very large.

We address these issues by combining aspects of model-free and model-based approaches (Fig. 2). In particular, we learn to reason into the future by directly costing trajectories without explicit model rollouts, resulting in more efficient inference. In addition to using real experience from the environment, our trajectory output representation allows us to learn from imagined (*i.e.*, counterfactual) experiences collected from an approximate world model, greatly improving sample efficiency.

3.1 Preliminaries on RL

The goal of an SDV is to make safe decisions sequentially. This can be modeled as a Markov Decision Process (MDP): $\mathcal{M} = (\mathcal{S}, \mathcal{A}, P, R, \gamma)$, where $\mathcal{S}$ and $\mathcal{A}$ represent state and action spaces respectively, such as raw observations of the scene and control signals for the ego-vehicle. $P(s'|s,a)$ and $R(s,a,s')$ represent the transition dynamics and reward functions respectively, and $\gamma \in (0,1)$ is the discount factor. We are interested in learning an optimal policy that maximizes the expected discounted return,

$$\pi^*(a|s) = \arg\max_{\pi} \mathbb{E}_{\pi,P}\left[\sum_{t=0}^{T} \gamma^t R(s^t, a^t, s^{t+1})\right]$$

Off-policy RL algorithms are popular solutions due to their high data efficiency since they are agnostic to the data collecting policy and thus do not constantly require fresh data. A general form of the off-policy learning process can be described as iteratively alternating between *policy evaluation* and *policy improvement* steps. Specifically, one can maintain a learnable Q-function $Q^k(s,a) = \mathbb{E}_{\pi,P}\left[\sum_{t=0}^{T} \gamma^t R(s^t, a^t, s^{t+1})\right]$, which captures the expected future return when executing $\pi^k(a|s)$, with k being the learning iteration. In the policy evaluation step, the Bellman operator $\mathcal{B}_\pi Q := R + \gamma \mathcal{P}_\pi Q$ is applied to update Q based on simulated data samples. Here, the transition matrix $\mathcal{P}_\pi$ is a matrix coupled with the policy π, *i.e.*, $\mathcal{P}_\pi Q(s,a) = \mathbb{E}_{s' \sim P(s'|s,a), a' \sim \pi(a'|s')}[Q(s',a')]$. We can then improve the policy π to favor selecting actions that maximize the expected Q-value. However, both steps require evaluations on all possible (s,a) pairs, and thus this is intractable for large state and action spaces. In practice, one can instead apply empirical losses over a replay buffer, *e.g.*, minimizing the empirical ℓ_2 loss between the left and right hand side of the Bellman operator. The replay buffer is defined as the set $\mathcal{D} = \{(s,a,r,s')\}$ holding past experiences sampled from $P(s'|s,a)$ and $\pi(a|s)$. Putting this together, the updating rules can be described as follows,

$$Q^{k+1} \leftarrow \arg\min_{Q} \mathbb{E}_{s,a,r,s' \sim \mathcal{D}}\left[\left((r + \gamma \mathbb{E}_{a' \sim \pi^k}[Q^k(s',a')]) - Q(s,a)\right)^2\right] \text{ (evaluation)},$$
$$\pi^{k+1} \leftarrow (1-\epsilon) \arg\max_{\pi} \mathbb{E}_{s \sim \mathcal{D}, a \sim \pi}[Q^{k+1}(s,a)] + \epsilon U(a), \quad \text{(improvement)} \qquad (1)$$

where U is the uniform distribution and ϵ is introduced for epsilon-greedy exploration, *i.e.*, making a greedy action under Q with probability $1-\epsilon$ and otherwise randomly exploring other actions with probability ϵ. Note that Eq. 1 reduces to standard Q-learning when we use $\pi^{k+1}(s) = \arg\max_a Q^{k+1}(s,a)$ as the policy improvement step instead.

3.2 Planning with TRAVL

Our goal is to design a driving model that can perform long term reasoning into the future by *planning.* To this end, we define our action as a trajectory $\tau = \{(x^0, y^0), (x^1, y^1), \cdots, (x^T, y^T)\}$, which navigates the ego-vehicle for the next T timesteps. Here (x^t, y^t) is the spatial location in birds eye view (BEV) at timestep t. Inspired by humans, we decompose the cost of following a trajectory into a short-term *cost-to-come*, $R_\theta(s, \tau)$, defined over the next T timesteps, and a long-term *cost-to-go* $V_\theta(s, \tau)$ that operates beyond that horizon. The final Q-function is defined as

$$Q_\theta(s, \tau) = R_\theta(s, \tau) + V_\theta(s, \tau)$$

Note that both R_θ and V_θ are predicted with a neural network. In the following, we describe our input state representation, the backbone network and cost predictive modules used to predict Q_θ follow by our planning inference procedure.

Input Representation: Following [3,32,57], our state space $\mathcal{S}$ contains an HD map as well as the motion history of the past T' seconds of both the ego-vehicle and other actors. To make the input data amenable to standard convolutional neural networks (CNNs), we rasterize the information into a BEV tensor, where for each frame within the history horizon T', we draw bounding boxes of all actors as 1 channel using a binary mask. The ego-vehicle's past positions are also rasterized similarly into T' additional channels. We utilize an M channel tensor to represent the HD map, where each channel encodes a different map primitive, such as centerlines or the target route. Finally, we include two more channels to represent the (x, y) coordinates of BEV pixels [42]. This results in a input tensor of size $\mathbb{R}^{H \times W \times (2T'+M+2)}$, where H and W denotes the size of our input region around the SDV.

Backbone Network: To extract useful contextual information, we feed the input tensor to a backbone network. As the input modality is a 2D image, we employ a CNN backbone adapted from ResNet [28]. Given an input tensor of size $\mathbb{R}^{H \times W \times (2T'+M+2)}$, the backbone performs downsampling and computes a final feature map $\mathbf{F} \in \frac{H}{8} \times \frac{W}{8} \times C$, where C is the feature dimension. More details on the architecture are provided in the supplementary.

Cost Predictive Header: We use a cost predictive header that takes an arbitrary trajectory τ and backbone feature map $\mathbf{F}$ as inputs, and outputs two scalar values representing the cost-to-come and cost-to-go of executing τ. As τ is represented by a sequence of 2D waypoints $\{(x^0, y^0), (x^1, y^1), \cdots, (x^T, y^T)\}$, we can extract context features of τ by indexing the t channel of the backbone feature $\mathbf{F}$ at position (x^t, y^t) for each timestep t. We then concatenate the features from all timesteps into a single feature vector $\mathbf{f}_\tau$. Note that the backbone feature $\mathbf{F}$ encodes rich information about the environment, and thus such an indexing operation is expected to help reason about the goodness of a trajectory, *e.g.*, if it is collision-free and follows the map. We also include kinematic information of τ into $\mathbf{f}_\tau$ by concatenating the position (x^t, y^t), velocity (v^t), acceleration (a^t), orientation (θ_t) and curvature $(\kappa_t, \dot{\kappa}_t)$. We use two shallow (3 layer) multi-layer perceptrons (MLPs) to regress the cost-to-come and cost-to-go from $\mathbf{f}_\tau$ before finally summing them to obtain our estimate of $Q(s, \tau)$.

Efficient Inference: Finding the optimal policy $\tau^* := \arg\max_\tau Q(s, \tau)$ given a Q-function over trajectories is difficult as τ lies in a continuous and high-dimensional space that has complex structures (*e.g.*, dynamic constraints). To make inference efficient, we approximate such an optimization problem using a sampling strategy [61,63,77,79]. Towards this goal, we first sample a wide variety of trajectories $\mathcal{T}$ that are physically feasible for the ego-vehicle, and then pick the one with maximum Q-value

$$\tau^* = \arg\max_{\tau \in \mathcal{T}} Q(s, \tau).$$

To obtain a set of trajectory candidates $\mathcal{T}$, we use a map-based trajectory sampler, which samples a set of lane following and lane changing trajectories following a bicycle model [54]. Inspired by [61], our sampling procedure is in the Frenet frame of the road, allowing us to easily sample trajectories which consider map priors, *e.g.*, follow curved lanes. Specifically, longitudinal trajectories are obtained by fitting quartic splines to knots corresponding to varying speed profiles, while lateral trajectories are obtained by first sampling sets of various lateral offsets (defined with respect to reference lanes) at different longitudinal locations and then fitting quintic splines to them. In practice, we find embedding map priors in this manner can greatly improve the model performance. In our experiments we sample roughly 10k trajectories per state. Note that despite outputting an entire trajectory as the action, for inference we use an MPC style execution [10] where the agent only executes an initial segment of the trajectory before replanning with the latest observation. Further discussion on this method of planning can be found in the supplementary.

3.3 Efficient Learning with Counterfactual Rollouts

The most straightforward way to learn our model is through classical RL algorithms. We can write the policy evaluation step in Eq. 1 as

$$Q^{k+1} \leftarrow \arg\min_{Q_\theta} \mathbb{E}_{\mathcal{D}} \left[\left(Q_\theta - \mathcal{B}_\pi^k Q^k \right)^2 \right], \quad s.t. \;\; Q_\theta = R_\theta + V_\theta, \tag{2}$$

where $\mathcal{B}_\pi Q := R + \gamma \mathcal{P}_\pi Q$ is the Bellman operator. However, as we show in our experiments and also demonstrated in other works [76], such model-free RL algorithms learn very slowly. Fortunately, our trajectory-based formulation and decomposition of Q_θ into R_θ and V_θ allow us to design a more efficient learning algorithm that follows the spirit of model-based approaches.

One of the main benefits of model-based RL [72] is the ability to efficiently sample imagined data through the world dynamics model, as this helps bypass the need for unrolling the policy in simulation which can be computationally expensive. However, for complex systems the learned model is likely to be also expensive, *e.g.*, neural networks. Therefore, we propose a simple yet effective world model where we assume the actors are not intelligent and do not react to different SDV trajectories. Suppose we have a replay buffer $\mathcal{D} = \{(s^t, \tau^t, r^t, s^{t+1})\}$ collected by interacting our current policy π with the simulator. To augment the training data with cheaply generated imagined data (s^t, τ', r', s'), we consider a counterfactual trajectory τ' that is different from τ. The resulting counterfactual state s' simply modifies s^{t+1} such that the ego-vehicle follows τ', while keeping the actors' original states the same. The counterfactual reward can then be computed as $r' = R(s^t, \tau', s')$. We exploit our near limitless source of counterfactual data for dense supervision on the short term predicted cost-to-come R_θ. Efficiently learning R_θ in turn benefits the learning of the Q-function overall. Our final learning objective (policy evaluation) is then

$$Q^{k+1} \leftarrow \arg\min_{Q_\theta} \mathbb{E}_{\mathcal{D}} \left[\underbrace{\left(Q_\theta(s,\tau) - \mathcal{B}_\pi^k Q^k(s,\tau)\right)^2}_{\text{Q-learning}} + \alpha_k \underbrace{\mathbb{E}_{\tau' \sim \mu(\tau'|s)} \left(R_\theta(s,\tau') - r'\right)^2}_{\text{Counterfactual Reward Loss}} \right],$$
$$s.t. \quad Q_\theta = R_\theta + V_\theta, \quad V_\theta = \gamma \mathcal{P}_\pi^k Q^k. \tag{3}$$

Here, μ is an arbitrary distribution over possible trajectories and characterizes the exploration strategy for counterfactual supervision. We use a uniform distribution over the sampled trajectory set $\mathcal{T}$ in all our experiments for simplicity. To perform an updating step in Eq. 3, we approximate the optimal value of Q_θ by applying a gradient step of the empirical loss of the objective. In practice, the gradients are applied over the parameters of R_θ and V_θ, whereas Q_θ is simply obtained by $R_\theta + V_\theta$. We also find that simply using the Q-learning policy improvement step suffices in practice. We provide more implementation details in the supplementary material. Note that we use counterfactual rollouts, and thus the simplified non-reactive world dynamics, only as supervision for short term reward/cost-to-come component. This can help avoid compounding errors of using such an assumption for long-term imagined simulations.

Theoretical Analysis: The counterfactual reward loss component can introduce modeling errors due to our approximated world modeling. As the Q-learning loss term in Eq. 3 is decreasing when k is large, such errors can have significant effects, hence it is non-obvious whether our learning procedure Eq. 3 can satisfactorily converge. We now show that iteratively applying policy evaluation in Eq. 3 and policy update in Eq. 1 indeed converges under some mild conditions.

Lemma 1. *Assuming R is bounded by a constant R_{max} and α_k satisfies*

$$\alpha_k < \left(\frac{1}{\gamma^k C} - 1\right)^{-1} \left(\frac{\pi^k}{\mu}\right)_{min}, \tag{4}$$

with C an arbitrary constant, iteratively applying Eq. 3 and the policy update step in Eq. 1 converges to a fixed point.

Furthermore, it converges to the optimal Q-function, Q^*. We refer the reader to the supplementary material for detailed proofs.

Theorem 1. *Under the same conditions as Lemma 1, our learning procedure converges to Q^*.*

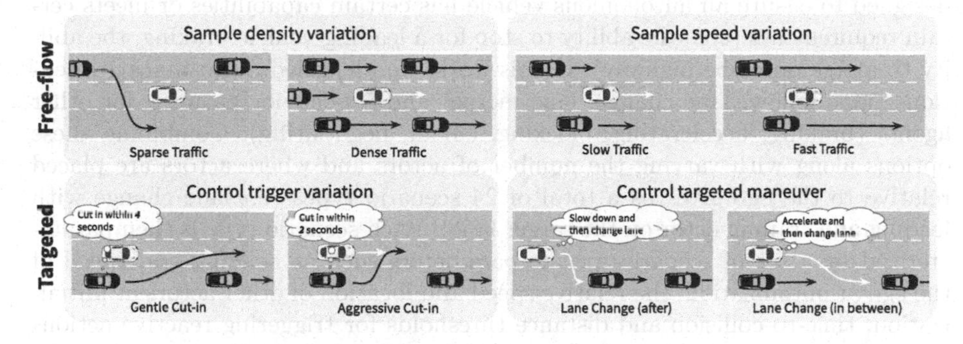

Fig. 3. Top row: Free-flow scenarios generated by sampling parameters such as density and actor speed. **Bottom row:** Targeted scenarios generated by enacting fine-grained control on the actors to target specific traffic situations.

4 Large Scale Closed-Loop Benchmark Dataset

We now describe how we design our large scale closed-loop benchmark (Fig. 3). In our simulator the agent drives on a diverse set of highways, which contain standard, on-ramp, merge and fork map topologies with varying curvature and number of lanes. We use IDM [74] and MOBIL [34] policies to control actors. As our work is focused on learning neural planners in closed-loop, we simulate bounding boxes and trajectory information of actors, and leave sensor simulation for future work.

There are two popular ways to testing an autonomous driving system: 1) uncontrolled traffic environments and 2) controlled scenarios that test certain self-driving capabilities such as reacting to a cut-in. However, there has been no analysis in the literature of the effects of *training* in these different environments. Towards this goal, we construct two training benchmarks based on these two different paradigms.

Free-Flow Scenario Set: Free-flow scenarios are similar to what we observe in real-world data, where we do not enact any fine-grained control over other actors. We define a generative model which samples from a set of parameters which define a scenario. Parameters which vary the initial conditions of actors include density, initial speed, actor class, and target lane goals. Parameters which vary actor driving behavior include actor target speed, target gap, speed limit, maximum acceleration and maximum deceleration. We also vary the map topology, road curvature, geometry, and the number of lanes. We use a mixture of truncated normal distributions for continuous properties and categorical distribution for discrete properties. More details are provided in the supplementary.

Targeted Scenario Set: Targeted scenarios are those which are designed to test autonomous vehicles in specific traffic situations. These scenarios are designed to ensure an autonomous vehicle has certain capabilities or meets certain requirements (*e.g.*, the ability to stop for a leading vehicle braking, the ability to merge onto the highway). In this work, we identified 3 ego-routing intentions (lane follow, lane change, lane merge) and 5 behavior patterns for other agents (braking, accelerating, blocking, cut-in, negotiating). Combining these options along with varying the number of actors and where actors are placed relative to the ego gives us a total of 24 scenario types (*e.g.*lane change with leading and trailing actor on the target lane). Each scenario type is then parameterized by a set of scenario-specific parameters such as heading and speed of the ego at initialization, the relative speed and location of other actors at initialization, time-to-collision and distance thresholds for triggering reactive actions

Table 1. We compare our approach against several baselines. Here C is using the standard control setting and T is using our proposed trajectory-based architecture and formulation. We see that trajectory-based approaches outperforms their control-based counterparts. We also see that our proposed method is able to learn more efficiently and outperform baselines.

Method		Pass Rate ↑	Col. Rate ↓	Prog. ↑	MinTTC↑	MinDist↑
Imit. Learning	C	0.545	0.177	240	0.00	2.82
PPO [66]		0.173	0.163	114	0.00	5.56
A3C [46]		0.224	0.159	284	0.03	4.65
RAINBOW[a] [30]		0.435	0.270	234	0.00	1.38
Imit. Learning	T	0.617	0.261	**286**	0.00	1.49
PPO [66]		0.273	0.249	200	0.00	1.73
A3C [46]		0.362	0.137	135	0.30	6.14
RAINBOW[a] [30]		0.814	0.048	224	0.45	9.70
TRAVL (ours)		**0.865**	**0.026**	230	**0.82**	**12.62**

[a] We only include prioritized replay and multistep learning as they were found to be the most applicable and important in our setting.

(*e.g.*when an actor performs a cut-in), IDM parameters of other actors as well as map parameters. We then procedurally generate variations of these scenarios by varying the values of these parameters, which result in diverse scenario realizations with actor behaviors that share similar semantics.

Benchmarking: We briefly explain how we sample parameters for scenarios. As the free-flow scenarios aim to capture nominal traffic situations, we simply sample *i.i.d* random parameter values and hold out a test set. In contrast, each targeted scenario serves for benchmarking a specific driving capability, and thus we should prevent training and testing on the same (or similar) scenarios. To this end, we first generate a test set of scenarios aiming to provide thorough evaluations over the entire parameterized spaces. Because enumerating all possible combinations of parameter is intractable, we employ an all-pairs generative approach [50] which provides a much smaller set that contains all possible combinations for any pair of discrete parameters. This is expected to provide efficient testing while covering a significant amount of failure cases [45]. More details on this approach are in the supplementary. Finally, we hold out those test parameters when drawing random samples for the training and validation set.

Table 2. We train RAINBOW[a] and TRAVL on different sets and evaluate on different sets. We see that training on targeted scenarios performs better than training on free-flow scenarios, even when evaluated on free-flow scenarios.

			Test					
			Pass Rate ↑		Collision Rate ↓		Progress ↑	
			Free-flow	Targeted	Free-flow	Targeted	Free-flow	Targeted
Train	RB[a]+T	Free-flow	0.783	0.453	0.198	0.228	146	173
		Targeted	0.885	0.815	0.104	0.048	231	224
	TRAVL	Free-flow	0.784	0.696	0.198	0.177	229	219
		Targeted	0.903	0.865	0.089	0.026	172	230

[a] We only include prioritized replay and multistep learning as they were found to be the most applicable and important in our setting.

5 Experiments

In this section, we showcase the benefits of our proposed learning method TRAVL by comparing against several baselines. We empirically study the importance of using targeted scenarios compared to free-flow scenarios for training and evaluation. Finally, we further study the effect of data diversity and scale and show that large scale, behaviorally diverse data is crucial in learning good policies.

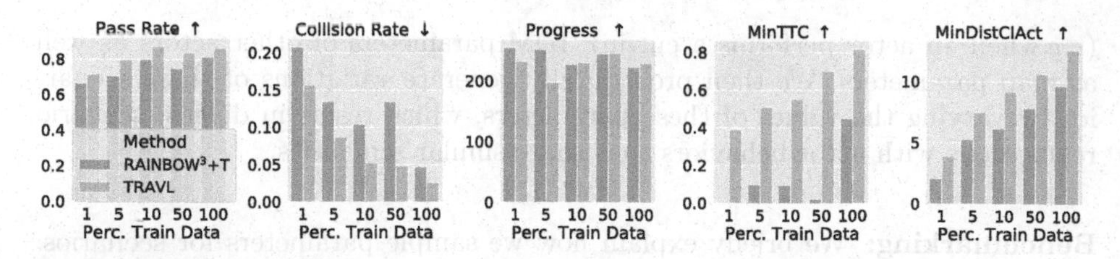

Fig. 4. Increasing scenario diversity improves performance across the board.

Datasets and Metrics: The free-flow dataset contains 834 training and 274 testing scenarios. The targeted dataset contains 783 training and 256 testing scenarios. All scenarios last for 15 s on average.

We use a number of autonomy metrics for evaluating safety and progress. *Scenario Pass Rate* is the percentage of scenarios that pass, which is defined as reaching the goal (e.g., maintain a target lane, reach a distance in a given time) without collision or speeding violations. *Collision Rate* computes the percentage of scenarios where ego collides. *Progress* measures the distance traveled in meters before the scenario ends. *Minimum Time to Collision (MinTTC)* measures the time to collision with another actor if the ego state were extrapolated along its future executed trajectory, with lower values meaning closer to collision. *Minimum Distance to the Closest Actor (MinDistClAct)* computes the minimum distance between the ego and any other actor in the scene. We use the median when reporting metrics as they are less sensitive to outliers.

Closed-Loop Benchmarking: We train and evaluate TRAVL and several baselines on our proposed targeted scenario set. We evaluate A3C [46], PPO [66] and a simplified RAINBOW[3] [30] as baseline RL algorithms, and experiment with control (C) and trajectory (T) output representations. We also include imitation learning (IL) baselines supervised from a well-tuned auto-pilot. In the control setting, the action space consists of 9 possible values for steering angle and 7 values for throttle, yielding a total of 63 discrete actions. In the trajectory setting, each trajectory sample is treated as a discrete action. All methods use the same backbone and header architecture, trajectory sampler, and MPC style inference when applicable. Our reward function consists of a combination of progress, collision and lane following terms. More details on learning, hyperparameters, reward function, and baselines are provided in the supplementary material.

[3] We only include prioritized replay and multistep learning as they were found to be the most applicable and important in our setting.

As shown in Table 1, trajectory-based methods outperform their control-based counterparts, suggesting our proposed trajectory-based formulation and architecture allow models to better learn long-term reasoning and benefit from the trajectory sampler's inductive bias. We also see that RAINBOW(see footnote 3)+T outperforms other RL trajectory-based baselines. This suggests that trajectory *value learning* is easier compared to learning a policy directly over the trajectory set. Furthermore, its off-policy nature allows the use of a replay buffer for more efficient learning. In contrast, on-policy methods such as A3C and PPO have difficulty learning, *e.g.*, it takes 10x longer to overfit to a single scenario compared to off-policy approaches. This aligns with similar findings in [22]. Additionally, IL baselines outperform weaker RL baselines that suffer from issues of sample complexity, but falls short to more efficient RL baselines due to the problem of distribution shift. Finally, TRAVL outperforms the baselines. We believe this is because our model-based counterfactual loss provides denser supervisions and thus reduces noisy variances during learning. This is empirically validated in Fig. 5 as our method has the least variance and converges much faster.

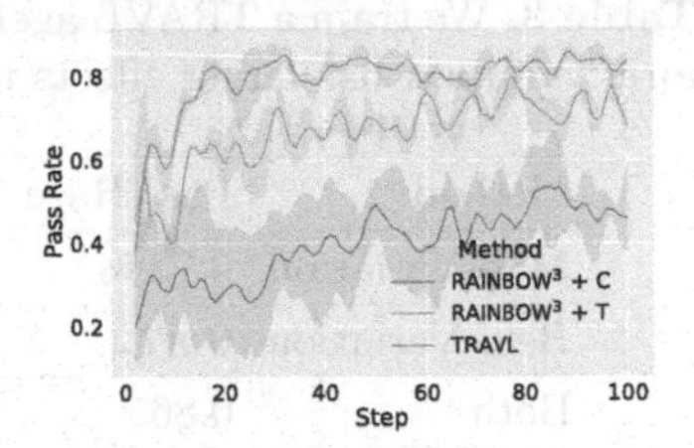

Fig. 5. Training curves for 3 runs. TRAVL has the least variance and converges faster.

Targeted vs Free-Flow: We now study the effect of using different types of scenarios for training by comparing models learned on our *targeted* vs *free-flow* set. As shown in Table 2, a model learned on the targeted set performs the best. Notably this is *even true when evaluating on the free-flow test set*, which is closer in distribution to the free-flow train set. The effectiveness of the targeted set can be due to two reasons. Firstly, as scenarios and actors are carefully designed, each targeted scenario is more likely to provide interesting interactions, resulting in stronger learning signals. On the contrary, many of the free-flow scenarios can be relatively monotonous, with fewer interactions among actors. Secondly, the design of each targeted scenario is driven by autonomous driving capabilities which where determined with expert prior knowledge and are specifically meant to capture what is necessary to be able to drive in nearly all scenarios. As a result, each type of targeted scenario can be viewed as a basis over the scenario space. Sampling behavioral variations results in a scenario set that provides wide coverage.

Behavioral Scale and Diversity: We now study how many scenarios we need for learning robust policies. Standard RL setups use only a single environment (*e.g.*, a fixed set of behavioral parameters for non-ego agents) and rely on the stochastic nature of the policy to collect diverse data. However, we find this is not enough. We train models on datasets with varying amount of scenario variations while keeping the total number of training simulation steps constant.

Table 3. We train a TRAVL agent on datasets with different axes of variation. Behavioral variation has larger effects than map for learning robust driving policies.

Method	Pass Rate ↑	Col. Rate ↓	Prog. ↑	MinTTC ↑	MinDist ↑
Map Variation	0.738	0.070	230	0.53	8.97
Beh. Variation	**0.872**	**0.022**	228	0.60	9.82
Both	0.865	0.026	**231**	**0.82**	**12.6**

As shown in Fig. 4, models trained with more diverse scenario variations exhibit better performance. In particular, we see that while metrics like pass rate and progress saturate quickly, safety-related metrics improve as we increase the number of variations. This suggests that adding in data diversity allows the model to be better prepared for safety-critical situations. We also study which axis of variation has the largest effect. Specifically, we disentangle map (*i.e.*, geolocation) variations (road curvature, number of lanes, topologies) and behavioral variation (actor triggers, target speeds) and construct datasets that only contain one source of diversity while keeping the total number of scenarios the same. Table 3 shows that TRAVL is able to perform well even without map variations as our trajectory-based formulation allows us to embed strong map priors into the model. However, the behavioral variation component is crucial in learning more robust policies.

6 Conclusion

We have studied how to design traffic scenarios and scale training environments in order to create an effective closed-loop benchmark for autonomous driving. We have proposed a new method to efficiently learn driving policies which can perform long-term reasoning and planning. Our method reasons in trajectory space and can efficiently learn in closed-loop by leveraging additional imagined experiences. We provide theoretical analysis and empirically demonstrate the advantages of our method over the baselines on our new benchmark.

References

1. Apollo simulation (2021)
2. Balmer, M., Rieser, M., Meister, K., Charypar, D., Lefebvre, N., Nagel, K.: MATSim-T: architecture and simulation times. In: Multi-Agent Systems for Traffic and Transportation Engineering (2009)
3. Bansal, M., Krizhevsky, A., Ogale, A.: Chauffeurnet: learning to drive by imitating the best and synthesizing the worst. arXiv (2018)
4. Ben-Akiva, M., et al.: Traffic simulation with MITSIMLab. In: Fundamentals of Traffic Simulation (2010)
5. Bergamini, L., et al.: SimNet: learning reactive self-driving simulations from real-world observations. In: ICRA (2021)

6. Bernhard, J., Esterle, K., Hart, P., Kessler, T.: Bark: open behavior benchmarking in multi-agent environments. In: IROS (2020)
7. Brockman, G., et al.: OpenAI gym. arXiv (2016)
8. Buckman, J., Hafner, D., Tucker, G., Brevdo, E., Lee, H.: Sample-efficient reinforcement learning with stochastic ensemble value expansion. In: NeurIPS (2018)
9. Caesar, H., et al.: NuPlan: a closed-loop ML-based planning benchmark for autonomous vehicles. arXiv (2021)
10. Camacho, E.F., Alba, C.B.: Model predictive control (2013)
11. Casas, J., Ferrer, J.L., Garcia, D., Perarnau, J., Torday, A.: Traffic simulation with aimsun. In: Fundamentals of Traffic Simulation (2010)
12. Chen, C., Seff, A., Kornhauser, A., Xiao, J.: Deepdriving: learning affordance for direct perception in autonomous driving. In: ICCV (2015)
13. Chen, D., Koltun, V., Krähenbühl, P.: Learning to drive from a world on rails. In: ICCV (2021)
14. Chen, D., Zhou, B., Koltun, V., Krähenbühl, P.: Learning by cheating. CoRL (2020)
15. Chua, K., Calandra, R., McAllister, R., Levine, S.: Deep reinforcement learning in a handful of trials using probabilistic dynamics models. In: NeurIPS (2018)
16. Codevilla, F., Müller, M., López, A., Koltun, V., Dosovitskiy, A.: End-to-end driving via conditional imitation learning. In: ICRA (2018)
17. Codevilla, F., Santana, E., López, A.M., Gaidon, A.: Exploring the limitations of behavior cloning for autonomous driving. In: ICCV (2019)
18. De Haan, P., Jayaraman, D., Levine, S.: Causal confusion in imitation learning. In: NeurIPS (2019)
19. Deng, J., Dong, W., Socher, R., Li, L.J., Li, K., Fei-Fei, L.: Imagenet: a large-scale hierarchical image database. In: CVPR (2009)
20. Ding, W., Chen, B., Li, B., Eun, K.J., Zhao, D.: Multimodal safety-critical scenarios generation for decision-making algorithms evaluation. IEEE Robot. Autom. Lett. **6**(2), 1551–1558 (2021)
21. Ding, W., Xu, C., Lin, H., Li, B., Zhao, D.: A survey on safety-critical scenario generation from methodological perspective. arXiv preprint arXiv:2202.02215 (2022)
22. Dosovitskiy, A., Ros, G., Codevilla, F., Lopez, A., Koltun, V.: Carla: an open urban driving simulator. CoRL (2017)
23. Feinberg, V., Wan, A., Stoica, I., Jordan, M.I., Gonzalez, J.E., Levine, S.: Model-based value estimation for efficient model-free reinforcement learning. arXiv (2018)
24. Feng, S., Yan, X., Sun, H., Feng, Y., Liu, H.X.: Intelligent driving intelligence test for autonomous vehicles with naturalistic and adversarial environment. Nat. Commun. **12**(1), 1–14 (2021)
25. Finn, C., Levine, S.: Deep visual foresight for planning robot motion. In: ICRA (2017)
26. Hafner, D., et al.: Learning latent dynamics for planning from pixels. In: ICML (2019)
27. Halkias, J., Colyar, J.: Model-predictive policy learning with uncertainty regularization for driving in dense traffic. FHWA-HRT-06-137, Washington, DC, USA (2006)
28. He, K., Zhang, X., Ren, S., Sun, J.: Deep residual learning for image recognition. In: CVPR (2016)
29. Henaff, M., Canziani, A., LeCun, Y.: Model-predictive policy learning with uncertainty regularization for driving in dense traffic. arXiv (2019)
30. Hessel, M., et al.: Rainbow: combining improvements in deep reinforcement learning. In: AAAI (2018)

31. Ho, J., Ermon, S.: Generative adversarial imitation learning. In: NeurIPS (2016)
32. Houston, J., et al.: One thousand and one hours: self-driving motion prediction dataset. arXiv (2020)
33. Kendall, A., et al.: Learning to drive in a day. In: ICRA (2019)
34. Kesting, A., Treiber, M., Helbing, D.: General lane-changing model MOBIL for car-following models. Transp. Res. Rec. **1999**(1), 86–94 (2007)
35. Koren, M., Kochenderfer, M.J.: Efficient autonomy validation in simulation with adaptive stress testing. In: 2019 IEEE Intelligent Transportation Systems Conference (ITSC), pp. 4178–4183. IEEE (2019)
36. Kuefler, A., Morton, J., Wheeler, T., Kochenderfer, M.: Imitating driver behavior with generative adversarial networks. In: 2017 IEEE Intelligent Vehicles Symposium (IV) (2017)
37. Levine, S., Kumar, A., Tucker, G., Fu, J.: Offline reinforcement learning: tutorial, review, and perspectives on open problems. arXiv (2020)
38. Li, Y., Song, J., Ermon, S.: Infogail: interpretable imitation learning from visual demonstrations. In: NeurIPS (2017)
39. Liang, X., Wang, T., Yang, L., Xing, E.: Cirl: controllable imitative reinforcement learning for vision-based self-driving. In: ECCV (2018)
40. Lillicrap, T.P., et al.: Continuous control with deep reinforcement learning. arXiv (2015)
41. Lin, T.-Y., et al.: Microsoft COCO: common objects in context. In: Fleet, D., Pajdla, T., Schiele, B., Tuytelaars, T. (eds.) ECCV 2014. LNCS, vol. 8693, pp. 740–755. Springer, Cham (2014). https://doi.org/10.1007/978-3-319-10602-1_48
42. Liu, R., et al.: An intriguing failing of convolutional neural networks and the coordconv solution. In: NeurIPS (2018)
43. Lopez, P.A., et al.: Microscopic traffic simulation using sumo. In: 2018 21st International Conference on Intelligent Transportation Systems (ITSC) (2018)
44. Manivasagam, S., et al.: LiDARsim: realistic lidar simulation by leveraging the real world. In: CVPR (2020)
45. MIRA, H., Hillman, R.: Test methods for interrogating autonomous vehicle behaviour-findings from the humandrive project (2019)
46. Mnih, V., et al.: Asynchronous methods for deep reinforcement learning. In: ICML (2016)
47. Muller, U., Ben, J., Cosatto, E., Flepp, B., Cun, Y.: Off-road obstacle avoidance through end-to-end learning. In: NeurIPS (2005)
48. Nagabandi, A., Kahn, G., Fearing, R.S., Levine, S.: Neural network dynamics for model-based deep reinforcement learning with model-free fine-tuning. In: ICRA (2018)
49. Najm, W.G., Smith, J.D., Yanagisawa, M., et al.: Pre-crash scenario typology for crash avoidance research. Technical report (2007)
50. Nie, C., Leung, H.: A survey of combinatorial testing. ACM Comput. Surv. (CSUR) (2011)
51. Oh, J., Singh, S., Lee, H.: Value prediction network. In: NeurIPS (2017)
52. Pan, X., Chen, X., Cai, Q., Canny, J., Yu, F.: Semantic predictive control for explainable and efficient policy learning. In: ICRA (2019)
53. Pan, Y., et al.: Agile autonomous driving using end-to-end deep imitation learning. arXiv (2017)
54. Polack, P., Altché, F., d'Andréa Novel, B., de La Fortelle, A.: The kinematic bicycle model: a consistent model for planning feasible trajectories for autonomous vehicles? In: 2017 IEEE Intelligent Vehicles Symposium (IV) (2017)

55. Pomerleau, D.A.: Alvinn: an autonomous land vehicle in a neural network. In: NeurIPS (1988)
56. Prakash, A., Chitta, K., Geiger, A.: Multi-modal fusion transformer for end-to-end autonomous driving. In: CVPR (2021)
57. Rhinehart, N., et al.: Contingencies from observations: tractable contingency planning with learned behavior models. arXiv (2021)
58. Riedmaier, S., Ponn, T., Ludwig, D., Schick, B., Diermeyer, F.: Survey on scenario-based safety assessment of automated vehicles. IEEE Access **8**, 87456–87477 (2020)
59. Ross, S., Gordon, G., Bagnell, D.: A reduction of imitation learning and structured prediction to no-regret online learning. In: AAAI (2011)
60. Sadat, A., Casas, S., Ren, M., Wu, X., Dhawan, P., Urtasun, R.: Perceive, predict, and plan: safe motion planning through interpretable semantic representations. In: Vedaldi, A., Bischof, H., Brox, T., Frahm, J.-M. (eds.) ECCV 2020. LNCS, vol. 12368, pp. 414–430. Springer, Cham (2020). https://doi.org/10.1007/978-3-030-58592-1_25
61. Sadat, A., Ren, M., Pokrovsky, A., Lin, Y.C., Yumer, E., Urtasun, R.: Jointly learnable behavior and trajectory planning for self-driving vehicles. In: IROS (2019)
62. Sauer, A., Savinov, N., Geiger, A.: Conditional affordance learning for driving in urban environments. CoRL (2018)
63. Schlechtriemen, J., Wabersich, K.P., Kuhnert, K.D.: Wiggling through complex traffic: planning trajectories constrained by predictions. In: 2016 IEEE Intelligent Vehicles Symposium (IV) (2016)
64. Schrittwieser, J., et al.: Mastering atari, go, chess and shogi by planning with a learned model. Nature **588**(7839), 604–609 (2020)
65. Schulman, J., Levine, S., Abbeel, P., Jordan, M., Moritz, P.: Trust region policy optimization. In: ICML (2015)
66. Schulman, J., Wolski, F., Dhariwal, P., Radford, A., Klimov, O.: Proximal policy optimization algorithms. arXiv (2017)
67. Shah, S., Dey, D., Lovett, C., Kapoor, A.: Airsim: high-fidelity visual and physical simulation for autonomous vehicles. In: Field and Service Robotics (2018)
68. Shi, T., Chen, D., Chen, K., Li, Z.: Offline reinforcement learning for autonomous driving with safety and exploration enhancement. arXiv (2021)
69. Silver, D., et al.: A general reinforcement learning algorithm that masters chess, shogi, and go through self-play. Science **362**(6419), 1140–1144 (2018)
70. Srinivas, A., Jabri, A., Abbeel, P., Levine, S., Finn, C.: Universal planning networks: learning generalizable representations for visuomotor control. In: ICML (2018)
71. Suo, S., Regalado, S., Casas, S., Urtasun, R.: Trafficsim: learning to simulate realistic multi-agent behaviors. In: CVPR (2021)
72. Sutton, R.S.: Integrated architectures for learning, planning, and reacting based on approximating dynamic programming. In: Machine Learning Proceedings 1990 (1990)
73. Toromanoff, M., Wirbel, E., Moutarde, F.: End-to-end model-free reinforcement learning for urban driving using implicit affordances. In: CVPR (2020)
74. Treiber, M., Hennecke, A., Helbing, D.: Congested traffic states in empirical observations and microscopic simulations. Phys. Rev. E **62**(2), 1805 (2000)
75. Wang, J., et al.: Advsim: generating safety-critical scenarios for self-driving vehicles. In: Proceedings of the IEEE/CVF Conference on Computer Vision and Pattern Recognition, pp. 9909–9918 (2021)
76. Wang, T., et al.: Benchmarking model-based reinforcement learning. arXiv (2019)

77. Werling, M., Ziegler, J., Kammel, S., Thrun, S.: Optimal trajectory generation for dynamic street scenarios in a frenet frame. In: ICRA (2010)
78. Wymann, B., Espié, E., Guionneau, C., Dimitrakakis, C., Coulom, R., Sumner, A.: TORCS, the open racing car simulator (2000). http://torcs.sourceforge.net/
79. Zeng, W., et al.: End-to-end interpretable neural motion planner. In: CVPR (2019)
80. Zeng, W., Wang, S., Liao, R., Chen, Y., Yang, B., Urtasun, R.: DSDNet: deep structured self-driving network. In: Vedaldi, A., Bischof, H., Brox, T., Frahm, J.-M. (eds.) ECCV 2020. LNCS, vol. 12366, pp. 156–172. Springer, Cham (2020). https://doi.org/10.1007/978-3-030-58589-1_10
81. Zhong, Z., Tang, Y., Zhou, Y., Neves, V.D.O., Liu, Y., Ray, B.: A survey on scenario-based testing for automated driving systems in high-fidelity simulation. arXiv preprint arXiv:2112.00964 (2021)
82. Zhou, M., et al.: Smarts: scalable multi-agent reinforcement learning training school for autonomous driving. arXiv (2020)

SLiDE: Self-supervised LiDAR De-snowing Through Reconstruction Difficulty

Gwangtak Bae[1], Byungjun Kim[1], Seongyong Ahn[2], Jihong Min[2], and Inwook Shim[3](✉)

[1] Seoul National University, Seoul, South Korea
[2] Agency for Defense Development, Daejeon, South Korea
[3] Inha University, Incheon, South Korea
iwshim@inha.ac.kr

Abstract. LiDAR is widely used to capture accurate 3D outdoor scene structures. However, LiDAR produces many undesirable noise points in snowy weather, which hamper analyzing meaningful 3D scene structures. Semantic segmentation with snow labels would be a straightforward solution for removing them, but it requires laborious point-wise annotation. To address this problem, we propose a novel self-supervised learning framework for snow points removal in LiDAR point clouds. Our method exploits the structural characteristic of the noise points: low spatial correlation with their neighbors. Our method consists of two deep neural networks: Point Reconstruction Network (PR-Net) reconstructs each point from its neighbors; Reconstruction Difficulty Network (RD-Net) predicts point-wise difficulty of the reconstruction by PR-Net, which we call *reconstruction difficulty*. With simple post-processing, our method effectively detects snow points without any label. Our method achieves the state-of-the-art performance among label-free approaches and is comparable to the fully-supervised method. Moreover, we demonstrate that our method can be exploited as a pretext task to improve label-efficiency of supervised training of de-snowing.

Keywords: LiDAR de-snowing · Vision for adverse weather · Self-supervised de-snowing

1 Introduction

Robust and accurate 3D scene measurement is an essential component of outdoor machine perceptions, e.g., autonomous vehicles. LiDAR is a commonly used 3D measurement sensor that gives reliable 3D point clouds in favorable weather conditions. However, in snowy weather conditions, LiDAR frequently generates

Supplementary Information The online version contains supplementary material available at https://doi.org/10.1007/978-3-031-19842-7_17.

S. Avidan et al. (Eds.): ECCV 2022, LNCS 13699, pp. 283–300, 2022.
https://doi.org/10.1007/978-3-031-19842-7_17

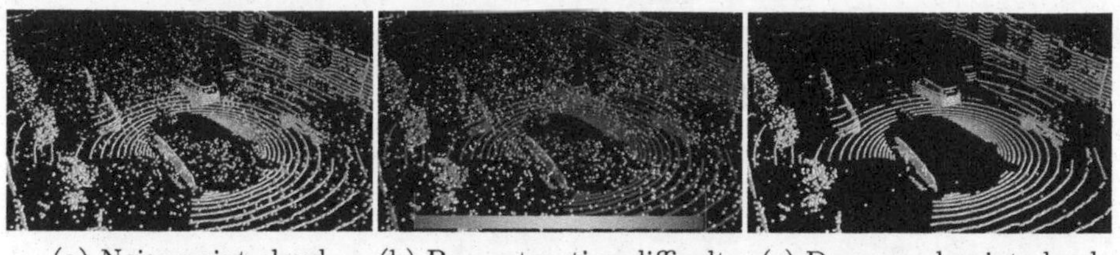

(a) Noisy point cloud (b) Reconstruction difficulty (c) De-snowed point cloud

Fig. 1. An example of the proposed LiDAR de-snowing process estimating how difficult to reconstruct each point from neighboring points. In (b), the closer to the red, the more difficult to reconstruct. (Color figure online)

a large number of particle noise points by detecting solid snowflakes [42]. These noise points could have a fatal impact on point cloud applications for outdoor systems [14,22,50].

Conventional filter-based approaches [5,40,47] have been presented for the LiDAR de-noising task to alleviate this problem. They attempt to remove the noise points by evaluating their spatial vicinity, but these approaches often suffer from misclassification since they only rely on simple spatial sparsity. Following the success of deep learning in various 3D point cloud applications (e.g., classification [43,44], detection [49,59], segmentation [27,38], etc.), a deep learning-based LiDAR de-noising approach, WeatherNet [16], is recently introduced. It takes advantage of deep learning-based semantic segmentation methods to detect point-wise LiDAR noise points.

While WeatherNet [16] outperforms the conventional approaches by a significant margin, it requires point-wise annotations. Even though there were attempts on efficient 3D point annotation [32,33,41], manual annotation of 3D point cloud is still laborious and time-consuming [10]. Moreover, labeling snow noise points requires even more efforts. Labeling the snow points is hard to take advantage of the assistance of camera view, which is the convention in point cloud labeling process [4,11,52], since camera images are also heavily degraded, e.g., snowflakes on the lens and in the air. Besides, in most cases, the degradation is not consistent with the noise points of LiDAR.

To address this issue, we propose a self-supervised learning framework for de-noising LiDAR point clouds in snowy weather, i.e., *LiDAR de-snowing*. We focus on the structural characteristic of the noise points in snowy weather that they have low spatial correlations with their neighbors. Therefore, they are difficult to be reconstructed from their neighboring points. Based on this insight, we propose a novel self-supervised learning approach for snow points removal in LiDAR point clouds. Our method consists of two deep neural networks: Point Reconstruction Network (PR-Net) reconstructs randomly selected target points from their neighbors; Reconstruction Difficulty Network (RD-Net) predicts point-wise error of PR-Net reconstruction, which we call *reconstruction difficulty*. An example of the estimated reconstruction difficulty and the de-snowing result is visualized in Fig. 1. A set of reconstruction target points for training is selected from the point cloud itself, which leads to a self-supervised training scheme. The two deep neural networks are jointly trained with a shared loss function, and then only RD-Net, followed by simple post-processing, is used to detect the noise points in the inference step. Our model trained without any labeled data outperforms previous label-free approaches with a large margin and achieves a comparable performance to the fully supervised model.

While our self-supervised method is basically label-free, it can be extended to a semi-supervised training scheme where training data consist of a small amount of labeled data and a large amount of unlabeled data. We demonstrate that an extension of our method as a pretext task enables supervised training of de-snowing in a label-efficient manner. Our semi-supervised extension yields better performance with fewer labeled data, which shows that our method as a pretext task is well-aligned with supervised training of de-snowing.

Our contributions can be summarized as follows:

- To the best of our knowledge, we are the first to introduce a self-supervised approach for LiDAR de-noising under snowy weather, i.e., *LiDAR de-snowing.* Our method identifies noise points that have low spatial correlations with their neighboring points.
- Our method outperforms previous label-free approaches by a large margin and achieves comparable results to the supervised method.
- We demonstrate that our self-supervised approach can be exploited as a pretext task for supervised de-snowing, which largely improves label-efficiency.

2 Related Work

2.1 LiDAR Point Cloud De-noising

Conventional filter-based methods classify noise points by their spatial sparsity [5,40,46,47,58]. Radius outlier removal (ROR) [47] finds the number of neighbors within a fixed radius and classifies points as noises if the number of neighbors is lower than a predefined threshold. ROR often fails at far-distant points because the density of the LiDAR point cloud decreases as the distance increases. Dynamic radius outlier removal (DROR) [5] employs varying search radii according to the distance to overcome the limitation of the previous works and successfully removes snow noise points. While those filter-based methods are easy to use and have a low computational burden, they still cannot handle point cloud's irregularity well, which often results in a significantly worse performance.

Recently, a supervised learning-based approach, WeatherNet [16], is introduced. It formulates the LiDAR de-noising problem as 2D semantic segmentation on the range image representation. Despite WeatherNet outperforms the filter-based methods, it requires expensive point-wise annotated training data.

2.2 Self-supervised Image De-noising

Self-supervised deep learning approaches have accomplished remarkable advances in the image de-noising task. Noise2Noise [31] is a pioneering work that restores a noisy image using another noisy image generated from the same clean image source. Since it is difficult to obtain a pair of noisy images in dynamic scenes, following works [1,25] improved the idea to work with a single noisy image by predicting a de-noised version of each pixel without depending on the pixel itself.

Despite the success of self-supervised de-noising in the image domain [1,23, 25,31,45] and dense point cloud domain [17,34,35], there are problems in utilizing those methods for the LiDAR de-noising task in snowy weather, i.e., the *de-snowing* task. First, the image de-noising cannot produce sufficiently reliable results when a pixel value is difficult to be precisely predicted by its neighboring pixel information [25]. Second, the objective of LiDAR de-noising [5,40,46,47] mainly focuses on removing noise points, not restoring the original points. For many applications that use LiDAR data (e.g., grasping an object, avoiding obstacles), rather than restoring the clean version of every elements, which is the main objective of the image de-noising tasks, it is essential to discard unreliable measurements while preserving clean measurements.

2.3 Semi-supervised Learning

Semi-supervised learning is a training scheme to learn from both labeled data and unlabeled data [54]. How to design an unsupervised loss function for leveraging sufficient unlabeled data is the main concern of semi-supervised learning. The categories of semi-supervised learning methods include unsupervised pretraining followed by fine-tuning [7,19,28,56], consistency regularization [26,48,53], pseudo labeling [18,29,55], and combination of these methods [2,51,60]. While our method is label-free, i.e., self-supervised, we demonstrate that our method can be extended to a semi-supervised training scheme. Combining our self-supervised loss with a supervised loss from a limited number of labeled data, the large performance gain shows that our method as a pretext task is well-aligned with supervised learning, which leads to an effective semi-supervised training scheme.

3 Proposed Method

Section 3.1 describes the representation of input LiDAR data. Section 3.2 explains our self-supervised learning framework for detecting noise points. Section 3.3 and Sect. 3.4 give detailed information on multi-hypothesis point reconstruction and post-processing process, respectively. Section 3.5 explains the extension of our self-supervised method into a semi-supervised training scheme.

3.1 Input Representation

The input to the proposed method is a 2D range image representation, which is the raw data structure of rotating LiDAR [9,37,38], e.g., Velodyne LiDAR. Such representation simplifies the 3D position reconstruction problem into 1D depth

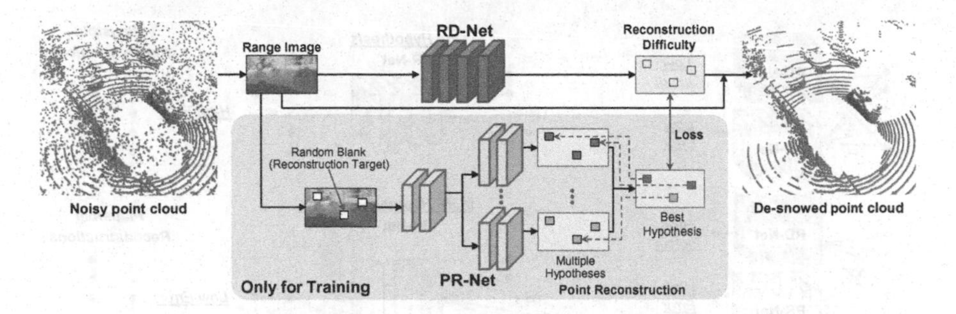

Fig. 2. The overall structure of our proposed LiDAR de-snowing method.

reconstruction along the LiDAR rays. The range image is generated as follows: Let P be a finite LiDAR point cloud, which contains K points: $P = \{\mathbf{p}_1, ..., \mathbf{p}_K\}$. Each point $\mathbf{p} = (p_x, p_y, p_z)$ is projected onto the image plane as $\mathbf{u} = (u, v)$ via a mapping function $\Pi : \mathbb{R}^3 \mapsto \mathbb{R}^2$. By following [16,37], column u is defined by $u = (\pi - arctan(p_y, p_x))/\delta_h$, where δ_h is the horizontal resolution of the LiDAR we used. Row v represents the laser id of $\mathbf{p}$, which corresponds to one of the sender/receiver modules in LiDAR. The projected point at (u, v) has a range value $r = (p_x^2 + p_y^2 + p_z^2)^{1/2}$. For each scan of LiDAR, we generate a corresponding range image $\mathrm{R} \in \mathbb{R}^{\mathrm{n} \times \mathrm{m}}$.

3.2 Self-supervised Learning Framework

We propose a self-supervised approach of LiDAR de-snowing that does not require point-wise annotations. To detect noise points that have low spatial correlations to their neighboring points, we designed a point reconstruction task and utilized errors from the task as guidance for training our de-snowing network. As shown in Fig. 2, reconstruction target points are randomly selected in the range image for every iteration of training. The Point Reconstruction Network (PR-Net) then predicts depth values of the target points by aggregating information of their neighboring points. At the same time, the Reconstruction Difficulty Network (RD-Net) estimates the extent of errors of the points reconstructed by PR-Net, which we call ***reconstruction difficulty***.

The two deep neural networks, PR-Net and RD-Net, are jointly trained with a shared loss function. We design a loss function with two objectives. First, as seen in the left figure in Fig. 3, the loss function should guide RD-Net to produce high output for noise points on which PR-Net has high error. Second, when training the point reconstruction task, the loss function should attenuate the loss contribution of noise points because they are extremely difficult to be reconstructed. It is to make PR-Net concentrate on reconstructing non-noise points. Our baseline loss function is given as follows:

$$\mathcal{L}_{self} = \sum I \odot \left[\sqrt{2} \frac{\left| \theta(\widetilde{R}) - R \right|}{\exp(\phi(R))} + \phi(R) \right], \tag{1}$$

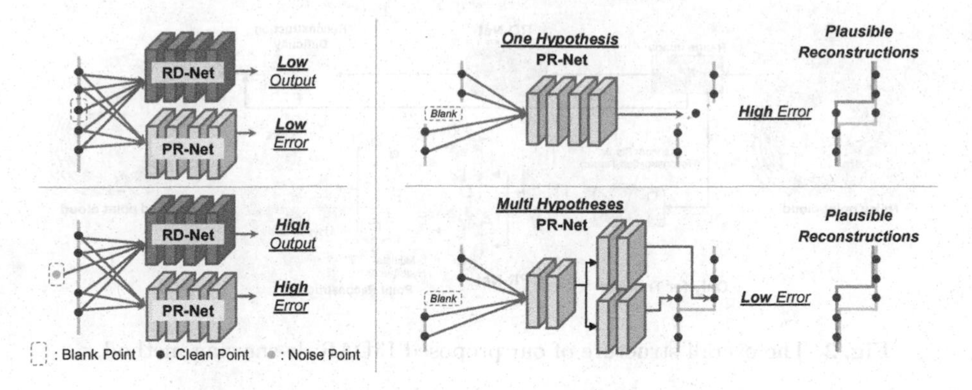

Fig. 3. The left image explains how PR-Net and RD-Net infer differently depending on whether a clean point or a noise point is blanked. The right image presents an example of ambiguity inherent in the point reconstruction task. PR-Net with multiple hypotheses can infer both of the multiple plausible reconstructions.

where R is a range image of LiDAR data and $\widetilde{R}$ is a randomly blanked range image. I is a binary mask that only selects loss from the blanked points. $\odot$ is an element-wise multiplication. $\theta(\cdot)$ and $\phi(\cdot)$ denote PR-Net and RD-Net, respectively. The blanked points are selected randomly for each iteration of training. The structure of Eq. (1) is inspired by the negative log-likelihood of a Laplacian distribution [21,24].

PR-Net's reconstruction error guides RD-Net to learn to estimate the reconstruction difficulty of each point. PR-Net takes the blanked range image $\widetilde{R}$ as an input and predicts the depth value of the blanked points, and then we calculate $L1$ loss. Equation (1) guides $\exp(\phi(R))$ to be high when the reconstruction error $L1$ is expected to be high, and to be low for the opposite. In other words, RD-Net is trained to predict the expected reconstruction quality of PR-Net for each point by taking the original range image R, which is not blanked, as an input. Consequently, we can identify the noise points by only utilizing the prediction of RD-Net. The regularization term $\phi(R)$ is added to prevent RD-Net from predicting an infinite value.

Equation (1) also allows training of PR-Net to be robust to noise points. Trying to minimize the loss on noise points could decrease the reconstruction ability for clean points. Therefore, in Eq. (1), gradients from noise points are attenuated by $\exp(\phi(R))$ as a weighting factor for the loss given to PR-Net. Points with high output from RD-Net have a smaller effect on the loss of PR-Net.

At the test time, only RD-Net is used to detect noise points. It takes a range image R as an input and predicts how difficult it will be to reconstruct each point. If the output of RD-Net $\phi(R)$ is higher than a certain threshold, the point is classified as a noise point.

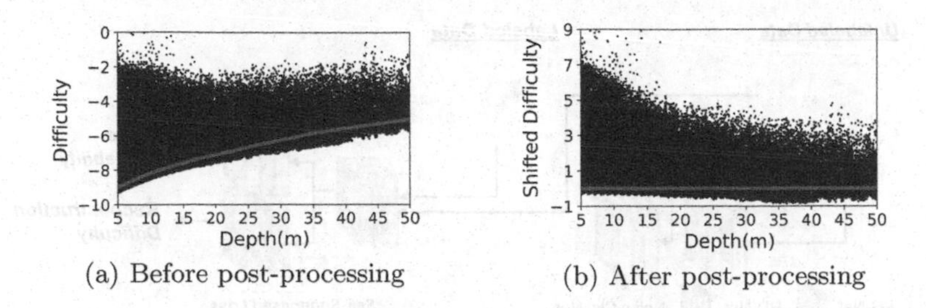

(a) Before post-processing (b) After post-processing

Fig. 4. RD-Net's output of each point in order of depth before (a) and after post-processing (b). Red lines indicate the 20th percentile of each meter of depth. (Color figure online)

3.3 Point Reconstruction with Multiple Hypotheses

PR-Net is trained to reconstruct randomly selected target points and their reconstruction errors are computed. However, the point reconstruction task has an inherent ambiguity, where some clean points may have multiple plausible answers for reconstruction. For example, in the right of Fig. 3, when a point on the object boundary is selected as a target, there can be two plausible answers: the object boundary or background. Since PR-Net with a single output cannot cover both answers precisely, the output is collapsed to the mean of them, which leads to an undesirable increase of the reconstruction error for clean points.

In order to distinguish the ambiguous clean points and noise, PR-Net is modified to have multiple outputs as shown in Fig. 2, which is inspired by multi-hypothesis learning. Following [30,57], only predictions with a minimum error are used for loss calculation. Such prediction for each point is given as follows:

$$C_i = \min_k \left| \theta^k(\tilde{R}) - R \right|_i , \tag{2}$$

where $\theta^k(\cdot)$ is the kth output, i.e., hypothesis, of PR-Net and C_i indicates the output of point i with a minimum error. Then, our reconstruction loss in Eq. (1) can be modified as follows:

$$\mathcal{L}_{self,mhl} = \sum I \odot \left[\sqrt{2} \frac{C}{\exp(\phi(R))} + \phi(R) \right] . \tag{3}$$

The modified loss guides PR-Net to have multiple plausible predictions rather than to be collapsed to the mean. If at least one of the multiple predictions is well-reconstructed, the reconstruction error will be low. However, since noise points are still difficult to reconstruct, albeit with a finite number of multiple predictions, reconstruction errors will be high.

3.4 Post-processing

As the sensing distance increases, the minimum output of RD-Net gradually increases, as shown in Fig. 4a. This can be seen as the effects of the decreased

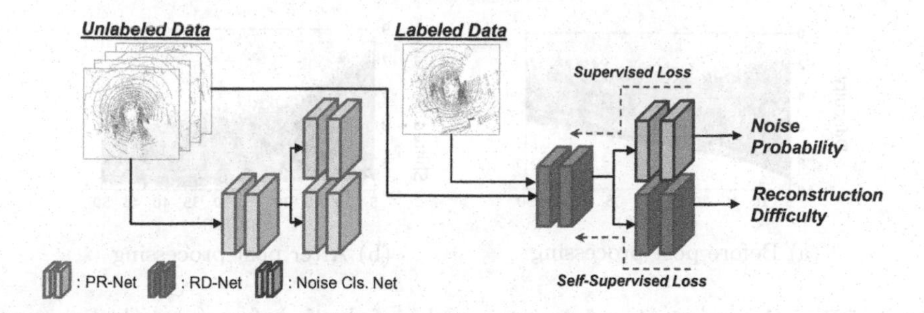

Fig. 5. Architecture design of our semi-supervised extension. The feature encoder of RD-Net and the noise classification network is shared.

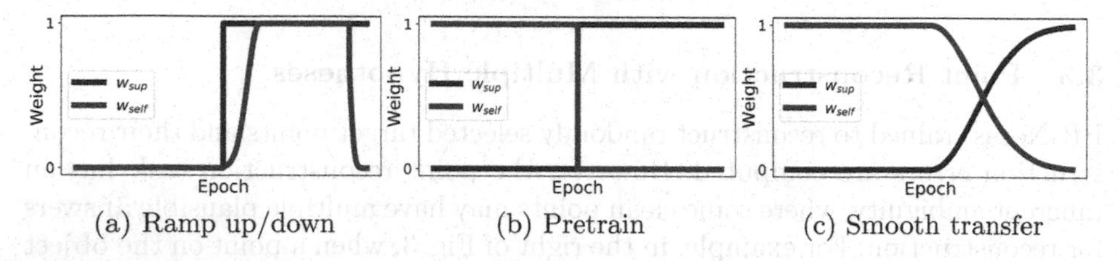

(a) Ramp up/down (b) Pretrain (c) Smooth transfer

Fig. 6. Weighting functions for combining the supervised and self-supervised loss.

density of the point cloud at a far distance, which makes the reconstruction more difficult. To compensate this effect, with partitioning the points by $1m$ depth interval, RD-Net's output is shifted downward in the amount of the shifting parameter for each depth interval. For example, in Fig. 4b, the shifting parameter is defined as the 20th percentile of RD-Net's output value. It compensates the depth-dependent bias on reconstruction difficulty. We then detect the noise points based on a certain threshold that is empirically determined. The ablation studies on the shifting parameter are described in Sect. 4.1.

3.5 Semi-supervised Learning

While our self-supervised method is able to remove noise points without any labeled data, we found out that it can also be extended to a semi-supervised training scheme. Since RD-Net is trained to regress the reconstruction difficulty, which is one of the characteristics to distinguish noise points from others, we can expect that the backbone network of RD-Net extracts features suitable for noise classification. As shown in Fig. 5, the feature encoder of RD-Net is shared for the feature encoder of a noise classification network. All of the networks are jointly trained with the weighted sum of two loss functions as follows,

$$\mathcal{L} = w_{self} * \mathcal{L}_{self,mhl} + w_{sup} * \mathcal{L}_{sup}, \tag{4}$$

where L_{sup} is the cross entropy loss by following WeatherNet [16]. w_{self} and w_{sup} are the weights for the self-supervised loss and the supervised loss, respectively.

Three different weighting functions are proposed and evaluated. Figure 6a indicates weighting functions that are widely used in consistency regularization methods for semi-supervised learning [20,26,36,53]. Weighting functions in Fig. 6b are inspired by self-supervised learning [6,8,12,13,39]. Our self-supervised method is regarded as a pretext task. The feature encoder of the noise classification network is initialized with the feature encoder of RD-Net. We further extend Fig. 6b into Fig. 6c. The pretrained features for estimating reconstruction difficulty are smoothly transferred into the noise classification network. The equations for the weighting functions are explained in the supplementary material.

4 Experimental Result

4.1 Point-wise Evaluation on Synthetic Data

Dataset. A number of scene data with point-wise annotations should be provided for a fair quantitative evaluation, regardless of whether training is done in a supervised or an unsupervised manner. However, to the best of our knowledge, there is no published dataset that satisfies it. We collect real-snow noise points using our stationary data-capturing system and synthesize the collected noise points into various clean weather road scenes.

Since a background point cannot be detected if a noise point is on the same LiDAR ray, we can directly pick out noises by comparing range images of noisy point cloud sets (*Noise-Set*) and clean point cloud sets (*Clean-Set*) as follows:

$$L^N_{(u,v)} \leftarrow \begin{cases} \mathbf{N}\ (\text{Noise}), & \text{if } R^F_{(u,v)} \geq R^N_{(u,v)} + \tau \\ \mathbf{C}\ (\text{Clean}), & \text{else} \end{cases} \tag{5}$$

where R^F is the reference range image generated from *Clean-Set*, R^N is a range image of *Noise-Set*, L^N indicates a label map of R^N and τ is a margin for sensing errors of LiDAR. In the case of scanning an empty space (e.g., sky), since no valid background information is projected to $R^F_{(u,v)}$, we always assign the 'Noise' label to $L^N_{(u,v)}$. The collected noise points are then injected into road scene point cloud sets (*Base-Set*) taken in clean weather as follows:

$$R^S_{(u,v)} = \begin{cases} R^N_{(u,v)}, & \text{if } R^N_{(u,v)} \leq R^{max}_{(u,v)} \text{ and } L^N_{(u,v)} = \mathbf{N}\ (\text{Noise}) \\ R^B_{(u,v)}, & \text{else} \end{cases} \tag{6}$$

where R^B is a range image in *Base-Set*, and R^S is a synthesized range image. R^{max} is the maximum detectable range which is introduced for a realistic synthesis [3,16] by considering scene structures of R^B as follows:

$$R^{max}_{(u,v)} = \min\left(\frac{-ln(\frac{n}{I^B_{(u,v)}+g})}{2 * \beta}, R^B_{(u,v)}\right), \tag{7}$$

Table 1. Quantitative results on the synthesized snow noise data. *Labeled data* indicates the number of labeled training data used.

Method	Labeled data	IoU	Precision	Recall
ROR [47]	0	17.82	17.86	98.65
DROR [5]	0	33.68	33.87	98.37
Ours w/o MHL	0	65.75	70.38	90.89
Ours	0	79.62	85.69	91.83
WeatherNet [16]	239	41.40	76.78	47.32
Ours (Semi.sup.)	239	82.44	96.39	85.07
WeatherNet [16]	23,908	84.04	97.48	85.90
Ours (Semi.sup.)	23,908	84.24	97.24	86.30

given the received laser intensity $I^B_{(u,v)}$, the adaptive laser gain g, the atmospheric extinction coefficient β, and the detectable noise floor n. In this paper, Nuscenes dataset [4] is used as *Base-Set* to synthesize snowy scenes. Please see the supplementary material for more details of the dataset generation process.

Implementation Details. Our self-supervised model is trained with the synthesized snow noise dataset. We split a total of 34, 139 scans into training, validation, and test sets at a ratio of 70 : 15 : 15. PR-Net and RD-Net consist of residual blocks [15] for the self-supervised model. For the semi-supervised model, PR-Net and RD-Net use the same backbone with WeatherNet to ensure a fair comparison. Separated layers in Fig. 5 consist of a *LiLaBlock* [41] and a convolution layer by following WeatherNet. Each training step encounters an independently selected random set of target points in order to learn various cases of the reconstruction. Horizontal flipping is randomly performed to augment training data. At the test time, points with an RD-Net output higher than the threshold are classified as noise points. Throughout experiments in this paper, the threshold is determined as 2.9, which achieves the highest performance for the validation dataset. Quantitative performance is evaluated using the Intersection-over-Union (IoU) metric, following WeatherNet [16].

Quantitative Comparisons. In Table 1, our proposed methods are compared with previous LiDAR de-noising approaches: ROR [47], DROR [5], and WeatherNet [16]. Table 1 shows that our proposed method yields a significantly higher IoU than the state-of-the-art label-free method, DROR. Notably, our approach achieves a comparable IoU to the supervised method without using any labeled data. Applying multi-hypothesis learning in Eq. (3) improved the baseline method in all metrics. For the case where only 1% of labeled data provided, the semi-supervised extension yields significantly higher performance than the supervised method, WeatherNet. Even when 100% labeled data provided, our method performs better than the supervised method without using any addi-

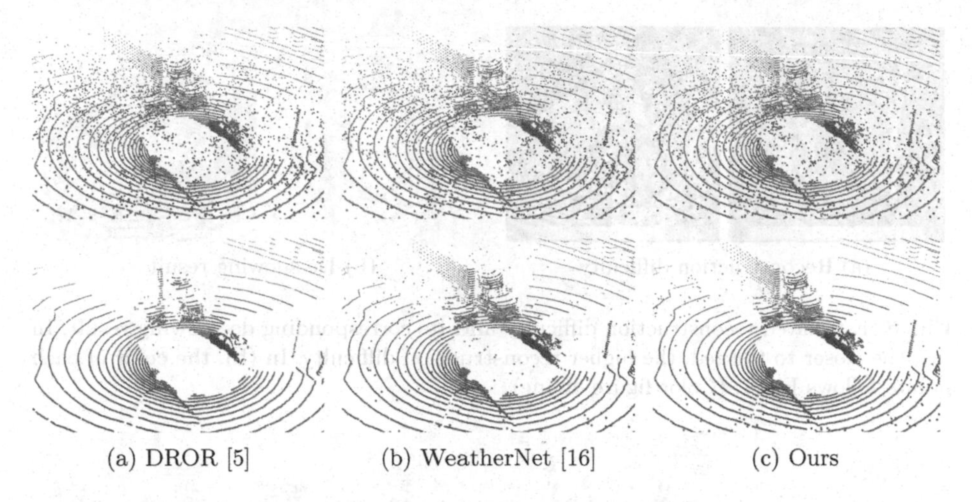

(a) DROR [5] (b) WeatherNet [16] (c) Ours

Fig. 7. Qualitative comparisons on the synthesized snow noise data. The first row shows all points with their prediction results (red: true positive, green: false positive, gray: true negative, yellow: false negative). Classification as noise corresponds to *positive*. The second row shows de-snowed point clouds. DROR misclassifies many sparsely distributed points at the side facade of the truck. (Color figure online)

tional unlabeled data, which shows better exploitation of the same given data. *Ours (Semi-sup)* refers to the semi-supervised extension with *Smooth transfer* in Fig. 6, which is described in Sect. 2.3.

Table 2. Analyses on the performance changes according to the noise level.

Method	Metric	Noise level			
		Light	Medium	Heavy	Extreme
DROR [5]	IoU	12.76	22.57	32.99	52.59
WeatherNet [16]	IoU	82.29	83.57	83.57	84.00
Ours(Semi-sup)	IoU	82.35	83.87	83.72	84.83
Ours	IoU	60.81	71.48	79.37	85.69
Ours	Precision	64.35	77.81	85.48	85.69
Ours	Recall	91.70	89.78	91.74	92.42

Table 2 shows an analysis on the performance changes of our method according to the noise level. The noise level is decided by following Canadian Adverse Driving Condition Dataset [42]. While WeatherNet and our semi-supervised model generate relatively consistent performances when the noise level changes, all of the label-free methods including ours have lower IoU as the noise level decreases. Since unsupervised methods do not use direct supervision from point-wise labeled data, noise-like points among clean points can be misclassified

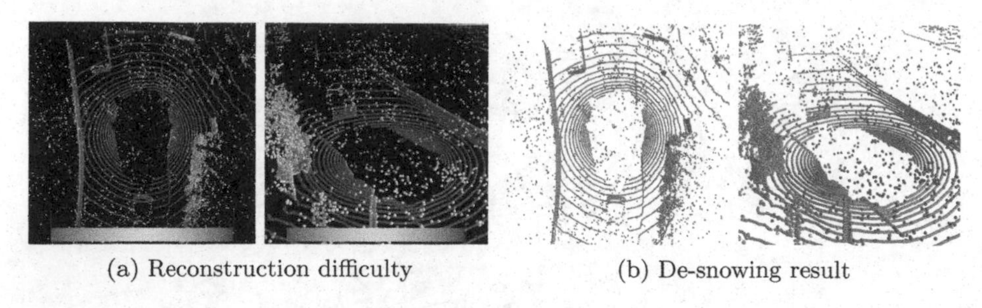

Fig. 8. Estimated reconstruction difficulty and its corresponding de-snowing result. In (a), the closer to the red, the higher reconstruction difficulty. In (b), the color of each points follows Fig. 7. (Color figure online)

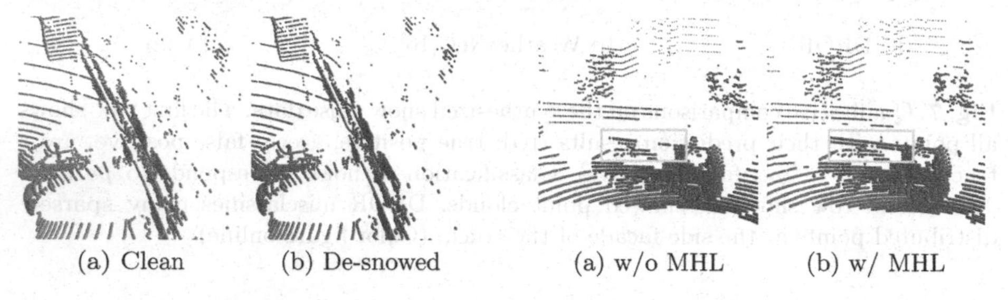

Fig. 9. A de-snowing result of only-clean points scene. Some of floating clean points are similar to noises and eliminated by our method.

Fig. 10. Comparison between the single hypothesis model and the multi hypotheses model.

as false positives. Figure 9 depicts the case where noise-like clean points exist. Although only clean points are displayed, clean points floating in the air look very similar to noise points, and those are removed in Fig. 9b.

In Table 3, we evaluate three weighting functions proposed for our semi-supervised method in Fig. 6. All of the weighting functions have better performances than the supervised method when limited data are given. Compared to the supervised method, *Ramp up/down*, which is widely used in semi-supervised methods, shows higher IoU when 1% and 10% labels are given but lower when sufficient labels are given, 100%. *Pretrain* yields better IoU than the supervised method even when 100% of labels are given. *Smooth transfer* shows the highest performance when 1% data are labeled and higher than the supervised method even when 100% of data are available for both of methods.

Ablation Studies. In Table 4, we investigate performance changes of our self-supervised method according to variations in configurations. First, multi hypotheses learning yields significantly better performance than our baseline method that has a single hypothesis. Among the different number of hypotheses, the model predicting three hypotheses achieves the highest IoU. Second,

Table 3. De-snowing performances of the supervised method and our semi-supervised extension when limited labeled data are provided.

Method	IoU (1% labels)	IoU (10% labels)	IoU (100% labels)
WeatherNet [16]	41.40	78.75	84.04
Ramp up/down	79.39	81.73	82.57
Pretrain	62.46	82.87	84.86
Smooth transfer	82.44	83.70	84.24

Table 4. Ablation experiments of the proposed self-supervised method.

# Hypotheses	1	2	3	4
IoU	65.75	76.17	79.62	76.52
Blank ratio	10%	30%	50%	70%
IoU	76.64	78.93	79.62	76.57
Shifting param.	min	10th	20th	30th
IoU	72.63	79.36	79.62	79.22

we analyze the effects of the blank ratio for training. Our method yields the highest performance when 50% of points are blanked. Third, we look into the impact of different shifting parameters in the post-processing step. The experiment with the minimum RD-Net output value for each depth interval as the shifting parameter shows a slightly lower result while other settings achieve similar performance.

Qualitative Comparisons. Figure 7 demonstrates a qualitative analysis of our method with DROR and WeatherNet. First, DROR misclassifies clean points if they are sparsely distributed. Assuming the same distance, as an angle between a LiDAR ray and its hitting surface gets farther from perpendicular, a point density of the surface gets lower. As depicted in Fig. 7a, it leads to the failure case of DROR, which solely depends on sparsity for detecting noise. For example, in spite of strong semantic consistency of the side facade of the truck, DROR incorrectly removes the points on it based on sparsity. In contrast, our method preserves clean points if they can be reconstructed from neighboring points. Second, WeatherNet has the smallest number of false positives, which are marked as green points in Fig. 7. It still has remaining noise points around the LiDAR sensor. More results are in the supplementary material.

Figure 8 visualizes the reconstruction difficulty estimated by RD-Net and its evaluation. Clean points have low reconstruction difficulty as they have high spatial correlations with neighbors. On the contrary, RD-Net assigns high difficulty to noise points. In Fig. 8a, points on trees have relatively high RD-Net output than the points on the road and cars. It shows that RD-Net successfully learns to reflect the reconstruction difficulty of each point. Although points on trees are inferred as more difficult points than other clean points, they still have lower difficulty than noise points and correctly classified as seen in Fig. 8b.

Figure 10 demonstrates the effects of multi-hypothesis learning. When PR-Net infers a single hypothesis, many points on the upper side of a car are misclassified and removed, as shown in the yellow box in Fig. 10a. This is the case explained in Fig. 3. On the contrary, as seen in Fig. 10b, many of those points are preserved when multi-hypothesis outputs are inferred by PR-Net.

4.2 Qualitative Evaluation on Real-world Data

We also evaluate our self-supervised method in real snowy weather scenarios captured by our mobility platform, equipped with a Velodyne 'VLS-128'. A total of 9,000 scans were captured in snowy weather. We qualitatively evaluate our method with DROR, which also does not require point-wise labels. Other experimental settings follow Sect. 4.1, except for the height of range images which is set to 128.

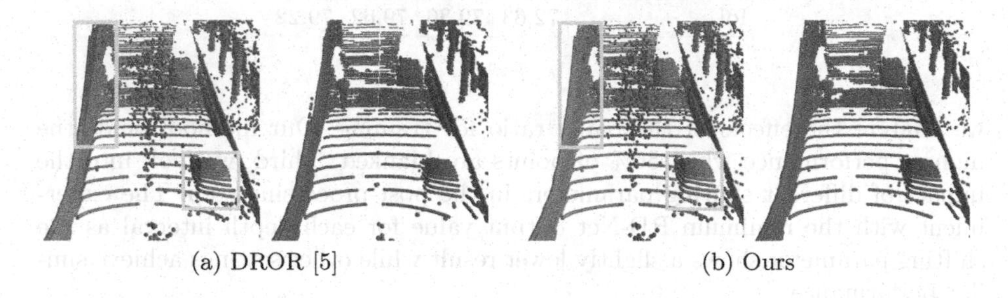

(a) DROR [5] (b) Ours

Fig. 11. Qualitative comparisons on the real-world snowy weather data (red: positive, gray and blue: negative). (Color figure online)

Figure 11 shows a de-snowing result of DROR and ours. As 'VLS-128' LiDAR generates dense point clouds due to its high vertical resolution, the sparsity-based de-snowing method, DROR, generates fewer false positives than in Sect. 4.1. However, a number of clean points are still misclassified. For example, as shown in the yellow boxes in Fig. 11a, clean points on the car and the wall are filtered out when they have low spatial density. On the other hand, in Fig. 11b, since our method estimates point-wise reconstruction difficulty, clean points on the surface of objects are well preserved. More results are presented in the supplementary material.

5 Conclusion

In this work, we proposed a novel self-supervised method for de-snowing LiDAR point clouds in snowy weather conditions. By utilizing the characteristic of noise points that they have low spatial correlations with their neighboring points, our method is designed to detect noise points that are difficult to reconstruct from their neighboring points. Our proposed self-supervised approach outperforms the state-of-the-art label-free methods and yields comparable results to the supervised approach without using any annotation. Furthermore, we present that our self-supervised method can be exploited as a pretext task for the supervised training, which significantly improves the label-efficiency.

Acknowledgement. This work was supported by the Agency for Defense Development (ADD) and by the National Research Foundation of Korea (NRF) grant funded by the Korea government (MSIT) (No. NRF-2022R1F1A1073505).

References

1. Batson, J., Royer, L.: Noise2self: blind denoising by self-supervision. In: International Conference on Machine Learning, pp. 524–533 (2019)
2. Berthelot, D., Carlini, N., Goodfellow, I., Papernot, N., Oliver, A., Raffel, C.A.: MixMatch: a holistic approach to semi-supervised learning. In: Advances in Neural Information Processing Systems, vol. 32 (2019)
3. Bijelic, M., et al.: Seeing through fog without seeing fog: deep multimodal sensor fusion in unseen adverse weather. In: IEEE Conference on Computer Vision and Pattern Recognition, pp. 11682–11692 (2020)
4. Caesar, H., et al.: nuScenes: a multimodal dataset for autonomous driving. In: IEEE Conference on Computer Vision and Pattern Recognition, pp. 11621–11631 (2020)
5. Charron, N., Phillips, S., Waslander, S.L.: De-noising of LiDAR point clouds corrupted by snowfall. In: Conference on Computer and Robot Vision, pp. 254–261 (2018)
6. Chen, X., He, K.: Exploring simple siamese representation learning. In: IEEE Conference on Computer Vision and Pattern Recognition, pp. 15750–15758 (2021)
7. Devlin, J., Chang, M.W., Lee, K., Toutanova, K.: BERT: pre-training of deep bidirectional transformers for language understanding. arXiv preprint arXiv:1810.04805 (2018)
8. Doersch, C., Gupta, A., Efros, A.A.: Unsupervised visual representation learning by context prediction. In: IEEE International Conference on Computer Vision, pp. 1422–1430 (2015)
9. Fan, L., Xiong, X., Wang, F., Wang, N., Zhang, Z.: RangeDet: in defense of range view for LiDAR-based 3D object detection. In: IEEE International Conference on Computer Vision, pp. 2918–2927 (2021)
10. Gao, B., Pan, Y., Li, C., Geng, S., Zhao, H.: Are we hungry for 3D LiDAR data for semantic segmentation? A survey of datasets and methods. IEEE Trans. Intell. Transp. Syst. (2021)
11. Geiger, A., Lenz, P., Urtasun, R.: Are we ready for autonomous driving? The KITTI vision benchmark suite. In: IEEE Conference on Computer Vision and Pattern Recognition, pp. 3354–3361 (2012)
12. Gidaris, S., Singh, P., Komodakis, N.: Unsupervised representation learning by predicting image rotations. arXiv preprint arXiv:1803.07728 (2018)
13. Grill, J.B., et al.: Bootstrap your own latent-a new approach to self-supervised learning. In: Advances in Neural Information Processing Systems, vol. 33, pp. 21271–21284 (2020)

14. Gruber, T., Bijelic, M., Heide, F., Ritter, W., Dietmayer, K.: Pixel-accurate depth evaluation in realistic driving scenarios. In: International Conference on 3D Vision, pp. 95–105. IEEE (2019)
15. He, K., Zhang, X., Ren, S., Sun, J.: Deep residual learning for image recognition. In: IEEE Conference on Computer Vision and Pattern Recognition, pp. 770–778 (2016)
16. Heinzler, R., Piewak, F., Schindler, P., Stork, W.: CNN-based LiDAR point cloud de-noising in adverse weather. IEEE Robot. Autom. Lett. **5**(2), 2514–2521 (2020)
17. Hermosilla, P., Ritschel, T., Ropinski, T.: Total denoising: Unsupervised learning of 3D point cloud cleaning. In: IEEE International Conference on Computer Vision, pp. 52–60 (2019)
18. Iscen, A., Tolias, G., Avrithis, Y., Chum, O.: Label propagation for deep semi-supervised learning. In: IEEE Conference on Computer Vision and Pattern Recognition, pp. 5070–5079 (2019)
19. Jarrett, K., Kavukcuoglu, K., Ranzato, M., LeCun, Y.: What is the best multi-stage architecture for object recognition? In: IEEE International Conference on Computer Vision, pp. 2146–2153. IEEE (2009)
20. Ke, Z., Wang, D., Yan, Q., Ren, J., Lau, R.W.: Dual student: breaking the limits of the teacher in semi-supervised learning. In: IEEE International Conference on Computer Vision, pp. 6728–6736 (2019)
21. Kendall, A., Gal, Y.: What uncertainties do we need in Bayesian deep learning for computer vision? In: Advances in Neural Information Processing Systems, pp. 5580–5590 (2017)
22. Kilic, V., et al.: LiDAR light scattering augmentation (LISA): physics-based simulation of adverse weather conditions for 3D object detection. arXiv preprint arXiv:2107.07004 (2021)
23. Kim, K., Ye, J.C.: Noise2score: tweedie's approach to self-supervised image denoising without clean images. In: Advances in Neural Information Processing Systems, vol. 34 (2021)
24. Klodt, M., Vedaldi, A.: Supervising the new with the old: learning SFM from SFM. In: Ferrari, V., Hebert, M., Sminchisescu, C., Weiss, Y. (eds.) ECCV 2018. LNCS, vol. 11214, pp. 713–728. Springer, Cham (2018). https://doi.org/10.1007/978-3-030-01249-6_43
25. Krull, A., Buchholz, T.O., Jug, F.: Noise2Void-learning denoising from single noisy images. In: IEEE Conference on Computer Vision and Pattern Recognition, pp. 2129–2137 (2019)
26. Laine, S., Aila, T.: Temporal ensembling for semi-supervised learning. arXiv preprint arXiv:1610.02242 (2016)
27. Landrieu, L., Simonovsky, M.: Large-scale point cloud semantic segmentation with superpoint graphs. In: IEEE Conference on Computer Vision and Pattern Recognition, pp. 4558–4567 (2018)
28. Le, Q.V.: Building high-level features using large scale unsupervised learning. In: International Conference on Acoustics, Speech and Signal Processing, pp. 8595–8598. IEEE (2013)
29. Lee, D.H., et al.: Pseudo-label: the simple and efficient semi-supervised learning method for deep neural networks. In: Workshop on Challenges in Representation Learning, ICML, vol. 3, p. 896 (2013)
30. Lee, S., Prakash, S.P.S., Cogswell, M., Ranjan, V., Crandall, D., Batra, D.: Stochastic multiple choice learning for training diverse deep ensembles. In: Advances in Neural Information Processing Systems, pp. 2119–2127 (2016)

31. Lehtinen, J., et al.: Noise2noise: learning image restoration without clean data. In: International Conference on Machine Learning, pp. 2965–2974 (2018)
32. Luo, H., et al.: Patch-based semantic labeling of road scene using colorized mobile LiDAR point clouds. IEEE Trans. Intell. Transp. Syst. **17**(5), 1286–1297 (2015)
33. Luo, H., et al.: Semantic labeling of mobile LiDAR point clouds via active learning and higher order MRF. IEEE Trans. Geosci. Remote Sens. **56**(7), 3631–3644 (2018)
34. Luo, S., Hu, W.: Differentiable manifold reconstruction for point cloud denoising. In: ACM International Conference on Multimedia, pp. 1330–1338 (2020)
35. Luo, S., Hu, W.: Score-based point cloud denoising. In: IEEE International Conference on Computer Vision, pp. 4583–4592 (2021)
36. Luo, Y., Zhu, J., Li, M., Ren, Y., Zhang, B.: Smooth neighbors on teacher graphs for semi-supervised learning. In: IEEE Conference on Computer Vision and Pattern Recognition, pp. 8896–8905 (2018)
37. Meyer, G.P., Laddha, A., Kee, E., Vallespi-Gonzalez, C., Wellington, C.K.: LaserNet: an efficient probabilistic 3D object detector for autonomous driving. In: IEEE Conference on Computer Vision and Pattern Recognition, pp. 12677–12686 (2019)
38. Milioto, A., Vizzo, I., Behley, J., Stachniss, C.: RangeNet++: fast and accurate LiDAR semantic segmentation. In: IEEE/RSJ International Conference on Intelligent Robots and Systems, pp. 4213–4220. IEEE (2019)
39. Noroozi, M., Favaro, P.: Unsupervised learning of visual representations by solving Jigsaw puzzles. In: Leibe, B., Matas, J., Sebe, N., Welling, M. (eds.) ECCV 2016. LNCS, vol. 9910, pp. 69–84. Springer, Cham (2016). https://doi.org/10.1007/978-3-319-46466-4_5
40. Park, J.I., Park, J., Kim, K.S.: Fast and accurate desnowing algorithm for LiDAR point clouds. IEEE Access **8**, 160202–160212 (2020)
41. Piewak, F., et al.: Boosting LiDAR-based semantic labeling by cross-modal training data generation. In: Leal-Taixé, L., Roth, S. (eds.) ECCV 2018. LNCS, vol. 11134, pp. 497–513. Springer, Cham (2019). https://doi.org/10.1007/978-3-030-11024-6_39
42. Pitropov, M., et al.: Canadian adverse driving conditions dataset. Int. J. Robot. Res. (2020)
43. Qi, C.R., Su, H., Mo, K., Guibas, L.J.: PointNet: deep learning on point sets for 3D classification and segmentation. In: IEEE Conference on Computer Vision and Pattern Recognition, pp. 652–660 (2017)
44. Qi, C.R., Yi, L., Su, H., Guibas, L.J.: PointNet++: deep hierarchical feature learning on point sets in a metric space. In: Advances in Neural Information Processing Systems, vol. 30 (2017)
45. Quan, Y., Chen, M., Pang, T., Ji, H.: Self2self with dropout: learning self-supervised denoising from single image. In: IEEE Conference on Computer Vision and Pattern Recognition, pp. 1890–1898 (2020)
46. Roriz, R., Campos, A., Pinto, S., Gomes, T.: DIOR: a hardware-assisted weather denoising solution for LiDAR point clouds. IEEE Sens. J. (2021)
47. Rusu, R.B., Cousins, S.: 3D is here: point cloud library (PCL). In: IEEE International Conference on Robotics and Automation, pp. 1–4 (2011)
48. Sajjadi, M., Javanmardi, M., Tasdizen, T.: Regularization with stochastic transformations and perturbations for deep semi-supervised learning. In: Advances in Neural Information Processing Systems, vol. 29 (2016)
49. Shi, S., Wang, X., Li, H.: PointRCNN: 3D object proposal generation and detection from point cloud. In: IEEE Conference on Computer Vision and Pattern Recognition, pp. 770–779 (2019)

50. Shim, I., et al.: Vision system and depth processing for DRC-HUBO+. In: IEEE International Conference on Robotics and Automation, pp. 2456–2463 (2016)
51. Sohn, K., et al.: FixMatch: simplifying semi-supervised learning with consistency and confidence. In: Advances in Neural Information Processing Systems, vol. 33, pp. 596–608 (2020)
52. Sun, P., et al.: Scalability in perception for autonomous driving: Waymo open dataset. In: IEEE Conference on Computer Vision and Pattern Recognition, pp. 2446–2454 (2020)
53. Tarvainen, A., Valpola, H.: Mean teachers are better role models: weight-averaged consistency targets improve semi-supervised deep learning results. In: Advances in Neural Information Processing Systems, vol. 30 (2017)
54. Van Engelen, J.E., Hoos, H.H.: A survey on semi-supervised learning. Mach. Learn. **109**(2), 373–440 (2020)
55. Xie, Q., Luong, M.T., Hovy, E., Le, Q.V.: Self-training with noisy student improves ImageNet classification. In: IEEE Conference on Computer Vision and Pattern Recognition, pp. 10687–10698 (2020)
56. Xie, S., Gu, J., Guo, D., Qi, C.R., Guibas, L., Litany, O.: PointContrast: unsupervised pre-training for 3D point cloud understanding. In: Vedaldi, A., Bischof, H., Brox, T., Frahm, J.-M. (eds.) ECCV 2020. LNCS, vol. 12348, pp. 574–591. Springer, Cham (2020). https://doi.org/10.1007/978-3-030-58580-8_34
57. Yang, G., Hu, P., Ramanan, D.: Inferring distributions over depth from a single image. In: IEEE/RSJ International Conference on Intelligent Robots and Systems, pp. 6090–6096 (2019)
58. Zhou, H., Chen, K., Zhang, W., Fang, H., Zhou, W., Yu, N.: DUP-Net: denoiser and upsampler network for 3D adversarial point clouds defense. In: IEEE Conference on Computer Vision and Pattern Recognition, pp. 1961–1970 (2019)
59. Zhou, Y., Tuzel, O.: VoxelNet: end-to-end learning for point cloud based 3D object detection. In: IEEE Conference on Computer Vision and Pattern Recognition, pp. 4490–4499 (2018)
60. Zoph, B., et al.: Rethinking pre-training and self-training. In: Advances in Neural Information Processing Systems, vol. 33, pp. 3833–3845 (2020)

Generative Meta-Adversarial Network for Unseen Object Navigation

Sixian Zhang[1,2(✉)], Weijie Li[1,2], Xinhang Song[1,2], Yubing Bai[1,2], and Shuqiang Jiang[1,2]

[1] Key Lab of Intelligent Information Processing Laboratory of the Chinese Academy of Sciences (CAS), Institute of Computing Technology (ICT), Beijing, China
{sixian.zhang,weijie.li,xinhang.song,yubing.bai}@vipl.ict.ac.cn, sqjiang@ict.ac.cn

[2] University of Chinese Academy of Sciences (UCAS), Beijing, China

Abstract. Object navigation is a task to let the agent navigate to a target object. Prevailing works attempt to expand navigation ability in new environments and achieve reasonable performance on the seen object categories that have been observed in training environments. However, this setting is somewhat limited in real world scenario, where navigating to unseen object categories is generally unavoidable. In this paper, we focus on the problem of navigating to unseen objects in new environments only based on limited training knowledge. Same as the common ObjectNav tasks, our agent still gets the egocentric observation and target object category as the input and does not require any extra inputs. Our solution is to let the agent "imagine" the unseen object by synthesizing features of the target object. We propose a generative meta-adversarial network (GMAN), which is mainly composed of a feature generator and an environmental meta discriminator, aiming to generate features for unseen objects and new environments in two steps. The former generates the initial features of the unseen objects based on the semantic embedding of the object category. The latter enables the generator to further learn the background characteristics of the new environment, progressively adapting the generated features to approximate the real features of the target object. The adapted features serve as a more specific representation of the target to guide the agent. Moreover, to fast update the generator with a few observations, the entire adversarial framework is learned in the gradient-based meta-learning manner. The experimental results on AI2THOR and RoboTHOR simulators demonstrate the effectiveness of the proposed method in navigating to unseen object categories. The code is available at https://github.com/sx-zhang/GMAN.git.

Keywords: Object navigation · Unseen object · Adversarial learning

---

Supplementary Information The online version contains supplementary material available at https://doi.org/10.1007/978-3-031-19842-7_18.

S. Avidan et al. (Eds.): ECCV 2022, LNCS 13699, pp. 301–320, 2022.
https://doi.org/10.1007/978-3-031-19842-7_18

1 Introduction

Visual object navigation is a task that requires an agent to navigate to the target object depending on the visual environment information. Recent works are typically trained by reinforcement learning (RL) to predict actions in real-time, with the input of observed visual information and target object embedding. Existing object navigation methods achieve reasonable performance on seen objects (observed in training environments) [5,56,60,62]. However, these seen objects make up only a subset of all real world objects. As illustrated in Fig. 1, imagine a situation where the agent is trained to find several indoor furnitures (e.g. chair, sofa, shelf), when adapting to more practical applications, it's unavoidable that the user may require it to find the unseen objects (e.g. spray bottle). Since the actions of the agent are mainly driven by the correlation between visual representation and target semantic embedding, while the unseen object categories have not been trained to correlate with such visual information. Thus, the navigation ability of the agent to unseen objects is significantly limited.

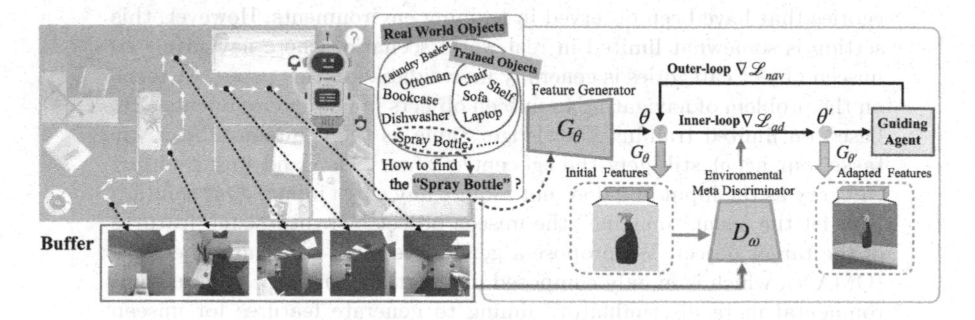

Fig. 1. Overview. We focus on navigating to the unseen objects in unseen environments. Given an unseen object category, our method is to firstly generate the initial features, and then adapt the generator to be compatible with the environment. The adapted features serve as a more specific goal to guide the agent.

Now that the agent is unfamiliar with the unseen target, a great challenge here is how to associate the unseen target with the limited "knowledge" learned from training. Some previous works rely on the helpful visual semantic information (e.g. object detection or instance segmentation) to establish semantic SLAM [5,7] or relation knowledge graph [10,62]. These methods may be unsuitable for unseen object navigation since unseen object detection or segmentation is not supported in those works. For other works focusing on visual embedding [35] or policy learning [33,56], although they can process the unseen objects (similar to seen objects) by using the same pipelines as seen objects, the performance is still limited, for that the visual representation and semantic embedding of unseen objects are neither modeled nor correlated during training. Therefore, an intuitive idea is to introduce some priors about the unseen targets, e.g. the object relationships [60] (involving seen and unseen objects) from the external dataset. Such way may be helpful in guiding agent to get close to unseen objects according to the relationships with other seen objects. However, only getting close may not be capable of locating the precise location of the target, since the visual characteristics of unseen objects are essentially unknown to the agent (i.e. the agent does not know what unseen objects look like).

In order to precisely locate the unseen objects, the agent may be required to "imagine" what the unseen objects look like. In some generative methods [13,25,58,59], the model learns the mapping from the semantic embedding to the visual features on seen object categories. Then given the semantic embedding of the unseen object, the model could "imagine" its visual features by analogy. Those works focus on generating the representations of the object itself, which are effective in static tasks (e.g. object recognition and fine-grained classification) that do not require many environment descriptions. However, in the navigation task, the visual observations typically contain objects and background. In this case, only generating objects features is not enough, and predicting precise navigation actions also requires considering the current environmental background.

In general, the key challenge of unseen object navigation is how to generate the unseen object representations the current working environment. In particular, the visual characteristics of both objects and environment (foreground and background) are critical to the navigation. Motivated by the challenges of generating comprehensive representations of unseen objects in new environments, we investigate our researches mainly from the following two aspects: 1) generating initial visual features of the unseen object from its semantic embedding; 2) proposing the generative adversarial learning model within the meta-learning structure to fast adapt the generator to the current working environment.

In this paper, we propose a generative meta-adversarial network (GMAN) for unseen object navigation, which consists of a feature generator (FG) and an environmental meta discriminator (EMD). The FG is pre-trained in advance to generate the unseen object features by learning the mapping from semantic embedding to the visual features on seen objects. Furthermore, to obtain the background information of current working environment, the EMD is proposed to adapt the FG to fit the environment, with an adversarial loss between the real-time observation features and generated object features. Significantly, a gradient-based meta-learning method is implemented to rapidly adapt the FG based on a few observations, which is shown as Fig. 1, where the adaptation of the FG the current environment can be regarded as the inner-loop, and maximizing the navigation reward serves as the outer-loop. The experimental results on AI2THOR [29] and RoboTHOR [8] simulators demonstrate the effectiveness of our GMAN on unseen object navigation.

2 Related Work

Visual Object Navigation. Goal-driven visual navigation can be categorized [2] into PointGoal [6,19,50,55], AreaGoal [30,57] and ObjectGoal. Several ObjectGoal works set an image as the goal [7,49,63], which contains more environment information, while our work sets the object category as the target (following most works) and our agent does not know about the unseen environments at all. Previous map-based methods typically construct a map in advance or in real-time [11,26,49,53]. Recently, learning-based methods are mainly composed of visual embedding and policy learning. The basic visual representation generally utilizes the ResNet [21]. [60] extracts a knowledge graph from external dataset and uses GCN [28] to embed the egocentric view and the knowledge graph. [35] proposes the attention probability module. Some works use more visual semantic information (e.g. object detection, instance segmentation) to establish semantic SLAM [5], spatial layout [39], prior knowledge graph [61,62] and scene memory [12]. [5] projects the segmentation of first-person view into a top-down semantic map. [61,62] utilize object detection to construct prior objects relationships. [39] takes both semantic segmentation and object detection to jointly train models on real and simulated data to realize sim-to-real transfer. [12] proposes a memory-based transformer policy to embed the RGB-D observation and segmentation. As to the policy learning, [56] adopts meta-reinforcement learning so that the agent can dynamically adapt to an unseen scene. These works mainly focus on navigating to seen objects in the unseen environments, while our work focuses on unseen objects in the unseen environments. So far, there are few researches on navigating to unseen objects. [60] first proposes the unseen objects (namely novel objects) and builds a knowledge graph (from the external dataset) which provides the spatial and visual relationships between seen and unseen objects. Only the seen objects are used for training. During testing, the agent can infer the location of unseen objects according to the prior spatial relationships with seen objects. [61] employs similar settings. These two works both provide the strong prior knowledge to correlate the unseen objects with seen objects. Our work transfers the knowledge (i.e. mapping the semantic embedding to the visual features and navigating with the generated features) from seen to unseen objects without such strong priors. Therefore, our task is more general and challenging.

Feature Generation. Feature generation for unseen objects has been widely studied in zero-shot recognition tasks [1,16,17,23,46,47,58]. Early works [23,31] learn attribute classifiers to associate seen and unseen categories. Some works [1,16,47] learn matching functions between visual representation and semantic representation. Recently, generative adversarial network [18] has been used in the generalized zero-shot learning (GZSL) to synthesize unseen category features to train GZSL classifiers [13,32,58,59]. Our idea of generating features for the unseen object based on its semantic embedding is motivated by [58], while we focus on the navigation task more challenging than the static classification task.

Meta-learning. Meta-learning (learning to learn) adapts to new tasks efficiently through experience learned from multiple tasks. The previous methods are as follows: 1) *gradient-based* methods [3,4,14,15,22,42,45] optimize through gradient updates. 2) *metric-based* methods [51,52,54] adapt to new tasks through significant distance metrics. 3) *memory-based* methods [37,40,43,48] store the past experience as memory to learn efficiently. [14] proposes a model-agnostic

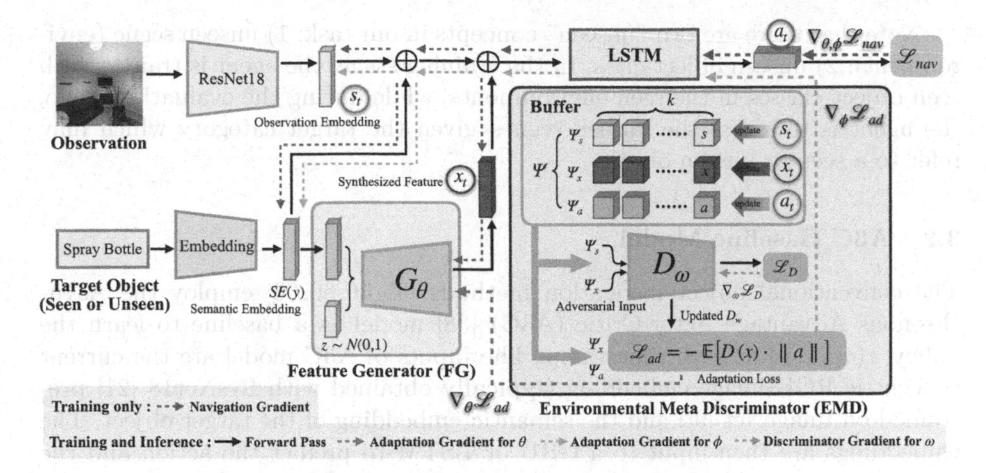

Fig. 2. Framework. Our generative meta-adversarial network (GMAN) mainly consists of a feature generator (FG) and an environmental meta discriminator (EMD). The FG synthesizes initial features of the target, then the EMD adversarially optimizes the FG to incorporate environmental features into object features. During navigation, the observation embedding s_t, generated features x_t and the action a_t are continually saved into a buffer Ψ, which is used to fast adapt the FG with meta-learning.

algorithm MAML, which learns the parameters initialization that can perform well in new tasks within a few gradients updates. Recently, some object navigation works use the gradient-based meta-learning for adapting in new scenes [35,56] and multi-task [33]. Our work also adopts the gradient-based meta-learning, while we aim at fast adaptation in adversarial learning and effectively initial parameters learning for both feature generator and policy.

3 Unseen Object Navigation

3.1 Task Definition

The prevailing object navigation task requires the agent to navigate to the seen target objects in new environments. In our work, the target categories in evaluation involve both seen and unseen object classes.

Formally, let $\mathcal{Y}^s = \{y_1^s, \ldots, y_M^s\}$ denote the set of M seen target object classes. Let $\mathcal{Y}^u = \{y_1^u, \ldots, y_N^u\}$ denote the set of N unseen classes. These two sets have no intersection. Considering a set of scenes, in each navigation episode, the agent is initialized at a random location p in an environment $e \in Env$ given the target object y ($y \in \mathcal{Y}^s \cup \mathcal{Y}^u$). The agent captures an egocentric RGB image (embedded as s_t) at the timestamp t and is trained to learn a policy $\pi(a_t|s_t, y)$ that predicts an action $a_t \in \mathcal{A}$. At each time t, the agent takes action a_t until executing the termination action. **The successful episode** is defined as the situation, where the agent finally gets close to the target object within a threshold of distance and the target is visible in agent's egocentric view.

Note that there are two "unseen" concepts in our task: 1) unseen scene (environment); 2) unseen object class. In the training stage, the agent is trained with seen object classes in the seen environments, while during the evaluation stage, the agent is tested in the unseen scenes given the target category which may refer to a seen or unseen object.

3.2 A3C Baseline Model

The conventional object navigation methods [35,56,60,63] employ the Asynchronous Advantage Actor-Critic (A3C) [38] model as a baseline to learn the policy $\pi(a_t|s_t, y)$ at each timestamp. The inputs of A3C model are the current egocentric RGB image embedding (typically obtained with ResNet18 [21] pre-trained on ImageNet [9]) and the semantic embedding of the target object. The embeddings are then input to a GRU or LSTM to predict the action and the value. Generally, the agent is trained to minimize the supervised actor-critic navigation loss $\mathcal{L}_{nav}$ [36,56,63], which is used to optimize the whole model. In this paper, our GMAN follows the framework of the A3C model and additionally synthesizes the target object features to guide the unseen object navigation.

4 Generative Meta-Adversarial Network

4.1 Feature Generator

The Generative Meta-Adversarial Network (GMAN) is illustrated in Fig. 2. The feature generator (FG) module contains a generator G that synthesizes features of the unseen objects based on their semantic embedding and a random noise. The generator is formulated as $G\left(SE\left(y\right), z\right)$, where $y \in \mathcal{Y}^s \cup \mathcal{Y}^u$ is the target category, z is a random Gaussian noise and $SE\left(\cdot\right)$ is the embedding module that converts the object category into a class-specific semantic vector. The generator is pre-trained with the dataset $D_{train} = \{(x^s, SE\left(y^s\right)) | x^s \in \mathcal{X}^s, y^s \in \mathcal{Y}^s\}$, where $\mathcal{X}^s$ is the set of seen object features and $\mathcal{Y}^s$ is the seen object labels. We collect the dataset D_{train} in the training scenes of the AI2THOR and RoboTHOR simulators, where the agent collects several egocentric RGB images for each object $y^s \in \mathcal{Y}^s$. The images are then extracted by ResNet18 pre-trained on ImageNet to obtain object feature $x^s \in \mathcal{X}^s$. The pre-training process teaches the generator to learn the mapping from the semantic embedding to the visual features. Therefore, the pre-trained generator could synthesize the unseen objects features based on their semantic embedding. This pre-training process is similar with [58] and detailed in supplements.

4.2 Environmental Meta Discriminator

The pre-trained G initially generates the class-specific object features according to semantic embedding. However, such features imply a general representation of all seen environments, rather than the current specific unseen environment. As

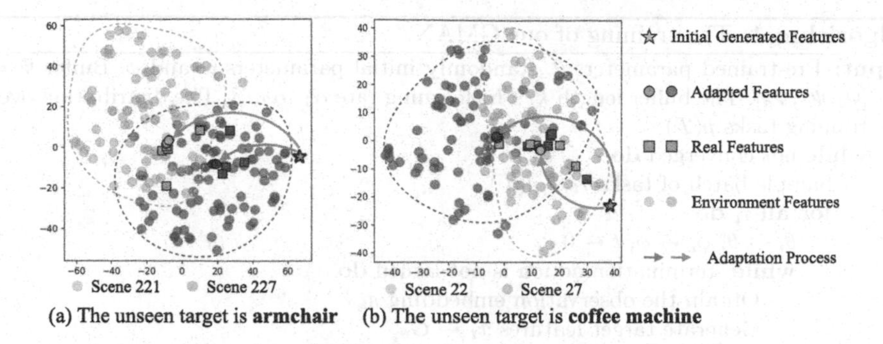

(a) The unseen target is **armchair** (b) The unseen target is **coffee machine**

Fig. 3. The T-SNE [34] visualization of initial features and adapted features. The orange and blue colors mean different environments. The light colored circles represent the environment features (sampled by the random agent in the egocentric view). Given an unseen target, the squares represent the real features (sampled from 5 different views around the target object). The arrows represent the process of features adaptation, where the initial features are generated by the initial generator, and the adapted features are generated by the optimized generator that is adapted by our environmental meta discriminator. (Color figure online)

shown in Fig. 3, the initial generated features are far from the real features of the target (different in different scenes). These initial features are not informative to guide the agent due to the lack of current environment information. Therefore, we propose the environmental meta discriminator (EMD) to optimize the generator G to learn the environment information during navigation. The EMD consists of a navigation buffer $\Psi = [\Psi_s, \Psi_x, \Psi_a]$ and a discriminator D, where Ψ_s records the embedded observation s_t during navigation, Ψ_x records the generated target features $x_t = G(SE(y), z)$ and Ψ_a records the action a_t output by the policy π. The capacity of Ψ_s, Ψ_s and Ψ_s are all set to k. Set that θ denotes the parameters of the pre-trained generator G, ω denotes the parameters of the discriminator D and ϕ denotes the remaining parameters of our model.

The generator G and discriminator D are a pair of adversarial learners and trained in a self-supervised way. We adopt the classical WGAN [20] to optimize G and D. The D aims to accurately distinguish the embedded observation features and the generated features, which is optimized by maximizing the following

$$\mathcal{L}_D = \mathbb{E}[D(s)] - \mathbb{E}[D(x)] - \lambda \mathbb{E}[(\|\nabla_{\tilde{x}} D(\tilde{x})\|_2 - 1)^2] \tag{1}$$

where $s \in \Psi_s$, $x \in \Psi_x$, λ is the penalty coefficient, $\tilde{x} = \varepsilon s + (1-\varepsilon)x$ with $\varepsilon \in U(0,1)$, and the $\mathbb{E}(\cdot)$ represents the mathematical expectation. The generator tries to generate realistic object features that are close to the environment features. The optimization objective of generator is to minimize the following

$$\mathcal{L}_G = -\mathbb{E}[D(x)] \tag{2}$$

As illustrated in Fig. 3, the adversarial learning (see the arrows) narrows the gap between the generated features and the real features by reducing the distance

Algorithm 1. The training of our GMAN.

Input: Pre-trained parameters θ. Randomly initial parameters ω and ϕ. Buffer $\Psi = [\Psi_s, \Psi_x, \Psi_a]$. The buffer length k. The learning rate α_1, α_2, β. The distribution over training tasks $p(\mathcal{T})$.

1: **while** not converged **do**
2: Sample batch of tasks $\tau_i \sim p(\mathcal{T})$
3: **for all** τ_i **do**
4: $\theta_i \leftarrow \theta$, $\phi_i \leftarrow \phi$, $t \leftarrow 0$
5: **while** termination action is not issued **do**
6: Obtain the observation embedding s_t
7: Generate target features $x_t \leftarrow G_{\theta_i}$
8: Take action a_t from $\pi_{\theta_i,\omega,\phi_i}$
9: $t \leftarrow t+1$
10: **if** t is not divisible by k **then**
11: Update Ψ_s by $\Psi_s \cup \{s_t\}$
12: Update Ψ_x by $\Psi_x \cup \{x_t\}$
13: Update Ψ_a by $\Psi_a \cup \{a_t\}$
14: **if** t is divisible by k **then**
15: Calculate $\mathcal{L}_D$ with Ψ_s, Ψ_x (Eq. 1)
16: $\omega \leftarrow \omega + \alpha_1 \nabla_\omega \mathcal{L}_D(\omega, \theta_i)$
17: Calculate $\mathcal{L}_{ad}$ with Ψ_x, Ψ_a (Eq. 3)
18: $\theta_i \leftarrow \theta_i - \alpha_2 \nabla_{\theta_i} \mathcal{L}_{ad}(\omega, \theta_i, \phi_i)$
19: $\phi_i \leftarrow \phi_i - \alpha_2 \nabla_{\phi_i} \mathcal{L}_{ad}(\omega, \theta_i, \phi_i)$
20: Empty Ψ_s, Ψ_x and Ψ_a
21: $\theta \leftarrow \theta - \beta \sum_{\tau_i \sim p(\mathcal{T})} \nabla_\theta \mathcal{L}_{nav}(\omega, \theta_i, \phi_i)$
22: $\phi \leftarrow \phi - \beta \sum_{\tau_i \sim p(\mathcal{T})} \nabla_\phi \mathcal{L}_{nav}(\omega, \theta_i, \phi_i)$

Output: θ, ω, ϕ

between the generated features and the environment features. The adversarial learning makes the generator capture the feature distributions of current environment, thus helping generate more informative features of the unseen objects.

Furthermore, since the navigation trajectory is limited and the navigation process is dynamic compared to various static images in classification tasks [41, 58], a fast adaptation of the generator, with reference to only a few observations, is also necessary to be considered. The MAML [14] provides an algorithm to find the optimal initial parameters, which could fit the sub-tasks within only a small number of adaptation steps. Inspired by MAML, we introduce the meta-learning into the adversarial learning. The MAML and its variants consist of an inner-loop and an outer-loop. The inner-loop updates the initial parameters through a few gradient steps to achieve great performance on a specific task. The outer-loop is to minimize the total loss on all tasks. The inner-loop is executed in both training and inference, while the outer-loop is only conducted in training.

In our case, we regard each episode in navigation as a new task. In the inner-loop, we expect to obtain a well-adapted generator and a wise policy which takes the generated features as the input. Therefore, we propose the adaptation loss $\mathcal{L}_{ad}$ to optimize the generator and the policy by minimizing the following

$$\mathcal{L}_{ad} = -\mathbb{E}[D(x) \cdot \|a\|] \tag{3}$$

where $a \in \Psi_a$ represents the action output by the policy function π. Each dimension of $a \in \mathbb{R}^{1\times 6}$ denotes the probability of each action. The adaptation loss $\mathcal{L}_{ad}$ is based on $\mathcal{L}_G$ (Eq. 2) with additional optimization to the policy function π rather than only to the generator. The intuition behind multiplying $-D(x)$ and $\|a\|$ is to encourage those policies whose decisions are made based on more realistic features (i.e. $-D(x)$ is lower). When calculating the gradient $\nabla_\phi \mathcal{L}_{ad}$ for the policy (line 19 in Algorithm 1), $-D(x)$ is regarded as the weight of $\nabla_\phi \mathbb{E}[\|a\|]$ (i.e. the policy with more realistic features will have a lower loss). Considering $p(\mathcal{T})$ as the distribution of training episodes, in every training sample $\tau_i \sim p(\mathcal{T})$, the generator parameters and policy parameters are firstly optimized by adaptation loss $\mathcal{L}_{ad}$ and updated to θ_i and ϕ_i. Then the outer-loop minimizes the total loss over all episodes. The training objective of the outer-loop is given by

$$\min_{\theta,\phi} \sum_{\tau_i \sim p(\mathcal{T})} \mathcal{L}_{nav}(\omega, \theta_i, \phi_i) \tag{4}$$

where $\mathcal{L}_{nav}$ is the navigation loss. Algorithm 1 summarizes the details of our method for training. Note that the D (with parameter ω) is optimized over all training episodes and the G is optimized in each episode with only a few iterations different from many iterations in prevailing works [41,58,59]. The inference process is similar to the training, except that the line 21 and 22 are removed.

5 Experiments

5.1 Experiment Setup

Datasets. We employ two editable simulators AI2THOR [29] and RoboTHOR [8], which are detailed in supplements. To guarantee that unseen objects do not appear in the seen (training) scenes, we edit the simulators to remove the unseen objects from the training scenes. We choose 24 types seen objects and 12 types unseen objects in AI2THOR, and 10 types seen objects and 4 types unseen objects in RoboTHOR. The split of the seen and unseen object categories is detailed in supplements. For both simulators, each validation set is used to select the best model, which is then respectively evaluated for 1000 episodes on seen and unseen objects. All experimental results are repeated three times and presented with the mean and standard deviation (in small gray font). Additionally, although our method focuses on the navigation task of unseen objects, the evaluations for seen objects (typically in the conventional navigation task) are also included.

Semantic Embedding. Semantic embedding serves as a bridge that transfers the knowledge of visual feature generation from seen to unseen objects.

Therefore, selecting an informative semantic embedding for object categories is critical for generating discriminative object features. Previous object navigation works [10,56,60] typically utilize the Glove [44] or FastText [24] as the semantic embedding for the object category, while zero-shot learning works [41,58] generally employ the attribute vector as the semantic embedding. Each dimension of the attribute vector represents the probability of containing such an attribute. We evaluate two semantic embeddings: Glove and attribute vector (detailed in the supplements), and find that the attribute vector achieves better navigation performance. Therefore, we adopt the attribute vector as semantic embedding. Every input object category is converted into an attribute vector through the semantic embedding module. Note that such converting process is pre-defined and does not require the user's involvement. More details of the attribute vector are also introduced in the supplements.

Rewards. We experiment with the following four rewards to train the agent.

Navigation Reward R_n. The previous works [10,35,56,60,62] typically employ the navigation reward R_n that penalizes each step with -0.01 and rewards agent for successfully finding the target object with 5.

Distance Reward R_{n+d}. The distance reward R_{n+d} is based on the R_n with an additional reward $r_d = max\left(0.1 \cdot \left(Dis\left(s_t, y\right) - Dis\left(s_{t+1}, y\right)\right), 0\right)$ for each step, where $Dis\left(s_t, y\right)$ computes the Euclidean distance from state s_t to target y, and s_{t+1} is the transferred state after performing the action a_t.

Similarity Reward R_{n+s}. To enhance the transfer ability from seen to unseen objects, the similarity reward R_{n+s} is designed to add an additional reward r_s to R_n, whenever the object (e.g. orange) found by the agent at last is similar to the target (e.g. tangerine) in some semantic aspects. The r_s is defined as $r_s = 0.1 S_{jaccard}\left(SE\left(y\right), SE\left(V\left(s_{done}\right)\right)\right)$, where $V\left(s_{done}\right)$ represents the visible object categories when the agent executes the action *Done*, $SE\left(\cdot\right)$ represents the semantic embedding (i.e. the attribute vector) of an object, and $S_{jaccard}\left(\cdot\right)$ computes the Jaccard similarity. When $V\left(s_{done}\right)$ contains multiple objects, we only consider the one that maximizes the $S_{jaccard}$.

Mixed Reward R_{n+d+s}. We also combine the distance reward R_{n+d} and the similarity reward R_{n+s} for the experiment.

Evaluation Metrics. We evaluate models using Success Rate (SR), Success weighted by Path Length (SPL), Exploration Area (EPA) and Distance To Goal in meters (DTS). These metrics are detailed in the supplements and the "Success" (i.e. successful episode) is defined in Sect. 3.1. In the following results, ↑ indicates that the larger value is better, while ↓ indicates the opposite.

Implementation Details. Following primary recommendation of [35, 56,60], the action set is defined as $\mathcal{A} = \{MoveAhead, RotateLeft, RotateRight, LookUp, LookDown, Done\}$. The horizontal rotation angle is set as 45 °C while the pitch angle is 30 °C. The action *Done* is decided by the agent rather than the simulator. The generator G and discriminator D are implemented as 2 fully connected layers. The G following [58,59] has 4096 hidden

Table 1. The impact of different rewards. We compare the effect of the navigation reward R_n, distance reward R_{n+d}, similarity reward R_{n+s} and the mixed reward R_{n+d+s}.

Reward	Unseen Objects				Seen Objects			
	SR↑ (%)	SPL↑ (%)	EPA (%)	DTS↓ (m)	SR↑ (%)	SPL↑ (%)	EPA (%)	DTS↓ (m)
R_n	14.50 1.15	6.95 0.52	5.65 0.74	1.24 0.02	27.03 0.72	12.46 0.66	7.88 0.08	1.20 0.01
R_{n+d}	18.43 0.50	8.23 0.11	5.92 0.61	1.20 0.04	30.83 0.85	14.14 0.56	9.08 0.25	1.11 0.02
R_{n+s}	21.43 0.75	**9.60** 0.38	6.51 0.45	1.16 0.01	29.53 0.35	13.14 0.61	9.24 0.29	1.11 0.01
R_{n+d+s}	**21.50** 1.23	9.36 0.66	8.56 0.15	**1.09** 0.01	**31.37** 1.00	**14.19** 0.30	9.36 0.18	**1.06** 0.01

units, while D is adjusted to have 1024 hidden units. We train our model using reinforcement learning with 12 asynchronous workers. The inner-loop is updated by SGD, while the outer-loop is optimized by Adam [27]. The learning rates (α_1, α_2, β) are all set to 10^{-4}. The penalty coefficient is $\lambda = 10$. The buffer length is set to $k = 20$.

Table 2. The ablation studies on different components (FG and EMD) with two baselines (A3C and A3C†(i.e. A3C with the PS module)).

PS	FG	EMD	Unseen Objects				Seen Objects			
			SR↑ (%)	SPL↑ (%)	EPA (%)	DTS↓ (m)	SR↑ (%)	SPL↑ (%)	EPA (%)	DTS↓ (m)
			21.50 1.23	9.36 0.66	8.56 0.15	1.09 0.01	31.37 1.00	14.19 0.30	9.36 0.18	1.06 0.01
	✓		23.20 1.82	8.87 0.41	8.68 0.03	1.09 0.03	34.20 1.42	15.11 0.72	8.85 0.27	1.06 0.01
	✓	✓	28.03 0.91	13.02 0.14	8.71 0.52	1.08 0.02	39.30 0.87	15.61 0.14	10.46 0.17	0.98 0.02
✓			32.70 0.75	20.30 0.38	8.02 0.07	1.17 0.01	48.80 0.17	26.85 0.42	10.09 0.06	1.02 0.01
✓	✓		34.20 1.56	19.64 0.24	10.99 0.06	1.13 0.02	50.73 0.38	26.38 1.32	12.03 0.48	0.99 0.01
✓	✓	✓	**48.83** 0.60	**25.09** 0.37	10.04 0.09	**0.93** 0.01	**57.80** 0.78	**28.41** 0.66	14.12 0.15	**0.91** 0.03

5.2 Methods for Comparison

We compare the following methods: 1) **Random**: The agent adopts a random action at each step. 2) **A3C**: The baseline model described in Sect. 3.2. 3) **SP** [60]: The agent navigates using scene priors knowledge graph extracted from the external dataset. 4) **SAVN** [56]: The agent minimizes the self-supervised loss to optimize the policy function with MAML for fast adaptation in unseen environments. 5) **EOTP** [35]: The agent is based on SAVN with additional attention probability module, which encodes semantic and spatial information.

Inspired by the effectiveness of the similarity rewards R_{n+s} on unseen objects (discussed in Sect. 5.3), an intuitive idea is that the similarity of the semantic embedding between the target object $SE(y)$ and the objects appearing in current observation $SE(V(s_t))$ may be beneficial for navigating to unseen objects.

Table 3. Comparisons on different EMD variants. "CosSim" replaces the discriminator with the cosine similarity. "w/o meta" removes the meta-learning from our EMD.

EMD variants	Unseen Objects				Seen Objects			
	SR↑ (%)	SPL↑ (%)	EPA (%)	DTS↓ (m)	SR↑ (%)	SPL↑ (%)	EPA (%)	DTS↓ (m)
CosSim	46.47 $_{0.31}$	24.20 $_{0.26}$	14.58 $_{0.28}$	0.98 $_{0.02}$	53.63 $_{0.58}$	24.66 $_{0.56}$	15.73 $_{1.07}$	0.92 $_{0.03}$
w/o meta	41.93 $_{0.80}$	21.13 $_{0.40}$	10.73 $_{0.41}$	1.01 $_{0.01}$	53.47 $_{0.92}$	25.68 $_{0.15}$	13.09 $_{0.21}$	0.93 $_{0.03}$
EMD (ours)	**48.83** $_{0.60}$	**25.09** $_{0.37}$	10.04 $_{0.09}$	**0.93** $_{0.01}$	**57.80** $_{0.78}$	**28.41** $_{0.66}$	14.12 $_{0.15}$	**0.91** $_{0.03}$

Therefore, a simple module PS is proposed to take the current egocentric view as the input and Predicts the Semantic embedding of all contained objects. The PS is pre-trained using collected images and the semantic embedding ground truth $\mathbb{E}_{y' \in V(I)}(SE(y'))$, where $V(I)$ is the set of visible objects in the image I. The PS is implemented with the ResNet pre-trained on ImageNet. The pre-training only uses the seen objects in the seen environments. Thus, there is another experimental group. 6) **A3C**†: The output of the PS is concatenated with the semantic embedding of the target object, together input to the LSTM. The A3C† (the A3C model equipped with the PS) is defined as another baseline. 7) **SP**†, **SAVN**†, **EOTP**†: All original methods are equipped with the PS.

There are other methods [5,7,10,61] for object navigation. However, these methods require pre-trained visual clues such as object detection or instance segmentation to construct object relation graphs [10,61] or semantic maps [5,7]. These methods are inapplicable to unseen objects because they require unseen object detection or segmentation, so that we modify them by fairly adding the PS module and the mixed reward. The comparisons are detailed in the supplements.

5.3 Evaluation Results

The Impact of Rewards. We use the A3C baseline to investigate the effect of rewards R_n, R_{n+d}, R_{n+s} and R_{n+d+s}, as shown in Table 1. Compared to the navigation reward R_n, the distance reward R_{n+d} provides the distance information of the target. Thus, the R_{n+d} improves the efficiency of RL training, thereby significantly improving the performance on both seen and unseen objects. Since the similarity reward R_{n+s} gives additional rewards which encourage the agent to find semantically similar objects. Thereby R_{n+s} enhances the transfer of navigation ability from seen to unseen objects (correlated through semantic embedding) and achieves more improvement on the unseen objects. Additionally, combining the advantages of R_{n+d} and R_{n+s}, the mixed reward R_{n+d+s} achieves the best performance on both seen and unseen objects. As a result, we choose the mixed reward R_{n+d+s} to train all models, including our method and the related works.

Ablation Studies on Different Modules. We choose two baselines (A3C and A3C†) for ablation studies as shown in Table 2. Directly using the FG brings a slight improvement for both two baselines, which demonstrates that directly employing feature generation methods without considering current environmental background is not enough for the unseen object navigation task. Comparatively, combining FG with EMD gains significant improvement especially on unseen objects, indicating that the continuous adaptation of the FG to obtain more environment information is necessary. Furthermore, the baseline A3C† also significantly outperforms A3C, which shows the effectiveness of the PS module, further indicating that our semantic embedding of the unseen object is meaningful so that the semantic similarity in the PS module can play its value. Besides, combining FG, EMD and PS obtains the best performance on both seen and unseen objects, again indicating that these modules could complement each other.

Comparisons on Some EMD Variants. To further explore the optimal structure of the proposed EMD, based on the GMAN† (line 6 in Table 2), some EMD variants are considered as shown in Table 3. The first variant (line 1) replaces the discriminator D with cosine similarity to calculate the similarity between the generated features x_t and the environment features s_t. The results indicate that the cosine similarity does bring some improvements compared with that of no discriminator (line 5 in Table 2), while the improvement is less than the proposed EMD. Compared to the fixed similarity measurement (cosine distance), the discriminator D seems to be a better "learnable measurement". The second variant without meta-learning (only through adversarial learning) can also obtain improvements than that of no discriminator (line 5 in Table 2), while is still inferior to our method. The results indicate that meta-learning is indeed a powerful tool to improve performance.

Table 4. Comparisons with the related works for navigation in unseen environments on AI2THOR simulator. The "†" indicates the combination with the PS module.

Method	Unseen Objects				Seen Objects			
	SR↑ (%)	SPL↑ (%)	EPA (%)	DTS↓ (m)	SR↑ (%)	SPL↑ (%)	EPA (%)	DTS↓ (m)
Random	$6.70_{1.01}$	$3.58_{0.67}$	$4.01_{0.03}$	$1.57_{0.02}$	$6.23_{0.60}$	$3.63_{0.30}$	$3.89_{0.06}$	$1.53_{0.04}$
A3C	$21.50_{1.23}$	$9.36_{0.66}$	$8.56_{0.15}$	$1.09_{0.01}$	$31.37_{1.00}$	$14.19_{0.30}$	$9.36_{0.18}$	$1.06_{0.01}$
SP [60]	$22.43_{1.71}$	$9.60_{1.25}$	$7.21_{0.35}$	$1.16_{0.01}$	$34.00_{0.87}$	$13.23_{0.87}$	$8.33_{0.11}$	$1.13_{0.02}$
SAVN [56]	$17.63_{0.59}$	$4.69_{0.61}$	$9.76_{1.74}$	$1.19_{0.04}$	$35.87_{1.48}$	$13.47_{0.58}$	$10.99_{0.37}$	$1.02_{0.02}$
EOTP [35]	$19.90_{0.40}$	$4.36_{0.41}$	$11.89_{0.07}$	$\mathbf{0.99}_{0.01}$	$36.97_{0.39}$	$14.56_{0.32}$	$11.03_{0.16}$	$0.98_{0.02}$
GMAN (ours)	$\mathbf{28.03}_{0.91}$	$\mathbf{13.02}_{0.14}$	$8.71_{0.52}$	$1.08_{0.02}$	$\mathbf{39.30}_{0.87}$	$\mathbf{15.61}_{0.14}$	$10.46_{0.17}$	$\mathbf{0.98}_{0.02}$
A3C†	$32.70_{0.75}$	$20.30_{0.38}$	$8.02_{0.07}$	$1.17_{0.01}$	$48.80_{0.17}$	$26.85_{0.42}$	$10.09_{0.06}$	$1.02_{0.01}$
SP†	$38.00_{0.40}$	$21.77_{1.38}$	$8.98_{0.45}$	$1.06_{0.01}$	$49.40_{0.61}$	$26.84_{0.49}$	$9.81_{0.04}$	$1.02_{0.02}$
SAVN†	$41.47_{1.26}$	$18.97_{0.90}$	$15.64_{0.14}$	$1.00_{0.01}$	$53.63_{0.68}$	$23.31_{0.43}$	$16.22_{0.07}$	$\mathbf{0.89}_{0.01}$
EOTP†	$38.57_{1.19}$	$15.44_{0.25}$	$11.41_{0.06}$	$1.03_{0.01}$	$52.87_{1.19}$	$\mathbf{29.50}_{0.58}$	$10.34_{0.10}$	$0.95_{0.02}$
GMAN† (ours)	$\mathbf{48.83}_{0.60}$	$\mathbf{25.09}_{0.37}$	$10.04_{0.09}$	$\mathbf{0.93}_{0.01}$	$\mathbf{57.80}_{0.78}$	$28.41_{0.66}$	$14.12_{0.15}$	$0.91_{0.03}$

5.4 Comparisons with the Related Works

The experimental results in AI2THOR and RoboTHOR are shown in Table 4 and 5. The SP attempts to navigate to unseen objects, which is task-related to our GMAN. Both SAVN and EOTP are structured in MAML-liked reinforcement learning, which is framework-related to our GMAN. However, SAVN and EOTP focus on seen objects and achieve poor performance on unseen objects. For a fair comparison, we enhance all related works from two aspects. 1) **The mixed reward**. All related works are implemented with the mixed reward that is conducive to unseen objects. Therefore, the SAVN and EOTP can improve the navigation ability on unseen objects although the performances are still lower than the SP. The SP benefits from its object relation graph and outperforms the A3C, SAVN and EOTP on unseen objects. 2) **The PS module**. All related works are equipped with the PS (i.e. under the A3C† baseline), which are denoted with the superscript †. Since the PS compares the semantic similarity of current view and the target object, all methods gain a significant improvement on navigating to unseen objects. Besides, the MAML-liked methods SAVN† and EOTP† outperform the SP†, which indicates that the MAML-liked methods get more benefit from the semantic similarity information.

Comparing our GMAN with the related works on unseen objects, both GMAN and GMAN† outperform the related works with a large margin. Significantly, under A3C† baseline, the GMAN† outperforms the related works by 7.36% in SR, 3.32% in SPL, and −0.07 m in DTS in AI2THOR and 3.70% in SR, 1.27% in SPL, and −0.05 m in DTS in RoboTHOR. The results reveal the great advantage of our GMAN on unseen objects. As for the seen objects, since our GMAN is based on the MAML-liked methods with a generative module that generates the features of the target objects, both GMAN and GMAN† can also improve the performance on seen objects, despite that the improvement is less

Table 5. Comparisons with the related works for navigation in unseen environments on RoboTHOR simulator.

Method	Unseen Objects				Seen Objects			
	SR↑ (%)	SPL↑ (%)	EPA (%)	DTS↓ (*m*)	SR↑ (%)	SPL↑ (%)	EPA (%)	DTS↓ (*m*)
Random	2.12 $_{0.91}$	1.03 $_{0.35}$	6.53 $_{0.09}$	2.26 $_{0.02}$	2.30 $_{0.26}$	1.11 $_{0.12}$	6.58 $_{0.10}$	2.22 $_{0.08}$
A3C	9.73 $_{0.06}$	5.06 $_{0.61}$	7.69 $_{0.08}$	2.11 $_{0.01}$	11.33 $_{0.64}$	5.39 $_{0.33}$	9.46 $_{0.09}$	2.18 $_{0.01}$
SP [60]	11.37 $_{0.76}$	5.52 $_{0.37}$	8.76 $_{0.19}$	2.08 $_{0.05}$	10.33 $_{0.64}$	5.03 $_{0.38}$	9.76 $_{0.52}$	2.17 $_{0.02}$
SAVN [56]	11.00 $_{0.35}$	4.32 $_{0.30}$	9.86 $_{0.42}$	1.98 $_{0.02}$	13.93 $_{0.38}$	6.02 $_{0.50}$	10.79 $_{0.18}$	**1.97** $_{0.10}$
EOTP [35]	11.30 $_{0.62}$	4.39 $_{0.33}$	10.56 $_{0.16}$	1.99 $_{0.01}$	14.53 $_{0.91}$	6.25 $_{0.71}$	11.12 $_{0.17}$	2.04 $_{0.05}$
GMAN (ours)	**13.27** $_{0.32}$	**5.60** $_{0.35}$	10.39 $_{0.14}$	**1.96** $_{0.05}$	**15.07** $_{0.15}$	**6.45** $_{0.38}$	11.06 $_{0.12}$	2.02 $_{0.05}$
A3C†	10.87 $_{0.51}$	7.26 $_{0.38}$	8.97 $_{0.08}$	2.17 $_{0.02}$	17.23 $_{0.06}$	11.78 $_{0.05}$	10.49 $_{0.09}$	2.11 $_{0.01}$
SP†	12.93 $_{0.93}$	7.33 $_{0.45}$	10.82 $_{0.14}$	2.11 $_{0.01}$	18.67 $_{0.46}$	11.89 $_{0.47}$	11.00 $_{0.12}$	2.09 $_{0.01}$
SAVN†	23.97 $_{0.30}$	13.02 $_{0.80}$	12.23 $_{0.23}$	1.73 $_{0.03}$	31.90 $_{0.70}$	17.01 $_{0.78}$	13.45 $_{0.10}$	1.82 $_{0.01}$
EOTP†	21.53 $_{0.45}$	10.38 $_{0.41}$	14.32 $_{0.03}$	1.87 $_{0.03}$	36.43 $_{1.06}$	18.94 $_{0.75}$	14.28 $_{0.09}$	1.81 $_{0.03}$
GMAN† (ours)	**27.67** $_{0.67}$	**14.29** $_{0.37}$	12.47 $_{0.02}$	**1.68** $_{0.02}$	**37.10** $_{0.61}$	**19.12** $_{0.12}$	13.65 $_{0.04}$	**1.78** $_{0.02}$

outstanding than that on unseen objects. The results indicate that the generated features bring limited improvements to those well-trained seen objects.

Note that the reported results of SP, SAVN and EOTP is basically consistent with [35,56] while different from [60]. Because our setting (24 types seen objects and 12 types unseen objects) has huge difference with [60] (46 types seen objects and 11 types unseen objects) but is similar to [35,56] (18 types seen objects).

6 Conclusions

In this paper, we propose a generative meta-adversarial network (GMAN) for unseen object navigation. Our method is composed of a feature generator (FG) and an environmental meta discriminator (EMD). The FG synthesizes the object features by learning the mapping from semantic embeddings to features. The EMD adapts the FG with adversarial learning to let FG learn the background information of the navigation scene. Besides, meta-learning is introduced to the adversarial learning for fast adaptation. Experimental results on AI2THOR and RoboTHOR show the effectiveness of our method on unseen object navigation.

Acknowledgement. This work was supported by National Key Research and Development Project of New Generation Artificial Intelligence of China, under Grant 2018AAA0102500, in part by the National Natural Science Foundation of China under Grant 62125207, 62032022, 61902378 and U1936203, in part by Beijing Natural Science Foundation under Grant Z190020, in part by the National Postdoctoral Program for Innovative Talents under Grant BX201700255.

References

1. Akata, Z., Perronnin, F., Harchaoui, Z., Schmid, C.: Label-embedding for image classification. CoRR abs/1503.08677 (2015)
2. Anderson, P., et al.: On evaluation of embodied navigation agents. arXiv preprint arXiv:1807.06757 (2018)
3. Cao, T., Xu, Q., Yang, Z., Huang, Q.: Task-distribution-aware meta-learning for cold-start CTR prediction. In: MM 2020: The 28th ACM International Conference on Multimedia, Virtual Event/Seattle, WA, USA, 12–16 October 2020, pp. 3514–3522 (2020)
4. Cao, T., Xu, Q., Yang, Z., Huang, Q.: Meta-wrapper: differentiable wrapping operator for user interest selection in CTR prediction. IEEE Trans. Pattern Anal. Mach. Intell. (2021)
5. Chaplot, D.S., Gandhi, D., Gupta, A., Salakhutdinov, R.R.: Object goal navigation using goal-oriented semantic exploration. In: Advances in Neural Information Processing Systems 33: Annual Conference on Neural Information Processing Systems 2020, NeurIPS 2020, December 6–12, 2020, virtual (2020)
6. Chaplot, D.S., Gandhi, D., Gupta, S., Gupta, A., Salakhutdinov, R.: Learning to explore using active neural SLAM. In: 8th International Conference on Learning Representations, ICLR 2020, Addis Ababa, Ethiopia, 26–30 April 2020. OpenReview.net (2020)

7. Chaplot, D.S., Salakhutdinov, R., Gupta, A., Gupta, S.: Neural topological SLAM for visual navigation. In: 2020 IEEE/CVF Conference on Computer Vision and Pattern Recognition, CVPR 2020, Seattle, WA, USA, 13–19 June 2020, pp. 12872–12881 (2020)
8. Deitke, M., et al.: RoboTHOR: an open simulation-to-real embodied AI platform. In: 2020 IEEE/CVF Conference on Computer Vision and Pattern Recognition, CVPR 2020, Seattle, WA, USA, 13–19 June 2020, pp. 3161–3171 (2020)
9. Deng, J., Dong, W., Socher, R., Li, L., Li, K., Li, F.: ImageNet: a large-scale hierarchical image database. In: 2009 IEEE Computer Society Conference on Computer Vision and Pattern Recognition (CVPR 2009), Miami, Florida, USA, 20–25 June 2009, pp. 248–255 (2009)
10. Du, H., Yu, X., Zheng, L.: Learning object relation graph and tentative policy for visual navigation. In: Vedaldi, A., Bischof, H., Brox, T., Frahm, J.-M. (eds.) ECCV 2020, Part VII. LNCS, vol. 12352, pp. 19–34. Springer, Cham (2020). https://doi.org/10.1007/978-3-030-58571-6_2
11. Elfes, A.: Using occupancy grids for mobile robot perception and navigation. Computer **22**(6), 46–57 (1989)
12. Fang, K., Toshev, A., Li, F., Savarese, S.: Scene memory transformer for embodied agents in long-horizon tasks. In: IEEE Conference on Computer Vision and Pattern Recognition, CVPR 2019, Long Beach, CA, USA, 16–20 June 2019, pp. 538–547. Computer Vision Foundation/IEEE (2019)
13. Felix, R., Vijay Kumar, B.G., Reid, I., Carneiro, G.: Multi-modal cycle-consistent generalized zero-shot learning. In: Ferrari, V., Hebert, M., Sminchisescu, C., Weiss, Y. (eds.) ECCV 2018, Part VI. LNCS, vol. 11210, pp. 21–37. Springer, Cham (2018). https://doi.org/10.1007/978-3-030-01231-1_2
14. Finn, C., Abbeel, P., Levine, S.: Model-agnostic meta-learning for fast adaptation of deep networks. In: Proceedings of the 34th International Conference on Machine Learning, ICML 2017, Sydney, NSW, Australia, 6–11 August 2017, pp. 1126–1135 (2017)
15. Frans, K., Ho, J., Chen, X., Abbeel, P., Schulman, J.: Meta learning shared hierarchies. In: 6th International Conference on Learning Representations, ICLR 2018, Vancouver, BC, Canada, 30 April–3 May 2018, Conference Track Proceedings. OpenReview.net (2018)
16. Frome, A., et al.: Devise: a deep visual-semantic embedding model. In: Burges, C.J.C., Bottou, L., Ghahramani, Z., Weinberger, K.Q. (eds.) Advances in Neural Information Processing Systems 26: 27th Annual Conference on Neural Information Processing Systems 2013. Proceedings of a Meeting Held, Lake Tahoe, Nevada, United States, 5–8 December 2013, pp. 2121–2129 (2013)
17. Fu, Y., Hospedales, T.M., Xiang, T., Gong, S.: Transductive multi-view zero-shot learning. IEEE Trans. Pattern Anal. Mach. Intell. **37**(11), 2332–2345 (2015)
18. Goodfellow, I.J., et al.: Generative adversarial nets. In: Ghahramani, Z., Welling, M., Cortes, C., Lawrence, N.D., Weinberger, K.Q. (eds.) Advances in Neural Information Processing Systems 27: Annual Conference on Neural Information Processing Systems 2014, Montreal, Quebec, Canada, 8–13 December 2014, pp. 2672–2680 (2014)
19. Gupta, S., Davidson, J., Levine, S., Sukthankar, R., Malik, J.: Cognitive mapping and planning for visual navigation. In: 2017 IEEE Conference on Computer Vision and Pattern Recognition, CVPR 2017, Honolulu, HI, USA, 21–26 July 2017, pp. 7272–7281. IEEE Computer Society (2017)

20. Guyon, I., et al. (eds.): Advances in Neural Information Processing Systems 30: Annual Conference on Neural Information Processing Systems 2017, Long Beach, CA, USA, 4–9 December 2017 (2017)
21. He, K., Zhang, X., Ren, S., Sun, J.: Deep residual learning for image recognition. In: 2016 IEEE Conference on Computer Vision and Pattern Recognition, CVPR 2016, Las Vegas, NV, USA, 27–30 June 2016, pp. 770–778 (2016)
22. Hochreiter, S., Younger, A.S., Conwell, P.R.: Learning to learn using gradient descent. In: Dorffner, G., Bischof, H., Hornik, K. (eds.) ICANN 2001. LNCS, vol. 2130, pp. 87–94. Springer, Heidelberg (2001). https://doi.org/10.1007/3-540-44668-0_13
23. Jayaraman, D., Grauman, K.: Zero-shot recognition with unreliable attributes. In: Ghahramani, Z., Welling, M., Cortes, C., Lawrence, N.D., Weinberger, K.Q. (eds.) Advances in Neural Information Processing Systems 27: Annual Conference on Neural Information Processing Systems 2014, Montreal, Quebec, Canada, 8–13 December 2014, pp. 3464–3472 (2014)
24. Joulin, A., Grave, E., Bojanowski, P., Mikolov, T.: Bag of tricks for efficient text classification. CoRR abs/1607.01759 (2016)
25. Karras, T., Laine, S., Aila, T.: A style-based generator architecture for generative adversarial networks. In: IEEE Conference on Computer Vision and Pattern Recognition, CVPR 2019, Long Beach, CA, USA, 16–20 June 2019, pp. 4401–4410 (2019)
26. Kidono, K., Miura, J., Shirai, Y.: Autonomous visual navigation of a mobile robot using a human-guided experience. Robotics Auton. Syst. **40**(2–3), 121–130 (2002)
27. Kingma, D.P., Ba, J.: Adam: a method for stochastic optimization. In: 3rd International Conference on Learning Representations, ICLR 2015, San Diego, CA, USA, May 7–9 2015, Conference Track Proceedings (2015)
28. Kipf, T.N., Welling, M.: Semi-supervised classification with graph convolutional networks. In: 5th International Conference on Learning Representations, ICLR 2017, Toulon, France, 24–26 April 2017, Conference Track Proceedings (2017)
29. Kolve, E., Mottaghi, R., Gordon, D., Zhu, Y., Gupta, A., Farhadi, A.: AI2-THOR: an interactive 3D environment for visual AI. CoRR abs/1712.05474 (2017)
30. Kumar, A., Gupta, S., Malik, J.: Learning navigation subroutines from egocentric videos. In: Kaelbling, L.P., Kragic, D., Sugiura, K. (eds.) 3rd Annual Conference on Robot Learning, CoRL 2019, Osaka, Japan, 30 October– 1 November 2019, Proceedings. Proceedings of Machine Learning Research, vol. 100, pp. 617–626. PMLR (2019)
31. Lampert, C.H., Nickisch, H., Harmeling, S.: Attribute-based classification for zero-shot visual object categorization. IEEE Trans. Pattern Anal. Mach. Intell. **36**(3), 453–465 (2014)
32. Li, J., Jing, M., Lu, K., Ding, Z., Zhu, L., Huang, Z.: Leveraging the invariant side of generative zero-shot learning. In: IEEE Conference on Computer Vision and Pattern Recognition, CVPR 2019, Long Beach, CA, USA, 16–20 June 2019, pp. 7402–7411. Computer Vision Foundation/IEEE (2019)
33. Li, J., et al.: Unsupervised reinforcement learning of transferable meta-skills for embodied navigation. In: 2020 IEEE/CVF Conference on Computer Vision and Pattern Recognition, CVPR 2020, Seattle, WA, USA, 13–19 June 2020, pp. 12120–12129 (2020)
34. Van der Maaten, L., Hinton, G.: Visualizing data using t-SNE. J. Mach. Learn. Res. **9**(11) (2008)

35. Mayo, B., Hazan, T., Tal, A.: Visual navigation with spatial attention. In: IEEE Conference on Computer Vision and Pattern Recognition, CVPR 2021, virtual, 19–25 June 2021, pp. 16898–16907 (2021)
36. Mirowski, P., et al.: Learning to navigate in complex environments. In: 5th International Conference on Learning Representations, ICLR 2017, Toulon, France, 24–26 April 2017, Conference Track Proceedings (2017)
37. Mishra, N., Rohaninejad, M., Chen, X., Abbeel, P.: A simple neural attentive meta-learner. In: 6th International Conference on Learning Representations, ICLR 2018, Vancouver, BC, Canada, 30 April– 3 May 2018, Conference Track Proceedings. OpenReview.net (2018)
38. Mnih, V., et al.: Asynchronous methods for deep reinforcement learning. In: Proceedings of the 33nd International Conference on Machine Learning, ICML 2016, New York City, NY, USA, 19–24 June 2016, pp. 1928–1937 (2016)
39. Mousavian, A., Toshev, A., Fiser, M., Kosecká, J., Wahid, A., Davidson, J.: Visual representations for semantic target driven navigation. In: International Conference on Robotics and Automation, ICRA 2019, Montreal, QC, Canada, 20–24 May 2019, pp. 8846–8852. IEEE (2019)
40. Munkhdalai, T., Yu, H.: Meta networks. In: Precup, D., Teh, Y.W. (eds.) Proceedings of the 34th International Conference on Machine Learning, ICML 2017, Sydney, NSW, Australia, 6–11 August 2017. Proceedings of Machine Learning Research, vol. 70, pp. 2554–2563. PMLR (2017)
41. Narayan, S., Gupta, A., Khan, F.S., Snoek, C.G.M., Shao, L.: Latent embedding feedback and discriminative features for zero-shot classification. In: Vedaldi, A., Bischof, H., Brox, T., Frahm, J.-M. (eds.) ECCV 2020, Part XXII. LNCS, vol. 12367, pp. 479–495. Springer, Cham (2020). https://doi.org/10.1007/978-3-030-58542-6_29
42. Nichol, A., Achiam, J., Schulman, J.: On first-order meta-learning algorithms. CoRR abs/1803.02999 (2018)
43. Oreshkin, B.N., López, P.R., Lacoste, A.: TADAM: task dependent adaptive metric for improved few-shot learning. In: Bengio, S., Wallach, H.M., Larochelle, H., Grauman, K., Cesa-Bianchi, N., Garnett, R. (eds.) Advances in Neural Information Processing Systems 31: Annual Conference on Neural Information Processing Systems 2018, NeurIPS 2018, Montréal, Canada, 3–8 December 2018, pp. 719–729 (2018)
44. Pennington, J., Socher, R., Manning, C.D.: Glove: Global vectors for word representation. In: Proceedings of the 2014 Conference on Empirical Methods in Natural Language Processing, EMNLP 2014, Doha, Qatar, 25–29 October 2014, A meeting of SIGDAT, a Special Interest Group of the ACL, pp. 1532–1543 (2014)
45. Ravi, S., Larochelle, H.: Optimization as a model for few-shot learning. In: 5th International Conference on Learning Representations, ICLR 2017, Toulon, France, 24–26 April 2017, Conference Track Proceedings. OpenReview.net (2017)
46. Rohrbach, M., Ebert, S., Schiele, B.: Transfer learning in a transductive setting. In: Burges, C.J.C., Bottou, L., Ghahramani, Z., Weinberger, K.Q. (eds.) Advances in Neural Information Processing Systems 26: 27th Annual Conference on Neural Information Processing Systems 2013. Proceedings of a meeting held 5–8 December 2013, Lake Tahoe, Nevada, United States, pp. 46–54 (2013)
47. Romera-Paredes, B., Torr, P.H.S.: An embarrassingly simple approach to zero-shot learning. In: Bach, F.R., Blei, D.M. (eds.) Proceedings of the 32nd International Conference on Machine Learning, ICML 2015, Lille, France, 6–11 July 2015. JMLR Workshop and Conference Proceedings, vol. 37, pp. 2152–2161. JMLR.org (2015)

48. Santoro, A., Bartunov, S., Botvinick, M., Wierstra, D., Lillicrap, T.P.: Meta-learning with memory-augmented neural networks. In: Balcan, M., Weinberger, K.Q. (eds.) Proceedings of the 33nd International Conference on Machine Learning, ICML 2016, New York City, NY, USA, 19–24 June 016. JMLR Workshop and Conference Proceedings, vol. 48, pp. 1842–1850. JMLR.org (2016)
49. Savinov, N., Dosovitskiy, A., Koltun, V.: Semi-parametric topological memory for navigation. In: 6th International Conference on Learning Representations, ICLR 2018, Vancouver, BC, Canada, 30 April–3 May 2018, Conference Track Proceedings. OpenReview.net (2018)
50. Savva, M., et al.: Habitat: A platform for embodied AI research. In: 2019 IEEE/CVF International Conference on Computer Vision, ICCV 2019, Seoul, Korea (South), 27 October– 2 November 2019, pp. 9338–9346. IEEE (2019)
51. Snell, J., Swersky, K., Zemel, R.S.: Prototypical networks for few-shot learning. In: Guyon, I., et al. (eds.) Advances in Neural Information Processing Systems 30: Annual Conference on Neural Information Processing Systems 2017, Long Beach, CA, USA, 4–9 December 2017, pp. 4077–4087 (2017)
52. Sung, F., Yang, Y., Zhang, L., Xiang, T., Torr, P.H.S., Hospedales, T.M.: Learning to compare: Relation network for few-shot learning. In: 2018 IEEE Conference on Computer Vision and Pattern Recognition, CVPR 2018, Salt Lake City, UT, USA, 18–22 June 2018, pp. 1199–1208. IEEE Computer Society (2018)
53. Thrun, S.: Learning metric-topological maps for indoor mobile robot navigation. Artif. Intell. **99**(1), 21–71 (1998)
54. Vinyals, O., Blundell, C., Lillicrap, T., Kavukcuoglu, K., Wierstra, D.: Matching networks for one shot learning. In: Lee, D.D., Sugiyama, M., von Luxburg, U., Guyon, I., Garnett, R. (eds.) Advances in Neural Information Processing Systems 29: Annual Conference on Neural Information Processing Systems 2016, Barcelona, Spain, 5–10 December 2016, pp. 3630–3638 (2016)
55. Wijmans, E., et al.: DD-PPO: learning near-perfect pointgoal navigators from 2.5 billion frames. In: 8th International Conference on Learning Representations, ICLR 2020, Addis Ababa, Ethiopia, 26–30 April 2020. OpenReview.net (2020)
56. Wortsman, M., Ehsani, K., Rastegari, M., Farhadi, A., Mottaghi, R.: Learning to learn how to learn: self-adaptive visual navigation using meta-learning. In: IEEE Conference on Computer Vision and Pattern Recognition, CVPR 2019, Long Beach, CA, USA, 16–20 June 2019, pp. 6750–6759 (2019)
57. Wu, Y., Wu, Y., Tamar, A., Russell, S.J., Gkioxari, G., Tian, Y.: Bayesian relational memory for semantic visual navigation. In: 2019 IEEE/CVF International Conference on Computer Vision, ICCV 2019, Seoul, Korea (South), 27 October–2 November 2019, pp. 2769–2779 (2019)
58. Xian, Y., Lorenz, T., Schiele, B., Akata, Z.: Feature generating networks for zero-shot learning. In: 2018 IEEE Conference on Computer Vision and Pattern Recognition, CVPR 2018, Salt Lake City, UT, USA, 18–22 June 2018, pp. 5542–5551 (2018)
59. Xian, Y., Sharma, S., Schiele, B., Akata, Z.: F-VAEGAN-D2: A feature generating framework for any-shot learning. In: IEEE Conference on Computer Vision and Pattern Recognition, CVPR 2019, Long Beach, CA, USA, 16–20 June 2019, pp. 10275–10284 (2019)
60. Yang, W., Wang, X., Farhadi, A., Gupta, A., Mottaghi, R.: Visual semantic navigation using scene priors. In: 7th International Conference on Learning Representations, ICLR 2019, New Orleans, LA, USA, 6–9 May 2019 (2019)

61. Ye, X., Yang, Y.: Hierarchical and partially observable goal-driven policy learning with goals relational graph. In: IEEE Conference on Computer Vision and Pattern Recognition, CVPR 2021, virtual, 19–25 June 2021, pp. 14101–14110 (2021)
62. Zhang, S., Song, X., Bai, Y., Li, W., Chu, Y., Jiang, S.: Hierarchical object-to-zone graph for object navigation. In: Proceedings of the IEEE/CVF International Conference on Computer Vision (ICCV), pp. 15130–15140, October 2021
63. Zhu, Y., et al.: Target-driven visual navigation in indoor scenes using deep reinforcement learning. In: 2017 IEEE International Conference on Robotics and Automation, ICRA 2017, Singapore, Singapore, 29 May– 3 June 2017, pp. 3357–3364 (2017)

Object Manipulation via Visual Target Localization

Kiana Ehsani[1(✉)], Ali Farhadi[2], Aniruddha Kembhavi[1,2], and Roozbeh Mottaghi[1,2]

[1] Allen Institute for AI, Seattle, USA
kianae@allenai.org
[2] University of Washington, Seattle, USA

Abstract. Object manipulation is a critical skill required for Embodied AI agents interacting with the world around them. Training agents to manipulate objects, poses many challenges. These include occlusion of the target object by the agent's arm, noisy object detection and localization, and the target frequently going out of view as the agent moves around in the scene. We propose Manipulation via Visual Object Location Estimation (m-VOLE), an approach that explores the environment in search for target objects, computes their 3D coordinates once they are located, and then continues to estimate their 3D locations even when the objects are not visible, thus robustly aiding the task of manipulating these objects throughout the episode. Our evaluations show a massive $3x$ improvement in success rate over a model that has access to the same sensory suite but is trained without the object location estimator, and our analysis shows that our agent is robust to noise in depth perception and agent localization. Importantly, our proposed approach relaxes several assumptions about idealized localization and perception that are commonly employed by recent works in navigation and manipulation – an important step towards training agents for object manipulation in the real world. Our code and data are available at prior.allenai.org/projects/m-vole.

Keywords: Object manipulation · Embodied AI · Mobile manipulation

1 Introduction

In recent years the computer vision community has made steady progress on a variety of Embodied AI tasks including navigation [1,4,6,58], object manipulation [11,45,46,50] and language-based tasks [2,9,14,42] within interactive worlds in simulation [12,24,39,49]. Performances of state-of-the-art systems on

---

Supplementary Information The online version contains supplementary material available at https://doi.org/10.1007/978-3-031-19842-7_19.

S. Avidan et al. (Eds.): ECCV 2022, LNCS 13699, pp. 321–337, 2022.
https://doi.org/10.1007/978-3-031-19842-7_19

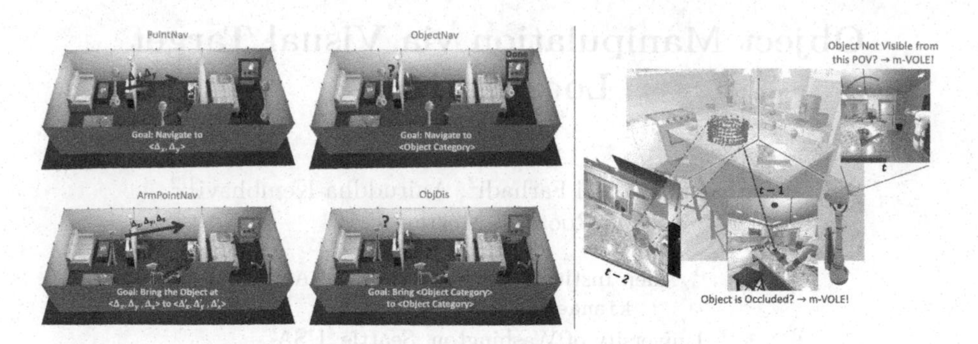

Fig. 1. We propose to solve a manipulation task, Object Displacement (ObjDis), where the goal is to bring a source object towards a destination object (e.g., bring a bowl to sink). In previous popular Embodied AI tasks shown in the left panel, either the goal is defined by the relative 3D coordinates (PointNav [39] and ArmPointNav [11]) or there is no manipulation involved (ObjectNav [4]). In contrast, we estimate the goal location from visual observations and manipulate objects across the scenes. The right panel shows an example that the agent robustly estimates the relative object location despite the occlusion by the arm and being out of view.

these tasks vary greatly depending on the complexity of the task, the assumptions made about the agent and environment, and the sensors employed by the agent. While robust and reliable methods have emerged for some of these tasks, developing generalizable and scalable solutions remains a topic of research. A challenging problem in this domain is visual object manipulation, a critical skill that enables the agents to interact with objects and change the world around them. Avoiding the collision of the arm with other objects in the scene, inferring the state of the scene using noisy visual observations, and planning an efficient path for the agent and its arm towards objects are a few of many interesting challenges in this domain.

One of the main obstacles in training capable and generalizable agents for visual object manipulation is the sparsity of training signals. Early works in visual navigation in the Embodied AI (EAI) research community (e.g., [58]) suffered from this training signal problem. To alleviate this issue, the EAI community provided the agent with a powerful suite of sensors. The first incarnations of the Point Goal Navigation (PointNav) task (navigating to a specified X-Y coordinate within an environment) relied on perfect sensory information that included GPS localization for the target along with compass and GPS sensors for the agent [39]. As a result, the agent was able to access accurate relative coordinates of its goal at every time step, which acted as a dense supervisory signal during training and inference. Under these assumptions, models with minimal inductive biases achieve near-perfect accuracy [47] in unseen test environments given enough training. Mirroring this success story, researchers have also been able to train effective agents for visual object manipulation [11].

While these are promising steps towards building embodied agents, the strong sensory assumptions employed by these models limit their applicability in the real world. Their task definition requires specifying the target via 3D coordinates, which can be extremely hard to determine, if not impossible, in indoor environments. Moreover, they rely on a perfect compass and GPS sensory information, which enables the agent to localize itself with respect to the target. In this paper, we take a step closer to a more realistic task definition for object manipulation by specifying the goal via a representative image of a particular object category (instead of their 3D coordinates) and propose a method to localize the target object without relying on any perfect localization sensory information.

We introduce Object Displacement (ObjDis), the task of bringing an object to a target location (e.g., *bring a bowl to the sink*). This involves searching for an object, navigating to it, picking it up, and placing it at the desired target location. Figure 1(left) contrasts ObjDis with other popular navigation and manipulation tasks. We propose Manipulation via Visual Object Location Estimation (m-VOLE), a model for ObjDis that continually estimates the 3D relative location of the target to the agent via visual observations and learns a policy to perform the desired manipulation task. m-VOLE predicts a segmentation mask for the objects of interest and leverages the depth sensor to estimate the relative 3D coordinates for these objects. However, target segmentations are not always available to the agent due to several reasons including: objects may be out of view due to the agent being in a different location, objects may be occluded by the arm of the agent, and masks may be unavailable due to imperfect object segmentation models; and these noisy observations of the target can lead to the failure of the agent. To alleviate these issues, m-VOLE aggregates estimates for the objects' 3D coordinates over time, leverages previous estimates when the object mask is unavailable, and can seamlessly re-localize the object in its coordinate frame when the agent observes it again – rendering the model robust to noise in movement and perception. Figure 1(right) shows a schematic of the agent's object localization in the presence of occlusion.

We conduct our experiments using the ManipulaTHOR [11] framework which provides visually rich, physics-enabled environments with a variety of objects. We show that:

(a) Our model achieves 3x success rate compared to a model that is trained without target localization but with an identical set of sensory suites.
(b) Our model is robust against noise in perception and agent movements compared to the baselines.
(c) Our method allows for zero-shot manipulation of novel objects in unseen environments.

2 Related Works

Robotic Manipulation. Manipulation and rearrangement of objects is a long-lasting problem in robotics [20,22,26]. Various approaches assume the full visibility of the environment and operate based on the assumption that a perception module provides the perfect observation of the environment [13,19,23,25]. In contrast to these approaches, we focus on the visual perception problem and solve the task when the agent has partial and noisy observations. Several approaches have been proposed (e.g., [8,27,53,55]) that address visual perception as well. However, the mentioned works focus on the tabletop rearrangement of objects. In contrast, we consider mobile manipulation of objects, which is a more general problem. Mobile manipulation has been explored in the literature as well. However, they typically use a single environment to develop and evaluate the models [5,33,34,44]. One of the important problems we address in this paper is the generalization to unseen environments and configurations.

Manipulation in Virtual Environments. Recently, various works have addressed the problem of object manipulation and rearrangement in virtual environments. These works typically abstract away grasping and, unlike the works mentioned above, mostly focus on visual perception for manipulation, learning-based planning, and generalization to unseen environments and objects. Robosuite [59], RLBench [21] and Meta-world [54] provide a set of benchmarks for tabletop manipulation. ManiSkill benchmark [31] built upon the Sapien framework [51] is designed to benchmark manipulation skills over diverse objects using a static robotic arm. In contrast to these works, we focus on object displacement (i.e., navigation and manipulation) in unseen scenes (e.g., a kitchen). [50] and [45] propose a set of tasks in the iGibson [40] and Habitat 2.0 [45] frameworks. However, the same environment and objects are used for train and test. In contrast, our focus is on generalization to unseen environments. ManipulaTHOR [11] is an object manipulation framework that employs a robotic arm and introduces a task and benchmark for mobile manipulation. Their task highly depends on various accurate sensory information, which renders the task unrealistic. In contrast to this work, we relax most supervisory signals to better mimic real-world scenarios.

Relaxing Supervisory Sensors. There have been some attempts to relax the assumptions about perfect sensory information using visual odometry for the navigation task [10,57]. While these are effective approaches for PointGoal navigation, the same approach does not apply to manipulation as they still require the goal's location in the agent's initial coordinate frame. In contrast, we have no access to the target or goal location at any time step.

Embodied Interactive Tasks. Navigation is a crucial skill for embodied agents. Various recent techniques have addressed the navigation problem [1,4,6,16,36,38,47,48,52,58]. These tasks mostly assume the environments are static, and objects do not move. We consider joint manipulation and navigation, which poses different challenges. [49,56] assume objects can move during navigation, but the set of manipulation actions is restricted to push actions. [2,30,42] propose instruction following to navigate and manipulate objects. They are designed either for static environments [2] or abstract object manipulation [30,42] (e.g., objects snap to the agent). Works such as [28,29] use auxiliary tasks to overcome the issues related to sparse training signals for navigation. We focus on manipulation, which deals with a different set of challenges. [43] uses high-level primitives such as `placeOnTop`, which uses privileged information to plan a path for the arm. Similarly, [12] uses `Go to Grasp (object)` action that moves the agent towards the object and grasps the object using privileged information provided by the environment. [35] uses `grab` ⟨`id`⟩ action, which ignores arm movements. [3,46] propose room rearrangement tasks using mobile robots. However, their object manipulation is unrealistic in that they assume a magic pointer abstraction, i.e., no arm movement is involved. In contrast, we manipulate objects using an arm, which introduces unique challenges.

3 Object Displacement

We introduce the task of Object Displacement (ObjDis), which requires an agent to navigate and manipulate objects in visually rich, cluttered environments. Given two object references, a source object O_S and a destination object O_D, the goal is to locate and pick up O_S and move it to the proximity of O_D (e.g., *bring an egg to a pot*). The objects of interest are specified to the agent via query images of an instance of each of the categories. Referring to objects via query images as opposed to object names enables us to task the agent with manipulating objects that it has not been trained for. Note that the query images *do not match* the appearance of objects within the scene but are canonical object images (we obtain the images from simulation, but they can also be obtained via other sources such as an image search engine). This enables the user to easily specify the task without the knowledge of object instances in the environment.

The ObjDis task consists of multiple implicit stages. To be successful, an agent must (1) explore its environment until it finds the source object, (2) navigate to it, (3) move its arm to the object so that its gripper may pick it up, (4) locate the destination object within the environment (5) navigate to this object and (6) place the object within its gripper in the proximity of the destination object. In visually rich and cluttered environments such as the ones present in AI2-THOR [24], this poses several challenges to the agent. *Firstly*, the agent must learn to move its body (navigate) as well as its arm (manipulate) effectively towards objects of interest to complete the desired task. *Secondly*, the agent must avoid collisions with other objects in the scene, which may occur with its body, its arm, or the source object once it has been picked up. *Thirdly*, the agent must be able to plan its actions over long horizons as it involves multiple objects, exploration, and manipulation. *Finally*, the agent must overcome noisy perception caused by frequent obstruction of its view due to its occluding arm, noisy depth sensors, and imperfect visual processing such as object detection.

We situate our experiments in AI2-THOR [24], a simulated set of environments built in Unity with a powerful physics engine, and adopt the agent provided by the ManipulaTHOR [11] framework. This agent has a rigid body with a 6DOF arm. The agent is capable of moving its body and arm in the environment while satisfying the kinematic constraints of the arm. Grasping is abstracted to a magnetic gripper (i.e., the object can be grasped if the gripper touches the object). While object grasping is a rich and challenging problem, using a magnetic gripper enables us to focus our efforts on other challenges, including exploration, navigation, arm manipulation, and long-horizon planning.

The action space for the agent consists of 11 actions. There are three agent movement actions (*MoveAhead*, *RotateRight*, and *RotateLeft*), two arm base movements (*MoveUp*, *MoveDown*), which move the base of the arm up and down with respect to the body of the agent, and six arm movements that move the gripper in x, y, z directions in agent's coordinate frame. Similar to [11], the agent movement actions move/rotate the agent's body by 0.2 m/45 °C, and the arm movements move the base of the arm or the gripper by 5 cm. The framework uses inverse kinematic to calculate the final position of each joint.

4 Manipulation via Visual Object Location Estimation

One of the main challenges of the ObjDis task is finding and localizing the object of interest in the scenarios that the object is out of the field of view, is not detected, or is occluded by the arm. We propose a model to find the target, estimate its location, keep track of the location over time, and plan a path to accomplish the task.

Agents tackling the task of ObjDis accept as input ego-centric RGB and Depth observations along with query images for the source and destination objects. They must choose one of 11 actions at each step and follow a long sequence of steps to succeed. In the absence of perfect perception and localization, this task proves quite hard to learn, and models fail to generalize to new environments (as shown in Sect. 5).

4.1 Estimating Relative 3D Object Location

We first explain how the relative location of the object of interest is estimated. This is not possible for an unobserved object because its location remains unknown. However, once the object comes into view (while exploring the environment using the policy described later in Sect. 4.3), the agent can start this estimation. Note that this is in contrast to the ArmPointNav [11] task where the groundtruth location of the target is provided at all time steps.

At time step t, if the desired objects are in view, one can generate their segmentation masks (we explain in Sect. 4.2 how we obtain these masks). We refer to these segmentations as M_S and M_D for the source and destination objects, respectively. Given a segmentation mask and the depth observation, the object's center in 3D is estimated by backprojecting the depth map within the segmentation mask into the 3D local coordinate of the agent. Formally, $\hat{d}_O = \pi(D[M_O], K)$, where $\hat{d}_O$ is the estimated 3D distance from the agent to the center of the object O in the agent's coordinate frame, M_O is the segmentation mask of the object in the current observation, $D[M_O]$ is the observed depth frame masked by the object's segmentation (i.e., depth values for the visible regions of the object), K is the intrinsic matrix of the camera, and π is a function that projects pixels into the 3D space based on the depth values and camera intrinsics.

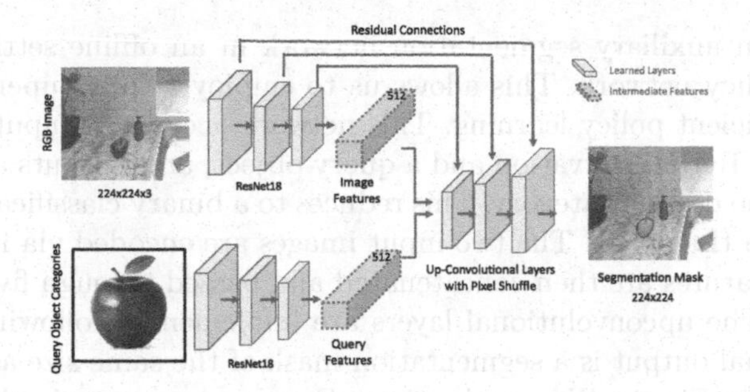

Fig. 2. Conditional Segmentation Architecture. We use a conditional segmentation model to estimate the segmentation mask. The network receives an RGB image and a query image (representing an object category) as input and outputs the segmentation mask for an instance of that category.

There are various sources of noise in object instance segmentation that make it challenging to have a reliable estimate of the 3D object location. First, segmentation models are far from perfect. They often have false positives and false negatives. Policies that rely on these masks or a quantity derived from these masks would not be able to produce a reliable sequence of actions. Second, the arm may obstruct the view of the agent as it moves in the scene – this causes the segmentation mask to periodically disappear. Third, the object might go outside the agent's camera frame after it is observed once and as a result not visible. To overcome these issues, we propose to aggregate this information over time.

To sequentially aggregate the relative coordinates of the object, a weighted average of the previous and the current estimated 3D distance is used. At each time step (after observing the target once), m-VOLE calculates the distance $\hat{d}_O$ in the agent's coordinate frame at time t. However, as the agent takes action, its coordinate frame keeps shifting. At each time step that the target object is visible, the agent re-localizes the goal in its current observation frame and can readjust the coordinates. To accurately convert all the past estimates to the current agent frame, we must keep track of the agent's relative location L_t^r with respect to its starting location. Since we aim to reduce the reliance on perfect sensing, we evaluate the effect of noise in the location estimation (see Sect. 5).

4.2 Conditional Segmentation

Estimating the 3D location of an observed object requires the agent to estimate its segmentation mask. We refer to this segmentation as 'conditional' since the goal is to obtain a mask for an object instance from the object category shown in the conditioning query image. Formulating the problem as a conditional segmentation problem as opposed to the traditional instance segmentation enables the network to focus on generating the mask for the target object[1], and the task reduces to a simpler task of category matching.

[1] We use *target object* to refer either the source or destination object.

We train an auxiliary segmentation network in an offline setting, independent of the policy network. This allows us to employ a fully supervised setting resulting in efficient policy learning. This network accepts as input two images, the ego-centric RGB observation and a query object, and outputs a mask for all objects from the desired category. This reduces to a binary classification problem at each pixel in the image. The two input images are encoded via ResNet18 [18] models. The features are then concatenated and passed through five upconvolutional layers. The upconvolutional layers are implemented following PixelShuffle [41]. The final output is a segmentation mask of the same size as the original image (details in Fig. 2). We use the Cross-Entropy loss to train this model.

We train and evaluate this model using an offline dataset collected from the train scenes in AI2-THOR [24], where we randomize object locations, textures, and scene lighting. We initialize the ResNet18 models with ImageNet pre-training and finetune them on our offline dataset. We compare this method with using a state-of-the-art instance segmentation model trained on the same data in Sect. 5.3.

4.3 Policy Network

Figure 3 shows a schematic of our proposed model m-VOLE. m-VOLE receives the RGBD observation I_t at the current timestep t, and the query images O_S and O_D. The conditional segmentation module estimates the segmentation masks for the target objects M_S and M_D and the visual object location estima-

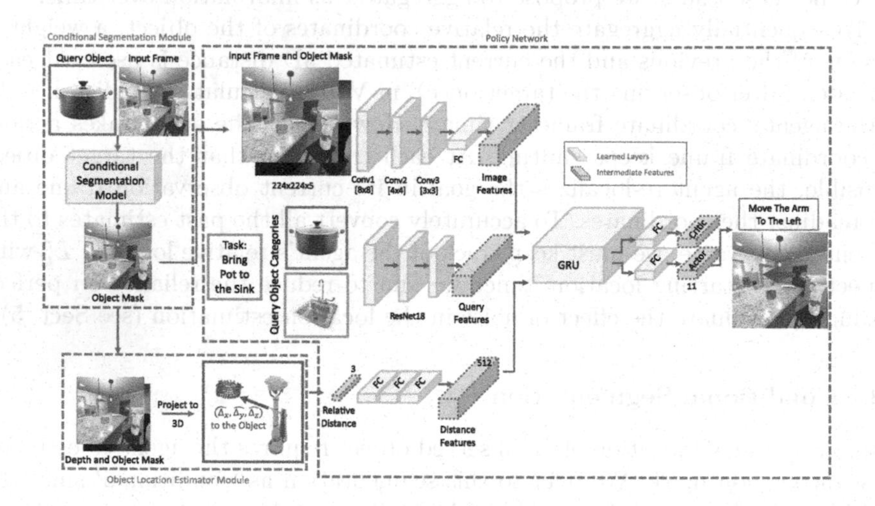

Fig. 3. Model architecture. Our model, m-VOLE, uses the RGBD observation and a canonical query image of the object to 1) predict the *target object*'s mask (Conditional Segmentation Module), 2) estimate the relative distance of the *target object* in agent's coordinate frame (Object Location Estimator Module), and 3) predict the next action (Policy Network). Note that we use *target object* to refer to either the source or destination object.

Table 1. Rewards. In addition to standard RL reward for embodied AI, we include reward components encouraging exploration and efficient manipulation.

Standard RL reward		Motivating exploration		Efficient manipulation	
Failed action	−0.03	Object observed	+1	Pick Up O_S	+5
Step	−0.01	Visit new state	+0.1	Δ distance to obj	$-\delta$
Episode success	+10				

tor calculates the relative object coordinates $\hat{d}_o$. The policy network then uses $I_t, O_S, O_D, M_S, M_D, \hat{d}_o$ to sample actions.

The RGBD observation and segmentation masks of the target objects in the current observation M_S, M_D are concatenated depth-wise and embedded using three convolutional layers. This visual encoding is combined with ResNet18 [18] (pre-trained on ImageNet [37]) features of the query images of target classes O_S, O_D and the embedding of object's relative location $\hat{d}_o$. The resulting feature vector is provided to an LSTM, and a final linear layer generates a distribution over the possible actions. We use DD-PPO [47] to train the model. Since our method does not have access to the object's location, and the object is not necessarily initially in sight (only in 13.9% of episodes the object is initially visible to the agent), our agent is required to explore the environment until the object is found. We use rewards shown in Table 1 to encourage exploration and more efficient manipulation. In the appendix, we ablate the effects of these rewards on the agent's performance.

5 Experiments

We present our experiments comparing m-VOLE with baselines (Sect. 5.1), a robustness analysis of our model with regards to noise in the agent's motions and sensors (Sect. 5.2) and ablations of m-VOLE's design choices (Sect. 5.3). Finally we evaluate m-VOLE for displacing novel objects, not used during training (Sect. 5.4).

Baselines: We consider the following baselines:

- **No Mask Baseline** – This model uses the RGBD observation and the query image as input and directly predicts a policy for completing the task. It does not have a segmentation module and does not have access to the agent's location relative to the starting point.
- **Mask Driven Model (MDM)** – This network uses a very similar architecture as our model with RGBD observation, query images, and the segmentation mask of the target objects as inputs. However, it does not have access to the agent's location relative to the starting point.
- **Location-aware Mask Driven Model (Loc-MDM)** – This baseline shares the same architecture as MDM. However, it also uses the agent's relative location from the start of the episode. The inputs available to Loc-MDM

Table 2. Quantitative Results. Rows (1)–(4) present the results with the predicted segmentation masks. Rows (5)–(8) present the results for models when provided with ground truth segmentation. Our model m-VOLE outperforms all the baselines across all metrics. The No Mask Baseline is simply unable to train well, in spite of repeated attempts by us to vary hyperparameters and other design choices.

Model	Segmentation	Additional input	PU	SR	SRwD
(1) No Mask Baseline	Prediction	N/A	0.0	0.0	0.0
(2) Mask Driven (MDM)	Prediction	N/A	16.6	1.62	0.09
(3) Loc-MDM	Prediction	Agent's Relative Loc	20.3	3.24	1.08
(4) m-VOLE (Ours)	Prediction	Agent's Relative Loc	**38.7**	**11.6**	**4.59**
(5) Mask Driven (MDM)	GT mask	N/A	50.5	17.1	8.74
(6) Loc-MDM	GT mask	Agent's Relative Loc	57.3	21.6	9.01
(7) ArmPointNav [11]	N/A	Compass + GPS	76.6	58.3	28.5
(8) m-VOLE (Ours)	GT mask	Agent's Relative Loc	**81.2**	**59.6**	**31.0**

are the same as our proposed model (m-VOLE). The difference is that Loc-MDM uses the agent's location naively, whereas m-VOLE uses it for target localization.

In addition to these baselines, we also train the model from [11] on our task ObjDis. Note that this model uses the perfect location of the agent and the target objects at each time step, so it is not a fair comparison to our model. However, it is a useful point of reference.

- **ArmPointNav** [11] – This method is the same architecture as the one introduced in [11] with the addition of the query images as input. The network is provided with the perfect location information for the agent and source and destination objects.

Dataset. We use the APND dataset proposed in [11] to define the tasks' configurations. APND consists of 12 object categories (Apple, Bread, Tomato, Lettuce, Pot, Mug, Potato, Pan, Egg, Spatula, Cup, and SoapBottle), 30 kitchen scenes, 130 agent initial locations per scene, and 400 initial locations per object-pair per room. For each task, we choose two objects, place them in two locations (randomly chosen from APND) and choose one as the source object O_S and the other as the destination O_D. We also randomize the initial location of the agent. We split the 30 scenes into 20 for training, 5 for validation, and 5 for testing. We generate 132 object pairs per training scene and 1600 pairs of initial locations of objects per training scene. We select a fixed set of 1100 tasks evenly distributed across scenes and object categories for validation and test. For more details on the dataset used for training the conditional segmentation model refer to the appendix.

Metrics. We use the same evaluation metrics as [11]: 1) Success rate **(SR)** - Fraction of episodes in which the agent successfully picks up the source object (O_S) and brings it to the destination object (O_D). 2) Success rate without disturbance **(SRwD)** - Fraction of episodes in which the task is successful, and the

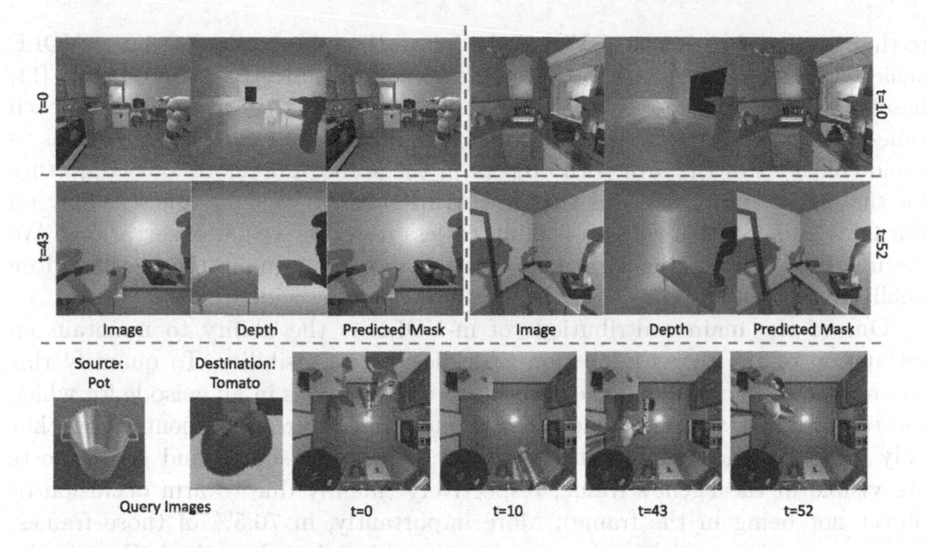

Fig. 4. Qualitative Results. The figure presents a successful episode of *Bringing a Pot to Tomato*. Despite the errors in Pot segmentation and with only a few pixels of the object being segmented, our agent successfully completes the task. Moreover, towards the end of the episode ($t = 52$), the pot in hand occludes the target object (tomato), but the agent is able to remember the target location despite the occlusion. Note that the topdown view (bottom row) is only shown for visualization purposes and is not an input to the network.

agent does not move other objects in the scene. 3) Pick up success rate **(PU)** - Fraction of episodes where the agent successfully picks up the object.

5.1 How Well Does m-VOLE Work?

Table 2 presents our primary quantitative findings on the test environments. Rows 1–4 provide results for m-VOLE compared to our baselines. Rows 5–8 evaluate these models when using ground truth segmentation masks in order to assess these models independent of the segmentation inaccuracies.

Our model m-VOLE, row 4, outperforms all of the baselines (rows 1–3). Directly learning a policy from input images results in a 0 success rate, showing the advantage of using a separate segmentation and target localization network. This result is despite trying various hyperparameters and designs to get this model off the ground. The improvement of the location-aware methods (rows 3 and 4) over the Mask-driven model (row 2) that does not encode relative location demonstrates that the agents benefit from encoding localization information. Our method m-VOLE provides a massive 3x improvement over the baseline in row 3 in terms of Success Rate (SR). This result shows that our approach that estimates the object's distance from the agent is quite effective.

Rows 5–8 evaluate models when employing ground truth segmentation masks. m-VOLE outperforms all baselines significantly, including Loc-MDM with access to the same information as m-VOLE. An interesting observation is that m-VOLE achieves better performance in unseen scenes compared to the ArmPointNav [11] baseline (row 7), which receives the relative location of the target object at each time step. Object Displacement requires attending to the visual observations to avoid collisions of the arm with objects in the scene. Therefore, our conjecture for the higher performance is that the ArmPointNav baseline relies heavily on the GPS and location sensors leading to less focus on visual observations. We discuss this further in the Appendix. Figure 4 shows our qualitative results. More qualitative results are presented in the supplementary video.

One of the main contributions of m-VOLE is the ability to maintain an estimation of the target's location regardless of its visibility. To quantify this contribution, we calculated the percentage of the frames in an episode for which the target is not visible during the inference time. Our experiments show that only in 43.9% and 10.5% of the observed frames the source and goal objects are visible in the agent's frame, respectively (mainly due to arm occlusion or object not being in the frame). More importantly, in 70.5% of those frames, the segmentation model fails to segment the object (misdetection). Despite the low percentage of object visibility, m-VOLE successfully completes the task by aggregating the temporal information.

5.2 How Robust Is m-VOLE to Noise?

There are two primary sources of noise that can impact our model, noise in the agent's movements and noise in depth perception. In the following experiments, we evaluate the effect of each one.

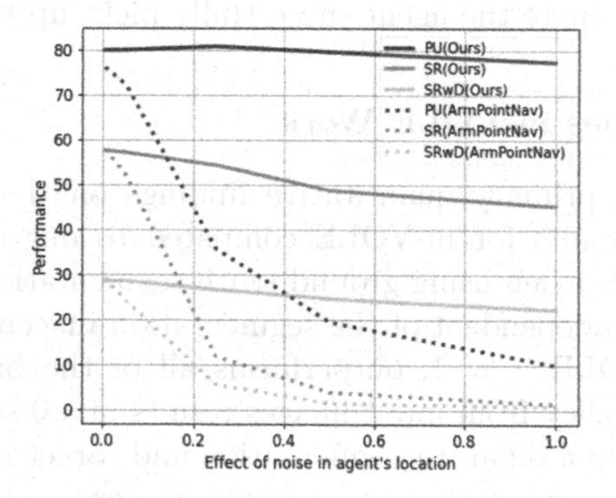

Fig. 5. Robustness to noise in agent's movements. We use the noise model from [32]. The x axis shows the noise multiplier and the y axis shows the performance on the Object Displacement task. Performance without any noise is shown at $x = 0$. $x = 1$ corresponds to a noise as large as the agent step size (20 cm).

Noise in Agent Movements. We use the motion model presented in [32], which models the noise in the traveled distance and angle of rotation. The details of the noise model can be found in the supplementary material. Figure 5 shows the robustness of our method to motion noise in comparison with ArmPointNav [11] modified to incorporate the same noise model. Our approach m-VOLE is far more robust, whereas ArmPointNav degrades significantly. We believe our method better leverages visual information to recover from noisy estimates.

Table 3. Robustness to noise in depth. Our model achieves the best performance on all metrics and the lowest relative performance drop compared to the baselines (shown in parentheses).

Model	PU	SR	SRwD
MDM @100M	20.5 (−40.0%)	3.5 (−71.5%)	2.07 (−72.6%)
Loc-MDM @114M	25.3 (−27.3%)	4.05 (−64.8%)	1.89 (−75.1%)
m-VOLE (Ours) @20M	**36.3 (−6.2%)**	**6.13 (−47.7%)**	**2.52 (−45.0%)**

Table 4. Instance segmentation ablation. We ablate the performance of our model using different instance segmentation methods. To do so we train a MaskRCNN [17] model on our data and show that our conditional segmentation helps our policy achieve better performance on the ObjDis task.

Model	PU	SR	SRwD
Our Policy with MaskRCNN [17] detection	31.3	8.7	3.87
Our Policy with Conditional Segmentation	**38.7**	**11.6**	**4.59**

Noise in Depth Perception. We use the Redwood depth noise model presented in [7]. Table 3 shows the performance of models in the presence of this noise. Note that models are trained with no noise, but inference takes place in the presence of noise. In this experiment, we are not only interested in evaluating models' absolute performance in the presence of noise but also in the relative degradation of each model. With this in mind, to have a fair comparison, we train all models until they reach the same success rate on the training set as m-VOLE. We then evaluate them with and without depth noise and calculate the relative performance drop in all metrics. As seen in Table 3, m-VOLE is more robust to noise in depth compared to our baselines (in absolute terms) and also shows a smaller relative performance degradation. We hypothesize that the aggregated information throughout the episode helps our model recover from the locally observed noise in depth.

5.3 Why Conditional Segmentation?

Simultaneously learning to segment objects and learning to plan is challenging. Hence, we isolate the segmentation branch of our network. Most approaches to ObjectNav [4] (navigating towards a specified object) do not use external object detectors and learn to detect and plan jointly, but that is not effective for our task. As we showed in Table 2, the baseline that does not use a separate segmentation network (row 1) generalizes poorly and achieves 0 success rate.

There are two advantages to our conditional segmentation method compared to the standard state-of-the-art detection/segmentation models such as MaskR-CNN [17]. First, our task is more straightforward since we are interested in simply segmenting an instance of the specified object category. Thus, we can afford to train a much smaller network. Second, as shown in Table 4, the model that uses conditional segmentation obtains better performance in all metrics. For this experiment, we use the MaskRCNN [17] model pre-trained on the LVIS dataset [15] and finetuned on AI2-THOR images. We obtain the highest confidence prediction for the target class from MaskRCNN and use that as the input to our policy. This is effective because there is typically just one object instance of the specified category in view.

5.4 Can m-VOLE Do Zero-Shot Manipulation?

So far, we have evaluated our model on generalization to unseen scenes. We evaluate whether the model can manipulate novel objects not used for training. This is a challenging task since the variation in object size and shape can result in different types of collisions and occlusions. For this evaluation, we train our model on six object categories (Apple, Bread, Tomato, Lettuce, Pot, Mug) and evaluate on the held-out classes (Potato, Pan, Egg, Spatula, Cup, SoapBottle). Note that our conditional segmentation model that provides the segmentation input to the policy has been trained on all categories. However, our policy has not been exposed to the test objects. Table 5 shows the results. Our model generalizes well to the novel categories, which is challenging as the novel objects' shapes and sizes can result in unseen patterns of collision.

Table 5. Zero-shot manipulation results.

Model	Object set	PU	SR	SRwD
Loc-MDM	NovelObj	9.52	0.9	0
m-VOLE	NovelObj	**26.2**	**5.24**	**3.33**
Loc-MDM	SeenObj	29.3	5.67	2.33
m-VOLE	SeenObj	49.3	18.7	10.7

6 Conclusion

We propose a method for visual object manipulation, where the goal is to displace an object between two locations in a scene. Our proposed approach learns to estimate target object location via aggregating noisy observations caused by missed detection, view occlusion by the arm, and noisy depth. We show that our approach provides a 3x improvement in success rate over a baseline without this auxiliary information, and it is more robust against noise in depth and agent movements.

References

1. Anderson, P., et al.: On evaluation of embodied navigation agents. arXiv (2018)
2. Anderson, P., et al.: Vision-and-language navigation: interpreting visually-grounded navigation instructions in real environments. In: CVPR (2018)
3. Batra, D., et al.: Rearrangement: a challenge for embodied AI. arXiv (2020)
4. Batra, D., et al.: ObjectNav revisited: on evaluation of embodied agents navigating to objects. arXiv (2020)
5. Chang, L.Y., Srinivasa, S.S., Pollard, N.S.: Planning pre-grasp manipulation for transport tasks. In: ICRA (2010)
6. Chaplot, D.S., Gandhi, D.P., Gupta, A., Salakhutdinov, R.R.: Object goal navigation using goal-oriented semantic exploration. In: NeurIPS (2020)
7. Choi, S., Zhou, Q.Y., Koltun, V.: Robust reconstruction of indoor scenes. In: CVPR (2015)
8. Danielczuk, M., Mousavian, A., Eppner, C., Fox, D.: Object rearrangement using learned implicit collision functions. In: ICRA (2021)
9. Das, A., Datta, S., Gkioxari, G., Lee, S., Parikh, D., Batra, D.: Embodied question answering. In: CVPR (2018)
10. Datta, S., Maksymets, O., Hoffman, J., Lee, S., Batra, D., Parikh, D.: Integrating egocentric localization for more realistic point-goal navigation agents. In: CoRL (2020)
11. Ehsani, K., et al.: ManipulaTHOR: a framework for visual object manipulation. In: CVPR (2021)
12. Gan, C., et al.: ThreeDWorld: a platform for interactive multi-modal physical simulation. In: NeurIPS (dataset track) (2021)
13. Garrett, C.R., Lozano-Perez, T., Kaelbling, L.P.: FFRob: leveraging symbolic planning for efficient task and motion planning. Int. J. Robot. Res. **37**, 104–136 (2018)
14. Gordon, D., et al.: IQA: visual question answering in interactive environments. In: CVPR (2018)
15. Gupta, A., Dollar, P., Girshick, R.: LVIS: a dataset for large vocabulary instance segmentation. In: CVPR (2019)
16. Gupta, S., Davidson, J., Levine, S., Sukthankar, R., Malik, J.: Cognitive mapping and planning for visual navigation. In: CVPR (2017)
17. He, K., Gkioxari, G., Dollár, P., Girshick, R.B.: Mask R-CNN (2017)
18. He, K., Zhang, X., Ren, S., Sun, J.: Deep residual learning for image recognition. In: CVPR (2016)
19. Huang, E., Jia, Z., Mason, M.T.: Large-scale multi-object rearrangement. In: ICRA (2019)
20. Hwang, Y.K., Ahuja, N.: Gross motion planning-a survey. ACM Comput. Surv. (CSUR) **24**, 219–291 (1992)
21. James, S., Ma, Z., Arrojo, D.R., Davison, A.J.: RLBench: the robot learning benchmark & learning environment. IEEE Robot. Autom. Lett. **5**, 3019–3026 (2020)

22. Kaelbling, L.P., Lozano-Pérez, T.: Integrated task and motion planning in belief space. Int. J. Robot. Res. **32**, 1194–1227 (2013)
23. King, J.E., Cognetti, M., Srinivasa, S.S.: Rearrangement planning using object-centric and robot-centric action spaces. In: ICRA (2016)
24. Kolve, E., et al.: AI2-THOR: an interactive 3D environment for visual AI. arXiv (2017)
25. Krontiris, A., Bekris, K.E.: Dealing with difficult instances of object rearrangement. In: RSS (2015)
26. Lozano-Pérez, T., Jones, J.L., Mazer, E., O'Donnell, P.A.: Task-level planning of pick-and-place robot motions. Computer **22**, 21–29 (1989)
27. Mahler, J., Goldberg, K.: Learning deep policies for robot bin picking by simulating robust grasping sequences. In: CoRL (2017)
28. Marza, P., Matignon, L., Simonin, O., Wolf, C.: Teaching agents how to map: spatial reasoning for multi-object navigation. arXiv (2021)
29. Mirowski, P.W., et al.: Learning to navigate in complex environments. In: ICLR (2017)
30. Misra, D., Bennett, A., Blukis, V., Niklasson, E., Shatkhin, M., Artzi, Y.: Mapping instructions to actions in 3D environments with visual goal prediction. In: EMNLP (2018)
31. Mu, T., et al.: ManiSkill: generalizable manipulation skill benchmark with large-scale demonstrations. In: NeurIPS (dataset track) (2021)
32. Murali, A., et al.: PyroBot: an open-source robotics framework for research and benchmarking. arXiv (2019)
33. Nedunuri, S., Prabhu, S., Moll, M., Chaudhuri, S., Kavraki, L.E.: SMT-based synthesis of integrated task and motion plans from plan outlines. In: ICRA (2014)
34. Nieuwenhuisen, M., et al.: Mobile bin picking with an anthropomorphic service robot. In: ICRA (2013)
35. Puig, X., et al.: Watch-and-help: a challenge for social perception and human-AI collaboration. In: International Conference on Learning Representations (2021). https://openreview.net/forum?id=w_7JMpGZRh0
36. Ramakrishnan, S.K., Al-Halah, Z., Grauman, K.: Occupancy anticipation for efficient exploration and navigation. In: Vedaldi, A., Bischof, H., Brox, T., Frahm, J.-M. (eds.) ECCV 2020. LNCS, vol. 12350, pp. 400–418. Springer, Cham (2020). https://doi.org/10.1007/978-3-030-58558-7_24
37. Russakovsky, O., et al.: ImageNet large scale visual recognition challenge. IJCV **115**, 211–252 (2015)
38. Savinov, N., Dosovitskiy, A., Koltun, V.: Semi-parametric topological memory for navigation. In: ICLR (2018)
39. Savva, M., et al.: Habitat: a platform for embodied AI research. In: ICCV (2019)
40. Shen, B., et al.: iGibson, a simulation environment for interactive tasks in large realistic scenes. In: IROS (2021)
41. Shi, W., et al.: Real-time single image and video super-resolution using an efficient sub-pixel convolutional neural network. In: CVPR (2016)
42. Shridhar, M., et al.: Alfred: a benchmark for interpreting grounded instructions for everyday tasks. In: CVPR (2020)
43. Srivastava, S., et al.: Behavior: benchmark for everyday household activities in virtual, interactive, and ecological environments. arXiv:abs/2108.03332 (2021)
44. Stilman, M., Schamburek, J.U., Kuffner, J., Asfour, T.: Manipulation planning among movable obstacles. In: ICRA (2007)
45. Szot, A., et al.: Habitat 2.0: training home assistants to rearrange their habitat. In: NeurIPS (2021)

46. Weihs, L., Deitke, M., Kembhavi, A., Mottaghi, R.: Visual room rearrangement. In: CVPR (2021)
47. Wijmans, E., et al.: DD-PPO: learning near-perfect PointGoal navigators from 2.5 billion frames. In: ICLR (2020)
48. Wortsman, M., Ehsani, K., Rastegari, M., Farhadi, A., Mottaghi, R.: Learning to learn how to learn: self-adaptive visual navigation using meta-learning. In: CVPR (2019)
49. Xia, F., et al.: Interactive Gibson benchmark: a benchmark for interactive navigation in cluttered environments. IEEE Robot. Autom. Lett. **5**, 713–720 (2020)
50. Xia, F., et al.: ReLMoGen: leveraging motion generation in reinforcement learning for mobile manipulation. In: ICRA (2021)
51. Xiang, F., et al.: SAPIEN: a simulated part-based interactive environment. In: CVPR (2020)
52. Yang, W., Wang, X., Farhadi, A., Gupta, A., Mottaghi, R.: Visual semantic navigation using scene priors. In: ICLR (2019)
53. Yen-Chen, L., Zeng, A., Song, S., Isola, P., Lin, T.Y.: Learning to see before learning to act: visual pre-training for manipulation. In: ICRA (2020)
54. Yu, T., et al.: Meta-world: a benchmark and evaluation for multi-task and meta reinforcement learning. In: CoRL (2019)
55. Zeng, A., et al.: Robotic pick-and-place of novel objects in clutter with multi-affordance grasping and cross-domain image matching. In: ICRA (2018)
56. Zeng, K.H., Weihs, L., Farhadi, A., Mottaghi, R.: Pushing it out of the way: interactive visual navigation. In: CVPR (2021)
57. Zhao, X., Agrawal, H., Batra, D., Schwing, A.G.: The surprising effectiveness of visual odometry techniques for embodied PointGoal navigation. arXiv (2021)
58. Zhu, Y., et al.: Target-driven visual navigation in indoor scenes using deep reinforcement learning. In: ICRA (2017)
59. Zhu, Y., Wong, J., Mandlekar, A., Martín-Martín, R.: Robosuite: a modular simulation framework and benchmark for robot learning. arXiv (2020)

MoDA: Map Style Transfer for Self-supervised Domain Adaptation of Embodied Agents

Eun Sun Lee, Junho Kim, SangWon Park, and Young Min Kim(✉)

Seoul National University, Seoul 08826, Republic of Korea
{eunsunlee,82magnolia,paulmoguri,youngmin}@snu.ac.kr

Abstract. We propose a domain adaptation method, MoDA, which adapts a pretrained embodied agent to a new, noisy environment without ground-truth supervision. Map-based memory provides important contextual information for visual navigation, and exhibits unique spatial structure mainly composed of flat walls and rectangular obstacles. Our adaptation approach encourages the inherent regularities on the estimated maps to guide the agent to overcome the prevalent domain discrepancy in a novel environment. Specifically, we propose an efficient learning curriculum to handle the visual and dynamics corruptions in an online manner, self-supervised with pseudo clean maps generated by style transfer networks. Because the map-based representation provides spatial knowledge for the agent's policy, our formulation can deploy the pretrained policy networks from simulators in a new setting. We evaluate MoDA in various practical scenarios and show that our proposed method quickly enhances the agent's performance in downstream tasks including localization, mapping, exploration, and point-goal navigation.

Keywords: Domain adaptation · Self-supervised learning · Image translation · Embodied agent · Visual navigation

1 Introduction

The absence of ground-truth labels is a critical bottleneck for training embodied agents in complex 3D world. A widely-used alternative is to train the agents in interactive simulators [41,44] which can load various 3D indoor scenes [7, 41,44]. Yet, when the agent optimized for a simulator is deployed in the real world, it fails to persist its performance due to the various unseen environmental noises [31]. The domain gap between simulators and the real world may be diminished by modeling the environmental noises [9,30], but it is not possible to obtain the correct noise model for the countless combinations of practical set-ups. Nonetheless, the visual agents collect an enormous amount of unlabelled data

Supplementary Information The online version contains supplementary material available at https://doi.org/10.1007/978-3-031-19842-7_20.

S. Avidan et al. (Eds.): ECCV 2022, LNCS 13699, pp. 338–354, 2022.
https://doi.org/10.1007/978-3-031-19842-7_20

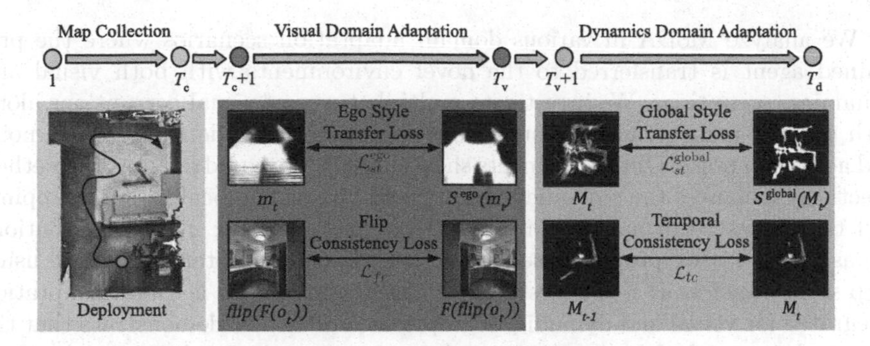

Fig. 1. MoDA suggests an integrated domain adaptation method for visual and dynamics corruptions, which fine-tunes an agent in an efficient learning curriculum. The agent collects a new map dataset to learn map style transfer networks (*green*). The agent is then trained for the new visual domain with ego style transfer loss and flip consistency loss (*yellow*). Lastly, the dynamics domain adaptation transfers the agent with global style transfer loss and temporal consistency loss (*blue*). (Color figure online)

from 3D scenes over spatial movement. If an agent can transfer its performance utilizing such data, the adaptation scheme can serve as a generic solution for an embodied agent to serve in diverse real environments.

Many studies in embodied agents have shown how the map-based memory aids an agent for robust visual navigation [10,32,48]. The agent aligns its egocentric visual observations to generate a top-down map representing the environment's layout. The allocentric understanding helps the agent to localize itself and plan for various navigation tasks efficiently. Furthermore, the map provides a domain-agnostic representation, disentangling the agent's perceptual module from its planning. In a scenario where the pretrained agent is transferred to a new, noisy environment with various visual and dynamics corruptions, we suggest a domain adaptation method which only fine-tunes the domain-agnostic map memory, rather than the overall pipeline of the embodied agent.

To compensate for the absence of ground-truth in the real world, we propose a self-supervised domain adaptation method, MoDA. The proposed method transfers the pretrained agent to the new, noisy environment by learning style transfer networks on maps. Our objective is to learn the structural regularities of indoor scenes from the clean maps obtained from the noiseless simulator. We then transfer the style onto the new maps generated amidst visual and dynamics noises. Our self-supervision loss compares the generated maps with the style-transferred maps. More specifically, our method suggests a learning curriculum as shown in Fig. 1. First, the agent is deployed to collect the noisy maps and learns two style transfer networks for egocentric and global maps. We then transfer the agent for the visual corruptions through the ego style loss, followed by compensating for the dynamics corruptions with the global style loss. To stabilize the training, we additionally encourage the flip consistency on RGB observations and temporal consistency over the agent's movement. MoDA provides an integrated self-supervised solution for both visual and dynamics corruptions and enables online adaptation in an environment where the ground-truth is unavailable.

We analyze MoDA in various domain adaptation scenarios where the pre-trained agent is transferred to the novel environments with both visual and dynamics corruptions. We investigate multiple types of visual corruptions along with the two main dynamics corruptions, which are the odometry sensor noise and actuation noise. Our experiments show that the proposed adaptation method effectively enhances the embodied agents' performance in localization, mapping, and the downstream navigation tasks. To summarize, our main contributions are as follows: i) we propose a self-supervised domain adaptation method using map style transfer, ii) we suggest an efficient curriculum to learn an adaptation integrated for visual and dynamics corruptions, and iii) we demonstrate that the proposed approach enhances the agent's performance in localization, mapping, and the final downstream visual navigation tasks in novel, unseen environments.

2 Related Work

In this section, we describe existing approaches for our main task, visual navigation and simulation-to-reality adaptation (Sim2Real), along with methods for image translation which is the key technique of our work.

Visual Navigation. The objective of visual navigation is to devise vision-based mapping and planning policies for solving a designated task. Classical approaches generate a map of the environment using SLAM techniques [5,18,19], and apply planning algorithms [26,35,46] using the generated map. On the other hand, recent learning-based approaches often train agents end-to-end with integrated mapping and planning [9–11,39]. These agents have shown competitive performance in a wide variety of tasks such as embodied QA [15] and goal-oriented navigation [10,39].

Maintaining a dedicated spatial memory unit is a key to such learning-based navigation agents. The spatial structure of the surrounding environment is often implicitly encoded with LSTM or GRU [14,27], or using graph structures that embed keyframes as graph nodes [11,13,16,40]. Nevertheless, map-based memories that depict spatial information on occupancy grid maps [8–10,39] efficiently aid embodied agents for tasks requiring long-range tracking and spatially-grounded planning. We mainly retain our focus on map-based spatial memory and propose a self-supervised task formulated on grid maps for effective domain adaptation.

Sim2Real. Simulators enable training an embodied agent with ground-truth poses or labels. While recent simulators [41,44] can realistically model the world to a certain degree, there are non-idealities in agent and object dynamics. More importantly, domain gap is inevitable when deploying an agent trained in simulation to real-world environments. As a result, agents trained on simulators often fail to generalize in real-word settings [31]. To alleviate Sim2Real gap, domain randomization [12,45] proposes to train the agent in various dynamics and visual simulations, which in turn allows the agent to observe a wide range of domains before actual deployment. Furthermore, representation learning techniques are used to improve the generalization performance in the context of embodied agent studies [17,20,25,42,43]. Alternatively, Hansen et al. [24] proposes to adapt the policy during real-world deployment, using self-supervised objectives such as inverse dynamics and rotation prediction. Recently, Lee et al. [36] introduced a self-supervised domain adaptation algorithm that is formulated upon occupancy maps, which showed performance enhancement in various deployment scenarios. However, Lee et al. [36] requires multiple agent round trips for successful adaptation, which limits the practical usage of the algorithm.

We compare MoDA against existing approaches for adapting agents to Sim2Real deployment and demonstrate that MoDA can perform effective adaptation without mandating fixed agent trajectories such as round trips.

Image-to-Image Translation. The goal of image-to-image translation is to transfer an image from a source domain into the style of target domain but to maintain its key contents. Early approaches such as Pix2Pix [29] propose to use generative adversarial networks (GANs) [22] for paired image-to-image translation. CycleGAN [50] aims to solve a more challenging problem of unpaired translation, where only a group of source and target domain images are provided for training. To accommodate for the lack of paired data, CycleGAN [50] proposes a novel cycle consistency loss that learns a forward and backward mapping simultaneously, leading to realistic image transfer. We leverage CycleGAN for transforming occupancy grid maps to the target domain, which allows for effective map generation under new, unseen environments. While recent advances in image-to-image translation enable high-resolution or multi-modal synthesis [28,47,51], we find that CycleGAN [50] is sufficient for transferring between occupancy grid maps that is not as diverse as real-world images.

3 Method

Given a pretrained agent from a noiseless simulator, MoDA transfers the agent to unseen, noisy environments with visual and dynamics corruptions. We first describe the overall pipeline of visual navigation with map-based memory in Sect. 3.1. Then Sect. 3.2 describes our map-to-map style transfer network which serves as the self-supervision signal. Lastly, Sect. 3.3 provides the learning curriculum of our online domain adaptation.

3.1 Visual Navigation with Spatial Map Memory

MoDA builds on conventional map-based navigation agents, where the action policy is planned based on map representation as shown in Fig. 2. The global map is estimated from the mapping and localization models. At each step, the RGB observation o_t is given as an input to the mapping model f_M which predicts the egocentric map m_t:

$$f_M(o_t) = m_t. \tag{1}$$

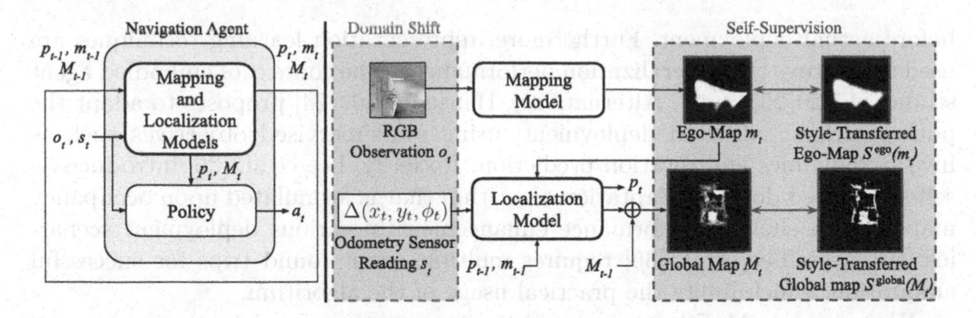

Fig. 2. The structure of map-based navigation agent decouples the mapping and localization models from the policy with the intermediate map memory (*left*). We suggests a domain adaptation method which transfers an agent to a shifted domain with visual and dynamics corruptions (*right*). Given the corrupted sensory inputs, o_t and s_t, our method only fine-tunes the agent's mapping f_M and localization f_L models with the self-supervision loss, encouraging the generated ego-map m_t and global map M_t to be similar to the style-transferred maps.

Given the odometry sensor measurement s_t, the localization model f_L predicts the 2D pose, $p_t = (x, y, \phi)$, where (x, y) indicates the 2D coordinate and ϕ denotes the 1D orientation. The localization model f_L is represented as

$$f_L(p_{t-1}, s_t, \{m_{t-1}, m_t\}) = p_t. \tag{2}$$

Note that the predicted egocentric maps from the previous and current timesteps $\{m_{t-1}, m_t\}$ are given to obtain the pose prediction $\Delta p_t = p_t - p_{t-1}$. Then the egocentric maps from the mapping model are transformed by the estimated poses from the localization model and accumulated as a global map M_t:

$$M_{t-1} \oplus \mathcal{T}_{p_t}(m_t) = M_t, \tag{3}$$

where $\oplus$ represents the fusion of 2D grid maps. The representation of transformation is simplified in Fig. 2. Then the policy module π_ψ plans an action $a_t = (u_x, u_y, u_\phi) \in \mathcal{A}$. The policy is derived from the sensory inputs o_t, s_t, the agent's current pose p_t and the global map M_t,

$$\pi_\psi(o_t, s_t, p_t, M_t) = a_t. \tag{4}$$

Nonetheless, many studies have shown that the policy is most dependent on the map-based memory M_t, which is useful for long-range or complex navigation, rather than the sensory inputs providing partial observability [4,23,48].

The agent performs various tasks including mapping and localization using the policy module trained in an ideal environment with ground-truth poses and maps. However, in realistic environments, the overall architecture is disturbed by two main types of corruption: visual and dynamics corruptions. The visual corruptions on RGB observations make it difficult to predict egocentric maps while the dynamics corruptions on odometry readings and actuation degrade the pose estimation. As the modular pipeline of map-based agents decouples the policy from mapping and localization models with the intermediate spatial memory, MoDA only fine-tunes the perceptual models, namely localization and mapping, to learn the perturbations in measurements and generate a domain-agnostic map for the subsequent policy.

Visual Corruptions. Visual corruptions affect the RGB observation, which is the input of the mapping model shown in Fig. 2. As a result, the egocentric perception of the embodied agent suffers from the domain discrepancy. While our adaptation does not assume any particular form of visual corruption, we test our adaptation in scenarios that properly represent the wide varieties of possible visual variations in the real world [12]. The tested scenarios are described in Sect. 4.

Dynamics Corruptions. Dynamics corruptions degrade the accuracy of the agent's pose estimated by the localization model, as shown in Fig. 2. The two main sources of dynamics corruptions are the actuation and odometry sensor noises. The actuation noise interrupts an agent from reaching the target location provided by the control commands. After an action, the agent's ground-truth movement $\Delta p_t = \Delta(x_t, y_t, \phi_t)$ is defined as

$$\Delta(x_t, y_t, \phi_t) = (u_x, u_y, u_\phi) + \epsilon_{\text{act}}, \tag{5}$$

where (u_x, u_y, u_ϕ) and ϵ_{act} indicate the intended action control and the actuation noise, respectively. Additionally, the odometry sensor noise ϵ_{sen} disturbs the agent from accurately perceiving its own movement. The final pose reading with the odometry sensor noise ϵ_{sen} is erroneously measured as

$$s_t = \Delta(x_t, y_t, \phi_t) + \epsilon_{\text{sen}}. \tag{6}$$

The actuation and odometry sensor noises are expected in all realistic settings and many studies present realistic models for both types of dynamics corruptions [12,21,33,41].

3.2 Unpaired Map-to-Map Translation Network

The main objective of our self-supervised domain adaptation is to translate maps observed from the new, noisy domain such that it recovers the noiseless structure in the absence of paired data. The neural network learns to capture the structural regularities of indoor scenes from the collection of ground-truth maps D_{gt} and translate the learned characteristics into the collection of noisy maps D_{noisy}. The set of ground-truth maps D_{gt} is collected from a noiseless simulator. When the pretrained agent is deployed in a new environment, it collects another set of map data D_{noisy} which is generated amidst visual and dynamics corruptions. We adopt the unpaired image-to-image translation network suggested in [50] on our maps. Since the formulation does not require one-to-one correspondences

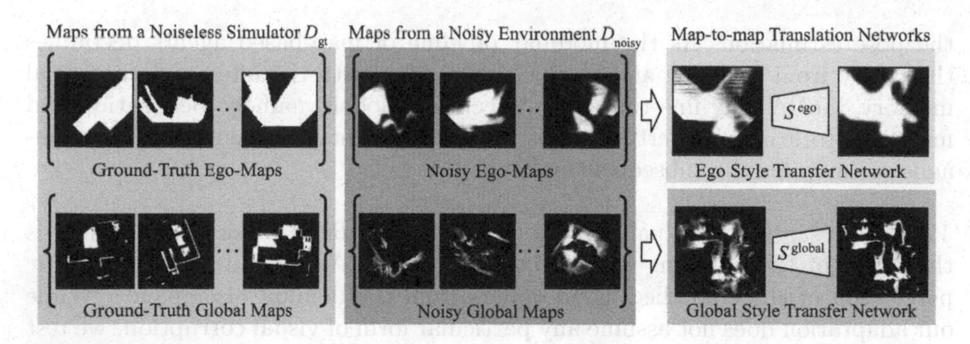

Fig. 3. Given the set of ground-truth maps D_{gt}(*grey*) obtained from the noiseless simulator, the set of noisy maps D_{noisy}(*green*) is collected by the pretrained agent deployed in a novel environment amidst visual and dynamics corruptions. We then learn two map-to-map translation networks for the egocentric map S^{ego} (*yellow*) and global map S^{global} (*blue*) to translate the maps in D_{noisy} into the style of the maps in D_{gt} (Color figure online)

between the two sets, our domain adaptation is completely self-supervised and successfully reasons about the stylistic difference between the two collections.

More specifically, we learn two map-to-map translation networks for the egocentric map S^{ego} and the global map S^{global} as shown in Fig. 3. The egocentric maps observe magnified views of the environment, and the style transfer network S^{ego} can help the agent to learn the detailed visual structure of the indoor scene to train the mapping model f_M against visual corruptions. The global style transfer network S^{global}, on the other hand, enforces a globally coherent structure over long-horizon navigation. The self-supervised loss on global maps restricts the localization model f_L from making erroneous predictions due to dynamics corruptions.

The success of the style transfer loss is tightly coupled with the clear structural regularities within the desired set of map data, which are represented as simple gray-scale images. We use different representations that contain more information in respective scenarios; the network learns over the explored area for egocentric maps, whereas the obstacle maps are used for global maps. While the two different representations are easily converted from one to the other given the pose of the agent, the obstacle maps are much more sparse than the explored area. Because the ego-maps cover smaller region, the obstacle maps cannot produce meaningful prior, consisting only about 6% of nonzero values on average within the images. On the other hand, the large overlapping regions in explored areas can be challenging in noisy global maps, whereas obstacle maps exhibit clearly distinguished structure.

3.3 Curriculum Learning for Domain Adaptation

We design a sequential curriculum using the hierarchical structure of map-based models as shown in Fig. 1. Once the agent collects the map data during its initial deployment up to time T_c, it then learns the two style transfer networks for the new environment. In the next step, our self-supervised adaptation method fine-tunes the perceptual module to robustly handle unknown corruptions. The style transfer loss on the egocentric maps provides signals to adapt the mapping model f_M against visual corruptions. Additionally, we enforce flip consistency on the RGB observations. Next, the global style transfer loss fine-tunes the localization model f_L up to time T_d against dynamics corruptions, in addition to the temporal consistency in the global map. The transferred agent then stably performs various navigation tasks in the new, noisy environment.

Visual Domain Adaptation. According to the learning curriculum, we first adapt the mapping model for unknown visual changes in the new environments. The style transfer network for the egocentric map S^{ego} converts the predicted egocentric map m_t into noiseless style. The ego style transfer loss minimizes the discrepancy between the two maps:

$$\mathcal{L}_{st}^{\text{ego}} = \sum_{t=T_c+1}^{T_v} \|m_t - S^{\text{ego}}(m_t)\|_2, \tag{7}$$

with T_v indicating the ending time of visual domain adaptation. In addition, we fine-tune the feature extractor F of the mapping model f_M along with a consistency loss to stabilize the training. The flip consistency loss $\mathcal{L}_{fc}$ assumes that the feature extractor should make consistent predictions over the flipped observations [3,37]. Specifically, when a horizontally flipped RGB observation, flip(o_t), and a non-flipped original observation, o_t, are given as inputs, the estimated egocentric maps should be equal but flipped. We, therefore, define our flip consistency loss as

$$\mathcal{L}_{fc} = \sum_{t=T_c+1}^{T_v} \|flip(F(o_t)) - F(flip(o_t))\|_2, \tag{8}$$

Together the visual domain adaptation transfers the mapping model of the pre-trained agent with the visual domain loss

$$\mathcal{L}_V = \lambda_{st}^{\text{ego}} \mathcal{L}_{st}^{\text{ego}} + \lambda_{fc}\mathcal{L}_{fc}. \tag{9}$$

The values for hyper-parameters are provided in the supplementary material.

Dynamics Domain Adaptation. In the next stage, we adapt the localization model f_L to the dynamics corruptions that are present in the new environment. The agent generates a global map over its trajectory and encodes the predicted pose information onto the map. Thus, by learning the structural priors from the style transfer network, the agent can inversely learn to estimate more accurate pose. Given the estimated and style-transferred global maps, the global style transfer loss, $\mathcal{L}_{st}^{\text{global}}$ is formulated as

$$\mathcal{L}_{st}^{\text{global}} = \sum_{t=T_v+1}^{T_d} \|M_t - S^{\text{global}}(M_t)\|_2. \tag{10}$$

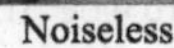

Fig. 4. Visualization of RGB observations with visual corruptions: speckle noise, low-lighting and large scene scale

where T_d denotes the ending time of dynamics domain adaptation. Moreover, we encourage the agent to generate consistent global maps over time. As the pose error accumulates over time, the global map generated in the earlier step encodes more accurate pose information over the global map generated in the later step. The temporal consistency loss $\mathcal{L}_{tc}$ compares the generated global map to the map from the previous time-step, and it is defined as

$$\mathcal{L}_{tc} = \sum_{t=T_v+1}^{T_d} \|M_t - M_{t-1}\|_2. \tag{11}$$

Therefore, the full objective of dynamics domain adaptation becomes

$$\mathcal{L}_{\mathcal{D}} = \lambda_{st}^{\text{global}} \mathcal{L}_{st}^{\text{global}} + \lambda_{tc} \mathcal{L}_{tc}, \tag{12}$$

completing the final stage of suggested learning curriculum.

4 Experiments

We show the validity of MoDA using the navigation agent from Active Neural SLAM [9] which is widely adapted for other navigation models [10,39]. Nonetheless, MoDA is applicable to various navigation agents with map-like memory. We use the Habitat simulator [41]. The pretrained agent is trained on the standard train split [41] of Gibson dataset [49] with ground-truth supervision. We split the unseen scenes of Gibson and Matterport3D [7] for adaptation and evaluation. The scenes for each split are listed in the supplementary material. MoDA is implemented using Pytorch [38] and accelerated with an RTX 2080 GPU.

Visual and Dynamics Corruptions. We evaluate the proposed method in three environments where visual and dynamics corruptions are present. Each environment is distinguished by the three visual variations: speckle noise, low-lighting, and scene scale change. Specifically, our experiments transfer the pretrained agent in three types of variation visualized in Fig. 4. First, we apply image quality degradation, which may be caused by the physical condition of the mounted camera. We generate low-quality RGB observations with additive speckle noises. The second type of perturbation is the low-lighting condition to reflect the common light variations in the real world. We show if our agent can be transferred to low-lighting scenes by adjusting the contrast and brightness of the input RGB image. Lastly, we evaluate if our pretrained agent from Gibson scenes can generalize to different scene scale by transferring the agent to the scenes in Matterport3D. While the Gibson scenes consist of scans collected from offices, the scenes in Matterport3D generally consist of large-scale homes. As the dynamics corruptions are seen in all realistic environments, we add the odometry sensor and actuation noise models to all three scenarios. In our experiment, we use the noise parameters generated from the actual physical deployment of LoCobot [1] in previous work [9,31] and draw from a Gaussian Mixture Model at each step.

Baselines. We extensively compare our agent, referred as "**MoDA**" to various baselines. "No adaptation (**NA**)" reflects the performance degradation of the un-adapted, pretrained model due to the domain gap. "Domain Randomization (**DR**)" adapts the pretrained model with ground-truth supervision in a randomized domain with various combinations of visual and dynamics corruptions. "Policy Adaptation during Deployment (**PAD**)" proposed by Hansen et al. [24] performs visual domain adaptation using an auxiliary task, namely rotation prediction. As the original method mainly targets visual adaptation, we further extend PAD for dynamics corruption by additionally training with our dynamics adaptation method. Lastly, "Global Map Consistency (**GMC**)" from Lee et al. [36] imposes global map consistency loss on round trip trajectories to adapt to dynamics corruptions. Further details of baseline implementation are explained in the supplementary material.

Tasks and Evaluation Metric. We report the transferred agent's localization and mapping performance in the new, noisy environment. For fair comparison, we evaluate each adaptation method on an identical set of trajectories obtained from the un-adapted agent in each environment. Following [6,34], localization is evaluated with the median translation (x, y) and rotation (ϕ) error. Mapping performance is evaluated with the mean squared error (MSE) of the generated occupancy grid maps compared to the ground-truth.

We also demonstrate our adapted agent's performance in downstream navigation tasks. Following Chaplot et al. [9], we report exploration performance using the explored area and explored area ratio after letting the agent to explore for a fixed number of steps. In addition, we report the collision ratio, which is the percentage of collisions from the agent's total steps. We include the collision ratio to distinguish simple random policy, which often results in undesirable collisions coupled with sliding along the walls, and eventually explore large areas. We further evaluate our agent on point-goal navigation(PointNav) as suggested in [2]. Here, we report the success rate and Success weighted by Path Length(SPL) [2].

We investigate MoDA in two settings, generalization and specialization, following [8]. In our main experiment, *generalization* adapts the agent in a set of unseen scenes with unknown noises (Sect. 4.1). The trained agent is then evaluated in a different set of novel scenes but with the same visual and dynamics corruptions. For *specialization*, the agent is fine-tuned and evaluated in the

Table 1. Generalization performance in the three new environments with speckle noise, low-lighting, and scene scale change. All three scenes contain dynamic corruptions not present in the original training setup of the pretrained agent

(a) Gibson Scenes Containing Speckle Noise and Dynamics Corruptions

	Pose		Map(MSE)		Exploration			PointNav	
	x, y(m)	$\theta(°)$	ego	global	area	ratio	collision	success	SPL
NA	0.16	15.02	1.11	0.32	28.45	0.82	0.40	0.12	0.10
DR	0.13	8.74	1.14	0.27	29.35	**0.88**	0.43	0.20	0.13
PAD	**0.04**	**1.08**	1.35	0.32	22.83	0.66	0.51	0.08	0.07
GMC	0.06	5.10	1.11	0.28	**29.73**	0.85	**0.35**	0.36	0.29
MoDA	**0.04**	2.61	**1.08**	**0.25**	28.63	0.82	0.36	**0.56**	**0.47**

(b) Gibson Scenes under Low-Lighting and Dynamics Corruptions

	Pose		Map(MSE)		Exploration			PointNav	
	x, y(m)	$\theta(°)$	ego	global	area	ratio	collision	success	SPL
NA	0.18	15.78	0.90	0.32	30.31	0.87	0.34	0.22	0.17
DR	0.14	7.60	0.96	0.26	29.95	0.88	0.41	0.20	0.14
PAD	**0.05**	**2.17**	1.02	0.27	26.88	0.78	0.37	0.22	0.18
GMC	0.06	5.39	0.90	0.28	**32.45**	**0.91**	0.32	0.46	0.37
MoDA	**0.05**	2.87	**0.89**	**0.25**	31.56	**0.91**	**0.26**	**0.56**	**0.45**

(c) MatterPort3D Scenes with Large Scene Scale and Dynamics Corruptions

	Pose		Map(MSE)		Exploration			PointNav	
	x, y(m)	$\theta(°)$	ego	global	area	ratio	collision	success	SPL
NA	0.17	14.42	1.07	0.39	52.16	0.45	0.44	0.04	0.02
DR	0.14	9.17	1.10	0.35	59.22	0.50	0.43	0.08	0.06
PAD	**0.05**	3.12	1.26	0.41	41.66	0.35	0.49	0.02	0.02
GMC	0.10	6.05	1.07	0.40	54.73	0.47	0.36	0.10	0.08
MoDA	**0.05**	**2.34**	**1.02**	**0.31**	**63.68**	**0.54**	**0.28**	**0.22**	**0.18**

same set of unseen scenes, but it starts from a different initial pose for evaluation (Sect. 4.2).

4.1 Task Adaptation to Noisy Environments: Generalization

In generalization, we test whether the adaptation method properly transfers the agent to the existing visual and dynamics corruptions, and avoids over-fitting the agents to the particular scenes. We compare our agent transferred to the three new environments with MoDA, as shown in Table 1. MoDA shows significant performance improvements in all three environments. By effectively fine-tuning pretrained agents using style-transfer in the map domain, MoDA improves localization and mapping, further enhancing the downstream task performance in pointNav and exploration. We additionally report the collision ratio to investigate our agent's stability in exploration. Compared to the baselines, agent trained by MoDA distinctively reports the low number of collision ratio during its exploration steps. While NA and DR agents also show competitive exploration performance, they also exhibit high collision ratios, which indicate instability. Further, while GMC shows competitive performance against MoDA, it mandates

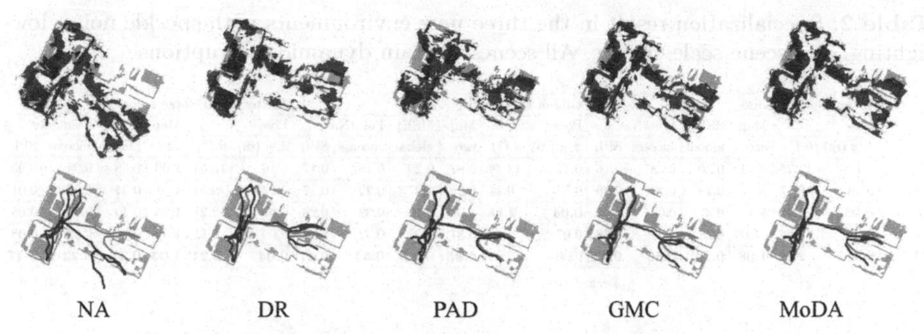

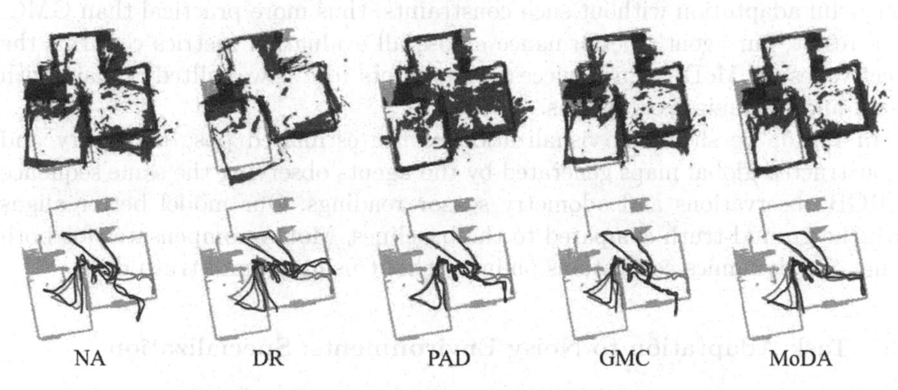

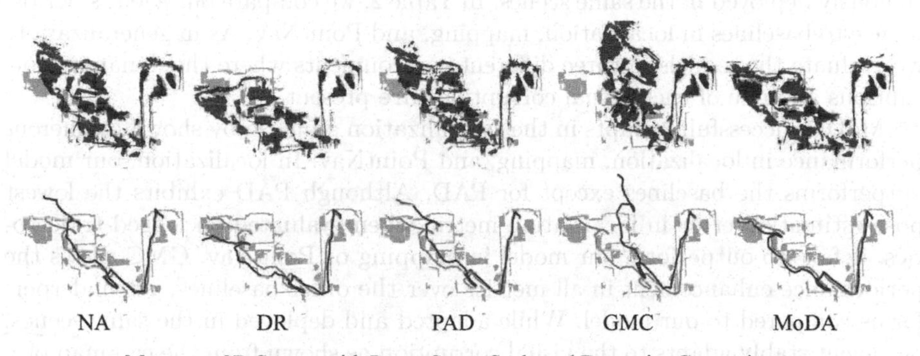

Fig. 5. Qualitative result of mapping (*top*) and localization (*bottom*) obtained from agents observing the identical sequence of RGB observations and odometry sensor readings. The reconstructed maps (*blue*) are aligned on the ground-truth maps (*grey*), and the estimated pose trajectories (*blue line*) are compared to the ground-truth trajectories (*red line*) (Color figure online)

Table 2. Specialization result in the three new environments with speckle noise, low-lighting, and scene scale change. All scenes contain dynamics corruptions

	Gibson speckle noise						Gibson low-lighting						Matterport3D large scene scale					
	Pose		Map (MSE)		PointNav		Pose		Map (MSE)		PointNav		Pose		Map (MSE)		PointNav	
	x, y (m)	θ (°)	ego	global	success	SPL	x, y (m)	θ (°)	ego	global	success	SPL	x, y (m)	θ (°)	ego	global	success	SPL
NA	0.17	15.25	1.10	0.30	0.18	0.16	0.17	14.44	0.87	0.29	0.18	0.17	0.16	15.04	1.09	0.38	0.04	0.03
DR	0.13	9.17	1.14	0.28	0.20	0.16	0.14	9.11	0.92	0.27	0.22	0.17	0.13	9.54	1.11	0.34	0.02	0.01
PAD	**0.03**	**1.09**	1.33	0.31	0.06	0.05	**0.04**	**2.88**	1.01	0.26	0.32	0.28	0.05	3.21	1.25	0.40	0.06	0.05
GMC	0.06	5.66	1.09	0.28	0.44	0.38	0.07	7.82	0.87	0.28	0.42	0.37	0.09	6.51	1.09	0.38	0.12	0.08
MoDA	0.04	2.54	**1.08**	**0.25**	**0.56**	**0.47**	0.06	3.24	**0.85**	**0.25**	**0.54**	**0.47**	**0.04**	**2.21**	**1.03**	**0.31**	**0.22**	**0.17**

the agent to navigate in round trip trajectories for adaptation. MoDA performs successful adaptation without such constraints, thus more practical than GMC. As a result, our agent's performance across all evaluation metrics confirms the effectiveness of MoDA which successfully adapts to a new, shifted domain with visual and dynamics corruptions.

In Fig. 5, we show the visualization of the estimated pose trajectory and reconstructed global maps generated by the agents observing the same sequence of RGB observations and odometry sensor readings. Our model better aligns with the ground-truth compared to the baselines. MoDA compensates for both visual and dynamics corruptions online without using ground-truth data.

4.2 Task Adaptation to Noisy Environments: Specialization

The specialization setting reflects the practical scenario where the agent is continuously deployed in the same scenes. In Table 2, we compare our agent's performance to baselines in localization, mapping, and PointNav. As in generalization, we evaluate the models in three different environments where the dynamics corruptions and one of each visual corruptions are present.

MoDA successfully adapts in the specialization scenario by showing coherent performance in localization, mapping, and PointNav. In localization, our model outperforms the baselines except for PAD. Although PAD exhibits the lowest pose estimation error in localization metric when evaluated on logged trajectories, it fails to outperform our model in mapping or PointNav. GMC shows the performance enhancement in all metrics over the other baselines, yet underperforms compared to our model. While adapted and deployed in the same scenes, our agent stably adapts to the visual corruption as shown from the ego-map prediction result in all three environments. We also observe that our agent estimates distinctively accurate poses, leading to generating an accurate global map. The performance improvement in mapping and localization, which is tightly coupled to the corruptions added to the observation, aids our agent to generate a more accurate intermediate spatial map. Then, the enhanced domain-agnostic representation further leads our model to outperform the baselines for the PointNav task. Therefore we conclude that MoDA provides a powerful, integrated adaption method for closing the domain gap from visual and dynamics corruptions present in a realistic environment.

Table 3. Ablation study on various loss functions employed in MoDA

	Pose		Map (MSE)		PointNav	
	x, y (m)	θ (°)	Ego	Global	Success	SPL
NA	0.16	15.02	1.11	0.32	0.12	0.10
NA + $\mathcal{L}_{fc}$	0.16	15.01	1.12	0.32	0.20	0.16
NA + $\mathcal{L}_{fc}$ + $\mathcal{L}_{st}^{\text{ego}}$	0.16	14.98	**1.08**	0.31	0.20	0.17
NA + $\mathcal{L}_{fc}$ + $\mathcal{L}_{st}^{\text{ego}}$ + $\mathcal{L}_{tc}$	**0.04**	3.16	**1.08**	0.25	0.42	0.34
NA + $\mathcal{L}_{fc}$ + $\mathcal{L}_{st}^{\text{ego}}$ + $\mathcal{L}_{tc}$ + $\mathcal{L}_{st}^{\text{global}}$	**0.04**	**2.61**	**1.08**	**0.25**	**0.56**	**0.47**

4.3 Ablation Study

In this section, we verify the effectiveness of each loss function in visual and dynamics domain adaptation. In Table 3, we report the adaptation result of the agent transferred to the environment with speckle noise visual corruption and dynamics corruptions. The overall experiment setup is the same as generalization. Beginning from the pretrained agent, referred as "NA", we gradually add the four losses mentioned in Sect. 3.3. We first train the pretrained agent only with the flip consistency loss $\mathcal{L}_{fc}$, which improves the agent's performance in localization and PointNav, but not in mapping. However, the model trained with both flip consistency loss $\mathcal{L}_{fc}$ and ego style transfer loss $\mathcal{L}_{st}^{\text{ego}}$ results in the performance enhancement in all evaluation metrics. This ablated model with $\mathcal{L}_{fc}$ and $\mathcal{L}_{st}^{\text{ego}}$ indicates that style transfer loss is more effective in adapting the agent during the visual domain adaptation stage. The adaptation for visual perturbations, which targets at predicting more accurate egocentric maps, also leads to the improvement in the subsequent evaluation tasks, localization, global map prediction and PointNav performance. The visually adapted agent is then trained with the temporal consistency loss $\mathcal{L}_{tc}$. The addition of $\mathcal{L}_{tc}$ effectively transfers the agent for the dynamics corruptions. Nonetheless, our full model, jointly trained with the global style transfer loss $\mathcal{L}_{st}^{\text{global}}$, exhibits the best performance in all metrics compared to all versions of ablated model.

5 Conclusion

In conclusion, we propose MoDA, a self-supervised domain adaptation method which provides an integrated solution to adapt the pretrained embodied agents to visual and dynamics corruptions. By transferring the noisy maps into clean-style maps, the agent can successfully adapt to the new environment with additional assistance with consistency loss. Our evaluation in generalization and specialization proves that MoDA is a powerful and practical domain adaptation method, showing its applicability in the noisy real world in the absence of ground-truth.

Acknowledgements. This work was partly supported by the National Research Foundation of Korea (NRF) grant funded by the Korea government(MSIT) (No. 2020R1C1C1008195), Creative-Pioneering Researchers Program through Seoul National University, and Institute of Information & communications Technology Planning & Evaluation (IITP) grant funded by the Korea government(MSIT) (No. 2021-0-02068, Artificial Intelligence Innovation Hub).

References

1. LoCoBot: an open source low cost robot (2021). http://www.locobot.org/
2. Anderson, P., et al.: On evaluation of embodied navigation agents. arXiv preprint arXiv:1807.06757 (2018)
3. Araslanov, N., Roth, S.: Self-supervised augmentation consistency for adapting semantic segmentation. In: Proceedings of the IEEE/CVF Conference on Computer Vision and Pattern Recognition (CVPR), pp. 15384–15394, June 2021
4. Bhatti, S., Desmaison, A., Miksik, O., Nardelli, N., Siddharth, N., Torr, P.H.: Playing doom with slam-augmented deep reinforcement learning. arXiv preprint arXiv:1612.00380 (2016)
5. Cadena, C., et al.: Past, present, and future of simultaneous localization and mapping: toward the robust-perception age. IEEE Trans. Robot. **32**(6), 1309–1332 (2016)
6. Campbell, D., Petersson, L., Kneip, L., Li, H.: Globally-optimal inlier set maximisation for camera pose and correspondence estimation. IEEE Trans. Pattern Anal. Mach. Intell. **42**(2), 328–342 (2018)
7. Chang, A., et al.: Matterport3D: learning from RGB-D data in indoor environments. arXiv preprint arXiv:1709.06158 (2017)
8. Chaplot, D.S., Dalal, M., Gupta, S., Malik, J., Salakhutdinov, R.R.: Seal: Self-supervised embodied active learning using exploration and 3D consistency. In: Advances in Neural Information Processing Systems 34 (2021)
9. Chaplot, D.S., Gandhi, D., Gupta, S., Gupta, A., Salakhutdinov, R.: Learning to explore using active neural slam. arXiv preprint arXiv:2004.05155 (2020)
10. Chaplot, D.S., Gandhi, D.P., Gupta, A., Salakhutdinov, R.R.: Object goal navigation using goal-oriented semantic exploration. In: Advances in Neural Information Processing Systems 33 (2020)
11. Chaplot, D.S., Salakhutdinov, R., Gupta, A., Gupta, S.: Neural topological slam for visual navigation. In: Proceedings of the IEEE/CVF Conference on Computer Vision and Pattern Recognition, pp. 12875–12884 (2020)
12. Chattopadhyay, P., Hoffman, J., Mottaghi, R., Kembhavi, A.: RobustNav: towards benchmarking robustness in embodied navigation. arXiv:abs/2106.04531 (2021)
13. Chen, K., et al.: A behavioral approach to visual navigation with graph localization networks. arXiv preprint arXiv:1903.00445 (2019)
14. Chung, J., Gulcehre, C., Cho, K., Bengio, Y.: Empirical evaluation of gated recurrent neural networks on sequence modeling (2014)
15. Das, A., Datta, S., Gkioxari, G., Lee, S., Parikh, D., Batra, D.: Embodied question answering. In: Proceedings of the IEEE Conference on Computer Vision and Pattern Recognition, pp. 1–10 (2018)
16. Deng, Z., Narasimhan, K., Russakovsky, O.: Evolving graphical planner: contextual global planning for vision-and-language navigation. arXiv preprint arXiv:2007.05655 (2020)

17. Du, Y., Gan, C., Isola, P.: Curious representation learning for embodied intelligence. In: Proceedings of the IEEE/CVF International Conference on Computer Vision, pp. 10408–10417 (2021)
18. Durrant-Whyte, H., Bailey, T.: Simultaneous localization and mapping: Part I. IEEE Robot. Autom. Mag. **13**(2), 99–110 (2006)
19. Engel, J., Schöps, T., Cremers, D.: LSD-SLAM: large-scale direct monocular SLAM. In: Fleet, D., Pajdla, T., Schiele, B., Tuytelaars, T. (eds.) ECCV 2014. LNCS, vol. 8690, pp. 834–849. Springer, Cham (2014). https://doi.org/10.1007/978-3-319-10605-2_54
20. Fan, L., Wang, G., Huang, D.A., Yu, Z., Fei-Fei, L., Zhu, Y., Anandkumar, A.: Secant: Self-expert cloning for zero-shot generalization of visual policies. arXiv preprint arXiv:2106.09678 (2021)
21. Gonçalves, J., Lima, J., Oliveira, H., Costa, P.: Sensor and actuator modeling of a realistic wheeled mobile robot simulator. In: 2008 IEEE International Conference on Emerging Technologies and Factory Automation, pp. 980–985. IEEE (2008)
22. Goodfellow, I., et al.: Generative adversarial nets. In: Ghahramani, Z., Welling, M., Cortes, C., Lawrence, N., Weinberger, K.Q. (eds.) Advances in Neural Information Processing Systems, vol. 27. Curran Associates, Inc. (2014). https://proceedings.neurips.cc/paper/2014/file/5ca3e9b122f61f8f06494c97b1afccf3-Paper.pdf
23. Gupta, S., Davidson, J., Levine, S., Sukthankar, R., Malik, J.: Cognitive mapping and planning for visual navigation. In: Proceedings of the IEEE Conference on Computer Vision and Pattern Recognition, pp. 2616–2625 (2017)
24. Hansen, N., et al.: Self-supervised policy adaptation during deployment. arXiv preprint arXiv:2007.04309 (2020)
25. Hansen, N., Wang, X.: Generalization in reinforcement learning by soft data augmentation. In: 2021 IEEE International Conference on Robotics and Automation (ICRA), pp. 13611–13617. IEEE (2021)
26. Hart, P., Nilsson, N., Raphael, B.: A formal basis for the heuristic determination of minimum cost paths. IEEE Trans. Syst. Sci. Cybern. **4**(2), 100–107 (1968). https://doi.org/10.1109/tssc.1968.300136
27. Hochreiter, S., Schmidhuber, J.: Long short-term memory. Neural Comput. **9**(8), 1735–1780 (1997)
28. Huang, X., Liu, M.-Y., Belongie, S., Kautz, J.: Multimodal unsupervised image-to-image translation. In: Ferrari, V., Hebert, M., Sminchisescu, C., Weiss, Y. (eds.) ECCV 2018. LNCS, vol. 11207, pp. 179–196. Springer, Cham (2018). https://doi.org/10.1007/978-3-030-01219-9_11
29. Isola, P., Zhu, J.Y., Zhou, T., Efros, A.A.: Image-to-image translation with conditional adversarial networks. In: CVPR (2017)
30. Kadian, A., et al.: Are we making real progress in simulated environments? Measuring the Sim2Real gap in embodied visual navigation (2019)
31. Kadian, A., et al.: Sim2real predictivity: does evaluation in simulation predict real-world performance? IEEE Robot. Autom. Lett. **5**(4), 6670–6677 (2020)
32. Karkus, P., Cai, S., Hsu, D.: Differentiable slam-net: Learning particle slam for visual navigation. In: Proceedings of the IEEE/CVF Conference on Computer Vision and Pattern Recognition, pp. 2815–2825 (2021)
33. Khosla, P.K.: Categorization of parameters in the dynamic robot model. IEEE Trans. Robot. Autom. **5**(3), 261–268 (1989)
34. Kim, J., Choi, C., Jang, H., Kim, Y.M.: PICCOLO: point cloud-centric omnidirectional localization. In: Proceedings of the IEEE/CVF International Conference on Computer Vision (ICCV), pp. 3313–3323, October 2021

35. Koenig, S., Likhachev, M.: D* lite. In: AAAI/IAAI, vol. 15 (2002)
36. Lee, E.S., Kim, J., Kim, Y.M.: Self-supervised domain adaptation for visual navigation with global map consistency. In: Proceedings of the IEEE/CVF Winter Conference on Applications of Computer Vision (WACV), pp. 1707–1716, January 2022
37. Li, B., Hu, M., Wang, S., Wang, L., Gong, X.: Self-supervised visual-LiDAR odometry with flip consistency. In: Proceedings of the IEEE/CVF Winter Conference on Applications of Computer Vision, pp. 3844–3852 (2021)
38. Paszke, A., et al.: PyTorch: an imperative style, high-performance deep learning library. In: Advances in Neural Information Processing Systems, vol. 32, pp. 8026–8037 (2019)
39. Ramakrishnan, S.K., Al-Halah, Z., Grauman, K.: Occupancy anticipation for efficient exploration and navigation. In: Vedaldi, A., Bischof, H., Brox, T., Frahm, J.-M. (eds.) ECCV 2020. LNCS, vol. 12350, pp. 400–418. Springer, Cham (2020). https://doi.org/10.1007/978-3-030-58558-7_24
40. Savinov, N., Dosovitskiy, A., Koltun, V.: Semi-parametric topological memory for navigation. arXiv preprint arXiv:1803.00653 (2018)
41. Savva, M., et al..: Habitat: a platform for Embodied AI Research. In: Proceedings of the IEEE/CVF International Conference on Computer Vision (ICCV) (2019)
42. Shah, R., Kumar, V.: RRL: Resnet as representation for reinforcement learning. arXiv preprint arXiv:2107.03380 (2021)
43. Srinivas, A., Laskin, M., Abbeel, P.: CURL: contrastive unsupervised representations for reinforcement learning. arXiv preprint arXiv:2004.04136 (2020)
44. Szot, A., et al.: Habitat 2.0: training home assistants to rearrange their habitat. arXiv preprint arXiv:2106.14405 (2021)
45. Tobin, J., Fong, R., Ray, A., Schneider, J., Zaremba, W., Abbeel, P.: Domain randomization for transferring deep neural networks from simulation to the real world. In: 2017 IEEE/RSJ International Conference on Intelligent Robots and Systems (IROS), pp. 23–30 (2017). https://doi.org/10.1109/IROS.2017.8202133
46. Wang, H., Yu, Y., Yuan, Q.: Application of Dijkstra algorithm in robot path-planning. In: 2011 Second International Conference on Mechanic Automation and Control Engineering, pp. 1067–1069. IEEE (2011)
47. Wang, T.C., Liu, M.Y., Zhu, J.Y., Tao, A., Kautz, J., Catanzaro, B.: High-resolution image synthesis and semantic manipulation with conditional GANs. In: Proceedings of the IEEE Conference on Computer Vision and Pattern Recognition (2018)
48. Wani, S., Patel, S., Jain, U., Chang, A.X., Savva, M.: MultiON: benchmarking semantic map memory using multi-object navigation. arXiv preprint arXiv:2012.03912 (2020)
49. Xia, F., Zamir, A.R., He, Z., Sax, A., Malik, J., Savarese, S.: Gibson env: real-world perception for embodied agents. In: Proceedings of the IEEE Conference on Computer Vision and Pattern Recognition, pp. 9068–9079 (2018)
50. Zhu, J.Y., Park, T., Isola, P., Efros, A.A.: Unpaired image-to-image translation using cycle-consistent adversarial networks. In: 2017 IEEE International Conference on Computer Vision (ICCV) (2017)
51. Zhu, J.Y., et al.: Toward multimodal image-to-image translation. In: Guyon, I., et al.(eds.) Advances in Neural Information Processing Systems 30, pp. 465–476. Curran Associates, Inc. (2017). http://papers.nips.cc/paper/6650-toward-multimodal-image-to-image-translation.pdf

Housekeep: Tidying Virtual Households Using Commonsense Reasoning

Yash Kant[1,2(✉)], Arun Ramachandran[2], Sriram Yenamandra[2], Igor Gilitschenski[1], Dhruv Batra[2,3], Andrew Szot[2], and Harsh Agrawal[2]

[1] University of Toronto, Toronto, Canada
ysh.kant@gmail.com
[2] Georgia Tech, Atlanta, USA
[3] Meta AI, Menlo Park, USA

Abstract. We introduce **Housekeep**, a benchmark to evaluate commonsense reasoning in the home for embodied AI. In Housekeep, an embodied agent must tidy a house by rearranging misplaced objects *without explicit instructions specifying which objects need to be rearranged.* Instead, the agent must learn from and is evaluated against human preferences of which objects *belong* where in a tidy house. Specifically, we collect a dataset of where humans typically place objects in tidy and untidy houses constituting 1799 objects, 268 object categories, 585 placements, and 105 rooms. Next, we propose a modular baseline approach for Housekeep that integrates planning, exploration, and navigation. It leverages a fine-tuned large language model (LLM) trained on an internet text corpus for effective planning. We find that our baseline planner generalizes to some extent when rearranging objects in unknown environments. See our webpage for code, data and more details: https://yashkant.github.io/housekeep/.

1 Introduction

Imagine your house after a big party: there are dirty dishes on the dining table, cups left on the couch, and maybe a board game lying on the coffee table. Wouldn't it be nice for a household robot to clean up the house *without needing explicit instructions specifying which objects are to be rearranged*?

Building AI reasoning systems that can perform such housekeeping tasks is an important scientific goal that has seen a lot of recent interest from the embodied AI community. The community has recently tackled various problems such as navigation [3,7,21,33,46,69], interaction and manipulation [19,64], instruction following [4,62], and embodied question answering [17,22,71]. Each of these

Y. Kant—Work done partially when visiting Georgia Tech.
A. Szot and H. Agrawal—Equal Contribution.

Supplementary Information The online version contains supplementary material available at https://doi.org/10.1007/978-3-031-19842-7_21.

S. Avidan et al. (Eds.): ECCV 2022, LNCS 13699, pp. 355–373, 2022.
https://doi.org/10.1007/978-3-031-19842-7_21

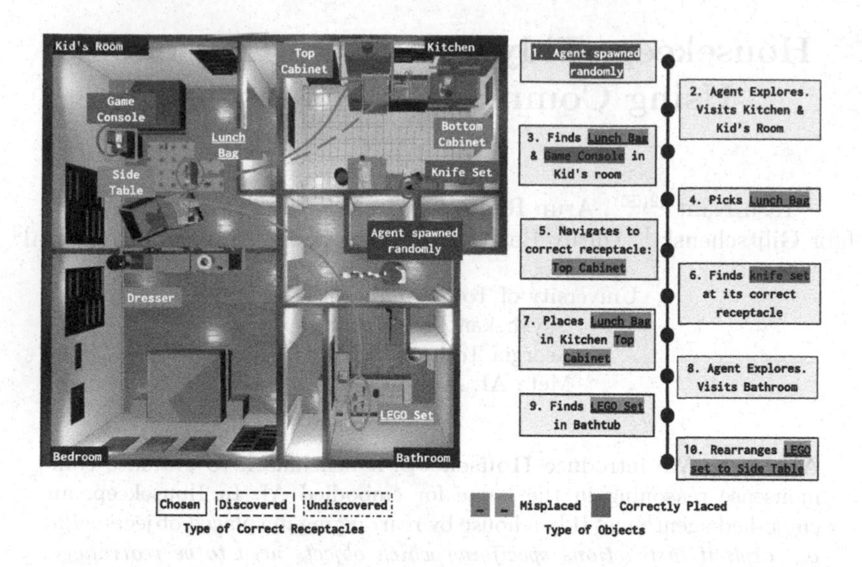

Fig. 1. In Housekeep, an agent is spawned in an untidy environment and tasked with rearranging objects to suitable locations without explicit instructions. The agent explores the scene and discovers misplaced objects, correctly placed objects, and receptacles where objects belong. The agent rearranges a misplaced object (like a lunch box on the floor in the kid's room) to a better receptacle like the top cabinet in the kitchen.

tasks defines a goal, e.g. navigating to a given location, moving objects to correct locations, or answering a question correctly. However, defining a goal for tidying a messy house is more tedious – one will have to write down a rule for where every object can or cannot be kept. Previous works in semantic reasoning frameworks for physical and relational commonsense [1,9,10,16,38,39] are often limited to specific settings (*e.g.* evaluating multi-relational embeddings) without instantiating these tasks in a physically plausible scenario, or by not capturing the full context of a complete household (*e.g.* table-top organization). We believe the time may be right to bridge the gap between the above two lines of research.

We introduce the Housekeep task to benchmark the ability of embodied AI agents to use physical commonsense reasoning and infer rearrangement goals that mimic human-preferred placements of objects in indoor environments. Figure 1 illustrates our task, where the Fetch robot is randomly spawned in an unknown house that contains unseen objects. Without explicit instructions, the agent must then discover objects placed in the house, classify the misplaced ones (LEGO set and lunch bag in Fig. 1), and finally rearrange them to one of many suitable receptacles (matching color-coded square). We collect a dataset of human preferences of object placements in tidy and untidy homes and use this dataset for: a) generating semantically meaningful initializations of unclean houses, and b) defining evaluation criteria for what constitutes a clean house. This dataset contains rearrangement preferences for 1799 objects, in 585 placements, in 105 rooms, constituting 1500+ hours of effort from 372 total annotators with 268 object categories curated from the Amazon-Berkeley [29], YCB objects [12],

Google Scanned Objects [53], and iGibson [61] datasets. Housekeep evaluates how an agent is able to rearrange novel objects not seen during training.

Housekeep is a challenging task for several reasons. First, agents need to reason about the correct placement of *novel* objects. Second, agents in Housekeep must operate in unseen environments using only egocentric visual observations. In the absence of any goal specification, the agent must *explore* areas that get cluttered frequently (*e.g.* coffee table, kitchen counter) for discovering potentially misplaced objects, and also find their suitable receptacles. Finally, since the environment is partially observable, the agent must continuously re-plan for when and where to rearrange objects via commonsense reasoning. For instance, on discovering a toy on the coffee table in the living room, the agent may choose to not rearrange it immediately if it hasn't discovered a more suitable receptacle such as the closet in the kid's room yet. The agent also has to reason about multiple potentially correct receptacles for any given object. For example, a toy could go in the closet in the master bedroom or in the kid's room.

We propose a modular baseline and demonstrate that embodied (physical) commonsense extracted from large language models (LLMs) [11,40] or traditional GloVe [49] vectors serves as an effective planner. Specifically, we find that finetuning these embeddings on a subset of human preferences generalizes well, and helps to reason about correct rearrangements for novel objects never seen during training. We integrate this planning module into a hierarchical policy that coordinates navigation, exploration, and planning as a baseline approach to Housekeep. Our hierarchical approach with the aid of few perfect sensors achieves an object success rate of 0.23 for unseen (versus 0.30 on seen objects). We also qualitatively analyze different failure cases of our baseline.

2 Related Work

Embodied AI Tasks. In recent times, we have seen a proliferation of Embodied AI tasks. Benchmarks on indoor navigation use point-goal specification [24,60], object-goal [7,69], room navigation [46], and language-guided navigation [4,66]. Some interactive tasks study the agent's ability to follow natural language instruction such as ALFRED [62] and TEACh [48] while others focus on rearranging objects following a geometric goal or predicate based specification [21,63,64,70]. [6] provides a summary of rearrangement tasks. All these tasks require an explicit goal specification lifting the burden of learning semantic compatibility of objects and their locations in the house from the agent. In contrast, in this work, we argue that agents shouldn't require an explicit goal specification to perform household tasks such as tidying up the house. Instead, it should use its common sense knowledge to infer the human-preferred goal state.

Capturing Human Preferences. Several works (summarized in Appendix A) in robotics model human preferences for assistive robots. Some [31] looked at furniture rearrangement based on surrounding human activities (*e.g.* standing by the kitchen shelf) while others [1,32] looked at table-top or a shelf rearrangement conditioned on a user. We differ from these works because we are interested in tidying up *entire houses* instead of a particular shelf or a table-top. In addition, the agent needs to operate with partial observations, and generalize to unseen environments and object types. [65] comes closest to our work. They learn a spatial model of object placements in a tidy environment. Our benchmark has a larger scale (1799 objects spanning 268 categories vs $\leq$55 object instances; 100+ room configurations vs 1 scene in [65]). Our benchmark also tests generalization to *unseen* objects, utilizing a dataset of human preferences instead of learning from a small set of tidy house instances. Dealing with unseen objects is important for real applications since humans can bring new objects into the home.

Commonsense Reasoning. Prior work in Natural Language Processing has studied the problem of imbuing commonsense knowledge in AI systems, from social common-sense knowledge [10,35,56,58,59,73] to understand the likely intents, goals, and social dynamics of people, abductive commonsense reasoning [8], next event prediction [74,75], to temporal common sense knowledge about temporal order, duration, and frequency of events [2,23,44,76]. Most similar to our work is the study of physical commonsense knowledge [9] about object affordances, interaction, and properties (such as flexibility, curvature, porousness). However, these benchmarks are static in nature (as a dataset of textual or visual prompts). Our task, on the other hand, is instantiated in an embodied interactive environment and more realistic – the environment is partially observed, and the agent has to explore unseen regions, discover misplaced objects and use common-sense reasoning to infer compatibility between objects and receptacles.

Application of Large Language Models. With the introduction of Transformer [67] style architectures, we have seen a diverse range of applications of large language models (LLMs) pre-trained on web-scale textual data. They have not only performed well on natural language processing tasks [40,67], but the implicit knowledge learned by these models have shown to be effective for other unrelated tasks [42]. LLMs has had a lot of success in vision-and-language tasks like Visual Question Answering (VQA) [41,68] and image captioning [27,37], external knowledge-based question answering [11,54] and construction [10]. They have also been shown to improve performance on Embodied AI tasks like vision-and-language navigation [43,45], instruction following [25], and planning for embodied tasks [28,36]. In our work, we explore if language models can display common-sense knowledge of how humans prefer to tidy up their homes.

3 Housekeep: Task and Dataset

Here, we define the Housekeep task and its instantiation in the Habitat [60,64].

3.1 Task Specification

Definition: Recall, in Housekeep an embodied agent is required to clean up the house by rearranging misplaced objects to their correct location within a limited number of time steps. The agent is spawned randomly in an unseen environment and has to explore the environment to find misplaced objects and put them in their correct locations (receptacles).

Scenes and Rooms: We use 14 interactive and realistic iGibson scenes [61]. These scenes span 17 room types (*e.g.* living room, garage) and contain multiple rooms with an average of 7.5 rooms per scene. We remove one scene from the original iGibson dataset (`benevolence_0_int`) because it's unfurnished.

Receptacles: We define *receptacles* as flat horizontal surfaces in a household (furniture, appliances) where objects can be found – misplaced or correctly placed. We remove assets that are neither objects nor receptacles (*e.g.* windows, paintings, etc.) and end up with 395 unique receptacles spread over 32 categories. An iGibson scene can contain between 19–78 receptacles. Notice that a valid object-receptacle placement requires the additional context of what room the receptacle is situated in. For example, a counter in the kitchen is a suitable receptacle to place a fruit basket, however, a counter in the bathroom may not be. Hence, we care about the diversity in combinations of room-receptacle occurrences for Housekeep. Overall, there are 128 distinct room-receptacles in the iGibson scenes.

Objects: We collect objects from four popular asset repositories – Amazon Berkeley Objects [29], Google Scanned Objects [53], ReplicaCAD [64], and YCB Objects [12]. We filter out objects with large dimensions (*e.g.* ladders, televisions), and objects that do not usually move in a household (*e.g.* garbage cans). After filtering, we have 1799 unique objects spread across 268 categories. We further categorize these objects into 19 high-level semantic categories such as stationery, food, electronics, toys, etc. More details about the filtering, semantic classes, and high/low-level object categories are in the Appendix B.

Agent: We simulate a Fetch robot [55], which has a wheeled base with a 7-DoF arm manipulator, parallel-jaw gripper, and an RGBD camera (90° FoV, 128×128 pixels) on the robot's head. The robot moves its base and head through five discrete actions – move forward by 0.25 m, rotate base right or left by 10°, rotate head camera up or down (pitch) by 10°. The robot interacts with objects through a "magic pointer abstraction" [6] where at any step the robot can select a discrete "interact" action. We provide more details in Appendix E.2.

3.2 Human Preferences Dataset: Where Do Objects Belong?

The central challenge of Housekeep is understanding how humans prefer to put everyday household objects in an organized and disorganized house. We want to capture where objects are typically found in an unorganized house (before tidying the house), and in a tidy house where objects are kept in their correct position (after the person has tidied the house). To this end, we run a study on Amazon MTurk [15, 57] with 372 participants. Each participant is shown an object (*e.g.* salt-shaker), a room (*e.g.* kitchen) for context, and asked to classify all the receptacles present in the room into the following categories:

- `misplaced`: subset of receptacles where object is found *before* housekeeping.
- `correct`: subset of receptacles where object is found *after* housekeeping.

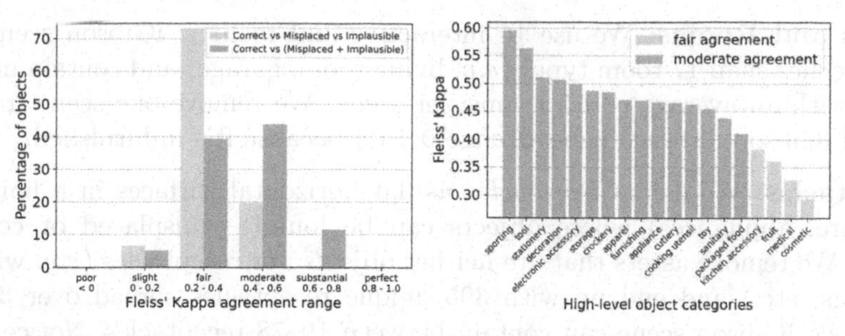

(a) Object Category Agreement (b) High-Level Category Agreement

Fig. 2. Analysis of agreement between reviewer ratings in the Housekeep human rearrangement preferences dataset.

- `implausible`: subset of receptacles where object is unlikely to be found either in a clean or an untidy house.

We also ask each participant to rank receptacles classified under `misplaced` and `correct`. For example, given a can of food, someone may prefer placing it in kitchen cabinets while others will rank pantry over the kitchen cabinet.

For each object-room pair (268×17), we collect 10 human annotations. We collect human annotations through multiple batches of smaller annotation tasks. In a single annotation task, we ask participants to classify-then-rank receptacles for 10 randomly sampled object-room pairs. On average a participant took 21 min to complete one annotation task. Overall, participants spent 1633 h hours doing our study. Appendix C provides more details about the instructions page, user interface, training videos, and FAQs provided in the beginning of the task.

Agreement Analysis. We evaluate the quality of our human annotations, using the Fleiss' kappa (`FK`) metric [20], which is widely used to assess the reliability of agreement between raters when classifying items. Recall that we collect 10 annotations to classify receptacles for each object-room pair into `correct`, `misplaced`, or `incompatible` bins. In Fig. 2a, we report `FK` agreement per object across all room-receptacle pairs (269×128) after keeping 8/10 annotations with the highest inter-human agreement. We use the agreement ranges proposed by [34] to interpret the `FK` scores. We also show agreement when combining the `misplaced` and `implausible` categories. Figure 2a demonstrates about 90% of our collected data has fair to moderate agreement between annotators. Figure 2b shows the mean agreement for high-level semantic categories. The agreement is higher for sporting, tool, and stationery categories because they go to specific places (office desks, garage, etc.). The agreement is low for objects like fruits, medicines, packaged foods because people differ in where they like to keep these objects (packaged food can go in cabinets, shelves, kitchen counters, refrigerators). Overall, these results indicate that our data defines a high-quality source of ground truth rearrangement preferences agreed upon by the majority of annotators.

3.3 Episodes

Each Housekeep episode is created by instantiating 7–10 objects within a scene, out of which 3–5 objects are misplaced and the remaining are placed correctly. Next, we concretely define the notions of *correct* and *misplaced* objects. For a given scene, let $\mathcal{R}$ be the set of receptacles available, and $\mathcal{O}$ be the set of all the objects which could be instantiated on these. Given an object $o \in \mathcal{O}$, let c_{or}, m_{or} respectively be the ratio of annotators who placed receptacle $r \in \mathcal{R}$ in `correct` and `misplaced` bins respectively. We call an object *correctly placed* if $c_{or} > 0.5$, and *misplaced* if $m_{or} > 0.5$, where both cannot be simultaneously true.

Splits: We create three non-overlapping sets of objects – `seen` (fork, gloves, etc.), `val-unseen` (chopping board, dishtowel, etc.), and `test-unseen` (banana, scissors, etc.). `seen`, `val-unseen` and `test-unseen` contains 8, 2 and 9 high-level object categories respectively. Note that *only* 40% of all objects are provided for training, making Housekeep a strong benchmark to test generalization to unseen objects.

We also split the 14 scenes into `train`, `val` and `test` with 8:2:4 scenes each respectively. We provide five different splits to test agents on a wide array of commonsense reasoning and rearrangement capabilities.

- **train**: 9K episodes with `seen` objects and `train` scenes
- **val-seen**: 200 episodes with `seen` objects and `val` scenes
- **val-unseen**: 200 episodes with `unseen` objects and `val` scenes
- **test-seen**: 800 episodes with `seen` objects and `test` scenes
- **test-unseen**: 800 episodes with `unseen` objects and `test` scenes

More details on episode statistics, and generation are in Appendix D.

3.4 Evaluation

We evaluate agents in three different dimensions of rearrangement quality, efficiency, and exploration. All metrics are reported per episode and then aggregated across multiple episodes to report averages and standard errors. While we only describe these metrics informally here, a more nuanced discussion with formal definitions for these can be found in Appendix D.3.

Metrics for Rearrangement. These metrics evaluate the relative change in the placement of objects between start and end states of the episode.

- **Episode Success (`ES`)**: Strict binary (*all* or *none*) metric that is one if and only if all objects (irrespective of whether initially misplaced or correctly placed) in the episode are correctly placed at the end of the episode.
- **Object Success (`OS`)**: Fraction of the objects placed correctly.
- **Soft Object Success (`SOS`)**: The ratio of reviewers that agree that an object is placed correctly.

- **Rearrange Quality (RQ)**: A normalized value in [0, 1] (via mean reciprocal rank [14]) is given to each object-receptacle based on the ranking collected from human preferences, 0 is given if misplaced.

Metrics OS, SOS and RQ are averaged across objects that are *initially misplaced* or *ever picked up* by the agent during the episode.

Exploration and Efficiency Metrics: We also study how well the agent explores an unseen environment as well as efficiency at rearranging objects.

- **Map Coverage (MC)**: The % of the navigable map area explored.
- **Misplaced Objects Coverage (MOC)**: The fraction of misplaced objects discovered. Agent discovers an object when it appears in FoV at any point.
- **Pick and Place Efficiency (PPE)**: The minimum number of picks and places required to solve the episode divided by the number of picks and places made by agent in the episode.

4 Methods

In this section, we describe our hierarchical baseline for the Housekeep benchmark. Our baseline breaks the multi-stage rearrangement into three natural components: a) exploration and mapping, b) planning, and c) navigation and rearrangement. The planning module communicates with all the other modules and determines what the agent does (explore or rearrange). Before we dive into the details of our baseline, we discuss some additional sensors that our baseline has access to.

Additional Sensors: In the Housekeep specification the agent operates from an RGBD sensor. However, to scope the problem and focus on the planning and commonsense reasoning we allow access to the following:

- *semantic* and *instance* sensor: Provides two pixel-wise masks aligned with egocentric RGB observations. The semantic segmentation mask maps every pixel to an object or receptacle category (*e.g.* bowl, cabinet). The instance mask maps every pixel to a unique instance ID, which helps to disambiguate between instances of the same object/receptacle category.
- *relationship* sensor: Given instance IDs of an object and a receptacle in the egocentric view, the relationship sensor predicts a binary value if the object is on top of the receptacle or not.
- *receptacle-room* map: Receptacles are static within a scene, so we also assume access to a mapping that provides us with the room name for any receptacle discovered (*e.g.* an oven maps to the kitchen).

In the future, these sensors can be easily swapped with their learned counterparts. [13,30] demonstrate it is possible to learn a segmentation sensor for indoor scenes, and [5] shows it is possible to learn to infer relationships between 3D objects.

4.1 Mapping and Exploration

Mapping: At the start of an episode, this module initializes an empty top-down allocentric map. As the agent navigates through the environment, it continuously updates the map at each step using egocentric observations and camera projection matrix. We further use the RGBD-aligned pixel-wise instance and semantic masks to localize objects and receptacles and update our allocentric map with them. Finally, the mapping module also keeps track of the room and relationship information of discovered objects and receptacles via the *relationship sensor* and known *receptacle-room map.*

Exploration: To discover misplaced objects as well as suitable receptacles to place them on, our exploration module aims to maximize the area on the map it has seen. This module only requires the hyperparameter n_e—the number of exploration steps—as input and executes low-level actions via the navigation module. We use frontier-based exploration [72] (FRT) for our main experiments, which iteratively visits unexplored frontiers, which are the edges between visited and unvisited space. We keep our implementation details same as those in [52].

4.2 Planning

Our planner communicates with all the modules to build a high-level rearrangement plan that the agent follows. It consists of:

Rearrange Submodule: Stores a list of locations of discovered objects and receptacles. From this list, it produces a list of object-receptacle pairs indicating the order of rearrangements to perform. There are 3 key decisions the rearrange submodule needs to make to create this list: 1) what objects are misplaced, 2) what order to arrange misplaced objects, and 3) what receptacle to place each misplaced object on. It makes these decisions via a **Ranker** submodule which ranks potential object-receptacle pairings by modeling the joint distribution $\mathbb{P}(\text{receptacle}, \text{room}|\text{object})$. To solve (3), for a given object the agent picks the receptacle in the room with the highest joint probability. We model the joint distribution of the receptacle and room because the context of a receptacle will change based on the room. For example, a plate belongs on the counter in the kitchen, but not a counter in the bathroom. Section 4.3 describes how we compute $\mathbb{P}(\text{receptacle}, \text{room}|\text{object})$, and also how we solve (1). To solve (2), we evaluate 4 heuristic orderings which are described in Appendix G.2.

Planner Submodule: At any given step, the planner decides to explore only if there are no more pending rearrangements. The agent explores for a fixed number of steps (n_e). Intuitively, higher values of n_e will encourage the agent to explore the environment at the beginning of the episode whereas lower values of n_e will encourage the agent to rearrange as soon as a better receptacle is found. While exploring, the planner ensures that map and rearrange modules are synchronized at each step. At the end of the exploration phase, the planner uses the rank (L) module to update compatibility scores by considering newly discovered objects and receptacles. We provide the planner pseudocode in Algorithm 2.

Navigation and Pick-Place: Please see Appendix E for details.

4.3 Extracting Embodied Commonsense from LLMs

One of the main goals of Housekeep is to equip the agent with commonsense knowledge to reason about the compatibility of an object with different receptacles present across different rooms. Large Language Models (LLMs) trained on unstructured web-corpora have been shown to work well for several embodied AI tasks like navigation [25,26,28,36,43]. We study whether we can use LLMs to extract physical (embodied) common sense about how humans prefer to rearrange objects to tidy a house. For this, we build a ranking module (`L`) which takes as input a list of objects and a list of receptacles in rooms and then outputs a sequence of desired rearrangements based on which object receptacle pairings are most likely. We select the rearrangements that maximize $\mathbb{P}(\text{receptacle}, \text{room}|\text{object})$. We decompose computing this probability into a product of two probabilities:

- Object Room `[OR]--` $\mathbb{P}$(`room|object`): Generate compatibility scores for rooms for a given object.
- Object Room Receptacle `[ORR]--` $\mathbb{P}$(`receptacle|object,room`): Generate compatibility scores for receptacles within a given room and for a given object.

Both of these are learned from the human rearrangement preferences dataset. From the compatibility scores in the `ORR` task, we first determine which objects in our list of objects are misplaced and which are correctly placed. To do this, we compute a hyperparameter s_L—the score threshold—from our `val` episodes using a grid search. Receptacles whose scores are above s_L for a given object-room pair are marked as correct, while those whose scores are below s_L are marked as incorrect. We then treat this as a classification task and pick s_L that maximizes the F1 score on the `val` episodes.

Next, to determine the ranking of receptacles for a given misplaced object, we use the probabilities from both the `OR` and `ORR` tasks. For a given object, we first rank the rooms in descending order of $\mathbb{P}(\text{room}|\text{object})$. Then, for each object-room pair in the ranked room list, we rank the *correct* receptacles in the room in descending order of $\mathbb{P}(\text{receptacle}|\text{object}, \text{room})$. Finally, we place the *incorrect* receptacles at the end of our list.

To learn the probability scores in the `OR` and `ORR` tasks, we start by extracting word embeddings from a pretrained RoBERTa LLM [40] of all objects, receptacles. We experiment with various contextual prompts [50,51] for extracting embeddings of paired room-receptacle (*e.g.* "`<receptacle> of <room>`") and object-room (*e.g.* "`<object> in <room>`") combinations. Next, we implemented the following 2 methods of using these embeddings to get the final compatibility scores:

Zero-Shot Ranking via MLM (`ZS-MLM`). Masked Language Modeling (MLM) is used extensively for pretraining LLMs [18,40], which involves predicting a masked word (*i.e.* `[mask]`) given the surrounding context words. This objective can be extended for zero-shot ranking using various contextual prompts. We use a frozen LLM to compute log-likelihood scores of prompts and use these scores to rank rooms and receptacles for the OR and ORR tasks. For ORR, we use the prompt "in <room>, usually you put <object> <spatial-preposition> [mask]" to rank receptacles given an object, a room, and a spatial preposition (*e.g.* in or on). For OR, we use the prompt "in a household, it is likely that you can find <object> in the room called [mask]".

Finetuning by Contrastive Matching (CM). Apart from using prompts in a zero-shot manner, we also train a 3-layered MLP on top of language embeddings generated by the LLM used in ZS-MLM and compute pairwise cosine similarity between any two embeddings. Embeddings are trained using objects from seen split. We train separate models for ORR and OR. For ORR, we match an object-room pair to the receptacle with the best average rank across annotators. We use contrastive loss [47] to promote similarity between an object-room pair and the matching receptacle. For OR, we match an object with all rooms that have at least one correct receptacle for it. In this case, we use the binary cross entropy (BCE) loss to handle multiple rooms per object.

We compare these ranking approaches with other baselines in Sect. 5.1. We provide training details of our ranking module in Appendix F.

5 Experiments

We first test whether LLMs can capture the embodied commonsense reasoning needed for planning in Housekeep. Then we deploy our modular agent equipped with this LLM-based planner to benchmark its ability to generalize to unseen environments cluttered with novel objects from seen (*i.e.* test-seen) and unseen (*i.e.* test-unseen) categories. Finally, we perform a thorough qualitative analysis of its failure modes and highlight directions for further improvements.

5.1 Language Models Capture Embodied Commonsense

Methods. We evaluate CM and ZS-MLM using RoBERTa [40] as our base LLM. We also compare these with GloVe-based [49] embeddings, and a baseline that randomly ranks rooms (for OR task) and receptacles (for ORR task).

Evaluation. We evaluate mean average precision (mAP) across objects to compare the ranked list of rooms/receptacles obtained from our ranking module to the list of rooms/receptacles deemed correct by the human annotators. Recall from Sect. 3.3, for a given object, a receptacle is considered correct when at least 6 annotators vote for

Table 1. We report mAP scores on train, and unseen objects splits of val and test for both OR and ORR matching tasks. The fine-tuning with CM objective is performed using objects *only* from train split

#	Method	ORR			OR		
		train	val-u	test-u	train	val-u	test-u
1	RoBERTa+CM	0.81	**0.79**	**0.81**	**1.0**	**0.65**	0.65
2	GloVe+CM	**0.88**	0.76	0.76	**1.0**	**0.65**	**0.66**
3	ZS-MLM	0.43	0.46	0.42	0.51	0.54	0.52
4	Random	0.47	0.47	0.46	0.58	0.52	0.59

Table 2. Results using our modular baseline on Housekeep test-seen and test-unseen splits. OR: Oracle, LM: LLM-based ranking, FT: Frontier exploration. GLV: GloVe

	Modules			Rearrange		Soft-Score		Explore		Efficiency
	#	Rank	Explore	ES ↑	OS ↑	SOS ↑	RQ ↑	MC ↑	OC ↑	PPE ↑
t-seen	1	OR	OR	1.00 ± 0.00	1.00 ± 0.00	0.65 ± 0.00	0.63 ± 0.00	–	1.00 ± 0.00	1.00 ± 0.00
	2	OR	FTR	0.35 ± 0.02	0.64 ± 0.01	0.49 ± 0.01	0.41 ± 0.01	73 ± 1	0.73 ± 0.01	1.00 ± 0.00
	3	LM	OR	0.04 ± 0.01	0.44 ± 0.01	0.46 ± 0.00	0.30 ± 0.01	–	1.00 ± 0.00	0.57 ± 0.01
	4	LM	FTR	0.01 ± 0.00	0.30 ± 0.01	0.39 ± 0.00	0.19 ± 0.01	77 ± 1	0.76 ± 0.01	0.41 ± 0.01
	5	GLV	FTR	0.01 ± 0.00	0.29 ± 0.01	0.36 ± 0.00	0.19 ± 0.01	71 ± 1	0.73 ± 0.01	0.39 ± 0.01
t-unseen	6	OR	OR	1.00 ± 0.00	1.00 ± 0.00	0.64 ± 0.00	0.61 ± 0.00	–	1.00 ± 0.00	1.00 ± 0.00
	7	OR	FTR	0.35 ± 0.02	0.65 ± 0.01	0.49 ± 0.01	0.40 ± 0.01	74 ± 1	0.74 ± 0.01	1.00 ± 0.00
	8	LM	OR	0.02 ± 0.00	0.32 ± 0.01	0.42 ± 0.00	0.20 ± 0.01	–	1.00 ± 0.00	0.42 ± 0.01
	9	LM	FTR	0.01 ± 0.00	0.23 ± 0.01	0.36 ± 0.00	0.14 ± 0.01	73 ± 1	0.74 ± 0.01	0.35 ± 0.01
	10	GLV	FTR	0.00 ± 0.00	0.23 ± 0.01	0.34 ± 0.00	0.15 ± 0.01	72 ± 1	0.74 ± 0.01	0.26 ± 0.01

it, and a room is considered correct if it has at least one correct receptacle within it. Higher AP score indicates correct items are likely to ranked higher than the incorrect items.

Results. Table 1 shows that RoBERTa+CM outperforms ZS-MLM by a large margin even when fintuned on a relatively small-sized training set (40% of total data, see Sect. 3.4). We find good transfer of results from val to test splits by RoBERTa+CM method on both tasks demonstrating the better generalization capabilities of LLMs. On the other hand, GloVe+CM does not seem to transfer well for the ORR task. Also, ZS-MLM performs worse than the Random baseline. We found that predictions of ZS-MLM baseline are biased towards certain receptacles (e.g. chair, and carpet are in top-4 most frequent choices). This bias is frequently not aligned with human preferences. We hypothesize this is likely an artifact of the original training data. Finally, notice that Random baseline performs relatively well on room-matching (OR) task, which is expected since there are ample of rooms with at least one correct receptacle for any given object.

5.2 Main Results for Housekeep

We use RoBERTa+CM as scoring function in Ranker module to continuously rerank (thus replan) discovered rooms and receptacles while exploring Housekeep episodes.

Oracle Modules. We show oracle agent's performance, by swappping Ranker and Explore modules with their oracle (perfect) counterparts. Oracle ranker uses the ground truth human preferences to rank the objects and receptacles found. Oracle exploration gives a complete map of the environment, *i.e.* agent knows all objects, receptacles and their respective locations.

Upper Bounds. In Table 2, we show results on both test-seen and test-unseen splits. Rows 1, 6 with oracle ranking and exploration denote the

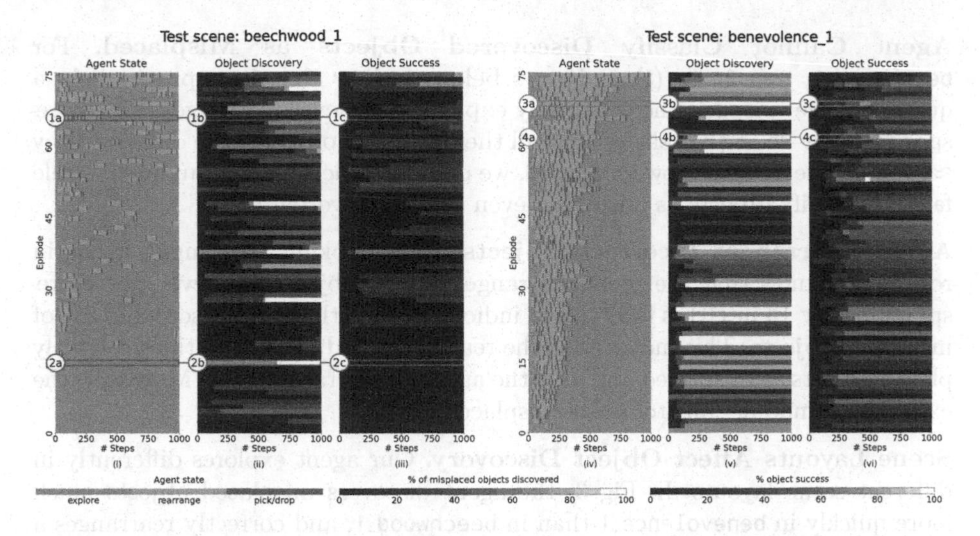

Fig. 3. Visually depicting agent's progress on 75 randomly-sampled episodes from two test scenes, beechwood_1 and benevolence_1. Plots (i) and (iv) depict Agent's state, (ii) and (v) show % of objects discovered, (iii) and (vi) show % object success, and x-axis is the timestep. All 3 plots of same scenes are aligned, *i.e.* show same episodes on y-axis.

upper bounds achievable across all metrics. Note that Soft Object Success (SOS) and Rearrangement Quality (RQ) are not perfect since human agreement across correct receptacles is not 100%.

Frontier Exploration, Full baseline. Using Frontier exploration (rows 1,2), OS drops by 47%. This drop in performance signifies the importance of task-driven exploration needed for Housekeep to find misplaced objects or correct receptacles quickly. Finally, we evaluate the fully non-oracle baseline (row 4) which achieves a 30% object success rate. From rows 4 and 9, we see that OS drops by 7%, but SOS drops only by 3% across seen vs unseen objects, which demonstrates some level of generalization capability to unseen environments. We also evaluate our baseline agent with a GloVe-based ranker (rows 5, 10) and observe similar OS performance to the LLM ranker.

We put additional experiments analyzing the effect of exploration steps (n_e), exploration strategies in Appendix G, and qualitative results in Appendix H.

5.3 Qualitative Analysis

Figure 3 visually depicts the baseline agent's progress across episodes on two test scenes. Agent State plots show the *module currently being executed*: explore (blue), rearrange (orange), or pick/place (red). Object Discovery plots show the *percentage of misplaced objects discovered* until any given time step. Object Success plots show the *object success* at any given time step. Dark to light shade corresponds to an increasing number of misplaced objects found/increasing object success. Each row corresponds to one episode, and the x-axis denotes time step.

Agent Cannot Classify Discovered Objects as Misplaced. For beechwood_1, row 2a in **(i)** and rows below it show that in approximately a quarter of the episodes, the agent only explores and never rearranges. The corresponding row 2b in **(ii)** tells us that all the misplaced objects were discovered by $\approx$700 time steps. From row 2a and 2b, we can conclude that the ranking module fails to identify objects as misplaced even after discovering them.

Agent Rearranges Incorrect Objects. Next, looking at orange regions in row 1a, we know that the agent rearranges several objects. However, the corresponding row 1b in **(ii)** is fully black, indicating that the agent discovered 0% of misplaced objects. This means that the reasoning module misidentifies correctly placed objects as misplaced and asks the agent to rearrange them. Moreover, the exploration module fails to locate misplaced objects.

Scene Layouts Affect Object Discovery. Our agent explores differently in different scene layouts. In Fig. 3, the agent discovers misplaced objects much more quickly in benevolence_1 than in beechwood_1, and correctly rearranges a higher fraction of them. Rows 3a and 3b show this trend – all objects are discovered within the first 200 steps of the episode in stark contrast to beechwood_1 episodes. Row 4c even shows an episode with 100% object success. This is explained by the fact that benevolence_1 is a smaller home with just one partitioning wall (4 rooms) versus beechwood_1 (8 rooms), making exploration and object discovery easier. We provide top-down maps of both scenes in Appendix H.1.

6 Conclusion

We presented the Housekeep benchmark to evaluate commonsense reasoning in the home for embodied AI. We collected a dataset of human preferences of where objects go in tidy and untidy houses, and used it to generate episodes and evaluate agent performance. Then we proposed a modular baseline that plans using commonsense reasoning extracted from a large language model. Housekeep is a challenging task, and the overall episode success rate remains low despite the use of additional sensors (*e.g.* segmentation, relationship) needed for planning and commonsense reasoning. Two areas of improvement are exploration module and reasoning module. A learned exploration module can visit areas that get cluttered more frequently, and optimize object coverage instead of map coverage. Improving the reasoning module's recall and precision at identifying misplaced objects can increase performance on our task. Finally, replacing additional sensors related with their learned counterparts will make baselines more realistic.

Acknowledgements. We thank the Habitat team for their support. The Georgia Tech effort was supported in part by NSF, AFRL, DARPA, ONR YIPs, ARO PECASE, Amazon. The views and conclusions contained herein are those of the authors and should not be interpreted as necessarily representing the official policies or endorsements, either expressed or implied, of the U.S. Government, or any sponsor.

References

1. Abdo, N., Stachniss, C., Spinello, L., Burgard, W.: Robot, organize my shelves! tidying up objects by predicting user preferences. In: 2015 IEEE International Conference on Robotics and Automation (ICRA) (2015)
2. Agrawal, H., Chandrasekaran, A., Batra, D., Parikh, D., Bansal, M.: Sort story: sorting jumbled images and captions into stories. In: Proceedings of the 2016 Conference on Empirical Methods in Natural Language Processing (2016)
3. Anderson, P., et al.: On evaluation of embodied navigation agents. arXiv preprint arXiv:1807.06757 (2018)
4. Anderson, P., et al.: Vision-and-language navigation: interpreting visually-grounded navigation instructions in real environments. In: 2018 IEEE Conference on Computer Vision and Pattern Recognition, CVPR 2018, Salt Lake City, UT, USA, 18–22 June 2018 (2018)
5. Armeni, I., et al.: 3D scene graph: a structure for unified semantics, 3D space, and camera. In: 2019 IEEE/CVF International Conference on Computer Vision, ICCV 2019, Seoul, Korea (South), 27 October- 2 November 2019 (2019)
6. Batra, D., et al.: Rearrangement: a challenge for embodied AI (2020)
7. Batra, D., et al.: ObjectNav revisited: on evaluation of embodied agents navigating to objects. arXiv preprint arXiv:2006.13171 (2020)
8. Bhagavatula, C., et al.: Abductive commonsense reasoning. In: 8th International Conference on Learning Representations, ICLR 2020, Addis Ababa, Ethiopia, 26–30 April 2020 (2020)
9. Bisk, Y., Zellers, R., LeBras, R., Gao, J., Choi, Y.: PIQA: reasoning about physical commonsense in natural language. In: The Thirty-Fourth AAAI Conference on Artificial Intelligence, AAAI 2020, The Thirty-Second Innovative Applications of Artificial Intelligence Conference, IAAI 2020, The Tenth AAAI Symposium on Educational Advances in Artificial Intelligence, EAAI 2020, New York, NY, USA, 7–12 February 2020 (2020)
10. Bosselut, A., Rashkin, H., Sap, M., Malaviya, C., Celikyilmaz, A., Choi, Y.: COMET: commonsense transformers for automatic knowledge graph construction. In: Proceedings of the 57th Annual Meeting of the Association for Computational Linguistics (2019)
11. Brown, T.B., et al.: Language models are few-shot learners. In: Advances in Neural Information Processing Systems 33: Annual Conference on Neural Information Processing Systems 2020, NeurIPS 2020, 6–12 December 2020, virtual (2020)
12. Calli, B., Singh, A., Walsman, A., Srinivasa, S., Abbeel, P., Dollar, A.M.: The YCB object and model set: towards common benchmarks for manipulation research. In: 2015 International Conference on Advanced Robotics (ICAR). IEEE (2015)
13. Cartillier, V., Ren, Z., Jain, N., Lee, S., Essa, I., Batra, D.: Semantic MapNet: building allocentric semanticmaps and representations from egocentric views. arXiv preprint arXiv:2010.01191 (2020)
14. Craswell, N.: Mean reciprocal rank. In: Encyclopedia of Database Systems (2009)
15. Crowston, K.: Amazon mechanical turk: a research tool for organizations and information systems scholars. In: Bhattacherjee, A., Fitzgerald, B. (eds.) IS&O 2012. IAICT, vol. 389, pp. 210–221. Springer, Heidelberg (2012). https://doi.org/10.1007/978-3-642-35142-6_14
16. Daruna, A., Liu, W., Kira, Z., Chernova, S.: RoboCSE: robot common sense embedding (2019)

17. Das, A., Datta, S., Gkioxari, G., Lee, S., Parikh, D., Batra, D.: Embodied question answering. In: 2018 IEEE Conference on Computer Vision and Pattern Recognition, CVPR 2018, Salt Lake City, UT, USA, 18–22 June 2018 (2018)
18. Devlin, J., Chang, M.W., Lee, K., Toutanova, K.: BERT: pre-training of deep bidirectional transformers for language understanding. In: Proceedings of the 2019 Conference of the North American Chapter of the Association for Computational Linguistics: Human Language Technologies, Volume 1 (Long and Short Papers) (2019)
19. Ehsani, K., et al.: ManipulaTHOR: a framework for visual object manipulation. In: Proceedings of the IEEE/CVF Conference on Computer Vision and Pattern Recognition (2021)
20. Fleiss, J., et al.: Measuring nominal scale agreement among many raters. Psychol. Bull. **76**(5), 378 (1971)
21. Gan, C., et al.: Threedworld: a platform for interactive multi-modal physical simulation. In: NeurIPS, abs/2007.04954 (2020)
22. Gordon, D., Kembhavi, A., Rastegari, M., Redmon, J., Fox, D., Farhadi, A.: IQA: visual question answering in interactive environments. In: 2018 IEEE Conference on Computer Vision and Pattern Recognition, CVPR 2018, Salt Lake City, UT, USA, 18–22 June 2018 (2018)
23. Granroth-Wilding, M., Clark, S.: What happens next? Event prediction using a compositional neural network model. In: Proceedings of the Thirtieth AAAI Conference on Artificial Intelligence, Phoenix, Arizona, USA, 12–17 February 2016 (2016)
24. Habitat: Habitat Challenge (2021). https://aihabitat.org/challenge/2021/
25. Hill, F., Mokra, S., Wong, N., Harley, T.: Human instruction-following with deep reinforcement learning via transfer-learning from text. arXiv:abs/2005.09382 (2020)
26. Hong, Y., Wu, Q., Qi, Y., Rodriguez-Opazo, C., Gould, S.: A recurrent vision-and-language BERT for navigation. In: ECCV (2021)
27. Hu, X., et al.: Vivo: surpassing human performance in novel object captioning with visual vocabulary pre-training. arXiv:abs/2009.13682 (2020)
28. Huang, W., Abbeel, P., Pathak, D., Mordatch, I.: Language models as zero-shot planners: extracting actionable knowledge for embodied agents. ArXiv abs/2201.07207 (2022)
29. Jasmine, C., et al.: ABO: dataset and benchmarks for real-world 3D object understanding. arXiv preprint arXiv:2110.06199 (2021)
30. Jiang, J., Zheng, L., Luo, F., Zhang, Z.: RedNet: residual encoder-decoder network for indoor RGB-D semantic segmentation. arXiv preprint arXiv:1806.01054 (2018)
31. Jiang, Y., Lim, M., Saxena, A.: Learning object arrangements in 3D scenes using human context. In: Proceedings of the 29th International Conference on Machine Learning, ICML 2012, Edinburgh, Scotland, UK, 26 June- 1 July 2012 (2012)
32. Kapelyukh, I., Johns, E.: My house, my rules: learning tidying preferences with graph neural networks. In: CoRL (2021)
33. Kolve, E., et al.: AI2-THOR: an interactive 3D environment for visual AI. arXiv (2017)
34. Landis, J.R., Koch, G.G.: The measurement of observer agreement for categorical data. Biometrics (1977)
35. Levesque, H.J., Davis, E., Morgenstern, L.: The winograd schema challenge. In: KR (2011)
36. Li, S., et al.: Pre-trained language models for interactive decision-making. arXiv:abs/2202.01771 (2022)

37. Li, X., et al.: Oscar: object-semantics aligned pre-training for vision-language tasks. In: Vedaldi, A., Bischof, H., Brox, T., Frahm, J.-M. (eds.) ECCV 2020. LNCS, vol. 12375, pp. 121–137. Springer, Cham (2020). https://doi.org/10.1007/978-3-030-58577-8_8
38. Liu, W., Bansal, D., Daruna, A., Chernova, S.: Learning instance-level n-ary semantic knowledge at scale for robots operating in everyday environments. In: Proceedings of Robotics: Science and Systems (2021)
39. Liu, W., Paxton, C., Hermans, T., Fox, D.: StructFormer: learning spatial structure for language-guided semantic rearrangement of novel objects. arXiv preprint arXiv:2110.10189 (2021)
40. Liu, Y., et al.: RoBERTa: a robustly optimized BERT pretraining approach. arXiv:1907.11692 (2019)
41. Lu, J., Batra, D., Parikh, D., Lee, S.: ViLBERT: pretraining task-agnostic visiolinguistic representations for vision-and-language tasks. In: Advances in Neural Information Processing Systems 32: Annual Conference on Neural Information Processing Systems 2019, NeurIPS 2019, Vancouver, BC, Canada, 8–14 2019 December (2019)
42. Lu, K., Grover, A., Abbeel, P., Mordatch, I.: Pretrained transformers as universal computation engines. arXiv:abs/2103.05247 (2021)
43. Majumdar, A., Shrivastava, A., Lee, S., Anderson, P., Parikh, D., Batra, D.: Improving vision-and-language navigation with image-text pairs from the web. arXiv:abs/2004.14973 (2020)
44. Mostafazadeh, N., et al.: A corpus and cloze evaluation for deeper understanding of commonsense stories. In: Proceedings of the 2016 Conference of the North American Chapter of the Association for Computational Linguistics: Human Language Technologies (2016)
45. Moudgil, A., Majumdar, A., Agrawal, H., Lee, S., Batra, D.: SOAT: a scene-and object-aware transformer for vision-and-language navigation. In: Advances in Neural Information Processing Systems 34 (2021)
46. Narasimhan, M., et al.: Seeing the un-scene: learning amodal semantic maps for room navigation. CoRR abs/2007.09841 (2020)
47. van den Oord, A., Li, Y., Vinyals, O.: Representation learning with contrastive predictive coding. arXiv preprint arXiv:1807.03748 (2018)
48. Padmakumar, A., et al.: Teach: task-driven embodied agents that chat. arXiv:abs/2110.00534 (2021)
49. Pennington, J., Socher, R., Manning, C.: GloVe: global vectors for word representation. In: Proceedings of the 2014 Conference on Empirical Methods in Natural Language Processing (EMNLP) (2014)
50. Petroni, F., et al.: How context affects language models' factual predictions. In: Automated Knowledge Base Construction (2020)
51. Petroni, F., et al.: Language models as knowledge bases? In: Proceedings of the 2019 Conference on Empirical Methods in Natural Language Processing and the 9th International Joint Conference on Natural Language Processing (EMNLP-IJCNLP) (2019)
52. Ramakrishnan, S.K., Jayaraman, D., Grauman, K.: An exploration of embodied visual exploration. Int. J. Comput. Vis. **129**(5), 1616–1649 (2021). https://doi.org/10.1007/s11263-021-01437-z
53. Google Research: Google Scanned Objects (2020). https://app.ignitionrobotics.org/GoogleResearch/fuel/collections/Google%20Scanned%20Objects. Accessed Feb 2022

54. Roberts, A., Raffel, C., Shazeer, N.: How much knowledge can you pack into the parameters of a language model? In: Proceedings of the 2020 Conference on Empirical Methods in Natural Language Processing (EMNLP) (2020)
55. Fetch robotics: Fetch (2020). http://fetchrobotics.com/
56. Sakaguchi, K., Le Bras, R., Bhagavatula, C., Choi, Y.: WinoGrande: an adversarial winograd schema challenge at scale. In: AAAI (2020)
57. Salganik, M.J.: Bit by Bit: Social Research in the Digital Age. Open review edition (2017)
58. Sap, M., et al.: ATOMIC: an atlas of machine commonsense for if-then reasoning. In: The Thirty-Third AAAI Conference on Artificial Intelligence, AAAI 2019, The Thirty-First Innovative Applications of Artificial Intelligence Conference, IAAI 2019, The Ninth AAAI Symposium on Educational Advances in Artificial Intelligence, EAAI 2019, Honolulu, Hawaii, USA, 27 January–1 February 2019 (2019)
59. Sap, M., Rashkin, H., Chen, D., Le Bras, R., Choi, Y.: Social IQa: commonsense reasoning about social interactions. In: Proceedings of the 2019 Conference on Empirical Methods in Natural Language Processing and the 9th International Joint Conference on Natural Language Processing (EMNLP-IJCNLP) (2019)
60. Savva, M., et al.: Habitat: a platform for embodied AI research. In: 2019 IEEE/CVF International Conference on Computer Vision, ICCV 2019, Seoul, Korea (South), 27 October–2 November 2019 (2019)
61. Shen, B., et al.: iGibson, a simulation environment for interactive tasks in large realistic scenes. arXiv preprint arXiv:2012.02924 (2020)
62. Shridhar, M., et al.: ALFRED: a benchmark for interpreting grounded instructions for everyday tasks. In: 2020 IEEE/CVF Conference on Computer Vision and Pattern Recognition, CVPR 2020, Seattle, WA, USA, 13–19 June 2020 (2020)
63. Srivastava, S., et al.: Behavior: benchmark for everyday household activities in virtual, interactive, and ecological environments. In: CoRL (2021)
64. Szot, A., et al.: Habitat 2.0: training home assistants to rearrange their habitat. In: Advances in Neural Information Processing Systems 34 (2021)
65. Taniguchi, A., Isobe, S., Hafi, L.E., Hagiwara, Y., Taniguchi, T.: Autonomous planning based on spatial concepts to tidy up home environments with service robots. Adv. Robot. **35**, 471–489 (2021)
66. Thomason, J., Murray, M., Cakmak, M., Zettlemoyer, L.: Vision-and-dialog navigation. In: CoRL (2019)
67. Vaswani, A., et al.: Attention is all you need. In: Advances in Neural Information Processing Systems 30: Annual Conference on Neural Information Processing Systems 2017, Long Beach, CA, USA, 4–9 December 2017 (2017)
68. Wang, W., Bao, H., Dong, L., Wei, F.: VLMO: unified vision-language pre-training with mixture-of-modality-experts. arXiv:abs/2111.02358 (2021)
69. Wani, S., Patel, S., Jain, U., Chang, A.X., Savva, M.: MultiON: benchmarking semantic map memory using multi-object navigation. In: NeurIPS (2020)
70. Weihs, L., Deitke, M., Kembhavi, A., Mottaghi, R.: Visual room rearrangement. In: Proceedings of the IEEE/CVF Conference on Computer Vision and Pattern Recognition (2021)
71. Wijmans, E., et al.: Embodied question answering in photorealistic environments with point cloud perception. In: IEEE Conference on Computer Vision and Pattern Recognition, CVPR 2019, Long Beach, CA, USA, 16–20 June 2019 (2019)
72. Yamauchi, B.: A frontier-based approach for autonomous exploration. In: CIRA, vol. 97 (1997)

73. Zellers, R., Bisk, Y., Farhadi, A., Choi, Y.: From recognition to cognition: Visual commonsense reasoning. In: IEEE Conference on Computer Vision and Pattern Recognition, CVPR 2019, Long Beach, CA, USA, 16–20 June 2019 (2019)
74. Zellers, R., Bisk, Y., Schwartz, R., Choi, Y.: SWAG: a large-scale adversarial dataset for grounded commonsense inference. In: Proceedings of the 2018 Conference on Empirical Methods in Natural Language Processing (2018)
75. Zellers, R., Holtzman, A., Bisk, Y., Farhadi, A., Choi, Y.: HellaSwag: can a machine really finish your sentence? In: Proceedings of the 57th Annual Meeting of the Association for Computational Linguistics (2019)
76. Zhou, B., Khashabi, D., Ning, Q., Roth, D.: "Going on a vacation" takes longer than "going for a walk": a study of temporal commonsense understanding. In: Proceedings of the 2019 Conference on Empirical Methods in Natural Language Processing and the 9th International Joint Conference on Natural Language Processing (EMNLP-IJCNLP) (2019)

Domain Randomization-Enhanced Depth Simulation and Restoration for Perceiving and Grasping Specular and Transparent Objects

Qiyu Dai[1], Jiyao Zhang[2], Qiwei Li[1], Tianhao Wu[1], Hao Dong[1], Ziyuan Liu[3], Ping Tan[3,4], and He Wang[1](✉)

[1] Peking University, Beijing, China
{qiyudai,lqw,hao.dong,hewang}@pku.edu.cn, thwu@stu.pku.edu.cn
[2] Xi'an Jiaotong University, Xi'an, China
zhangjiyao@stu.xjtu.edu.cn
[3] Alibaba XR Lab, Beijing, China
[4] Simon Fraser University, Burnaby, Canada
pingtan@sfu.ca

Abstract. Commercial depth sensors usually generate noisy and missing depths, especially on specular and transparent objects, which poses critical issues to downstream depth or point cloud-based tasks. To mitigate this problem, we propose a powerful RGBD fusion network, SwinDRNet, for depth restoration. We further propose Domain Randomization-Enhanced Depth Simulation (DREDS) approach to simulate an active stereo depth system using physically based rendering and generate a large-scale synthetic dataset that contains 130K photorealistic RGB images along with their simulated depths carrying realistic sensor noises. To evaluate depth restoration methods, we also curate a real-world dataset, namely STD, that captures 30 cluttered scenes composed of 50 objects with different materials from specular, transparent, to diffuse. Experiments demonstrate that the proposed DREDS dataset bridges the sim-to-real domain gap such that, trained on DREDS, our SwinDRNet can seamlessly generalize to other real depth datasets, e.g. ClearGrasp, and outperform the competing methods on depth restoration. We further show that our depth restoration effectively boosts the performance of downstream tasks, including category-level pose estimation and grasping tasks. Our data and code are available at https://github.com/PKU-EPIC/DREDS.

Keywords: Depth sensor simulation · Specular and transparent objects · Domain randomization · Pose estimation · Grasping

Q. Dai and J. Zhang— Equal contributions.

Supplementary Information The online version contains supplementary material available at https://doi.org/10.1007/978-3-031-19842-7_22.

S. Avidan et al. (Eds.): ECCV 2022, LNCS 13699, pp. 374–391, 2022.
https://doi.org/10.1007/978-3-031-19842-7_22

1 Introduction

With the emerging depth-sensing technologies, depth sensors and 3D point cloud data become more and more accessible, rendering many applications in VR/AR and robotics. Compared with RGB images, depth images or point clouds contain the true 3D information of the underlying scene geometry, thus depth cameras have been widely deployed in many robotic systems, *e.g.* for object grasping [3,11] and manipulation [17,18,32], that care about the accurate scene geometry. However, an apparent disadvantage of accessible depth cameras is that they may carry non-ignorable sensor noises more significant than usual noises in colored images captured by commercial RGB cameras. A more drastic failure case of depth sensing would be on objects that are either transparent or their surfaces are highly specular, where the captured depths would be highly erroneous and even missing around the specular or transparent region. It should be noted that specular and transparent objects are indeed ubiquitous in our daily life, given most of the metallic surfaces are specular and many man-made objects are made of glasses and plastics which can be transparent. The existence of so many specular and transparent objects in our real-world scenes thus poses severe challenges to depth-based vision systems and limits their application scenarios to well-controlled scenes and objects made of diffuse materials.

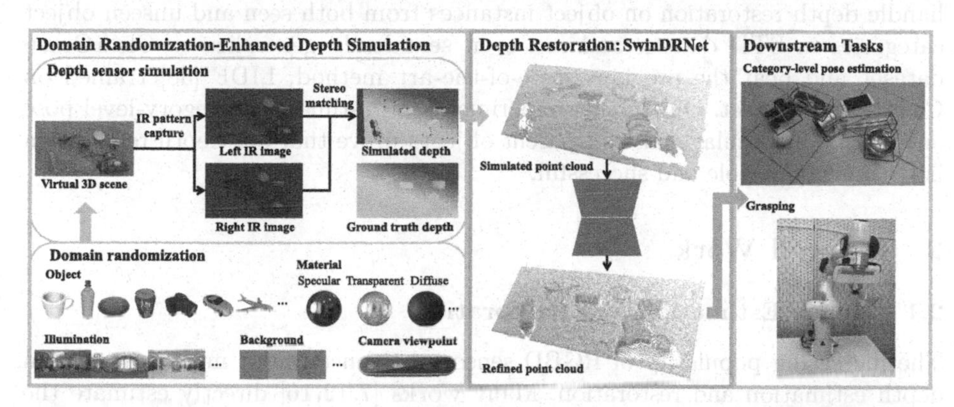

Fig. 1. Framework overview. From the left to right: we leverage domain randomization-enhanced depth simulation to generate paired data, on which we can train our depth restoration network SwinDRNet, and the restored depths will be fed to downstream tasks and improve estimating category-level pose and grasping for specular and transparent objects.

In this work, we devise a two-stream Swin Transformer [15] based RGB-D fusion network, SwinDRNet, for learning to perform depth restoration. However, it is a lack of real data composed of paired sensor depths and perfect depths to train such a network. Previous works on depth completion for transparent objects, like ClearGrasp [26] and LIDF [38], leverage synthetic perfect depth image for network training. They simply remove the transparent area in the perfect depth and their methods then learn to complete the missing depths in a feedforward way or further combines with depth optimization. We argue that both the methods can only access incomplete depth images during training and never see a depth with realistic sensor noises, leading to suboptimality when directly deployed on real sensor depths. Also, these two works only consider a small number of similar objects with little shape variations and all being transparent and hence fail to demonstrate their usefulness when adopted in scenes with completely novel object instances. Given material specularity or transparency forms a continuous spectrum, it is further questionable whether their methods can handle objects of intermediate transparency or specularity.

To mitigate the problems in the existing works, we thus propose to synthesize depths with realistic sensor noise patterns by simulating an active stereo depth camera resembling RealSense D415. Our simulator is built on Blender and leverages raytracing to mimic the IR stereo patterns and compute the depths from them. To facilitate generalization, we further adopt domain randomization techniques that randomize the object textures, object materials (from specular, transparent, to diffuse), object layout, floor textures, illuminations along camera poses. This domain randomization-enhanced depth simulation method, or in short DREDS, leads to 130K photorealistic RGB images and their corresponding simulated depths. We further curate a real-world dataset, STD dataset, that contains 50 objects with specular, transparent, and diffuse material. Our extensive experiments demonstrate that our SwinDRNet trained on DREDS dataset can handle depth restoration on object instances from both seen and unseen object categories in STD dataset and can even seamlessly generalize to ClearGrasp dataset and beat the previous state-of-the-art method, LIDF [38] trained on ClearGrasp dataset. Our further experiments on estimating category-level pose and grasping specular and transparent objects prove that our depth restoration is both generalizable and successful.

2 Related Work

2.1 Depth Estimation and Restoration

The increasing popularity of RGBD sensors has encouraged much research on depth estimation and restoration. Many works [7,12,16] directly estimate the depth from a monocular RGB image, but fail to restore accurate geometries of the point cloud because of the few geometric constraints of the color image. Other studies [19,25,33] restore the dense depth map given the RGB image and the sparse depth from LiDAR, but the estimated depth still suffers from low quality due to the limited geometric guidance of the sparse input. Recent research focuses on commercial depth sensors, trying to complete and refine the depth values from the RGB and noisy dense depth images. Sajjan *et al.* [26] proposed a two-stage method for transparent object depth restoration, which firstly estimates surface normals, occlusion boundaries, and segmentations from RGB images, and then calculates the refined depths via global optimization. However, the optimization is time-consuming, and heavily relies on the previous network predictions. Zhu *et al.* [38] proposed an implicit transparent object depth completion model, including the implicit representation learning from ray-voxel pairs and the self-iterating refinement, but voxelization of the 3D space results in heavy geometric discontinuity of the refined point cloud. Our method falls into this category and outperforms those methods, ensuring fast inference time and better geometries to improve the performance of downstream tasks.

2.2 Depth Sensor Simulation

To close the sim-to-real gap, the recent research focuses on generating simulated depth maps with realistic noise distribution. [14] simulated the pattern projection and capture system of Kinect to obtain simulated IR images and perform stereo matching, but could not simulate the sensor noise caused by object materials and scene environments. [22] proposed an end-to-end framework to simulate the mechanism of various types of depth sensors. However, the rasterization method limits the photorealistic rendering and physically correct simulation. [21] presented a new differentiable structure-light depth sensor simulation pipeline, but cannot simulate the transparent material, limited by the renderer. Recently, [37] proposed a physics-grounded active stereovision depth sensor simulator for various sim-to-real applications, but focused on instance-level objects and the robot arm workspace. Our DREDS pipeline generates realistic RGBD images for various materials and scene environments, which can generalize the proposed model to category-level unseen object instances and novel categories.

2.3 Domain Randomization

Domain randomization bridges the sim-to-real gap in the way of data augmentation. Tobin *et al.* [27] first explore transferring to real environments by generating training data through domain randomization. Subsequent works [23,28,35] generate synthetic data with sufficient variation by manually setting randomized features. Other studies [36] perform randomization using the neural networks. These works have verified the effectiveness of domain randomization on the tasks such as robotic manipulation [20], object detection and pose estimation [13], *etc.*. In this work, we combine the depth sensor simulation pipeline with domain randomization, which, for the first time, enables direct generalization to unseen diverse real instances on specular and transparent object depth restoration.

3 Domain Randomization-Enhanced Depth Simulation

3.1 Overview

In this work, we propose a simulated RGBD data generation pipeline, namely Domain Randomization Enhanced Depth Simulation (DREDS), for tasks of depth restoration, object perception, and robotic grasping. We build a depth sensor simulator, modeling the mechanism of the active stereo vision depth camera system based on the physically based rendering, along with the domain randomization technique to handle real-world variations.

Leveraging domain randomization and active stereo sensor simulation, we present DREDS, the large-scale simulated RGBD dataset, containing photorealistic RGB images and depth maps with the real-world measurement noise

Table 1. Comparisons of specular and transparent depth restoration dataset. S, T, and D refer to specular, transparent, and diffuse materials, respectively. #Objects refers to the number of objects. SN+CG means the objects are selected from ShapeNet and ClearGrasp (the number are not mentioned).

Dataset	Type	#Objects	Type of material	Size
ClearGrasp-Syn [26]	Syn	9	T	50K
Omniverse [38]	Syn	SN+CG	T+D	60K
ClearGrasp-Real [26]	Real	10	T	286
TODD [34]	Real	6	T	1.5K
DREDS	Sim	1,861	S+T+D	130K
STD	Real	50	S+T+D	27K

and error, especially for the hand-scale objects with specular and transparent materials. The proposed DREDS dataset bridges the sim-to-real domain gap, and generalizes the RGBD algorithms to unseen objects. DREDS dataset's comparison to the existing specular and transparent depth restoration datasets is summarized in Table 1.

3.2 Depth Sensor Simulation

A classical active stereo depth camera system contains an infrared (IR) projector, left and right IR stereo cameras, and a color camera. To measure the depth, the projector emits an IR pattern with dense dots to the scene. Subsequently, the two stereo cameras capture the left and right IR images, respectively. Finally, the stereo matching algorithm is used to calculate per-pixel depth values based on the discrepancy between the stereo images, to get the final depth scan. Our depth sensor simulator follows this mechanism, containing light pattern projection, capture, and stereo matching. The simulator is mainly built upon Blender [1].

Light Pattern Capture via Physically Based Rendering. For real-world specular and transparent objects, the IR light from the projector may not be received by the stereo cameras, due to the reflection on the surface or the refraction through the transparent objects, resulting in inaccurate and missing depths. To simulate the physically correct IR pattern emission and capture process, we thus adopt physically based ray tracing, a technique that mimics the real light transportation process, and supports various surface materials especially specular and transparent materials.

Specifically, the textured spotlight projects a binary pattern image into the virtual scene. Sequentially, the binocular IR images are rendered from the stereo cameras. We manage to simulate IR images via visible light rendering, where both the light pattern and the reduced environment illumination contribute to the IR rendering. From the perspective of physics, the difference between IR and visible light is the reflectivity and refractive index of the object. We note that the wavelength (850 nm) of IR light used in depth sensors, *e.g.* RealSense D415, is close to the visible light (400–800 nm). So the resulting effects have already been well-covered by the randomization in object reflectivity and refractive index used in DREDS, which constructs a superset of real IR images. To mimic the portion of IR in environmental light, we reduce its intensity. Finally, all RGB values are converted to intensity, which is our final IR image.

Stereo Matching. We perform stereo matching to obtain the disparity map, which can be transferred to the depth map leveraging the intrinsic parameters of the depth sensor. In detail, we compute a matching cost volume over the left and right IR images along the epipolar line and find the matching results with minimum matching cost. Then we perform sub-pixel detection to generate a more accurate disparity map using the quadratic curve fitting method. To generate a more realistic depth map, we perform post-processing, including left/right consistency check, uniqueness constraint, median filtering, *etc.*.

3.3 Simulated Data Generation with Domain Randomization

Based on the proposed depth sensor simulator, we formulate the simulated RGBD data generation pipeline as $D = Sim(\mathcal{S}, \mathcal{C})$, where $\mathcal{S} = \{\mathcal{O}, \mathcal{M}, \mathcal{L}, \mathcal{B}\}$ denotes scene-related simulation parameters in the virtual environment, including $\mathcal{O}$ the setting of the objects with random categories, poses, arrangements, and scales, $\mathcal{M}$ the setting of random object materials from specular, transparent, to diffuse, $\mathcal{L}$ the setting of environment lighting from varying scenes with different intensities, $\mathcal{B}$ the setting of background floor with diverse materials. $\mathcal{C}$ is the cameras' statue parameters, consisting of intrinsic and extrinsic parameters, the pattern image, baseline distance, *etc.*. Taking these settings as input, the proposed simulator Sim generates the realistic RGB and depth images D.

To construct scenes with sufficient variations so that the proposed method can generalize to the real, we adopt domain randomization to enhance the generation, considering all these aspects. See supplementary materials for more details.

3.4 Simulated Dataset: DREDS

Making use of domain randomization and depth simulation, we construct the large-scale simulated dataset, DREDS. In total, DREDS dataset consists of two subsets: 1) **DREDS-CatKnown**: 100,200 training and 19,380 testing RGBD images made of 1,801 objects spanning 7 categories from ShapeNetCore [5], with randomized specular, transparent, and diffuse materials, 2) **DREDS-CatNovel**: 11,520 images of 60 category-novel objects, which is transformed from GraspNet-1Billion [8] that contains CAD models and annotates poses, by changing their object materials to specular or transparent, to verify the ability of our method to generalize to new object categories. Examples of paired simulated RGBD images of DREDS-Catknown and DREDS-CatNovel datasets are shown in Fig. 2.

4 STD Dataset

4.1 Real-world Dataset: STD

To further examine the proposed method in real scenes, we curate a real-world dataset, composed of Specular, Transparent, and Diffuse objects, which we call it STD dataset. Similar to DREDS dataset, STD dataset contains 1) **STD-CatKnown**: the subset with category-level objects, for the evaluation of depth restoration and category-level pose estimation tasks, and 2) **STD-CatNovel**: the subset with category-novel objects for evaluating the generalization ability of the proposed SwinDRNet method. Figure 3 shows the scene examples and annotations of STD dataset.

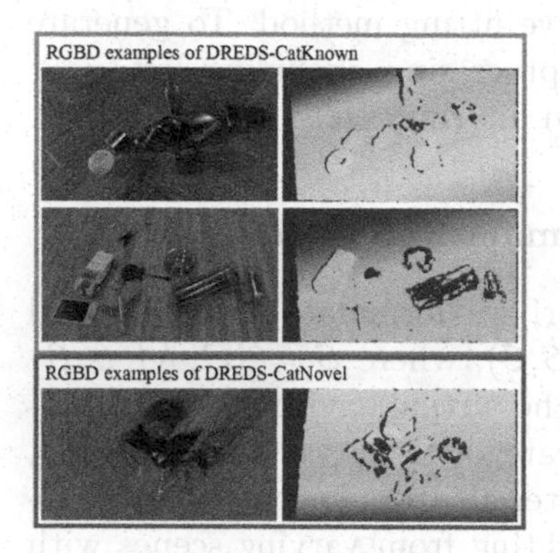

Fig. 2. RGBD examples of DREDS dataset.

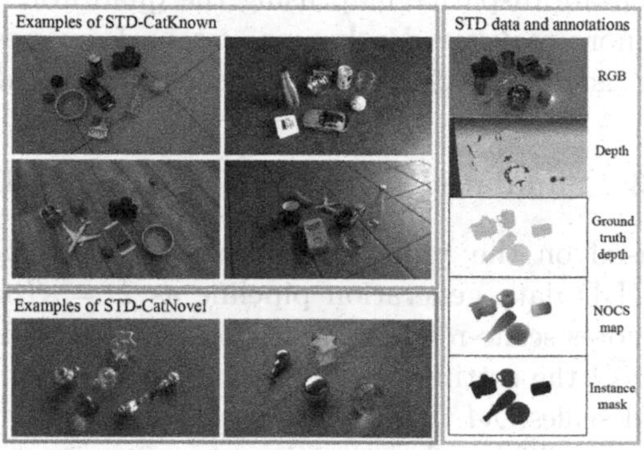

Fig. 3. Scene examples and annotations of STD dataset.

4.2 Data Collection

We collect an object set, covering specular, transparent, and diffuse materials. Specifically, for STD-CatKnown dataset, we collect 42 instances from 7 known ShapeNetCore [5] categories, and several category-unseen objects from the YCB dataset [4] and our own as the distractors. For STD-CatNovel dataset, we pick 8 specular and transparent objects from unseen categories. For each object except the distractors, we utilize the photogrammetry-based reconstruction tool, Object Capture API [2], to obtain its clean and accurate 3D mesh for ground truth poses annotation, so that we can yield ground truth depth and object masks.

We capture data from 30 different scenes (25 for STD-CatKnown, 5 for STD-CatNovel) with various backgrounds and illuminations, using RealSense D415. In each scene, over 4 objects with random arrangements are placed in a cluttered way. The sensor moves around the objects in an arbitrary trajectory. In total, we take 22,500 RGBD frames for STD-CatKnown, and 4,500 for STD-CatNovel.

Overall, the proposed real-world STD dataset consists of 27K RGBD frames, 30 diverse scenes, and 50 category-level and category-novel objects, making it facilitate the further generalizable object perception and grasping research.

5 Method

In this section, we introduce our network for depth restoration in Sect. 5.1 and then introduce the methods we used for downstream tasks, *i.e.* category-level 6D object pose estimation and robotic grasping, in Sect. 5.2.

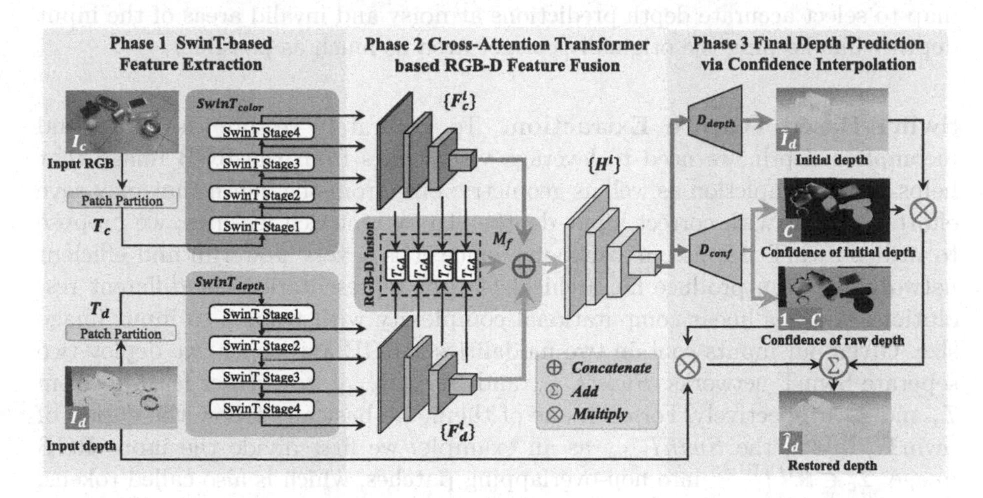

Fig. 4. Overview of our proposed depth restoration network SwinDRNet. We first extract the multi-scale features of RGB and depth image in phase 1, respectively. Next, in phase 2, our network fuse features of different modalities. Finally, we generate the initial depth map and confidence maps via two decoders, respectively, and fuse the raw depth and initial depth using the predicted confidence map.

5.1 SwinDRNet for Depth Restoration

Overview. To restore the noisy and incomplete depth, we propose a SwinTransformer [15] based depth restoration network, namely **SwinDRNet**.

SwinDRNet takes as input a RGB image $\mathcal{I}_c \in \mathbb{R}^{H\times W\times 3}$ along with its aligned depth image $\mathcal{I}_d \in \mathbb{R}^{H\times W}$ and outputs a refined depth $\hat{\mathcal{I}}_d \in \mathbb{R}^{H\times W}$ that restores the error area of the depth image and completes the invalid area, where H and W are the input image sizes.

We notice that prior works, *e.g.* PVN3D [9], usually leverage a heterogeneous architecture that extracts CNN features from RGB and extracts PointNet++ [24] features from depth. We, for the first time, devise a homogeneous and mirrored architecture that only leverages SwinT to extract and hierarchically fuse the RGB and depth features.

As shown in Fig. 4, the architecture of SwinDRNet is a two-stream fused encoder-decoder and can be further divided into three phases: in the first phase of feature extraction, we leverage two separate SwinT backbones to extract hierarchical features $\{\mathcal{F}_c^i\}$ and $\{\mathcal{F}_d^i\}$ from the input RGB image $\mathcal{I}_c$ and depth $\mathcal{I}_d$, respectively; In the second stage of RGBD feature fusion, we propose a fusion module M_f that utilizes cross-attention transformers to combine the features from the two streams and generate fused hierarchical features $\{\mathcal{H}^i\}$; and finally in the third phase, we propose two decoder modules, the depth decoder module D_{depth} decodes the fused feature into a raw depth and the confidence decoder module D_{conf} outputs a confidence map of the predicted raw depth, and from the outputs we can compute the final restored depth by using the confidence map to select accurate depth predictions at noisy and invalid areas of the input depth while keeping the originally correct area as much as possible.

SwinT-Based Feature Extraction. To accurately restore the noisy and incomplete depth, we need to leverage visual cues from the RGB image that helps depth completion as well as geometric cues from the depth that may save efforts at areas with correct input depths. To extract rich features, we propose to utilize SwinT [15] as our backbone, since it is a very powerful and efficient network that can produce hierarchical feature representations at different resolutions and has linear computational complexity with respect to input image size. Given our inputs contain two modalities – RGB and depth, we deploy two seperate SwinT networks, $SwinT_{\text{color}}$ and $SwinT_{\text{depth}}$, to extract features from $\mathcal{I}_c$ and $\mathcal{I}_d$, respectively. For each one of them, we basically follow the design of SwinT. Taking the $SwinT_{\text{color}}$ as an example: we first divide the input RGB image $\mathcal{I}_c \in \mathbb{R}^{H \times W \times 3}$ into non-overlapping patches, which is also called tokens, $\mathcal{T}_c \in \mathbb{R}^{\frac{H}{4} \times \frac{W}{4} \times 48}$; we then pass $\mathcal{T}_c$ through the four stages of SwinT to generate the multi-scale features $\{\mathcal{F}_c^i\}$, which are especially useful for dense depth prediction thanks to the hierarchical structure. The encoder process can be formulated as:

$$\{\mathcal{F}_c^i\}_{i=1,2,3,4} = SwinT_{\text{color}}(\mathcal{T}_c), \tag{1}$$

$$\{\mathcal{F}_d^i\}_{i=1,2,3,4} = SwinT_{\text{depth}}(\mathcal{T}_d). \tag{2}$$

where $\mathcal{F}^i \in \mathbb{R}^{\frac{H}{4i} \times \frac{W}{4i} \times iC}$ and C is the output feature dimension of the linear embedding layer in the first stage of SwinT.

Cross-Attention Transformer Based RGB-D Feature Fusion. Given the hierarchical features $\{\mathcal{F}_c^i\}$ and $\{\mathcal{F}_d^i\}$ from the two-stream SwinT backbone, our RGB-D fusion module M_f leverages cross-attention transformers to fuse the corresponding $\mathcal{F}_c^i$ and $\mathcal{F}_d^i$ into $\mathcal{H}^i$. For attending feature $\mathcal{F}_\mathcal{A}$ to $\mathcal{F}_\mathcal{B}$, a common cross-attention transformer T_{CA} first calculates the query vector Q_A from $\mathcal{F}_A$ and the key K_B and value V_B vectors from feature $\mathcal{F}_B$:

$$Q_A = \mathcal{F}_A \cdot W_q, \quad K_B = \mathcal{F}_B \cdot W_k, \quad V_B = \mathcal{F}_B \cdot W_v, \tag{3}$$

where Ws are the learnable parameters, and then computes the cross-attention feature $\mathcal{H}_{\mathcal{F}_A \to \mathcal{F}_B}$ from $\mathcal{F}_A$ to $\mathcal{F}_B$:

$$\mathcal{H}_{\mathcal{F}_A \to \mathcal{F}_B} = T_{CA}(\mathcal{F}_A, \mathcal{F}_B) = \text{softmax}\left(\frac{Q_A \cdot K_B^T}{\sqrt{d_K}}\right) \cdot V_B, \tag{4}$$

where d_K is the dimension of Q and K.

In our module M_f, we leverage bidirectional cross-attention by deploying two cross-attention transformers to obtained the cross-attention features from both directions, and then concatenates them with the original features to form the fused hierarchical features $\{\mathcal{H}^i\}$, as shown below:

$$\mathcal{H}^i = \mathcal{H}_{\mathcal{F}_c^i \to \mathcal{F}_d^i} \bigoplus \mathcal{H}_{\mathcal{F}_d^i \to \mathcal{F}_c^i} \bigoplus \mathcal{F}_c^i \bigoplus \mathcal{F}_d^i, \tag{5}$$

where $\bigoplus$ represents concatenation along the channel axis.

Final Depth Prediction via Confidence Interpolation. The credible area of the input depth map (*e.g.*, the edges of specular or transparent objects in contact with background or diffusive objects) plays a critical role in providing information about spatial arrangement. Inspired by the previous works [10, 30], we make use of a confidence map between the raw and predicted depth maps. However, unlike [10, 30] predicting the confidence map between the multi-modality, we focus on preserving the correct original value to generate more realistic depth maps with less distortion. The final depth map can be formulated as:

$$\hat{\mathcal{I}}_d = C \bigotimes \tilde{\mathcal{I}}_d + (1 - C) \bigotimes \mathcal{I}_d \tag{6}$$

where $\bigotimes$ represents elementwise multiplication, and $\hat{\mathcal{I}}_d$ and $\tilde{\mathcal{I}}_d$ denote the final restored depth and the output of depth decoder head, respectively.

Loss Functions. For SwinDRNet training, we supervise both the final restored depth $\hat{\mathcal{I}}_d$ and the output of depth decoder head $\tilde{\mathcal{I}}_d$, which is formulated as:

$$\mathcal{L} = \omega_{\tilde{\mathcal{I}}_d} \mathcal{L}_{\tilde{\mathcal{I}}_d} + \omega_{\hat{\mathcal{I}}_d} \mathcal{L}_{\hat{\mathcal{I}}_d}, \tag{7}$$

where $\mathcal{L}_{\hat{\mathcal{I}}_d}$ and $\mathcal{L}_{\tilde{\mathcal{I}}_d}$ are the losses of $\hat{\mathcal{I}}_d$ and $\tilde{\mathcal{I}}_d$, respectively. $\omega_{\hat{\mathcal{I}}_d}$ and $\omega_{\tilde{\mathcal{I}}_d}$ are weighting factors. Each of the two loss can be formulated as:

$$\mathcal{L}_i = \omega_n \mathcal{L}_n + \omega_d \mathcal{L}_d + \omega_g \mathcal{L}_g, \tag{8}$$

where $\mathcal{L}_n$, $\mathcal{L}_d$ and $\mathcal{L}_g$ are the L1 losses between the predicted and ground truth surface normal, depth and the gradient map of depth image, respectively. ω_n, ω_d and ω_g are the weights for different losses. We further add higher weight to the loss within the foreground region, to push the network to concentrate more on the objects.

5.2 Downstream Tasks

Category-Level 6D Object Pose Estimation. Inspired by [31], we use the same backbone with SwinDRNet, and add two decoder heads to predict coordinates of the NOCS map and semantic segmentation mask. Then we follow the method [31], perform pose fitting between the restored object point clouds in the world coordinate space and the predicted object point clouds in the normalized object coordinate space, and perform pose fitting to get the 6D object pose.

Robotic Grasping. By combining SwinDRNet to the object grasping task, we can analyze the performance of depth restoration on the robotic manipulation. We adopt the end-to-end network, GraspNet-baseline [8], to predict the 6-DoF grasping poses directly from the scene point cloud. Given the restored depth map from SwinDRNet, the scene point cloud is transformed and sent to GraspNet-baseline. Then the model predicts the grasp candidates. Finally, the gripper of the parallel-jaw robot arm executes the target rotation and position selected from those candidates.

6 Tasks, Benchmarks and Results

In this section, we train our SwinDRNet on the train split of DREDS-CatKnown dataset and deploy it on the tasks including category-level 6D object pose estimation and robotic grasping.

6.1 Depth Restoration

Evaluation Metrics. We follow the metrics of transparent objects depth completion in [38]: 1) **RMSE**: the root mean squared error, 2) **REL**: the mean absolute relative difference, 3) **MAE**: the mean absolute error, 4) the percentage of d_i satisfying $max(\frac{d_i}{d_i^*}, \frac{d_i^*}{d_i}) < \delta$, where d_i denotes the predicted depth, d_i^* is GT and $\delta \in \{1.05, 1.10, 1.25\}$. We resize the prediction and GT to 126×224 resolution for fair comparisons, and evaluate in all objects area and challenging area (specular and transparent objects), respectively.

Baselines. We compare our method with several state-of-the-art methods, including LIDF [38], the SOTA method for depth completion of transparent objects, and NLSPN [19], the SOTA method for depth completion on NYUv2 [29] dataset. All baselines are trained on the train split of DREDS-CatKnown and evaluated on four types of testing data: 1) the test split of DREDS-CatKnown: simulated images of category-known objects. 2) DREDS-CatNovel: simulated images of category-novel objects. 3) STD-CatKnown: real images of category-known objects; 4) STD-CatNovel. real images of category-novel objects.

Results. The quantitative results reported in Table 2 show that we achieve the best performance compared to other methods on DREDS and STD datasets, and have a powerful generalization ability to transfer to not only novel category objects in the simulation environment but also in the real world. Moreover, SwinDRNet (30 FPS) is significantly faster than LIDF (13 FPS) and the two-branch baseline that uses PointNet++ on depth (6 FPS). Although it is a little slower than NLSPN (35 FPS) because the code still has room for further optimization and acceleration, SwinDRNet is real-time for downstream tasks. The methods are all evaluated on an NVIDIA RTX 3090 GPU.

Sim-to-Real and Domain Transfer. We perform sim-to-real and domain transfer experiments to verify the generalization ability of the DREDS dataset. For sim-to-real experiments, SwinDRNet is trained on DREDS-CatKnown, but takes different depth images as input of training (one follow [38] and takes the cropped synthetic depth image as input, and another takes the simulated depth image). The results evaluated on STD in Table 3 reveal the powerful potential of our depth simulation pipeline, which can significantly close the sim-to-real gap and generalize to the new categories. For domain transfer experiments, we train SwinDRNet on the train split of DREDS-CatKnown dataset and evaluate on Cleargrasp dataset. The results reported in Table 4 testify that model only trained on DREDS-CatKnown can easily generalize to the new domain Claer-Grasp and outperform the previous results directly trained on ClearGrasp and Omniverse [38] (LIDF train on Omniverse and ClearGrasp), which verifies the generalization ability of our dataset.

Table 2. Quantitative comparison to state-of-the-art methods on DREDS and STD. ↓ means lower is better, ↑ means higher is better. The left of '/' shows the results evaluated on all objects, and the right of '/' shows the results evaluated on specular and transparent objects. Note that only one result is reported on STD-CatNovel, because all the objects are specular or transparent.

Methods	RMSE↓	REL↓	MAE↓	$\delta_{1.05}$ ↑	$\delta_{1.10}$ ↑	$\delta_{1.25}$ ↑
	DREDS-CatKnown (Sim)					
NLSPN	**0.010**/0.011	0.009/0.011	0.006/0.007	97.48/96.41	99.51/99.12	99.97/99.74
LIDF	0.016/0.015	0.018/0.017	0.011/0.011	93.60/94.45	98.71/98.79	99.92/99.90
Ours	**0.010/0.010**	**0.008/0.009**	**0.005/0.006**	**98.04/97.76**	**99.62/99.57**	**99.98/99.97**
	DREDS-CatNovel (Sim)					
NLSPN	0.026/0.031	0.039/0.054	0.015/0.021	78.90/69.16	89.02/83.55	97.86/96.84
LIDF	0.082/0.082	0.183/0.184	0.069/0.069	23.70/23.69	42.77/42.88	75.44/75.54
Ours	**0.022/0.025**	**0.034/0.044**	**0.013/0.017**	**81.90/75.27**	**92.18/89.15**	**98.39/97.81**
	STD-CatKnown (Real)					
NLSPN	0.114/0.047	0.027/0.031	0.015/0.018	94.83/89.47	98.37/97.48	99.38/99.32
LIDF	0.019/0.022	0.019/0.023	0.013/0.015	93.08/90.32	98.39/97.38	99.83/99.62
Ours	**0.015/0.018**	**0.013/0.016**	**0.008/0.011**	**96.66/94.97**	**99.03/98.79**	**99.92/99.85**
	STD-CatNovel (Real)					
NLSPN	0.087	0.050	0.025	**81.95**	90.36	96.06
LIDF	0.041	0.060	0.031	53.69	79.80	99.63
Ours	**0.025**	**0.033**	**0.017**	81.55	**93.10**	**99.84**

Table 3. Quantitative results for Sim-to-Real. *Synthetic* means taking the cropped synthetic depth images for training, and *Simulated* means taking the simulated depth images from the train split of DREDS-CatKnown for training.

Trainset	RMSE↓	REL↓	MAE↓	$\delta_{1.05}$ ↑	$\delta_{1.10}$ ↑	$\delta_{1.25}$ ↑
	STD-CatKnown (Real)					
Synthetic	0.0467/0.056	0.0586/0.070	0.0377/0.047	49.12/39.42	86.50/79.85	98.98/97.66
Simulated	**0.015/0.018**	**0.013/0.016**	**0.008/0.011**	**96.66/94.97**	**99.03/98.79**	**99.92/99.85**
	STD-CatNovel (Real)					
Synthetic	0.065	0.101	0.053	21.04	55.87	96.96
Simulated	**0.025**	**0.033**	**0.017**	**81.55**	**93.10**	**99.84**

Table 4. Quantitative results for domain transfer. *The previous best results* means that the best previous method is trained on ClearGrasp and Omniverse, and evaluated on ClearGrasp. *Domain transfer* means that SwinDRNet is trained on DREDS-CatKnown and evaluated on ClearGrasp.

Model	RMSE↓	REL↓	MAE↓	$\delta_{1.05}$ ↑	$\delta_{1.10}$ ↑	$\delta_{1.25}$ ↑
	ClearGrasp real-known					
The previous best results	0.028	0.033	0.020	82.37	92.98	98.63
Domain transfer	**0.022**	**0.017**	**0.012**	**91.46**	**97.47**	**99.86**
	ClearGrasp real-novel					
The previous best results	0.025	0.036	0.020	79.5	94.01	99.35
Domain transfer	**0.016**	**0.008**	**0.005**	**96.73**	**98.83**	**99.78**

6.2 Category-Level Pose Estimation

Evaluation Metrics. We use two aspects of metrics to evaluate: 1) **3D IoU.** It computes the intersection over union of ground truth and predicted 3D bounding boxes. We choose the threshold of 25% (IoU25), 50%(IoU50) and 75%(IoU75) for this metric. 2) **Rotation and translation errors.** It computes the rotation and translation errors between the ground truth pose and predicted pose. We choose 5°2cm, 5°5cm, 10°2cm, 10°5cm, 10°10cm for this metric.

Baselines. We choose two models as baselines to show the usefulness of the restored depth for category-level pose estimation and the effectiveness of SwinDRNet+NOCSHead: 1) **NOCS** [31]. It takes a RGB image as input to predict the per-pixel normalized coordinate map and obtain the pose with the help of the depth map. 2) **SGPA** [6]. The state-of-the-art method. It leverages one object and its corresponding category prior to dynamically adapting the prior to the observed object. Then the prior adaptation is used to reconstruct the 3D canonical model of the specific object for pose fitting.

Results. To verify the usefulness of the restored depth, we report the results of three methods using raw or restored (output of SwinDRNet) depth in Table 5. *-only* means using raw depth in the whole experiment, *Refined depth+* means using restored depth for pose fitting in NOCS and SwinDRNet+NOCSHead. Due to the fact that SGPA deforms the point cloud to get the results which are sensitive to depth, we use restored depth for both training and inference. We observe that restored depth improves the performance of three methods by large margins under all the metrics on both dataset. These performance gains suggest that depth restoration is truly useful for category-level pose estimation. Moreover, SwinDRNet+NOCSHead outperforms NOCS and SGPA under all the metrics.

Table 5. Quantitative results for category-level pose estimation. *only* means using raw depth in the whole experiment,*Refined* means using restored depth for training and inference in SGPA and for pose fitting in NOCS and our method.

Methods	IoU25	IoU50	IoU75	5°2cm	5°5cm	10°2cm	10°5cm	10°10cm
	DREDS-CatKnown (Sim)							
NOCS-only	85.7	66.0	23.0	21.3	25.4	40.0	47.9	49.0
SGPA-only	79.5	66.7	49.1	29.5	32.5	48.7	54.7	55.7
Refined depth + NOCS	86.7	73.2	40.7	30.4	31.8	54.1	57.5	57.6
Refined depth + SGPA	82.3	72.0	60.5	45.9	46.8	66.4	68.4	68.5
Ours-only	94.3	82.5	57.9	34.5	37.6	55.7	62.6	63.2
Refined depth + Ours	**94.7**	**84.8**	**68.0**	**49.1**	**50.1**	**69.8**	**72.4**	**72.5**
	STD-CatKnown (Real)							
NOCS-only	83.2	66.9	16.9	20.4	26.0	37.9	52.5	53.5
SGPA-only	77.6	67.1	46.6	30.0	32.3	47.7	53.3	53.9
Refined depth + NOCS	82.6	72.6	35.6	28.5	30.0	54.4	57.6	57.7
Refined depth + SGPA	78.8	71.6	62.8	49.3	49.7	70.5	71.5	71.6
Ours-only	**92.4**	87.4	61.7	37.9	42.6	57.8	70.6	71.0
Refined depth + Ours	**92.4**	**88.0**	**75.9**	**52.9**	**53.8**	**77.1**	**79.1**	**79.1**

6.3 Robotic Grasping

Experiments Setting. We conduct real robot experiments to evaluate the depth restoration performance on robotic grasping tasks. In our physical setup, we use a 7-DoF Panda robot arm from Franka Emika with a parallel-jaw gripper. RealSense D415 depth sensor is mounted on the tripod in front of the arm. We set 6 rounds of table clearing experiments. For each round, 4 to 5 specular and transparent objects are randomly picked from STD objects to construct a cluttered scene. For each trial, the robot arm executes the grasping pose with the highest score, and removes the grasped object until the workspace is cleared, or 10 attempts are reached.

Evaluation Metrics. Real grasping performance is measured using the following metrics: 1) **Success Rate**: the ratio of grasped object number and attempt number, 2) **Completion Rate**: the ratio of successfully removed object number and the original object number in a scene.

Baselines. We follow the 6-DoF grasping pose prediction network GraspNet-baseline, using the released pretrained model. *GraspNet* means GraspNet-baseline directly takes the captured raw depth as input, while *SwinDR-Net+GraspNet* means the network receives the refined point cloud from SwinDRNet that is trained only on DREDS-CatKnown dataset.

Table 6. Results of real robot experiments. *#Objects* denotes the sum of grasped object numbers in all rounds. *#Attempts* denotes the sum of robotic grasping attempt numbers in all rounds.

Methods	#Objects	#Attempts	Success rate	Completion rate
GraspNet	19	49	38.78%	40%
SwinDRNet+GraspNet	25	26	**96.15%**	**100%**

Results. Table 6 reports the performance of real robot experiments. *SwinDR-Net+GraspNet* obtains high success rate and completion rate, while *GraspNet* is lower. Without depth restoration, it is difficult for a robot arm to grasp specular and transparent objects due to the severely incomplete and inaccurate raw depth. The proposed SwinDRNet significantly improves the performance of specular and transparent object grasping.

7 Conclusions

In this work, we propose a powerful RGBD fusion network, SwinDRNet, for depth restoration. Our proposed framework, DREDS, synthesizes a large-scale RGBD dataset with realistic sensor noises, so as to close the sim-to-real gap for specular and transparent objects. Furthermore, we collect a real dataset STD, for real-world performance evaluation. Evaluations on depth restoration, category-level pose estimation, and object grasping tasks demonstrate the effectiveness of our method.

References

1. Blender. https://www.blender.org/
2. Object capture API on macos. https://developer.apple.com/augmented-reality/object-capture/
3. Breyer, M., Chung, J.J., Ott, L., Roland, S., Juan, N.: Volumetric grasping network: Real-time 6 DOF grasp detection in clutter. In: Conference on Robot Learning (2020)

4. Calli, B., et al.: Yale-CMU-berkeley dataset for robotic manipulation research. Int. J. Robot. Res. **36**(3), 261–268 (2017)
5. Chang, A.X., et al.: ShapeNet: An information-rich 3d model repository. arXiv preprint arXiv:1512.03012 (2015)
6. Chen, K., Dou, Q.: SGPA: structure-guided prior adaptation for category-level 6D object pose estimation. In: Proceedings of the IEEE/CVF International Conference on Computer Vision, pp. 2773–2782 (2021)
7. Eigen, D., Puhrsch, C., Fergus, R.: Depth map prediction from a single image using a multi-scale deep network. In: Advances in Neural Information Processing Systems 27 (2014)
8. Fang, H.S., Wang, C., Gou, M., Lu, C.: Graspnet-1billion: a large-scale benchmark for general object grasping. In: Proceedings of the IEEE/CVF Conference on Computer Vision and Pattern Recognition, pp. 11444–11453 (2020)
9. He, Y., Sun, W., Huang, H., Liu, J., Fan, H., Sun, J.: PVN3D: a deep point-wise 3D keypoints voting network for 6DoF pose estimation. In: Proceedings of the IEEE/CVF Conference on Computer Vision and Pattern Recognition, pp. 11632–11641(2020)
10. Hu, M., Wang, S., Li, B., Ning, S., Fan, L., Gong, X.: PENET: towards precise and efficient image guided depth completion. In: 2021 IEEE International Conference on Robotics and Automation (ICRA), pp. 13656–13662. IEEE (2021)
11. Jiang, Z., Zhu, Y., Svetlik, M., Fang, K., Zhu, Y.: Synergies between affordance and geometry: 6-DoF grasp detection via implicit representations. Robot.: Sci. Syst. (2021)
12. Jiao, J., Cao, Y., Song, Y., Lau, R.: Look deeper into depth: monocular depth estimation with semantic booster and attention-driven loss. In: Proceedings of the European conference on computer vision (ECCV), pp. 53–69 (2018)
13. Khirodkar, R., Yoo, D., Kitani, K.: Domain randomization for scene-specific car detection and pose estimation. In: 2019 IEEE Winter Conference on Applications of Computer Vision (WACV), pp. 1932–1940. IEEE (2019)
14. Landau, M.J., Choo, B.Y., Beling, P.A.: Simulating kinect infrared and depth images. IEEE Trans. Cybernet. **46**(12), 3018–3031 (2015)
15. Liu, Z., et al.: Swin transformer: Hierarchical vision transformer using shifted windows. In: Proceedings of the IEEE/CVF International Conference on Computer Vision, pp. 10012–10022 (2021)
16. Long, X., et al.: Adaptive surface normal constraint for depth estimation. In: Proceedings of the IEEE/CVF International Conference on Computer Vision, pp. 12849–12858 (2021)
17. Mo, K., Guibas, L.J., Mukadam, M., Gupta, A., Tulsiani, S.: Where2Act: from pixels to actions for articulated 3D objects. In: Proceedings of the IEEE/CVF International Conference on Computer Vision, pp. 6813–6823 (2021)
18. Mu, T., et al.: ManiSkill: generalizable manipulation skill benchmark with large-scale demonstrations. In: Annual Conference on Neural Information Processing Systems (NeurIPS) (2021)
19. Park, J., Joo, K., Hu, Z., Liu, C.-K., So Kweon, I.: Non-local spatial propagation network for depth completion. In: Vedaldi, A., Bischof, H., Brox, T., Frahm, J.-M. (eds.) ECCV 2020. LNCS, vol. 12358, pp. 120–136. Springer, Cham (2020). https://doi.org/10.1007/978-3-030-58601-0_8
20. Peng, X.B., Andrychowicz, M., Zaremba, W., Abbeel, P.: Sim-to-real transfer of robotic control with dynamics randomization. In: 2018 IEEE International Conference on Robotics and Automation (ICRA), pp. 3803–3810. IEEE (2018)

21. Planche, B., Singh, R.V.: Physics-based differentiable depth sensor simulation. In: Proceedings of the IEEE/CVF International Conference on Computer Vision, pp. 14387–14397 (2021)
22. Planche, B., et al.: DepthSynth: real-time realistic synthetic data generation from cad models for 2.5 D recognition. In: 2017 International Conference on 3D Vision (3DV), pp. 1–10. IEEE (2017)
23. Prakash, A., et al.: Structured domain randomization: bridging the reality gap by context-aware synthetic data. In: 2019 International Conference on Robotics and Automation (ICRA), pp. 7249–7255. IEEE (2019)
24. Qi, C.R., Yi, L., Su, H., Guibas, L.J.: PointNet++: deep hierarchical feature learning on point sets in a metric space. In: Advances in Neural Information Processing Systems 30 (2017)
25. Qu, C., Liu, W., Taylor, C.J.: Bayesian deep basis fitting for depth completion with uncertainty. In: Proceedings of the IEEE/CVF International Conference on Computer Vision, pp. 16147–16157 (2021)
26. Sajjan, S., et al.: Clear grasp: 3D shape estimation of transparent objects for manipulation. In: 2020 IEEE International Conference on Robotics and Automation (ICRA), pp. 3634–3642. IEEE (2020)
27. Tobin, J., Fong, R., Ray, A., Schneider, J., Zaremba, W., Abbeel, P.: Domain randomization for transferring deep neural networks from simulation to the real world. In: 2017 IEEE/RSJ International Conference on Intelligent Robots and Systems (IROS), pp. 23–30. IEEE (2017)
28. Tremblay, J., et al.: Training deep networks with synthetic data: bridging the reality gap by domain randomization. In: Proceedings of the IEEE Conference on Computer Vision and Pattern Recognition Workshops, pp. 969–977 (2018)
29. Uhrig, J., Schneider, N., Schneider, L., Franke, U., Brox, T., Geiger, A.: Sparsity invariant CNNs. In: 2017 International Conference on 3D Vision (3DV), pp. 11–20. IEEE (2017)
30. Van Gansbeke, W., Neven, D., De Brabandere, B., Van Gool, L.: Sparse and noisy lidar completion with RGB guidance and uncertainty. In: 2019 16th International Conference on Machine Vision Applications (MVA), pp. 1–6. IEEE (2019)
31. Wang, H., Sridhar, S., Huang, J., Valentin, J., Song, S., Guibas, L.J.: Normalized object coordinate space for category-level 6D object pose and size estimation. In: Proceedings of the IEEE/CVF Conference on Computer Vision and Pattern Recognition, pp. 2642–2651 (2019)
32. Weng, Y., et al.: CAPTRA: category-level pose tracking for rigid and articulated objects from point clouds. In: Proceedings of the IEEE/CVF International Conference on Computer Vision, pp. 13209–13218 (2021)
33. Xiong, X., Xiong, H., Xian, K., Zhao, C., Cao, Z., Li, X.: Sparse-to-dense depth completion revisited: sampling strategy and graph construction. In: Vedaldi, A., Bischof, H., Brox, T., Frahm, J.-M. (eds.) ECCV 2020. LNCS, vol. 12366, pp. 682–699. Springer, Cham (2020). https://doi.org/10.1007/978-3-030-58589-1_41
34. Xu, H., Wang, Y.R., Eppel, S., Aspuru-Guzik, A., Shkurti, F., Garg, A.: Seeing glass: joint point-cloud and depth completion for transparent objects. In: 5th Annual Conference on Robot Learning (2021)
35. Yue, X., Zhang, Y., Zhao, S., Sangiovanni-Vincentelli, A., Keutzer, K., Gong, B.: Domain randomization and pyramid consistency: Simulation-to-real generalization without accessing target domain data. In: Proceedings of the IEEE/CVF International Conference on Computer Vision, pp. 2100–2110 (2019)

36. Zakharov, S., Kehl, W., Ilic, S.: DeceptionNet: network-driven domain randomization. In: Proceedings of the IEEE/CVF International Conference on Computer Vision, pp. 532–541 (2019)
37. Zhang, X., et al.: Close the visual domain gap by physics-grounded active stereovision depth sensor simulation. arXiv preprint arXiv:2201.11924 (2022)
38. Zhu, L., et al.: RGB-D local implicit function for depth completion of transparent objects. In: Proceedings of the IEEE/CVF Conference on Computer Vision and Pattern Recognition, pp. 4649–4658 (2021)

Resolving Copycat Problems in Visual Imitation Learning via Residual Action Prediction

Chia-Chi Chuang[1], Donglin Yang[1], Chuan Wen[1], and Yang Gao[1,2(✉)]

[1] Institute for Interdisciplinary Information Sciences, Tsinghua University, Beijing, China
{zhuangjq19,ydl18,cwen20}@mails.tsinghua.edu.cn,
gaoyangiiis@tsinghua.edu.cn
[2] Shanghai Qi Zhi Institute, Shanghai, China

Abstract. Imitation learning is a widely used policy learning method that enables intelligent agents to acquire complex skills from expert demonstrations. The input to the imitation learning algorithm is usually composed of both the current observation and historical observations since the most recent observation might not contain enough information. This is especially the case with image observations, where a single image only includes one view of the scene, and it suffers from a lack of motion information and object occlusions. In theory, providing multiple observations to the imitation learning agent will lead to better performance. However, surprisingly people find that sometimes imitation from observation histories performs worse than imitation from the most recent observation. In this paper, we explain this phenomenon from the information flow within the neural network perspective. We also propose a novel imitation learning neural network architecture that does not suffer from this issue by design. Furthermore, our method scales to high-dimensional image observations. Finally, we benchmark our approach on two widely used simulators, CARLA and MuJoCo, and it successfully alleviates the copycat problem and surpasses the existing solutions.

Keywords: Imitation learning · Autonomous driving · Copycat problem

1 Introduction

Learning to control in complex environments is a challenging task. Imitation learning is a powerful technique that learns useful skills from a pre-collected expert demonstration [1,16,29,31,42]. Compared with reinforcement

C.-C. Chuang and D. Yang—Equal contribution.

Supplementary Information The online version contains supplementary material available at https://doi.org/10.1007/978-3-031-19842-7_23.

S. Avidan et al. (Eds.): ECCV 2022, LNCS 13699, pp. 392–409, 2022.
https://doi.org/10.1007/978-3-031-19842-7_23

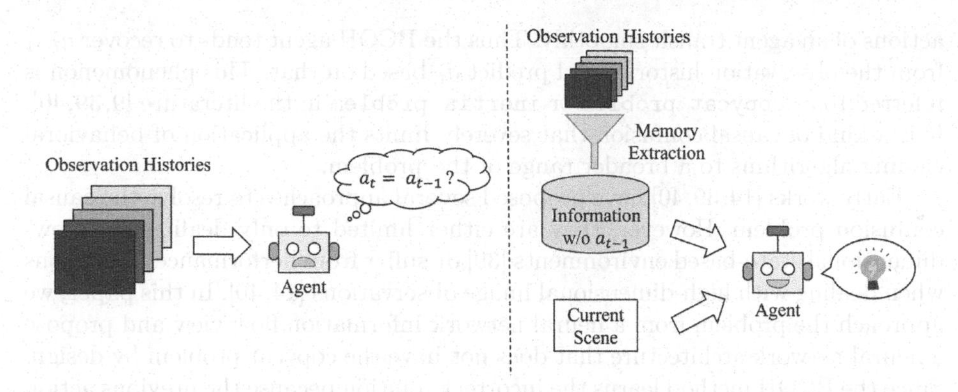

Fig. 1. The copycat problem & our solution: Left: Behavioral cloning from observation histories might learn a shortcut that directly outputs its previous action as the current action. Unfortunately, this leads to inferior performance during the test time. Right: We propose to solve the problem by a copycat-free memory extraction module. The history information is carefully extracted such that the shortcut no longer exists.

learning [24,32,33], imitation learning is generally more data-efficient and does not require destructive exploration. Behavioral cloning (BC) [4,13,19,25,30] is one of the most widely used imitation algorithms. It directly mimics the expert behavior by learning the mapping from the observations to the expert actions with supervised learning. Imitation learning has achieved many successes in the past in domains like autonomous driving [4,30] and drone flying [21]. Imitation learning has also become an essential component in many other policy learning algorithms [20,31].

In an Markov Decision Process (MDP), the input to the behavioral cloning algorithm can be the current state (Behavioral Cloning from Single Observation, i.e., BCSO). However, in practice, the agent's observation might be far from Markovian. This is the case, especially with visual observations. The most recent visual frame usually misses essential information, such as the objects' motion and appearance that are occluded in the current frame but visible in previous frames. Thus in many applications, it is more reasonable to provide the behavioral cloning algorithm with the observation histories (Behavioral Cloning from Observation Histories, i.e., BCOH) instead of a single observation. This allows the agent to access the state of the problem better.

However, recent works [14,38] find that behavioral cloning with observation histories can sometimes perform even worse than behavioral cloning from the most recent observation. This phenomenon is counter-intuitive: the more complete input features (observation history) should lead to better downstream performance, but in practice, it won't! [14,39–41] explain this phenomenon as BCOH is prone to learn an action prediction shortcut instead of the correct concept. BCOH predicts the current action a_t from the observation histories $o_t, o_{t-1}, \cdots$. However, the observation histories are actually obtained by executing previous expert actions $a_{t-1}, a_{t-2}, \cdots$ in the environments. Usually, the actions of an agent transit smoothly. Thus the BCOH agent tends to recover a_{t-1} from the observation histories and predict a_t based on that. This phenomenon is referred to as `copycat problem` or `inertia problem` in the literature [9,39,40]. It is a kind of causal confusion that severely limits the application of behavioral cloning algorithms to a broader range of the problem.

Early works [14,39,40] have proposed several approaches to resolve the causal confusion problem. However, they are either limited to only dealing with low-dimensional state-based environments [39] or suffer from performance limitations when dealing with high-dimensional image observations [14,40]. In this paper, we approach the problem from a neural network information flow view and propose a neural network architecture that does not have the copycat problem by design. Since the BCOH method learns the incorrect solution because the previous action a_{t-1} is a shortcut solution, we propose to design the neural network architecture such that it cannot learn to act based on the previous actions. Figure 1 illustrates the concept of why copycat problem occurs and our solution. Unlike previous approaches, which achieve similar effects by using adversarial training [39] or using per sample re-weighting [40], our policy avoids the incorrect solution more thoroughly by cutting the information flow of the wrong clue. Thus, our method trains more easily and achieves better task performances.

We benchmark our method in a challenging autonomous driving environment, CARLA [10]. CARLA is a visually realistic driving simulator. It is widely used as a benchmark for autonomous driving and imitation learning [8,9,15]. Besides CARLA, we evaluate the performance in a commonly used robotics simulator, MuJoCo [37], with image-based observation. Our method outperforms all previous approaches on these challenging benchmarks.

2 Related Work

Imitation Learning and Copycat Problems: Imitation learning [1,29,42] is a powerful technology to learn complex policies from expert demonstrations. Among the different imitation learning methods, behavioral cloning is a simple but effective paradigm that directly regresses the expert actions from the observations. However, like other imitation methods, behavioral cloning suffers from the distributional shift that small errors will accumulate over time during testing and finally make the imperfect imitators encounter out-of-distribution states [31]. We focus on a specific phenomenon arising under the distributional shift in the partially observed settings – the copycat problem [39–41]. Although environmental interactions [5,14,16] or a queryable expert [18,31,34–36] can resolve it, purely offline methods [2,39–41] are understudied. The specific definition and existing solutions for the copycat problem will be introduced in detail in Sect. 3.3. In this paper, we propose a new neural network architecture for imitation policies to solve the copycat problem more thoroughly.

Shortcut Learning: While the numerous success stories of deep neural networks (DNN) have rapidly spread over science, industry, and society, its limitations are coming into focus. For example, in computer vision, DNN image classifier tends to rely on texture rather than shape [12] and the background rather than objects [3,41]. And in NLP, some language models turn out to depend on spurious features like word length without understanding the content of a sentence [23,27]. [11] summarize this phenomenon and name it "shortcut learning," i.e., the DNNs prefer to learn the more straightforward solution (shortcut) rather than taking more effort to understand the intended solution. We regard the copycat problem as an instance of shortcut learning: when the scenes and labels change smoothly, the neural networks tend to cheat by extrapolating from the historical information, a shortcut solution. Beyond imitation learning, the copycat problem also exists in other sequential tasks, such as robotics manipulation [22], tracking [44], forecasting [17], etc. Our method may generalize to these tasks, and we leave this to our future work.

3 Preliminaries

3.1 Partially Observed Markov Decision Process

The decision process of an agent equipped with sensors can be best described by a partially observed Markov decision process (POMDP) since the equipped sensors such as cameras or LIDARs can only perceive part of the environment at each time step. A POMDP can be formalized as a tuple $(\mathbf{S}, \mathbf{A}, \mathbf{T}, r, \mathbf{O})$, where $\mathbf{S}$ is the state space of the environment, $\mathbf{A}$ is the action space of the agent, $\mathbf{T}$ is the transition probabilities between states with a given action, $r : \mathbf{S} \times \mathbf{A} \rightarrow \mathbb{R}$ is the reward function and $\mathbf{O}$ is the observation space of the agent. In practice, the dimension of $\mathbf{S}$ is much greater than the dimension of $\mathbf{O}$. At each time step t, the agent has no access to the underlying true state s_t and has to take the action a_t according to the observation o_t. To deal with partial observation, it is common to take the historical information into account [2,24,26,39,40], constructing observation history $\tilde{o}_t = [o_t, o_{t-1}, \cdots, o_{t-H}]$ where H is the history length.

In the imitation learning setup, we are given a demonstration dataset composed of observation-action tuples: $\{(o_t, a_t)\}$. They represent good behaviors, i.e., the trajectories that achieve high accumulated rewards. However, the reward is not provided in the imitation learning setup. Instead, the goal of imitation learning is to learn a function that maps observation histories $\tilde{o}_t$ to the expert action a_t that leads to high accumulated rewards.

3.2 Behavioral Cloning

Among all the variants of imitation learning, we focus on behavioral cloning (BC), a straightforward but powerful approach to mimic the expert behaviors from a pre-collected demonstration $\mathcal{D} = \{(o_i, a_i)\}_{i=1}^N$, where the N is the number of samples. BC reduces the complex policy learning into supervised learning by maximizing the log-likelihood of the action a_t conditioned on the observation history $\tilde{o}_t$ so that the agent can behave as similarly to the expert as possible, i.e., $\theta^* = \arg\max_\theta \mathbb{E}_\mathcal{D}[\log P(a_t|\tilde{o}_t; \theta)]$. $H = 0$ is BC from single observation (BCSO), and $H > 0$ is BC from observational histories (BCOH).

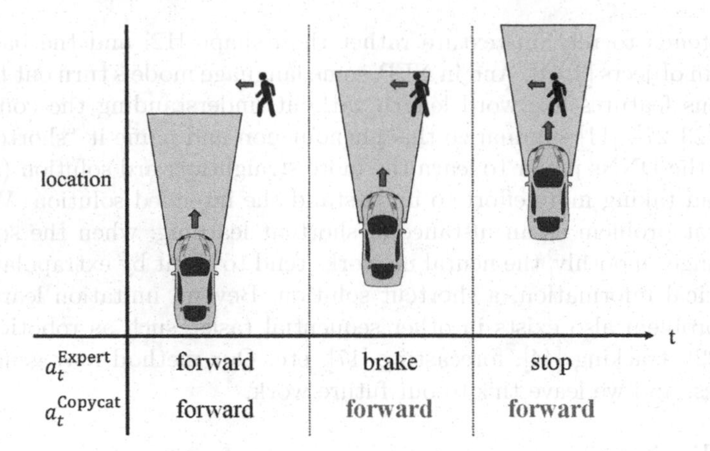

Fig. 2. An illustrative example of the copycat problem: The autonomous vehicle encounters a pedestrian. The trapezoidal area indicates the region can be observed by the driving agent. The expert agent would brake as soon as it sees the pedestrian, but the copycat agent tends to repeat previous actions and move forward.

3.3 Copycat Problem

The copycat problem is a phenomenon that the BC agent tends to infer a "copycat" action similar to the previous action. When the copycat problem happens, both the training and the validation loss are low, but the in-environment evaluation has poor performance. It is not an instance of the over-fitting problem. Prior works [2,9] showed using only a single observation o_t can achieve better performance than using multiple observations on their tasks. In addition, [14,39,40] show that the agent can completely fail when the input contains information about the previous actions. The copycat problem is a widely existing phenomenon in many downstream tasks when using BC, such as autonomous driving [38,40] and robot control [14,39].

Figure 2 displays a concrete example of the copycat problem. In a driving scenario, the expert brakes immediately when it observes a person crossing the road so it can stop timely to prevent a traffic accident from happening. On the other hand, a copycat agent tends to keep its own previous action and usually brakes late or doesn't brake at all. Similar failure cases have been confirmed by multiple previous works [38,40].

The reason why copycat problems happen in BC has been investigated in previous works [14,39,40]. It happens because there exists a shortcut from the input observation history $\tilde{o}_t$ to the target current action a_t. The input observation history is obtained by executing the expert actions $a_{t-H}, a_{t-H+1}, \cdots, a_{t-1}$ in the environment. Since most sequential decision problems exhibit a smooth transition property, expert actions also usually change smoothly. This means the previous action a_{t-1} is very close to the prediction target a_t. In many cases, the agent predicts the action a_t either completely or partially based on the information of the previous action a_{t-1}. However, decisions based on an agent's own previous action could be wrong in many cases, such as the vehicle stopping for the pedestrian case discussed above.

4 Methodology

Since the copycat problem is caused by the model learning the spurious prediction pathway from the previous action a_{t-1} to the current action a_t, in this section, we propose a neural network architecture that doesn't have this pathway by design. I.e., we would like our neural network to be unable to access the previous action a_{t-1} information according to the neural network architecture. We note that similar ideas have been explored by [39]. However, they propose to achieve this with adversarial networks, and we show that our solution has a much better scaling property than the previous one.

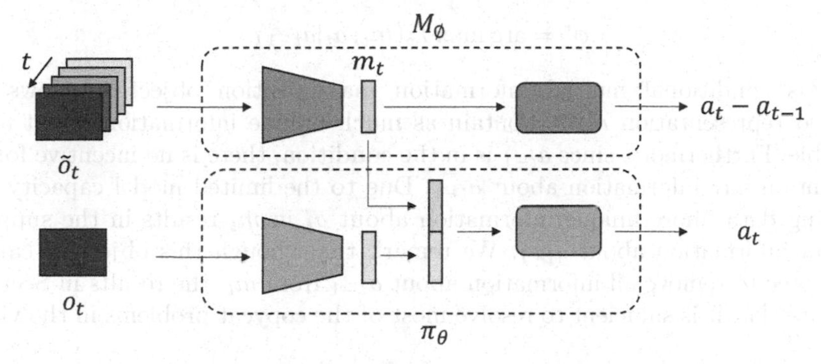

Fig. 3. Our framework: We propose to solve the problem with one memory extraction stream (upper stream M_ϕ) and a policy stream (lower stream π_θ). The memory extraction stream extracts copycat-free historical features, and the policy stream fuse the current observation and the history to output final decisions. Furthermore, blocks with the same color indicate they are updated in the same optimization step.

4.1 Model Architecture

Since the source of copycat problem is the nuisance information about a_{t-1} implied in the temporal information in the observation history $\tilde{o}_t$, the principle of designing the imitation model architecture should be to extract as much information as possible from the observations to predict the action a_t while ensuring that most information about a_{t-1} is removed. Only the observation history $\tilde{o}_t$ contains information about a_{t-1}, and the current observation o_t does not, so we split the neural network into two pathways: the $\tilde{o}_t$ pathway and the o_t pathway. We propose a **memory extraction module** to extract history-related features m_t from the $\tilde{o}_t$ input, such that it contains little information about a_{t-1}. Then we concatenate the feature m_t and the current observation o_t to predict the final action. We name the second network **policy network** since it fuses the history information and the current information to make the decision. Figure 3 shows our framework, and we describe each of the components as follows.

Memory Extraction Module. The memory extraction module M_ϕ aims to get the additional copycat-free feature from $\tilde{o}_t$ and compensate for the missing information due to the partial observation when predicting the action a_t. We extract the intermediate embedding of M_ϕ, named memory feature m_t, to represent the historical information. To make m_t free from the copycat problem, we design a specific prediction objective for the memory extraction module M_ϕ. As discussed in Sect. 3.3, the copycat problem is mainly caused by the excessive information about a_{t-1} in $\tilde{o}_t$. Motivated by this, m_t is expected to contain the essential information for predicting the action a_t while avoiding the copycat problem by retaining as little information about a_{t-1} as possible. I.e. M_ϕ should be trained to maximize the mutual information between m_t and a_t while minimizing that between m_t and a_{t-1}. This goal can be formalized as a maximizing conditional mutual information objective:

$$\phi^* = \arg\max_\phi I_\phi(m_t; a_t|a_{t-1}). \tag{1}$$

This conditional mutual information maximization objective allows the learned representation m_t to contain as much unique information about a_t as possible. Furthermore, since a_{t-1} is in the condition, there is no incentive for m_t to contain any information about a_{t-1}. Due to the limited model capacity and training data, more unique information about a_t in m_t results in the suppression of information about a_{t-1}. We remark that though this objective cannot guarantee to remove all information about a_{t-1} from m_t, the results in Sect. 5.5 indicate that it is sufficient to resolve most of the copycat problems in the visual tasks.

In practice, Eq. (1) is challenging to implement and unstable to optimize. Therefore, we propose to optimize over a lower bound on the conditional mutual information to learn the representation m_t better.

Theorem 1 (lower bound of Eq. (1)). *Let $r_t := a_t - a_{t-1}$ be the action residual, then we have $I_\phi(m_t, a_t|a_{t-1}) \geq \mathbf{H}(r_t|a_{t-1}) - \mathbf{H}_\phi(r_t|m_t)$, where $\mathbf{H}$ represents the Shannon entropy.*

We can maximize this lower bound to approximate Eq. (1). Because the first term is not correlated with the parameter ϕ, the objective to optimize M_ϕ can be rewrited as $\phi^* = \arg\min_\phi H_\phi(r_t|m_t)$. Furthermore, minimizing the conditional mutual information is equivalent to the maximum likelihood estimation, so we propose an **action residual prediction** objective for the memory extraction module M_ϕ:

$$\phi = \arg\max_\phi \mathbb{E}_\mathcal{D}[\log P(a_t - a_{t-1}|\tilde{o}_t; \phi)]. \tag{2}$$

Intuitively, this approximated objective function aims to predict the change of actions. Since the expert actions in a decision process usually transit smoothly, predicting the action change approximately predicts the critical differences in the environments that need expert action change.

Trained with Eq. (2), the memory extraction module M_ϕ optimizes over the objective in Eq. (1) as well, i.e. extracting the sufficient information about a_t and removing the shortcut information about a_{t-1}. Therefore, the memory feature m_t provides additional information from history and cuts off the shortcut path $a_{t-1} \rightarrow a_t$ by removing most information about a_{t-1}. With the proposed architecture, the policy model π_θ will take the advantage of both the visual clues in o_t and the complementary information provided by m_t to learn the correct way for predicting a_t.

Policy Module. The policy module π_θ takes the current observation o_t and the memory feature m_t as input and predicts a_t, where m_t is extracted from $M_\phi(\tilde{o}_t)$ and contains additional and copycat-free historical information about predicting a_t. In practice, we process the observation with a convolutional neural network and fuse the mid-layer representation with the memory extraction module by concatenation. The objective of the policy model is:

$$\theta^* = \arg\max_{\theta} \mathbb{E}_{\mathcal{D}}[\log P(a_t|o_t, m_t; \theta)]. \tag{3}$$

4.2 Implementation Details

In our implementation, we feed o_t and $\tilde{o}_t$ into π_θ and M_ϕ respectively. We take the intermediate feature of M_ϕ as m_t and then fuse it into π_θ. Specifically, we directly concatenate m_t with the intermediate feature of π_θ through a stop-gradient layer and feed them into the following layers of π_θ. The respective targets of π_θ and M_ϕ are a_t and $a_t - a_{t-1}$. π_θ and M_ϕ are trained jointly with the objectives Eq. (3) and Eq. (2). We will introduce experiment setup and other implementation details in Sect. 5.1 and Appendix respectively.

5 Experiments

In this section, we aim to answer the following questions. 1) How well does our method performs? Does it outperform the previously proposed methods? 2) We would like to understand the role of the memory extraction module and provide empirical evidence that it indeed excludes information about a_{t-1}, thus resolving the copycat problem. 3) We would like to understand in what circumstances the proposed model is better qualitatively.

To answer these questions and evaluate our method, we conduct experiments in the CARLA [10] autonomous driving simulator and three standard OpenAI Gym MuJoCo [37] continuous control environments: Hopper, HalfCheetah and Walker2D.

5.1 Experiment Setup

CARLA [10]: CARLA is a photo-realistic autonomous driving simulator that features realistic driving scenarios. CARLA is also the most challenging and widely used environment to benchmark how well the copycat problem is resolved in the literature [40]. We mostly follow the experimental setting from [9], but withhold the ego-agent velocity from the observation for all methods following [40] to ensure the environment is partially observed. The expert demonstration dataset is collected by a scripted expert. Under the hood, the expert knows the scene layout, including information such as the location and speed of other vehicles and pedestrians.

Each expert driving trajectory can be described by a list of (o_t, c_t, a_t) tuples, where t indexes the timestep. Here o_t is the observation in the form of an RGB image. In our setup, the image is resized to 200×88. c_t is the driving command (one of the following commands: follow the lane, turn left, turn right, and go straight). The driving command is necessary because one has to command the autonomous driving agent about where to go. Finally, $a_t \in [-1, 1]^2$ is a two-dimensional vector for controlling the steer and acceleration. Here the positive acceleration means throttle, and negative means brake. We use the CARLA100 [9] dataset that consists of 100 h of driving data. Table 1 shows actions are smooth most of the time, so we can effectively verify whether the method can deal with the copycat problem. The BC agent aims to learn a function $f(\tilde{o}_t, c_t) \rightarrow a_t$. We follow previous setups [40] and use $H = 6$ for BCOH.

Table 1. The distribution of $||a_t - a_{t-1}||_2^2$ in CARLA100

$[0, 10^{-3})$	$[10^{-3}, 10^{-2})$	$[10^{-2}, 10^{-1})$	$\geq 10^{-1}$
68.8%	7.8%	14.3%	9.0%

CARLA *NoCrash* benchmark [9]: *NoCrash* benchmark [9] is a series of navigation tasks consisting of 25 routes over 4 kinds of weather. Finishing a route means following the given route within the time constraint without collisions. There are three traffic conditions: *Empty*, *Regular*, and *Dense* from easy to difficult. We focus on the *Dense* and *Regular* to ensure the agent learns the information about pedestrians or vehicles on the road. In addition to the same weathers and town as the training data, we evaluate the performance in the new town and the new weather. For each experiment, we count the number of successful episode denoted as *#SUCCESS* to be the main performance metric. We will describe model detail and other important metrics in Appendix.

MuJoCo-Image(Hopper,Halfcheetah,Walker2D) [37]: Following the same setting in [40], we set the observation o_t at time step t to be a 128×128 RGB image of the rendering environment and $H = 1$ for BCOH. These control tasks have various state and action spaces, transition dynamics, and reward functions. Expert datasets are collected by a TRPO agent [32] with access to fully observed states. Each dataset is a list of (o_t, a_t) tuples, where t denotes the timestep. We collected 1k samples for HalfCheetah, and 20k for Hopper and Walker2D. The BC agent is required to learn a function $f(\tilde{o}_t) \rightarrow a_t$. The performance metric is the average reward from the environment. We will report the details about the model architectures in Appendix.

5.2 Previous Methods

We compare our method to previous methods that are proposed to solve the copycat problem. We introduce those methods as follows.

BCSO & BCOH: For CARLA, We use the model proposed in [9] as the baseline model, and it was proposed along with CARLA *Nocrash* benchmark. For MuJoCo, we follow the model in [40].

DAGGER [31]: **DAGGER** is a widely used online method in BC to deal with the causal confusion by querying experts to label new trajectories generated by the agent and then updating it. We note that this is considered an oracle method since DAGGER requires extra online interactions and expert labels. Therefore, we set the number of online queries to 150k for CARLA and 1k for MuJoCo.

Historical-Dropout (HD) [2]: **HD** randomly dropouts historical frames to cope with the copycat problem. In practice, we add a dropout layer with a probability of 0.5 on past observations $o_{t-1}, o_{t-2}, \cdots, o_{t-H}$.

Fighting Copycat Agent (FCA) [39]: **FCA** is an adversarial training approach to remove the information about the previous action in the observation histories. It encodes the observation histories to a latent vector, asks it to predict the current action with supervised loss, and asks it not to predict the previous action using adversarial loss.

Keyframe-Focused (Keyframe) [40]: **Keyframe** re-weights the samples according to a weighting defined by how fast action changes. The places where action changes unexpectedly are considered as important timesteps.

Note that there are some other methods [6,7,28,43] that achieve better performance on the CARLA benchmark. However, our goal is to resolve the copycat problem, instead of achieving the best results on CARLA. Following prior works [40], we take [9] as the base algorithm. We expect our techniques will also benefits more complicated imitation methods.

5.3 Results

We report the mean and standard deviation of the performance metric for all entries and analyze the experiment results in this section.

CARLA *Nocrash* benchmark results are shown in the Table 2. We conclude that our method achieves the best performance compared to all previous approaches, including the oracle approach DAGGER. Besides our superior performance, we would also like to describe the results qualitatively. BCSO in CARLA severely fails because the agent can not observe the motion of the surrounding agent, as well as that of itself. Thus, BCOH is necessary in this case. We observe severe copycat problems with BCOH. A lot of failure cases for BCOH are when it is stopped for some obstacle, and it cannot restart again due to the copycat issue. This leads to a high timeout failure rate. Figure 4 shows the copycat problem of BCOH and these are the main reasons for the poor performance. The oracle method DAGGER is better than baseline in the *Dense* traffic

Table 2. The *#SUCCESS* on CARLA *Nocrash* benchmark. For each method, we train 3 policies from different initial seeds.

Environment	Train town & Weather		New weather		New town	
Traffic	*Regular*	*Dense*	*Regular*	*Dense*	*Regular*	*Dense*
BCSO	37.6 ± 6.1	13.1 ± 1.8	18.3 ± 1.7	5.7 ± 5.7	10.3 ± 2.9	2.0 ± 0.8
BCOH	67.1 ± 10.8	34.1 ± 7.5	26.0 ± 6.5	6.0 ± 2.9	27.0 ± 5.0	15.3 ± 2.9
OURS	$\mathbf{78.1 \pm 0.9}$	$\mathbf{52.0 \pm 2.3}$	$\mathbf{39.3 \pm 5.8}$	$\mathbf{19.0 \pm 2.7}$	$\mathbf{40.7 \pm 7.0}$	$\mathbf{25.7 \pm 4.2}$
DAGGER	69.7 ± 8.4	42.7 ± 5.7	24.7 ± 4.8	14.7 ± 4.2	34.0 ± 4.3	12.0 ± 4.5
HD	70.1 ± 4.0	35.6 ± 3.5	28.0 ± 6.5	16.7 ± 6.2	32.0 ± 4.3	11.3 ± 2.5
FCA	58.0 ± 8.0	31.2 ± 5.2	20.0 ± 11.5	8.7 ± 4.0	21.3 ± 3.1	8.3 ± 2.9
Keyframe	74.4 ± 7.3	41.9 ± 6.2	34.0 ± 5.4	13.7 ± 5.4	33.3 ± 7.3	16.3 ± 3.9

case but similar to baseline in *Regular* traffic. We hypothesize that the *Dense* traffic evaluation has a significant distributional shift compared to the training data. HD performs similarly to BCOH, indicating it can not solve the copycat problem in this challenging case. FCA fails to outperform the BCOH baseline because it can not handle the large complex input space with the adversarial training mechanism which was also mentioned in [40]. Keyframe significantly outperforms the BCOH baseline but is still worse than our method.

MuJoCo-Image results are shown in Table 3. BCOH yields higher rewards than BCSO due to access to motion information, and the performance can be further improved by addressing the copycat problem. DAGGER performs well on MuJoCo and gets the best results in Hopper. HD fails in these settings and has a similar performance to BCSO, indicating it cannot resolve the copycat problem. FCA yields a lower reward than BCOH because it is difficult to handle high-dimensional input with adversarial learning. For the other offline method Keyframe, though it significantly improves the performance, ours outperforms it in Walker2D and HalfCheetah and achieves a comparable performance in Hopper. Results on more MuJoCo environments are shown in Appendix.

Table 3. The average reward in MuJoCo-Image. For each method and task, we train 5 policies from different initial seeds.

Method	BCSO	BCOH	OURS	DAGGER	HD	FCA	Keyframe
Hopper	601 ± 168	740 ± 35	894 ± 38	$\mathbf{1034 \pm 45}$	617 ± 111	735 ± 106	951 ± 117
Walker2D	481 ± 40	614 ± 107	$\mathbf{800 \pm 58}$	699 ± 111	594 ± 61	534 ± 99	769 ± 91
HalfCheetah	4 ± 5	615 ± 41	$\mathbf{914 \pm 115}$	822 ± 186	96 ± 40	270 ± 168	819 ± 96

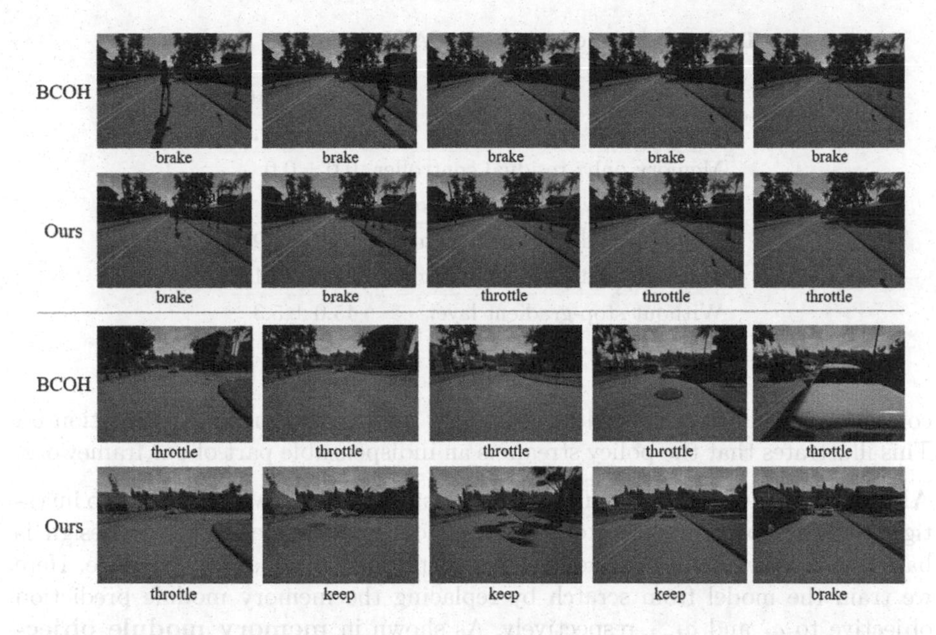

Fig. 4. Visualizations of BCOH and our method in CARLA: Each row is a video sequence. The first two rows show a case where BCOH is stopped for a pedestrian and never restarts, while our method handles that successfully. The last two rows show a case where BCOH passes an intersection and then fails to stop behind a vehicle while our method executes the proper behavior. The "throttle/keep/brake" indicates the acceleration of the agent.

5.4 Ablation Studies

To evaluate the effect of each module and technique we propose in Sect. 4.1, we conduct ablation study experiments in CARLA *Nocrash Dense*. The results are shown in Table 4, and we provide detailed analysis for each ablation, respectively.

The memory extracted module only: We want to show both parts of our framework are necessary. From the BCSO result, we know the policy module alone achieves inferior performance. To evaluate the memory extraction module alone, we set up two experiments that only uses the well-trained memory extracted module to derive the action: one uses the residual directly (**Memory only: residual controller**), i.e., it uses the output of the memory extracted module plus the previous action to get the current action $a_t = a_{t-1} + r_t$; the other uses the extracted feature (**Memory only: learned controller**) to control the vehicle with the same controller as the baseline model, i.e., the controller is trained by using m_t from memory extracted module to fit a_t. However, the imitators in both setups fail in every episode. It indicates the memory extracted feature m_t alone contains insufficient information for the task. The residual prediction not only removes the information about a_{t-1} but also erases some necessary information about the current scene. Therefore, the policy module

Table 4. Ablation results in CARLA *Nocrash Dense*

Architecture	#*Success*(↑)
Ours	**52.0 ± 2.3**
Memory only: residual controller	0.0 ± 0.0
Memory only: learned controller	0.0 ± 0.0
Memory module objective: a_t	41.3 ± 1.9
Memory module objective: a_{t-1}	47.0 ± 5.1
Without stop-gradient layer	45.0 ± 5.3

completes information related to control by adding the current observation o_t. This illustrates that the policy stream is an indispensable part of our framework.

Alternative memory module objectives: Moreover, we would like to investigate the performance of alternative memory module designs. Our design is based on a lower bound of the conditional mutual information objective. Here we train the model from scratch by replacing the memory module prediction objective to a_t and a_{t-1} respectively. As shown in **memory module objective:** a_t and **memory module objective:** a_{t-1}, both of the objective changes lead to deterioration of the performances. This indicates that other features are less informative or more likely to suffer from the copycat problem.

Without stop-gradient layer: The stop-gradient is applied to the memory feature m_t to ensure the policy module doesn't directly learn shortcut information from the observation histories. After removing it, the result shows the performance degradation, indicating that the imitation learner might learn some copycat information from the observation histories. Though the action residual prediction objective removes some shortcut information about a_{t-1} and improves the performance compared to BCOH, the stop-gradient is indispensable for the policy module to cut off the nuisance information flow from observation histories.

5.5 Analysis

The performance of our framework is clearly above that of alternative methods on the CARLA *Nocrash* benchmark. In this subsection, we prove that the training objective proposed in Eq. (2) can successfully resolve the copycat problem. We design two analytical experiments to evaluate our method qualitatively. The first analysis directly proves our method has fewer copycat behaviors by an intervention experiment. The second one validates our training objective indeed reduces the mutual information between a_{t-1} and m_t. We choose BCOH and Keyframe as the baselines for these two analytical experiments.

Table 5. The analysis results on the validation set

Method	BCOH	Keyframe	OURS
Change(%)(↓)	40.88	22.96	**16.18**
MSE loss $\times 10^{-2}$(↑)	4.83	5.04	**7.50**

Intervention on Observation Histories. To verify whether our model learns the causal effect of the observation histories $\tilde{o}_t$, we propose an intervention analysis. We replace the original $\tilde{o}_t$ with a counterfactual history $do(\tilde{o}_t) = [o_t, o_t, \cdots, o_t]$, i.e., all histories are the current observation o_t. For a copycat agent, then with the counterfactual observation histories $do(\tilde{o}_t)$ as the input, it will have an illusion that the vehicle is in the stationary state and will merely repeat the previous motion of staying still. A causally correct agent should perform according to the content of the scene. We test our model and the baseline models under the cases where the vehicles have positive speed and the latest action is non-stop (denoted by $\pi(\cdot) > 0$ with a slight abuse of notation). A causally correct agent should drive forward in this case, i.e. $\pi(do(\tilde{o}_t)) > 0$. We count the percentage where the model stopped on the intervened observation, i.e., $\frac{N(v>0, \pi(\tilde{o}_t)>0, \pi(do(\tilde{o}_t))=0)}{N(v>0, \pi(\tilde{o}_t)>0)}$ where $N(.)$ denotes the counting function. The behavior change percentage can be regarded as the error rate of this experiment.

The 2nd row of Table 5 shows the intervention results. As shown in the table, 40.88% of samples change in BCOH, even there is no need for the agent to stop in these scenes (e.g., no vehicles or pedestrians ahead; the traffic signal is not red). It illustrates that BCOH makes decisions by merely repeating the previous action but ignoring the perception of the current scene in nearly half of the cases. The metric of Keyframe is relatively lower than that of BCOH, which indicates Keyframe does not rely just on the previous action and partially mitigates the copycat problem. Meanwhile, OURS is only 16.18%, showing that our approach significantly alleviates the copycat problem.

Mutual Information. In Sect. 4, we set up a conditional mutual information objective, i.e. maximize $I(m_t; a_t | a_{t-1})$. And we use an approximation to this objective during the optimization process. In this analysis, we empirically evaluate how much information about a_{t-1} is left in m_t. This would help to justify both the objective function as well as the approximation. More concretely, we regress a_{t-1} from the representation with a 2-layer MLP. We calculate the validation loss of a_{t-1} and use it as a proxy to the negative of the mutual information between the representation and a_{t-1}. The less the information, the less likely the method would suffer from the copycat issue. We evaluate our method's representation m_t and corresponding layers of the BCOH and the Keyframe methods.

The results in the 3rd row of Table 5 show that the feature in BCOH is more predictive of the previous action. This conclusion matches the preliminary results that BCOH is more prone to copycat problems and learns the spurious temporal correlation. The MSE of Keyframe is slightly larger than that of BCOH, which shows that this method removes part of the information about the previous action and mitigates the copycat problem. Meanwhile, the MSE of OURS is the largest, which indicates that the extracted feature m_t in our model contains the smallest amount of mutual information with a_{t-1} and cuts off the shortcut of simply imitating the previous action from observation histories.

6 Conclusion

Behavioral cloning is a fundamental algorithm that helps the agent to imitate complex behaviors. However, it suffers from causal issues due to the temporal structure of the problem. We view the causal issue from an information flow perspective and propose a simple yet effective method to improve performance drastically. Our method outperforms previous solutions to the copycat problem. In the future, we would like to investigate casual issues in a broader range of algorithms, e.g., reinforcement learning and other sequential decision problems.

Acknowledgement. We thank Jiaye Teng, Renhao Wang and Xiangyue Liu for insightful discussions and comments. This work is supported by the Ministry of Science and Technology of the People's Republic of China, the 2030 Innovation Megaprojects "Program on New Generation Artificial Intelligence" (Grant No. 2021AAA0150000), and a grant from the Guoqiang Institute, Tsinghua University.

References

1. Argall, B.D., Chernova, S., Veloso, M.M., Browning, B.: A survey of robot learning from demonstration. Robot. Auton. Syst. **57**(5), 469–483 (2009)
2. Bansal, M., Krizhevsky, A., Ogale, A.S.: Chauffeurnet: Learning to drive by imitating the best and synthesizing the worst. In: Bicchi, A., Kress-Gazit, H., Hutchinson, S. (eds.) Robotics: Science and Systems XV, University of Freiburg, Freiburg im Breisgau, Germany, 22–26 June 2019 (2019)
3. Beery, S., Van Horn, G., Perona, P.: Recognition in terra incognita. In: Ferrari, V., Hebert, M., Sminchisescu, C., Weiss, Y. (eds.) ECCV 2018. LNCS, vol. 11220, pp. 472–489. Springer, Cham (2018). https://doi.org/10.1007/978-3-030-01270-0_28
4. Bojarski, M., et al.: End to end learning for self-driving cars. CoRR abs/1604.07316 (2016)
5. Brantley, K., Sun, W., Henaff, M.: Disagreement-regularized imitation learning. In: 8th International Conference on Learning Representations, ICLR 2020, Addis Ababa, Ethiopia, 26–30 April 2020. OpenReview.net (2020)
6. Chen, D., Koltun, V., Krähenbühl, P.: Learning to drive from a world on rails. CoRR abs/2105.00636 (2021)
7. Chen, D., Zhou, B., Koltun, V., Krähenbühl, P.: Learning by cheating. In: Kaelbling, L.P., Kragic, D., Sugiura, K. (eds.) 3rd Annual Conference on Robot Learning, CoRL 2019, Osaka, Japan, October 30–November 1, 2019, Proceedings. Proceedings of Machine Learning Research, vol. 100, pp. 66–75. PMLR (2019)

8. Codevilla, F., Müller, M., López, A.M., Koltun, V., Dosovitskiy, A.: End-to-end driving via conditional imitation learning. In: 2018 IEEE International Conference on Robotics and Automation, ICRA 2018, Brisbane, Australia, 21–25 May 2018, pp. 1–9. IEEE (2018)
9. Codevilla, F., Santana, E., López, A.M., Gaidon, A.: Exploring the limitations of behavior cloning for autonomous driving. In: 2019 IEEE/CVF International Conference on Computer Vision, ICCV 2019, Seoul, Korea (South), October 27–November 2 2019, pp. 9328–9337. IEEE (2019)
10. Dosovitskiy, A., Ros, G., Codevilla, F., López, A.M., Koltun, V.: CARLA: an open urban driving simulator. In: 1st Annual Conference on Robot Learning, CoRL 2017, Mountain View, California, USA, 13–15 November 2017, Proceedings. Proceedings of Machine Learning Research, vol. 78, pp. 1–16. PMLR (2017)
11. Geirhos, R., et al.: Shortcut learning in deep neural networks. Nat. Mach. Intell. **2**(11), 665–673 (2020)
12. Geirhos, R., Rubisch, P., Michaelis, C., Bethge, M., Wichmann, F.A., Brendel, W.: Imagenet-trained CNNs are biased towards texture; increasing shape bias improves accuracy and robustness. In: 7th International Conference on Learning Representations, ICLR 2019, New Orleans, LA, USA, 6–9 May 2019. OpenReview.net (2019)
13. Giusti, A., et al.: A machine learning approach to visual perception of forest trails for mobile robots. IEEE Robot. Autom. Lett. **1**(2), 661–667 (2016)
14. de Haan, P., Jayaraman, D., Levine, S.: Causal confusion in imitation learning. In: Wallach, H.M., Larochelle, H., Beygelzimer, A., d'Alché-Buc, F., Fox, E.B., Garnett, R. (eds.) Advances in Neural Information Processing Systems, vol. 32, Annual Conference on Neural Information Processing Systems 2019, NeurIPS 2019, 8–14 December 2019, Vancouver, BC, Canada, pp. 11693–11704 (2019)
15. Heinze-Deml, C., Meinshausen, N.: Conditional variance penalties and domain shift robustness. Mach. Learn. **110**(2), 303–348 (2020). https://doi.org/10.1007/s10994-020-05924-1
16. Ho, J., Ermon, S.: Generative adversarial imitation learning. In: Lee, D.D., Sugiyama, M., von Luxburg, U., Guyon, I., Garnett, R. (eds.) Advances in Neural Information Processing Systems, vol. 29, Annual Conference on Neural Information Processing Systems 2016, December 5–10, 2016, Barcelona, Spain, pp. 4565–4573 (2016)
17. Hu, P., Huang, A., Dolan, J.M., Held, D., Ramanan, D.: Safe local motion planning with self-supervised freespace forecasting. In: IEEE Conference on Computer Vision and Pattern Recognition, CVPR 2021, virtual, 19–25 June 2021, pp. 12732–12741. Computer Vision Foundation/IEEE (2021)
18. Laskey, M., Lee, J., Fox, R., Dragan, A.D., Goldberg, K.: DART: noise injection for robust imitation learning. In: 1st Annual Conference on Robot Learning, CoRL 2017, Mountain View, California, USA, 13–15 November 2017, Proceedings. Proceedings of Machine Learning Research, vol. 78, pp. 143–156. PMLR (2017)
19. LeCun, Y., Muller, U., Ben, J., Cosatto, E., Flepp, B.: Off-road obstacle avoidance through end-to-end learning. In: Advances in Neural Information Processing Systems, vol. 18 [Neural Information Processing Systems, NIPS 2005, 5–8 December 2005, Vancouver, British Columbia, Canada], pp. 739–746 (2005)
20. Levine, S., Koltun, V.: Guided policy search. In: Proceedings of the 30th International Conference on Machine Learning, ICML 2013, Atlanta, GA, USA, 16–21 June 2013. JMLR Workshop and Conference Proceedings, vol. 28, pp. 1–9. JMLR.org (2013)

21. Loquercio, A., Kaufmann, E., Ranftl, R., Dosovitskiy, A., Koltun, V., Scaramuzza, D.: Deep drone racing: From simulation to reality with domain randomization. IEEE Trans. Robot. **36**(1), 1–14 (2020)
22. Mandlekar, A., et al.: What matters in learning from offline human demonstrations for robot manipulation. In: Faust, A., Hsu, D., Neumann, G. (eds.) Conference on Robot Learning, 8–11 November 2021, London, UK. Proceedings of Machine Learning Research, vol. 164, pp. 1678–1690. PMLR (2021)
23. McCoy, T., Pavlick, E., Linzen, T.: Right for the wrong reasons: Diagnosing syntactic heuristics in natural language inference. In: Korhonen, A., Traum, D.R., Màrquez, L. (eds.) Proceedings of the 57th Conference of the Association for Computational Linguistics, ACL 2019, Florence, Italy, July 28–August 2, 2019, Volume 1: Long Papers, pp. 3428–3448. Association for Computational Linguistics (2019)
24. Minh, V., et al.: Human-level control through deep reinforcement learning. Nature **518**(7540), 529–533 (2015)
25. Mülling, K., Kober, J., Kroemer, O., Peters, J.: Learning to select and generalize striking movements in robot table tennis. Int. J. Robot. Res. **32**(3), 263–279 (2013)
26. Murphy, K.: A survey of POMDP solution techniques. Environment **2**, 1–12 (2000)
27. Niven, T., Kao, H.: Probing neural network comprehension of natural language arguments. In: Korhonen, A., Traum, D.R., Màrquez, L. (eds.) Proceedings of the 57th Conference of the Association for Computational Linguistics, ACL 2019, Florence, Italy, July 28–August 2 2019, Volume 1: Long Papers, pp. 4658–4664. Association for Computational Linguistics (2019)
28. Ohn-Bar, E., Prakash, A., Behl, A., Chitta, K., Geiger, A.: Learning situational driving. In: 2020 IEEE/CVF Conference on Computer Vision and Pattern Recognition, CVPR 2020, Seattle, WA, USA, 13–19 June 2020, pp. 11293–11302. Computer Vision Foundation/IEEE (2020)
29. Osa, T., Pajarinen, J., Neumann, G., Bagnell, J.A., Abbeel, P., Peters, J.: An algorithmic perspective on imitation learning. CoRR abs/1811.06711 (2018)
30. Pomerleau, D.: ALVINN: an autonomous land vehicle in a neural network. In: Touretzky, D.S. (ed.) Advances in Neural Information Processing Systems, vol. 1, [NIPS Conference, Denver, Colorado, USA, 1988], pp. 305–313. Morgan Kaufmann (1988)
31. Ross, S., Gordon, G.J., Bagnell, D.: A reduction of imitation learning and structured prediction to no-regret online learning. In: Gordon, G.J., Dunson, D.B., Dudík, M. (eds.) Proceedings of the Fourteenth International Conference on Artificial Intelligence and Statistics, AISTATS 2011, Fort Lauderdale, USA, 11–13 April 2011. JMLR Proceedings, vol. 15, pp. 627–635. JMLR.org (2011)
32. Schulman, J., Levine, S., Moritz, P., Jordan, M.I., Abbeel, P.: Trust region policy optimization. CoRR abs/1502.05477 (2015)
33. Schulman, J., Wolski, F., Dhariwal, P., Radford, A., Klimov, O.: Proximal policy optimization algorithms. CoRR abs/1707.06347 (2017)
34. Spencer, J.C., Choudhury, S., Venkatraman, A., Ziebart, B.D., Bagnell, J.A.: Feedback in imitation learning: The three regimes of covariate shift. CoRR abs/2102.02872 (2021)
35. Sun, W., Bagnell, J.A., Boots, B.: Truncated horizon policy search: Combining reinforcement learning & imitation learning. In: 6th International Conference on Learning Representations, ICLR 2018, Vancouver, BC, Canada, April 30–May 3 2018, Conference Track Proceedings. OpenReview.net (2018)

36. Sun, W., Venkatraman, A., Gordon, G.J., Boots, B., Bagnell, J.A.: Deeply aggregated: differentiable imitation learning for sequential prediction. In: Precup, D., Teh, Y.W. (eds.) Proceedings of the 34th International Conference on Machine Learning, ICML 2017, Sydney, NSW, Australia, 6–11 August 2017. Proceedings of Machine Learning Research, vol. 70, pp. 3309–3318. PMLR (2017)
37. Todorov, E., Erez, T., Tassa, Y.: MuJoCo: a physics engine for model-based control. In: 2012 IEEE/RSJ International Conference on Intelligent Robots and Systems, IROS 2012, Vilamoura, Algarve, Portugal, 7–12 October 2012, pp. 5026–5033. IEEE (2012)
38. Wang, D., Devin, C., Cai, Q., Krähenbühl, P., Darrell, T.: Monocular plan view networks for autonomous driving. In: 2019 IEEE/RSJ International Conference on Intelligent Robots and Systems, IROS 2019, Macau, SAR, China, 3–8 November 2019, pp. 2876–2883. IEEE (2019)
39. Wen, C., Lin, J., Darrell, T., Jayaraman, D., Gao, Y.: Fighting copycat agents in behavioral cloning from observation histories. In: Larochelle, H., Ranzato, M., Hadsell, R., Balcan, M., Lin, H. (eds.) Advances in Neural Information Processing Systems, vol. 33, Annual Conference on Neural Information Processing Systems 2020, NeurIPS 2020, 6–12 December 2020, virtual (2020)
40. Wen, C., Lin, J., Qian, J., Gao, Y., Jayaraman, D.: Keyframe-focused visual imitation learning. In: Meila, M., Zhang, T. (eds.) Proceedings of the 38th International Conference on Machine Learning, ICML 2021, 18–24 July 2021, Virtual Event. Proceedings of Machine Learning Research, vol. 139, pp. 11123–11133. PMLR (2021)
41. Wen, C., Qian, J., Lin, J., Teng, J., Jayaraman, D., Gao, Y.: Fighting fire with fire: avoiding DNN shortcuts through priming. In: International Conference on Machine Learning, pp. 23723–23750. PMLR (2022)
42. Widrow, B., Smith, F.W.: Pattern-recognizing control systems (1964)
43. Zhang, Z., Liniger, A., Dai, D., Yu, F., Gool, L.V.: End-to-end urban driving by imitating a reinforcement learning coach. CoRR abs/2108.08265 (2021)
44. Zhou, X., Koltun, V., Krähenbühl, P.: Tracking objects as points. In: Vedaldi, A., Bischof, H., Brox, T., Frahm, J.-M. (eds.) ECCV 2020. LNCS, vol. 12349, pp. 474–490. Springer, Cham (2020). https://doi.org/10.1007/978-3-030-58548-8_28

OPD: Single-View 3D Openable Part Detection

Hanxiao Jiang(✉), Yongsen Mao, Manolis Savva, and Angel X. Chang

Simon Fraser University, Burnaby, Canada
{Shawn_Jiang,sam_mao,msavva,angelx}@sfu.ca
https://3dlg-hcvc.github.io/OPD/

Abstract. We address the task of predicting what parts of an object can open and how they move when they do so. The input is a single image of an object, and as output we detect what parts of the object can open, and the motion parameters describing the articulation of each openable part. To tackle this task, we create two datasets of 3D objects: OPDSynth based on synthetic objects, and OPDReal based on RGBD reconstructions of real objects. We then design OpdRcnn, a neural architecture that detects openable parts and predicts their motion parameters. Our experiments show that this is a challenging task especially when considering generalization across object categories, and the limited amount of information in a single image. Our architecture outperforms baselines and prior work especially for RGB image inputs.

Keywords: Articulated parts · Single-view motion parameter estimation

1 Introduction

There is increasing interest in training embodied AI agents that interact with the world based on visual perception. Recently, Batra et al. [2] introduced rearrangement as a challenge bringing together researchers in machine learning, vision, and robotics. Common household tasks such as "put dishes in the cupboard" or "get cup from the cabinet" can be viewed as object rearrangement. A key challenge in such tasks is identifying which parts can be opened, and how they move to open and close.

To address this problem, we introduce the *openable part detection* (OPD) task, where the goal is to identify openable parts, and their motion type and parameters (see Fig. 1). We focus on predicting the openable parts ("door", "drawer", and "lid") of an object, and their motion parameters from a single RGB image input. More specifically, we focus on container objects (e.g. cabinets, ovens, etc.). Containers need to be opened to look for hidden objects to reach, or

Supplementary Information The online version contains supplementary material available at https://doi.org/10.1007/978-3-031-19842-7_24.

S. Avidan et al. (Eds.): ECCV 2022, LNCS 13699, pp. 410–426, 2022.
https://doi.org/10.1007/978-3-031-19842-7_24

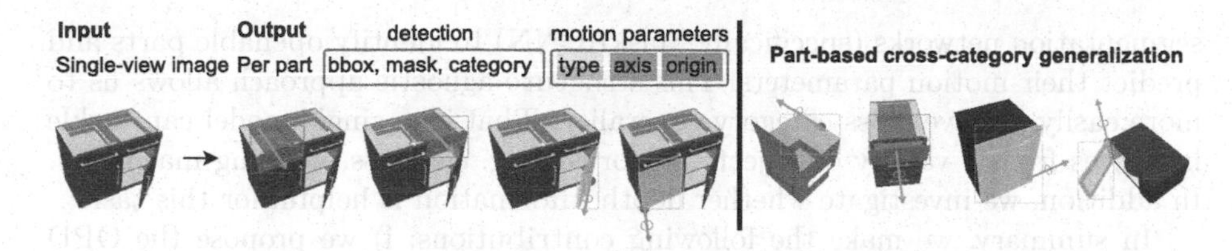

Fig. 1. In the openable part detection (OPD) task the input is a single view image, and the outputs are detected openable parts as well as their motion parameters. We design OPDRCNN: a neural architecture for this task that operates at the part level, allowing generalization across diverse object categories.

to place away objects. Methods that can identify what can be opened and how offer a useful abstraction that can be used in more complex tasks [18,21–23,31].

Prior work [7,11,36] has looked at predicting motion types and parameters from 3D point clouds. Point-cloud based methods rely on depth information, often sampled from a reconstructed mesh that aggregates information from multiple views. In addition, Li et al. [11] assumes that the kinematic chain and the part structure of the objects are given. To handle different kinematic structures, they train a separate model for each structure. This approach is not scalable to the wide variety of structures found in the real world, even for objects in the same category. For example, cabinets may have two, three or four drawers.

We study the task of identifying openable parts and motion parameters from a single-view image, and investigate whether a structure-agnostic image-based approach can be effective at this task. Our approach OPDRCNN extends instance

Table 1. Summary of prior work on motion parameter estimation (motion type, axis, and rotation origin). We indicate the input modality, whether the method has cross-category generalization (CC), whether part segmentations are predicted (Seg) as opposed to using ground-truth parts, whether object pose (OP) and part state (PS) are predicted. Most prior work takes point cloud (PC) inputs. In contrast, our input is single-view images (RGB, D, or RGB-D).

Method	Input	CC	Seg	OP	PS	#cat	#obj	#part
Snapshot [6]	3D mesh						368	368
Shape2Motion [30]	PC	✓	✓			45	2440	6762
RPMNet [36]	PC	✓	✓				969	1420
DeepPartInduction [37]	Pair of PCs		✓			16	16881	
MultiBodySync [8]	Multiple PCs		✓			16		
ScrewNet [9]	Depth video	✓				9	4496	4496
Liu et al. [14]	RGB video	✓	✓			3		
ANCSH [11]	Single-view PC		✓	✓	✓	5	237	343
Abbatematteo et al. [1]	RGB-D		✓			5	6	8
VIAOP [39]	RGB-D	✓				6	149	618
OPDRCNN (ours)	Single-view RGB(-D)	✓	✓	✓	✓	11	683	1343

segmentation networks (specifically MaskRCNN) to identify openable parts and predict their motion parameters. This structure-agnostic approach allows us to more easily achieve cross-category generality. That is, a single model can tackle instances from a variety of object categories (e.g. cabinets, washing machines). In addition, we investigate whether depth information is helpful for this task.

In summary, we make the following contributions: i) we propose the OPD task for predicting openable parts and their motion parameters from single-view images; ii) we contribute two datasets of objects annotated with their openable parts and motion parameters: OPDSynth and OPDReal; and iii) we design OPDRCNN: a simple image-based, structure-agnostic neural architecture for openable part detection and motion estimation across diverse object categories. We evaluate our approach with different input modalities and compare against baselines to show which aspects of the task are challenging. We show that depth is not necessary for accurate part detection and that we can predict motion parameters from RGB images. Our approach significantly outperforms baselines especially when requiring both accurate part detection and part motion parameter estimation.

2 Related Work

Part Segmentation and Analysis. Part segmentation has been widely studied in the vision and graphics communities for both 2D images and 3D representations. Much of the prior work relies on part annotations in 2D images [4,40], or 3D CAD models with semantically labelled part segmentations [19,29,30,36,38]. Annotated 3D part datasets have been used to study part segmentation by rendering images [24] or directly in 3D (e.g. meshes or point clouds) [19,38]. These datasets have fostered the development of different methods to address part segmentation on images [27,28,33]. Many of these focus on human/animal part segmentation, and use hierarchical methods to parse both objects and the parts. Beyond part segmentation, [15] further enhanced the PASCAL VOC Part dataset [4] with state information. In contrast, we directly detect and segment the parts of interest using standard object instance segmentation methods [3,5].

Unlike prior work, we focus on a small set of openable part categories. Part datasets differ in what parts they focus on (e.g., human body parts, or fine-grained vs coarse-grained object parts). Determining the set of parts of interest can be tricky, as the set of possible parts can be large and ill-defined, with what constitutes a part varying across object categories. Therefore, we focus on a small set of openable parts in common household objects. We believe this set of parts is a small, practical set that is important for object interaction.

Articulated Object Motion Prediction. Part mobility analysis is a long-standing problem in 3D computer graphics. Early work [17] has focused on learning the part mobility in mechanical assemblies. Xu et al. [35] used a mobility tree formalism to further explore object and part mobility in indoor 3D scenes.

More recent work [20] proposed a joint-aware deformation framework based on shape analysis and optimization to predict motion joint parameters. Part mobility analysis has also been performed on sequences of RGBD scans [10]. More recently, there has been increasing interest in data-driven methods for studying articulated objects and estimating motion parameters [11,30]. To support these data-driven approaches, there has been concurrent development of datasets of annotated part articulations for synthetic [30,34,36] and reconstructed [13,16] 3D objects.

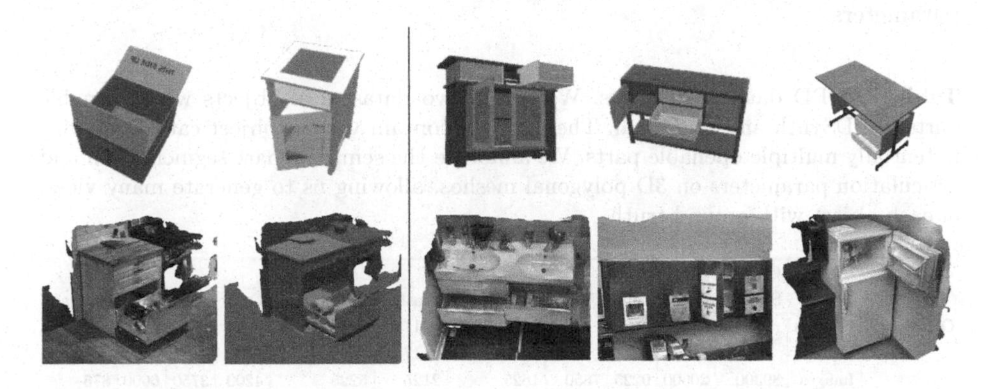

Fig. 2. Example articulated objects from the OPDSynth dataset and the OPDReal dataset. The first row is from OPDSynth. Left: different openable part categories (`lid`, in orange, `drawer` in green, `door` in red). Right: Cabinet objects with different kinematic structures and varying numbers of openable parts. The second row is from our OPDReal dataset. Left: reconstructed cabinet and its semantic part segmentation. Right: example reconstructed objects from different categories. (Color figure online)

Table 1 summarizes the part mobility prediction tasks defined by this and other recent work. Hu et al. [6] predict joint parameters given pre-segmented 3D objects. Other work predicts segmentation together with motion parameters, for 2.5D inputs [36,37], 3D point clouds [30] or for sequences of RGBD scans [10]. Abbatematteo et al. [1], Li et al. [11] have the most similar task setting with our work, predicting the part segmentation, part pose, and joint parameters from a single view image. However, both require depth as input and knowledge of the kinematic chain of each object. They require training a separate model for each object category, where the object category is defined as having the same structure (i.e. same kinematic chain). This means that different models need to be trained for cabinets with 3 drawers and cabinets with 4 drawers. In contrast, we identify all openable parts of an object in an input RGB image, without assuming a specific structure with given number of parts. This allows us to train a single model that generalizes across categories. Note that Abbatematteo et al. [1] also use MaskRCNN for segmentation, but they do not analyze or report the part segmentation and detection performance of their model.

More recent work has started to explore training of single models for motion prediction across categories and structures [8,9,9,14]. Zeng et al. [39] proposed an optical flow-based approach on RGB-D images given segmentation masks of the moving part and fixed part. They evaluate only on ground truth segmentation and do not investigate how part segmentation and detection influences the accuracy of motion prediction. Others have proposed to predict articulated part pose from depth sequences [9], image video sequences [14], or synchronizing multiple point clouds [8]. In contrast, we focus on single-view image input and show that even without depth information, we can accurately predict motion parameters.

Table 2. OPD dataset statistics. We create two datasets of objects with openable parts: OPDSynth and OPDReal. The datasets contain various object categories with potentially multiple openable parts. We annotate the semantic part segmentation and articulation parameters on 3D polygonal meshes, allowing us to generate many views of each object with ground truth.

		Category										
		Storage	Table	Bin	Fridge	Microwave	Washer	Dishwasher	Oven	Safe	Box	Suitcase
OPDSynth	Objects	366	76	69	42	13	17	41	24	30	28	7
	Parts	809	187	75	68	13	17	42	36	30	59	7
	Images	89300	20600	9225	7850	1625	2125	5225	4200	3750	6600	875
OPDReal	Objects	231	16	7	12	7	3	3	5	-	-	-
	Parts	787	35	7	27	7	3	3	6	-	-	-
	Images	27394	1175	474	1321	570	159	186	253	-	-	-

3 Problem Statement

Our input is a single RGB image I of a single articulated object and the output is the set of openable parts $P = \{p_1 \dots p_k\}$ (i.e. drawers, doors and lids) with their motion parameters $\Phi = \{\phi_1 \dots \phi_k\}$. Figure 1 illustrates the input and output for our task. For each part $p_i = \{b_i, m_i, l_i\}$, we predict the 2D bounding box b_i, the segmentation mask m_i, and the semantic label $l_i \in \{\texttt{drawer}, \texttt{door}, \texttt{lid}\}$. The motion parameters ϕ_i of each openable part specify the motion type $c_i \in \{\texttt{prismatic}, \texttt{revolute}\}$, motion axis direction $a_i \in R^3$ and motion origin $o_i \in R^3$. Specifically, we have $\phi_i = [c_i, a_i, o_i]$ for revolute joints (e.g., door rotating around a hinge), and $\phi_i = [c_i, a_i]$ for prismatic joints (e.g., drawer sliding out). For simplicity, we assume that each part has only one motion.

4 OPD Dataset

To study our task, we collate two datasets of objects with openable parts (Fig. 2), a dataset of synthetic 3D models (OPDSynth) and a dataset of real objects scanned with RGB-D sensors (OPDReal). For OPDSynth, we select objects with openable parts from an existing dataset of articulated 3D models PartNet-Mobility [34]. For OPDReal, we reconstruct 3D polygonal meshes for articulated objects in real indoor environments and annotate their parts and articulation information. Table 2 shows summary statistics of the number of objects and openable parts for each object category (see supplement for more detailed statistics).

OPDSynth. The PartNet-Mobility dataset contains 14K movable parts across 2,346 3D objects from 46 common indoor object categories [34]. We canonicalize the part names to 'drawer', 'door', or 'lid'. We select object categories with openable parts that can serve as containers. For each category, we identify the openable parts and label them as `drawer`, `door`, or `lid` (see supplement). Overall, we collated 683objects with 1343 parts over 11 categories (see Table 2). We then render several views of each object to produce RGB, depth, and semantic part mask frames. Specifically, for each articulated object we render different motions for each part. In one motion state all parts are set to the minimum value in their motion range. Then for each part, we pick 3 random motion states and the maximum of the motion range, while the other parts remain at the minimum value. We render five images for each motion state from different camera viewpoints, and each image is composited on four different randomly selected background images (see supplement for rendering details).

OPDReal. To construct a dataset of real objects, we take 863 RGB-D video scans of indoor environments with articulated objects (residences, campus lounges, furniture showrooms) using iPad Pro 2021 devices. We obtain polygonal mesh reconstructions from these scans using the Open3D [41] implementation of RGB-D integration and Waechter et al. [25]'s implementation of texturing. We follow an annotate-validate strategy to filter and annotate this scanned data. Specifically, we: 1) Annotate the model quality and filter the scans with bad quality (insufficient geometry to annotate articulations); 2) Annotate the semantic part segmentation; 3) Validate the semantic segmentation; 4) Annotate the articulation parameters for articulated parts; 5) Validate the articulations through animating the moving parts; 6) Annotate a 'semantic OBB' (center at origin, semantic 'up' and 'front' axis direction) for each object; 7) Calculate consistent object pose (i.e. transformation between camera coordinates and object coordinates) based on semantic OBB. After filtering and annotation, we have a total of 763 polygonal meshes for 284 different objects across 8 object categories. Table 2 shows the distribution over different object categories.

We then project the 3D annotations back to the original RGB and depth frames from the scan videos. We select around 100 frames from each video and project the segmentation mask in 3D back to 2D. The articulation parameters in world coordinate are also projected to the camera coordinates of each frame. When selecting frames, we sample one frame every second ensuring that at least 1% of pixels belong to an openable part and at least 20% of parts are visible.

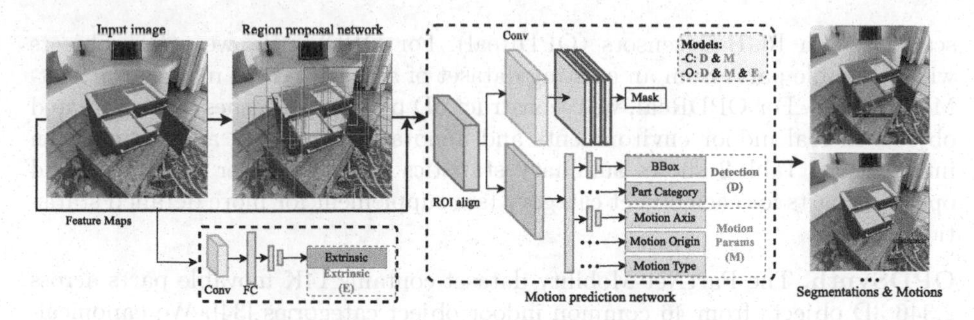

Fig. 3. Illustration of the network structure for our OPDRCNN-C and OPDRCNN-O architectures. We leverage a MaskRCNN backbone to detect openable parts. Additional heads are trained to predict motion parameters for each part.

5 Approach

To address openable part detection, we leverage a instance segmentation network to identify openable parts by treating each part as an 'object instance'. Specifically, we augment Mask-RCNN [5] with heads for predicting the motion parameters. Mask-RCNN uses a backbone network for feature extraction and a region proposal network (RPN) to propose regions of interest (ROI) which are fed into branches that refine the bounding box and predict the mask and category label. By training Mask-RCNN on our OPD dataset, we can detect and segment openable parts. We attach additional branches to the output of the `RoiAlign` module to predict motion parameters. We consider two variants: i) OPDRCNN-C directly predicts the motion parameters in camera coordinates; and ii) OPDRCNN-O predicts camera extrinsics (object pose in our single-object setting) and motion parameters in world coordinates (canonical object coordinates). Figure 3 shows the overall architecture. All models use a cross-entropy loss for categorical prediction and a smooth L1 loss with $\beta = 1$ when regressing real values (see supplement).

OPDRCNN-C. For OPDRCNN-C, we add separate fully-connected layers to the `RoiAlign` module to directly predict the motion parameters. The original MaskRCNN branches predict the part label l_i and part bounding box delta δ_i for each part p_i in the box head. The box delta δ_i is combined with the box proposal from the RPN module to calculate the final output bounding box b_i. We add additional branches to predict the motion parameters $\phi_i = [c_i, a_i, o_i]$ (motion category, motion axis, motion origin) in the camera coordinate frame. We use the smooth L1 loss for the motion origin and motion axis, and the cross-entropy loss for the part joint type. Note that we only apply the motion origin loss for revolute joints, since the motion origin is not meaningful for prismatic joints.

OPDRCNN-O. For OPDRCNN-O, additional layers predict the 6 DoF object pose parameters to establish an object coordinate frame within which to predict motion axes and origins for openable parts. Following prior work [26], the object coordinate frame has consistent up and front orientations. The motivation is that motion origins and axes in object coordinates are more consistent than in camera coordinates. We are only dealing with a single object per image and consistent poses are available for each annotated object, so predicting the object pose is equivalent to predicting the extrinsic parameters of the camera pose. We regress the object pose parameters from image features using convolution and fully-connected layers (see Fig. 3). OpdRcnn-O is trained with the same motion parameter loss as OpdRcnn-C, but with motion axes and origins expressed in object coordinates. We treat the extrinsics matrix as a vector of length 12 (9 for rotation, 3 for translation) and use the smooth L1 loss.

Implementation Details. We implement our architecture with Detectron2 [32]. We initialize weights from a ResNet50 pretrained on ImageNet, and optimize with SGD. Unless otherwise specified, we use a learning rate of 0.001 with linear warmup for 500 steps and decay by 0.1 at 0.6 and 0.8 of the total steps.

We first train only on the detection and segmentation task with a learning rate of 0.0005 for 30000 steps. Then we pick the best weights for RGB, depth, and RGBD independently. Because our OpdRcnn-C and OpdRcnn-O have the same structure for detection and segmentation, we load the weights from the best detection and segmentation models and fully train with all losses. In both OpdRcnn-C and OpdRcnn-O we use the ratios $[1, 8, 8]$ to weigh the motion category loss, motion axis loss and motion origin loss respectively. In OpdRcnn-O we use 15 as the weight for the object pose loss.

During training we employ image space data augmentation to avoid overfitting (random flip, random brightness, and random contrast). During inference, we use a greedy non-maximum suppression (NMS) with IoU threshold 0.5 and choose the predicted bounding box with highest score. We use a confidence threshold of 0.05, and allow for 100 maximum part detections per image.

6 Experiments

6.1 Metrics

Part Detection. To evaluate part detection and segmentation we use standard object detection and segmentation metrics over the part category, as implemented for MSCOCO [12]. In the main paper, we report mAP@IoU=0.5 for the predicted part label and 2D bounding box (**PDet**).

Part Motion. To evaluate motion parameter estimation and measure the influence of the motion type we compute detection metrics over the motion type (motion-averaged mAP). This is in contrast with part-averaged mAP which is over the part category. A 'match' for motion-averaged mAP considers the predicted motion type and the 2D bounding box (**MDet**), and error thresholds for the motion axis and motion origin. We set the thresholds to 10° for axis error and 0.25 of the object's diagonal length for the origin distance (predicted origin to

GT axis line). Motion parameters are evaluated in the camera coordinate frame. Motion origins for translation joints are all considered to match. We report several variants of part-averaged mAP: **PDet**, **PDet** with motion type matched (+**M**), **PDet** with motion type and motion axis matched (+**MA**), and **PDet** with motion type, motion axis and motion origin matched (+**MAO**). Motion-averaged mAP has the same variants with **MDet** instead of **PDet**. When evaluating motion parameters, prior work [11,30] has focused on the average error only for correctly detected parts. In contrast, our metrics incorporate whether the part was successfully detected or not (in addition to error thresholds).

6.2 Baselines

As noted by Li et al. [11], knowing the canonical object coordinates can assist with predicting the motion axes and origin. This is because the motion axes are often one of the $\{x, y, z\}$ axes in object coordinates. Similarly, the motion origin for revolute joints is often at the edge of the object. So we design baselines that select (randomly or using the most frequent heuristic) from the three axes and the edges and corners of the normalized object bounding box. These baselines rely on known canonical object coordinates, so we take the part and object pose predictions from OPDRCNN-O. For origin prediction we use ground truth object sizes to convert from normalized to canonical coordinates, so these are strong baselines with access to additional information not available to OPDRCNN.

RANDMOT. A lower bound on performance that randomly predicts the part type, motion type, motion axis and motion origin. The motion origin is picked randomly from 19 points in the corner, edge center, face center and center of the object bounding box. The motion axis is picked randomly from three axis-aligned directions in the canonical object coordinates.

MOSTFREQ. Selects the most frequent motion type, axis, and origin in object coordinates conditioned on the predicted part category (statistics from train set).

ANCSH. Li et al. [11]'s ANCSH predicts motion parameters (motion axis, motion origin) and part segmentation for single-view point clouds. This work assumes a fixed kinematic chain for all objects in a category. In other words, the number of parts, part labels, and part motion types are given as input. We re-implement ANCSH in PyTorch matching the results reported by the authors. Since ANCSH requires a fixed kinematic chain, we choose the most frequent kinematic chain: objects with one rotating door part (243 out of 683 total objects in OPDSynth). We train ANCSH on this 'one-door dataset'. The main evaluation is on the complete validation and test set (same as other baselines and approaches). The supplement provides additional comparisons on only one-door objects.

OPDPN. Baseline using a PointNet++ backbone to predict instance segmentation and motion parameters directly from an input point cloud. This baseline predicts the part category, part instance id, and motion parameters for each point. This architecture operates on a fixed number of parts so we train on objects with less than 5 parts. See the supplement for details on the architecture.

Table 3. Evaluation of openable part detection and part motion parameter estimation on the OPDSynth test set. The RandMot and MostFreq baselines use detections and extrinsic parameters from OpdRcnn-O. Both variants of OpdRcnn outperform baselines and prior work especially for RGB-only inputs and on metrics accounting for part motion parameter estimation.

Input	Model	Part-averaged mAP % ↑				Motion-averaged mAP % ↑		
		PDet	+M	+MA	+MAO	**MDet**	+MA	+MAO
RGBD	RandMot	5.0	1.3	0.2	0.1	6.2	0.7	0.3
	MostFreq	69.4	66.1	49.2	27.8	73.6	61.6	38.8
	OpdRcnn-C	**69.2**±0.26	**67.3**±0.25	42.6±0.25	40.9±0.31	**75.3**±0.09	55.3±0.23	53.9±0.20
	OpdRcnn-O	68.5±0.31	66.6±0.38	**52.6**±0.26	**49.0**±0.21	74.9±0.13	**65.3**±0.27	**62.9**±0.24
D (PC)	ANCSH [11]	2.7	2.7	2.3	2.1	3.9	3.1	2.8
	OpdPN	20.4	19.3	14.0	13.6	22.0	18.1	17.6
D	OpdRcnn-C	**68.2**±0.39	**66.5**±0.42	41.0±0.21	39.2±0.20	**73.0**±0.09	53.0±0.18	51.5±0.19
	OpdRcnn-O	67.3±0.43	65.6±0.28	**51.2**±0.44	**47.7**±0.24	72.2±0.14	**62.3**±0.25	**60.0**±0.23
RGB	OpdRcnn-C	**67.4**±0.26	**66.2**±0.18	40.9±0.21	38.0±0.19	**75.0**±0.14	53.4±0.20	51.4±0.18
	OpdRcnn-O	66.6±0.28	65.5±0.27	**50.7**±0.23	**46.9**±0.26	74.5±0.26	**63.8**±0.32	**61.5**±0.26

6.3 Results

Model Comparisons. Table 3 shows the performance of the baselines and the two variants of OpdRcnn. We can make a number of observations. The RandMot baseline performs quite poorly for both part detection and part motion estimation metrics on all input scenarios, indicating the challenge of detecting openable parts and estimating their motion parameters. The MostFreq baseline is quite competitive if we only look at part detection and motion type detection (**PDet** and **MDet** metrics). This is not surprising as the MostFreq baseline leverages detections from OpdRcnn-O and the most frequent motion heuristic is relatively strong. However, when we look at precision of motion axis and motion origin predictions (+**MAO** and related metrics) we see that OpdRcnn-O significantly outperforms simpler baselines, especially for the motion-averaged metrics (higher than 60% mAP). It also outperforms the camera-centric OpdRcnn-C, showing the benefit of object-centric representation relative to camera-centric representation when accurate estimation of motion axes and origins is important.

Effect of Input Modalities. Comparing different input modalities in Table 3 we see that overall, methods do best with RGBD, while D (depth, or point cloud) input, and RGB-only input are more challenging. For the depth input modality we compare against ANCSH [11] and the OPDPN baseline. Note that ANCSH requires separate models for different kinematic chains so it is severely disadvantaged when evaluating on the OPDSynth dataset that includes objects with varying number of parts and differing motion types. We observe that ANCSH is outperformed by OPDPN and both variations of OPDRCNN. One of the limitations of ANCSH is that it requires a prespecified kinematic chain (with fixed number of parts, part categories and motion types). Therefore, we also evaluated ANCSH in a setting constrained to 'single door' objects, where it performs more competitively (see supplement). These observations demonstrate the increased generality of our approach over baselines in terms of handling arbitrary object categories with changing numbers of moving parts and motion types. Moreover, both OPDRCNN variants perform well in the RGB-only setting.

Error Metrics. We also compute error metrics as in Li et al. [11]. Prior work computes these metrics only for detected parts that match a ground truth part, thus including different number of instances for different methods. ANCSH and OPDPN have average motion axis error of 10.36° (for 6975 instances) and 6.25° (for 9862 instances) respectively. Both have average motion origin error of 0.09. Our proposed methods have comparable errors but over more detected parts. OPDRCNN-C and OPDRCNN-O have axis error of 9.06° and 6.67° for ~22000 instances, and origin error of 0.11. These error metrics are restricted only to matched predictions and fail to capture detection performance differences.

Real-World Images. We demonstrate that we can apply our method on real-world images to detect openable parts and their motion parameters. We finetune the RGB and RGBD models trained on OPDSynth on the OPDReal training set. All hyperparameters are the same except we set the learning rates to 0.001 for RGB and RGBD, and 0.0001 for depth-only (D) models. Table 4 summarizes performance on our OPDReal dataset. Overall, the task is much more challenging with real data as seen by lower performance across all metrics for all methods. We see that OPDRCNN-O has the best performance overall across most metrics, and in particular for metrics that measure motion parameter estimation.

Qualitative Visualizations. Figure 4 shows qualitative results from both the OPDSynth and OPDReal datasets. We see the overall trend that OPDRCNN-O outperforms OPDRCNN-C in terms of motion parameter estimation. We also see that both datasets have hard cases, with OPDReal being particularly challenging due to real-world appearance variability and the limited field of view of single-view images resulting in many partially observed parts.

Table 4. Results on the OPDReal test set. Overall, the task is more challenging on real objects. OPDRCNN-O has the highest performance across most metrics, and in particular for motion-averaged metrics including motion estimation (+**MAO**).

Input	Model	Part-averaged mAP % ↑				Motion-averaged mAP % ↑		
		PDet	+**M**	+**MA**	+**MAO**	**MDet**	+**MA**	+**MAO**
RGBD	RANDMOT	5.3	1.4	0.1	0.1	7.1	0.5	0.3
	MOSTFREQ	56.6	54.6	34.0	21.6	71.6	50.4	32.2
	OPDRCNN-C	54.7	53.3	21.8	21.3	**73.4**	32.3	31.5
	OPDRCNN-O	**56.6**	**54.3**	**33.8**	**32.4**	73.3	**50.0**	**48.1**
D	OPDPN	15.4	15.3	12.1	11.5	22.8	17.9	17.0
	OPDRCNN-C	49.4	46.6	12.2	11.6	**61.1**	17.7	17.0
	OPDRCNN-O	**49.2**	**47.3**	**18.1**	**16.1**	60.3	**26.7**	**23.9**
RGB	OPDRCNN-C	**58.0**	**57.0**	22.2	21.3	73.6	32.6	31.4
	OPDRCNN-O	57.8	56.4	**33.0**	**30.8**	**74.0**	**48.7**	**45.7**

6.4 Analysis

Are Some Parts More Challenging than Others? Table 6 shows the performance of OPDRCNN-O broken down by openable part category. The `drawer` parts exhibit translational motion, `lid` parts have rotational motions, and `door` parts exhibit both. We see that `lid` is more challenging than `drawer` and `door`, with considerably lower part detection mAP and significantly lower part motion estimation performance (+**MA** and +**MAO** metrics). This may be caused by fewer `lid` in the training data (89 `lid` vs 508 `door` and 363 `drawer` parts).

How Does Ground Truth Part and Pose Affect Motion Prediction? To understand how part detection influences motion prediction, we use ground truth (GT) for the part label, part 2D bounding box, and object pose with OPDRCNN-O on the OPDSynth validation set (see Table 5. As expected, when the GT part label and box are provided **PDet** is close to 100%. Surprisingly, having just the GT part label and box does not improve motion prediction. The ground truth object pose is more important for predicting the motion axis and origin correctly. Even with GT pose and 2D part information, the motion prediction is still imperfect. See the supplement for additional analysis.

Does Depth Information Help? Table 3 shows that depth only (D) models are outperformed by RGB and RGBD models. Depth information is helpful in conjuction with RGB information as seen by the small performance boost between RGB and RGBD across all metrics. Depth-only models perform worse than RGB-only models across all motion-averaged metrics. We suspect that this is because most openable parts have minimal difference in depth values along their edges, and thus color is more helpful than depth for predicting openable part segmentation masks. Table 6 provides some insight. We see that for the `drawer` and `lid` category where detection is overall more challenging, depth

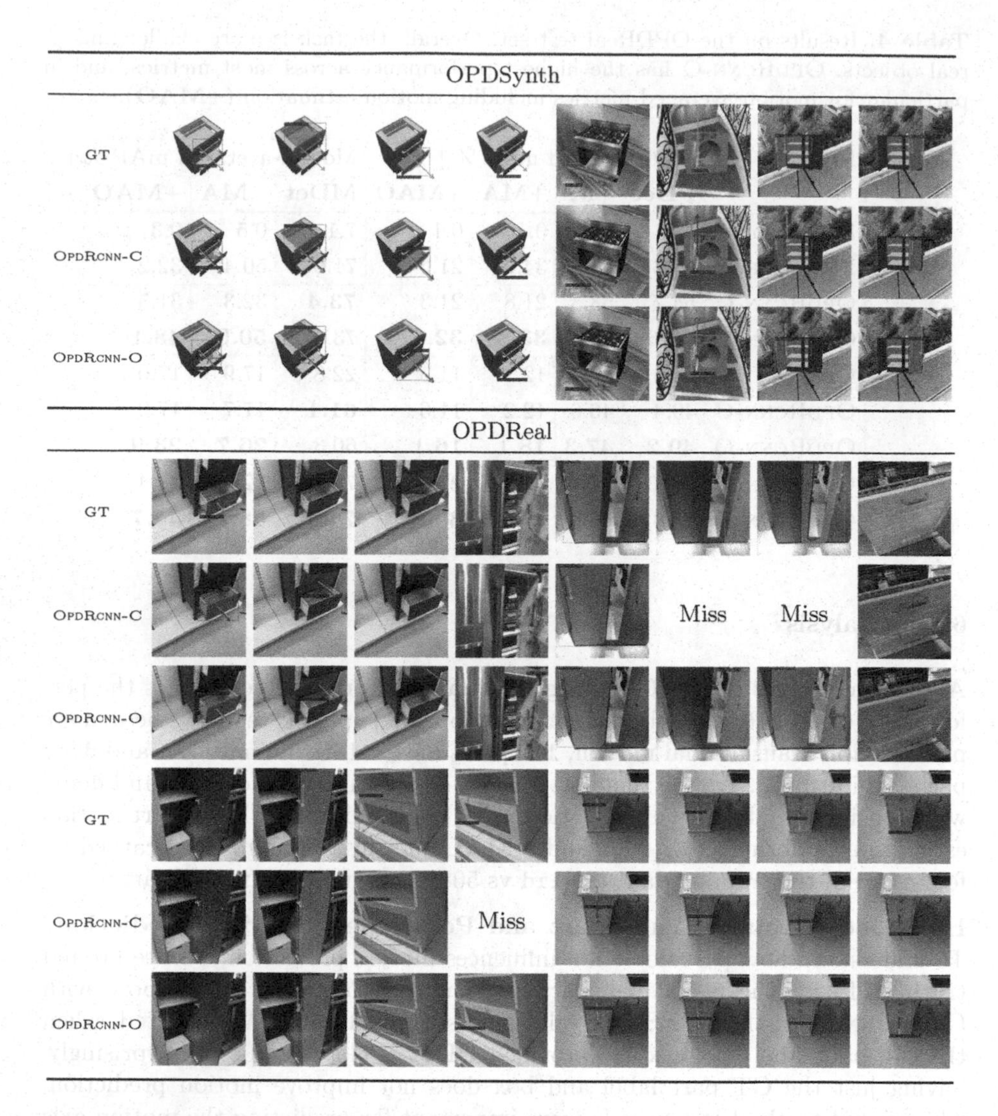

Fig. 4. Qualitative results from the OPDSynth and OPDReal val sets. The first row in each triplet shows the ground truth (GT) with each openable part mask, and the translational or rotational axis indicated in green. Other rows are predictions using OPDRcnn-C and OPDRcnn-O on RGBD inputs. The predicted axis is green if within 5° of the ground truth (also shown, in blue), in orange if within 10°, and in red if more than 10°. The first few OPDSynth examples show that translational `drawer` openable parts are relatively easy. Rotating `door` parts are more challenging (see high error predictions by OPDRcnn-C in the second row from the top). The OPDReal data is more challenging with many high axis error cases and entirely undetected parts (Miss). Particularly hard examples include unusual rotating doors that look like translating drawers (last column in top set), and openable parts only partly visible in the image frame (fourth column). (Color figure online)

Table 5. Analysis of performance given ground truth part category, 2D bounding box, and object pose. Results for OpdRcnn-O are on the OPDSynth val set.

	Part-averaged mAP % ↑				Motion-averaged mAP % ↑		
	PDet	+M	+MA	+MAO	**MDet**	+MA	+MAO
RGBD OpdRcnn-O	72.5±0.34	70.6±0.29	51.7±0.62	47.1±0.59	75.4±0.07	61.6±0.32	59.0±0.32
GT Box2DPart	**99.0**±0.00	**90.9**±0.16	50.6±0.36	45.4±0.27	**89.7**±0.15	58.1±0.32	54.7±0.28
GT Pose	73.1±0.10	71.0±0.05	60.5±0.06	59.4±0.05	75.2±0.08	67.0±0.14	66.2±0.09
GT Box2DPartPose	**99.0**±0.00	90.6±0.37	**65.5**±0.24	**63.8**±0.17	89.5±0.19	**73.3**±0.26	**72.0**±0.30

Table 6. Per-category evaluation ofOpdRcnn-O model on the OPDSynth val set. All metrics use part-averaged mAP. The `drawer` openable parts are easiest overall and do not benefit much from depth information. In contrast `door` and in particular `lid` parts are more challenging and do benefit from depth in the input.

Input	`drawer`				`door`				`lid`			
	PDet	+M	+MA	+MAO	**PDet**	+M	+MA	+MAO	**PDet**	+M	+MA	+MAO
RGB	**81.4**	**80.9**	**71.5**	**71.5**	**86.0**	**81.1**	61.6	57.0	58.7	58.4	27.7	17.8
D	70.3	70.1	63.4	63.4	83.3	78.6	61.7	57.4	57.8	57.5	30.0	18.3
RGBD	71.9	71.4	65.9	65.9	85.9	79.9	**63.7**	**59.8**	**62.2**	**61.8**	**31.4**	**19.7**

information does not help and the motion prediction results are also bad. For the `door` category where detection results are higher depth offers additional information that improves motion parameter estimation (higher +**MA** and +**MAO** metrics for depth-only (D) and RGBD models). See the supplement for an additional analysis based on breaking down performance across detection mAP ranges.

Can We Infer Part Motion States? A discrete notion of 'motion state' is often useful (e.g., "the fridge door is open" vs "the fridge door is closed"). We manually annotate a binary open vs closed motion state and continuous distance or angle offset from the closed state for all objects. We then add an MLP to the box head to predict binary motion state under a smooth L1 loss. We finetune the OpdRcnn-O model to predict the binary motion state along with part detection and motion parameter estimation. Finetuning is done for 5000 mini-batches. To evaluate motion state prediction we compute mAP values for 'match' or 'no match' of the binary motion state. Here, we define the mAP over motion type to evaluate the motion state prediction. Part-averaged mAP of the OpdRcnn-O model on the OPDSynth validation set is 62.8% for RGBD, 62.1% for RGB and 60.7% for D inputs. Motion-averaged mAP values are 62.0%, 64.2% and 59.9% respectively. These values should be contrasted with **PDet** and **MDet** in Table 3. We see that overall we can predict binary part motion states fairly well, though this too is a non-trivial task. We hypothesize that learning a 'threshold' for a binary notion of open vs closed is challenging (e.g., "fridge door cracked open").

6.5 Limitations

We focused on single objects with openable parts. The objects are fairly simple household objects without complex kinematic chains or complex motions (e.g., we cannot handle bifold doors). We also did not consider image inputs with multiple objects, each potentially possessing openable parts. A simple strategy to address this limitation would be to first detect distinct objects and then apply our approach on each object. Lastly, we focused on estimating the translation or rotation axis and rotation origin parameters but we do not estimate the range of motion for each part. This would be required to estimate a full part pose and to track the motion of an openable part from RGB video data.

7 Conclusion

We proposed the task of openable part detection and motion parameter estimation for single-view RGB images. We created a dataset of images from synthetic 3D articulated objects (OPDSynth) and of real objects reconstructed using RGBD sensors (OPDReal). We used these datasets to systematically study the performance of approaches for the openable part detection and motion estimation task, and investigate what aspects of the task are challenging. We found that the openable part detection task from RGB images is challenging especially when generalization across object and part categories is important. Our work is a first step, and there is much potential for future work in better understanding of articulated objects from real-world RGB images and RGB videos.

Acknowledgements. This work was funded in part by a Canada CIFAR AI Chair, a Canada Research Chair and NSERC Discovery Grant, and enabled in part by support from WestGrid and Compute Canada. We thank Sanjay Haresh for help with scanning and video narration, Yue Ruan for scanning and data annotation, and Supriya Pandhre, Xiaohao Sun, and Qirui Wu for help with data annotation.

References

1. Abbatematteo, B., Tellex, S., Konidaris, G.: Learning to generalize kinematic models to novel objects. In: Proceedings of the 3rd Conference on Robot Learning (2019)
2. Batra, D., et al.: Rearrangement: A challenge for embodied AI. arXiv preprint arXiv:2011.01975 (2020)
3. Bochkovskiy, A., Wang, C.Y., Liao, H.Y.M.: YOLOV4: optimal speed and accuracy of object detection. arXiv preprint arXiv:2004.10934 (2020)
4. Chen, X., Mottaghi, R., Liu, X., Fidler, S., Urtasun, R., Yuille, A.: Detect what you can: detecting and representing objects using holistic models and body parts. In: Proceedings of the IEEE Conference on Computer Vision and Pattern Recognition (CVPR), pp. 1971–1978 (2014)

5. He, K., Gkioxari, G., Dollár, P., Girshick, R.: Mask R-CNN. In: Proceedings of the IEEE International Conference on Computer Vision (ICCV), pp. 2961–2969 (2017)
6. Hu, R., Li, W., Van Kaick, O., Shamir, A., Zhang, H., Huang, H.: Learning to predict part mobility from a single static snapshot. ACM Trans. Graph. (TOG) **36**(6), 227 (2017)
7. Hu, R., Savva, M., van Kaick, O.: Functionality representations and applications for shape analysis. Comput. Graph. Forum **37**(2), 603–624 (2018)
8. Huang, J., et al.: MultiBodySync: multi-body segmentation and motion estimation via 3D scan synchronization. arXiv preprint arXiv:2101.06605 (2021)
9. Jain, A., Lioutikov, R., Niekum, S.: ScrewNet: Category-independent articulation model estimation from depth images using screw theory. arXiv preprint arXiv:2008.10518 (2020)
10. Li, H., Wan, G., Li, H., Sharf, A., Xu, K., Chen, B.: Mobility fitting using 4D RANSAC. Comput. Graph. Forum **35**(5), 79–88 (2016)
11. Li, X., Wang, H., Yi, L., Guibas, L., Abbott, A.L., Song, S.: Category-level articulated object pose estimation. In: Proceedings of the IEEE Conference on Computer Vision and Pattern Recognition (CVPR) (2020)
12. Lin, T.-Y., et al.: Microsoft COCO: common objects in context. In: Fleet, D., Pajdla, T., Schiele, B., Tuytelaars, T. (eds.) ECCV 2014. LNCS, vol. 8693, pp. 740–755. Springer, Cham (2014). https://doi.org/10.1007/978-3-319-10602-1_48
13. Liu, L., Xu, W., Fu, H., Qian, S., Han, Y., Lu, C.: AKB-48: a real-world articulated object knowledge base. arXiv preprint arXiv:2202.08432 (2022)
14. Liu, Q., Qiu, W., Wang, W., Hager, G.D., Yuille, A.L.: Nothing but geometric constraints: a model-free method for articulated object pose estimation. arXiv preprint arXiv:2012.00088 (2020)
15. Lu, C., et al.: Beyond holistic object recognition: enriching image understanding with part states. In: Proceedings of the IEEE Conference on Computer Vision and Pattern Recognition (CVPR), pp. 6955–6963 (2018)
16. Martín-Martín, R., Eppner, C., Brock, O.: The RBO dataset of articulated objects and interactions. Int. J. Robot. Res. **38**(9), 1013–1019 (2019)
17. Mitra, N.J., Yang, Y.L., Yan, D.M., Li, W., Agrawala, M.: Illustrating how mechanical assemblies work. ACM Trans. Graph.-TOG **29**(4), 58 (2010)
18. Mittal, M., Hoeller, D., Farshidian, F., Hutter, M., Garg, A.: Articulated object interaction in unknown scenes with whole-body mobile manipulation. arXiv preprint arXiv:2103.10534 (2021)
19. Mo, K., et al.: PartNet: a large-scale benchmark for fine-grained and hierarchical part-level 3D object understanding. In: Proceedings of the IEEE Conference on Computer Vision and Pattern Recognition (CVPR), pp. 909–918 (2019)
20. Sharf, A., Huang, H., Liang, C., Zhang, J., Chen, B., Gong, M.: Mobility-trees for indoor scenes manipulation. Comput. Graph. Forum **33**(1), 2–14 (2014)
21. Shridhar, M., et al.: Alfred: a benchmark for interpreting grounded instructions for everyday tasks. In: Proceedings of the IEEE/CVF Conference on Computer Vision and Pattern Recognition, pp. 10740–10749 (2020)
22. Srivastava, S., et al.: BEHAVIOR: Benchmark for everyday household activities in virtual, interactive, and ecological environments. In: Conference on Robot Learning, pp. 477–490. PMLR (2022)
23. Szot, A., et al.: Habitat 2.0: training home assistants to rearrange their habitat. Adv. Neural. Inf. Process. Syst. **34**, 251–266 (2021)
24. Varol, G., et al.: Learning from synthetic humans. In: Proceedings of the IEEE Conference on Computer Vision and Pattern Recognition, pp. 109–117 (2017)

25. Waechter, M., Moehrle, N., Goesele, M.: Let there be color! Large-scale texturing of 3d reconstructions. In: Fleet, D., Pajdla, T., Schiele, B., Tuytelaars, T. (eds.) ECCV 2014. LNCS, vol. 8693, pp. 836–850. Springer, Cham (2014). https://doi.org/10.1007/978-3-319-10602-1_54
26. Wang, H., Sridhar, S., Huang, J., Valentin, J., Song, S., Guibas, L.J.: Normalized object coordinate space for category-level 6D object pose and size estimation. In: Proceedings of the IEEE Conference on Computer Vision and Pattern Recognition (CVPR), pp. 2642–2651 (2019)
27. Wang, J., Yuille, A.L.: Semantic part segmentation using compositional model combining shape and appearance. In: Proceedings of the IEEE Conference on Computer Vision and Pattern Recognition (CVPR), pp. 1788–1797 (2015)
28. Wang, P., Shen, X., Lin, Z., Cohen, S., Price, B., Yuille, A.L.: Joint object and part segmentation using deep learned potentials. In: Proceedings of the IEEE International Conference on Computer Vision (ICCV), pp. 1573–1581 (2015)
29. Wang, X., Zhou, B., Fang, H., Chen, X., Zhao, Q., Xu, K.: Learning to group and label fine-grained shape components. ACM Trans. Graph. (TOG) **37**(6), 1–14 (2018)
30. Wang, X., Zhou, B., Shi, Y., Chen, X., Zhao, Q., Xu, K.: Shape2Motion: joint analysis of motion parts and attributes from 3D shapes. In: Proceedings of the IEEE Conference on Computer Vision and Pattern Recognition (CVPR), pp. 8876–8884 (2019)
31. Weihs, L., Deitke, M., Kembhavi, A., Mottaghi, R.: Visual room rearrangement. In: Proceedings of the IEEE Conference on Computer Vision and Pattern Recognition (CVPR) (2021)
32. Wu, Y., Kirillov, A., Massa, F., Lo, W.Y., Girshick, R.: Detectron2 (2019). https://github.com/facebookresearch/detectron2
33. Xia, F., Wang, P., Chen, X., Yuille, A.L.: Joint multi-person pose estimation and semantic part segmentation. In: Proceedings of the IEEE Conference on Computer Vision and Pattern Recognition (CVPR), pp. 6769–6778 (2017)
34. Xiang, F., et al.: SAPIEN: a simulated part-based interactive environment. In: Proceedings of the IEEE Conference on Computer Vision and Pattern Recognition (CVPR), pp. 11097–11107 (2020)
35. Xu, W., et al.: Joint-aware manipulation of deformable models. ACM Trans. Graph. (TOG) **28**(3), 1–9 (2009)
36. Yan, Z., et al.: RPM-Net: recurrent prediction of motion and parts from point cloud. ACM Trans. Graph. (TOG) **38**(6), 240 (2019)
37. Yi, L., Huang, H., Liu, D., Kalogerakis, E., Su, H., Guibas, L.: Deep part induction from articulated object pairs. ACM Trans. Graph. (TOG) **37**(6), 209 (2019)
38. Yi, L., et al.: A scalable active framework for region annotation in 3D shape collections. ACM Trans. Graph. (TOG) **35**(6), 1–12 (2016)
39. Zeng, V., Lee, T.E., Liang, J., Kroemer, O.: Visual identification of articulated object parts. arXiv preprint arXiv:2012.00284 (2020)
40. Zhou, B., Zhao, H., Puig, X., Fidler, S., Barriuso, A., Torralba, A.: Scene parsing through ADE20K dataset. In: Proceedings of the IEEE Conference on Computer Vision and Pattern Recognition (CVPR) (2017)
41. Zhou, Q.Y., Park, J., Koltun, V.: Open3D: a modern library for 3D data processing. arXiv:1801.09847 (2018)

AirDet: Few-Shot Detection Without Fine-Tuning for Autonomous Exploration

Bowen Li[1,2], Chen Wang[1], Pranay Reddy[1,3], Seungchan Kim[1], and Sebastian Scherer[1(✉)]

[1] Robotics Institute, Carnegie Mellon University, Pittsburgh, USA
{bowenli2,seungch2,basti}@andrew.cmu.edu
[2] School of Mechanical Engineering, Tongji University, Shanghai, China
[3] Electronics and Communication Engineering, IIITDM Jabalpur, Jabalpur, India
2018033@iiitdmj.ac.in

Abstract. Few-shot object detection has attracted increasing attention and rapidly progressed in recent years. However, the requirement of an exhaustive offline fine-tuning stage in existing methods is time-consuming and significantly hinders their usage in online applications such as autonomous exploration of low-power robots. We find that their major limitation is that the little but valuable information from a few support images is not fully exploited. To solve this problem, we propose a brand new architecture, AirDet, and surprisingly find that, by learning *class-agnostic relation* with the support images in all modules, including cross-scale object proposal network, shots aggregation module, and localization network, AirDet without fine-tuning achieves comparable or even better results than many fine-tuned methods, reaching up to **30–40%** improvements. We also present solid results of onboard tests on real-world exploration data from the DARPA Subterranean Challenge, which strongly validate the feasibility of AirDet in robotics. To the best of our knowledge, AirDet is the first feasible few-shot detection method for autonomous exploration of low-power robots. The code and pre-trained models are released at https://github.com/Jaraxxus-Me/AirDet.

Keywords: Few-shot object detection · Online · Robot exploration

1 Introduction

Few-shot object detection (FSOD) [36,39–41,44] aims at detecting objects out of base training set with few support examples per class. It has received increasing attention from the robotics community due to its vital role in autonomous exploration, since robots are often expected to detect novel objects in an unknown environment but only a few examples can be provided online. For example, in a mission of rescue shown in Fig. 1 (a), the robots are required to detect some uncommon objects such as drill, rope, helmet, vent.

Supplementary Information The online version contains supplementary material available at https://doi.org/10.1007/978-3-031-19842-7_25.

S. Avidan et al. (Eds.): ECCV 2022, LNCS 13699, pp. 427–444, 2022.
https://doi.org/10.1007/978-3-031-19842-7_25

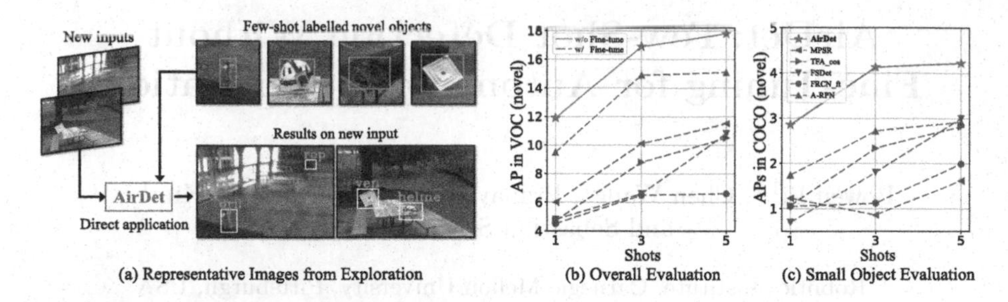

Fig. 1. Representative images from robots' exploration and performance comparison of the state-of-the-arts [6,28,36,39,40] and proposed AirDet. Solid lines denote results with no fine-tuning and dashed lines indicate results fine-tuned on few-shot data. Without further updating, AirDet can outperform prior work. Besides, unlike the fine-tuned models meeting bottleneck in small objects, AirDet has set an outstanding new level.

Despite its recent promising developments, most of existing methods [2,7,16,19,24,31,36,37,39,42] require a careful *offline* fine-tuning stage on novel images before inference. However, the requirement of the fine-tuning process is infeasible for robotic *online* applications, since **(1)** new object categories can be dynamically added during exploration, thus re-fine-tuning the model with limited onboard computational resources for novel classes is extremely inefficient for the time-starved tasks such as search and rescue [8,33–35]. **(2)** to save human effort, only very few samples can be provided online[1], thus the fine-tuning stage [7,16,19,24,31,36,37,39,42] needs careful *offline* hyper-parameter tuning to avoid over-fitting, which is infeasible for *online* exploration, and **(3)** fine-tuned models usually perform well for in-domain test [19,24,28,36,39,40], while suffer from cross-domain test, which is unfavourable for robotic applications.

Therefore, we often expect a few-shot detector that is able to inference without fine-tuning such as [6]. However, the performance of [6] is still severely hampered for challenging robotics domain due to **(1)** ineffective multi-scale detection; **(2)** ineffective feature aggregation from multi-support images; and **(3)** inaccurate bounding box location prediction. Surprisingly, in this paper, we find that all three problems can be effectively solved by learning *class-agnostic relation* with support images. We name the new architecture AirDet, which can produce promising results even without abominable fine-tuning as shown in Fig. 1, which, to the best of our knowledge, is the first feasible few-shot detection model for autonomous robotic exploration. Specifically, the following three modules are proposed based on *class-agnostic relation*, respectively.

Support-Guided Cross-Scale Fusion (SCS) for Object Proposal. One reason for performance degradation in multi-scale detection is that the region proposals are not effective for small scale objects, even though some existing

[1] Since online annotation is needed during mission execution, only 1–5 samples can be provided in most of the robotic applications, which is the main focus of this paper.

works adopt multiple scale features from query images [42,44]. We argue that the proposal network should also include cross-scale information from the support images. To this end, we present a novel SCS module, which explicitly extracts multi-scale features from cross-scale relations between support and query images.

Global-Local Relation (GLR) for Shots Aggregation. Most prior work [6,39,41] simply average multi-shot support feature to obtain a class prototype for detection head. However, this cannot fully exploit the little but valuable information from every support image. Instead, we construct a shots aggregation module by learning the relationship between the multi-support examples, which achieves significant improvements with more shots.

Prototype Relation Embedding (PRE) for Location Regression. Some existing works [6,42] introduced a relation network [32] into the classification branch; however, the location regression branch is often neglected. To settle this, we introduce cross-correlation between the proposals from SCS and support features from GLR into the regression branch. This results in the PRE module, which explicitly utilize support images for precise object localization.

In summary, AirDet is a fully relation-based few-shot object detector, which can be applied directly to the novel classes without fine-tuning. It surprisingly produces comparable or even better results than exhaustively fine-tuned SOTA methods [6,28,36,39,40], as shown in Fig. 1(b). Besides, as shown in Fig. 1(c), AirDet maintains high robustness in small objects due to the SCS module, which fully takes advantage of the multi-scale support feature. Note that in this paper, fine-tuning is undesired because it cannot satisfy the online responsive requirement for robots, but it can still improve the performance of AirDet.

2 Related Works

2.1 General Object Detection

The task of object detection [9,10,13,23,25,28] is to find out all the pre-defined objects in an image, predicting their categories and locations, which is one of the core problems in the field of computer vision. Object detection algorithms are mainly divided into: two-stage approaches [9,10,13,28] and one-stage approaches [23,25–27]. R-CNN [10], and its variants [9,10,13,28] serve as the foundation of the former branch; among them, Faster R-CNN [28] used region proposal network (RPN) to generate class-agnostic proposals from the dense anchors, which greatly improved the speed of object detection based on R-CNN [9]. On the other hand, YOLO series [25–27] fall into the second branch, which tackles object detection as an end-to-end regression problem. Besides, the known SSD series [20,23] propose to utilize pre-defined bounding boxes to adjust to various object scales inspired by [28].

One shortcoming of the above methods is that they require abundant labeled data for training. Moreover, the types and number of object categories are fixed after training (80 classes in COCO, for instance), which is not applicable to robot's autonomous exploration, where unseen, novel objects often appear online.

2.2 Few-Shot Object Detection

Trained with abundant data for base classes, few-shot object detectors can learn to generalize only using a few labeled novel image shots. Two main branches leading in FSOD are meta-learning-based approaches [6,12,39–41] and transfer-learning-based approaches [24,31,36,38,44].

Transfer-learning approaches seek for the best learning strategy of general object detectors [28] on a few novel images. Wang *et al.* [36] proposed to fine-tune only the last layer with a cosine similarity-based classifier. Using manually defined positive refinement branch, MPSR [39] mitigated the scale scarcity issue. Recent works have introduced semantic relations between novel-base classes [44] and contrastive proposal encoding [31].

Aiming at training meta-models on episodes of individual tasks, meta-learning approaches [6,12,15,40–42] generally contain two branches, one for extracting support information and the other for detection on the query image. Among them, Meta R-CNN [41], and FSDet [40] target at support guided query channel attention. With novel attention RPN and multi-relation classifier, A-RPN [6] has set the current SOTA. Very recent works also cover support-query mutual guidance [42], aggregating context information [15], and constructing heterogeneous graph convolutional networks on proposals [12].

2.3 Relation Network for Few-Shot Learning

In few-shot image classification, relation network [32], also known as learning to compare, has been introduced to train a classifier by modeling the class-agnostic relation between a query image and the support images. Once trained and provided with a few novel support images, inference on novel query images can be implemented without further updating.

For few-shot object detection, such relation has only been utilized for the classification branch so far in very few works. For example, Fan *et al.* proposed a multi-relation classification network, which consists of global, local, and patch relation branches [6]. Zhang *et al.* leveraged general relation network [32] architecture to build multi-level proposal scoring, and support weighting modules [42]. In this work, we thoroughly explore such relation in few-shot detection and propose a fully relation-based architecture.

2.4 Multi-scale Feature Extraction

Multi-scale features have been exhaustively exploited for multi-scale objects in general object detection [17,20,21,23,26,30]. For example, FSSD [20] proposed to fuse multi-scale feature and implement detection on the fused feature map. Lin *et al.* constructed the feature pyramid network (FPN) [21], which builds a top-down architecture and employs multi-scale feature map for detection. For few-shot detection, standard FPN [21] has been widely adopted in prior transfer-learning-based methods [31,36,39,44]. In meta-learning, existing meta-learner [42] employs all scales from FPN and implement detection on each scale in parallel, which is computationally inefficient.

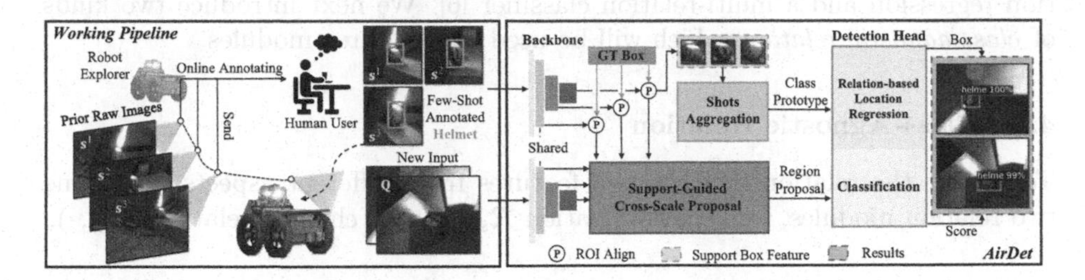

Fig. 2. The pipeline of the autonomous exploration task and the framework of AirDet. During exploration, a few prior raw images that potentially contain novel objects (helmet) are sent to a human user first. Provided with online annotated few-shot data, the robot explorer is able to detect those objects by observing its surrounding environment. AirDet includes 4 modules, *i.e.*, the shared backbone, support-guided cross-scale (SCS) feature fusion module for region proposal, global-local relation (GLR) module for shots aggregation, and relation-based detection head, which are visualized by different colors. (Color figure online)

3 Preliminary

In few-shot object detection [4,39–41], the classes are divided into B base classes $\mathcal{C}_\mathrm{b}$ and N novel ones $\mathcal{C}_\mathrm{n}$, satisfying that $\mathcal{C}_\mathrm{b} \cap \mathcal{C}_\mathrm{n} = \varnothing$. The objective is to train a model that can detect novel classes in $\mathcal{C}_\mathrm{n}$ by only providing k-shot labeled samples for $\mathcal{C}_\mathrm{n}$ and abundant images from base classes $\mathcal{C}_\mathrm{b}$.

During training, we adopt the episodic paradigm [41]. Basically, images from the base classes $\mathcal{C}_\mathrm{b}$ are split into query images $\mathbf{Q}_\mathrm{b}$ and support images $\mathbf{S}_\mathrm{b}$. Given all support images $\mathbf{S}_\mathrm{b}$, the model learns to detect objects in query images $\mathbf{Q}_\mathrm{b}$. During test, the model is to detect objects in novel query images $\mathbf{Q}_\mathrm{n}$ by only providing few (1–5) labeled novel support images $\mathbf{S}_\mathrm{n}$.

Remark 1. Most existing methods [2,24,31,36,38–41,44] have to be fine-tuned on $\mathbf{S}_\mathrm{n}$ due to the class-specific model design, while AirDet can be applied directly to $\mathbf{Q}_\mathrm{n}$ by providing $\mathbf{S}_\mathrm{n}$ without fine-tuning.

4 Methodology

Since only a few shots are given during model test, information from the support images is little but valuable. We believe that the major limitation of the existing algorithms is that such information from support images is not fully exploited.

Therefore, we propose to learn *class-agnostic relation* with the support images in all the modules of AirDet. As exhibited in Fig. 2, the structure of AirDet is simple: except for the shared backbones, it only consists of three modules, *i.e.*, a support-guided cross-scale fusion (SCS) module for regional proposal, a global-local relation (GLR) module for shots aggregation, and a relation-based detection head, containing prototype relation embedding (PRE) module for location regression and a multi-relation classifier [6]. We next introduce two kinds of *class-agnostic relation*, which will be used by the three modules.

4.1 Class-Agnostic Relation

To exploit the relation between two features from different aspects, we define two relation modules, *i.e.*, spatial relation $\mathcal{R}_{\mathrm{s}}(\cdot,\cdot)$ and channel relation $\mathcal{R}_{\mathrm{c}}(\cdot,\cdot)$.

1. Spatial Relation: Object features from the same category are often correlated along the spatial dimension, thus we define the spatial relation features $\mathcal{R}_{\mathrm{s}}$ in (1) leveraging on the regular and depth-wise convolution.

$$\mathcal{R}_{\mathrm{s}}(\mathbf{A},\mathbf{B}) = \mathbf{A} \odot \mathrm{MLP}\Big(\mathrm{Flatten}\big(\mathrm{Conv}(\mathbf{B})\big)\Big), \tag{1}$$

where inputs $\mathbf{A},\mathbf{B} \in \mathbb{R}^{C\times W\times H}$ denote 2 general tensors. Flatten means flatten the features in spatial (image) domain and MLP denotes multilayer perceptron (MLP), so that $\mathrm{MLP}\Big(\mathrm{Flatten}\big(\mathrm{Conv}(\mathbf{B})\big)\Big) \in \mathbb{R}^{\mathrm{C}\times 1\times 1}$. $\odot$ indicates depth-wise convolution [6]. Note that we use convolution to calculate correlation since both operators are composed of inner products.

2. Channel Relation: Inspired by the phenomenon that features of different classes are often stored in different channels [18], we propose a simple but effective channel relation $\mathcal{R}_{\mathrm{c}}(\cdot,\cdot)$ in (2) to extract the cross-class relation features.

$$\mathcal{R}_{\mathrm{c}}(\mathbf{A},\mathbf{B}) = \mathrm{Conv}\big(\mathrm{Cat}(\mathbf{A},\mathbf{B})\big) + \mathrm{Cat}\big(\mathrm{Conv}(\mathbf{A}),\mathrm{Conv}(\mathbf{B})\big), \tag{2}$$

where $\mathrm{Cat}(\cdot,\cdot)$ is to concatenate features along the channel dimension.

Remark 2. The two simple but effective *class-agnostic relation* learners are fundamental building blocks of AirDet, which, to the best of our knowledge, is the first attempt towards a fully relation-based structure in few-shot detection.

4.2 Support-guided Cross-Scale Fusion (SCS) for Object Proposal

As mentioned earlier, existing works generate object proposals only using single scale information from query images [16,37,39,40], while such strategy may not be effective for small scale novel objects.

Differently, we propose support-guided cross-scale fusion (SCS) in AirDet to introduce multi-scale features and take the relation between query and support images for region proposal. As shown in Fig. 3(a), SCS takes support and query features from different backbone blocks (ResNet2, 3, and 4 block) as input. We first apply *spatial relation*, where the query and support features from the same backbone block are $\mathbf{A}, \mathbf{B}$ in (1), respectively. Then we use *channel relation* to fuse the ResNet2 and ResNet3 block features, which are $\mathbf{A}, \mathbf{B}$ in (2), respectively. The fused channel relation feature is later merged with the spatial relation feature from ResNet4 block. The final merged feature is sent to the region proposal network (RPN) [28] to generate region proposals.

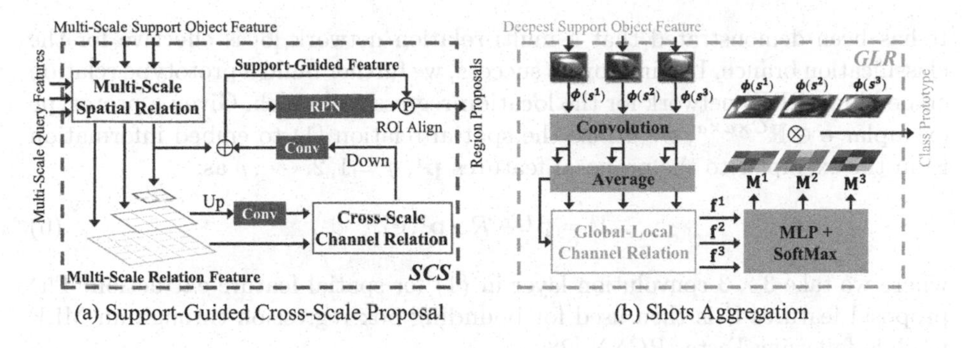

Fig. 3. Network architecture of SCS for region proposal and GLR for shots aggregation.

4.3 Global-Local Relation (GLR) for Shots Aggregation

In prior attempts [16,39–41], the support object features from multiple shots are usually averaged to represent the class prototype, which is then used for regression and classification. Although it can be effective with fine-tuning, we argue that a simple average cannot fully utilize the information from few-shot data. To this end, we built global-local relation (GLR) module in Fig. 3(b), which leverages the features from every shot to construct the final prototype.

Suppose the k-shot deepest support features are $\phi(\mathbf{s}^i)$, $i = 1, \cdots, k$, our final class prototype $\mathbf{e}$ can be expressed as a weighted average of the features:

$$\mathbf{e} = \sum_{i=1}^{k} (\phi(\mathbf{s}^i) \otimes \mathbf{M}^i), \tag{3}$$

where $\otimes$ is the element-wise multiplication, and $\mathbf{M}^i$ is a confidence map:

$$\mathbf{M}^i = \text{SoftMax}\Big(\text{MLP}(\mathbf{f}^i)\Big), \tag{4}$$

where $\mathbf{f}^i$ is the output from the channel relation extractor:

$$\mathbf{f}^i = \mathcal{R}_c\left(\text{Conv}(\phi(\mathbf{s}^i)), \frac{1}{k}\sum_{i=1}^{k} \text{Conv}(\phi(\mathbf{s}^i))\right). \tag{5}$$

Note that to include both "global" (all shots) and "local" (single shot) features, the inputs of the channel relation extractor in (5) include both the feature from that shot and the averaged features over all shots.

Remark 3. Unlike prior work [16,39–41] relying on fine-tuning with more support data for performance gain, AirDet can extract stronger prototype to achieve improvement on more shots without fine-tuning.

4.4 Prototype Relation Embedding (PRE) for Location Regression

It has been demonstrated that a multi-relation network [6] is effective for the classification branch. Inspired by its success, we further build a prototype relation embedding (PRE) network for the location regression branch. Given a prototype exemplar $\mathbf{e} \in \mathbb{R}^{C \times a \times a}$, we utilize the spatial relation (1) to embed information from the exemplar to the proposal features $\mathbf{p}^j$, $j = 1, 2, \cdots, p$ as:

$$\mathbf{l}^j = \mathbf{p}^j + \mathcal{R}_\mathrm{s}(\mathbf{p}^j, \mathbf{e}), \tag{6}$$

where we take 3×3 convolution layer in (1) for spatial feature extraction. The proposal features $\mathbf{l}^j$ is then used for bounding box regression through an MLP module following Faster-RCNN [28].

Remark 4. Class-related feature $\mathbf{l}^j$ contains information from support objects, which turns out more effective for location regression even if the objects have never been seen in the training set.

5 Experiments

5.1 Implementation

We adopt the training pipeline from [6]. To maintain a fair comparison with other methods [28,36,39,40], we mainly adopt ResNet101 [14] pre-trained on ImageNet [4] as backbone. The performance of other backbones is presented in Appendix B. For fair comparison [6,28,36,39,40], we utilized their official implementation, support examples, and models (if provided) in all the experiments. AirDet and the baseline [6] take the **same** supports in all the settings. We use 4 NVIDIA GeForce Titan-X Pascal GPUs for experiments. Detailed configuration of AirDet can be found in Appendix A and the source code.

Remark 5. To save human effort, only very few support examples (1–5 samples per class) can be provided during online exploration. Therefore, we mainly focused on $k = 1, 2, 3, 5$-shot evaluation. Since the objects from exploration are usually unseen, we only test novel classes throughout the experiments.

5.2 In-Domain Evaluation

We first present the in-domain evaluation on COCO benchmark [22], where the models are trained and tested on the same dataset. Following prior works [2,6,7,16,31,36,38–41,44], the 80 classes are split into 60 non-VOC base classes and 20 novel classes. During training, the base class images from COCO trainval2014 are considered available. With few-shot samples per novel class, the models are evaluated on 5,000 images from COCO val2014 dataset.

Overall Performance. As shown in Table 1, AirDet achieves significant performance gain on the baseline [6]. AirDet without fine-tuning amazingly also achieves comparable or even better results than many fine-tuned methods. With fine-tuning, AirDet outperformed existing SOTAs [2,6,28,36,39,40,43]. Since the results without fine-tuning may be sensitive to support images, we report the averaged performance, and the standard deviation on 3–5 randomly sampled support images, where we surprisingly find AirDet more robust to the variance of support images compared with the baseline [6].

Multi-scale Objects. We next report the performance of methods [6,28,36,39,40] and AirDet on multi-scale objects in Table 2. Thanks to SCS, AirDet achieves the highest performance for multi-scale objects among all the SOTAs. Especially for small objects, given 5-shots, AirDet can achieve a surprising **4.22** AP_s, nearly doubling the fine-tuned methods with multi-scale FPN features [36,39].

Comparison of 10-Shot. For a more thorough comparison, we present the 10-shot evaluation on the COCO validation dataset in Table 3. Without fine-tuning, AirDet can surprisingly achieve comparable performance against recent

Table 1. Performance comparison on COCO validation dataset. In each setting, red and green fonts denote the best and second-best performance, respectively. AirDet achieves significant performance gain on baseline without fine-tuning. With fine-tuning, AirDet sets a new SOTA performance. †We randomly sampled 3–5 different groups of support examples and reported the average performance and their standard deviation.

Shots		1			2			3			5		
Method	Fine-tune	AP	AP_{50}	AP_{75}	AP	AP_{50}	AP_{75}	AP	AP_{50}	AP_{75}	AP	AP_{50}	AP_{75}
A-RPN [6]†	✗	4.32 ±0.7	7.62 ±1.3	4.3 ±0.7	4.67 ±0.3	8.83 ±0.5	4.49 ±0.3	5.28 ±0.6	9.95 ±0.8	5.05 ±0.6	6.08 ±0.3	11.17 ±0.4	5.88 ±0.3
AirDet (Ours)†	✗	5.97 **±0.4**	10.52 **±0.9**	5.98 **±0.2**	6.58 **±0.2**	12.02 **±0.4**	6.33 **±0.2**	7.00 **±0.5**	12.95 **±0.8**	6.71 ±0.7	7.76 **±0.3**	14.28 **±0.4**	7.31 ±0.4
FRCN [28]	✓	3.26	6.66	3.04	3.73	7.79	3.22	4.59	9.52	4.07	5.32	11.20	4.54
TFA_{fc} [36]	✓	2.78	5.39	2.36	4.14	7.98	4.01	6.33	12.10	5.94	7.92	15.58	7.29
TFA_{cos} [36]	✓	3.09	5.24	3.21	4.21	7.70	4.35	6.05	11.48	5.93	7.61	14.56	7.17
FSDetView [40]	✓	2.20	6.20	0.90	3.40	10.00	1.50	5.20	14.70	2.10	8.20	21.60	4.70
MPSR [39]	✓	3.34	6.11	3.25	5.41	9.68	5.52	5.70	10.54	5.50	7.20	13.55	6.89
A-RPN [6]	✓	4.59	8.85	4.37	6.15	12.05	5.76	8.24	15.52	7.92	9.02	17.29	8.53
W. Zhang *et al.* [43]	✓	4.40	7.50	4.90	5.60	9.90	5.90	7.20	13.30	7.40	–	–	–
FADI [2]	✓	5.70	10.40	6.00	7.00	13.01	7.00	8.60	15.80	8.30	10.10	18.60	11.90
AirDet (Ours)	✓	6.10	11.40	6.04	8.73	16.24	8.35	9.95	19.39	9.09	10.81	20.75	10.27

Table 2. Performance evaluation on multi-scale objects from COCO. Highest-ranking and second-best scores are marked out with red and green, respectively. Without fine-tuning, AirDet can avoid over-fitting and shows robustness on small-scale objects. By virtue of the SCS module, AirDet can achieve higher results than those with FPN.

Shots			1			2			3			5		
Method	FPN	Fine-tune	AP_s	AP_m	AP_l	AP_s	AP_m	AP_l	AP_s	AP_m	AP_l	AP_s	AP_m	AP_l
A-RPN [6]†	✗	✗	2.43 ±0.4	5.00 ±1.0	6.74 ±1.1	2.67 ±0.3	5.01 ±0.3	7.18 ±0.4	3.42 ±0.2	6.15 ±0.5	8.77 ±0.8	3.54 ±0.3	6.73 ±0.03	9.97 ±0.2
AirDet (Ours)†	✗	✗	2.85 **±0.3**	6.33 **±0.7**	9.00 **±0.8**	4.00 **±0.3**	6.84 **±0.1**	9.94 **±0.3**	4.13 **±0.1**	7.95 **±0.5**	11.30 ±0.9	4.22 **±0.2**	8.24 ±0.04	12.90 ±0.5
FRCN [28]	✓	✓	1.05	3.68	5.41	0.94	4.39	6.42	1.12	5.11	7.83	1.99	5.30	8.84
TFA_{fc} [36]	✓	✓	1.06	2.71	4.38	1.17	4.02	7.05	1.97	5.48	11.09	2.40	6.86	12.86
TFA_{cos} [36]	✓	✓	1.07	2.78	5.12	1.64	4.12	7.27	2.34	5.48	10.43	2.82	6.70	12.21
FSDetView [36]	✗	✓	0.70	2.70	3.70	0.60	4.00	4.20	1.80	5.10	8.00	3.00	9.70	12.30
MPSR [39]	✓	✓	1.23	3.82	5.58	1.89	5.69	8.73	0.86	4.60	9.96	1.62	6.78	11.66
A-RPN [6]	✗	✓	1.74	5.17	6.96	2.20	7.55	10.49	2.72	9.51	14.74	2.92	10.67	16.08
AirDet (Ours)	✗	✓	3.05	6.40	10.03	4.00	9.65	13.91	3.46	11.44	16.04	3.27	11.20	18.64

work [36,39,41], while all of them require a careful fine-tuning stage. Moreover, our fine-tuned model outperforms most prior methods [2,3,6,12,15,16,19,31,36–41,44] in most metrics, especially average recall rate (AR). Besides, the performance superiority on small objects (AP_s and AR_s) further demonstrates the effectiveness of AirDet on multi-scale, especially the small-scale objects.

Efficiency Comparison. We report the fine-tuning and inference time of AirDet, and the SOTA methods [6,36,39,40] in a setting of 3-shot one class

Table 3. Performance comparison with 10-shot on COCO validation dataset. Red and green fonts indicate best and second-best scores, respectively. AirDet achieves comparable results without fine-tuning and outperforms most methods with fine-tuning, which strongly demonstrates its effectiveness.

Method	Venue	Fine-tune	AP	AP_{50}	AP_{75}	AP_s	AP_m	AP_l	AR_1	AR_{10}	AR_{100}	AR_s	AR_m	AR_l
LSTD [3]	AAAI 2018	✓	3.2	8.1	2.1	0.9	2.0	6.5	7.8	10.4	10.4	1.1	5.6	19.6
MetaDet [37]	ICCV 2019	✓	7.1	14.6	6.1	1.0	4.1	12.2	11.9	15.1	15.5	1.7	9.7	30.1
FSRW [16]	ICCV 2019	✓	5.6	12.3	4.6	0.9	3.5	10.5	10.1	14.3	14.4	1.5	8.4	28.2
Meta RCNN [41]	ICCV 2019	✓	8.7	19.1	6.6	2.3	7.7	14.0	12.6	17.8	17.9	7.8	15.6	27.2
TFA_{fc} [36]	ICML 2020	✓	9.1	17.3	8.5	–	–	–	–	–	–	–	–	–
TFA_{cos} [36]	ICML 2020	✓	9.1	17.1	8.8	–	–	–	–	–	–	–	–	–
FSDetView [40]	ECCV 2020	✓	12.5	27.3	9.8	2.5	13.8	19.9	20.0	25.5	25.7	7.5	27.6	38.9
MPSR [39]	ECCV 2020	✓	9.8	17.9	9.7	3.3	9.2	16.1	15.7	21.2	21.2	4.6	19.6	34.3
A-RPN [6]	CVPR 2020	✓	11.1	20.4	10.6	–	–	–	–	–	–	–	–	–
SRR-FSD [44]	CVPR 2021	✓	11.3	23.0	9.8	–	–	–	–	–	–	–	–	–
FSCE [31]	CVPR 2021	✓	11.9	–	10.5	–	–	–	–	–	–	–	–	–
DCNet [15]	CVPR 2021	✓	12.8	23.4	11.2	4.3	13.8	21.0	18.1	26.7	25.6	7.9	24.5	36.7
Y. Li *et al.* [19]	CVPR 2021	✓	11.3	20.3	–	–	–	–	–	–	–	–	–	–
FADI [2]	NIPS 2021	✓	12.2	22.7	11.9	–	–	–	–	–	–	–	–	–
QA-FewDet [12]	ICCV 2021	✓	11.6	23.9	9.8	–	–	–	–	–	–	–	–	–
$FSOD^{up}$ [38]	ICCV 2021	✓	11.0	–	10.7	4.5	11.2	17.3	–	–	–	–	–	–
AirDet	**Ours**	✗	8.7	15.3	8.8	4.3	9.7	14.8	19.1	33.8	34.8	13.0	37.4	52.9
AirDet	**Ours**	✓	13.0	23.9	12.4	4.5	15.2	22.8	20.5	33.7	34.4	9.6	36.4	55.0

Table 4. Efficiency comparison with official source code. We adopt the pre-trained models provided by [36], so their fine-tuning time is unavailable.

Method	**AirDet**	A-RPN [6]	FSDet [40]	MPSR [39]	TFA_{fc} [36]	TFA_{cos} [36]	$FRCN_{ft}$ [36]
Fine-tuning (min)	**0**	21	11	3	–	–	–
Inference (s/img)	**0.081**	0.076	0.202	0.109	0.085	0.094	0.091

in Table 4, in which the official code and implementation with ResNet101 as the backbone are adopted. Without fine-tuning, AirDet can make direct inferences on novel objects with a comparable speed, while the others methods [6,36,39,40] require a fine-tuning time of about 3–30 minutes, which cannot meet the requirements of online exploration. Note that the fine-tuning time is measured on TITAN X GPU, while such computational power is often unavailable on robots.

Remark 6. Many methods [2,3,6,12,15,16,19,31,36–41,44] also require an offline process to fine-tune hyper-parameters for different shots. While such *offline* tuning is infeasible for robotic *online* exploration. Instead, AirDet can adopt **the same** base-trained model without fine-tuning for implementation.

5.3 Cross-Domain Evaluation

Robots are often deployed to novel environments that have never been seen during training, thus cross-domain test is crucial for robotic applications. In this section, we adopt the same model trained on COCO, while test on PASCAL VOC [5] and LVIS [11] to evaluate the model generalization capability.

PASCAL VOC. We report the overall performance on PASCAL VOC-2012 [5] for all methods in Table 5. In the cross-domain setting, even without fine-tuning, AirDet achieves better performance than methods [6,28,36,39,40] that

Table 5. Cross-domain performance on VOC-2012 validation dataset. Red and green fonts denote the first and second place, respectively. AirDet has been demonstrated strong generalization capability, maintaining obvious superiority against others.

Shots		1			2			3			5		
Method	Fine-tune	AP	AP_{50}	AP_{75}	AP	AP_{50}	AP_{75}	AP	AP_{50}	AP_{75}	AP	AP_{50}	AP_{75}
A-RPN [6]†	✗	10.45 ±0.1	18.10 ±0.1	10.32 ±0.1	13.10 ±0.2	22.60 ±0.4	13.17 ±0.2	14.05 ±0.2	24.08 ±0.2	14.24 ±0.2	14.87 ±0.08	25.03 ±0.07	15.26 ±0.1
AirDet (Ours)†	✗	11.92 **±0.06**	21.33 **±0.08**	11.56 **±0.08**	15.80 **±0.08**	26.80 **±0.3**	16.08 **±0.05**	16.89 **±0.1**	28.61 **±0.1**	17.36 **±0.1**	17.83 **±0.03**	29.78 **±0.03**	18.38 **±0.1**
FRCN [28]	✓	4.49	9.44	3.85	5.20	11.92	3.84	6.50	14.39	5.11	6.55	14.48	5.09
TFA_{cos} [36]	✓	4.66	7.97	5.14	6.59	11.91	6.49	8.78	17.09	8.15	10.46	20.93	9.53
TFA_{fc} [36]	✓	4.40	8.60	4.21	7.02	13.80	6.21	9.24	18.48	8.03	11.11	22.83	9.78
FSDetView [40]	✓	4.80	14.10	1.40	3.70	11.60	0.60	6.60	22.00	1.20	10.80	26.50	5.50
MPSR [39]	✓	6.01	11.23	5.74	8.20	15.08	8.22	10.08	18.29	9.99	11.49	21.33	11.06
A-RPN [6]	✓	9.49	17.41	9.42	12.71	23.66	12.44	14.89	26.30	14.76	15.09	28.08	14.17
AirDet (Ours)	✓	13.33	24.64	12.68	17.51	30.35	17.61	17.68	32.05	17.34	18.27	33.02	17.69

Table 6. Cross-domain performance of A-RPN [6] and AirDet on LVIS dataset. We report the results for 5-shot without fine-tuning on 4 random splits.

Split	1				2				3				4			
Metrict	AP	AP_{50}	AP_{75}	AR_{10}	AP	AP_{50}	AP_{75}	AR_{10}	AP	AP_{50}	AP_{75}	AR_{10}	AP	AP_{50}	AP_{75}	AR_{10}
AirDet	**6.71**	**12.31**	**6.51**	**27.57**	**9.35**	**14.23**	**9.98**	**25.42**	**9.09**	**15.64**	**8.82**	**34.64**	**11.07**	**16.90**	**12.30**	**25.76**
A-RPN	5.49	10.04	5.27	26.59	8.85	13.41	9.46	24.45	7.49	12.34	8.13	33.85	10.80	15.46	12.24	25.05

perform relatively well in in-domain test. This means AirDet has a much stronger generalization capability than most fine-tuned prior methods.

LVIS. We randomly sample LVIS [11] to form 4 splits of classes, each of which contains 16 different classes. To provide valid evaluation, the classes that have 20 to 200 images are taken for the test. More details can be found in Appendix E. The averaged performance with 5-shot without fine-tuning is presented in Table 6, where AirDet outperforms the baseline [6] in every split under all metrics. Since the novel categories in the 4 LVIS splits are more (64 classes in total) and rarer (many of them are uncommon) than the VOC 20 classes, the superiority of AirDet in Table 6 highly demonstrate its robustness under class variance.

5.4 Ablation Study and Deep Visualization

In this section, we address the effectiveness of the proposed three modules via quantitative results and qualitative visualization using Grad-Cam [29].

Quantitative Evaluation. We report the overall performance on 3-shot and 5-shot for the baseline [6] and AirDet by enabling the three modules, respectively. It can be seen in Table 7 that AirDet outperforms the baseline in all cases. With the modules enabled one by one, the results get gradually higher, which strongly demonstrates the necessity and effectiveness of SCS, GLR, and PRE.

Table 7. Ablation study of the three modules, *i.e.*, PRE, GLR, and SCS in AirDet. With each module enabled, the performance is improved step by step on our baseline. With the full modules, AirDet can amazingly achieve up to **35%** higher results.

Module			3						5					
PRE	GLR	SCS	AP	Δ%	AP_{50}	Δ%	AP_{75}	Δ%	AP	Δ%	AP_{50}	Δ%	AP_{75}	Δ%
Baseline [6]			4.80	0.00	9.24	0.00	4.49	0.00	5.73	0.00	10.68	0.00	5.53	0.00
✓			5.15	+7.29	10.11	+9.41	4.71	+4.90	5.94	+3.66	11.54	+8.05	5.34	-3.43
✓	✓		5.59	+16.46	10.61	+14.83	5.12	+14.03	6.44	+12.39	12.08	+13.11	6.06	+9.58
✓	✓	✓	**6.50**	**+35.41**	**12.30**	**+33.12**	**6.11**	**+36.08**	**7.27**	**+26.78**	**13.63**	**+27.62**	**6.71**	**+21.34**

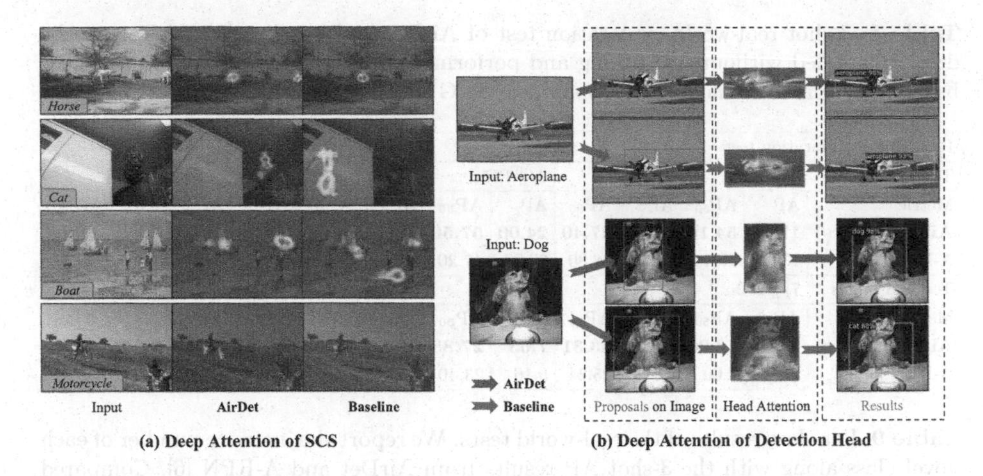

Fig. 4. Deep visualization comparison between AirDet and baseline [6]. In (a), By virtue of SCS, AirDet is capable of finding given support objects effectively. In (b), with similar proposals (red boxes), AirDet can focus on the entire object (aeroplane) and notice the most representative parts (dog), resulting in more precise regression box and correct classification results. More examples are presented in Appendix D. (Color figure online)

How Effective is SCS? Given 2-shot per class, we first take the highest ranking proposal from RPN [28] to backpropagate the objectiveness score and resize the gradient map to the original image. Figure 4 (a) exhibits the heat map from both AirDet and the baseline. We observe that AirDet generally concentrates on objects more precisely than the baseline. Moreover, AirDet can focus better on objects belonging to the support class and is not distracted by other objects (2nd and 3rd row). This means that AirDet can generate novel object proposals more effectively.

How Effective is GLR and Detection Head? In Fig. 4 (b), we observe that with similar proposal boxes, AirDet head can better focus on the entire object, *e.g.*, aeroplane is detected with a precise regression box, *e.g.*, the dog is correctly classified with high score. This again demonstrates the effectiveness of our GLR and detection head.

5.5 Real-World Test

Real-world tests are conducted for AirDet and our baseline [6] with 12 sequences that were collected from the DARPA Subterranean (SubT) challenge [1]. Due to the requirements of *online* response during the mission, the models can only be evaluated **without fine-tuning**, which makes existing methods [2,3,6,12, 15,16,19,31,36–41,44] impractical. The environments of SubT challenge also poses extra difficulties, *e.g.*, a lack of lighting, thick smoke, dripping water, and

Table 8. 3-shot real-world exploration test of AirDet and baseline [6]. AirDet can be directly applied without fine-tuning and performs considerably more robust than the baseline by virtue of the newly proposed SCS, GLR, and PRE modules.

Real-world exploration test												
Test/#Frames	1/#248		2/#146		3/#127		4/#41		5/#248		6/#46	
Metric	AP	AP_{50}	AP	AP_{50}	AP	AP_{50}	AP	AP_{50}	AP	AP_{50}	AP	AP_{50}
AirDet (Ours)	**17.10**	**54.10**	**17.90**	**47.40**	**24.00**	**57.50**	**26.94**	**48.20**	**11.28**	**38.17**	**20.40**	**70.63**
A-RPN [6]	13.56	40.40	14.30	38.80	20.20	47.20	22.41	40.14	6.75	24.10	14.70	59.38
Test/#Frames	7/#212		8/#259		9/#683		10/#827		11/#732		12/#50	
Metric	AP	AP_{50}	AP	AP_{50}	AP	AP_{50}	AP	AP_{50}	AP	AP_{50}	AP	AP_{50}
AirDet (Ours)	**5.90**	**16.00**	**15.26**	**43.31**	**7.63**	**27.88**	**13.55**	**23.92**	**15.74**	**34.43**	**21.45**	**45.83**
A-RPN [6]	2.39	7.60	11.27	25.24	6.16	23.40	8.10	14.85	11.54	27.28	18.20	33.98

Table 9. Per class results of the real-world tests. We report the instance number of each novel class along with the 3-shot AP results from AirDet and A-RPN [6]. Compared with the baseline, AirDet achieves higher results for all classes.

Class	Backpack	Helmet	Rope	Drill	Vent	Extinguisher	Survivor
Instances	626	674	723	587	498	1386	205
AirDet	**32.3**	**9.7**	**13.9**	**10.8**	**16.2**	**10.5**	**10.7**
Baseline [6]	26.6	9.7	6	9	14.4	5.6	9.1

cluttered or irregularly shaped environments, *etc.*. To test the generalization capabilities, we adopt the same models of AirDet and the baseline as those evaluated in Sect. 5.2 and Sect. 5.3. The performance of 3-shot for each class is exhibited in Table 8, where AirDet is proved better. The robot is equipped with an NVIDIA Jetson AGX Xavier, where our method runs at 1–2 FPS without TensorRT acceleration or other optimizations.

In Table 9, we present the number of instances and the performance on each novel class. To our excitement, AirDet shows smaller variance and higher precision cross different classes. We also present the support images and representative detected objects in Fig. 5. Note that AirDet can detect the novel objects accurately in the query images even if they have distinct scales and different illumination conditions with the supports. We regard this capability to the carefully designed SCS in AirDet. More visualization are presented in Appendix C. The robustness and strong generalization capability of AirDet in the real-world tests demonstrated its promising prospect and feasibility for autonomous exploration.

6 Limitation and Future Work

Despite the promising prospect and outstanding performance, AirDet still has several limitations. (1) Since abundant base classes are needed to generalize, AirDet needs a relatively large base dataset to train before inference on novel classes. (2) Second, AirDet relies on the quality of support images to work well

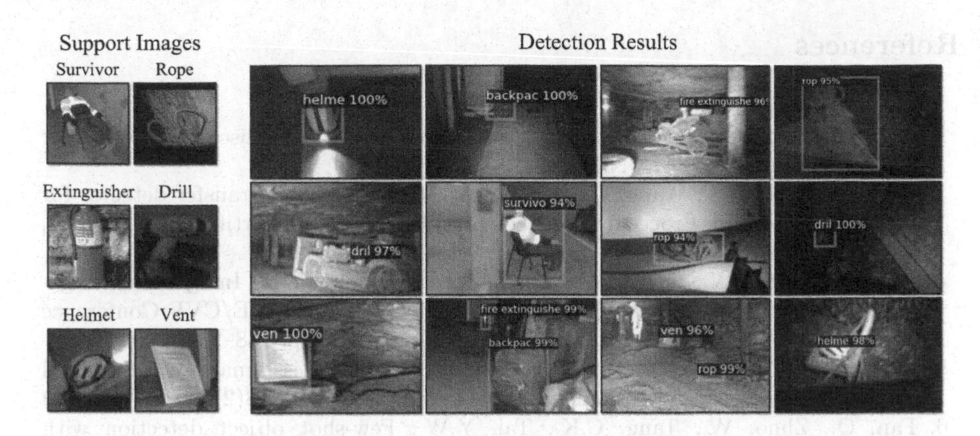

Fig. 5. The provided support images and examples of detection results in the real-world tests. AirDet is robust to distinct object scales and different illumination conditions.

without fine-tuning. This is because the provided few support images are the only information for the unseen classes. (3) We observe that the failure cases of AirDet are mainly due to false classification, resulting in a high result variance among different classes in COCO and VOC. (4) Since SCS and the detection head run in loops for multiple novel classes, the efficiency of AirDet will suffer from a large number of novel classes. We provide quantitative results for limitation (1), (2), and (3) in Appendix F.

7 Conclusion

This paper presents a brand new few-shot detector, AirDet, which consists of 3 newly proposed *class-agnostic relation*-based modules and is free of fine-tuning. Specifically, with proposed spatial relation and channel relation, we construct support guided cross-scale feature fusion for region proposals, global-local relation network for shots aggregation, and prototype relation embedding for precise localization. With the strong capability to extract *class-agnostic relation*, AirDet can work comparably or even better than those exhaustively fine-tuned methods in both in-domain and cross-domain evaluation. AirDet is also tested on real-world data with a robotic platform, where its feasibility for autonomous exploration is demonstrated.

Acknowledgement. This work was sponsored by ONR grant #N0014-19-1-2266 and ARL DCIST CRA award W911NF-17-2-0181. The work was done when Bowen Li and Pranay Reddy were interns at The Robotics Institute, Carnegie Mellon University. The authors would like to thank all members of the Team Explorer for providing data collected from the DARPA Subterranean Challenge.

References

1. https://subtchallenge.com
2. Cao, Y., et al.: Few-shot object detection via association and discrimination. Adv. Neural Inf. Process. Syst. (NIPS) **34**, 1–12 (2021)
3. Chen, H., Wang, Y., Wang, G., Qiao, Y.: LSTD: a low-shot transfer detector for object detection. In: Proceedings of the AAAI Conference on Artificial Intelligence, vol. 32 (2018)
4. Deng, J., Dong, W., Socher, R., Li, L.J., Li, K., Fei-Fei, L.: Imagenet: a large-scale hierarchical image database. In: Proceedings of the IEEE/CVF Conference on Computer Vision and Pattern Recognition (CVPR), pp. 248–255 (2009)
5. Everingham, M., Van Gool, L., Williams, C.K., Winn, J., Zisserman, A.: The pascal visual object classes (VOC) challenge. Int. J. Comput. Vision **88**(2), 303–338 (2010)
6. Fan, Q., Zhuo, W., Tang, C.K., Tai, Y.W.: Few-shot object detection with attention-RPN and multi-relation detector. In: Proceedings of the IEEE/CVF Conference on Computer Vision and Pattern Recognition (CVPR), pp. 4013–4022 (2020)
7. Fan, Z., Ma, Y., Li, Z., Sun, J.: Generalized few-shot object detection without forgetting. In: Proceedings of the IEEE/CVF Conference on Computer Vision and Pattern Recognition (CVPR), pp. 4527–4536 (2021)
8. Farooq, N., Ilyas, U., Adeel, M., Jabbar, S.: Ground robot for alive human detection in rescue operations. In: Proceedings of the International Conference on Intelligent Informatics and Biomedical Sciences (ICIIBMS), vol. 3, pp. 116–123 (2018)
9. Girshick, R.: Fast R-CNN. In: Proceedings of the IEEE/CVF Conference on Computer Vision and Pattern Recognition (CVPR), pp. 1440–1448 (2015)
10. Girshick, R., Donahue, J., Darrell, T., Malik, J.: Rich feature hierarchies for accurate object detection and semantic segmentation. In: Proceedings of the IEEE/CVF Conference on Computer Vision and Pattern Recognition (CVPR), vol. 3, pp. 580–587 (2014)
11. Gupta, A., Dollar, P., Girshick, R.: LVIS: a dataset for large vocabulary instance segmentation. In: Proceedings of the IEEE/CVF Conference on Computer Vision and Pattern Recognition (CVPR), pp. 5356–5364 (2019)
12. Han, G., He, Y., Huang, S., Ma, J., Chang, S.F.: Query adaptive few-shot object detection with heterogeneous graph convolutional networks. In: Proceedings of the IEEE/CVF International Conference on Computer Vision (ICCV), pp. 3263–3272 (2021)
13. He, K., Gkioxari, G., Dollar, P., Girshick, R.: Mask R-CNN. In: Proceedings of the IEEE/CVF International Conference on Computer Vision (ICCV), pp. 2961–2969 (2017)
14. He, K., Zhang, X., Ren, S., Sun, J.: Deep residual learning for image recognition. In: Proceedings of the IEEE/CVF Conference on Computer Vision and Pattern Recognition (CVPR), pp. 770–778 (2016)
15. Hu, H., Bai, S., Li, A., Cui, J., Wang, L.: Dense relation distillation with context-aware aggregation for few-shot object detection. In: Proceedings of the IEEE/CVF Conference on Computer Vision and Pattern Recognition (CVPR), pp. 10185–10194 (2021)
16. Kang, B., Liu, Z., Wang, X., Yu, F., Feng, J., Darrell, T.: Few-shot object detection via feature reweighting. In: Proceedings of the IEEE/CVF International Conference on Computer Vision (ICCV), pp. 8420–8429 (2019)

17. Kong, T., Yao, A., Chen, Y., Sun, F.: HyperNet: towards accurate region proposal generation and joint object detection. In: Proceedings of the IEEE/CVF Conference on Computer Vision and Pattern Recognition (CVPR) (2016)
18. Li, B., Wu, W., Wang, Q., Zhang, F., Xing, J., Yan, J.: Siamrpn++: evolution of siamese visual tracking with very deep networks. In: Proceedings of the IEEE/CVF Conference on Computer Vision and Pattern Recognition (CVPR), pp. 4282–4291 (2019)
19. Li, Y., Zhu, H., et al.: Few-shot object detection via classification refinement and distractor retreatment. In: Proceedings of the IEEE/CVF Conference on Computer Vision and Pattern Recognition (CVPR), pp. 15395–15403 (2021)
20. Li, Z., Zhou, F.: FSSD: feature fusion single shot multibox daetector. arXiv preprint arXiv:1712.00960 (2017)
21. Lin, T.Y., Dollár, P., Girshick, R., He, K., Hariharan, B., Belongie, S.: Feature pyramid networks for object detection. In: Proceedings of the IEEE/CVF Conference on Computer Vision and Pattern Recognition (CVPR), pp. 2117–2125 (2017)
22. Lin, T.-Y., et al.: Microsoft COCO: common objects in context. In: Fleet, D., Pajdla, T., Schiele, B., Tuytelaars, T. (eds.) ECCV 2014. LNCS, vol. 8693, pp. 740–755. Springer, Cham (2014). https://doi.org/10.1007/978-3-319-10602-1_48
23. Liu, W., et al.: SSD: single shot multibox detector. In: Leibe, B., Matas, J., Sebe, N., Welling, M. (eds.) ECCV 2016. LNCS, vol. 9905, pp. 21–37. Springer, Cham (2016). https://doi.org/10.1007/978-3-319-46448-0_2
24. Qiao, L., Zhao, Y., Li, Z., Qiu, X., Wu, J., Zhang, C.: Defrcn: decoupled faster r-cnn for few-shot object detection. In: Proceedings of the IEEE/CVF International Conference on Computer Vision (ICCV), pp. 8681–8690 (2021)
25. Redmon, J., Divvala, S., Girshick, R., Farhadi, A.: You only look once: unified, real-time object detection. In: Proceedings of the IEEE/CVF Conference on Computer Vision and Pattern Recognition (CVPR), pp. 779–788 (2016)
26. Redmon, J., Farhadi, A.: YOLO9000: better, faster, stronger. In: Proceedings of the IEEE/CVF Conference on Computer Vision and Pattern Recognition (CVPR), pp. 7263–7271 (2017)
27. Redmon, J., Farhadi, A.: YOLOv3: an incremental improvement, pp. 1–6. arXiv preprint arXiv:1804.02767 (2018)
28. Ren, S., He, K., Girshick, R., Sun, J.: Faster R-CNN: towards real-time object detection with region proposal networks. IEEE Trans. Pattern Anal. Mach. Intell. **39**(6), 1137–1149 (2016)
29. Selvaraju, R.R., Cogswell, M., Das, A., Vedantam, R., Parikh, D., Batra, D.: Grad-Cam: visual explanations from deep networks via gradient-based localization. In: Proceedings of the IEEE/CVF International Conference on Computer Vision (ICCV), pp. 618–626 (2017)
30. Shen, Z., Liu, Z., Li, J., Jiang, Y.G., Chen, Y., Xue, X.: DSOD: learning deeply supervised object detectors from scratch. In: Proceedings of the IEEE International Conference on Computer Vision (ICCV) (2017)
31. Sun, B., Li, B., Cai, S., Yuan, Y., Zhang, C.: FSCE: few-shot object detection via contrastive proposal encoding. In: Proceedings of the IEEE/CVF Conference on Computer Vision and Pattern Recognition (CVPR), pp. 7352–7362 (2021)
32. Sung, F., Yang, Y., Zhang, L., Xiang, T., Torr, P.H., Hospedales, T.M.: Learning to compare: relation network for few-shot learning. In: Proceedings of the IEEE/CVF Conference on Computer Vision and Pattern Recognition (CVPR), pp. 1199–1208 (2018)

33. Tariq, R., Rahim, M., Aslam, N., Bawany, N., Faseeha, U.: Dronaid: a smart human detection drone for rescue. In: Proceedings of the 15th International Conference on Smart Cities: Improving Quality of Life Using ICT & IoT (HONET-ICT), pp. 33–37 (2018)
34. Wang, C., Qiu, Y., Wang, W., Hu, Y., Kim, S., Scherer, S.: Unsupervised online learning for robotic interestingness with visual memory. IEEE Trans. Rob. **34**, 1–15 (2021)
35. Wang, C., Wang, W., Qiu, Y., Hu, Y., Scherer, S.: Visual memorability for robotic interestingness via unsupervised online learning. In: Vedaldi, A., Bischof, H., Brox, T., Frahm, J.-M. (eds.) ECCV 2020. LNCS, vol. 12347, pp. 52–68. Springer, Cham (2020). https://doi.org/10.1007/978-3-030-58536-5_4
36. Wang, X., Huang, T.E., Darrell, T., Gonzalez, J.E., Yu, F.: Frustratingly simple few-shot object detection. In: Proceedings of the International Conference on Machine Learning (ICML), pp. 1–12 (2020)
37. Wang, Y.X., Ramanan, D., Hebert, M.: Meta-learning to detect rare objects. In: Proceedings of the IEEE/CVF International Conference on Computer Vision (ICCV), pp. 9925–9934 (2019)
38. Wu, A., Han, Y., Zhu, L., Yang, Y.: Universal-prototype enhancing for few-shot object detection. In: Proceedings of the IEEE/CVF International Conference on Computer Vision (ICCV), pp. 9567–9576 (2021)
39. Wu, J., Liu, S., Huang, D., Wang, Y.: Multi-scale positive sample refinement for few-shot object detection. In: Vedaldi, A., Bischof, H., Brox, T., Frahm, J.-M. (eds.) ECCV 2020. LNCS, vol. 12361, pp. 456–472. Springer, Cham (2020). https://doi.org/10.1007/978-3-030-58517-4_27
40. Xiao, Y., Marlet, R.: Few-shot object detection and viewpoint estimation for objects in the wild. In: Vedaldi, A., Bischof, H., Brox, T., Frahm, J.-M. (eds.) ECCV 2020. LNCS, vol. 12362, pp. 192–210. Springer, Cham (2020). https://doi.org/10.1007/978-3-030-58520-4_12
41. Yan, X., Chen, Z., Xu, A., Wang, X., Liang, X., Lin, L.: Meta R-CNN: towards general solver for instance-level low-shot learning. In: Proceedings of the IEEE/CVF International Conference on Computer Vision (ICCV), pp. 9577–9586 (2019)
42. Zhang, L., Zhou, S., Guan, J., Zhang, J.: Accurate few-shot object detection with support-query mutual guidance and hybrid loss. In: Proceedings of the IEEE/CVF Conference on Computer Vision and Pattern Recognition (CVPR), pp. 14424–14432 (2021)
43. Zhang, W., Wang, Y.X.: Hallucination improves few-shot object detection. In: Proceedings of the IEEE/CVF Conference on Computer Vision and Pattern Recognition (CVPR), pp. 13008–13017 (2021)
44. Zhu, C., Chen, F., Ahmed, U., Shen, Z., Savvides, M.: Semantic relation reasoning for shot-stable few-shot object detection. In: Proceedings of the IEEE/CVF Conference on Computer Vision and Pattern Recognition (CVPR), pp. 8782–8791 (2021)

TransGrasp: Grasp Pose Estimation of a Category of Objects by Transferring Grasps from Only One Labeled Instance

Hongtao Wen, Jianhang Yan, Wanli Peng(✉), and Yi Sun

Dalian University of Technology, Dalian, China
{wenht,yjh97,1136558142}@mail.dlut.edu.cn, lslwf@dlut.edu.cn

Abstract. Grasp pose estimation is an important issue for robots to interact with the real world. However, most of existing methods require exact 3D object models available beforehand or a large amount of grasp annotations for training. To avoid these problems, we propose TransGrasp, a category-level grasp pose estimation method that predicts grasp poses of a category of objects by labeling only one object instance. Specifically, we perform grasp pose transfer across a category of objects based on their shape correspondences and propose a grasp pose refinement module to further fine-tune grasp pose of grippers so as to ensure successful grasps. Experiments demonstrate the effectiveness of our method on achieving high-quality grasps with the transferred grasp poses. Our code is available at https://github.com/yanjh97/TransGrasp.

Keywords: Grasp pose estimation · Dense shape correspondence · Grasp transfer · Grasp refinement

1 Introduction

Grasp pose estimation refers to the problem of finding a grasp configuration for the grasping task given an object. It's one of the most important issues for robots to interact with the real world and has great potential to replace human hands to execute various tasks, such as helping elderly or disabled people carry out everyday tasks. However, predicting a stable grasp of an object is a challenging problem as it requires modeling physical contact between a robot hand and various objects in real scenarios. Currently, most existing works rely on grasp pose annotations on each object or exact 3D object models available beforehand. However, annotating 6D grasp pose is not easy as the rotation, translation and width of the gripper have to be adjusted to ensure a stable grasp. In this paper,

H. Wen and J. Yan—Equal contributions.

Supplementary Information The online version contains supplementary material available at https://doi.org/10.1007/978-3-031-19842-7_26.

S. Avidan et al. (Eds.): ECCV 2022, LNCS 13699, pp. 445–461, 2022.
https://doi.org/10.1007/978-3-031-19842-7_26

we deal with the problems from the perspective of computer vision to develop a category-level grasp pose prediction model. Our approach aims to estimate the grasp poses of all instances by labeling the grasp poses on only one instance of the category and transferring the grasp to other instances, as shown in Fig. 1. As we have known, the study on category-level grasp pose estimation has not been fully explored so far.

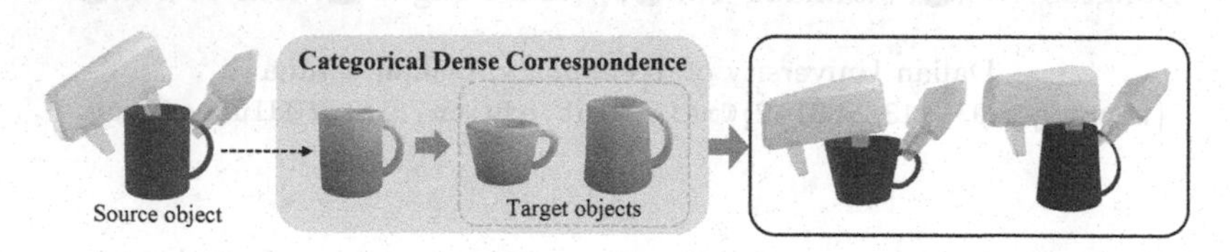

Fig. 1. Our method for Grasp Pose Estimation with **only one labeled instance**. We transfer grasps from one labeled source object to target objects, using the learned categorical dense correspondence among a category of objects

Generally, grasp pose estimation can be divided into model-based methods [10,12,15,26,31,33] and data-driven (model-free) methods [2,8,13,16–18,22,30]. Both methods have their own limitations. Model-based methods generate grasp poses of an object regarding the criteria such as force closure. It requires precise geometric and complete 3D model of an object which is not always available in real scenarios, while data-driven methods estimate grasp poses by experience that is gathered during grasp execution. It requires dense grasp pose annotations which are extremely time-consuming and is also uncertain whether the experience can be generalized to novel objects. In this paper, we propose a novel grasp pose estimation approach, named TransGrasp, to solve the above problems.

Our TransGrasp is based on the fact that a category of object instances have consistent topologies and grasping affordances, such as mugs with grasping handles for drinking. This motivates us to align grasp contact points within a category of objects by first building their shape correspondences and then transfer labeled grasps on one object to the others, finally these initial grasps on all objects are fine-tuned continuously. In this work, we first learn the categorical shape latent space and a template that captures common structure of the category from the given object instances. The variance of each object instance is represented by the difference of its shape relative to the template, and dense correspondence across instance shapes can be established by deforming their surfaces to the template. Then we design a novel model to transfer grasp points on one object instance across a category of objects based on their shape deformation correspondence. To avoid extreme distortions of grasp points during correspondence deformation, the grasp pose refinement module is introduced to further alter the gripper, allowing the grasp points on the surface to form a stable grasp. Note that the correction on gripper is crucial to guarantee a successful grasp. Finally, grasp points of gripper on any object surface can be aligned across an entire category of objects.

Compared with previous works, the proposed model predicts grasp poses of an entire category of objects with grasp pose annotations of only one object instance, which greatly reduces the amount of 6DoF grasp pose annotations of objects in data-driven methods. Furthermore, category-level grasp pose estimation model learns a categorical shape latent space shared by all instances, which enables shape reconstruction and grasp pose estimation within a category. We additionally estimate rotation R, translation t, and scale s given object observable point cloud as input to transform grasp poses from Object Coordinates to Camera Coordinates and subsequently conduct simulation and real robot experiments. The experiment results verify that our estimated grasp poses can align well across the instances of a category.

The contributions of this work are as follows:

- We propose a category-level grasp pose estimation model to predict grasp poses of a category of objects by only labeling one object instance, which does not heavily relying on a large amount of grasp pose annotations and 3D model on every object.
- We exploit consistent structure within a category of objects to perform grasp pose transfer across a category of objects. In addition, the grasp pose refinement module further corrects grasp poses to ensure successful grasps.
- We set up simulated and real robot system and conduct sufficient experiments that verify the effectiveness of the proposed method on achieving high-quality grasps with the transferred grasp poses.

2 Related Work

In this section, we briefly review the literatures on 6DoF grasp pose estimation, which can be approximately divided into model-based methods and data-driven (model-free) methods.

Model-Based Methods: A common pipeline for model-based methods is to generate the grasp pose on the known object models using the criteria such as force closure [12,15], and then transform annotated grasps from Object Coordinates to Camera Coordinates with the estimated object 6D pose. Zeng *et al.* [33] segment objects with fully convolutional network from multiple views and fit pre-scanned 3D models with ICP [1] to obtain object pose. Others apply learning-based and instance-level object pose estimation methods such as [10,26,31] for grasp transformation. However, these model-based methods require exact 3D models beforehand, which cannot be used in general settings because the target objects may be different from the existing 3D models. To estimate the 6D pose of unseen objects, researches for category-level object pose estimation [4,5,25,27] have drawn more and more attention, but they have hardly been applied to the grasp estimation. In this work, we explore the challenging problem for category-level grasp pose estimation by transferring grasps from one labeled instance to the others.

Model-Free Methods: To estimate the grasp pose on unseen objects, some model-free methods [2,8,13,16–18,22,30] that analyze the 3D geometry of the object surface directly using deep neural network are proposed. GPD [18] and PointNetGPD [13] generate a large set of grasp candidates firstly and then build a CNN or PointNet [20] to identify whether each candidate is a grasp or not. Following this pipeline, 6-DOF GraspNet [16] builds a variational auto-encoder (VAE) for efficiently candidates sampling and introduces a grasp evaluator network for grasp refinement. CollisionNet [17] builds a collision evaluation extension based on [16] to check the cluttered environment for collisions. Since analyzing every candidate costs significantly, some end-to-end methods [2,8,22,30] are proposed. Among them, S^4G [22] builds a single-shot grasp proposal network based on PointNet++ [21] to predict 6DoF grasp pose and quality score of each point directly. Fang *et al.* [8] build a large scale grasp pose dataset called GraspNet-1Billion and propose an 3D grasp detection network that learns approaching direction, operation parameters and grasp robustness end-to-end. However, all these methods need a sufficient amount of dense grasp pose annotations for all the objects in the scene for training, which is time-consuming and laborious. To avoid this problem, our method estimates grasp poses of a category of objects by labeling only one object instance, which greatly reduces the cost of grasp pose annotations.

Closely related to our work, some methods transfer grasps to other objects by building dense correspondence. DGCM-Net [19] transfers grasps by predicting view-dependent normalized object coordinate (VD-NOC) values between pairs of depth images. CaTGrasp [29] maps the input point cloud to a Non-Uniform Normalized Object Coordinate Space (NUNOCS) where the spatial correspondence is built. DON [9] learns pixel-wise descriptors in a self-supervised manner to transfer grasp points between 2D images. NDF [24] builds 3D dense correspondence using Neural Descriptor Field, through which the grasp poses are transferred directly. Different from these methods, we construct dense surface correspondence between the complete shape of objects to transfer the grasp points on the surface to obtain coarse grasps, and further refine these grasps using a refinement module to ensure reliable and successful grasps for the gripper.

3 Method

Our ultimate goal is to estimate grasp poses of a category of objects by labeling only one object instance, rather than heavily relying on a large amount of grasp pose annotations on every object. Our TransGrasp establishes point-to-point dense correspondence among intra-class shapes (Sect. 3.1) to transfer grasp points on one source object instance to other target object instances by aligning grasp poses on the object surface (Sect. 3.2). Furthermore, to avoid extreme distortions during grasp transfer, we introduce a grasp pose refinement module to further adjust the pose of gripper, encouraging the grasps to satisfy antipodal principle (Sect. 3.3). Our overall architecture is illustrated in Fig. 2. In the remainder of this section, we will discuss each component in detail.

3.1 Learning Categorical Dense Correspondence

Objects can be classified according to shape, function, *etc.*. Objects classified by shape usually have consistent topologies and grasping affordances and they are likely to be successfully grasped in a similar way. Based on this prior, we can build the categorical dense correspondence across object instances so that grasp poses could be transferred from one object to others. The categorical correspondence across object instances needs to handle the intra-class shape variation and most recent works model the variation explicitly from a pre-learned template [6,25,34]. The variance of each object instance is represented by the difference of its shape relative to the template, and dense correspondence among instance shapes can be established through the template. In this paper, we build an encoder-decoder framework to reconstruct object model given partial point cloud and provide dense shape correspondence between the reconstructed model and the template. The encoder-decoder framework is illustrated at the top of Fig. 2.

Firstly, a latent shape space representing various 3D shapes and a template of a class for shape correspondence are learned by a decoder-only network. Here

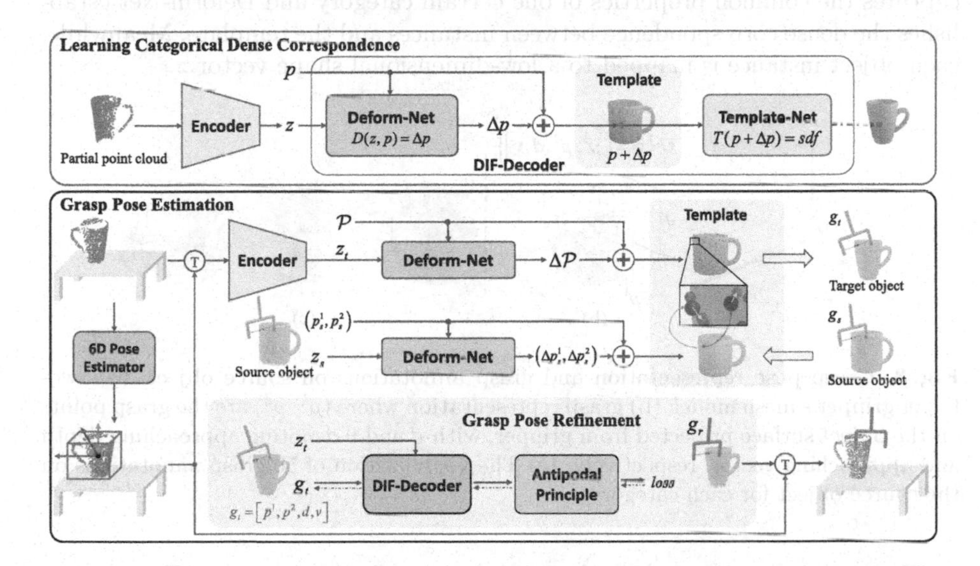

Fig. 2. Overview of the proposed TransGrasp. Top: The network composed of Shape Encoder and DIF-Decoder is trained to learn categorical dense correspondence by deforming object points to the template. Bottom: During inference process, point clouds are first transformed from Camera Coordinates to Object Coordinates and then serve as the input of our trained network, generating the object instance's deformation to template. Meanwhile, the source model with grasp annotation g_s is fed into Deform-Net to obtain its deformation. The dense correspondence established by their common template subsequently guides the transfer from g_s to g_t by aligning grasp points on object surface. The transferred grasp pose g_t is then refined to g_r by the grasp pose refinement module. Finally, g_r is transformed from Object Coordinates to Camera Coordinates so as to conduct a grasp. (Color figure online)

we adopt Deformed Implicit Field [6] decoder (DIF-Decoder), with a continuous random variable z as input to adapt to the intra-class shape variation. The DIF-Decoder can be divided into two parts, a Template-Net (T) and a Deform-Net (D). The Template-Net can map a 3D point p to a scalar SDF value sdf:

$$T(p) = sdf,\ p \in \mathbb{R}^3,\ sdf \in \mathbb{R}. \tag{1}$$

The network weights of T are shared across the whole class, therefore it is enforced to learn common patterns within the class. For each object, the Deform-Net learns a deformation offset to the template $D(z, p) = \Delta p$ for each query point p with the latent code z. And then, the SDF value of point p can be obtained via Template-Net $T(p + \Delta p) = sdf$. This process establishes dense correspondence between an object instance and the template. In other words, the implicit template is "warped" according to the latent codes to model different SDFs. Therefore, dense correspondence among instance shapes can be established through the template. This provides a good reference for grasp transfer across object instances. The DIF-Decoder is trained separately for each category using the CAD models selected from ShapeNetCore [3]. After training, the Template-Net captures the common properties of one certain category and Deform-Net establishes the dense correspondence between instances and the template. Meanwhile, each object instance is mapped to a low-dimensional shape vector z.

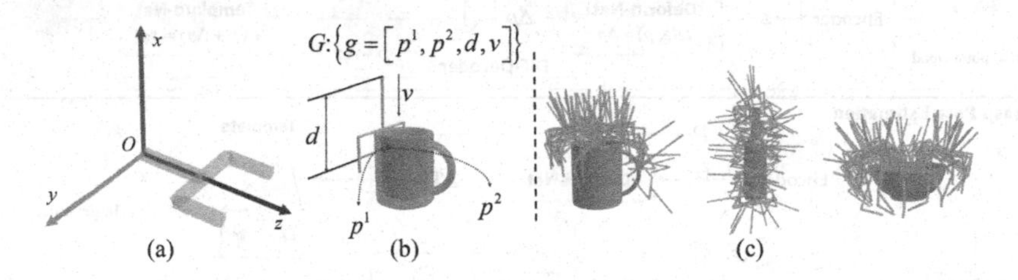

Fig. 3. Grasp pose representation and grasp annotations on source object. (a) two-finger gripper's mesh model; (b) grasp representation where (p^1, p^2) are the grasp points on the object surface projected from gripper, with d and v denoting approaching depth and approaching vector, respectively; (c) The visualization of 50 grasp annotations on the source object for each category

However, input data are usually partial point clouds in practice. To complete the shape information, a Shape Encoder learns to map each input to a latent shape code z so that given the shape code a complete shape can be reconstructed using the priors learned by the DIF-Decoder. The Shape Encoder is trained by taking a partial point cloud rendered from its CAD model as input and shape vector z of each object instance as output.

3.2 Grasp Pose Estimation

After learning the categorical dense correspondence among different objects, we then focus on estimating grasp poses for every object instance by grasp pose transfer from only one labeled object. This transfer process is illustrated in the middle of Fig. 2. The source object is the only one object instance with grasp pose annotations, and its grasp poses on the object surface will be transferred to other target object instances by its established correspondence to other ones. We first introduce a novel grasp pose representation that establishes the connection between grasp pose of the gripper and object surface points, and then implement the grasp pose transfer from the source object surface to the target object surface.

Grasp Pose Representation: To easily implement grasp transfer using dense correspondence between objects, we seek a novel suitable grasp representation for the gripper which follows the two-finger gripper's mesh model shown in Fig. 3(a). Unlike previous works [16,17,32] that define the grasp pose as $(R,\ t,\ width)$ of the gripper, we introduce a novel grasp representation which connects grasp pose with the object surface points as illustrated in Fig. 3(b). Formally, we denote the two grasp points on the object surface projected from gripper as p^1, p^2, the vertical distance from gripper's origin O to the line linking p^1 and p^2 as approaching depth d, and the approaching vector as v. These four parameters form the grasp pose of the gripper where v determines the approaching direction and d the approaching depth along this direction. The width of gripper is determined by the distance between the two grasp points. In this section of Grasp Pose Estimation, only grasp points $(p^1,\ p^2)$ are transferred from the labeled source object to other target objects keeping the d and v unchanged, while in the next section of Grasp Pose Refinement (Sect. 3.3) all pose parameters $(p^1,\ p^2,\ d,\ v)$ are simultaneously refined to obtain optimal stable grasp poses.

Labeling One Instance: To realize grasp transfer among the instances of a class, it is essential to provide a known instance with grasp pose annotations as the source object. Therefore, for each category, one of the instances needs to be labeled in advance. For convenience, we randomly select one instance for each category from the existing large-scale grasp dataset ACRONYM [7] with pose annotations. Meanwhile, to ensure their robustness and avoid inaccuracy, all the chosen grasps are tested in the simulation environment IsaacGym [14], and only the grasps that can successfully grasp the corresponding source objects in the simulation platform are regarded as the source for grasp transfer. We have about 1k grasp poses for each categorical source object. Figure 3(c) shows 50 grasp annotations on the source object for each category.

Grasp Transfer: With our grasp pose representation and one labeled source object, the grasp poses could then be transferred to other object instances of the same category during the inference procedure of our approach shown in the center of Fig. 2. As the encoder and decoder have been learned previously in Sect. 3.1, the shape dense correspondence between the source and target object can be established through the template immediately. Given grasp points p_s^1, p_s^2 on the source object, their deformation offsets Δp_s^1, Δp_s^2 to the template are predicted by Deform-Net, then its grasp poses on the template will be obtained by this deformation accordingly. To further transfer the grasp points on the template to any target object, the input partial point cloud of the target object is first transformed from Camera Coordinates to Object Coordinates using off-the-shelf 6D object pose estimator such as [25], then encoded to the latent code z_t from which its complete shape is reconstructed using DIF-Decoder. And then query point set $\mathcal{P}$ on the surface of reconstructed model is deformed to the template by Deform-Net. Finally, the grasp poses on the template transferred from source object are naturally aligned to the target object through their shared template. It is noted that although the deformed source and target object shape have similar shapes on the template, the grasp points on the source object cannot be accurately matched to surface points on the target. Here we apply the nearest point matching to transfer the grasp points on the source (blue points) to the target (gold points), as shown in the enlarged detail in Fig. 2.

3.3 Grasp Pose Refinement

Although point-to-point dense correspondence can realize the grasp transfer from a source object with grasp pose annotations to any target object, when handling some outlier objects with strange shapes, poor deformation could be estimated by Deform-Net, which in turn leads to the failure of grasp transfer. Therefore, we propose a grasp pose refinement module to improve performance after grasp transfer, as shown at the bottom of Fig. 2. To refine grasp pose, we utilize spatial relationship between gripper and object by calculating the distances from the points on gripper to the object surface. Specifically, the transferred grasp points (p^1, p^2) and the sampled points p^g from the gripper, along with the latent code of target object z_t, are input to the DIF-Decoder to get the SDF values of these points. The grasp pose is iteratively refined following the refinement principles for parallel-jaw gripper, until a stable grasp pose g_r is achieved. The gripper poses before and after adjustment are shown as the silver and gold grippers in Fig. 2 respectively.

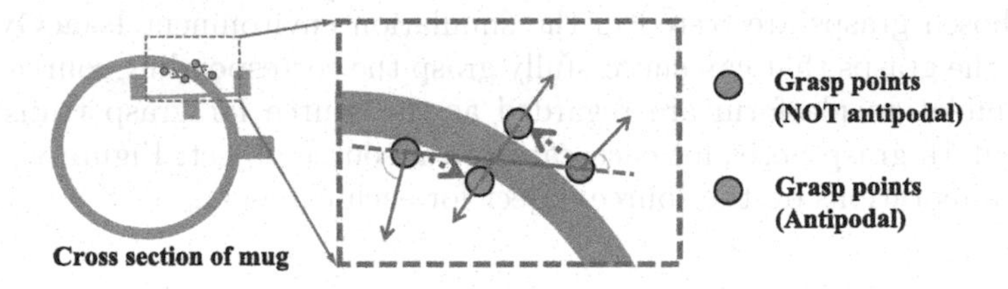

Fig. 4. The refinement process of grasp points based on antipodal principle

We first encourage the two grasp points to satisfy the antipodal principle (a simplified version of force closure for parallel-jaw gripper) on the object surface, as shown in Fig. 4. In mathematics, antipodal points of a sphere are those diametrically opposite to each other. For the two-finger grasping problem, the grasp points are antipodal when their normals are in opposite directions and stay in one line with the grasp points. Let n^1 and n^2 be the normal vectors at p^1 and p^2 respectively, then they need to satisfy:

$$\cos\langle p^2 - p^1, n^1\rangle = 1, \cos\langle p^1 - p^2, n^2\rangle = 1, \tag{2}$$

where n^1 and n^2 can be easily computed by taking the derivative of sdf with respect to the grasp point: $\partial sdf / \partial p$, using the back-propagation of DIF-Decoder.

Therefore, we define the antipodal loss that encourages the grasp points to satisfy the antipodal principle:

$$L_{anti} = -\cos\langle p^2 - p^1, n^1\rangle - \cos\langle p^1 - p^2, n^2\rangle. \tag{3}$$

In addition, we constrain the SDF value of the grasp points (p^1, p^2) to zero, making sure that the grasp points are on the object surface using the touch loss:

$$L_{touch} = \|sdf^1\|_2 + \|sdf^2\|_2, \tag{4}$$

where sdf^1 and sdf^2 are SDF values of the transferred two grasp points (p^1, p^2).

Moreover, to penalize the interpenetration between the gripper and object, we build a collision avoidance loss to constrain the gripper outside the object, that is, the SDF values of the gripper points are greater than 0 via:

$$L_{collision} = -\frac{1}{N}\sum_{i}^{N} \min(0, sdf_i^g), \tag{5}$$

where N is the number of the sampled points p^g of the gripper at its grasp pose, and sdf^g are the SDF values of the these sampled points. Based on such a loss function, the sampled points of the gripper inside the object, of which the SDF values are less than 0, are pushed out of the object surface, so as to avoid the collision between the gripper and the target. Note that this loss function can adjust all the items (p^1, p^2, d, v) of the grasp pose.

Besides, we apply regularization loss to ensure that the grasp pose does not change too much, preventing excessive deviation from stable grasp poses:

$$L_{reg} = \|\Delta p^1\|_2 + \|\Delta p^2\|_2 + \|\Delta d\|_2 + \|\Delta v\|_2, \tag{6}$$

where $\Delta *$ denotes the grasp pose offset during the refinement process.

Among all the losses above, L_{anti} and L_{touch} fine-tune the grasp poses by directly changing the projected grasp points (p^1, p^2) while the effect of $L_{collision}$ and L_{reg} is performed by changing all the items of the grasp pose (p^1, p^2, d, v). Our final refinement loss is defined as:

$$L_r = \lambda_1 L_{anti} + \lambda_2 L_{collision} + \lambda_3 L_{touch} + \lambda_4 L_{reg}, \tag{7}$$

where λ_1, λ_2, λ_3 and λ_4 are the trade-off parameters to balance each loss. We set λ_1, λ_2, λ_3 and λ_4 as 100, 10, 20 and 200 respectively in practice. During refinement, the weights of the DIF-Decoder and the latent shape code z_t are fixed and L_r is back-propagated through the DIF-Decoder to the grasp pose g_t, so that g_t can be refined to a more stable grasp using Adam algorithm [11].

4 Experiments

In this section, we first introduce the implementation details, along with the dataset and evaluation metrics we adopt. Then we conduct experiments to verify the effectiveness of each part of TransGrasp in Sect. 4.1 and make comparisons with other methods in Sect. 4.2. Finally, real robot experiments are conducted to explore the practical application of our proposed method in Sect. 4.3.

Implementation Details: DIF-Decoder is trained following [34]. For PointNet-like [20] shape encoder, we use the Adam optimizer [11] with an initial learning rate of 0.0001 for training which takes 30 epochs for each category. For the trade-off between the transfer success rate and optimization speed, we set the refine iteration number as 10 in practice.

Dataset: To evaluate our proposed TransGrasp, we select three household categories of objects including 100 mugs, 218 bottles and 107 bowls from ShapeNetCore [3]. Meanwhile, considering the various shapes of objects within a class, we introduce data augmentation which enables the learned categorical shape latent space to cover possible object instances of a class. For instance, mugs of different shapes and sizes are generated from the existing mug models by randomly enlarging or shrinking their top or bottom. Please refer to the supplementary for augmentation examples. The training/test distribution of augmented data is 765/135 for mug, 940/150 for bottle and 801/162 for bowl, respectively. We use only one object instance in each class of objects as the source object which are labeled hundreds of possible grasp poses (802 for mug, 1000 for both bottle and bowl, respectively). From this labeled source object, its grasp poses will be transferred to other target objects of the same class. To train Shape Encoder and 6D Object Pose Estimator, We render partial point clouds from 100 random views for each model in training set using Pytorch3D [23].

Evaluation Metrics: In this section, we evaluate the effectiveness of grasp pose estimation for the target objects of each category. As there are no ground truth grasp pose annotations on the target objects for comparison, we utilize IsaacGym [14] to build a parallel-jaw simulation platform where the success rate of grasp transfer is replaced by the success rate of grasp. Our simulation platform is shown in Fig. 6(a). All simulations are done using a Franka Panda manipulator. Different from other grasp generation methods that are interested in the score of one best grasp among all predicted grasps, we focus on every transferred grasp pose and aim to evaluate the effect of our grasp pose transfer. Therefore, for each instance, we count the ratio of successful grasps among all transferred grasps as transfer success rate. And then the transfer success rate for one category is calculated as the average of all its instances' ones:

$$s_{trans} = \frac{1}{M} \sum_{i}^{M} \frac{N_i^{suc}}{N_i^{all}}, \tag{8}$$

where M denotes the number of object instances while N^{suc} and N^{all} denote the number of successful grasps and total grasps respectively. Specifically, the platform consists of free-floating grippers and objects with gravity, and the grippers are placed according to their grasp poses. A grasp is considered as successful one if the object is held for more than 15 s.

4.1 Ablation Study

To verify the effectiveness of each part of TransGrasp and avoid other influencing factors, we first conduct the experiments using ground-truth 6D poses and scales of objects. Specifically, the following four experiments are conducted. ***Direct Mapping:*** The grasp points are directly obtained from the source labeled model and then scaled according to the size of target model while both depth d and approaching vector v remain unchanged. ***Direct Mapping w/ Refine:*** In addition to *Direct Mapping*, it contains our grasp refinement module that fine-tunes grasp poses by changing grasp points (p^1, p^2), depth d and approaching vector v. ***Grasp Transfer:*** The grasp points of target model are transferred from the source labeled model by our method and also scaled according to the size of target model while both depth d and approaching vector v remain unchanged. ***Grasp Transfer w/ Refine:*** With both grasp transfer and refinement module, it's the complete version of our proposed method. The quantitative results are shown in Table 1.

Table 1. Performance of grasp transfer for each category

Method	Transfer success rate (%)			
	Mug	Bottle	Bowl	Average
Direct mapping	60.68	79.81	33.27	57.92
Direct mapping w/Refine	59.26	79.86	34.83	57.98
Grasp transfer	82.67	86.93	46.24	71.95
Grasp transfer w/Refine	**87.77**	**87.24**	**61.52**	**78.84**

Since grasp points after direct mapping may not be on the surface of target object, which provides a poor initialization for refinement, refinement after direct mapping barely has improvement (57.92% to 57.98% in average). Compared to *Direct Mapping*, *Grasp Transfer* greatly increases grasp success rate with 14.03% for the average success rate. This is because through deformation, the transferred grasp points can adapt to object instances with different shapes, which largely avoids the collision of the gripper with the object. It benefits from the categorical shape space learned by TransGrasp which establishes dense correspondences. Besides, *Grasp Transfer w/ Refine* with the proposed grasp pose refinement module further increases the grasp transfer success rates with 6.89% for the average success rate, indicating that our grasp pose refinement module effectively optimizes the transferred grasp poses and improves the stability of grasping. It is worth noting that each method has a relatively low grasp success rate for the bowl. The first reason is that about 28% of the bowl models are single-layer meshes which are not suitable for grasping. The second reason is that the grasp poses of the bowl are always surrounding the rim of bowl, far away from the center of mass of the bowl, which makes the external force insufficient to produce friction that can resist the gravity of the bowl.

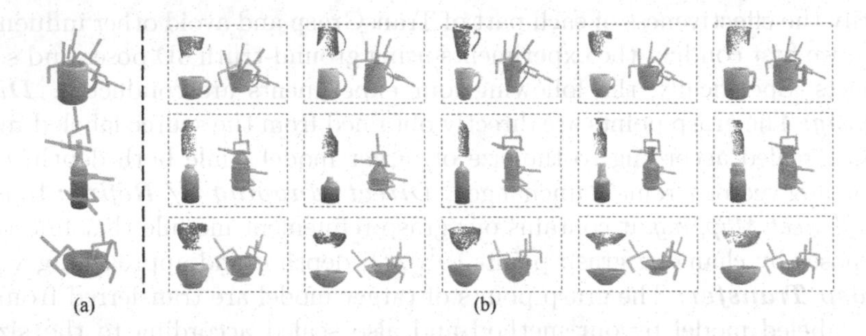

Fig. 5. Qualitative results of three categories. (a) shows source labeled models and some grasp pose annotations. (b) illustrates the transferred grasp poses. For every instance represented inside a dashed box, we show its input partial point cloud (upper left), reconstructed model (bottom left) and transferred grasp poses (right)

Figure 5 shows the qualitative results of the transferred grasps after refinement on the test set of each category. We only show a few grasps for clarity. It can be observed that our method can efficiently estimate high-quality grasps even if the input data is incomplete observed point cloud of unseen instance. Figure 6(b) shows the grasp results on the simulation platform [14]. It can be seen that the grippers can grasp objects stably using our transferred grasp poses, even for those objects with slender handle of the mug and thin edge of the bowl.

4.2 Comparisons with Other Methods in Simulation

The above ablation experiments verify the effectiveness of the proposed TransGrasp, we continue to compare our method with others in this section. Among the grasping methods, GPD [18] and 6-DOF GraspNet [16] have similar grippers and object samples as ours, and the two methods are also representative grasping methods at present. Therefore, we make comparison with them. The major difference from the above two methods is that TransGrasp only labels one instance in grasp pose estimation of an entire category of objects, while GPD and 6-DOF GraspNet require grasp pose annotations for every object.

In this experiment, we evaluate grasp success rate on test set of three categories: *mug*, *bottle* and *bowl*. We employ [25] as the 6D Object Pose Estimator to estimate 6D pose of the input point cloud of a target object, as shown in the left of Fig. 2, by which the grasp pose is transformed from Object Coordinates to

Table 2. Quantitative comparison of grasp success rate and inference time

Method	Grasp success rate (%)				Inference time (s)
	Mug	Bottle	Bowl	Average	(NVIDIA 1080 GPU)
GPD [18]	34.81	74.00	44.44	51.08	1.14
6-DOF GraspNet [16]	76.30	78.00	**75.93**	76.74	1.82
TransGrasp (Ours)	**86.67**	**88.00**	72.84	**82.50**	**0.53**

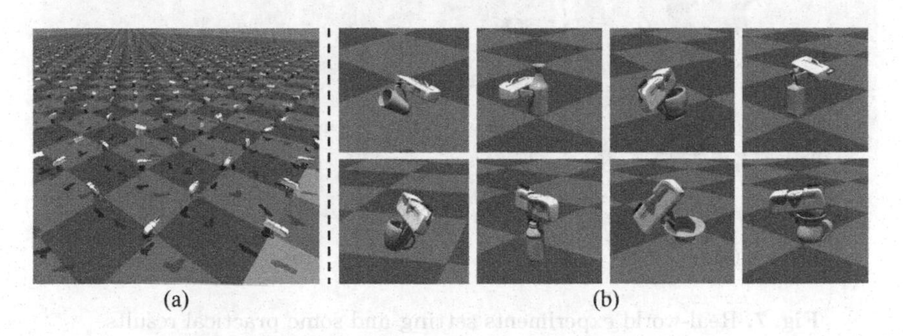

Fig. 6. Simulation platform and some grasp results

Camera Coordinates for the next step. We choose the grasp pose with minimal refinement loss as the final one. For GPD and 6-DOF GraspNet, their final grasp poses are chosen according to the grasp scores computed by their own evaluation networks. To be fair, all the methods generate 100 grasp samples for selection.

Table 2 shows the quantitative results of the three methods. Compared with GPD and 6-DOF GraspNet, TransGrasp obtains the highest success rate for the mug and bottle category (88.67% and 88.00%), and achieves comparable performance with 6-DOF GraspNet for the bowl category although our method only labels one instance of an entire category of objects. One dominant factor is that our method reconstructs the complete shapes of objects, rather than directly estimate grasp poses from partial input point clouds like GPD and 6-DOF GraspNet, which introduces more feasible grasps. And benefiting from the reconstructed complete shape, our refinement module can avoid the interpenetration between gripper and unseen parts of object. Furthermore, even our method contains the refinement module, it takes the least inference time with 0.53s among these methods, which satisfies the requirement of grasping household objects.

4.3 Real Robot Experiments

To verify the practical application of our method, we further conduct grasping experiments on a real robot. All experiments are done using a JACO manipulator with an Intel RealSense D435 mounted on its gripper. We choose 5 objects per category with weights from 135.6 g to 436.2 g for grasping. Figure 7(a) shows

Table 3. Quantitative comparison of grasp success rate in real robot experiments

Method	Mug	Bottle	Bowl	Average
GPD [18]	7/25(28%)	19/25(76%)	11/25(44%)	37/75(49.3%)
6-DOF GraspNet [16]	16/25(64%)	17/25(68%)	**19/25(76%)**	52/75(69.3%)
TransGrasp (Ours)	**19/25(76%)**	**21/25(84%)**	**19/25(76%)**	**59/75(78.7%)**

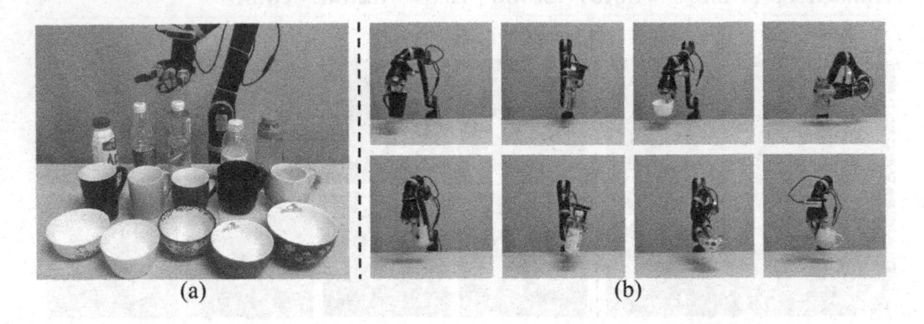

Fig. 7. Real-world experiments setting and some practical results

the hardware setup and object test set. We make comparison with GPD and 6-DOF GraspNet in our experiments. During the grasping process, we first classify and segment each instance in the scene using off-the-shelf instance segmentation method [28] and then estimate 6D object pose of each instance using [25]. We transfer the grasps with the models trained on our synthetic dataset and test in real scenario directly for our method. For GPD and 6-DOF GraspNet, we use the trained models given by authors to generate grasp poses. After grasps are generated or transferred, we choose the grasp using the grasp selection strategy consistent with the simulation experiments in Sect. 4.2. If the gripper can lift the object, the grasp is marked as success. We run 5 trials per object and totally 25 trials for each category. As shown in Table 3, our method outperforms GPD and 6-DOF GraspNet on success rate for mug and bottle, and achieves the same performance with 6-DOF GraspNet for bowl. Figure 7(b) shows some of our grasping results in the real-world setting. It can be observed that the grippers can successfully grasp objects with different shapes in real scenes, which indicates that our method can be applied to practical applications. More experiment results can be seen in the supplementary material.

5 Limitations and Future Work

Although the experiments have demonstrated the effectiveness of our method, there is still a limitation to be addressed. As shown in Fig. 8, the grasps are difficult to be transferred across objects with large shape variances. When the grasps on a cylindrical bottle are transferred to a flat bottle or a square bottle, some of the transferred grasps (in red) penetrate the object. We can further improve the transfer success rate by classifying these objects into fine-grained subclass in the future.

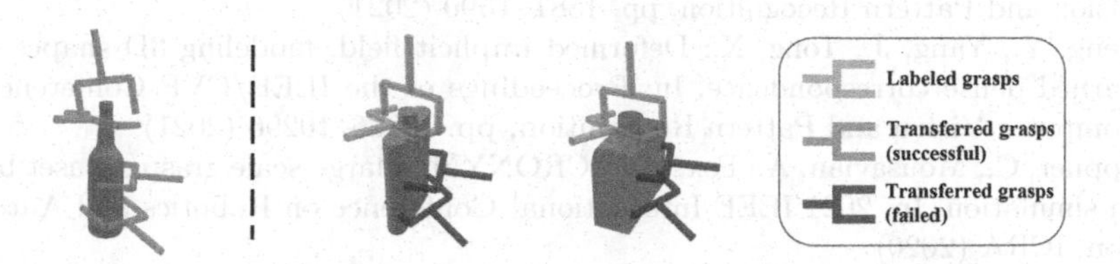

Fig. 8. Grasp Transfer with large shape variances for the bottle category (Color figure online)

6 Conclusion

In this work, we present a category-level grasp pose estimation method to predict grasp poses of a category of objects by labeling only one object instance, which does not rely on a large amount of grasp annotations or exact 3D object models available beforehand. We firstly establish dense shape correspondence among an entire category of objects to transfer the grasp poses from the one source object to other instances, and then build a grasp pose refinement module to improve performance after grasp transfer. Sufficient experiments show that our estimated grasp poses can align well across the instances of one category and achieve high-quality grasp performance compared with existing methods.

Acknowledgments. This research was supported by the National Natural Science Foundation of China (No.U1708263 and No.61873046).

References

1. Besl, P.J., McKay, N.D.: Method for registration of 3-d shapes. In: Sensor Fusion IV: Control Paradigms and Data Structures, vol. 1611, pp. 586–606. International Society for Optics and Photonics (1992)
2. Breyer, M., Chung, J.J., Ott, L., Siegwart, R., Nieto, J.: Volumetric grasping network: real-time 6 dof grasp detection in clutter. arXiv preprint arXiv:2101.01132 (2021)
3. Chang, A.X., et al.: ShapeNet: an information-rich 3D model repository. Technical Report arXiv:1512.03012 [cs.GR]. Stanford University – Princeton University – Toyota Technological Institute at Chicago (2015)
4. Chen, D., Li, J., Wang, Z., Xu, K.: Learning canonical shape space for category-level 6D object pose and size estimation. In: Proceedings of the IEEE/CVF Conference on Computer Vision and Pattern Recognition, pp. 11973–11982 (2020)

5. Chen, W., Jia, X., Chang, H.J., Duan, J., Shen, L., Leonardis, A.: Fs-net: fast shape-based network for category-level 6D object pose estimation with decoupled rotation mechanism. In: Proceedings of the IEEE/CVF Conference on Computer Vision and Pattern Recognition, pp. 1581–1590 (2021)
6. Deng, Y., Yang, J., Tong, X.: Deformed implicit field: modeling 3D shapes with learned dense correspondence. In: Proceedings of the IEEE/CVF Conference on Computer Vision and Pattern Recognition, pp. 10286–10296 (2021)
7. Eppner, C., Mousavian, A., Fox, D.: ACRONYM: a large-scale grasp dataset based on simulation. In: 2021 IEEE International Conference on Robotics and Automation, ICRA (2020)
8. Fang, H.S., Wang, C., Gou, M., Lu, C.: Graspnet-1billion: a large-scale benchmark for general object grasping. In: Proceedings of the IEEE/CVF Conference on Computer Vision and Pattern Recognition, pp. 11444–11453 (2020)
9. Florence, P.R., Manuelli, L., Tedrake, R.: Dense object nets: learning dense visual object descriptors by and for robotic manipulation. arXiv preprint arXiv:1806.08756 (2018)
10. He, Y., Sun, W., Huang, H., Liu, J., Fan, H., Sun, J.: PVN3D: a deep point-wise 3D keypoints voting network for 6dof pose estimation. In: Proceedings of the IEEE/CVF Conference on Computer Vision and Pattern Recognition, pp. 11632–11641 (2020)
11. Kingma, D.P., Ba, J.: Adam: a method for stochastic optimization. arXiv preprint arXiv:1412.6980 (2014)
12. León, B., et al.: OpenGRASP: a toolkit for robot grasping simulation. In: Ando, N., Balakirsky, S., Hemker, T., Reggiani, M., von Stryk, O. (eds.) SIMPAR 2010. LNCS (LNAI), vol. 6472, pp. 109–120. Springer, Heidelberg (2010). https://doi.org/10.1007/978-3-642-17319-6_13
13. Liang, H., et al.: Pointnetgpd: detecting grasp configurations from point sets. In: 2019 International Conference on Robotics and Automation (ICRA), pp. 3629–3635. IEEE (2019)
14. Makoviychuk, V., et al.: Isaac gym: high performance gpu-based physics simulation for robot learning (2021)
15. Miller, A.T., Allen, P.K.: Graspit! a versatile simulator for robotic grasping. IEEE Rob. Autom. Mag. **11**(4), 110–122 (2004)
16. Mousavian, A., Eppner, C., Fox, D.: 6-dof graspnet: variational grasp generation for object manipulation. In: Proceedings of the IEEE/CVF International Conference on Computer Vision, pp. 2901–2910 (2019)
17. Murali, A., Mousavian, A., Eppner, C., Paxton, C., Fox, D.: 6-dof grasping for target-driven object manipulation in clutter. In: 2020 IEEE International Conference on Robotics and Automation (ICRA), pp. 6232–6238. IEEE (2020)
18. ten Pas, A., Gualtieri, M., Saenko, K., Platt, R.: Grasp pose detection in point clouds. Int. J. Rob. Res. **36**(13–14), 1455–1473 (2017)
19. Patten, T., Park, K., Vincze, M.: Dgcm-net: dense geometrical correspondence matching network for incremental experience-based robotic grasping. Front. Rob. AI **7**, 120 (2020)
20. Qi, C.R., Su, H., Mo, K., Guibas, L.J.: Pointnet: deep learning on point sets for 3D classification and segmentation. In: Proceedings of the IEEE Conference on Computer Vision and Pattern Recognition, pp. 652–660 (2017)
21. Qi, C.R., Yi, L., Su, H., Guibas, L.J.: Pointnet++: deep hierarchical feature learning on point sets in a metric space. arXiv preprint arXiv:1706.02413 (2017)

22. Qin, Y., Chen, R., Zhu, H., Song, M., Xu, J., Su, H.: S4g: amodal single-view single-shot se (3) grasp detection in cluttered scenes. In: Conference on Robot Learning, pp. 53–65. PMLR (2020)
23. Ravi, N., et al.: Accelerating 3D deep learning with pytorch3d. arXiv:2007.08501 (2020)
24. Simeonov, A., et al.: Neural descriptor fields: Se (3)-equivariant object representations for manipulation. arXiv preprint arXiv:2112.05124 (2021)
25. Tian, M., Ang, M.H., Lee, G.H.: Shape prior deformation for categorical 6d object pose and size estimation. In: Vedaldi, A., Bischof, H., Brox, T., Frahm, J.-M. (eds.) ECCV 2020. LNCS, vol. 12366, pp. 530–546. Springer, Cham (2020). https://doi.org/10.1007/978-3-030-58589-1_32
26. Wang, C., et al.: Densefusion: 6D object pose estimation by iterative dense fusion. In: Proceedings of the IEEE/CVF Conference on Computer Vision and Pattern Recognition, pp. 3343–3352 (2019)
27. Wang, H., Sridhar, S., Huang, J., Valentin, J., Song, S., Guibas, L.J.: Normalized object coordinate space for category-level 6D object pose and size estimation. In: The IEEE Conference on Computer Vision and Pattern Recognition (CVPR) (2019)
28. Wang, X., Kong, T., Shen, C., Jiang, Y., Li, L.: SOLO: segmenting objects by locations. In: Vedaldi, A., Bischof, H., Brox, T., Frahm, J.-M. (eds.) ECCV 2020. LNCS, vol. 12363, pp. 649–665. Springer, Cham (2020). https://doi.org/10.1007/978-3-030-58523-5_38
29. Wen, B., Lian, W., Bekris, K., Schaal, S.: Catgrasp: learning category-level task-relevant grasping in clutter from simulation. arXiv preprint arXiv:2109.09163 (2021)
30. Wu, C., et al.: Grasp proposal networks: an end-to-end solution for visual learning of robotic grasps. arXiv preprint arXiv:2009.12606 (2020)
31. Xiang, Y., Schmidt, T., Narayanan, V., Fox, D.: Posecnn: a convolutional neural network for 6D object pose estimation in cluttered scenes. arXiv preprint arXiv:1711.00199 (2017)
32. Yang, D., Tosun, T., Eisner, B., Isler, V., Lee, D.: Robotic grasping through combined image-based grasp proposal and 3D reconstruction. In: 2021 IEEE International Conference on Robotics and Automation (ICRA), pp. 6350–6356. IEEE (2021)
33. Zeng, A., et al.: Multi-view self-supervised deep learning for 6D pose estimation in the amazon picking challenge. In: 2017 IEEE International Conference on Robotics and Automation (ICRA), pp. 1386–1383. IEEE (2017)
34. Zheng, Z., Yu, T., Dai, Q., Liu, Y.: Deep implicit templates for 3D shape representation. In: Proceedings of the IEEE/CVF Conference on Computer Vision and Pattern Recognition, pp. 1429–1439 (2021)

StARformer: Transformer with State-Action-Reward Representations for Visual Reinforcement Learning

Jinghuan Shang(✉), Kumara Kahatapitiya, Xiang Li, and Michael S. Ryoo

Stony Brook University, Stony Brook, NY 11794, USA
{jishang,kkahatapitiy,xiangli8,mryoo}@cs.stonybrook.edu

Abstract. Reinforcement Learning (RL) can be considered as a sequence modeling task: given a sequence of past state-action-reward experiences, an agent predicts a sequence of next actions. In this work, we propose **St**ate-**A**ction-**R**eward Transformer (**StAR**former) for visual RL, which explicitly models *short-term* state-action-reward representations (StAR-representations), essentially introducing a Markovian-like inductive bias to improve *long-term* modeling. Our approach first extracts StAR-representations by self-attending image state patches, action, and reward tokens within a short temporal window. These are then combined with pure image state representations—extracted as convolutional features, to perform self-attention over the whole sequence. Our experiments show that StARformer outperforms the state-of-the-art Transformer-based method on image-based Atari and DeepMind Control Suite benchmarks, in both offline-RL and imitation learning settings. StARformer is also more compliant with longer sequences of inputs. Our code is available at https://github.com/elicassion/StARformer.

Keywords: Reinforcement learning · Transformer · Sequence modeling

1 Introduction

Reinforcement Learning (RL) naturally operates sequentially: an agent observes a state from the environment, takes an action, observes the next state, and receives a reward from the environment. In the past, RL problems have been usually modeled as Markov Decision Processes (MDP). It enables us to take an action solely based on the current state, which is assumed to represent the whole history. With this scheme, sequences are broken into single steps so that

Supplementary Information The online version contains supplementary material available at https://doi.org/10.1007/978-3-031-19842-7_27.

S. Avidan et al. (Eds.): ECCV 2022, LNCS 13699, pp. 462–479, 2022.
https://doi.org/10.1007/978-3-031-19842-7_27

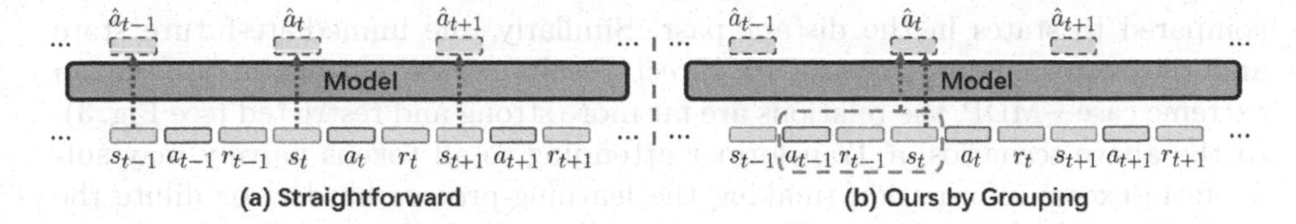

Fig. 1. Illustration of RL as sequence modeling using Transformer: (a) A straightforward approach and, (b) Our proposed improvement. The intuition is to explicitly model local features (green boundary) to help long-term sequence modeling. (Color figure online)

algorithms like TD-learning [58] can be mathematically derived via Bellman Equation to solve RL problems. Recent advances such as [9,26] formulate (offline-)RL differently—as a sequence modeling task, and Transformer [64] architectures have been adopted as generative trajectory models to solve it, i.e., given past experiences of an agent composed of a sequence of state-action-reward triplets, a model iteratively generates an output sequence of action predictions.

This new scheme softens the MDP assumption, where an action is predicted considering multiple steps in history. To implement this, methods such as [9,26] process the input sequence *plainly* through self-attention (with a causal attention mask) using Transformers [64]. This way, a given state, action, or reward token may attend to any of the (previous) tokens in the sequence, which allows the model to capture long-term relations. Moreover, each image state is usually encoded with convolutional networks (CNNs) *as-a-whole* prior to self-attention.

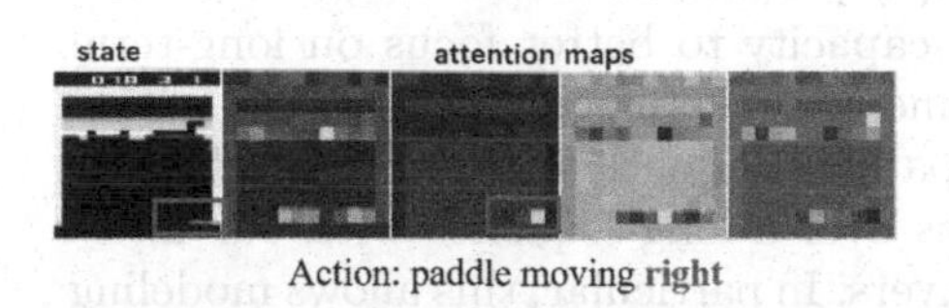

Fig. 2. Attention maps between action token and pixel state patches in our method. In the second attention map from the left, weights in the paddle region are directed towards right (highlighted in red), corresponding to the semantic meaning of "right" action. (Color figure online)

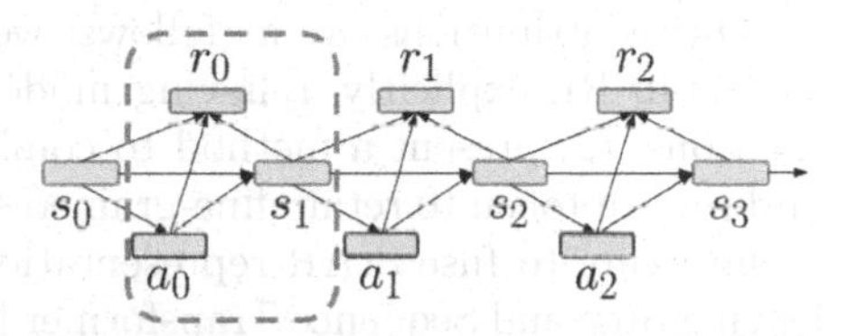

Fig. 3. MDP view of an RL process. Only the connected pairs (directed arrows) are causally-related, whereas others are independent of each other. Green boundary highlights our motivation: explicitly considering a single transition helps long-term modeling. (Color figure online)

However, if we consider states, actions, and rewards within adjacent time-steps, they generally have strong connections due to potential causal relations. For instance, states in the recent past have a stronger effect on the next action, compared to states in the distant past. Similarly, the immediate-future state and the corresponding reward are direct results of the current action. In an extreme case—MDP, the relations are far more strong and restricted (see Fig. 3). In the above scenarios, a Transformer attending to all tokens naively may suffer from excess information (making the learning-process harder) or dilute the truly-essential relation priors. This is especially critical when input sequences are quite large, either in spatial [68] or temporal [26] dimension, and when Transformer models become heavy, i.e., contain a large number of layers [62]. Learning Markovian-like dependencies between tokens from scratch is hard and may waste computations [26], as the rest of the dependencies are possibly weaker. Moreover, tokenizing image states as-a-whole based on CNNs further prohibits Transformer models from capturing detailed spatial relations. Such loss of information can be critical especially in RL tasks with fine-grained regions-of-interest.

To alleviate such issues, we propose to explicitly model single-step transitions, introducing a Markovian-like inductive bias and relieving the capacity to be used for long sequence modeling. We introduce **St**ate-**A**ction-**R**eward Transformer (**StAR**former) for visual RL, which consists of two interleaving components: a Step Transformer and a Sequence Transformer. The Step Transformer learns local representations (i.e., **StAR**-representations) by self-attending state-action-reward tokens *within a window of single time-step*. Here, image states are encoded as ViT-like [18] patches, retaining fine-grained spatial information. The Sequence Transformer then combines StAR-representations with pure image state representations (extracted as convolutional features) *from the whole sequence* to make action predictions. Our experiments validate the benefits of StARformer over prior work in both offline-RL and imitation learning settings, while also being more compliant with longer input sequences.

Our contributions are as follows: we (1) propose to model single-step transitions in RL explicitly, relieving model capacity to better focus on long-term relations, (2) present a method to combine ViT-like image patches with action and reward token to retain fine-grained spatial information, and (3) introduce an architecture to fuse StAR-representations over a long-sequence with our interleaving Step and Sequence Transformer layers. In particular, this allows modeling sequences of state-action feature representations at multiple different levels.

2 Related Work

2.1 Reinforcement Learning to Sequence Modeling

Reinforcement Learning (RL) is usually modeled as a Markov Decision Process (MDP). Based on this, single-step value-estimation methods have been derived from the Bellman equation, including Q-learning [66] and Temporal Difference (TD) learning [30,53,58,60], along with their extensions [24,43,71].

More recent directions [9,26] formulate RL a different way—as a sequence modeling task, i.e., given a sequence of recent experiences including state-actions-reward triplets, a model predicts a sequence of next actions. This approach can be trained in a supervised learning manner, being more compliant with offline RL [36] and imitation learning settings [25,56,61]. Zheng et al. [72] adapt this formulation to online settings. Furuta et al. [20] extend DT [9] to match given hindsight information. Reed et al. [52] train a single agent that performs a wide range of RL and language tasks. Sequence modeling can be also viewed as solving RL by learning trajectory representations. Other than methods learning visual representations only [34,35,38,55,69–71], our approach combines visual and trajectory representations together, thanks to the power of Transformer.

2.2 Transformers

Transformer architectures [64] have been first introduced in language processing tasks [17,50,51], to model interactions between a sequence of word embeddings, or more generally, unit representations or *tokens*. Recently, Transformers have been adopted in vision tasks with the key idea of breaking down images/videos into tokens [5,7,10,18,23,28,45], often outperforming convolutional networks (CNNs) in practice. Inspired by designs from both Transformers and CNNs, combining the two [15,42] shows further improvements. Transformers also found to be useful in handling sensory information [59] and doing one-shot imitation learning [16]. Chen et al. [9] explore how GPT [51] can be applied to RL under the sequence modeling setting.

Sequence modeling in visual RL is similar to learning from videos in terms of input data, which are composed of sequences of observed images (i.e. states). One challenge of applying Transformers to videos is the large number of input tokens and quadratic computation. These problems have been investigated in multiple directions, including attention approximation [12,29,65], separable attention in different dimensions [5,7], reducing the number of tokens using local windows [40, 41], adaptively generating a small amount of tokens [54] or using a CNN-stem to come up with a small amount of high-level tokens [14,46,67].

StARformer shares a similar concept to performing spatial and temporal attention separately as in [5,7]. In contrast to such methods designed to reduce attention computation, our primary target is introducing inductive bias: modeling short-term and long-term contexts separately. Our method also operates on different sets of tokens, in short-term (s-a-r tokens) and in long-term (learned intermediate StAR-representation), which deviates from previous methods.

3 Preliminary

3.1 Transformer

Transformer [64] architectures have shown diverse applications in language [17] and vision tasks [5,18]. Given a sequence of input tokens $X = \{x_1, x_2, ..., x_n\}$, where $x_i \in \mathbb{R}^d$, a Transformer layer maps it to an output sequence of tokens $Z = \{z_1, z_2, ..., z_n\}$, where $z_i \in \mathbb{R}^d$. A Transformer model is obtained by stacking multiple such layers. We denote the mapping for each layer (l) as $F(\cdot)$: $Z^l = F(Z^{l-1})$. We use $F(\cdot)$ to represent a Transformer layer in the remaining sections.

Self-attention [11,39,48,64] is the core component of Transformers, which models pairwise relations between tokens. As introduced in [64], an input token representation X is linearly mapped into query, key and value representations, i.e., $\{Q, K, V\} \in \mathbb{R}^{n \times d}$ respectively, to compute self-attention as follows:

$$\text{Attention}(Q, K, V) = \text{softmax}(\frac{QK^T}{\sqrt{d}})V. \quad (1)$$

Vision Transformer (ViT) [18] extends the same idea of self-attention to the image domain. Given an input image $s \in \mathbb{R}^{H \times W \times C}$, a set of n non-overlapping local patches $P = \{p_i\} \in \mathbb{R}^{h \times w \times C}$ is extracted, flattened and linearly mapped to a sequence of tokens $\{x_i\} \in \mathbb{R}^d$. We extend ViT [18] so that action, reward, and state patches can jointly attend, where we find semantic meanings could be learned within action-patch attention in RL tasks (Fig. 1).

3.2 RL as Sequence Modeling

We consider a Markov Decision Process (MDP), described by tuple $(\mathcal{S}, \mathcal{A}, P, \mathcal{R})$, where $s \in \mathcal{S}$ represents the state, $a \in \mathcal{A}$, the action, $r \in \mathcal{R}$, the reward, and P, the transition dynamics given by $P(s'|s, a)$. In MDP, a trajectory (τ) is defined as the past experience of an agent, which is a sequence composed of states, actions, and rewards in the following temporal order:

$$\tau = \{s_1,\ a_1,\ r_1,\ s_2,\ a_2,\ r_2,\ \ldots,\ s_t,\ a_t,\ r_t\}. \quad (2)$$

Sequence modeling for RL is making action predictions from past experience [9,26]:

$$Pr(\hat{a}_t) = p(a_t|\ s_{1:t},\ a_{1:t-1},\ r_{1:t-1}). \quad (3)$$

Recent work [9,26] try to adopt an existing Transformer architecture [51] for RL with the formulation as above. In [9,26], states (s), actions (a), and rewards (r) are considered as input tokens (see Fig. 1a), while using a causal mask to ensure an autoregressive output sequence generation (i.e. following Eq. 3), where a token can access any of its preceding tokens through self-attention.

In contrast, our formulation attends tokens with (potentially) strong causal relations *explicitly*, while attending to long-term relations as well. To do this, in this work, we break a trajectory into small groups of state-action-reward tuples (i.e., s, a, r). It learns local relations within the tokens of each group through self-attention (see Fig. 1b, and Fig. 2), followed by long-term sequence modeling.

4 StARformer

4.1 Overview

StARformer consists of two basic components: Step Transformer and Sequence Transformer, together with interleaving connections (see Fig. 4b). Step Transformer learns StAR-representations from strongly-connected local tokens *explicitly*, which are then fed into the Sequence Transformer along with pure state

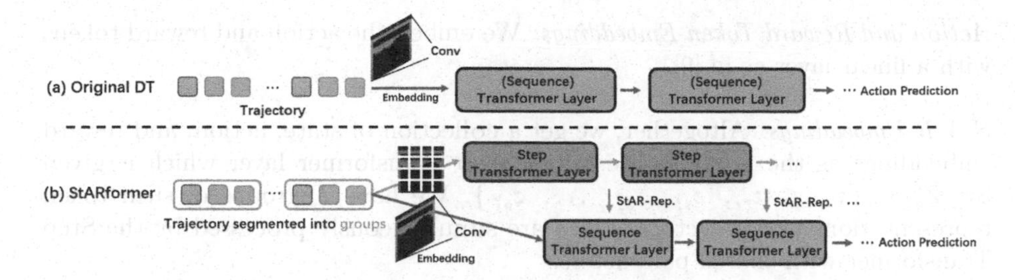

Fig. 4. (a) Structure summary of original DT [9], where the Transformer Layer acts similar as our Sequence Transformer. (b) StARformer consists of Step Transformer and Sequence Transformer, to separately model a single-step and the sequence as-a-whole, respectively. Two types of layers are connected at each level via learned StAR-representations. In terms of state embedding methods, DT uses only convolution, while StARformer uses ViT-like [18] embeddings (patches) in Step Transformer and convolution in Sequence Transformer separately. (Color figure online)

representations to model the whole input trajectory. At the output of the final Sequence Transformer layer, we make action predictions via a prediction head. In the following subsections, we will introduce the two Transformer components, and their corresponding token embeddings in detail.

4.2 Step Transformer

Grouping State-Action-Reward: Our intuition of grouping is to model strong local relations explicitly. To do so, we first segment a trajectory (τ) into a set of groups, where each group consists of previous action (a_{t-1}), reward (r_{t-1}) and current state (s_t) (Fig. 5). Each element within a group has a strong causal relation with the others.

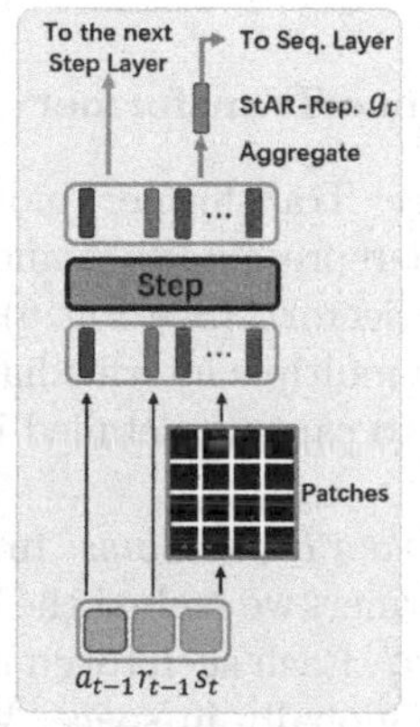

Fig. 5. Overview of Step Transformer. Output tokens are (1) sent to the next Step Transformer layer and (2) aggregated to produce StAR-representation.

Patch-Wise State Token Embeddings: In Step Transformer, we tokenize each input image state by dividing it into a set of non-overlapping spatial patches z_{s_t} along its spatial dimensions, following ViT [18] (Fig. 5). Our motivation for using patch embeddings is to create fine-grained state embeddings. This allows the Step Transformer to model the relations of actions and rewards with local-regions of state (Fig. 2). Such local correspondences provide more information compared to highly-abstracted convolutional features in this single-step modeling, which is empirically validated in our ablation studies (Sect. 6.4).

Action and Reward Token Embeddings: We embed the action and reward tokens with a linear layer as in [9].

S-A-R Embeddings: Altogether, we get a collection of state, action, and reward embeddings as the input to the initial Step Transformer layer which is given by: $Z_t^0 = \{z_{a_{t-1}}, z_{r_t}, z_{s_t^1}, z_{s_t^2}, \ldots, z_{s_t^n}\}$. We have T groups of such token representations per trajectory, which are simultaneously processed by the Step Transformer with shared parameters.

Step Transformer Layer: We adopt the conventional Transformer design from [64] (Sect. 3.1) as our Step Transformer layer. Each group of tokens from the previous layer Z_t^{l-1} is transformed to Z_t^l by a Step Transformer layer with the mapping F_{step}^l: $Z_t^l = F_{\text{step}}^l(Z_t^{l-1})$.

StAR-representation: At the output of each Step Transformer layer l, we further obtain a State-Action-Reward-representation (StAR-representation) $g_t^l \in \mathbb{R}^D$ by aggregating output tokens $Z_t^l \in \mathbb{R}^{n \times d}$ (see green flows in Fig. 4b):

$$g_t^l = \text{FC}([Z_t^l]) + e_t^{\text{temporal}}. \tag{4}$$

Here $[\cdot]$ represents concatenation of the tokens within each group and $e_t^{\text{temporal}} \in \mathbb{R}^D$, the temporal positional embeddings for each timestep. Finally, the output StAR-representation g_t^l is fed into the corresponding Sequence Transformer layer for long-term sequence modeling.

4.3 Sequence Transformer

Our Sequence Transformer models long-term sequences by looking at the learned StAR-representations and the *pure state tokens* (introduced below) over the whole trajectory (See Fig. 6). Notice that, as illustrated in Fig. 4 (b), this happens with multiple intermediate StAR representations, allowing the Sequence Transformer to capture detailed information.

Pure State Token Embeddings: In addition to the patch-wise token embeddings in Step Transformer, we embed the input image state s_t as-a-whole, to create pure state tokens h_t^0. Each such token represents a single state representation, describing the state globally in space. We do this by processing each state through a CNN encoder, since the convolutional layers mix features spatially:

$$h_t^0 = \text{Conv}(s_t) + e_t^{\text{temporal}}, \tag{5}$$

where $e_t^{\text{temporal}} \in \mathbb{R}^D$ represents the temporal positional embeddings exactly the same as we add to g_t for each timestep. The convolutional encoder is from [44].

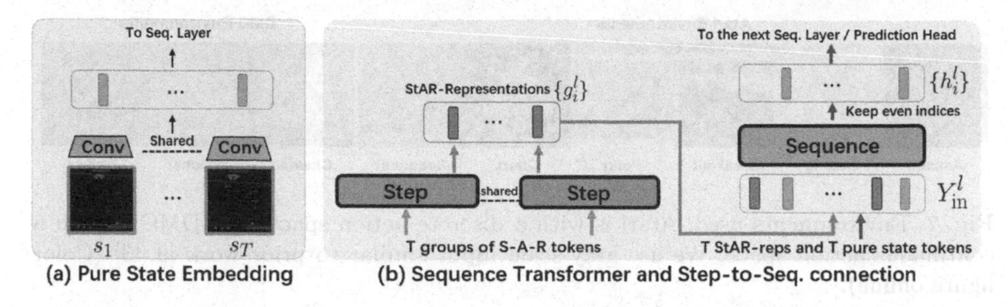

Fig. 6. (a) Pure state embeddings are learned from shared convolutional layers. (b) Sequence Transformer takes StAR-representation and the pure state tokens and generate output tokens.

Sequence Transformer Layer: Similar to Step Transformer, we use the conventional Transformer layer design from [64] for our Sequence Transformer. The input to the Sequence Transformer layer l consists of representations from two sources: (1) the learned StAR-representations $g_i^l \in \mathbb{R}^D$ from the corresponding Step Transformer layer, and (2) $h_t^{l-1} \in \mathbb{R}^D$ from the outputs of the previous Sequence Transformer layer. Here, as mentioned above, we set h_t^0 to be the pure state representation. The two types of token representations are merged to form a single sequence, preserving their temporal order (as elaborated below):

$$Y_{\text{in}}^l = \{g_1^l,\ h_1^{l-1},\ g_2^l,\ h_2^{l-1},\ldots,\ g_T^l,\ h_T^{l-1}\}. \tag{6}$$

We place g_t^l before h_t^{l-1}—which originates from s_t—because g_t^l contains information of the *previous* action a_{t-1}, which comes prior to s_t in the trajectory. We also apply a causal mask in the Sequence Transformer to ensure that the tokens at time t cannot attend any future tokens (i.e., $> t$).

Here, Sequence Transformer takes StAR-representations generated from each intermediate Step Transformer layer, rather than taking the final StAR-represent-ations after all Step Transformer layers. In this way, the model gains an ability to look at StAR-representations at multiple abstraction levels. In Sect. 6.4, we empirically validate the benefit of this layer-wise fusion.

Sequence Transformer computes an intermediate set of output tokens as in: $Y_{\text{out}}^l = F_{\text{sequence}}^l(Y_{\text{in}}^l)$. We then select the tokens at even indices of Y_{out}^l (where indexing starts from 1) to be the pure state tokens $h_i^l := y_{\text{out};2i}^l$, which are then fed into the next Sequence Transformer layer.

Action Prediction: The output of the last Sequence Transformer layer is used to make action predictions, based on a linear head: $\hat{a}_t = \phi(h_t^l)$.

4.4 Training and Inference

StARformer is a drop-in replacement of DT [9], as training and inference procedures remain the same. StARformer can easily operate on step-wise reward without a performance drop (detailed discussed in Sect. 6.4). In contrast, it is critical to design a Return-to-go (RTG, target return) carefully in DT, which needs more trials and tuning to find the best value.

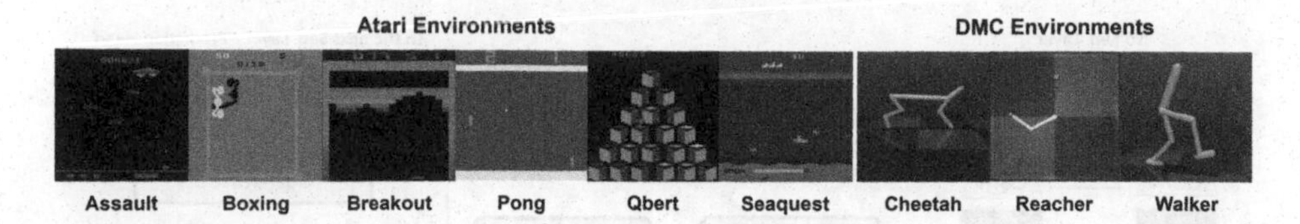

Fig. 7. Environments used: Atari is with a discrete action space, and DMC is with a continuous action space. We use gray-scale input similar to prior work [9,43] (Color figure online).

5 Experimental Setup

5.1 Settings

We consider offline RL [36] and imitation learning (behavior cloning) in our experiments. In offline RL, we have a fixed memory buffer of sub-optimal trajectory rollouts. Offline RL is generally more challenging compared to conventional RL due to the shifted distribution problem [36].

In our Imitation learning setting, the agent is not exposed to reward signals and online-collected data from the environment. This is an even harder problem due to provided trajectories being sub-optimal, compared to traditional imitation learning that could collect new data and do Inverse Reinforcement Learning [1, 47]. We simply remove the rewards in the dataset used in offline RL to come up with this setting. Both DT [9] and proposed method can operate without reward, by simply removing T reward tokens in DT [9] (T is trajectory length), or removing the reward token in Step Transformer in our model.

5.2 Environments and Datasets

We consider image-based Atari [6] (discrete action space) and DeepMind Control Suite (DMC) [63] (continuous action space) to evaluate our model in different types of tasks, listed in Fig. 7 with image examples. We pick 6 games in Atari: Assault, Boxing, Breakout, Pong, Qbert, and Seaquest. Similar to [9] we use 1% (500k steps) of the DQN replay buffer dataset [2] to perform a thorough and fair comparison. We select 3 continuous control tasks in DMC [63]: Cheetah-run, Reacher-easy, and Walker-walk. In DMC, we collect a replay buffer (i.e. sub-optimal trajectories) generated by training a SAC [21] agent from scratch for 500k steps for each task. Note that these continuous control tasks are with image inputs, which previous work [9] does not cover (originally using Gym [8]).

We report the absolute value of episodic returns (i.e., cumulative rewards). Results are averaged across 7 random seeds in Atari and 10 seeds in DMC, each seed is evaluated by 10 randomly initialized episodes.

5.3 Baselines

We select Decision-Transformer (DT) [9], a SOTA Transformer-based sequence modeling method for RL. We notice there is also Trajectory-Transformer [26], which however, is not designed for image inputs. We use most of the same hyper-parameters as in DT [9] for Atari environments without extra tuning (details in Supplementary Table 4 and 5). As for DMC environments, since they are not covered by DT [9], we carefully tune the baseline first and then use the same set of hyper-parameters in our method. We also compare with SOTA non-Transformer offline-RL methods including CQL [33], QR-DQN [13], REM [3], and BEAR [31]. For imitation (behavior cloning), we only compare with DT [9] and straightforward behaviour cloning with ViT (referred to as BC-ViT).

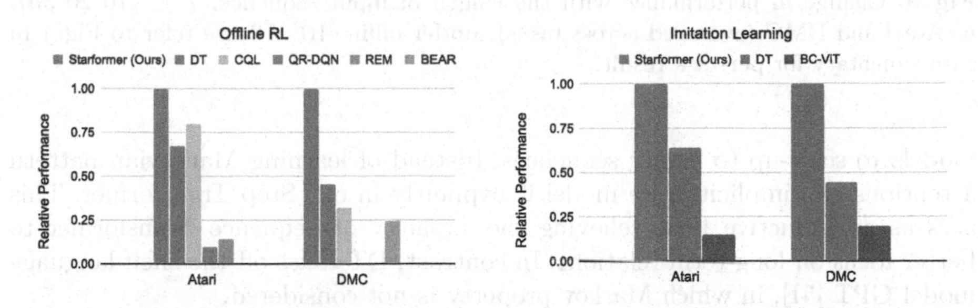

Fig. 8. Relative performance of episodic returns. The results are averaged across all environments and random seeds (same in later experiments), and normalized w.r.t. the performance of StAR. Please refer Table 1 in supplementary for details.

6 Results

6.1 Improving Sequence Modeling for RL

We first compare our StARformer (StAR) with the state-of-the-art Transformer-based RL method in Atari and image-based DMC environments, under both offline RL and imitation learning settings. We select the Decision-Transformer proposed in [9], (referred as DT) as our baseline. Here, we keep $T = 30$ for all environments, which is the number of time-steps (length) of each input trajectory. We also compare our method to CQL [33], a SOTA non-Transformer offline-RL method. Figure 8 shows that our method outperforms baselines, in both offline RL and imitation learning settings, suggesting that our method can better model reinforcement sequences with images.

6.2 Scaling-Up to Longer Sequences

In this experiment, we evaluate how StARformer and DT perform with different input sequence lengths, specifically $T = \{10, 20, 30\}$, under offline-RL setting. In Fig. 9, we see that StARformer gains performance with longer trajectories, whereas DT [9] saturates as early as $T = 10$. This validates our claim that considering short-term and long-term relations separately (and then fusing) help

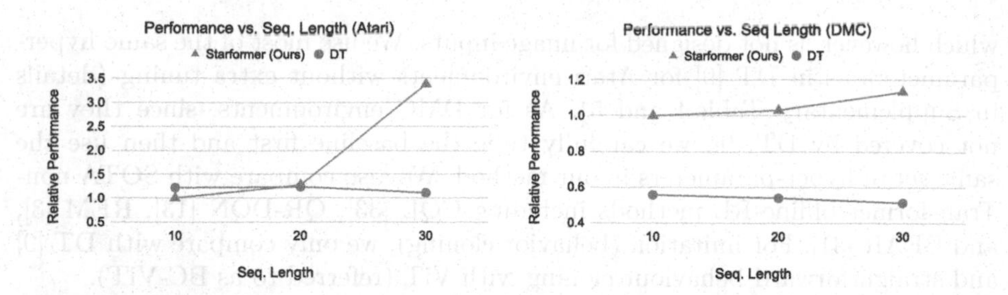

Fig. 9. Change in performance with the length of input sequence, $T \in \{10, 20, 30\}$, in Atari and DMC (averaged across tasks), under offline-RL. Please refer to Fig. 1 in supplementary for per-task result.

models to scale-up to longer sequences. Instead of learning Markovian pattern attentions [26] implicitly, we model it explicitly in our Step Transformer. This acts as an inductive bias, relieving the capacity of Sequence Transformer to better focus on long-term relations. In contrast, DT takes off-the-shelf language model GPT [51], in which Markov property is not considered.

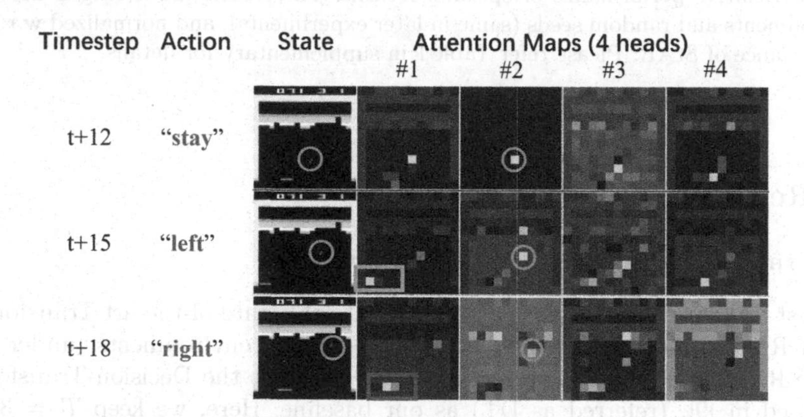

Fig. 10. Visualization of attention maps in our Step Transformer, extracted for Breakout game. Attention weights are computed between the action token and state patch tokens. We highlight the ball (orange circle) in the input for convenience. Please find out more visualizations in our supplementary material (Color figure online).

6.3 Visualization

We show attention maps between action and state patches in Step Transformer at several timesteps extracted from a trajectory in Breakout (see Fig. 10). In this game, the agent should move the paddle to bounce the ball back from the bottom, after the ball falls down while breaking the bricks on the top. In the presented attention maps, the regions with a high attention score (highlighted) mainly overlap with the locations of the ball, paddle, and potential target bricks. We find the attention maps in head #1 to be particularly interesting. Here, the focused regions corresponding to the paddle show a directional pattern, corresponding to the semantic meaning of actions "moving the paddle right", "left" or "stay". This validates that Step Transformer captures essential spatial relations between actions and state patches, which is important for decision making. Moreover, in head #2, we observe that the focused regions correspond to the locations of the ball, except when the ball is out of the boundary, too-close to the paddle or indistinguishable within bricks. Overall, these attention maps suggest how our model can show a basic understanding of the Breakout game.

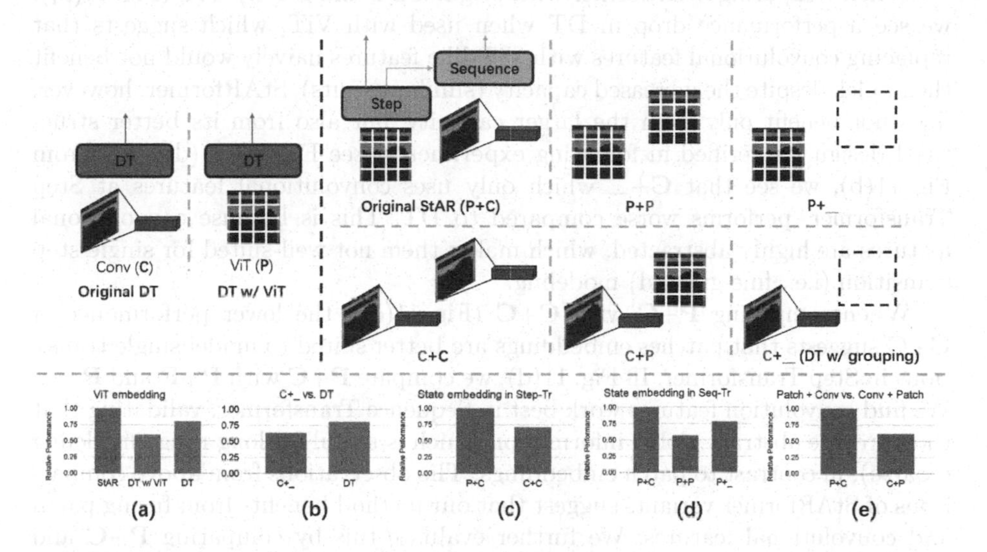

Fig. 11. (Top): Embedding methods used in original DT, StARformer (StAR), and their variants. We label ViT (patches) as **P**, convolution as **C**, and None (not using the corresponding embedding) as "_". (Bottom): (a–e) Performance comparisons between variants. Per-task results are in Supplementary Table 2.

6.4 Ablations

StARformer has three design differences compared to baseline DT [9]: it (1) learns StAR-representation from single-step transitions (grouping), (2) uses both ViT-like [18] patch embeddings and convolutions for state representation, and (3) merges these two types of embeddings in Step Transformer *layer-wise.*

StAR-representationand State Representations: We first discuss designs of StAR-representation and state representation methods ((1) and (2) mentioned above) jointly, as they can be unified into variants shown in Fig. 11. We vary state embedding methods used to learn s_t in Step Transformer and h_t in Sequence Transformer. Namely, we consider: (1) ViT features (patch embeddings, labeled as **P**), (2) Convolutional features (labeled as **C**), or (3) None (not having the corresponding embedding, labeled as "__"). The original StARformer can be represented by **P+C** (patch embeddings for s_t and convolutional embeddings for h_t). Other variants include: **P+P**, **P+__**, **C+P**, **C+C**, and **C+__**. We note variant **C+__** could be viewed as DT + grouping, where we simply adapt DT to our framework, and learn StAR-representation from convolutional features only. We also implement a variant of DT using ViT for state embedding (noted as DT w/ ViT), to match our method in terms of having a similar embedding method and capacity (13M parameters vs. 14M parameters in ours).

When comparing StARformer with original DT and DT w/ ViT (Fig. 11(a)), we see a performance drop in DT when used with ViT, which suggests that replacing convolutional features with ViT-like features naively would not benefit the model, despite the increased capacity (similar to ours). StARformer, however, does not benefit only from the larger capacity, but also from its better structural design, as verified in following experiments (see Fig. 11(c) (d) (e)). From Fig. 11(b), we see that **C+__** which only uses convolutional features at Step Transformer, performs worse compared to DT. This is because convolutional features are highly abstracted, which makes them not well-suited for single-step transition (i.e., fine-grained) modeling.

When comparing **P+C** with **C+C** (Fig. 11(c)), the lower performance of **C+C** suggests that patches embeddings are better suited to model single transitions in Step Transformer. In Fig. 11(d), we compare **P+C** with **P+P** and **P+__**. We find convolution features work best in Sequence Transformer, validating that they provide abstract global information which is useful for long range modeling (coarse), in contrast to patch embeddings. The observations from above comparisons of StARformer variants suggest that our method benefits from fusing patch and convolutional features. We further evaluate this by comparing **P+C** and **C+P** (Fig. 11(e)), where **P+C** performs better, confirming this fusion method of "fine-grained (patches) to high-level (conv)" best matches with our sequence modeling scheme of "single-transition followed-by long-range-context".

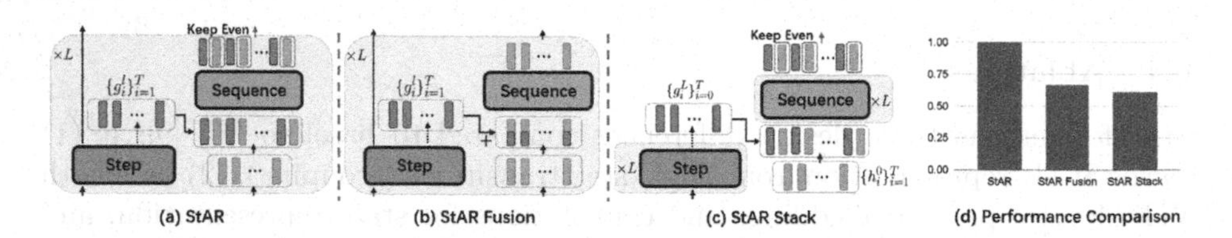

Fig. 12. Variants of our model w/ different connections. Two variants, (b) StAR Fusion and (c) StAR Stack are shown in comparison to our original (a) StAR model. (d) Experiments (offline-RL) show original structure, which is a layer-wise fusion, works best. Please refer to Supplementary Table 3 for per-task results.

In our model, we model whole trajectory using representations from two sources: StAR-representations g from Step Transformer, and pure state representation h from previous layer of Sequence Transformer. We combine g and h in a layer-wise manner (i.e., at each corresponding layer). We investigate two other variants: (1) g_t^l is fused with h_t^l by summation (referred as StAR Fusion, see Fig. 12(b)), and (2) the Sequence Transformer is "stacked" on-top of the Step Transformer (referred as StAR Stack, see Fig. 12(c)). Results of these configurations are shown in Fig. 12(d) and StAR works the best. We see that attending to all tokens is better than token summation at Sequence Transformer. Also, having StAR-representations from different abstraction levels is beneficial compared to having one.

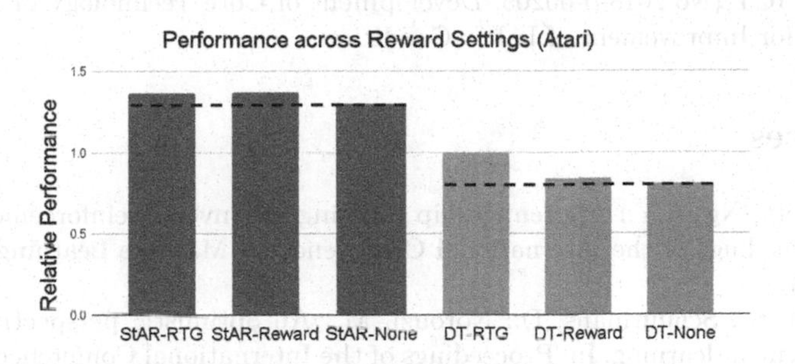

Fig. 13. Performance in different reward settings: return-to-go (RTG), stepwise reward, or no reward at-all (labeled as 'None') settings.

Reward Setting: Return-to-go, Stepwise Reward, or No Reward At-All? We investigate how different reward settings affect sequence modeling, specifically, return-to-go (RTG) [9], stepwise reward, and no reward at-all. Decision-Transformer [9] originally uses RTG $\hat{R}_t$, which is defined as the sum of future step-wise rewards: $\hat{R}_t = \sum_{t'=t}^{T} r_{t'}$, widely used in [4,19,27,32,37,40,57] Stepwise reward r_t is the immediate reward generated by an environment in each step, which is generally used in most RL algorithms [24,43]. StARformer uses r_t by default, guided by the motivation of modeling single-step transitions. No reward at-all corresponds to imitation (behavior cloning).

Results are shown in Fig. 13. StARformer and DT behaves differently when reward settings are varied. Both methods show performance gains with reward. Also, StARformer performs similarly regardless of RTG or stepwise reward, whereas DT relies more on RTG, and StARformer-None can still outperform DT-RTG, even without reward. These observations tell sequence modeling can even work on state-action-only trajectories when the model has enough capacity. Such observation is consistent with Dreamerv2 [22], where no-reward setting performs as well as having reward due to the strong dynamics model.

7 Conclusion

In this work, we introduce StARformer, which models strong local relations explicitly (Step Transformer) to help improve the long-term sequence modeling (Sequence Transformer) in Visual RL. Our extensive empirical results show how the learned StAR-representations help our model to outperform the baseline. We further demonstrate that our method successfully models trajectories, with an emphasis on long sequences.

Acknowledgements. We thank members in Robotics Lab at Stony Brook for valuable discussions. This work was supported by Institute of Information & communications Technology Planning & Evaluation (IITP) grant funded by the Ministry of Science and ICT (No.2018-0-00205, Development of Core Technology of Robot Task-Intelligence for Improvement of Labor Condition.

References

1. Abbeel, P., Ng, A.Y.: Apprenticeship learning via inverse reinforcement learning. In: Proceedings of the International Conference on Machine Learning (ICML), p. 1 (2004)
2. Agarwal, R., Schuurmans, D., Norouzi, M.: An optimistic perspective on offline reinforcement learning. In: Proceedings of the International Conference on Machine Learning (ICML), pp. 104–114, July 2020
3. Agarwal, R., Schuurmans, D., Norouzi, M.: An optimistic perspective on offline reinforcement learning. In: Proceedings of the International Conference on Machine Learning (ICML), pp. 104–114. PMLR (2020)
4. Andrychowicz, M., et al.: Hindsight experience replay. In: Advances in Neural Information Processing Systems (NeurIPS) (2017)
5. Arnab, A., Dehghani, M., Heigold, G., Sun, C., Lučić, M., Schmid, C.: ViViT: a video vision transformer. In: Proceedings of the International Conference on Computer Vision (ICCV), October 2021
6. Bellemare, M.G., Naddaf, Y., Veness, J., Bowling, M.: The arcade learning environment: an evaluation platform for general agents. J. Artif. Intell. Res. **47**(1), 253–279 (2013)
7. Bertasius, G., Wang, H., Torresani, L.: Is space-time attention all you need for video understanding? In: Proceedings of the International Conference on Machine Learning (ICML), July 2021
8. Brockman, G., et al.: OpenAI gym (2016). arXiv:1606.01540
9. Chen, L., et al.: Decision transformer: reinforcement learning via sequence modeling. In: Advances in Neural Information Processing Systems (NeurIPS), December 2021
10. Chen, M., et al.: Generative pretraining from pixels. In: Proceedings of the International Conference on Machine Learning (ICML), pp. 1691–1703, July 2000
11. Cheng, J., Dong, L., Lapata, M.: Long short-term memory-networks for machine reading (2016). arXiv:1601.06733
12. Choromanski, K., et al.: Rethinking attention with performers. In: Proceedings of the International Conference on Learning Representations (ICLR), April 2020

13. Dabney, W., Rowland, M., Bellemare, M., Munos, R.: Distributional reinforcement learning with quantile regression. In: Proceedings of the AAAI Conference on Artificial Intelligence (AAAI), vol. 32 (2018)
14. Dai, R., Das, S., Kahatapitiya, K., Ryoo, M.S., Bremond, F.: MS-TCT: multi-scale temporal convtransformer for action detection. In: Proceedings of the IEEE/CVF Conference on Computer Vision and Pattern Recognition (CVPR), pp. 20041–20051 (2022)
15. Dai, Z., Liu, H., Le, Q.V., Tan, M.: CoAtNet: marrying convolution and attention for all data sizes. In: Advances in Neural Information Processing Systems (NeurIPS), December 2021
16. Dasari, S., Gupta, A.: Transformers for one-shot visual imitation. In: Conference on Robot Learning (CoRL) (2020)
17. Devlin, J., Chang, M.W., Lee, K., Toutanova, K.: BERT: pre-training of deep bidirectional transformers for language understanding (2019). arXiv:1810.04805
18. Dosovitskiy, A., et al.: An image is worth 16x16 words: transformers for image recognition at scale. In: Proceedings of the International Conference on Learning Representations (ICLR), April 2020
19. Eysenbach, B., Geng, X., Levine, S., Salakhutdinov, R.: Rewriting history with inverse RL: hindsight inference for policy improvement. In: Advances in Neural Information Processing Systems (NeurIPS) (2020)
20. Furuta, H., Matsuo, Y., Gu, S.S.: Distributional decision transformer for hindsight information matching. In: Proceedings of the International Conference on Learning Representations (ICLR) (2022)
21. Haarnoja, T., Zhou, A., Abbeel, P., Levine, S.: Soft actor-critic: off-policy maximum entropy deep reinforcement learning with a stochastic actor. In: Proceedings of the International Conference on Machine Learning (ICML), pp. 1861–1870, July 2018
22. Hafner, D., Lillicrap, T., Norouzi, M., Ba, J.: Mastering atari with discrete world models. arXiv preprint arXiv:2010.02193 (2020)
23. Han, K., Xiao, A., Wu, E., Guo, J., Xu, C., Wang, Y.: Transformer in transformer. In: Advances in Neural Information Processing Systems (NeurIPS), December 2021
24. Hessel, M., et al.: Rainbow: combining improvements in deep reinforcement learning. In: Proceedings of the AAAI Conference on Artificial Intelligence (AAAI) (2018)
25. Ho, J., Ermon, S.: Generative adversarial imitation learning. In: Advances in Neural Information Processing Systems (NeurIPS), December 2016
26. Janner, M., Li, Q., Levine, S.: Offline reinforcement learning as one big sequence modeling problem. In: Advances in Neural Information Processing Systems (NeurIPS), December 2021
27. Kaelbling, L.P.: Learning to achieve goals. In: International Joint Conference on Artificial Intelligence (IJCAI) (1993)
28. Kahatapitiya, K., Ryoo, M.S.: Swat: spatial structure within and among tokens. arXiv preprint arXiv:2111.13677 (2021)
29. Kitaev, N., Kaiser, L., Levskaya, A.: Reformer: the efficient transformer. In: Proceedings of the International Conference on Learning Representations (ICLR), May 2019
30. Konda, V.R., Tsitsiklis, J.N.: Actor-critic algorithms. In: Advances in Neural Information Processing Systems (NeurIPS), December 2000
31. Kumar, A., Fu, J., Soh, M., Tucker, G., Levine, S.: Stabilizing off-policy Q-learning via bootstrapping error reduction. In: Advances in Neural Information Processing Systems (NeurIPS), vol. 32 (2019)

32. Kumar, A., Peng, X.B., Levine, S.: Reward-conditioned policies. arXiv preprint arXiv:1912.13465 (2019)
33. Kumar, A., Zhou, A., Tucker, G., Levine, S.: Conservative q-learning for offline reinforcement learning. Adv. Neural Inf. Process. Syst. (NeurIPS) **33**, 1179–1191 (2020)
34. Laskin, M., Srinivas, A., Abbeel, P.: Curl: contrastive unsupervised representations for reinforcement learning. In: Proceedings of the International Conference on Machine Learning (ICML), pp. 5639–5650. PMLR (2020)
35. Laskin, M., Lee, K., Stooke, A., Pinto, L., Abbeel, P., Srinivas, A.: Reinforcement learning with augmented data. Adv. Neural. Inf. Process. Syst. **33**, 19884–19895 (2020)
36. Levine, S., Kumar, A., Tucker, G., Fu, J.: Offline reinforcement learning: tutorial, review, and perspectives on open problems (2020). arXiv:2005.01643
37. Li, A.C., Pinto, L., Abbeel, P.: Generalized hindsight for reinforcement learning. In: Advances in Neural Information Processing Systems (NeurIPS) (2020)
38. Li, X., Shang, J., Das, S., Ryoo, M.S.: Does self-supervised learning really improve reinforcement learning from pixels? arXiv preprint arXiv:2206.05266 (2022)
39. Lin, Z., et al.: A structured self-attentive sentence embedding (2017). arXiv:1703.03130
40. Liu, Z., et al.: Swin transformer: hierarchical vision transformer using shifted windows. In: Proceedings of the International Conference on Computer Vision (ICCV), pp. 10012–10022, October 2021
41. Liu, Z., et al.: Video swin transformer (2021). arXiv:2106.13230
42. Liu, Z., Mao, H., Wu, C.Y., Feichtenhofer, C., Darrell, T., Xie, S.: A convnet for the 2020s. arXiv preprint arXiv:2201.03545 (2022)
43. Mnih, V., et al.: Playing atari with deep reinforcement learning (2013). arXiv:1312.5602
44. Mnih, V., et al.: Human-level control through deep reinforcement learning. Nature **518**(7540), 529–533 (2015)
45. Neimark, D., Bar, O., Zohar, M., Asselmann, D.: Video transformer network (2021). arXiv:2102.00719
46. Neimark, D., Bar, O., Zohar, M., Asselmann, D.: Video transformer network. In: Proceedings of the International Conference on Computer Vision (ICCV), pp. 3163–3172 (2021)
47. Ng, A.Y., Russell, S.J., et al.: Algorithms for inverse reinforcement learning. In: Proceedings of the International Conference on Machine Learning (ICML), vol. 1, p. 2 (2000)
48. Parikh, A.P., Täckström, O., Das, D., Uszkoreit, J.: A decomposable attention model for natural language inference (2016). arXiv:1606.01933
49. Pong, V., Gu, S., Dalal, M., Levine, S.: Temporal difference models: model-free deep RL for model-based control. In: Proceedings of the International Conference on Learning Representations (ICLR) (2018)
50. Radford, A., Narasimhan, K., Salimans, T., Sutskever, I.: Improving language understanding by generative pre-training (2018)
51. Radford, A., Wu, J., Child, R., Luan, D., Amodei, D., Sutskever, I.: Language models are unsupervised multitask learners. OpenAI Blog **1**(8), 9 (2019)
52. Reed, S., et al.: A generalist agent. arXiv preprint arXiv:2205.06175 (2022)
53. Rummery, G.A., Niranjan, M.: On-line Q-learning using connectionist systems, vol. 37. Citeseer (1994)

54. Ryoo, M.S., Piergiovanni, A., Arnab, A., Dehghani, M., Angelova, A.: TokenLearner: adaptive space-time tokenization for videos. In: Advances in Neural Information Processing Systems (NeurIPS), December 2021
55. Shang, J., Das, S., Ryoo, M.S.: Learning viewpoint-agnostic visual representations by recovering tokens in 3D space. arXiv preprint arXiv:2206.11895 (2022)
56. Shang, J., Ryoo, M.S.: Self-supervised disentangled representation learning for third-person imitation learning. In: IEEE/RSJ International Conference on Intelligent Robots and Systems (IROS), pp. 214–221. IEEE (2021)
57. Srivastava, R.K., Shyam, P., Mutz, F., Jaśkowski, W., Schmidhuber, J.: Training agents using upside-down reinforcement learning. arXiv preprint arXiv:1912.02877 (2019)
58. Sutton, R.S.: Learning to predict by the methods of temporal differences. Mach. Learn. **3**(1), 9–44 (1988)
59. Tang, Y., Ha, D.: The sensory neuron as a transformer: permutation-invariant neural networks for reinforcement learning. arXiv preprint arXiv:2109.02869 (2021)
60. Tesauro, G., et al.: Temporal difference learning and TD-Gammon. Commun. ACM **38**(3), 58–68 (1995)
61. Torabi, F., Warnell, G., Stone, P.: Generative Adversarial Imitation from Observation (2019). arXiv:1807.06158
62. Touvron, H., Cord, M., Sablayrolles, A., Synnaeve, G., Jégou, H.: Going deeper with image transformers. In: Proceedings of the International Conference on Computer Vision (ICCV), pp. 32–42, October 2021
63. Tunyasuvunakool, S., et al.: dm_control: software and tasks for continuous control. Softw. Impacts **6**, 100022 (2020)
64. Vaswani, A., et al.: Attention is all you need. In: Advances in Neural Information Processing Systems (NeurIPS), December 2017
65. Wang, S., Li, B.Z., Khabsa, M., Fang, H., Ma, H.: Linformer: self-attention with linear complexity (2020). arXiv:2006.04768
66. Watkins, C.J., Dayan, P.: Q-learning. Mach. Learn. **8**(3–4), 279–292 (1992)
67. Xiao, T., Singh, M., Mintun, E., Darrell, T., Dollár, P., Girshick, R.: Early convolutions help transformers see better. Adv. Neural Inf. Process. Syst. (NeurIPS) **34**, 30392–30400 (2021)
68. Yang, J., et al.: Focal self-attention for local-global interactions in vision transformers (2021). arXiv:2107.00641
69. Yarats, D., Fergus, R., Lazaric, A., Pinto, L.: Mastering visual continuous control: improved data-augmented reinforcement learning. arXiv preprint arXiv:2107.09645 (2021)
70. Yarats, D., Kostrikov, I., Fergus, R.: Image augmentation is all you need: regularizing deep reinforcement learning from pixels. In: Proceedings of the International Conference on Learning Representations (ICLR) (2020)
71. Yarats, D., Zhang, A., Kostrikov, I., Amos, B., Pineau, J., Fergus, R.: Improving sample efficiency in model-free reinforcement learning from images. In: Proceedings of the AAAI Conference on Artificial Intelligence (AAAI), pp. 10674–10681, May 2021
72. Zheng, Q., Zhang, A., Grover, A.: Online decision transformer (2022)

TIDEE: Tidying Up Novel Rooms Using Visuo-Semantic Commonsense Priors

Gabriel Sarch[1(✉)], Zhaoyuan Fang[1], Adam W. Harley[1], Paul Schydlo[1], Michael J. Tarr[1], Saurabh Gupta[2], and Katerina Fragkiadaki[1]

[1] Carnegie Mellon University, Pittsburgh, USA
gsarch@andrew.cmu.edu
[2] University of Illinois at Urbana-Champaign, Champaign, USA

Abstract. We introduce TIDEE, an embodied agent that tidies up a disordered scene based on learned commonsense object placement and room arrangement priors. TIDEE explores a home environment, detects objects that are out of their natural place, infers plausible object contexts for them, localizes such contexts in the current scene, and repositions the objects. Commonsense priors are encoded in three modules: i) visuo-semantic detectors that detect out-of-place objects, ii) an associative neural graph memory of objects and spatial relations that proposes plausible semantic receptacles and surfaces for object repositions, and iii) a visual search network that guides the agent's exploration for efficiently localizing the receptacle-of-interest in the current scene to reposition the object. We test TIDEE on tidying up disorganized scenes in the AI2THOR simulation environment. TIDEE carries out the task directly from pixel and raw depth input without ever having observed the same room beforehand, relying only on priors learned from a separate set of training houses. Human evaluations on the resulting room reorganizations show TIDEE outperforms ablative versions of the model that do not use one or more of the commonsense priors. On a related room rearrangement benchmark that allows the agent to view the goal state prior to rearrangement, a simplified version of our model significantly outperforms a top-performing method by a large margin. Code and data are available at the project website: https://tidee-agent.github.io/.

1 Introduction

For robots to operate in home environments and assist humans in their daily lives, they need to be more than step-by-step instruction followers: they need to proactively take action in circumstances that violate expectations, priors, and norms, and effectively interpret incomplete or noisy instructions by human users. Consider Fig. 1. A robot should realize the remote is out-of-place, should be able to infer alternative plausible repositions, and tidy-up the scene by rearranging

---

Supplementary Information The online version contains supplementary material available at https://doi.org/10.1007/978-3-031-19842-7_28.

S. Avidan et al. (Eds.): ECCV 2022, LNCS 13699, pp. 480–496, 2022.
https://doi.org/10.1007/978-3-031-19842-7_28

Fig. 1. TIDEE is an embodied agent that tidies up disorganized scenes using commonsense knowledge of object placements and room arrangements. (a) It explores the scene to detect out-of-place (OOP) objects (in this case the remote control). (b) It then infers plausible receptacles (the coffee table) through graph inference over a neural graph memory of objects and relations. (c–d) It then searches for the inferred receptacle (the coffee table) guided by a visual search network and repositions the object.

the objects to their regular locations. Such understanding would also permit the robot to follow incomplete instructions from human users, such as *"put the remote away"*. For this, a robot needs to have commonsense knowledge regarding contextual, object-object, and object-room spatial relations.

What is the form of this commonsense knowledge and how can it be acquired? There are two sources of commonsense knowledge: i) communication of such knowledge via natural language, for example, *"the lamp should be placed on the bed stand"*, and ii) acquisition of such knowledge via visually observing the world and encoding statistical relationships between objects and places. These two sources are complementary. Commonsense in natural language is easy to specify and modify through instruction, while commonsense through visual observation is scalable and often more expressive. Consider, for example, tall yellow IKEA lamps that are often placed on the floor, while shorter lamps are usually placed on bed stands and are appropriately centered and oriented towards the bed. In this example, object contextual relationships depend on more than the category label "lamp"; they depend on sub-categorical information, which is easily encoded in the visual features of the objects [25].

We introduce Teachable Interactive Decluttering Embodied Explorer (TIDEE), which combines semantic and visual commonsense knowledge with embodied components to tidy up disorganized home environments it has never seen before, from raw RGB-D input. TIDEE explores a home environment to detect objects that are not in their normal locations (that therefore need to be repositioned), as shown in Fig. 1(a). When an out-of-place (OOP) object is detected, TIDEE infers plausible receptacles for the object to be placed onto, through graph inference over the union of a neural memory graph of objects and spatial relations and the scene graph of the room at hand (Fig. 1(b)). It then actively explores the scene to find instances of the predicted receptacle category guided by a visual search network, and repositions the detected out-of-place objects (Fig. 1(c–d)).[1] TIDEE uses both visual features and semantic information to encode commonsense knowledge. This knowledge is encoded in the weights of the out-of-place detectors, the neural memory graph weights, and the visual search network weights, and is learned end-to-end to optimize objectives of the rearrangement task, such as classifying out-of-place objects, inferring plausible repositions, and efficiently locating an object of interest. To the best of our knowledge, this is the first work that attempts to tidy up novel room environments directly from pixel and depth input, without any explicit instructions for object placements, relying instead on learned prior knowledge to solve the task.

We test TIDEE in tidying up kitchens, living rooms, bathrooms and bedrooms in the AI2THOR simulation environment [23]. We generate untidy scenes by applying random forces that push or pull objects within each room. We show that human evaluators prefer TIDEE's rearrangements more often than those obtained by baselines or ablative versions of our model that do not use semantics for out-of-place detection, do not use a learnable graph memory (defaulting instead to most common placement), or do not have neural guidance during object search. We further show that TIDEE can be adapted to respect preferences of users by fine-tuning its out-of-place visuo-semantic object classifier based on individual instructions. Finally, we test a reduced version of TIDEE on the recent scene rearrangement benchmark [3,37], where an AI agent is tasked to reposition the objects to bring the scene to a desirable target configuration. TIDEE outperforms the current state of the art. We attribute TIDEE's excellent performance to the modular organization of its architecture and the object-centric scene representation TIDEE uses to reason about rearrangements.

2 Related Work

Embodied AI. The development of learning-based embodied AI agents has made significant progress across a wide variety of tasks, including: scene rearrangement [3,17,37], object-goal navigation [1,6,8,19,40,42], point-goal navigation [1,19,30,31,39], scene exploration [7,10], embodied question answering [12, 18], instructional navigation [2,34], object manipulation [14,43], home task completion with explicit instructions [27,34,35], active visual learning [9,15,20,38], and collaborative task completion with agent-human conversations [29]. While these works have driven much progress in embodied AI, ours is the first agent to tackle the task of tidying up rooms, which requires commonsense reasoning about whether or not an object is out of place, and inferring where it belongs in the context of the room. Progress in embodied AI has been accelerated tremendously through the availability of high visual fidelity simulators, such as, Habitat [31], GibsonWorld [33], ThreeDWorld [16], and AI2THOR [23]. Our work builds upon

[1] We follow the terminology from AI2THOR [23] and define a receptacle as a type of object that can contain or support other objects. Sinks, refrigerators, cabinets, and tabletops are some examples of receptacles.

AI2THOR by relying on the (approximate) dynamic manipulation the simulator enables for household objects.

Representing Commonsense Knowledge Regarding Object Spatial Relations. Visual commonsense knowledge is often represented in terms of a knowledge graph, namely, a graph of visual entity nodes (objects, parts, attributes) where edge types represent pairwise relationships between entities. Knowledge graphs have been successfully used in visual classification and detection [11,26], zero-shot classification of images [36], object goal navigation [42], and image retrieval [22].

Closest to our work is the work of Yang et al. [42] where a knowledge graph is used to help an agent navigate to semantic object goals. While in the knowledge graph of Yang et al. [42] each node stands for an object category described by its semantic embedding, in our case each node is an object instance described by both semantic and visual features, similar to the earlier work of Malisiewicz and Efros on visual Memex [25]. Moreover, we consider tidying up rooms, where navigation to semantic goals is one submodule of what the agent needs to do. Lastly, while [42] maps images to actions directly trained with reinforcement learning, and graph indexing provides simply an additional embedding to concatenate to the agent's state, our model is modular and hierarchical, using a "theory" of out-of-place objects, inferring regular object placements, exploration to localize placements in the scene, and then taking actions to achieve the inferred object re-arrangement. We show that TIDEE outperforms non-modular image-to-action mapping agents in the scene re-arrangement benchmark in Sect. 4.5.

3 Teachable Interactive Decluttering Embodied Explorer (TIDEE)

The architecture of TIDEE is illustrated in Fig. 2. The agent navigates a home environment and receives RGB-D images at each time step alongside egomotion information. We consider both groundtruth depth and egomotion, as well as noisy versions of both, and estimated depth in our experimental section. The agent builds geometrically consistent spatial 2D and 3D maps of the environment by fusing RGB-D input, following prior works [7] (Sect. 3.1). TIDEE detects objects and classifies them as in or out-of-place (OOP) using a combination of visual and semantic features (Sect. 3.2). When an OOP object is detected, the agent infers plausible object context (i.e., plausible receptacle categories for the OOP object to be repositioned on) through inference over a memory graph of objects and relations (Memex) and the current scene graph (Sect. 3.3). The agent then searches the current scene to find instances of the receptacle category and a visual search network guides its exploration by proposing locations in the scene to visit (Sect. 3.4). Once the receptacle is detected, the agent places the OOP object on it. Navigation actions move the agent in discrete steps. For picking up and placing objects, the agent must specify an object to interact with via a relative coordinate (x, y) in the (ego-centric) frame.

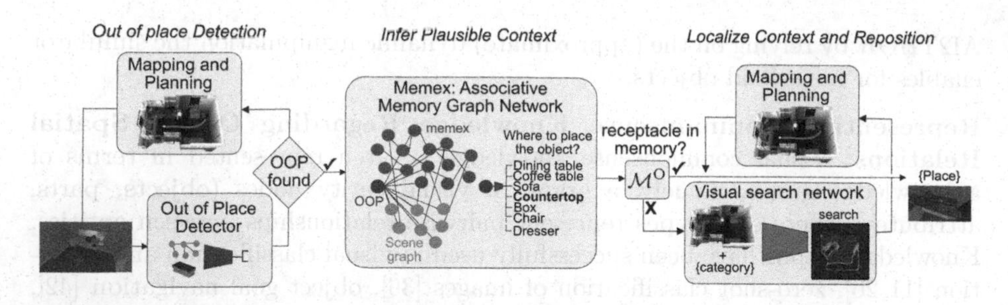

Fig. 2. Architecture of TIDEE. TIDEE explores the scene, detects objects and classifies whether they are in-place or out-of-place. If an object is out-of-place, TIDEE uses graph inference in its joint external graph memory and scene graph to infer plausible receptacle categories. It then explores the scene guided by a visual search network that suggests where instances of a receptacle category may be found, given the scene spatial semantic map. TIDEE iterates the steps above until it cannot detect any more OOP objects, in which case it concludes that the room has been tidied up.

3.1 Background: Semantic 3D Mapping

TIDEE builds 3D semantic maps of the home environment it visits augmented with 3D object detection centroids. These maps are used to infer spatial relations among objects and to guide exploration to objects-of-interest. Specifically, TIDEE maintains two spatial visual maps of the environment that it updates at each time step from the input RGB-D stream, similar to previous works [8]: i) a 2D overhead occupancy map $\mathbf{M}_t^{\text{2D}} \in \mathbb{R}^{H \times W}$ and, ii) a 3D occupancy and semantics map $\mathbf{M}_t^{\text{3D}} \in \mathbb{R}^{H \times W \times D \times K}$, where K is the number of semantic object categories; we use $K = 116$. The $\mathbf{M}^{\text{2D}}$ maps is used for exploration and navigation in the environment. More details on our exploration and planning strategy can be found in the supplementary.

We detect objects from K semantic object categories in each input RGB image using the state-of-the-art d-DETR detector [45], pretrained on the MS-COCO datasets [24] and finetuned on images from the AI2THOR training houses. We obtain 3D object centroids by using the depth input to map detected 2D object bounding boxes into a 3D box centroids. We add these in the 3D semantic map with one channel per semantic class, similar to Chaplot et al. [9], but in 3D as opposed to a 2D overhead map. We did not use 3D object detectors directly because we found that 2D object detectors are more reliable than 3D ones likely because of the tremendous pretraining in large-scale 2D object detection datasets, such as MS-COCO [24]. Finally, to create the 3D maps $\mathbf{M}^{\text{3D}}$, we concatenate the 3D occupancy maps with the 3D semantic maps.

We further maintain an object memory $\mathcal{M}^{\text{O}}$ as a list of object detection 3D centroids and their predicted semantic category labels $\mathcal{M}^{\text{O}} = \{[(X, Y, Z)_i, \ell_i \in \{1 \dots K\}], i = 1 \dots N\}$, where N is the number of objects detected thus far. The object centroids are expressed with respect to the coordinate system of the agent, and, similar to the semantic maps, are updated over time using egomotion.

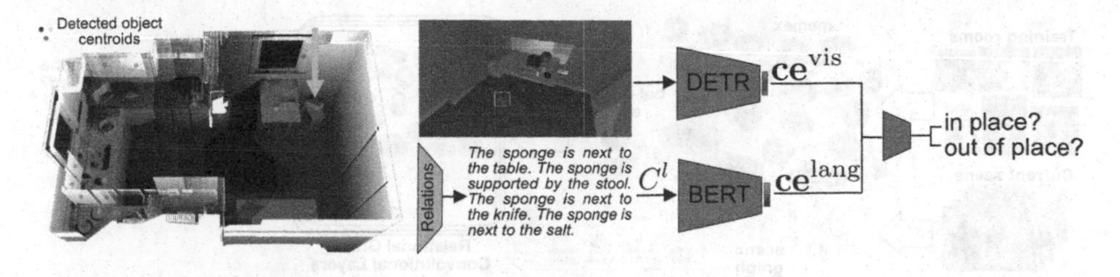

Fig. 3. Out-of-place objection classification using spatial language description features $\mathbf{ce}^{\text{lang}}$ and visual features $\mathbf{ce}^{\text{vis}}$.

3.2 Detecting Out-of-Place Objects

TIDEE detects objects and classifies whether each one is in or out-of-place (OOP) using both visual object features and language descriptions of the object's spatial relations with its surrounding objects, such as *"The alarm clock is on the sofa. The alarm clock is next to the coffee table."* We train three OOP classifiers: one that relies only on visual features, one that relies only on language descriptions of the relations of the object with its surroundings that can more easily adapt to user preferences, and one that fuses both visual and language features, as shown in Fig. 3.

The visual OOP classifier (`dDETR-OOP`) builds upon our d-DETR detector. Specifically, we augment our d-DETR detector with a second decoding head and jointly train it under the tasks of localizing objects and predicting their semantic categories, as well as their in or out-of-place status. We consider the query embedding of the d-DETR decoder as relevant visual features $\mathbf{ce}^{\text{vis}}$ for OOP classification.

The language OOP classifier (`BERT-OOP`) infers the relations of the detected object to surrounding objects and describes them in language form. We consider the following spatial relations: (i) *A supported-by B*, where B is a receptacle class, (ii) *A next-to B*, *A closest-to B*. We detect these pairwise relations using Euclidean distances on detected 3D object centroids in the object memory $\mathcal{M}^{\text{O}}$. For more details on our object spatial relation detection, please see the supplementary. We represent all detected pairwise relations as sentences of the form "The {detection class} is {relation} the {related class}", and concatenate the sentences to form a paragraph, as shown in Fig. 3. We map this object spatial context description paragraph into a neural vector $\mathbf{ce}^{\text{lang}}$ for the relation set given by the [CLS] token from the BERT model [13] pretrained on a language masking task and then trained for plausible/non-plausible classification in our training set. A benefit of the language OOP classifier is that it can adapt to user's specifications without any visual exemplars of plausible/implausible object arrangements. Consider, for example, the instruction *"I want my alarm clock on the bed stand"*. Using such instruction, we generate positive and negative descriptions of in and out-of-place alarm clocks by adapting the preference into a positive sample (e.g. "Alarm clock supported-by the bed stand"), and

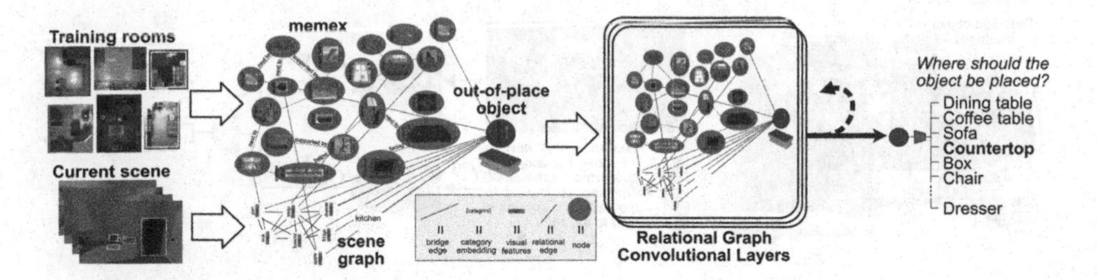

Fig. 4. Graph inference over the union of the Memex graph and the current scene graph infers plausible receptacle categories for an out-of-place object.

taking relations in the training set that include the alarm clock and a different receptacle class as negative samples ("Alarm clock supported-by the desk").

The multimodal classifier (`dDETR+BERT-OOP`) concatenates $\mathbf{ce}^{\text{vis}}$ and $\mathbf{ce}^{\text{lang}}$ as input to predict OOP classification labels for the detected object.

3.3 Inferring Plausible Object Contexts with a Neural Associative Graph Memory

Once an OOP object is detected and picked up, TIDEE infers a plausible placement location for the object in the current scene. As shown in Fig. 4, TIDEE includes a neural graph module which is trained to predict plausible object placement proposals of OOP objects by passing information between the OOP object to be placed, a memory graph encoding plausible contextual relations from training scenes, and a scene graph encoding the object-relation configuration in the current scene. Message passing is trained end-to-end to predict one of the possible receptacle classes in AI2THOR to place the OOP object on.

We instantiate an OOP node, denoted n_{OOP}, consisting of the detected OOP object for which we want to infer a plausible receptacle category by concatenating the ROI-pooled detector backbone features and a category embedding of the predicted object category.

The structure of the memory graph (nodes and edges) is instantiated from 5 out of 20 training houses. Each object in the scene is given a node in the graph that consists of a category embedding and ROI-pooled detector backbone features using the bounding box of the object at a nearby egocentric viewpoint. Edge weights in the memory graph correspond to spatial relations detected between pairs of object instances that are within a distance threshold. We consider six spatial relations and corresponding edge types: *above*, *below*, *next to*, *supported by*, *aligned with*, and *facing* [21]. We infer these using spatial relation classifiers that operate on ground-truth 3D oriented bounding boxes. Though the graph may contain noisy, non-important edges between object instances, for example, *"the coffee table is next to the bed"* which may introduce a spurious dependence between a bed and a coffee table instance, the edge kernel weights are trained end-to-end to infer plausible receptacles for OOP objects, and thus graph inference can learn to ignore such spurious edges. We call our memory

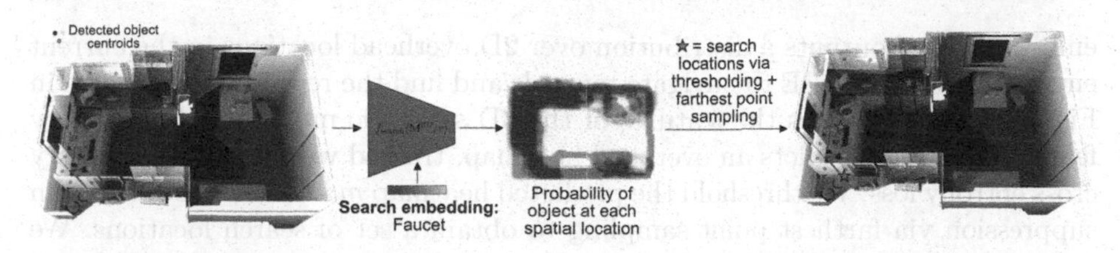

Fig. 5. The visual search network conditions on an object category of interest, and proposes locations for the agent to visit in the scene to find instances of that category.

graph "Memex" to highlight that nodes represent object instances, similar to [25], and not object categories as in previous works [42].

The structure of the scene graph [22] is instantiated from observations obtained while mapping the current scene, as in Sect. 3.1. Nodes in the scene graph represent ROI-pooled features and category embeddings of objects detected by the agent in $\mathcal{M}^{\mathrm{O}}$. We include an additional node for the room type. We fully-connect all nodes within the scene graph. Compared to the Memex graph, we do not include separate edge weights for relations as most of the Memex relations require accurate 3D bounding boxes that we do not have access to at inference time.

We add "bridge edges", as additional learnable edge weights, between nodes in the scene graph and Memex nodes with the same category, following [44], to allow information to flow between the current scene and the memory graph. We further connect n_{OOP} to all current scene nodes and to the room type node. After message passing, we pass the updated n_{OOP} through an MLP to get logits for each possible receptacle class in AI2THOR.

The network is trained for predicting plausible receptacles for OOP objects in 15 training houses. We use 15/20 houses to train the weights so as to not overlap with the houses used for the memory graph. Relation-specific edge weights are learned end-to-end by Relational Graph Convolutions (rGCN) [32]. We supervise the network via a cross-entropy loss using ground-truth receptacle categories for each "pickupable" object from the AI2THOR original scene configurations. More details of our graph inference can be found in the supplementary.

3.4 Intelligent Exploration Using a Visual Search Network

After inferring a target receptacle category, TIDEE localizes it in the scene and places the OOP object on top of it. In the case that instances of the target receptacle category have already been detected in the scene, our agent navigates to the corresponding instance using its navigation path planning controllers from Sect. 3.1. In the case that the target receptacle category has not yet been detected, our model predicts plausible locations to search for the receptacle using a category-conditioned visual search network $f_{\text{search}}(\mathrm{M}^{3\mathrm{D}}, r)$.

The visual search network $f_{\text{search}}(\mathrm{M}^{3\mathrm{D}}, r)$ takes as input a 3D spatial semantic map $\mathbf{M}^{3\mathrm{D}}$ and a receptacle category label r represented by a learned category embedding and outputs a distribution over 2D overhead locations in the current environment for TIDEE to navigate towards and find the receptacle, as shown in Fig. 5. f_{search} convolves the features of the 3D semantic map with the category features of r and predicts an overhead heatmap, trained with a standard binary cross entropy loss. We threshold the predicted heatmap m and use non-maximum suppression via farthest point sampling to obtain a set of search locations. We rank the search locations based on their score and visit them sequentially until the target receptacle category is detected with high probability. Further architectural details for f_{search} can be found in the supplementary.

4 Experiments

We test TIDEE on reorganizing untidy rooms in the test houses of the AI2THOR simulation environment. Our experiments aim to answer the following questions:

(i) How well does TIDEE perform in tidying up scenes? Sect. 4.2
(ii) How much does the combination of visual and semantic features help in detecting out-of-place objects over visual features alone? Sect. 4.3
(iii) How much does exploration guided by the proposed visual search network improve upon random exploration for detecting objects of interest? Sect. 4.4
(iv) How well does TIDEE perform in the task of scene rearrangement [3]—which requires memorization of a specific prior scene configuration? Sect. 4.5
(v) How well can TIDEE adapt zero-shot to human instructions and alter placement priors accordingly? Sect. 4.6

4.1 Tidying-Up Task Definition

Dataset. We create untidy scenes by selecting a subset of "pickupable" objects[2]. We displace each object from its default location by moving the object to a random location in the scene and either dropping the object or applying a force in a random direction and allowing the AI2THOR physics engine to resolve the object's end location. We consider all available room types, namely bedrooms, living rooms, kitchens and bathrooms. We generate 8000 training, 200 validation, and 100 testing messy configurations. The goal of the agent is to manipulate the messy objects back to plausible locations within the room. An episode ends once the agent executes the "done" action or a maximum of 1000 steps have been taken. For more details on the task and dataset, please see the supplementary.

4.2 Object Repositioning Evaluation

We have TIDEE and all baselines perform the tidy task to detect out-of-place objects and reposition them within the scene.

[2] Pickupable objects are a predefined set of 62 object classes in AI2THOR [23] that are able to be picked up and repositioned by the agent, such apple, book, and laptop.

Evaluation Metrics. Quantitative evaluation of object repositioning is difficult: an object may have multiple plausible locations in a scene, and therefore measuring the distance from a single initial ground-truth 3D location is usually not reflective of performance. We thus evaluate the plausibility of object repositions of our model from those of baseline models by querying human evaluators in Amazon Mechanical Turk (AMT). Given two candidate repositions by for the same object TIDEE and a baseline, we ask human evaluators to select the one they find most plausible. We include the AMT interface we used in the supplementary.

Table 1. Percent of human evaluators that prefer TIDEE **object repositions versus baselines.** Reported is mean and standard error across subjects ($n=5$). All preferences are significantly above chance (*$p<0.05$, **$p<0.01$, Binomial test). Bold indicates higher preference for TIDEE.

TIDEE vs `CommonMemory`	**54.30 ± 3.32***
TIDEE vs `WithoutMemex`	**54.32 ± 4.67***
TIDEE vs `3DSmntMap2Place`	**57.69 ± 1.29****
TIDEE vs `RandomReceptacle`	**64.59 ± 2.94****
TIDEE vs `MessyPlacement`	**92.06 ± 1.57****
TIDEE vs `AI2THORPlacement`	34.00 ± 3.13**

Table 2. Evaluating visual search performance for finding objects of interest in test scenes for TIDEE and an exploration baseline that uses our 2D overhead occupancy maps to propose random search locations [41].

	% Success ↑	Time steps ↓
TIDEE	**72.4**	**88.8**
w/o VSN	64.8	100.9

Baselines. We compare TIDEE against baselines that vary in their way of inferring plausible receptacle categories for repositioning of out-of-place objects. All baselines use the same mapping and planning for navigation, the same multimodal classifier for detecting out-of-place objects (`dDETR+BERT-OOP`), and the visual search network for localizing receptacle instances of a category. We compare placements from TIDEE against the following baselines: (i) `CommonMemory`: A model that considers the most common receptacle in the training set for the out-of-place object category. (ii) `WithoutMemex`: A model that uses the scene graph but not the Memex for graph inference. (iii) `3DSmntMap2Place`: A model that proposes repositioning locations within the current scene by conditioning the visual search network on the category label of the out-of-place object. We threshold all predicted map locations and do farthest point sampling to obtain a set of diverse object placement proposals. The proposals are sorted by confidence value and visited sequentially until any receptacle is found within the local region of the proposed location. (iv) `RandomReceptacle`: A model that selects as the target receptacle the first receptacle detected by a random exploration agent. (v) `AI2THORPlacement`: The location of the OOP object in the original (tidy) AITHOR scene. The default object positions usually follow commonsense priors of scene arrangements. (vi) `MessyPlacement`: The location of the OOP object in the messy scene.

We report human preferences for OOP object repositions for our model versus each of the baselines in Table 1. TIDEE is preferred 54.3% of the time over `CommonMemory`, the most competitive of the baselines. `CommonMemory` does not consider the visual features of the out-of-place object, rather, only its semantic category, and thus cannot reason using sub-categorical information regarding object placements. TIDEE is still preferred 34% of the time over the `AI2THORPlacement` placements indicating that its re-placements are plausible and competitive with an oracle. We note that a perfect model would at best obtain a (50-50) preference compared to these placements provided by the AITHOR environment designers.

4.3 Out-of-Place Detector Evaluation

In this section, we evaluate TIDEE's accuracy for detecting objects in and out-of-place from images collected from the test home environments. An in-place object is one in its default location in the AITHOR scene, while an out-of-place object is one moved out-of-place as defined at the beginning of Sect. 4. We compute average precision (AP) at IOU thresholds of 0.25 and 0.5 for in-place (*IP*) and out-of-place (*OOP*) objects, as well as the meanAP (mAP) for visual only (`dDETR-OOP`), language only (`BERT-OOP`) and multimodal (`dDETR+BERT-OOP`) classifiers described in Sect. 3.2. We also compare against an oracle BERT classifier that assumes access to ground-truth 3D object centroids, bounding boxes, and category labels to detect relations and form descriptive utterances of in and out-of-place objects, which we call `oracle-BERT-OOP`.

We show quantitative comparisons in Table 3. Combining language and visual features performs slightly better than using language or visual features alone for out-of-place object detection. The benefit of the language classifier is that it can be re-trained on-demand to adjust to human instructions without any visual training data, as we explain in Sect. 4.6. The good performance of the oracle BERT classifier suggests that simple relations inferred from accurate 3D centroids likely suffice to classify in- and out-of-place objects in AI2THOR scenes if perception is perfect.

Table 3. Average precision (AP) for in and out-of-place object detection. Combining vision and language features helps detection performance. IP = in place; OOP = out of place.

	$mAP_{0.25}$	$AP^{IP}_{0.25}$	$AP^{OOP}_{0.25}$	$mAP_{0.5}$	$AP^{IP}_{0.5}$	$AP^{OOP}_{0.5}$
`dDETR+BERT-OOP`	**51.09**	**58.41**	**43.78**	**46.26**	**53.64**	**38.88**
`dDETR-OOP`	49.98	57.60	42.37	44.98	52.79	37.17
`BERT-OOP`	31.71	41.13	22.30	25.25	33.79	16.71
`oracle-BERT-OOP`	–	–	–	90.70	96.24	85.16

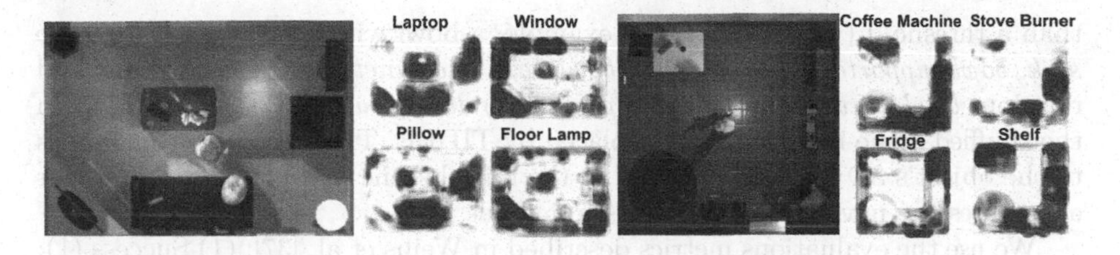

Fig. 6. Visual Search Network predictions encode object location priors for different object categories.

4.4 Visual Search Network Evaluation

In this section, we compare exploration for finding objects of interest in test scenes (one category of the possible 116 per episode) guided by TIDEE's visual search network against an exploration agent that uses the 2D overhead occupancy map and samples unvisited locations to visit, similar to Yamauchi [41]. We adopt the success criteria similar to the object goal navigation [4] and define a successful trial as one where the agent is within a radius of *any* target object category instance and the object is visible within view. We report the percentage of successful episodes performed by the agent and average number of time steps across all episodes in Table 2. If an agent fails an episode, the number of time steps defaults to the maximum allowable steps for each episode (200). TIDEE outperforms the exploration baseline. We show visualizations of the network predictions in Fig. 6, and also in the supplementary.

4.5 Scene Rearrangement Challenge

We test TIDEE to generalize to the recent scene rearrangement benchmark of [37], which considers an AI agent tasked with repositioning objects in a scene in order to match the prior configuration of an identical scene. We consider the two-phase rearrangement setup where in the first "walkthrough" phase, the agent observes a room in its initial configuration, and in the second so called "unshuffle" phase, observes the same room with some objects in new configurations and is tasked to rearrange the room back to its initial configuration. While the challenge considers both rearranging objects to different locations within a room and changing their open/close states, we only consider repositioning of objects because our current model does not handle opening and closing receptacles.

We simplify TIDEE's architecture and only maintain the 2D & 3D occupancy map for navigation and the object memory $\mathcal{M}^O$ for keeping track of objects and their labels over time. We start each phase by exploring the scene and detecting objects. As in Sect. 3.2, we infer the relations for all pickupable objects in the object memory $\mathcal{M}^O$ in the initial and shuffled scenes. We consider an object of the initial scene displaced if its category label has been detected in the shuffled scene and the proportion of inferred relations that are different across the two scenes ({# same relations}/{# different relations}), initial and shuffled, is less than a threshold (we use 0.35). For example, a bowl with relations *bowl next to sink, bowl supported by countertop, bowl next to cabinet* in the initial scene, and relations *bowl next to chair, bowl supported by dining table, bowl next to lamp* in the shuffled scene is considered misplaced by TIDEE. Then, our agent navigates to the object's 3D location detected in the initial scene and places it there. Our agent uses the navigation controllers from Sect. 3.1.

We use the evaluations metrics described in Weihs et al. [37]: (1) Success (↑): the trial is a success if the initial configuration is fully recovered in the unshuffle phase; (2) % FixedStrict (↑): the proportion of objects that were misplaced initially but ended in the correct configuration (if a single in-place object is moved out-of-place, this metric is set to 0); (3) % Energy (↓): the energy is a measure for the similarity of the rearranged scene and the original scene, the lower the more similar (for more details, refer to Weihs et al. [37]); (4) % Misplaced (↓): this metric equals the number of misplaced objects at the end of the episode divided by the number of misplaced objects at the start.

We report TIDEE's performance compared to the top performing methods for the two-phase re-arrangement in Table 4. The model from Weihs et al. [37] trains a reinforcement learning (RL) agent with proximal policy optimization (PPO) and imitation learning (IL) given RGB images as input and includes a semantic mapping component adapted from the Active Neural SLAM model [7]. We additionally show the robustness of TIDEE to realistic sensor measurements. We consider three different versions of TIDEE depending on the source of egomotion and depth information: **(i)** TIDEE uses ground-truth egomotion and depth. **(ii)** TIDEE*+noisy pose* uses ground-truth depth and egomotion from the LocoBot agent in AI2THOR with Gaussian movement noise added to each movement based on measurements of the real LocoBot robot [28] (forward movement $\sigma = 0.005$ m; rotation $\sigma = 0.5°$). **(iii)** TIDEE*+est. depth* uses ground-truth egomotion and depth obtained from the depth prediction model of Blukis et al. [5], which takes in egocentric RGB images. The model is pre-trained and then finetuned on the training scenes of ALFRED [34].

4.6 Updating Placement Priors by Instruction

In this section, we test whether we can alter the OOP classifier on-demand using language specifications for in and out-of-place. Since alarm clocks are often found on desks in AI2THOR, we tested whether augmenting training by pairing the sentence *"alarm clock is supported by desk"* with the out-of-place label would allow us to alter the OOP classifier's output. As shown in Table 5, across three test scenes where alarm clocks are found on desks, the initial OOP object classifier gives us low probability that the alarm clock on the desk is out-of-place. We then add in the language description *"alarm clock is supported by desk"* for a small amount of additional iterations. As shown in Table 5, we find that our procedure suffices to alter the priors of the classifier. We provide additional examples using various object-relation pairings in the supplementary.

Table 4. Test set performance on 2-Phase Rearrangement Challenge (2022). TIDEE outperforms the baseline of [37] even with realistic noise.

	% FixedStrict ↑	% Success ↑	% Energy ↓	% Misplaced ↓
TIDEE	**11.6**	**2.4**	**93**	**94**
TIDEE *+noisy pose*	7.7	1.2	101	101
TIDEE *+est. depth*	5.9	0.6	97	97
TIDEE *+noisy depth*	11.4	2.0	94	95
Weihs *et al.* [37]	0.5	0.0	110	110

Limitations. TIDEE has the following two limitations: i) It does not consider open and closed states of objects, or their 3D pose as part of the messy and reorganization process, which are direct avenues for future work. ii) The messy rooms we create by randomly misplacing objects may not match the messiness in human environments.

Table 5. Altering priors with instructions. The confidence of the out-of-place classifier for clocks found on desks in three test scenes increases when the additional spatial description for indicating out-of-place clocks.

	Before	After
Clock #1	.08	.73
Clock #2	.10	.62
Clock #3	.12	.76

5 Conclusion

We have introduced TIDEE, an agent that tidies up rooms in home environments using commonsense priors encoded in visuo-semantic out of place detectors, visual search networks that guide exploration to objects, and a Memex neural graph memory of objects and relations that infers plausible object context. We evaluate with human evaluators, and find that TIDEE outperforms agents that lack it's modular architecture, as well as modular agents that lack TIDEE's commonsense priors. TIDEE can be instructed in natural language to follow on-demand specifications for object placement. Finally, we establish a new state-of-the-art for the scene rearrangement challenge of Weihs et al. [37] by simplifying TIDEE's architecture to memorize a single scene as opposed to using a prior learned across multiple environments. We believe TIDEE takes an important step towards embodied visuomotor commonsense reasoning.

Acknowledgements. This material is based upon work supported by National Science Foundation grants GRF DGE1745016 & DGE2140739 (GS), a DARPA Young Investigator Award, a NSF CAREER award, an AFOSR Young Investigator Award, and DARPA Machine Common Sense. Any opinions, findings and conclusions or recommendations expressed in this material are those of the authors and do not necessarily reflect the views of the United States Army, the National Science Foundation, or the United States Air Force.

References

1. Anderson, P., et al.: On evaluation of embodied navigation agents. arXiv preprint arXiv:1807.06757 (2018)
2. Anderson, P., et al.: Vision-and-language navigation: interpreting visually-grounded navigation instructions in real environments. In: Proceedings of the IEEE Conference on Computer Vision and Pattern Recognition, pp. 3674–3683 (2018)
3. Batra, D., et al.: Rearrangement: a challenge for embodied AI. arXiv abs/2011.01975 (2020)
4. Batra, D., et al.: Objectnav revisited: on evaluation of embodied agents navigating to objects. arXiv preprint arXiv:2006.13171 (2020)
5. Blukis, V., Paxton, C., Fox, D., Garg, A., Artzi, Y.: A persistent spatial semantic representation for high-level natural language instruction execution. In: Conference on Robot Learning, pp. 706–717. PMLR (2022)
6. Chang, M., Gupta, A., Gupta, S.: Semantic visual navigation by watching Youtube videos. Adv. Neural. Inf. Process. Syst. **33**, 4283–4294 (2020)
7. Chaplot, D.S., Gandhi, D., Gupta, S., Gupta, A., Salakhutdinov, R.: Learning to explore using active neural slam. In: International Conference on Learning Representations (ICLR) (2020)
8. Chaplot, D.S., Gandhi, D.P., Gupta, A., Salakhutdinov, R.R.: Object goal navigation using goal-oriented semantic exploration. In: Advances in Neural Information Processing Systems, vol. 33 (2020)
9. Chaplot, D.S., Jiang, H., Gupta, S., Gupta, A.: Semantic curiosity for active visual learning. In: Vedaldi, A., Bischof, H., Brox, T., Frahm, J.-M. (eds.) ECCV 2020. LNCS, vol. 12351, pp. 309–326. Springer, Cham (2020). https://doi.org/10.1007/978-3-030-58539-6_19
10. Chen, T., Gupta, S., Gupta, A.: Learning exploration policies for navigation. In: International Conference on Learning Representations (2019). https://openreview.net/pdf?id=SyMWn05F7
11. Chen, X., Li, L.J., Fei-Fei, L., Gupta, A.: Iterative visual reasoning beyond convolutions. In: Proceedings of the IEEE Conference on Computer Vision and Pattern Recognition, pp. 7239–7248 (2018)
12. Das, A., Datta, S., Gkioxari, G., Lee, S., Parikh, D., Batra, D.: Embodied question answering. In: Proceedings of the IEEE Conference on Computer Vision and Pattern Recognition, pp. 1–10 (2018)
13. Devlin, J., Chang, M.W., Lee, K., Toutanova, K.: BERT: pre-training of deep bidirectional transformers for language understanding. In: Proceedings of the 2019 Conference of the North American Chapter of the Association for Computational Linguistics: Human Language Technologies, Volume 1 (Long and Short Papers), Minneapolis, Minnesota, pp. 4171–4186. Association for Computational Linguistics, June 2019. https://doi.org/10.18653/v1/N19-1423. https://aclanthology.org/N19-1423
14. Fan, L., et al.: Surreal: open-source reinforcement learning framework and robot manipulation benchmark. In: Conference on Robot Learning, pp. 767–782. PMLR (2018)
15. Fang, Z., Jain, A., Sarch, G., Harley, A.W., Fragkiadaki, K.: Move to see better: self-improving embodied object detection. In: The British Machine Vision Conference (2021)
16. Gan, C., et al.: Threedworld: a platform for interactive multi-modal physical simulation. arXiv preprint arXiv:2007.04954 (2020)

17. Gan, C., et al.: The threedworld transport challenge: a visually guided task-and-motion planning benchmark towards physically realistic embodied AI. In: 2022 International Conference on Robotics and Automation (ICRA), pp. 8847–8854 (2022). https://doi.org/10.1109/ICRA46639.2022.9812329
18. Gordon, D., Kembhavi, A., Rastegari, M., Redmon, J., Fox, D., Farhadi, A.: IQA: visual question answering in interactive environments. In: Proceedings of the IEEE Conference on Computer Vision and Pattern Recognition, pp. 4089–4098 (2018)
19. Gupta, S., Davidson, J., Levine, S., Sukthankar, R., Malik, J.: Cognitive mapping and planning for visual navigation. In: Proceedings of the IEEE Conference on Computer Vision and Pattern Recognition (2017)
20. Haber, N., Mrowca, D., Fei-Fei, L., Yamins, D.L.: Learning to play with intrinsically-motivated self-aware agents. In: 32nd Conference on Neural Information Processing Systems (2018)
21. Hayward, W.G., Tarr, M.J.: Spatial language and spatial representation. Cognition **55**, 39–84 (1995)
22. Johnson, J., et al.: Image retrieval using scene graphs. In: Proceedings of the IEEE Conference on Computer Vision and Pattern Recognition, pp. 3668–3678 (2015)
23. Kolve, E., et al.: AI2-THOR: An Interactive 3D Environment for Visual AI. arXiv (2017)
24. Lin, T.-Y., et al.: Microsoft COCO: common objects in context. In: Fleet, D., Pajdla, T., Schiele, B., Tuytelaars, T. (eds.) ECCV 2014. LNCS, vol. 8693, pp. 740–755. Springer, Cham (2014). https://doi.org/10.1007/978-3-319-10602-1_48
25. Malisiewicz, T., Efros, A.A.: Beyond categories: the visual memex model for reasoning about object relationships. In: NIPS, December 2009
26. Marino, K., Salakhutdinov, R., Gupta, A.: The more you know: using knowledge graphs for image classification. In: 2017 IEEE Conference on Computer Vision and Pattern Recognition (CVPR), pp. 20–28 (2017). https://doi.org/10.1109/CVPR.2017.10
27. Min, S.Y., Chaplot, D.S., Ravikumar, P., Bisk, Y., Salakhutdinov, R.: Film: following instructions in language with modular methods (2021)
28. Murali, A., et al.: Pyrobot: an open-source robotics framework for research and benchmarking. arXiv preprint arXiv:1906.08236 (2019)
29. Padmakumar, A., et al.: Teach: task-driven embodied agents that chat (2021)
30. Ramakrishnan, S.K., Al-Halah, Z., Grauman, K.: Occupancy anticipation for efficient exploration and navigation. In: Vedaldi, A., Bischof, H., Brox, T., Frahm, J.-M. (eds.) ECCV 2020. LNCS, vol. 12350, pp. 400–418. Springer, Cham (2020). https://doi.org/10.1007/978-3-030-58558-7_24
31. Savva, M., et al.: Habitat: a platform for embodied AI research. In: Proceedings of the IEEE/CVF International Conference on Computer Vision, pp. 9339–9347 (2019)
32. Schlichtkrull, M., Kipf, T.N., Bloem, P., van den Berg, R., Titov, I., Welling, M.: Modeling relational data with graph convolutional networks. In: Gangemi, A., et al. (eds.) ESWC 2018. LNCS, vol. 10843, pp. 593–607. Springer, Cham (2018). https://doi.org/10.1007/978-3-319-93417-4_38
33. Shen, B., et al.: iGibson 1.0: a simulation environment for interactive tasks in large realistic scenes. In: 2021 IEEE/RSJ International Conference on Intelligent Robots and Systems. IEEE (2021)
34. Shridhar, M., et al.: Alfred: a benchmark for interpreting grounded instructions for everyday tasks. In: Proceedings of the IEEE/CVF Conference on Computer Vision and Pattern Recognition, pp. 10740–10749 (2020)

35. Suglia, A., Gao, Q., Thomason, J., Thattai, G., Sukhatme, G.S.: Embodied bert: a transformer model for embodied, language-guided visual task completion. In: EMNLP 2021 Workshop on Novel Ideas in Learning-to-Learn through Interaction (2021). https://www.amazon.science/publications/embodied-bert-a-transformer-model-for-embodied-language-guided-visual-task-completion
36. Wang, X., Ye, Y., Gupta, A.: Zero-shot recognition via semantic embeddings and knowledge graphs. In: Proceedings of the IEEE Conference on Computer Vision and Pattern Recognition, pp. 6857–6866 (2018)
37. Weihs, L., Deitke, M., Kembhavi, A., Mottaghi, R.: Visual room rearrangement. In: IEEE/CVF Conference on Computer Vision and Pattern Recognition (CVPR), June 2021
38. Weihs, L., et al.: Learning generalizable visual representations via interactive gameplay. In: International Conference on Learning Representations (2021)
39. Wijmans, E., et al.: DD-PPO: learning near-perfect pointgoal navigators from 2.5 billion frames. In: ICLR (2020)
40. Wortsman, M., Ehsani, K., Rastegari, M., Farhadi, A., Mottaghi, R.: Learning to learn how to learn: self-adaptive visual navigation using meta-learning. In: Proceedings of the IEEE/CVF Conference on Computer Vision and Pattern Recognition, pp. 6750–6759 (2019)
41. Yamauchi, B.: A frontier-based approach for autonomous exploration. In: Proceedings 1997 IEEE International Symposium on Computational Intelligence in Robotics and Automation CIRA 1997. Towards New Computational Principles for Robotics and Automation, pp. 146–151. IEEE (1997)
42. Yang, W., Wang, X., Farhadi, A., Gupta, A., Mottaghi, R.: Visual semantic navigation using scene priors. In: Proceedings of (ICLR) International Conference on Learning Representations, May 2019
43. Yu, T., et al.: Meta-world: a benchmark and evaluation for multi-task and meta reinforcement learning. In: Conference on Robot Learning, pp. 1094–1100. PMLR (2020)
44. Zareian, A., Karaman, S., Chang, S.-F.: Bridging knowledge graphs to generate scene graphs. In: Vedaldi, A., Bischof, H., Brox, T., Frahm, J.-M. (eds.) ECCV 2020. LNCS, vol. 12368, pp. 606–623. Springer, Cham (2020). https://doi.org/10.1007/978-3-030-58592-1_36
45. Zhu, X., Su, W., Lu, L., Li, B., Wang, X., Dai, J.: Deformable DETR: deformable transformers for end-to-end object detection. In: International Conference on Learning Representations (2021). https://openreview.net/forum?id=gZ9hCDWe6ke

Learning Efficient Multi-agent Cooperative Visual Exploration

Chao Yu[1], Xinyi Yang[1], Jiaxuan Gao[2], Huazhong Yang[1], Yu Wang[1(✉)], and Yi Wu[2,3(✉)]

[1] Department of Electronic Engineering, Tsinghua University, Beijing, China
yu-wang@tsinghua.edu.cn
[2] Institute for Interdisciplinary Information Sciences, Tsinghua University, Beijing, China
jxwuyi@gmail.com
[3] Shanghai Qi Zhi Institute, Shanghai, China

Abstract. We tackle the problem of cooperative visual exploration where multiple agents need to jointly explore unseen regions as fast as possible based on visual signals. Classical planning-based methods often suffer from expensive computation overhead at each step and a limited expressiveness of complex cooperation strategy. By contrast, reinforcement learning (RL) has recently become a popular paradigm for tackling this challenge due to its modeling capability of arbitrarily complex strategies and minimal inference overhead. In this paper, we propose a novel RL-based multi-agent planning module, *Multi-agent Spatial Planner* (MSP). MSP leverages a transformer-based architecture, *Spatial-TeamFormer*, which effectively captures spatial relations and intra-agent interactions via hierarchical spatial self-attentions. In addition, we also implement a few multi-agent enhancements to process local information from each agent for an aligned spatial representation and more precise planning. Finally, we perform policy distillation to extract a meta policy to significantly improve the generalization capability of final policy. We call this overall solution, *Multi-Agent Active Neural SLAM* (MAANS). MAANS substantially outperforms classical planning-based baselines for the first time in a photo-realistic 3D simulator, Habitat. Code and videos can be found at https://sites.google.com/view/maans.

Keywords: Multi-agent reinforcement learning · Visual exploration

C. Yu, X. Yang, and J. Gao—Equal contribution.

This research is supported by NSFC (U20A20334, U19B2019 and M-0248), Tsinghua-Meituan Joint Institute for Digital Life, Tsinghua EE Independent Research Project, Beijing National Research Center for Information Science and Technology (BNRist), and Beijing Innovation Center for Future Chips.

Supplementary Information The online version contains supplementary material available at https://doi.org/10.1007/978-3-031-19842-7_29.

S. Avidan et al. (Eds.): ECCV 2022, LNCS 13699, pp. 497–515, 2022.
https://doi.org/10.1007/978-3-031-19842-7_29

1 Introduction

Visual exploration [41] is an important task for building intelligent embodied agents and has been served as a fundamental building block for a wide range of applications, such as scene reconstruction [1,21], autonomous driving [3], disaster rescue [26] and planetary exploration [50]. In this paper, we consider a multi-agent exploration problem, where multiple homogeneous robots simultaneously explore an unknown spatial region via visual and sensory signals in a cooperative fashion. The existence of multiple agents enables complex cooperation strategies to effectively distribute the workload among different agents, which could lead to remarkably higher exploration efficiency than the single-agent counterpart.

Planning-based solutions have been widely adopted for navigation problems for both single-agent and multi-agent scenarios [4,45,53]. Planning-based methods require little training and can be directly applied to different scenarios. However, these methods often suffer from limited expressiveness capability on coordination strategies, require non-trivial hyper-parameter tuning for each test scenario, and are particularly time-consuming due to repeated re-planning at each decision step. By contrast, reinforcement learning (RL) has been promising solution for a wide range of decision-making problems [28,33], including various visual navigation tasks [6,9,45]. Once a policy is well trained by an RL algorithm, the robot can capture arbitrarily complex strategies and produce real-time decisions with efficient inference computation (i.e., a single forward-pass of neural network).

However, training effective RL policies can be particularly challenging. This includes two folds: (1) learning a cooperative strategy over multiple agents in an end-to-end manner becomes substantially harder thanks to an exponentially large action space and observation space when tackling the exploration task based on visual signals; (2) RL policies often suffer from poor generalization ability to different scenarios or team sizes compared with classical planning-based approaches. Hence, most RL-based visual exploration methods focus on the single-agent case [6,9,45] or only consider a relatively simplified multi-agent setting (like maze or grid world [55]) of a fixed number of agents [30].

In this work, we develop *Multi-Agent Active Neural SLAM* (MAANS), the first RL-based solution for cooperative multi-agent exploration that substantially outperforms classical planning-based methods in a photo-realistic physical simulator, Habitat [46]. In MAANS, an agent consists of 4 components, a neural SLAM module, a planning-based local planner, a local policy for control, and the most critical one, a novel *Multi-agent Spatial Planner (MSP)*. Which is an RL-trained planning module that can capture complex intra-agent interactions via a self-attention-based architecture, *Spatial-TeamFormer*, and produce effective navigation targets for a varying number of agents.

We also implement a map refiner to align the spatial representation of each agent's local map, and a map merger, which enables the local planner to perform more precise sub-goal generation over a manually combined approximate 2D map. Finally, instead of directly running multi-task RL over all the training scenes, we first train a single policy on each individual scene and then use policy distillation to extract a meta policy, leading to a much improved generalization capability. We compare MAANS with a collection of classical planning-based methods and RL-based variants. Empirical results show that MAANS has a 20.56% and 7.99% higher exploration efficiency on training and testing scenes than the best planning-based competitor. The learned policy can further generalize to novel team sizes in a zero-shot manner as well.

2 Related Work

2.1 Visual Exploration

In classical visual exploration solutions, a search-based planning algorithm could be adopted to generate valid exploration trajectories. Representative variants include frontier-based methods [53,62,65], which always choose navigation targets from the explored region, and sampling-based methods [27], which generate goals via a stochastic process. In addition to the expensive search computation for planning, these methods do not involve learning and thus have limited representation capabilities for particularly challenging scenarios. Hence, RL-based methods have been increasingly popular for their training flexibility and strong expressiveness power. Early methods simply train navigation policies in a purely end-to-end fashion [9,22] while recent works start to incorporate the inductive bias of a spatial map structure into policy representation by introducing a differentiable spatial memory [16,34,37], semantic prior knowledge [30] or learning a topological scene graph [2,8,63].

The Active Neural SLAM (ANS) method [6] is the state-of-the-art framework for single-agent visual exploration (details in Sect. 3.2). There are also follow-up enhancements based on the ANS framework, such as improving map reconstruction with occupancy anticipation [40] and incorporating semantic signals into the reconstructed map for semantic exploration [7]. Our MAANS can be viewed as a multi-agent extension of ANS with a few multi-agent-specific components.

2.2 Multi-agent Cooperative Exploration

There have been works extending planning-based visual exploration solutions to the multi-agent setting by introducing handcraft planning heuristics over a shared reconstructed 2D map [5,11,12,18,35,38,60]. However, due to the lack of learning, these methods may have the limited potential of capturing non-trivial multi-agent interactions in challenging domains. By contrast, multi-agent reinforcement learning (MARL) has shown its strong performances in a wide range of domains [36], so many works have been adopting MARL to solve challenging cooperative problems. Representative works include value decomposition for approximating credit assignment [42,49], learning intrinsic rewards to tackle sparse rewards [20,29,57] and curriculum learning [31,58].

However, jointly optimizing multiple policies makes multi-agent RL training remarkably more challenging than its single-agent counterpart. Hence, these end-to-end RL methods either focus on much simplified domains, like grid world or particle world [55], or still produce poor exploration efficiency compared with classical planning-based solutions. Our MAANS framework adopts a modular design and is the first RL-based solution that significantly outperforms classical planning-based baselines in a photo-realistic physical environment.

Finally, we remark that MAANS utilizes a centralized global planner MSP, which assumes perfect communication between agents. There are also works on multi-agent cooperation with limited or constrained communication [15,22,24, 39,48,56,69], which are parallel to our focus.

2.3 Size-Invariant Representation Learning

There has been rich literature in deep learning studying representation learning over an arbitrary number of input entities in deep learning [67,68]. In MARL, the self-attention mechanism [54] has been the most popular policy architecture to tackle varying input sizes [14,23,44,59] or capture high-order relations [19,32,63,66]. A concurrent work [56] also considers the zero-shot team-size adaptation in the photo-realistic environment by learning a simple attention-based communication channel between agents. By contrast, our works develop a much expressive network architecture, Spatial-TeamFormer, which adopts a hierarchical self-attention-based architecture to capture both intra-agent and spatial relationships and results in substantially better empirical performance (see Sect. 5.4). Besides, parameter sharing is another commonly used technique in MARL for team-size generalization, which has been also shown to help reduce nonstationarity and accelerate training [10,52]. Our work follows this paradigm as well.

3 Preliminary

3.1 Task Setup

We consider a multi-agent coordination indoor active SLAM problem, in which a team of agents needs to cooperatively explore an unknown indoor scene as fast as possible. At each timestep, each agent performs an action among *Turn Left*, *Turn Right* and *Forward*, and then receives an RGB image through a camera and noised pose change through a sensor, which is provided from the Habitat environment. We consider a decision-making setting by assuming perfect communication between agents. The objective of the task is to maximize the accumulative explored area within a limited time horizon.

3.2 Active Neural SLAM

The ANS framework [6] consists of 4 parts: a neural SLAM module, a RL-based global planner, a planning-based local planner and a local policy. The neural SLAM module takes an RGB image, the pose sensory signals, and its past outputs as inputs, and outputs an updated 2D reconstructed map and a current pose estimation. Note that in ANS, the output 2D map only covers a neighboring region of the agent location and always keeps the agent at the egocentric position. For clarification, we call this raw output map from the SLAM module a *agent-centric local map*. The global planner in ANS takes in an augmented agent-centric *local map* as its input, and outputs two real numbers from two Gaussian distributions denoting the coordinate of the long-term goal. The local planner performs classical planning, i.e., Fast Marching Method (FMM) [47], over the agent-centric local map towards a given long-term goal, and outputs a trajectory of short-term sub-goals. Finally, the local policy produces actions given an RGB image and a sub-goal and is trained by imitation learning.

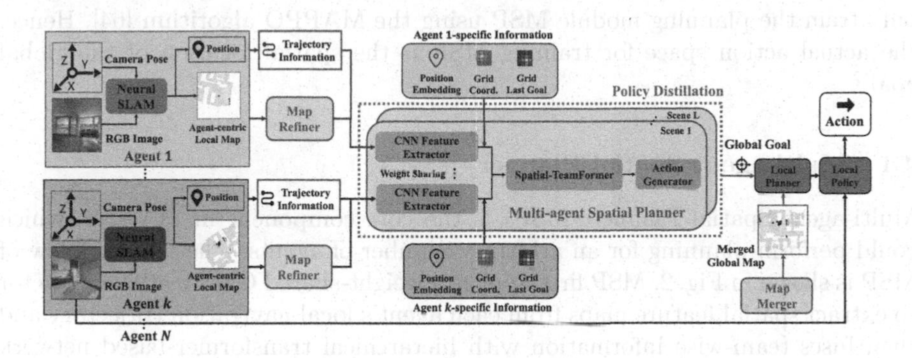

Fig. 1. Overview of *Multi-Agent Active Neural SLAM* (MAANS).

With the advantage of the modeling capability of arbitrarily complex strategy in RL, an RL-based global planner which determines the global goals encourages exploration faster. To apply RL training, we model the problem as a decentralized partially observable Markov decision process (Dec-POMDP). Dec-POMDP is parameterized by $\langle S, A, O, R, P, n, \gamma, h \rangle$. n is the number of agents. S is the state space, A is the joint action space. $o^{(i)} = O(s; i)$ is agent i's observations at state s. $P(s'|s, a)$ defines the transition probability from state s to state s' via joint action a. $R(s, A)$ is the shared reward function. γ is the discount factor. The objective function is $J(\theta) = \mathbb{E}_{a,s}[\sum_t \gamma^t R(s^t, a^t)]$. In this task, the policy π_θ generates a global goal for each agent every decision-making step. The shared reward function is defined as the accumulative environment reward every global goal planning step.

4 Methodology

The overview of MAANS is demonstrated in Fig. 1, where each agent is presented in a modular structure. The *Neural SLAM* module corrects the sensor error and performs SLAM in order to build a top-down 2D occupancy map. Then we use a *Map Refiner* to rotate each agent's egocentric local map to a global coordinate system. We augment these refined maps with each agent's trajectory information and feed these spatial inputs along with other agent-specific information to our core planning module, *Multi-agent Spatial Planner (MSP)* to generate a global goal as the long-term navigation target for each individual agent. We remark that only estimated geometric information is utilized in this map fusion process. To effectively reach a global goal, the agent first plans a path to this long-term goal in a manually merged global map using FMM and generates a sequence of short-term sub-goals. Finally, given a short-term sub-goal, a *Local Policy* outputs the final navigation action based on the visual input and the relative spatial distance as well as the relative angle to the sub-goal. Note that the Neural SLAM module and the Local Policy do not involve multi-agent interactions, so we directly reuse these two modules from ANS [45]. We fix these modules throughout training and only train the planning module MSP using the MAPPO algorithm [64]. Hence, the actual action space for training MSP is the spatial location of the global goal.

4.1 Multi-agent Spatial Planner

Multi-agent Spatial Planner (MSP) is the core component in MAANS, which could perform planning for an arbitrary number of agents. The full workflow of MSP is shown in Fig. 2. MSP first applies a weight-shared CNN feature extractor to extract spatial feature maps from each agent's local navigation trajectory and then fuses team-wise information with hierarchical transformer-based network architecture, Spatial-TeamFormer. Finally, an action generator will generate a spatial global goal based on the features from Spatial-TeamFormer. Suppose there are a total of N agents and the current decision-making agent has ID k. We will describe how agent k generates its long-term goal via the 3 parts in MSP

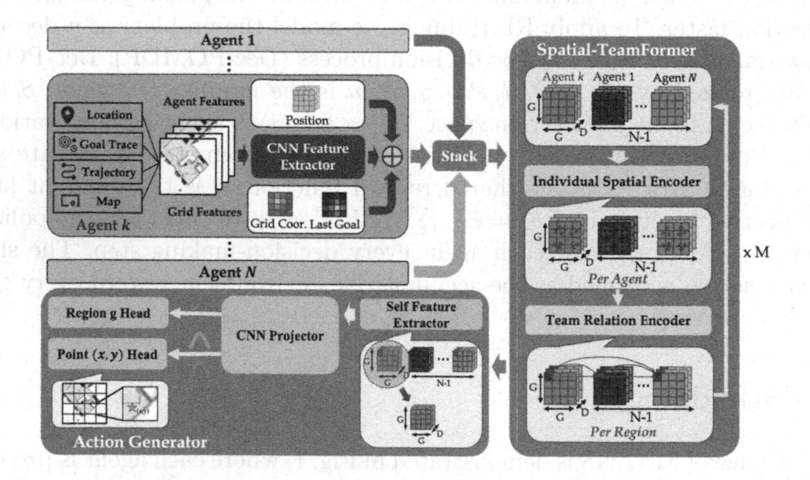

Fig. 2. Workflow of *Multi-agent Spatial Planner* (MSP), including a CNN-based Feature Extractor, a Spatial-TeamFormer for representation learning and an Action Generator.

in the following content. Note that due to space constraints, we only present the main ideas while more computation details can be found in Appendix A.4.

(1) CNN Feature Extractor. For every single agent, we use its current location, movement trajectory, previous goal, goal history, self-occupancy map and obstacle map as inputs and convert them to a 480×480 2D map with 6 channels over a global coordinate system. We adopt a weight-shared CNN network with 5 layers to process each agent's input map, which produces a $G \times G$ feature map with $D = 32$ channels. G corresponds to the discretization level of the scene. We choose $G = 8$ in our work, leading to G^2 grids corresponding to different spatial regions in the underlying indoor scene.

Besides CNN spatial maps, we also introduce additional features, including agent-specific embeddings of its current position and grid features, i.e., the embeddings of the relative coordinate of each grid to the agent position as well as the embedding of the previous global goal.

(2) Spatial-TeamFormer. With a total of N extracted $G \times G$ feature maps, we aim to learn a team-size-invariant spatial representation over all the agents. Transformer has been a particularly popular candidate for learning invariant representations, but it may not be trivially applied in this case. Standard Transformer model in NLP [54] tackles 1-dimensional text inputs, which ignores the spatial structure of input features. Visual transformers [13] capture spatial relations well by performing spatial self-attention. However, we have a total of N spatial inputs from the entire team.

Hence, we present a specialized architecture to jointly leverage intra-agent and spatial relationships in a hierarchical manner, which we call *Spatial-TeamFormer*. A Spatial-Teamformer block consists of two layers, i.e., an *Individual Spatial Encoder* for capturing spatial features for each agent, and a *Team Relation Encoder* for reasoning cross agents. Similar to visual transformer [13], *Individual Spatial Encoder* focuses only on spatial information by performing a spatial self-attention over each agent's own $G \times G$ spatial map without any cross-agent computation. By contrast, *Team Relation Encoder* completely focuses on capturing team-wise interactions without leveraging any spatial information. In particular, for each of the $G \times G$ grid, *Team Relation Encoder* extracts the features w.r.t. that grid from the N agents and performs a standard transformer over these N features. We can further stack multiple Spatial-TeamFormer blocks for even richer interaction representations.

We remark that another possible alternative to Spatial-TeamFormer is to simply use a big transformer over the aggregated $N \times G \times G$ features. Such a naive solution is substantially more expensive to compute ($O(N^2G^4)$ time complexity) than Spatial-TeamFormer ($O(N^2G^2 + NG^4)$ time complexity), which may also incur significant learning difficulty in practice (see Sect. 5.4).

(3) Action Generator. The Action Generator is the final part of MSP, which outputs a long-term global goal over the reconstructed map. Since spatial-TeamFormer produces a total of N rich spatial representation, which can be denoted as $N \times G \times G$, we take the first $G \times G$ grid, which is the feature map of the current agent, to derive a single team-size-invariant representation.

In order to produce accurate global goals, we adopt a spatial action space with two separate action heads, i.e., a discrete region head for choosing a region g from the $G \times G$ discretized grids, and a continuous point head for outputing a coordinate (x, y), indicating the relative position of the global goal within the selected region g. To compute the action probability for g, we compute a spatial softmax operator over all the grids while to ensure the scale of (x, y) is bounded between 0 and 1, we apply a sigmoid function before outputting the value of (x, y). We remark that such a spatial design of action space is beneficial since it alleviates the problem of multi-modal issue of modeling potential "good" goals, which could not be simply represented by a simple normal distribution as used in [9] (see Sect. 5.4).

4.2 Map Refiner for Aligned 2D Maps

We develop a map refiner to ensure all the maps from the neural SLAM module are within the same coordinate system. The workflow is shown as the blue and green part in Fig. 3. The map refiner first composes all the past *agent-centric local maps* to recover the agent-centric *global* map. Then, we transform the coordinate system based on the pose estimates to normalize the global maps from all the agents w.r.t. the same coordinate system. To ensure the feature extractor in MSP concentrates only on the viable part and also induce a more focused spatial action space, we crop the unexplorable boundary of the normalized map and enlarge the house region as our final refined map.

4.3 Map Merger for Improved Local Planning

The local planner from ANS plans sub-goals on the agent-centric local map, while in our setting, we can also leverage the information from other agents to plan over a more accurate map. The diagram of map merger is shown in Fig. 3. After obtaining N enlarged global maps via the map refiner, the map merger simply integrates all these maps by applying a max-pooling operator for each pixel location. We remark that the artificial merged global map is only utilized in the local planner, but not in the global planner MSP. We empirically observe that having a coarse merged map produces better short-term local goal while such an artificial map is not sufficient for accurate global planning (see Sect. 5.4).

4.4 Policy Distillation for Improved Generalization

The common training paradigm for visual exploration is multi-task learning, i.e., at each training episode, a random training scene or team size is sampled and all collected samples are aggregated for policy optimization [6,9]. However, we empirically observe that different Habitat scenes and team sizes may lead to drastically different exploration difficulties. During training, gradients from

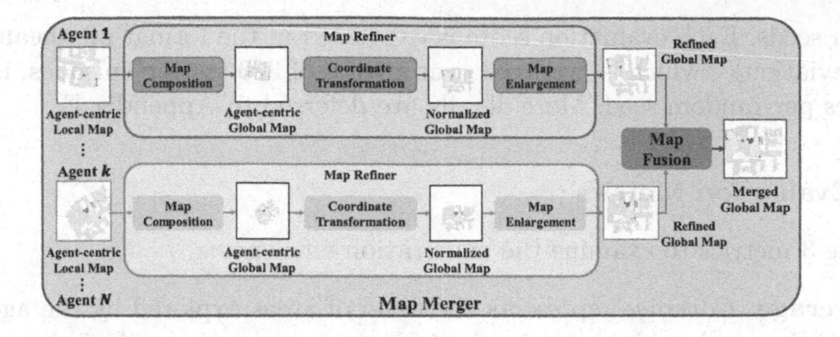

Fig. 3. Computation workflow of *map refiner* (blue and green) and *map merger* (purple). (Color figure online)

different configurations may negatively impact each other. Similar observations have been also reported in the existing literature [17,51]. We use policy distillation to tackle this problem. Therefore, we adopt a two-phase distillation-based solution: in the first phase, we train separate policies for representative training scenes with a fixed team size, i.e., we choose $N = 2$ in our experiments. in the second phase, we learn another policy with $N = 2$ agents to distill the collection of pretrained policies over different training scenes and directly measure the generalization ability of this distillation policy to novel scenes and different team sizes.

More specifically, for the i-th training scene, we first learn a specialized teacher policy $\pi(g, x, y|s, \theta_i)$ given state s with parameter θ_i, where g denotes the region output and (x, y) is the point head output. Then we train another distillation policy $\pi(g, x, y|s, \theta)$ by simply running a dagger-style [43] imitation learning, i.e., randomly rollout trajectories w.r.t. the distillation policy $\pi(s, \theta)$ and imitate the output from the specific teacher policy.

Since the region action g is discrete, we adopt a KL-divergence-based loss function while for the continuous point action (x, y), a squared difference loss between the teacher policy and distillation policy is optimized.

5 Experiment Results

5.1 Experiment Setting

We adopt scene data from the Gibson Challenge dataset [61] while the visual signals and dynamics are simulated by the Habitat simulator [46]. Although Gibson Challenge dataset provides 72 training and 14 validation scenes, we discard scenes that are not appropriate for our task, such as scenes that have large disconnected regions or multiple floors so that the agents are not possible to achieve 90% coverage of the entire house. Then we categorize the remaining scenes into 23 training scenes and 10 testing scenes. We consider $N = 2, 3, 4$ agents in our experiments. Every RL training is performed with 10^4 training episodes over 3 random seeds. Each evaluation score is expressed in the format of "mean (standard deviation)", which is averaged over a total of 300 testing episodes, i.e., 100 episodes per random seed. More details are deferred to Appendix E.

5.2 Evaluation Metrics

We take 3 metrics to examine the exploration efficiency:

1. **Coverage**: *Coverage* represents the ratio of areas explored by the agents to the entire explorable space in the indoor scene at the end of the episode. Higher *Coverage* implies more effective exploration.
2. **Steps**: *Steps* is the number of timesteps used by agents to achieve a coverage ratio of 90% within an episode. Fewer *Steps* implies faster exploration.
3. **Mutual Overlap**: For effective collaboration, each agent should visit regions different from those explored by its teammates. We report the average overlapping explored area over each pair of agents when the coverage ratio reaches 90%. *Mutual Overlap* denotes the normalized value of this metric. Lower *Mutual Overlap* suggests better multi-agent coordination.

5.3 Baselines

We first adapt 3 single-agent planning-based methods, namely *Nearest* [62], *Utility* [25], and *RRT* [53], to our problems by planning on the merged global map. The 3 planning-based baselines are frontier-based, i.e., they choose long-term navigation goals from the boundary between currently explored and unexplored area using different heuristics. Note that though these are originally single-agent methods and are adapted to multi-agent settings by planning on the merged global map. When choosing global goals, each agent performs computation based on the merged global map, its current position and its past trajectory.

For multi-agent baselines, we compare MAANS with 3 planning-based methods, namely *Voronoi* [18], *APF* [65] and *WMA-RRT* [35]. *APF* [65] computes artificial potential field over clustered frontiers and plans a potential-descending path with maximum information gain. APF introduces resistance force among multiple agents to avoid repetitive exploration. *WMA-RRT* [35] is a multi-agent variant of RRT, in which agents cooperatively maintain a single tree and follow a formal locking-and-search scheme. *Voronoi*-based method partitions the map into different parts using a voronoi partition and each agent only searches unexplored area in its own partition. More details can be found in Appendix B.

5.4 Ablation Study

We report the training performances of multiple RL variants on 2 selected scenes, *Colebrook* and *Dryville*, and measure the *Steps* and the *Mutual Overlap* over these 2 scenes.

(1) Comparison with ANS Variants. We first consider 2 ANS variants:

1. **ANS-idv.** We train N ANS agents to explore individually, i.e., without any communication, in the environment.
2. **ANS-stack.** We directly stack all the agent-centric local maps from the neural SLAM module as the input representation to the global planner, and retrain the ANS global planner under our multi-agent task setting.

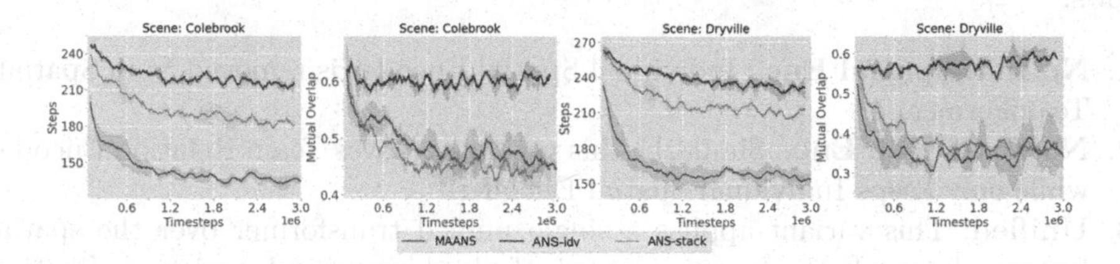

Fig. 4. Comparison between MAANS (red) and other ANS variants. (Color figure online)

We demonstrate the training curves in Fig. 4. Regarding *Steps*, both ANS variants perform consistently worse than MAANS on each map. Regarding *Mutual Overlap*, the idv variant fails to cooperate completely while the stack variant produces comparable *Mutual Overlap* to MAANS despite its low exploration efficiency. We remark that ANS-stack performs global and local planning completely on the agent-centric *local* map while the local map is a narrow sub-region over the entire house, which naturally leads to a much conservative exploration strategy and accordingly helps produce a lower *Mutual Overlap.*

(2) Ablation Study on MSP. We consider 3 additional MSP variants:

1. **MSP w.o. TeamFormer**: We completely substitute Spatial-TeamFormer with a simple average pooling layer over the extracted spatial features from CNN extractors.
2. **MSP w.o. Act. Gen.** We remove the region head from the spatial action generator, so that the global goal is directly generated over the entire refined global map via two Gaussian action distributions. We remark that such an action space design follows the original ANS paper [45].
3. **MSP-merge.** We consider another MSP variant that applies a single CNN feature extractor over the manually merged global map from the map merger, instead of forcing the network to learn to fuse each agent's information.

As shown in Fig. 5, the full MSP module produces the lowest *Steps* and *Mutual Overlap.* Among all the MSP variants, *MSP w.o. Act. Gen.* produces the highest *Steps.* This suggests that a simple Gaussian representation of actions may not be able to fully capture the distribution of good long-term goals, which can be highly multi-modal in the early exploration stage. In scene *Dryville*, *MSP w.o. TeamFormer* performs much worse and shows larger training unstability than the full model, showing the importance of jointly leveraging intra-agent and spatial relationships in a hierarchical manner. In addition, *MSP-merge* produces a very high *Mutual Overlap* in scene *Dryville*. We hypothesis that this is due to the fact that many agent-specific information are lost in the manually merged maps while MSP can learn to utilize these features implicitly.

(3) Ablation Study on Spatial-TeamFormer. We consider the following variants of MAANS by altering the components of Spatial-TeamFormer as follows:

1. **No Ind. Spatial Enc.**: Individual Spatial Encoder is removed from Spatial-TeamFormer
2. **No Team Rel. Enc.**: Similarly, this variant removes Team Relation Encoder while only keeps Individual Spatial Encoder.
3. **Unified**: This variant applies a single unified transformer over the spatial features from all the agents instead of the hierarchical design in Spatial-TeamFormer. In particular, we directly feed all the $N \times G \times G$ features into a big transformer model to generate an invariant representation.
4. **Flattened**: In this variant, we do not keep the spatial structure of feature maps. Instead, we first convert the CNN extracted feature into a flatten vector for each agent and then simply feed these N flattened vectors to a standard transformer model for feature learning. We remark that this variant is exactly the same as [56].

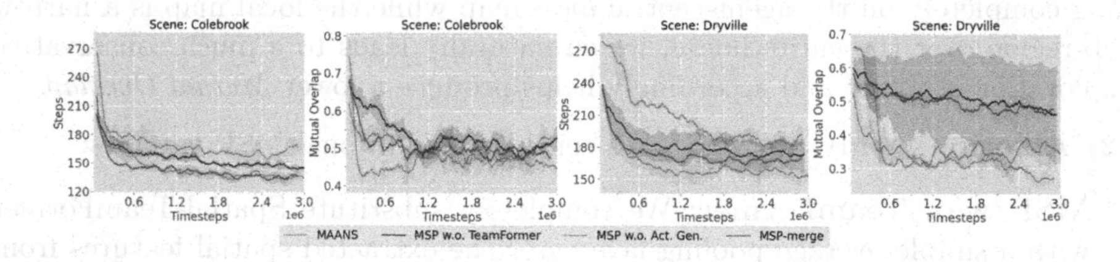

Fig. 5. Ablation studies on MSP components.

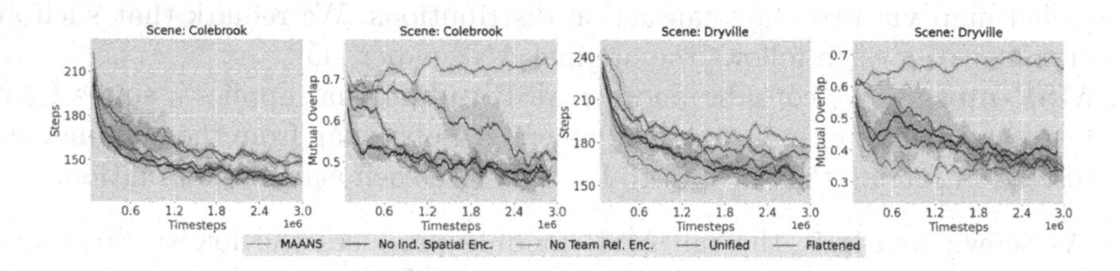

Fig. 6. Ablation studies on Spatial-TeamFormer.

We report training curves in Fig. 6. Compared with Spatial-TeamFormer, *No Team Rel. Enc.* has the highest *Mutual Overlap* and worst *Steps* on both scenes, which suggests that lacking partners' relationship attention significantly lowers the cooperation efficiency. We remark that *No Team Rel. Enc.* is indeed a single-agent variant of Spatial-TeamFormer: each agent plans global goal using its individual information while doing path planning still with the merged global map. The variant using *Flattened* features is also performing much worse than the full model with a clear margin, showing that the network architecture without utilizing spatial inductive bias could hurt final performance. When individual spatial encoder is removed (*No Ind. Spatial Enc.*), the sample efficiency drops greatly in scene *Dryville* and the method achieves a higher *Mutual Overlap* than MAANS. *Unified* also has worse sample efficiency that the full model. Note that *Unified* shows greater performance than *Flattened*, again confirming the importance of utilizing spatial inductive bias.

5.5 Main Results

Due to space constraints, we defer the full results to Appendix F.

(1) Comparison with Planning-Based Baselines and RL Baseline

a. Training with a Fixed Team Size: We first report the performance of MAANS and selected the baseline methods with a fixed team size of $N = 2$ agents on both representative training scenes and testing scenes in Table 1. We remark that only 9 policies of representative training scenes are used to do policy distillation(PD) since it takes a lot of work to train a separated policy for each scene. Except for the *Mutual Overlap* on the testing scenes where the performance is slightly worse than *Voronoi*, MAANS still outperforms all planning-based baselines in *Steps* and *Coverage* metrics on training and testing scenes. More concretely, MAANS reduces 20.56% exploration steps on training scenes and 7.99% exploration steps on testing scenes than the best planning-based competitor. We also compare with an RL baseline *MAANS w.o. PD*, which is trained by randomly sampling all training scenes instead of policy distillation. *MAANS w.o. PD* performs much worse than MAANS on both training and testing scenes, and slightly worse than the best single-agent planning-based method *RRT* and multi-agent planning-based method *Voronoi* on testing scenes, indicating the necessity of introducing policy distillation. Further illustration and analysis could be found in Appendix F.3.

Table 1. Performance of MAANS and selected *planning-based* baselines and RL baseline with a fixed size of $N = 2$ agents on both training and testing scenes.

Sce.	Metrics	Utility	RRT	APF	WMA-RRT	Voronoi	MAANS w.o PD	MAANS
Train	*Mut. Over.* ↓	0.68(0.01)	0.53(0.02)	0.61(0.01)	0.61(0.01)	0.44(0.01)	0.46(0.01)	**0.42(0.01)**
	Steps ↓	236.15(3.61)	199.59(3.27)	251.41(3.15)	268.20(2.24)	237.04(2.95)	180.25(2.35)	**158.55(2.25)**
	Coverage ↑	0.92(0.01)	0.96(0.00)	0.90(0.01)	0.87(0.01)	0.93(0.00)	0.96(0.00)	**0.97(0.00)**
Test	*Mut. Over.*↓	0.69(0.01)	0.57(0.01)	0.57(0.01)	0.64(0.01)	**0.51(0.01)**	0.57(0.01)	0.54(0.02)
	Steps ↓	161.28(2.32)	157.29(2.59)	181.18(4.17)	198.92(3.83)	156.68(3.21)	159.53(2.73)	**144.16(2.52)**
	Coverage ↑	0.95(0.00)	0.95(0.01)	0.93(0.01)	0.91(0.01)	**0.96(0.01)**	**0.96(0.00)**	**0.96(0.00)**

b. Zero-Shot Transfer to Different Team Sizes: In this part, we directly apply the policies trained with $N = 2$ agents to the scenes of $N = 3, 4$ agents respectively. The zero-shot generalization performance of MAANS compared with the best single-agent baseline *RRT* and the best multi-agent baseline *Voronoi* on both training and testing scenes is shown in Table 2. Note that experiments on testing scenes are extremely challenging since MAANS is never trained with $N = 3, 4$ team sizes on testing scenes. MAANS achieves much better performance than the best planning-based methods with every novel team size on training scenes (17.56% fewer *Steps* with $N = 3$ and 18.36% fewer *Steps* with $N = 4$) and comparable performance on testing scenes (<3 more *Steps* with $N = 3, 4$).

Table 2. Generalization performance of MAANS and selected *planning-based methods* to novel fixed and varying team sizes on training and testing scenes. Note that MAANS has the best performance on training scenes and comparable results on testing scenes.

# Agent	Metrics	Training scenes			Testing scenes		
		RRT	Voronoi	MAANS	RRT	Voronoi	MAANS
3	*Mut. Over.* ↓	0.44(0.01)	**0.37(0.01)**	0.42(0.01)	0.45(0.01)	**0.43(0.01)**	0.53(0.01)
	Steps ↓	155.13(3.26)	180.27(2.51)	**127.88(1.91)**	128.33(1.66)	**119.98(2.31)**	122.48(2.22)
	Coverage ↑	0.95(0.01)	0.95(0.00)	**0.97(0.00)**	0.95(0.01)	**0.96(0.00)**	**0.96(0.00)**
4	*Mut. Over.* ↓	0.36(0.01)	**0.34(0.01)**	0.42(0.01)	0.41(0.01)	**0.39(0.01)**	0.50(0.01)
	Steps ↓	140.57(1.78)	147.01(2.38)	**114.75(1.69)**	111.30(1.58)	**101.90(2.36)**	109.07(2.02)
	Coverage ↑	0.92(0.01)	0.93(0.00)	**0.96(0.00)**	0.93(0.01)	**0.95(0.00)**	0.94(0.00)
2 ⇒ 3	*Mut. Over.*↓	0.36(0.01)	0.35(0.01)	**0.30(0.01)**	0.43(0.01)	**0.32(0.02)**	0.46(0.01)
	Steps↓	185.94(1.83)	200.91(2.32)	**148.82(2.01)**	136.42(2.41)	148.12(6.69)	**134.11(2.88)**
	Coverage↑	0.94(0.00)	0.92(0.00)	**0.96(0.00)**	**0.96(0.01)**	0.92(0.05)	**0.96(0.00)**
3 ⇒ 2	*Mut. Over.*↓	0.35(0.01)	**0.33(0.01)**	0.41(0.01)	**0.39(0.01)**	0.42(0.01)	0.43(0.01)
	Steps↓	187.93(1.98)	206.94(2.50)	**145.14(2.83)**	139.52(3.74)	**133.77(2.83)**	145.43(3.44)
	Coverage↑	0.91(0.00)	0.89(0.01)	**0.95(0.00)**	0.94(0.01)	**0.95(0.01)**	0.94(0.01)

c. Varying Team Size within an Episode. We further consider the setting where the team size varies within an episode in Table 2. We use "$N_1 \Rightarrow N_2$" to denote that each episode starts with N_1 agents and the team size immediately switches to N_2 after half of episode length. In cases where the team size increases, MAANS produces substantially better performances w.r.t. every metric. In particular, MAANS achieves 33 fewer *Steps* in training scenes and lower *Steps* than other methods in testing scenes, which suggests that MAANS has the capability to adaptively adjust its strategy. Regarding the cases where the team size decreases, MAANS consumes over 40 fewer *Steps* in 3 ⇒ 2 than RRT in training scenes.

(2) Learned Strategy

Figure 7 demonstrates two 2-agent trials of MAANS and RRT, the most competitive planning-based method, with the same birth place. The merged global map are shown in keep timesteps. As shown in Fig. 7, MAANS's coverage ratio goes up faster than RRT, indicating higher exploration efficiency. At timestep around 90, MAANS produces global goals successfully allocate the agents towards two distant unexplored area while RRT guides the agents towards the same part of the map. And at timestep around 170 when MAANS reaches 90% coverage ratio, RRT still stuck in previous explored area though there is obviously another large open space. Notice that at this key timestep RRT selects two frontiers that are marked unexplored but with no actual benefit, which an agent utilizing prior knowledge about room structures would certainly avoid.

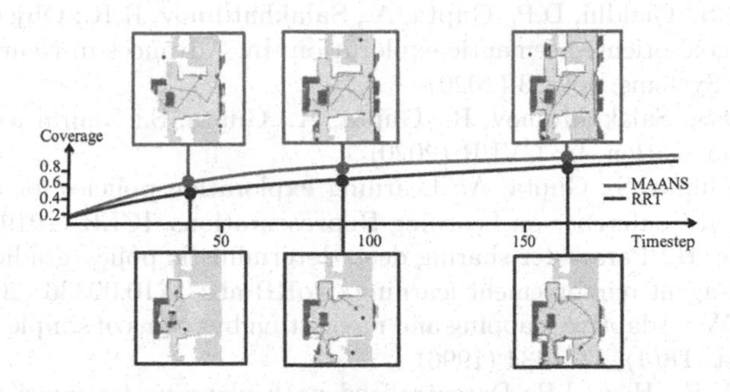

Fig. 7. Learned strategy on scene *Colebrook* of MAANS vs. RRT, where the red line with arrow represents the trajectory, the explored area shows in blue and the obstacle shows in green. MAANS achieves much higher and faster coverage ratio than RRT. (Color figure online)

6 Conclusion

We propose the first multi-agent cooperative exploration framework, *Multi-Agent Active Neural SLAM* (MAANS) that outperforms planning-based competitors in a photo-realist physical environment. The key component of MAANS is the RL-based planning module, *Multi-agent Spatial Planner (MSP)*, which leverages a transformer-based architecture, Spatial-TeamFormer, to capture team-size-invariant representation with strong spatial structures. We also implement a collection of multi-agent-specific enhancements and policy distillation for better generalization. Experiments on Habitat show that MAANS achieves better training and testing performances than all the baselines. We hope MAANS can inspire more powerful multi-agent methods in the future.

References

1. Anguelov, D., et al.: Google street view: capturing the world at street level. Computer **43**(6), 32–38 (2010)
2. Bhatti, S., Desmaison, A., Miksik, O., Nardelli, N., Siddharth, N., Torr, P.H.: Playing doom with slam-augmented deep reinforcement learning. arXiv preprint arXiv:1612.00380 (2016)

3. Bresson, G., Alsayed, Z., Yu, L., Glaser, S.: Simultaneous localization and mapping: a survey of current trends in autonomous driving. IEEE Trans. Intell. Veh. **2**(3), 194–220 (2017)
4. Burgard, W., Moors, M., Stachniss, C., Schneider, F.E.: Coordinated multi-robot exploration. IEEE Trans. Rob. **21**(3), 376–386 (2005)
5. Čáp, M., Novák, P., Vokřínek, J., Pěchouček, M.: Multi-agent RRT*: sampling-based cooperative pathfinding. arXiv preprint arXiv:1302.2828 (2013)
6. Chaplot, D.S., Gandhi, D., Gupta, S., Gupta, A., Salakhutdinov, R.: Learning to explore using active neural slam. In: International Conference on Learning Representations. ICLR (2020)
7. Chaplot, D.S., Gandhi, D.P., Gupta, A., Salakhutdinov, R.R.: Object goal navigation using goal-oriented semantic exploration. In: Advances in Neural Information Processing Systems, vol. 33 (2020)
8. Chaplot, D.S., Salakhutdinov, R., Gupta, A., Gupta, S.: Neural topological slam for visual navigation. In: CVPR (2020)
9. Chen, T., Gupta, S., Gupta, A.: Learning exploration policies for navigation. In: International Conference on Learning Representations. ICLR (2019)
10. Chu, X., Ye, H.: Parameter sharing deep deterministic policy gradient for cooperative multi-agent reinforcement learning. CoRR abs/1710.00336 (2017)
11. Cohen, W.W.: Adaptive mapping and navigation by teams of simple robots. Robot. Auton. Syst. **18**(4), 411–434 (1996)
12. Desaraju, V.R., How, J.P.: Decentralized path planning for multi-agent teams in complex environments using rapidly-exploring random trees. In: 2011 IEEE International Conference on Robotics and Automation, pp. 4956–4961. IEEE (2011)
13. Dosovitskiy, A., et al.: An image is worth 16x16 words: transformers for image recognition at scale. arXiv preprint arXiv:2010.11929 (2020)
14. Duan, Y., et al.: One-shot imitation learning. In: NIPS (2017)
15. Foerster, J.N., Assael, Y.M., De Freitas, N., Whiteson, S.: Learning to communicate with deep multi-agent reinforcement learning. arXiv preprint arXiv:1605.06676 (2016)
16. Henriques, J.F., Vedaldi, A.: Mapnet: an allocentric spatial memory for mapping environments. In: proceedings of the IEEE Conference on Computer Vision and Pattern Recognition, pp. 8476–8484 (2018)
17. Hessel, M., Soyer, H., Espeholt, L., Czarnecki, W., Schmitt, S., van Hasselt, H.: Multi-task deep reinforcement learning with popart. In: Proceedings of the AAAI Conference on Artificial Intelligence, vol. 33, pp. 3796–3803 (2019)
18. Hu, J., Niu, H., Carrasco, J., Lennox, B., Arvin, F.: Voronoi-based multi-robot autonomous exploration in unknown environments via deep reinforcement learning. IEEE Trans. Veh. Technol. **69**(12), 14413–14423 (2020)
19. Iqbal, S., Sha, F.: Actor-attention-critic for multi-agent reinforcement learning. In: International Conference on Machine Learning, pp. 2961–2970. PMLR (2019)
20. Iqbal, S., Sha, F.: Coordinated exploration via intrinsic rewards for multi-agent reinforcement learning. arXiv preprint arXiv:1905.12127 (2019)
21. Isler, S., Sabzevari, R., Delmerico, J., Scaramuzza, D.: An information gain formulation for active volumetric 3D reconstruction. In: 2016 IEEE International Conference on Robotics and Automation (ICRA), pp. 3477–3484. IEEE (2016)
22. Jain, U., et al.: Two body problem: collaborative visual task completion. In: Proceedings of the IEEE/CVF Conference on Computer Vision and Pattern Recognition, pp. 6689–6699 (2019)
23. Jiang, J., Dun, C., Huang, T., Lu, Z.: Graph convolutional reinforcement learning. arXiv preprint arXiv:1810.09202 (2018)

24. Jiang, J., Lu, Z.: Learning attentional communication for multi-agent cooperation. Adv. Neural. Inf. Process. Syst. **31**, 7254–7264 (2018)
25. Juliá, M., Gil, A., Reinoso, O.: A comparison of path planning strategies for autonomous exploration and mapping of unknown environments. Auton. Robot. **33**(4), 427–444 (2012)
26. Kleiner, A., Prediger, J., Nebel, B.: RFID technology-based exploration and slam for search and rescue. In: 2006 IEEE/RSJ International Conference on Intelligent Robots and Systems, pp. 4054–4059. IEEE (2006)
27. Li, A.Q.: Exploration and mapping with groups of robots: recent trends. Curr. Robot. Rep. **1**(4), 1–11 (2020)
28. Lillicrap, T.P., et al.: Continuous control with deep reinforcement learning. arXiv preprint arXiv:1509.02971 (2015)
29. Liu, I.J., Jain, U., Yeh, R.A., Schwing, A.: Cooperative exploration for multi-agent deep reinforcement learning. In: International Conference on Machine Learning, pp. 6826–6836. PMLR (2021)
30. Liu, X., Guo, D., Liu, H., Sun, F.: Multi-agent embodied visual semantic navigation with scene prior knowledge. arXiv preprint arXiv:2109.09531 (2021)
31. Long, Q., Zhou, Z., Gupta, A., Fang, F., Wu, Y., Wang, X.: Evolutionary population curriculum for scaling multi-agent reinforcement learning. In: International Conference on Learning Representations (2020)
32. Malysheva, A., Sung, T.T., Sohn, C.B., Kudenko, D., Shpilman, A.: Deep multi-agent reinforcement learning with relevance graphs. arXiv preprint arXiv:1811.12557 (2018)
33. Mnih, V., et al.: Playing atari with deep reinforcement learning. arXiv preprint arXiv:1312.5602 (2013)
34. Mousavian, A., Toshev, A., Fišer, M., Košecká, J., Wahid, A., Davidson, J.: Visual representations for semantic target driven navigation. In: 2019 International Conference on Robotics and Automation (ICRA), pp. 8846–8852. IEEE (2019)
35. Nazif, A.N., Davoodi, A., Pasquier, P.: Multi-agent area coverage using a single query roadmap: a swarm intelligence approach. In: Bai, Q., Fukuta, N. (eds.) Advances in Practical Multi-Agent Systems, pp. 95–112. Springer, Heidelberg (2010). https://doi.org/10.1007/978-3-642-16098-1_7
36. Nguyen, T.T., Nguyen, N.D., Nahavandi, S.: Deep reinforcement learning for multiagent systems: a review of challenges, solutions, and applications. IEEE Trans. Cybern. **50**(9), 3826–3839 (2020)
37. Parisotto, E., Salakhutdinov, R.: Neural map: structured memory for deep reinforcement learning. In: International Conference on Learning Representations. ICLR (2018)
38. Patel, S., et al.: Multi-agent ergodic coverage in urban environments (2021)
39. Peng, P., et al.: Multiagent bidirectionally-coordinated nets for learning to play starcraft combat games. CoRR abs/1703.10069 (2017). https://arxiv.org/abs/1703.10069
40. Ramakrishnan, S.K., Al-Halah, Z., Grauman, K.: Occupancy anticipation for efficient exploration and navigation. In: Vedaldi, A., Bischof, H., Brox, T., Frahm, J.-M. (eds.) ECCV 2020. LNCS, vol. 12350, pp. 400–418. Springer, Cham (2020). https://doi.org/10.1007/978-3-030-58558-7_24
41. Ramakrishnan, S.K., Jayaraman, D., Grauman, K.: An exploration of embodied visual exploration. Int. J. Comput. Vision **129**(5), 1616–1649 (2021)

42. Rashid, T., Samvelyan, M., Schroeder, C., Farquhar, G., Foerster, J., Whiteson, S.: Qmix: monotonic value function factorisation for deep multi-agent reinforcement learning. In: International Conference on Machine Learning, pp. 4295–4304. PMLR (2018)
43. Ross, S., Gordon, G., Bagnell, D.: A reduction of imitation learning and structured prediction to no-regret online learning. In: Proceedings of the Fourteenth International Conference on Artificial Intelligence and Statistics, pp. 627–635. JMLR Workshop and Conference Proceedings (2011)
44. Ryu, H., Shin, H., Park, J.: Multi-agent actor-critic with hierarchical graph attention network. In: Proceedings of the AAAI Conference on Artificial Intelligence, vol. 34, pp. 7236–7243 (2020)
45. Savinov, N., et al.: Episodic curiosity through reachability. In: International Conference on Learning Representations. ICLR (2019)
46. Savva, M., et al.: Habitat: a platform for embodied AI research. In: Proceedings of the IEEE/CVF International Conference on Computer Vision, pp. 9339–9347 (2019)
47. Sethian, J.A.: A fast marching level set method for monotonically advancing fronts. Proc. Natl. Acad. Sci. **93**(4), 1591–1595 (1996)
48. Sukhbaatar, S., Fergus, R., et al.: Learning multiagent communication with backpropagation. Adv. Neural. Inf. Process. Syst. **29**, 2244–2252 (2016)
49. Sunehag, P., et al.: Value-decomposition networks for cooperative multi-agent learning based on team reward. In: Proceedings of the 17th International Conference on Autonomous Agents and MultiAgent Systems, pp. 2085–2087 (2018)
50. Tagliabue, A., Schneider, S., Pavone, M., Agha-mohammadi, A.: Shapeshifter: a multi-agent, multi-modal robotic platform for exploration of titan. CoRR abs/2002.00515 (2020)
51. Teh, Y.W., et al.: Distral: robust multitask reinforcement learning. In: NIPS (2017)
52. Terry, J.K., Grammel, N., Hari, A., Santos, L., Black, B., Manocha, D.: Parameter sharing is surprisingly useful for multi-agent deep reinforcement learning. CoRR abs/2005.13625 (2020)
53. Umari, H., Mukhopadhyay, S.: Autonomous robotic exploration based on multiple rapidly-exploring randomized trees. In: 2017 IEEE/RSJ International Conference on Intelligent Robots and Systems (IROS), pp. 1396–1402 (2017). https://doi.org/10.1109/IROS.2017.8202319
54. Vaswani, A., et al.: Attention is all you need. In: Guyon, I., et al. (eds.) Advances in Neural Information Processing Systems, vol. 30. Curran Associates, Inc. (2017)
55. Wakilpoor, C., Martin, P.J., Rebhuhn, C., Vu, A.: Heterogeneous multi-agent reinforcement learning for unknown environment mapping. arXiv preprint arXiv:2010.02663 (2020)
56. Wang, H., Wang, W., Zhu, X., Dai, J., Wang, L.: Collaborative visual navigation. arXiv preprint arXiv:2107.01151 (2021)
57. Wang, T., Wang, J., Wu, Y., Zhang, C.: Influence-based multi-agent exploration. In: International Conference on Learning Representations (2020)
58. Wang, W., et al.: From few to more: large-scale dynamic multiagent curriculum learning. In: Proceedings of the AAAI Conference on Artificial Intelligence, vol. 34, pp. 7293–7300 (2020)
59. Wang, X., Girshick, R., Gupta, A., He, K.: Non-local neural networks. In: Proceedings of the IEEE Conference on Computer Vision and Pattern Recognition, pp. 7794–7803 (2018)

60. Wurm, K.M., Stachniss, C., Burgard, W.: Coordinated multi-robot exploration using a segmentation of the environment. In: 2008 IEEE/RSJ International Conference on Intelligent Robots and Systems, pp. 1160–1165. IEEE (2008)
61. Xia, F., Zamir, A.R., He, Z., Sax, A., Malik, J., Savarese, S.: Gibson ENV: real-world perception for embodied agents. In: Proceedings of the IEEE Conference on Computer Vision and Pattern Recognition, pp. 9068–9079 (2018)
62. Yamauchi, B.: A frontier-based approach for autonomous exploration. In: Proceedings 1997 IEEE International Symposium on Computational Intelligence in Robotics and Automation CIRA 1997. Towards New Computational Principles for Robotics and Automation, pp. 146–151. IEEE (1997)
63. Yang, W., Wang, X., Farhadi, A., Gupta, A., Mottaghi, R.: Visual semantic navigation using scene priors. arXiv preprint arXiv:1810.06543 (2018)
64. Yu, C., Velu, A., Vinitsky, E., Wang, Y., Bayen, A., Wu, Y.: The surprising effectiveness of mappo in cooperative, multi-agent games. arXiv preprint arXiv:2103.01955 (2021)
65. Yu, J., et al.: SMMR-explore: submap-based multi-robot exploration system with multi-robot multi-target potential field exploration method. In: 2021 IEEE International Conference on Robotics and Automation (ICRA) (2021)
66. Zambaldi, V., et al.: Relational deep reinforcement learning. arXiv preprint arXiv:1806.01830 (2018)
67. Zhang, C., Song, D., Huang, C., Swami, A., Chawla, N.V.: Heterogeneous graph neural network. In: Proceedings of the 25th ACM SIGKDD International Conference on Knowledge Discovery & Data Mining, pp. 793–803 (2019)
68. Zhang, Y., Hare, J., Prugel-Bennett, A.: Deep set prediction networks. Adv. Neural. Inf. Process. Syst. **32**, 3212–3222 (2019)
69. Zhu, F., et al.: Main: a multi-agent indoor navigation benchmark for cooperative learning (2021)

Zero-Shot Category-Level Object Pose Estimation

Walter Goodwin[1(✉)], Sagar Vaze[2], Ioannis Havoutis[1], and Ingmar Posner[1]

[1] Oxford Robotics Institute, University of Oxford, Oxford, UK
{walter.goodwin,ioannis.havoutis,ingmar.posner}@robots.ox.ac.uk
[2] Visual Geometry Group, University of Oxford, Oxford, UK
sagar.vaze@robots.ox.ac.uk

Abstract. Object pose estimation is an important component of most vision pipelines for embodied agents, as well as in 3D vision more generally. In this paper we tackle the problem of estimating the pose of novel object categories in a zero-shot manner. This extends much of the existing literature by removing the need for pose-labelled datasets or category-specific CAD models for training or inference. Specifically, we make the following contributions. First, we formalise the zero-shot, category-level pose estimation problem and frame it in a way that is most applicable to real-world embodied agents. Secondly, we propose a novel method based on semantic correspondences from a self-supervised vision transformer to solve the pose estimation problem. We further re-purpose the recent CO3D dataset to present a controlled and realistic test setting. Finally, we demonstrate that all baselines for our proposed task perform poorly, and show that our method provides a *six-fold improvement* in average rotation accuracy at 30 °C. Our code is available at https://github.com/applied-ai-lab/zero-shot-pose.

1 Introduction

Consider a young child who is presented with two toys of an object category they have never seen before: perhaps, two toy aeroplanes. Despite having never seen examples of 'aeroplanes' before, the child has the ability to understand the spatial relationship between these related objects, and would be able to align them if required. This is the problem we tackle in this paper: the zero-shot prediction of pose offset between two instances from an object category, without the need for any pose annotations. We propose this as a challenging task which removes many assumptions in the current pose literature, and which more closely resembles the setting encountered by embodied agents in the real-world. To substantiate this claim, consider the information existing pose recognition algorithms have

W. Goodwin and S. Vaze—These authors contributed equally

Supplementary Information The online version contains supplementary material available at https://doi.org/10.1007/978-3-031-19842-7_30.

S. Avidan et al. (Eds.): ECCV 2022, LNCS 13699, pp. 516–532, 2022.
https://doi.org/10.1007/978-3-031-19842-7_30

Fig. 1. Zero-shot category-level pose estimation enables the alignment of different instances of the same object category, without any pose labels for that category or any other. For each category, the estimated pose of the first object, relative to the second, is visualised through transformation of the first object's point cloud.

access to. Current methods make one (or more) of the following assumptions: that evaluation is performed at the *instance-level* (i.e there is no intra-category variation between objects) [41]; that we have access to *labelled pose datasets* for all object categories [3,9,18,27,38,42,45]; and/or that we have access to a realistic *CAD model* for each object category the model will encounter [8,17,44].

Meanwhile, humans can understand pose without access to any of this information. How is this possible? Intuitively, we suggest humans use an understanding of *object parts*, which generalise across categories, to correspond related objects. This process can be followed by using *basic geometric primitives* to understand the spatial relationship between objects. Humans typically also have a coarse *depth estimate* and can inspect the object from *multiple viewpoints*.

In this paper, we use these intuitions to build a solution to estimate the pose offset between two instances of a given category. We perform 'zero-shot' pose-estimation in the sense that our models have never seen pose-labelled examples of the test categories, and neither do they rely on category-specific CAD models. We first make use of features extracted from a vision transformer (ViT [13]), trained in a self-supervised manner on large scale data [6], to establish semantic correspondences between two object instances of the same category. Prior work has demonstrated that self-supervised ViTs have an understanding of object parts which can transfer to novel instances and categories [4,36]. Next, using a weighting of the semantic correspondences, we obtain a coarse estimate of the pose offset by select an optimal viewpoint for one of the object instances. Having obtained semantic correspondences and selected the best view, we use depth maps to create sparse point clouds for each object at the corresponding semantic locations. Finally, we align these point clouds with a rigid-body transform using a robust least squares estimation [35] to give our final pose estimate.

We evaluate our method on the CO3D dataset [28], which provides high-resolution imagery of diverse object categories, with substantial intra-category variance between instances. We find that this allows us to reflect a realistic setting while performing quantitative evaluation in a controlled manner. We consider a range of baselines which could be applied to this task, but find that they perform poorly and often fail completely, demonstrating the highly challenging nature of the problem.

In summary, we make the following contributions:

- We formalise a new and challenging setting for pose estimation, which is an important component of most 3D vision systems. We suggest our setting closely resembles those encountered by real-world embodied agents (Sect. 3).
- We propose a novel method for zero-shot, category-level pose estimation, based on semantic correspondences from self-supervised ViTs (Sect. 4).
- Through rigorous experimentation on a devised CO3D benchmark, we demonstrate that our method facilitates zero-shot pose alignment when the baselines often fail entirely (Sect. 5).

2 Related Work

2.1 Category-Level Pose Estimation

While estimating pose for a single object instances has a long history in robotics and computer vision [41], in recent years there has been an increased interest in the problem of *category-level* pose estimation, alongside the introduction of several category-level datasets with labelled pose [2,38–40]. Approaches to category-level pose estimation can be broadly delineated into: those defining pose explicitly through the use of reference CAD models [17,29,31,31,44]; those which learn category-level representations against which test-time observations can be in some way matched to give relative pose estimates [7,8,11,27,33,37,38,45]; and those that learn to directly predict pose estimates for a category from observations [3,9,18,42].

Most methods (e.g. [8,24,27,33,38]) treat each object category distinctly, either by training a separate model per category, or by using different templates (e.g. CAD models) for each category. A few works (e.g. [42,45]) attempt to develop category-agnostic models or representations, and several works consider the exploitation of multiple views to enhance pose estimation [20,21]. In contrast to existing works in category-level pose estimation, we do not require any pose-labelled data or CAD models in order to estimate pose for a category, and tackle pose estimation for unseen categories.

2.2 Few-Shot and Self-supervised Pose Estimation

There has been some recent work that notes the difficulty of collecting large, labelled, in-the-wild pose datasets, and thus seeks to reduce the data burden by employing few-shot approaches. For instance, Pose-from-Shape [44] exploits existing pose-labelled RGB datasets, along with CAD models, to train an object-agnostic network that can predict the pose of an object in an image, with respect to a provided CAD model. Unlike this work, we seek to tackle an in-the-wild setting in which a CAD model is not available for the objects encountered. Self-supervised, embodied approaches for improving pose estimation for given object instances have been proposed [12], but require extensive interaction and still do not generalise to the category level. Few-shot approaches that can quickly fine-tune to previously unseen categories exist [34,42], but still require a non-trivial number of labelled examples to fine-tune to unseen categories, while in contrast we explore the setting in which no prior information is available. Furthermore, recent work has explored the potential for unsupervised methods with equivariant inductive biases to infer category-level canonical frames without labels [23], and to thus infer 6D object pose given an observed point cloud. This method, while avoiding the need for pose labels, only works on categories for which it has been trained. Finally, closest in spirit to the present work is [14], who note that the minimal requirement to make zero-shot pose estimation a well-posed problem is to provide an implicit canonical frame through use of a reference image, and formulate pose estimation as predicting the relative viewpoint from this view. However, this work can only predict pose for single object instances, and does not extend to the category level.

2.3 Semantic Descriptor Learning

A key component of the presented method to zero-shot category level pose estimation is the ability to formulate semantic keypoint correspondences between pairs of images within an object category, in a zero-shot manner. There has been much interest in semantic correspondences in recent years, with several works proposing approaches for producing these without labels [1,4,6,22]. Semantic correspondence is particularly well motivated in robotic settings, where problems such as extending a skill from one instance of an object to any other demand the ability to relate features across object instances. Prior work has considered learning dense descriptors from pixels [15] or meshes [32] in a self-supervised manner, learning skill-specific keypoints from supervised examples [26], or robust matching at the whole object level [16]. The descriptors in [15,26,32] are used to infer the relative pose of previously unseen object instances to instances seen in skill demonstrations. In contrast to these robotics approaches, in our method we leverage descriptors that are intended to be category-agnostic, allowing us to formulate a zero-shot solution to the problem of pose estimation.

3 Zero-Shot Category-Level Pose Estimation

In this section, we formalise and motivate our proposed zero-shot pose estimation setting. To do this, we first outline the generic problem of object pose estimation. 6D pose estimation is the regression problem of, given an image of the object, regressing to the offset (translation and rotation) of the object with respect to some frame of reference. This frame of reference can be defined implicitly (e.g in the supervised setting, the labels are all defined with respect to some 'canonical' frame) or explicitly (e.g with a reference image). In either case, pose estimation is fundamentally a relative problem. In the zero-shot setting we consider, the frame of reference cannot be implicitly defined by labels: we do not have labelled pose for any objects. Therefore, the pose estimation problem is that of aligning (computing the pose offset between) two instances of a given category.

In our proposed setting, we assume access to N views of a target object, as well as depth information for both objects, and suggest that these constraints reflect practical settings. For objects in the open-world, we are unlikely to have realistic CAD models or labelled pose training sets. On the other hand, many embodied agents are fitted with depth cameras or can recover depth (up to a scale) from structure from motion or stereo correspondence. Furthermore, real-world agents are able to interact with the object and hence gather images from multiple views.

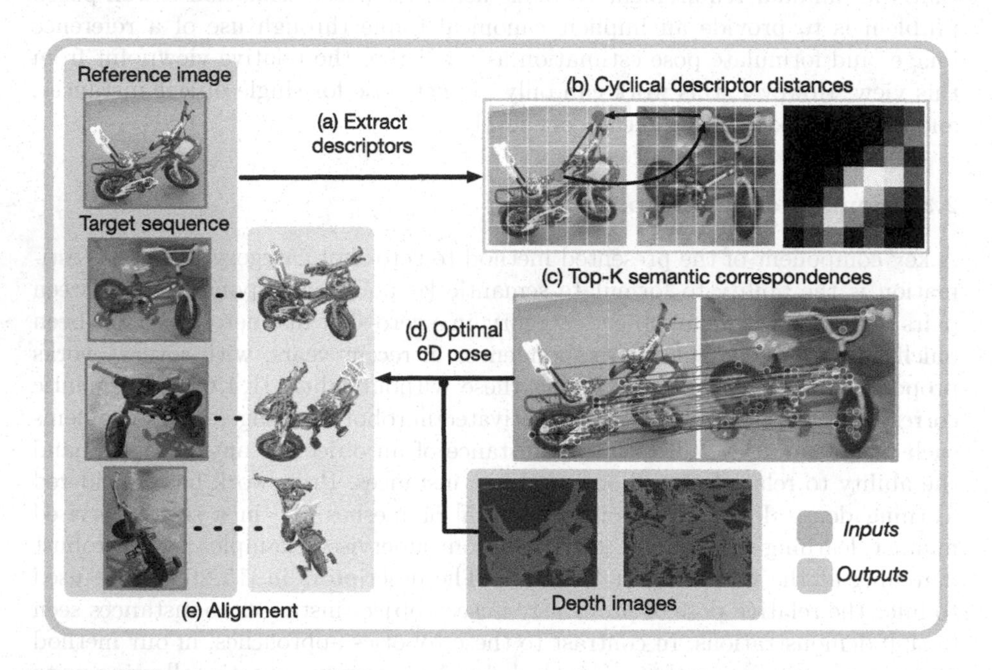

Fig. 2. Our method for zero-shot pose estimation between two instances of an object, given a reference image and a sequence of target images. In our method, we: **(a)** Extract spatial feature descriptors for all images with a self-supervised vision transformer (ViT). **(b)** Compare the reference image to all images in the target sequence by building a set of cyclical distance maps (Sect. 4.1). **(c)** Use these maps to establish K semantic correspondences between compared images and select a suitable view from the target sequence (Sect. 4.2). **(d)** Given the semantic correspondences and a suitable target view, we use depth information to compute a rigid transformation between the reference and target objects (Sect. 4.3). **(e)** Given relative pose transformations between images in the target sequence, we can align the point cloud of the *reference image* with *the entire target sequence.*

Formally, we consider a reference image, $I_{\mathcal{R}}$, and a set of target images $I_{\mathcal{T}_{1:N}} = \{I_{\mathcal{T}_1}...I_{\mathcal{T}_N}\}$, where $I_i \in \mathbb{R}^{H \times W \times 3}$. We further have access to depth maps, $D_i \in$

$\mathbb{R}^{H \times W}$ for all images. Given this information, we require a model, $\mathcal{M}$, to output a single 6D pose offset between the object in the reference image and the object in the target sequence, as:

$$T^* = \mathcal{M}(I_{\mathcal{R}}, I_{\mathcal{T}_{1:N}} | \quad D_{\mathcal{R}}, D_{\mathcal{T}_{1:N}}) \tag{1}$$

Finally, we note that, in practice, the transformations between the target views must be known for the predicted pose offset to be most useful. These transformations are easily computed by an embodied agent and can be used to, given an alignment between $I_{\mathcal{R}}$ and *any* of the target views, align the reference instance with the entire target sequence.

4 Methods

In this section, we detail our method for zero-shot pose estimation. First, semantic correspondences are obtained between the reference and target object (Sect. 4.1). These correspondences are used to select a suitable view for pose estimation from the N images in the target sequence (Sect. 4.2). Finally, using depth information, the correspondences' spatial locations are used to estimate the pose offset between the reference and target object instances (Sect. 4.3).

4.1 Self-supervised Semantic Correspondence with Cyclical Distances

The key insight of our method is that semantic, parts-based correspondences generalise well between different object instances within a category, and tend to be spatially distributed in similar ways for each such object. Indeed, a parts-based understanding of objects can also generalise between categories; for instance, 'eyes', 'ears' and 'nose' transfer between many animal classes. Recent work has demonstrated that parts-based understanding emerges naturally from self-supervised vision transformer (ViT) features [4,6,36], and our solution leverages such a network with large scale pre-training [6]. The ViT is trained over ImageNet-1K, and we assume that it carries information about a sufficiently large set of semantic object parts to generalise to arbitrary object categories.

As described in Sect. 3, the proposed setting for pose estimation considers a relative problem, between a reference object (captured in a single image) and a target object (with potentially multiple views available). We compare two images (for now referred to as I_1, I_2) by building a 'cyclical distance' map for every pixel location in I_1 using feature similarities. Formally, consider $\Phi(I_i) \in \mathbb{R}^{H' \times W' \times D}$ as the normalised spatial feature of an image extracted by a ViT. Letting u be an index into $\Phi(I_1)$ as $u \in \{1...H'\} \times \{1...W'\}$, we find it's cyclical point u' as:

$$u' = \operatorname*{argmin}_w d(\Phi(I_1)_w, \Phi(I_2)_v) \quad | \quad v = \operatorname*{argmin}_w d(\Phi(I_1)_u, \Phi(I_2)_w) \tag{2}$$

Here $d(\cdot,\cdot)$ is the L2-distance, and a cyclical distance map is constructed as $C \in \mathbb{R}^{H' \times W'}$ with $C_u = -d(u, u')$. Using the top-K locations in C, we take features from $\Phi(I_1)$ and their nearest neighbours in $\Phi(I_2)$ as correspondences. This process is illustrated in Fig. 2b.

The cyclical distance map can be considered as a soft mutual nearest neighbours assignment. Mutual nearest neighbours [4] between I_1 and I_2 return a cyclical distance of zero, while points in I_1 with a small cyclical distance can be considered to 'almost' have a mutual nearest neighbour in I_2. The proposed cyclical distance metric has two key advantages over the hard constraint. Firstly, while strict mutual nearest neighbours gives rise to an unpredictable number of correspondences, the soft measure allows us to ensure K semantic correspondences are found for every pair of images. We find having sufficient correspondences is critical for the downstream pose estimation. Secondly, the soft constraint adds a spatial prior to the correspondence discovery process: features belonging to the same object part are likely to be close together in pixel space.

Finally, following [4], after identifying an initial set of matches through our cyclical distance method, we use K-Means clustering on the selected features in the reference image to recover points which are spatially well distributed on the object. We find that well distributed points result in a more robust final pose estimate (see supplementary). In practise, we select the top-$2K$ correspondences by cyclical distance, and filter to a set of K correspondences with K-Means.

4.2 Finding a Suitable View for Alignment

Finding semantic correspondences between two images which view (two instances of) an object from very different orientations is challenging. For instance, it is possible that images from the front and back of an object have no semantic parts in common. To overcome this, an agent must be able to choose a suitable view from which to establish semantic correspondences. In the considered setting, this entails selecting the best view from the N target images. We do this by constructing a correspondence score between the reference image, $I_{\mathcal{R}}$, and each image in the target sequence, $I_{\mathcal{T}_{1:N}}$. Specifically, given the reference image and an image from the target sequence, the correspondence score is the sum the of the feature similarities between their K semantic correspondences. Mathematically, given a set of K correspondences between the jth target image and the reference, $\{(u_k^j, v_k^j)\}_{k=1}^K$, this can be written as:

$$j^* = \underset{j \in 1:N}{\operatorname{argmax}} \sum_{k=1}^{K} -d(\Phi(I_{\mathcal{R}})_{u_k^j}, \Phi(I_{\mathcal{T}_j})_{v_k^j}) \tag{3}$$

4.3 Pose Estimation from Semantic Correspondences and Depth

The process described in Sect. 4.1 gives rise to a set of corresponding points in 2D pixel coordinates, $\{(u_k, v_k)\}_{k=1}^K$. Using depth information and camera intrinsics, these are unprojected to their corresponding 3D coordinates, $\{(\mathbf{u}_k, \mathbf{v}_k)\}_{k=1}^K$, where $\mathbf{u}_k, \mathbf{v}_k \in \mathbb{R}^3$. In the pose estimation problem, we seek a single 6D pose that describes the orientation and translation of the target object, relative to the frame defined by the reference object. Given a set of corresponding 3D points, there are a number of approaches for solving for this rigid body transform. As we assume our correspondences are both noisy and likely to contain outliers, we use a fast least-squares method based on the singular value decomposition [35], and use RANSAC to handle outliers. We run RANSAC for up to 10,000 iterations, with further details in supplementary. The least squares solution recovers a 7-dimensional transform: rotation $\mathbf{R}$, translation $\mathbf{t}$, and a uniform scaling parameter λ, which we found crucial for dealing with cross-instance settings. The least-squares approach minimises the residuals and recovers the predicted 6D pose offset, T^* as:

$$T^* = (\mathbf{R}^*, \mathbf{t}^*) = \underset{(\mathbf{R},\mathbf{t})}{\operatorname{argmin}} \sum_{k=1}^{K} \mathbf{v}_k - (\lambda \mathbf{R} \mathbf{u}_k + \mathbf{t}) \tag{4}$$

5 Experiments

5.1 Evaluation Setup

Dataset, CO3D [28]*:* To evaluate zero-shot, category-level pose estimation methods, a dataset is required that provides images of multiple object categories, with a large amount of intra-category instance variation, and with varied object viewpoints. The recently released Common Objects in 3D (CO3D) dataset fulfils these requirements with 1.5 million frames, capturing objects from 50 categories, across nearly 19k scenes [28]. For each object instance, CO3D provides approximately 100 frames taken from a 360° viewpoint sweep with handheld cameras, with labelled camera pose offsets. The proposed method makes use of depth information, and CO3D provides estimated object point clouds, and approximate depth maps for each image, that are found by a Structure-from-Motion (SfM) approach applied over the sequences [30]. We note that, while other pose-oriented datasets exist [2,40,41], we find them to either be lacking in necessary meta-data (e.g no depth information), have little intra-category variation (e.g be instance level), contain few categories, or only provide a single image per object instance. We expand on dataset choice in the supplementary.

Labels for Evaluation: While the proposed pose estimation method requires no pose-labelled images for training, we label a subset of sequences across the CO3D categories for quantitative evaluation. We do this by assigning a category-level canonical frame to each selected CO3D sequence. We exclude categories that have infinite rotational symmetry about an axis (e.g 'apple') or have an insufficient number of instances with high quality point clouds (e.g 'microwave'). For the remaining 20 categories, we select the top-10 sequences based on a point cloud quality metric. Point clouds are manually aligned within each category with a rigid body transform. As CO3D provides camera extrinsics for every frame in a sequence with respect to its point cloud, these alignments can be propagated to give labelled category-canonical pose for every frame in the chosen sequences. Further details are in the supplementary.

Evaluation Setting: For each object category, we sample 100 combinations of sequence pairs, between which we will compute pose offsets. For the first sequence in each pair, we sample a single reference frame, $I_{\mathcal{R}}$, and from the second we sample N target frames, $I_{\mathcal{T}_{1:N}}$. We take $N = 5$ as our standard setting, with results for different numbers of views in Table 2 and the supplementary. For each pair of sequences, we compute errors in pose estimates between the ground truth and the predictions. For the rotation component, following standard practise in the pose estimation literature, we report the median error across samples, as well as the accuracy at 15° and 30°, which are given by the percentage of predictions with an error less than these thresholds. Rotation error is given by the geodesic distance between the ground truth predicted rotations [19].

'Zero-shot' Pose Estimation: In this work, we leverage models with large-scale, self-supervised pre-training. The proposed pose estimation method is 'zero-shot' in the sense that it does not use labelled examples (either pose labels or category labels) for any of the object categories it is tested on. The self-supervised features, though, may have been trained on images containing unlabelled instances of some object categories considered. To summarise, methods in this paper **do not require** labelled pose training sets or CAD models for the categories they encounter during evaluation. They **do require** large-scale unsupervised pre-training, depth estimates, and multiple views of the target object. We assert that these are more realistic assumptions for embodied agents (see Sect. 3).

5.2 Baselines

We find very few baselines in the literature which can be applied to the highly challenging problem of pose-detection on unseen categories. Though some methods have tackled the zero-shot problem before, they are difficult to translate to our setting as they require additional information such as CAD models for the test objects. We introduce the baselines considered.

PoseContrast [42]: This work seeks to estimate 3D pose (orientation only) for previously unseen categories. The method trains on pose-labelled images and assumes unseen categories will have both sufficiently similar appearance and geometry, and similar category-canonical frames, to those seen in training. We adapt this method for our setting and train it on all 100 categories in the ObjectNet3D dataset [43]. During testing, we extract global feature vectors for the reference and target images with the model, and use feature similarities to select a suitable view. We then run the PoseContrast model on the reference and selected target image, with the model regressing to an Euler angle representation of 3D pose. PoseContrast estimates pose for each image independently, implicitly inferring the canonical frame for the test object category. We thus compute the difference between the predicted the pose predictions for the reference and chosen target image to arrive at a relative pose estimate.

Iterative Closest Point (ICP): ICP is a point cloud alignment algorithm that assumes no correspondences are known between two point clouds, and seeks an optimal registration. We use ICP to find a 7D rigid body transform (scale, translation and rotation, as in Sect. 4.3) between the reference and target objects. We use the depth estimates for each image to recover point clouds for the two instances, aggregating the N views in the target sequence for a maximally complete target point cloud. We use these point clouds with ICP. As ICP is known to perform better with good initialisation, we also experiment with initialising it from the coarse pose-estimate given by our 'best view' method (see Sect. 4.2) which we refer to as 'ICP + BV'.

Image Matching: Finally, we experiment with other image matching techniques. In the literature, cross-instance correspondence is often tackled by learning category-level keypoints. However, this usually involves learning a different model for each category, which defeats the purpose of our task. Instead, we use category-agnostic features and obtain matches with mutual nearest neighbours between images, before combining the matches' spatial locations with depth information to compute pose offsets (similarly to Sect. 4.3). We experiment both with standard SIFT features [25] and deep features extracted with an ImageNet self-supervised ResNet-50 (we use SWaV features [5]). In both cases, we select the best view using the strength of the discovered matches between the reference and target images (similarly to Sect. 4.2).

5.3 Implementation Details

In this work we use pre-trained DINO ViT features [6] to provide semantic correspondences between object instances. Specifically, we use ViT-Small with a patch size of 8, giving feature maps at a resolution of 28×28 from square 224×224 images. Prior work has shown that DINO ViT features encode information on generalisable object parts and correspondences [4,36]. We follow [4] for feature processing and use 'key' features from the 9th ViT layer as our feature representation, and use logarithmic spatial binning of features to aggregate local context to at each ViT patch location. Furthermore, the attention maps in the ViT provide a reasonable foreground segmentation mask. As such, when computing cyclical distances, we assign infinite distance to any point which lands off the foreground at any stage in the reference-target image cycle (Sect. 4.1), to ensure that all correspondences are on the objects of interest. We refer to the supplementary for further implementation details on our method and baselines.

5.4 Main Results

We report results averaged over the 20 considered categories in CO3D in the leftmost columns of Table 1. We first highlight that the baselines show poor performance across the reported metrics. ICP and SIFT perform most poorly, which we attribute to them being designed for within-instance matching. Alignment with the SWaV features, which contain more semantic information, fares slightly better, though still only reports a 7.5% accuracy at 30°. Surprisingly, we also found PoseContrast to give low accuracies in our setting. At first glance, this could simply be an artefact of different canonical poses—between those inferred by the model, and those imposed by the CO3D labels. However, we note that we *take the difference* between the reference and target poses as our pose prediction, which cancels any constant-offset artefacts in the canonical pose.

Table 1. We report median error and accuracy at 30°, 15° averaged across all 20 categories. We also report Accuracy at 30° broken down by class for an illustrative subset of categories. We provide full, per category breakdowns in the supplementary.

	All categories			Per category (Acc30↑)				
	Med. Err (↓)	Acc30 (↑)	Acc15 (↑)	Bike	Hydrant	M'cycle	Teddy	Toaster
ICP	111.8	3.8	0.7	3.0	6.0	1.0	1.0	7.0
SIFT	129.4	4.0	1.5	3.0	11.0	1.0	1.0	0.0
SWaV	123.1	7.5	3.3	13.0	9.0	8.0	5.0	7.0
ICP+BV	109.3	5.4	1.2	6.0	5.0	8.0	5.0	5.0
PoseContrast	111.5	6.9	1.1	2.0	4.0	13.0	4.0	12.0
Ours (K=30)	60.2	43.5	29.0	63.0	24.0	80.0	33.0	39.0
Ours (K=50)	**53.8**	**46.3**	**31.1**	**71.0**	**26.0**	**82.0**	**39.0**	**42.0**

Meanwhile, our method shows substantial improvements over all implemented baselines. Our method reports less than half the Median Error aggregated over all categories, and further demonstrates a *six-fold increase* in accuracy at 30°. We also note that this improvement cannot solely be attributed to the scale of DINO's ImageNet pre-training: the SWaV-based baseline also uses self-supervised features trained on ImageNet [5], and PoseContrast is initialised with MoCo-v2 [10] weights, again from self-supervision on ImageNet.

We find that performance varies substantially according to the specific geometries and appearances of individual categories. As such, in the rightmost columns of Table 1, we show per-category results for an illustrative subset of the selected classes in CO3D. We find that textured objects, which induce high quality correspondences, exhibit better results (e.g 'Bike' and 'Motorcycle'). Meanwhile, objects with large un-textured regions (e.g 'Toaster') proved more challenging.

The results for 'Hydrant' are illustrative of a challenging case. In principle, a hydrant has a clearly defined canonical frame, with faucets appearing on only three of its four 'faces' (see Fig. 3). However, if the model fails to identify all three faucets as salient keypoints for correspondence, the object displays a high degree of rotational symmetry. In this case, SIFT, which focuses exclusively on appearance (i.e it does not learn semantics), performs higher than its average, as the hydrant faucets are consistently among the most textured regions on the object. Meanwhile, our method, which focuses more on semantics, performs worse than its own average on this category.

5.5 Making Use of Multiple Views

The Number of Target Views : A critical component of our setting is the availability of multiple views of the target object. We argue that this is important for the computation of zero-shot pose offset between two object instances, as a single image of the target object may not contain anything in common with the reference image. An important factor, therefore, is the number of images available in the target sequence. In principle, if one had infinite views of the target sequence, and camera transformations between each view, the pose estimation problem collapses to that of finding the best view. However, we note that this is unrealistic. Firstly, running inference on a set of target views is expensive, with the computational burden generally scaling linearly with the number of views. Secondly, collecting and storing an arbitrarily large number of views is also expensive. Finally, the number of views required to densely and uniformly sample viewpoints of an object is very high, as it requires combinatorially sampling with respect to three rotation parameters.

In this work we experiment with the realistic setting of a 'handful' of views of the target object. In Table 2, we experiment with varying N in $\{1, 3, 5\}$ instances in the target sequence. In the bottom three rows, we show the performance of our full method as N is varied and find that, indeed, the performance increases with the number of available views (further results are in supplementary). However, we find that even from a *single view*, our method can outperform the baselines with access to five views.

Only Finding Best Target View : We disambiguate the 'coarse' and 'fine' pose-estimation steps of our method (Sect. 4.2 and Sect. 4.3 respectively). Specifically, we experiment with our method's performance if we assume the reference image is perfectly aligned with the selected best target view. We show these figures as 'Ours-BV' in the top rows of Table 2. It can be seen that this part of our method alone can substantially outperform the strongest baselines. However, we also show that the subsequent fine alignment step using the depth information (Sect. 4.3) provides an important improvement in performance. For instance, when $N = 5$, this component of our method boosts Acc30 from 35.4% to 46.3%.

How to Pick the Best Target View : Here, we explore the importance of our particular recipe for arriving at the optimal target view. First, we experiment with a standard baseline of choosing the target view which maximises the similarity with respect to the ViT's global feature vector (termed *'GlobalSim'*). We also try maximising the Intersection-over-Union of the foreground masks, as provided by the ViT attention maps, of the reference and target frames (*'SaliencyIoU'*). Finally, we try maximising the IoU between the foreground mask of a target object and its cyclical distance map with respect to the reference image. The intuition here is to recover a target view where a large proportion of the foreground object pixels have a unique nearest neighbour in the reference image (*'CyclicalDistIoU'*). We present the results of these findings in Table 3.

5.6 Qualitative Results

In Fig. 3 we provide qualitative alignment results for four object categories, including for 'Hydrant', which we include as a failure mode. The images show the reference image and the best view from the target sequence, along with the semantic correspondences discovered between them. We further show the point cloud for the reference image aligned with the target sequence using our method. Specifically, we first compute a relative pose offset between the reference image and the best target view, and then propagate this pose offset using camera extrinsics to the other views in the target sequence. Finally, we highlight the practical utility of this setting. Consider a household robot wishing to tidy away a 'Teddybear' (top row) into a canonical pose (defined by a reference image). Using this method, the agent is able to view the toy from a number of angles (in the target sequence), align the reference image to an appropriate view, and thus understand the pose of the toy *from any other angle.*

Table 2. We experiment with varying numbers of images available in the target sequence (N). Even with only one view, our method substantially outperforms existing baselines with access to multiple views. We further show the utility of pose alignment from the best view ('Ours') over simply choosing the best view with our method ('Ours-BV').

Method	Med. Err	Acc30	Acc15
Ours-BV (N=1)	92.8	12.6	3.7
Ours-BV (N=3)	69.5	26.1	8.0
Ours-BV (N=5)	61.1	35.4	10.6
Ours (N=1)	97.3	23.8	13.4
Ours (N=3)	63.9	38.9	26.2
Ours (N=5)	**53.8**	**46.3**	**31.1**

Table 3. We ablate different methods for selecting the best view from the target sequence, from which we perform our final pose computation. Compared to a other intuitive options for this task, we demonstrate the importance of our proposed best view selection pipeline for downstream performance.

Method	Med. Err	Acc30	Acc15
CyclicalDistIOU	94.7	22.3	13.1
GlobalSim	71.9	36.8	23.7
SaliencyIOU	87.0	29.1	18.1
CorrespondSim	**53.8**	**46.3**	**31.1**

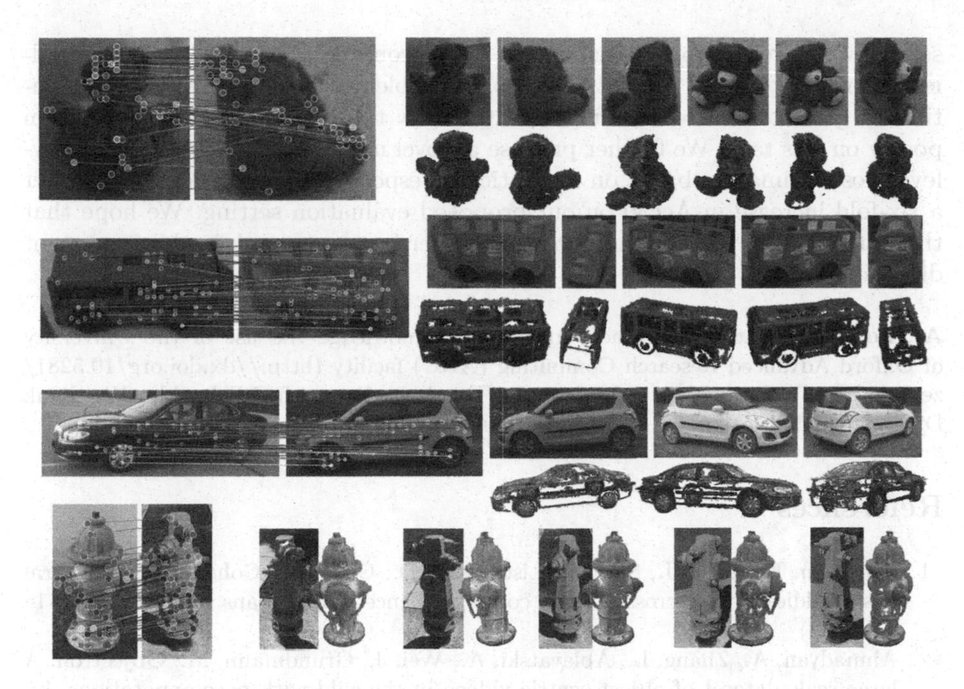

Fig. 3. Example results for the categorics **Teddybear**, **Toybus**, **Car**, **Hydrant**. Depicted are the correspondences found between the reference image and the best-matching frame from the target sequence found following Sect. 4.2. To the right, the estimated pose resulting from these correspondences is shown as an alignment between the reference object (shown as a rendered point cloud) and the target sequence. **Hydrant** depicts a failure mode—while the result looks visually satisfying, near rotational symmetry (about the vertical axis) leads to poor alignment.

6 Conclusion

Consideration of Limitations: We have proposed a model which substantially outperforms existing applicable baselines for the task of zero-shot category-level pose detection. However, absolute accuracies remain low and we suggest fall far short of human capabilities. Firstly, our performance across the considered classes is 46.3% Acc30 with 5 views available. We imagine these accuracies to be substantially lower than the a human baseline for this task. Secondly, though single view novel category alignment is highly challenging for machines, humans are capable of generalising highly abstract concepts to new categories, and thus would likely be able to perform reasonably in a single view setting.

Final Remarks: In this paper we have proposed a highly challenging (but realistic) setting for object pose estimation, which is a critical component in most 3D vision pipelines. In our proposed setting, a model is required to align two instances of an object category without having any pose-labelled data for training. We further re-purpose the recently released CO3D dataset and devise a test setting which reasonably resembles the one encountered by a real-world embodied agent. Our setting presents a complex problem which requires both semantic and geometric understanding, and we show that existing baselines perform poorly on this task. We further propose a novel method for zero-shot, category-level pose estimation based on semantic correspondences and show it can offer a six-fold increase in Acc30 on our proposed evaluation setting. We hope that this work will serve as a spring-board to foster future research in this important direction.

Acknowledgment. The authors gratefully acknowledge the use of the University of Oxford Advanced Research Computing (ARC) facility (http://dx.doi.org/10.5281/zenodo.22558). Sagar Vaze is funded by a Facebook Research Scholarship. We thank Dylan Campbell for many useful discussions.

References

1. Aberman, K., Liao, J., Shi, M., Lischinski, D., Chen, B., Cohen-Or, D.: Neural best-buddies: sparse cross-domain correspondence. ACM Trans. Graph. **37**(4), 1–14 (2018)
2. Ahmadyan, A., Zhang, L., Ablavatski, A., Wei, J., Grundmann, M.: Objectron: A large scale dataset of object-centric videos in the wild with pose annotations. In: CVPR, pp. 7822–7831 (2021)
3. Akizuki, S.: ASM-Net : Category-level pose and shape estimation using parametric deformation. In: BMVC (2021)
4. Amir, S., Gandelsman, Y., Bagon, S., Dekel, T.: Deep ViT Features as Dense Visual Descriptors (2021)
5. Caron, M., Misra, I., Mairal, J., Goyal, P., Bojanowski, P., Joulin, A.: Unsupervised learning of visual features by contrasting cluster assignments. NeurIPS **33**, 9912–9924 (2020)
6. Caron, M., et al.: Emerging properties in self-supervised vision transformers. In: ICCV, pp. 9650–9660 (2021)
7. Chen, D., Li, J., Wang, Z., Xu, K.: Learning canonical shape space for category-level 6D object pose and size estimation. In: CVPR, pp. 11973–11982 (2020)
8. Chen, K., Dou, Q.: SGPA: Structure-guided prior adaptation for category-level 6D object pose estimation. In: ICCV, pp. 2773–2782 (2021)
9. Chen, W., Jia, X., Chang, H.J., Duan, J., Shen, L., Leonardis, A.: FS-Net: fast shape-based network for category-level 6d object pose estimation with decoupled rotation mechanism. In: CVPR, pp. 1581–1590 (2021)
10. Chen, X., Fan, H., Girshick, R.B., He, K.: Improved baselines with momentum contrastive learning (2020). https://arxiv.org/abs/2003.04297
11. Chen, X., Dong, Z., Song, J., Geiger, A., Hilliges, O.: Category level object pose estimation via neural analysis-by-synthesis. In: Vedaldi, A., Bischof, H., Brox, T., Frahm, J.-M. (eds.) ECCV 2020. LNCS, vol. 12371, pp. 139–156. Springer, Cham (2020). https://doi.org/10.1007/978-3-030-58574-7_9
12. Deng, X., Xiang, Y., Mousavian, A., Eppner, C., Bretl, T., Fox, D.: Self-supervised 6D object pose estimation for robot manipulation. In: ICRA, pp. 3665–3671 (2020)
13. Dosovitskiy, A., et al.: An image is worth 16 × 16 words: transformers for image recognition at scale. In: ICLR (2021)

14. El Banani, M., Corso, J.J., Fouhey, D.F.: Novel object viewpoint estimation through reconstruction alignment. In: CVPR, pp. 3113–3122 (2020)
15. Florence, P.R., Manuelli, L., Tedrake, R.: Dense object nets: learning dense visual object descriptors by and for robotic manipulation. In: CoRL (2018)
16. Goodwin, W., Vaze, S., Havoutis, I., Posner, I.: Semantically grounded object matching for robust robotic scene rearrangement. In: ICRA, pp. 11138–11144 (2021)
17. Grabner, A., Roth, P.M., Lepetit, V.: 3D pose estimation and 3D model retrieval for objects in the wild. In: CVPR, pp. 3022–3031 (2018)
18. Gupta, S., Arbeláez, P., Girshick, R., Malik, J.: Inferring 3D object pose in RGB-D images (2015)
19. Huynh, D.Q.: Metrics for 3D rotations: comparison and analysis. J. Math. Imaging Vis. **35**(2), 155–164 (2009)
20. Kanezaki, A., Matsushita, Y., Nishida, Y.: RotationNet: joint object categorization and pose estimation using multiviews from unsupervised viewpoints. In: CVPR, pp. 5010–5019 (2018)
21. Kundu, J.N., Rahul, M.V., Ganeshan, A., Babu, R.V.: Object pose estimation from monocular image using multi-view keypoint correspondence. In: Leal-Taixé, L., Roth, S. (eds.) ECCV 2018. LNCS, vol. 11131, pp. 298–313. Springer, Cham (2019). https://doi.org/10.1007/978-3-030-11015-4_23
22. Lee, J., Kim, D., Ponce, J., Ham, B.: SFNET: learning object-aware semantic correspondence. In: CVPR, pp. 2278–2287 (2019)
23. Li, X., et al.: Leveraging SE(3) equivariance for self-supervised category-level object pose estimation. NeurIPS **34**, 15370–15381 (2021)
24. Lin, Y., Tremblay, J., Tyree, S., Vela, P.A., Birchfield, S.: Single-stage keypoint-based category-level object pose estimation from an RGB image. In: 2022 International Conference on Robotics and Automation (ICRA), pp. 1547–1553 (2021)
25. Lowe, D.G.: Distinctive image features from scale-invariant keypoints. Int. J. Comput. Vis. **60**(2), 91–110 (2004)
26. Manuelli, L., Gao, W., Florence, P., Tedrake, R.: kPAM: KeyPoint Keypoint affordances for category-level robotic manipulation. In: International Symposium on Robotics Research (ISRR), pp. 132–157 (2019)
27. Pavlakos, G., Zhou, X., Chan, A., Derpanis, K.G., Daniilidis, K.: 6-DoF object pose from semantic keypoints. In: ICRA, pp. 2011–2018 (2017)
28. Reizenstein, J., Shapovalov, R., Henzler, P., Sbordone, L., Labatut, P., Novotny, D.: Common objects in 3D: large-scale learning and evaluation of real-life 3D category reconstruction. In: ICCV, pp. 10901–10911 (2021)
29. Sahin, C., Kim, T.K.: Category-level 6D object pose recovery in depth images. In: ECCV (2019)
30. Schonberger, J.L., Frahm, J.M.: Structure-from-motion revisited. In: CVPR, pp. 4104–4113 (2016)
31. Shi, J., Yang, H., Carlone, L.: Optimal pose and shape estimation for category-level 3D object perception. Robot. Sci. Syst. XVII (2021)
32. Simeonov, A., et al.: Neural descriptor fields: SE(3)-equivariant object representations for manipulation. In: 2022 International Conference on Robotics and Automation (ICRA), pp. 6394–6400 (2021)
33. Tian, M., Ang, M.H., Lee, G.H.: Shape prior deformation for categorical 6D object pose and size estimation. In: Vedaldi, A., Bischof, H., Brox, T., Frahm, J.-M. (eds.) ECCV 2020. LNCS, vol. 12366, pp. 530–546. Springer, Cham (2020). https://doi.org/10.1007/978-3-030-58589-1_32

34. Tseng, H.Y., et al.: Few-shot viewpoint estimation. In: BMVC (2020)
35. Umeyama, S.: Least-squares estimation of transformation parameters between two point patterns. IEEE Trans. Pattern Anal. Mach. Intell. **13**(4), 376–380 (1991)
36. Vaze, S., Han, K., Vedaldi, A., Zisserman, A.: Generalized category discovery. In: CVPR, pp. 7492–7501 (2022)
37. Wang, A., Kortylewski, A., Yuille, A.: NeMo: neural mesh models of contrastive features for robust 3D pose estimation. In: ICLR (2021)
38. Wang, H., Sridhar, S., Huang, J., Valentin, J., Song, S., Guibas, L.: Normalized object coordinate space for category-level 6D object pose and size estimation. In: CVPR, pp. 2642–2651 (2019)
39. Xiang, Y., et al.: ObjectNet3D: a large scale database for 3D object recognition. In: Leibe, B., Matas, J., Sebe, N., Welling, M. (eds.) ECCV 2016. LNCS, vol. 9912, pp. 160–176. Springer, Cham (2016). https://doi.org/10.1007/978-3-319-46484-8_10
40. Xiang, Y., Mottaghi, R., Savarese, S.: Beyond PASCAL: a benchmark for 3D object detection in the wild. In: WACV, pp. 75–82 (2014)
41. Xiang, Y., Schmidt, T., Narayanan, V., Fox, D.: PoseCNN: a convolutional neural network for 6D object pose estimation in cluttered scenes. In: Robotics: Science and Systems XIV (2018)
42. Xiao, Y., Du, Y., Marlet, R.: Posecontrast: class-agnostic object viewpoint estimation in the wild with pose-aware contrastive learning. In: 3DV, pp. 74–84 (2021)
43. Xiao, Y., Marlet, R.: Few-shot object detection and viewpoint estimation for objects in the wild. In: Vedaldi, A., Bischof, H., Brox, T., Frahm, J.-M. (eds.) ECCV 2020. LNCS, vol. 12362, pp. 192–210. Springer, Cham (2020). https://doi.org/10.1007/978-3-030-58520-4_12
44. Xiao, Y., Qiu, X., Langlois, P.A., Aubry, M., Marlet, R.: Pose from shape: deep pose estimation for arbitrary 3D objects. In: BMVC (2019)
45. Zhou, X., Karpur, A., Luo, L., Huang, Q.: StarMap for category-agnostic keypoint and viewpoint estimation. In: ECCV, pp. 318–334 (2018)

Sim-to-Real 6D Object Pose Estimation via Iterative Self-training for Robotic Bin Picking

Kai Chen[1], Rui Cao[1], Stephen James[2], Yichuan Li[1], Yun-Hui Liu[1], Pieter Abbeel[2], and Qi Dou[1](✉)

[1] The Chinese University of Hong Kong, Hong Kong, China
qidou@cuhk.edu.hk

[2] University of California, Berkeley, US

Abstract. 6D object pose estimation is important for robotic bin-picking, and serves as a prerequisite for many downstream industrial applications. However, it is burdensome to annotate a customized dataset associated with each specific bin-picking scenario for training pose estimation models. In this paper, we propose an iterative self-training framework for sim-to-real 6D object pose estimation to facilitate cost-effective robotic grasping. Given a bin-picking scenario, we establish a photo-realistic simulator to synthesize abundant virtual data, and use this to train an initial pose estimation network. This network then takes the role of a teacher model, which generates pose predictions for unlabeled real data. With these predictions, we further design a comprehensive adaptive selection scheme to distinguish reliable results, and leverage them as pseudo labels to update a student model for pose estimation on real data. To continuously improve the quality of pseudo labels, we iterate the above steps by taking the trained student model as a new teacher and re-label real data using the refined teacher model. We evaluate our method on a public benchmark and our newly-released dataset, achieving an ADD(-S) improvement of 11.49% and 22.62% respectively. Our method is also able to improve robotic bin-picking success by 19.54%, demonstrating the potential of iterative sim-to-real solutions for robotic applications. Project homepage: www.cse.cuhk.edu.hk/~kaichen/sim2real_pose.html.

Keywords: Sim-to-real adaptation · 6D object pose estimation · Iterative self-training · Robotic bin picking

1 Introduction

6D object pose estimation aims to identify the position and orientation of objects in a given environment. It is a core task for robotic bin-picking and plays an

Supplementary Information The online version contains supplementary material available at https://doi.org/10.1007/978-3-031-19842-7_31.

S. Avidan et al. (Eds.): ECCV 2022, LNCS 13699, pp. 533–550, 2022.
https://doi.org/10.1007/978-3-031-19842-7_31

increasingly important role in elevating level of automation and reducing production cost for various industrial applications [10,28,29,49,50]. Though recent progress has been made, this task still remains highly challenging, given that industrial objects are small and always cluttered together, leading to heavy occlusions and incomplete point clouds. Early methods have relied on domain-specific knowledge and designed descriptors based on material textures or shapes of industrial objects [11,20]. Recently, learning-based models [19,34,41,48] that directly regress 6D pose from RGB-D inputs have emerged as more general solutions with promising performance for scalable application in industry.

However, training deep networks requires a large amount of annotated data. Different from labeling for ordinary computer vision tasks such as object detection or segmentation, annotations of 6D object pose is extremely labor-intensive (if possible), as it manages RGB-D data and involves several cumbersome steps, such as camera localization, scene reconstruction, and point cloud registration [2,15,32]. Without an easy way to get pose labels for real environments, the practicality of training object pose estimation networks lessens.

Perhaps then, the power of simulation can be leveraged to virtually generate data with labels of object 6D pose to train deep learning models for this task [22]. However, due to the simulation-to-reality gap, if we exclusively train a model with synthetic data, it is not guaranteed to also work well on real data. Sim-to-real adaptation methods are therefore required. Some methods use physical-based simulation [23] or mixed reality [53] to make simulators as realistic as possible. Other works [37,39,45] use domain randomization [24,43] to vary the simulated environments with different textures, illuminations or reflections. Some recent methods resort to features that are less affected by the sim-to-real gap for object pose estimation [14,30]. Alternatively, some methods initially train a model on synthetic data, and then refine the model by self-supervised learning on unlabeled real data [7,52]. Unfortunately, these sim-to-real methods either rely on a customized simulation to produce high-quality synthetic data or need a customized network architecture for extracting domain-invariant signal for sim-to-real pose estimation. Given that a robot would be applied to various upcoming bin-picking scenarios, customizing the method to fit for various different real scenes could also be labor-intensive. In addition, since the adaptation to complex bin-picking scenarios is not sufficient, obvious performance gap between simulation and real evaluation is still present in existing sim-to-real methods.

These current limitations encourage us resorting to a new perspective to tackle this challenging task. In general, a model trained on synthetic data should have acquired essential knowledge about how to estimate object poses. If we directly expose it to real data, its prediction is not outrageously wrong. If the model could directly self-train itself leveraging the successful practices, it is highly likely to efficiently adapt to reality, just with unlabeled data. Collecting a large amount of unlabeled real data is easy to achieve for a robot. More importantly, this adaptation process is scalable and network-agnostic. In any circumstances, the robot can be deployed with the most suitable deep learning network for its target task, record sufficient unlabeled real data, and then leverage the above self-training adaptation idea for accurate object pose estimation in its own scenario. Promisingly, similar self-training paradigms have been recently explored for some other computer vision applications such as self-driving and medical image diagnosis, for handling distribution shift towards real-world use [33,36,55]. However, the question of how it can be used for reducing annotation cost while keeping up real-data performance of 6D object pose estimation for robots is unknown.

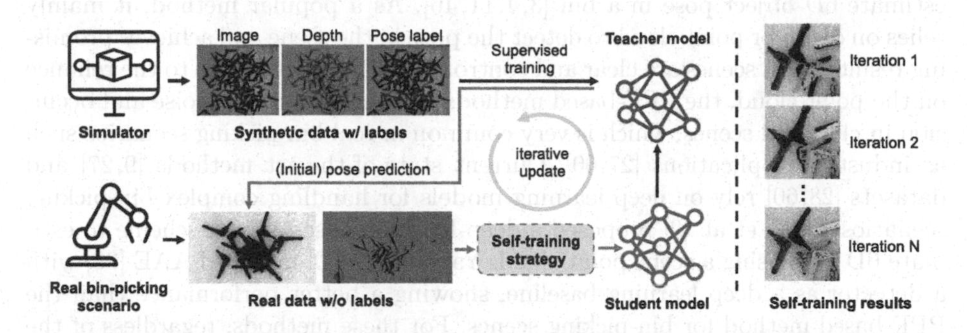

Fig. 1. For industrial bin-picking, we build a simulator to generate synthetic data for learning a teacher model. Starting from it, our novel iterative self-training method can select reliable pseudo-labels and progressively improve the 6D object pose accuracy on real data, without the need for any manual annotation.

In this paper, we propose the first sim-to-real iterative self-training framework for 6D object pose estimation. Fig. 1 depicts an overview of our framework. First, we establish a photo-realistic simulator with multiple rendering attributes to synthesize abundant data, on which a pose estimation network is initially well-trained. Then, it is taken as a teacher model to generate pose predictions for unlabeled real data. From these results, we design a new pose selection method, which comprehensively harnesses both 2D appearance and 3D geometry information of objects to carefully distinguish reliable predictions. These high-quality pose predictions are then leveraged as pseudo labels for the corresponding real data to train the pose estimation network as a student model. Moreover, to progressively improve the quality of pseudo labels, we consider the updated student model as a new teacher, refine pose predictions, and re-select the highest-quality pseudo labels again. Such an iterative self-training scheme ensures to simultaneously improve the quality of pseudo labels and model performance on real data. According to extensive experiments on a public benchmark dataset and our constructed *Sim-to-real Industrial Bin-Picking (SIBP)* dataset, our proposed iterative self-training method significantly outperforms the state-of-the-art sim-to-real method [39] by 11.49% and 22.62% in terms of the ADD(-S) metric.

2 Related Works

2.1 6D Object Pose Estimation for Bin-Picking

Though recent works [8,23,34,37] show superior performance on household object datasets (*e.g.*, LineMOD [20] and YCB Video [56]), 6D pose estimation for bin-picking still remains a challenging task for highly cluttered and complex scenes. Conventional methods typically leveraged Point Pair Features (PPF) to estimate 6D object pose in a bin [3,4,11,46]. As a popular method, it mainly relies on depth or point cloud to detect the pose in the scene and achieves promising results when scenes are clear and controlled [23]. However, due to the reliance on the point cloud, the PPF-based methods are sensitive to the noise and occlusion in cluttered scenes, which is very common in real bin-picking scenarios such as industrial applications [27,60]. Current state-of-the-art methods [9,27] and datasets [28,60] rely on deep learning models for handling complex bin-picking scenarios. Dong et al. [9] proposed a deep-learning-based voting scheme to estimate 6D pose using a scene point cloud. Yang et al. [60] modified AAE [38] with a detector as a deep learning baseline, showing a better performance than the PPF-based method for bin-picking scenes. For these methods, regardless of the particular network design, a large amount of real-world labeled data is always required in order to achieve a high accuracy.

2.2 Visual Adaptation for Sim-to-Real Gap

Sim-to-real transfer is a crucial topic to tackle the bottleneck for deep learning-based robotic applications, which is essential to bridge the domain gap between simulated data and real-world observation through robotic visual systems [31, 52]. A promising technique for achieving sim-to-real transfer is domain randomization [26,31,45], which samples a huge diversity of simulation settings (e.g. camera position, lighting, background texture, etc.) in a simulator to force the trained model to learn domain-invariant attributes and to enhance generalization on real-world data. Visual domain adaptation [37–39] has also had recent success for sim-to-real transfer, where the source domain is usually the simulation, and the target domain is the real world [1,25]. Recently, some works [8,23,44,45,52] leverage physically-based renderer (PBR) data to narrow the sim-to-real gap by simulating more realistic lighting and texture in the simulator. However, with the additional synthetic data, current works on 6D pose estimation still mainly rely on annotated real data to ensure performance.

2.3 Self-training via Iterative Update

A typical self-training paradigm [16,61] first trains a teacher model based on the labeled data, then utilizes the teacher model to generate pseudo labels for unlabeled data. After that, the unlabeled data with their pseudo labels are used to train a student model. As a typical label-efficient methodology, self-training techniques have recently gained attention in many fields. Some methods apply self-training to image classification [47,59]. They use the teacher model to generate soft pseudo labels for unlabeled images, and train the student model with a standard cross-entropy loss. The input consistency regularization [57] and noisy student training [58] are widely used techniques to further enhance the student model with unlabeled data. Some works adopt self-training in semantic segmentation [33,63]. Instead of training the student model with merely pseudo labels, they train the student by jointly using a large amount of data with pseudo labels and a small number of data with manual labels. Some other works resort to self-training to perform unsupervised domain adaptation [36,64], which train the teacher and student models on different domains. Recently, some methods study pseudo label selection. They pick out reliable pseudo labels based on the probability outputs or the uncertainty measurements derived from the network [13,35]. Though the effectiveness of self-training has been revealed in many fields, we find that self-training has not been investigated for 6D object pose estimation.

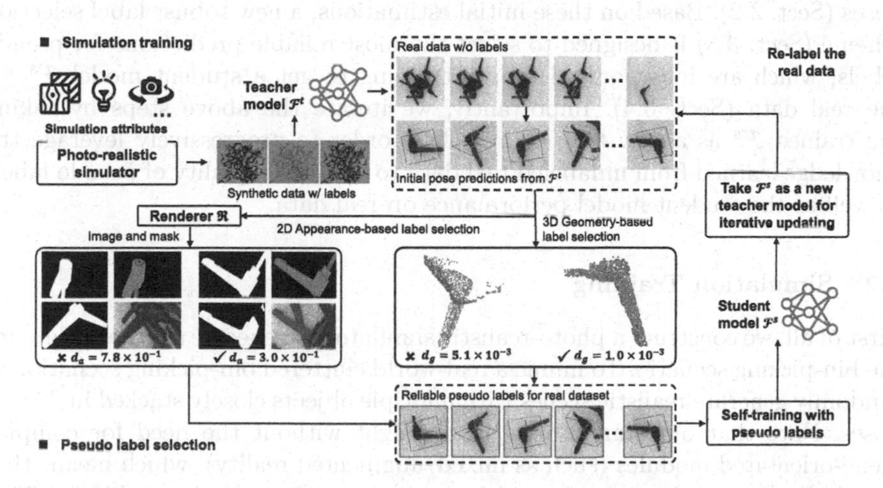

Fig. 2. Overview of our proposed iterative self-training method for 6D object pose estimation. We first build a photo-realistic simulation with multiple rendering attributes to generate synthetic data. We then train a teacher model on the synthetic data, which is used to generate initial pose predictions for the unlabeled real data. Subsequently, a pseudo label selection scheme with both 2D appearance and 3D geometry metrics are adopted to select data with reliable pose predictions, which are used to train a student model for the real data. We iterate this self-training scheme multiple times by taking the trained student model as a new teacher for the next iteration.

3 Method

In this paper, we take advantage of advanced self-training to solve the problem of sim-to-real object pose estimation. Our method is designed to be scalable and network-agnostic, and aims to dramatically improve pose estimation accuracy on real-world data without need of any manual annotation.

3.1 Overview of Sim-to-Real Object Pose Estimation

Without loss of generality, let $\mathcal{F}$ denote an arbitrary 6D object pose estimation network, which takes RGB-D data of an object as input, and outputs its corresponding 6D pose with respect to the camera coordinate frame. Let I be the RGB image and G denote the 3D point cloud recovered from the depth map, the 6D object pose p is predicted by $\mathcal{F}$ as:

$$p = [R|t] = \mathcal{F}(I, G), \tag{1}$$

where $R \in SO(3)$ is the rotation and $t \in \mathbb{R}^3$ is the translation. Fig. 2 summarizes our network-agnostic, sim-to-real pose estimation solution. Given a target bin-picking scenario, we first create a photo-realistic simulation to generate abundant synthetic data with a low cost, which is used to initially train a teacher model $\mathcal{F}^t$ with decent quality. We then apply $\mathcal{F}^t$ on real data to predict their object poses (Sect. 3.2). Based on these initial estimations, a new robust label selection scheme (Sect. 3.3) is designed to select the most reliable predictions for pseudo labels, which are incorporated by self-training to get a student model $\mathcal{F}^s$ for the real data (Sect. 3.4). Importantly, we iterate the above steps by taking the trained $\mathcal{F}^s$ as a new teacher model, in order to progressively leverage the knowledge learned from unlabeled real data to boost the quality of pseudo labels as well as the student model performance on real data.

3.2 Simulation Training

First of all, we construct a photo-realistic simulator to generate synthetic data for the bin-picking scenario. To mimic a real-world cluttered bin-picking scenario, we randomly generate realistic scenes with multiple objects closely stacked in diverse poses. Note that our simulator is light-weight without the need for complex scene-orientated modules (such as mixed/augmented reality), which means that once built, the simulator can be widely applied to different industrial bin-picking scenarios. Using the simulator, we generate a large amount of data with precise labels, and use it to train a teacher model $\mathcal{F}^t$ with acceptable initial performance. After that, when we expose $\mathcal{F}^t$ to real data (I_i, G_i), we can obtain its pose prediction as $\widetilde{p}_i = \mathcal{F}^t(I_i, G_i)$. As shown in Fig. 2, due to the unavoidable gap between virtual and real data, $\widetilde{p}_i$ is not guaranteed to always be correct. In order to self-train the model on unlabeled real data, we need to carefully pick out reliable predictions of $\mathcal{F}^t$. This step is known as pseudo label selection in a typical self-training paradigm.

3.3 Pose Selection for Self-training

Given unlabeled real data $\{(I_i, G_i)\}_{i=1}^m$ and their initial pose predictions $\{\widetilde{p}_i = [R_i|t_i]\}_{i=1}^m$, pose selection aims to find out reliable pose results. Existing pseudo label selection methods [13,35,58] are limited to classification-based tasks, such as image recognition and semantic segmentation. But for object pose estimation, which is typically formulated as a regression learning process, it is unclear how to select pseudo labels. In order to address this problem, we propose a comprehensive pose selection method. The core idea is to first virtually generate the observation data that corresponds to the predicted pose. Subsequently, we compare the generated data with the real collected one to determine whether the predicted pose is reliable or not. Note that for a robust pose selection, both the 2D image and the 3D point cloud are important. As shown in Fig. 3, although 2D image provides rich features for discriminating object poses, they could be ambiguous due to the information loss when projecting the 3D object onto the 2D image plane. Using 3D point cloud to assess the object pose can avoid this issue. But different from 2D images, the 3D point cloud could be quite noisy in the bin-picking scenario, which affects the assessment result. Our proposed method therefore leverages the complementary advantages of 2D image and 3D point cloud for a comprehensive pose selection for self-training.

2D Appearance-Based Pose Selection. To obtain the image that corresponds to the predicted pose, we leverage an off-the-shelf renderer [6], which is denoted as $\Re$. The renderer takes an object pose and the corresponding object CAD model as inputs, and will virtually generate an RGB image I^r and a binary object mask M^r that correspond to the predicted pose[1]:

$$\{I^r, M^r\} = \Re(\widetilde{p}). \tag{2}$$

The binary mask exhibits clear object contour, while the RGB image contains richer texture and semantic information (see Fig. 2). We make combined use of them for appearance-based pose selection on the 2D image plane.

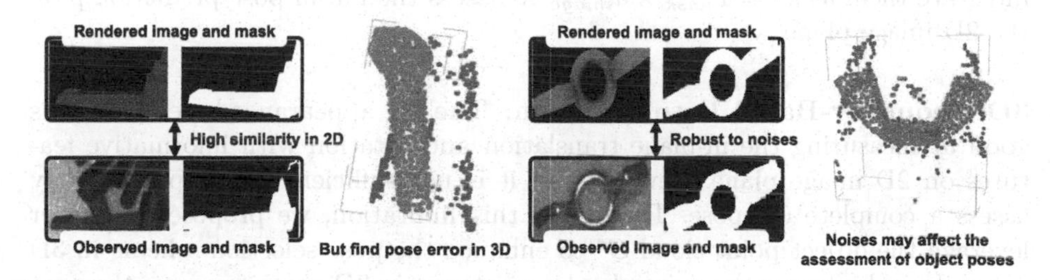

Fig. 3. Both 2D appearance and 3D geometry are important for reliable pseudo label selection. On the one hand, the 3D geometry metric can notice pose errors that are hard to detect in 2D. On the other hand, 3D geometry metric gets unreliable for noisy point cloud, but the 2D appearance metric is robust to these noises.

Let M^o denote the object mask for the observed RGB image, which can be accurately acquired by an existing segmentation network [18] trained on synthetic data. Rather than directly measuring the contour similarity [31] and to

[1] For clarity, we will omit the subscript i for formulations in this section.

ensure that the mask-based metric is robust to segmentation noise, we evaluate the pose quality based on the pixel-wise overlap of M^r and M^o, which is less affected by imperfect segmentation:

$$s(M^r, M^o) = \frac{1}{2} \times (\frac{1}{N_+} \sum_{q \in M_+^o} \mathbb{1}(q \in M_+^r) + \frac{1}{N_-} \sum_{q \in M_-^r} \mathbb{1}(q \in M_-^o)), \quad (3)$$

where $M_+^{\{r,o\}}$ denotes the foreground region of the mask and $M_-^{\{r,o\}}$ denotes the background region of the mask. N_+ and N_- are the number of pixels of M_+^o and M_-^r, respectively. Given the mask overlap, the corresponding mask-based pose distance metric could be computed as $d_{mask} = 1 - s(M^r, M^o)$. This mask-based metric coarsely leverages the object contour information for pose evaluation, but neglects the detailed texture and semantic information of the object. As a consequence, d_{mask} would be sensitive to occlusion and not reliable enough for objects with a complex shapes.

In this regard, we further leverage the rendered RGB image to enhance the pose assessment. In order to extract representative features from the RGB image, we use a pre-trained CNN Φ to transform I^r and I^o into multi-level high-dimension features, based on which we measure the perceptual distance [62] between I^r and I^o as:

$$d_{image} = \sum_{l=1}^{L} \frac{1}{N_l} \sum_{q} \|\Phi_l^r(q) - \Phi_l^o(q)\|_2, \quad (4)$$

where $\Phi_l^{\{r,o\}}$ denotes the l-th level normalized feature of Φ. N_l is the number of pixels of the corresponding feature map. The perceptual distance d_{image} complements d_{mask} with low-level texture and high-level semantic features. We then integrate them as $d_a = d_{mask} \times d_{image}$ to assess the initial pose prediction $\widetilde{p}$ on the 2D image plane.

3D Geometry-Based Pose Selection. The 2D appearance-based metric is good at measuring the in-plane translation and rotation with informative features on 2D image plane. Nevertheless, it is not sufficient to comprehensively assess a complete 6D pose. To address this limitation, we propose to further leverage the object point cloud G^o to enhance the pose selection scheme in 3D space. In order to use the point cloud consistency in 3D space to assess the pose quality, we need to generate the object point cloud that corresponds to the predicted pose. With the object CAD model $\mathbb{C}$ and the predicted pose $\widetilde{p}$, we perform the following 2 steps. First, we apply $\widetilde{p}$ to the CAD model: $\mathbb{C}^r = \widetilde{R} \times \mathbb{C} + \widetilde{t}$. It transforms the original CAD model $\mathbb{C}$ in the object coordinate frame to a model $\mathbb{C}^r$ in the camera coordinate frame. After that, since $\mathbb{C}^r$ is a complete point cloud while G^o only contains the surface point cloud that could be observed from a view point, directly evaluating the point cloud consistency between $\mathbb{C}^r$ and G^o cannot precisely reveal the quality of $\widetilde{p}$. To mitigate this problem, we further apply a projection operator proposed by [17] to the CAD model in camera coordinate frame: $G^r = \mathbb{P}(\mathbb{C}^r)$. This generates the surface points of $\mathbb{C}^r$ that can be observed from the predicted pose. Intuitively, if $\widetilde{p}$ is precise, G^r should be well aligned with the observed point cloud G^o in 3D space. We use the Chamfer Distance (CD) between G^r and G^o as a 3D metric to quantify the quality of $\widetilde{p}$:

$$d_g = \frac{1}{N_1} \sum_{x \in G^r} \min_{y \in G^o} \|x - y\|_2 + \frac{1}{N_2} \sum_{y \in G^o} \min_{x \in G^r} \|x - y\|_2, \tag{5}$$

where x and y denote 3D points from G^r and G^o, respectively. N_1 and N_2 are the total numbers of points for G^r and G^o.

Comprehensive Selection. Given the complementary advantages of d_a and d_g, we leverage both of them to single out reliable pose predictions. Pose predictions are designated as pseudo labels and included in the next iteration of self-training when $d_a < \tau_a$ and $d_g < \tau_g$. These two thresholds τ_a and τ_g can be determined flexibly according to the metric value distribution on unlabeled real data. In this paper, we adaptively set $\tau_{(a,g)} = \mu_{(a,g)} + \sigma_{(a,g)}$, where $\mu_{(a,g)}$ and $\sigma_{(a,g)}$ are the mean and standard deviation of d_a and d_g on real data.

3.4 Iterative Self-training on Real Data

After label selection, we incorporate data with reliable pseudo labels to the 6D object pose estimation network to train a student model $\mathcal{F}^s$. Our self-training scheme aims to enforce the prediction of $\mathcal{F}^s$ to be consistent with pose pseudo labels. In this regard, we define the self-training loss function as $\mathcal{L}_{\text{self-training}} = \frac{1}{M}\sum_{i=1}^{M} \ell(\widetilde{p}_i, p'_i)$, where p'_i is the pose prediction from the student model, ℓ could be any loss function of object pose estimation and is depended on the architecture of $\mathcal{F}$. Through self-training, we transfer the knowledge that the teacher model has learned from the synthetic data to the student model on real data.

To further improve the capability of the student model, we iterate the above process for multiple runs. With progressive training on the real data, the pseudo labels generated by the updated teacher model will have a higher quality than the previous iteration's labels. Our experiments demonstrate that this iterative optimization scheme can effectively boost the student model on real data.

4 Experiments

In this section, we will answer the following questions: (1) Does the proposed self-training method improve the performance of sim-to-real object pose estimation? (2) Do the appearance-based and geometry-based pose selection metrics improve the student model performance? (3) Could the proposed self-training framework be applied to different backbone architectures? (4) How many iterations are required to self-train the pose estimation network? (5) Does the proposed self-training indeed improve the bin-picking success rate on real robots? To answer question (1)–(4), we evaluate our method and compare it with state-of-the-art sim-to-real methods on both the ROBI [60] dataset and our proposed SIBP dataset. To answer question (5), we deploy our pose estimation model after self-training on a real robot arm and conduct real-world bin-picking experiments.

4.1 Experiment Dataset

ROBI Dataset. ROBI [60] is a recent public dataset for industrial bin-picking. It contains seven reflective metal objects, with two of them being composed of different materials[2]. Since ROBI only contains real RGB-D data, we use the simulator presented in Sect. 3.2 to extend ROBI by randomly synthesizing 1000 virtual scenes for each object. For real data, we use 3129 real RGB-D scenes of ROBI. The 3129 real scenes are in two cluttered levels. 1288 scenes are low-bin scenarios, in which a bin contains only 5–10 objects and corresponds to easy cases. 1841 sceres are full-bin scenarios, in which a bin is full with tens of objects and correspond to hard cases. We use 2310 scenes (910 low-bin and 1400 full-bin scenarios) without using their annotations for self-training and use the remaining scenes for evaluation. Please refer to [60] and the supplementary for more detailed information of ROBI.

Sim-to-real Industrial Bin-Picking (SIBP) Dataset. We further build a new SIBP dataset. It provides both virtual and real RGB-D data for six texture-less objects in industrial bin-picking scenarios. Compared with ROBI, objects in SIBP are more diverse in color, surface material (3 plastic, 2 metallic and 1 alloyed) and object size (from a few centimeters to one decimeter). We use SIBP for a more comprehensive evaluation. For synthetic data, we use the same simulator to generate 6000 synthetic RGB-D scenes. For real data, we use a Smarteye Tech HD-1000 industrial stereo camera to collect 2743 RGB-D scenes in a real-world bin-picking environment. We use 2025 scenes without using their annotations for self-training, and use the remaining scenes for evaluation. Please refer to supplementary for more detailed information of SIBP.

4.2 Evaluation Metrics

We follow [37,40,52] and compare pose estimation results of different methods *w.r.t.* the ADD metric [20], which measures whether the average distance between models transformed by the predicted pose and the ground-truth pose is smaller than 10% of object diameter. For symmetric objects, we adopted the ADD-S metric [21] when computing the average distance of models. We report the average recall of ADD(-S) for quantitative evaluation.

4.3 Comparison with State-of-the-Art Methods

We compared our self-training model with a baseline DC-Net [42] trained exclusively with synthetic data, and two state-of-the-art sim-to-real object pose esti-

[2] Din-Connector and D-Sub Connector.

mation methods: MP-AAE [37] and AAE [39]. Note that all methods are evaluated based on the same segmentation. Table 1 and Table 2 presents the quantitative evaluation results. In comparison with the baseline model, on ROBI, our self-training model outperforms DC-Net by 16.73%. On SIBP, it also exceeds DC-Net by 14.94%. These results demonstrate the effectiveness of our self-training method for mitigating the sim-to-real gap for object pose estimation.

Moreover, we compared our method with two SOTA sim-to-real methods. AAE first trained an Augmented Autoencoder on synthetic data and then used the embedded feature of Autoencoder for object pose estimation on real data. MP-AAE further resorted to multi-path learning to learn a shared embedding space for pose estimation of different objects on real data. Our self-training method consistently outperforms these two SOTA methods for object pose estimation. On ROBI, our model outperforms MP-AAE by 24.5% and exceeds AAE by 11.49%. On SIBP, our self-training model achieves an average recall of 68.72%, which is 32.79% higher than MP-AAE and 22.62% higher than AAE. Fig. 4 presents corresponding qualitative comparisons on ROBI. MP-AAE and AAE have difficulties in adapting the model to a real bin-picking environment. The occlusions caused by the container or by neighboring objects may affect their pose estimation results. In comparison, our method can directly adapt the pose estimation model to real data with self-training. It can smoothly handle complex industrial bin-picking scenarios and achieve high pose estimation accuracy. Please refer to the supplementary for more qualitative results.

Table 1. Quantitative comparison with state-of-the-art methods on ROBI dataset. The average recall (%) of the ADD(-S) metric is reported. * indicates asymmetric objects.

Object	Zigzag*	Chrome screw	Gear	Eye Bolt	Tube fitting	Din* connector	D-Sub* connector	Mean
MP-AAE [37]	22.76	47.49	66.22	50.85	69.54	15.92	4.70	39.64
AAE [39]	25.80	64.02	90.64	67.31	91.58	21.83	7.35	52.65
DC-Net [42]	30.96	67.74	77.56	53.52	74.84	18.72	10.60	47.41
Ours	**45.83**	**79.88**	**97.81**	**89.21**	**97.45**	**24.21**	**14.62**	**64.14**

Table 2. Quantitative comparison with state-of-the-art methods on SIBP dataset. The average recall (%) of the ADD(-S) metric is reported. * indicates asymmetric objects.

Object	Cosmetic	Flake	Handle*	Corner	Screw head	T-Shape connector	Mean
MP-AAE [37]	58.04	21.56	27.26	39.59	36.67	32.44	35.93
AAE [39]	58.93	53.82	21.67	55.93	39.04	47.23	46.10
DC-Net [42]	68.54	58.74	43.44	53.96	44.53	53.47	53.78
Ours	**79.59**	**69.94**	**75.24**	**62.57**	**56.98**	**68.02**	**68.72**

4.4 Validation of the Pose Selection Strategy

We study the performance of our self-training method with different pose selection strategies. Specifically, we separately remove the appearance-based metric d_a and geometry-based metric d_g from the framework, and evaluate their performances on ROBI.

Table 3. Quantitative evaluation of the proposed self-training method with different pose selection settings. d_a denotes the appearance-based metric and d_g denotes the geometry-based metric. LB denotes the lower bound of our model, which is trained with only synthetic data. UB denotes the upper bound of our model, which additionally uses pose labels of real data for training.

d_a	d_g	Zigzag*	Chrome screw	Gear	Eye bolt	Tube fitting	Din* connector	D-Sub* connector	Mean
✓		38.54	78.00	91.29	85.09	95.52	22.83	12.06	60.48
	✓	43.10	**79.98**	95.20	87.84	96.91	21.61	13.95	62.66
✓	✓	**45.83**	79.88	**97.81**	**89.21**	**97.45**	**24.21**	**14.62**	**64.14**
LB		30.96	67.74	77.56	53.52	74.84	18.72	10.60	47.41
UB		56.90	93.71	98.90	96.12	99.46	44.16	33.86	74.73

Table 3 reports the comparative results. Compared with the lower-bound model (LB), both the proposed appearance metric and geometry metric help to significantly improve the pose accuracy by self-training. Among three different settings, using both appearance and geometry information for pose selection achieves the best pose accuracy. Removing d_g results in a larger pose accuracy drop than removing d_a. These experimental results indicate that merely assessing the pose label in 2D is not sufficient for self-training of 6D object pose. Using both 2D appearance and 3D geometry information provides the most reliable scheme for pose selection, and achieves the best pose accuracy.

4.5 Self-training with Different Backbone Networks

In contrary to other self-supervised/self-training methods [7,13,33,52], our proposed self-training framework can be easily applied to different networks for sim-to-real object pose estimation. In this section, we change the backbone network to DenseFusion [51], another popular RGB-D based object pose estimation network. Compared with DC-Net, DenseFusion is a two-stage method which is composed of an initial *estimator* and a pose *refiner*. We then use the same synthetic data and real data with previous experiments for self-training and evaluation. Table 4 reports the experiment results. Even with a different backbone network, the proposed self-training method can consistently adapt the model to real data and significantly improve the pose accuracy on real data.

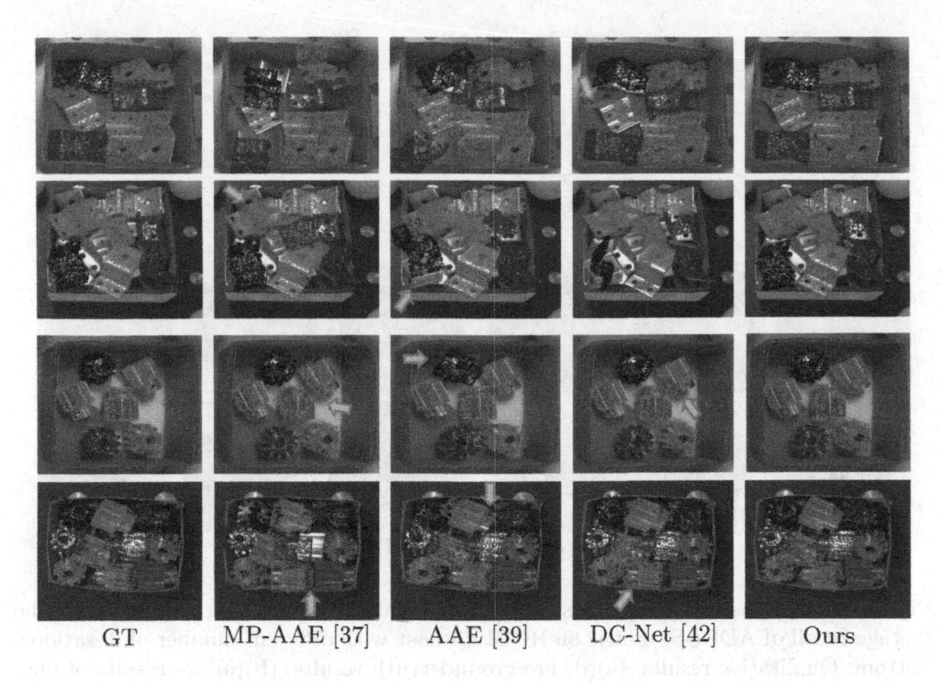

Fig. 4. Qualitative comparison with state-of-the-arts. The 6D object pose estimation results are depicted by projecting the object model onto the image plane using the object pose and camera intrinsic parameters. DC-Net is our baseline model w/o our proposed self-training. Colored points denote the projected model (best viewed in color). (Color figure online)

4.6 Continuous Improvement With Iterative Self-training

In order to study the effect of iteration numbers on the final pose accuracy, we perform iterative self-training with a maximal number of 10 iterations, and compared the pose accuracy after each iteration. Figure 5 depicts the experiment results. In general, the pose error reduces as the number of self-training iterations increase. In most cases, the first iteration usually accounts for the most improvement of the pose accuracy. After that, the second and third iteration can continuously enhance the pose accuracy, with a relatively smaller improvement. After about 5 iterations, the performance slowly saturates without further improvement. Figure 5 further presents qualitative comparisons between one-time iteration model and five-time iteration model. With the increase of iterations, the pose estimation results get visually closer to the ground truth.

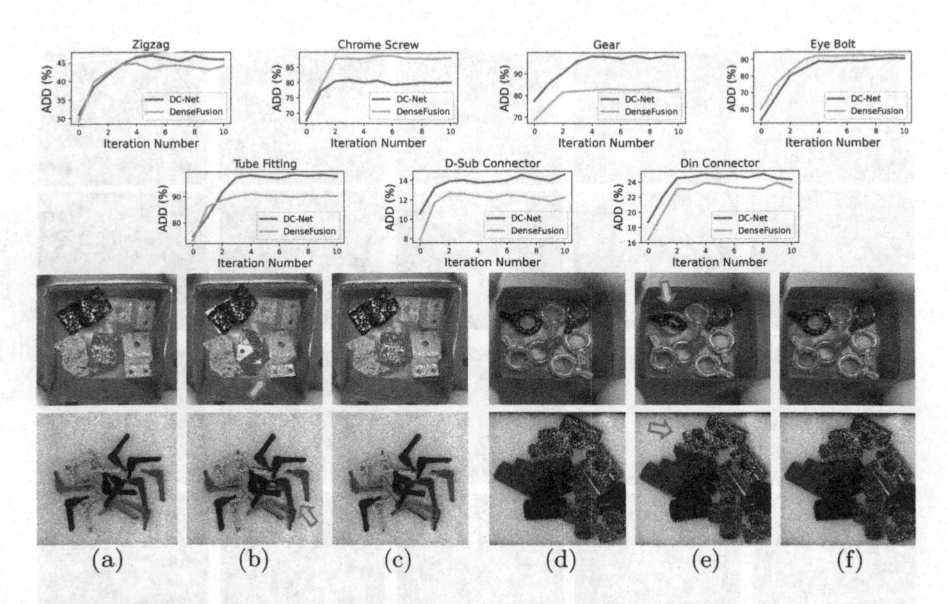

Fig. 5. Object pose estimation results of our iterative self-training model. Top: The average recall of ADD(-S) metric on ROBI dataset with different number of iterations. Bottom: Qualitative results. (a)(d) are ground-truth results. (b)(e) are results of one-time self-training. (c)(f) are results of five-time iterative self-training.

Table 4. Results of the proposed self-training method with a different backbone network [51]. w/o ST and w/ ST denote models without and with proposed self-training.

	Zigzag*	Chrome screw	Gear	Eye Bolt	Tube fitting	Din* connector	D-Sub* connector	Mean
w/o ST	29.29	69.61	68.59	60.07	73.29	16.42	7.92	46.46
w/ ST	**44.46**	**87.87**	**82.73**	**92.36**	**90.88**	**23.88**	**12.66**	**62.12**

4.7 Effectiveness Demonstration on Robots

In order to further demonstrate the effectiveness of proposed self-training object pose estimation method, we deploy the trained pose estimation model on a Franka Emika Panda robot arm and conduct real-world bin-picking experiments on the *'Handle'* object from SIBP. The grasp configuration is generated offline based on the object CAD model [12], and then projected to the camera coordinate system according to estimated object pose. This means that overall grasping success rate is highly reliant on pose accuracy. Table 5 presents the experiment results. Following self-training of the pose estimation network, the grasping success rate dramatically improves. These results demonstrate the effectiveness of our proposed self-training method for practical industrial bin-picking.

Table 5. Results of the robot bin-picking experiments. The *left* picture illustrates the robot bin-picking environment. **Inst.** denotes the total number of objects to be grasped. **Tria.** denotes the total number of trials needed to access all of the objects.

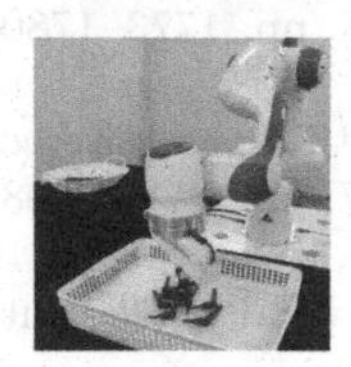

	Inst. / Tria.					Mean
	01	02	03	04	05	
w/o ST	14/24	14/20	14/18	14/22	14/19	67.96%
w/ ST	14/14	14/16	14/18	14/16	14/16	**87.50%**

5 Conclusion

With the goal of achieving accurate 6D object pose estimation in real-world scenes, we have presented a sim-to-real method that first trains a 6D pose estimation on high-fidelity simulation data, and then performs our iterative self-training method. Our method provides an efficient solution for pose estimation where labeled data is hard (or even impossible) to collect. Extensive results demonstrate that our approach can significantly improve the predicted pose quality, with great potential to be applied to industrial robotic bin-picking scenarios.

Currently, our method assumes access to object models. Although this is acceptable in the industrial bin-picking domain and is indeed common in the pose estimation literature, it would nonetheless be interesting to extend this iterative self-training method to pose estimation of previously unseen objects [5, 54]. Our method exploits object mask predicted by a segmentation network trained with synthetic data. It would be meaningful to extend our iterative self-training method to joint instance segmentation and object pose estimation.

Acknowledgement. The work was supported by the Hong Kong Centre for Logistics Robotics.

References

1. Bousmalis, K., et al.: Using simulation and domain adaptation to improve efficiency of deep robotic grasping. In: ICRA, pp. 4243–4250 (2018)
2. Brachmann, E., Rother, C.: Visual camera re-localization from RGB and RGB-D images using DSAC. TPAMI **44**(9), 5847–5865 (2020)
3. Buch, A.G., Kiforenko, L., Kraft, D.: Rotational subgroup voting and pose clustering for robust 3D object recognition. In: ICCV, pp. 4137–4145 (2017)
4. Buch, A.G., Kraft, D., Robotics, S., Odense, D.: Local point pair feature histogram for accurate 3D matching. In: BMVC (2018)
5. Chen, K., Dou, Q.: SGPA: structure-guided prior adaptation for category-level 6D object pose estimation. In: ICCV, pp. 2773–2782 (2021)
6. Chen, W., et al.: Learning to predict 3D objects with an interpolation-based differentiable renderer. NeurIPS (2019)
7. Deng, X., Xiang, Y., Mousavian, A., Eppner, C., Bretl, T., Fox, D.: Self-supervised 6D object pose estimation for robot manipulation. In: ICRA, pp. 3665–3671 (2020)

8. Di, Y., Manhardt, F., Wang, G., Ji, X., Navab, N., Tombari, F.: So-pose: exploiting self-occlusion for direct 6D pose estimation. In: ICCV, pp. 12396–12405 (2021)
9. Dong, Z., et al.: PPR-Net: point-wise pose regression network for instance segmentation and 6D pose estimation in bin-picking scenarios. In: IROS, pp. 1773–1780 (2019)
10. Drost, B., Ulrich, M., Bergmann, P., Hartinger, P., Steger, C.: Introducing mvtec itodd-a dataset for 3D object recognition in industry. In: ICCVW, pp. 2200–2208 (2017)
11. Drost, B., Ulrich, M., Navab, N., Ilic, S.: Model globally, match locally: efficient and robust 3D object recognition. In: CVPR, pp. 998–1005 (2010)
12. Fang, H.S., Wang, C., Gou, M., Lu, C.: Graspnet-1billion: a large-scale benchmark for general object grasping. In: CVPR, pp. 11444–11453 (2020)
13. Gal, Y., Ghahramani, Z.: Dropout as a Bayesian approximation: representing model uncertainty in deep learning. In: ICML, pp. 1050–1059 (2016)
14. Georgakis, G., Karanam, S., Wu, Z., Kosecka, J.: Learning local rgb-to-cad correspondences for object pose estimation. In: ICCV, pp. 8967–8976 (2019)
15. Gojcic, Z., Zhou, C., Wegner, J.D., Guibas, L.J., Birdal, T.: Learning multiview 3D point cloud registration. In: CVPR, pp. 1759–1769 (2020)
16. Grandvalet, Y., Bengio, Y.: Semi-supervised learning by entropy minimization. In: NeurIPS (2004)
17. Gu, J., et al.: Weakly-supervised 3D shape completion in the wild. In: Vedaldi, A., Bischof, H., Brox, T., Frahm, J.-M. (eds.) ECCV 2020. LNCS, vol. 12350, pp. 283–299. Springer, Cham (2020). https://doi.org/10.1007/978-3-030-58558-7_17
18. He, K., Gkioxari, G., Dollár, P., Girshick, R.: Mask r-cnn. In: ICCV (2017)
19. He, Y., Sun, W., Huang, H., Liu, J., Fan, H., Sun, J.: Pvn3d: a deep point-wise 3D keypoints voting network for 6dof pose estimation. In: CVPR, pp. 11632–11641 (2020)
20. Hinterstoisser, S., Lepetit, V., Ilic, S., Holzer, S., Bradski, G., Konolige, K., Navab, N.: Model based training, detection and pose estimation of texture-less 3D objects in heavily cuttered scenes. In: Lee, K.M., Matsushita, Y., Rehg, J.M., Hu, Z. (eds.) ACCV 2012. LNCS, vol. 7724, pp. 548–562. Springer, Heidelberg (2013). https://doi.org/10.1007/978-3-642-37331-2_42
21. Hodaň, T., Matas, J., Obdržálek, Š: On evaluation of 6D object pose estimation. In: Hua, G., Jégou, H. (eds.) ECCV 2016. LNCS, vol. 9915, pp. 606–619. Springer, Cham (2016). https://doi.org/10.1007/978-3-319-49409-8_52
22. Hodan, T., et al.: Bop: benchmark for 6D object pose estimation. In: ECCV, pp. 19–34 (2018)
23. Hodaň, T., et al.: BOP challenge 2020 on 6D object localization. In: Bartoli, A., Fusiello, A. (eds.) ECCV 2020. LNCS, vol. 12536, pp. 577–594. Springer, Cham (2020). https://doi.org/10.1007/978-3-030-66096-3_39
24. James, S., Davison, A.J., Johns, E.: Transferring end-to-end visuomotor control from simulation to real world for a multi-stage task. In: CoRL, pp. 334–343 (2017)
25. James, S., et al.: Sim-to-real via sim-to-sim: data-efficient robotic grasping via randomized-to-canonical adaptation networks. In: CVPR, pp. 12627–12637 (2019)
26. Kehl, W., Manhardt, F., Tombari, F., Ilic, S., Navab, N.: SSD-6D: making RGB-based 3D detection and 6D pose estimation great again. In: ICCV, pp. 1521–1529 (2017)
27. Kleeberger, K., Huber, M.F.: Single shot 6D object pose estimation. In: ICRA, pp. 6239–6245 (2020)
28. Kleeberger, K., Landgraf, C., Huber, M.F.: Large-scale 6D object pose estimation dataset for industrial bin-picking. In: IROS, pp. 2573–2578 (2019)

29. Li, X., et al.: A sim-to-real object recognition and localization framework for industrial robotic bin picking. RAL **7**(2), 3961–3968 (2022)
30. Li, Z., Hu, Y., Salzmann, M., Ji, X.: SD-pose: Semantic decomposition for cross-domain 6D object pose estimation. In: AAAI, vol. 35, no. 3, pp. 2020–2028 (2021)
31. Manhardt, F., Kehl, W., Navab, N., Tombari, F.: Deep model-based 6D pose refinement in RGB. In: ECCV, pp. 800–815 (2018)
32. Murez, Z., et al.: Atlas: end-to-end 3D scene reconstruction from posed images. In: Vedaldi, A., Bischof, H., Brox, T., Frahm, J.-M. (eds.) ECCV 2020. LNCS, vol. 12352, pp. 414–431. Springer, Cham (2020). https://doi.org/10.1007/978-3-030-58571-6_25
33. Pastore, G., Cermelli, F., Xian, Y., Mancini, M., Akata, Z., Caputo, B.: A closer look at self-training for zero-label semantic segmentation. In: CVPR, pp. 2693–2702 (2021)
34. Peng, S., Zhou, X., Liu, Y., Lin, H., Huang, Q., Bao, H.: Pvnet: pixel-wise voting network for 6dof object pose estimation. TPAMI, pp. 4561–4570 (2020)
35. Rizve, M.N., Duarte, K., Rawat, Y.S., Shah, M.: In defense of pseudo-labeling: an uncertainty-aware pseudo-label selection framework for semi-supervised learning. In: ICLR (2020)
36. RoyChowdhury, A., et al.: Automatic adaptation of object detectors to new domains using self-training. In: CVPR, pp. 780–790 (2019)
37. Sundermeyer, M., et al.: Multi-path learning for object pose estimation across domains. In: CVPR, pp. 13916–13925 (2020)
38. Sundermeyer, M., Marton, Z.C., Durner, M., Brucker, M., Triebel, R.: Implicit 3D orientation learning for 6d object detection from RGB images. In: ECCV, pp. 699–715 (2018)
39. Sundermeyer, M., Marton, Z.C., Durner, M., Triebel, R.: Augmented autoencoders: Implicit 3d orientation learning for 6D object detection. IJCV **128**(3), 714–729 (2020)
40. Thalhammer, S., Leitner, M., Patten, T., Vincze, M.: Pyrapose: feature pyramids for fast and accurate object pose estimation under domain shift. In: ICRA, pp. 13909–13915 (2021)
41. Tian, M., Ang, M.H., Lee, G.H.: Shape prior deformation for categorical 6D object pose and size estimation. In: Vedaldi, A., Bischof, H., Brox, T., Frahm, J.-M. (eds.) ECCV 2020. LNCS, vol. 12366, pp. 530–546. Springer, Cham (2020). https://doi.org/10.1007/978-3-030-58589-1_32
42. Tian, M., Pan, L., Ang, M.H., Lee, G.H.: Robust 6D object pose estimation by learning RGB-D features. In: ICRA, pp. 6218–6224 (2020)
43. Tobin, J., Fong, R., Ray, A., Schneider, J., Zaremba, W., Abbeel, P.: Domain randomization for transferring deep neural networks from simulation to the real world. In: IROS, pp. 23–30 (2017)
44. Tremblay, J., To, T., Birchfield, S.: Falling things: a synthetic dataset for 3D object detection and pose estimation. In: CVPRW, pp. 2038–2041 (2018)
45. Tremblay, J., To, T., Sundaralingam, B., Xiang, Y., Fox, D., Birchfield, S.: Deep object pose estimation for semantic robotic grasping of household objects. In: CoRL (2018)
46. Tuzel, O., Liu, M.-Y., Taguchi, Y., Raghunathan, A.: Learning to rank 3D features. In: Fleet, D., Pajdla, T., Schiele, B., Tuytelaars, T. (eds.) ECCV 2014. LNCS, vol. 8689, pp. 520–535. Springer, Cham (2014). https://doi.org/10.1007/978-3-319-10590-1_34

47. Veit, A., Alldrin, N., Chechik, G., Krasin, I., Gupta, A., Belongie, S.: Learning from noisy large-scale datasets with minimal supervision. In: CVPR, pp. 839–847 (2017)
48. Wada, K., Sucar, E., James, S., Lenton, D., Davison, A.J.: Morefusion: multi-object reasoning for 6D pose estimation from volumetric fusion. In: CVPR, pp. 14540–14549 (2020)
49. Wada, K., James, S., Davison, A.J.: Reorientbot: learning object reorientation for specific-posed placement. ICRA (2022)
50. Wada, K., James, S., Davison, A.J.: Safepicking: learning safe object extraction via object-level mapping. ICRA (2022)
51. Wang, C., et al.: Densefusion: 6D object pose estimation by iterative dense fusion. In: CVPR, pp. 3343–3352 (2019)
52. Wang, G., Manhardt, F., Shao, J., Ji, X., Navab, N., Tombari, F.: Self6D: self-supervised monocular 6D object pose estimation. In: Vedaldi, A., Bischof, H., Brox, T., Frahm, J.-M. (eds.) ECCV 2020. LNCS, vol. 12346, pp. 108–125. Springer, Cham (2020). https://doi.org/10.1007/978-3-030-58452-8_7
53. Wang, H., Sridhar, S., Huang, J., Valentin, J., Song, S., Guibas, L.J.: Normalized object coordinate space for category-level 6D object pose and size estimation. In: CVPR, pp. 2642–2651 (2019)
54. Wang, J., Chen, K., Dou, Q.: Category-level 6d object pose estimation via cascaded relation and recurrent reconstruction networks. arXiv:2108.08755 (2021)
55. Wang, X., Chen, H., Xiang, H., Lin, H., Lin, X., Heng, P.A.: Deep virtual adversarial self-training with consistency regularization for semi-supervised medical image classification. Med. Image Anal. **70**, 102010 (2021)
56. Xiang, Y., Schmidt, T., Narayanan, V., Fox, D.: Posecnn: a convolutional neural network for 6D object pose estimation in cluttered scenes. In: RSS (2018)
57. Xie, Q., Dai, Z., Hovy, E., Luong, T., Le, Q.: Unsupervised data augmentation for consistency training. NeurIPS **33**, 6256–6268 (2020)
58. Xie, Q., Luong, M.T., Hovy, E., Le, Q.V.: Self-training with noisy student improves imagenet classification. In: CVPR, pp. 10687–10698 (2020)
59. Yalniz, I.Z., Jégou, H., Chen, K., Paluri, M., Mahajan, D.: Billion-scale semi-supervised learning for image classification. arXiv:1905.00546 (2019)
60. Yang, J., Gao, Y., Li, D., Waslander, S.L.: Robi: a multi-view dataset for reflective objects in robotic bin-picking. In: IROS, pp. 9788–9795 (2021)
61. Yarowsky, D.: Unsupervised word sense disambiguation rivaling supervised methods. In: ACL, pp. 189–196 (1995)
62. Zhang, R., Isola, P., Efros, A.A., Shechtman, E., Wang, O.: The unreasonable effectiveness of deep features as a perceptual metric. In: CVPR, pp. 586–595 (2018)
63. Zhu, Y., et al.: Improving semantic segmentation via self-training. arXiv:2004.14960 (2020)
64. Zou, Y., Yu, Z., Kumar, B., Wang, J.: Unsupervised domain adaptation for semantic segmentation via class-balanced self-training. In: ECCV, pp. 289–305 (2018)

Active Audio-Visual Separation of Dynamic Sound Sources

Sagnik Majumder[1(✉)] and Kristen Grauman[1,2]

[1] UT Austin, Austin, TX, USA
{sagnik,grauman}@cs.utexas.edu
[2] Facebook AI Research, Austin, TX, USA

Abstract. We explore active audio-visual separation for dynamic sound sources, where an embodied agent moves intelligently in a 3D environment to *continuously* isolate the *time-varying* audio stream being emitted by an object of interest. The agent hears a mixed stream of multiple audio sources (e.g., multiple people conversing and a band playing music at a noisy party). Given a limited time budget, it needs to extract the target sound accurately at *every step* using egocentric audio-visual observations. We propose a reinforcement learning agent equipped with a novel transformer memory that learns motion policies to control its camera and microphone to recover the dynamic target audio, using self-attention to make high-quality estimates for current timesteps and also simultaneously improve its past estimates. Using highly realistic acoustic SoundSpaces [13] simulations in real-world scanned Matterport3D [11] environments, we show that our model is able to learn efficient behavior to carry out continuous separation of a dynamic audio target. Project: https://vision.cs.utexas.edu/projects/active-av-dynamic-separation/.

1 Introduction

Our daily lives are full of *dynamic audio-visual events*, and the activity and physical space around us affect how well we can perceive them. For example, an assistant trying to respond to his boss's call in a noisy office might miss the initial part of her comment but then visually spot other noisy actors nearby and move to a quieter corner to hear better; to listen to a street musician playing a violin at a busy intersection, a pedestrian might need to veer around other onlookers and shift a few places to clearly hear.

These examples show how *smart sensor motion* is necessary for accurate audio-visual understanding of dynamic events. For audio sensing in an environment full of distractor sounds, variations in acoustic attributes like volume, pitch, and semantic content of a sound source—in concert with a listener's proximity and direction from it—determine how well the listener is able to hear a

Supplementary Information The online version contains supplementary material available at https://doi.org/10.1007/978-3-031-19842-7_32.

S. Avidan et al. (Eds.): ECCV 2022, LNCS 13699, pp. 551–569, 2022.
https://doi.org/10.1007/978-3-031-19842-7_32

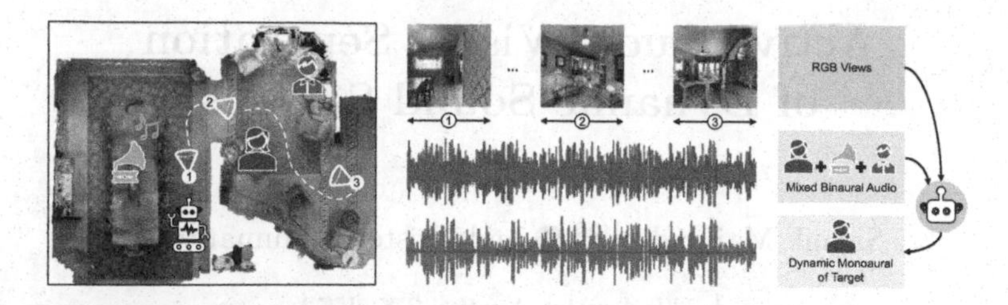

Fig. 1. Active audio-visual separation of dynamic sources. Given multiple *dynamic* (time-varying, non-periodic) audio sources S, all mixed together, the proposed agent separates the audio signal by actively moving in the 3D environment on the basis of its egocentric audio-visual input, so that it is able to accurately retrieve the target signal at every step of its motion.

target sound. However, just getting close or far is not always enough: effective hearing might require the listener to move around by *visually sensing* competing sound sources and surrounding obstacles, and discovering locations favorable to listening (e.g., an intersection of two walls that reinforces listening through audio reflections, the front of a tall cabinet that can dull acoustic disturbances).

In this work, we investigate how to induce such intelligent behaviors in autonomous agents through audio-visual learning. Specifically, we propose the new task of *active audio-visual separation of dynamic sources*: given a stream of egocentric audio-visual observations, an agent must determine how to move in an environment with multiple sounding objects in order to *continuously* retrieve the *dynamic* (temporally changing) sounds being emitted by some object of interest, under the constraint of a limited time budget. See Fig. 1. This task is relevant for augmented reality (AR) and mobile robotics applications, where a user equipped with an assistive hearing device or a service robot needs to better understand a sound source of interest in a busy environment.

Our task is distinct from related efforts in the literature. Whereas traditional audio-visual separation models extract sounds passively from pre-recorded videos [1,15,18,21,24,27,29,30,37,45,58,67,69], our task requires active placement of the agent's camera and microphones over time. Whereas embodied audio-visual navigation [4,12–14,26] entails moving towards a sound source, our task requires recovering the sounds of a target object. The recent Move2Hear model performs active source separation [38] but is limited to *static* (*i.e.*, periodic or constant) sound sources, such as a ringing phone or fire alarm—where recovering one timestep of the sound is sufficient. In contrast, our task involves *dynamic* audio sources and calls for extracting the target audio at *every step* of the agent's movement. Variations in the observed audio arise not only from the room acoustics and audio spatialization, but also from the temporally-changing and fleeting nature of the target and distractor sounds. This means the agent must recover a new audio segment at every step of its motion, which it hears only once. The proposed task is thus both more realistic and more difficult than existing active audio-visual separation settings.

To address active dynamic audio-visual source separation, we introduce a reinforcement learning (RL) framework that trains an agent how to move to continuously listen to the dynamic target sound. Our agent receives a stream of egocentric audio-visual observations in the form of RGB images and mixed binaural audio, along with the target category of interest (human voice, musical instrument, etc.) and decides its next action (translation or rotation of its camera and microphones) at every time step.

The technical design of our approach aims to meet the challenges outlined above. First, the proposed motion policy accumulates its observations with a recurrent network. Second, a transformer [60]-based acoustic memory leverages self-attention to simultaneously produce an estimate of the current segment of the dynamic target audio while refining estimates of past timesteps. The bi-directionality of the transformer's attention mechanism helps capture regularities in acoustic attributes (e.g., volume, pitch, timbre) and semantic content (words in a speech, musical notes, etc.) and model their variations over a long temporal range. Third, the motion policy is trained using a reward that encourages movements conducive to the best-possible separation of the target audio at every step, thus forcing both the policy and the acoustic memory to learn about patterns in the acoustic and semantic characteristics of the target.

We validate our ideas with realistic audio-visual simulations from SoundSpaces [13] together with 53 real-world Matterport3D [11] environment scans, along with non-periodic sounds from many diverse human speakers, music, and other common background sources. Our agent successfully learns a motion policy for hearing its dynamic target more clearly in unseen scenes, outperforming the state-of-the-art Move2Hear [38] approach in addition to several baselines.

2 Related Work

Audio(-Visual) Source Separation. Substantial prior work uses audio alone to separate sounds in a passive (non-embodied) way, whether from single-channel (monaural) audio [35,55,57,62] or multi-channel audio that better surfaces spatial cues [19,20,28,42,63,66,68]. Bringing vision together with audio, approaches based on signal processing and matrix factorization [23,27,34,47,49,54,56], visual tracking [7,36], and deep learning [25,29,65,70] have all been explored, with particular recent emphasis on audio-visual separation of speech [1,2,18, 21,24,30,45]. Other methods explore isolating a target audio source of interest [31,32,41,44,59].

Active models for hearing better include sound source localization methods in robotics that steer a microphone towards a localized source (e.g., [5,9,43]) and audio-visual methods that detect when people are speaking [3,6,61],[?], which could help attend to one at a time. Most recently, Move2Hear [38] trains an embodied agent to move in a 3D environment using audio and vision in order to hear a target object. Although it is an important first step in tackling active separation, Move2Hear deals only with static sounds, as discussed above. This severely limits its real-world scope, where almost all sounds, including those of interest like human speech, music, etc., are time-varying. Addressing dynamic sounds is decidedly more complex: the model must leverage not only the agent's surrounding geometry (to shut out undesirable interfering sounds) but also predict a continuously evolving audio sequence while hearing it just once within the mixed input sound. Aside from substantially generalizing the task, compared to [38] our core technical contributions are a novel transformer [60]-based acoustic memory that does bi-directional self-attention to not only produce high-quality current separations on the basis of past separations but also refine past separations by taking useful cues from current separations, and a dense reward that promotes accurate dynamic separation at every step of the agent's motion. We demonstrate our model's advantages over [38] in results.

Transformer Memory. Transformers [60] have been shown to do very well on long-horizon embodied tasks like navigation and exploration [10,12,16,22,39, 40,50]. The performance gains arise from a transformer's ability to effectively leverage past experiences of the agent [10,22,39,40,50] and do cross-modal reasoning [12,16]. Different from these methods, our idea is to use a transformer as a memory model for capturing long-range acoustic correlations for audio-visual separation. Unlike prior approaches that show the successful use of transformers for passive separation [15,37,58,67,69], we train an agent to actively position itself in a 3D environment so that the transformer memory can improve current and past estimates of the target audio through bi-directional self-attention.

3 Task Formulation

We introduce a novel embodied task: active audio-visual separation of dynamic sound sources. In this task, an autonomous agent hears multiple *dynamic* audio sources of different types placed at various locations in an unmapped 3D environment, where the raw audio emitted by each source varies with time. The agent's objective is to move around intelligently to isolate the *target source's*[1] dynamic audio from the observed audio mixture at every step of its motion.

Our agent relies on both acoustic and visual cues to actively listen to the ever-changing target audio. The audio stream conveys cues about the spatial placement of the audio sources relative to the agent. Besides letting the agent see sounding objects within its field of view, the visual signal can help it avoid obstacles and go towards more closed spots in the 3D scene to suppress undesired acoustic interference. Both audio and vision can reveal how far the agent can venture in search of possibly favorable locations to cut out distractor sounds—without failing to follow the target. The temporal variations in the heard audio can also inform the agent about patterns in the fluctuations of the sound quality, intensity, and semantics (e.g., a voice raising and falling), letting it anticipate those attributes for future timesteps and move accordingly to maximize separation.

[1] e.g., a human speaker or instrument the agent wishes to listen to.

Task Definition. An episode of the proposed task begins by randomly instantiating multiple audio sources in a 3D environment by sampling their locations, types, and the dynamic audio tracks that play at the source positions. At every timestep, the agent receives a mixed binaural audio waveform, which depends on the surrounding scene's geometric and material composition, the relative spatial arrangement of the agent and the sources, and the audio types (e.g., speech, musical instrument, etc.). One of the sources is the target. The agent's goal is to extract the latent monaural signal playing at the target at every timestep.[2]

To succeed, the agent needs to leverage egocentric audio-visual cues to move intelligently so that it can predict the dynamic audio sequence. The agent cannot re-experience sounds from past timesteps, but it is free to revise its past predictions.

Episode Specification. We formally specify an episode by the tuple $(\mathcal{E}, p_0, S_1(t), S_2(t), \ldots S_k(t), G^c)$. Here, $\mathcal{E}$ refers to the 3D scene in which the episode is executed, $p_0 = (l_0, o_0)$ denotes the initial agent pose determined by its location l and orientation o, and $S_i(t) = (S_i^w(t), S_i^l, S_i^c)$ specifies an audio source by its dynamic monaural waveform S_i^w as a function of the episode step t, location S_i^l and audio type S_i^c. k is the total number of audio sources of distinct types $(S_1^c \neq S_2^c \neq \cdots \neq S_k^c)$ present in the scene, and G^c is the target audio type, such that $G \in \{S_i\}$. At each step, the agent's objective is to listen to the binaural mixture of all sources and predict the dynamic signal $G^w(t)$ associated with the target audio type G^c. Note that distinct human voices are considered distinct audio types. We limit the episode length to $\mathcal{T}$ steps, thus assigning the agent a fixed time budget to retrieve the entire audio clip.

Action Space. At each timestep, the agent samples an action a_t from its action space $\mathcal{A} = \{MoveForward, TurnLeft, TurnRight\}$ and moves on an unobserved *navigability graph* of the environment. While turns are always valid, $MoveForward$ is not valid unless there is an edge from the current node to the next in the direction the agent is facing (i.e., no walls or obstacles exist there).

3D Environment and Audio-Visual Simulation. Following the state-of-the-art [12,13,38] in embodied audio-visual learning, we use the AI-Habitat simulator [53] with SoundSpaces [13] audio simulations and Matterport3D scenes [11]. Matterport3D provides dense 3D meshes and image scans of real-world houses and other indoor environments. SoundSpaces contains room impulse responses (RIR) to render realistic spatial audio at a resolution of 1 m for Matterport3D. It models how sound waves from each source travel in the 3D environment and

[2] The *monaural* source is the ground truth target, since it is devoid of all material and spatial effects due to the environment and the mutual positioning of the agent and the sources. Using a spatialized (e.g., binaural) audio as ground truth would permit undesirable shortcut solutions: the agent could move somewhere in the environment where the target is inaudible, and technically return the right answer of silence [38].

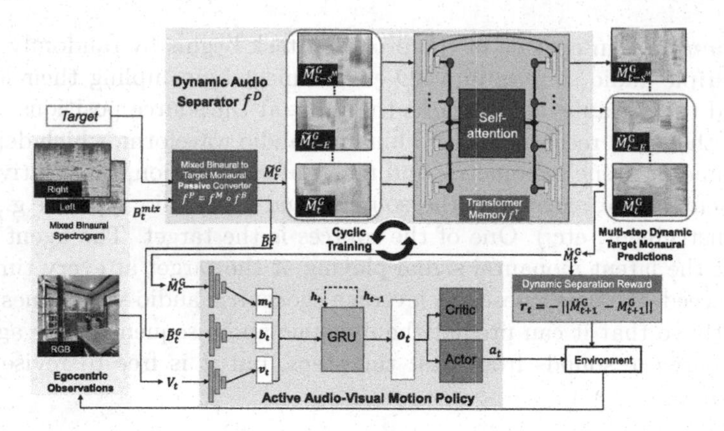

Fig. 2. Our model addresses active audio-visual separation of dynamic sources by leveraging a synergy between a *dynamic audio separator* and an *active audio-visual motion policy*. The dynamic separator f^D uses self-attention to continuously isolate a target signal M_t^G from its received mixed binaural B_t^{mix} on the basis of its past and current initial estimates $\{\tilde{M}_{t-s^{\mathcal{M}}}^G, \ldots, \tilde{M}_t^G\}$, while also using its current initial separation $\tilde{M}_t^G$ to refine past final separations $\{\ddot{M}_{t-E}^G, \ldots, \ddot{M}_{t-1}^G\}$. The motion policy uses the separator's outputs, $\ddot{M}_t^G$ and $\ddot{B}_t^G$ and egocentric RGB images V_t to guide the agent to areas suitable for separating future dynamic targets.

interact with the surrounding materials and geometry. In particular, the state-of-the-art RIRs simulate most real-world acoustic phenomena: direct sounds, early specular/diffuse reflections, reverberations, binaural spatialization, and frequency-dependent effects of materials and air absorption. See [13] for details. By using these highly perceptually realistic simulators together with real-world scanned environments and real-world audio data (cf. Sect. 5), we minimize the sim2real gap; we leave exploring transfer to a real robot for future work.

In every episode, we place k point objects as the monaural sources in a Matterport 3D scene and render a binaural mixture B_t^{mix} of the audio coming to the agent from all the source locations, by convolving the monaural waveforms at each step with the corresponding RIRs and taking their mean. Our agent moves through the 3D spaces while receiving real-time egocentric visual and audio observations that are a function of the 3D scene and its surface materials, the agent's current pose, and the sources' positions.

4 Approach

We approach the problem with reinforcement learning. We train a motion policy to make sequential movement decisions on the basis of egocentric audio-visual observations, guided by dynamic audio separation quality. Our model has two main components (see Fig. 2): 1) an audio separator network and 2) an active audio-visual (AV) motion policy.

The separator network serves three functions at every step: 1) it passively separates the current target audio segment from its heard mixture, 2) it improves its past separations by exploiting their correlations in acoustic attributes like volume, pitch, quality, content semantics, etc. with the current separation, and 3) it uses the current separation to guide the motion policy towards locations favorable for accurate separation of future audio segments. The motion policy is trained to repeatedly maximize dynamic separation quality. It moves the agent in the 3D scene to get the best possible estimate of the *complete* target signal.

These two components share a symbiotic relationship, each feeding off of useful learning cues from the other while training. This allows the agent to learn the complex links between separation quality and the dynamic acoustic attributes of the target source, its location relative to the agent, the scene's material layout (walls, floors, furniture, etc.), and the inferred spatial arrangement of distractor sources in the 3D environment.

4.1 Dynamic Audio Separator Network

The dynamic audio separator network f^D receives the mixed binaural sound B_t^{mix} coming from all sources in the environment and the target audio type G^c at every step. Using these inputs, it produces estimates $\ddot{M}^G$ of the monaural target at both the current and earlier E steps, i.e., $f^D(B_t^{mix}, G^c) = \{\ddot{M}_{t-E}^G, \dots, \ddot{M}_t^G\}$ (Fig. 2 top). We use short-time Fourier transform (STFT) to represent the monaural M and binaural B as audio magnitude spectrograms of dimensionality $F \times N$ and $2 \times F \times N$, respectively, where F is the number of frequency levels, N is the number of overlapping temporal windows, and B has 2 channels (left and right).

The audio separation takes place by using two modules, such that $f^D = f^P \circ f^T$. First, the passive audio separator module f^P predicts the current monaural target $\tilde{M}_t^G$, given the audio goal category G^c and the binaural mixture B_t^{mix}. Next, the transformer memory f^T uses self-attention to combine cues from the current monaural prediction $\tilde{M}_t^G$ and an external memory bank $\mathcal{M}$ of f^P's past predictions to both produce an enhanced version $\ddot{M}_t^G$ of the current monaural prediction and also refine the past E predictions $\{\ddot{M}_{t-E}^G, \dots, \ddot{M}_{t-1}^G\}$. Our multi-step prediction is motivated by the time-varying and fleeting nature of dynamic audio; we leverage semantic and acoustic structure in the target signal to both anticipate future targets and correct errors in past estimates. Next, we describe each of these modules in detail.

Passive Audio Separator. We adopt a factorized architecture [38] for the passive separator f^P, which comprises a binaural extractor f^B and a monaural predictor f^M. Given the mixed binaural and the target audio type, f^B predicts the target binaural $\tilde{B}_t^G$ by estimating a real-valued ratio mask [29,65,70]. f^M converts B_t^G into an initial estimate $\tilde{M}_t^G$ of the current monaural target. Both f^B and f^M use U-Net like architectures [51] with ReLU activations. We enhance these initial monaural predictions with a transformer memory, as we will describe next.

Transformer Memory. The proposed memory is a transformer encoder [60] that encodes the current and past $s^{\mathcal{M}}$ monaural predictions from f^M. Every transformer block has a pre-norm architecture that has been found to work particularly well for audio separation [58,69]. For storing past estimates from f^M, we maintain an external memory bank of size $s^{\mathcal{M}}$. At every step, we encode the past and current estimates using a multi-layer convolutional network to output a lower-dimensional monaural feature set $\{e_{t-s^{\mathcal{M}}}, \ldots, e_t\}$. We feed these monaural features to the transformer along with sinusoidal positional encodings.

The transformer encoder uses the monaural features for its keys, values, and queries to build a joint representation through self-attention, such that all features across the temporal range $[t - s^{\mathcal{M}}, t]$ can share information with each other. Further, the positional encoding for every feature informs the transformer about the feature's temporal position in its context, thus helping it capture both short- and long-range similarities in the target dynamic signal.

On the transformer's output side, we sample the feature set $\{d_{t-E}, \ldots, d_t\}$ associated with the timesteps in the range $[t - E, t]$, where $E \leq s^{\mathcal{M}}$, and decode them using another multi-layer convolution to obtain monaural estimates $\{\ddot{M}^G_{t-E}, \ldots, \ddot{M}^G_t\}$. We also connect the convolutional encoder and decoder with additive skip connections for faster training and improved generalization [33].

This multi-step prediction is particularly useful in the dynamic setting, where at each step, the agent hears a new segment of the target sound. Consequently, the agent can benefit from an error-correction mechanism that uses the current estimates of the target to amend for mistakes in past estimates. Our results show that unlike RNN-based alternatives, the transformer memory naturally provides the agent with this option as it can exploit temporally local and global patterns in the dynamic audio—such as regularities in pitch and volume in human speech, and notes and timbre in music—in a bi-directional manner. By doing self-attention on an explicit storage of memory entries in a non-sequential fashion, it can accurately separate the current audio target and also refine past separations. The high quality of the current separations allows the agent to search for separation-friendly locations in its environment more extensively, by providing robustness in separation even when the agent temporarily passes through areas with high distractor interference.

Both modules f^P and f^T are trained in a supervised manner using the ground truth of the target binaural and monaural signals, as detailed in Sect. 4.3.

4.2 Active Audio-Visual Motion Policy

The audio-visual (AV) motion policy guides the agent in the 3D environment to predict the best possible estimate of the dynamic audio target at every timestep (Fig. 2 bottom). On the basis of real-time egocentric visual and audio inputs, the AV motion policy sequentially emits actions a_t that help continuously maximize the separation quality of the dynamic separator network f^D (Sect. 4.1). It has two modules: an observation encoder and a policy network.

Observation Space and Encoding. At each step, the AV motion policy receives the egocentric RGB image V_t, the current binaural separation $\tilde{B}_t^G$ from f^B, and the current monaural prediction $\ddot{M}_t^G$ from f^T.

The audio and visual streams provide the agent with useful complementary information for continuous separation of the dynamic target. $\tilde{B}^G$ informs the agent about the relative spatial position (both distance and direction) of the target and also the scene's materials and geometry, allowing it to anticipate how these factors might affect its future movements and separation quality. The binaural cue also prevents the agent from venturing too far away from the target, where it would be unable to retrieve the current dynamic target, even with the help of its transformer memory. Besides informing the agent about the relation between separation quality and its pose relative to the target, as implicitly inferred from $\tilde{B}^G$, $\ddot{M}^G$ also allows the agent to track acoustic and semantic similarities in $\ddot{M}^G$ over time by using its recurrent memory module (see below) and choose an action that could help maximize the next separation.

The visual signal V_t informs the policy about the 3D scene's geometric configuration and helps it avoid collisions with obstacles. The synergy of vision and audio lets the agent learn about relations between the 3D scene's geometry, semantics, and surface materials on the one hand, and the expected separation quality for different agent poses on the other hand.

We learn three separate CNN encoders to represent these inputs: $v_t = \mathcal{F}^V(V_t)$, $b_t = \mathcal{F}^B(\tilde{B}_t^G)$ and $m_t = \mathcal{F}^M(\ddot{M}_t^G)$. We concatenate v_t, b_t and m_t to obtain the complete audio-visual representation o_t.

Policy Network. The policy network in our model is a gated recurrent neural network (GRU) [17] followed by an actor-critic architecture. The GRU receives the current audio-visual representation o_t and its accumulated history of states from the last step h_{t-1} to produce an updated history h_t and also emit the current state feature s_t. The actor-critic module receives s_t and h_{t-1}, and predicts the policy distribution $\pi_\theta(a_t|s_t, h_{t-1})$ and the current state's value $V_\theta(s_t, h_{t-1})$, where θ are the policy parameters. The agent samples actions a_t from $\mathcal{A}$ as per the policy distribution to interact with the 3D environment.

4.3 Training

Dynamic Audio Separator Training. As discussed in Sect. 4.1, the separator network has two modules: the passive separator f^P that outputs $\tilde{B}$ and $\tilde{M}$, and the transformer memory f^T that outputs $\ddot{M}$. We train both modules using the respective target spectrogram ground truths (B^G for binaural and M^G for monaural), which are provided by the simulator.

We train f^P with the passive separation loss:

$$\mathcal{L}_t^P = ||\tilde{B}_t^G - B_t^G||_1 + ||\tilde{M}_t^G - M_t^G||_1. \tag{1}$$

The training loss $\mathcal{L}_t^T$ for the transformer memory aims to maximize the quality of both past and current monaural estimates, $\{\ddot{M}_{t-E}^G, \dots, \ddot{M}_{t-1}^G\}$ and $\ddot{M}_t^G$, respectively:

$$\mathcal{L}_t^T = \frac{1}{E+1} \sum_{u=t-E}^{t} ||\ddot{M}_u^G - M_u^G||_1. \tag{2}$$

Note f^P outputs step-wise predictions ($\tilde{B}_t^G$ and $\tilde{M}_t^G$), unlike f^T that uses past separation history to produce its separation output for the current step and its current output to correct past errors. Hence, we pretrain f^P using $\mathcal{L}^P$ and a static dataset pre-collected from the training scenes by randomly instantiating the number of audio sources, their locations, the target audio type, the static monaural segments for the sources, and the agent pose for each data point. Besides reducing the online computation cost by lowering the amount of trainable parameters, preliminary experiments showed that pretraining leads to stationarity in the distribution of $\tilde{M}^G$ during the early phases of on-policy updates, thus stabilizing the training of the transformer.

Having pretrained f^P, we jointly train f^T and the AV motion policy with on-policy data while keeping the parameters of f^P frozen, since the agent's trajectories directly affect the distribution of inputs for both f^T and the policy and hence, their training.

Audio-Visual Motion Policy Training. The dynamic separation task calls for a continuous retrieval of the time-varying target audio by the agent. Towards this goal, we train our motion policy with a dense RL reward:

$$r_t = -||\ddot{M}_{t+1}^G - M_{t+1}^G||_1. \tag{3}$$

At each step, r_t encourages the agent to act to maximize the next separation.

We train the policy using Decentralized Distributed PPO (DD-PPO) [64] with trajectory rollouts of 20 steps. The DD-PPO loss consists of a value loss, policy loss, and entropy loss to promote exploration (see Supp. Sect. 11).

Joint Training. We train the dynamic separator and AV motion policy by switching training between them after every parameter update.

5 Experiments

Experimental Setup. We instantiate each episode with $k = 2$ audio sources that are placed randomly in the scene at least 8 m apart and specify one source as the target (Supp. details the dataset construction (Sect. 8) and shows results for $k > 2$ (Sect. 6)). To clearly distinguish our task from audio-navigation, the agent is initially co-located with the target source at the episode start and is expected to fine-tune its position to isolate the dynamic audio target. We set the time budget $\mathcal{T}$ to 20 for all episodes. We split 53 large Matterport3D scenes into non-overlapping train/val/test with 731K/100/1K episodes; we always evaluate the agent in unseen, unmapped environments.

Our dataset consists of 102 distinct types of sounds from three broad categories: speech, music, and background sounds. For speech, we use 100 different speakers from LibriSpeech [46]. For music, we combine multiple instruments from the MUSIC [70] dataset. For background sounds, we use non-speech and non-music sounds from ESC-50 [48], like running water, dog barking, etc. One of the 100 speakers (all speakers are considered distinct types) or music is the target. The distractors can be either background sounds, a speaker, or music. All sources play unique audio types. Our diverse dataset lets us evaluate different separation scenarios of varying difficulty: speech vs. speech, speech vs. music, or subtracting assorted background sounds.

There are 3,292 clips in total across all categories for use as monaural audio, each at least 20 s long. This helps avoid repetition in the monaural audio at each source in an episode with $\mathcal{T} = 20$. For unheard sounds, we split the clips into non-overlapping train/val/test in 18:3:4. For pretraining f^P, we generate 200K 1 s segments from the long clips and split it into train/val in 47:3, where the segments for each split are sampled from the corresponding split for long clips.

Existing Method and Baselines. We compare against the following methods:

- **Stand In-Place:** audio-only agent that does not change its pose throughout the episode, thereby separating the target in a completely passive way.
- **Rotate In-Place:** audio-only agent that rotates clockwise at its starting location at all steps to sample the mixed audio from all possible directions
- **DoA:** audio-only agent that takes one step away from the target, turns around, and samples the direct sound by facing its direction of arrival (DoA) [43] until the episode ends.
- **Random:** audio-only agent that randomly selects actions from the action space $\mathcal{A}$ at all steps.
- **Proximity Prior:** an agent that selects random actions but stays inside a radius of 2 m (selected through validation) of the target so that it does not lose track of the dynamic goal audio. Unlike our agent, this agent assumes access to the privileged information of the ground truth distance to the target.
- **Novelty** [8]: a visual exploration agent trained with RL to maximize its coverage of novel locations in the environment within the episode budget. This helps sample diverse audio cues which can potentially aid dynamic separation.
- **Move2Hear** [38]: a state-of-the-art audio-visual RL agent for active source separation that was originally proposed to tackle static audio sources. We retrain this model on our task by using the dynamic separation reward (Sect. 4.3).

For fair comparison, we use the same dynamic audio separator f^D for all baselines and our agent, while the acoustic stream that the separator for each individual model receives depends on the its agent's actions. While f^P shares its pre-trained parameters across all agents, f^T is trained separately for each agent by collecting on-policy data as per the individual agent's policy. This helps disentangle the contributions to the separation quality stemming from the separator network and the policy designs, respectively. We set $s^{\mathcal{M}}$ to 19 and E to 14 on the basis of validation (see Supp. Sect. 4 for additional analysis on the choice of E). These choices allow f^T to do accurate separation by leveraging the maximum amount of context possible in an episode ($s^{\mathcal{M}} = \mathcal{T} - 1$) while also refining past predictions. This way there is a balance between the amount of forward and backward information flow through f^T's bi-directional self-attention.

Evaluation. We evaluate all models in terms of mean separation quality of the dynamic monaural target over all $\mathcal{T}$ steps, averaged over 1,000 test episodes and 3 random seeds. We use standard metrics: **STFT distance**, the error between separated and ground truth spectrograms, and **SI-SDR** [52], a scale-invariant measure of the amount of distortion present in the separated waveform.

Table 1. Active audio-visual dynamic source separation.

Model	*Heard*		*Unheard*	
	SI-SDR ↑	STFT ↓	SI-SDR ↑	STFT ↓
Stand In-Place	2.49	0.328	2.03	0.343
Rotate In-Place	2.50	0.327	2.04	0.343
DoA	2.78	0.313	1.88	0.342
Random	2.81	0.314	1.95	0.343
Proximity Prior	2.92	0.309	2.05	0.338
Novelty [8]	1.68	0.358	1.44	0.366
Move2Hear [38]	2.31	0.331	2.06	0.339
Ours	**3.93**	**0.273**	**2.57**	**0.318**
Ours w/o transformer memory f^T	2.32	0.330	2.02	0.340
Ours w/o multi-step predictions (*i.e.*, $E = 0$)	2.69	0.317	2.29	0.33
Ours w/o visual input V_t	3.61	0.284	2.40	0.324

We provide more details about the data, spectrogram computation, baseline implementation, network architectures, training and evaluation in Supp Sect. 8–11.

5.1 Source Separation Results

Table 1 shows the separation quality of all models. The passive baselines Stand and Rotate In-Place fare worse than the ones that move and sample more diverse audio cues, like Random and Proximity Prior. However, despite being able to move, Novelty [8] performs poorly; in its effort to maximize coverage of the environment, it wanders too far from the target and fails to hear it in certain phases of its motion. DoA improves over the stationary baselines, since standing a step away from the target source allows it to sample a cleaner audio cue.

On the unheard sounds setting, Move2Hear [38] outperforms some baselines (e.g., DoA, Random, Novelty [8], etc.) but fares comparably against others (e.g., stationary baselines and Proximity Prior). This behavior can be attributed to the absence of the transformer memory f^T in Move2Hear. Although it is trained to maximize the separation quality at every step, its acoustic memory refiner is unable to handle the time-varying and transient nature of audio. To further illustrate our task complexity, we plot the distribution of separation performance (SI-SDR) of Move2Hear for both static and dynamic sources in Fig. 3a. The switch from dynamic to static audio results in a substantial degradation of separation quality (compare dotted and solid red curves: the mode of the SI-SDR distribution shifts leftward upon going from dynamic to static). This clearly shows the distinct new challenges posed by the dynamic source separation setting.

Our model outperforms all baselines and Move2Hear by a statistically significant margin ($p \leq 0.05$). It overcomes the drop in Move2Hear's separation quality induced by dynamic sources (compare solid red and green curves in Fig. 3a: the mode for dynamic audio shifts rightward upon using our method). Its performance demonstrates the impact of our active policy and long-term memory. Simply staying at or close to the source to be able to continuously hear the target (as done by Stand or Rotate In-Place, DoA, and Proximity Prior) is not enough. Our improvement over Move2Hear [38] further emphasizes the advantage of our transformer memory f^T in dealing with dynamic audio, both in terms of boosting separation when the agent is able to sample a cleaner signal, and providing robustness to the separator when the agent is passing through zones that are relatively less suitable for separation.

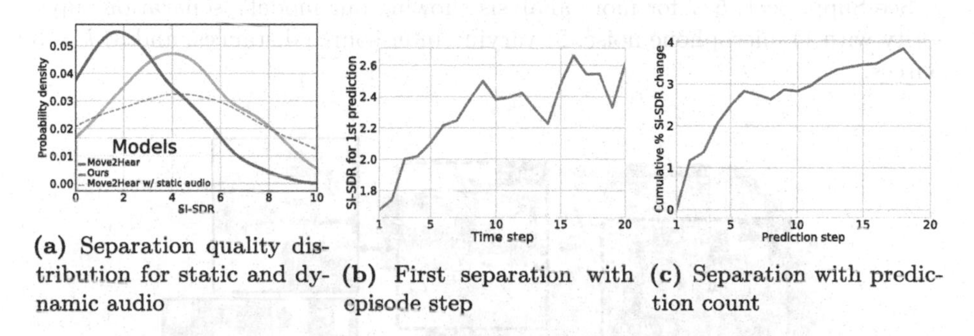

(a) Separation quality distribution for static and dynamic audio

(b) First separation with episode step

(c) Separation with prediction count

Fig. 3. (a) Distribution of separation quality for static and dynamic audio (higher SI-SDR is better). (b) Quality at first separation as a function of time. (c) Separation quality for a target as a function of prediction count.

5.2 Model Analysis

Ablations. We report the ablation of our model components in Table 1 bottom. Our model suffers a severe drop in performance upon removing our transformer memory (f^T). This shows that f^T plays a significant role in boosting dynamic separation quality. f^T learns joint representations of past and present separation estimates through self-attention, successfully leveraging long-range temporal relations in the audio signal for high-quality separation. The drop when E is set to 0 shows that the proposed multi-step predictions are also important in the dynamic setting, where the agent hears each segment of the time-varying signal only once and might not be able to separate a target very well at its first attempt. Our multi-step predictions let the agent improve past separations over time by extrapolating acoustic and semantic cues present in the current estimate. The visual cue V_t also contributes to our agent's separation performance; without it, the agent is prone to collisions and has to solely rely on the separated binaural B_t^G to understand how the surrounding geometry and materials affect separation.

Transformer Memory. Next we analyze different aspects of our transformer memory f^T to understand how they boost our agent's separation performance.

Figure 3b shows how a longer memory leads to better initial estimates of the target audio segments. With more memory, f^T is better able to find consistencies in past estimates to improve the current estimate.

Figure 3c shows how making more estimates of the same target using the multi-step prediction mechanism of f^T improves separation. Our model keeps refining all its predictions until late in the episode. This shows how f^T uses self-attention for enhancing past estimates on the basis of the current estimate through backward flow of useful information shared across multiple audio segments.

See Supp. Sect. 5–7 for more analysis showing our model's separation superiority with 1) microphone noise, 2) varying inter-source distances, and 3) $k = 3$ sources.

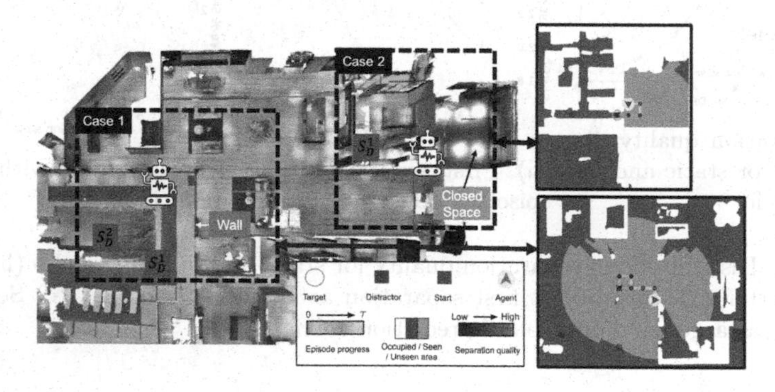

Fig. 4. Sample trajectories of our model. Our model not only uses audio-visual cues to find locations in the 3D scene that help dampen the distractors, but also utilizes its transformer memory to provide separation robustness and stability, while non-myopically looking for sweet spots with the potential to improve overall separation in the future.

Qualitative Results. In Fig. 4, we show two successful dynamic separation episodes of our method. In case 1, the agent starts outdoors and its motion choices are limited by the lack of potential acoustic barriers between the two distractors and the agent. Hence, it ventures away from the target in order to go indoors and make use of a wall to place itself in the acoustic shadow of the distractors, while still being able to sample the target audio. Such behavior is possible with our transformer memory, which maintains reasonable separation quality while the agent crosses regions of low separation quality looking for places that could improve future separation. In case 2, the agent makes short movements to find a closed space in its vicinity that not only blocks the distractor, but also reinforces its hearing of the target through acoustic reflections. See Supp. Sect. 2 and 3 for a discussion on our method's failure cases and our model's separation quality as a function of agent pose, respectively.

6 Conclusion

We introduced the task of active audio-visual separation of dynamic sources, where an agent must see and hear to move in an environment to continuously isolate the sounds of the time-varying source of interest. Our proposed approach outperforms the previous state-of-the-art in active audio-visual separation, as well as strong baselines and a popular exploration policy from prior work. In future work, we aim to extend our model to tackle sim2real transfer and moving sources, e.g., with audio source motion anticipation.

Acknowledgements. Thank you to Ziad Al-Halah for very valuable discussions. Thanks to Tushar Nagarajan, Kumar Ashutosh, and David Harwarth for feedback on paper drafts. UT Austin is supported in part by DARPA L2M, NSF CCRI, and the IFML NSF AI Institute. K.G. is paid as a research scientist by Meta.

References

1. Afouras, T., Chung, J.S., Zisserman, A.: The conversation: deep audio-visual speech enhancement. arXiv preprint arXiv:1804.04121 (2018)
2. Afouras, T., Chung, J.S., Zisserman, A.: My lips are concealed: audio-visual speech enhancement through obstructions. arXiv preprint arXiv:1907.04975 (2019)
3. Alameda-Pineda, X., Horaud, R.: Vision-guided robot hearing. Int. J. Robot. Res. **34**(4–5), 437–456 (2015)
4. Yu, Y., Huang, W., Sun, F., Chen, C., Wang, Y., Liu, X.: sound adversarial audio-visual navigation. In: Submitted to The Tenth International Conference on Learning Representations (2022). https://openreview.net/forum?id=NkZq4OEYN-
5. Asano, F., Goto, M., Itou, K., Asoh, H.: Real-time sound source localization and separation system and its application to automatic speech recognition. In: Eurospeech (2001)
6. Ban, Y., Li, X., Alameda-Pineda, X., Girin, L., Horaud, R.: Accounting for room acoustics in audio-visual multi-speaker tracking. In: IEEE International Conference on Acoustics, Speech and Signal Processing (ICASSP) (2018)
7. Barzelay, Z., Schechner, Y.Y.: Harmony in motion. In: 2007 IEEE Conference on Computer Vision and Pattern Recognition, pp. 1–8 (2007). https://doi.org/10.1109/CVPR.2007.383344

8. Bellemare, M.G., Srinivasan, S., Ostrovski, G., Schaul, T., Saxton, D., Munos, R.: Unifying count-based exploration and intrinsic motivation. arXiv preprint arXiv:1606.01868 (2016)
9. Bustamante, G., Danès, P., Forgue, T., Podlubne, A., Manhès, J.: An information based feedback control for audio-motor binaural localization. Auton. Robots **42**(2), 477–490 (2017). https://doi.org/10.1007/s10514-017-9639-8
10. Campari, T., Eccher, P., Serafini, L., Ballan, L.: Exploiting scene-specific features for object goal navigation. In: Bartoli, A., Fusiello, A. (eds.) ECCV 2020. LNCS, vol. 12538, pp. 406–421. Springer, Cham (2020). https://doi.org/10.1007/978-3-030-66823-5_24
11. Chang, A., et al.: Matterport3d: learning from RGB-D data in indoor environments. In: International Conference on 3D Vision (3DV) (2017). matterPort3D dataset license. http://kaldir.vc.in.tum.de/matterport/MP_TOS.pdf
12. Chen, C., Al-Halah, Z., Grauman, K.: Semantic audio-visual navigation. In: CVPR (2021)
13. Chen, C., et al.: SoundSpaces: audio-visual navigation in 3d environments. In: Vedaldi, A., Bischof, H., Brox, T., Frahm, J.-M. (eds.) ECCV 2020. LNCS, vol. 12351, pp. 17–36. Springer, Cham (2020). https://doi.org/10.1007/978-3-030-58539-6_2
14. Chen, C., Majumder, S., Al-Halah, Z., Gao, R., Ramakrishnan, S.K., Grauman, K.: Learning to set waypoints for audio-visual navigation. In: International Conference on Learning Representations (2021). https://openreview.net/forum?id=cR91FAodFMe
15. Chen, J., Mao, Q., Liu, D.: Dual-path transformer network: direct context-aware modeling for end-to-end monaural speech separation. arXiv preprint arXiv:2007.13975 (2020)
16. Chen, K., Chen, J.K., Chuang, J., Vázquez, M., Savarese, S.: Topological planning with transformers for vision-and-language navigation. In: Proceedings of the IEEE/CVF Conference on Computer Vision and Pattern Recognition, pp. 11276–11286 (2021)
17. Chung, J., Kastner, K., Dinh, L., Goel, K., Courville, A.C., Bengio, Y.: A recurrent latent variable model for sequential data. In: NeurIPS (2015)
18. Chung, S.W., Choe, S., Chung, J.S., Kang, H.G.: Facefilter: audio-visual speech separation using still images. arXiv preprint arXiv:2005.07074 (2020)
19. Deleforge, A., Horaud, R.: The cocktail party robot: sound source separation and localisation with an active binaural head. In: HRI 2012–7th ACM/IEEE International Conference on Human Robot Interaction, pp. 431–438. ACM, Boston, United States, March 2012. https://doi.org/10.1145/2157689.2157834,https://hal.inria.fr/hal-00768668
20. Duong, N.Q., Vincent, E., Gribonval, R.: Under-determined reverberant audio source separation using a full-rank spatial covariance model. IEEE Trans. Audio Speech Lang. Process. **18**(7), 1830–1840 (2010)
21. Ephrat, A., et al.: Looking to listen at the cocktail party: a speaker-independent audio-visual model for speech separation. arXiv preprint arXiv:1804.03619 (2018)
22. Fang, K., Toshev, A., Fei-Fei, L., Savarese, S.: Scene memory transformer for embodied agents in long-horizon tasks. In: Proceedings of the IEEE/CVF Conference on Computer Vision and Pattern Recognition, pp. 538–547 (2019)
23. Fisher III, J.W., Darrell, T., Freeman, W., Viola, P.: Learning joint statistical models for audio-visual fusion and segregation. In: Leen, T., Dietterich, T., Tresp, V. (eds.) Advances in Neural Information Processing Systems, vol. 13,

pp. 772–778. MIT Press (2001). https://proceedings.neurips.cc/paper/2000/file/11f524c3fbfeeca4aa916edcb6b6392e-Paper.pdf
24. Gabbay, A., Shamir, A., Peleg, S.: Visual speech enhancement. arXiv preprint arXiv:1711.08789 (2017)
25. Gan, C., Huang, D., Zhao, H., Tenenbaum, J.B., Torralba, A.: Music gesture for visual sound separation. In: Proceedings of the IEEE/CVF Conference on Computer Vision and Pattern Recognition, pp. 10478–10487 (2020)
26. Gan, C., Zhang, Y., Wu, J., Gong, B., Tenenbaum, J.B.: Look, listen, and act: towards audio-visual embodied navigation. In: ICRA (2020)
27. Gao, R., Feris, R., Grauman, K.: Learning to separate object sounds by watching unlabeled video. In: Proceedings of the European Conference on Computer Vision (ECCV), pp. 35–53 (2018)
28. Gao, R., Grauman, K.: 2.5d visual sound. In: CVPR (2019)
29. Gao, R., Grauman, K.: Co-separating sounds of visual objects. In: Proceedings of the IEEE/CVF International Conference on Computer Vision, pp. 3879–3888 (2019)
30. Gao, R., Grauman, K.: Visualvoice: audio-visual speech separation with cross-modal consistency. arXiv preprint arXiv:2101.03149 (2021)
31. Gu, R., et al.: Neural spatial filter: target speaker speech separation assisted with directional information. In: Kubin, G., Kacic, Z. (eds.) Interspeech 2019, 20th Annual Conference of the International Speech Communication Association, Graz, Austria, 15–19 September 2019, pp. 4290–4294. ISCA (2019). https://doi.org/10.21437/Interspeech.2019-2266
32. Gu, R., Zou, Y.: Temporal-spatial neural filter: direction informed end-to-end multi-channel target speech separation. arXiv preprint arXiv:2001.00391 (2020)
33. He, K., Zhang, X., Ren, S., Sun, J.: Deep residual learning for image recognition. In: Proceedings of the IEEE Conference on Computer Vision and Pattern Recognition, pp. 770–778 (2016)
34. Hershey, J.R., Movellan, J.R.: Audio vision: Using audio-visual synchrony to locate sounds. In: NeurIPS (2000)
35. Huang, P., Kim, M., Hasegawa-Johnson, M., Smaragdis, P.: Deep learning for monaural speech separation. In: 2014 IEEE International Conference on Acoustics, Speech and Signal Processing (ICASSP), pp. 1562–1566 (2014). https://doi.org/10.1109/ICASSP.2014.6853860
36. Li, B., Dinesh, K., Duan, Z., Sharma, G.: See and listen: score-informed association of sound tracks to players in chamber music performance videos. In: 2017 IEEE International Conference on Acoustics, Speech and Signal Processing (ICASSP), pp. 2906–2910 (2017). https://doi.org/10.1109/ICASSP.2017.7952688
37. Lu, W.T., Wang, J.C., Won, M., Choi, K., Song, X.: Spectnt: a time-frequency transformer for music audio. arXiv preprint arXiv:2110.09127 (2021)
38. Majumder, S., Al-Halah, Z., Grauman, K.: Move2Hear: active audio-visual source separation. In: ICCV (2021)
39. Mezghani, L., et al.: Memory-augmented reinforcement learning for image-goal navigation. arXiv preprint arXiv:2101.05181 (2021)
40. Mezghani, L., Sukhbaatar, S., Szlam, A., Joulin, A., Bojanowski, P.: Learning to visually navigate in photorealistic environments without any supervision. arXiv preprint arXiv:2004.04954 (2020)
41. Žmolíková, K., et al.: Speakerbeam: speaker aware neural network for target speaker extraction in speech mixtures. IEEE J. Sel. Top. Sign. Proces. **13**(4), 800–814 (2019). https://doi.org/10.1109/JSTSP.2019.2922820

42. Nakadai, K., Hidai, K.i., Okuno, H.G., Kitano, H.: Real-time speaker localization and speech separation by audio-visual integration. In: Proceedings 2002 IEEE International Conference on Robotics and Automation (Cat. No. 02CH37292), vol. 1, pp. 1043–1049. IEEE (2002)
43. Nakadai, K., Lourens, T., Okuno, H.G., Kitano, H.: Active audition for humanoid. In: AAAI (2000)
44. Ochiai, T., et al.: Listen to what you want: neural network-based universal sound selector. In: Meng, H., Xu, B., Zheng, T.F. (eds.) Interspeech 2020, 21st Annual Conference of the International Speech Communication Association, Virtual Event, Shanghai, China, 25–29 October 2020, pp. 1441–1445. ISCA (2020). https://doi.org/10.21437/Interspeech.2020-2210, https://doi.org/10.21437/Interspeech.2020-2210
45. Owens, A., Efros, A.A.: Audio-visual scene analysis with self-supervised multisensory features. In: Proceedings of the European Conference on Computer Vision (ECCV), pp. 631–648 (2018)
46. Panayotov, V., Chen, G., Povey, D., Khudanpur, S.: Librispeech: An ASR corpus based on public domain audio books. In: 2015 IEEE International Conference on Acoustics, Speech and Signal Processing (ICASSP), pp. 5206–5210 (2015). https://doi.org/10.1109/ICASSP.2015.7178964
47. Parekh, S., Essid, S., Ozerov, A., Duong, N.Q.K., Pérez, P., Richard, G.: Motion informed audio source separation. In: 2017 IEEE International Conference on Acoustics, Speech and Signal Processing (ICASSP), pp. 6–10 (2017). https://doi.org/10.1109/ICASSP.2017.7951787
48. Piczak, K.J.: ESC: dataset for environmental sound classification. In: Proceedings of the 23rd Annual ACM Conference on Multimedia, pp. 1015–1018. ACM Press. https://doi.org/10.1145/2733373.2806390, http://dl.acm.org/citation.cfm?doid=2733373.2806390
49. Pu, J., Panagakis, Y., Petridis, S., Pantic, M.: Audio-visual object localization and separation using low-rank and sparsity. In: 2017 IEEE International Conference on Acoustics, Speech and Signal Processing (ICASSP), pp. 2901–2905. IEEE (2017)
50. Ramakrishnan, S.K., Nagarajan, T., Al-Halah, Z., Grauman, K.: Environment predictive coding for embodied agents. arXiv preprint arXiv:2102.02337 (2021)
51. Ronneberger, O., Fischer, P., Brox, T.: U-net: convolutional networks for biomedical image segmentation. In: Navab, N., Hornegger, J., Wells, W.M., Frangi, A.F. (eds.) MICCAI 2015. LNCS, vol. 9351, pp. 234–241. Springer, Cham (2015). https://doi.org/10.1007/978-3-319-24574-4_28
52. Roux, J.L., Wisdom, S., Erdogan, H., Hershey, J.R.: SDR - half-baked or well done? In: ICASSP 2019–2019 IEEE International Conference on Acoustics, Speech and Signal Processing (ICASSP), pp. 626–630 (2019). https://doi.org/10.1109/ICASSP.2019.8683855
53. Savva, M., et al.: Habitat: a platform for embodied AI research. In: ICCV (2019)
54. Sedighin, F., Babaie-Zadeh, M., Rivet, B., Jutten, C.: Two multimodal approaches for single microphone source separation. In: 2016 24th European Signal Processing Conference (EUSIPCO), pp. 110–114 (2016). https://doi.org/10.1109/EUSIPCO.2016.7760220
55. Smaragdis, P., Raj, B., Shashanka, M.: Supervised and semi-supervised separation of sounds from single-channel mixtures. In: Davies, M.E., James, C.J., Abdallah, S.A., Plumbley, M.D. (eds.) ICA 2007. LNCS, vol. 4666, pp. 414–421. Springer, Heidelberg (2007). https://doi.org/10.1007/978-3-540-74494-8_52

56. Smaragdis, P., Smaragdis, P.: Audio/visual independent components. In: Proceedings of International Symposium on Independant Component Analysis and Blind Source Separation (2003)
57. Spiertz, M., Gnann, V.: Source-filter based clustering for monaural blind source separation. In: Proceedings of International Conference on Digital Audio Effects DAFx'09 (2009)
58. Subakan, C., Ravanelli, M., Cornell, S., Bronzi, M., Zhong, J.: Attention is all you need in speech separation. In: ICASSP 2021–2021 IEEE International Conference on Acoustics, Speech and Signal Processing (ICASSP), pp. 21–25. IEEE (2021)
59. Tzinis, E., et al.: Into the wild with audioscope: unsupervised audio-visual separation of on-screen sounds. arXiv preprint arXiv:2011.01143 (2020)
60. Vaswani, A., et al.: Attention is all you need. In: Advances in Neural Information Processing Systems, pp. 5998–6008 (2017)
61. Viciana-Abad, R., Marfil, R., Perez-Lorenzo, J., Bandera, J., Romero-Garces, A., Reche-Lopez, P.: Audio-visual perception system for a humanoid robotic head. Sensors **14**(6), 9522–9545 (2014)
62. Virtanen, T.: Monaural sound source separation by nonnegative matrix factorization with temporal continuity and sparseness criteria. IEEE Trans. Audio Speech Lang. Process. **5**(3), 1066–1074 (2007)
63. Weiss, R.J., Mandel, M.I., Ellis, D.P.: Source separation based on binaural cues and source model constraints. In: Ninth Annual Conference of the International Speech Communication Association, vol. 2008 (2009)
64. Wijmans, E., et al.: DD-PPO: learning near-perfect pointgoal navigators from 2.5 billion frames. arXiv preprint arXiv:1911.00357 (2019)
65. Xu, X., Dai, B., Lin, D.: Recursive visual sound separation using minus-plus net. In: Proceedings of the IEEE/CVF International Conference on Computer Vision, pp. 882–891 (2019)
66. Özgür Yılmaz, Rickard, S.: Blind separation of speech mixtures via time-frequency masking. In: IEEE Transactions on Signal Processing (2002) Submitted (2004)
67. Zadeh, A., Ma, T., Poria, S., Morency, L.P.: Wildmix dataset and spectro-temporal transformer model for monoaural audio source separation. arXiv preprint arXiv:1911.09783 (2019)
68. Zhang, X., Wang, D.: Deep learning based binaural speech separation in reverberant environments. IEEE/ACM Trans. Audio Speech Lang. Process. **25**(5), 1075–1084 (2017)
69. Zhang, Z., He, B., Zhang, Z.: Transmask: a compact and fast speech separation model based on transformer. In: ICASSP 2021–2021 IEEE International Conference on Acoustics, Speech and Signal Processing (ICASSP), pp. 5764–5768. IEEE (2021)
70. Zhao, H., Gan, C., Rouditchenko, A., Vondrick, C., McDermott, J., Torralba, A.: The sound of pixels. In: Proceedings of the European Conference on Computer Vision (ECCV), pp. 570–586 (2018)

DexMV: Imitation Learning for Dexterous Manipulation from Human Videos

Yuzhe Qin, Yueh-Hua Wu, Shaowei Liu, Hanwen Jiang, Ruihan Yang, Yang Fu, and Xiaolong Wang(✉)

University of California San Diego, La Jolla 92093, USA
xlw012@ucsd.edu

Abstract. While significant progress has been made on understanding hand-object interactions in computer vision, it is still very challenging for robots to perform complex dexterous manipulation. In this paper, we propose a new platform and pipeline DexMV (**Dex**terous **M**anipulation from **V**ideos) for imitation learning. We design a platform with: (i) a simulation system for complex dexterous manipulation tasks with a multi-finger robot hand and (ii) a computer vision system to record large-scale demonstrations of a human hand conducting the same tasks. In our novel pipeline, we extract 3D hand and object poses from videos, and propose a novel demonstration translation method to convert human motion to robot demonstrations. We then apply and benchmark multiple imitation learning algorithms with the demonstrations. We show that the demonstrations can indeed improve robot learning by a large margin and solve the complex tasks which reinforcement learning alone cannot solve. Code and videos are available at https://yzqin.github.io/dexmv

Keywords: Dexterous manipulation · Learning from human demonstration · Reinforcement learning

1 Introduction

Dexterous manipulation of objects is the primary means for humans to interact with the physical world. Humans perform dexterous manipulation in everyday tasks with diverse objects. To understand these tasks, in computer vision, there is significant progress on 3D hand-object pose estimation [28,78] and affordance reasoning [13,71]. While computer vision techniques have greatly advanced, it is still very challenging to equip robots with human-like dexterity. Recently,

Y. Qin and Y.-H. Wu—Equal Contribution.

Supplementary Information The online version contains supplementary material available at https://doi.org/10.1007/978-3-031-19842-7_33.

S. Avidan et al. (Eds.): ECCV 2022, LNCS 13699, pp. 570–587, 2022.
https://doi.org/10.1007/978-3-031-19842-7_33

there has been a lot of effort on using reinforcement learning (RL) for dexterous manipulation with an anthropomorphic robot hand [51]. However, given the high Degree-of-Freedom joints and nonlinear tendon-based actuation of the multi-finger robot hand, it requires a *large amount* of training data with RL. Robot hands trained using only RL will also adopt *unnatural* behavior. Given these challenges, can we leverage humans' experience in the interaction with the physical world to guide robots, with the help of computer vision techniques?

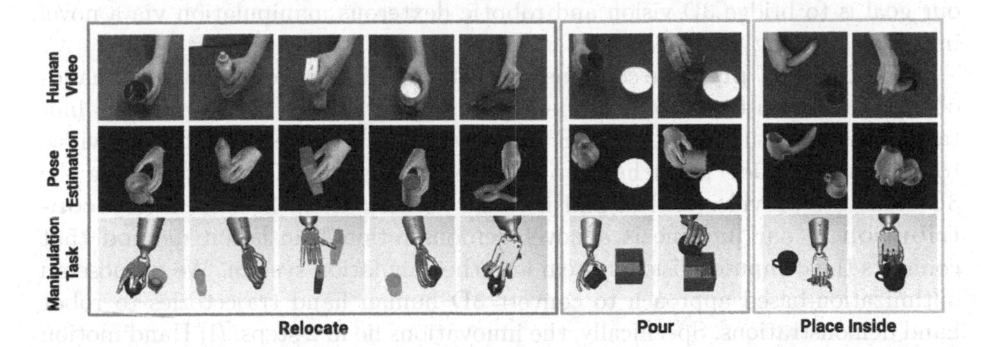

Fig. 1. We record human videos on manipulation tasks (1st row) and perform 3D hand-object pose estimations from the videos (2nd row) to construct the demonstrations. We have a paired simulation system providing the same dexterous manipulation tasks for the multi-finger robot (3rd row), including *relocate*, *pour*, and *place inside*, which we can solve using imitation learning with the inferred demonstrations.

One promising avenue is imitation learning from human demonstrations [59, 64]. Particularly, for dexterous manipulation, Rajeswaran *et al.* [59] introduces a simulation environment with four different manipulation tasks and paired sets of human demonstrations collected with a Virtual Reality (VR) headset and a motion capture glove. However, data collection with VR is relatively high-cost and not scalable, and there are only 25 demonstrations collected for each task. It also limits the complexity of the task: It is shown in [57] that RL can achieve similar performance with or without the demonstrations in most tasks proposed in [59]. Instead of focusing on a small scale of data, we look into increasing the difficulty and complexity of the manipulation tasks with diverse daily objects. This requires large-scale human demonstrations which are hard to obtain with VR but are much more available from human videos.

In this paper, we propose **a new platform and a novel imitation learning pipeline** for benchmarking complex and generalizable dexterous manipulation, namely DexMV (**Dex**terous **M**anipulation from **V**ideos). We introduce new tasks with the multi-finger robot hand (Adroit Robotic Hand [41]) on diverse objects in simulation. We collect real human hand videos performing the same tasks as demonstrations. By using human videos instead of VR, it *largely reduces the cost* for data collection and allows humans to perform more *complex and diverse* tasks. While the video demonstrations might not be optimal for perfect imitation (e.g., behavior cloning) to learn successful policies, the diverse dataset is beneficial for augmenting the training data for RL, which can learn from both successful and unsuccessful trials.

Our DexMV platform contains a paired systems with: (i) A computer vision system which records the videos of human performing manipulation tasks (1st row in Fig. 1); (ii) A physical simulation system which provides the interactive environments for dexterous manipulation with a multi-finger robot (3rd row in Fig. 1). The two systems are aligned with the same tasks. With this platform, our goal is to bridge 3D vision and robotic dexterous manipulation via a novel imitation learning pipeline.

Our DexMV pipeline contains three stages. First, we extract the 3D hand-object poses from the recorded videos (2nd row in Fig. 1). Unlike previous imitation learning studies with 2-DoF grippers [69,79], we need the human video to guide the 30-DoF robot hand to move each finger in 3D space. Parsing the 3D structure provides critical and necessary information. Second, a **key contribution** in our pipeline is a novel demonstration translation method that connects the computer vision system and the simulation system. We propose an optimization-based approach to convert 3D human hand trajectories to robot hand demonstrations. Specifically, the innovations lie in 2 steps: (i) Hand motion retargeting approach to obtain robot hand states; and (ii) Robot action estimation to obtain the actions for learning. Third, given the robot demonstrations, we perform imitation learning in the simulation tasks. We investigate and benchmark algorithms which augment RL objectives with state-only [57] and state-action [31,59] demonstrations.

We experiment with three types of challenging tasks with the YCB objects [15]. The first task is to *relocate* an object to a goal position. Instead of relocating a single ball as [59], we increase the task difficulty by using diverse objects (first 5 columns in Fig. 1). The second task is *pour*, which requires the robot to pour the particles from a mug into a container (Fig. 1 from column 6). The third task is *place inside*, where the robot hand needs to place an object into a container (last 2 columns in Fig. 1). In our experiments, we benchmark different imitation learning algorithms and show human demonstrations improve task performance by a large margin. More interestingly, we find the learned policy with our demonstrations can even generalize to unseen instances within the same category or outside the category. We highlight our contributions as follows:

- DexMV platform for learning dexterous manipulation using human videos. It contains paired computer vision and simulation systems with multiple dexterous complex manipulation tasks.
- DexMV pipeline to perform imitation learning, with a key innovation on demonstration translation to convert human videos to robot demonstrations.
- With DexMV platform and pipeline, we largely improve the dexterous manipulation performance over multiple complex tasks, and its generalization ability to unseen object instances.

2 Related Work

Dexterous Manipulation. Manipulation with multi-finger hands is one of the most challenging robotics tasks, which has been actively studied with optimization and planning [3,7,10,20,49,62]. Recently, researchers have started exploring reinforcement learning for dexterous manipulation [50,51]. However, training with only RL requires huge data samples. Different from a regular 2-DoF robot gripper, the Adroit Robotic Hand used in our experiment has 30 DoFs, which greatly increases the optimization space.

Imitation Learning from Human Demonstrations. Imitation learning is a promising paradigm for robot learning. It is not limited to behavior cloning [8,11, 54,61,75] and inverse reinforcement learning [1,5,23,31,44,48,63,76], but also augmenting the RL training with demonstrations [21,53,57,59,77]. For example, Rajeswaran *et al.* [59] propose to incorporate demonstrations with on-policy RL. However, all these approaches rely on expert demonstrations collected using expert policies or VR, which is not scalable and generalizable to complex tasks. In fact, it is shown in [57] that RL can achieve similar performance and sample efficiency with or without the demonstrations in most dexterous manipulation tasks in [59] besides *relocate*. Going beyond a VR setup, researchers have recently explored imitation learning and RL with videos [17,64,65,68,69,79]. However, all tasks are relatively simple (e.g., pushing a block) with a parallel gripper. In this paper, we propose new challenging dexterous manipulation tasks. We leverage pose estimation for providing demonstrations that largely improve imitation learning performance, while RL alone fails to solve these tasks.

Hand-Object Interaction. Hand-object interaction is a widely studied topic. One line of work focus on object pose estimation [29,32,38,52,56,72,78] and hand pose estimation [6,12,25,28,33,40,45,70,80]. For example, Berk *et al.* [15] propose YCB dataset with real objects and their 3D scans. Beyond object poses, datasets and methods for joint estimation of the hand-object poses are also proposed [18,24,26,45]. Another line of work focus on hand-object contact reasoning [13,71], which can be used for functional grasps [14,34,46]. Recently, researchers explored to use hand pose estimation to control a robot hand [4,27,43]. They require motion retargeting to map the hand pose into robot motion. These work focus on teleoperation where no objects are involved. In DexMV, we consider the demonstration translation setting, which converts a sequence of hand-object poses into robot demonstrations. We show that the translated demonstration is beneficial for imitation learning for dexterous manipulation.

3 Overview

We propose DexMV for **Dex**terous **M**anipulation with imitation learning from human **V**ideos including:

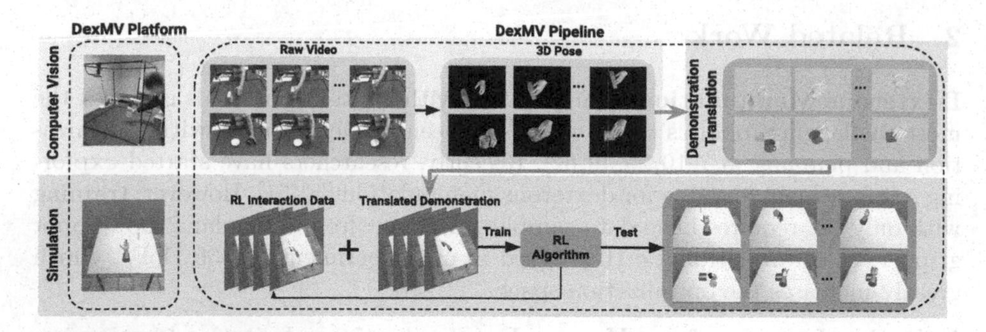

Fig. 2. DexMV platform and pipeline overview. Our platform is composed of a computer vision system (yellow), a simulation system (blue), and a demonstration translation module (colored with green). In computer vision system, we collect human manipulation videos. In simulation, we design the same tasks for the robot hand. We apply 3D hand-object pose estimation from videos, followed by demonstration translation to generate robot demonstrations, which are then used for imitation learning. (Color figure online)

(i) **DexMV platform**, which offers a paired computer vision system recording human manipulation videos and physical simulators conducting the same tasks. We design three challenging tasks in the simulator and use computer vision system to collect human demonstrations for these tasks in real world. The videos can be efficiently collected with around *100 demonstrations per hour*.

(ii) **DexMV pipeline**, a new pipeline for imitation learning from human videos. Given the recorded videos from the computer vision system, we perform pose estimation and motion retargeting to translate human video to robot hand demonstrations. These demonstrations are then used for policy learning where multiple imitation learning and RL algorithms are investigated. By learning from demonstrations, our policy can be deployed in environments with *different goals and object configurations*, instead of just following one trajectory.

4 DexMV Platform

Our DexMV platform is shown in Fig. 2. It is composed of a computer vision system and a simulation system.

Computer Vision System. The computer vision system is used to collect human demonstration videos on manipulating diverse real objects. In this system, we build a cubic frame (35 in.^{3}) and attach two RealSense D435 cameras (RGBD cameras) on the top front and top left, as shown on the top row of Fig. 2. During data collection, a human will perform manipulation tasks inside the frame, e.g., relocate sugar box. The manipulation videos will be recorded using the two cameras (from a front view and a side view).

Simulation System. Our simulation system is built on MuJoCo [73] with the Adroit Hand [41]. We design multiple dexterous manipulation tasks aligned with human demonstrations. As shown in the bottom row of Fig. 2, we perform imitation learning by augmenting the Reinforcement Learning with the collected demonstrations. Once the policy is trained, it can be tested on the same tasks with different goals and object configurations.

Task Description. We propose 3 types of manipulation tasks with different objects. We select the YCB objects [15] with a reasonable size for human to manipulate. For all tasks, the state is composed of robot joint readings, object pose. For *relocate*, we also include target position. The action is 30-d control command of the position actuator for each joint.

Relocate. It requires the robot hand to pick up an object on the table to a target location. It is inspired by the hardest task in [59] where the robot hand needs to grasp a sphere and move it to the target position. We further increase the task difficulty by using 5 complex objects as shown in the table above and visualized in the first 5 columns of Fig. 1. The transparent green shape represents the goal, which can change during training and testing (goal-conditioned). The task is successfully solved if the object reaches the goal, without the need for a specific orientation. We train one policy for relocating each object.

Pour. It requires the robot hand to reach the mug and pour the particles inside into a container (Fig. 1 column 6). The robot needs to grasp the mug and then manipulate it precisely. The evaluation criteria is based on the percentage of particles poured inside the container.

Place Inside. It requires the robot to pick up an object (e.g., a banana) and place it into a container. The robot needs to rotate the object to a suitable orientation and approach the container carefully to avoid collisions. The evaluation is based on how much percentage of the object mesh volume is inside the container.

Besides testing on the trained objects, we also evaluate the generalization ability of the trained policies on unseen object instances, within the training categories like can, bottle, mug, and also outside the training categories such as camera from ShapeNet dataset [16].

5 Pose Estimation

5.1 Object Pose Estimation

We use the 6-DoF pose to represent the location and orientation of objects, which contains translation $T \in \mathbf{R}^3$ and rotation $R \in \mathbf{SO(3)}$. For each frame t in the video, we use the PVN3D [29] model trained on the YCB dataset [15] to detect objects and estimate 6-DoF poses. By taking both the RGB image and the point clouds as inputs, the model first estimates the instance segmentation mask. With dense voting on the segmented point clouds, the model then predicts the 3D location of object key points. The 6-DoF object pose is optimized by minimizing the PnP matching error.

5.2 Hand Pose Estimation

We utilize the MANO model [60] to represent the hand that consists of hand pose parameter θ_t for 3D rotations of 15 joints and root global pose r_t, and shape parameters β_t for each frame t. The 3D hand joints can be computed using hand kinematics function $j_t^{3d} = \mathbf{J}(\theta_t, \beta_t, r_t)$.

Given a video, we use the off-the-shelf skin segmentation [39] and hand detection [67] models to obtain a hand mask M_t. We use the trained hand pose estimation models in [45] to predict the 2D hand joints j_t^{2d} and the MANO parameters θ_t and β_t for every frame t using RGB image. We estimate the root joint r_t using the center of the depth image masked by M_t. Given the initial estimation θ_t, β_t, r_t of each frame t and the camera pose Π, we formulate the 3D hand joint estimation as an optimization problem,

$$\theta_t^*, \beta_t^*, r_t^* = \mathop{\arg\min}_{\theta_t, \beta_t, r_t} ||\Pi J(\theta_t, \beta_t, r_t) - j_t^{2d}||^2 + \lambda||M_t \cdot (\mathbf{R}(\theta_t, \beta_t, r_t) - D_t)||^2 \quad (1)$$

where $\lambda = 0.001$ and $\mathbf{R}$ is a depth rendering function [37] and D_t is the corresponding depth map in frame t. We minimize the re-projection error in 2D and optimize the hand 3D locations from the depth map. Equation 1 can be further extended to a multi-camera setting by minimizing the objective from different cameras with calibrated extrinsics. We use two cameras by default to handle occlusion. Additionally, we deploy a post-processing procedure with minimizing the difference of j_t^{3d} and β_t between frames.

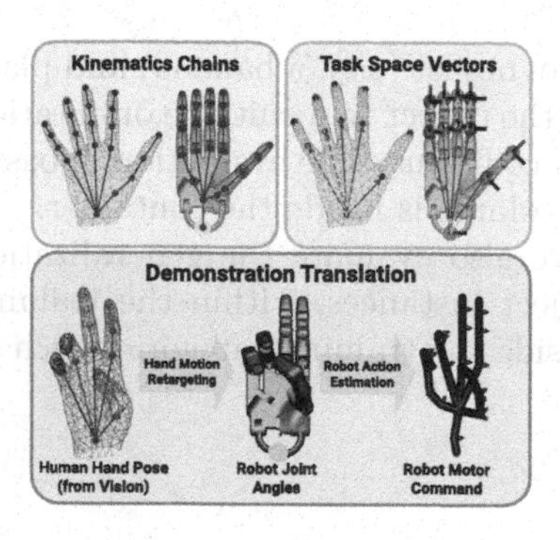

Fig. 3. Top row: Kinematic Chains and Task Space Vectors (TSV). The TSV (dash arrows) are ten vectors to be matched between both hands. **Bottom row**: (i) Hand motion retargeting; (ii) Action Estimation.

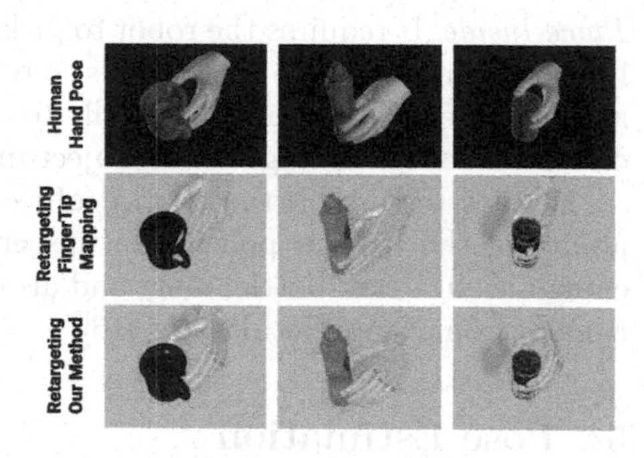

Fig. 4. Visualization of hand motion retargeting results. **Top row:** 3D hand-object poses. **Mid row:** Retargeting results using Finger Tip Mapping as optimization objective. **Bottom row:** Retargeting results using the proposed optimization objective.

6 Demonstration Translation

Common imitation learning algorithm consumes state-action pairs from expert demonstrations. It requires the state of the robot and the action of the motor as training data but not directly using the human hand pose. As shown in the kinematics chains in the blue box of Fig. 3, although both robot and human hands share a similar five-finger morphology, their kinematics chains are different. The finger length of each knuckle is also different.

In this paper, we propose a novel method to translate demonstrations from *human-centric pose estimation* result into *robot-centric imitation data*. Specifically, there are two steps in demonstration translation (as shown in the red box of Fig. 3): (i) **Hand motion retargeting** to align the human hand motion to robot hand motion, which are with different DoF and geometry; (ii) **Predicting robot action**, i.e. torque of robot motor: Without any wired sensors, we need to recover the action from only pose estimation results.

6.1 Hand Motion Retargeting

In computer graphics and animation, motion retargeting is used to retarget a performer's motion to a virtual character for applications such as movies and cartons [2,30]. While the animation community emphasizes realistic visual appearance, we consider more on the physical effect of the retargeted motion in the context of robotics manipulation. That is, the robot motion should be executable, respecting the physical limit of motors, e.g. acceleration and torque.

Formally, given a sequence of human hand pose estimated $\{(\theta_t, \beta_t, r_t)\}_{t=0}^{T}$ from a video, the hand pose retargeting can be defined as computing the sequence of robot joint angles $\{q_t\}_{t=0}^{T}$, where t is the step number in a sequence of length $T + 1$. We formulate the hand pose retargeting as an optimization problem.

Optimization with Task Space Vectors. In most manipulation tasks, human and multi-finger robot hand contacts the object using finger tips. Thus in fingertip-based mapping [4,27], retargeting is solved by preserving the Task Space Vectors (TSV) defined from finger tips position to the hand palm position (green arrow) as shown in the green box of Fig. 3, so that the human and robot hands will have same finger tip position relative to the palm. However, only considering finger tip may results in unexpected optimization result as shown in Fig. 4. Although the tip position is preserved, the bending information of hand fingers are lost, leading to penetrating the object after retargeting. It becomes more severe when the joint angles are closed to the robot singularity [47]. To solve this issue, we include the TSV from palm to middle phalanx (blue arrow) as shown in the green box of Fig. 3. The optimization objective is,

$$\min_{q_t} \sum_{i=0}^{N} ||\mathbf{v}_i^{\mathbf{H}}(\theta_t, \beta_t, r_t) - \mathbf{v}_i^{\mathbf{R}}(q_t)||^2 + \alpha ||q_t - q_{t-1}||^2, \tag{2}$$

where the forward kinematics function for human $\mathbf{v}_i^{\mathbf{H}}(\theta, \beta, r)$ computes the i-th TSV for human hand and the robot forward kinematics function $\mathbf{v}_i^{\mathbf{R}}(q)$ computes

i-th TSV for the robot. The first term in Eq. 2 is to find the best $\{q_t\}_{t=0}^T$ that matches the finger tip-palm TSV and finger tip-phalanx TSV from both hands. Note the human hand TSV $\{v_i^H(\theta_0, \beta_0, r_0)\}_{t=0}^T$ is first processed by a low pass filter before optimization for better smoothness. We also add an L2 normalization (second term in Eq. 2) to improve the temporal consistency. For optimization, we use Sequential Least-Squares Quadratic Programming in NLopt [35], where $\alpha = 8e-3$ in implementation.

When retargeting hand pose, the objective in Eq. 2 is optimized for each t from $t = 0$ to $t = T$. For $t \geq 1$, we initialize q_t using the optimization results q_{t-1} from last step to further improve the temporal smoothness.

6.2 Robot Action Estimation

Hand motion retargeting provides temporal-consistent translation from human hand poses to robot joint angles $\{q_t\}_{t=0}^T$ in different time steps. But the action, i.e. joint torque or control command, is unknown. In robotics, the joint torque τ can be computed via robot inverse dynamics function $\tau = f_{inv}(q, q', q'')$. To use this function, we first fit the sequence of joint angles into a continuous joint trajectory function $\mathbf{q}(t)$. Then, the first and second order derivative $\mathbf{q}'(t)$ and $\mathbf{q}''(t)$ can be used to compute the torque $\tau(t) = f_{inv}(q(t), q'(t), q''(t))$.

One Important Property of $\mathbf{q}(t)$ is: The absolute value of third-order derivative $\mathbf{q}'''(t)$, which also refers as jerk, should be as small as possible. We call it minimum-jerk requirement. The reason is two-fold: (i) First, physiological study [22] shows that the motion of human hand is a minimum jerk trajectory, where the lowest effort is required during the movement. This requirement will resemble the behavior of human, which in turn leads to more natural robot motion. (ii) Second, minimizing jerk can ensure low joint position errors for motors [42] and limit excessive wear on the physical robot [19]. Thus we also expect the fitted trajectory function $\{q_t\}_{t=0}^T$ to be minimum-jerk. In implementation, we use the model proposed in [74] to achieve this.

7 Imitation Learning

We perform imitation learning using the translated demonstrations. Instead of using behavior cloning, we adopt imitation learning algorithms that incorporate the demonstrations into RL.

Background. We consider the Markov Decision Process (MDP) represented by $\langle S, A, P, R, \gamma \rangle$, where S and A are state and action space, $P(s_{t+1}|s_t, a_t)$ is the transition density of state s_{t+1} at step $t+1$ given action a_t. $R(s, a)$ is the reward function, and γ is the discount factor. The goal of RL is to maximize the expected reward under policy $\pi(a|s)$. We incorporate RL with imitation learning. Given trajectories $\{(s_i, a_i)\}_{i=1}^n$ from demonstrations π_{D}, we optimize the agent policy π_θ with $\{(s_i, a_i)\}_{i=1}^n$ and reward R. We evaluate imitation learning under two settings: state-action imitation and state-only imitation.

7.1 State-Action Imitation Learning

We will introduce two algorithms. The first one is the Generative Adversarial Imitation Learning (GAIL) [31], which is the SOTA IL method that performs occupancy measure matching to learn policy. Occupancy measure ρ_π [55] is a density measure of the state-action tuple on policy π. GAIL aims to minimize the objective $d(\rho_{\pi_\mathrm{E}}, \rho_{\pi_\theta})$, where d is a distance function, π_θ is the policy parameterized by θ. GAIL use generative adversarial training to estimate the distance and minimize it with the objective as,

$$\min_\theta \max_w \mathop{\mathbb{D}}_{s,a\sim\rho_{\pi_\theta}} [\log D_w(s,a)] + \mathop{\mathbb{E}}_{s,a\sim\rho_{\pi_\mathrm{D}}} [\log(1 - D_w(s,a))], \tag{3}$$

where D_w is a discriminator. To equip GAIL with reward function, we adopt the approach proposed by [36], denoted as GAIL+ in this paper.

The second algorithm is Demo Augmented Policy Gradient (DAPG) [58]. The objective function for policy optimization at each iteration k is as follow.

$$g_{aug} = \sum_{(s,a)\in\rho_{\pi_\theta}} \nabla_\theta \ln \pi_\theta(a|s) A^{\pi_\theta}(s,a) + \sum_{(s,a)\in\rho_{\pi_\mathrm{D}}} \nabla_\theta \ln \pi_\theta(a|s)\lambda_0\lambda_1^k, \tag{4}$$

where A^{π_θ} is the advantage function [9] of policy π_θ, and λ_0 and λ_1 are hyper-parameters to determine the advantage of state-action pair in demonstrations.

7.2 State-Only Imitation Learning

Besides state-action imitation, we also evaluate the State-Only Imitation Learning (SOIL) [57] algorithm which does using action information from demonstrations. SOIL extends DAPG to the state-only imitation setting by learning an inverse model h_ϕ with the collected trajectories when running the policy. The inverse model predicts the missing actions in demonstrations and the policy is trained as DAPG. The policy and inverse model are optimized simultaneously.

8 Experiment

Experiment Settings. We adopt TRPO [66] as our RL baseline. We benchmark imitation learning algorithms SOIL, GAIL+, and DAPG. All these algorithms incorporate demonstrations with TRPO of same hyper-parameters. Thus all the algorithms are comparable. All policies are evaluated with the same three random seeds. By default, the hand pose is estimated using two cameras.

8.1 Experiments with Relocate

Main Comparisons. We benchmark four methods: SOIL, GAIL+, DAPG, and RL on the *relocate* tasks. The *success rate* is shown in Table 1 and training curves are in Fig. 5. A trial is counted as success only when the final position of the object (after 200 steps) is within 0.1 unit length to the target. The initial object position and target position are randomized. In Fig. 5, the x-axis is training iterations and the y-axis is normalized average-return over three seeds.

Both Table 1 and Fig. 5 show that the imitation learning methods outperform RL baseline for all five tasks. For mustard bottle and sugar box, a pure RL agent is not able to learn anything. SOIL performs the best for sugar box while DAPG achieves comparable or better performance on mug, mustard bottle, and tomato soup can. *Relocate* with sugar box is most challenging since it is tall and thin with a flat surface. Once it drops on the table, it is hard to grasp again. We conjecture training the inverse dynamics model online in SOIL helps obtain more accurate action for *Relocate*.

Ablation on Motion Retargeting Method. Figure 6 (a) illustrates the performance of SOIL with respect to demonstrations generated by different retargeting method, on relocating a tomato soup can. It shows that our retargeting design (blue curve) can improve the imitation learning performance over fingertip mapping (red curve) based retargeting baseline. Besides, without L2 norms in Eq. 2 (green curve), the performance shows a huge drop. It indicates the importance of smooth robot motion when used as demonstrations.

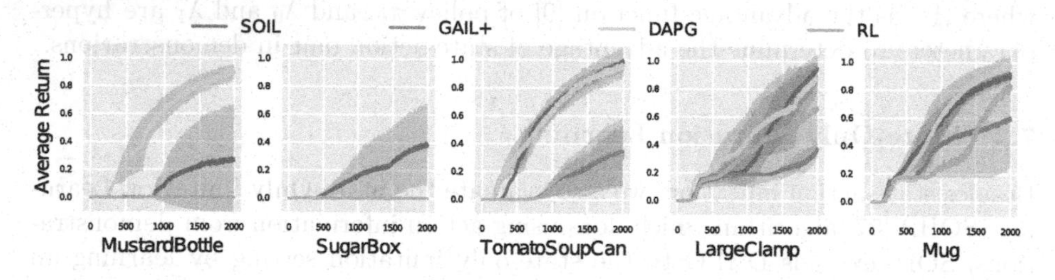

Fig. 5. Learning curves of the four methods on the relocate task with respect to five different objects. The x-axis is training iterations. The shaded area indicates standard error and the performance is evaluated with three individual random seeds. (Color figure online)

Table 1. Success rate of the evaluated methods on *Relocate* with five different objects. Success is defined based on the distance between object and target, evaluated via 100 trials for three seeds.

Model	Task - Relocate				
	Mustard	Sugar Box	Tomato Can	Clamp	Mug
SOIL	0.33 ± 0.42	$\mathbf{0.67 \pm 0.47}$	0.98 ± 0.02	0.89 ± 0.15	0.71 ± 0.35
GAIL+	0.06 ± 0.01	0.00 ± 0.00	0.66 ± 0.47	0.52 ± 0.39	0.53 ± 0.37
DAPG	$\mathbf{0.93 \pm 0.05}$	0.00 ± 0.00	$\mathbf{1.00 \pm 0.00}$	$\mathbf{1.00 \pm 0.00}$	$\mathbf{1.00 \pm 0.00}$
RL	0.06 ± 0.01	0.00 ± 0.00	0.67 ± 0.47	0.51 ± 0.37	0.49 ± 0.36

Ablation on the Number of Demonstrations. Figure 6 (b) shows the performance of SOIL with respect to different number of demonstrations, on relocating a tomato soup can. SOIL can achieve better sample efficiency, performance and the variance is reduced when more demonstrations are given. Table 2 illustrates the success rate of the different policies trained with a different number of demonstrations. *This shows the importance of our efficient platform to scale up the number and diversity of demonstrations.*

Ablation on Environmental Conditions. Figure 6 (c) and (d) illustrates that our demonstrations can transfer to different physical environments. Given the same demonstrations on *Relocate* the tomato soup can, we train policies with objects scaling into different size from $\times 0.75$ to $\times 1.125$ (Fig. 6 (c)) and applying different frictions (Fig. 6 (d)) from $\times 0.8$ to $\times 1.2$. With same demonstrations, we can still perform imitation learning in different environments and achieve consistent results. All policies achieve much better performance than pure RL. *This experiment shows the robustness of our pipeline against the gap between simulation (environment) and real (demonstration).*

Ablation on Hand Pose Estimation. We choose 4 different settings for hand pose estimation from video captured by: (i) single camera; (ii) single camera plus post-processing (iii) dual camera; (iv) dual cameras plus the post-processing

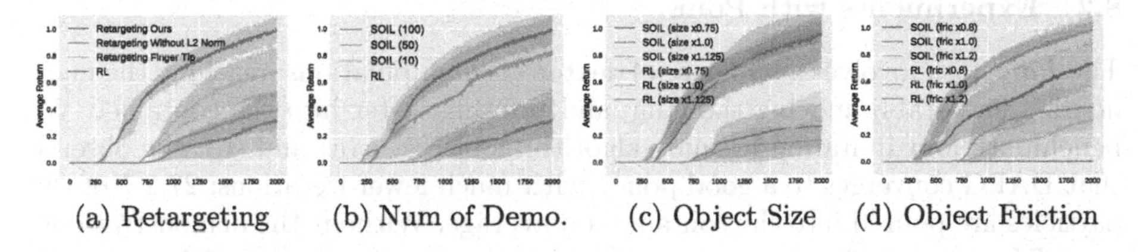

(a) Retargeting (b) Num of Demo. (c) Object Size (d) Object Friction

Fig. 6. Ablation Study: Learning curves of SOIL on *Relocate* with tomato soup can. The x-axis is training iterations. We ablate: (a) hand pose estimation methods; (b) number of demonstrations used to train SOIL; (c) scaling of object size; (d) friction of the relocated object. The demonstrations are kept the same for all conditions.

Table 2. Success rate with different number of demonstrations on *Relocate* task with a tomato soup can, evaluated via 100 trials for three random seeds at 400, 600, and 800 training iterations.

#Demo	400 Iter.	600 Iter.	800 Iter.
SOIL (100)	$\mathbf{0.36 \pm 0.13}$	$\mathbf{0.70 \pm 0.25}$	$\mathbf{1.00 \pm 0.00}$
SOIL (50)	0.19 ± 0.23	0.40 ± 0.43	0.59 ± 0.39
SOIL (10)	0.04 ± 0.05	0.17 ± 0.11	0.36 ± 0.21
RL (0)	0.00 ± 0.00	0.00 ± 0.00	0.13 ± 0.19

Table 3. Hand pose error and the success rate of learned policy for 4 hand pose estimation settings. Smaller MPJPE means better hand pose estimation performance.

Setting	MPJPE	Success (%)
1 Cam	41.7	66.7 ± 57.7
1 Cam Post	36.2	69.7 ± 33.3
2 Cam	36.6	84.7 ± 25.7
2 Cam Post	**32.5**	$\mathbf{93.3 \pm 11.5}$

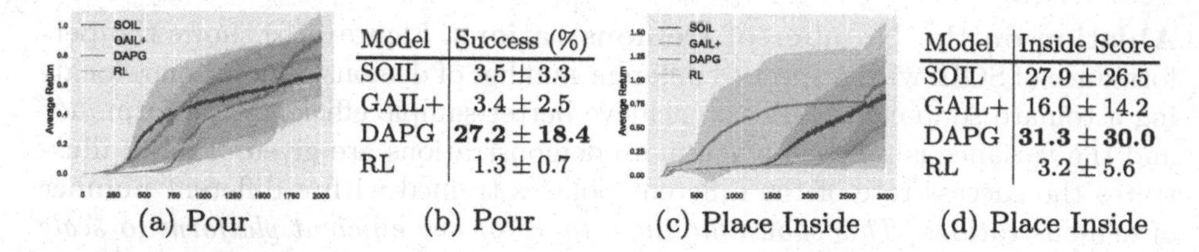

(b) Pour

Model	Success (%)
SOIL	3.5 ± 3.3
GAIL+	3.4 ± 2.5
DAPG	**27.2 ± 18.4**
RL	1.3 ± 0.7

(d) Place Inside

Model	Inside Score
SOIL	27.9 ± 26.5
GAIL+	16.0 ± 14.2
DAPG	**31.3 ± 30.0**
RL	3.2 ± 5.6

Fig. 7. *Pour and Place Inside.* (a) learning curves of *Pour*; (b) success rate of *Pour*; (c) learning rate of *Place Inside*; (d) Inside score of *Place Inside*.

(mentioned in Sect. 5.2) Since we have no ground-truth pose annotation in our dataset, we evaluate the performance of hand pose estimation approaches on the DexYCB [18] dataset, which provides the pose ground-truths. We follow DexYCB to use Mean Per Joint Position Error (MPJPE, smaller the better). We experiment with SOIL on relocating a tomato soup can. The object pose remains the same for all four settings. Table 3 shows better pose estimation in general corresponds to better imitation, except the comparison on 1-Cam Post and 2-Cam. It seems 2-Cam are better than 1-Cam in imitation, even pose estimation result is close. We conjecture 2-Cam can provide more smooth trajectories.

8.2 Experiments with Pour

The *Pour* task involves a sequence of dexterous manipulations: reaching the mug, holding and stably moving the mug, and pouring water into the container. We benchmark four imitation learning algorithms in Fig. 7 (a) and (b). We observe that DAPG converges to a good policy with much fewer iterations: 27.2% of the particles are poured into the container on average. Without the demonstrations, pure RL will only have a very small chance to pour a few particles into the container. State-action method DAPG performs better than state-only method SOIL. It is challenging to learn inverse model with water particles in this task for

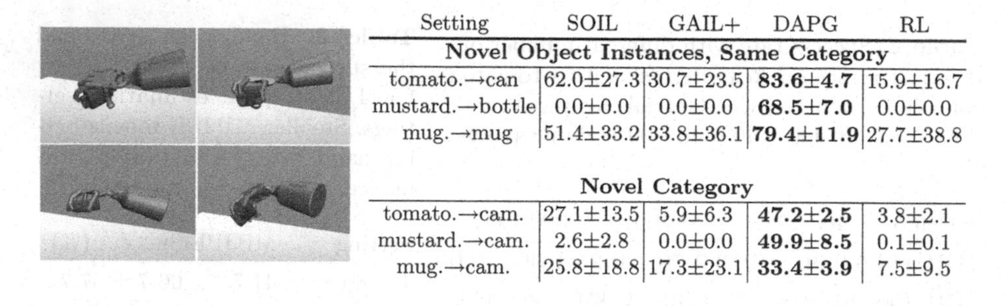

Setting	SOIL	GAIL+	DAPG	RL
Novel Object Instances, Same Category				
tomato.→can	62.0±27.3	30.7±23.5	**83.6±4.7**	15.9±16.7
mustard.→bottle	0.0±0.0	0.0±0.0	**68.5±7.0**	0.0±0.0
mug.→mug	51.4±33.2	33.8±36.1	**79.4±11.9**	27.7±38.8
Novel Category				
tomato.→cam.	27.1±13.5	5.9±6.3	**47.2±2.5**	3.8±2.1
mustard.→cam.	2.6±2.8	0.0±0.0	**49.9±8.5**	0.1±0.1
mug.→cam.	25.8±18.8	17.3±23.1	**33.4±3.9**	7.5±9.5

Fig. 8. *Relocate* generalization. Left: visualization of *Relocate* with unseen ShapeNet objects from 4 categories. Right: success rate of *Relocate*. A→B denotes using the policy trained on YCB object A to evaluate the performance on ShapeNet category B.

(a) Visualization of Two Tasks

Model	Success (%)
SOIL	3.8 ± 3.2
GAIL+	2.1 ± 0.8
DAPG	$\mathbf{23.3 \pm 10.4}$
RL	0.5 ± 0.3

(b) Pour

Category	Inside Score
SOIL	21.6 ± 16.1
GAIL+	10.9 ± 4.6
DAPG	$\mathbf{26.9 \pm 17.3}$
RL	1.8 ± 1.5

(c) Place Inside

Fig. 9. Generalization with *Pour* and *Place Inside*. (a) visualization of *Pour* (top row) with ShapeNet mug, and *Place Inside* with ShapeNet cellphone; (b) success rate of *Pour* with ShapeNet Mug; (c) inside score of *Place Inside* with ShapeNet cellphone.

SOIL, while the analytical inverse dynamics function still provides reasonable actions for DAPG. See our videos and website for visualization.

8.3 Experiments with Place Inside

The *Place Inside* task requires the robot hand to first pick up a banana, rotate it to the appropriate orientation, and place it inside the mug. We define a metric *Inside Score*, which is computed based on the volume percentage of the banana inside of the mug. We benchmark the imitation learning algorithms in Fig.7 (c) and (d). We find that DAPG outperforms other approaches whereas RL hardly learns to manipulate the object. Between state-only method (SOIL) and state-action method (DAPG), we again observe that computing action offline analytically achieves better results, due to the complexity of the task.

8.4 Generalization on Novel Objects and Category

Generalization with Relocate. We benchmark the generalization ability of the trained policy with different imitation learning algorithms. We directly deployed the trained policies for the *Relocate* task using our demonstrations on: (i) novel unseen objects within the same category; (ii) object instances from a new category. For evaluation on unseen object instances within the same category, we choose the can, bottle, and mug categories from ShapeNet [16] dataset, which corresponds to the tomato soup can, mustard bottle, and mug in our demonstrations. We omit evaluation on the clamp and sugar box policies since there are no corresponding categories in ShapeNet. For the evaluation on new category, we choose the camera category from ShapeNet. We select 100 object instances for each category and pre-process the objects to fit the size of robot hand. More details about data pre-processing can be found in supplementary. We report the results in Fig. 8. The success rate is computed as a mean accuracy over all object instances within a category. We find DAPG generalizes well on unseen objects in both same category and novel category experiments, significantly outperforming other approaches. We can also observe generalization to novel categories brings a larger challenge.

Generalization with Pour. We benchmark the generalization performance of the trained policies from all four methods on new mug instances for *Pour* task. We choose 100 object instances from the ShapeNet mug category for deployment. The top row of Fig. 9 (a) visualizes the task with different mugs and Fig. 9 (b) shows the success rate. In this benchmark, DAPG still performs the best across all methods. Interestingly, the performance is not much worse than *Pour* with the trained object (by comparing Fig. 7 (b)). This shows the robustness of imitation learning using our pipeline.

Generalization with Place Inside. For *Place Inside*, we replace the YCB banana using the objects from cellphone category in ShapeNet as visualized in the bottom row of Fig. 9 (a). Figure 9 (c) shows the generalization performance when directly deploying the policy trained with banana. We find the policy performance is also close to Place Inside with banana (by comparing Fig. 7 (d)). This indicates the robustness of policy against small shape variations and human demonstration on single object can also benefit the tasks on new objects.

9 Conclusion

To the best of our knowledge, DexMV is the first work to provide a platform on computer vision/simulation systems and a pipeline on learning dexterous manipulation from human videos. We propose a novel demonstration translation module to bridge the gap between these two systems. We hope DexMV opens new research opportunities for benchmarking imitation learning algorithms for dexterous manipulation.

Acknowledgement. This work was supported, in part, by grants from DARPA LwLL, NSF CCF-2112665 (TILOS), NSF 1730158 CI-New: Cognitive Hardware and Software Ecosystem Community Infrastructure (CHASE-CI), NSF ACI-1541349 CC*DNI Pacific Research Platform, and gifts from Meta, Google, Qualcomm and Picsart.

References

1. Abbeel, P., Ng, A.Y.: Apprenticeship learning via inverse reinforcement learning (2004)
2. Aberman, K., Wu, R., Lischinski, D., Chen, B., Cohen-Or, D.: Learning character-agnostic motion for motion retargeting in 2d. arXiv preprint arXiv:1905.01680 (2019)
3. Andrews, S., Kry, P.G.: Goal directed multi-finger manipulation: control policies and analysis. Comput. Graph. **37**(7), 830–839 (2013)
4. Antotsiou, D., Garcia-Hernando, G., Kim, T.K.: Task-oriented hand motion retargeting for dexterous manipulation imitation. In: ECCV Workshops (2018)
5. Aytar, Y., Pfaff, T., Budden, D., Paine, T., Wang, Z., de Freitas, N.: Playing hard exploration games by watching youtube. In: NeurIPS (2018)
6. Baek, S., Kim, K.I., Kim, T.K.: Pushing the envelope for RGB-based dense 3d hand pose estimation via neural rendering. In: CVPR (2019)

7. Bai, Y., Liu, C.K.: Dexterous manipulation using both palm and fingers (2014)
8. Bain, M., Sammut, C.: A framework for behavioural cloning. In: Machine Intelligence (1995)
9. Baird III, L.C.: Advantage updating. Technical Report (1993)
10. Bicchi, A.: Hands for dexterous manipulation and robust grasping: a difficult road toward simplicity. IEEE Trans. Robot. Autom. **16**(6), 652–662 (2000)
11. Bojarski, M., et al.: End to end learning for self-driving cars. arXiv (2016)
12. Boukhayma, A., Bem, R.D., Torr, P.H.: 3d hand shape and pose from images in the wild. In: CVPR (2019)
13. Brahmbhatt, S., Ham, C., Kemp, C.C., Hays, J.: Contactdb: analyzing and predicting grasp contact via thermal imaging. In: CVPR (2019)
14. Brahmbhatt, S., Handa, A., Hays, J., Fox, D.: Contactgrasp: functional multi-finger grasp synthesis from contact. arXiv (2019)
15. Calli, B., Walsman, A., Singh, A., Srinivasa, S., Abbeel, P., Dollar, A.M.: Benchmarking in manipulation research: the YDB object and model set and benchmarking protocols. arXiv (2015)
16. Chang, A.X., et al.: Shapenet: an information-rich 3d model repository. arXiv preprint arXiv:1512.03012 (2015)
17. Chang, M., Gupta, A., Gupta, S.: Semantic visual navigation by watching Youtube videos. In: NIPS (2020)
18. Chao, Y.W., et al.: Dexycb: a benchmark for capturing hand grasping of objects. In: CVPR (2021)
19. Craig, J.J.: Introduction to Robotics: Mechanics and Control, 3/E. Pearson Education India, Noida (2009)
20. Dogar, M.R., Srinivasa, S.S.: Push-grasping with dexterous hands: mechanics and a method (2010)
21. Duan, Y., Chen, X., Houthooft, R., Schulman, J., Abbeel, P.: Benchmarking deep reinforcement learning for continuous control (2016)
22. Flash, T., Hogan, N.: The coordination of arm movements: an experimentally confirmed mathematical model. J. Neurosci. **5**(7), 1688–1703 (1985)
23. Fu, J., Luo, K., Levine, S.: Learning robust rewards with adversarial inverse reinforcement learning. arXiv (2017)
24. Garcia-Hernando, G., Yuan, S., Baek, S., Kim, T.K.: First-person hand action benchmark with RGB-D videos and 3d hand pose annotations. In: CVPR (2018)
25. Ge, L., et al.: 3d hand shape and pose estimation from a single RGB image. In: CVPR (2019)
26. Hampali, S., Rad, M., Oberweger, M., Lepetit, V.: Honnotate: a method for 3d annotation of hand and object poses. In: CVPR (2020)
27. Handa, A., et al.: Dexpilot: vision-based teleoperation of dexterous robotic hand-arm system. In: ICRA (2020)
28. Hasson, Y., et al.: Learning joint reconstruction of hands and manipulated objects. In: CVPR (2019)
29. He, Y., Sun, W., Huang, H., Liu, J., Fan, H., Sun, J.: Pvn3d: a deep point-wise 3d keypoints voting network for 6dof pose estimation. In: CVPR (2020)
30. Hecker, C., Raabe, B., Enslow, R.W., DeWeese, J., Maynard, J., van Prooijen, K.: Real-time motion retargeting to highly varied user-created morphologies. ACM Trans. Graph. (TOG) **27**(3), 1–11 (2008)
31. Ho, J., Ermon, S.: Generative adversarial imitation learning. In: NeurIPS (2016)
32. Hu, Y., Hugonot, J., Fua, P., Salzmann, M.: Segmentation-driven 6d object pose estimation. In: CVPR (2019)

33. Iqbal, U., Molchanov, P., Breuel Juergen Gall, T., Kautz, J.: Hand pose estimation via latent 2.5 d heatmap regression. In: ECCV (2018)
34. Jiang, H., Liu, S., Wang, J., Wang, X.: Hand-object contact consistency reasoning for human grasps generation. arXiv (2021)
35. Johnson, S.G.: The nlopt nonlinear-optimization package (2014)
36. Kang, B., Jie, Z., Feng, J.: Policy optimization with demonstrations. In: ICML (2018)
37. Kato, H., Ushiku, Y., Harada, T.: Neural 3d mesh renderer. In: CVPR (2018)
38. Kehl, W., Manhardt, F., Tombari, F., Ilic, S., Navab, N.: Ssd-6d: making RGB-based 3d detection and 6d pose estimation great again. In: ICCV (2017)
39. Khaled, S.M., et al.: Combinatorial color space models for skin detection in sub-continental human images. In: IVIC (2009)
40. Kulon, D., Guler, R.A., Kokkinos, I., Bronstein, M.M., Zafeiriou, S.: Weakly-supervised mesh-convolutional hand reconstruction in the wild. In: CVPR (2020)
41. Kumar, V., Xu, Z., Todorov, E.: Fast, strong and compliant pneumatic actuation for dexterous tendon-driven hands. In: ICRA (2013)
42. Kyriakopoulos, K.J., Saridis, G.N.: Minimum jerk path generation. In: Proceedings. 1988 IEEE International Conference on Robotics and Automation, pp. 364–369. IEEE (1988)
43. Li, S., et al.: Vision-based teleoperation of shadow dexterous hand using end-to-end deep neural network. In: ICRA (2019)
44. Liu, F., Ling, Z., Mu, T., Su, H.: State alignment-based imitation learning. In: ICLR (2020)
45. Liu, S., Jiang, H., Xu, J., Liu, S., Wang, X.: Semi-supervised 3d hand-object poses estimation with interactions in time. In: CVPR (2021)
46. Mandikal, P., Grauman, K.: Dexterous robotic grasping with object-centric visual affordances. arXiv (2020)
47. Nakamura, Y., Hanafusa, H.: Inverse kinematic solutions with singularity robustness for robot manipulator control (1986)
48. Ng, A.Y., Russell, S.J., et al.: Algorithms for inverse reinforcement learning (2000)
49. Okamura, A.M., Smaby, N., Cutkosky, M.R.: An overview of dexterous manipulation. In: ICRA (2000)
50. Akkaya, I., et al.: Solving rubik's cube with a robot hand. OpenAI, arXiv (2019)
51. Andrychowicz, M., et al.: Learning dexterous in-hand manipulation. OpenAI, arXiv (2018)
52. Peng, S., Liu, Y., Huang, Q.X., Bao, H., Zhou, X.: Pvnet: pixel-wise voting network for 6dof pose estimation. In: CVPR (2019)
53. Peters, J., Schaal, S.: Reinforcement learning of motor skills with policy gradients. Neural Netw. **21**(4), 682–697 (2008)
54. Pomerleau, D.A.: Alvinn: an autonomous land vehicle in a neural network. In: NeurIPS (1989)
55. Puterman, M.L.: Markov Decision Processes: Discrete Stochastic Dynamic Programming. Wiley, Hoboken (1994)
56. Rad, M., Lepetit, V.: Bb8: a scalable, accurate, robust to partial occlusion method for predicting the 3d poses of challenging objects without using depth. In: ICCV (2017)
57. Radosavovic, I., Wang, X., Pinto, L., Malik, J.: State-only imitation learning for dexterous manipulation. In: IROS (2021)
58. Rajeswaran, A., et al.: Learning complex dexterous manipulation with deep reinforcement learning and demonstrations. arXiv (2017)

59. Rajeswaran, A., et al.: Learning complex dexterous manipulation with deep reinforcement learning and demonstrations (2018)
60. Romero, J., Tzionas, D., Black, M.J.: Embodied hands: modeling and capturing hands and bodies together. In: ToG (2017)
61. Ross, S., Bagnell, D.: Efficient reductions for imitation learning. In: AISTATS (2010)
62. Rus, D.: In-hand dexterous manipulation of piecewise-smooth 3-d objects. Int. J. Robot. Res. **18**(4), 355–381 (1999)
63. Russell, S.: Learning agents for uncertain environments (1998)
64. Schmeckpeper, K., Rybkin, O., Daniilidis, K., Levine, S., Finn, C.: Reinforcement learning with videos: combining offline observations with interaction. arXiv (2020)
65. Schmeckpeper, K., et al.: Learning predictive models from observation and interaction. arXiv (2019)
66. Schulman, J., Levine, S., Abbeel, P., Jordan, M., Moritz, P.: Trust region policy optimization. In: ICML (2015)
67. Shan, D., Geng, J., Shu, M., Fouhey, D.: Understanding human hands in contact at internet scale. In: CVPR (2020)
68. Shao, L., Migimatsu, T., Zhang, Q., Yang, K., Bohg, J.: concept2robot: learning manipulation concepts from instructions and human demonstrations. In: RSS (2020)
69. Song, S., Zeng, A., Lee, J., Funkhouser, T.: Grasping in the wild: learning 6dof closed-loop grasping from low-cost demonstrations. Robot. Autom. Lett. **5**(3), 4978–4985 (2020)
70. Spurr, A., Song, J., Park, S., Hilliges, O.: Cross-modal deep variational hand pose estimation. In: CVPR (2018)
71. Taheri, O., Ghorbani, N., Black, M.J., Tzionas, D.: Grab: a dataset of whole-body human grasping of objects. In: ECCV (2020)
72. Tekin, B., Sinha, S.N., Fua, P.: Real-time seamless single shot 6d object pose prediction. In: CVPR (2018)
73. Todorov, E., Erez, T., Tassa, Y.: Mujoco: a physics engine for model-based control. In: IROS (2012)
74. Todorov, E., Jordan, M.I.: Smoothness maximization along a predefined path accurately predicts the speed profiles of complex arm movements. J. Neurophysiol. **80**(2), 696–714 (1998)
75. Torabi, F., Warnell, G., Stone, P.: Behavioral cloning from observation. arXiv (2018)
76. Torabi, F., Warnell, G., Stone, P.: Generative adversarial imitation from observation. arXiv (2018)
77. Večerík, M., et al.: Leveraging demonstrations for deep reinforcement learning on robotics problems with sparse rewards. arXiv (2017)
78. Xiang, Y., Schmidt, T., Narayanan, V., Fox, D.: Posecnn: a convolutional neural network for 6d object pose estimation in cluttered scenes. arXiv (2018)
79. Young, S., Gandhi, D., Tulsiani, S., Gupta, A., Abbeel, P., Pinto, L.: Visual imitation made easy. arXiv (2020)
80. Zimmermann, C., Brox, T.: Learning to estimate 3d hand pose from single RGB images. In: CVPR (2017)

Sim-2-Sim Transfer for Vision-and-Language Navigation in Continuous Environments

Jacob Krantz(✉) and Stefan Lee

Oregon State University, Corvallis, USA
krantzja@oregonstate.edu
https://jacobkrantz.github.io/sim-2-sim

Abstract. Recent work in Vision-and-Language Navigation (VLN) has presented two environmental paradigms with differing realism – the standard VLN setting built on topological environments where navigation is abstracted away [3], and the VLN-CE setting where agents must navigate continuous 3D environments using low-level actions [21]. Despite sharing the high-level task and even the underlying instruction-path data, performance on VLN-CE lags behind VLN significantly. In this work, we explore this gap by transferring an agent from the abstract environment of VLN to the continuous environment of VLN-CE. We find that this sim-2-sim transfer is highly effective, improving over the prior state of the art in VLN-CE by +12% success rate. While this demonstrates the potential for this direction, the transfer does not fully retain the original performance of the agent in the abstract setting. We present a sequence of experiments to identify what differences result in performance degradation, providing clear directions for further improvement.

Keywords: Vision-and-Language Navigation (VLN) · Embodied AI

1 Introduction

Vision-and-Language Navigation (VLN) is a popular instruction-guided navigation task where a vision-equipped agent must navigate to a goal location in an never-before-seen environment by following a path described by a natural language instruction. In the standard VLN setting, the environment is abstracted as a topology of interconnected panoramic images (called a nav-graph) such that navigation amounts to traversing the graph by iteratively selecting among a small set of neighboring node locations at each time step. In effect, this induces a prior of possible locations for the agent and assumes perfect navigation between nodes. Recent work has identified that these assumptions do not reflect the challenges a deployed system would experience in a real environment. To reduce this

Supplementary Information The online version contains supplementary material available at https://doi.org/10.1007/978-3-031-19842-7_34.

S. Avidan et al. (Eds.): ECCV 2022, LNCS 13699, pp. 588–603, 2022.
https://doi.org/10.1007/978-3-031-19842-7_34

gap, Krantz et al. [21] introduced the VLN in Continuous Environments (VLN-CE) setting which instantiates VLN in a 3D simulator and drops the nav-graph assumptions – requiring agents to navigate with low-level actions.

So far, VLN-CE has proven to be significantly more challenging than its more abstract counterpart, with published work reporting episode success rates less than half of those reported in standard VLN. Typically, models for VLN-CE have been end-to-end systems trained to directly predict low-level actions (or nearby waypoints) from language and observations. As such, the lower performance compared to VLN is commonly ascribed to the challenge of learning navigation and language grounding jointly in a long-horizon task. This suggests two avenues towards improvement – developing more capable models directly in VLN-CE or exploring transfer of knowledge from VLN to VLN-CE. Given the significant differences between the abstract and continuous settings, the potential of sim-2-sim transfer has so far been under-explored and uncertain.

In this work, we explore the sim-2-sim transfer of a VLN agent to VLN-CE. To support this transfer, we build a modular harness such that any VLN agent can be run in VLN-CE. Analogous to sim-2-real experiments from [2], this harness includes a local navigation policy and a subgoal generation module to mimic nav-graph provided candidates. We present a sequence of systematic experiments to quantify what differences between the two task settings are most impactful. Our final model demonstrates that transfer can lead to significant improvements over end-to-end VLN-CE-trained methods – achieving an increase of +12% success rate over prior published state-of-the-art (32% vs 44%) on the VLN-CE test set.

Along the way, our analysis quantifies the effect of (1) differences between VLN and VLN-CE data subsets, (2) the visual domain gap between real panoramas and simulated renders in VLN-CE, (3) error induced by navigation policies, and (4) subgoal candidate generation – identifying fruitful directions for future work. Through an analysis of our model's errors, we identify that episodes containing significant elevation change (i.e. stairs) are a challenge for our model.

Overall, our results demonstrate an alternative research direction for improving instruction-guided navigation agents in continuous environments. We show transfer with compelling results and present the community with clear remaining challenges. We release our model-agnostic VLN-2-VLNCE harness code[1] to facilitate research in this area – helping to unify progress between both settings.

Contributions. To summarize the contributions of this work:

- We present a first-of-its-kind demonstration that instruction-following agents trained in the abstract VLN setting can effectively transfer to VLN-CE — setting a new state of the art in VLN-CE by +12% success rate.
- We quantify performance loss for key steps of the transfer process – providing insight to existing gaps in our transfer paradigm that could be improved.
- We develop and release a modular VLN-2-VLNCE transfer harness to spur innovation on transfer techniques and help progress from VLN to VLN-CE.

[1] github.com/jacobkrantz/Sim2Sim-VLNCE.

2 Related Work

VLN Agents. The VLN task [3] has received significant attention in recent years with the continual improvement of benchmark models. A higher-level panoramic action space was introduced by Fried et al. [11] and widely adopted in follow-up works. Early efforts proposed cross-modal grounding of vision and language [27]. Pre-training and data augmentation methods can mitigate issues regarding limited data [11,26] and challenges of jointly processing vision, language, and action [14]. Recently, transformer-based architectures and multi-step training routines have demonstrated superior performance [8,16,22]. VLN↻BERT [16] is one such model that introduced a recurrent history context to a transformer-based agent. We adopt VLN↻BERT as the core VLN agent in this work. With continual interest in improving topological instruction followers, we establish expectations and infrastructure for their transfer to continuous environments.

VLN-CE Agents. Recognizing the unrealistic affordances of navigation graphs, Krantz et al. [21] proposed replacing the topological definition of VLN with lower-level actions in continuous environments. Initial end-to-end baselines that directly predict lower-level actions were modeled after sequence-to-sequence [3] and cross-modal [27] VLN methods and demonstrated significantly lower performance than in VLN. Raychaudhuri et al. [25] proposed language-aligned path supervision to better guide off-the-path decision-making. Krantz et al. [20] explored a higher-level waypoint action space trained with deep reinforcement learning, finding increased performance but still far below VLN. Other works exploited continuous environments to study topological map generation [7] and semantic mapping for navigation [18]. In the context of a velocity-based continuous action space, Irshad et al. [17] paired a discrete-action model with a learned control policy to demonstrate the benefit of hierarchy. Rather than train agents from scratch in VLN-CE as these prior works do, we analyze the transfer of pre-trained VLN agents to continuous environments.

Sim-2-Real Transfer of Instruction Following. Transferring skills learned in simulation to reality is a critical step for embodied agents [4,10,12,19]. The most similar to our work is that of Anderson et al. [2], which directly transferred VLN agents to reality. We analyze the intermediate step: sim-2-sim transfer from high-level simulation to more realistic, lower-level simulation. The study of Anderson et al. [2] emulated the VLN action space with subgoal candidate prediction and performed navigation with the Robot Operating System (ROS) [24]. They found a more than 2x drop in performance when evaluating agents in reality, showing that effective VLN sim-2-real transfer remains an open research question. We identify similar challenges in sim-2-sim transfer as Anderson et al. [2] did in sim-2-real. We diagnose and begin mitigating these issues in VLN-CE using methods infeasible in VLN and impractical in reality.

3 Task Descriptions

In this section, we describe both the vision-and-language navigation task (VLN) and its continuous-environment counterpart, VLN-CE.

Vision-and-Language Navigation (VLN). VLN was proposed by Anderson, et al. [3], in which agents navigate previously-unseen indoor environments. The task is to follow a path described in natural language and stop at the goal. Instructions are unstructured and crowd-sourced in English to form the Room-to-Room (R2R) dataset. VLN takes place in a topological simulator, where a pre-defined nav-graph dictates allowable navigation through edge connections.

At each time step t, a VLN agent receives the instruction $\mathcal{I}$, a panoramic RGB observation $\mathcal{O}_t$, and a set of n subgoal candidates $\mathcal{C}_t^{(1 \dots n)}$. The instruction $\mathcal{I}$ is a sequence of L words $(w_0, w_1, \dots, w_L)$. The panoramic observation $\mathcal{O}_t$ is represented as 36 views such that 12 equally-spaced horizontal headings (0°, 30°, ..., 360°) are represented at each of 3 elevation angles (–30°, 0°, +30°). Each image frame in $\mathcal{O}_t$ has a resolution of 480 × 640 with a horizontal field of view of 75°. Each subgoal candidate $\mathcal{C}_t^i$ is represented by a relative heading angle θ_t^i, an elevation angle ϕ_t^i, and the observation view frame $\mathcal{O}_t^j$ that most closely centers the candidate: $\mathcal{C}_t^i = (\theta_t^i, \phi_t^i, \mathcal{O}_t^j)$. The VLN agent then navigates by selecting a candidate. The environment then transitions and produces observations for time $t+1$. This repeats until the agent issues the `STOP` action.

VLN in Continuous Environments (VLN-CE). VLN-CE as proposed by Krantz et al. [21] converts the topologically-defined VLN task into a continuous-environment task more representative of real-world navigation. Instead of selecting and navigating by environment-provided subgoal candidates, agents in VLN-CE must navigate a continuous-valued 3D mesh by selecting actions from a lower-level action space (`FORWARD` 0.25 m, `TURN-LEFT` 15°, `TURN-RIGHT` 15°, `STOP`). In this work, we seek to port VLN agents into continuous environments. Thus, we adopt what we can of the VLN observation space: the instruction $\mathcal{I}$ and the panoramic observation $\mathcal{O}_t$. We introduce a 2D laser scan which we describe in Sect. 4.3. This provides our subgoal generation module with 2D occupancy. Such a sensor is commonly used for both localization and mapping in real-world navigation stacks [24]. In our analysis below, we perform intermediate experiments that may allow oracle navigation or topological subgoal candidates but note that these experiments do not result in admissible VLN-CE agents.

Room-to-Room Dataset. The Room-to-Room dataset (R2R) consists of 7,189 shortest-path trajectories across all train, validation, and test splits [3]. The VLN-CE dataset is a subset of R2R consisting of 5,611 trajectories traversible in continuous environments (78% of R2R) [21]. For both tasks, we report performance on several validation splits. Val-Seen contains episodes with novel paths and instructions but from scenes observed in training. Val-Unseen contains novel paths, instructions, and scenes. Train* is a random subset of the training split derived by Hong et al. [16] to support further analysis.

Matterport3D Scene Dataset. Both VLN and VLN-CE use the Matterport3D (MP3D) Dataset [5] which consists of 90 scenes, 10,800 panoramic RGBD images, and 90 3D mesh reconstructions. VLN agents interact with MP3D through a connectivity graph that queries panoramic images. VLN-CE agents interact with the 3D mesh reconstructions through the Habitat Simulator [23].

Evaluation Metrics. We report evaluation metrics standard to the public leaderboards of both VLN and VLN-CE. These include trajectory length (TL), navigation error (NE), oracle success rate (OS), success rate (SR), and success weighted by inverse path length (SPL) as described in [1,3]. We consider SR the primary metric for comparison and SPL for evaluating path efficiency. Both SR and SPL determine success using a 3.0 m distance threshold between the agent's terminal location and the goal. We report the same metrics across both VLN and VLN-CE to enable empirical diagnosis of performance differences. We note that distance is computed differently between VLN and VLN-CE resulting in subtle differences in evaluation for the same effective path. In VLN, the distance between nodes is Euclidean[2], whereas in VLN-CE, distances are geodesic as computed on the 3D navigation mesh. This tends to result in longer path lengths in VLN-CE and a drop in SPL.

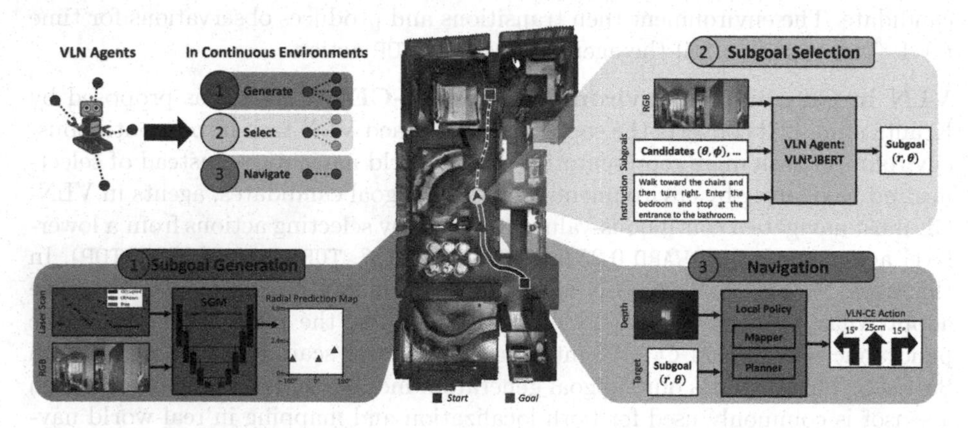

Fig. 1. We diagnose and quantify the VLN to VLN-CE gap by operating a topological VLN agent in continuous environments. We emulate the VLN observation space by generating subgoal candidates from egocentric observation (Sect. 4.3) and emulate the action space by calling a map-based navigation policy (Sect. 4.2).

4 Porting a VLN Agent to Continuous Environments

To perform a sim-2-sim transfer, we emulate the VLN action space in VLN-CE by developing a harness consisting of a lower-level navigation policy and a subgoal generation module. An overview can be seen in Fig. 1. Our harness follows the structure used in sim-2-real experiments by Anderson et al. [2]. We develop

[2] As defined by the Matterport3D Simulator used in VLN.

our harness such that subgoal generation modules, VLN agents, and navigation policies are all modular components for drop-in replacement and evaluation. This not only enables the analysis in this work, but the released codebase will ease sim-2-sim transfer of future VLN agents to continuous environments.

4.1 Core VLN Agent

We adopt VLN↻BERT [16], a VLN agent that performs near the state-of-the-art[3] on R2R Test. This agent uses observation and action spaces typical of recent work in VLN. VLN↻BERT has a recurrence-modified transformer architecture and encodes RGB vision with a ResNet-152 trained on Places365 [15,28]. We use VLN↻BERT as our core VLN agent for all experiments but note that our harness and related design choices are agnostic to the choice of VLN agent.

4.2 Navigation Policies

With the abstracted action space in VLN, navigation between subgoal locations is effectively performed by teleportation. In contrast, continuous settings require a navigation policy to convey the agent between subgoals. This navigation subtask can be considered a short-range version of PointGoal navigation [1] given that the average distance between nodes in the VLN nav-graph is 2.25 m. To evaluate the impact of navigation between subgoals in VLN-CE, we first establish upper bounds with two oracle methods: teleportation and an oracle policy. We then evaluate with a local policy admissible in VLN-CE. We describe these navigation policies as follows.

Teleportation. A teleportation action involves translation to the target location and rotation to face away from the previous location. In our teleportation policy, the heading is snapped to the nearest global 30° increment to match the heading discretization in VLN. This policy matches the navigation assumption made in VLN and serves as both an upper bound on navigation performance and a confirmation that the VLN agent is operating as expected in VLN-CE.

Oracle Policy. An oracle navigation policy has access to the navigation mesh used by the simulator to take optimal actions from the VLN-CE action space. The input to our oracle policy is 3D coordinates that exist on the navigation mesh. Analytical search using the A* algorithm then determines an action plan. We set a stopping distance threshold of 0.15 m to the goal which we empirically determine results in minimal navigation error and minimal action jitter near the goal. The Oracle Policy serves as an upper bound on navigation performance to policies operating under the same action space.

Local Policy. We employ a mapping and planning navigation policy based on [6] to convey the agent to selected subgoal locations as shown in Fig. 1 "Navigation". The Local Policy receives a target location specified by a distance and relative heading (r, θ). At each time step, a local occupancy map is aggregated from the geometric projection of an egocentric depth observation. The unexplored map

[3] eval.ai/web/challenges/challenge-page/97.

area is assumed to be free space. We then plan a path to the target location using the Fast Marching Method (FMM) which produces position coordinates along the shortest path lying 0.25 m away from the agent. Following [13], we greedily decode a discrete action to approach this point. We adopt the VLN-CE action space and stop once the agent is within 0.15 m of the target or 40 actions have been taken. The occupancy map is re-initialized for each new target.

4.3 Subgoal Candidate Generation

The VLN action space requires a set of subgoal candidates to select from. Where VLN uses neighboring nodes in the nav-graph, such oracle information does not exist in VLN-CE. As shown in Fig. 1 "Subgoal Generation", we emulate the presence of a nav-graph by having a learned module predict subgoal candidates from observations. Anderson et al. [2] developed a subgoal generation module (SGM) for VLN sim-2-real transfer. We adopt SGM and modify it for VLN-CE.

The observation space of the SGM is panoramic RGB vision (same as the VLN agent), supplemented with a 2D laser scanner for range-finding. We emulate a 360° laser scanner in the Habitat Simulator mounted at 0.24 m above ground level. We note that such scanners are commercially available. As shown in Fig. 1 "Laser Scan", the laser scan is represented as a radial obstacle map limited to a 4.8 m range. The obstacle map size is 24×48 with 24 discrete range bins of size 0.2 m and 48 heading bins of size 7.5°. The RGB panorama frames are encoded with the same ResNet-152 used by the VLN agent.

A U-Net architecture fuses these RGB features with the obstacle map to predict a radial map of subgoal candidates. This map follows the same range and heading discretization as the input. Gaussian non-maximum suppression filters localized predictions. The $k = 5$ candidates with the highest probability mass are converted into (r, θ) candidates consumed by the VLN agent.

Following Anderson et al. [2], we train the SGM to minimize Sinkhorn divergence [9] to ground-truth subgoal candidates on the VLN nav-graph. Nodes from Val-Unseen are held out for validation. To better match the visual domain of 3D-reconstructed environments, we train the SGM with panoramas rendered using the Habitat Simulator. We provide an ablation analysis motivating our changes to 360° laser scanning and reconstructed RGB vision in the supplementary.

5 Results and Analysis

We evaluate the transferability of VLN agents to continuous environments using the harness described above. We identify potential sources of performance degradation and evaluate their isolated impacts. Specifically, our analysis starts in VLN and moves toward VLN-CE, covering differences in dataset episodes (Sect. 5.1), replacing MP3D panoramas with reconstructed vision (Sect. 5.2), using a navigation policy to reach selected subgoals (Sect. 5.3), and removing nav-graph subgoals (Sect. 5.4). After demonstrating favorable performance over previous agents trained purely in VLN-CE (Sect. 5.5), we diagnose a remaining failure mode to be addressed in both future VLN-2-VLNCE transfer and sim-2-real transfer efforts (Sect. 5.6).

5.1 How Different Is the VLN-CE Subset of R2R?

We find the VLN-CE subset is slightly more difficult but comparable to the full R2R dataset. This suggests that performance in VLN is accurate to, but slightly over-estimates, the upper bound on performance transferred to VLN-CE. We make this comparison in Table 1 by evaluating VLN↻BERT on the topological VLN task for the VLN and VLN-CE subsets of R2R. Differences in these datasets arise from conversion errors which cause the VLN-CE subset of R2R to contain 22% fewer trajectories than in VLN. VLN↻BERT performs worse on the VLN-CE subset by 3 SR in Val-Seen and 1 SR in Val-Unseen.

Table 1. Difficulty of Room-to-Room (R2R) episodes in the VLN-CE subset vs. all episodes of R2R. We find performance on the VLN-CE subset is slightly lower than full R2R; Val-Seen success drops by 3 points and Val-Unseen success drops by 1 point.

Task: VLN		Train*					Val-Seen					Val-Unseen				
#	Dataset	TL ↓	NE ↓	OS ↑	SR ↑	SPL ↑	TL ↓	NE ↓	OS ↑	SR ↑	SPL ↑	TL ↓	NE ↓	OS ↑	SR ↑	SPL ↑
1	Room-to-Room [3]	9.98	0.93	95	93	90	10.81	3.00	76	72	68	11.85	4.24	67	61	56
2	VLN-CE subset [21]	9.96	1.00	94	93	89	10.94	3.30	74	69	65	11.79	4.43	66	60	54

5.2 What Is the Visual Domain Gap from VLN to VLN-CE?

We find a significant visual domain gap from VLN to VLN-CE in previously-seen environments of 10–13 SR and a moderate gap in novel environments of 3 SR (Table 2). This suggests that VLN agents may be overfitting to visual representation in training. This visual domain gap exists despite VLN and VLN-CE being derived from the same real-world scenes and sharing paths. We demonstrate characteristic examples in Fig. 2. VLN observations rendered from Matterport3D panoramas (MP3D panos) are high quality and captured directly from Matterport Pro2 3D cameras [5]. On the other hand, VLN-CE observations are rendered from 3D scene reconstructions in Habitat-Sim (Habitat renders) which introduce reconstruction errors and a domain gap in lighting. We demonstrate that training the agent with Habitat renders eliminates this gap almost entirely.

Experiment Setup. We evaluate VLN↻BERT in VLN with MP3D panos and Habitat renders. To compute vision features from Habitat, we map each node in VLN to a matching location in VLN-CE using known camera poses. We capture a panorama in the scene reconstruction matching the camera height, angle, and resolution expected in VLN. We then encode the panorama with the same feature encoder used by VLN↻BERT, a ResNet-152 trained on Places365 [15,28]. These features are then used as a drop-in replacement for MP3D pano features.

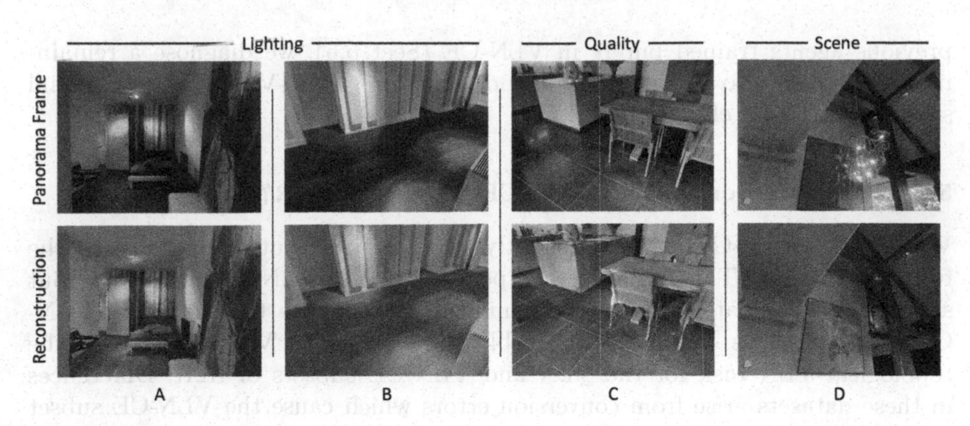

Fig. 2. Visual observation differences between VLN and VLN-CE. VLN renders observations from high-quality Matterport3D panoramas (top) and VLN-CE renders observations from 3D scene reconstructions (bottom). Examples of domain differences include lighting (A, B), furniture resolution (C), and object presence in the scene (D).

Zero-Shot Generalization. We present zero-shot generalization performance in Table 2, row 1 vs. 2. Switching to Habitat renders results in success rate drops of 13 points in Train*, 10 points in Val-Seen, and 3 points in Val-Unseen. The significant performance degradation in seen environments (Train* and Val-Seen) demonstrates the impact of the visual domain gap between VLN and VLN-CE. The relatively smaller drop in Val-Unseen suggests that this gap is not as catastrophic when generalizing to novel environments and that VLN agents may be overfitting to visual representation in training environments.

Domain Transfer. We show in row 4 that re-training VLN↻BERT directly on reconstructed vision can entirely recover task performance in seen environments and recover all by 1 point in success in Val-Unseen (row 4 vs. 1). This reconstruction-trained model achieves a success rate of 60 in Val-Unseen regardless of using MP3D panos or Habitat renders (rows 3 and 4). This suggests that training VLN agents on Habitat renders may result in less visual overfitting.

Table 2. Effect of the visual domain gap between Matterport3D panoramas (MP3D) and reconstruction renders (Recon.) on VLN performance. Reconstructed vision decreases success rate in Val-Unseen by 3 points (rows 1 vs. 2). This drop is mitigated by training VLN agents directly on reconstructed images (row 4).

Task: VLN	Camera		Train*					Val-Seen					Val-Unseen				
#	Train	Eval	TL ↓	NE ↓	OS ↑	SR ↑	SPL ↑	TL ↓	NE ↓	OS ↑	SR ↑	SPL ↑	TL ↓	NE ↓	OS ↑	SR ↑	SPL ↑
1	MP3D	MP3D	9.98	0.93	95	93	90	10.81	3.00	76	72	68	11.85	4.24	67	61	56
2		Recon.	11.17	2.20	84	80	76	11.66	4.02	68	62	58	11.69	4.56	64	58	53
3	Recon.	MP3D	10.63	1.18	92	90	86	11.35	3.41	72	66	61	11.72	4.12	66	60	54
4		Recon.	9.76	0.48	98	98	96	10.56	2.88	77	72	69	11.16	4.14	66	60	56

5.3 What Is the Navigation Gap Between VLN and VLN-CE?

Conveying the VLN agent to selected subgoals using a realistic navigator causes a decrease in performance across all splits, highlighted by a 5 point drop in Val-Unseen success. The VLN task, which effectively teleports between subgoals, makes navigation assumptions of perfect obstacle avoidance and stopping precision. These challenges must be contended with in VLN-CE. In the worst case, navigation failure causes episode failure regardless of the decisions made by the VLN agent. In the best case, an unexpected pose and observation are produced at the next time step, a situation VLN agents do not encounter in training.

Experiment Setup. We evaluate our reconstruction-trained VLN↻BERT in VLN-CE with each navigation policy from Sect. 4.2 and provide known subgoal candidates from the VLN nav-graph. We note that navigation errors introduce ambiguity in graph localization which must be resolved to provide the next subgoal candidates. In our experiments, we keep the agent at its navigation-terminated location but provide subgoal candidates by assuming it is located at the nearest node on the nav-graph by geodesic distance. This reflects the consequences of poor navigation where a large navigation error may cause the VLN agent to observe candidates from a different graph node than intended. We first evaluate the impact of using the VLN-CE action space instead of teleportation. We then evaluate the effect of a realistic navigation policy. Results are in Table 3.

Teleportation vs. VLN-CE Actions. We use the Oracle Policy to represent ideal actions within the VLN-CE action space. We find that performance under the Oracle Policy matches performance under Teleportation in Val-Unseen (row 2 vs. 1). This suggests that the navigation gap between VLN and VLN-CE can be minimized with performant navigation policies. This is an expected result due to the navigation verification that originally culled episodes for the VLN-CE dataset (see [21]). We additionally record the navigation error of the Oracle Policy for each short-range navigation performed, finding an average navigation error of 0.11 m in Val-Seen and 0.08 m in Val-Unseen. This demonstrates that the VLN agent is robust to position jitter about official graph nodes.

Optimal Navigation vs. Realistic Navigation. In row 3, we present performance under our Local Policy. Compared to the Oracle Policy, we find a drop in success in both Val-Seen (3 points) and Val-Unseen (5 points). The SPL drop is identical, suggesting that path efficiency remains high in successful episodes. The average short-range navigation error is 0.26 m in Val-Seen and 0.40 m in Val-Unseen, which are higher than that of the Oracle Policy. We observe that most calls to the Local Policy perform similarly to the Oracle Policy. However, the Local Policy has a longer tail of significant navigation failures (error >0.5 m). Specifically, the Local Policy has a >0.5 m failure rate that is 15x higher than the Oracle Policy. We expand on this error comparison in the supplementary.

5.4 What Is the Subgoal Candidate Gap?

In the previous experiments, VLN agents were selecting from subgoal candidates provided by the nav-graph, but these are not available in VLN-CE. Here, we

Table 3. Impact of navigation policies on VLN agents operating in VLN-CE. We assume a known nav-graph but require lower-level actions between nodes. The Oracle Policy results in better performance than the Local Policy, with a 5 SR Val-Unseen gap.

Task: VLN-CE		Train*					Val-Seen					Val-Unseen				
#	Navigator	TL ↓	NE ↓	OS ↑	SR ↑	SPL ↑	TL ↓	NE ↓	OS ↑	SR ↑	SPL ↑	TL ↓	NE ↓	OS ↑	SR ↑	SPL ↑
1	Teleportation	10.04	0.58	97	97	88	11.28	3.24	75	70	63	11.98	4.06	66	60	52
2	Oracle Policy	10.09	0.72	97	96	87	11.19	3.10	75	69	61	12.04	4.07	68	60	52
3	Local Policy	10.58	1.57	90	88	78	11.37	3.49	72	66	58	12.28	4.51	63	55	47

Table 4. Comparing VLN-CE performance when subgoal candidates are provided by the nav-graph (rows 1,2) vs. predicted by a subgoal generation module (SGM)(rows 3,4). Rows 3 and 4 demonstrate a complete harness for a VLN agent admissible in VLN-CE, consisting of both subgoal candidate generation and lower-level navigation.

Task: VLN-CE			Train*					Val-Seen					Val-Unseen				
#	Subgoal Candidates	F-tune	TL ↓	NE ↓	OS ↑	SR ↑	SPL ↑	TL ↓	NE ↓	OS ↑	SR ↑	SPL ↑	TL ↓	NE ↓	OS ↑	SR ↑	SPL ↑
1	Nav-Graph	-	10.58	1.57	90	88	78	11.37	3.49	72	66	58	12.28	4.51	63	55	47
2	Optimal SGM	-	11.40	2.07	86	81	72	12.69	3.78	71	63	55	14.56	4.96	61	49	41
3	SGM	-	13.12	3.54	71	65	55	12.69	4.51	60	51	44	13.74	5.83	51	41	35
4	SGM	✓	11.15	3.77	73	66	56	11.18	4.67	61	52	44	10.69	6.07	52	43	36

predict subgoal candidates online (at each time step) and observe significant performance degradation (14–23 SR). Fine-tuning the VLN agent on the online distribution of subgoal candidates slightly mitigates this drop, recovering 1–2 SR.

Experiment Setup. We replace the nav-graph with our subgoal generation module (SGM: Sect. 4.3) as the source of subgoal candidates. We join our SGM with reconstruction-trained VLN↻BERT and our Local Policy as navigator. In Table 4, we evaluate the impact of the prediction space and the SGM predictions.

Optimal SGM Predictions. We first consider the case where SGM predictions optimally match the nav-graph (Table 4, row 2). We convert nav-graph subgoals into a radial prediction map matching the SGM output space. This discretizes subgoals into polar bins with a 0.20 m range and 7.5° resolution. This prediction space accounts for a non-trivial drop in performance: a 3 point drop in Val-Seen success rate and a 6 point drop in Val-Unseen success rate (row 2 vs. 1).

Performance with SGM. Row 3 demonstrates subgoal predictions from the SGM with no nav-graph reliance. This experiment is compliant with the VLN-CE task definition; subgoal candidates are predicted from egocentric observations, a VLN agent selects a candidate, and a local policy conveys the agent using VLN-CE actions. We find a decrease of 12 SR in Val-Seen and 8 SR in Val-Unseen. These drops indicate a large domain gap between the nav-graph and the SGM. This result parallels findings in sim-2-real transfer by Anderson et al. [2] that point to an alignment problem between the VLN nav-graph and the SGM.

Fine-Tuning VLN↻BERT with SGM. We seek to determine if fine-tuning a VLN agent on the SGM candidates can reclaim nav-graph performance. To motivate, consider how learned priors in VLN could result in detrimental behavior under a different distribution of subgoals. For example, a slight change in the distribution of subgoal distances could break a learned time horizon prior – the VLN agent may choose to stop either too early or too late. Such a prior can be very strong in VLN since all paths require 5–7 actions. We fine-tune our reconstruction-trained VLN↻BERT via imitation learning in VLN-CE. Specifically, we train with teacher forcing to maximize the probability of predicting the subgoal candidate that greedily minimizes the geodesic distance to the goal. We set a batch size of 12, a learning rate of 1e-7, and train with cross-entropy loss and early stopping. In Table 4 row 4, we present the result of fine-tuning on SGM predictions. Performance increases slightly across all validation splits, highlighted by a 2-point increase in Val-Unseen success rate. This still leaves a 6 point gap in success rate to predicting optimal subgoal candidates and a 12 point gap in success rate to candidates directly from the nav-graph. Fine-tuning experiments such as this one cannot be performed on the rigid topology of VLN. Neither can they be performed practically in reality, making VLN-CE a promising setting for generalizing VLN agents for sim-2-sim or sim-2-real transfer.

5.5 Comparison with Previous VLN-CE Models

We compare our best agent against previously published methods on the VLN-CE Challenge Leaderboard[4]. This agent includes subgoal candidates from the SGM, reconstruction-trained VLN↻BERT fine-tuned on SGM candidates, and Local Policy navigation. We label our submission VLN-CE↻BERT. As shown in Table 5, VLN-CE↻BERT surpasses the success rate of all previous models on all splits including Test. Notably, all previous methods trained exclusively in VLN-CE, with the best amongst them requiring extensive compute (7000+ GPU-hours) to achieve a 32 SR. With just 12 GPU-hours of fine-tuning, our model surpasses this by 12 SR to achieve a 44 SR in Test (a relative improvement of 38%). This result shows that VLN to VLN-CE transfer is a highly promising avenue for making progress on the VLN-CE task and closing the gap to VLN.

5.6 Failures of Subgoal Candidate Generation

Starting from VLN↻BERT operating in VLN, we diagnosed a 1 SR drop from dataset differences, a 1 SR drop from the visual domain gap, a 5 SR drop from navigation policies, and a 12 SR drop from subgoal candidate generation. We now look into what failure modes constitute the subgoal candidate drop. We find that half of these failures (6/12 SR) are caused by episodes that require the agent to gain or lose at least 1.0 m of elevation (i.e. traversing stairs). In such episodes, an oracle VLN agent (a greedy-optimal subgoal selector) paired with the Oracle Policy for navigation is successful less than 50% of the time. This

[4] eval.ai/web/challenges/challenge-page/719.

Table 5. Results on the VLN-CE Challenge Leaderboard. Our submission outperforms previously-published results, demonstrating a 12 point improvement over the next best model in success rate on Test (a 38% relative improvement).

Task: VLN-CE	Val-Seen					Val-Unseen					Test				
# Model	TL ↓	NE ↓	OS ↑	SR ↑	SPL ↑	TL ↓	NE ↓	OS ↑	SR ↑	SPL ↑	TL ↓	NE ↓	OS ↑	SR ↑	SPL ↑
1 **VLN-CE↻BERT** (ours)	11.18	**4.67**	**61**	**52**	**44**	10.69	**6.07**	**52**	**43**	**36**	11.43	**6.17**	**52**	**44**	**37**
2 HPN+DN [20]	8.54	5.48	53	46	43	7.62	6.31	40	36	34	8.02	6.65	37	32	30
3 WPN+DN [20]	9.52	6.23	45	37	33	9.86	6.93	40	33	29	9.68	7.49	36	29	25
4 LAW [25]	9.34	6.35	49	40	37	8.89	6.83	44	35	31	9.67	7.69	38	28	25
5 CMA+PM+DA+Aug [21]	9.06	7.21	44	34	32	8.27	7.60	36	29	27	8.85	7.91	36	28	25

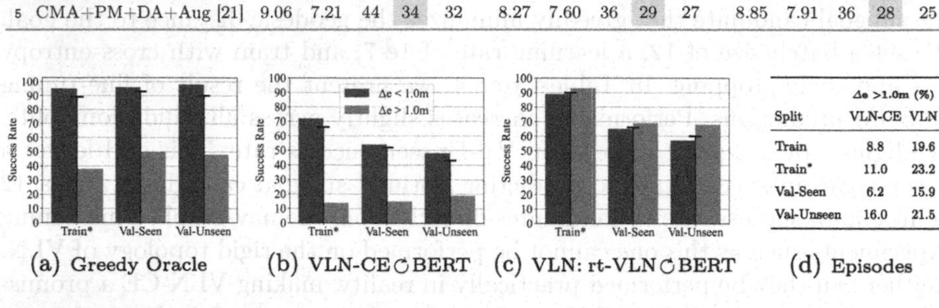

Split	Δe >1.0m (%) VLN-CE	VLN
Train	8.8	19.6
Train*	11.0	23.2
Val-Seen	6.2	15.9
Val-Unseen	16.0	21.5

(a) Greedy Oracle (b) VLN-CE↻BERT (c) VLN: rt-VLN↻BERT (d) Episodes

Fig. 3. Success rates in episodes with high (>1.0 m) vs. low (<1.0 m) elevation delta. Black bars indicate success rate irrespective of elevation. (a,b) use SGM subgoal candidates in VLN-CE and perform worse with elevation. However in VLN (c), our reconstruction-trained agent performs better with elevation. (d) shows the percentage of episodes that have a >1.0 m delta.

indicates that the SGM has a recall problem in generating elevation-changing subgoals necessary for success. Details and experimental support are below.

Oracle Selection and Navigation. We first consider whether candidates produced by the SGM can lead to episode success independent of failures stemming from selection or navigation issues. We define a greedy oracle VLN agent that selects the subgoal candidate with minimal geodesic distance to the goal. STOP is called when no candidate is closer to the goal than the current position. We run this oracle over SGM candidates and navigate via the Oracle Policy. We find success rates of 89% in Train*, 93% in Val-Seen, and 90% in Val-Unseen (Fig. 3(a)). However, failures disproportionately stem from episodes requiring elevation change. This is exemplified in Train* where episodes requiring at least a 1.0 m elevation change result in 38% success – a highly restrictive upper bound.

Elevation-Based Performance of VLN-CE↻BERT. We compare success rates for VLN-CE↻BERT in episodes requiring a high (>1.0 m) vs. low (<1.0 m) elevation delta (Fig. 3(b)). In Val-Unseen, there is a 29 SR gap (19 SR vs. 48 SR). This is unsurprising given the poor high-elevation performance of the oracle VLN agent. The failure of the SGM to predict high-elevation subgoals (a recall problem) may be related to training data; subgoal candidate locations are highly correlated with free-space in the 2D laser scan. However, 2D free-space is a poor predictor of elevation subgoals, which are relatively rare in training (Fig. 3(d)).

Elevation-Based Performance in VLN. Elevation-based VLN evaluation provides an upper bound on VLN-CE performance in elevation episodes. Surprisingly, we find in Fig. 3(c) that our reconstruction-trained VLN↻BERT has a higher success rate in episodes requiring a high elevation delta (68 SR vs. 57 SR in Val-Unseen). We suspect this result stems from episodes that feature stairs have a smaller search space; nav-graph nodes on stairs have a smaller degree. From this result, we estimate that solving the elevation problem in VLN-CE can mitigate 6 points of the 12 SR drop attributable to subgoal candidate generation, leaving a cumulative 12 SR gap between VLN and VLN-CE.

5.7 Qualitative Example

We present a qualitative example of VLN-CE↻BERT successfully navigating in a novel environment within VLN-CE (Fig. 4). This example demonstrates all components of our agent: subgoal candidate generation, subgoal selection, and navigation. In each step, subgoal candidates (large, yellow triangles) enable navigation in primary directions. However, not all can be reached, like in Step 1 where a candidate is centered on the kitchen countertop. The SGM candidates diverge from nav-graph node locations. This leads the VLN agent to observe the environment from poses off the nav-graph, yet proper candidate selections are still made over the course of 5 actions. Between steps 0 and 1, the selected subgoal cannot be navigated to in a straight line. Our Local Policy demonstrates obstacle avoidance while planning and executing a path that traverses around the countertop. This example demonstrates that through modular construction, VLN agents can operate successfully and efficiently in continuous environments.

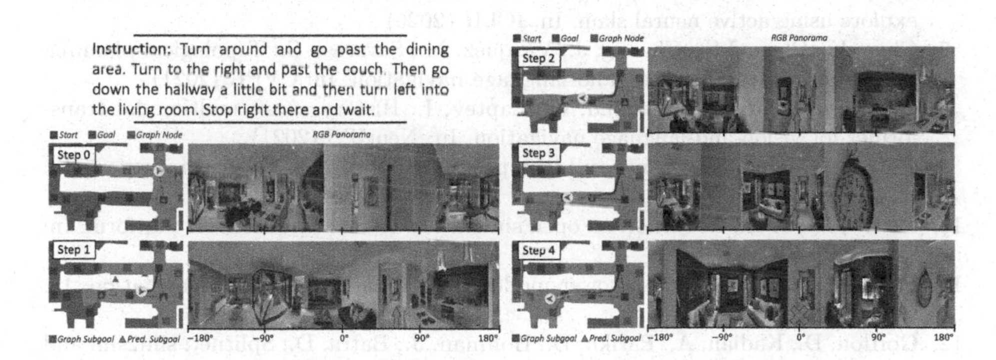

Fig. 4. Qualitative example of our VLN-CE↻BERT agent operating in VLN-CE.

6 Conclusion

In summary, we explore the sim-to-sim transfer of instruction-following agents from topological VLN to unconstrained VLN-CE. Our transfer results in an absolute improvement of 12% over the prior state-of-the-art in VLN-CE. We diagnose the remaining VLN-to-VLNCE gap, identifying subgoal candidate generation as a primary hindrance to transfer. We outline the problem of generating candidates in multi-floor environments to guide future work. Operating VLN agents in continuous environments enables a new interplay between language and navigation topologies. This can lead not only to higher performance in realistic environments, but also to the development of more robust topological navigators.

Acknowledgements. This work was supported in part by the DARPA Machine Common Sense program. The views and conclusions contained herein are those of the authors and should not be interpreted as necessarily representing the official policies or endorsements, either expressed or implied, of the U.S. Government, or any sponsor.

References

1. Anderson, P., et al.: On evaluation of embodied navigation agents. arXiv preprint arXiv:1807.06757 (2018)
2. Anderson, P., et al.: Sim-to-real transfer for vision-and-language navigation. In: CoRL (2020)
3. Anderson, et al.: Vision-and-language navigation: interpreting visually-grounded navigation instructions in real environments. In: CVPR (2018)
4. Blukis, V., Terme, Y., Niklasson, E., Knepper, R.A., Artzi, Y.: Learning to map natural language instructions to physical quadcopter control using simulated flight. In: CoRL (2020)
5. Chang, A., et al.: Matterport3d: learning from RGB-D data in indoor environments. In: 3DV (2017), MatterPort3D dataset license. http://kaldir.vc.in.tum.de/matterport/MP_TOS.pdf
6. Chaplot, D.S., Gandhi, D., Gupta, S., Gupta, A., Salakhutdinov, R.: Learning to explore using active neural slam. In: ICLR (2020)
7. Chen, K., Chen, J.K., Chuang, J., Vázquez, M., Savarese, S.: Topological planning with transformers for vision-and-language navigation. In: CVPR (2021)
8. Chen, S., Guhur, P.L., Schmid, C., Laptev, I.: History aware multimodal transformer for vision-and-language navigation. In: NeurIPS (2021)
9. Cuturi, M.: Sinkhorn distances: Lightspeed computation of optimal transport. In: NeurIPS (2013)
10. Deitke, M., et al.: Robothor: an open simulation-to-real embodied AI platform. In: CVPR (2020)
11. Fried, D., et al.: Speaker-follower models for vision-and-language navigation. In: NeurIPS (2018)
12. Gordon, D., Kadian, A., Parikh, D., Hoffman, J., Batra, D.: Splitnet: sim2sim and task2task transfer for embodied visual navigation. In: CVPR (2019)
13. Hahn, M., Chaplot, D.S., Tulsiani, S., Mukadam, M., Rehg, J.M., Gupta, A.: No RL, no simulation: learning to navigate without navigating. In: NeurIPS (2021)
14. Hao, W., Li, C., Li, X., Carin, L., Gao, J.: Towards learning a generic agent for vision-and-language navigation via pre-training. In: CVPR (2020)
15. He, K., Zhang, X., Ren, S., Sun, J.: Deep residual learning for image recognition. In: CVPR (2016)
16. Hong, Y., Wu, Q., Qi, Y., Rodriguez-Opazo, C., Gould, S.: VLN BERT: a recurrent vision-and-language bert for navigation. In: CVPR (2021)

17. Irshad, M.Z., Ma, C.Y., Kira, Z.: Hierarchical cross-modal agent for robotics vision-and-language navigation. In: Proceedings of the IEEE International Conference on Robotics and Automation (ICRA) (2021)
18. Irshad, M.Z., Mithun, N.C., Seymour, Z., Chiu, H.P., Samarasekera, S., Kumar, R.: Sasra: semantically-aware spatio-temporal reasoning agent for vision-and-language navigation in continuous environments. In: International Conference on Pattern Recognition (ICPR) (2022)
19. Kadian, A., et al.: Are we making real progress in simulated environments? measuring the sim2real gap in embodied visual navigation. In: IROS (2020)
20. Krantz, J., Gokaslan, A., Batra, D., Lee, S., Maksymets, O.: Waypoint models for instruction-guided navigation in continuous environments. In: ICCV (2021)
21. Krantz, J., Wijmans, E., Majumdar, A., Batra, D., Lee, S.: Beyond the nav-graph: vision-and-language navigation in continuous environments. In: ECCV (2020)
22. Majumdar, A., Shrivastava, A., Lee, S., Anderson, P., Parikh, D., Batra, D.: Improving vision-and-language navigation with image-text pairs from the web. In: ECCV (2020)
23. Savva, M., et al.: Habitat: a platform for embodied AI research. In: ICCV (2019)
24. Quigley, M., et al.: Ros: an open-source robot operating system. In: ICRA Workshop on Open Source Software (2009)
25. Raychaudhuri, S., Wani, S., Patel, S., Jain, U., Chang, A.X.: Language-aligned waypoint (law) supervision for vision-and-language navigation in continuous environments. In: EMNLP (2021)
26. Tan, H., Yu, L., Bansal, M.: Learning to navigate unseen environments: back translation with environmental dropout. In: NAACL HLT (2019)
27. Wang, X., et al.: Reinforced cross-modal matching and self-supervised imitation learning for vision-language navigation. In: CVPR (2019)
28. Zhou, B., Lapedriza, A., Khosla, A., Oliva, A., Torralba, A.: Places: a 10 million image database for scene recognition. In: TPAMI (2017)

Style-Agnostic Reinforcement Learning

Juyong Lee, Seokjun Ahn, and Jaesik Park(✉)

Pohang University of Science and Technology (POSTECH), Pohang, South Korea
{joy.lee,sdeveloper,jaesik.park}@postech.ac.kr

Abstract. We present a novel method of learning style-agnostic representation using both style transfer and adversarial learning in the reinforcement learning framework. The style, here, refers to task-irrelevant details such as the color of the background in the images, where generalizing the learned policy across environments with different styles is still a challenge. Focusing on learning style-agnostic representations, our method trains the actor with diverse image styles generated from an inherent adversarial style perturbation generator, which plays a min-max game between the actor and the generator, without demanding expert knowledge for data augmentation or additional class labels for adversarial training. We verify that our method achieves competitive or better performances than the state-of-the-art approaches on Procgen and Distracting Control Suite benchmarks, and further investigate the features extracted from our model, showing that the model better captures the invariants and is less distracted by the shifted style. The code is available at https://github.com/POSTECH-CVLab/style-agnostic-RL.

Keywords: Reinforcement learning · Domain generalization · Neural style transfer · Adversarial learning

1 Introduction

Learning visual representation in reinforcement learning (RL) framework incorporated with deep convolutional neural networks enabled achieving remarkable performances in various control tasks, including video games [29,41], robot manipulation [23,38], and autonomous driving [46]. Unfortunately, however, generalization of the learned policies to unseen environments often results in failures, even with slight changes in the backgrounds [4,15,44].

Several methods have been proposed to overcome this limitation of RL agents, such as having an encoder with generative models [6,8,12,21] or training with auxiliary tasks [19,24,27]. Methods using generative models are designed to train

J. Lee and S. Ahn—These authors contributed equally to this work.

Supplementary Information The online version contains supplementary material available at https://doi.org/10.1007/978-3-031-19842-7_35.

S. Avidan et al. (Eds.): ECCV 2022, LNCS 13699, pp. 604–620, 2022.
https://doi.org/10.1007/978-3-031-19842-7_35

the agents to understand the world environment, and auxiliary tasks enable the agent to extract better features that will lead to better performances. Due to its simplicity, the latter technique is gaining interest. For example, recent works have shown that representation learning with self-supervision objectives [9,19], data randomization with feature matching [22], and data augmentation with additional regularization [10,11,17] result in high success.

The central concept of these approaches is to diversify training data so that the RL agents can learn invariants to the *different styles of environments*. Here, the style of the environment indicates too detailed or irrelevant elements in the observation. In an autonomous driving situation, for instance, detecting the road or pedestrians is key to success, while the texture of the road, the colors of the other cars, or the weather condition can be regarded as different styles, which distract the agent from abstract and understand the situation. Data augmentation, thus, might lead to better generalization capacity by mimicking natural style changes of observations. However, the results are inefficient or unstable without a careful choice of augmentation type and timing [16,20]. To tackle this issue, sounder training methods of adding more regularization terms can be applied [10,11,17,32], but this makes the training objectives much more complex.

In this work, we focus on learning style-agnostic representations and propose **SAR**: **S**tyle-**A**gnostic **R**L, which adopts the concept of both style transfer and adversarial learning. Style transfer has been applied in many computer vision tasks, including domain generalization in RL [14,42,45]. Here, we further examine how style transfer is used to train the agents via generating images of new styles. The generator module in our model generates *never-seen* styles and helps the actor generalize its learned policy to the unseen styles with various background images, including realistic images, without any heuristics or explicit environment class labels. Notably, the generator is trained with adversarial loss to perform adaptive style perturbation to the encoded feature representation. To our best knowledge, this attempt and success have not been presented anywhere before. An overview of our model is described in Fig. 1.

In summary, the contributions of this paper are as follows:

- First, we introduce SAR, a novel method of learning style-agnostic representation for domain generalization in RL.
- Second, we conduct extensive empirical evaluations showing that the model better captures invariants between different styles of environment.
- Finally, we show that the SAR agents achieve competitive or better results on the Procgen [3] and Distracting Control Suite [36] benchmarks than the previous state-of-the-art algorithms.

2 Related Work

2.1 Domain Generalization in RL

The main target of the domain generalization in RL can be summarized as training an agent to learn a robust policy that can be generalized to unseen environments. This allows RL algorithms to be applied in more realistic situations because agents are often tested in different environments from the training stage. One example is deploying a policy learned from the simulation to the real world in the robot manipulation task.

Data randomization is a promising technique for such domain generalization in many cases [2,38]. However, it is difficult to build an accurate and practical simulator that enables using data randomization. Visual augmentation, on the other hand, is much easier to apply as it is based on simple image transformations. Laskin and Lee et al. [20], for example, demonstrated that simply using data augmentation, such as random cropping or gray scaling, is indeed helpful in improving the generalization capacity of RL agents. Also, Yarats & Kostrikov et al. [17] suggested using regularization terms for stabilizing the model training when using data augmentation.

However, although data augmentation is potentially effective, it has several limitations. For example, a naïve choice of the augmentation type may degrade the generalization performance [20]. Applying cropping to an essential part of the image may confuse the agent, or training the model to produce the same action from a rotated image may be unreasonable. Here, we present a method for domain generalization by diversifying the training examples without requiring a complex strategy for data augmentation. The generator in the SAR model generates new feature examples having different styles and helps the agents with learning style-agnostic representations.

2.2 Adversarial Feature Learning

Adversarial feature learning has become popular for domain generalization in computer vision tasks [18,25,30,35,43]. Li et al. [25] showed that adversarial objectives help a model learn universal feature representations across different domains. Furthermore, Nam & Lee et al. [30] proposed a method of reducing the style gap for domain generalization in the image classification task. Inspired by this work, we investigate the adversarial feature learning for RL agents, but with a simpler training procedure, i.e., without dividing training phases or considering the environment style's *class*es.

We note that adopting adversarial training for RL is not new [24,31]. To our best knowledge, however, exploiting adversarial learning to the latent features in RL framework and the min-max game scheme is not presented before. Especially, our method can be interpreted as domain randomization beyond pixel space. Mixing styles with linear interpolation for representation learning in RL setting has been proposed in the earlier work [45]. However, unlike in the previous study, the style perturbation generator in SAR produces new synthetic styles that will not be seen with a simple interpolation. The adversarial examples help the actor extract style-agnostic embeddings without any label of styles and, finally, learn a robust policy for unseen environments.

3 Backgrounds

3.1 Deep Reinforcement Learning

RL agents interact and get trained with the world environment within a Markov decision process, which is defined as a tuple of (state space $\mathcal{S}$, action space $\mathcal{A}$, transition probability P, reward space $\mathcal{R}$, and discount factor γ); at every timestep t, the agent observes a state $s_t \in \mathcal{S}$ and takes an action $a_t \in \mathcal{A}$ from its policy $\pi(a_t|s_t)$ [1]. Then, the agent is rewarded with $r_t \in \mathcal{R}$, and moves to the next state s_{t+1} sampled from the transition probability $P(s_{t+1}|s_t, a_t)$.

The policy of the agent is optimized to maximize the discounted sum of rewards $G_t = \sum_{k=t}^{\infty} \gamma^k r_k$. With given state s_t, the value of the state $V(s_t)$ is estimated as $\mathbb{E}_{\tau\sim\pi}[G_t|s_t]$ and the value of the state-action $Q(s_t, a_t)$ is computed as $\mathbb{E}_{\tau\sim\pi}[G_t|s_t, a_t]$, with trajectory τ sampled from the policy π.

With deep RL algorithms, the policy π gets parameterized by a set of learnable parameters ψ, and value function V or Q is optimized with network parameter ϕ. Also, especially for visual-based RL, since the images only offer partial observations, Mnih et al. [28] has proposed that defining the state s_t as a stacked consecutive image frames $(o_{t-k}, o_{t-k+1}, \ldots, o_t)$, where $\mathcal{O}$ is a high-dimensional image space and $o \in \mathcal{O}$, is effective.

Proximal Policy Optimization. (PPO) [33] is a state-of-the-art on-policy RL algorithm that is used for, in our setting, discrete control tasks. Here, on-policy refers to a situation in which the model is trained with trajectories collected from the current policy. With PPO, the actor is updated using policy gradients, where the gradients are computed by using (i) action-advantages A_t to reduce the gradient variances and (ii) clipped-ratio loss to constraint the update region. The critic estimates the state-value V_ϕ, and gets trained with mean-squared error loss toward a target state-value V_t^{target} using generalized advantage estimation [33]. So, the objectives for the actor and critic network can be written as follows:

$$A_t = Q_\phi(s_t, a_t) - V_\phi(s_t) \tag{1}$$

$$\mathcal{L}_{\text{actor}}(\psi) = -\mathbb{E}_{s_t, a_t\sim\pi}\Big[\min\Big(\frac{\pi_\psi(a_t|s_t)}{\pi_{\psi_{\text{old}}}(a_t|s_t)}A_t, \text{clip}(\frac{\pi_\psi(a_t|s_t)}{\pi_{\psi_{\text{old}}}(a_t|s_t)}, \epsilon)A_t\Big)\Big] \tag{2}$$

$$\mathcal{L}_{\text{critic}}(\phi) = \mathbb{E}_{s_t\sim\pi}\Big[(V_\phi(s_t) - V_t^{target})^2\Big], \tag{3}$$

where ϵ is a coefficient for clipping function $\text{clip}(\cdot) \rightarrow [1-\epsilon, 1+\epsilon]$.

Soft Actor-Critic. (SAC) [7] is an off-policy RL algorithm for continuous control tasks. Since off-policy algorithms can train the agent with trajectories collected from the different policies, other than the current one, it appears to be more flexible to alternative routes but may get slower. With SAC, the actor learns a policy π_ψ, with the guide of critic estimating the state-action value Q_ϕ to maximize an objective as a sum of the reward and the policy entropy

$\mathbb{E}_{s_t,a_t\sim\pi}[\sum_t r_t + \alpha H(\pi(a_t|s_t))]$. Here, α is an entropy coefficient determining the priority of exploration over exploitation.

The actor, then, is trained by maximizing the expected return of its sampled actions where the objective can be denoted as follows:

$$\mathcal{L}_{\text{actor}}(\psi) = -\mathbb{E}_{a_t\sim\pi}\Big[Q_\phi(s_t,a_t) - \alpha\log\pi_\psi(a_t|s_t)\Big]. \tag{4}$$

The critic is updated to minimize the temporal difference. The objectives for the critic, with the estimated target value of the next state, are as follows:

$$V(s_{t+1}) = \mathbb{E}_{a_t\sim\pi}\Big[Q_\phi(s_{t+1},a_t) - \alpha\log\pi_\psi(a_t|s_{t+1})\Big] \tag{5}$$

$$\mathcal{L}_{\text{critic}}(\phi) = \mathbb{E}_{s_t,a_t,r_t,s_{t+1}\sim\mathcal{D}}\Big[\Big(Q_\phi(s_t,a_t) - \big(r_t + \gamma V(s_{t+1})\big)\Big)^2\Big] \tag{6}$$

where $\mathcal{D}$ is the replay buffer.

In this work, we show that our method can be attached to *both on-policy and off-policy* RL algorithms, namely PPO and SAC. Also, our method can be applied to *both continuous and discrete* control tasks as tested with the Procgen and Distracting Control Suite benchmark.

Style Transfer via Instance Normalization. For style transfer, many recent works adopt a method of using instance normalization (IN) [5,13,39,40,45]. The underlying idea is that the mean and standard deviations of feature maps, computed across the spatial dimension within each feature channel, reflect the images' style. For example, the color or texture of an image can be captured with these statistics, which may be irrelevant features for classifying or detecting an object. By using IN, the effect of styles can be normalized with the formula:

$$\text{IN}(z) = \gamma\cdot\frac{z-\mu(z)}{\sigma(z)} + \beta \tag{7}$$

where $z \in \mathbb{R}^{C\times H\times W}$ is a feature map with channel C, height H and width W, and $\beta,\gamma \in \mathbb{R}^C$ refers to the affine transformation parameters.

Note that $\mu(z) \in \mathbb{R}^C$ and $\sigma(z) \in \mathbb{R}^C$ are denoted as:

$$\mu(z)_c = \frac{1}{HW}\sum_{h=1}^{H}\sum_{w=1}^{W} z_{c,h,w},\quad \sigma(z)_c = \sqrt{\frac{1}{HW}\sum_{h=1}^{H}\sum_{w=1}^{W}(z_{c,h,w}-\mu(z)_c)^2} \tag{8}$$

with $c \in \{1,\ldots,C\}$.

Moreover, Huang and Belongie [13] proposed the method of adaptive instance normalization (AdaIN), which can be understood as replacing the style statistics of a target content image with those of a source style image with the definition below:

$$\text{AdaIN}(z,z') = \sigma(z')\cdot\frac{z-\mu(z)}{\sigma(z)} + \mu(z') \tag{9}$$

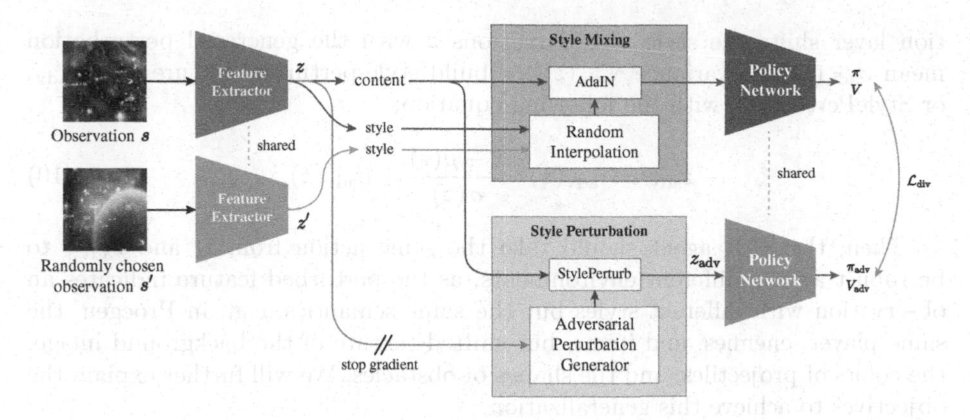

Fig. 1. Overview of the proposed Style-Agnostic Reinforcement learning (SAR) with the base model of PPO. The upper *Style Mixing* module makes the policy network focus on the critical content in the observations by mixing styles from randomly chosen states s'. We newly employ our *Style Perturbation* module, helping the agent with learning a robust policy by adversarially perturbing latent features.

where z' is the feature map extracted from the source style image.

This idea can be used for mixing styles between images within a mini-batch. Especially in the domain adaptation for image classification, this has been proved to be successful [30]. Zhou et al. [45] adopted style mixing for domain generalization in RL. However, the scope of mixing styles is restricted only to the training mini-batch as AdaIN is an interpolation. Here, our method enables the agents to observe unseen styles by generating new adversarial feature examples.

4 Method

Overview. SAR is composed of an actor-critic module with RL objectives and a style perturbation generator helping the agents to observe more diverse styles of observations. While the generator is updated to produce more substantial perturbations for style transfer by *maximizing* the difference between the action predictions, the actor learns a more robust policy to the attack from the generator by *minimizing* the gap between predicted action distributions.

To perform this min-max game between actor and generator, we present a **style perturbation layer**, shown in Fig. 1. Unlike the conventional approach using only style mixing within the mini-batch [45], the model in the training phase generates new styles and observes a broader range of feature examples. Note that this does not require explicit data augmentation that potentially degrades performance without a cautious choice of augmentation type.

4.1 Style Perturbation Layer

Our method is based on the concept of style transfer, which was proven to be successful in generating images with new styles [5,14]. The style perturbation layer shifts the style of observations z with the generated perturbation mean $\beta_{\text{adv}}(z)$ and variance $\gamma_{\text{adv}}(z)$, to build style-perturbed feature map z_{adv}, or StylePerturb(z), with the following equation:

$$z_{\text{adv}} = \gamma_{\text{adv}}(z) \cdot \frac{z - \mu(z)}{\sigma(z)} + \beta_{\text{adv}}(z). \tag{10}$$

Then, the SAR agent should take the same action from z_t and $z_{\text{adv},t}$ to be robust among different environments, as the perturbed feature indicates an observation with different styles but the same semantics, e.g., in Procgen, the same player, enemies, and items, but shifted texture of the background image, the colors of projectiles, and the shapes of obstacles. We will further explain the objectives to achieve this generalization.

4.2 SAR Objectives

Primarily, the policy network is updated via PPO or SAC objectives. Thus, the actor loss of SAR is adopted from Eq. 2 with PPO baseline or from Eq. 4 when using SAC. We will denote this loss be $\mathcal{L}^{\circ}_{\text{actor}}$. Also, for the critic loss, as suggested in RAD [20], we adopt the critic objective of PPO or SAC interchangeably, denoted as $\mathcal{L}^{\circ}_{\text{critic}}$.

Another big goal of SAR is to be robust to different environments. Therefore, the agent should learn its policy by minimizing the difference between the distributions of actions from the style-perturbed features $z_{\text{adv},t}$ and the original ones z_t. By leveraging KL-divergence, we can calculate the objective as $\mathcal{L}_{\text{div}} = \text{KL}[\pi(\cdot|z_t)||\pi(\cdot|z_{\text{adv},t})]$. Integrating this with a weight coefficient λ, the objective for the SAR actor module can be written as:

$$\mathcal{L}_{\text{actor}}(\psi) = \mathcal{L}^{\circ}_{\text{actor}}(\psi) + \lambda \cdot \mathcal{L}_{\text{div}} \tag{11}$$

On the other hand, the generator participates in the min-max game in another manner: to maximize the differences between the action distributions. This module is trained with the objective of the same $\mathcal{L}_{\text{div}}$ but with a converted sign. Unlike the previous works using class *label* information of the environment style [24] or additional heavy background images [10], the objectives for the robust policy (i.e., adversarial loss) do not demand any secondary labors. Hence, the overall goals for the generator can be formalized as:

$$\mathcal{L}_{\text{gen}}(\theta) = -\lambda' \cdot \mathcal{L}_{\text{div}}, \tag{12}$$

where λ' can be different coefficient from that of actor objective.

Finally, the critic gets updated to guide the actor to optimize its policy to maximize the value function. Meanwhile, we observed that the sharing critic network, for predicting the value for both style-perturbed features and the original ones, does not bring a huge difference in the performance from decoupling the critic network but lighter training computation. Instead, we add a regularization term G_{critic} for the value function, to minimize the difference between the value predicted from the adversarial example, i.e., $(V_\phi(z_t) - V_\phi(z_{\text{adv},t}))^2$, which helps stabilization. Thus, the critic's objectives can be computed as follows:

$$\mathcal{L}_{\text{critic}}(\phi) = \mathcal{L}^{\circ}_{\text{critic}}(\phi) + \kappa \cdot G_{\text{critic}}, \tag{13}$$

with hyperparameter κ[1].

On Convergence. When the SAR agents learn the optimal policy π^*, the KL divergence term, or $\mathcal{L}_{\text{div}}$, becomes zero. This is the situation where the actors infer the same actions from the features with different styles. This might be one of the two cases: (i) the generator produces the same style statistics for all images in the mini-batch, or more possibly (ii) the actor well focuses on the invariant part of all observations.

Since the model should learn an additional generator module, the training procedure indeed demands more computations. However, the sample efficiency is not highly degraded even with limited training timesteps, e.g., the usual 25M timesteps in Procgen. Although the agent may not learn the optimal policy due to the limited number of epochs, we also empirically observed that the performances of the SAR agents converge as shown in Fig. 3.

4.3 Pseudo-code

Here, we present the pseudo-code of the SAR algorithm. As depicted in Fig. 1, to maximize the effect of style transfer, we design the z_t to pass a *Style Mixing* module and a *Style Perturbation* module with two divided branches. In the *Style Mixing* module, the styles of observations in the mini-batch get interpolated with Eq. 9. In *Style Perturbation* module, on the other hand, the styles of observations are shifted with new styles generated from the generator network with Eq. 10.

With two different features z_t and $z_{\text{adv},t}$, the SAR agent predicts two different action distributions π_t and $\pi_{\text{adv},t}$. The difference between these predictions $\mathcal{L}_{\text{div}}$ is computed, and it gets interpreted in two different ways: by the generator to produce more unfamiliar styles and by the actor to make its policy more robust.

5 Results

5.1 Setup

In this section, we exhibit the experiment results for the generalization performance of our SAR model on Procgen [3] and Distracting Control Suite [36] benchmarks. Recently, these benchmarks have become a standard for measuring the generalization performance of visual-based RL algorithms [10,11,17,20,32]. These contain reasonably challenging and diverse tasks, which are highly relevant to real-world robot learning.

[1] The values used for each hyperparameters λ, λ', κ in the experiment are described in the supplementary material.

Algorithm 1. SAR algorithm

1: Initialize rollout or replay buffer $\mathcal{D}$
2: Initialize parameters for policy ψ, generator θ, and critic ϕ
3: **for** every epoch **do**
4:     **for** every environment step **do**
5:         Sample (s_t, a_t, r_t, s_{t+1})
6:         Update $\mathcal{D} \leftarrow \mathcal{D} \cup \{(s_t, a_t, r_t, s_{t+1})\}$
7:     **end for**
8:     **for** each mini-batch sampled from $\mathcal{D}$ **do**
9:         $z_t \leftarrow \text{Encoder}(s_t)$ ▷ Encoder in the actor network
10:         Generate $\beta_{\text{adv}}(z_t), \gamma_{\text{adv}}(z_t)$ ▷ From the generator network
11:         $z_{\text{adv},t} \leftarrow \text{StylePerturb}(z_t)$ ▷ Use Eq. 10
12:         $z_t \leftarrow \text{AdaIN}(z_t, z'_t)$ ▷ z'_t is permuted from z_t within mini-batch
13:         Compute $\mathcal{L}_{\text{div}}$ from $z_t, z_{\text{adv},t}$
14:         Compute $\mathcal{L}_{\text{actor}}$, $\mathcal{L}_{\text{gen}}$, and $\mathcal{L}_{\text{critic}}$
15:         Update ψ, θ, and ϕ
16:     **end for**
17: **end for**

While the Procgen benchmark is with a *discrete* action space, the Distracting Control Suite presents *continuous* control tasks. We exploited PPO as the basic baseline on the Procgen, and SAC as the basic baseline on the Distracting Control Suite, showing that the SAR algorithm can be applied to both on-policy and off-policy algorithms. Figure 2 visualizes some examples of training and test environments in the two different benchmarks.

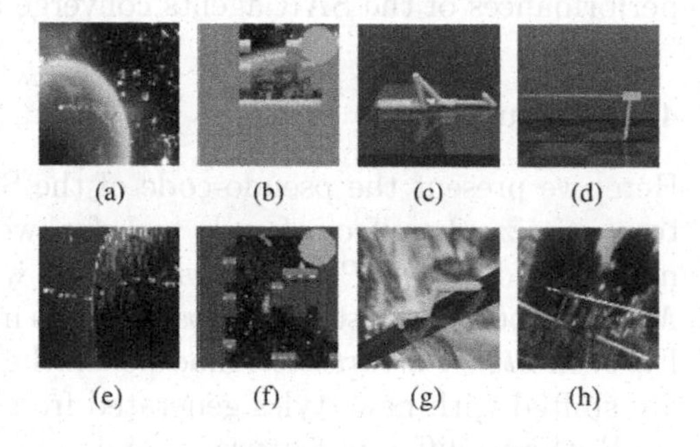

Fig. 2. Examples of seen training environments from (a) `starpilot` and (b) `jumper` in Procgen (c) `walker:walk` and (d) `cartpole:balance` task in Distracting Control Suite, with examples of unseen test environments from (e) `starpilot` and (f) `jumper` in Procgen (g) `walker:walk` and (h) `cartpole:balance` task in Distracting Control Suite.

OpenAI Procgen. One key reason for choosing this benchmark is that this presents different styles between test and training environments. We train the agents on the first 200 levels in the Procgen environment. Then, we test the generalization performance of the agents on the environment levels sampled from the full distribution of unseen levels, with *easy* distribution mode. Among 16 tasks, we selected four tasks demonstrating comparably more considerable differences (`starpilot, climber, jumper, ninja`) and four tasks showing comparably less significant differences (`coinrun, maze, bigfish, dodgeball`) between the training and test environments style.

Distracting Control Suite. DeepMind Control Suite [37] presents various continuous control tasks where RL agents can be tested. On top of the DMC, Stone et al. [36] proposed Distracting Control Suite that distracts the agents by applying a color shift, changing the background images into videos, and rotating the camera angle. We test our model and other baselines with different noise coefficient values and show how these models generalize to unseen situations.

5.2 Generalization Performance

Procgen. First, Table 1 shows the result of the generalization test of SAR with six other baselines. The SAR agent achieved high and robust performances in the zero-shot generalization test: 3 *top-1 scores* and 7 *top-3 scores* out of 8 tasks.

The baselines are six visual-learning RL algorithms showing state-of-the-art results on Procgen. **PPO** [33] is the vanilla on-policy RL baseline, and **RAD** [20] uses data augmentation on top of PPO. We performed random translation (denoted as 'trans') and random color cutout (denoted as 'color') for RAD, as they are reporting the best performance. Among many advanced algorithms on RAD, **UCB DrAC**[32], and **Meta DrAC** [32] are chosen to be compared with our method among three variants of DrAC; the former one presents the best performance among the variants. **Mixstyle** [45] exploits the style mixing, and **DARL** [24] uses an adversarial objective for regularization with style *labels*.[2]

Table 1. The generalization scores of SAR and baseline methods on Procgen. The results are averaged over three runs with 100M training timesteps without smoothing. The ranking stands for the average rank among all tasks. The *top-1 score* is bold.

	PPO [33]	RAD [20] (trans)	RAD [20] (color)	UCB DrAC [32]	Meta DrAC [32]	MixStyle [45]	DARL [24]	**SAR** (Ours)
Starpilot	30.37 ±11.14	29.57 ±7.52	27.03 ±7.51	33.17 ±6.37	29.40 ±4.61	25.70 ±8.13	21.97 ±10.66	**35.87 ±9.13**
Climber	6.73 ±1.27	4.87 ±1.31	7.23 ±2.05	**9.43 ±1.35**	7.77 ±0.68	7.37 ±2.71	7.03 ±1.37	7.93 ±1.10
Jumper	6.00 ±2.00	4.67 ±0.58	5.67 ±1.53	5.67 ±0.58	7.33 ±2.52	6.00 ±2.65	**7.67 ±1.53**	6.33 ±1.15
Ninja	6.00 ±2.83	5.33 ±2.52	5.33 ±2.08	6.33 ±1.53	7.33 ±0.58	**8.67 ±1.53**	7.33 ±0.58	8.33 ±1.15
Coinrun	8.67 ±1.15	8.33 ±1.15	**9.33 ±1.15**	9.00 ±1.00	8.33 ±0.58	**9.33 ±0.58**	**9.33 ±1.15**	9.00 ±1.00
Maze	4.67 ±0.58	5.33 ±1.53	5.33 ±0.58	**7.33 ±1.53**	4.67 ±0.58	5.33 ±0.58	3.67 ±1.15	5.00 ±1.00
Bigfish	10.37 ±3.27	6.03 ±2.06	10.13 ±1.84	9.37 ±3.16	12.03 ±4.38	9.00 ±2.94	9.07 ±4.05	**13.20 ±6.16**
Dodgeball	4.13 ±1.75	4.93 ±1.53	3.20 ±1.56	**8.13 ±1.33**	2.40 ±2.46	3.60 ±2.31	4.47 ±2.73	3.60 ±2.23
Avg. Rank	4.9	5.8	4.8	3.1	4.5	3.9	4.5	**2.9**

Distracting Control Suite. As Table 2 demonstrates, SAR again showed robust performances in selected four tasks in Distracting Control suite compared

[2] We reproduced all the results of the baselines. The results showed better than the reported performance in several tasks as more training steps [20,32,45].

to the baselines. This experiment implies that our method can also be applied in continuous control tasks and is attachable to the off-policy RL algorithms.

In this experiment, we purposely tested different baselines from Procgen to compare SAR with various algorithms. **SAC** [7] is the vanilla off-policy RL algorithm, and **CURL** [19] uses a contrastive objective for representation learning on top of SAC. **DrQ** [17] was chosen as the representative baseline using the data augmentation with additional regularization terms. **PAD** [9] adapts to a new test environment using self-supervision.[3],[4]

Table 2. The generalization results on Distracting Control Suite after training 500k timesteps. The models are evaluated in two distraction settings: `moderate` setup with the noise coefficient $\beta_{\text{cam}} = \beta_{\text{rgb}} = 0.3$ and 60 background videos, and `hard` setup with the noise coefficient $\beta_{\text{cam}} = \beta_{\text{rgb}} = 0.5$ and 60 background videos, where β_{cam} and β_{rgb} mean camera angle noise intensity and color noise intensity. The results are averaged over 3 runs with different seeds, and rank is calculated within distracted environments.

		SAC [7]	CURL [19]	DrQ [17]	PAD [9]	SAR (Ours)
`walker:walk`	zero noise	373±89	828±99	**930±23**	838±47	325±57
	`moderate`	96±10	88±11	126±33	125±27	**139±19**
	`hard`	85±8	57±7	80±11	71±8	**112±15**
`cartpole:balance`	zero noise	**996±1**	995±3	**996±3**	992±6	990±5
	`moderate`	262±20	215±57	246±15	236±17	**266±26**
	`hard`	251±12	216±62	240±26	238±22	**261±17**
`reacher:easy`	zero noise	197±7	**960±24**	844±63	671±285	177±51
	`moderate`	88±11	79±11	83±10	75±19	**98±13**
	`hard`	72±11	67±12	78±3	71±8	**93±10**
`cheetah:run`	zero noise	316±159	280±12	**332±21**	285±29	304±80
	`moderate`	**55±10**	46±8	47±8	49±5	49±11
	`hard`	**53±15**	41±8	33±13	41±10	46±13
Avg. Rank		2.125	4.625	3.125	3.625	**1.25**

Model Behavior. Figure 3 provides the learning curve of the SAR agents. They exhibit competitive sample efficiency compared to the baselines. A quantitative comparison of the models' computational complexity is in Table 3. Although the SAR model requires more parameters, it does not sacrifice much training and test time in comparison with methods using data augmentation.

[3] We reproduced all the results of the baselines and applied 'trans' to DrQ. The results with zero noise well match the reported performances in most cases [7,17,19].

[4] The performance of PAD differs from the reported value because of the simultaneous application of natural video backgrounds, color noise, and camera angle noise.

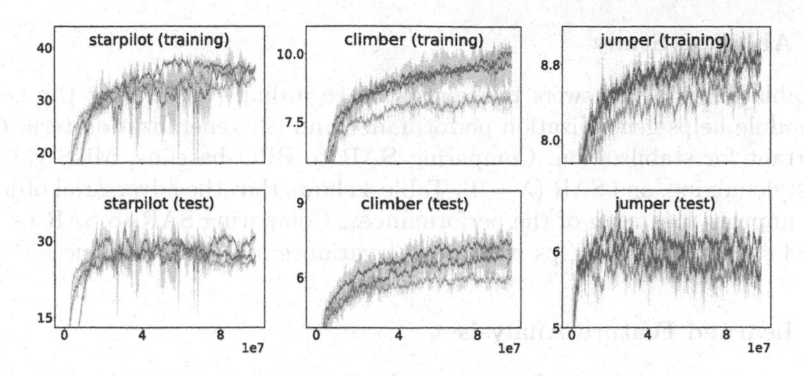

Fig. 3. The learning curve of SAR with baselines. For better visualization, we selected three models and three tasks: PPO (blue), RAD (orange), and SAR (red). Here, we applied exponential moving average smoothing with a coefficient value of 0.98. (Color figure online)

Table 3. A comparison between the number of parameters, training time, and test time. The training time refers to the time consumed for 256 timesteps and an update, and the test time is for running ten episodes in Procgen.

	PPO [33]	RAD [20](color)	UCB DrAC[32]	MixStyle [45]	DARL [24]	SAR (ours)
Parameters ($\times 10^6$)	0.626	0.626	0.626	0.626	0.678	1.151
Training time (s)	6.507	11.605	12.841	6.735	6.542	13.377
Test time (s)	2.983	2.656	3.154	2.349	2.521	3.969

With Augmentation. Training the SAR agents can be integrated with other techniques. For example, Table 4 presents the result of the SAR agents with data augmentation. Both the use of random translation and color cutout improved the performance. This result implies that the SAR agents can potentially be improved using other auxiliary tasks or regularization terms.

On Curriculum Learning. Choice of timing for adopting the min-max game, i.e., curriculum learning, can improve final generalization performances for SAR. See supplementary for the results of the experiment.

Table 4. Results on generalization performances of SAR with the application of data augmentation and ablation study in `starpilot`. SAR ($\lambda = 0$) refers to the setting without adversarial loss, and SAR ($\kappa = 0$) refers to the setting without regularization loss. We apply two different data augmentation: trans and color. The results are averaged over three runs.

	PPO [33]	MixStyle [45]	SAR ($\lambda = 0$)	SAR ($\kappa = 0$)	SAR SAR	SAR (trans)	SAR (color)
`starpilot`	27.09 ± 0.83	26.81 ± 0.89	**27.44**±2.59	**29.28** ± 7.79	28.92 ± 4.60	**30.76** ± 0.90	**33.72** ± 1.16

5.3 Ablation Study

This ablation study answers two questions regarding (1) whether the generator module helps generalization performance and (2) generalization term G_V is important for stabilization. Comparing SAR to PPO baseline, MixStyle using only style mixing, and SAR ($\lambda = 0$), Table 4 shows that the adversarial objective helps improve the mean of the performances. Comparing SAR to SAR ($\kappa = 0$), Table 4 shows that G_V helps stabilize the variance of the performances. [5]

5.4 Learned Feature Analysis

Furthermore, we qualitatively examine the features extracted from the encoder learned with the SAR objectives. The feature z we analyzed is from encoded features before entering the AdaIN layer to exclude the effect of explicit style mixing.

We demonstrate three analyses on the embedding:

- GradCAM [34] visualization for the high-level understanding interpretation.
- Reconstruction images from the feature maps.
- t-SNE [26] for analyzing the latent representations.

Visualization of Model Decision. We use GradCAM [34] to visualize where the trained agents are focusing with respect to the decisions. GradCAM can be computed by averaging the activation scores across the channels of the target convolutional layer and weighting by their gradients. Both the agent trained by the vanilla PPO and our agents predict their actions as focused on similar objects in the training environment; in the case of `starpilot`, they are focusing on the shooter and the projectiles from enemies. In the unseen test environment shown in Fig. 4, however, the vanilla PPO agent gets more distracted by the changed backgrounds and focuses on irrelevant areas in the images.

Image Reconstruction from Embedded Features. Reconstructing images from the feature maps displays a more straightforward visualization of the characteristics of the learned features. We trained a new decoder network that converts the feature maps into the original images from training environments. In Fig. 4, we show the reconstructed and original images. While the meaningful semantics, e.g., shooters or enemies, are remained, the reconstructed background seems invariant to the different original styles.

t-SNE Analysis. Li et al. [24] addressed that the distance between the embedding in the latent space may reflect the dissimilarities between the features. Thus, by observing the t-SNE [26] of the embedding from different environments,

[5] Note that the results in Table 4 are slightly different from Table 1, as we applied exponential moving average smoothing before averaging with coefficient value 0.98.

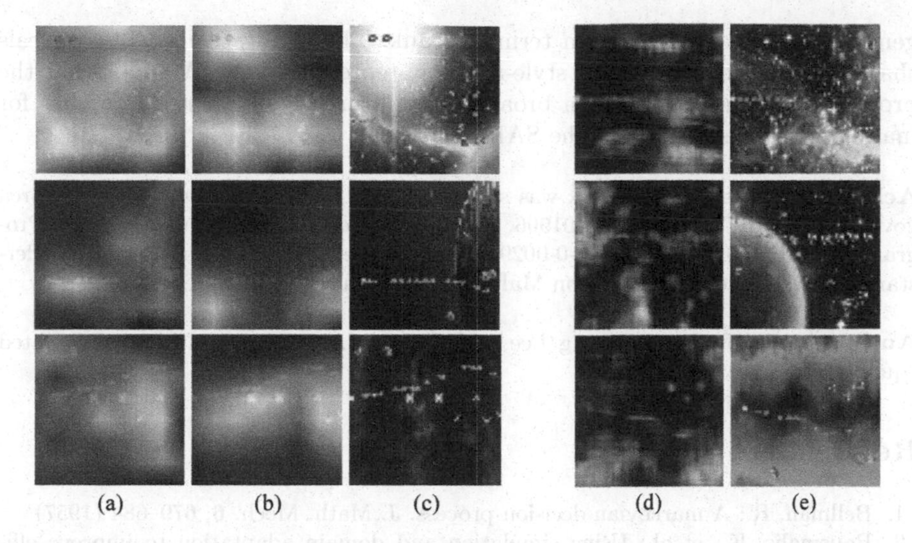

Fig. 4. GradCAM results of (a) PPO and (b) SAR, overlaid on (c) the original images from `starpilot` in Procgen. The highlighted regions represent where the agent is focusing. The SAR model better focuses on what is important with the style shifts. (d) Image reconstruction results from features extracted with the SAR agents, and (e) the original observations of `starpilot` in Procgen are displayed.

how the feature maps are correlated with the style of images can be visualized. While the features extracted from the PPO encoder are patterned with respect to the level *labels*, i.e., the styles, the SAR encoder extracts invariant embedding regardless of them. The visualization result can be seen in supplementary materials.

6 Limitation

We address the limitation of the SAR agents, mainly shown in the noise-free setting in Distracting Control Suite in Table 2, although they could well adapt to heavy noise. The additional terms in learning objectives may negatively affect the performance when there is zero noise in the test environment. Not enough styles of training environments would have also affected the actors, as they could not observe a sufficient amount of styles of training features to compete well with the generator. The generator would have taken the wrong direction for generating the new styles, and a failure in the min-max game may happen. Curriculum learning may help alleviate such concerns.

7 Conclusion

The SAR agents learn style-agnostic representations by observing features with a wide range of styles by (i) mixing with style randomization and (ii) producing from an adversarial style perturbation generator. In both Procgen and Distracting Control Suite benchmark experimentation, the SAR agents show the best generalization performances in terms of rank. The qualitative analysis reveals that the model helps to learn style-agnostic representations. We hope that the progress made here provides a broader view bringing out more techniques for many other tasks as well, as the SAR agents do.

Acknowledgement. This work was supported by IITP grant funded by the Korea government(MSIT) (No.2019-0-01906, Artificial Intelligence Graduate School Program(POSTECH) and No.2022-0-00290, Visual Intelligence for Space-Time Understanding and Generation based on Multi-layered Visual Common Sense).

Author contributions. Juyong Lee and Seokjun Ahn–These authors contributed equally to this work.

References

1. Bellman, R.: A markovian decision process. J. Math. Mech. **6**, 679–684 (1957)
2. Bousmalis, K., et al.: Using simulation and domain adaptation to improve efficiency of deep robotic grasping. In: Proceedings of the International Conference on Robotics and Automation (ICRA) (2018)
3. Cobbe, K., Hesse, C., Hilton, J., Schulman, J.: Leveraging procedural generation to benchmark reinforcement learning. In: Proceedings of the International Conference on Machine Learning (ICML) (2020)
4. Cobbe, K., Klimov, O., Hesse, C., Kim, T., Schulman, J.: Quantifying generalization in reinforcement learning. In: Proceedings of the International Conference on Machine Learning (ICML) (2019)
5. Dumoulin, V., Shlens, J., Kudlur, M.: A learned representation for artistic style. In: Proceedings of the International Conference on Learning Representations (ICLR) (2017)
6. Ha, D., Schmidhuber, J.: World models. arXiv preprint arXiv:1803.10122 (2018)
7. Haarnoja, T., Zhou, A., Abbeel, P., Levine, S.: Soft actor-critic: Off-policy maximum entropy deep reinforcement learning with a stochastic actor. In: Proceedings of the International Conference on Machine Learning (ICML) (2018)
8. Hafner, D., Lillicrap, T., Ba, J., Norouzi, M.: Dream to control: Learning behaviors by latent imagination. Proceedings of the International Conference on Learning Representations (ICLR) (2020)
9. Hansen, N., et al.: Self-supervised policy adaptation during deployment. In: Proceedings of the International Conference on Learning Representations (ICLR) (2021)
10. Hansen, N., Su, H., Wang, X.: Stabilizing deep q-learning with convnets and vision transformers under data augmentation. In: Advances in Neural Information Processing Systems (NIPS) (2021)
11. Hansen, N., Wang, X.: Generalization in reinforcement learning by soft data augmentation. In: Proceedings of the International Conference on Robotics and Automation (ICRA) (2021)
12. Higgins, I., et al.: Darla: Improving zero-shot transfer in reinforcement learning. In: Proceedings of the International Conference on Machine Learning (ICML) (2017)
13. Huang, X., Belongie, S.: Arbitrary style transfer in real-time with adaptive instance normalization. In: Proceedings of the International Conference on Computer Vision (ICCV) (2017)

14. Karras, T., Laine, S., Aila, T.: A style-based generator architecture for generative adversarial networks. In: Proceedings of the IEEE International Conference on Computer Vision and Pattern Recognition (CVPR) (2019)
15. Kirk, R., Zhang, A., Grefenstette, E., Rocktäschel, T.: A survey of generalisation in deep reinforcement learning. arXiv preprint arXiv:2111.09794 (2021)
16. Ko, B., Ok, J.: Time matters in using data augmentation for vision-based deep reinforcement learning. arXiv preprint arXiv:2102.08581 (2021)
17. Kostrikov, I., Yarats, D., Fergus, R.: Image augmentation is all you need: Regularizing deep reinforcement learning from pixels. In: Proceedings of the International Conference on Learning Representations (ICLR) (2021)
18. Laidlaw, C., Feizi, S.: Functional adversarial attacks. In: Advances in Neural Information Processing Systems (NIPS) (2019)
19. Laskin, M., Srinivas, A., Abbeel, P.: CURL: Contrastive unsupervised representations for reinforcement learning. In: Proceedings of the International Conference on Machine Learning (ICML) (2020)
20. Laskin, M., Lee, K., Stooke, A., Pinto, L., Abbeel, P., Srinivas, A.: Reinforcement learning with augmented data. In: Advances in Neural Information Processing Systems (NIPS) (2020)
21. Lee, A.X., Nagabandi, A., Abbeel, P., Levine, S.: Stochastic latent actor-critic: Deep reinforcement learning with a latent variable model. In: Advances in Neural Information Processing Systems (NIPS) (2020)
22. Lee, K., Lee, K., Shin, J., Lee, H.: Network randomization: A simple technique for generalization in deep reinforcement learning. In: Proceedings of the International Conference on Learning Representations (ICLR) (2020)
23. Levine, S., Finn, C., Darrell, T., Abbeel, P.: End-to-end training of deep visuomotor policies. J. Mach. Learn. Res. **17**, 1–40 (2016)
24. Li, B., François-Lavet, V., Doan, T., Pineau, J.: Domain adversarial reinforcement learning. arXiv preprint arXiv:2102.07097 (2021)
25. Li, H., Pan, S.J., Wang, S., Kot, A.C.: Domain generalization with adversarial feature learning. In: Proceedings of the IEEE International Conference on Computer Vision and Pattern Recognition (CVPR) (2018)
26. van der Maaten, L., Hinton, G.: Visualizing data using t-SNE. J. Mach. Learn. Res. **9**, 2579–2605 (2008)
27. Mazoure, B., Ahmed, A.M., MacAlpine, P., Hjelm, R.D., Kolobov, A.: Cross-trajectory representation learning for zero-shot generalization in rl. arXiv preprint arXiv:2106.02193 (2021)
28. Mnih, V., et al.: Playing atari with deep reinforcement learning. arXiv preprint arXiv:1312.5602 (2013)
29. Mnih, V., et al.: Human-level control through deep reinforcement learning. Nature **518**, 529–533 (2015)
30. Nam, H., Lee, H., Park, J., Yoon, W., Yoo, D.: Reducing domain gap by reducing style bias. In: Proceedings of the IEEE International Conference on Computer Vision and Pattern Recognition (CVPR) (2021)
31. Pinto, L., Davidson, J., Sukthankar, R., Gupta, A.: Robust adversarial reinforcement learning. In: Proceedings of the International Conference on Machine Learning (ICML) (2017)
32. Raileanu, R., Goldstein, M., Yarats, D., Kostrikov, I., Fergus, R.: Automatic data augmentation for generalization in deep reinforcement learning. In: Advances in Neural Information Processing Systems (NIPS) (2021)
33. Schulman, J., Wolski, F., Dhariwal, P., Radford, A., Klimov, O.: Proximal policy optimization algorithms. arXiv preprint arXiv:1707.06347 (2017)

34. Selvaraju, R.R., Cogswell, M., Das, A., Vedantam, R., Parikh, D., Batra, D.: Grad-cam: Visual explanations from deep networks via gradient-based localization. In: Proceedings of the International Conference on Computer Vision (ICCV) (2017)
35. Song, Y., Shu, R., Kushman, N., Ermon, S.: Constructing unrestricted adversarial examples with generative models. In: Advances in Neural Information Processing Systems (NIPS) (2018)
36. Stone, A., Ramirez, O., Konolige, K., Jonschkowski, R.: The distracting control suite - a challenging benchmark for reinforcement learning from pixels. arXiv preprint arXiv:2101.02722 (2021)
37. Tassa, Y., et al.: Deepmind control suite. arXiv preprint arXiv:1801.00690 (2018)
38. Tobin, J., Fong, R., Ray, A., Schneider, J., Zaremba, W., Abbeel, P.: Domain randomization for transferring deep neural networks from simulation to the real world. In: Proceedings of the IEEE/RSJ International Conference on Intelligent Robots and Systems (IROS) (2017)
39. Ulyanov, D., Vedaldi, A., Lempitsky, V.: Instance normalization: The missing ingredient for fast stylization. arXiv preprint arXiv:1607.08022 (2016)
40. Ulyanov, D., Vedaldi, A., Lempitsky, V.: Improved texture networks: Maximizing quality and diversity in feed-forward stylization and texture synthesis. In: Proceedings of the IEEE International Conference on Computer Vision and Pattern Recognition (CVPR) (2017)
41. Vinyals, O., et al.: Grandmaster level in starcraft ii using multi-agent reinforcement learning. Nature (2019)
42. Wang, Y., Qi, L., Shi, Y., Gao, Y.: Feature-based style randomization for domain generalization. arXiv preprint arXiv:2106.03171 (2021)
43. Xu, Q., Tao, G., Cheng, S., Zhang, X.: Towards feature space adversarial attack. arXiv preprint arXiv:2004.12385 (2020)
44. Zhang, C., Vinyals, O., Munos, R., Bengio, S.: A study on overfitting in deep reinforcement learning. arXiv preprint arXiv:1804.06893 (2018)
45. Zhou, K., Yang, Y., Qiao, Y., Xiang, T.: Domain generalization with mixstyle. In: Proceedings of the International Conference on Learning Representations (ICLR) (2021)
46. Zhu, Y., et al.: Target-driven visual navigation in indoor scenes using deep reinforcement learning. In: Proceedings of the International Conference on Robotics and Automation (ICRA) (2017)

Self-supervised Interactive Object Segmentation Through a Singulation-and-Grasping Approach

Houjian Yu(✉) and Changhyun Choi

Department of Electrical and Computer Engineering, University of Minnesota, Minneapolis, USA
{yu000487,cchoi}@umn.edu

Abstract. Instance segmentation with unseen objects is a challenging problem in unstructured environments. To solve this problem, we propose a robot learning approach to actively interact with novel objects and collect each object's training label for further fine-tuning to improve the segmentation model performance, while avoiding the time-consuming process of manually labeling a dataset. Given a cluttered pile of objects, our approach chooses pushing and grasping motions to break the clutter and conducts object-agnostic grasping for which the Singulation-and-Grasping (SaG) policy takes as input the visual observations and imperfect segmentation. We decompose the problem into three subtasks: (1) the object singulation subtask aims to separate the objects from each other, which creates more space that alleviates the difficulty of (2) the collision-free grasping subtask; (3) the mask generation subtask obtains the self-labeled ground truth masks by using an optical flow-based binary classifier and motion cue post-processing for transfer learning. Our system achieves 70% singulation success rate in simulated cluttered scenes. The interactive segmentation of our system achieves 87.8%, 73.9%, and 69.3% average precision for toy blocks, YCB objects in simulation, and real-world novel objects, respectively, which outperforms the compared baselines. Please refer to our project page for more information: https://z.umn.edu/sag-interactive-segmentation.

Keywords: Interactive segmentation · Reinforcement learning · Robot manipulation

1 Introduction

Instance segmentation is one of the most informative inputs to visual-based robot manipulation systems. It greatly accelerates the robotic learning process while improves the motion efficiency for target-oriented tasks [12,22,25,38]. However,

---

Supplementary Information The online version contains supplementary material available at https://doi.org/10.1007/978-3-031-19842-7_36.

S. Avidan et al. (Eds.): ECCV 2022, LNCS 13699, pp. 621–637, 2022.
https://doi.org/10.1007/978-3-031-19842-7_36

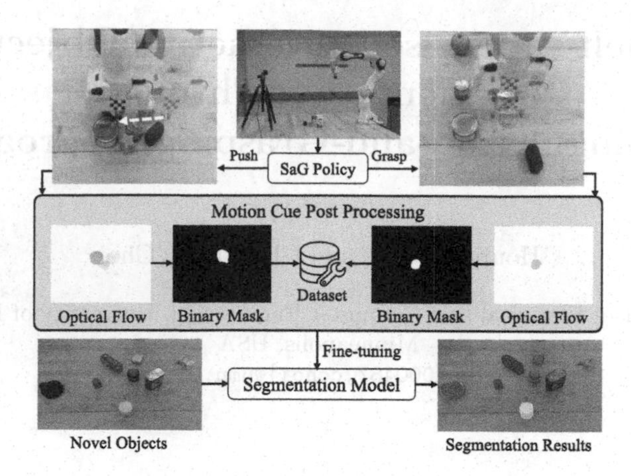

Fig. 1. The robot agent learns a Singulation-and-Grasping (SaG) policy via deep Q-learning in simulation. We collect the RGB images before and after applying the actions and use coherent motion to create pseudo ground truth masks for the segmentation transfer learning.

in real cases, robot agents frequently encounter novel objects in unstructured environments, exacerbating the accuracy of object segmentation [36,37]. In such a situation, humans often employ multiple interactions with the unknown objects and perceive object segments having consistent motions. This allows us to eventually get familiar with the novel objects and understand their shapes and contours [34]. Our work aims to enable robots to perform the same task. Given an imperfect segmentation model and unseen objects, our robot agent learns to obtain object segment labels in a self-supervised manner via object pushing and grasping interactions and then improves its segmentation model to perceive the novel objects more effectively.

Classical learning-based segmentation methods require a large amount of human-labeled training annotation, such as ImageNet [31] and MS COCO dataset [26]. While these methods have shown generalization to novel objects to some extent, they underperform when objects are out of the distribution of the trained objects. Interactive segmentation approach has taken an orthogonal avenue by actively collecting labels for novel objects using a robotic manipulator. Pathak et al. uses picking-and-placing [29] and Eitel et al. adopts pushing for singulation [10] to generate single object location displacement and obtain the ground truth label. However, these methods are limited because the simple frame difference method in [29] is noisy in label annotations and inefficient when multiple objects move simultaneously due to grasping collisions and failures. The work in [10] requires a relatively accurate segmentation method and a large amount of hand-labelled pushing actions to train their push proposal network [11] beforehand and cannot be directly applied to unseen scenes.

To address the limitations above and obtain high quality object annotations with minimal human intervention, we propose a Singulation-and-Grasping (SaG) pipeline free of laborious manual annotation to improve the segmentation results through robot-object interaction. Figure 1 shows our solution to the problem. The main contributions of our work are as follows:

- We train the Singulation-and-Grasping (SaG) policy in an end-to-end learning of a Deep Q-Network (DQN) without human annotations.
- We propose a data collection pipeline combining both the pushing and grasping motions to generate high-quality pseudo ground truth masks for unseen objects. The segmentation results after transfer learning show that our method can be used for unseen object segmentation in highly cluttered scenes.
- We evaluate our system in a real-world setting without fine-tuning the DQN, which shows the system generalization capability.

2 Related Work

Interactive Segmentation. Previous works dealing with the interactive perception problem focus on generating interactions with the environment based on objectness hypotheses and obtaining feedback after applying actions to update the segmentation results in recognition, data collection, and pose estimation tasks [2,13,20,23]. Many methods use a robot manipulator to distinguish one object from the others by applying pre-planned non-prehensile actions to specific object hypothesis [5,24,33]. However, the non-prehensile action, such as pushing, for a specific object is challenging in cluttered environments due to inevitable collisions with other surrounding objects. Our work utilizes both pushing and grasping actions to facilitate object isolation from a clutter.

The SE3-Net [3] learns to segment distinct objects from raw scene point clouds and predicts an object's rigid motion, but it only considers up to 3 objects in a less dense clutter. The closest works to our approach are Pathak et al. [29] and Eitel et al. [10] that use grasping and pushing, respectively. However, both of them require collision-free interactions to work effectively. [1] exploits motion cue to differentiate the grasped novel objects from the manipulator and background and gets single object annotation. However, in real cases, the model trained on such data will not reach high performance in heavy clutter. Our Singulation-and-Grasping (SaG) policy manages to solve the collision problem during grasping and even obtains the pseudo ground truth annotations during the singulation phase.

Pushing and Grasping Collaboration. Synergistic behaviors between pushing and grasping have been well explored in [6,9,17,39,40]. The visual pushing for grasping (VPG) [40] provides a model-free deep Q-learning framework to jointly learn pushing and grasping policies, where the pushing action is applied to facilitate future grasps. Both [17] and [39] use robust foreground segmentation methods, which track object location through interaction. In such a case, the ground truth transformation for each object can be matched, and the reward from measurements such as border occupancy ratio [9] can be designed accordingly. [9] and [6] conduct grasps by actively exploring and making rearrangements of the environment until the rule-based grasp detect algorithm or the

DQN decide whether the goal object is suitable for grasping. Analogous to these methods, the visual system in our work cannot provide robust tracking information before and after the interaction, especially when the objects are previously unseen. In such a case, the reward design for our DQN is much more challenging. Our system instead collects high quality data annotation rather than achieve a simple object removal task.

Object Singulation. Previous work [11] effectively solves the singulation problem but uses human-labeled pushing actions to train a push proposal network. On the other hand, [16] selects pushing actions to verify if visible edges correspond to proposed object boundaries without learning features. [32] and [21] focus on the target-oriented object singulation problem, however, it is much more challenging to train a singulation policy that separates all objects than a target-oriented singulation policy that separates only one target object from a clutter. As such, our approach focuses on an object-agnostic singulation problem.

3 Problem Formulation

We formulate the interactive object segmentation problem as follows:

Definition 1. *Given multiple novel objects on a planar surface, the manipulator executes pushing or grasping motion primitives based on the potentially noisy segmentation results (e.g., under- or over-segments). The goal is to improve the segmentation performance via fine-tuning with the data collected during robot-object interactions.*

To solve the interactive data collection problem, we divide the problem into three subtasks:

Subtask 1. *Given a pile of novel objects, the robot executes objects singulation motions to separate them from each other to increase free space, facilitating grasping actions later. We define this task as the* ***object singulation*** *task.*

Subtask 2. *Given a well-singulated scene where the pairwise distances of object segments are above a threshold, the robot grasps and removes objects from the scene. We define this task as the* ***collision-free grasping*** *task.*

Subtask 3. *Given the RGB images collected from the previous two subtasks, the binary segmentation masks are generated by using a learned classifier and a motion cue post-processing. We define this task as the* ***mask generation*** *task.*

4 Method

We model the problem as a discrete Markov Decision Process (MDP) as in [40] and [39]. Given a state s_t, the agent executes an action a_t according to the trained policy $\pi(s_t)$ and obtains the new state s_{t+1} receiving a current reward $R_{a_t}(s_t, s_{t+1})$. The goal of our network is to obtain an action-value function $Q_\pi(s_t, a_t)$ that approximates the expected future return for each motion a_t. We also introduce the Singulation-and-Grasping (SaG) pipeline, an interactive data collection process, from which objects annotations are self-generated.

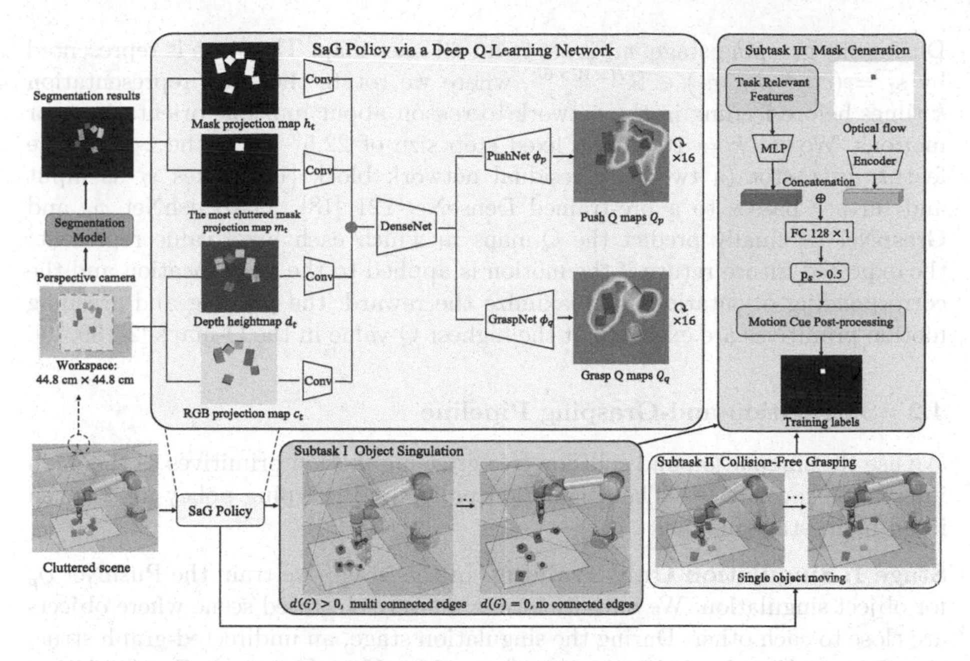

Fig. 2. The SaG pipeline for interactive data collection. The deep Q-network takes as input the state representation s_t, which consists of the orthographically projected RGB-D images (c_t, d_t) and object segmentation masks (h_t, m_t). The initially cluttered objects are singulated and grasped via the SaG policy π. Both the scenes of interaction and the task-relevant features are recorded to obtain object segment annotations. (Color figure online)

4.1 System Overview

As illustrated in Fig. 2, an RGB-D camera is affixed to the environment to provide visual information of the workspace. The original RGB-D image I_t at time t is first segmented by a segmentation model to provide objectness hypotheses. In this work, we use the UOIS segmentation model [37] taking the RGB and depth images to make inferences. By treating each segmented connected component $\{s^1, s^2, ..., s^n\}$ as a single object instance, we relabel the segmentation results and get 2D instance center locations $\{c^1, c^2, ..., c^m\}$ based on their axis-aligned bounding box coordinates.

We orthographically project the RGB, depth, segmentation hypotheses, and the most cluttered mask in the gravity direction with known camera parameters to get the color projection map $c_t \in \mathbb{R}^{H \times W \times 3}$, depth heightmap $d_t \in \mathbb{R}^{H \times W \times 1}$, mask projection map $h_t \in \mathbb{R}^{H \times W \times 1}$, and the most cluttered mask projection map $m_t \in \mathbb{R}^{H \times W \times 1}$ (see Sect. 4.2 for the details of m_t). The mask projection map h_t introduces the global clutter distribution information to the system, while the most cluttered mask projection map m_t highlights the possible target object that requires most effort to be singulated. This additional input m_t intuitively suggests removing the most cluttered area during singulation.

During the grasping stage, m_t is set to an all-ones map. The state is represented by $s_t = (c_t, d_t, h_t, m_t) \in \mathbb{R}^{H \times W \times 6k}$, where we rotate the state representation k times before feeding in the network to reason about multiple orientations for motions. We set $k = 16$ with a fixed step size of 22.5° w.r.t. the z-axis. The feature extractor (a two-layer residual network block [15]) takes s_t as input and further passes to a pre-trained DenseNet-121 [18]. The PushNet ϕ_p and GraspNet ϕ_g finally predict the Q-maps in which each pixel value represents the expected future return if the motion is applied to the pixel location and the corresponding orientation. To maximize the reward, the pushing and grasping motion primitives are executed at the highest Q-value in the Q-maps [27,39,40].

4.2 Singulation-and-Grasping Pipeline

We use the singulation and collision-free grasping motion primitives as the main interaction mechanism. The singulation policy and grasping policy are trained in an multi-stage manner:

Stage I: Singulation Only Training. In this stage, we train the PushNet ϕ_p for object singulation. We initially form a densely-cluttered scene where objects are close to each other. During the singulation stage, an undirected-graph structure $G = (V, E)$ is formed for each state s_t, where $V := \{1, ..., m\}$, $E \subset V \times V$ and each node $i \in V$ is represented by c^i. The edge E is constructed by the Euclidean distance between nodes. When the pairwise distance is under a threshold p, an edge connects two nodes. We then find the most cluttered mask from the segmentation hypothesis that corresponds to the largest number of connected edges.

To effectively train the singulation policy via the PushNet ϕ_p, we need to carefully design a reward function R_p. Existing target-oriented methods have a strong assumption that a target object can be robustly detected in the course of interactions [38,39]. We relax that assumption since the segmentation hypotheses in h_t is possibly noisy (i.e., over- or under-segments may exist) in the presence of novel objects. In that case, it is challenging to robustly segment/track objects. Instead, we employ a set of surrogate measures. To represent the degree of singulation of the scene, we obtain the graph density value [7] as

$$d(G) = \frac{2|E|}{|V|(|V|-1)} \tag{1}$$

where $|E|$ represents the number of the edges and $|V|$ is the number of the vertices. For a cluttered scene, the vertices in the graph are highly connected, and hence the density value $d(G)$ is close to one. In contrast, when objects are well singulated, the density value $d(G)$ is close to zero.

A good singulation motion is supposed to create an end-effector trajectory across the segmentation masks but also separate the under-segmented clutter, resulting in a non-decreasing number of object masks. Additionally, effective motions should increase average pairwise distance between objects and decrease the graph density $d(G)$. We also consider a two-dimensional multivariate Gaussian distribution $\mathcal{N}$ fitted to the center locations of the object segments c^i. The determinant of the covariance matrix Σ of $\mathcal{N}$ indicates the sparsity of the spatial distribution. Therefore, we design the pushing reward function as:

$$R_p = \begin{cases} -0.5, & d(G) \text{ increases} \\ 0.25, & \text{pushing passes mask } h_t \text{ and} \\ & |\{c^1, ..., c^m\}|_{t+1} \text{ non-decreases} \\ 0.5, & a_d \text{ or } a_{var} \text{ increases} \\ 1.0, & d(G) \text{ decreases or } |\Sigma| \text{ increases} \end{cases} \quad (2)$$

where a_d and a_{var} indicate the average and variance of the pairwise center location distance, respectively. $|\{c^1, ..., c^m\}|_{t+1}$ represents the number of instance masks at time $t+1$. $|\Sigma|$ represents the determinant of the covariance matrix Σ of $\mathcal{N}$.

Stage II: Grasping only Training. In this stage, the parameters of the pre-trained PushNet ϕ_p are fixed, and we mainly train the GraspNet ϕ_g with a relatively scattered scene to simulate the scenarios where objects have already been well singulated. Inspired by [40], we conduct object-agnostic grasping tasks and use the reward function as follows:

$$R_g = \begin{cases} 1.5, & \text{if grasping successfully} \\ 0, & \text{otherwise} \end{cases} \quad (3)$$

Since the previous stage has created enough space for object-agnostic grasping, training a grasp-only policy maximizes the grasping success thanks to the stage I.

Stage III: Coordination. In this stage, we combine the stage I and II as the SaG policy π for pushing and grasping collaboration. The pushing action is executed iteratively until the graph density value $d(G)$ reaches zero or grasp trials reaches the maximum pushing number, while the grasping action dominates when the objects are well singulated.

Algorithm 1 summarizes the details of the SaG policy learning in Supplementary Sect. A.1, and the training and implementation details can be found at Supplementary Sect A.2.

4.3 Mask Generation

Through SAG interactions, we self-generate object annotations to be used to improve the segmentation model. Prior work [10] has explored the similar idea, but it cannot filter out multi-object moving cases as it often generates inaccurate training labels that negatively affects the transfer learning.

We propose a learning-based binary classifier to identify single object moving cases using optical flow and apply a motion cue post-processing method on them. The classifier takes optical flow and task relevant features as input and outputs the single object moving probability. The task relevant features consist of graph density $d(G)$, average and variance of pairwise center location distance a_d and

a_{var}, and target border occupancy ratio r_b as defined in [39]. Since the training data for the flow classifier is collected from simulation only and the real robot setting has a domain gap from simulation, we consider such task relevant features to help the classification. We obtain object's ground truth location directly from the simulation (V-REP [30]) and by comparing the object locations change. We set the probability of single object movement to be 1 as the ground truth label when only one object was moving and 0 otherwise.

We compute the optical flow using the FlowNet2 [19] with images I_t and I_{t+1} before and after executing the motion primitive a_t, respectively, and feed the optical flow together with task relevant features to the classifier. Inspired by [10], we use normalized graph cut on each optical flow to obtain a set of segments in binary mask format $L_t = \{l_t^1, ..., l_t^N\}$ for frame I_t, and we select segment $l_t^n \in L_t$ that satisfies related constraints (e.g., location, size). We add the RGB image I_t and its corresponding binary mask l_t^n as a ground truth label into the training dataset $D = \{(I_0, l_0^{n_0}), ..., (I_t, l_t^{n_t})\}$ for transfer learning.

4.4 Mask R-CNN Transfer Learning

We use the Mask R-CNN [14] model pre-trained on COCO instance segmentation dataset with the ResNet-50-FPN backbone implemented by Detectron2 [35]. It is a common practice to use a baseline model pre-trained on a well-annotated standard image dataset, for instance ImageNet [31], where the backbone serves as a universal feature extractor in the network. Moreover, the Feature Pyramid Network (FPN) [14] type backbone extracts image features from different scales, which provides better anchors prediction in various levels. We fine-tune the segmentation model with the self-generated dataset D. The details of segmentation results can be found at Sect. 5.

5 Experiments

In this section, we conduct multiple experiments to evaluate the proposed **SaG** approach. The goals of the experiments are 1) to compare our **SaG** with several baselines in both singulation and segmentation performances and 2) to show whether the fine-tuned **SaG** segmentation model is effective and further applicable to in other downstream robot manipulation tasks (e.g., grasping).

5.1 Datasets and Evaluation Metrics

Singulation. We evaluate our singulation performance in simulation with 6 basic shape toy blocks. We conduct 200 test trials with various object arrangements for the singulation task and record $d(G)$ values. A trial is considered to be successful when the pairwise distances between all objects are above a threshold p and $d(G)$ reaches zero within 8 pushes.

Segmentation. We collect 404 and 200 testing images for toy blocks and YCB [4] objects in simulation. For default test setting, we randomly drop 10 toy blocks and 8 YCB objects in the workspace. Additionally, for cluttered test setting, we increase the number of objects to 18 toy blocks and 15 YCB objects where all objects are located in a dense pile. For real robot default testing, we manually labeled 100 images with 6 to 8 objects in the workspace. In addition, 50 images are labeled with up to 16 objects in a clutter for the cluttered test cases.

We evaluate the instance segmentation performance with the standard MS COCO evaluation metric, average precision (AP). We also use another evaluation metric as defined in [8] to compare current state-of-the-art non-interactive segmentation method, where scores for segmentation instances are not provided and cannot be evaluated with COCO AP. To compute the overlap precision, recall, and F-measure (P/R/F), the Hungarian matching method is used for the predicted and ground truth masks. Given the matching, the P/R/F are computed by $P = \frac{\sum_i |a_i \cap g(a_i)|}{\sum_i |a_i|}$, $R = \frac{\sum_i |a_i \cap g(a_i)|}{\sum_j |g_j|}$, $F = \frac{2PR}{P+R}$, where a_i denotes the set of pixels belonging to predicted object i, $g(a_i)$ is the ground truth matched to each predicted region, and g_j denotes ground truth pixels of object j.

5.2 Singulation Performance

We utilize the simulation environment in V-REP [30] running a UR5 arm with an RG2 gripper. Five baselines are compared with our approach: 1) **SaG-maskall**, the baseline without the most cluttered object mask projection map m_t and filters the Q maps by the binary mask projection map h_t, 2) **SaG-maskone**, the baseline without the m_t input and filters the Q maps with m_t, 3) **VPG**, target agnostic pushing and grasping to clean the cluttered objects [40]. 4) **SaG no m_t**, our proposed method without m_t, and 5) **SaG no h_t**, our proposed method without h_t.

Figure 3 shows the average singulation success rate with different pairwise distance thresholds p from 6 cm to 10 cm versus the number of pushes. The pushing motions achieve about 80% singulation success rate with the small threshold of $6cm$ and over 50% with the large threshold of 10 cm after eight pushes.

To show that **SaG** is effective, we further conduct the five push-only baseline comparisons mentioned above. We reuse the test cases when evaluating **SaG** singulation success for each method. The average singulation success rate for each baseline combines performance measurement with thresholds from 6 cm to 10 cm. Figure 4 demonstrates that the **VPG** push-only method barely has the object singulation effect. On the other hand, our proposed approach improves the performance by a large margin about 60% after eight pushes. Although the singulation task is target-agnostic, the results of **SaG no m_t** and **SaG no h_t** show that providing the network with global and local clutter information helps improve the overall performance. **SaG-maskall** may push objects that have already been well singulated, resulting in an ineffective pushing policy. While **SaG-maskone** always pushes the most cluttered mask, it lacks the global object arrangement information, resulting in unsatisfactory performance as well.

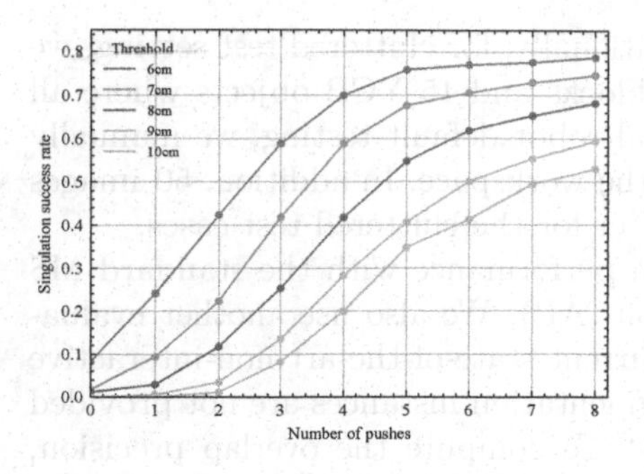

Fig. 3. Singulation policy performance with different distance thresholds.

Fig. 4. Singulation success rate for different baselines.

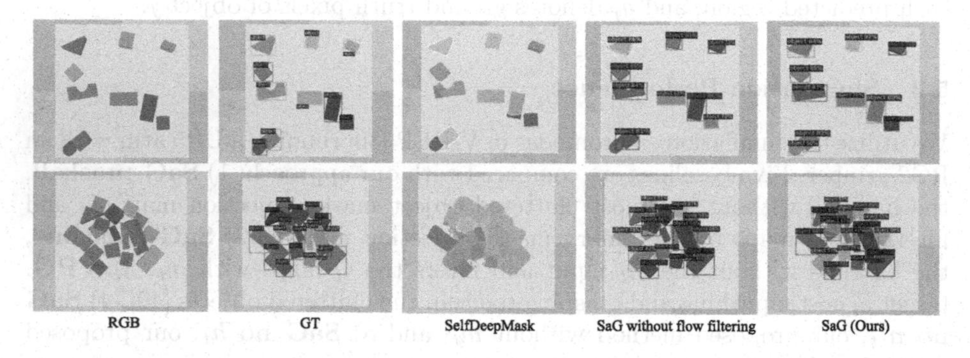

Fig. 5. Basic toy blocks segmentation qualitative results. The top row is related to a default test case and the bottom row is the highly cluttered test case.

5.3 Interactive Segmentation

We evaluate the instance segmentation performance with transfer learning in simulation. The Detectron2 COCO instance segmentation model with the ResNet-50-FPN backbone [35] is used in our experiments. We fine-tune the model for 200, 250, and 150 iterations with 2000 toy blocks interactions, 2000 YCB objects interactions, and 1000 real robot novel objects interactions. Our models are trained with the initial learning rate of 0.0005 with SGD for optimization. Weight decay and momentum are set as 0.0001 and 0.9. All our models are trained on a single NVIDIA RTX 2080 Ti.

Note that the push proposal network in **SelfDeepMask** [10] and the complete data collection pipeline of **Seg-by-Interaction (SBI)** [29] are not released. We instead prepared the training data with our SaG policy and fine-tuned their corresponding segmentation models. While the baselines comparison in such a way can be slightly unfair since we could not use their data collection methods, we followed their hyperparameter setting and the loss function

Table 1. Segmentation results in simulation with toy blocks. Ablation study on different architectures of SaG pipeline.

Method	*OPF*	Default			Cluttered		
		AP_{50}	AP_{75}	$AP_{50:95}$	AP_{50}	AP_{75}	$AP_{50:95}$
SaG (ours)	✓	**98.7**	**96.6**	**87.8**	**88.6**	**81.0**	**73.0**
SaG (ours)		96.8	94.7	81.5	87.8	80.6	72.1
SaG no m_t	✓	98.3	95.9	81.6	86.1	78.2	69.1
SaG no h_t	✓	93.2	90.5	79.0	82.6	73.1	65.4
SaG grasp	✓	97.1	94.5	83.4	86.8	79.7	71.7
SaG push	✓	96.7	94.1	80.1	85.7	77.0	69.7
SaG push		94.2	92.0	77.7	85.5	76.9	68.2
SaG-maskone	✓	98.0	95.7	86.0	87.3	77.9	70.5
SaG-maskall	✓	97.9	95.7	86.0	85.6	77.1	69.5
VPG [40]	✓	89.2	86.3	69.4	81.4	72.1	65.0
SelfDeepMask [10]	✓	77.4	66.4	53.7	52.0	32.0	29.9
SelfDeepMask [10]		74.6	62.1	50.4	43.9	26.9	24.6
DeepMask [28]	✓	71.9	48.0	41.1	50.4	31.4	29.3
SBI [29]		72.1	54.2	45.6	52.8	29.1	27.3

selections. For **DeepMask** [28] method, we fine-tune the pre-trained ResNet-50 *DeepMask* model for 10 epochs.

Toy Blocks Segmentation. The quantitative results are in Table 1. We use the standard COCO instance segmentation average precision (AP) for segmentation evaluation. In Table 1, *OPF* denotes the use of optical flow filtering classifier. Our proposed approach combining pushing and grasping interactions provides the optimal performance of 87.8% in default setting and 73.0% in highly cluttered setting, both in $AP_{50:95}$. The **SaG push** method has relatively low performance since the singulation policy often moves multiple objects simultaneously, creating

Table 2. Segmentation results in simulation with YCB objects [4].

Method	*OPF*	Default			Cluttered		
		AP_{50}	AP_{75}	$AP_{50:95}$	AP_{50}	AP_{75}	$AP_{50:95}$
SaG (ours)	✓	**92.9**	**82.8**	**73.9**	**84.7**	**66.8**	**61.7**
SaG (ours)		91.7	82.7	73.1	81.5	65.1	58.7
SelfDeepMask [10]	✓	72.0	37.8	38.7	52.6	22.8	25.6
SelfDeepMask [10]		70.4	39.1	38.6	51.6	23.3	25.6
DeepMask [28]	✓	69.8	28.2	33.6	51.3	18.9	23.6
SBI [29]		68.3	26.5	32.2	47.7	14.9	20.8

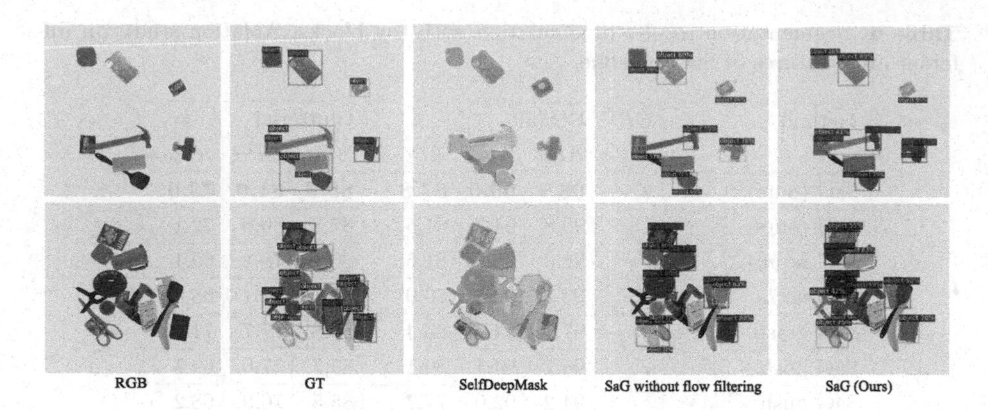

Fig. 6. YCB objects segmentation qualitative results. The top row shows default test setting and bottom row is the cluttered test case.

noisy labels. Our approach outperforms the compared baselines [10,28,29,40] by large margins. Figure 5 shows the visualization results.

YCB Object Segmentation. We also compare baselines with more challenging objects in heavy clutter. The quantitative results are in Table 2. Our method achieves 73.9% and 61.7% $AP_{50:90}$ in default and cluttered test sets, respectively, and outperforms other baselines by large margins. Segmentation visualization can be found in Fig. 6.

5.4 SaG Downstream Robot Task Application

We evaluate the robotic top-down grasping task as one of the downstream tasks in simulation. The grasping performance is defined as the grasping success rate (%) over the last 1000 attempts. We compare the grasping success rate with VPG [40] (no segmentation input at all), UOIS-VPG, and SaG-VPG, where UOIS-VPG uses UOIS [37] as the segmentation model and SaG-VPG uses the fine-tuned Mask-RCNN to provide object binary masks information as an additional input. All three methods execute the grasping actions only and were trained from scratch for 1000 epochs with 6 toy blocks. The experiment results in Table 3 show that the SaG based segmentation model improves the grasping performance more effectively.

Table 3. Top-down grasping success rate in simulation

Settings	SaG-VPG	UOIS-VPG	VPG
6 toy blocks	**85.5**	78.9	70.7
10 toy blocks	**78.4**	73.8	61.3

Table 4. Segmentation results on real robot.

Method	*OPF*	Default			Cluttered		
		AP_{50}	AP_{75}	$AP_{50:95}$	AP_{50}	AP_{75}	$AP_{50:95}$
SaG (ours)	✓	**94.6**	**87.0**	**69.3**	84.3	**78.4**	**63.2**
SaG (ours)		88.1	80.0	61.1	**85.3**	78.3	62.5
SelfDeepMask [10]	✓	79.9	57.7	51.3	70.7	51.9	43.8
SelfDeepMask [10]		74.1	53.3	47.1	69.3	50.6	43.1
DeepMask [28]	✓	72.8	39.5	38.8	66.5	50.0	39.6

Table 5. SOTA comparison with the non-interactive approach on highly cluttered unseen object segmentation.

Method	Overlap			Boundary		
	P	R	F	P	R	F
SaG (ours)	**91.4**	**89.5**	**90.4**	**79.3**	**81.2**	**80.1**
UOIS [37]	70.8	76.7	73.6	38.5	73.6	50.3

5.5 Real Robot Experiments

Fig. 7. Training and default testing object sets in real-robot experiments. The testing objects are never seen during training process.

We collect the training data via our SaG pipeline with a Franka Emika Panda robot. The **SaG** policy is only trained in simulation and not fine-tuned in real-robot setting. There are 41 different objects in the training set and 25 novel objects in the default testing set as shown in Fig. 7.

The quantitative segmentation results with different AP thresholds are in Table 4. Our approach outperforms **SelfDeepMask** and **DeepMask** by 14.7 and 21.8 average precision points (AP_{50}) on default test cases, respectively. We also compare our method with pre-trained UOIS [37] non-interactive approach in Table 5.

The qualitative visualizations are in Fig. 8 and Fig. 9. The results show that our segmentation model can be generalized to novel objects. Even in a cluttered

Fig. 8. Highly cluttered test set visualization in real robot experiments.

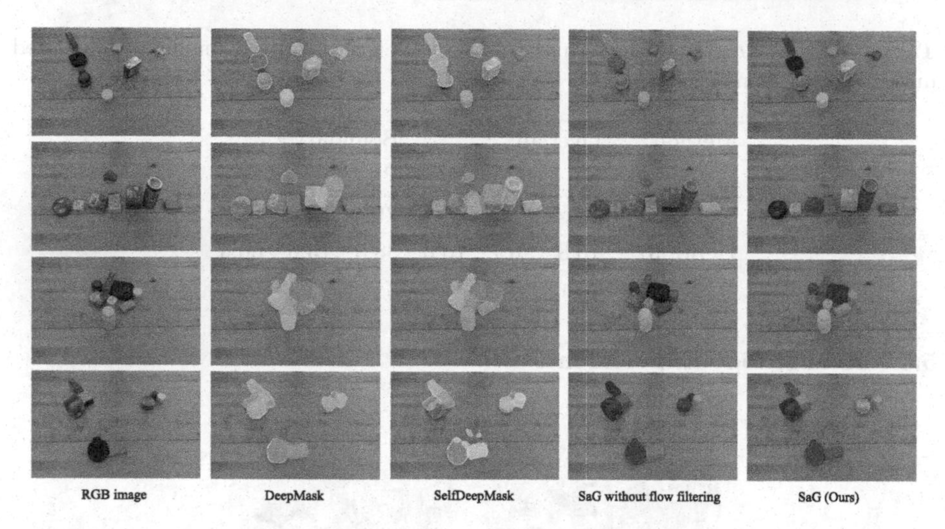

Fig. 9. Visualizations of test cases (sparsely distributed, a cluttered scene, and piles of objects).

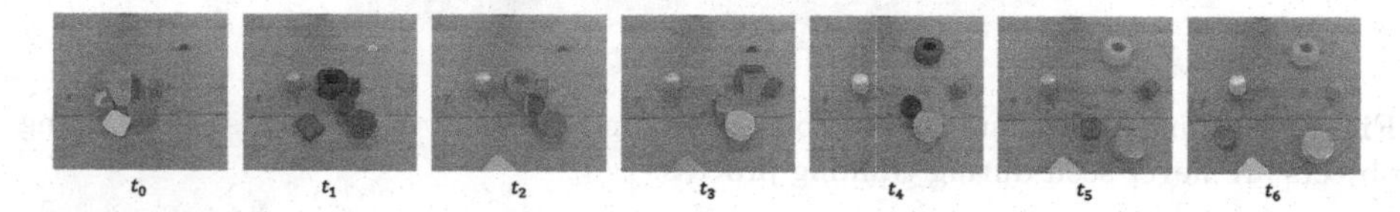

Fig. 10. Instance segmentation visualization with increasing number of pushes. Our singulation policy helps improve the segmentation results by breaking the clutter.

scene, our model manages to segment individual objects. Figure 10 shows the interactive segmentation in a push experiment, where the singulation motion separating objects further contributes to the better segmentation performance.

6 Conclusions

We presented an interactive object segmentation method through the SaG policy learning in the end-to-end deep Q-learning. The robot interacted with unseen objects using pushing and grasping actions and automatically generated pseudo ground truth annotations for further transfer learning. We showed our approach outperforms all the compared baselines by large margins in both simulation and real-robot experiments. Additionally, the proposed approach was applied to a downstream robot manipulation task, object grasping.

Although effective, the current optical-flow based classifier lowers the data collection efficiency. A future direction would learn a motion grouping method that directly provides multiple ground truth masks from optical flow.

Acknowledgements. This work was supported in part by the Sony Research Award Program and NSF Award 2143730.

References

1. Boerdijk, W., Sundermeyer, M., Durner, M., Triebel, R.: Self-supervised object-in-gripper segmentation from robotic motions. arXiv preprint arXiv:2002.04487 (2020)
2. Bohg, J.: Interactive perception: Leveraging action in perception and perception in action. IEEE Trans. Rob. **33**(6), 1273–1291 (2017)
3. Byravan, A., Fox, D.: Se3-nets: Learning rigid body motion using deep neural networks. In: 2017 IEEE International Conference on Robotics and Automation (ICRA), pp. 173–180. IEEE (2017)
4. Calli, B., Walsman, A., Singh, A., Srinivasa, S., Abbeel, P., Dollar, A.M.: Benchmarking in manipulation research: Using the yale-cmu-berkeley object and model set. IEEE Robotics Autom. Mag. **22**(3), 36–52 (2015). https://doi.org/10.1109/MRA.2015.2448951
5. Chaudhary, K., et al.: Retrieving unknown objects using robot in-the-loop based interactive segmentation. In: 2016 IEEE/SICE International Symposium on System Integration (SII), pp. 75–80. IEEE (2016)
6. Chen, Y., Ju, Z., Yang, C.: Combining reinforcement learning and rule-based method to manipulate objects in clutter. In: 2020 International Joint Conference on Neural Networks (IJCNN), pp. 1–6. IEEE (2020)
7. Coleman, T.F., Moré, J.J.: Estimation of sparse Jacobian matrices and graph coloring problems. SIAM J. Numer. Anal. **20**(1), 187–209 (1983)
8. Dave, A., Tokmakov, P., Ramanan, D.: Towards segmenting anything that moves. In: Proceedings of the IEEE/CVF International Conference on Computer Vision (ICCV) Workshops, Oct 2019
9. Deng, Y., et al.: Deep reinforcement learning for robotic pushing and picking in cluttered environment. In: 2019 IEEE/RSJ International Conference on Intelligent Robots and Systems (IROS), pp. 619–626. IEEE (2019)
10. Eitel, A., Hauff, N., Burgard, W.: Self-supervised transfer learning for instance segmentation through physical interaction. In: 2019 IEEE/RSJ International Conference on Intelligent Robots and Systems (IROS), pp. 4020–4026. IEEE (2019)
11. Eitel, A., Hauff, N., Burgard, W.: Learning to singulate objects using a push proposal network. In: Amato, N.M., Hager, G., Thomas, S., Torres-Torriti, M. (eds.) Robotics Research. SPAR, vol. 10, pp. 405–419. Springer, Cham (2020). https://doi.org/10.1007/978-3-030-28619-4_32

12. Fang, K., Bai, Y., Hinterstoisser, S., Savarese, S., Kalakrishnan, M.: Multi-task domain adaptation for deep learning of instance grasping from simulation. In: 2018 IEEE International Conference on Robotics and Automation (ICRA), pp. 3516–3523. IEEE (2018)
13. Fitzpatrick, P.: First contact: an active vision approach to segmentation. In: Proceedings 2003 IEEE/RSJ International Conference on Intelligent Robots and Systems (IROS 2003) (Cat. No. 03CH37453), vol. 3, pp. 2161–2166. IEEE (2003)
14. He, K., Gkioxari, G., Dollár, P., Girshick, R.: Mask r-cnn. In: Proceedings of the IEEE International Conference on Computer Vision, pp. 2961–2969 (2017)
15. He, K., Zhang, X., Ren, S., Sun, J.: Deep residual learning for image recognition. In: Proceedings of the IEEE Conference on Computer Vision and Pattern Recognition, pp. 770–778 (2016)
16. Hermans, T., Rehg, J.M., Bobick, A.: Guided pushing for object singulation. In: 2012 IEEE/RSJ International Conference on Intelligent Robots and Systems, pp. 4783–4790. IEEE (2012)
17. Huang, B., Han, S.D., Boularias, A., Yu, J.: Dipn: Deep interaction prediction network with application to clutter removal. In: 2021 IEEE International Conference on Robotics and Automation (ICRA), pp. 4694–4701. IEEE (2021)
18. Huang, G., Liu, Z., Van Der Maaten, L., Weinberger, K.Q.: Densely connected convolutional networks. In: Proceedings of the IEEE Conference on Computer Vision and Pattern Recognition, pp. 4700–4708 (2017)
19. Ilg, E., Mayer, N., Saikia, T., Keuper, M., Dosovitskiy, A., Brox, T.: Flownet 2.0: Evolution of optical flow estimation with deep networks. In: Proceedings of the IEEE Conference on Computer Vision and Pattern Recognition, pp. 2462–2470 (2017)
20. Kenney, J., Buckley, T., Brock, O.: Interactive segmentation for manipulation in unstructured environments. In: 2009 IEEE International Conference on Robotics and Automation, pp. 1377–1382. IEEE (2009)
21. Kiatos, M., Malassiotis, S.: Robust object grasping in clutter via singulation. In: 2019 International Conference on Robotics and Automation (ICRA), pp. 1596–1600. IEEE (2019)
22. Kurenkov, A., et al.: Visuomotor mechanical search: Learning to retrieve target objects in clutter. In: 2020 IEEE/RSJ International Conference on Intelligent Robots and Systems (IROS), pp. 8408–8414. IEEE (2020)
23. Kuzmič, E.S., Ude, A.: Object segmentation and learning through feature grouping and manipulation. In: 2010 10th IEEE-RAS International Conference on Humanoid Robots, pp. 371–378. IEEE (2010)
24. Le Goff, L.K., Mukhtar, G., Le Fur, P.H., Doncieux, S.: Segmenting objects through an autonomous agnostic exploration conducted by a robot. In: 2017 First IEEE International Conference on Robotic Computing (IRC), pp. 284–291. IEEE (2017)
25. Liang, H., Lou, X., Yang, Y., Choi, C.: Learning visual affordances with target-orientated deep q-network to grasp objects by harnessing environmental fixtures. In: 2021 IEEE International Conference on Robotics and Automation (ICRA), pp. 2562–2568. IEEE (2021)
26. Lin, T.-Y., et al.: Microsoft coco: Common objects in context. In: Fleet, D., Pajdla, T., Schiele, B., Tuytelaars, T. (eds.) ECCV 2014. LNCS, vol. 8693, pp. 740–755. Springer, Cham (2014). https://doi.org/10.1007/978-3-319-10602-1_48
27. Mnih, V., et al.: Human-level control through deep reinforcement learning. Nature **518**(7540), 529 (2015)
28. O Pinheiro, P.O., Collobert, R., Dollár, P.: Learning to segment object candidates. In: Advances in Neural Information Processing Systems, vol. 28 (2015)

29. Pathak, D., et al.: Learning instance segmentation by interaction. In: Proceedings of the IEEE Conference on Computer Vision and Pattern Recognition Workshops, pp. 2042–2045 (2018)
30. Rohmer, E., Singh, S.P., Freese, M.: V-rep: A versatile and scalable robot simulation framework. In: 2013 IEEE/RSJ International Conference on Intelligent Robots and Systems, pp. 1321–1326. IEEE (2013)
31. Russakovsky, O., et al.: Imagenet large scale visual recognition challenge. Int. J. Comput. Vision **115**(3), 211–252 (2015)
32. Sarantopoulos, I., Kiatos, M., Doulgeri, Z., Malassiotis, S.: Split deep q-learning for robust object singulation. In: 2020 IEEE International Conference on Robotics and Automation (ICRA), pp. 6225–6231. IEEE (2020)
33. Schiebener, D., Ude, A., Asfour, T.: Physical interaction for segmentation of unknown textured and non-textured rigid objects. In: 2014 IEEE International Conference on Robotics and Automation (ICRA), pp. 4959–4966. IEEE (2014)
34. Spelke, E.S.: Principles of object perception. Cogn. Sci. **14**(1), 29–56 (1990)
35. Wu, Y., Kirillov, A., Massa, F., Lo, W.Y., Girshick, R.: Detectron2. https://github.com/facebookresearch/detectron2 (2019)
36. Xie, C., Xiang, Y., Mousavian, A., Fox, D.: The best of both modes: Separately leveraging rgb and depth for unseen object instance segmentation. In: Conference on Robot Learning, pp. 1369–1378. PMLR (2020)
37. Xie, C., Xiang, Y., Mousavian, A., Fox, D.: Unseen object instance segmentation for robotic environments. IEEE Trans. Robot. 1–17 (2021)
38. Xu, K., Yu, H., Lai, Q., Wang, Y., Xiong, R.: Efficient learning of goal-oriented push-grasping synergy in clutter. IEEE Robot. Autom. Lett. **6**(4), 6337–6344 (2021)
39. Yang, Y., Liang, H., Choi, C.: A deep learning approach to grasping the invisible. IEEE Robot. Autom. Lett. **5**(2), 2232–2239 (2020)
40. Zeng, A., Song, S., Welker, S., Lee, J., Rodriguez, A., Funkhouser, T.: Learning synergies between pushing and grasping with self-supervised deep reinforcement learning. In: 2018 IEEE/RSJ International Conference on Intelligent Robots and Systems (IROS), pp. 4238–4245. IEEE (2018)

Learning from Unlabeled 3D Environments for Vision-and-Language Navigation

Shizhe Chen[1(✉)], Pierre-Louis Guhur[1], Makarand Tapaswi[2], Cordelia Schmid[1], and Ivan Laptev[1]

[1] Inria, École normale supérieure, CNRS, PSL Research University, Paris, France
shizhe.chen@inria.fr
[2] IIIT Hyderabad, Hyderabad, India
https://cshizhe.github.io/projects/hm3d_autovln.html

Abstract. In vision-and-language navigation (VLN), an embodied agent is required to navigate in realistic 3D environments following natural language instructions. One major bottleneck for existing VLN approaches is the lack of sufficient training data, resulting in unsatisfactory generalization to unseen environments. While VLN data is typically collected manually, such an approach is expensive and prevents scalability. In this work, we address the data scarcity issue by proposing to automatically create a large-scale VLN dataset from 900 unlabeled 3D buildings from HM3D [45]. We generate a navigation graph for each building and transfer object predictions from 2D to generate pseudo 3D object labels by cross-view consistency. We then fine-tune a pretrained language model using pseudo object labels as prompts to alleviate the cross-modal gap in instruction generation. Our resulting HM3D-AutoVLN dataset is an order of magnitude larger than existing VLN datasets in terms of navigation environments and instructions. We experimentally demonstrate that HM3D-AutoVLN significantly increases the generalization ability of resulting VLN models. On the SPL metric, our approach improves over state of the art by 7.1% and 8.1% on the unseen validation splits of REVERIE and SOON datasets respectively.

Keywords: Vision-and-language · Navigation · 3D Environments

1 Introduction

Having a robot carry out various chores has been a common vision from science fiction. Such a long-term goal requires an embodied agent to understand our human language, navigate in the physical environment and interact with objects. As an initial step towards this goal, the vision-and-language navigation task (VLN) [3] has emerged and attracted growing research attention. Early

---

Supplementary Information The online version contains supplementary material available at https://doi.org/10.1007/978-3-031-19842-7_37.

S. Avidan et al. (Eds.): ECCV 2022, LNCS 13699, pp. 638–655, 2022.
https://doi.org/10.1007/978-3-031-19842-7_37

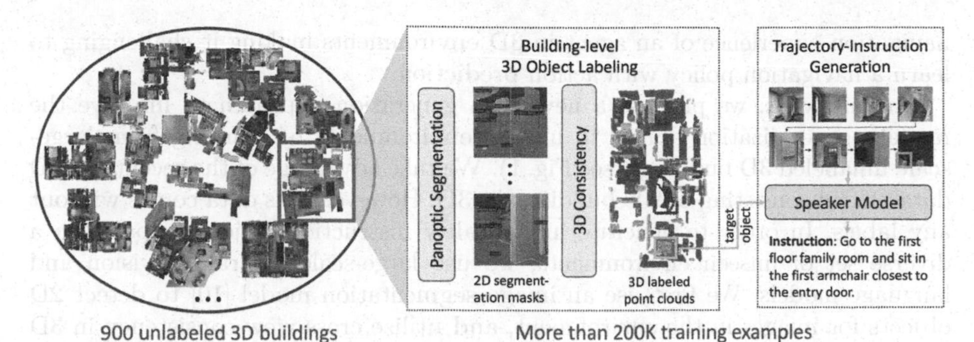

Fig. 1. HM3D-AutoVLN dataset. We use 900 unlabeled 3D buildings from the HM3D [45] dataset. We improve labels obtained with a 2D segmentation model based on 3D cross-view consistency (middle, see Fig. 2). Then we rely on these pseudo labels to generate instructions via a speaker model (right, see Fig. 3). We automatically create over 200K realistic training samples for VLN.

VLN approaches [3,28] provide agents with step-by-step navigation instructions to arrive at a target location, such as *"Walk out of the bedroom. Turn right and walk down the hallway. At the end of the hallway turn left. Walk in front of the couch and stop"*. While these detailed instructions reduce the difficulty of the task, they lower the practical value for people when commanding robots in real life. Thus, more recent VLN methods [43,57] focus on high-level instructions and request an agent to find a specific object at the target location, *e.g.*, *"Go to the living room and bring me the white cushion on the sofa closest to the lamp"*. The agent needs to explore 3D environments on its own to find the *"cushion"*.

Compared to step-by-step instructions, following such high-level goal-driven instructions is more challenging. As there is no detailed guidance, an agent needs to learn about the structure of the environment for effective exploration. However, most existing VLN tasks such as REVERIE [43] or SOON [57] are based on 3D scans from the Matterport3D (MP3D) dataset [4], and contain less than 60 buildings and approximately 10K training trajectories. The limited amount of training data makes VLN models overfit to seen environments, resulting in less generalizable navigation policies in unseen environments. Manually collecting more VLN data is however expensive and not scalable. To address this data scarcity issue, previous works have investigated various data augmentation methods, such as synthesizing more instructions and trajectories in seen environments by a speaker model [16], environment dropout [49] or editing [30], and mixing up seen environments [33]. Nevertheless, these approaches are still based on a small amount of 3D environments that cannot cover a wide range of objects and scenes. To address visual diversity, VLN-BERT [36] utilizes image-caption pairs from the web to improve generalization ability, while Airbert [18] shows that pairs from indoor environments, the BnB dataset, are more beneficial for VLN tasks. However, it is hard for image-caption pairs to mimic the real navigation experience of an agent in 3D environments making it challenging to learn a navigation policy with action prediction.

In this work, we propose a new data generation approach to improve the model's generalization ability to unseen environments by learning from large-scale unlabeled 3D buildings (see Fig. 1). We take advantage of the recent HM3D dataset [45] consisting of 900 buildings in 3D. However, this data comes without any labels. In order to generate high-quality instruction-trajectory pairs on a diverse set of unseen environments, we use large-scale pretrained vision and language models. We first use an image segmentation model [10] to detect 2D objects for images in the environment, and utilize cross-view consistency in 3D to increase the accuracy of pseudo 3D object annotations. Then, we fine-tune a language model, GPT-2 [44], with pseudo object labels as prompts to generate high-level navigation instructions to this object. In this way, we construct the HM3D-AutoVLN dataset that uses 900 3D buildings, and consists of 36,562 navigable nodes, 172,000 3D objects and 217,703 object-instruction-trajectory triplets for training – an order of magnitude larger than prior VLN datasets. We train multiple state-of-the-art VLN models [8,9,22,49] with the generated HM3D-AutoVLN data and show significant gains. Specifically, we improve over the state-of-the-art DUET model [9] on REVERIE and SOON datasets by 7.1% and 8.1% respectively. In summary, our contributions are as follows:

- We introduce an automatic approach to construct a large-scale VLN dataset, HM3D-AutoVLN, from unlabeled 3D buildings. We rely on 2D image models to obtain pseudo 3D object labels and on pretrained language models to generate instructions.
- We carry out extensive experiments on two challenging VLN tasks, REVERIE and SOON. The training on HM3D-AutoVLN dataset significantly improves the performance for multiple state-of-the-art VLN models.
- We provide insights on data collection and challenges inherent to leveraging unlabeled environments. It suggests that the diversity of environments is more important than the number of training samples alone.

2 Related Work

Vision-and-Language Navigation. The VLN tasks have recently been popularized with the emergence of various supportive datasets [3,6,23,27,28,39,48,50]. Different instructions define the variations of VLN tasks. Step-by-step instructions [3,6,28] require an agent to strictly follow the path, whereas goal-driven high-level instructions [43,57] mainly describe the target place and object and command the agent to retrieve a particular remote object. The embodied question answering task asks an agent to navigate and answer a question [12,55], and vision-and-language dialog [13,38,39,50] tasks use interactive communications with agents to guide the navigation. Due to the inherent multimodal nature, works on VLN tasks provide appealing and creative model architectures, such as cross-modal attention mechanism [16,34,35,49,52], awareness of objects [20,37,42], sequence modeling using transformers [8,22,40], Bayesian state tracking [1], and graph-based structures for better exploration [9,14,51]. The models are usually pretrained in a teacher forcing manner [8,19] and then fine-tuned using student forcing [3,9], stochastic sampling [31], or reinforcement learning [8,22,49,53]. While most existing VLN works focus on discrete environment with predefined navigation graphs, VLN in continuous environments [26,27] is more practical in real world [2]. The discrete environments also prove to be beneficial for continuous VLN such as providing waypoint supervision [46] and enabling hierarchical modeling of low-level and high-level actions [7,21].

Data-Centric VLN Approaches. One of the major challenges of VLN remains the scarcity of training data, leading to a large gap between seen and unseen environments. Data augmentation is one effective approach to address over-fitting, such as to dropout environment features [49], change image styles [30], mixup rooms in different environments [33], generate more path-instruction pairs from speaker models [16,17,49] and new images from GANs [24]. However, those data augmentation methods are still based on a limited number of environments, abating generalization ability on unseen environments. Contrary to the above, VLNBert [36] and Airbert [18] exploit abundant web image-captions to improve generalization. Nevertheless, such data do not fully resemble real navigation experience and thus they can only be used to train a compatibility model to measure path-instruction similarity instead of learning a navigation policy. In our work, photo-realistic 3D environments are used to learn a navigation policy.

3D Environments. 3D environments and simulation platforms [25,29,41,47] are a basic foundation to promote research in embodied intelligence. There are artificial environments based on video game engines such as iThor [25] and VirtualHome [41], which utilize synthetic scenes and allow interactions with objects. However, synthetic scenes have limited visual diversity and do not fully reflect the real world. Thus, we focus on photo-realistic 3D environments for VLN. The MP3D [4] dataset is a collection of 90 labeled environments and is most widely adopted in existing VLN datasets such as R2R [3], RxR [28], REVERIE [43] and SOON [57]. Though possessing high-quality reconstructions, the number of houses in MP3D is limited. The Gibson [54] dataset contains 571 scenes, however they have poor 3D reconstruction quality. The recent HM3D dataset [45] has the largest number of 3D environments though with no labels. It contains 1,000 high-quality and diverse building-scale reconstructions (900 publicly released) around the world such as multi-floor residences, stores, offices, *etc.* In this work, we employ the unlabeled HM3D dataset to scale up VLN datasets.

3 Generating VLN Dataset from Unlabeled 3D Buildings

In this section, we present our approach to automatically generate large-scale VLN data from unlabeled 3D environments, specifically HM3D [45]. We consider a discrete navigation environment [3,43,57], which is the mainstream setting for VLN. The discrete setup treats each environment as an undirected graph $\mathcal{G} = \{\mathcal{V}, \mathcal{E}\}$, where $\mathcal{V}$ denotes navigable nodes, and $\mathcal{E}$ denotes connectivity edges. Each node $V_i \in \mathcal{V}$ corresponds to a panoramic RGB-D image captured at its location (x, y, z). An agent is equipped with a camera and GPS sensors and moves between connected nodes on this navigation graph.

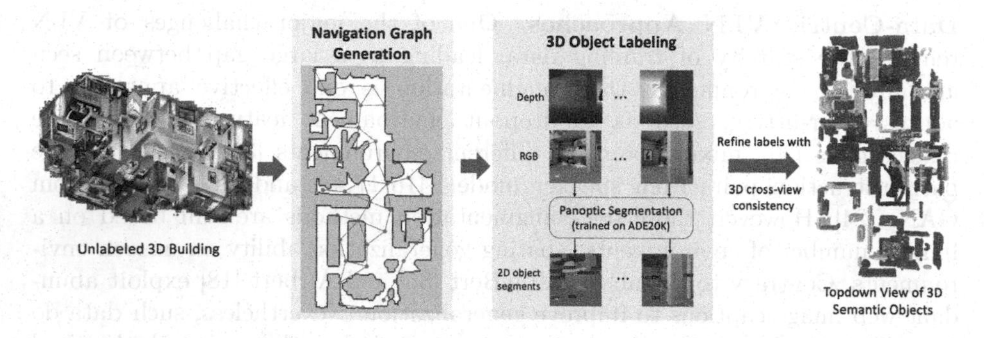

Fig. 2. We automatically construct the navigation graph and label 3D objects for each unlabeled 3D building.

We aim to generate VLN data with high-level instructions (similar to [43, 57]), which mainly describe the target scene and specific object instance. The success of the agent is defined as arriving at a node where the target object is close and visible and correctly localizing the object. Therefore, the training data should consist of a language instruction, an initial position and orientation of the agent, a ground-truth trajectory to the target location and the position of the target object. In order to generate such training data, it is essential to know object locations in the 3D space instead of bounding boxes in 2D images. Otherwise, it can be hard to infer relationships between objects in different views leading to unreliable target locations. Moreover, a sentence that refers to a specific object should be generated since the object name alone is not specific. Generation of such data is challenging because the original 3D buildings come without any labels. Therefore, we propose to use image and text models pretrained on large-scale external datasets for pseudo labeling. This involves two stages: (i) building-level pre-processing to obtain navigation graphs and 3D object annotations (Sect. 3.1); and (ii) object-trajectory-instruction triplet generation (Sect. 3.2).

3.1 Building-Level Pre-processing

We utilize the Habitat simulator [47] to load 3D buildings in HM3D [45] such that agents can navigate. This is a continuous navigation environment, thus we first convert each environment into a discrete navigation graph and then extract 3D pseudo object annotations for each building as illustrated in Fig. 2.

Constructing Navigation Graphs. We follow the characteristics of navigation graphs in the MP3D simulator [3] to create new graphs for continuous HM3D. For example, the average distance between connected nodes is around 2 m and connectivity is defined by whether two nodes are visible and navigable from each other. However, the MP3D simulator relies on human efforts to create such graphs where the nodes are selected and navigability is inspected manually. Instead, we build the graph in a fully automatic way. We first randomly sample 20,000 navigable locations in the 3D environment and then greedily add locations as new nodes into the graph. The newly added location is the nearest one to existing nodes in the graph among all remaining candidates that are more than 2 m away from the existing nodes. After collecting all the nodes, we use two criteria to connect different nodes: (i) the geodesic distance between nodes is below 3 m; and (ii) the average distance of the depth image captured from one point to another should be larger than 2 m. Figure 2 (left) presents an example of navigation graph generation on one floor of the environment. For each node, we extract 36 RGB-D images from different orientations to represent the panoramic view, following the VLN setting [3].

Labeling 3D Objects from 2D Predictions. Since existing 3D datasets such as Scannet [11] are small and contain a limited number of object classes, pretrained 3D object detectors are incapable of providing high-quality 3D object labels for HM3D which covers a diverse set of scenes and objects. Therefore, we propose to use an existing 2D image model to label 3D objects. Specifically, we use a panoptic segmentation model Mask2Former [10] trained on ADE20K [56] to generate instance masks for all images in the constructed graph. We project 2D pixel-wise semantic predictions into 3D point clouds using the camera intrinsic parameters and depth information, and thus obtain 3D bounding boxes. However, due to the domain gap between existing 2D labeled images and images from HM3D, the predictions of 2D models can be noisy as shown in Fig. 5. Even worse, a 3D object is often partially observed from one view, making the estimated 3D bounding box from a single location less accurate. In order to further improve 3D object labeling, we take advantage of cross-view consistency of 3D objects and merge 2D predictions from multiple views. To ease computation and reduce label noise, we downsample the extracted point clouds with class probabilities into voxels of size $0.1 \times 0.1 \times 0.1$ m^3 in a similar way as semantic map construction [5]. Hence, 2D predictions from different views of an object can be merged together. We average class probabilities of all points inside each voxel and take the label with maximum probability to refine semantic prediction. The neighboring voxels of the same class are grouped together as a complete 3D object.

In this way, we generate 3D objects for the whole building and also map 2D objects of different views into the unified 3D objects. This enables us to create reliable goal locations for target objects in the VLN task. Figure 2 (right) shows example predictions of 3D semantic objects from a top-down view.

3.2 Data Generation: VLN Training Triplets

Based on the pseudo 3D object labels of the building, we generate object-trajectory-instruction triplets as shown in Fig. 3. For each 3D object, we obtain the corresponding goal locations in the graph where the object is visible and located within d_o meters. Then we randomly select a start node that is 4 to 9 steps away from the target location, and use the navigation graph to compute the shortest path as the expert trajectory. The final panoramic image in the trajectory is used to generate high-level VLN instructions.

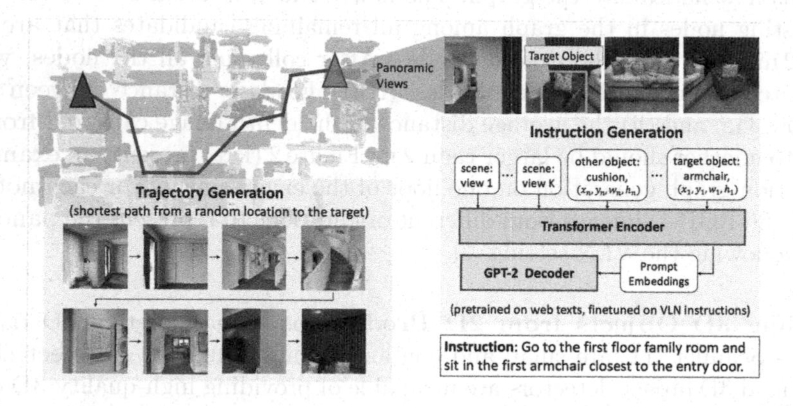

Fig. 3. To generate VLN training triplets, we first select a target object using the pseudo 3D object labels and then sample a trajectory using the navigation graph. Finally, we use a trained speaker model to generate high-level VLN instructions.

Generating Instructions with Object Prompts. Previous works [16,49] train speaker models from scratch on VLN datasets to generate instructions from an image trajectory. Due to the small amount of training examples in VLN datasets, such speaker models may suffer from inaccurate classification and are also unable to generate novel words beyond the training vocabulary. As a result, they even perform worse than template-based instructions on unseen images with diverse objects [18]. We alleviate these problems by (i) using object labels obtained from the 3D object labeling as prompts for the speaker model; and (ii) fine-tuning a pretrained large language model (GPT-2 [44]).

To be specific, our speaker model consists of a transformer encoder that produces prompt embeddings and a language decoder that generates sentences conditioning on the prompts. The encoder is fed with three types of tokens: target object token, other object tokens in the panorama, and view image tokens. The object token embedding encodes the object label, location, size, and visual representation, while the view image embedding is obtained from the visual representation and orientation embedding. A multi-layer transformer is adopted to learn relations of different tokens in order to generate more accurate referring expression for the target object. We average the output embeddings for each type of tokens separately, resulting in three tokens (target object, other objects, and view image) as prompts to the decoder. The decoder is initialized with pretrained GPT-2 [44], a state-of-the-art language model. It takes the above prompts as the first three tokens and sequentially predicts words. We finetune the speaker model end-to-end on the REVERIE dataset [43].

Table 1. Statistics of training data on different VLN datasets. (gt.obj denotes whether the data relies on groundtruth object annotations)

Dataset	gt.obj	#env.	#objs	#instr	#vocab	Inst. length
REVERIE [43]	✓	60	12,029	10,466	1,140	18.64
REVERIE-Speaker [9]	✓	64	12,062	19,636	399	15.20
SOON [57]	✓	34	4,358	2,780	735	44.09
HM3D-AutoVLN	×	900	172,000	217,703	1,696	20.52

3.3 HM3D-AutoVLN: A Large-Scale VLN Dataset

We employ the above procedure to process all unlabeled buildings in HM3D and generate a large-scale VLN dataset with high-level instructions, named HM3D-AutoVLN. Table 1 compares HM3D-AutoVLN with existing VLN datasets that also use high-level instructions. REVERIE and SOON are manually annotated datasets and REVERIE-Speaker is an augmented dataset generated by a speaker model trained with groundtruth room and object annotations. Our HM3D-AutoVLN is an order of magnitude larger than these datasets, in terms of environments, objects, and instructions. Due to the pretrained language model, our dataset also features a diverse vocabulary[1] in the generated instructions.

4 VLN Model and Training

In this section, we briefly introduce a state-of-the-art VLN model DUET [9] and then describe its training procedure.

[1] In Table 1 vocab, we only count words that occur more than 5 times in the dataset.

4.1 DUET VLN Model

The main idea of the DUal scalE graph Transformer (DUET) [9] model is to build a topological map during navigation, which allows the model to make efficient long-term navigation plans over all navigable nodes in the graph instead of being limited to the neighboring nodes. The model consists of two modules: topological mapping and global action planning. The topological mapping module gradually adds newly observed locations to the map and updates node representations. The global action planning module is a dual-scale graph transformer to predict a next location in the map or a stop action at each step. The dual-scale architecture enables the model to jointly use fine-grained language grounding with fine-scale representations of the current location and global graph reasoning with coarse-scale representation of the map. The DUET model is the winning entry in ICCV 2021 REVERIE and SOON navigation challenge. We use the publicly released code[2] in our experiments.

4.2 Training

Stage I: Pretraining on HM3D-AutoVLN. We use three proxy tasks to pretrain DUET [9], including Masked Language Modeling (MLM), Single-step Action Prediction (SAP) and Object Grounding (OG). For MLM, the inputs are a masked instruction and a full expert trajectory $(V_1, \cdots, V_T)$. The goal is to recover the masked words from the cross-modal visual representation. For SAP, we ask the model to predict the next action given the instruction and its navigation history. We randomly select the navigation history as: (i) $(V_1, \cdots, V_T)$ with target action *STOP*; (ii) $(V_1, \cdots, V_t)$ where $1 \leq t < T$ with target action V_{t+1}; and (iii) a random node V_i in the graph, and choose a target node V_{i+1} with shortest overall distance from V_i to V_{i+1} and from V_{i+1} to V_T among all nodes in the constructed map. Finally, for OG, the model is fed with an instruction and expert trajectory $(V_1, \cdots, V_T)$ to predict the ground-truth object in V_T.

Stage II: Fine-Tuning on Downstream VLN Datasets. We fine-tune a pretrained DUET on each downstream task by using the pseudo interactive demonstrator algorithm [9] – a special case of student forcing that addresses exposure bias. We suggest two strategies for fine-tuning: (i) fine-tuning on the downstream dataset only; or (ii) balancing the size of HM3D-AutoVLN and the downstream dataset, and combining the two datasets for joint training. The latter strategy is successful at preventing catastrophic forgetting of the previously learned policy as shown in our experiments in Table 2.

[2] https://github.com/cshizhe/VLN-DUET.

5 Experiments

5.1 Experiment Setup

Datasets. We evaluate models on two VLN tasks with high-level instructions, REVERIE [43] and SOON [57]. Each dataset is divided into training, val seen, val unseen and a hidden test unseen split. The statistics of training splits are presented in Table 1. The REVERIE dataset provides groundtruth object bounding boxes at each location and only requires an agent to select one of the bounding boxes. The shortest path from the agent's initial location to the target location is between 4 to 7 steps. The SOON dataset instead does not give groundtruth bounding boxes, and asks an agent to directly predict the orientation of the object's center. The path lengths are also longer than REVERIE with 9.5 steps on average. Due to the increased task difficulty and smaller training dataset size, it is more challenging to achieve high performance on SOON than on REVERIE.

Evaluation Metrics. We measure two types of metrics for the VLN task, navigation-only and remote object grounding. The navigation-only metrics focus solely on whether an agent arrives at the target location. We use standard navigation metrics [3] including **Success Rate (SR)**, the percentage of paths with the *final* location near any target locations within 3 m; **Oracle Success Rate (OSR)**, the percentage of paths with *any* location close to target within 3 m; and **SR weighted by Path Length (SPL)** which multiplies SR with the ratio between the length of shortest path and the agent's predicted path. As SPL takes into account navigation accuracy and efficiency, it is the primary metric in navigation. The remote object grounding metrics [43] consider both navigation and object grounding performance. The standard metrics are **Remote Grounding Success (RGS)** and **RGS weighted by Path Length (RGSPL)**. RGS in REVERIE is defined as correctly selecting the object among groundtruth bounding boxes, while in SOON as predicting a center point that is inside of the groundtruth polygon. RGSPL penalizes RGS by path length similar to SPL. For all these metrics, higher is better.

Implementation Details. We adopt ViT-B/16 [15] pretrained on ImageNet to extract view and object features. For SOON dataset, we use the Mask2Former trained on ADE20K [10] to extract object bounding boxes and transfer the prediction in the same way as in REVERIE. We use the same hyper-parameters in modeling as the DUET model [9]. All experiments were run on a single Nvidia RTX8000 GPU. The best epoch is selected based on SPL on the val unseen split.

5.2 Ablation Studies

The main objective of our work is to explore how much VLN agents can benefit from large-scale synthesized dataset. In this section, we carry out extensive ablations on the REVERIE dataset to study the effectiveness of our HM3D-AutoVLN dataset and key design choices.

Table 2. DUET performance on REVERIE (RVR) val unseen split using different training data. † denotes manual object annotations are used to synthesize data

	Training data		Navigation			Grounding	
	Pretrain	Finetune	OSR	SR	SPL	RGS	RGSPL
R1	RVR	RVR	48.74	44.36	30.79	30.30	21.08
R2	RVR+Speaker†	RVR	51.07	46.98	33.73	32.15	23.03
R3	HM3D	RVR	54.81	50.87	39.36	34.65	**26.79**
R4	HM3D	RVR+HM3D	**62.14**	**55.89**	**40.85**	**36.58**	26.76

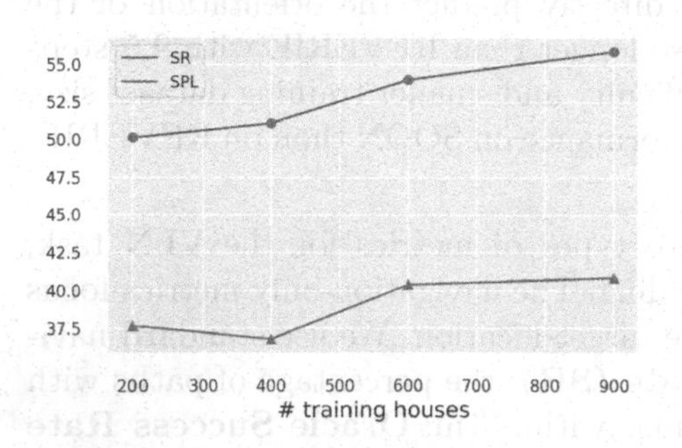

(a) Navigation performance with respect to the number of training environments in HM3D-AutoVLN dataset.

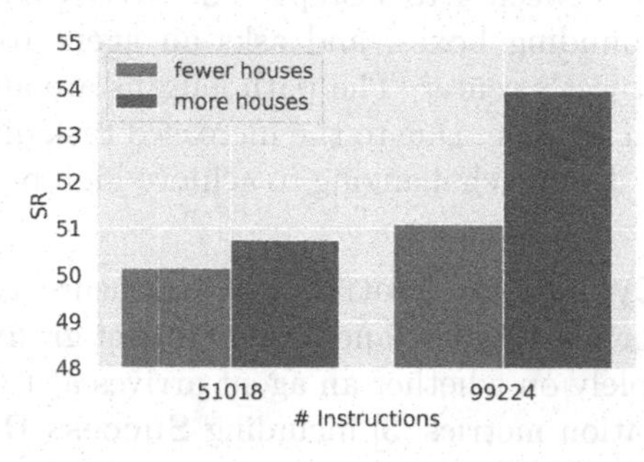

(b) Given the same number of training examples, collecting data from more environments (blue) performs better than fewer environments.

Fig. 4. Influence of the number of environments on DUET performance.

Contributions from HM3D-AutoVLN Dataset. In Table 2, we compare the impact of VLN data generated from different sources for training DUET. Row 1 (R1) only relies on manually annotated instructions on 60 buildings in MP3D. Row 2 (R2) utilizes groundtruth room and object annotations to generate instructions for all objects in the 60 seen buildings. Such data is of high-quality and significantly improves VLN performance, *e.g.* 3% on SPL. In R3, we only use our generated HM3D-AutoVLN for pretraining, while still finetuning on the REVERIE train split. Due to the visual diversity of the 900 additional buildings, it brings a larger boost than R2. Compared to R1, the improvement of the SPL metric is 8.6%, whereas other metrics like SR improve by 6.5%. This suggests that the pretraining on large-scale environments enables the model to learn efficient exploration, even without any additional manual annotations. Moreover, when the model is fine-tuned jointly on the HM3D-AutoVLN and REVERIE datasets (R4), the SR metric is further improved by 5% over R3. This indicates that fine-tuning on the downstream dataset alone may suffer from forgetting and lead to worse generalization performance. Compared to the navigation metrics, the remote object grounding metrics see modest improvements. We hypothesize that though our generated instructions often describe the object and scene accurately, they are not discriminative enough to refer to a specific object in the environment (*e.g.* unable to discriminate between multiple pillows placed on a sofa) or confuse predicting relationships between objects (*e.g.* second pillow from the left). We see improving RGS and RGSPL as a promising future direction that would need to take into account relations between objects.

Impact of the Number of Training Environments. Here, we evaluate the impact of the number of training environments. As pretraining on HM3D-AutoVLN mostly improves navigation performance, we mainly show navigation metrics SR and SPL. The full result table is provided in the supplementary material. As shown in Fig. 4a, more environments continuously improve the navigation performance. We observe that even with the full 900 environments in HM3D, the gains are not saturated but increase gradually. We also evaluate the impact of the number of environments given a fixed budget of VLN training examples. Figure 4b shows that using more environments improves the performance. On the **left**, the red bar uses 200 environments with 51,018 instructions, and the blue bar usc the same amount of instructions but covers all 900 environments in HM3D. The **right** part is the same, but uses 400 environments for the red bar. The results suggest that it is beneficial to increase the number of environments rather than just increasing the trajectories for a limited number of environments.

Table 3. DUET performance when using a fraction of the supervised data

HM3D Pretrain	#environments	#instructions	Navigation			Grounding	
			OSR	SR	SPL	RGS	RGSPL
×	60	10,466	48.74	44.36	30.79	30.30	21.08
✓	0	0	43.08	36.81	25.28	20.82	13.74
✓	1	449	50.78	42.12	29.55	25.02	17.26
✓	10	1,404	50.47	43.79	33.61	26.30	20.24
✓	30	5,244	60.81	53.71	39.26	34.42	25.11
✓	60	10,466	62.14	55.89	40.85	36.58	26.76

Finetuning with a Small Amount of Supervised Data. A large-scale automatically generated dataset improves pretraining and hence can reduce the supervised data requirement in downstream tasks. We verify the effectiveness of HM3D-AutoVLN on a few-shot learning setting [18] that compares the impact of finetuning on a variable number of REVERIE environments, see Table 3. Even without finetuning on any supervised instructions from REVERIE, our pretrained model achieves fair performance in comparison to a baseline that is only finetuned on all supervised data. On the SPL metric, we outperform the baseline DUET model when finetuning on 10 environments (1/6 of the supervised data). Moreover, finetuning with half the original data (30 environments) achieves significant boosts on all metrics compared to the baseline model. A similar trend can be observed on the SOON dataset, see supplementary material.

Comparing Distance to Objects. As mentioned in Sect. 3.2, we select some of the visible objects that are close to the agent, within d_o meters, to generate instructions. In Table 4, we compare the influence of different distances to the objects on the VLN performance. Larger d_o allows us to generate more instructions. We can see that while including additional remote objects increases the number of instructions, it leads to a small drop in performance as the model struggles to identify objects that are small and far away.

Table 4. DUET performance on instructions where the visible objects are at a different distance d_o from the agent location

d_o	#instrs	Navigation			Grounding	
		OSR	SR	SPL	RGS	RGSPL
2	217,703	**62.14**	**55.89**	**40.85**	**36.58**	**26.76**
3	396,401	57.37	53.25	40.38	34.28	25.65
∞	544,606	59.98	53.37	38.03	35.70	25.50

Table 5. Comparison of different speaker models in terms of manual captioning evaluation and the followup navigation performance

	Captioning			Navigation	
	Room	Obj	Rel	SR	RGS
Template	0.13	0.76	0.05	52.20	32.75
LSTM	0.58	0.65	0.27	49.59	32.29
GPT2 (Ours)	**0.73**	**0.78**	**0.35**	**55.89**	**36.58**

Evaluating Quality of the HM3D-AutoVLN Dataset. We validate each annotation procedure in our automatic dataset construction. For *navigation graph generation*, we measure whether the graph covers the whole building. Assuming each navigation node covers a circle with a radius of 2 m, the graph achieves a high coverage rate of 93.4% on average. For *3D object labeling*, we randomly select 300 bounding boxes and manually annotate semantic labels for them. We observe that 37.4% of them are correctly predicted with 2D predictions, whereas 58.3% of them were correctly predicted when applying cross-view consistency, showing an absolute improvement of 21%. Examples in Fig. 5 highlight the advantages of using cross-view consistency. Due to the distorted view, it's reasonable that the 2D models wrongly predict the `mirror` and `wardrobe` to be a `window` and `door` respectively. Cross-view consistency also helps to integrate scene context, swapping a `table` to a `desk` and a normal `chair` to a `swivel chair`. For *instruction generation*, we further compare our GPT2-based speaker model with template-based methods and a LSTM baseline [9]. We manually evaluate 100 randomly selected instructions by measuring whether the instruction correctly mentions the target room, object class and object instance (with correct relations). We also measure the VLN performance using the generated instructions. Our model performs best on manual evaluation and on downstream VLN tasks as shown in Table 5.

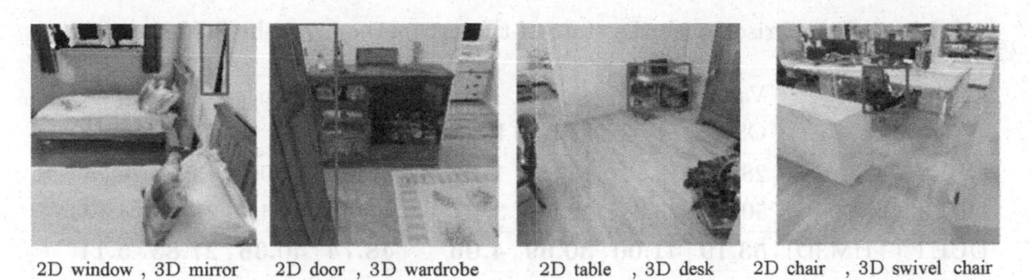

Fig. 5. Qualitative examples of pseudo labelling

5.3 Comparison with State of the Art

In Table 6, we compare with the state of the art on the REVERIE dataset. To demonstrate the contributions of our automatically constructed dataset, we further train additional VLN agents with the augmentation of the HM3D-AutoVLN dataset, including EnvDrop [49], RecBert [22] and HAMT [8]. Our proposed dataset improves results of all methods and gives a particularly large boost to high-capacity models. When pretraining DUET on HM3D-AutoVLN, the increase in performance over DUET with pretraining is significant for all metrics on both the val unseen and the test unseen splits. For example, the SPL measure increases by 7.1% and 2.8% on val unseen and test unseen splits of REVERIE. Table 7 provides results for the SOON dataset. Note that instructions from the SOON dataset are not used to train our speaker model (*c.f.* Sect. 3.2) and are somewhat different from the REVERIE instructions. Nevertheless, the

Table 6. Comparison with the state-of-the-art methods on REVERIE dataset

Methods	Val unseen					Test unseen				
	Navigation			Grounding		Navigation			Grounding	
	OSR	SR	SPL	RGS	RGSPL	OSR	SR	SPL	RGS	RGSPL
Human	–	–	–	–	–	86.83	81.51	53.66	77.84	51.44
Seq2Seq [3]	8.07	4.20	2.84	2.16	1.63	6.88	3.99	3.09	2.00	1.58
RCM [53]	14.23	9.29	6.97	4.89	3.89	11.68	7.84	6.67	3.67	3.14
SMNA [34]	11.28	8.15	6.44	4.54	3.61	8.39	5.80	4.53	3.10	2.39
FAST-MAttNet [43]	28.20	14.40	7.19	7.84	4.67	30.63	19.88	11.61	11.28	6.08
SIA [32]	44.67	31.53	16.28	22.41	11.56	44.56	30.80	14.85	19.02	9.20
Airbert [18]	34.51	27.89	21.88	18.23	14.18	34.20	30.28	23.61	16.83	13.28
EnvDrop [49]	26.3	23.0	19.9	–	–	–	–	–	–	–
+HM3D-AutoVLN	29.3	25.2	20.6	–	–	–	–	–	–	–
RecBERT (oscar) [22]	27.66	25.53	21.06	14.20	12.00	26.67	24.62	19.48	12.65	10.00
+HM3D-AutoVLN	33.23	29.20	23.12	18.12	14.18	–	–	–	–	–
HAMT [8]	36.84	32.95	30.20	18.92	17.28	33.41	30.40	26.67	14.88	13.08
+HM3D-AutoVLN	42.09	37.80	31.35	23.03	18.88	–	–	–	–	–
DUET [9]	51.07	46.98	33.73	32.15	23.03	56.91	52.51	36.06	31.88	22.06
+HM3D-AutoVLN	**62.14**	**55.89**	**40.85**	**36.58**	**26.76**	**62.3**	**55.17**	**38.88**	**32.23**	**22.68**

Table 7. Comparison with the state-of-the-art methods on the SOON dataset

Methods	Val unseen				Test unseen			
	OSR	SR	SPL	RGSPL	OSR	SR	SPL	RGSPL
GBE [57]	28.54	19.52	13.34	1.16	21.45	12.90	9.23	0.45
DUET [9]	50.91	36.28	22.58	3.75	43.00	33.44	21.42	4.17
DUET (+HM3D)	**53.19**	**41.00**	**30.69**	**4.06**	**48.74**	**40.36**	**27.83**	**5.11**

large performance increase demonstrates cross-domain benefits of pretraining on an automatically collected large-scale dataset. Finally, we can also observe that the gain for object grounding is less significant, as discussed before this can be explained by the confusion between objects of the same categories and erroneous spatial relations.

6 Conclusion

This work addresses the lack of training data for VLN tasks. We propose to automatically generate pseudo 3D object labels and VLN instructions for a collection of large-scale unlabeled 3D environments. Training on our new dataset HM3D-AutoVLN significantly improves VLN performance due to the large-scale pretraining or co-training. It also provides insights on the importance of dataset collection and the challenges inherent to leveraging unlabeled environments. In particular, it shows that the diversity of navigation environments is more important than the number of training samples alone.

Acknowledgements. This work was granted access to the HPC resources of IDRIS under the allocation 101002 made by GENCI. This work is funded in part by the French government under management of Agence Nationale de la Recherche as part of the "Investissements d'avenir" program, reference ANR19-P3IA-0001 (PRAIRIE 3IA Institute) and by Louis Vuitton ENS Chair on Artificial Intelligence.

References

1. Anderson, P., Shrivastava, A., Parikh, D., Batra, D., Lee, S.: Chasing ghosts: instruction following as bayesian state tracking. NeurIPS **32** (2019)
2. Anderson, P., et al.: Sim-to-real transfer for vision-and-language navigation. In: CoRL, pp. 671–681. PMLR (2021)
3. Anderson, P., et al.: Vision-and-language navigation: interpreting visually-grounded navigation instructions in real environments. In: CVPR, pp. 3674–3683 (2018)
4. Chang, A., et al.: Matterport3d: learning from rgb-d data in indoor environments. In: 3DV, pp. 667–676. IEEE (2017)
5. Chaplot, D.S., Gandhi, D.P., Gupta, A., Salakhutdinov, R.R.: Object goal navigation using goal-oriented semantic exploration. NeurIPS **33**, 4247–4258 (2020)

6. Chen, H., Suhr, A., Misra, D., Snavely, N., Artzi, Y.: Touchdown: natural language navigation and spatial reasoning in visual street environments. In: CVPR, pp. 12538–12547 (2019)
7. Chen, K., Chen, J.K., Chuang, J., Vázquez, M., Savarese, S.: Topological planning with transformers for vision-and-language navigation. In: Proceedings of the IEEE/CVF Conference on Computer Vision and Pattern Recognition, pp. 11276–11286 (2021)
8. Chen, S., Guhur, P.L., Schmid, C., Laptev, I.: History aware multimodal transformer for vision-and-language navigation. In: NeurIPS (2021)
9. Chen, S., Guhur, P.L., Tapaswi, M., Schmid, C., Laptev, I.: Think global, act local: Dual-scale graph transformer for vision-and-language navigation. In: CVPR (2022)
10. Cheng, B., Misra, I., Schwing, A.G., Kirillov, A., Girdhar, R.: Masked-attention mask transformer for universal image segmentation. arXiv (2021)
11. Dai, A., Chang, A.X., Savva, M., Halber, M., Funkhouser, T., Nießner, M.: Scannet: Richly-annotated 3d reconstructions of indoor scenes. In: CVPR, pp. 5828–5839 (2017)
12. Das, A., Datta, S., Gkioxari, G., Lee, S., Parikh, D., Batra, D.: Embodied question answering. In: CVPR, pp. 1–10 (2018)
13. De Vries, H., Shuster, K., Batra, D., Parikh, D., Weston, J., Kiela, D.: Talk the walk: navigating New York city through grounded dialogue. arXiv preprint arXiv:1807.03367 (2018)
14. Deng, Z., Narasimhan, K., Russakovsky, O.: Evolving graphical planner: contextual global planning for vision-and-language navigation. In: NeurIPS, vol. 33 (2020)
15. Dosovitskiy, A., et al.: An image is worth 16× 16 words: transformers for image recognition at scale. In: ICLR (2020)
16. Fried, D., et al.: Speaker-follower models for vision-and-language navigation. In: NeurIPS, pp. 3318–3329 (2018)
17. Fu, T.-J., Wang, X.E., Peterson, M.F., Grafton, S.T., Eckstein, M.P., Wang, W.Y.: Counterfactual vision-and-language navigation via adversarial path sampler. In: Vedaldi, A., Bischof, H., Brox, T., Frahm, J.-M. (eds.) ECCV 2020. LNCS, vol. 12351, pp. 71–86. Springer, Cham (2020). https://doi.org/10.1007/978-3-030-58539-6_5
18. Guhur, P.L., Tapaswi, M., Chen, S., Laptev, I., Schmid, C.: Airbert: in-domain pretraining for vision-and-language navigation. In: ICCV, pp. 1634–1643 (2021)
19. Hao, W., Li, C., Li, X., Carin, L., Gao, J.: Towards learning a generic agent for vision-and-language navigation via pre-training. In: CVPR, pp. 13137–13146 (2020)
20. Hong, Y., Rodriguez, C., Qi, Y., Wu, Q., Gould, S.: Language and visual entity relationship graph for agent navigation. NeurIPS **33**, 7685–7696 (2020)
21. Hong, Y., Wang, Z., Wu, Q., Gould, S.: Bridging the gap between learning in discrete and continuous environments for vision-and-language navigation. In: CVPR, pp. 15439–15449 (2022)
22. Hong, Y., Wu, Q., Qi, Y., Rodriguez-Opazo, C., Gould, S.: Vln BERT: a recurrent vision-and-language BERT for navigation. In: CVPR, pp. 1643–1653 (2021)
23. Irshad, M.Z., Ma, C., Kira, Z.: Hierarchical cross-modal agent for robotics vision-and-language navigation. In: ICRA, pp. 13238–13246 (2021)
24. Koh, J.Y., Lee, H., Yang, Y., Baldridge, J., Anderson, P.: Pathdreamer: a world model for indoor navigation. In: ICCV, pp. 14738–14748 (2021)
25. Kolve, E., et al.: AI2-THOR: an interactive 3d environment for visual ai. ArXiv abs/1712.05474 (2017)

26. Krantz, J., Gokaslan, A., Batra, D., Lee, S., Maksymets, O.: Waypoint models for instruction-guided navigation in continuous environments. In: ICCV, pp. 15162–15171 (2021)
27. Krantz, J., Wijmans, E., Majumdar, A., Batra, D., Lee, S.: Beyond the nav-graph: vision-and-language navigation in continuous environments. In: Vedaldi, A., Bischof, H., Brox, T., Frahm, J.-M. (eds.) ECCV 2020. LNCS, vol. 12373, pp. 104–120. Springer, Cham (2020). https://doi.org/10.1007/978-3-030-58604-1_7
28. Ku, A., Anderson, P., Patel, R., Ie, E., Baldridge, J.: Room-across-room: multilingual vision-and-language navigation with dense spatiotemporal grounding. In: EMNLP, pp. 4392–4412 (2020)
29. Li, C., et al.: igibson 2.0: object-centric simulation for robot learning of everyday household tasks. In: CoRL, pp. 455–465. PMLR (2022)
30. Li, J., Tan, H., Bansal, M.: Envedit: environment editing for vision-and-language navigation. In: CVPR, pp. 15407–15417 (2022)
31. Li, X., et al.: Robust navigation with language pretraining and stochastic sampling. In: EMNLP, pp. 1494–1499 (2019)
32. Lin, X., Li, G., Yu, Y.: Scene-intuitive agent for remote embodied visual grounding. In: CVPR, pp. 7036–7045 (2021)
33. Liu, C., Zhu, F., Chang, X., Liang, X., Ge, Z., Shen, Y.D.: Vision-language navigation with random environmental mixup. In: ICCV, pp. 1644–1654 (2021)
34. Ma, C.Y., et al.: Self-monitoring navigation agent via auxiliary progress estimation. In: ICLR (2019)
35. Ma, C.Y., Wu, Z., AlRegib, G., Xiong, C., Kira, Z.: The regretful agent: Heuristic-aided navigation through progress estimation. In: CVPR, pp. 6732–6740 (2019)
36. Majumdar, A., Shrivastava, A., Lee, S., Anderson, P., Parikh, D., Batra, D.: Improving vision-and-language navigation with image-text pairs from the web. In: Vedaldi, A., Bischof, H., Brox, T., Frahm, J.-M. (eds.) ECCV 2020. LNCS, vol. 12351, pp. 259–274. Springer, Cham (2020). https://doi.org/10.1007/978-3-030-58539-6_16
37. Moudgil, A., Majumdar, A., Agrawal, H., Lee, S., Batra, D.: Soat: a scene-and object-aware transformer for vision-and-language navigation. NeurIPS 34 (2021)
38. Nguyen, K., Daumé III, H.: Help, ANNA! visual navigation with natural multimodal assistance via retrospective curiosity-encouraging imitation learning. arXiv preprint arXiv:1909.01871 (2019)
39. Padmakumar, A., et al.: Teach: task-driven embodied agents that chat. In: AAAI, vol. 36, pp. 2017–2025 (2022)
40. Pashevich, A., Schmid, C., Sun, C.: Episodic transformer for vision-and-language navigation. In: ICCV, pp. 15942–15952 (2021)
41. Puig, X., et al.: Virtualhome: simulating household activities via programs. In: Proceedings of the IEEE Conference on Computer Vision and Pattern Recognition, pp. 8494–8502 (2018)
42. Qi, Y., Pan, Z., Hong, Y., Yang, M.H., van den Hengel, A., Wu, Q.: The road to know-where: an object-and-room informed sequential bert for indoor vision-language navigation. In: ICCV, pp. 1655–1664 (2021)
43. Qi, Y., et al.: Reverie: remote embodied visual referring expression in real indoor environments. In: CVPR, pp. 9982–9991 (2020)
44. Radford, A., et al.: Language models are unsupervised multitask learners. OpenAI Blog **1**(8), 9 (2019)
45. Ramakrishnan, S.K., et al.: Habitat-matterport 3d dataset (hm3d): 1000 large-scale 3d environments for embodied ai. arXiv preprint arXiv:2109.08238 (2021)

46. Raychaudhuri, S., Wani, S., Patel, S., Jain, U., Chang, A.: Language-aligned waypoint (law) supervision for vision-and-language navigation in continuous environments. In: EMNLP, pp. 4018–4028 (2021)
47. Savva, M., et al.: Habitat: a platform for embodied ai research. In: ICCV, pp. 9339–9347 (2019)
48. Shridhar, M., et al.: Alfred: a benchmark for interpreting grounded instructions for everyday tasks. In: CVPR, pp. 10740–10749 (2020)
49. Tan, H., Yu, L., Bansal, M.: Learning to navigate unseen environments: back translation with environmental dropout. In: NAACL, pp. 2610–2621 (2019)
50. Thomason, J., Murray, M., Cakmak, M., Zettlemoyer, L.: Vision-and-dialog navigation. In: CoRL, pp. 394–406. PMLR (2020)
51. Wang, H., Wang, W., Liang, W., Xiong, C., Shen, J.: Structured scene memory for vision-language navigation. In: Proceedings of the IEEE/CVF Conference on Computer Vision and Pattern Recognition, pp. 8455–8464 (2021)
52. Wang, H., Wang, W., Shu, T., Liang, W., Shen, J.: Active visual information gathering for vision-language navigation. In: Vedaldi, A., Bischof, H., Brox, T., Frahm, J.-M. (eds.) ECCV 2020. LNCS, vol. 12367, pp. 307–322. Springer, Cham (2020). https://doi.org/10.1007/978-3-030-58542-6_19
53. Wang, X., et al.: Reinforced cross-modal matching and self-supervised imitation learning for vision-language navigation. In: CVPR, pp. 6629–6638 (2019)
54. Xia, F., Zamir, A.R., He, Z., Sax, A., Malik, J., Savarese, S.: Gibson env: real-world perception for embodied agents. In: CVPR, pp. 9068–9079 (2018)
55. Yu, L., Chen, X., Gkioxari, G., Bansal, M., Berg, T.L., Batra, D.: Multi-target embodied question answering. In: CVPR, pp. 6309–6318 (2019)
56. Zhou, B., Zhao, H., Puig, X., Fidler, S., Barriuso, A., Torralba, A.: Scene parsing through ade20k dataset. In: CVPR, pp. 633–641 (2017)
57. Zhu, F., Liang, X., Zhu, Y., Yu, Q., Chang, X., Liang, X.: Soon: scenario oriented object navigation with graph-based exploration. In: CVPR, pp. 12689–12699 (2021)

BodySLAM: Joint Camera Localisation, Mapping, and Human Motion Tracking

Dorian F. Henning[1(✉)], Tristan Laidlow[1], and Stefan Leutenegger[1,2]

[1] Imperial College London, London, UK
{d.henning,t.laidlow15}@imperial.ac.uk
[2] Technische Universität München, Munich, Germany
stefan.leutenegger@tum.de

Abstract. Estimating human motion from video is an active research area due to its many potential applications. Most state-of-the-art methods predict human shape and posture estimates for individual images and do not leverage the temporal information available in video. Many "in the wild" sequences of human motion are captured by a moving camera, which adds the complication of conflated camera and human motion to the estimation. We therefore present BodySLAM, a monocular SLAM system that jointly estimates the position, shape, and posture of human bodies, as well as the camera trajectory. We also introduce a novel human motion model to constrain sequential body postures and observe the scale of the scene. Through a series of experiments on video sequences of human motion captured by a moving monocular camera, we demonstrate that BodySLAM improves estimates of all human body parameters and camera poses when compared to estimating these separately.

Keywords: Human pose estimation · Human motion model · Camera tracking · Dynamic SLAM

1 Introduction

Estimating the orientation and translation, shape, and posture of human bodies from a sequence of images is an active research area in computer vision due to its many potential applications in domains such as safe human-robot interaction, biomechanical analysis, action recognition, and augmented and virtual reality (AR/VR). To fully realise these types of applications at scale, it is desirable that this estimation can be done "in the wild" from images captured by a moving

Research presented in this paper has been supported by Dyson Technology Ltd. and the Technical University of Munich.

Supplementary Information The online version contains supplementary material available at https://doi.org/10.1007/978-3-031-19842-7_38.

S. Avidan et al. (Eds.): ECCV 2022, LNCS 13699, pp. 656–673, 2022.
https://doi.org/10.1007/978-3-031-19842-7_38

monocular camera without relying on any special markers or motion-capture systems. This would even enable human motion analysis on image sequences that were not captured with these applications in mind, such as from videos found on YouTube or other online repositories. State-of-the-art estimation methods for human shape and posture involve training powerful deep neural networks (DNNs), but these methods require large amounts of annotated training data. As 3D manual annotations are extremely difficult to obtain, in-the-wild datasets are only labelled in 2D, with 3D annotations only being available in indoor settings where a motion capture system was used to produce ground truth labels. These restrictions have led to most methods only providing shape and posture estimates for individual frames and not leveraging the temporal information that is available from image sequences. Another critical limitation is that most of these training datasets only feature stationary cameras, and a true in-the-wild system will need to operate in environments where the camera is moving along an unknown trajectory. This further complicates the estimation procedure, as with a moving camera, the camera motion and human motion become conflated.

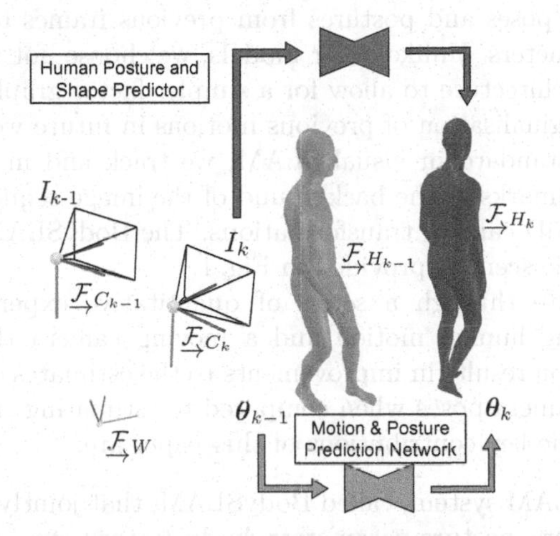

Fig. 1. BodySLAM Problem Definition and Example Scene. The human body postures of the previous n frames are used to predict its motion; and a Human Mesh Regressor [28] is used for human centre pose, posture, and shape prediction. Then, a factor graph of camera trajectory, human states and natural landmarks are jointly optimised.

The standard method to incrementally estimate the position of a moving monocular camera in real time is with a visual simultaneous localisation and mapping (SLAM) system. While many visual SLAM methods have been shown to be both accurate and robust, image sequences containing people in motion pose a particular challenge, since most state-of-the-art visual SLAM methods assume a static environment. It is possible to use a separate network to segment and mask-out humans from the visual SLAM system, but in many sequences humans take up a large proportion of the frame leaving too few points for the

SLAM system to track robustly. Importantly, even in static scenes with many observable points, classic monocular SLAM systems are not able to observe the scale of a scene, introducing ambiguity to the absolute poses of the cameras and any human motion observed by them.

With these limitations in mind, we propose a method, called BodySLAM, for jointly estimating the orientation and translation, shape, and posture of human bodies, as well as the trajectory of the moving monocular camera that observes them, in a single factor graph optimisation. We represent 3D human bodies using the shape and posture parameters of the SMPL mesh model [32], and the orientation and translation of the root of the corresponding kinematic tree underlying the SMPL model. In each frame, we obtain a measurement of these human body parameters using an off-the-shelf neural network. The body shape measurements are fused together in the factor graph optimisation as these should remain constant for a single individual over the image sequence. This constraint also helps to resolve the scale ambiguity from the monocular SLAM formulation.

Posture parameters are constrained between frames through the use of a novel motion model. The motion model uses body shape parameters and estimates of the human centre poses and postures from previous frames to form a prior on future body parameters. Unlike other models, we choose not to use a recurrent neural network architecture to allow for a simpler factor graph formulation and to enable the marginalisation of previous motions in future work.

Finally, as is standard in visual SLAM, we track and map a sparse set of feature-based landmarks in the background of the image sequences to help estimate the relative 6D camera transformations. The BodySLAM problem definition for an example scene is provided in Fig. 1.

We demonstrate through a series of quantitative experiments on image sequences featuring human motion and a moving camera that our approach of joint optimisation results in improvements to the estimates of all human body parameters and camera poses when compared to estimating these separately.[1]

In summary, the key contributions of this paper are:

- a monocular SLAM system, called BodySLAM, that jointly optimises human shape parameters, posture parameters, body centre poses, camera poses, and a set of 3D landmarks,
- a novel motion model that robustly predicts future human body parameters from previous ones, and
- an experimental evaluation of our joint estimation approach on our own dataset consisting of more than 8k frames with ground truth human body parameters and camera poses obtained by a Vicon motion tracking system, demonstrating that we can accurately recover human body parameters and camera trajectories at metric scale.

[1] We encourage the reader to view our supplementary video available at: https://youtu.be/0-SL3VeWEvU.

2 Related Work

2.1 3D Human Body Representations

3D representations for the human body generally fall into three broad categories: sparse sets of independent body joints, articulated body skeletons with rigidity constraints between joints, and (parametric) human mesh models.

Both of the first two categories involve the estimation of a set of 3D joint landmarks from 2D images. By using datasets such as Human3.6m [10] with ground truth 3D labels generated by motion capture systems, it is possible to directly regress from the input images to 3D joint positions using DNNs [39,40]. Another option is to use the abundance of 2D ground truth labels to learn to detect 2D joint keypoints [9,13], and then "lift" these to 3D using methods such as dictionary of skeletons [2,43,48] or DNNs [14,34,51]. While the direct approach tends to overfit to indoor lab environments, the lifting strategy discards a lot of potentially useful information from the captured images.

More recently, parametric 3D human body models [3,32,36] have been successfully used as output targets for human body shape and posture estimation. Human body models encapsulate statistics of body shape and other priors to reduce the ambiguity of the 3D estimation problem, and can provide high-level details such as facial expressions and hand or foot articulation [38]. It is for these advantages that we choose to use the SMPL parametric human mesh model [32] as the 3D representation for human bodies in BodySLAM.

These parametric mesh models can be fitted end-to-end using either an optimisation-based top-down approach, a regression-based bottom-up approach, or a combination of both [8,23,28]. SMPLify [8] proposes an optimisation-based fitting routine where the 3D mesh model is fitted to predicted 2D keypoints. While top-down approaches such as SMPLify require a 2-step process of keypoint prediction and mesh optimisation, regression-based bottom-up approaches start directly from pixel-level information. [23] develops a direct parametric mesh regressor trained on 2D and 3D data, predicting mesh and camera parameters from an image input. [28] uses an optimisation routine in the training loop of a deep convolutional neural network to improve the supervisory signal and quality of the predictions. In our work, we use the method of [28] to initialise the human shape and posture parameters, but then refine these estimates through our joint factor graph optimisation procedure.

2.2 Human Motion Models

While there are many approaches to estimate human motion from videos looking only at joint locations [11,18,41], our focus is on methods that use parametric human body models like SMPL.

A common approach is to enforce a smoothness prior on the posture parameters rather than define an explicit motion model. For example, [4] does this to include temporal information in their extension of the previously introduced SMPLify routine. By adding smoothness constraints on posture parameters and enforcing consistent body shape, the task of human shape and posture estimation from video is essentially formulated as a bundle adjustment problem. [19] uses multiple views of the same scene and silhouette constraints to improve single-frame fitting. While smoothness constraints help improve estimates of the human body parameters, they are agnostic to the direction of motion and do not help to estimate scale. As scale recovery is an important part of our work, we use a model to predict future body position and posture parameters from previous estimates, in which smoothness is implicitly included.

There are other examples that use DNNs to learn a human motion model [24,26,31,44]. In [24], a network learns human kinematics by prediction of past and future image frames. Recently, [26] introduced both a temporal encoder with a gated recurrent unit (GRU) to estimate temporally consistent motion and shapes, and a training procedure with a motion discriminator that forces the network to generate feasible human motion. While such a recurrent network is capable of generating very accurate human motion predictions, including it in a factor graph optimisation would be difficult as each hidden state would need to become a state variable. Instead, we opt to use a multi-layer perceptron that takes the body shape parameters and the past n body posture estimates as input to predict the current posture parameters. This formulation allows us to greatly simplify the factor graph and will allow for much easier marginalisation of past states as part of future work.

2.3 Visual SLAM with Dynamic Objects

Most visual SLAM systems assume that the scene they are mapping and tracking against is static (e.g. ORB-SLAM [35]). Many of these systems are still capable of handling some small dynamic regions of the scene as any keypoints in these areas will be treated as outliers when using robust methods; however, they will fail if the dynamic objects occupy large segments of the frame as is often the case in image sequences capturing human motion. Some systems are designed to handle dynamic environments through the use of elimination strategies, where moving objects are detected and removed from the tracking and mapping processes (e.g. [5,7,12,20,21,47]). While this strategy offers an improvement over the static scene assumption in standard SLAM systems, it still requires that enough static background is visible for robust tracking and throws away any information contained in the dynamic parts of the environment. Alternative methods explicitly model motion with techniques such as rigid object-level maps [6,45,46,50].

More recently, there have been some approaches that estimate the camera motion and dynamic scenes simultaneously, without any *a priori* knowledge of the environment (e.g. [15,16,22]). While these systems are limited to estimating rigid-body motion, AirDOS [42], the work most closely related to our own, extends this line of work to include articulated objects such as humans. AirDOS uses a combination of rigidity and motion constraints to create articulated models for humans and vehicles within a factor graph framework. By jointly optimising for the object motion, object 3D structure and the camera trajectory, improved camera tracking over an elimination strategy baseline is shown.

While AirDOS presents impressive results, we are able to make a number of key improvements over their approach. Firstly, whereas AirDOS depends on stereo input, BodySLAM works with only a monocular camera, making it suitable to more "in the wild" applications. Unlike stereo SLAM systems, standard monocular SLAM systems are unable to observe the scale of the scene; however, we demonstrate that by simultaneously optimising for the human posture, shape and centre poses, BodySLAM is able to recover the metric scale. Secondly, since the focus of AirDOS is on improved camera tracking and not human motion estimation, it uses a simple skeleton-based model of human rigidity. It furthermore does not use a learned temporal model, but uses motion constraints for the rigid body segments. BodySLAM, on the other hand, uses the SMPL parametric mesh model, allowing for full human body reconstructions, and uses a temporal human motion model to predict translation, orientation, and posture change. Finally, while AirDOS provides quantitative camera tracking results, only a few qualitative results for the object tracking and structure estimation are provided. In our experiments on real-world datasets, we quantitatively demonstrate that BodySLAM improves not only camera tracking, but the estimation of the human body centre poses, shape and posture parameters.

3 Methodology

3.1 Preliminaries and Notation

We use the following notation throughout this work: a reference frame is denoted as $\underrightarrow{\mathcal{F}}_A$, and vectors expressed in it are written as ${}_A\mathbf{r}$ or translations ${}_A\mathbf{r}_{OP}$ with O and P as start and endpoints, respectively. The homogeneous transformation from reference frame $\underrightarrow{\mathcal{F}}_B$ to $\underrightarrow{\mathcal{F}}_A$ is denoted as $\boldsymbol{T}_{AB}$. Its rotation matrix is denoted as $\mathbf{C}_{AB}$, whose minimal representation we parameterise as an axis-angle rotation vector in the respective frame, ${}_A\boldsymbol{\alpha}_{AB}$. Commonly used reference frames are the static world frame $\underrightarrow{\mathcal{F}}_W$, the camera centric frame $\underrightarrow{\mathcal{F}}_C$, and the human body centric frame $\underrightarrow{\mathcal{F}}_H$. Camera frames are indexed by time steps k, and body joints by j. Measurements of quantity $\mathbf{z}$ are denoted with a tilde, $\tilde{\mathbf{z}}$.

3.2 System Overview

An overview of the BodySLAM framework is provided in Fig. 2. BodySLAM can be split into two main components: a front-end that performs the pre-processing of the input image stream, and the back-end that performs the graph optimisation (bundle adjustment).

In the front-end, BRISK keypoints are extracted from each image for use in camera tracking [29]. The monocular bundle adjustment problem is then initialised with static landmarks and camera poses obtained by applying RANSAC on the keypoints. Each image is also processed by an instance segmentation network [49] to detect human bounding boxes, and by OpenPose [9] to extract human keypoints. From these, human mesh parameters are estimated using a

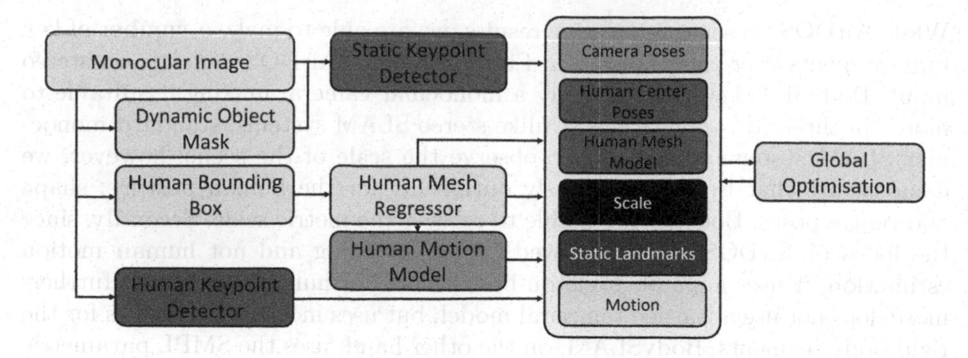

Fig. 2. The BodySLAM framework.

Human Mesh Regressor (HMR) network [28]. Finally, human motion and posture estimates are obtained from our deep motion model (see Sect. 3.5).

Many state-of-the-art methods perform bundle adjustment only on keyframes, a selected subset of the total frames; however, to avoid complications to the training and inference of the motion model by having a variable frame rate, we assume a constant frame rate of 30 FPS and estimate the camera pose, human centre pose, and human shape and posture parameters at every frame.

3.3 3D Representation of the Human Body

The SMPL model [32] is a parametric human mesh model that supplies a function, $\mathcal{M}(\boldsymbol{\beta}, \boldsymbol{\theta})$, of the shape parameters, $\boldsymbol{\beta}$, and posture parameters, $\boldsymbol{\theta}$, and returns a mesh, ${}_H\mathbf{M} \in \mathbb{R}^{N\times 3}$, as a set of $N = 6890$ vertices in the human-centric frame $\underrightarrow{\mathcal{F}}_H$. The body shape is parameterised as a 10-dimensional vector, $\boldsymbol{\beta} \in \mathbb{R}^{10}$, where the components are determined by a principal component analysis to capture as much of the variability in human body shape as possible. The body posture parameters, $\boldsymbol{\theta}$, represent each of the 23 joint orientations in the respective parent frame in axis-angle representation with dimensionality $\dim(\boldsymbol{\theta}) = 23 \times 3 = 69$. With this kinematic tree representation, it is easy to encode the same posture for different human shapes, as the joint positions can be defined as a linear combination of a set of mesh vertices. For J joints, a linear regressor $\mathbf{W} \in \mathbb{R}^{J\times N}$ is pre-trained to regress the J 3D body joints as ${}_H\mathbf{X} = \mathbf{W}\,{}_H\mathbf{M}$ with ${}_H\mathbf{X} \in \mathbb{R}^{J\times 3}$.

The 6D human centre pose is defined as $\mathbf{x}^{\text{human}} = [{}_W\mathbf{r}_{WH}, {}_W\boldsymbol{\alpha}_{WH}] \in \mathbb{R}^6$, where ${}_W\mathbf{r}_{WH}$ is the body centre position with respect to the static world frame, and ${}_W\boldsymbol{\alpha}_{WH}$ is the body centre orientation represented as a rotation vector.

3.4 Measurements of Human Body Parameters

SMPL Parameter Prediction. The state-of-the-art method from Kolotouros *et al.* [28] is used to generate measurements of the SMPL parameters $\{\tilde{\boldsymbol{\beta}}, \tilde{\boldsymbol{\theta}}\}$ in each frame. This method uses an HMR network to regress the SMPL parameters from a normalised and cropped image of a human. The cropped human image is obtained by detecting human bounding boxes using Detectron2 [49].

Human Centre Pose Prediction. We also use the network in [28] to generate measurements of the 6D transformation between the camera and human centre frames. In addition to the SMPL parameters, the HMR network also predicts a weak perspective camera model. As detailed in [17], after rectifying and removing focal distortion from the images of a calibrated camera, we can use the weak perspective camera parameters and the rotation between the camera centre and the centre of the detected bounding box to derive the transformation $\tilde{\boldsymbol{T}}_{CH}$.

3.5 Human Motion Model

Formulation. For the motion model, we use a deep neural network. Its design follows the architecture from Martinez *et al.* [34], but instead of a recurrent unit used for the temporal encoding, we use a sequential backbone only consisting of a multi-layer perceptron. The motion model uses the past n body posture estimates $\boldsymbol{\Theta} = [\boldsymbol{\theta}_{k\text{-}n}, \ldots, \boldsymbol{\theta}_{k\text{-}1}]$ and the (constant) body shape $\boldsymbol{\beta}$ as inputs, to predict the current body posture, $\boldsymbol{\theta}_k^{\text{pred}}$, and relative human centre translation, ${}_{H_{k\text{-}1}}\mathbf{r}^{\text{pred}}_{H_k H_{k\text{-}1}}$. This prediction is used as a motion measurement in the optimisation problem discussed in Sect. 3.6. The model is trained using the loss:

$$\mathcal{L} = c_p + \lambda c_\theta, \tag{1}$$

where $c_p = \|{}_{H_{k\text{-}1}}\mathbf{r}^{\text{true}}_{H_{k\text{-}1}H_k} - {}_{H_{k\text{-}1}}\mathbf{r}^{\text{pred}}_{H_{k\text{-}1}H_k}\|^2$ is the error on the predicted relative position, and $c_\theta = \|\boldsymbol{\theta}_k^{\text{true}} - \boldsymbol{\theta}_k^{\text{pred}}\|^2$ is the error on the human body posture. The global orientation change of the human centre frame is predicted as part of the 24 body joint angles. The relative weighting between the two error terms λ is set to 1/24, weighting the position error as much as each rotation in the kinematic tree. Since the input and output values of the network were normalised with the respective means and standard deviations of the dataset, the loss values quickly converge. No robust cost function was used for the optimisation loss.

Architecture. For the human posture and shape encoder, we use a 3-layer linear encoder, with an input layer of size $(n \times 3 \times 24) + 10$ for the n previous human body posture and shape estimates, and two fully connected layers of size 1024. We use 2 separate decoders for the human position change and human body posture, each with two fully connected layers, and a final layer of size 3 and 72, respectively. The human posture decoder has an additional residual connection to the most recent posture $\boldsymbol{\theta}_k = \boldsymbol{\theta}_{k\text{-}1} + \boldsymbol{\theta}^{\text{pred}}$.

Training Procedure. The human motion model is trained on a subset of the AMASS dataset [33], only considering sequences with walking, running, or jogging motions. Since we predict the pose change between two consecutive camera frames, we processed the data of the AMASS dataset to match the most common camera frame rate of 30 FPS. The network uses a fixed number of input frames for its prediction. To avoid overlapping sequences, we processed the dataset into chunks containing the required number of input frames and a target frame. We experimented with different sequence lengths of $n = \{2, 4, 8, 16\}$, but did not notice a drastic change in performance with longer sequences, and so trained the model with $n = 2$. This left us with more than 340,000 sequences, from which we reserved 20% for validation. We trained the network until convergence (100 epochs), with a batch size of 4096 on a single Nvidia GeForce GTX1080Ti GPU. We used the Adam optimiser [25] with a learning rate of 1×10^{-3}.

3.6 Factor Graph Formulation

Overview. During bundle adjustment, we jointly optimise camera poses, $\mathbf{x}^{\text{cam}} = [{}_W\mathbf{r}'_{WC_k}, {}_W\boldsymbol{\alpha}_{WC_k}]$, human centre poses, $\mathbf{x}^{\text{human}} = [{}_W\mathbf{r}_{WH_k}, {}_W\boldsymbol{\alpha}_{WH_k}]$, body shape, $\boldsymbol{\beta}$, postures, $\boldsymbol{\theta}_k$, and L static landmarks, ${}_W\boldsymbol{l}_l$ with $l \in 1, \ldots, L$. For faster convergence, we over-parameterise the camera position as ${}_W\mathbf{r}_{WC_k} = s\ {}_W\mathbf{r}'_{WC_k}$, where the scale s is estimated along with the (scaled) position vector $\mathbf{r}'_{WC_k}$. Thus we estimate $\mathbf{x} = [s, \boldsymbol{\beta}, \mathbf{x}_1^{\text{cam}}, \mathbf{x}_1^{\text{human}}, \boldsymbol{\theta}_1, \ldots, \mathbf{x}_K^{\text{cam}}, \mathbf{x}_K^{\text{human}}, \boldsymbol{\theta}_K, {}_W\boldsymbol{l}_1, \ldots, {}_W\boldsymbol{l}_L]$.

Figure 3 shows the respective factor graphs for three different SLAM formulations. The variables to be estimated are drawn as circles and the measurements are drawn as boxes. The measurements include 3D landmarks, OpenPose keypoint measurements and relative motion factors. The leftmost factor graph describes the traditional monocular SLAM problem without temporal constraints between the camera poses. The middle factor graph shows the naïve formulation, where human reprojection error terms are included in blue, but there are still no temporal constraints. The rightmost factor graph includes measurements from our learned motion model, creating temporal constraints between the estimated camera and human centre poses, rendering scale observable. Through our experimental results, we will demonstrate that using the rightmost factor graph leads to the most accurate estimates of both human body parameters and camera poses. Therefore, the total cost function optimised by BodySLAM is given by:

$$c(\mathbf{x}) = \lambda^{\text{lm}} \sum_{k=1}^{K} \sum_{l=1}^{L} \mathbf{e}_{k,l}^{\text{lm}\,\text{T}} \mathbf{e}_{k,l}^{\text{lm}} + \sum_{k=1}^{K} \sum_{j=1}^{J} \lambda_{k,j}^{\text{joints}} \mathbf{e}_{k,j}^{\text{joints}\,\text{T}} \mathbf{e}_{k,j}^{\text{joints}} + \lambda^{\text{mm}} \sum_{k=2}^{K} \mathbf{e}_k^{\text{mm}\,\text{T}} \mathbf{e}_k^{\text{mm}}$$
$$+ \lambda^{\text{posture}} \sum_{k=1}^{K} \mathbf{e}_k^{\text{posture}\,\text{T}} \mathbf{e}_k^{\text{posture}} + \lambda^{\text{shape}} \sum_{k=1}^{K} \mathbf{e}_k^{\text{shape}\,\text{T}} \mathbf{e}_k^{\text{shape}},$$

where $\mathbf{e}_{k,l}^{\text{lm}}$, $\mathbf{e}_{k,j}^{\text{joints}}$, $\mathbf{e}_k^{\text{mm}}$, $\mathbf{e}_k^{\text{posture}}$, and $\mathbf{e}^{\text{shape}}$ are the error terms for the structural landmarks, human joint landmarks, human motion, body posture, and body shape, respectively. Each of these error terms is discussed in detail below. The structural landmarks are weighted by $\lambda^{\text{lm}} = 64/b^2$, where b is the size of the BRISK keypoint ($b = 2.25$px in our implementation). The human joint landmarks are weighted by $\lambda^{\text{joints}}_{k,j} = \sigma^{-2}_{k,j}$, where $\sigma_{k,j}$ is confidence predicted by OpenPose. Finally, in our implementation, we use $\lambda^{\text{mm}} = 100$, $\lambda^{\text{posture}} = 1$, and $\lambda^{\text{shape}} = 1$, the values of which were determined experimentally.

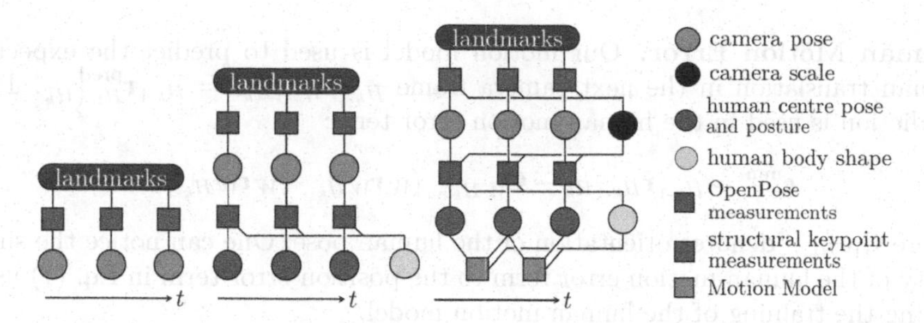

Fig. 3. Baseline and proposed Factor Graphs of the visual SLAM problem. Without human bodies, the monocular SLAM Factor Graph has no relation between successive camera poses (left). A naive formulation of SLAM problem with human 2D keypoint measurements as dynamic keypoints fails to create a relation between successive frames (middle). Our proposed BodySLAM formulation adds a motion model that predicts human centre pose and posture, and creates a temporal constraint for camera and human center poses (right). Additionally, we overparameterise the camera pose with an additional scale factor s, to achieve faster convergence of our Bundle Adjustment.

Reprojection Error for Structural Landmarks. For the structural landmarks, ${}_W\boldsymbol{l}_l$, we use the standard reprojection error term (as described in [30]):

$$\mathbf{e}^{\text{lm}}_{k,l} = \tilde{\mathbf{z}}_{k,l} - \mathbf{u}(\boldsymbol{T}^{-1}_{WC_k}\ {}_W\boldsymbol{l}_l), \tag{2}$$

where $\mathbf{u}(\cdot)$ denotes the camera projection model and $\tilde{\mathbf{z}}_{k,l}$ is the 2D keypoint measurement of the static landmark ${}_W\boldsymbol{l}_l$ in frame k. The measurements are BRISK keypoints filtered for outliers by 3D-2D RANSAC. To handle any remaining outliers, we use the robust Cauchy loss function on the error term.

Reprojection Error for Human Joint Landmarks. As done in [8,28], we use OpenPose keypoints and their corresponding confidence values as measurements in the joint reprojection error term, employing the same standard reprojection error term as for the structural keypoints:

$$\mathbf{e}^{\text{joints}}_{k,j} = \tilde{\mathbf{z}}_{k,j} - \mathbf{u}(\boldsymbol{T}^{-1}_{WC_k}\ \boldsymbol{T}_{WH_k}\ {}_H\boldsymbol{l}_{j,k}(\boldsymbol{\theta}_k, \boldsymbol{\beta})), \tag{3}$$

where $\tilde{\mathbf{z}}_{k,j}$ denotes the OpenPose keypoint measurement, and ${}_B\boldsymbol{l}_{j,k}(\boldsymbol{\theta}_k, \boldsymbol{\beta})$ denotes the SMPL joint j in frame k expressed as homogeneous points in the human centered frame $\underrightarrow{\mathcal{F}}_H$ as a function of the posture $\boldsymbol{\theta}_k$ and shape $\boldsymbol{\beta}$. To increase robustness in particular with (self-)occluded joints and noisy keypoint detections, we use the Geman-McLure robust cost function in line with [8].

Note that the SMPL parameters $\{\boldsymbol{\theta}, \boldsymbol{\beta}\}$ can be directly initialised from the predictions as explained in Sect. 3.4. In turn, the human poses can be initialised as $\boldsymbol{T}_{WH_k} = \boldsymbol{T}_{WC_k} \tilde{\boldsymbol{T}}_{C_k H_k}$.

Human Motion Error. Our motion model is used to predict the expected human translation in the next camera frame ${}_{H_{k\text{-}1}}\tilde{\mathbf{r}}_{H_{k\text{-}1}H_k} := {}_{H_{k\text{-}1}}\mathbf{r}^{\text{pred}}_{H_{k\text{-}1}H_k}$. This prediction is used in the human motion error term:

$$\mathbf{e}_k^{\text{mm}} = {}_{H_{k\text{-}1}}\tilde{\mathbf{r}}_{H_{k\text{-}1}H_k} - \mathbf{C}^T_{WH_{k\text{-}1}} \left({}_W\mathbf{r}_{WH_k} - {}_W\mathbf{r}_{WH_{k\text{-}1}} \right), \tag{4}$$

where $\mathbf{C}_{WH_{k\text{-}1}}$ denotes orientation of the human pose. One can notice the similarity of the human motion error term to the position error term in Eq. (1) used during the training of the human motion model.

Human Body Posture Error. Our motion model predicts the expected human body posture in the current frame, $\tilde{\boldsymbol{\theta}}_k$. The posture error is formulated as:

$$\mathbf{e}_k^{\text{posture}} = \tilde{\boldsymbol{\theta}}_k - \boldsymbol{\theta}_k. \tag{5}$$

Human Shape Error. In each frame, we use an HMR network to predict the SMPL parameters $\{\tilde{\boldsymbol{\beta}}_k, \tilde{\boldsymbol{\theta}}_k\}$. As $\boldsymbol{\beta}$ should be constant over the image sequence, we constrain this to be a constant using the body shape error term:

$$\mathbf{e}_k^{\text{shape}} = \tilde{\boldsymbol{\beta}}_k - \boldsymbol{\beta}. \tag{6}$$

Optimisation Procedure and Implementation Details. To initialise the full BodySLAM optimisation, we perform a straightforward implementation of bundle adjustment using the BRISK keypoints, RANSAC-based pose initialisation and optimisation using the Google Ceres optimiser [1] to obtain estimates for the camera poses and landmark locations up to an arbitrary scale. After this initialisation, we use a first-order nonlinear least squares optimisation in PyTorch [37] for bundle adjustment. However, to increase convergence speed and performance, one could easily implement a second-order optimisation.

Since the first-order optimisation is prone to converge at local minima, we use a two-step optimisation schedule often used in human mesh optimisation [4, 8,28]. First, we optimise only the relative transformations between the cameras and the human body centres, the transformations between the cameras and the world frame, and the static landmarks, while keeping all other variables constant. The purpose of this optimisation step is to roughly estimate the scale of the given scene, and the camera and human centre trajectories. We optimise until convergence, finding approximately 200 iterations to be sufficient. In a second step, we optimise all variables for another 100 iterations after which we always reached convergence in our experiments.

4 Experimental Results

To evaluate BodySLAM, we collected a dataset of 10 video sequences of human motion captured by a moving monocular camera. A motion capture system was used to collect ground truth camera trajectories and human body parameters. This dataset is described in more detail in Sect. 4.1.

For evaluation, we consider the standard metric of average trajectory error for the camera trajectory (C-ATE), the human centre trajectory (H-ATE), and the human joint trajectories (J-ATE). Furthermore, we align the estimated camera trajectories in Sim(3), and interpret the aligned scale $s_{\mathrm{Sim}(3)}$ as scale error. With a perfect scale estimation, this estimated scale would be 1.0.

As one of the main contributions of BodySLAM is its ability to estimate metric scale from a monocular image sequence, we first evaluate how well it can recover the scale of a scene when it is given the scale-perturbed ground truth camera trajectory. The results of this preliminary study are provided in Sect. 4.2. In Sect. 4.3, we evaluate the ability of the full BodySLAM system to estimate the camera poses, human body centre poses, and human joint positions for all sequences in our dataset. Finally, in Sect. 4.4, we perform an ablation study on our motion model, showing that our model is able to provide meaningful estimates of human motion and improves estimates of the trajectories.

4.1 Custom Human Pose Estimation Dataset

Current real-world evaluation datasets are lacking at least one of the following properties: a moving camera, known camera intrinsics, ground truth camera motion, a moving human subject, or ground truth human body parameters. An extensive but incomplete list of these can be found in Table 1 in the Appendix. Due to these limitations, we perform the evaluation on our own custom dataset that captures the ground truth camera poses of a moving camera with known intrinsics, along with human centre poses and posture parameters for moving humans. Ground truth values were obtained using a Vicon motion capture system. The dataset consists of 10 separate sequences, each between 600 to 1000 frames, totalling 8174 images. The dataset is split into easy, medium, and hard sequences, categorised by the speed and complexity of the camera motion.

4.2 Study of Scale Recovery Through Motion Factors

To evaluate the ability of BodySLAM to recover metric scale, we perform an ablation study where structural landmark measurements are ignored and where we initialise the camera trajectory with the Vicon ground truth poses after scaling the camera positions with an arbitrary perturbation factor between 0.1 and 5.0. We then freeze the Sim(3) transformations of the camera poses in the graph, only allowing the scale of the camera trajectory, s, to be optimised along with the human body parameters. As shown in Fig. 4, after initial perturbation by $1/s'$, we are able to reduce the H-ATE to approximately the value it had been before perturbation. Furthermore, we show that the scale is correctly estimated with a maximum error of 26.5% for the most extreme perturbation factor of 5.0 (as found in Table 3 in the Appendix). For smaller initial perturbations, the estimated scale deviates by only roughly 5% from the true scale.

4.3 BodySLAM: Optimisation of Camera Poses and Human States

In Table 1, we present the main results for our full BodySLAM system. As a baseline for comparison, we apply classic monocular bundle adjustment with naïve human motion tracking (i.e. only considering the unary OpenPose measurements and no motion model). This baseline, represented by the middle factor graph in Fig. 3, also does not use the scale overparameterisation of BodySLAM.

Our results show that both the C-ATE and the H-ATE, after $SE(3)$ alignment, are significantly lower than the baseline method. The average improvement through the introduction of the motion model is 48.3% for the root trajectory, and 35.6% for all joints combined.

Furthermore, we show that the motion model factor is able to accurately recover the scale, with scale estimation errors in the range of 0.93 to 1.03, compared to errors between 0.35 to 0.74 for the baseline, as shown in Table 1. However, one must note that the values of the scale error are completely arbitrary, since they depend only on the random initialisation of the camera poses. The C-ATE with $SE(3)$ alignment for BodySLAM shows a huge improvement compared to the baseline results. This improvement is attributed to the correctly estimated scale, as the results after Sim(3) alignment clearly show. After Sim(3)

Table 1. Comparison of absolute trajectory errors: camera (C-ATE), human (H-ATE), and joints (J-ATE) plus camera trajectory scale on test sequences for the baseline without motion model factor ("no MM") and the full BodySLAM (ours). Position errors are computed after SE(3) alignment to account for the difference in the Vicon global frame and visual world frame $\underset{\rightarrow}{\mathcal{F}}_W$. We also aligned the scale-unaware baseline using Sim(3), which reveals that our method still outperforms it regarding human centre position and joint locations, and performs on-par regarding camera pose accuracy. Note, however, that **in the wild Sim(3) alignment is not possible**, since it requires access to the ground truth camera trajectory – illustrating the usefulness of our method.

	C-ATE [mm]				H-ATE [mm]				J-ATE [mm]				Scale error	
	SE(3)		Sim(3)		SE(3)		Sim(3)		SE(3)		Sim(3)			
Seq.	no MM	(ours)	no MM	(ours)	no MM	(ours)	no MM	(ours)	no MM	(ours)	no MM	(ours)	no MM	(ours)
E 1	108.1	**44.3**	42.4	43.5	114.0	**56.3**	106.5	56.0	154.8	**118.8**	149.8	115.9	0.671	**0.940**
E 2	193.0	**50.8**	50.7	49.1	178.6	**55.0**	182.8	53.0	210.9	**124.2**	214.6	120.5	0.581	**1.030**
E 3	122.3	**40.9**	40.8	38.9	220.8	**160.9**	217.1	160.4	247.7	**196.1**	244.3	192.8	0.738	**0.981**
M 1	314.7	**92.6**	93.3	91.0	326.9	**201.4**	314.6	200.0	368.8	**236.2**	355.3	236.2	0.502	**0.978**
M 2	241.6	**64.0**	57.9	61.1	368.1	**295.4**	353.5	297.1	408.6	**323.3**	392.0	325.4	0.507	**0.936**
M 3	392.1	**128.3**	128	128.1	311.2	**121.7**	305.9	122.2	335.5	**170.2**	326.4	171.9	0.354	**0.985**
M 4	203.1	**92.4**	91.3	90.9	153.7	**70.4**	142.8	71.7	184.2	**123.0**	175.0	125.3	0.635	**0.951**
D 1	310.4	**108.1**	107	103.5	303.7	**136.7**	293.9	135.9	330.6	**180.6**	321.1	181.1	0.445	**0.951**
D 2	316.0	**100.0**	98.9	95.2	297.6	**91.2**	283.5	90.6	320.6	**142.6**	306.8	143.8	0.430	**0.933**
D 3	246.9	**115.3**	113	114.0	235.4	**149.9**	228.8	149.5	264.5	**183.9**	258.0	185.0	0.494	**0.942**

alignment, the C-ATEs differ only by a few millimeters, which can be caused by small convergence differences during the first order optimisation.

We performed an additional experiment, ablating the necessity of the structural landmarks in the joint optimisation. For this, we performed an optimisation of the camera poses and human states, with and without structural landmarks to constrain the camera poses. The initial camera poses were randomly perturbed (±100 mm, ±0.01 rad) and optimised until convergence. The optimisation reduced the mean camera position error from 228.96 mm after perturbation to 155.5 mm with landmarks, and 158.6 mm without landmarks (1.94% error increase), highlighting BodySLAM's versatility. The complete results are found in Table 2 in the Appendix.

4.4 Motion Model Prediction Performance

To evaluate the contribution of our motion model, we analyse the performance of the model on the test split of our preprocessed AMASS dataset.

The mean translation prediction error is 3.4 mm for a mean per-frame position change of 18.4 mm. The distribution of the error vector components is shown in Fig. 5. Please note that in the SMPL body model, the human body coordinate frame $\underrightarrow{\mathcal{F}}_H$ is aligned to the sagittal axis (forward) in the z-direction, and the longitudinal axis (upward) in y-direction. Despite a relatively large standard deviation, our prediction errors are centered around zero, and, as previous results have shown, can significantly improve the estimation of the scale and the human centre and joint trajectories when included in the graph optimisation.

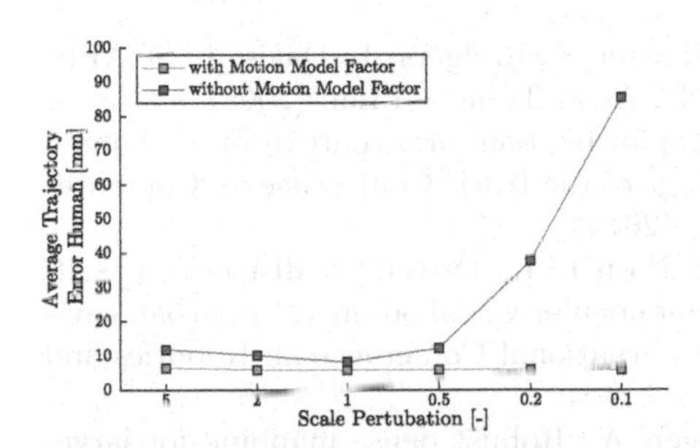

Fig. 4. Average Human Trajectory Error, $SE(3)$ aligned. Without Motion Model, the scale is not recovered.

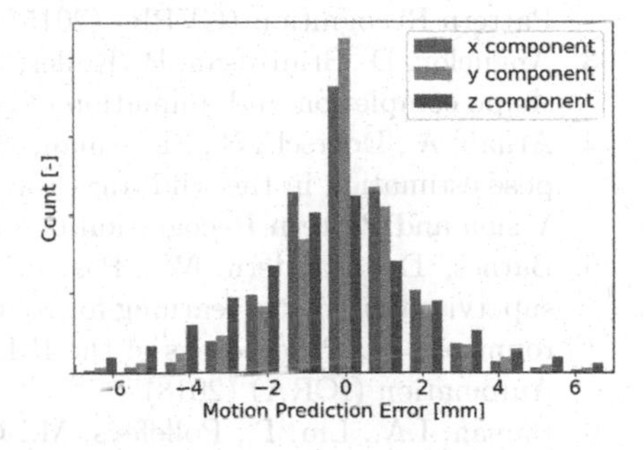

Fig. 5. Human motion prediction error histogram in x, y, and z direction.

5 Conclusions and Future Work

We have presented BodySLAM, a novel monocular SLAM system that jointly optimises a set of 6D camera poses, 3D landmarks, human body centre poses, and human body shape and posture parameters in a factor graph formulation.

To the best of our knowledge, this is the first system to jointly estimate for the configuration of a dense human body mesh model and the trajectory of a moving monocular camera. We have also introduced a human motion model to constrain sequential body postures and to help observe scale. To validate our contribution, we have collected a series of video sequences of human motion captured by a moving monocular camera, with ground truth camera trajectories and human body parameters from a motion capture system. Through a number of experiments, we have demonstrated that our joint formulation consistently results in improvements to both camera tracking and human body estimation.

In future work, we plan to extend the BodySLAM method into an incremental and real-time SLAM system. We would also like to explore more complex architectures for the human motion model and evaluate whether or not any gains in accuracy offset the increased complexity of the factor graph formulation that these architectures would require. Finally, one limitation of our method is that it assumes the intrinsic parameters of the camera are given, which may not always be the case. Some very recent work [27] has demonstrated the ability to estimate the parameters of a perspective camera from a single image. Adding the camera parameters to our factor graph formulation and initialising with this method may allow us to apply BodySLAM to true "in the wild" datasets.

References

1. Agarwal, S., Mierle, K., et al.: Ceres Solver. http://ceres-solver.org
2. Akhter, I., Black, M.J.: Pose-conditioned joint angle limits for 3D human pose reconstruction. In: Proceedings of the IEEE Conference on Computer Vision and Pattern Recognition (CVPR) (2015)
3. Anguelov, D., Srinivasan, P., Koller, D., Thrun, S., Rodgers, J., Davis, J.: SCAPE: shape completion and animation of people. ACM Trans. Graph. (2005)
4. Arnab, A., Doersch, C., Zisserman, A.: Exploiting temporal context for 3D human pose estimation in the wild. In: Proceedings of the IEEE Conference on Computer Vision and Pattern Recognition (CVPR) (2019)
5. Barnes, D., Maddern, W., Pascoe, G., Posner, I.: Driven to distraction: self-supervised distractor learning for robust monocular visual odometry in urban environments. In: Proceedings of the IEEE International Conference on Robotics and Automation (ICRA) (2018)
6. Bârsan, I.A., Liu, P., Pollefeys, M., Geiger, A.: Robust dense mapping for large-scale dynamic environments. In: Proceedings of the IEEE International Conference on Robotics and Automation (ICRA) (2018)
7. Bescos, B., Fácil, J.M., Civera, J., Neira, J.: DynaSLAM: tracking, mapping and inpainting in dynamic scenes. Technical report (2018)
8. Bogo, F., Kanazawa, A., Lassner, C., Gehler, P., Romero, J., Black, M.J.: Keep It SMPL: automatic estimation of 3D human pose and shape from a single image. In: Leibe, B., Matas, J., Sebe, N., Welling, M. (eds.) ECCV 2016. LNCS, vol. 9909, pp. 561–578. Springer, Cham (2016). https://doi.org/10.1007/978-3-319-46454-1_34
9. Cao, Z., Simon, T., Wei, S.E., Sheikh, Y.: Realtime multi-person 2D pose estimation using part affinity fields. In: Proceedings of the IEEE Conference on Computer Vision and Pattern Recognition (CVPR) (2017)

10. Catalin Ionescu Fuxin Li, C.S.: Latent structured models for human pose estimation. In: Proceedings of the International Conference on Computer Vision (ICCV) (2011)
11. Dabral, R., Mundhada, A., Kusupati, U., Afaque, S., Sharma, A., Jain, A.: Learning 3D human pose from structure and motion. In: Ferrari, V., Hebert, M., Sminchisescu, C., Weiss, Y. (eds.) ECCV 2018. LNCS, vol. 11213, pp. 679–696. Springer, Cham (2018). https://doi.org/10.1007/978-3-030-01240-3_41
12. Dai, W., Zhang, Y., Li, P., Fang, Z., Scherer, S.: RGB-D SLAM in dynamic environments using point correlations. IEEE Trans. Pattern Anal. Mach. Intell. (PAMI) **44**, 373–389 (2020)
13. Fang, H.S., Xie, S., Tai, Y.W., Lu, C.: RMPE: regional multi-person pose estimation. In: Proceedings of the International Conference on Computer Vision (ICCV) (2017)
14. Habibie, I., Xu, W., Mehta, D., Pons-Moll, G., Theobalt, C.: In the wild human pose estimation using explicit 2D features and intermediate 3D representations. In: Proceedings of the IEEE Conference on Computer Vision and Pattern Recognition (CVPR) (2019)
15. Henein, M., Kennedy, G., Mahony, R., Ila, V.: Exploiting rigid body motion for SLAM in dynamic environments. In: Proceedings of the IEEE International Conference on Robotics and Automation (ICRA) (2018)
16. Henein, M., Zhang, J., Mahony, R., Ila, V.: Dynamic SLAM: the need for speed. In: Proceedings of the IEEE International Conference on Robotics and Automation (ICRA) (2020)
17. Henning, D., Guler, A., Leutenegger, S., Zafeiriou, S.: HPE3D: human pose estimation in 3D (2020). https://github.com/dorianhenning/hpe3d
18. Hossain, M.R.I., Little, J.J.: Exploiting temporal information for 3D human pose estimation. In: Ferrari, V., Hebert, M., Sminchisescu, C., Weiss, Y. (eds.) ECCV 2018. LNCS, vol. 11214, pp. 69–86. Springer, Cham (2018). https://doi.org/10.1007/978-3-030-01249-6_5
19. Huang, Y., et al.: Towards accurate marker-less human shape and pose estimation over time. In: Proceedings of the International Conference on 3D Vision (3DV) (2017)
20. Jaimez, M., Kerl, C., Gonzalez-Jimenez, J., Cremers, D.: Fast odometry and scene flow from RGB-D cameras based on geometric clustering. In: Proceedings of the IEEE International Conference on Robotics and Automation (ICRA) (2017)
21. Ji, T., Wang, C., Xie, L.: Towards real-time semantic RGB-D SLAM in dynamic environments. In: Proceedings of the IEEE International Conference on Robotics and Automation (ICRA) (2021)
22. Judd, K.M., Gammell, J.D., Newman, P.: Multimotion Visual Odometry (MVO): simultaneous estimation of camera and third-party motions. In: Proceedings of the IEEE/RSJ International Conference on Intelligent Robots and Systems (IROS) (2018)
23. Kanazawa, A., Black, M.J., Jacobs, D.W., Malik, J.: End-to-end recovery of human shape and pose. In: Proceedings of the IEEE Conference on Computer Vision and Pattern Recognition (CVPR) (2018)
24. Kanazawa, A., Zhang, J.Y., Felsen, P., Malik, J.: Learning 3D human dynamics from video. In: Proceedings of the IEEE Conference on Computer Vision and Pattern Recognition (CVPR) (2019)
25. Kingma, D.P., Ba, J.L.: Adam: a method for stochastic optimization. In: 3rd International Conference on Learning Representations, ICLR 2015 - Conference Track Proceedings (2015)

26. Kocabas, M., Athanasiou, N., Black, M.J.: VIBE: video inference for human body pose and shape estimation. In: Proceedings of the IEEE Conference on Computer Vision and Pattern Recognition (CVPR) (2020)
27. Kocabas, M., Huang, C.H.P., Tesch, J., Müller, L., Hilliges, O., Black, M.J.: SPEC: seeing people in the wild with an estimated camera. In: Proceedings of the International Conference on Computer Vision (ICCV) (2021)
28. Kolotouros, N., Pavlakos, G., Black, M.J., Daniilidis, K.: Learning to reconstruct 3D human pose and shape via model-fitting in the loop. In: Proceedings of the International Conference on Computer Vision (ICCV) (2019)
29. Leutenegger, S., Chli, M., Siegwart, R.Y.: BRISK: binary robust invariant scalable keypoints. In: Proceedings of the International Conference on Computer Vision (ICCV) (2011)
30. Leutenegger, S., Lynen, S., Bosse, M., Siegwart, R., Furgale, P.: Keyframe-based visual-inertial odometry using nonlinear optimization. Int. J. Robot. Res. (IJRR) **34**, 314–334 (2015)
31. Ling, H.Y., Zinno, F., Cheng, G., van de Panne, M.: Character controllers using motion vaes. ACM Trans. Graph. **39**, 40 (2020)
32. Loper, M., Mahmood, N., Romero, J., Pons-Moll, G., Black, M.J.: SMPL: a skinned multi-person linear model. ACM Trans. Graph. **34**, 1–16 (2015)
33. Mahmood, N., Ghorbani, N., Troje, N.F., Pons-Moll, G., Black, M.J.: AMASS: archive of motion capture as surface shapes. In: Proceedings of the International Conference on Computer Vision (ICCV) (2019)
34. Martinez, J., Hossain, R., Romero, J., Little, J.J.: A simple yet effective baseline for 3d human pose estimation. In: Proceedings of the International Conference on Computer Vision (ICCV) (2017)
35. Mur-Artal, R., Tardos, J.D.: ORB-SLAM2: an open-source SLAM system for monocular, stereo, and RGB-D cameras. IEEE Trans. Robot. **33**, 1255–1262 (2017)
36. Osman, A.A.A., Bolkart, T., Black, M.J.: STAR: sparse trained articulated human body regressor. In: Vedaldi, A., Bischof, H., Brox, T., Frahm, J.-M. (eds.) ECCV 2020. LNCS, vol. 12351, pp. 598–613. Springer, Cham (2020). https://doi.org/10.1007/978-3-030-58539-6_36
37. Paszke, A., et al.: Automatic differentiation in PyTorch. In: Neural Information Processing Systems (NIPS) (2017)
38. Pavlakos, G., et al.: Expressive body capture: 3D hands, face, and body from a single image. In: Proceedings of the IEEE Conference on Computer Vision and Pattern Recognition (CVPR) (2019)
39. Pavlakos, G., Zhou, X., Daniilidis, K.: Ordinal depth supervision for 3D human pose estimation. In: Proceedings of the IEEE Conference on Computer Vision and Pattern Recognition (CVPR) (2018)
40. Pavlakos, G., Zhou, X., Derpanis, K.G., Daniilidis, K.: Coarse-to-fine volumetric prediction for single-image 3D human pose. In: Proceedings of the IEEE Conference on Computer Vision and Pattern Recognition (CVPR) (2017)
41. Pavllo, D., Feichtenhofer, C., Grangier, D., Auli, M.: 3D human pose estimation in video with temporal convolutions and semi-supervised training. In: Proceedings of the IEEE Conference on Computer Vision and Pattern Recognition (CVPR) (2018)
42. Qiu, Y., Wang, C., Wang, W., Henein, M., Scherer, S.: AirDOS: dynamic SLAM benefits from articulated objects. In: Proceedings of the IEEE International Conference on Robotics and Automation (ICRA) (2022)

43. Ramakrishna, V., Kanade, T., Sheikh, Y.: Reconstructing 3D human pose from 2D image landmarks. In: Fitzgibbon, A., Lazebnik, S., Perona, P., Sato, Y., Schmid, C. (eds.) ECCV 2012. LNCS, vol. 7575, pp. 573–586. Springer, Heidelberg (2012). https://doi.org/10.1007/978-3-642-33765-9_41
44. Rempe, D., Birdal, T., Hertzmann, A., Yang, J., Sridhar, S., Guibas, L.J.: HuMoR: 3D human motion model for robust pose estimation. In: Proceedings of the International Conference on Computer Vision (ICCV) (2021)
45. Rünz, M., Agapito, L.: Co-fusion: real-time segmentation, tracking and fusion of multiple objects. In: Proceedings of the IEEE International Conference on Robotics and Automation (ICRA) (2017)
46. Rünz, M., Buffier, M., Agapito, L.: MaskFusion: real-time recognition, tracking and reconstruction of multiple moving objects. In: Proceedings of the International Symposium on Mixed and Augmented Reality (ISMAR) (2018)
47. Scona, R., Jaimez, M., Petillot, Y.R., Fallon, M., Cremers, D.: StaticFusion: background reconstruction for dense RGB-D SLAM in dynamic environments. In: Proceedings of the IEEE International Conference on Robotics and Automation (ICRA) (2018)
48. Valmadre, J., Lucey, S.: Deterministic 3D human pose estimation using rigid structure. In: Daniilidis, K., Maragos, P., Paragios, N. (eds.) ECCV 2010. LNCS, vol. 6313, pp. 467–480. Springer, Heidelberg (2010). https://doi.org/10.1007/978-3-642-15558-1_34
49. Wu, Y., Kirillov, A., Massa, F., Lo, W.Y., Girshick, R.: Detectron2 (2019). https://github.com/facebookresearch/detectron2
50. Xu, B., Li, W., Tzoumanikas, D., Bloesch, M., Davison, A., Leutenegger, S.: MID-fusion: octree-based object-level multi-instance dynamic SLAM. In: Proceedings of the IEEE International Conference on Robotics and Automation (ICRA) (2019)
51. Zhao, R., Wang, Y., Martinez, A.: A simple, fast and highly-accurate algorithm to recover 3D shape from 2D landmarks on a single image. IEEE Trans. Pattern Anal. Mach. Intell. (PAMI) **40**, 3059–3066 (2016)

FusionVAE: A Deep Hierarchical Variational Autoencoder for RGB Image Fusion

Fabian Duffhauss[1,2(✉)], Ngo Anh Vien[1], Hanna Ziesche[1], and Gerhard Neumann[3]

[1] Bosch Center for Artificial Intelligence, Renningen, Germany
{Fabian.Duffhauss,AnhVien.Ngo,Hanna.Ziesche}@bosch.com
[2] University of Tübingen, Tübingen, Germany
[3] Karlsruhe Institute of Technology, Karlsruhe, Germany
gerhard.neumann@kit.edu

Abstract. Sensor fusion can significantly improve the performance of many computer vision tasks. However, traditional fusion approaches are either not data-driven and cannot exploit prior knowledge nor find regularities in a given dataset or they are restricted to a single application. We overcome this shortcoming by presenting a novel deep hierarchical variational autoencoder called FusionVAE that can serve as a basis for many fusion tasks. Our approach is able to generate diverse image samples that are conditioned on multiple noisy, occluded, or only partially visible input images. We derive and optimize a variational lower bound for the conditional log-likelihood of FusionVAE. In order to assess the fusion capabilities of our model thoroughly, we created three novel datasets for image fusion based on popular computer vision datasets. In our experiments, we show that FusionVAE learns a representation of aggregated information that is relevant to fusion tasks. The results demonstrate that our approach outperforms traditional methods significantly. Furthermore, we present the advantages and disadvantages of different design choices.

1 Introduction

Sensor fusion is a popular technique in computer vision as it allows to combine the advantages from multiple information sources. It is especially gainful in scenarios where a single sensor is not able to capture all necessary data to perform a task satisfactorily. Over the last years, we have seen many examples, where the accuracy of computer vision tasks was significantly improved by sensor fusion, e.g. in environmental perception for autonomous driving [6,25,59], for 6D pose estimation [14,15,53], and for robotic grasping [54,64]. However, traditional fusion methods usually focus more on the beneficial merging of multiple

---

Supplementary Information The online version contains supplementary material available at https://doi.org/10.1007/978-3-031-19842-7_39.

S. Avidan et al. (Eds.): ECCV 2022, LNCS 13699, pp. 674–691, 2022.
https://doi.org/10.1007/978-3-031-19842-7_39

modalities and less on teaching the model to obtain profound prior knowledge about the used dataset.

Our work tries to fill in this research gap by proposing a deep hierarchical variational autoencoder called FusionVAE that is able to perform both tasks: fusing information from multiple sources and supplementing it with prior knowledge about the data gained while training. As shown in Fig. 1, FusionVAE merges a varying number of input images for reconstructing the original target image using prior knowledge about the dataset. To the best of our knowledge, FusionVAE is the first approach that combines these two tasks. Therefore, we developed three challenging benchmarks based on well-known computer vision datasets to evaluate the performance of our approach. In addition, we perform comparisons to baselines by extending traditional approaches to perform the same tasks. FusionVAE outperforms all these traditional methods on all proposed benchmark tasks significantly. We show that FusionVAE can generate high-quality images given few input images with partial observability. We provide ablation studies to illustrate the impact of commonly used information aggregation operations and to prove the benefits of the employed posterior distribution.

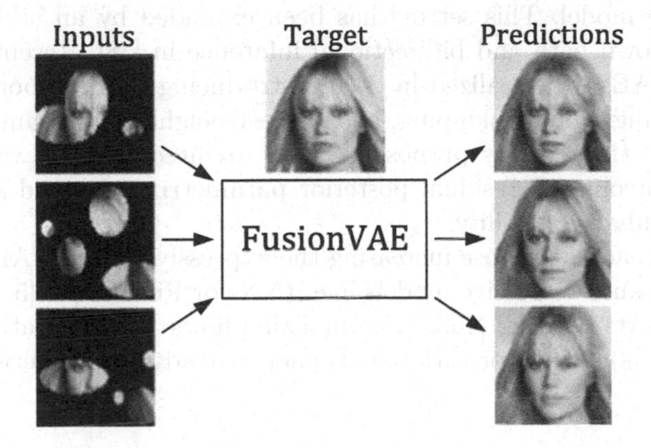

Fig. 1. Overview of our FusionVAE approach. The network receives up to three partly occluded input images, fuses them together with prior knowledge, and predicts different hypothesis of how the target images could look like.

We can summarize the four main contributions of our paper as follows: i) We create three challenging image fusion tasks for generative models. ii) We develop a deep hierarchical VAE called FusionVAE that is able to perform image-based data fusion while employing prior knowledge of the used dataset. iii) We show that FusionVAE produces high-quality fused output images and outperforms traditional methods by a large margin. iv) We perform ablation studies showing the benefits of our design choices regarding both the posterior distribution and commonly used aggregation methods.

2 Related Work

In this section, we present related work about image generation, image fusion, and image completion.

2.1 VAE-Based Image Generation

Variational autoencoders (VAE) are powerful networks that are able to compress the essence of datasets in a small latent space while being able to exploit it for generative tasks [22]. However, the standard VAE is limited in capacity and expressiveness and thus, when applied to image generation leads to oversmoothed results lacking fine-grained details. Over the last years much work has been invested into the effort of improving the generative performance of VAEs. One stream of work is based on introducing a hierarchy into the latent space of the VAE and scaling this hierarchy to greater and greater depth. First introduced in [46] many hierarchical VAEs are based on coupling the inference and generative processes by introducing a deterministic bottom-up path combined with a stochastic top-down process in the inference network and sharing the latter with the generative model. This setting has been extended by an additional deterministic top-down path and bidirectional inference in [38]. Recently, very deep hierarchical VAEs were realized in [7] by introducing residual bottlenecks with dedicated scaling, update skipping, and nearest neighbour up-sampling. Closest to our work is the recently proposed NVAE architecture [49], which relies on depth-wise convolution, residual posterior parametrization, and spectral regularization to enhance stability.

Other approaches propose increasing the expressiveness of VAEs by combining them with auto-regressive models like RNNs or PixelCNNs [5,9,11,43], conditioning contexts (CVAE) [45,52], normalizing flows [23], generative adverserial networks (GANs) [26,40], or variational generative adversarial networks (CVAE-GAN) [1].

2.2 Fusion of Multiple Images

Image fusion has long been dominated by classical computer vision. Only lately deep learning methods entered the domain with the CNN-based approach proposed by Liu et al. [33]. In a subsequent publication the authors extended their work to a multi-scale setting [32]. Shortly afterwards, Prabhakar et al. developed a fusion method based on a siamese network architecture, called DeepFuse [42] which was improved in subsequent work [28] by employing the DenseNet architecture [17]. Concurrently, Li et al. [30] proposed a fusion architecture based on VGG [44] and in order to scale to even greater depth another one [29] based on ResNet-50 [13]. The aforementioned methods use CNNs as feature extractors and as decoders, while the fusion operations themselves are restricted to classical methods like averaging or addition of feature maps or weighted source images. A fully CNN-based feature-map fusion mechanism was proposed in [20].

While all previous publications target only a specific fusion task (e.g. multi-focus fusion, multi-resolution fusion, etc.) or were limited to specific domains (e.g. medical images), two very recent works propose novel multi-purpose fusion networks, which are applicable to many fusion tasks and image types [56,63]. Very recently also GAN-based methods entered the domain of image fusion, starting with the work by Ma et al. on infrared-visible fusion [35,37] and with [36,55] on multi-resolution image fusion. Most recent are two publications on GAN-based multi-focus image fusion [12,18]. While GAN-based approaches can generate high-fidelity images, it is known that they suffer from the mode collapse problem. VAE-based methods in contrast are known to generate more faithful data distribution [49]. Different from previous work, this paper proposes a VAE-based multi-purpose fusion framework.

2.3 Image Completion

Similar to image fusion, also image completion has only recently become a playing field for deep learning methods. First approaches based on simple multilayer perceptrons (MLPs) [24] or CNNs [10] were targeted only to filling small holes in an image. However, with the introduction of GANs [8], the area quickly became dominated by GAN-based approaches, starting with the context encoders presented by Pathak et al. [41]. Many subsequent papers proposed extensions to this model in order to obtain fine-grained completions while preserving global coherence by introducing additional discriminators [19], searching for closest samples to the corrupted image in a latent embedding space conditioning on semantic labels [48], or designing additional specialized loss functions [31]. High resolution results were obtained recently by multi-scale approaches [58], iterative upsampling [62], and the application of contextual attention [47,57,60]. Another stream of current work focuses on multi-hypothesis image completion, leveraging probabilistic problem formulations [39,65].

3 Background

In this section, we outline the fundamentals of standard VAEs, conditional VAEs, and hierarchical VAEs upon which we build our approach. Another section is dedicated to aggregation methods for data fusion.

3.1 Standard VAE

A variational autoencoder (VAE) [22] is a neural network consisting of a probabilistic encoder $q(\boldsymbol{z}|\boldsymbol{y})$ and a generative model $p(\boldsymbol{y}|\boldsymbol{z})$. The generator models a distribution over the input data $\boldsymbol{y}$, conditioned on a latent variable $\boldsymbol{z}$ with prior distribution $p_\theta(\boldsymbol{z})$. The encoder approximates the posterior distribution $p(\boldsymbol{z}|\boldsymbol{y})$ of the latent variables $\boldsymbol{z}$ given input data $\boldsymbol{y}$ and is trained along with the generative model by maximizing the evidence lower bound (ELBO)

$$\mathrm{ELBO}(\boldsymbol{y}) = \mathbb{E}_{q(\boldsymbol{z}|\boldsymbol{y})}[\log p(\boldsymbol{y}|\boldsymbol{z})] - \mathrm{KL}(q(\boldsymbol{z}|\boldsymbol{y})||p(\boldsymbol{z})), \tag{1}$$

where KL is the Kullback-Leibler divergence and $\log p(\boldsymbol{y}) \geq \mathrm{ELBO}(\boldsymbol{y})$.

3.2 Conditional VAE

In VAEs, the generative model $p(\boldsymbol{y}|\boldsymbol{z})$ is unconditional. In contrast, conditional VAEs (CVAE) [45] consider a generative model for a conditional distribution $p(\boldsymbol{y}|\boldsymbol{x}, \boldsymbol{z})$ where $\boldsymbol{y}$ is the target data, $\boldsymbol{x}$ is the conditional input variable, and $\boldsymbol{z}$ is a latent variable. The prior of the latent variable is $p(\boldsymbol{z}|\boldsymbol{x})$, while its approximate posterior distribution is given by $q(\boldsymbol{z}|\boldsymbol{x}, \boldsymbol{y})$. The variational lower bound of the conditional log-likelihood can be written as follows

$$\log p(\boldsymbol{y}|\boldsymbol{x}) \geq \mathbb{E}_{q(\boldsymbol{z}|\boldsymbol{x},\boldsymbol{y})}[\log p(\boldsymbol{y}|\boldsymbol{x}, \boldsymbol{z})] - \mathrm{KL}(q(\boldsymbol{z}|\boldsymbol{x}, \boldsymbol{y})||p(\boldsymbol{z}|\boldsymbol{x})). \quad (2)$$

3.3 Hierarchical VAE

In hierarchical VAEs [21,46,49], the latent variables $\boldsymbol{z}$ are divided into L disjoint groups $\boldsymbol{z}_1, ..., \boldsymbol{z}_L$ in order to increase the expressiveness of both prior and approximate posterior which become

$$p(\boldsymbol{z}) = \prod_{l=1}^{L} p(\boldsymbol{z}_l|\boldsymbol{z}_{<l}) \quad \text{and} \quad q(\boldsymbol{z}|\boldsymbol{y}) = \prod_{l=1}^{L} q(\boldsymbol{z}_l|\boldsymbol{z}_{<l}, \boldsymbol{y}), \quad (3)$$

where $\boldsymbol{z}_{<l}$ denotes the latent variables in all previous hierarchies. All the conditionals in the prior $p(\boldsymbol{z}_l|\boldsymbol{z}_{<l})$ and in the approximate posterior $q(\boldsymbol{z}_l|\boldsymbol{z}_{<l}, \boldsymbol{y})$ are modeled by factorial Gaussian distributions. Under this modelling choice, the ELBO from Eq. (1) turns into

$$\mathrm{ELBO}(\boldsymbol{y}) = \mathbb{E}_{q(\boldsymbol{z}|\boldsymbol{y})}[\log p(\boldsymbol{y}|\boldsymbol{z})] - \sum_{l=1}^{L} \mathbb{E}_{q(\boldsymbol{z}_{<l}|\boldsymbol{y})}[\mathrm{KL}(q(\boldsymbol{z}_l|\boldsymbol{z}_{<l}, \boldsymbol{y})||p(\boldsymbol{z}_l|\boldsymbol{z}_{<l}))]. \quad (4)$$

3.4 Aggregation Methods

For fusing data within our approach, we consider different aggregation methods, such as mean aggregation, max aggregation, Bayesian aggregation [51] and pixel-wise addition. All described aggregation methods fuse a set of feature tensors $\boldsymbol{f}_1, ..., \boldsymbol{f}_K$, obtained by encoding K input images $\{\boldsymbol{x}_i\}_{i=1}^{K}$ in a permutation invariant way [61]. In mean aggregation, multiple feature vectors are fused by taking the pixel-wise average $\boldsymbol{f} = \frac{1}{K}\sum_{i=1}^{K} \boldsymbol{f}_i$. For max aggregation, we take the pixel-wise maximum instead $\boldsymbol{f} = \max_i(\boldsymbol{f}_i)$. Bayesian aggregation (BA) [51] considers an uncertainty estimate for the fused feature vectors. In order to obtain such an uncertainty estimate, the encoder has to predict means $\boldsymbol{\mu}_i$ and variances $\boldsymbol{\sigma}_i$ of a factorized Gaussian distribution over the latent feature vectors

$$\boldsymbol{f}_i = \mathcal{N}(\boldsymbol{\mu}_i, \mathrm{diag}(\boldsymbol{\sigma}_i)), \text{with } \boldsymbol{\mu}_i = \mathrm{enc}_{\mu}(\boldsymbol{x}_i, \boldsymbol{y}) \text{ and } \boldsymbol{\sigma}_i = \mathrm{enc}_{\sigma}(\boldsymbol{x}_i, \boldsymbol{y}), \quad (5)$$

instead of $\boldsymbol{f}_i$ directly. Here enc_{μ} and enc_{σ} represent the encoding process which generates means and variances respectively. The predicted distributions over latent feature vectors for multiple input images can be fused iteratively using the Bayes rule [2] (detailed derivation is given in Appendix B)

$$\boldsymbol{q}_i = \boldsymbol{\sigma}_{i-1}^2 \oslash (\boldsymbol{\sigma}_{i-1}^2 + \boldsymbol{\sigma}_i^2) \tag{6}$$

$$\text{where } \boldsymbol{\mu}_i = \boldsymbol{\mu}_{i-1} + \boldsymbol{q}_i \odot (\boldsymbol{\mu}_i - \boldsymbol{\mu}_{i-1}) \quad \text{and} \quad \boldsymbol{\sigma}_i^2 = \boldsymbol{\sigma}_{i-1}^2 \odot (1 - \boldsymbol{q}_i). \tag{7}$$

$\oslash$ and $\odot$ denote element-wise division and multiplication respectively.

4 Conditional Generative Models for Image Fusion

We propose a deep hierarchical conditional variational autoencoder, called FusionVAE *(Fusion Variational Auto-Encoder)*, that is able to fuse information from multiple sources and to infer the missing information in the images from a prior learned from the dataset. To the best of our knowledge, it is the first model that combines the generative ability of a hierarchical VAE to learn the underlying distribution of complex datasets with the ability to fuse multiple input images.

4.1 Problem Formulation

We consider image fusion problems that are concerned with generating the fused target image from multiple source images. Each source image contains partial information of the target image and the goal of the task is to recover the original target image given a finite set of source images. In particular, we denote the target image as $\boldsymbol{y}$ and the set of K source images as context $\boldsymbol{x} = \{\boldsymbol{x}_1, \boldsymbol{x}_2, \ldots, \boldsymbol{x}_K\}$, where each $\boldsymbol{x}_i$ is one source image. Given training sample $(\boldsymbol{x}, \boldsymbol{y})$, we aim to maximize the conditional likelihood $p(\boldsymbol{y}|\boldsymbol{x})$.

4.2 FusionVAE

Our approach is designed to maximize the conditional likelihood $p(\boldsymbol{y}|\boldsymbol{x})$. However, optimizing this objective directly is intractable. Therefore, we derive a variational lower bound as follows (detailed derivation can be found in Appendix B)

$$\begin{aligned} \log p(\boldsymbol{y}|\boldsymbol{x}) \geq\ & \mathbb{E}_{q_\phi(\boldsymbol{z}|\boldsymbol{y})}[\log p_\theta(\boldsymbol{y}|\boldsymbol{x}, \boldsymbol{z})] \\ & - \beta \sum_{l=1}^{L} \alpha_l \mathbb{E}_{q(\boldsymbol{z}_{<l}|\boldsymbol{y})}[\mathrm{KL}(q_\phi(\boldsymbol{z}_l|\boldsymbol{y}, \boldsymbol{z}_{<l})||p_\theta(\boldsymbol{z}_l|\boldsymbol{x}, \boldsymbol{z}_{<l}))], \end{aligned} \tag{8}$$

where we split the latent variables $\boldsymbol{z}$ into L disjoint groups $\{\boldsymbol{z}_1, \boldsymbol{z}_2, \ldots, \boldsymbol{z}_L\}$. β and α_l are annealing parameters that control the warming-up of the KL terms as in [49]. Inspired by [46], β is increased linearly from 0 to 1 during the first few training epochs to start training the reconstruction before introducing the KL term, which is increased gradually. α_l is a KL balancing coefficient [50] that is used during the warm-up period to encourage the equal use of all latent groups and to avoid posterior collapse.

FusionVAE consists of three main networks and a latent space as illustrated in Fig. 2: 1) a context encoder network which models the conditional prior $p_\phi(\boldsymbol{z}_l|\boldsymbol{x}, \boldsymbol{z}_{<l})$, 2) a target encoder which models the approximate posterior $q_\phi(\boldsymbol{z}|\boldsymbol{y})$, 3) a latent space comprising the L latent groups, and 4) a generator network $p_\theta(\boldsymbol{y}|\boldsymbol{x}, \boldsymbol{z})$ that aims to reconstruct the target image.

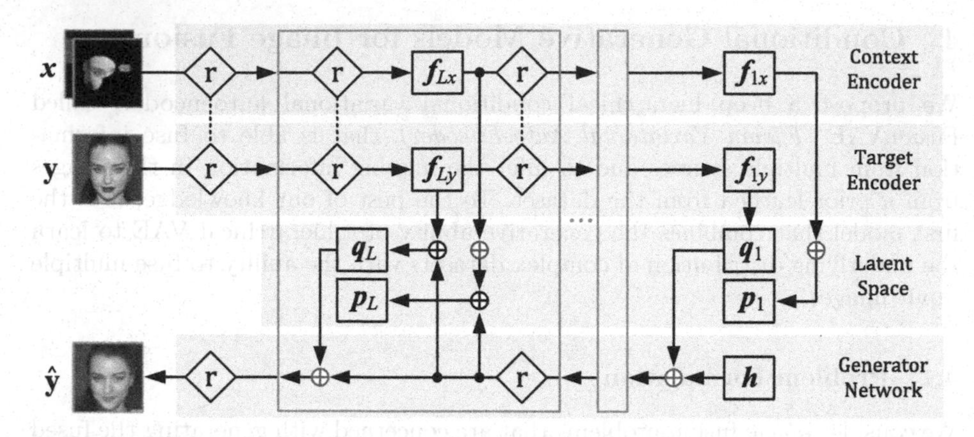

Fig. 2. Overview of the proposed network architecture. $\boldsymbol{h}$ is a trainable parameter vector, ⊕ denotes concatenation, ⊕ max aggregation, and ⊕ pixel-wise addition. ⓡ is a residual network like in [49]. The dotted lines between the residual networks indicate shared parameters. (Color figure online)

4.3 Network Architecture

Figure 2 illustrates the network architecture of our FusionVAE for training. It is built in a hierarchical way inspired by [49]. In each latent hierarchy $l \in 1, \ldots, L$ we have a set of feature maps $\boldsymbol{f}_{lx}$, $\boldsymbol{f}_{ly}$ and latent distributions p_l, q_l.

The first gray box contains the context encoder network that obtains a stack of source images $\boldsymbol{x}$ and employs residual cells [13] as in [49] to extract features $\boldsymbol{f}_{lx}$. The second gray box shows the target encoder network that encodes the target image $\boldsymbol{y}$ into the feature map $\boldsymbol{f}_{ly}$ using the same residual cells as the context encoder. The third gray box illustrates the latent space which contains the prior distributions $p_\phi(\boldsymbol{z}_l|\boldsymbol{x}, \boldsymbol{z}_{<l})$ (denoted p_1, ..., p_L in Fig. 2) and the approximate posterior distributions $q_\phi(\boldsymbol{z}_l|\boldsymbol{y}, \boldsymbol{z}_{<l})$ (denoted q_1, ..., q_L in Fig. 2). The fourth gray box contains the generator network which aims to create different output samples $\hat{\boldsymbol{y}}$. It employs a trainable parameter vector $\boldsymbol{h}$, concatenates the information from all hierarchies, and decodes them using residual cells.

In each latent hierarchy, we aggregate the context features $\boldsymbol{f}_{lx}$ using pixel-wise max aggregation. In all but the first hierarchy, we pixel-wisely add the corresponding feature map from the generator network to the aggregated context features and to the target image features f_{ly}. Using 2D convolutional layers, we learn the prior distributions p_l and the approximate posterior distributions q_l. We propose to use the approximate posterior distributions q_l as target distributions in order to learn good prior distributions p_l. Therefore, $q_\phi(\boldsymbol{z}_l|\boldsymbol{y}, \boldsymbol{z}_{<l})$ is created from the target image features f_{ly} as well as information from the generator network.

During training the generator network aims to create a prediction $\hat{\boldsymbol{y}}$ based on samples of the posterior distributions q_l and a trainable parameter vector $\boldsymbol{h}$.

For evaluation, we can omit the target image input $\boldsymbol{y}$ and sample from the prior distributions p_l In case no input image is given, we set p_1 to a standard normal distribution. Based on the samples and the trainable parameter vector $\boldsymbol{h}$, our FusionVAE can generate new output images.

5 Experimental Setup

To evaluate our approach, we conduct a series of experiments on three different datasets using data augmentation. Furthermore, we adapt traditional architectures for solving the same tasks in order to compare our results. Finally, we perform an ablation study to show the effects of specific design choices.

5.1 Datasets

For training and evaluating our approach, we create three novel fusion datasets based on MNIST [27], CelebA [34], and T-LESS [16] as described in the following.

FusionMNIST. Based on the dataset MNIST [27], we create an image fusion dataset called FusionMNIST. For each target image, it contains different noisy representations where only random parts of the target image are visible. The first three columns of Fig. 3 show different examples of FusionMNIST corresponding to the target images in the fourth column. To generate FusionMNIST, we applied zero padding to all MNIST images to obtain a resolution of 32×32. For creating a noisy representation, we generate a mask composed of the union of a varying number of ellipses with random size, shape and position. All parts of the given images outside the mask are blackened. Finally, we add Gaussian noise with a fixed variance and clip the pixel values afterwards to stay within $[0, 1]$.

FusionCelebA. We generate a similar fusion dataset based on the aligned and cropped version of CelebA [34] which we call FusionCelebA. Figure 4 depicts different example images in the first three columns which belong to the target image in the fourth column. To generate FusionCelebA, we center-crop the CelebA images to 148×148 before scaling them down to 64×64 as proposed by [26]. As in FusionMNIST, we create different representations by using masks based on random ellipses.

FusionT-LESS. A promising area of application for our FusionVAE is robot vision. Scenes in robotics settings can be very difficult to understand due to texture-less or reflective objects and occlusions. To examine the performance of our FusionVAE in this area, we create an object dataset with challenging occlusions based on T-LESS [16] which we call FusionT-LESS. To generate FusionT-LESS, we use the real training images of T-LESS and take all images of classes 19–24 as basis for the target images. This selection contains all objects with power sockets and therefore images with many similarities. Every tenth image is removed from the training set and used for evaluation. In order to create challenging occlusions, we cut all objects from images of other classes using a Canny edge detector [4] and overlay each target image with a random number between five and eight cropped objects. We select all images from classes 1, 2, 5–7, 11–14, and 25–27 as occluding objects for training and classes 3, 4, 8–10, 15–18, and 28–30 for evaluation.

5.2 Data Augmentation

During training we apply different augmentation methods on the datasets to avoid overfitting. For FusionMNIST, we apply the elliptical mask generation and the addition of Gaussian noise live during training so that we obtain an infinite number of different fusion tasks. For FusionT-LESS, almost the entire creation of occluded images is performed during training. We apply horizontal flips, rotations, scaling and movement of target and occluding images with random parameters before composing the different occluded representations. Solely the object cutting with the Canny edge detector is performed offline as a preprocessing step to keep the training time low. For FusionCelebA, we apply a horizontal flip of all images randomly in 50% of all occasions and also the elliptical mask generation is done live during training.

5.3 Architectures for Comparison

To the best of our knowledge, FusionVAE is the first fusion network for multiple images with a generative ability to fill areas without input information based on prior knowledge about the dataset under consideration. For lack of a suitable other model from the literature which would allow a fair comparison on our multi-image fusion tasks, we compare our approach with standard architectures that we adapted to support our tasks.

The first architecture for comparison is a CVAE with residual layers as employed in [49]. We use a shared encoder for processing the input images and applied max aggregation before the latent space as we did in our FusionVAE. The second architecture for comparison is a fully convolutional network (FCN) with shared encoder and max aggregation before the decoder.

For both baseline architectures, we created a version with skip connections (+S) and a version without. When using skip connections, we applied max aggregation at each shortcut for merging the features from the encoder with the decoder's features. To allow for a fair comparison, we designed all architectures so that they have a similar number of trainable parameters.

5.4 Training Procedure

We trained all networks in a supervised manner using the augmented target images $\boldsymbol{y}$ as described in Sect. 5.2. In order to teach the networks both to fuse information from a different number of input images and to learn prior knowledge about the dataset, we vary the number of input images $\boldsymbol{x}$ during the entire training. Specifically, we select a uniformly distributed random number between zero and three for each batch.

5.5 Evaluation Metrics

For evaluation, we estimate the negative log-likelihood (NLL) in bits per dimension (BPD) using weighted importance sampling [3]. We use 100 samples for all experiments with FusionCelebA as well as FusionT-LESS and 1000 samples for FusionMNIST. Since we cannot estimate the NLL of the FCN, we used the minimum over all samples of the mean squared error ($\mathrm{MSE}_{\mathrm{min}}$) as second metric.

6 Results

This section presents and discusses the quantitative and qualitative results of our research in comparison to the baseline methods mentioned in Sect. 5.3.

6.1 Quantitative Results

Tables 1, 2 and 3 show the NLL and the $\mathrm{MSE}_{\mathrm{min}}$ of all architectures on FusionMNIST, FusionCelebA, and FusionT-LESS respectively. The results are divided into the results based on zero to three input images and the average (avg) of it. We see that our FusionVAE outperforms all baseline methods on average. Regarding the NLL, our model surpasses the others additionally for 0 and 1 input images. For 2 and 3 images, CVAE+S reaches sometimes slightly better

Table 1. Results on FusionMNIST. The best results are printed in bold.

	NLL in 10^{-2} BPD					$\mathrm{MSE}_{\mathrm{min}}$ in 10^{-2}				
	0	1	2	3	Avg	0	1	2	3	Avg
FCN						10.99	5.81	5.78	5.79	7.25
FCN+S						5.80	3.74	2.54	1.78	3.56
CVAE	17.81	15.01	14.07	13.61	15.23	3.83	1.72	1.05	0.80	1.93
CVAE+S	18.43	14.57	**13.18**	**12.30**	14.77	3.62	1.75	1.19	0.97	1.95
FusionVAE	**15.93**	**14.17**	13.70	13.48	**14.39**	**3.14**	**0.99**	**0.74**	**0.65**	**1.45**

Table 2. Results on FusionCelebA. The best results are printed in bold.

	NLL in 10^{-2} BPD					MSE_{min} in 10^{-2}				
	0	1	2	3	Avg	0	1	2	3	Avg
FCN						13.77	14.82	13.10	11.24	13.23
FCN+S						12.56	8.96	6.06	4.09	7.92
CVAE	446.0	280.1	273.5	266.5	316.7	9.23	3.46	2.27	1.55	4.14
CVAE+S	525.0	270.1	233.5	**203.5**	308.3	11.08	5.49	3.66	2.57	5.71
FusionVAE	**248.1**	**227.6**	**231.2**	228.7	**233.9**	**5.11**	**0.88**	**0.86**	**0.84**	**1.93**

Table 3. Results on FusionT-LESS. The best results are printed in bold.

	NLL in 10^{-2} BPD					MSE_{min} in 10^{-2}				
	0	1	2	3	Avg	0	1	2	3	Avg
FCN						5.83	3.34	2.37	1.82	3.35
FCN+S						8.06	1.84	1.13	0.74	2.97
CVAE	25.24	23.73	22.70	23.13	23.71	5.57	1.54	0.77	0.37	2.08
CVAE+S	26.08	24.94	23.98	23.95	24.75	4.95	2.50	1.77	1.19	2.62
FusionVAE	**24.18**	**23.07**	**22.23**	**22.88**	**23.10**	**4.11**	**0.59**	**0.32**	**0.19**	**1.32**

NLL values. However, our approach reaches the best MSE_{min} values for each number of input images.

The outstanding performance of our architecture in comparison to the others is also obvious when looking at the qualitative results in Figs. 3, 4 and 5. For every row, these figures show the input, target, and up to three output predictions for all architectures. For the FCN, we depict just a single output prediction per row as all of them look almost identical.

In the first three rows when the network does not receive any input image, we see that our network provides very realistic images. This indicates that it is able to capture the underlying distribution of the used datasets very well and much better than the other architectures. Due to the difficulty of the FusionT-LESS dataset, none of the models is able to produce realistic images without any input. Still our model shows much better performance in generating object-like shapes. In case the models receive at least one input image (cf. rows 4–12), all architectures are able to extract the available information from the given input images. In addition, all VAE approaches, ours included, are able to complete the given input data based on prior knowledge. It is clearly visible, however, that the predictions of our model are much more realistic than the ones of the standard CVAE approaches especially for the more difficult datasets like FusionCelebA and FusionT-LESS.

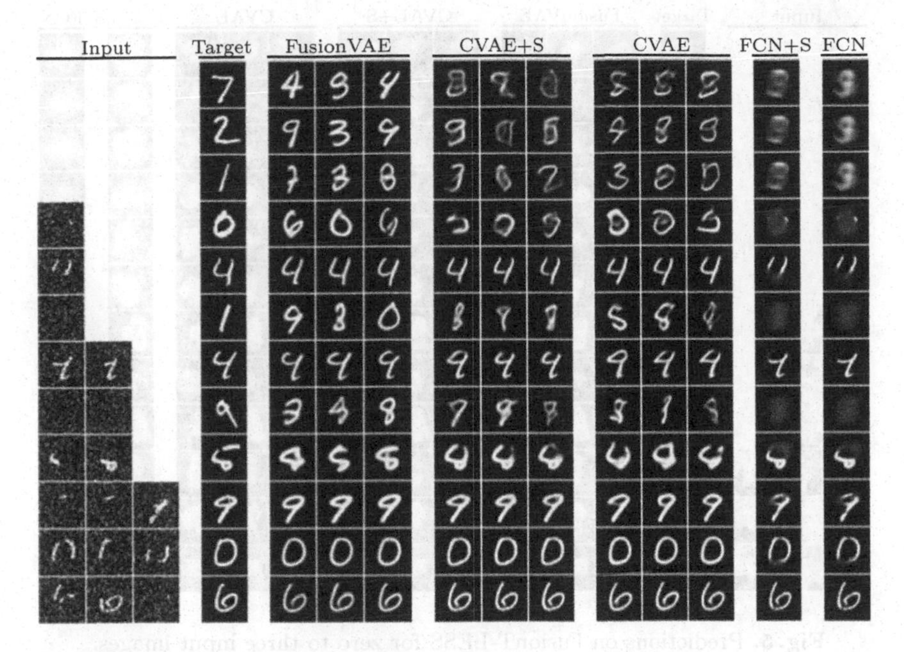

Fig. 3. Prediction results of the different architectures on FusionMNIST for zero to three input images.

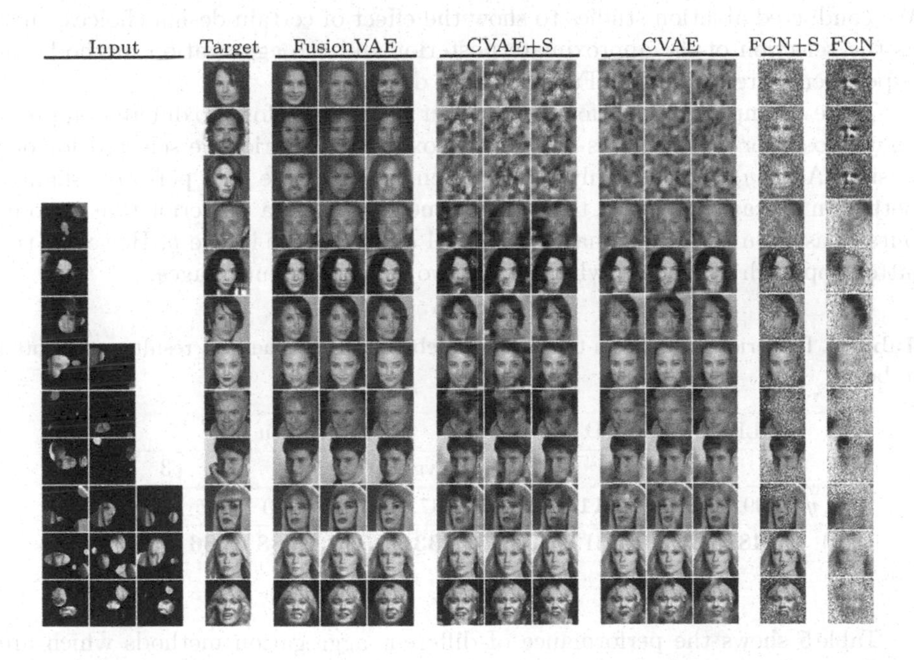

Fig. 4. Predictions on FusionCelebA for zero to three input images.

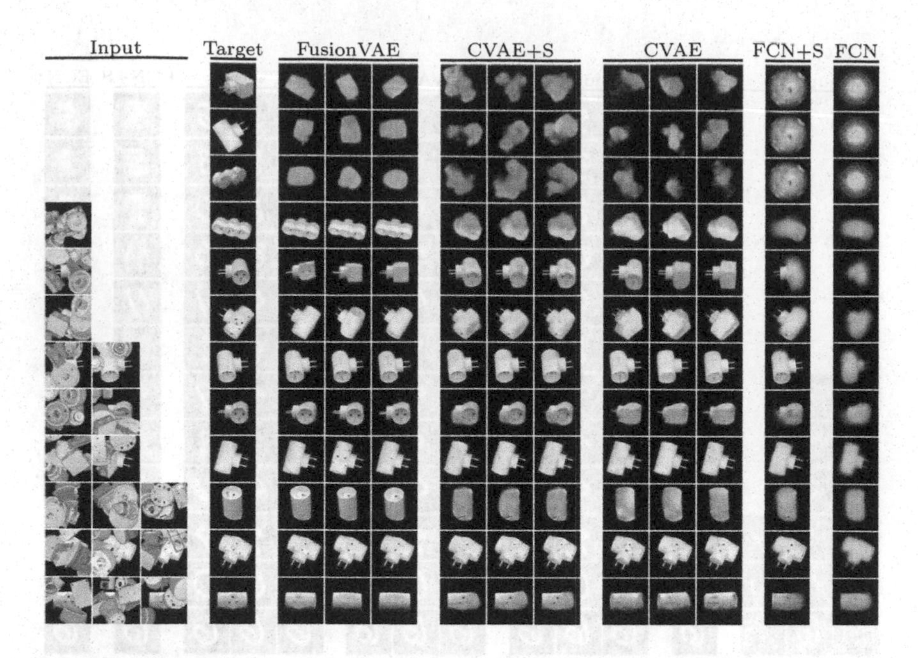

Fig. 5. Predictions on FusionT-LESS for zero to three input images.

6.2 Ablation Studies

We conducted ablation studies to show the effect of certain design choices, such as the selection of the approximate posterior and the aggregation method. All experiments are run on the FusionCelebA dataset.

Table 4 compares the performance of our FusionVAE for two different approximate posterior distributions q. The approximate posterior we selected for our FusionVAE $q(y)$, depends only on the given target image y. It performs slightly better on average compared to the same method using a posterior that is computed based on the input images x as well as the target image y. However, the latter approach is superior when fusing two or three input images.

Table 4. Posterior ablation on the FusionCelebA dataset. The best results are printed in bold.

	NLL in 10^{-2} BPD					MSE_{min} in 10^{-2}				
	0	1	2	3	Avg	0	1	2	3	Avg
$q(x, y)$	309.2	246.8	**211.1**	**180.5**	237.0	**5.04**	1.50	0.95	**0.66**	2.04
$q(y)$	**248.1**	**227.6**	231.2	228.7	**233.9**	5.11	**0.88**	**0.86**	0.84	**1.93**

Table 5 shows the performance of different aggregation methods which are applied to create the prior distributions p_l of every latent group. In our FusionVAE, the prior is created by fusing the input image features f_{lx} using max

Table 5. Aggregation ablation on the FusionCelebA dataset. The best results are printed in bold.

	NLL in 10^{-2} BPD					MSE_{min} in 10^{-2}				
	0	1	2	3	Avg	0	1	2	3	Avg
MaxAggAdd	**248.1**	227.6	231.2	228.7	233.9	5.11	**0.88**	0.86	0.84	**1.93**
MeanAggAdd	270.9	**223.3**	**216.4**	214.4	231.3	5.41	1.00	**0.79**	**0.70**	1.98
BayAggAdd	970.7	294.0	291.4	291.5	462.6	6.03	5.17	5.10	5.12	5.36
MaxAggAll	249.6	236.0	223.9	**212.7**	**230.6**	6.15	2.82	1.84	1.30	3.03
MeanAggAll	252.7	235.2	222.2	213.5	230.9	6.19	2.39	1.52	1.13	2.81
BayAggAll	255.6	568.3	414.7	1376.9	653.3	**5.10**	4.24	1.96	1.39	3.18

aggregation (MaxAgg) and adding them to the decoded features of the same latent group before applying a 2D convolution. We abbreviate that method with MaxAggAdd.

In addition to MaxAgg, we examined mean aggregation (MeanAgg) and Bayesian aggregation (BayAgg) [51] for comparison. For each aggregation principle, we tried two different versions: 1) aggregation of the input image features $\boldsymbol{f}_{lx}$ adding the corresponding information from the decoder in a pixel-wise manner (denoted by suffix Add), and 2) directly aggregating all features, i.e. both input image features $\boldsymbol{f}_{lx}$ and decoder features (denoted by suffix All).

For creating the prior p_i when using BayAgg, we moved the 2D convolutions before the aggregation in order to create the parameters μ and σ of a latent Gaussian distribution. Unlike MaxAgg and MeanAgg, BayAgg directly outputs a new Gaussian distribution that does not need to be processed any further by a convolution.

We can see that all variations of mean and max aggregation are significantly better than Bayesian aggregation. Also their training procedures are less often impaired due to numeric instabilities. Interestingly, the NLL is very similar independent of whether the aggregation is performed on all features or not. However, the MSE_{min} is much better for the aggregation with addition. Since the expressiveness of the metrics is limited, we provide additional visualizations of this ablation in Appendix D.

7 Conclusion

We have presented a novel deep hierarchical variational autoencoder for generative image fusion called FusionVAE. Our approach fuses multiple corrupted input images together with prior knowledge obtained during training. We created three challenging image fusion benchmarks based on common computer vision datasets. Moreover, we implemented four standard methods that we modified to support our tasks. We showed that our FusionVAE outperforms all other methods significantly while having a similar number of trainable parameters. The predicted images of our approach look very realistic and incorporate given input information almost perfectly. During ablation studies, we revealed the benefits of our design choices regarding the applied aggregation method and the used posterior distribution. In future work, our research could be extended by enabling the fusion of different modalities e.g. by using multiple encoders. Additionally, an explicit uncertainty estimation could be implemented that helps to weigh the impact of input information according to its uncertainty.

References

1. Bao, J., Chen, D., Wen, F., Li, H., Hua, G.: CVAE-GAN: fine-grained image generation through asymmetric training. In: ICCV, pp. 2745–2754 (2017)
2. Becker, P., Pandya, H., Gebhardt, G., Zhao, C., Taylor, C.J., Neumann, G.: Recurrent Kalman networks: factorized inference in high-dimensional deep feature spaces. In: ICML, pp. 544–552. PMLR (2019)
3. Burda, Y., Grosse, R., Salakhutdinov, R.: Importance weighted autoencoders. In: ICLR (2016)
4. Canny, J.: A computational approach to edge detection. IEEE TPAMI **8**(6), 679–698 (1986)
5. Chen, X., et al.: Variational lossy autoencoder. In: ICLR (2017)
6. Chen, X., Ma, H., Wan, J., Li, B., Xia, T.: Multi-view 3D object detection network for autonomous driving. In: CVPR, pp. 1907–1915 (2017)
7. Child, R.: Very deep VAEs generalize autoregressive models and can outperform them on images. In: ICLR (2021)
8. Goodfellow, I., et al.: Generative adversarial nets. In: NeurIPS, vol. 27, pp. 2672–2680 (2014)
9. Gregor, K., Danihelka, I., Graves, A., Rezende, D., Wierstra, D.: DRAW: a recurrent neural network for image generation. In: ICML. Proceedings of Machine Learning Research, vol. 37, pp. 1462–1471. PMLR, July 2015
10. Gu, J., et al.: Recent advances in convolutional neural networks. Pattern Recogn. **77**, 354–377 (2018)
11. Gulrajani, I., et al.: PixelVAE: a latent variable model for natural images. In: ICLR (2017)
12. Guo, X., Nie, R., Cao, J., Zhou, D., Mei, L., He, K.: FuseGAN: learning to fuse multi-focus image via conditional generative adversarial network. IEEE TMM **21**(8), 1982–1996 (2019)
13. He, K., Zhang, X., Ren, S., Sun, J.: Deep residual learning for image recognition. In: CVPR, pp. 770–778 (2016)
14. He, Y., Huang, H., Fan, H., Chen, Q., Sun, J.: FFB6D: a full flow bidirectional fusion network for 6D pose estimation. In: CVPR, pp. 3003–3013 (2021)
15. He, Y., Sun, W., Huang, H., Liu, J., Fan, H., Sun, J.: PVN3D: a deep point-wise 3D keypoints voting network for 6DoF pose estimation. In: CVPR, pp. 11632–11641 (2020)
16. Hodaň, T., Haluza, P., Obdržálek, Š., Matas, J., Lourakis, M., Zabulis, X.: T-LESS: an RGB-D dataset for 6D pose estimation of texture-less objects. In: WACV (2017)
17. Huang, G., Liu, Z., Van Der Maaten, L., Weinberger, K.Q.: Densely connected convolutional networks. In: CVPR, pp. 4700–4708 (2017)

18. Huang, J., Le, Z., Ma, Y., Mei, X., Fan, F.: A generative adversarial network with adaptive constraints for multi-focus image fusion. Neural Comput. Appl. **32**(18), 15119–15129 (2020). https://doi.org/10.1007/s00521-020-04863-1
19. Iizuka, S., Simo-Serra, E., Ishikawa, H.: Globally and locally consistent image completion. ACM TOG **36**(4), 1–14 (2017)
20. Jung, H., Kim, Y., Jang, H., Ha, N., Sohn, K.: Unsupervised deep image fusion with structure tensor representations. IEEE TIP **29**, 3845–3858 (2020)
21. Kim, J., Yoo, J., Lee, J., Hong, S.: SetVAE: learning hierarchical composition for generative modeling of set-structured data. In: CVPR, pp. 15059–15068 (2021)
22. Kingma, D.P., Welling, M.: Auto-encoding variational Bayes. In: ICLR (2014)
23. Kingma, D.P., Salimans, T., Jozefowicz, R., Chen, X., Sutskever, I., Welling, M.: Improving variational inference with inverse autoregressive flow. In: NeurIPS, vol. 29, pp. 4743–4751 (2016)
24. Köhler, R., Schuler, C., Schölkopf, B., Harmeling, S.: Mask-specific inpainting with deep neural networks. In: Jiang, X., Hornegger, J., Koch, R. (eds.) GCPR 2014. LNCS, vol. 8753, pp. 523–534. Springer, Cham (2014). https://doi.org/10.1007/978-3-319-11752-2_43
25. Ku, J., Mozifian, M., Lee, J., Harakeh, A., Waslander, S.L.: Joint 3D proposal generation and object detection from view aggregation. In: IROS, pp. 1–8. IEEE (2018)
26. Larsen, A.B.L., Sønderby, S.K., Larochelle, H., Winther, O.: Autoencoding beyond pixels using a learned similarity metric. In: ICML, pp. 1558–1566. PMLR (2016)
27. LeCun, Y.: The MNIST database of handwritten digits. http://yann.lecun.com/exdb/mnist (1998)
28. Li, H., Wu, X.J.: DenseFuse: a fusion approach to infrared and visible images. IEEE TIP **28**(5), 2614–2623 (2018)
29. Li, H., Wu, X.J., Durrani, T.S.: Infrared and visible image fusion with ResNet and zero-phase component analysis. Infrared Phys. Technol. **102**, 103039 (2019)
30. Li, H., Wu, X.J., Kittler, J.: Infrared and visible image fusion using a deep learning framework. In: ICPR, pp. 2705–2710. IEEE (2018)
31. Li, Y., Liu, S., Yang, J., Yang, M.H.: Generative face completion. In: CVPR, pp. 3911–3919 (2017)
32. Liu, Y., Chen, X., Cheng, J., Peng, H.: A medical image fusion method based on convolutional neural networks. In: International Conference on Information Fusion, pp. 1–7. IEEE (2017)
33. Liu, Y., Chen, X., Peng, H., Wang, Z.: Multi-focus image fusion with a deep convolutional neural network. Inf. Fusion **36**, 191–207 (2017)
34. Liu, Z., Luo, P., Wang, X., Tang, X.: Deep learning face attributes in the wild. In: ICCV, December 2015
35. Ma, J., et al.: Infrared and visible image fusion via detail preserving adversarial learning. Inf. Fusion **54**, 85–98 (2020)
36. Ma, J., Xu, H., Jiang, J., Mei, X., Zhang, X.P.: DDcGAN: a dual-discriminator conditional generative adversarial network for multi-resolution image fusion. IEEE TIP **29**, 4980–4995 (2020)
37. Ma, J., Yu, W., Liang, P., Li, C., Jiang, J.: FusionGAN: a generative adversarial network for infrared and visible image fusion. Inf. Fusion **48**, 11–26 (2019)
38. Maaløe, L., Fraccaro, M., Liévin, V., Winther, O.: BIVA: a very deep hierarchy of latent variables for generative modeling. In: NeurIPS, vol. 32, pp. 6551–6562 (2019)
39. Marinescu, R.V., Moyer, D., Golland, P.: Bayesian image reconstruction using deep generative models. arXiv:2012.04567 [cs.CV] (2020)

40. Parmar, G., Li, D., Lee, K., Tu, Z.: Dual contradistinctive generative autoencoder. In: CVPR, pp. 823–832 (2021)
41. Pathak, D., Krahenbuhl, P., Donahue, J., Darrell, T., Efros, A.A.: Context encoders: Feature learning by inpainting. In: CVPR, pp. 2536–2544 (2016)
42. Prabhakar, K.R., Srikar, V.S., Babu, R.V.: DeepFuse: a deep unsupervised approach for exposure fusion with extreme exposure image pairs. In: ICCV, pp. 4714–4722 (2017)
43. Sadeghi, H., Andriyash, E., Vinci, W., Buffoni, L., Amin, M.H.: PixelVAE++: improved PixelVAE with discrete prior. arXiv:1908.09948 [cs.CV] (2019)
44. Simonyan, K., Zisserman, A.: Very deep convolutional networks for large-scale image recognition. In: ICLR (2015)
45. Sohn, K., Lee, H., Yan, X.: Learning structured output representation using deep conditional generative models. In: NeurIPS, vol. 28, pp. 3483–3491 (2015)
46. Sønderby, C.K., Raiko, T., Maaløe, L., Sønderby, S.K., Winther, O.: Ladder variational autoencoders. In: NeurIPS, vol. 29, 3738–3746 (2016)
47. Song, Y., et al.: Contextual-based image inpainting: infer, match, and translate. In: Ferrari, V., Hebert, M., Sminchisescu, C., Weiss, Y. (eds.) ECCV 2018. LNCS, vol. 11206, pp. 3–18. Springer, Cham (2018). https://doi.org/10.1007/978-3-030-01216-8_1
48. Song, Y., Yang, C., Shen, Y., Wang, P., Huang, Q., Kuo, C.C.J.: SPG-Net: segmentation prediction and guidance network for image inpainting. In: BMVC (2018)
49. Vahdat, A., Kautz, J.: NVAE: a deep hierarchical variational autoencoder. In: NeurIPS, vol. 33, pp. 19667–19679 (2020)
50. Vahdat, A., Macready, W., Bian, Z., Khoshaman, A., Andriyash, E.: DVAE++: discrete variational autoencoders with overlapping transformations. In: ICML, pp. 5035–5044. PMLR (2018)
51. Volpp, M., Flürenbrock, F., Grossberger, L., Daniel, C., Neumann, G.: Bayesian context aggregation for neural processes. In: ICLR (2020)
52. Walker, J., Doersch, C., Gupta, A., Hebert, M.: An uncertain future: forecasting from static images using variational autoencoders. In: Leibe, B., Matas, J., Sebe, N., Welling, M. (eds.) ECCV 2016. LNCS, vol. 9911, pp. 835–851. Springer, Cham (2016). https://doi.org/10.1007/978-3-319-46478-7_51
53. Wang, C., et al.: DenseFusion: 6D object pose estimation by iterative dense fusion. In: CVPR, pp. 3343–3352 (2019)
54. Wang, W., Liu, W., Hu, J., Fang, Y., Shao, Q., Qi, J.: GraspFusionNet: a two-stage multi-parameter grasp detection network based on RGB-XYZ fusion in dense clutter. Mach. Vis. Appl. **31**(7), 1–19 (2020)
55. Xu, H., Liang, P., Yu, W., Jiang, J., Ma, J.: Learning a generative model for fusing infrared and visible images via conditional generative adversarial network with dual discriminators. In: IJCAI (2019)
56. Xu, H., Ma, J., Le, Z., Jiang, J., Guo, X.: FusionDN: a unified densely connected network for image fusion. In: AAAI, vol. 34, pp. 12484–12491, April 2020
57. Yan, Z., Li, X., Li, M., Zuo, W., Shan, S.: Shift-Net: image inpainting via deep feature rearrangement. In: Ferrari, V., Hebert, M., Sminchisescu, C., Weiss, Y. (eds.) Computer Vision – ECCV 2018. LNCS, vol. 11218, pp. 3–19. Springer, Cham (2018). https://doi.org/10.1007/978-3-030-01264-9_1
58. Yang, C., Lu, X., Lin, Z., Shechtman, E., Wang, O., Li, H.: High-resolution image inpainting using multi-scale neural patch synthesis. In: CVPR, pp. 6721–6729 (2017)

59. Yoo, J.H., Kim, Y., Kim, J., Choi, J.W.: 3D-CVF: generating joint camera and LiDAR features using cross-view spatial feature fusion for 3D object detection. In: Vedaldi, A., Bischof, H., Brox, T., Frahm, J.-M. (eds.) ECCV 2020. LNCS, vol. 12372, pp. 720–736. Springer, Cham (2020). https://doi.org/10.1007/978-3-030-58583-9_43
60. Yu, J., Lin, Z., Yang, J., Shen, X., Lu, X., Huang, T.S.: Generative image inpainting with contextual attention. In: CVPR, pp. 5505–5514 (2018)
61. Zaheer, M., Kottur, S., Ravanbakhsh, S., Póczos, B., Salakhutdinov, R., Smola, A.J.: Deep sets. In: NeurIPS, vol. 30, pp. 3391–3401 (2017)
62. Zeng, Yu., Lin, Z., Yang, J., Zhang, J., Shechtman, E., Lu, H.: High-resolution image inpainting with iterative confidence feedback and guided upsampling. In: Vedaldi, A., Bischof, H., Brox, T., Frahm, J.-M. (eds.) ECCV 2020. LNCS, vol. 12364, pp. 1–17. Springer, Cham (2020). https://doi.org/10.1007/978-3-030-58529-7_1
63. Zhang, H., Xu, H., Xiao, Y., Guo, X., Ma, J.: Rethinking the image fusion: a fast unified image fusion network based on proportional maintenance of gradient and intensity. In: AAAI, vol. 34, pp. 12797–12804, April 2020
64. Zhang, Q., Qu, D., Xu, F., Zou, F.: Robust robot grasp detection in multimodal fusion. In: MATEC Web of Conferences, vol. 139, p. 00060. EDP Sciences (2017)
65. Zheng, C., Cham, T.J., Cai, J.: Pluralistic image completion. In: CVPR, pp. 1438–1447 (2019)

Learning Algebraic Representation for Systematic Generalization in Abstract Reasoning

Chi Zhang[1(✉)], Sirui Xie[1], Baoxiong Jia[1], Ying Nian Wu[1], Song-Chun Zhu[1,2,3,4], and Yixin Zhu[2]

[1] University of California, Los Angeles, Los Angeles, CA 90095, USA
{chi.zhang,srxie,baoxiongjia}@ucla.edu, {ywu,sczhu}@stat.ucla.edu
[2] Institute for Artificial Intelligence, Peking University, Beijing 10080, China
yixin.zhu@pku.edu.cn
[3] Tsinghua University, Beijing 10080, China
[4] Beijing Institute for General Artificial Intelligence, Beijing 10080, China

Abstract. Is intelligence realized by connectionist or classicist? While connectionist approaches have achieved superhuman performance, there has been growing evidence that such task-specific superiority is particularly fragile in *systematic generalization*. This observation lies in the central debate between connectionist and classicist, wherein the latter continually advocates an *algebraic* treatment in cognitive architectures. In this work, we follow the classicist's call and propose a hybrid approach to improve systematic generalization in reasoning. Specifically, we showcase a prototype with algebraic representation for the abstract spatial-temporal reasoning task of Raven's Progressive Matrices (RPM) and present the ALgebra-Aware Neuro-Semi-Symbolic (ALANS) learner. The ALANS learner is motivated by abstract algebra and the representation theory. It consists of a neural visual perception frontend and an algebraic abstract reasoning backend: the frontend summarizes the visual information from object-based representation, while the backend transforms it into an algebraic structure and induces the hidden operator on the fly. The induced operator is later executed to predict the answer's representation, and the choice most similar to the prediction is selected as the solution. Extensive experiments show that by incorporating an algebraic treatment, the ALANS learner outperforms various pure connectionist models in domains requiring systematic generalization. We further show the generative nature of the learned algebraic representation; it can be decoded by isomorphism to generate an answer.

C. Zhang, S. Xie, B. Jia—Indicates equal contribution.

Supplementary Information The online version contains supplementary material available at https://doi.org/10.1007/978-3-031-19842-7_40.

S. Avidan et al. (Eds.): ECCV 2022, LNCS 13699, pp. 692–709, 2022.
https://doi.org/10.1007/978-3-031-19842-7_40

1 Introduction

"Thought is in fact a kind of Algebra."

—William James [25]

Imagine you are given two alphabetical sequences of "*c*, *b*, *a*" and "*d*, *c*, *b*", and asked to fill in the missing element in "*e*, *d*, ?". In nearly no time will one realize the answer to be *c*. However, more surprising for human learning is that, effortlessly and instantaneously, we can "freely generalize" [37] the solution to any partial consecutive ordered sequences. While believed to be innate in early development for human infants [39], such systematic generalizability has constantly been missing and proven to be particularly challenging in existing connectionist models [2,29]. In fact, such an ability to entertain a given thought and semantically related contents strongly implies an abstract algebra-like treatment [12]; in literature, it is referred to as the "language of thought" [11], "physical symbol system" [44], and "algebraic mind" [37]. However, in stark contrast, existing connectionist models tend only to capture statistical correlation [7,26,29], rather than providing any account for a structural inductive bias where systematic algebra can be carried out to facilitate generalization.

This contrast instinctively raises a question—what constitutes such an *algebraic* inductive bias? We argue that the foundation of modeling counterpart to the algebraic treatment in early human development [37,39] lies in algebraic computations set up on mathematical axioms, a form of formalized human intuition and the beginning of modern mathematical reasoning [17,34]. Of particular importance to the algebra's basic building blocks is the Peano Axiom [46]. In the Peano Axiom, the essential components of algebra, the algebraic set, and corresponding operators over it, are governed by three statements: (1) the existence of at least one element in the field to study ("zero" element), (2) a successor function that is recursively applied to all elements and can, therefore, span the entire field, and (3) the principle of mathematical induction. Building on such a fundamental axiom, we begin to form the notion of an algebraic set and induce the operator to construct an algebraic structure. We hypothesize that such an algebraic treatment set up on fundamental axioms is essential for a model's systematic generalizability, the lack of which will only make it sub-optimal.

To demonstrate the benefits of adopting such an algebraic treatment in systematic generalization, we showcase a prototype for Raven's Progressive Matrices (RPM) [48,49], an exemplar task for abstract spatial-temporal reasoning [51,68]. In this task, an agent is given an incomplete 3×3 matrix consisting of eight context panels with the last one missing, and asked to pick one answer from a set of eight choices that best completes the matrix. Human's reasoning capability of solving this abstract reasoning task has been commonly regarded as an indicator of "general intelligence" [4] and "fluid intelligence" [19,24,55,56]. In spite of the task being one that ideally requires abstraction, algebraization, induction, and generalization [4,48,49], recent endeavors unanimously propose pure connectionist models that attempt to circumvent such intrinsic cognitive requirements [21,51,58,64,68,70,74]. However, these methods' inefficiency is also evident in systematic generalization; they struggle to extrapolate to domains beyond training [51,70], shown also in this paper.

To address the issue, we introduce an ALgebra-Aware Neuro-Semi-Symbolic (ALANS) learner. At a high-level, the ALANS learner is embedded in a general neuro-symbolic architecture [14,36,66,67] but has the on-the-fly operator learnability, hence **semi-symbolic**. Specifically, it consists of a neural visual perception frontend and an algebraic abstract reasoning backend. For each RPM instance, the neural visual perception frontend first slides a window over each panel to obtain the object-based representation [26,63] for every object. A belief inference engine latter aggregates all object-based representation in each panel to produce the probabilistic *belief state*. The algebraic abstract reasoning backend then takes the belief states of the eight context panels, treats them as snapshots on an algebraic structure, lifts them into a matrix-based algebraic representation built on the Peano Axiom and the representation theory [23], and induces the hidden operator in the algebraic structure by solving an inner optimization problem [3,8,72]. The answer's algebraic representation is predicted by executing the induced operator: its corresponding set element is decoded by isomorphism, and the final answer is selected as the one most similar to the prediction.

The ALANS learner enjoys several benefits in abstract reasoning with an algebraic treatment:

1. Unlike previous monolithic models, the ALANS learner offers a more **interpretable** account of the entire abstract reasoning process: the neural visual perception frontend extracts object-based representation and produces belief states of panels by explicit probability inference, whereas the algebraic abstract reasoning backend induces the hidden operator in the algebraic structure. The final answer's representation is obtained by executing the induced operator, and the choice panel with minimum distance is selected. This process much resembles the top-down bottom-up strategy in human reasoning missed in recent literature [21,51,58,64,68,70,74]: humans reason by inducing the hidden relation, executing it to generate a feasible solution in mind, and choosing the most similar answer available [4].
2. While keeping the semantic interpretability and end-to-end trainability in existing neuro-symbolic frameworks [14,36,66,67], ALANS is **semi-symbolic** in the sense that the symbolic operator can be learned and concluded on the fly without manual definition for every one of them. Such an inductive ability also enables a greater extent of the desired generalizability.
3. By decoding the predicted representation in the algebraic structure, we can also generate an answer that satisfies the hidden relation in the context.

This work makes three major contributions. (1) We propose the ALANS learner, a neuro-semi-symbolic design, in contrast to existing monolithic models. (2) To demonstrate the efficacy of incorporating an algebraic treatment in reasoning, we show the superior systematic generalization ability of the proposed ALANS learner in various extrapolatory RPM domains. (3) We present analyses into both neural visual perception and algebraic abstract reasoning.

2 Related Work

2.1 Quest for Symbolized Manipulation

The idea to treat thinking as a mental language can be dated back to Augustine [1,62]. Since the 1970s, this school of thought has undergone a dramatic revival as the quest for symbolized manipulation in cognitive modeling, such as "language of thought" [11], "physical symbol system" [44], and "algebraic mind" [37]. In their study, connectionist's task-specific superiority and inability to generalize beyond training [7,26,51,68] have been hypothetically linked to a lack of such symbolized algebraic manipulation [7,29,38]. With evidence that an algebraic treatment adopted in early human development [39] can potentially address the issue [2,36,38], classicist [12] approaches for generalizable reasoning used in programs [40] and blocks world [61] have resurrected. As a hybrid approach to bridge connectionist and classicist, recent developments lead to neuro-symbolic architectures. In particular, the community of theorem proving has been one of the earliest to endorse the technique [13,50,53]: ∂ILP [10] and NLM [9] make inductive programming end-to-end, and DeepProbLog [35] connects learning and reasoning. Recently, Hudson and Manning [22] propose NSM for visual question answering where a probabilistic graph is used for reasoning. Yi *et al.* [67] demonstrate a neuro-symbolic prototype for the same task where a perception module and a language parsing module are separately trained, with the predefined logic operators associated with language tokens chained to process the visual information. Mao *et al.* [36] soften the predefined operators to afford end-to-end training with only question answers. Han *et al.* [14] use the hybrid architecture for metaconcept learning. Yi *et al.* [66] and Chen *et al.* [6] show how neuro-symbolic models can handle explanatory, predictive, and counterfactual questions in temporal and causal reasoning. Lately, NeSS [5] exemplifies an algorithmic stack machine that can be used to improve generalization in language learning. ALANS follows the classicist's call but adopts a neuro-*semi*-symbolic architecture: it is end-to-end trainable as opposed to Yi *et al.* [66,67] and the operator can be learned and concluded on the fly without manual specification.

2.2 Abstract Visual Reasoning

Recent works by Santoro *et al.* [51] and Zhang *et al.* [68] arouse the community's interest in abstract visual reasoning; the task of Raven's Progressive Matrices (RPM) is introduced as such a measure for intelligent agents. As an intelligence quotient test for humans [48,49], RPM is believed to be strongly correlated with human's general intelligence [4] and fluid intelligence [19,24,55,56]. Early RPM-solving systems employ symbolic representation based on hand-designed features and assume access to the underlying logics [4,31–33]. Another stream of research on RPM recruits similarity-based metrics to select the most similar answer from the choices [20,30,41–43,54]. However, these visual or semantic features are unable to handle uncertainty from imperfect perception, and directly assuming access to the logic operations simplifies the problem. Recently

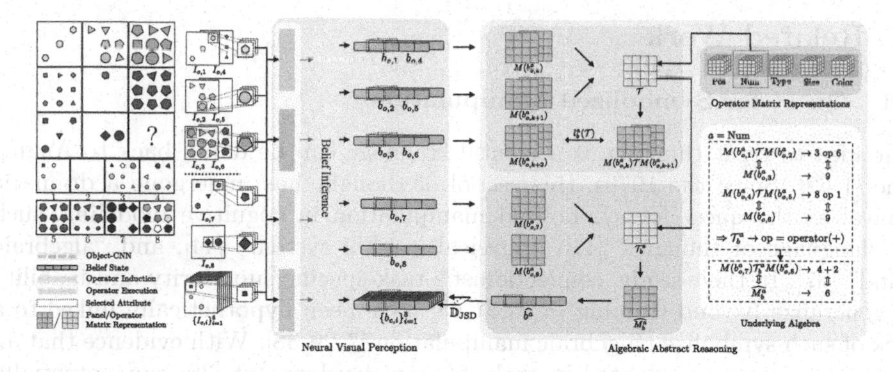

Fig. 1. An overview of the ALANS learner. For an RPM instance, the neural visual perception module produces the belief states for all panels: an object CNN extracts object attribute distributions for each image region, and a belief inference engine marginalizes them out to obtain panel attribute distributions. For each panel attribute, the algebraic abstract reasoning module transforms the belief states into matrix-based algebraic representation and induces hidden operators by solving inner optimizations. The answer representation is obtained by executing the induced operators, and the choice most similar to the prediction is selected as the solution. An example of the underlying discrete algebra and its correspondence is also shown on the right.

proposed data-driven approaches arise from the availability of large datasets: Santoro *et al.* [51] extend a pedagogical RPM generation method [59], whereas Zhang *et al.* [68] use a stochastic image grammar [75] and introduce structural annotations in it, which Hu *et al.* [21] further refine to avoid shortcut solutions by statistics in candidate panels. Despite the fact that RPM intrinsically requires one to perform abstraction, algebraization, induction, and generalization, existing methods bypass such cognitive requirements using a single feedforward pass in connectionist models: Santoro *et al.* [51] use a relational module [52], Steenbrugge *et al.* [57] augment it with a VAE [28], Zhang *et al.* [68] assemble a dynamic tree, Hill *et al.* [18] arrange the data in a contrastive manner, Zhang *et al.* [70] propose a contrast module, Zhang *et al.* [74] formulate it in a student-teacher setting, Wang *et al.* [58] build a multiplex graph network, Hu *et al.* [21] aggregate features from a hierarchical decomposition, and Wu *et al.* [64] apply a scattering transformation to learn objects, attributes, and relations. Recently, Zhang *et al.* [71] employ a neuro-symbolic design but requires full knowledge over the hidden relations to perform *abduction*. While our work adopts the visual perception module and employs a similar training strategy from Zhang *et al.* [71], the ALANS learner manages to *induce* the hidden relations, enabling on-the-fly relation induction and systematic generalization on relational learning. The recent work of Neural Interpreter (NI) [47] is a complementary neural approach to our method: Although both NI and ALANS decompose the reasoning process into sub-components and aggregate them, NI focuses more on compositionality, routing new input via different paths of learned modules to generalize, whereas ALANS more on induction, enabling a learned module to adapt on the fly.

3 The ALANS Learner

In this section, we introduce the ALANS learner for the RPM problem. In each RPM instance, an agent is given an incomplete 3×3 panel matrix with the last entry missing and asked to induce the operator hidden in the matrix and choose from eight choice panels one that follows it. Formally, let the answer variable be denoted as y, the context panels as $\{I_{o,i}\}_{i=1}^{8}$, and choice panels as $\{I_{c,i}\}_{i=1}^{8}$. Then the problem can be formulated as estimating $P(y \mid \{I_{o,i}\}_{i=1}^{8}, \{I_{c,i}\}_{i=1}^{8})$. According to the common design [4,51,68], there is one operator that governs each panel attribute. Hence, by assuming independence among attributes, we propose to factorize the probability of $P(y = n \mid \{I_{o,i}\}_{i=1}^{8}, \{I_{c,i}\}_{i=1}^{8})$ as

$$\prod_{a} \sum_{\mathcal{T}^a} P(y^a = n \mid \mathcal{T}^a, \{I_{o,i}\}_{i=1}^{8}, \{I_{c,i}\}_{i=1}^{8}) \times P(\mathcal{T}^a \mid \{I_{o,i}\}_{i=1}^{8}), \tag{1}$$

where y^a denotes the answer selection based only on attribute a and $\mathcal{T}^a$ the operator on a.

Overview. As shown in Fig. 1, the ALANS learner decomposes the process into perception and reasoning: the neural visual perception frontend is adopted from Zhang *et al.* [71] and extracts the *belief states* from each of the sixteen panels, whereas the algebraic abstract reasoning backend views an instance as an example in an abstract algebra structure, transforms belief states into *algebraic representation* by the representation theory, *induces* the hidden operators, and *executes* the operators to predict the representation of the answer. Therefore, in Eq. (1), the operator distribution is modeled by the fitness of an operator and the answer distribution by the distance between the predicted representation and that of a candidate.

3.1 Neural Visual Perception

We follow the design in Zhang *et al.* [71] and decompose visual perception into an object CNN and a belief state inference engine. Specifically, for each panel, we use a sliding window to traverse the spatial domain of the image and feed each image region into an object CNN. The CNN has four branches, producing for each region its object attribute distributions, including objectiveness (if the region contains an object), type, size, and color. The belief inference engine summarizes the panel attribute distributions (over position, number, type, size, and color) by marginalizing out all object attribute distributions (over objectiveness, type, size, and color). As an example, the distribution of the panel attribute of Number can be computed as such: for N image regions and their predicted objectiveness

$$P(\text{Number} = k) = \sum_{\substack{R^o \in \{0,1\}^N \\ \sum_j R^o_j = k}} \prod_{j=1}^{N} P(r^o_j = R^o_j), \tag{2}$$

where $P(r_j^o)$ denotes the jth region's estimated objectiveness distribution, and R^o is a binary sequence of length N that sums to k. All panel attribute distributions compose the *belief state* of a panel. In the following, we denote the belief state as b and the distribution of an attribute a as $P(b^a)$. For more details, please refer to Zhang *et al.* [71] and the Appendix.

3.2 Algebraic Abstract Reasoning

Given the belief states of both context and choice panels, the algebraic abstract reasoning backend concerns the induction of hidden operators and the prediction of answer representation for each attribute. The fitness of induced operators is used for estimating the operator distribution and the difference between the prediction and the choice panel for estimating the answer distribution.

Algebraic Underpinning. Without loss of generality, here we assume row-wise operators. For each attribute, under perfect perception, the first two rows in an RPM instance provide snapshots into an example of *group* [15] constrained to an integer-indexed set, a simple algebra structure that is closed under a binary operator. To see this, note that an accurate perception module would see each panel attribute as a deterministic set element. Therefore, RPM instances with unary operators, such as progression, are group examples with special binary operators where one operand is constant. Instances with binary operators, such as arithmetics, directly follow the group properties. Those with ternary operators are ones defined on a three-tuple set from rows.

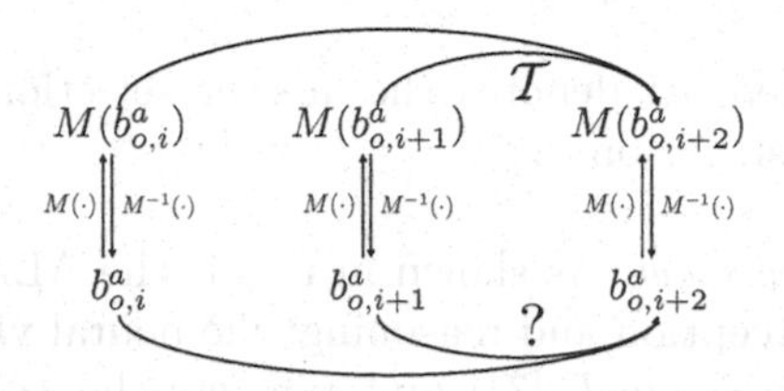

Fig. 2. Isomorphism between the abstract algebra and the matrix-based representation. In this view, operator induction is now reduced to solving for a matrix.

Algebraic Representation. A systematic algebraic view allows us to felicitously recruit ideas in the representation theory [23] to glean the hidden properties in the abstract structures: it makes abstract algebra amenable by reducing it onto linear algebra. Following the same spirit, we propose to lift both the set elements and the hidden operators to a learnable matrix space. To encode the set element, we employ the Peano Axiom [46]. According to the Peano Axiom, an integer-indexed set can be constructed by (1) a zero element ($\mathbf{0}$), (2) a successor function ($S(\cdot)$), and (3) the principle of mathematical induction, such that the kth element is encoded as $S^k(\mathbf{0})$. Specifically, we instantiate the zero element as a learnable matrix M_0 and the successor function as the matrix-matrix product parameterized by M. In an attribute-specific manner, the representation of an attribute taking the kth value is $(M^a)^k M_0^a$. For operators, we consider them to live in a learnable matrix group of a corresponding dimension, such that the action of an operator on a set can be represented as matrix multiplication.

Such algebraic representation establishes an isomorphism between the matrix space and the abstract algebraic structure: abstract elements on the algebraic structure have a bijective mapping to/from the matrix space, and inducing the abstract relation can be reduced to solving for a matrix operator. See Fig. 2 for a graphical illustration of the isomorphism.

Operator Induction. Operator induction concerns about finding a concrete operator in the abstract algebraic structure. By the property of closure, we formulate it as an inner-level regularized linear regression problem: a binary operator $\mathcal{T}_b^a$ for attribute a in a group minimizes $\ell_b^a(\mathcal{T})$ defined as

$$\ell_b^a(\mathcal{T}) = \sum_i \mathbb{E}\left[\|M(b_{o,i}^a)\mathcal{T}M(b_{o,i+1}^a) - M(b_{o,i+2}^a)\|_F^2\right] + \lambda_b^a \|\mathcal{T}\|_F^2, \tag{3}$$

where under visual uncertainty, we take the expectation w.r.t. the distributions in the belief states of context panels $P(b_{o,i}^a)$ in the first two rows, and denote its algebraic representation as $M(b_{o,i}^a)$. For unary operators, one operand can be treated as constant and absorbed into $\mathcal{T}$. Note that Eq. (3) admits a closed-form solution (please refer to the Appendix for details). Therefore, the operator can be learned and adapted for different instances of binary relations and concluded on the fly. Such a design also simplifies the recent neuro-symbolic approaches, where every single symbol operator needs to be hand-defined [14,36,66,67]. Instead, we only specify an inner-level optimization framework and allow symbolic operators to be quickly induced based on the neural observations, while keeping the semantic interpretability in the neuro-symbolic methods. Therefore, we term such a design semi-symbolic.

The operator probability in Eq. (1) is then modeled by each operator type's fitness, *e.g.*, for binary,

$$P(\mathcal{T}^a = \mathcal{T}_b^a \mid \{I_{o,i}\}_{i=1}^8) \propto \exp(-\ell_b^a(\mathcal{T}_b^a)). \tag{4}$$

Operator Execution. To predict the algebraic representation of the answer, we solve another inner-level optimization similar to Eq. (3), but now treating the representation of the answer as a variable:

$$\widehat{M_b^a} = \arg\min_M \ell_b^a(M) = \mathbb{E}[\|M(b_{o,7}^a)\mathcal{T}_b^a M(b_{o,8}^a) - M\|_F^2], \tag{5}$$

where the expectation is taken w.r.t. context panels in the last row. The optimization also admits a closed-form solution (please refer to the Appendix for details), which corresponds to the execution of the induced operator in Eq. (3).

The predicted representation is decoded probabilistically as the predicted belief state of the solution,

$$P(\widehat{b^a} = k \mid \mathcal{T}^a) \propto \exp(-\|\widehat{M^a} - (M^a)^k M_0^a\|_F^2). \tag{6}$$

Answer Selection. Based on Eqs. (1) and (4), estimating the answer distribution is now boiled down to estimating the conditional answer distributions for each attribute. Here, we propose to model it based on the Jensen-Shannon Divergence (JSD) of the predicted belief state and that of a choice,

$$P(y^a = n \mid \mathcal{T}^a, \{I_{o,i}\}_{i=1}^8, \{I_{c,i}\}_{i=1}^8) \propto \exp(-d_n^a), \tag{7}$$

where we define d_n^a as

$$d_n^a = \mathbb{D}_{\text{JSD}}(P(\widehat{b^a} \mid \mathcal{T}^a) \| P(b_{c,n}^a))). \tag{8}$$

Discussion. Comparing with the possible problem-solving process by humans [4], we argue that the proposed algebraic abstract reasoning module offers a computational and interpretable counterpart to human-like reasoning in RPM. Specifically, the induction component resembles fluid intelligence, where one quickly induces the hidden operator by observing the context panels. The execution component synthesizes an image by executing the induced operator, and the choice most similar to the image is selected as the answer.

We also note that by decoding the predicted representation in Eq. (6), a solution can be *generated*: by sequentially selecting the most probable operator and the most probable attribute value, a rendering engine can directly render the solution. The reasoning backend also enables end-to-end training: by integrating the belief states from neural perception, the module conducts both induction and execution in a soft manner, such that the gradients can be back-propagated and both the visual frontend and the reasoning backend jointly trained.

3.3 Training Strategy

We train the entire ALANS learner by minimizing the cross-entropy loss between the estimated answer distribution and the ground-truth selection and an auxiliary loss [51,58,68,71] that shapes the operator distribution from the reasoning engine, *i.e.*,

$$\min_{\theta, \{M_0^a\}, \{M^a\}} \ell(P(y \mid \{I_{o,i}\}_{i=1}^8, \{I_{c,i}\}_{i=1}^8), y_\star) + \sum_a \lambda^a \ell(P(\mathcal{T}^a \mid \{I_{o,i}\}_{i=1}^8), y_\star^a), \tag{9}$$

where $\ell(\cdot)$ denotes the cross-entropy loss, $y_\star$ the correct choice in candidates, and $y_\star^a$ the ground-truth operator selection for attribute a. The first part of the loss encourages the model to select the right choice for evaluation, while the second part motivates meaningful internal representation to emerge. Compared to Zhang *et al.* [71], the system requires joint operation from not only a trained perception module θ, but also the algebraic encodings from the zero elements $\{M_0^a\}$ and the successor functions $\{M^a\}$, and correspondingly, induced operators $\mathcal{T}$. We notice the three-stage curriculum in Zhang *et al.* [71] is crucial for such a neuro-semi-symbolic system. In particular, we use λ^a to balance the trade-off in the curriculum: in the first stage, we only train parameters regarding objectiveness; in the second stage, we freeze objectiveness parameters and cyclically train parameters involving type, size, and color; in the last stage, we fine-tune all parameters.

4 Experiments

A cognitive architecture with systematic generalization is believed to demonstrate the following three principles [12,37,38]: (1) systematicity, (2) productivity, and (3) localism. Systematicity requires an architecture to be able to entertain "semantically related" contents after understanding a given thought. Productivity states the awareness of a constituent implies that of a recursive application of the constituent; vice versa for localism.

To verify the effectiveness of an algebraic treatment in systematic generalization, we showcase the superiority of the proposed ALANS learner on the three principles in the abstract spatial-temporal reasoning task of RPM. Specifically, we use the generation methods proposed in Zhang *et al.* [68] and Hu *et al.* [21] to generate RPM problems and carefully split training and testing to construct the three regimes. The former generates candidates by perturbing only one attribute of the correct answer while the later modifies attribute values in a hierarchical manner to avoid shortcut solutions by pure statistics. Both methods categorize relations in RPM into three types, according to Carpenter *et al.* [4]: unary (Constant and Progression), binary (Arithmetic), and ternary (Distribution of Three), each of which comes with several instances. Grounding the principles into learning abstract relations in RPM, we fix the configuration to be 3×3Grid and generate the following data splits for evaluation (please refer to the Appendix for details):

- Systematicity: the training set contains only a subset of instances for each type of relation, while the test set all other relation instances.
- Productivity: as the binary relation results from a recursive application of the unary relation, the training set contains only unary relations, whereas the test set only binary relations.
- Localism: the training and testing sets in the productivity split are swapped to study localism.

We follow Zhang *et al.* [68] to generate $10,000$ instances for each split and assign 6 folds for training, 2 folds for validation, and 2 folds for testing.

4.1 Experimental Setup

We evaluate the systematic generalizability of the proposed ALANS learner on the above three splits, and compare the ALANS learner with other baselines, including ResNet [16], ResNet+DRT [68], WReN [51], CoPINet [70], MXGNet [58], LEN [74], HriNet [21], and SCL [64]. We use either official or public implementations that reproduce the original results. All models are implemented in PyTorch [45] and optimized using ADAM [27] on an Nvidia Titan Xp GPU. We validate trained models on validation sets and report performance on test sets.

4.2 Systematic Generalization

Table 1 shows the performance of various models on systematic generalization, *i.e.*, systematicity, productivity, and localism. Compared to results reported in existing works mentioned above, all pure connectionist models experience a devastating performance drop when it comes to the critical cognitive requirements on systematic generalization, indicating that pure connectionist models fail to perform abstraction, algebraization, induction, or generalization needed in solving the abstract reasoning task; instead, they seem to only take a shortcut to bypass them. In particular, MXGNet's [58] superiority is diminishing in systematic generalization. In spite of learning with structural annotations, ResNet+DRT [68] does not fare better than its base model. The recently proposed HriNet [21] slightly improves on ResNet [16] in this aspect, with LEN [74] being only marginally better. WReN [51], on the other hand, shows oscillating performance across three regimes. Evaluated under systematic generation, SCL [64] and CoPINet [70] also far deviate from "superior performance." These observations suggest that pure connectionist models highly likely learn from variation in visual appearance rather than the algebra underlying the problem.

Table 1. Model performance on different aspects of systematic generalization. The performance is measured by accuracy on the test sets. Results on datasets generated by Zhang *et al.* [68] (upper) and by Hu *et al.* [21] (lower).

Method	MXGNet	ResNet+DRT	ResNet	HriNet	LEN	WReN	SCL	CoPINet	ALANS	ALANS-Ind	ALANS-V
Systematicity	20.95%	33.00%	27.35%	28.05%	40.15%	35.20%	37.35%	59.30%	**78.45%**	52.70%	93.85%
Productivity	30.40%	27.95%	27.05%	31.45%	42.30%	56.95%	51.10%	60.00%	**79.95%**	36.45%	90.20%
Localism	28.80%	24.90%	23.05%	29.70%	39.65%	38.70%	47.75%	60.10%	**80.50%**	59.80%	95.30%
Average	26.72%	28.62%	25.82%	29.73%	40.70%	43.62%	45.40%	59.80%	**79.63%**	48.65%	93.12%
Systematicity	13.35%	13.50%	14.20%	21.00%	17.40%	15.00%	24.90%	18.35%	**64.80%**	52.80%	84.85%
Productivity	14.10%	16.10%	20.70%	20.35%	19.70%	17.95%	22.20%	29.10%	**65.55%**	32.10%	86.55%
Localism	15.80%	13.85%	17.45%	24.60%	20.15%	19.70%	29.95%	31.85%	**65.90%**	50.70%	90.95%
Average	14.42%	14.48%	17.45%	21.98%	19.08%	17.55%	25.68%	26.43%	**65.42%**	45.20%	87.45%

Embedded in a neural-semi-symbolic framework, the proposed ALANS learner improves on systematic generalization by a large margin. With an algebra-aware design, the model is considerably stable across different principles of systematic generalization. The algebraic representation learned in relations of either a constituent or a recursive composition naturally supports productivity and localism, while semi-symbolic inner optimization further allows various instances of an operator type to be induced from the algebraic representation and boosts systematicity. The importance of the algebraic representation is made more significant in the ablation study: ALANS-Ind, with algebraic representation replaced by independent encodings and the algebraic isomorphism broken, shows inferior performance. We also examine the performance of the learner with perfect visual annotations (denoted as ALANS-V) to see how the proposed algebraic reasoning module works: the gap despite of accurate perception indicates space for improvement for the inductive reasoning part of the model. In the next section, we further show that the neuro-semi-symbolic decomposition in ALANS's design enables diagnostic tests into its jointly learned perception module and reasoning module. This design is in stark contrast to black-box models.

4.3 Analysis into Perception and Reasoning

The neural-semi-symbolic design affords analyses into both perception and reasoning. To evaluate the neural perception and the algebraic reasoning modules, we extract region-based object attribute annotations from the datasets [21,68] and categorize all relations into three types, *i.e.*, unary, binary, and ternary.

Table 2 shows the perception module's performance on the test sets in the three regimes of systematic generalization. We note that in order for the ALANS learner to achieve the desired results shown in Table 1, ALANS learns to construct the concept of objectiveness perfectly. The model also shows fairly accurate prediction on the attributes of type and size. However, on the texture-related concept of color, ALANS fails to develop a reliable notion on it. Despite that, the general prediction accuracy of the perception module is still surprising, considering that the perception module is jointly learned with ground-truth annotations on answer selections. The relatively lower accuracy on color could be attributed to its larger space compared to other attributes.

Table 2. Perception accuracy of the proposed ALANS learner, measured by whether the module can correctly predict an attribute's value. Results on datasets generated by Zhang *et al.* [68] (left) and by Hu *et al.* [21] (right).

Object Attribute	Objectiveness	Type	Size	Color	Object Attribute	Objectiveness	Type	Size	Color
Systematicity	100.00%	99.95%	94.65%	71.35%	Systematicity	100.00%	96.34%	92.36%	63.98%
Productivity	100.00%	99.97%	98.04%	77.61%	Productivity	100.00%	94.28%	97.00%	69.89%
Localism	100.00%	95.65%	98.56%	80.05%	Localism	100.00%	95.80%	98.36%	60.35%
Average	100.00%	98.52%	97.08%	76.34%	Average	100.00%	95.47%	95.91%	64.74%

Table 3 lists the reasoning module's performance during testing for the three aspects. Note that on position, the unary operator (shifting) and binary operator (set arithmetics) do not systematically imply each other. Hence, we do not count them as probes into productivity and localism. In general, we notice that the better the perception accuracy on one attribute, the better the performance on reasoning. However, we also note that despite the relatively accurate perception of objectiveness, type, and size, near perfect reasoning is never guaranteed. This deficiency is due to the perception uncertainty handled by expectation in Eq. (3): in spite of correctness when we take arg max, marginalizing by expectation will unavoidably introduce noise into the reasoning process. Therefore, an ideal reasoning module requires the perception frontend to be not only correct but also certain. Computationally, one can sample from the perception module and optimize Eq. (9) using REINFORCE [60]. However, the credit assignment problem and variance in gradient estimation will further complicate training.

4.4 In-Distribution Performance

To further evaluate how models perform under the regular Identically Distributed (I.I.D.) setup, we train the models on the original datasets generated by Zhang *et al.* [68] and Hu *et al.* [21] and measure the model accuracy in the test splits. We compare ALANS with published baselines in Table 1.

Table 4 (left) shows the results on the RAVEN dataset [68]. With the jointly trained vision component, the ALANS learner does not fare better than the best connectionist approaches, making it on par with SCL only. As the dataset is known to have shortcut solutions, neural approaches like MXGNet and CoPINet could potentially find it easier to solve and hence achieve much superior results in this setup. However, ALANS-V, the variant with a perfect perception component reaches a level of much robustness and accuracy, attaining the best results in grid-like layouts, empirically believed to be the hardest in human evaluation [68].

Table 3. Reasoning accuracy of the proposed ALANS learner, measured by whether the module can correctly predict the type of a relation on an attribute. Results on datasets generated by Zhang *et al.* [68] (left) and by Hu *et al.* [21] (right).

Relation on	Position	Number	Type	Size	Color
Systematicity	72.04%	82.14%	81.50%	80.80%	40.40%
Productivity	-	98.75%	89.50%	72.10%	33.95%
Localism	-	74.70%	44.25%	56.40%	54.20%
Average	72.04%	85.20%	71.75%	69.77%	42.85%

Relation on	Position	Number	Type	Size	Color
Systematicity	69.96%	80.34%	83.50%	80.85%	28.85%
Productivity	-	99.10%	87.95%	68.50%	23.10%
Localism	-	70.55%	36.65%	42.30%	33.20%
Average	69.96%	83.33%	69.37%	63.88%	28.38%

Table 4. Model performance on RAVEN [68] (left) and I-RAVEN [21] (right) under the regular I.I.D. evaluation, measured by accuracy on the test sets.

Method	Acc	Center	2 × 2Grid	3 × 3Grid	L-R	U-D	O-IC	O-IG
WReN	34.0%/21.5%	58.4%/24.0%	38.9%/25.0%	37.7%/20.1%	21.6%/19.7%	19.8%/19.9%	38.9%/21.3%	22.6%/20.6%
ResNet	53.4%/18.4%	52.8%/22.6%	41.9%/15.5%	44.3%/18.1%	58.8%/19.0%	60.2%/19.6%	63.2%/17.5%	53.1%/16.6%
ResNet+DRT	59.6%/20.7%	58.1%/24.2%	46.5%/18.2%	50.4%/19.8%	65.8%/22.0%	67.1%/22.1%	69.1%/21.0%	60.1%/18.1%
LEN	71.6%/32.8%	79.1%/44.8%	56.1%/27.9%	60.3%/23.9%	80.5%/34.1%	76.4%/34.4%	79.3%/35.8%	69.9%/28.5%
HriNet	45.1%/60.8%	66.1%/78.2%	40.7%/50.1%	38.0%/42.4%	44.9%/70.1%	43.2%/70.3%	47.2%/68.2%	35.8%/46.3%
MXGNet	84.0%/33.1%	94.3%/40.7%	60.5%/27.9%	64.9%/24.7%	96.6%/35.8%	96.4%/34.5%	94.1%/36.4%	81.3%/31.6%
CoPINet	91.4%/46.1%	95.1%/54.4%	77.5%/36.8%	78.9%/31.9%	**99.1%**/51.9%	**99.7%**/52.5%	**98.5%**/52.2%	91.4%/42.8%
ALANS	74.4%/78.5%	69.1%/72.3%	80.2%/79.5%	75.0%/72.9%	72.2%/79.2%	73.3%/79.6%	76.3%/85.9%	74.9%/79.9%
SCL	74.2%/80.5%	82.8%/84.6%	70.4%/79.4%	64.1%/69.9%	77.6%/82.7%	78.4%/82.6%	84.2%/87.3%	62.2%/77.2%
ALANS-V	**94.4%/93.5%**	**98.4%/98.9%**	**91.5%/85.0%**	**87.0%/83.2%**	97.3%/**90.9%**	96.4%/**98.1%**	97.3%/**99.1%**	**93.2%/89.5%**

Table 4 (right) shows the results on the I-RAVEN dataset [21]. Apart from ALANS-V's realizing the best performance across all models, we also notice the consistency of performance of the proposed method across datasets, with or without the shortcut issues. All other methods show drastically varying performance, particularly for CoPINet and MXGNet, arguably because of the choice generation strategy that effectively prunes easy solution paths via statistics.

In summary, by analyzing the results from Tables 1 and 4 together, we notice that ALANS not only attains reasonable performance on the I.I.D. setup but also generalizes systematically.

4.5 Generative Potential

Compared to existing discriminative-only RPM-solving methods, the proposed ALANS learner is unique in its generative potential. As mentioned above, the final panel attribute can be decoded by sequentially selecting the most probable hidden operator and the attribute value. A solution can be generated when equipped with a rendering engine. In Fig. 3, we use the rendering program from Zhang *et al.* [68] to showcase the generative potential in the ALANS learner.

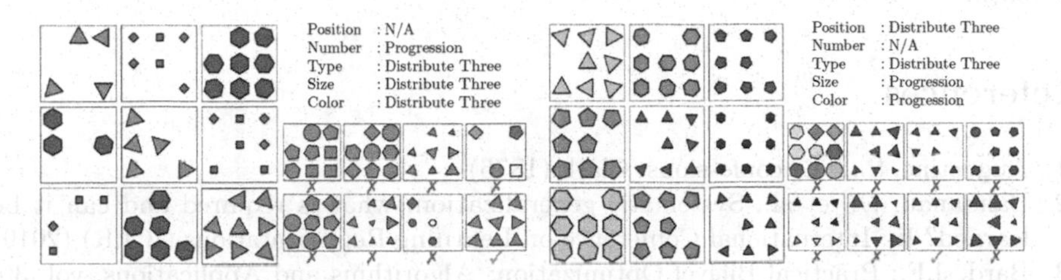

Fig. 3. Examples of RPM instances with the missing entries filled by solutions directly generated by the ALANS learner. Ground-truth relations are also listed. Note the generated results do not look exactly the same as the correct candidate choices due to random rotations during rendering, but they are semantically correct.

5 Conclusion and Limitation

In this work, we propose the ALgebra-Aware Neuro-Semi-Symbolic (ALANS) learner, echoing a normative theory in the connectionist-classicist debate that an algebraic treatment in a cognitive architecture should improve a model's systematic generalization ability. In particular, the ALANS learner employs a neural-semi-symbolic architecture, where the neural visual perception module is responsible for summarizing visual information and the algebraic abstract reasoning module transforms it into algebraic representation with isomorphism established by the Peano Axiom and the representation theory, conducts operator induction, and executes it to arrive at an answer. In three RPM domains reflective of systematic generalization, the proposed ALANS learner shows superior performance compared to other pure connectionist baselines.

The proposed ALANS learner also bears some limitations. For one thing, we make the assumption in our formulation that relations on different attributes are independent and can be factorized. This assumption is not universally correct and could potentially lead to failure in more complex reasoning scenarios when attributes are correlated. For another, we assume a fixed and known space for each attribute in the perception module, while in the real world the space for one attribute could be dynamically changing. In addition, the reasoning module is sensitive to perception uncertainty as has already been discussed in the experimental results. Besides, the gap between perfection and the status quo in reasoning remains to be filled. In this work, we only show how the hidden operator can be induced with regularized linear regression via the representation theory. However, more elaborate differentiable optimization problems can certainly be incorporated for other problems.

With the limitation in mind, we hope that this preliminary study could inspire more research on incorporating algebraic structures into current connectionist models and help address challenging modeling problems [65,69,73,76].

Acknowledgement. We thank Prof. Hongjing Lu and colleagues from UCLA for fruitful discussions. We would also like to thank anonymous reviewers for constructive feedback.

References

1. Augustine, S.: The confessions. Clark (1876)
2. Bahdanau, D., et al.: Systematic generalization: what is required and can it be learned? In: International Conference on Learning Representations (ICLR) (2019)
3. Bard, J.F.: Practical Bilevel Optimization: Algorithms and Applications, vol. 30. Springer, Dordrecht (2013)
4. Carpenter, P.A., Just, M.A., Shell, P.: What one intelligence test measures: a theoretical account of the processing in the raven progressive matrices test. Psychol. Rev. **97**(3), 404 (1990)
5. Chen, X., Liang, C., Yu, A.W., Song, D., Zhou, D.: Compositional generalization via neural-symbolic stack machines. In: Advances in Neural Information Processing Systems (2020)
6. Chen, Z., Mao, J., Wu, J., Wong, K.Y.K., Tenenbaum, J.B., Gan, C.: Grounding physical concepts of objects and events through dynamic visual reasoning. In: International Conference on Learning Representations (ICLR) (2020)
7. Chollet, F.: The measure of intelligence. arXiv preprint arXiv:1911.01547 (2019)
8. Colson, B., Marcotte, P., Savard, G.: An overview of bilevel optimization. Ann. Oper. Res. **153**(1), 235–256 (2007)
9. Dong, H., Mao, J., Lin, T., Wang, C., Li, L., Zhou, D.: Neural logic machines. In: International Conference on Learning Representations (ICLR) (2018)
10. Evans, R., Grefenstette, E.: Learning explanatory rules from noisy data. J. Artif. Intell. Res. (JAIR) **61**, 1–64 (2018)
11. Fodor, J.A.: The Language of Thought, vol. 5. Harvard University Press, Cambridge (1975)
12. Fodor, J.A., Pylyshyn, Z.W., et al.: Connectionism and cognitive architecture: a critical analysis. Cognition **28**(1–2), 3–71 (1988)
13. d'Avila Garcez, A.S. Broda, K.B., Gabbay, D.M.: Neural-Symbolic Learning Systems: Foundations and Applications. Springer, London (2012)
14. Han, C., Mao, J., Gan, C., Tenenbaum, J., Wu, J.: Visual concept-metaconcept learning. In: Advances in Neural Information Processing Systems (NeurIPS) (2019)
15. Hausmann, B.A., Ore, O.: Theory of quasi-groups. Am. J. Math. **59**(4), 983–1004 (1937)

16. He, K., Zhang, X., Ren, S., Sun, J.: Deep residual learning for image recognition. In: Conference on Computer Vision and Pattern Recognition (CVPR) (2016)
17. Heath, T.L., et al.: The Thirteen Books of Euclid's Elements. Courier Corporation (1956)
18. Hill, F., Santoro, A., Barrett, D.G., Morcos, A.S., Lillicrap, T.: Learning to make analogies by contrasting abstract relational structure. In: International Conference on Learning Representations (ICLR) (2019)
19. Hofstadter, D.R.: Fluid concepts and Creative Analogies: Computer Models of the Fundamental Mechanisms of Thought. Basic Books, New York (1995)
20. Holyoak, K.J., Ichien, N., Lu, H.: From semantic vectors to analogical mapping. Curr. Dir. Psychol. Sci. **31**, 09637214221098054 (2022)
21. Hu, S., Ma, Y., Liu, X., Wei, Y., Bai, S.: Hierarchical rule induction network for abstract visual reasoning. arXiv preprint arXiv:2002.06838 (2020)
22. Hudson, D., Manning, C.D.: Learning by abstraction: The neural state machine. In: Advances in Neural Information Processing Systems (2019)
23. Humphreys, J.E.: Introduction to Lie Algebras and Representation Theory, vol. 9. Springer, New York (2012)
24. Jaeggi, S.M., Buschkuehl, M., Jonides, J., Perrig, W.J.: Improving fluid intelligence with training on working memory. Proc. Natl. Acad. Sci. (PNAS) **105**(19), 6829–6833 (2008)
25. James, W.: The Principles of Psychology. Henry Holt and Company, New York (1891)
26. Kansky, K., et al.: Schema networks: zero-shot transfer with a generative causal model of intuitive physics. In: International Conference on Machine Learning (ICML) (2017)
27. Kingma, D.P., Ba, J.: Adam: a method for stochastic optimization. In: International Conference on Learning Representations (ICLR) (2014)
28. Kingma, D.P., Welling, M.: Auto-encoding variational Bayes. arXiv preprint arXiv:1312.6114 (2013)
29. Lake, B., Baroni, M.: Generalization without systematicity: On the compositional skills of sequence-to-sequence recurrent networks. In: International Conference on Machine Learning (ICML) (2018)
30. Little, D.R., Lewandowsky, S., Griffiths, T.L.: A Bayesian model of rule induction in raven's progressive matrices. In: Annual Meeting of the Cognitive Science Society (CogSci) (2012)
31. Lovett, A., Forbus, K.: Modeling visual problem solving as analogical reasoning. Psychol. Rev. **124**(1), 60 (2017)
32. Lovett, A., Forbus, K., Usher, J.: A structure-mapping model of Raven's progressive matrices. In: Annual Meeting of the Cognitive Science Society (CogSci) (2010)
33. Lovett, A., Tomai, E., Forbus, K., Usher, J.: Solving geometric analogy problems through two-stage analogical mapping. Cogn. Sci. **33**(7), 1192–1231 (2009)
34. Maddy, P.: Believing the axioms. I. J. Symb. Logic **53**(2), 481–511 (1988)
35. Manhaeve, R., Dumancic, S., Kimmig, A., Demeester, T., De Raedt, L.: DeepProbLog: neural probabilistic logic programming. In: Advances in Neural Information Processing Systems (NeurIPS) (2018)
36. Mao, J., Gan, C., Kohli, P., Tenenbaum, J.B., Wu, J.: The neuro-symbolic concept learner: Interpreting scenes, words, and sentences from natural supervision. In: International Conference on Learning Representations (ICLR) (2019)
37. Marcus, G.: The Algebraic Mind. MIT Press, Cambridge (2001)
38. Marcus, G.: The next decade in AI: four steps towards robust artificial intelligence. arXiv preprint arXiv:2002.06177 (2020)

39. Marcus, G.F., Vijayan, S., Rao, S.B., Vishton, P.M.: Rule learning by seven-month-old infants. Science **283**(5398), 77–80 (1999)
40. McCarthy, J.: Programs with common sense. RLE and MIT Computation Center (1960)
41. McGreggor, K., Goel, A.: Confident reasoning on raven's progressive matrices tests. In: AAAI Conference on Artificial Intelligence (AAAI) (2014)
42. McGreggor, K., Kunda, M., Goel, A.: Fractals and Ravens. Artif. Intell. **215**, 1–23 (2014)
43. Mekik, C.S., Sun, R., Dai, D.Y.: Similarity-based reasoning, Raven's matrices, and general intelligence. In: International Joint Conference on Artificial Intelligence (IJCAI) (2018)
44. Newell, A.: Physical symbol systems. Cogn. Sci. **4**(2), 135–183 (1980)
45. Paszke, A., et al.: Automatic differentiation in PyTorch. In: NIPS Autodiff Workshop (2017)
46. Peano, G.: Arithmetices principia: Nova methodo exposita. Fratres Bocca (1889)
47. Rahaman, N., et al.: Dynamic inference with neural interpreters. In: Advances in Neural Information Processing Systems (NeurIPS) (2021)
48. Raven, J.C.: Mental tests used in genetic studies: the performance of related individuals on tests mainly educative and mainly reproductive. Master's thesis, University of London (1936)
49. Raven, J.C., Court, J.H.: Raven's Progressive Matrices and Vocabulary Scales. Oxford Pyschologists Press, Oxford (1998)
50. Rocktäschel, T., Riedel, S.: End-to-end differentiable proving. In: Advances in Neural Information Processing Systems (NeurIPS) (2017)
51. Santoro, A., Hill, F., Barrett, D., Morcos, A., Lillicrap, T.: Measuring abstract reasoning in neural networks. In: International Conference on Machine Learning (ICML) (2018)
52. Santoro, A., et al.: A simple neural network module for relational reasoning. In: Advances in Neural Information Processing Systems (NeurIPS) (2017)
53. Serafini, L., d'Garcez, A.: Logic tensor networks: deep learning and logical reasoning from data and knowledge. arXiv preprint arXiv:1606.04422 (2016)
54. Shegheva, S., Goel, A.: The structural affinity method for solving the Raven's progressive matrices test for intelligence. In: AAAI Conference on Artificial Intelligence (AAAI) (2018)
55. Spearman, C.: The Nature of "Intelligence" and the Principles of Cognition. Macmillan, New York (1923)
56. Spearman, C.: The Abilities of Man, vol. 6. Macmillan, New York (1927)
57. Steenbrugge, X., Leroux, S., Verbelen, T., Dhoedt, B.: Improving generalization for abstract reasoning tasks using disentangled feature representations. arXiv preprint arXiv:1811.04784 (2018)
58. Wang, D., Jamnik, M., Lio, P.: Abstract diagrammatic reasoning with multiplex graph networks. In: International Conference on Learning Representations (ICLR) (2020)
59. Wang, K., Su, Z.: Automatic generation of Raven's progressive matrices. In: International Joint Conference on Artificial Intelligence (IJCAI) (2015)
60. Williams, R.J.: Simple statistical gradient-following algorithms for connectionist reinforcement learning. Mach. Learn. **8**(3–4), 229–256 (1992)
61. Winograd, T.: Procedures as a representation for data in a computer program for understanding natural language. Technical report, MIT Center for Space Research (1971)

62. Wittgenstein, L.: Philosophical Investigations. Philosophische Untersuchungen. Macmillan, New York (1953)
63. Wu, J., Tenenbaum, J.B., Kohli, P.: Neural scene de-rendering. In: Conference on Computer Vision and Pattern Recognition (CVPR) (2017)
64. Wu, Y., Dong, H., Grosse, R., Ba, J.: The scattering compositional learner: discovering objects, attributes, relationships in analogical reasoning. arXiv preprint arXiv:2007.04212 (2020)
65. Xu, M., Jiang, G., Zhang, C., Zhu, S.C., Zhu, Y.: EST: evaluating scientific thinking in artificial agents. arXiv preprint arXiv:2206.09203 (2022)
66. Yi, K., Gan, C., Li, Y., Kohli, P., Wu, J., Torralba, A., Tenenbaum, J.: CLEVRER: collision events for video representation and reasoning. In: International Conference on Learning Representations (ICLR) (2020)
67. Yi, K., Wu, J., Gan, C., Torralba, A., Kohli, P., Tenenbaum, J.: Neural-symbolic VQA: disentangling reasoning from vision and language understanding. In: Advances in Neural Information Processing Systems (NeurIPS) (2018)
68. Zhang, C., Gao, F., Jia, B., Zhu, Y., Zhu, S.C.: Raven: A dataset for relational and analogical visual reasoning. In: Conference on Computer Vision and Pattern Recognition (CVPR) (2019)
69. Zhang, C., Jia, B., Edmonds, M., Zhu, S.C., Zhu, Y.: ACRE: abstract causal reasoning beyond covariation. In: Conference on Computer Vision and Pattern Recognition (CVPR) (2021)
70. Zhang, C., Jia, B., Gao, F., Zhu, Y., Lu, H., Zhu, S.C.: Learning perceptual inference by contrasting. In: Advances in Neural Information Processing Systems (NeurIPS) (2019)
71. Zhang, C., Jia, B., Zhu, S.C., Zhu, Y.: Abstract spatial-temporal reasoning via probabilistic abduction and execution. In: Conference on Computer Vision and Pattern Recognition (CVPR) (2021)
72. Zhang, C., Zhu, Y., Zhu, S.C.: MetaStyle: three-way trade-off among speed, flexibility, and quality in neural style transfer. In: AAAI Conference on Artificial Intelligence (AAAI) (2019)
73. Zhang, W., Zhang, C., Zhu, Y., Zhu, S.C.: Machine number sense: a dataset of visual arithmetic problems for abstract and relational reasoning. In: AAAI Conference on Artificial Intelligence (AAAI) (2020)
74. Zheng, K., Zha, Z.J., Wei, W.: Abstract reasoning with distracting features. In: Advances in Neural Information Processing Systems (NeurIPS) (2019)
75. Zhu, S.C., Mumford, D., et al.: A stochastic grammar of images. Found. Trends® Comput. Graph. Vis. **2**(4), 259–362 (2007)
76. Zhu, Y., et al.: Dark, beyond deep: a paradigm shift to cognitive AI with humanlike common sense. Engineering **6**(3), 310–345 (2020)

Video Dialog as Conversation About Objects Living in Space-Time

Hoang-Anh Pham[1](✉), Thao Minh Le[1], Vuong Le[1], Tu Minh Phuong[2], and Truyen Tran[1]

[1] Applied Artificial Intelligence Institute, Deakin University, Geelong, Australia
{phamhoan,thao.le,vuong.le,truyen.tran}@deakin.edu.au
[2] Posts and Telecommunications Institute of Technology, Ho Chi Minh City, Vietnam
phuongtm@ptit.edu.vn

Abstract. It would be a technological feat to be able to create a system that can hold a meaningful conversation with humans about what they watch. A setup toward that goal is presented as a video dialog task, where the system is asked to generate natural utterances in response to a question in an ongoing dialog. The task poses great visual, linguistic, and reasoning challenges that cannot be easily overcome without an appropriate representation scheme over video and dialog that supports high-level reasoning. To tackle these challenges we present a new object-centric framework for video dialog that supports neural reasoning dubbed COST–which stands for *Conversation about Objects in Space-Time*. Here dynamic space-time visual content in videos is first parsed into object trajectories. Given this video abstraction, COST maintains and tracks object-associated *dialog states*, which are updated upon receiving new questions. Object interactions are dynamically and conditionally inferred for each question, and these serve as the basis for relational reasoning among them. COST also maintains a history of previous answers, and this allows retrieval of relevant object-centric information to enrich the answer forming process. Language production then proceeds in a step-wise manner, taking the context of the current utterance, the existing dialog, and the current question. We evaluate COST on the AVSD test splits (DSTC7 and DSTC8), demonstrating its competitiveness against state-of-the-arts.

1 Introduction

It is a hallmark of visual intelligence to build a system that can hold a meaningful conversation with humans about a video. A system of this capability would be a strong contender for passing the visual Turing test [11]. Posed as video dialog [1,12], this task challenges the current arts due to its sheer complexity on many fronts. A dialog has its natural flow through multiple turns, each of which builds upon the previous questions and answers. This demands deep linguistic understanding about, and keeping track of, what has been said and grounded on

Supplementary Information The online version contains supplementary material available at https://doi.org/10.1007/978-3-031-19842-7_41.

S. Avidan et al. (Eds.): ECCV 2022, LNCS 13699, pp. 710–726, 2022.
https://doi.org/10.1007/978-3-031-19842-7_41

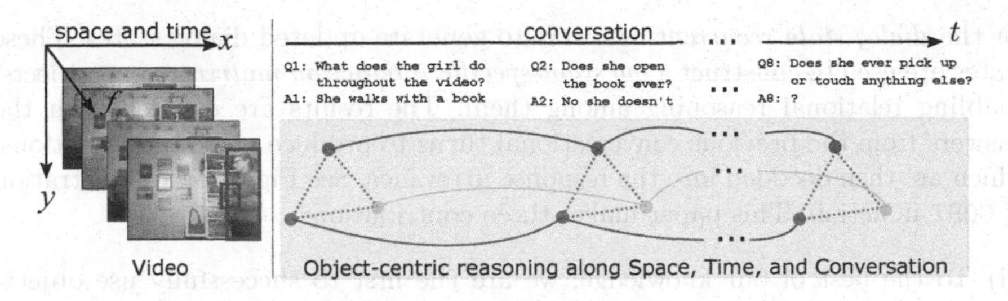

Fig. 1. We introduce COST, an object-centric reasoning framework that collects clues along with the Space and Time dimensions of the Video and the Conversation dimension of the dialog toward reliable QA.

the visual concepts found in the video, and then analyzing the new question in this newly established context. Given the question semantics and its constituent words, generating a linguistic answer necessitates symbolic grounding and visual reasoning through the complex space-time structure of the video, where in multiple objects interact in a dynamic manner.

Video dialog is inherently harder than the task of visual dialog over static images [8] due to the temporal dynamics of the scene [1]. It is also harder than the standard setting of video QA [28] since the next question may not be comprehensible without maintaining a history of, and referring to the previous answers. There have been several attempts to tackle these challenges. Early attempts [14,24,33,35] encode the dialog flow using recurrent neural networks. Later methods [22,29] resort to Transformers for better distant dependencies as well as cross-modality relations. More recent methods make use of graphs as a representation of dialog structures [12,23] and co-reference [18] which achieved the new state-of-the-art result for this task. However, we have only scratched the surface of what is possible and the principal challenges remain.

A plausible pathway toward solving the remaining video dialog challenges is through high-level, object-centric representation and reasoning as seen in human visual cognition: Humans see objects and agency as the core "living" constructs with natural compositionally, permanency, temporal dynamics, and n-body interactions [39]. Importantly, the object-centric approach has recently been found to be essential for reasoning in Visual QA [27] and Video QA [7], thanks to the ease of binding linguistic concepts to visual regions.

To this end, we propose a new object-centric framework dubbed COST (**C**onversation about **O**bjects in **S**pace-**T**ime) for video dialog. Video is first parsed into a set of object trajectories throughout the video across the spatial dimensions of the frame and the temporal spans of object lives. Central to COST is a model of the object states dynamics as the conversation progresses[1]. In particular, COST maintains a *recurrent system of dialog-induced object states* throughout the course of conversation. For each new question, the object lives will be consulted through semantic word-object grounding and selective frame attention, producing question-guided object representations. These serve as input

[1] This is related to, but distinct from, the dialog state tracking in typical task-oriented dialogs in NLP [10].

for the *dialog state recurrent networks* to generate updated dialog states. These states are used to construct a *question-specific interaction matrix* among objects, enabling relational reasoning among them. The results are coupled with the answers from the previous conversational turns to produce new representations, which are then decoded into the response utterance. See Fig. 1 for an illustration of COST in action. This paper makes three contributions:

(i) To the best of our knowledge, we are the first to successfully use object-centric reasoning for video dialog.
(ii) The inductive priors brought by our object-centric framework as *a recurrent system of dialog-induced object states* are *unique* and particularly beneficial for video dialog as empirically evident.
(ii) Experimental results on the AVSD dataset with two different test splits DSTC7 and DSTC8 shows that COST is highly competitive against rivals.

2 Related Work

Visual and Video Dialog. The visual dialog task and an accompanied dataset (VisDial) were first introduced in [8]. The task requires both conversational and visual reasoning ability over multiple turns. Like the precursor Visual QA task, visual dialog needs deep understanding of the visual concepts and relations in the images, and reasoning about them in response to the current question. A unique challenge in visual dialog is the co-referrence problem of linguistic information between dialog turns. The early works tried to resolve this by hierarchical encoding [37] or memory network based on attention [36]. However, these works mostly focus on history dialog reasoning. The work of [21] solved the co-reference on both visual and linguistic space using neural module networks. Recently, like the other vision-language tasks, visual dialog is also beneficial from cross-modal pre-training which was employed [13,32,43,53].

The video dialog task pushes the challenges further, thanks to the complexity of analyzing video. Video-grounded dialog system (VGDS) received more interested from community recently with the DSTC7 [51] and DSTC8 [19] challenge. Early attempts [14,24,33,35] employ the recurrent neural network to encode dialog history. Later methods [6,30,49] use the Attention mechanism, or [45] design a memory networks to extract the relationship between different modalities, [22,26,29] employ Transformer-based network to resolve the cross modality learning. More explicit relationships in dialogs are recently studied, showing promising results [12,23]. This extends to co-reference graph across visual and textual domains [18]. However, these approach all represent the video by frame features, just lacking the key concepts of of object permanence.

Object-Centric Visual Reasoning. Visual QA and visual dialog benefit greatly from object-centric representation of visual content as this fills the gap between low-level visual features and high-level linguistic semantics [7,27]. The early work proposed the object-centric representation [9] by leveraging the object detection for images and [4,17,44,47,48] combining with the tracking algorithm for video inputs. In problems that require further reasoning on objects, a relational network was introduced by [2], and [16,18,34,42,50,52] make use of graph-based method for transparent reasoning. However, these methods only use object's features as an additional input on a generic network without any prior reasoning structure in space-time, which is the key of success in further reasoning. The work in [7] introduces a generic reasoning unit through dynamically building interaction graphs between objects in space-time derived by the question, but this is limited to single question.

3 Method

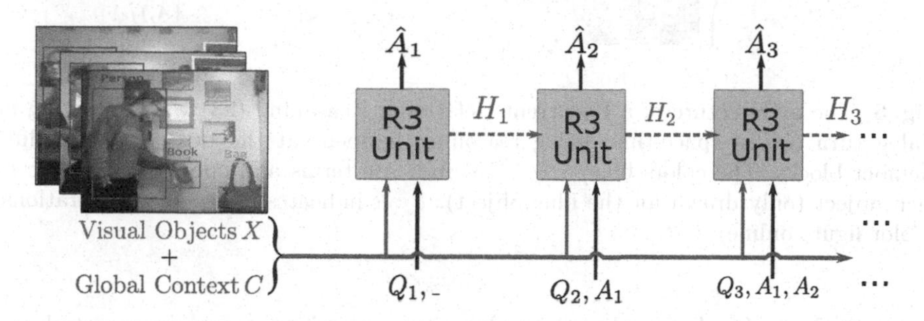

Fig. 2. The architecture of COST model features a chain of *Recurrent Relational Reasoning* (R3) Units which maintain the *dialog states* $\{H_i\}$ across the turns of the conversation.

Given a video V, the task of video dialog is to hold a smooth conversation of T turns. Each turn t is a textual question-answer pair (Q_t, A_t). We want to estimate a model parameterized by θ that returns the best answer for the corresponding question:

$$\hat{A}_t = \arg\max_{A} P\left(A \mid V, Q_{1:t-1}, A_{1:t-1}, Q_t; \theta\right), \tag{1}$$

for $t = 1, 2, ..., T$.

The main challenge lies in the coherent answering with respect to the existing conversation of length $t-1$, while reasoning over the space-time of video, which itself demands efficient and effective schemes of representation that supports a high-level dialog. With these constraints in mind, here we treat the video V as a collection of object lives, each of which is a trajectory of spatial positions equipped with object visual features. This object-oriented view enables ease of constructing reasoning paths in response to linguistic queries.

In the following we present COST, an *object-centric reasoning model* for video dialog. Figure 2 shows the overall architecture of COST.

3.1 Preliminaries

Following [7], we parse the video of F frames into N sequences of objects tracked over time. Objects at each frame are position-coded together with their appearance

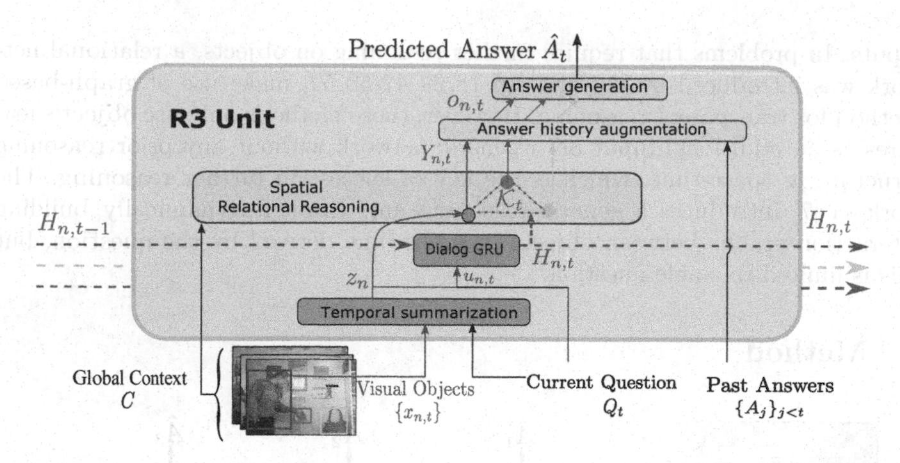

Fig. 3. The architecture of a Recurrent Relational Reasoning (R3) Unit operating at dialog turn t. The space-time-dialog reasoning happens at the three corresponding member blocks. The colors blue/red/yellow indicate terms and operations specific to each object (only drawn for the blue object). Pink indicates cross-object operations. (Color figure online)

features. Each *visual object live* throughout the video is therefore represented as a matrix $X_n \in \mathbb{R}^{F\times d}$ for $n = 1, 2, ..., N$. Moreover, each frame is represented by a holistic context vector $c \in \mathbb{R}^{1\times d}$, therefore the holistic context matrix representation of video is denoted as $C \in \mathbb{R}^{F\times d}$. Dialog sentences (questions and answers) are split into words and then embedded into a matrix $S = [w_{1:L}] \in \mathbb{R}^{L\times d}$ where L is sentence length. With a slight abuse of notation, we use d to denote the size of both visual feature vectors and linguistic vectors for ease of reading.

We also make use of the attention function [40] defined over the triplet of *query* $q \in \mathbb{R}^{1\times d}$, *keys* $K \in \mathbb{R}^{M\times d}$ and *values* $V \in \mathbb{R}^{M\times d}$:

$$\text{Attn}\left(q, K, V\right) := \sum_{m=1}^{M} \text{softmax}_m \left(\frac{K_m W_k \left(q W_q\right)^\top}{\sqrt{d}} \right) V_m W_v \in \mathbb{R}^{1\times d}. \tag{2}$$

where W_k, W_q, and W_v are learnable parameters. In essence, this function reads out the most relevant values whose keys match the query. Likewise, a linear projection $W \in \mathbb{R}^{d_1\times d_2}$ from $x \in \mathbb{R}^{1\times d_1}$ to $\mathbb{R}^{1\times d_2}$ is denoted as

$$\text{Linear}(x) := xW. \tag{3}$$

3.2 Recurrent Relational Reasoning over Space-Time and Turns

The COST is a recurrent system that keeps track of the evolution of the dialog states over turns. Each step is a reasoning unit R3 (short for Recurrent Relational Reasoning) which takes as input a question $Q_t \in \mathbb{R}^{S\times d}$ of the present turn t, the previous dialog states $H_{t-1} \in \mathbb{R}^{N\times d}$, a holistic context representation

$C \in \mathbb{R}^{F \times d}$, a set of N object sequences $X = \{X_n \mid X_n \in \mathbb{R}^{F \times d}\}_{n=1}^{N}$, and past answers $\{A_j\}_{j<t}$. The outputs of a R3 unit are the representations of object lives conditioned on the current query and information from past turns. Figure 3 illustrates the structure of R3 unit.

Query-Induced Temporal Summarization. At each *conversational turn* t, we produce query-specific object representation. Firstly, we generate a *query-specific* summary of each object sequence $X_n \in \mathbb{R}^{F \times d}$ into a vector $z_n \in \mathbb{R}^{1 \times d}$ using temporal attention over frames. The frame attention weights are driven by the words in the query $Q_t = \{Q_{s,t} \in \mathbb{R}^{1 \times d}\}_{s=1}^{S_t}$. These produce a query-specific *object resume* as follows:

$$z_n = \frac{1}{S_t} \sum_{s=1}^{S_t} \text{Attn}\left(Q_{s,t}, X_n, X_n\right) \in \mathbb{R}^{1 \times d}. \tag{4}$$

To handle the case where an object cannot be detected at particular frames, we place a binary mask in the appropriate place.

Next we generate an *object-specific* embedding of the question:

$$q_n = \text{Attn}\left(z_n, Q_t, Q_t\right) \in \mathbb{R}^{1 \times d},$$

Finally the object embedding is modulated by the question as:

$$u_{n,t} = \tanh\left(\left[z_n, q_n, q_n \odot z_n\right]\right) \in \mathbb{R}^{1 \times 3d}. \tag{5}$$

This embedding serves as input for the recurrent network, which is presented in the next subsection.

Recurrent Dialog States. In dialog, the question at a turn typically advances from, and co-refers to, the previous questions and answers. In video dialog, questions are semantically related to objects appearing in video. We hence maintain a **dialog state** at turn t in the form of a matrix $H_t \in \mathbb{R}^{N \times d}$, i.e., each row corresponds to an object. The state dynamics is modeled in a set of N parallel recurrent networks:

$$H_{n,t} = \text{GRU}\left(H_{n,t-1}, u_{n,t}\right) \in \mathbb{R}^{1 \times d}, \tag{6}$$

for $n = 1, 2, ..., N$, where GRU is a standard Gated Recurrent Unit [5], $u_{n,t}$ is turn-specific embedding of the object n at turn t calculated in Eq. (5).

As the dialog states are object-centric, co-references between questions are *indirectly* and *distributionally* captured into the current multi-object states through the integration of previous states $H_{n,t-1}$ (which contains information of the previous questions) and the current question as part of $u_{n,t}$. In what follows, we will show how H_t is used for relational reasoning.

Relational Reasoning Between Objects. Equipped with dialog states, we now model the inter-object interaction which describes the behavior of objects with their neighbors driven by the current question Q_t. We employ a spatial graph whose vertices are the object resumes $Z = \{z_n\}_{n=1}^{N}$ computed in Eq. (4), and the edges are represented by an adjacency matrix $\mathcal{K}_t \in \mathbb{R}^{N\times N}$ which is calculated dynamically as the *question-specific interaction matrix* between objects:

$$\mathcal{K}_t = \text{softmax}\left(\frac{H_t H_t^\top}{\sqrt{d}}\right), \tag{7}$$

where H_t is computed in Eq. (6). This matrix serves as a backbone for a Deep Graph Convolutional Network (DGCN) [27] to refine object representations by taking into account the relations with their neighboring nodes:

$$\bar{Z} = \text{DGCN}\left(Z; \mathcal{K}_t\right) \tag{8}$$

Details of DCGN $(\cdot;\cdot)$ is given in the Supplement.

Utilizing Visual Contex. To utilize the underlying background scene information and compensate for possible undetected objects, we augment the object representations with the holistic context information $C \in \mathbb{R}^{F\times d}$. We summarize the context sequence into a vector as follows:

$$\bar{c} = \frac{1}{S_t}\sum_{s=1}^{S_t} \text{Attn}\left(Q_{s,t}, C, C\right) \in \mathbb{R}^{1\times d}. \tag{9}$$

Finally, the final output of this component is

$$Y_{n,t} = \text{Linear}\left(\left[\bar{Z}[n]; \bar{c}\right]\right) \in \mathbb{R}^{1\times d}, \tag{10}$$

where $\bar{Z}[n]$ is the n-th row of $\bar{Z}$ in Eq. (8).

Leveraging Answer History. So far, the R3 unit has used the recurrent dialog states to generate the representation of objects for the current question, however, part of historical information has been forgotten after a long conversation. It is therefore necessary to maintain a dynamic history of previous turns, which will be queried when seeking an answer to a new question. This would help to mitigate the co-reference effect by borrowing parts of previous answers into the current answer when deemed relevant.

Recall that the reasoning by the R3 unit results in each object having a turn-specific representation $Y_{n,j} \in \mathbb{R}^{1\times d}$ for turn $j = 1, 2, ..., t-1$. Let $A_{j-1} \in \mathbb{R}^{L_{j-1}\times d}$ be the embedding the answer at turn $j-1$. The *object-specific answer embedding* is computed as:

$$a_{n,j-1} = \text{Attn}\left(Y_{n,j}, A_{j-1}, A_{j-1}\right) \in \mathbb{R}^{1\times d}. \tag{11}$$

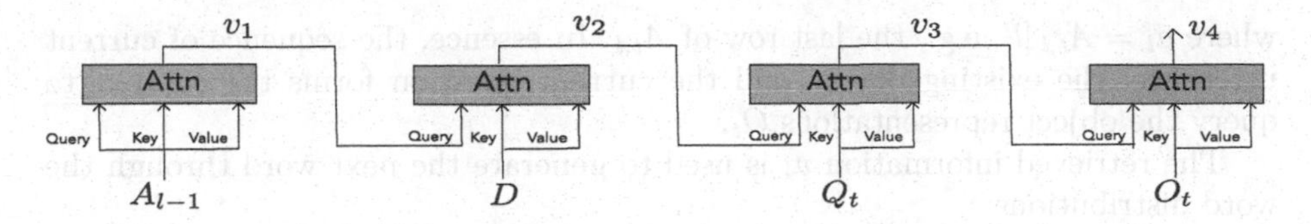

Fig. 4. Four-step Transformer decoder used in the Answer generator for the question at turn t at generation step l. A_{l-1}: embedding matrix of the unfinished utterance of length $l-1$; D: embedding matrix of the dialog history; Q_t : embedding of the question at turn t; O_t: output of COST.

This is combined with the object representation and turn position to generate a new object representation:

$$G_{n,j} = \text{Linear}\left([Y_{n,j}, a_{j-1}, p_j]\right) \in \mathbb{R}^{1\times d}, \tag{12}$$

where p_j is a positional encoding feature for each turn.

Thus the collection over turns $\mathfrak{M}_{n,t} = [G_{n,j}]_{j=1}^{t-1} \in \mathbb{R}^{(j-1)\times d}$ is a history of answer-guided representation for object n over previous turns. The answer history enables the retrieval of relevant pieces w.r.t. the current question at turn t:

$$\mathcal{H}_{n,t} = \text{Attn}\left(Y_{n,t}, \mathfrak{M}_{n,t}, \mathfrak{M}_{n,t}\right). \tag{13}$$

This is then augmented with the current object representation $Y_{n,t}$ in Eq. (10) to produce the final form:

$$O_{n,t} = \text{Linear}\left([\mathcal{H}_{n,t}, Y_{n,t}]\right); \qquad O_t = [O_{n,t}]_{n=1}^{N} \in R^{N\times d}. \tag{14}$$

This serves as input for the answer generation module, which we present next.

3.3 Answer Generation

To generate the response utterance, we employ the standard autoregressive framework to produce one word at a time by iteratively estimating the conditional word distribution $P\left(w \mid w_{t,1:l-1}; V, Q_{1:t-1}, A_{1:t-1}, Q_t\right)$ at generation step l. Inspired by the decoders in [25,26], we use a four-step Transformer decoder, as illustrated in Fig. 4.

Let $A_{t,l} \in \mathbb{R}^{l\times d}$ be the embedding matrix of the unfinished utterance of length l; $D = [Q_{1:t-1}, A_{1:t-1}] \in \mathbb{R}^{L_D\times d}$ the embedding of the dialog history of length L_D (words); $Q_t \in \mathbb{R}^{L_Q\times d}$ the embedding of the question, and $O_t \in \mathbb{R}^{N\times d}$ the output of generated by the COST in Eq. (14). The decoder generates a representation v_4 through a step-wise manner:

$$v_1 = \text{Attn}\left(a_l, A_{t,l-1}, A_{t,l-1}\right); \qquad v_2 = \text{Attn}\left(v_1, D, D\right);$$
$$v_3 = \text{Attn}\left(v_2, Q_t, Q_t\right); \qquad v_4 = \text{Attn}\left(v_3, O_t, O_t\right). \tag{15}$$

where $a_l = A_{t,l}[l]$, e.g., the last row of $A_{t,l}$. In essence, the sequence of current utterance, the existing dialog, and the current question forms the context to query the object representations O_t.

The retrieved information v_4 is used to generate the next word through the word distributions:

$$P_{vocab} = \text{softmax}\left(\text{Linear}(v_4)\right) \in \mathbb{R}^{1\times N_{vocab}}, \tag{16}$$

$$P_q = \text{Ptr}\left(Q_t, v_4\right) \in \mathbb{R}^{1\times N_{vocab}}, \tag{17}$$

$$P_l = \alpha P_q + (1-\alpha)\, P_{vocab}, \tag{18}$$

where $\alpha \in (0,1)$, and Ptr is a trainable Pointer Network [41] that "points" to all the tokens in question Q_t that are related to v_4, and P_l is a short-form for $P\left(w \mid w_{t,1:l-1}, V, Q_{1:t-1}, A_{1:t-1}, Q_t; \theta\right)$. Ptr seeks to reuse the relevant question words for the answer; and this is useful when facing rare words or word repetition is required. The gating function α is learnable (detailed in the Supplement).

3.4 Training

Given ground-truth answers $A^*_{1:T}$ of a full conversation of T turns, where $A^*_t = \left(w_1, w_2, ..., w_{L^a_t}\right)$. To make our generator more robust, we also add the log-likelihood of re-generated current turn's question $Q^*_t = \left(w_1, w_2, ..., w_{L^q_t}\right)$ to our loss. The network is trained with the cross-entropy loss w.r.t parameter θ, θ_q:

$$\mathcal{L} = \mathcal{L}\left(\theta\right) + \mathcal{L}\left(\theta_q\right), \text{where} \tag{19}$$

$$\mathcal{L}\left(\theta\right) = \sum_{t=1}^{T} \log P\left(A^*_t \mid V, Q_{1:t-1}, A^*_{1:t-1}, Q_t; \theta\right) \tag{20}$$

$$= \sum_{t=1}^{T} \sum_{l=1}^{L^a_t} \log P_l\left(w_l \mid w_{t,1:l-1}, V, Q_{1:t-1}, A^*_{1:t-1}, Q_t; \theta\right); \text{and} \tag{21}$$

$$\mathcal{L}\left(\theta_q\right) = \log P_{vocab}\left(Q^*_t \mid V, Q_t; \theta_q\right) \tag{22}$$

$$= \sum_{l=1}^{L^q_t} \log P_{vocab}\left(w^q_l \mid w^q_{t,1:l-1}, V, Q_t; \theta_q\right); \tag{23}$$

where $P_l\left(\cdot\right), P_{vocab}\left(\cdot\right)$ are computed in Eqs. (16–18).

4 Experiments

4.1 Experimental Settings

Dataset: We train our proposed method COST on the Audio-Visual Scene-Aware Dialogue (AVSD) [1], a large-scale video grounded dialog dataset. The dataset provides text-based dialogs visually grounded on untrimmed action videos from the popular Charades dataset [38]. Each annotated dialog consists of 10 rounds

Table 1. Statistics of the AVSD dataset and the test splits used at DSTC7 and DSTC8.

	Training	Validation	DSTC7 Test	DSTC8 test
No. videos	7,659	1,787	1,710	1,710
No. dialog turns	153,180	35,740	13,490	18,810

of question-answer about both static and dynamic scenes in a video, including objects, actions, and audio content. We benchmark against existing methods on video dialog using two different test splits used at the Seventh Dialog System Technology Challenges (DSTC7) [51] and the Eighth Dialog System Technology Challenges (DSTC8) [19]. See Table 1 for detailed statistics of the AVSD dataset and the two test splits at DSTC7 and DSTC8.

Often state-of-the-art methods in video dialog rely on different sources of information, including visual dynamic scenes, text description in the form of caption/video summary, and audio content. While the textual data containing high-level information of visual content can significantly attribute to the model's performance, they provide shortcuts due to linguistic biases that the models can exploit. As the ultimate goal of the video dialog task is to benchmark if a model can gather the visual clues to produce an appropriate response for a smooth conversation with humans, our experiments deliberately aim at challenging the visual reasoning capability of the models. In particular, we assume that the models only have access to the visual content and dialog history to answer a question at a specific turn while ignoring other additional textual data and audio data used by many other methods [25,26,33].

Implementation Details: All models are trained by optimizing the multi-label cross-entropy loss over generated tokens using Adam optimizer [20] with cosine learning rate scheduler [31]. At the training stage, we also use the auto-encoder loss function for the current question from [25]. We use a batch size of 128 samples distributed on 4 GPUs and train all the models for 50 epochs. Unless stated otherwise, each attention component in Eq. (15) is composed of a stack of 3 identical attention layers in Eq. (2). For each attention layer, we also use 4 parallel heads as suggested by [40]. Model parameters are selected based on the convergence of validation loss. At inference time, we adopt a beam search algorithm with a beam size of 3 for our answer generator.

Regarding object lives extraction, we strictly follow [7] and extract 30 object sequences per video. We further apply frame sub-sampling at an overall ratio of 4:1 to reduce the computational expensiveness. On average, each object live is composed of 176 time steps. For the context features C used in Eq. (9), we use I3D features [3] similar to other existing methods [22,26]. Pytorch implementation of our model is available online.[2]

[2] https://github.com/hoanganhpham1006/COST.

Evaluation Metrics: We adopt the same word-overlap-based metrics, including BLEU, METEOR, ROUGE-L, and CIDEr, as used by [51] to evaluate the effectiveness of the models.

4.2 Comparison Against SOTAs

We compare against the state-of-the-art methods, including MTN [25], FA+HRED [33], Student-Teacher [15], SCGA [18] and BiST [26] on both AVSD@DSTC7 and AVSD@DSTC8 test splits. For fair comparisons, all models only make use of the video content and dialog history. Results are shown in Table 2 and Table 3 for DSTC7 and DSTC8 test splits, respectively. In particular, COST consistently sets new SOTA performance on all evaluation metrics against existing methods on both test splits. The results strongly demonstrate the efficiency of our object-centric reasoning model with recurrent relational reasoning compared to methods that rely only on holistic visual features such as I3D [3] and ResNeXt [46].

Table 2. Experimental results on the AVSD@DSTC7 test split. All models only have access to video content and dialog history. †Models use visual features other than holistic video features such as I3D or ResNeXt. COST use object sequences and I3D features.

Methods	BLEU1	BLEU2	BLEU3	BLEU4	METEOR	ROUGE-L	CIDEr
FA+HRED [33]	0.648	0.505	0.399	0.323	0.231	0.510	0.843
MTN (I3D) [25]	0.654	0.521	0.420	0.343	0.247	0.520	0.936
MTN (ResNeXt) [25]	0.688	0.55	0.444	0.363	0.260	0.541	0.985
Student-Teacher [15]	0.675	0.543	0.446	0.371	0.248	0.527	0.966
BiST [26]	0.711	0.578	0.475	0.394	0.261	0.550	1.050
SCGA† [18]	0.702	0.588	0.481	0.398	0.256	0.546	1.059
COST (Ours)	**0.723**	**0.589**	**0.483**	**0.400**	**0.266**	**0.561**	**1.085**

Table 3. Experimental results on the AVSD@DSTC8 test split.

Methods	BLEU1	BLEU2	BLEU3	BLEU4	METEOR	ROUGE-L	CIDEr
MTN (I3D) [25]	0.611	0.496	0.404	0.336	0.233	0.505	0.867
MTN (ResNeXt) [25]	0.643	0.523	0.427	0.356	0.245	0.525	0.912
BiST [26]	0.684	0.548	0.457	0.376	0.273	0.563	1.017
SCGA† [18]	0.675	0.559	0.459	0.377	0.269	0.555	1.024
COST (Ours)	**0.695**	**0.559**	**0.465**	**0.382**	**0.278**	**0.574**	**1.051**

4.3 Model Analysis

Object-Centric Representation Facilitates Space-Time-Dialog Reasoning. To better highlight the effectiveness of our object-centric representation for reasoning over space-time to explore the semantic structure in videos, we design

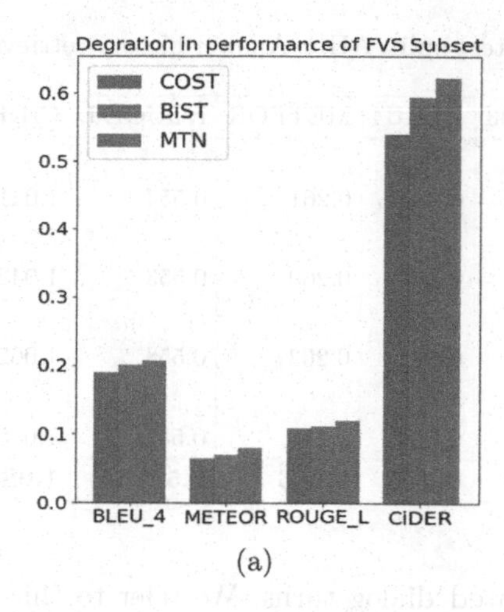

(a)

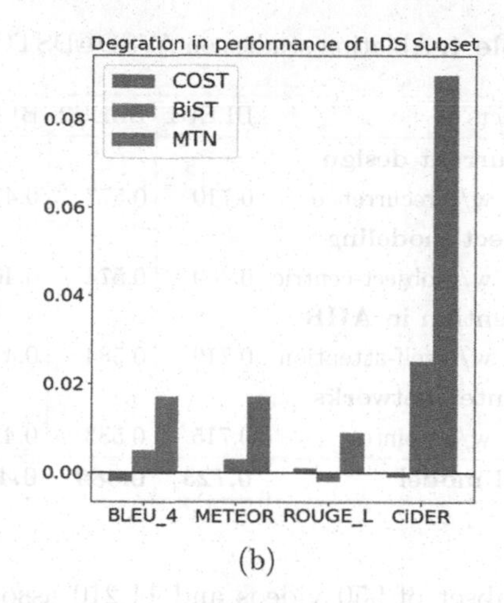

(b)

Fig. 5. Performance degradation points from DSTC7 full test splits to (a) FVS subset and (b) LDS subset. The FVS requires fine-grained visual understanding to answer questions while the LDS challenges the capability to handle long-distance dependencies questions of the models. The lower the better. Negative degradation points indicate improvements in performance. COST demonstrates its robustness against degradation in performance compared to BiST and MTN on these challenging subsets.

a subset of the AVSD@DSTC7 test split that poses challenges for models heavily relying on linguistic biases but undervaluing fine-grained visual information. First, we train a variant of the MTN model [25] with all the visual and audio components removed on the AVSD dataset. Next, we only pick dialog turns and their associated videos having BLEU4 score lower than 0.05 on the DSTC7 test split. This eliminates any questions whose answers can be guessed by the linguistic biases. Eventually, we obtain a subset of 1,062 videos with 8,782 associated dialog turns referred to as *Fine-grained Visual Subset (FVS)*. We evaluate the degradation of COST and the current state-of-the-art BiST and MTN on the FVS subset and report the results in Fig. 5(a). As shown, COST is more robust to questions in FVS while BiST, MTN struggle as it is degraded by larger margins on all the evaluation metrics. The results are clearly evident the benefits of the fine-grained video understanding brought by the object-centric representation compared to the holistic video representation used by MTN, BiST.

Recurrent Modeling Supports Long-Distance Dependencies. One of the main advantages of COST against SOTA methods is that it maintains a *recurrent system of dialog states,* which offers the better capability of handling long-distance dependencies questions. These questions require models to maintain and retrieve information appearing far in earlier turns. Methods without an explicit mechanism to propagate long-distance dependencies would struggle to generalize. In order to verify this, we design another subset of the AVSD@DSTC7 test split where we only collect questions at turns greater than 3. This results in

Table 4. Ablation studies on AVSD@DSTC7 test split. AHR: Answer history retrieval.

Effects of	BLEU1	BLEU2	BLEU3	BLEU4	METEOR	ROUGE-L	CIDEr
recurrent design							
w/o recurrence	0.710	0.577	0.471	0.388	0.261	0.554	1.041
object modeling							
w/o object-centric	0.709	0.574	0.467	0.385	0.260	0.553	1.042
attention in AHR							
w/o self-attention	0.719	0.584	0.477	0.395	0.263	0.558	1.062
pointer networks							
w/o pointer	0.715	0.583	0.477	0.394	0.260	0.557	1.045
Full model	**0.723**	**0.589**	**0.483**	**0.400**	**0.266**	**0.561**	**1.085**

a subset of 950 videos and 11,210 associated dialog turns. We refer to this as *Long-distance Dependencies Subset (LDS).* Figure 5(b) details the degradation in performance of COST, BiST, and MTN on the LDS. As shown, while COST achieves slight improvements in performance (negative degradation) thanks to it recurrent design, BiST, MTN experience consistent losses across the evaluation metrics. The results verify our hypothesis that maintaining the recurrent dialog object states is beneficial for handling long-distance dependencies between turns.

4.4 Ablation Studies

We conduct an extensive set of ablation studies to examine the contributions of each components in COST (See Table 4). These include ablating the recurrent design of COST, the use of object-centric video representation, and the use of self-attention layers for answer history retrieval for answer generation. We find that ablating any of these components would take the edge off the model's performance. The results are consistent with our analysis in Sect. 4.3 on the crucial effects of our recurrent design and the object-centric modeling by COST on the overall performance. We detail the effects as follows.

Effect of Recurrent Design: We remove the GRU in Eq. (6) and use query-specific object representation outputs of Eq. (5) as direct input to compute the adjacency matrix $\mathcal{K}_t$ in Eq. (7). By doing this, we ignore the effect of past dialog states on the current turn. As clearly seen, the model's performance considerably degrades on all evaluation metrics.

Effect of Object-Centric Modeling: In this experiment, we use only the context feature C (I3D features) and remove all the effects of the object representation. This leads to performance degradation by nearly 2% on all BLEU scores. The fine-grained object-centric representation clearly has an effect on improving the understanding of hidden semantic structure in video.

Effect of Self-attention-based Answer History Retrieval: This experiment ablates the role of prior generated answer tokens on information retrieval as in Eq. (13).

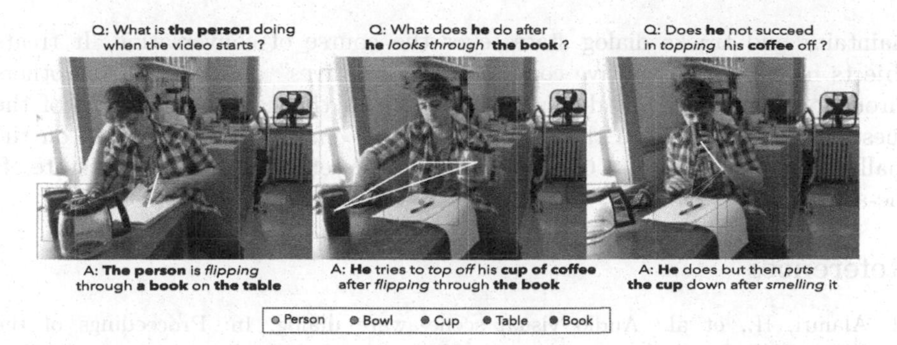

Q: What is the person doing when the video starts ?
Q: What does he do after he looks through the book ?
Q: Does he not succeed in topping his coffee off ?
A: The person is flipping through a book on the table
A: He tries to top off his cup of coffee after flipping through the book
A: He does but then puts the cup down after smelling it
Person
Bowl
Cup
Table
Book

Fig. 6. Visualization of the question-specific interaction matrices between objects $\mathcal{K}_t$ of Eq. (7). Each frame/question pair (left to right) represents a dialog turn, where the frame is selected to reflect the moment being queried. The visibility of edges denotes the relevance of visual objects' relations to the questions and desired answers. Detected objects are named by Faster RCNN. `COST` succeeds in constructing turn-specific graphs of relevant visual objects that facilitate answering questions. Sample is taken from DSTC7 test split - Best viewed in colors. (Color figure online)

We instead use the turn-specific visual representation as a direct input for the answer generator. The results show that this slightly affects the overall performance of the model.

Effect of Pointer Networks for Answer Generation: We remove the use of the pointer networks as in Eqs. (17 and 18) during answer generation. The results show that the removal of the pointer slightly hurt the model's performance.

4.5 Qualitative Analysis

We visualize a representative example from the AVSD@DSTC7 test split as a showcase to analyze the internal operation of the proposed method `COST`. We present the question-induced interaction matrix as it is a crucial component of our model design in Eq. (7). Figure 6 presents the evolution of the relationships between objects in video from turn to turn (left to right). `COST` succeeds in constructing turn-specific graphs of relevant visual objects that reflect the relationships of interest by respective questions and answers. The interpretability and strong quantitative results (Sect. 4.2 and 4.3) by `COST` are evident to the appropriateness of the object-centric representation towards solving video dialog task.

5 Conclusion

Addressing the highly challenging task of video dialog, we have proposed `COST`, a new *recurrent object-centric* system that learns to reason through multiple dialog turns, object dynamics, and interactions over space-time in video. `COST` maintains and tracks dialog states over the course of conversation. It treats objects in video as primitive constructs whose "lives" and relations to others throughout the video are dynamically examined through the guidance of the questions, conditioned on the *dialog states* and *answer history*. Tested on the challenging AVSD dataset, COST demonstrates its effectiveness against state-of-the-art models.

References

1. Alamri, H., et al.: Audio visual scene-aware dialog. In: Proceedings of the IEEE/CVF Conference on Computer Vision and Pattern Recognition, pp. 7558–7567 (2019)
2. Baradel, F., Neverova, N., Wolf, C., Mille, J., Mori, G.: Object level visual reasoning in videos. In: Ferrari, V., Hebert, M., Sminchisescu, C., Weiss, Y. (eds.) ECCV 2018. LNCS, vol. 11217, pp. 106–122. Springer, Cham (2018). https://doi.org/10.1007/978-3-030-01261-8_7
3. Carreira, J., Zisserman, A.: Quo Vadis, action recognition? A new model and the kinetics dataset. In: CVPR, pp. 6299–6308 (2017)
4. Chao, Y.W., Vijayanarasimhan, S., Seybold, B., Ross, D.A., Deng, J., Sukthankar, R.: Rethinking the faster R-CNN architecture for temporal action localization. In: Proceedings of the IEEE Conference on Computer Vision and Pattern Recognition, pp. 1130–1139 (2018)
5. Cho, K., Van Merriënboer, B., Bahdanau, D., Bengio, Y.: On the properties of neural machine translation: encoder-decoder approaches. arXiv preprint arXiv:1409.1259 (2014)
6. Chu, Y.W., Lin, K.Y., Hsu, C.C., Ku, L.W.: Multi-step joint-modality attention network for scene-aware dialogue system. arXiv preprint arXiv:2001.06206 (2020)
7. Dang, L.H., Le, T.M., Le, V., Tran, T.: Hierarchical object-oriented spatio-temporal reasoning for video question answering. In: IJCAI (2021)
8. Das, A., et al.: Visual Dialog. IEEE Trans. Pattern Anal. Mach. Intell. **41**(5), 1242–1256 (2019). https://doi.org/10.1109/TPAMI.2018.2828437
9. Desta, M.T., Chen, L., Kornuta, T.: Object-based reasoning in VQA. In: 2018 IEEE Winter Conference on Applications of Computer Vision (WACV), pp. 1814–1823. IEEE (2018)
10. Gao, S., Sethi, A., Agarwal, S., Chung, T., Hakkani-Tur, D.: Dialog state tracking: a neural reading comprehension approach. In: Proceedings of the 20th Annual SIGdial Meeting on Discourse and Dialogue, pp. 264–273 (2019)
11. Geman, D., Geman, S., Hallonquist, N., Younes, L.: Visual turing test for computer vision systems. Proc. Natl. Acad. Sci. **112**(12), 3618–3623 (2015)
12. Geng, S., et al.: Dynamic graph representation learning for video dialog via multi-modal shuffled transformers. In: Proceedings of AAAI Conference on Artificial Intelligence (2021)
13. Hong, Y., Wu, Q., Qi, Y., Rodriguez-Opazo, C., Gould, S.: VLN BERT: a recurrent vision-and-language BERT for navigation. In: Proceedings of the IEEE/CVF Conference on Computer Vision and Pattern Recognition (CVPR), pp. 1643–1653, June 2021
14. Hori, C., et al.: End-to-end audio visual scene-aware dialog using multimodal attention-based video features. In: ICASSP 2019–2019 IEEE International Conference on Acoustics, Speech and Signal Processing (ICASSP), pp. 2352–2356. IEEE (2019)

15. Hori, C., Cherian, A., Marks, T.K., Hori, T.: Joint student-teacher learning for audio-visual scene-aware dialog. In: INTERSPEECH, pp. 1886–1890 (2019)
16. Huang, D., Chen, P., Zeng, R., Du, Q., Tan, M., Gan, C.: Location-aware graph convolutional networks for video question answering. In: Proceedings of the AAAI Conference on Artificial Intelligence, vol. 34, pp. 11021–11028 (2020)
17. Kalogeiton, V., Weinzaepfel, P., Ferrari, V., Schmid, C.: Action tubelet detector for spatio-temporal action localization. In: Proceedings of the IEEE International Conference on Computer Vision, pp. 4405–4413 (2017)
18. Kim, J., Yoon, S., Kim, D., Yoo, C.D.: Structured co-reference graph attention for video-grounded dialogue. In: AAAI (2021)
19. Kim, S., et al.: The eighth dialog system technology challenge. arXiv preprint arXiv:1911.06394 (2019)
20. Kingma, D., Ba, J.: Adam: a method for stochastic optimization. In: International Conference on Learning Representations (ICLR) (2014)
21. Kottur, S., Moura, J.M.F., Parikh, D., Batra, D., Rohrbach, M.: Visual coreference resolution in visual dialog using neural module networks. In: Ferrari, V., Hebert, M., Sminchisescu, C., Weiss, Y. (eds.) ECCV 2018. LNCS, vol. 11219, pp. 160–178. Springer, Cham (2018). https://doi.org/10.1007/978-3-030-01267-0_10
22. Le, H., Chen, N.F.: Multimodal transformer with pointer network for the DSTC8 AVSD challenge. arXiv preprint arXiv:2002.10695 (2020)
23. Le, H., Chen, N.F., Hoi, S.C.: Learning reasoning paths over semantic graphs for video-grounded dialogues. arXiv preprint arXiv:2103.00820 (2021)
24. Le, H., Hoi, S., Sahoo, D., Chen, N.: End-to-end multimodal dialog systems with hierarchical multimodal attention on video features. In: DSTC7 at AAAI2019 Workshop (2019)
25. Le, H., Sahoo, D., Chen, N., Hoi, S.: Multimodal transformer networks for end-to-end video-grounded dialogue systems. In: Proceedings of the 57th Annual Meeting of the Association for Computational Linguistics, pp. 5612–5623 (2019)
26. Le, H., Sahoo, D., Chen, N., Hoi, S.C.: BiST: bi-directional spatio-temporal reasoning for video-grounded Dialogues. In: Proceedings of the 2020 Conference on Empirical Methods in Natural Language Processing (EMNLP), pp. 1846–1859 (2020)
27. Le, T.M., Le, V., Venkatesh, S., Tran, T.: Dynamic language binding in relational visual reasoning. In: IJCAI, pp. 818–824 (2020)
28. Le, T.M., Le, V., Venkatesh, S., Tran, T.: Hierarchical conditional relation networks for video question answering. In: Proceedings of the IEEE/CVF Conference on Computer Vision and Pattern Recognition, pp. 9972–9981 (2020)
29. Lee, H., Yoon, S., Dernoncourt, F., Kim, D.S., Bui, T., Jung, K.: DSTC8-AVSD: multimodal semantic transformer network with retrieval style word generator. arXiv preprint arXiv:2004.08299 (2020)
30. Lin, K.Y., Hsu, C.C., Chen, Y.N., Ku, L.W.: Entropy-enhanced multimodal attention model for scene-aware dialogue generation. arXiv preprint arXiv:1908.08191 (2019)
31. Loshchilov, I., Hutter, F.: SGDR: stochastic gradient descent with warm restarts. In: International Conference on Learning Representations (2017)
32. Murahari, V., Batra, D., Parikh, D., Das, A.: Large-scale pretraining for visual dialog: a simple state-of-the-art baseline. In: Vedaldi, A., Bischof, H., Brox, T., Frahm, J.-M. (eds.) ECCV 2020. LNCS, vol. 12363, pp. 336–352. Springer, Cham (2020). https://doi.org/10.1007/978-3-030-58523-5_20
33. Nguyen, D.T., Sharma, S., Schulz, H., Asri, L.E.: From film to video: multi-turn question answering with multi-modal context. In: DSTC7 Workshop at AAAI 2019 (2019)

34. Pan, B., et al.: Spatio-temporal graph for video captioning with knowledge distillation. In: Proceedings of the IEEE/CVF Conference on Computer Vision and Pattern Recognition, pp. 10870–10879 (2020)
35. Sanabria, R., Palaskar, S., Metze, F.: CMU Sinbad's submission for the DSTC7 AVSD challenge. In: DSTC7 at AAAI2019 Workshop, vol. 6 (2019)
36. Seo, P.H., Lehrmann, A., Han, B., Sigal, L.: Visual reference resolution using attention memory for visual dialog. arXiv preprint arXiv:1709.07992 (2017)
37. Serban, I., et al.: A hierarchical latent variable encoder-decoder model for generating dialogues. In: Proceedings of the AAAI Conference on Artificial Intelligence, vol. 31 (2017)
38. Sigurdsson, G.A., Varol, G., Wang, X., Farhadi, A., Laptev, I., Gupta, A.: Hollywood in homes: crowdsourcing data collection for activity understanding. In: Leibe, B., Matas, J., Sebe, N., Welling, M. (eds.) ECCV 2016. LNCS, vol. 9905, pp. 510–526. Springer, Cham (2016). https://doi.org/10.1007/978-3-319-46448-0_31
39. Spelke, E.S., Kinzler, K.D.: Core knowledge. Dev. Sci. **10**(1), 89–96 (2007)
40. Vaswani, A., et al.: Attention is all you need. In: Advances in Neural Information Processing Systems, pp. 5998–6008 (2017)
41. Vinyals, O., Fortunato, M., Jaitly, N.: Pointer networks. In: Advances in Neural Information Processing Systems 28 (2015)
42. Wang, X., Gupta, A.: Videos as space-time region graphs. In: Ferrari, V., Hebert, M., Sminchisescu, C., Weiss, Y. (eds.) ECCV 2018. LNCS, vol. 11209, pp. 413–431. Springer, Cham (2018). https://doi.org/10.1007/978-3-030-01228-1_25
43. Wang, Y., Joty, S., Lyu, M.R., King, I., Xiong, C., Hoi, S.C.: VD-BERT: a unified vision and dialog transformer with BERT. arXiv preprint arXiv:2004.13278 (2020)
44. Wojke, N., Bewley, A., Paulus, D.: Simple online and realtime tracking with a deep association metric. In: ICIP, pp. 3645–3649. IEEE (2017)
45. Xie, H., Iacobacci, I.: Audio visual scene-aware dialog system using dynamic memory networks. In: DSTC8 at AAAI2020 Workshop (2020)
46. Xie, S., Girshick, R., Dollár, P., Tu, Z., He, K.: Aggregated residual transformations for deep neural networks. In: CVPR, pp. 1492–1500 (2017)
47. Xie, S., Sun, C., Huang, J., Tu, Z., Murphy, K.: Rethinking spatiotemporal feature learning: speed-accuracy trade-offs in video classification. In: Ferrari, V., Hebert, M., Sminchisescu, C., Weiss, Y. (eds.) ECCV 2018. LNCS, vol. 11219, pp. 318–335. Springer, Cham (2018). https://doi.org/10.1007/978-3-030-01267-0_19
48. Yang, Z., Garcia, N., Chu, C., Otani, M., Nakashima, Y., Takemura, H.: BERT representations for video question answering. In: Proceedings of the IEEE/CVF Winter Conference on Applications of Computer Vision, pp. 1556–1565 (2020)
49. Yeh, Y.T., Lin, T.C., Cheng, H.H., Deng, Y.H., Su, S.Y., Chen, Y.N.: Reactive multi-stage feature fusion for multimodal dialogue modeling. arXiv preprint arXiv:1908.05067 (2019)
50. Yi, K., et al.: CLEVRER: collision events for video representation and reasoning. arXiv preprint arXiv:1910.01442 (2019)
51. Yoshino, K., et al.: Dialog system technology challenge 7. arXiv preprint arXiv:1901.03461 (2019)
52. Zeng, R., et al.: Graph convolutional networks for temporal action localization. In: Proceedings of the IEEE/CVF International Conference on Computer Vision, pp. 7094–7103 (2019)
53. Zhuge, M., et al.: Kaleido-BERT: vision-language pre-training on fashion domain. In: Proceedings of the IEEE/CVF Conference on Computer Vision and Pattern Recognition (CVPR), pp. 12647–12657, June 2021

Author Index

Printed in the United States
by Baker & Taylor Publisher Services